Wheat Breeding for Multifunctional Agriculture

Wheat Breeding for Multifunctional Agriculture

– Authors –

Dilip Kumar Verma

Principal Scientist (Genetics and Plant Breeding)

Anil Kumar Singh

Principal Scientist (Agricultural Extension),

Indian Agricultural Research Institute, Regional Station,
Indore – 452 001, Madhya Pradesh

Daya Nand Singh

Research and Development Head
M/s Eagle Seeds and Biotech Limited,
Stationed at Indore, Madhya Pradesh

BIOTECH

2021

BIOTECH BOOKS®

ISBN 978-81-7622-490-1

Published by: **BIOTECH BOOKS**®
4762-63/23, Ansari Road, Darya Ganj,
New Delhi - 110 002
Phone: 09899295259, 011-43003222
E-mail: biotechbooks@yahoo.co.in
Website: www.biotechbooks.in

Typeset at: **Classic Computer Services**
Delhi - 110 035

Printed at: **Replika Press Pvt. Ltd.**

PRINTED IN INDIA

डॉ० राजेन्द्र प्रसाद केन्द्रीय कृषि विश्वविद्यालय
पूसा (समस्तीपुर) - 848 125, बिहार
Dr. Rajendra Prasad Central Agricultural University
Pusa (Samastipur) - 848 125, Bihar

डॉ० रमेश चन्द्र श्रीवास्तव
कुलपति
Dr. R. C. Srivastava
M.Tech., Ph.D. (IIT, Kgp)
FNAAS, FIASWC, FISAE, FIE, FCHAI
Vice - Chancellor

Phone : 06274-240226 (O)
06274-240264 (R)
Fax : 06274-240255
Mob. : 09040033323
Email : vc@rpcau.ac.in
Website : www.rpcau.ac.in

No.__________/Dr RPCAU (VC) **Date :** 18/12/2021

Message

The challenges before the Agricultural Scientists of our country today are much more complex than ever before. Food production has to be increased to 240 MT within the next five years to respond to our ever-increasing population. There is a pressing need to increase crop productivity as well as quality in view of changing climatic scenarios. Wheat is one of the most important cereal crops grown all over the world. Though significant developments have been made in the past few years in wheat breeding, I am happy to see more relevant information is collected, compiled, edited and presented in this book titled "**Wheat Breeding for Multifunctional Agriculture**". This is a serious effort that will add a flag of knowledge of wheat breeding in a multifunctional way for the development and production of this important cereal in NEPZ (North Eastern Plain Zone) and CZ (Central Zone) of India in particular and the world in general.

This book provides a complete, comprehensive and broad subject based views through various chapters viz. Wheat Knowledge Management, Genetics and Breeding, Bread Wheat Breeding at Pusa, Bihar - Past, Present and Future, Potential Species of Gene Pool Enrichment, Breeding Wheat for Biotic and Abiotic Stresses, Breeding Wheat for Quality and Biofortification, Breeding wheat for Sustainable, Conservation, and Organic Agriculture and Wheat Seed production and maintenance. This book is monumental as well as an arduous piece of work being carried out by Dr. Dilip Kumar Verma and his team. I am hopeful the outcome of this book will serve the specific purpose of imparting completely comprehensive information about wheat breeding for the farmers, students, scholars and practitioners.

I wish for successful publication of this Book "**Wheat Breeding for Multifunctional Agriculture**".

RCSrivastava

(R. C. Srivastava)

BIHAR AGRICULTURAL UNIVERSITY, SABOUR
BHAGALPUR - 813210 (BIHAR)

Dr. Ajoy Kumar Singh
Vice Chancellor

Phone : 0641-2452606 (O)
: 0641-2452605 (R)
Fax : 0641-2452604
Patna : 0612-2222267 (O)
Fax : 0612-2225364
Mob. No. : +91-8902750693
Email ID : vcbausabour@gmail.com
URL : www.bausabour.ac.in

Ref. :...................... Date. :......................

Research is a creativity and exploring the creativity is the essence of scientific activity. The challenges before the agricultural scientists today are much more complex than were ever before. It is expected that by 2020 the world population will reach 9 billion mark and the food grain production needs to be almost doubled. Besides increase in population global food production system is also facing challenges like shrinking arable land, reduced availability of water for irrigation and reducing human interest in continuing farming activity.

Wheat is staple food crop of large majority of world's population across the continents. Wheat plays an important role in global food security as it provides approximately 20% of the protein for more than half of the world's population and it is t is the second most important food crop in the developing world after rice. With a predicted world population of 9 billion in 2050, the demand for wheat is expected to increase by 60%. To meet this demand, annual wheat yield increases must rise from the current level of below 1% to at least 1.6%.

Under these circumstances, wheat crop improvement remains an indispensable instrument for enhancing crop production and productivity. However, the traditional breeding approaches alone cannot achieve the target of doubling food production in few decades and it requires combination of advanced and traditional breeding approaches. Keeping this in mind, the important and relevant knowledgewere collected, compiled, collated, edited and presented in this book form titled **"Breeding Wheat for Multifunctional Agriculture"**.

This book provides complete, comprehensive and broad subject based views with addition of the wheat breeding in multifunctional ways viz., Wheat Knowledge Management; Breeder's Perspective of nanotechnology; Genetics, Cytogenetics, Breeding and Gene Regulation of Bread Wheat (*Triticum aestivum* L. *Triticum sativum*, Lamk); past, present and future prospects of Bread Wheat Breeding at various Research Station of Indian Agricultural Research Institute; Potential Species of Gene Pool Enrichment for Wheat Improvement; Breeding Wheat for abiotic and biotic stresses resistance; Breeding for Quality and Biofortification/Micronutrient-Dense Wheat; Breeding wheat for Sustainable, Conservation and Organic Agriculture; Wheat Seed, Maintenance Breeding, Basic Principles, Seed Priming, Electrophoresis, Seed Certification and Farmer Participatory Seed Production.

This book has been prepared with a specific purpose of bringing together comprehensive information about the wheat breeding in multifunctional ways and we hope that the readers will find this a useful guide for wheat breeding activity. We thank publishers/researchers and staffs of both IARI Research Station located at Pusa, and Indore, Madhya Pradesh for their valuable contribution.

Ajoy K. Singh

राजमाता विजयाराजे सिंधिया कृषि विश्वविद्यालय
राजा पंचम सिंह मार्ग, ग्वालियर (म.प्र.) – 474002
Rajmata Vijayaraje Scindia Krishi Vishwa Vidyalaya,
Raja Pancham Singh Marg, Gwalior (M.P.) – 474 002
(An ISO Certified 9001:2008)
Tel: 0751 – 2970502, Fax: 0751 – 2970504, E-mail: vcrvsaugwa@mp.gov.in

प्रो. एस. के. राव
कुलपति
Prof. S. K. Rao
Vice-Chancellor

No./VC/2020-21/4073
Date: 10/12/2020

MESSAGE

Research is creativity and exploring the creativity is our duty. The challenges before the Agricultural Scientists of our country today are much more complex than even before. Food production has to be increased to 240mt within the next five years. Wheat (***Triticum spp.***) is a worldwide cultivated grass grown in all the continents except Antarctica. To respond to an ever-increasing need of food by the growing world's population, standing today at 7 billion and expected to reach 9 billion by 2050, there is a pressing need to increase crop productivity. Under these circumstances, important and relevant information were collected, compiled, collated, edited and presented in a book form titled **"Breeding Wheat for Multifunctional Agriculture"**. Cereals are grown throughout the world to provide food for human consumption. The hopeful outcome depends on achieving an ambitions transformation involving widespread modernization in technology, based primarily on massive increase in inputs to agriculture.

This book provides complete, comprehensive and broad subject based views with addition of the wheat breeding in multifunctional ways comprising wide range of topics of wheat breeding and seed production This book has been prepared with a specific purpose of importing complete comprehensive information about wheat breeding in multifunctional ways and we hope that the students and readers will find this with much utility.

Though significant developments have been made in the development of wheat (bread wheat and durum) varieties, but this compilation and documentation is a very good effort which will add a flag of knowledge of wheat breeding.

(S. K. Rao)

Preface

"Everyone has inside of him or her, a piece of good news. The good news is that you don't know how great you can be!"

– Dr. Abdul Kalam

Science alone could solve the problems of hunger and poverty, instantiation and illiteracy, of superstition and deadening customs, of vast resources running to waste, of a rich country inhabited by starving people. The future belongs to science and those who make friends with science.

– Jawaharlal Nehru

"Our priority and our work is to find the best combination between the 100 000 genes of a plant".

"Life can only be understood backwards; but it must be lived forwards."

– Soren Kierkegaard, Danish philosopher, theologian, and psychologist

Sir Bent Skovmand mixes wry humor with biological wisdom. *"If the seeds disappear, so could your food – so could you.*

Salutation to God Prajápati Kasyapa. Agriculture may always flourish and the Goddess (may reside) in seeds and wealth. Channdavata he.

"Bent Skovmand is not exactly a household name, but he has more to do with the welfare of the earth's 5 billion people than many heads of state."

– Eugene Linden, Time Magazine, August 19, 1991

Research is a creativity and exploring the creativity is our duty. The challenges before the Agricultural Scientists of our country today are much more complex than even before. Food production has to be increased to 240 mt within the next five years. Wheat (*Triticum* spp.) is a worldwide cultivated grass grown in all the continents except Antarctica. To respond to an ever-increasing need of food by the growing world's population, standing today at 7 billion and expected to reach 9 billion by 2050, there is a pressing need to increase crop productivity. Under these circumstances, important and relevant informations were collected, compiled, collated, edited and presented in a book form entitled **"Wheat Breeding for Multifunctional Agriculture".** Cereals are grown throughout the world to provide food for human consumption. The hopeful outcome depends on achieving an ambitions transformation involving widespread modernization in technology, based primarily on massive increase in inputs to agriculture.

This book has been prepared with a specific purpose of importing complete comprehensive information about wheat breeding in multifunctional way and we hope that the students and readers will find this with much utility. We thank publishers/researchers and staffs of both IARI RS Pusa, Bihar [2006-2010] as well as IARI RS Indore, Madhya Pradesh [2010-2013 and 2018- till today], where I had served a significant piece of our scientific life to conduct, direct and follow up the wheat breeding for multifunctional agriculture in true spirit. We are sure that this book will serve as valuable text cum reference book to the graduate/post graduate researchers/students of agricultural institutions/universities. We profusely thank Dr R.C. Srivastava, Vice-Chancellor, Dr. Rajendra Prasad Central Agricultural University, Pusa, Bihar, India and Dr Ajay Kumar Singh, Vice-Chancellor, Bihar Agricultural University, Sabour, Bhagalpur, Bihar, India and Dr. S.K. Rao, Vice-Chancellor, Rajmata Vijayaraje Sciendia Krishi Vishwavidyalaya, Gwalior, Madhya Pradesh, India for his encouragement and for providing Foreword.

This book provides complete, comprehensive and broad subject based views with addition of the wheat breeding in multifunctional ways *viz.*, 1. Wheat Knowledge Management, Nano Dreams and Nightmares: A Breeder's Perspective; 2. Genetics, Cytogenetics, Breeding and Gene Regulation of Bread Wheat (*Triticum aestivum* L. *Triticum sativum*, Lamk); 3. Bread Wheat Breeding at IARI RS Pusa, Bihar - Past, Present (2006-2010) and Future; Bread Wheat and Durum Wheat Breeding at IARI RS Indore, Madhya Pradesh – Past, Present (2010-2013; 2018-2020) and Future; 5 *Leymus racemosus*: Potential Species of Gene Pool Enrichment for Wheat Improvement; 6. Breeding Wheat for Abiotic Stresses (Heat, Drought, Waterlogged, Cold) Resistance; 7. Breeding Wheat for Biotic Stresses (Disease, Insect pests and Nematodes) Resistance; 8. Breeding Wheat for Quality and Biofortification/ Micronutrient-Dense Wheat; 9. Breeding wheat for Sustainable, Conservation and Organic Agriculture; 10. Wheat Seed, Maintenance Breeding, Basic Principles, Seed Priming, Electrophoresis, Seed Certification and Farmer Participatory Seed Production.

We thank Drs S Srivastava; R Thakur; NB Singh; NK Singh; BC Saha; S Chowdhruy and A K Sinha and many more personnel for motivating us to take up this assignment. Sincere thanks are due to Dr (Mrs) Kiran Kumari, Er Heena Verma; DRx Somya Verma and Mr Jayant Verma for rendering help during the overall preparation of the manuscript in the form of a book. We acknowledge the photographs from various sources *viz.*, selfies, whatsapp, facebook *etc.* for giving the text and matter to more clarity of the readers. We particularly appreciate Mr Sunil Mittal jee of Biotech Book for his continued support. We sincerely hope that this book will serve as a milestone towards achieving meaningful wheat genetic improvement to meet the ever-increasing requirements of food of this region in particular and world in general.

Dilip Kumar Verma

Anil Kumar Singh

Daya Nand Singh

Glimpses of Wheat Breeding for Multifunctional Agriculture

Figure 1. Welcome to IARI

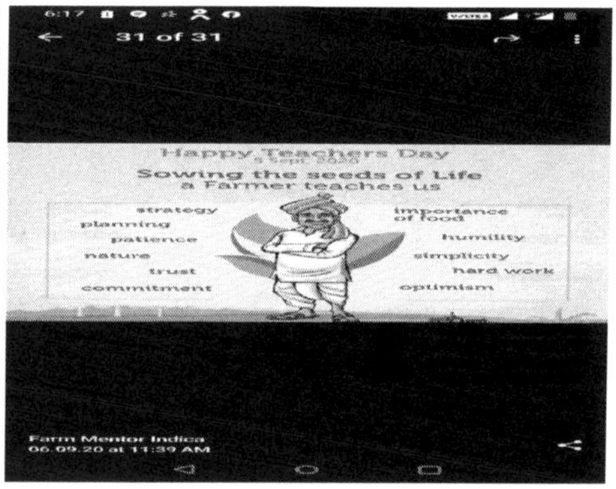

Figure 2. Sowing the seeds of life a farmer teaches us

Figure 3. Humanity is greater than status

Figure 4. Find a way not an excuse

Figure 5. Danielle Pierre, Your life is a reflection of your thoughts.

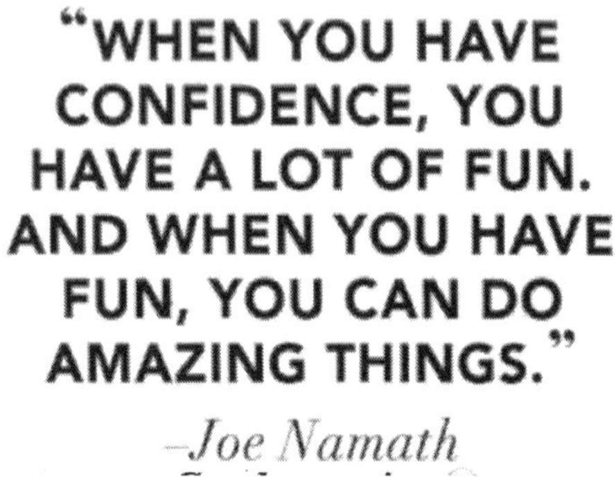

Figure 6. Joe Namath, When you have confidence, you have ....

Figure 7. The strongest people make time to help others, even if they are struggling...

Figure 8. Success speaks louder than promises

Figure 9. Pizza always confuses me....

Figure 11. Symbol of prosperity, green revolution and serving against food famine in India from past, present and future.

Figure 12. Dr Norman Borlaug (2009): Father of Green Revolution, Noble Prize Winner Peace

Figure 13. Dr MS Swaminathan, pioneered the Green Revolution in India

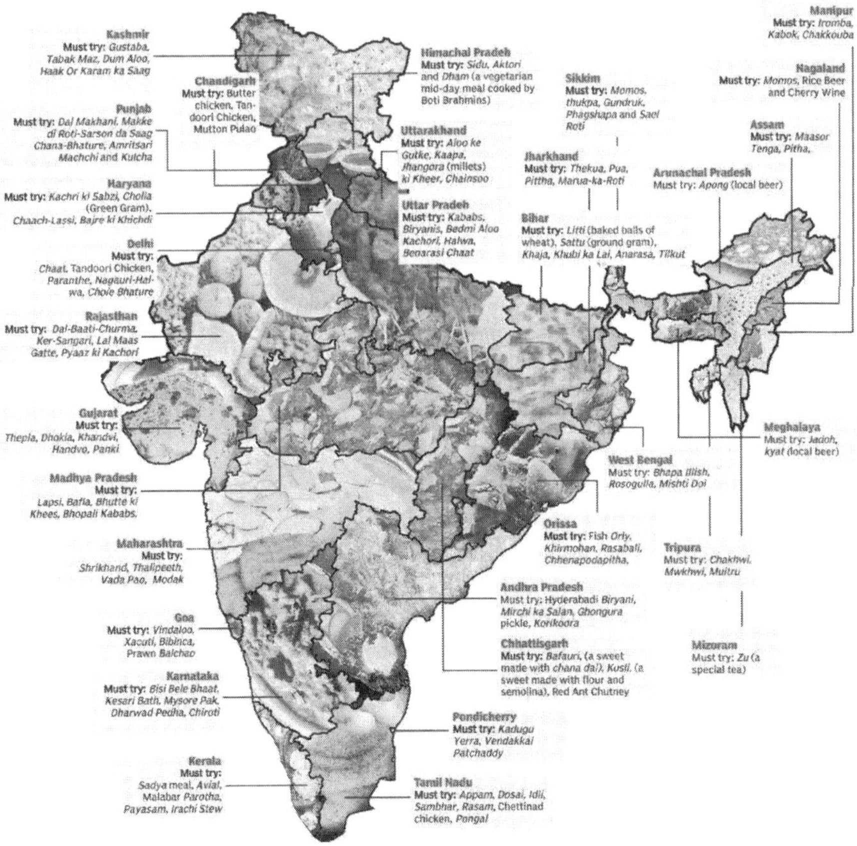

Figure 14. Various tastes of food prevalent in Indian states

Figure 15. A rising sun scene, stating good morning for all the human beings for food on this planet

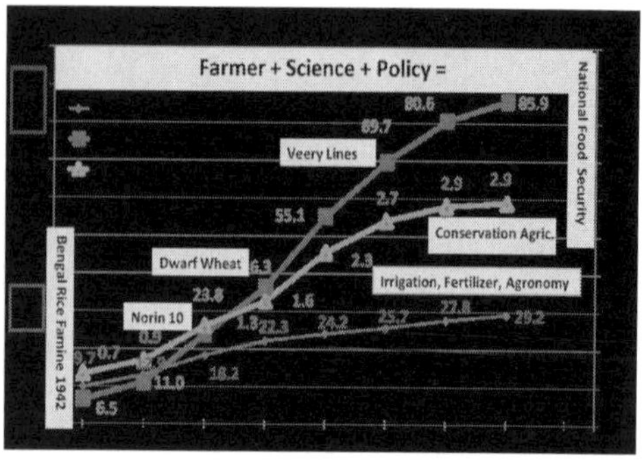

Figure 16. A graph deciphering the static policy issues from Bengal Famine 1942 to National Food Security

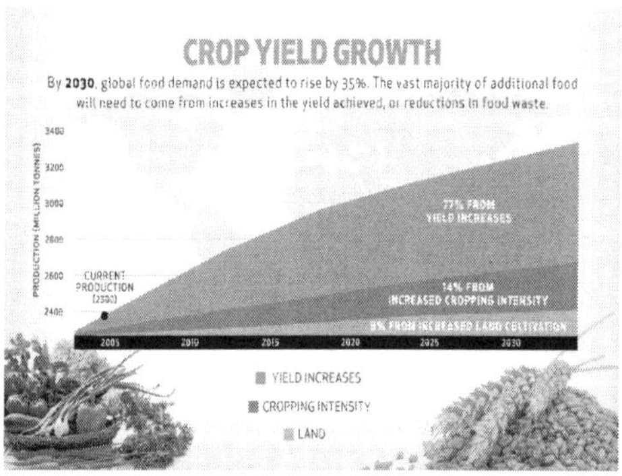

Figure 17. Crop Yield Growth

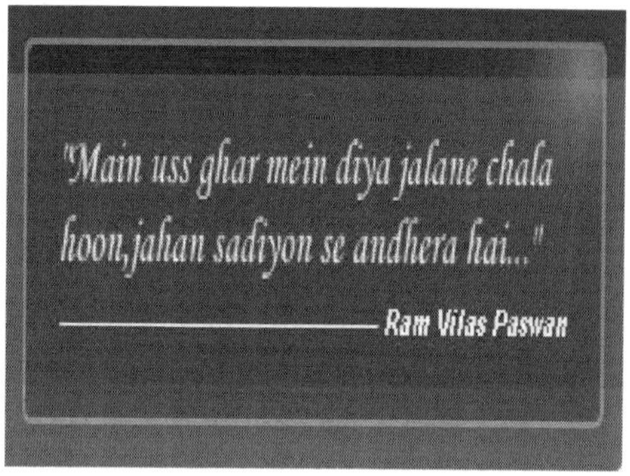

Figure 18. Quote by Honourable (Late) Ram Vilas Paswan, Minister of Public Distribution System, Govt of India

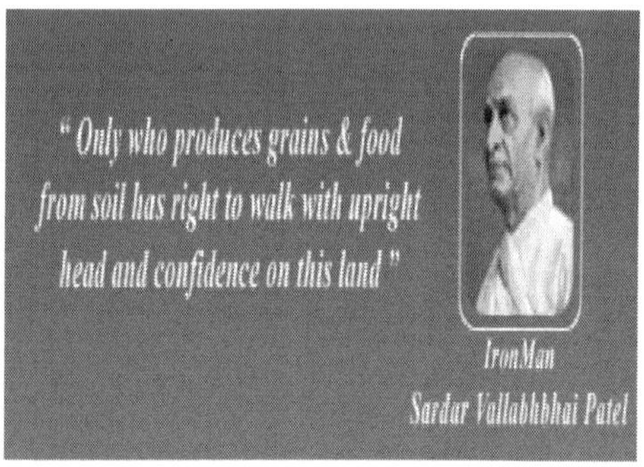

Figure 19. Iron Man Sardar Vallabhbhai Patel - Only who produces grains and food from soil has right to walk with upright head and confidence on this land.

Figure 20. Dr D K Verma addressing breeding for abiotic stress management during 101st Indian Science Congress-2014

Figure 21. Dr D K Verma and Dr Sheikh Md Sultan during 101st Indian Science Congress-2014

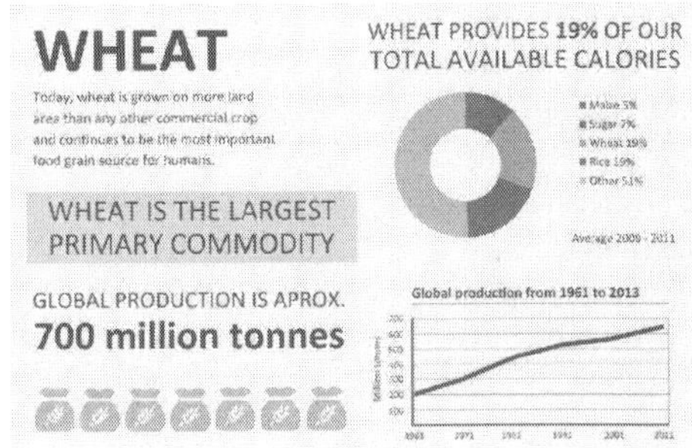

Figure 22. Wheat provides 19% of our total available calories

Figure 23. In India there are 823 TV Channel, but not even one channel for agriculture

Figure 24. What is the use of Figures, ...

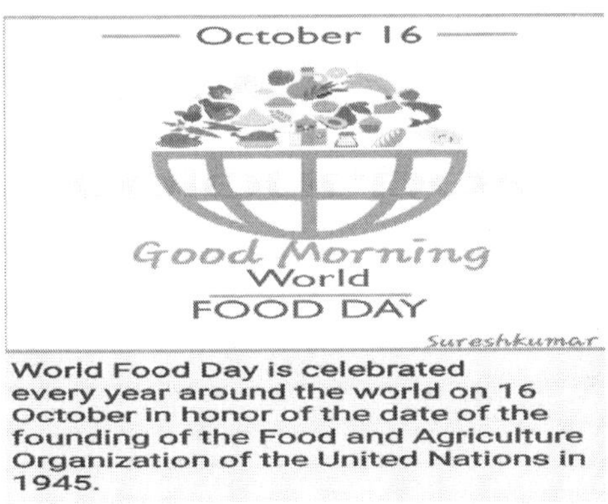

Figure 25. October 16: World Food Day

Figure 26. A symbol of prosperity in Bihar, Maan Annapurna

Figure 27. On the auspicious day of World Food Day, 16th October, 2020 at 11AM, address by our honourable Prime Minister, Govt of India, releasing coins of enumeration Rs 75/-

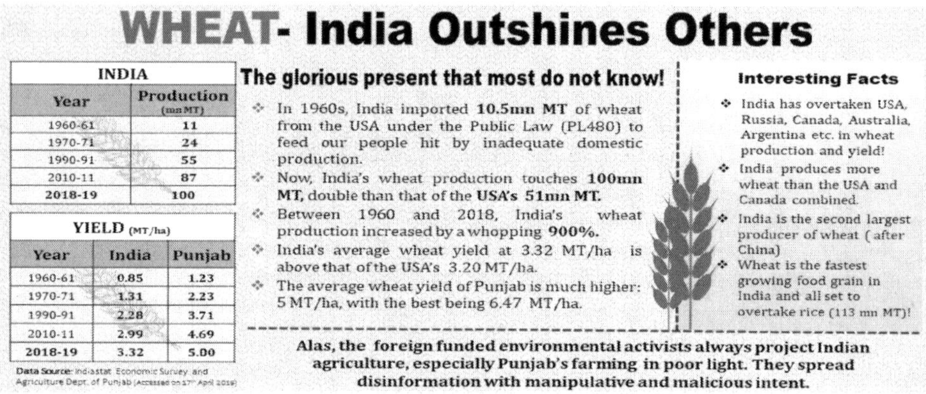

WHEAT- India Outshines Others

INDIA	
Year	Production (mn MT)
1960-61	11
1970-71	24
1990-91	55
2010-11	87
2018-19	100

YIELD (MT/ha)		
Year	India	Punjab
1960-61	0.85	1.23
1970-71	1.31	2.23
1990-91	2.28	3.71
2010-11	2.99	4.69
2018-19	3.32	5.00

Data Source: Indiastat Economic Survey and Agriculture Dept. of Punjab (Accessed on 17th April 2019)

The glorious present that most do not know!

- In 1960s, India imported **10.5mn MT** of wheat from the USA under the Public Law (PL480) to feed our people hit by inadequate domestic production.
- Now, India's wheat production touches **100mn MT**, double than that of the **USA's 51mn MT.**
- Between 1960 and 2018, India's wheat production increased by a whopping **900%.**
- India's average wheat yield at 3.32 MT/ha is above that of the USA's 3.20 MT/ha.
- The average wheat yield of Punjab is much higher: 5 MT/ha, with the best being 6.47 MT/ha.

Interesting Facts

- India has overtaken USA, Russia, Canada, Australia, Argentina etc. in wheat production and yield!
- India produces more wheat than the USA and Canada combined.
- India is the second largest producer of wheat (after China)
- Wheat is the fastest growing food grain in India and all set to overtake rice (113 mn MT)!

Alas, the foreign funded environmental activists always project Indian agriculture, especially Punjab's farming in poor light. They spread disinformation with manipulative and malicious intent.

Figure 28. Wheat – India Outshines Others

Food security mustn't lull us into complacency

Being one of the most populous nations in the world, India needs a food system so robust and renewable that even if another pandemic occurs, the impact on the economy in general and this system in particular is minimal. We have to find alternative ways of producing more food. The nation which will be able to fill the bellies of its citizens while preserving its natural resources will be the winner.

The Tribune Tribune Chandigarh 24-08-2020

Figure 29. Excerpts from Tribune, Chandigarh dt 24.08.2020: Food security must not lull us into complacency

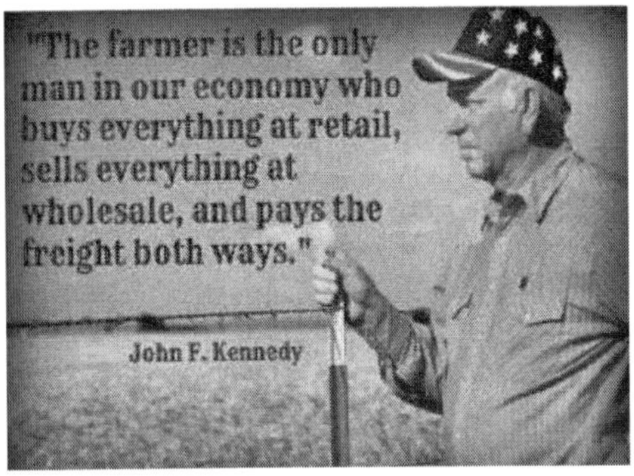

Figure 30. Quote, John F Kennedy

Figure 31. Centenary Celebration of ICAR-IARI, New Delhi in 2005 (erstwhile IARI Pusa, Bihar)

Figure 32. Centenary Celebration of ICAR-IARI New Delhi in 2005 (erstwhile IARI Pusa, Bihar)

Figure 33. Grain appearance of HD 2733

Figure 34. Grain appearance of HI 1563

Figure 35. Grain appearance of HI 1544 (Purna)

Figure 36. Grain appearance of HI 8777

Figure 37. Grain appearance of HI 8627 (Malavkirti)

Figure 38. Grain appearance of HI 8663 (Poshan)

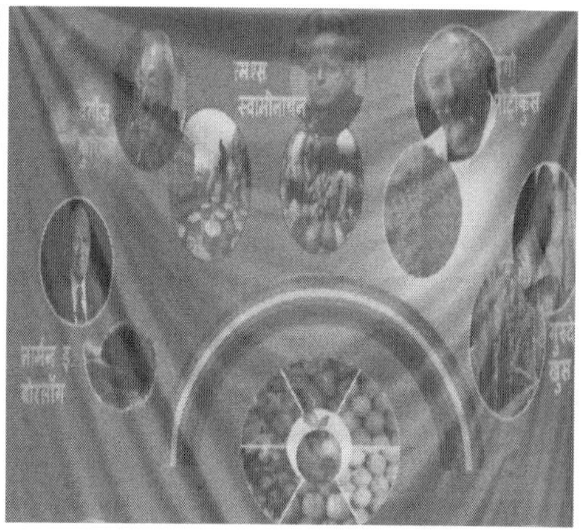

Figure 39. Rainbow revolution: Norman E Borlaug, Verguese Kurien, MS Swaminathan, Ingo Potriques, Gurudev Singh Khush

Figure 40. Glimpse of Bihar on the auspicious Republic day, Govt of India at Rajpath Delhi, 2020

Figure 41. Will the Budget Double Farmers' Income?

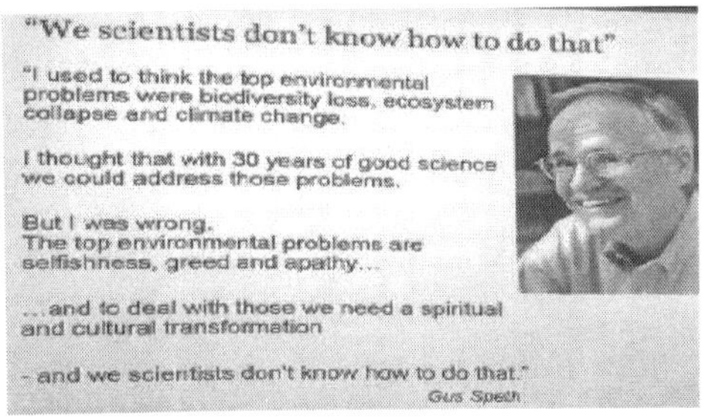

Figure 42. Quote by Gus Speth - We scientists do not know how to do that

Figure 43. Phipps Laboratory of Imperial Agricultural Research Institute, Pusa (Bihar) before 1934 Earthquake.

Figure 44. Replica of Phipps Laboratory of Imperial Agricultural Research Institute, Pusa (Bihar) at N Delhi

Figure 45a. Prime Minster, Govt of India, presenting Agriculture Minister's Krishi Karman Award for Progressive Farmers 2017-18 to Mrs Babita Dẹvi, Parari, Khagaria, Bihar

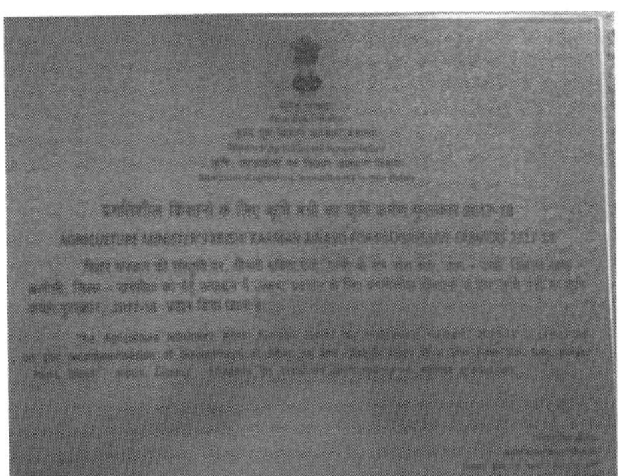

Figure 45b. Prime Minster, Govt of India, presenting Agriculture Minister's Krishi Karman Award for Progressive Farmers 2017-18 to Mrs Babita Devi, Parari, Khagaria, Bihar

Figure 46. A scene of ploughing field in North Eastern India (Loamy soil)

The Authors

Dr Dilip Kumar Verma, PhD, is a Principal Scientist (Plant Breeding and Genetics), Sr Scientist (Cytogenetics// Genetics) and Scientist (Plant Breeding), ICAR-IARI RRS Indore, Madhya Pradesh, India, with has *27+years' rich experience in Managing - Research Projects, Farm, Production and Product Development etc.* and teaching, guiding research scholars for post doctoral research programme, born on 12th December 1964 in village Rohiyama (a small helmet surrounded by rivulets/dhars), Khagaria district of Bihar, where he had came across with only two seasons, *viz.*, six-monthly *badh* (flood) and another six month *sukhad* (dry season) every year. He had completed primary schooling at his birth place, afterwards for middle schooling he has to walk on foot for 6kms encompassing six rivulets and ox-bow lakes (*dhar/chaur*) full of waist-deep water at *Middle Schhol, Beldour* and high schooling in the famine prone *Shastri Smarak High School Khara Budhama, Madhipura, Bihar*. He has faced the mighty sorrow of Kosi river of Bihar and several times his life was saved by the grace of God from the torrential rain and gutting water in the rivulets (*dhar*). He had faced food shortage from childhood, it is not exaggerated, the reality that upto Ph.D he had not taken the complete meals two times during early childhood, he had eaten *sorkhi/ghechul/ bhaint/koka/khubia etc.*

Dr Verma had got education at Intermediate Level from *TNB College, Bhagalpur, Bihar* and started graduation from Biology stream from *Sahibganj College, Sahibganj, Jharkhand* (erstwhile Bihar). In the meantime he did his graduation from *Tirhut College of Agriculture, Dholi, Bihar* and post graduation (M.Sc/Ph.D in Plant Breeidng) from *Dr Rajendra Prasad Central University, Pusa, Bihar* (erstwhile Rajendra Agricultrual University, Pusa, Bihar).

Dr Verma started his Professional Career as *Agricultural Farm Mechanics Instructor* at *Surya Narayan Singh and Deo Narayan Gurmaita, Higher Secondary School* (erstwhile Watson Zila School), *Madhubani, Bihar*. He started his scientific careers in ARS-1996 as Scientist (Plant Breeding) and served *ICAR RC NEH Region, Umiam, Meghalaya* upto August, 2005. From where he was transferred as Scientist (Sr Scale) at *Regional Maize Research and Seed Production Centre, Kushamahout, Begusarai, Bihar* [HQ; Indian Institute of Maize Research (erstwhile Directorate of Maize Research, New Delhi] from August 2005 to September 2006. Again his career was entrusted to work on wheat as Sr Scientist (Genetics/Cytogenetics) at *ICAR-IARI RS Pusa, Bihar*, w.e.f. August 2006 to October 2009. In October 2009 he was transferred to *ICAR-IARI RS Indore, Madhya Pradesh* and served upto July, 2013.

Dr Verma after serving a long span of Central India as wheat breeder, he got selected for the post of Principal Scientist (Genetics and Plant Breeding) and served *ICAR-IGFRI RRS Srinagar, J and K* on Temperate Grasses, Temperate Legumes, Grassland Breeding and Temperate Agroforestry, where there was no facility et all and lighted the lamp in the extremist prone areas of Srinagar, J and K. Sometimes to save his life he has to lock his residential accommodation from outside and follow the instruction not to light any items and even he had slept on the roof of Anand Dham, Husru, Budgam district, J and K. Dr Verma after serving five years, he was transferred to *ICAR-IARI New Delhi*, but honourable Director had given the place of duty at *ICAR-IARI RS Indore* in August, 2018. Dr Verma was recipient of several National/International Recognitions and as a team leader. He was associated and nominated with numbers of scientific socierty and committeees, who evolved for profesional and agricultural development in the country. *Indian Society of Genetics, New Delhi* felicitated him by Fellowship. Dr Verma visited *Thailand (CIMMYT-ARMP) Maize* and *Nepal (CIMMYT-BISA) Wheat*. He has published books on *Temperate Grasses and Legumes; Makhana (Black Diamond): A Miracle Crop for White Ball Revolution and Grasses, Grassland Breeding and Temperate Agroforestry*. He has written more than 12 boook chapters and more than 160 Research Papers.

Dr. Anil Kumar Singh is Principal Scientist (Agricultural Extension) at ICAR- IARI Regional Station Indore, Madhya Pradesh (HQ: ICAR-Indian Agricultural Research Institute, New Delhi, India). He obtained his B.Sc. (Ag.) and A.H. from GBPUA and T, Pantnagar, Uttarakhand (erstwhile Uttar Pradesh, India in 1988, M.Sc. Dairying (Dairy Extension) in 1990 and Ph.D. Dairying (Dairy Extension Education) in 1995 from ICAR-National Dairy Research Institute, Karnal, Haryana. He also obtained MBA (HRM) from Institute of Management Technology, Uttar Pradesh in 2013 (in-service). He started his career as Assistant Professor/Training Associate (Agricultural Extension) at GBPUA and T Pantnagar, Uttarakhand (Earstwhile Uttar Pradesh) in 1996. He joined ICAR-CRIJAF as Scientist (Agricultural Extension) in 1997 and worked for extension of Jute and Jute based cropping systems till 2001. He was transferred with a special assignment for Extension of wheat and wheat-based cropping systems in 2001 to ICAR- IARI Regional Station, Indore, Madhya Pradesh. Here he specialized in "Transfer of Technology" in Wheat based Cropping Systems in Central India and served as Scientist (Senior Scale) till 2001, Senior Scientist till 2006, Senior Scientist

(Senior Scale) till 2009 and Principal Scientist from 2012 at IARI Regional Station Indore, Madhya Pradesh.

Dr Singh served as I/C Media Cell and Principal Scientist (Agricultural Extension) at ICAR- IARI Delhi for a short period during 2015. Since 2015 he is providing his services to ICAR- IARI Regional Station, Indore, Madhya Pradesh as Principal Scientist (Agricultural Extension). He is continuing the job for almost two decades at Indore. During this period his major achievements were: Transfer, assessment, refinement and feedback of recommended technologies in wheat, organizing large scale wheat demonstrations of to induce varietal diversity and stabilize wheat productivity, introduce and popularize durum wheat to improve food; fodder and nutritional security, replace outdated wheat varieties with new varieties, develop "Wheat Seed Villages" to improve seed availability, identify desirable traits of wheat varieties as perceived by farmers and constraints in wheat productivity. He also organized more than 1200 short trainings for Capacity Building for different stakeholders, developed alternate livelihood strategies for climatically vulnerable farmers of tribal areas and introduced m-Krishi Mobile Sewa in collaboration with Tata Consultancy Services.

During this period Dr Singh published 4 books, 7 book chapters, 60 research papers, 111 popular articles, 11 technical bulletins and 21 extension literatures. He was also associated with release of various popular wheat varieties in different zones of India *viz.* HI 8627 (Malavkirti), HI 1531 (Harshita), HI 1544 (Purna), HI 8663 (Poshan), HI 1563 (Pusa Prachi), HI 8713 (Pusa Mangal), HI 8737 (Pusa Anmol), HI 8759 (Pusa Tejas), HI 1605 (Pusa Ujala), HI 1612, HI 8777, HI 1620, HI 1621, HI 1628, HI 8802, HI 8805, HI 1633 (Pusa Vani) and HI 1634 (Pusa Ahilya). During his tenure he was awarded by: ICAR Award for Outstanding Interdisciplinary Team Research in Agriculture and Allied Sciences 2009-10, Shri AK Singh Memorial "Gram Sanskriti Award 2012 for Transfer of Technology" "Fellow of Society for Scientific Development in Agriculture and Technology" SSDAT, Meerut. 2014, Outstanding Interdisciplinary Team Research for excellent team work in research achievement and special attainment in the field of agricultural science (Wheat Research) 2015 SSDAT, Meerut and "Nanaji Deshmukh ICAR Award for Outstanding Interdisciplinary Team Research in Agricultural and Allied Sciences 2015-16 for Social science" *etc.*

Dr. Daya Nand Singh is Research and Development Head at *M/s Eagle Seeds and Biotech Limited*, stationed at Indore, Madhya Pradesh. Dr Singh has *18+years' rich experience in Managing - Research Projects, Farm, Production and Product Development etc.* He obtained his B.Sc (Ag and AH) from GBPUAT, Pantanagar, Uttarakhand (erstwhile Uttar Pradesh, India in 1991, M.Sc. (Ag.) Plant Breeding in 1993 from *Dr Rajendra Prasad Central University, Pusa, Bihar* (erstwhile Rajendra Agricultrual University, Pusa, Bihar). He was awarded *Senior Research Fellowship* for Ph.D in Life Sciences from *Council of Scientific and Industrial Research, New Delhi*, at *Dr Rajendra Prasad Central University, Pusa, Bihar* (erstwhile Rajendra Agricultrual University, Pusa, Bihar). He started his

Figure 1.1. ICT based e-Plant Breeding

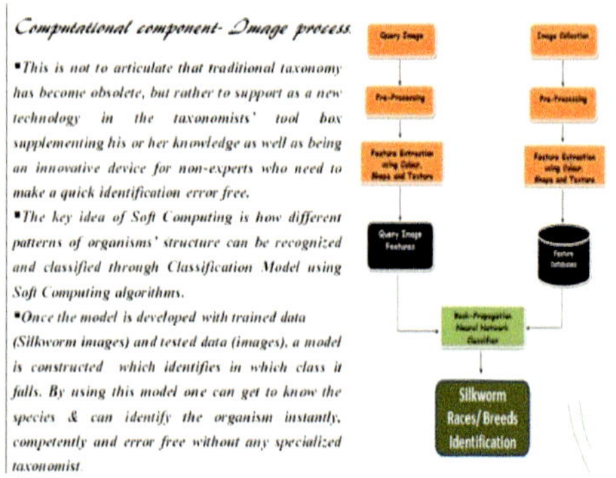

Figure 1.2. Computational component – Image process

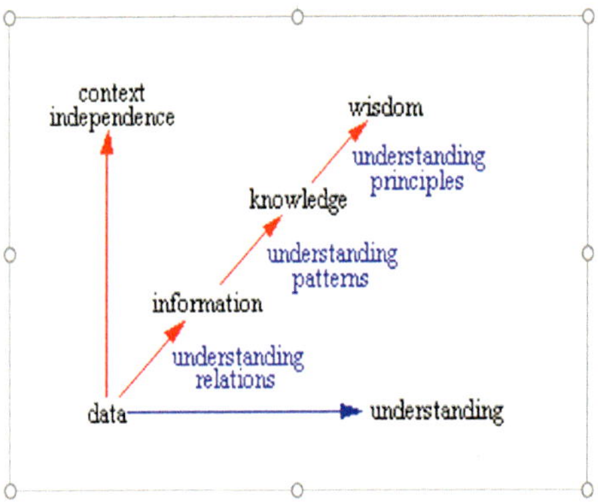

Figure 1.3. Context independence vs understanding

Figure 1.4. Intelligent without ambition is a bird without wings

Figure 1.5. Rivers never go reverse

Figure 1.6. Online participating the lecture delivered by Dr T Mohapatra, DG, ICAR, New Delhi

Figure 1.7. Dr D K Verma delivered lecture on 'Transform farmers livelihood through sustainable agriculture

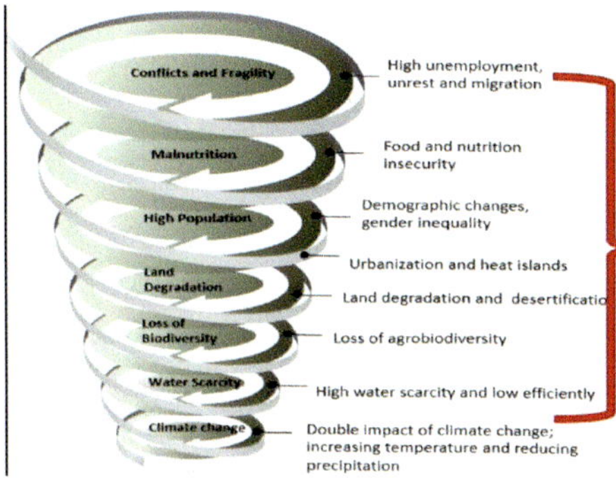

Figure 1.8. Dealing with "The Perfect Storm" Needs Business Unusual

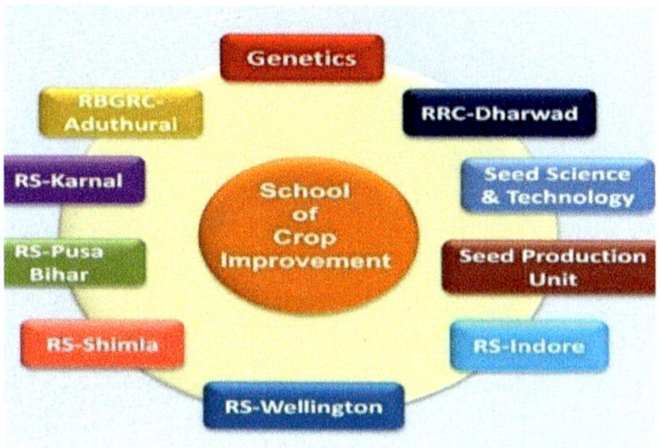

Figure 1.9. IARI and its regional stations under the auspices of School of Crop Improvement

Figure 1.10. Speed breeding

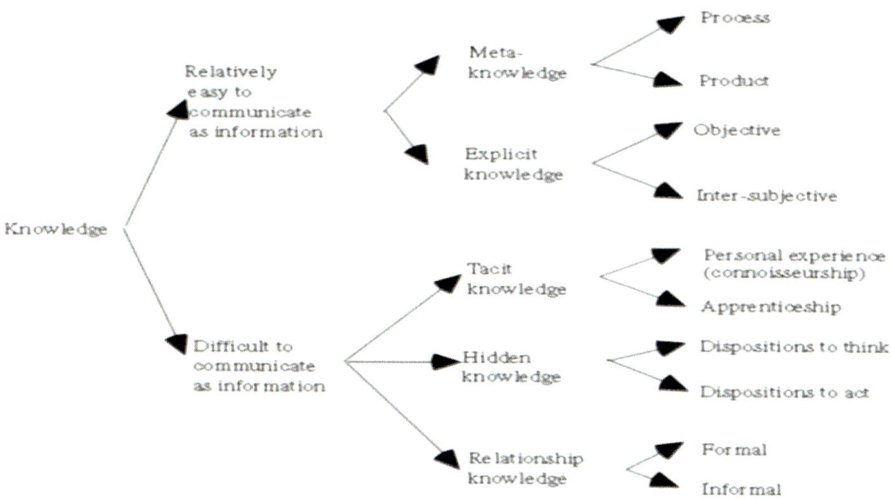

Figure 1.11. Knowledge Categorisation

Figure 1.12. Totipotency of plant cells

Figure 1.13. Dr Hans-Joachim Braun, DDG (Wheat), CIMMYT, Mexico

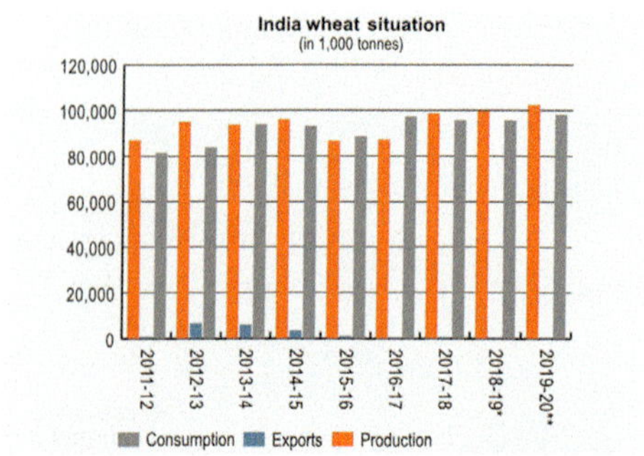

Figure 1.14. *Estimated **Projected // *Source*: US Department of Agriculture

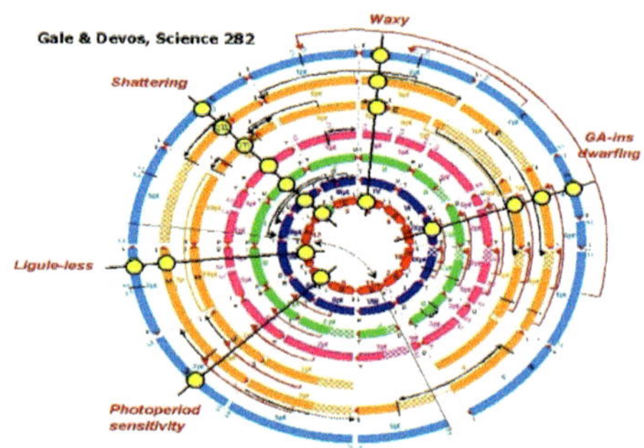

Figure 1.15. Wheat breeding

Figure 1.16. Cultivate a culture of innovation to ensure continued success

Figure 1.17. Teaching is the profession which creates all other professions

Cardinal temperatures for wheat sowing

Stage	Cardinal temperatures (in °C)	
	Minimum	Maximum
Germination	± 18	± 28
Growth	± or < 7	± 18
Development	± 16	± 30
Dough stage	± 20	± 35

Figure 1.18. Cardinal temperatures for wheat sowing

Figure 1.19. Research Methodology: The Design, Process and Application

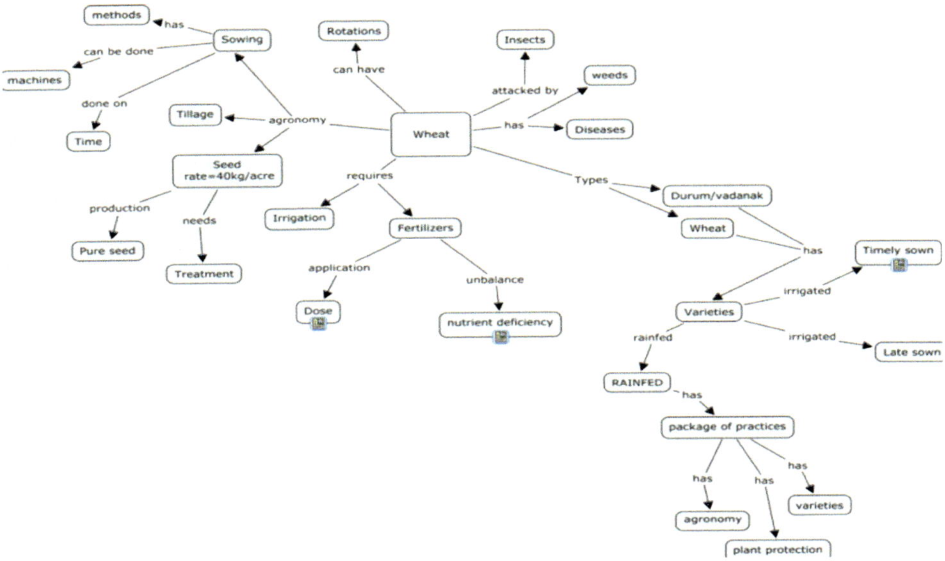

Figure 1.20. An ideal attempt at Concept map of wheat: package of practice

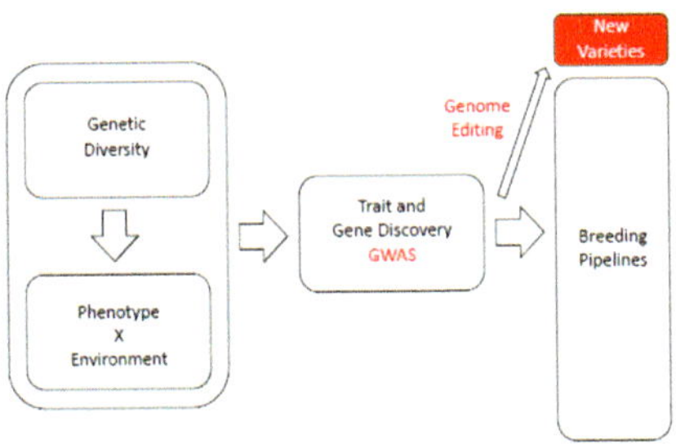

Figure 1.21. Trait and Gene Discovery is the need of the hour

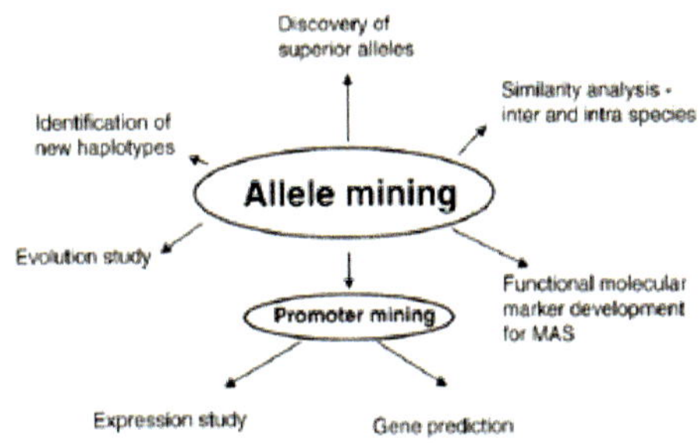

Figure 1.22. Allele mining

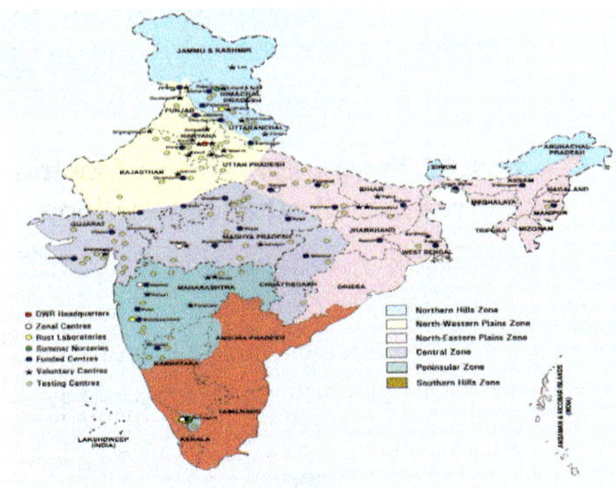

Figure 1.23. Wheat production Zone in India

Zones	States/Regions Covered	Approx Area (million ha)
Northern Hill Zone(NHZ)	Hilly areas of J&K(except Jammu, Kathua and Samba districts), Himachal Pradesh (except Una & Paonta valley),Uttarakhand(excluding Tarai region) & Sikkim	0.90
North Western Plains Zone(NWPZ)	Punjab,Haryana,Western UP(except Jhansi Div),Rajasthan (excluding Kota & Udaipur div),Delhi, Tarai region of Uttarakhand, Una & Paonta valley of HP, Jammu,Samba & Kathua districts of J&K and Chandigarh.	12.00
North Eastern Plains Zone(NEPZ)	Eastern UP(28 dist),Bihar,Jharkhand,West Bengal,Assam, Odisha and other NE states (except Sikkim)	9.12
Central Zone	MP, Gujarat, Chattisgarh, Kota & Udaipur Div of Rajasthan & Bundelkhand region of UP.	7.80
Peninsular Zone	Maharashtra, Tamil Nadu(except Nilgiris & Palani Hills),Karnataka & Andhra Pradesh	1.10
Southern Hill Zone(SHZ)	Nilgiris & Palani Hills of Tamil Nadu	0.01

Figure 1.24. Wheat growing areas of India

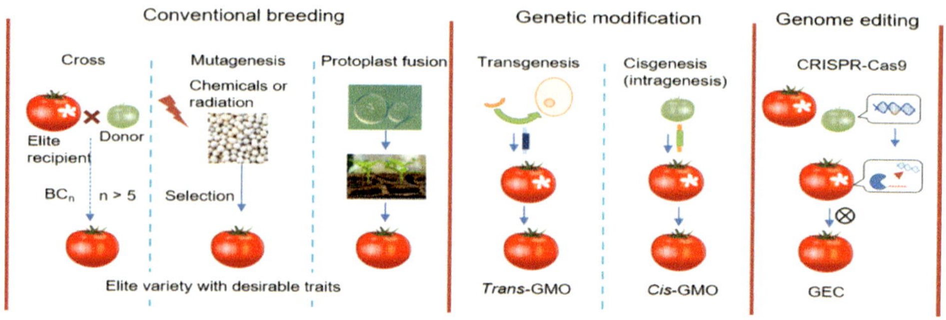

Figure 1.25. A comparison of genetic engineering techniques.
***Source*: Crop biotechnology 2020-2030.**

This bleak future was mostly avoided, largely thanks to a revolution in agritech that began in Mexico known as the Green Revolution. Using selective breeding, American biologist Norman Borlaug created a high yield strain of wheat that led to more grain per acre. Selective breeding is just one of a series of techniques for manipulating the DNA of plants that can be used to benefit agriculture. Over the past few decades, the genetic engineering tools available to scientists has expanded to include methods such as mutagenesis and transgenic breeding, the technique used to develop the crops widely known as "**genetically modified**". However, in recent years, technological advances such as next generation DNA sequencing and gene editing techniques such as CRISPR-Cas9 have vastly expanded the capabilities of genetic engineering. This has led to much excitement in the field of agricultural biotechnology, with proponents hoping that modern genetic technologies could help usher in a new Green Revolution of agricultural productivity. Plant architectural traits are important breeding targets to improve crops **yield potential** and **food security**. Wheat Knowledge Management with Nano Dreams and Nightmares is very much pertinent for a Breeder's Perspective. Knowledge can be defined as conclusion drawn from information that has been linked to other information and compared to what is already known. Knowledge means confident understanding of a subject with the ability to use it for a specific purpose if appropriate. Knowledge is basically using a set of data, in a particular context, and deriving some conclusion out of it. Agricultural Professionals, are involved in technology development, technology evaluation and technology transfer. These scientific activities are based on assessment of farmers' technology needs. Thereby linking technology sources to technology users. Every scientist, in our case breeder's, working today comes across large amount of data that he/she needs to analyse, not just that, during the process of technology development he generates large amount of data himself, all of that may not go into the development of single technology, nevertheless it may be needed at a later date. Knowledge management is an integrated systematic approach to identifying, managing and sharing all of an organisation's information asset, including databases, documents, policies, and procedures, as well as previously unarticulated expertise and experience held by individual workers. Fundamentally it is about making the collective information and experience of an organisation available to individual.

In a breeding programme large amount of lines evaluated every year for different traits thereby generating large amount of data, All the data when categorized and catalogued under different heads generates information regarding quality, resistance and agronomic traits. In the present world of high-speed internet with numerous facilities to connect to internet via broadband, Wi-Fi networks, mobile phone, landline phone, cables *etc.* the ease in internet usage and availability of internet has increased. The connectivity infrastructure in organisations-e-mail, voice mail, intranet and internet have made it far easier to create and distribute information very broadly. Increased internet usage and all-time availability of internet has triggered an explosion of information and knowledge on the user but nothing much has happened to expand our limited attention capacity, generally a user is flooded with lots of unrelated information with little use. It is further seen that very little has been done to help the receivers digest and use this sudden burst of information more effectively. To manage knowledge, different types of tools and technologies such as document management, content management *etc.* can be used. The management of knowledge is also critical at both the strategic and operational levels of an organization, though not much emphasis is being laid upon this in the public sector.

Knowledge can be broadly categorised into Explicit and Implicit or Tacit knowledge. **Explicit Knowledge** is any form of information, expertise, or experience that can be articulated in detail, codified, rendered persistent, and shared. This type of knowledge can be easily handled by today knowledge management systems. Explicit knowledge can take the form of a database, document, drawing, formula, patent, video, or presentation. **Tacit Knowledge** is much more subtle and hard to define, grab and communicate. Stephen Gourlay presents a clear definition of tacit knowledge as "a form of knowledge that is highly personal and context specific and deeply rooted in individual experiences, ideas, values and emotions". If explicit knowledge is an articulation of "what we know," then tacit knowledge is shared sense of "what we do with what we know." Tacit knowledge is experiential. We can call tacit knowledge the savoir-faire, the know-how, of employees. In most organizations, tacit knowledge is a common, unarticulated system of values, vision, purpose, and behaviours that directs the community's activities. In fact, the collective wisdom of the community is a valuable source of tacit knowledge. Nonaka and Takeuchi (1995) emphasise tacit knowledge as a main source creating new knowledge and continuous innovation. Jon-Arild (2008) further proposes categorization of knowledge on the basis of ease of communication and transfer. This model of categorization of knowledge includes five different types of knowledge's. The three additional types are Meta Knowledge, Hidden Knowledge and Relationship Knowledge. **Meta Knowledge** can be defined as "knowledge about knowledge". Meta-knowledge is the knowledgebase structuring explicit knowledge, *i.e.* know why. Meta-knowledge is also a sort of knowing how we know, appearing when reflections are made on the basis of our normative basis: meta-knowledge is both a process and a product. **Hidden Knowledge**, *i.e.* knowing how we know, is the premises, prerequisites and motives influencing our thoughts and action positions. Hidden knowledge influences the way we think and act, as a sort of personal paradigm, **Relationship Knowledge** is learned by interaction, and could be shared by systematic work in teams complementarily composed, both in the educational system and in the business world.

A successful strategy must be more than just a statement like 'To become a knowledge-enabled organisation'. The strategy must identify the key needs and issues within the organisation, and provide a framework for addressing these. An important aspect of knowledge management is the implementation of methods to share the idiosyncratic knowledge of expert practitioners within an organization. Concept Maps are used to form knowledge models by placing them in a hierarchical organization and appending elaborating media onto the nodes within each map. The entire knowledge model is linked together through a general, subsuming top-level map. Purpose of concept maps are note taking and summarizing key concepts, their relationships and hierarchy, to generate ideas (brainstorming), to design complex structures (long texts, hypermedia, large web sites), to aid learning by explicitly integrating new and old knowledge and to assess understanding or diagnose misunderstanding. As one puts ideas down on paper, the ideas become clearer and may trigger new associations leading to new ideas (***Creativity Tool).*** Writers must move from writing in linear fashion to creating hypertext documents with links to other documents. Concept mapping can be used to design the conceptual structure of hypertext (***Hypertext Design Tool).*** Concept mapping can be used to discuss concepts and the relationships between the concepts (***Communication Tool***). New knowledge should be integrated into existing structures in order to be remembered and receive meaning. Concept mapping stimulates learning process by drawing attention to the relationship between concepts (***Learning Tool).*** Concept maps drawn by students express their conceptions, as well as their misconceptions (***Assessment Tool).***

Concept Maps

Graphic displays of "concept spaces" visually indicate the relationships and important subsets of concepts, particularly subsets that constitute ontological commitments for representing given phenomena. In particular, such displays are intended to provide students with large-scale (and even global) views of the structure of concept spaces. The *conceptualization* operation allows users (instructors or students) to associate a given concept occurring in a presentation (such as *Stream Velocity*) with other concepts occurring in the presentation (such as *Depth, Slope, Roughness*), providing a network structure among the concepts. One of major uses can be in documenting research documents and data. The term nanotechnology was first used in 1974 to describe manufacturing that is precise down to the nanometre level. Today, it is **'the techno *buzzword de jour*'** (Mekel, 2006) and has become synonymous with futuristic breakthroughs. Indeed, in the late 20th century, the 'quest to master the nanoscale' was said to resemble a global race; however, it is far from a settled, uncontested area. Whilst scientific research at this scale is nothing new (UNESCO, 2006), nanotechnology can mostly simply be considered a *collective* term for the myriad research, engineering and technology development activities focussed on the application of a diverse, expanding set of techniques for manipulating physical and biological materials. It is often claimed that nanotechnology will be one the most significant technologies of the 21st century and that it will enable 'the next industrial revolution' (The White House, 2000). As Anthropologist Chris Toumey (2004) has noted, 'landscapes of nano-hyperbole' currently abound. While new technologies always stimulate visions of the future, "nano futures" are polarised between utopian dreams and apocalyptic nightmares. As Mekel (2006) puts it: the 'maelstrom of radical rhetoric swirling about nanotechnology includes two types of claims – those of "nanosalvation" (assertions based on the belief that nanotechnology is a technological magic bullet) and their flipside, those of "nanodamnation"', which focus on technological risk and the potential for unanticipated consequences. This provides an introduction to nanotechnology and the shaping of emerging technoscience, describing key narratives and the contestation of "nanofutures". With emerging technoscience becoming a central focus of futures thinking the paper concludes with some observations for practitioners and researchers in the futures field.

Technological Change, Expectations and Time Orientation of Wheat Breeding

The Science and Technology Studies field focusses our attention on the highly complex interactions between science and technology and society. Three key perspectives provide a broad theoretical framework: science and technology shape society (*i.e.* "technological determinism"), society shapes science and technology (*i.e.* "social determinism"), and the interactionist (two-way) perspective. Although technological determinism is still highly prevalent in discussions about technology, and 'often easiest to capture and analyse pronouncements made about emerging technology', overall, there has been a shift away from linear notions of technological progress and the "science-push" model. Newer, more complex pictures highlight the roles of a wide-range of factors including expectations, social choices, and path-dependency, taking a 'co-evolutionary' perspective. That said, Winner (2001) has also noted the 'glaring disconnect' between the academic critique of technological determinism and 'the visions of run-away technology that prevail in society at large'.

Table 1.1. Perspectives on Science, Technology and Society

	Science and Technology Shape Society	*Science Shapes Science and Technology*	*Interactive Approaches (co-evolutionary)*
Time	1950s-60s	1970s-80s	1990s-
Definition of technology	Cause	Consequence	Cause and Consequence
Independent variable	Technology	Society	Social group
Relation of actor to technology	Beneficiaries (or victims)	Negotiate interest	Seamless web
Role of policy	Protect or reject science and technology	Empower actors, create networks	Democratise
Power structure	Technological regime	Negotiation	Frames, discourses
Methodological approach	Study impact of technology	Follow the artefact	Follow the actors

Empirical studies demonstrate that expectations can play decisive roles in establishment of new technological fields – such as nanotechnology – at three levels: the *macro* (*e.g.* creation of government policy), *meso* (*e.g.* in innovation networks and industry sectors), and *micro* (*e.g.* in research groups). The final theoretical area considered is analysis of the future-orientation of modernity. Giddens contends that a central factor separating modern society from earlier forms is future-orientation. It sees the future 'as a territory to be conquered or colonized. Further, it is the most preoccupied with the future because, unlike traditional societies, it has no sense of control over the future. Similarly, Carvounas (2002) argues that modernity's unique temporality is fundamental to understanding it. He asserts that 'with the birth of modernity the past became undermined and the future valorised to such a degree that temporality became dominated by an open future'. Consequently, argues Carvounas, the problem of "temporal coordination" emerged creating the need for "temporal continuity" through 'new narratives connecting the modes of time'. The second aspect is how the future consequently must be managed. For Giddens (1999) this future-orientation creates the issue of *risk* and further asserts that 'our very attempts to control the future tend to rebound upon us, forcing us to look for different ways of relating to uncertainty'. More recently Adam and Grove (2007) have critically analysed Western society's relationship to the future, arguing that the West suffers from "structural irresponsibility", producing long-term futures through innovation and technology development and without the capacity for adequate forethought.

Emergence through Invention and Futurism

Nanotechnology's emergence stems from two contrasting developments. First, advances made in scientific instruments for scanning and probing at the "nanoscale" (the scale of nanometers, or one billionth of a metre), which later became commercially available in the 1980s, led to new efforts and capacities to understand and exploit it. Indeed, many of threads that came to be known as "nanotechnology" long pre-date the use of the term. In 1989 Donald Eigler famously arranged Xenom atoms to spell "IBM" and this achievement was given the cover of *Nature*. The IBM scientists involved claimed 'many new avenues of investigation are open to us. it should be possible to assemble or modify certain molecules in this way one can build novel structures that would otherwise be unobtainable'. Such moments give Schmidt's (2007b) definition of nanotechnology as 'domesticating atoms and harnessing them to serve our needs' meaning. Second, futuristic writings and pronouncements massively led tangible advancement. Early nanotechnology literature, is full of futuristic claims about its likely future development and consequences. Eric Drexler's seminal *Engines of Creation: the Coming Era of Nanotechnology*, published in 1986, outlined a controversial "molecular manufacturing" vision. This well-read book outlined a speculative, distant future technological vision of theoretic possibility which Drexler interchangeably termed "molecular technology", "molecular engineering" and "nanotechnology". Drexler forecast the engineering of tiny "nanomachines" and "nanosystems" that 'will allow the fabrication of complex objects to atomic precision'. This version of "nanotechnology" has ever-since been associated with futuristic concepts such as tiny medical nanobots in human arteries, or "nano-assemblers" creating materials from the "bottom-up" (*i.e.* atom-by-atom). Although this vision is usually discounted as ill-conceived and unlikely to ever be fully realised by scientists proponents of other nano-visions are 'not shy about saying that current research will inevitably generate a brave new world'. When former US President Bill Clinton introduced the National Nanotechnology Initiative (NNI) in 2000, he drew on the futuristic imagery that Nobel Laureate physicist Richard Feynman outlined in his now famous speeches 'There's plenty of room at the bottom' and 'Infinitesimal machinery'. It states that 'some scientists think they can combine carbon nanotubes with plastics to make composites that are far lighter, yet stronger than steel' and confidently asserts that 'properly structured gold nanoparticles. start absorbing light and can turn that light into heat, enough heat, in fact, to act like miniature thermal scalpels that can kill unwanted cells in the body, such as cancer cells'. Today, nanotechnology is considered to deserve special attention because *converging* and *emergent* technologies at the nanoscale are expected to have enormous consequences.

Nanotechnology's Complex

Nanotechnology is today generating a lot of attention, all around the world, and building great expectations in the academic community, and amongst governments, investors and industry. Lakhtakia (2004) argues that nanotechnology is increasingly being seen as 'the solution of any problem afflicting humanity' in the minds of many researchers as well as those who control research funds and is, thus, shaping up as a "**mega-ideology**". Much of the enthusiasm arguably stems from the lab observations of the properties of materials at the nanoscale, which is then extrapolated into potential applications. The only feature common to the diverse activities characterised as "nanotechnology" is the tiny dimensions at which they operate'. The broadest definition of nanotechnology is research conducted at the nanoscale'. To further explore nano's complex present it is necessary to first further identify nanotechnology's specificities. In this emergent area science, technology and society are tightly interconnected. 'Familiar distinctions between "applied" and "basic" research are troublesome' and there are dozens of definitions of what nanotechnology is or could be. Definitions typically focus on *control* (*i.e.* the ability to precisely manipulate matter at the nanoscale) and new tools and approaches which make it possible to *exploit objects at the nanoscale* such as the fabrication techniques used to make nanomaterials and/or nanostructures such as carbon-based nanotubes. Some define nanotechnology according to how it may be applied, *e.g.* as a *'platform technology'* (Walsh, 2007) or *'enabling set of technologies'* (PMSEIC, 2005). Loveridge, *et al.*, 2008) suggest three conditions distinguish the production of "nano-artefacts": *Dimensional scale* (focusing at the nano-range of 1-100 nanometres [nm]); Properties and behaviours of particles that come into effect when molecules attain a critical size, *i.e.* molecular disaggregation; and *System integration* of nano-artefacts (*e.g.* nanoparticles, fabricated nanomaterials and nano-engineered structures, *etc.*) to make new nanotechnologies and products. As part of this new production the enormous, and increasing, aspiration for ambitious, commercially-deployed application is an important development. This, and a mainstreaming view of "converging technologies" at the nanoscale, has led to a rapidly growing belief that nanotechnologies may be the beginning of a "new world" within a notional time horizon of 2030 (Loveridge *et al.*, 2008). Further consideration is informed by three main "types" of nanotechnology identified.

Table 1.2. Nanotechnology and Plant Breeding

Type	*Description*	*Examples*
Incremental nanotechnology	Scientific and technological developments that, essentially are a continuation of the research directions of the past 50 years	Includes much colloid science and materials that have superior/new properties as a result of their controlled nanoscale structure. Also includes bulk production of nanoparticles and their inclusion in consumer products
Evolutionary nanotechnology	The scaling down of existing technologies towards the nanoscale.	Includes development in information technology (*e.g.* new semiconductors, memory devices) and molecular delivery (drug delivery).
Radical nanotechnology	The creation of fully functional nanoscale machines and the engineering of systems (not simply materials and devices).	Embraces the proposals of Drexler and his followers for molecular manufacturing (*e.g.* new bottom up approaches) to fabricating materials and other approaches that may lead to this outcome (*e.g.* bionanotechnology).

Overall, there are two main sides of the nano-debate – the skeptics and the believers, which inform the two long-term visions of "**nano-optimism**" and "**nano-pessimism**". Similarly, Mody (2004) commented that 'for many practitioners, nano is still a bit of a put-on, a bandwagon whose content they do not quite understand but which they are trying to make the best of. sticking to relatively uncontroversial play'. For some Science and Technology Studies scholars nanotechnology is above-all a "political project" to raise finance and channel this to particular privileged projects. An American nanoscientist explains: 'the scientists and engineers are desperate to maintain funding levels, maintain government support.the stuff that people are doing today they have been doing for years'. Fuller (2009) bluntly describes nanotechnology as a "rebranding exercise for chemistry". Rosalyn Berne author of *Nanotalk: Conversations with Scientists and Engineers about Ethics, Meaning, and Belief in the Development of Nanotechnology,* believes it represents a new, still-forming, partnership between science and engineering. These goals are extremely far-reaching and, in their most radical forms, exhibit a large dose of techno-utopianism. It is to these "nano-imaginaries" now in turn.

From Experiments to Imaginaries

The future of nanotechnology has become overburdened and has overshadowed the present'. The intensity of "nano-hype" that has been quickly building over the past decade is well captured by *The Power of Small*. The initiative's supporting website asserts that 'nanotechnology will change your life' and, among the various forecasts of nanotechnology development, confidently predicts that:

Nanotechnology is certain to provide solutions for some of our worst environmental problems and lead the way to a more sustainable future. Nanotechnology will allow convenient sensor surveillance and tracking technologies anywhere in response to our desire for security. The solutions cover the spectrum.

new solar power technologies, lighter and stronger construction materials for cars and other vehicles for improved fuel efficiencies, and processes to inactivate and/ or remove a wide range of toxic substances from water and soil; all may be possible.

Similarly, *Nanologues*, featured scenarios looking out to 2015 where nanotechnology enables all sorts of amazing things such as: a 'nose filter. capturing impurities using nanofibre mesh'; computers '100 times as powerful as they were in 2005'; and 'unique packaging systems using nanosensors to change the colour of the packaging when the food inside is no longer edible, and alerts a networked monitoring system'. Nanotechnology even enables use of 'spray-on photovoltaic' solar power generation in one scenario. Likewise, in *Nanotalk* scientists discuss their ambitions: more efficient solar power via 'artificial photosynthesis'; the manufacturing of generic elements via 'nanoscale material manipulation', rather than via traditional mining and processing systems; and one nanoscientist would like have the encyclopaedia, or maybe all the books ever written, stored in a certain cubic centimetre of his brain. This desire for technical mastery is well expressed by one nanoscientist. Two goals are central to nano-imaginaries: solving the problems of the *industrial* age, such as climate change, and enhancing capacities gained during the *information* age. Indeed, Roco (2006), a major architect of the NNI, has asserted that 'nanotechnology promises to extend the limits of sustainable development.' The emerging technologies in particular nanotechnology, is argued to involve a related shift from a *physical* worldview to a *chemical* worldview: that is, a shift from seeking to 'discover the ultimate nature of things' to seeking to 'construct the most efficient means to our ends' – a focus on the "functional" and on matter in terms of instrumentality in bringing about humanly relevant ends. Further, utopianism permeates much of the nanotechnology literature. It can be seen as part of a shift from **'Liberation Theology'** to **'Liberation Technology'**, in which governments seek technological, rather than political, solutions to social problems. It can also be likened to the utopianism that accompanied computerisation. Just as Dunlop and Kling (1991) outlined a 'seductive equation' of computerization, 'Technological Progress = Social Progress', "nano-ization" suggests that humanity can solve its major challenges through accelerating technological advancement. Clearly there are various layers to the promised nano-future which need to be interpreted. A large amount of potential has come to reside in nanotechnology, with a wide range of governmental and non-governmental of actors creating hope for specific nanotechnologies and the emergence of nanotechnology as a "disruptive event".

Temporality and Narratives in Nanotechnology

Grounding discussions about nanotechnology in the present and in reality is a daunting task. Aside from the field's obviously intense future-orientation, this emerging area is characterised by a broader "non-presentism", integrating narratives, and multiple temporalities in actor networks. 'Nanoists often project a synthesis [of scientific and technological endeavour] far back into the past and the future'. The second key aspect of this appears to be a focus on the past, as 'the science of the nanoscale is acquiring its own histories' by reinterpreting past scientific and engineering developments. For example, the discovery in 1991 of a molecular shape known as the "carbon nanotube" is today widely viewed as a pivotal moment in nanotechnology's history. At the time, however, 'this work was not called nanotechnology, but simply chemistry'. As nanotechnology is currently comprised of a loose "community of communities", with each group having its own history, nanotechnologists can pick and choose to create histories and new narratives. Each framing it differently, is integrative, and aligns well with the different "types" of nanotechnology identified by Wood *et al.* (2008). They are: *'Understanding and utilising the properties of matter at the nanoscale'* (*i.e.* an *incremental* nanotechnology narrative); *'An extension of the miniaturisation imperative'* (*i.e.* an *evolutionary* narrative); *'The "clean and green" linchpin for solving societal problems'* (combining *incremental* and *evolutionary* nanotechnology narratives): *'Control over the structure of matter and transformative human mastery, through "technologies converging at the nano-scale"'* (*i.e.* a *radical* nanotechnology narrative).

Shaping the Emerging Technoscience

Social, cultural, moral, political and economic visions of promise, threat and governance have shaped and are shaping – in uneven and complicated ways – the research trajectories that will determine the eventual form of nanotechnologies. In this way, nanotechnology exemplifies the *political economy* of expectations (see Brown, 2003). "Hyper-expectations", fed by nano-hype, are fuelling counter-concerns. As *Nature* consulting editor Philip Ball has described, in 2003 nanoscientists started to be viewed as mad scientists, the 'new Victor Frankensteins, the modern Prometheuses, the contemporary Fausts, dabbling with dangerous forces they cannot control'. If concerns continue to grow, nano-futures could quickly become more uncertain and rhetoric far less deterministic. Consequently, government policy could become less supportive, investment dry up, additional precautionary risk assessments required, restrictive regulations then introduced, and so on and so forth. Public acceptability of products incorporating nanotechnologies could decline.15 Similar co-evolutionary interactions between science and society have recently been evident in biotechnology, another area of emerging science fuelled by implicit utopian imaginaries. Indeed, nanotechnology's novelty, complexity and publicity makes it fundamentally uncertain and "post-normal science". Certainly, a growing "meme" in Science and Technology Studies is *reflexiveness*, which is 'a process by which the broadened community of participants concerned about the direction and impacts of scientific advance and technological innovation gain a fuller understanding of the social context in which they operate'. This quality would be nurtured by greater futures-responsiveness.

This began with an outline of how two radically opposing long-term visions of a 'nano-enabled' future, "nano-optimism" and "nano-pessimism", appear set for heightened profile and conflict. In explicating and describing this, the paper also outlined the promises, narratives and temporalities mobilised in the recent intensification of expectations about nanotechnology and the political economy of "nano-futures". The discussion around nanotechnology was shown to epitomize contemporary processes 'making the future present' through nano-hype and hopes, new engagement and governance practices, and other emerging anticipatory strategies. Additionally, futures researchers and practitioners can take note of four key findings: Expectations have been found to play important performative roles in technological change. This is new evidence of the influence of images of the future and how they condition the present (Slaughter and Bussey, 2006); In considering the future trajectory and impact of emerging technologies (nano, bio and other technologies) it is necessary to critically consider increasing hype and related actor strategies and theories – *e.g.* the articulation of "promises" and the 'promise-requirement' cycles that emerge. Deterministic rhetoric in nanotechnology – articulated by both nano-advocates and nano-critics – is a stark contrast to its emergent status, fuzzy boundaries and unsettled make-up. It should also be noted that the label "emerging technologies" is 'effectively a hidden form of technological determinism'; These can inform futures work such as technology foresight, future scenario creation, *etc.*; and A useful analytical distinction can be made between 'looking into' and 'looking at' the future – between attempts to forecast futures and considering attempts made to define the future and create direction or other phenomena of future-orientation. The later grasps the *politics* of the future, analyses contestation. Ultimately, what we should primarily see in the emergence of "nanotechnology" is an important and fascinating attempt to define the future. As such, engagement with it must focus on this and reimagining of 'what is at stake – politically, culturally and socially – in the incredibleness of nanotechnology' (Kearnes, and Macnaughten, 2005) to enable greater reflexiveness.

Speed Breeding

University of Queensland's Dr Lee Hickey is working with UK researchers to speed breed a variety of important food crops. *With an ever-increasing human population, predicted to be 10 billion people by 2050, studies estimate that we need to double the rate of genetic gain in our crop improvement programs globally to meet this demand. Dr Lee Hickey is a plant breeder and crop geneticist. He and his team are seeking to develop plant breeding tools that could help to enable this genetic gain.* When one considers the other challenges of increased climatic fluctuations, more drought, more heat, and evolving pathogens; we begin to understand the challenges to food security we are facing globally. Without a doubt, improving the genetics of crop varieties needs to be done in parallel with improving farming practices to deliver more efficient food production systems. In recent years, we've had a massive boom in terms of genomics. The advances in the technology for genotyping have been huge. We're now able to genotype plants, at will, and at very low costs; and get sequence information or high-density DNA markers across the genome at very low costs. This has revolutionized plant breeding, for instance enabling plant breeders to track genes and even develop predictive breeding approaches such as genomic selection. All of this has transformed the science of plant breeding and has huge potential for crop improvement. However, one of the key limitations in the development of an improved crop variety is actually how long it takes to grow a plant generation. At the moment, using traditional approaches, it will take more than a decade from making the critical cross to releasing the improved variety. We need to develop more robust and productive crops faster than traditional approaches allow. The challenges we are facing call for an integration of technologies to enable us to develop crops faster than ever before. Speed breeding is all about growing plants quickly, efficiently and as cheaply as possible. Through speed breeding technology, we are trying to revolutionize how we grow plants to keep pace with the advances in genomic technology in recent years.

While speed breeding technology was originally developed and applied to wheat many years ago, more recently, we have adapted the approach to a range of important crops. We have published protocols that enable rapid cycling of other long-day and day-neutral crops like barley, canola, and chickpea. It has been a pleasure to collaborate with the John Innes Centre to assemble the data, information, and details needed to speed breed whatever crop researchers or breeders would like to grow around the world and make protocols for speed breeding available globally. There has been a limitation to speed breeding short-day crops, because of their requirements for flowering. However, most recently, we have been working with International Crops Research Institute for the Semi-Arid Tropics (ICRISAT), as part of a Bill and Melinda Gates Foundation project, to develop protocols suitable for short-day crops like sorghum, millet and pigeon pea. I believe that we can speed-breed any species. It is just a matter of optimizing the protocol and the conditions that are required for inducing early flowering and achieving rapid generation advance. At the University of Queensland, we're developing a system called Express Edit, which aims to take genome editing, out of the lab, to avoid bottlenecks and improve scalability. We want to make a very flexible tool that can be used on any genotype and avoid the whole tissue culture process. Our idea is to do this directly within the speed breeding process. If we can achieve this, it will be a flexible breeding tool that will transform plant breeding. It would mean that we can achieve up to six generations a year with genome editing incorporated into the whole breeding process.

Combining Genomic Selection with Speed Breeding

Genomic selection is the perfect match for speed breeding as genomic selection serves to reduce the breeding timeframe by removing the need to actually do the field testing to identify parents that can be used for the next cycle of breeding. This is done through algorithms that use sequence information, or DNA markers, to predict the performance of those plants. If you can accurately predict performance of a plant for any trait, like yield or quality, or disease, you don't need to go to the field for evaluations. This means that in a sense, speed breeding can become "detached" from the field testing operations and the predictive breeding cycles can go very, very fast. Of course, there's still a need to develop lines that are genetically stable through inbreeding and these must go into field trials as normal, but your actual engine room for genetic gain can go much, much faster, and this is a really exciting space to be experimenting with. We are very fortunate to be working with a breeding company in Australia which means we have access to massive datasets – we're talking training datasets comprising 700-yield trials and more than 50,000 breeding lines, fully genotyped. We're developing a totally new methodology where we combine speed breeding and genomic selection and we're calling it 'speed GS'. However, we're not just doing the standard genomic selection. We're taking it one step further. Collaborators at The University of Queensland, including Dr Kai Voss-Fels and one of the co-inventors of genomic selection, Professor Ben Hayes, are injecting some very innovative concepts into this project. We're using a branch of artificial intelligence to not only predict the best performance based on genotyping, but also predicting which are the best parents to use for crossing. In India, close to twenty-three million tonnes of leftover rice residues are annually burnt by farmers to get rid of the straw and prepare their fields for sowing wheat, which is the next crop, resulting in air pollution. Also, dry environments pose a challenge for the germination of wheat varieties with short coleoptile.

To overcome these problems, Scientists at Pune based Agharkar Research Institute (ARI), have mapped two alternative dwarfing genes *Rht14* and *Rht18* in wheat. These genes are associated with better seedling vigour and longer coleoptiles (sheath protecting the young shoot tip). ARI have mapped the dwarfing genes on chromosome 6A in durum wheat, and DNA-based markers were developed for a better selection of these genes in wheat breeding lines. The DNA-based markers will help wheat breeders to precisely select wheat lines carrying these alternative dwarfing genes from a massive pool of wheat breeding lines. These DNA based markers are being used at ARI for marker-assisted transfer of these genes in Indian wheat varieties, so as to make them suitable for sowing under rice stubble-retained conditions and dry environments. Wheat breeding lines with these alternative dwarfing genes are presently at an advanced stage. Wheat lines with these alternative dwarfing genes, apart from reducing crop residue burning, can allow deeper sowing of wheat seeds to avail advantage of residual moisture in the soil under dry environments. The presently available semi-dwarf wheat varieties, which were explored during the Green Revolution, carry conventional *Rht1* dwarfing alleles and produce optimum yields under high-fertility irrigated conditions. However, they are not well adapted for deeper sowing conditions in dry environments due to shorter coleoptiles, and low early vigor often results into reduced seedling emergence. Moreover, crop stands of *Rht1* wheat also remain poor where previous crop residues pose a barrier for seedling emergence due to the short coleoptiles. Burning of leftover rice crop residue has serious implications for the environment, soil, and human health. Therefore, there is a need to include alternative dwarfing genes in wheat improvement programs. Also, only two dwarfing alleles of *Rht1* are predominant in Indian wheat varieties; therefore, there is a need to diversify the genetic base of dwarfing genes considering diverse wheat growing zones in India. In genetic studies conducted at ARI, dwarfing genes *Rht14* and *Rht18* in wheat conferred a plant height reduction comparable to the *Rht1* alleles while retaining early vigour in wheat seedlings, but do not affect coleoptile length and seedling shoot length. These can, therefore, be utilized as an alternative dwarfing gene to *Rht1* for deep sowing conditions or in fields with retained stubble. These lines will also allow deeper sowing of wheat seeds to avail advantage of residual moisture in the soil, therefore, saving valuable water resources and reduce the cost of cultivation to farmers.

Wheat is a major source of energy, protein and dietary fiber in human nutrition since decades. Wheat is a major crop contributing importantly to the nutrient supply of the global population. Since the beginning of agriculture, ten thousand years ago, the importance of varietals improvement is well known. In the ancient time, selection and introduction were commonly used method, since knowledge about use of hybridization, mutation and polyploidy were not in practice. Selections were made on the basis of physical appearance and to fulfill the requirements. Till 15th century, most of the varieties were developed either through primary selection or introduction. Breeders had the advantages of variability until 1970s and due to intensive crossing programmes; green revolution took place during 1967-1968. The development and promotion of modern, high yielding varieties was the most important factor contributing to the enormous success of green revolution. During two generations leading up to the turn of the century the global population grew by 90 percent whilst food production expanded by 115 percent. The global food security is quite fragile, particularly when looking towards the middle of the century because of projected needs for human, animal and industrial uses. Global wheat production is expected to increase from nearly 600 million tons of present production level to around 760 million tons in 2020 with limited expansion of sown area. But now, due to non-existence of variability for the yield trait, it is not possible to develop a new variety either by selection or introduction. Selection based on knowing the location of the genes of interest gives the breeder a significant advantage, particularly for quantitative traits, where classical selection is done on the phenotype as a whole rather than on the underlying genetic determinants. Identification of significant QTL marker associations forms the baseline for MAS of quantitative traits. Furthermore, other breeding methods like hybridization, which reshuffle existing variability in the population and tools like polyploidy and mutation bring challenges for allele in the available traits and are not possible to introduce a new trait from unrelated species. Molecular markers have been introduced over last two decades, which has revolutionized the entire scenario of biological sciences. DNA based molecular markers have acted as versatile tools and have found there position in various fields like taxonomy, physiology, embryology, genetic engineering *etc.* PCR brought about a new class of DNA profiling markers that facilitated the development of marker-based gene tags, map-based cloning of agronomically important genes, variability studies, phylogenetic analysis, synteny mapping, marker assisted selection of various genotypes. Molecular markers are identifiable DNA sequences found at specific locations on the chromosomes and transmitted by the standard laws of inheritance from one generation to next and considered as landmarks in the chromosome maps that can be useful to monitor the transfer of specific chromosome segments known to carry useful agronomic traits. Molecular markers have also provided an excellent opportunity to develop saturated genetic maps and to integrate genetic, cytological and molecular maps. Molecular markers are being used to tag specific chromosome segments bearing the desired gene(s) to be transferred into the breeding lines.

Conventionally, plant breeding depends upon morphological/phenotypic markers for the identification of agronomic traits. With the development of methodologies for the analysis of plant gene structure and function, molecular markers have been utilized for identification of traits to locate the gene(s) for a trait of interest on a plant chromosome and are widely used to study the organization of plant genomes and for the construction of genetic linkage maps. Molecular markers are independent from environmental variables and can be scored at any stage in the life cycle of a plant. Over the last several years, there has thus been marked increase in the application of molecular markers in the breeding programmes of various crop plants. Molecular markers not only facilitate the development of new varieties by reducing the time required for the detection of specific traits in progeny plants, but also fasten the identification of desired genes and their corresponding molecular markers, thus accelerating efficient breeding of resistance traits into wheat cultivars by marker assisted selection (MAS). Breeders used molecular markers to increase the precision of selection for best trial combinations. Variety developed by MAS are not considered genetically modified organisms (GMOs) and accepted by local and international market. Molecular marker aided selection methods have resulted in significant improvement in breeding efficiency by reducing trial and error aspect of breeding process and by allowing for time and cost savings. Molecular marker systems will benefit from the constant increase in the integration of biotechnological production of segregating populations such as homozygous double haploids in wheat breeding cycles since, the major requirement of being co-dominant for molecular markers will disappear. Marker assisted selection (MAS) offers an opportunity to select desirable lines based on genotype rather than phenotype. Marker assisted selection (MAS) is based on the identification and use of markers, which are linked to the gene(s) controlling the trait of interest. By virtue of linkage, selection may be applied to the marker itself. The advantage consists in the opportunity of speeding up the application of the selection procedure. For instance, a character which is expressed only at the mature plant stage may be selected at the plantlet stage, if selection is applied to a molecular marker. Selection may be applied simultaneously to more than one character. Selection for a resistance gene may be carried out without needing to expose the plant to the pest, pathogen or deleterious agent. If linkage exists between a molecular marker and a quantitative trait locus (QTL), selection may become more efficient and rapid. The construction of detailed molecular and genetic maps of the genome of the species of interest is a prerequisite for most forms of MAS. However, the current cost of the application of these techniques is significant, and the choice of one cost technique rather than others may be dictated by factors. There are few examples of crop varieties in farmer's fields, which have developed through MAS.

RAPDs emerged as a convenient and effective technique for tracing alien chromosome segments in translocation lines (Williams *et al.*, 1990). RAPD markers provide a useful alternative to RFLP analysis for screening markers linked to a single trait within near isogenic lines and bulked segregants. He *et al.* (1992) reported the development of a DNA polymorphism detection method by combining RAPD with DGGE (denaturing gradient gel elecrophoresis) for pedigree analysis and fingerprinting of wheat cultivars. RAPD markers can be converted to more user-friendly Sequence Characterized Amplified Region (SCAR) markers, which display a less complex banding pattern. SCAR markers linked to resistance genes against fungal pathogens have been characterized in combination with RAPD and RFLP (Liu *et al.*, 1999). In recent years, RAPD and other PCR based markers like Sequence Characterized Amplified Regions (SCAR), Sequence Tagged Sites (STS) and Differential Display Reverse Transcriptase PCR (DDRT-PCR) are increasingly being used for identification of desirable traits in wheat and related genera. These markers have been used in particular for disease resistance against viral and fungal pathogens and also for insect and nematode pests and have the potential of pyramiding of resistance genes for effective breeding programs. PCR based markers have been extensively characterized for genes of resistance against common bunt, *Tilletia tritici* (Demeke *et al.*, 1996), powdery mildew, *Erysiphe graminis* (Qi *et al.*, 1996), leaf rust, *Puccinia recondita* (Seyfarth *et al.*, 1999), resistance against Hessian fly, *Mayetiola destructor* (Dweikat *et al.*, 1994) and Russian wheat aphid, *Diuraphis noxia* (Myburg *et al.*, 1998; Venter and Botha, 2000). SSRs or Microsatellites are more promising molecular markers for the identification and differentiation of genotypes within a species. The high level of polymorphism and easy handling has made microsatellites extremely useful for different applications in wheat breeding Roy *et al.*, 1999.). Microsatellites have also been used to identify resistance genes like *Pm6* from *Triticum timopheevii* (Tao *et al.*, 1999) and *Yr15* from breadwheat (Chague *et al.*, 1999).

Foldscope-Economically Feasible Research Tool

The Foldscope is an origami-based optical microscope that can be assembled from a flat sheet of paper in less than 10 minutes. Manu Prakash, an Associate Professor of Bioengineering at Stanford University and his team developed these flat microscopes by combining principles of optical design with origami (Cybulski *et al.*, 2014). The Foldscope performs most of the functions of a high-school lab microscope, but its parts cost less than a dollar. It consisted of a punched sheet of cardstock, a spherical glass lens, a light emitting diode and a diffuser panel, along with a watch battery that powers the LED. It can provide over 2,000 X magnification with submicron resolution, weighs less than two nickels (8.8g), is small enough to fit in a pocket, requires no external power, and can survive being dropped from a 3-story building or stepped on by a person. Prakash Lab at Stanford University, USA in associated with Department of Biotechnology (DBT), Government of India brought the Foldscope to India to encourage curiosity in science. This cost-effective and portable foldscope has good scope in agriculture field includes the study on viability and germination of pollens. These studies are important because the fate of pollen have a profound influence on crop yield as it has greater role in reproductive success. Our experimental results in garden plants, pulses and horticultural crops revealed that it can be used for pollen germination studies efficiently. Foldscope could differentiate the colour of viable and non-viable pollens very efficiently in our pollen viability studies too (unpublished). Moreover, it can be taken to field and observe the pollen viability at the field itself if we have lot many genotypes to observe at a time by engaging many people. The procedure to observe the pollen grains, its viability and germination by using fold scope should be explained.

References

Abd El-Mohsen, AA and Abo-Hegazy, SR (2013) Comparing the relative efficiency of two experimental designs in wheat field trials. *World Essays Journal*. **1**(3): 101-109.

Abd El-Shafi, MA (2014) Efficiency of classical complete and incomplete block designs in yield trial on bread wheat genotypes. *Research Journal of Agriculture and Biological Sciences*. **10**(1): 17-23.

Acquaah G (2009) Principles of Plant Genetics and Breeding. John Wiley and Sons.

Adam, B and Groves, C (2007) Future matters: Action, knowledge, ethics. Leiden, Netherlands: BRILL.

Ahlawat, IPS; Prakash, Om and Saini, GS (1998) Scientific Crop Production in India, Rama Publishing House, Meerut, India.

Anderson, B (2007) Hope for nanotechnology: anticipatory knowledge and the governance of affect. *Area,* **39**(2): 156-165.

Anderson, B; Matthew. K and Doubleday, R (2007) Geographies of nanotechnoscience. *Area,* **39**(2): 139-142.

Andrea, LS and Willis, JH (2012) Molecular evolution and genetics of postzygotic reproductive isolation in plants. *Biol Rep* **4**: 23.

Angus, RA; Hardwick, SW *and* Cline, GB (1988) A computer program that calculates various similarity coefficients between electropherograms. *J. Hered.* **79**: 74–75.

Ankele, E; Heberle, BE and Pfosser, MF (2005) Searching for mechanisms to albino plant formation in cereals. *Acta Physiol Plant* **27**(4B): 651–664.

Anon (2006) The Future of nanotechnology: We need to talk. Retrieved June 11, 2010, from http: //www.nanologue.net/.

Anonymous (2004) Bringing genomics to the wheat fields: implementation of molecular marker technologies in public wheat breeding programs. Report of stakeholder's workshop, February 22, 2004, Kansas City MO.

Anonymous (2004) *Report of the Coordinated Experiments.* Vol. III. 2003-04. Crop Protection Programme (Pathology, Nematology and Entomology), All India Coordinated Wheat and Barley Improvement Project, Directorate of Wheat Research, Karnal 132 001, India. pp. 184.

Appadurai, RR (1966) Field Crops of Ceylon. Lake House Colombo. pp. 147-176.

Arnall, A and Douglas, P (2005) Moving the nanoscience and technology (NST) debate forwards: Short-term impacts, long-term uncertainty and the social constitution. *Technology in Society,* **27**: 23-38.

Asiedu, R; Mujeeb-Kazi, A and ter Kuile, N (1989) Marker-assisted introgression of alien chromatin into wheat. In: Mujeeb-Kazi A, Sitch LA (eds) Review of advances in plant biotechnology, 1985-1988. 2nd Int Symp Genetic Manipulation of Crops. CIMMYT, Mexico City, and IRRI, Manila, pp. 133-144.

Ayachit SM (Tr) (2002) Kashyapiya Krishisukti (A treatise on Agriculture by Kashyapa). Agri – History Bulletin No.4. Asian Agri History Foundation, Secunderabad, India.

Baga, M; Chibbar, RN and Kartha, KK (1999a) Expression and regulation of transgenes for selection of transformants and modification of traits in cereals. In: Vasil I.K. ed. Advances in cellular and molecular biology of plants, Vol.5. Molecular Improvement of Cereal Crops. Kluwer Publ., Dordrecht, The Netherlands. pp. 83-132.

Balasubramaniyan, P and Palaniappan, SP (2002) Principles and Practices of Agronomy. Agrobios (India), Jodhpur, India.

Ball, P (2003) Nanotechnology in the firing line. Retrieved June 11, 2010, from http://nanotechweb.org/cws/article/indepth/18804.

Barben, D; Fisher, E; Selin, C and Guston, DH (2007) Anticipatory Governance of Nanotechnology: Foresight: Engagement and Integration. In Edward J. Hacket, Olga Amsterdamska, Michael Lynch and Judy Wajcman (Eds.), The Handbook of Science and Technologies, Third Edition: MIT Press.

Becker, D; Brettschneider, R and Lorz, H (1994) Fertile transgenic wheat from microprojectile bombardment of scutellar tissue. *Plant Journal* **5**: 299-307.

Berne, RW (2006) *Nanotalk*. New Jersey: Lawrence Erlbaum.

Biswas, TD and Mukherjee, SK (1994) Text Book of Soil Science, Tata Mc-Graw Hill Publishing Company Ltd., New Delhi. 667p.

Borup, M; Brown, N; Konrad, K and Lente, HV (2006) The sociology of expectations in science and technology. *Technology Analysis and Strategic Management*, **18**(3/4): 285-298.

Bowman, DM and Hodge, GA (2007) Nanotechnology and public interest dialogue: Some international observations. *Bulletin of Science, Technology and Society*, **27**(2): 118-132.

Bradbury, PJ; Zhang, Z; Kroon, DE; Casstevens, TM; Ramdoss, Y *and* Buckler, ES *(*2007*)* TASSEL: software for association mapping of complex traits in diverse samples. *Bioinformatics* **23***:* 2633–2635.

Brain, FB(1971) Crop Production: Cereals and Legumes, Academic Press, London and New York.

Braun, HJ; Atlin, G and Payne, T (2010) Multi-location testing as a tool to identify plant response to global climate change. In: Reynolds MP (ed) Climate change and crop production. CABI Climate Change Series, U.K., pp. 115–138.

Breseghello, F and Sorrells, ME *(*2006*)* Association analysis as a strategy for improvement of quantitative traits in plants. *Crop Sci.* **46***:* 1323–1330.

Briggs, FN and Allard, RW (1953) The Current Status of the Backcross Method of Plant Breeding. *Agronomy Journal* 45: 131-138.

Brown, N (2003) Hope against hype – accountability in biopasts, presents and futures. *Science Studies*, **16**(2): 3-21.

Century, K; Reuber, TL and Ratcliffe, OJ (2008) Regulating the regulators: the future prospects for transcription-factor-based agricultural biotechnology products. *Plant Physiology*. **147**: 20-29.

Chubb, JR and Bickmore, WA (2003) Considering nuclear compartmentalization in the light of nuclear dynamics. *Cell* **112 :** 403–406.

Collins, NC; Tardieu, F and Tuberosa, R (2008) Quantitative trait loci and crop performance under abiotic stresses: where do we stand? *Plant Physiol*. **147**: 469–486.

Colvin, V and Weisner, M (2005) Environmental implications of emerging nanotechnologies. In David Rejeski and Robert Olson (Eds.), Environmentalism and the technologies of tomorrow: Shaping the next industrial revolution (pp.41-52). Washington: Island.

Dahiya, BS and Srivastava, RB (2004) Emerging Issues in Wheat Production. Strategy Paper presented at 43rd All India Wheat and Barley Research meet, IARI, New Delhi, August 27-30, 2004.

Davenport, CB (1908) Degeneration, albinism and inbreeding. *Science* **28**(718): 454–455.

de Wit, CT; van Laar, HH and van Keulen, H (1979) Physiological potential of crop production. In *Plant Breeding Perspectives.* J Sneep and AJT Hendricksen, eds. Center for Agricultural Publishing and Documentation, Wageningen, pp. 47-82.

De, GC (1997) Fundamentals of Agronomy, Oxford and IBH Publishing Co. Pvt. Ltd., New Delhi.

Dixon, J (2007) The Economics of Wheat. Developments in Plant Breeding **12**: 9-22.

Drexler, E; Peterson, C and Pergamit, G (1991) Unbounding the future: The nanotechnology revolution. New York: William Morrow and Co.

Dunkley, RWS (2004) Nanotechnology: Social consequences and future implications. *Futures,* **36**(10): 1129-1132.

Dunlop, C and Kling, R (1991) Computerization and controversy: Value conflicts and social choice. Academic Press Inc.

Eberhart, SA and Russel, WA (1966) Stability Parameters for Comparing Varieties. *Crop Science,* **6:** 36-40.

Ethirajan, AS (1981) Some recent promising varieties from intensified breeding and selection programme. *Ind Fms Dig* **14**(2): 15-17.

Eudes, F and Chugh, A (2009) An overview of triticale doubled haploids. In: Touraev A, Forster B, Jain S (eds) Advances in haploid production in higher plants. Springer, Berlin, pp. 87–96.

Ewing, PM; Runck, BC; Kono, TYJ and Kantar, MB (2019) The home field advantage of modern plant breeding. *PLoS ONE* **14**(12): e0227079.

Fisher, RA (1926) The arrangement of field experiments. *Ministry of Agriculture of Great Britain.* **33**: 503-513.

Foisset, N and Delourme, R (1996) Segregation distortion in androgenic plants. In: Jain SM, Sopory SK SK, Veilleux RE (eds) *In vitro* haploid production in higher plants. Kluwer Academic Publishers, Netherlands, pp. 189–201.

Forster, BP and Thomas, WTB (2005) Double haploids in genetics and plant breeding. *Plant Breed Rev* **25**: 57–88.

Frankel, OH and Bennet, E (1970) Genetic resources in plants-their exploration and conservation. Davis FA. Philadelphia, USA.

Fu, DH; Xiao, ML and Hayward, A (2014) Utilization of crop heterosis: a review. *Euphytica* **197**: 161–173.

Fuglsang, L (2001) Three perspectives in STS in the policy context. In SH Cutcliffe and C Mitcham (Eds.), Visions of STS: Counterpoints in science, technology, and society studies (pp.33-49). State University Press.

Fuller, S (2009) Knowledge politics and new converging technologies: A social epistemological perspective. *Innovation - The European Journal of Social Science Research,* **22**(1): 7-34.

Gao, LF; Jing, RL; Huo, NX; Li, Y; Li, XP; Zhou, RH; Chang, XP; Tang, JF; Ma, ZY and Jia, JZ (2004) One hundred and one new microsatellite loci derived from ESTs (EST-SSRs) in bread wheat. *Theor Appl Genet* **108**: 1392-1400.

Gavelin, K; Richard, W and Doubleday, R (2007) Democratic technologies? The final report of the Nanotechnology Engagement Group. London: Involve.

Giesbrecht, FG and Gumpertz, ML (2004) Planning, Construction and Statistical Analysis of Comparative Experiments. Wiley, CT: New York.

Gill, KS and Gill, BS (1996) A PCR-based screening assay of *Ph1*, the chromosome pairing regulator gene of wheat. *Crop Sci* **36**: 719-722.

Gourlay, S (2002) Tacit Knowledge, Tacit Knowing or Behaving? 3rd European Organizational Knowledge, Learning, and Capabilities Conference, Athens, Greece, 5-6 April.2002.

Groszmann, M; Greaves, IK and Fujimoto, R (2013) The role of epigenetics in hybrid vigour. *Trends Genet* **29**(12): 684–690.

Gupta, PK; Balyan, HS; Kumar, J; Kulwal, PL; Kumar, N; Mir, RR; Kumar, A and Prabhu, KV (2008) QTL analysis and marker assisted selection in bread wheat. Procs of 11th International wheat genetics symposium (IWGS), August 24-29, 2008, Brisbane Australia.

Hans de Jong, J (1999) High resolution FISH in plants –techniques and applications. *Trends Plant Sci.* **4**: 258–263.

Hansen, G and Wright, MS (1999) Recent advances in the transformation of plants. *Trends in Plant Sciences* **4**: 226-231.

Harlan, JR and de Wet, JJ (1971) Toward a rational classification of cultivated plants, *Texas* **20** (1): 509-517.

He, S; Ohm, H and Mackenzie, S (1992). Detection of DNA sequence polymorphisms among wheat varieties. *Theor Appl Genet* **84**: 573-578.

He, Y and Amasino, RM (2005) Role of chromatin modification in flowering-time control. *Trends Plant Sci.***10**: 30–35.

Heisey, PW; Lantican, MA and Dubin, HJ (2002) Impacts of international wheat breeding research in developing countries. 1966–97. Mexico, D.F.: CIMMYT, 73 p.

Heszky, LE and Simon-Kiss, I (1992) "DAMA" the first plant variety of biotechnology origin in Hungary, registered in 1992. *Hung. Agric. Res.* **1**: 30–32.

Hinkelman, K and Kempthorne, O (2006) Design and Analysis of Experiments. Vols. 1 and 2. Wiley, CT: New York.

Hoisington, D; Bohorova, N; Fennell, S; Khairallah, M; Pellegrineschi, A and Ribaut, MJ (Undated) The application of biotechnology to wheat improvement. FAO Document Repository.

Idrees, N and Khan, MI (2009) Design improvement using uniformity trials experimental data. *Pakistan Journal Agricultural Sciences.* **46**(4): 315-320.

Imaizumi T and Kay SA (2006) Photoperiodic control of flowering: not only by coincidence. *Trends Plant Sci.***11**: 550–558.

Ishii, T and Yonezawa, K (2007b) Optimization of the marker-based procedures for pyramiding genes from multiple donor lines: II. Strategies for selecting the objective homozygous plant. Crop Sci. ***47**: 1878–1886.*

Jacquier, NMA; LM Gilles, DE Pyott, JP Martinant, PM Rogowsky and T Widiez (2020) Puzzling out plant reproduction by haploid induction for innovations in plant breeding, *Nature Plants* **6**: 610–619.

Jag Shoren, Singh, G; Tyagi, BS and Singh, SS (2009) Wheat Research in India: Past, present and future. Consolidating the productivity gain in wheat-An Outlook. A Souvenir. Indian Agricultural Research Institute, New Delhi, pp: 8-18.

Jain, KBL (1994) Wheat cultivars in India (Compiled). Directorate of Wheat Research, Karnal 132001, India. Research Bulletin No. 2: 66 p.

Johnson, DG and Wetmore, JM (2009) Technology and society: Building our sociotechnical future. Cambridge: The MIT.

Jon-Arild, J (2008) Organizational innovation as part of knowledge management. *International Journal of Information Management* **28**: 403– 412.

Jones, D (1995) Technical boundless optimism (Review of "Nano! Remaking the World Atom by Atom", by Ed Regis). *Nature,* **374**: 835-837.

Jorgensen, MS and Jorgensen, U (2009) Green technology foresight in high technology: A social shaping of technology approach to the analysis of hopes and hypes. *Technology Analysis and Strategic Management,* **21**(3): 363-379.

Jorgensen, MS; Jorgensen, U and Clausen, C (2009) The social shaping approach to technology foresight. *Futures,* **41**(2): 80-86.

Kaiser, D (2006) Notes toward a nanotech timeline. Office of Scientific and Technical Information (OSTI). Working Paper 6-06-001. Retrieved June 11, 2010.

Kapoor, R (Undated) Win-Win Strategies in agribusiness management for corporate and farmers. International Conference on Agribusiness and Food Industry in Developing Countries: Opportunities and Challenges.

Kasha, K and Maluszynski, M (2003) Production of doubled haploids in crop plants. An introduction. In: Maluszynski M, Kasha K, Forster BP, Szarejko I (eds) Doubled haploid production in crop plants. A manual. Kluwer Academic Publishers, Norwell, pp. 1–4.

Kashif, M; Khan, MI; Arif, M; Ahmad, M and Mahmood, K (2011) Experience in use alpha lattice design in Pakistan. *International Journal of Intelligent Technologies and Applied Statistics*. **4**(1): 133-146.

Kearnes, M; Robin, GW; Phil, M; James, W and Brian, W (2006) From bio to nano: Learning lessons from the UK agricultural biotechnology controversy. *Science as Culture,* **15**(4): 291-307.

Keiper, A (2003) The nanotechnology revolution. *The New Atlantis*, 17-34.

Keller, B and Feuillet, C (2000) Colinearity and gene density in grass genome. *Trends in Plant Science* **5**: 246-251.

Keller, KH (2007) Nanotechnology and society. *Journal of Nanoparticle Research,* **9**: 5-10.

Khetarpa, RK; Lal, A; Varaprasad, KS; Agarwal, PC; Shashi, B; Chalam, VC and Gupta, K (2006) Quarantine for safe exchange of plant genetic resources. In: *Hundred Years of Plant Genetic Resources Management in India*, Singh AK, Srinivasan K, Saxena S, Dhillon BS (eds)., National Bureau of Plant Genetic Resources, New Delhi, pp. 83–108.

Kikuchi, S; Saito, Y; Ryuto, H; Fukunishi, N; Abe, T; Tanaka, H and Tsujimoto, H (2009) Effects of heavy-ion beams on chromosomes of common wheat, *Triticum aestivum*. *Mutat Res/Fund Mol Mech Mutagen* **669**: 63–66.

Klein, TM; Wolf, ED; Wu, R and Sanford, JC (1987) High-velocity microprojectiles for delivering nucleic acids into living cells. *Nature* **327**: 70-73.

Korzun, V; Roder, M; Worland, AJ and Borner A (1997) Intrachromosomal mapping of genes for dwarfing (*Rht12*) and vernalization response (*Vrn1*) in wheat by using RFLP and microsatellite markers. *Plant Breeding* **116**: 227-232.

Kristensen, K (2012) Incomplete split-plot designs based on alpha-designs: a compromise between traditional split-plot designs and randomized complete block design. *Euphytica*. **183**(3): 401-413.

Kumar, J; Pratap, A and Kumar, S (2006) Phenomics in crop plants: Trends, Options and limitations, Springer India.

Kurz, A (1996) Active and inactive genes localize preferentially in the periphery of chromosome territories. *J. Cell Biol.* **135:** 1195–1205.

Lakhtakia, A (2004) Editorial - introduction. In Akhlesh Lakhtakia (Ed.), Nanometer structures: Theory, modeling, and simulation. Bellingham, WA, USA.

Liu, PY; Zhu, J; Lou, XY and Lu, Y (2003) A method for marker-assisted selection based on QTLs with epistatic effects. *Genetica* **119**: 75–86.

Liu, S; Banik, M; Yu, K; Park, SJ; Poysa, V and Guan, Y (2007) Marker-assisted election (MAS) in major cereal and legume crop breeding: current progress and future directions. Int. J. Plant Breed. **1**: 74–88.

Loveridge, D; Paul, D and Sally, R (2008) Converging technologies at the nanoscale: The making of a new world? *Technology Analysis and Strategic Management,* **20**(1): 29-43.

Lux, R (2007) The nanotech report: 5th Edition: Lux Research.

Mackay, TFC (2009) A-maize-ing diversity. *Science,* **325**: 688–689.

Makela, P; Muurinen, S and Pettonen-Sainio, P (2008) Spring cereals: From dynamic ideotype to cultivars in Northern Latitude. *Agric. Food Sci.* **17**: 289-306.

Masood, MA; Farooq, K; Mujahid, Y and Anwar, MZ (2008) Improvement in precision of agricultural field experiments through design and analysis. *Pakistan Journal of Life and Social Sciences.* **6**: 89-91.

Matthew, K and Macnaughten, P (2005) Introduction: (Re)imagining nanotechnology. *Science as Culture,* **15**(4): 279-290.

Matthew, K; Macnaughten, P and Wilsdon, J (2006) Governing at the nanoscale: People, policies and emerging technologies London.

McGrail, S (2010) Nano Dreams and Nightmares: Emerging Technoscience and the Framing and (re)Interpreting of the Future, Present and Past, *Journal of Futures Studies,* **14**(4): 23 – 48.

Mekel, M (2006) Nanotechnologies: Small science, big potential and bigger issues. *Development,* **49**(4): 47-53.

Meyer, P (1995) Understanding and controlling transgene expression. *Trends in Biotechnology* **13**: 332-337.

Miller, G (2008a) Nanotechnology – Don't believe the green hype. *Chain Reaction.*

Miller, G (2008b) Nanotechnology and the public interest: Repeating the mistakes of GM foods? *International Journal of Technology Transfer and Commercialization,* **7**(2/3).

Miller, G and Rye S (2008) Out of the laboratory and on to our plates: Nanotechnology in food and agriculture. Melbourne, Australia.

Miller, G; Lisa, A; Erich, P; Dick, B; Rye, S and Kimbrell, G (2006) Nanomaterials, sunscreens and cosmetics: small ingredients, big risks: Friends of the Earth Australia and Friends of the Earth United States.

Mir, RR and Varshney, RK (2013) Future prospects of molecular markers in plants. In: Henry RJ (ed) Molecular markers in plants. Blackwell Publishing Ltd., Oxford, pp. 169–190.

Mitchell, E (2007) Scientific promise: Reflections on nano-hype. In Nigel Cameron and Ellen Mitchell (Eds.), *Nanoscale: Issues and perspectives for the nano century* (pp.43-60). Hoboken, NJ: Wiley.

Mody, C (2004) Small, but determined: Technological determinism in nanoscience. *HYLE – International Journal for Philosophy in Chemistry,* **10**(2): 99-128.

Mohan, M; Nair, S; Bhagwat, A; Krishna, TG; Yano, M; Bhatia, CR and Sasaki, T (1997) Genome mapping, molecular markers and assisted selection in crop plants. *Mol Breeding* **3**: 87-103.

Mooney, P (2006) Hype and hope: A past and future perspective on new technologies for development. *Development,* **49**(4): 16-22.

Morachan, YB (1986) Crop Production and Management, Oxford and IBH Pub. Co. New Delhi.

Moreau, L; Charcosset, A; Hospital, F and Gallais, A (1998) Marker-assisted selection efficiency in populations of finite size. *Genetics* **148**: 1353–1365.

Munshi, D; Robert B; Kurian, P and Lakhtakia, A (2007) A map of the nanoworld: Sizing up the science, politics, and business of the infinitesimal. *Futures,* **39**(4): 432-452.

Murovec, J and Bohanec, B (2012) Haploids and doubled haploids in plant breeding. In: Abdurakhmonov I (ed) Plant breeding. INTECH.

Nagarajan, S (2000) Thought for the new millennium. Wheat production in India – a success story and future strategies. *Indian Farming.* **50**: 9-16.

Nagarajan, S and Mohan, D (1994) Shortage to surplus – Sustaining the wheat revolution in India through coordinated research. Directorate of Wheat Research, Karnal 132001, India. *Research Bulletin.* 1: 20 p.

Nagarajan, S; Singh, RP; Singh, R; Singh, S; Singh, A; Kumar, A and Chand, R (2001) Transfer of technologies in wheat through frontline demonstrations in India – A comprehensive report 1995 – 2000. Directorate of Wheat Research, Karnal 132001, India, No. 6: 21 p.

Nei, M (1973) Analysis of gene diversity in subdivided populations. *Proc. Natl Acad. Sci. USA* **70**: 3321–3323.

Nei, M and Li, WH (1979) Mathematical models for studying genetic variation in terms of restriction endonucleases. *Proc. Natl Acad. Sci. USA* **76**: 5269–5273.

Nonaka, I and Takeuchi, H (1995) The knowledge creating company: How Japanese companies create the dynamics of innovation. Oxford: Oxford University Press.

Nordborg, M and Tavare, S (2002) Linkage disequilibrium: what history has to tell us. *Trends Genet.* **18**: 83–90.

Patterson, HD and Silvey, V (1980) Statutory and recommended list trials of crop varieties in the United Kingdom. *Journal of the Royal Statistical Society,* A **143**: 219-252.

Payne, RW; Murray, DA; Harding, SA; Baird, DB and Soutar, DM (2011) An Introduction to GenStat for Windows (14th). VSN International, CT: Hemel Hempstead.

Phil, M; Kearnes, MB and Wynne, B (2005) Nanotechnology, governance and public deliberation: What role for the social sciences? *Science Communication,* **27**(2): 268-291.

Piepho, HP; Buchs, A and Truberg, B (2006) On the use of multiple lattice designs and á-designs in plant breeding trials. *Plant Breeding*. **125**: 523-528.

PMSEIC (2005) Nanotechnology: Enabling technologies for Australian innovative industries. Prime Minister's Science and Engineering and Innovation Council.

Poehlman, JM and Sleper, DA (1995) Breeding Field Crops. Iowa State University Press, Ames.

Porter, AL; Youtie, J; Shapira, P and Schoeneck, D (2008) Refining search terms for nanotechnology. *Journal of Nanoparticle Research,* **10**: 715-728.

Pritcharda, JK; Stephens, M and Donnelly, P (2000) Inference of population structure using multilocus genotype data. *Genetics,* **155**: 945–959.

Puja (2003) Introgression of productivity enhancing genes into durum wheat *(Triticum turgidum* L. var *durum)* from tetraploid progenitor species. *M.Sc. Thesis.* PAU, Ludhiana.

Rafalski, JA (2010) Association genetics in crop improvement. *Curr. Opin. Plant Biol.* **13**: 174–180.

Rai, B; Gupta, SK and Pratap, A (2011) Breeding methods, *Advances in Botanical Research* **45**: 21-48.

Rana, B; Rana, P; Yadav, MK and Kumar, S (Undated) Marker assisted selection: a strategy for wheat improvement, Personal Communication.

Randles, S; Paul D; Denis, L and Schmidt, JC (2008) Editorial: Nanoworlds as Schumpeterian emergence and Polanyian double-movements. *Technology Analysis and Strategic Management,* **20**(1): 1-11.

Roco, M (2006) Environmentally responsible development of nanotechnology. *Environmental Science and Technology,* **39**(5): 106A-112A.

Roco, M and Bainbridge, WS (2002) Converging technologies for improving human performance: Nanotechnology, biotechnology, information technology and cognitive science. Arlington, VA: U.S. National Science Foundation.

Rosenzweig, C and Hillel, D (1995) Potential impacts of climate change on agriculture and food supply. Consequences 1: 2. U.S. Global Change Research Information Office, Washington, D.C.

Salmenkaita, JP and Ahti, S (2002) Rationales for Government intervention in the commercialization of new technologies. *Technology Analysis and Strategic Management,* **14**(2): 183-200.

Sarawitz, D and Woodhouse, E (2003) Small is powerful. In Alan Lightman, Daniel Sarewitz, and Christina Desser (Eds.), Living with the genie: Essays on technology and the quest for human mastery. Washington, DC: Island.

Schmidt, KF (2007a) Green nanotechnology: It's easier than you think (PEN 8): Project on Emerging Nanotechnologies.

Schmidt, KF (2007b) Nanofrontiers: Visions for the future of nanotechnology (PEN 6): Project on Emerging Nanotechnologies.

Selin, C (2006a) Time matters: Temporal harmony and dissonance in nanotechnology networks. *Time and Society*, **15**(1): 121-139.

Selin, C (2006b) Volatile visions: Transactions in anticipatory knowledge. Unpublished PHD, Copenhagen Business School.

Selin, C (2007) Expectations and the emergence of nanotechnology. *Science, Technology and Human Values*, **32**(2): 196-220.

Senge, P and Goran, C (2001) Innovating our way to the next industrial revolution. *MIT Sloan Management Review* (Winter): 24-38.

Sharma SN, Bhatnagar VK, Mann MS, Shekhawat US, Sain RS (2002) Maximization of wheat yields with a unique variety in warmer areas. *Wheat Inform Serv* **95**: 11–16.

Singh, BD (2002) Plant Breeding. Kalyani Publishers, New Delhi.

Singh, C (1997) Modern techniques of raising field crops, Oxford and IBH Pub Co. Pvt. Ltd., New Delhi.

Singh, GP (2020) Wheat man of India: Dr G P Singh, Director, IIWBR, Karnal, India.

Singh, P (2001) Essentials of Plant Breeding. Kalyani Publishers.

Singh, SS (1999) Principles and Practices of Agronomy, Kalyani Publishers, New Delhi.

Singh, SS and Singh, BB (2001) IARI wheats for evergreen revolution, IARI, New Delhi, pp.92.

Slaughter, R (1996) Futures studies: From individual to social capacity. *Futures*, **28**(8): 751-762.

Slaughter, RA and Bussey, M (2006) Futures thinking for social foresight. Taipei, Taiwan: Tamkang Univ Press in association with Foresight International.

Smith JC, Ghosh A, Hijmans RJ (2019) Agricultural intensification was associated with crop diversification in India (1947-2014). *PLoS ONE* **14**(12): e0225555.

Stalker, HT (1980) Utilization of wild species for crop improvement. *Adv. Agron*.**33**: 111- 47.

Stilgoe, J (2007) Nanodialogues: Experiments in public engagement with science. London.

Swaminathan, MS (1969) The role of mutation breeding in a changing agriculture. Induced Mutations in Plants (Proc. Symp. Pullman, 1969), LAEA, Vienna.

Swaminathan, MS (1993) Wheat revolution – a dialogue. The MacMillan India Ltd. Madras. 164 p.

Tanksley, SD and Nelson, JC (1996) AB-QTL analysis: A method for simultaneous discovery and transfer of valuable QTLs from unadapted germplasm into elite breeding lines. *Thero Appl Genet* **92**: 191-203.

Tegart, G (2002) Nanotechnology: The technology for the 21st century (Volume 1 – summary report of a foresight project). APEC Center for Technology Foresight.

Tester, M and Langridge, P (2010) Breeding technologies to increase crop production in a changing world. *Science* **327**: 818–822.

Toumey, C (2004) Nano hyperbole and lessons from earlier technologies. *Nanotechnology Law and Business,* **1**(4): 397-405.

Tribe, D (1994) Feeding and Greening the World: The Role of International Agricultural Research, CAB International, Wallingford, UK.

Tyagi, BS; Singh, G; Shoran, G and Nagarajan, S (Undated) Enriching Wheat Germplasm Through Elite International Germplasm Nurseries, *DWR Newsletter.*

UNESCO (2006) The ethics and politics of nanotechnology. United Nations Educational Scientific and Cultural Organisation.

van Lente, H (1993) Promising technology: The dynamics of expectations in technological developments. Unpublished PhD, Univ of Twente, Enschede.

van Lente, H and Rip, A (1998) The rise of membrane technology: From rhetorics to social reality. *Social Studies of Science,* **28**(2): 221-254.

Varshney, RK; Langridge, P and Graner, A (2007) Application of genomics to molecular breeding of wheat and barley, *Advances of Genetics,* **58**: 121–155.

Verma, DK (1997) A Comprehensive List of Theses Submitted in the Department of Plant Breeding: Period: 1996-1976, Rajendra Agricultural University, Bihar, Pusa (Samastipur) – 848 125, India.

Walsh, B (2007) Environmentally beneficial nanotechnologies: Barriers and opportunities (A report for the UK Department for Environment, Food and Rural Affairs). Buckinghamshire, UK: Oakdene Hollins.

Wang, J; Chapman, SC; Bonnett, DG; Rebetzke, GJ and Crouch, J (2007) Application of population genetic theory and simulation models to efficiently pyramid multiple genes via marker-assisted selection. *Crop Sci.* **47**: 582–588.

Werr, W and Lorz, H (1986) Transient gene expression in a Gramineae cell line. *Molecular and General Genetics* **202**: 471-475.

Wickson, F (2008) Narratives of nature and nanotechnology. *Nature Nanotechnology,* **3**: 313-315.

Winter, P and Kahl, G (1995) Molecular marker technologies for plant improvement. *World Journal of Microbiology and Biotechnology,***11**: 438-448.

Wintle, B; Mark, B and Fiona, F (2007) How fast should nanotechnology advance? *Nature Nanotechnology,* **2**: 327.

Wolkenhauer, O (2001) Systems biology: The reincarnations of systems theory applied in biology? *Briefings in Bioinformatics.* **2:** 258-270.

Wood, S; Geldart, A and Jones, R (2008) Crystallizing the nanotechnology debate. *Technology Analysis and Strategic Management,* **20**(1): 13-27.

Xu, Y and Crouch, JH (2008) Marker assisted selection in Plant Breeding: From publications to practice. *Crop Science* **48**: 391-407.

Xu, Y; McCouch SR and Zhang, Q (2005) How can we use genomics to improve cereals with rice as a reference genome? *Plant Molecular Biology,* **59**: 7-26.

Yadav, R; Parashar, M; Jag Shoran; Kumar, J and Bhardwaj, SC (2007) Rust resistance in newly bred lines of wheat (*Triticum aestivum*). *Indian Journal of Genetics and Plant Breeding*. **67** (3): 291. - 293.

Yadav, R; Singh, SS; Jain, N; Singh, GP and Prabhu, KV (2010) Wheat Production in India: Technologies to face future challenges. *J. Agri. Sci.* **2** (2): Open access journal published from Canada.

Young, ND and Tanksley SD (1989) Restriction fragment length polymorphism maps and the concept of graphical genotypes. *Theoretical and Applied Genetics,* **77**: 95-101.

Chapter 2

Genetics, Cytogenetics, Breeding and Gene Regulation of Bread Wheat (*Triticum aestivum* L. *Triticum sativum*, Lamk)

Outlines

1. Nomenclature, Classification and Origin
2. Classification of Wheat
3. Classes of Wheat
4. Crop Growth Stages
5. Genetics of Some Morphological and Physiological Characters
6. Cytogenetics
   a. Relation of interphase chromosomal position with the domains of transcriptional activity
   b. Heterochromatin: indefinable, yet measurable
   c. DNA methylation directs changes in chromatin structure
   d. Modifications in histone tail: gene regulation through an epigenetic code
   e. Chromatin Imaging in living cells
   f. Pollen Viability Studies
   g. Pollen Germination Studies
   h. Wheat Pollination

13. Wheat Maize-Pollen Doubled Haploid
    a. Maize Pollen Donor Plants
    b. Wheat Parental Plants
14. Crossing
    a. Emasculation
    b. Pollination
    c. Hormone Spray
    d. Excision and Embryo Rescue
    e. Second Cold Treatment
    f. Transplanting Haploid Plants
15. Colchicine Treatment – Doubling
    a. Plant Preparation
    b. Colchicine Treatment and Rinse
16. RT-PCR Analysis
17. Generation of Transgenic Wheat Lines
18. Engineering Nuclear Male Sterility
19. Transgene Silencing in Wheat
20. Transposon Tagging in Wheat
21. Marker Assisted Selection in Wheat
22. Genomics and Future

Man obtains practically all his food directly or indirectly from plants. The cereal plants of the grass family producing large edible grains provide about one-half of man's food calories and a large part of his nutrient requirements. Wheat and rice are foremost among cereals and indeed among all crops, as direct sources of food for man. First of all bread making quality of all plant products only wheat flour contains the particular kinds and combinations of proteins necessary for the production of light porous leaves of leavened bread. Rye is the only other true 'bread grain' but its flour produces darker, heavier, coarser and less digestible bread than that of wheat and is therefore popular wheat is widely adopted to different soil and climatic conditions and its culture is relatively simple. All wheats, whether wild or cultivated belong to the genus *Triticum* of the family Gramineae, the grass family, *Triticum* is the only one of some 600 genera belonging to this great family, which itself comprises well over 5,000 species. In recent decades, the controlled breeding of wheat has added greatly to the range of variation of the crop in some directions and to a limitation of variation in others. Not every morphologic character is of equal value as a means for classification. Certain plant characters such as habit of growth (winter, intermediate or spring), time of heading, or plant height are determined genetically), but classes often are not distinct. For example, there is a more or less complete series of types from true winter wheat to true spring wheat and expression depends on temperature, length of day, or date of seeding. Leaf sheaths, ligules, and auricles very little from one variety or species to another. Root

Figure 2.1. Prof M S Swaminathan on the eve of MSSRF dt May 22, 2020

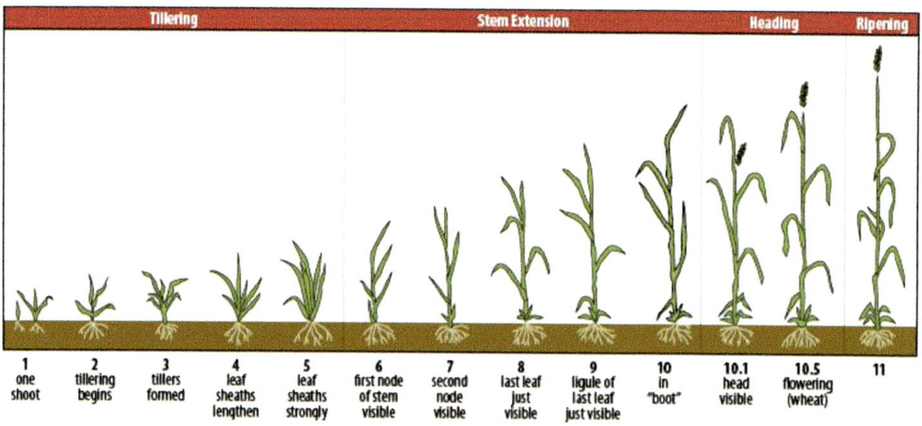

Figure 2.2. The Freeke scale of wheat development

Figure 2.3. Ears of einkorn, emmer and spelt wheat

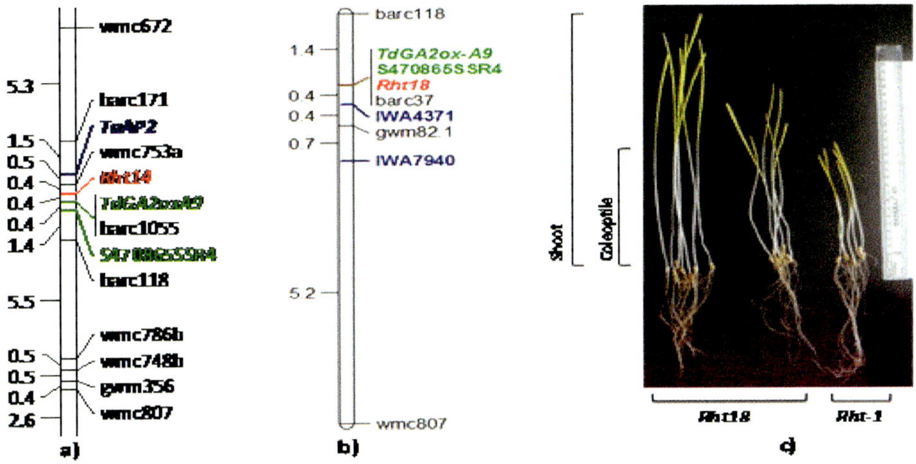

Figure 2.4. a & b) Alternative dwarfing genes Rht14 and Rht18 mapped on chromosome 6A in wheat. c) Rht18 introgression lines showing improved coleoptile length and seedling shoot length.

Figure 2.5. Black Wheat needs to be validated at different institutions for their grain content and productivity

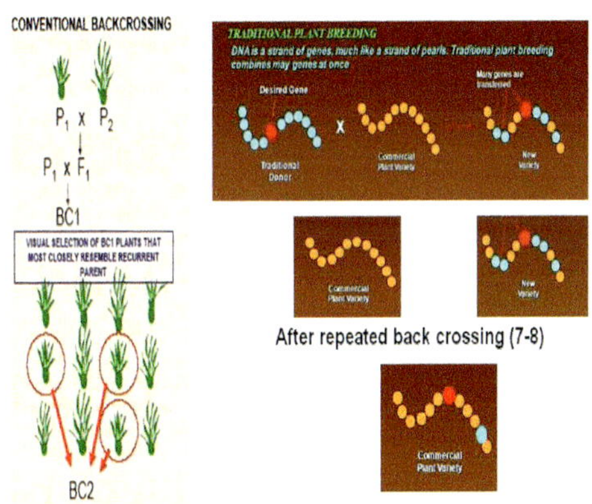

Figure 2.6. Plant breeders aim to produce superior crops...

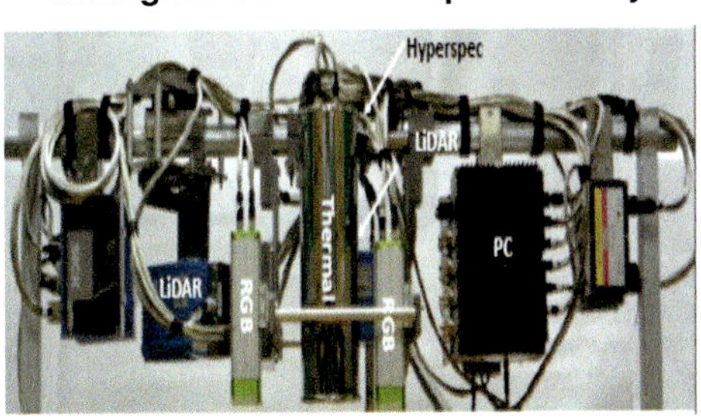

Figure 2.7a. High-resolution phenotyping

Figure 2.7b. High-throughput phenotyping

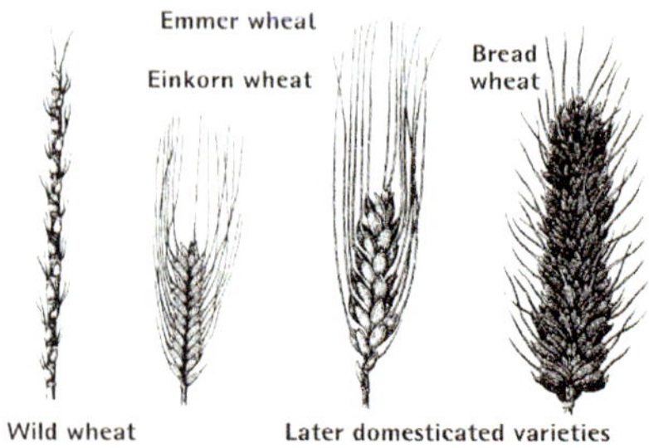

Figure 2.8. Wild wheat and later domesticated varieties

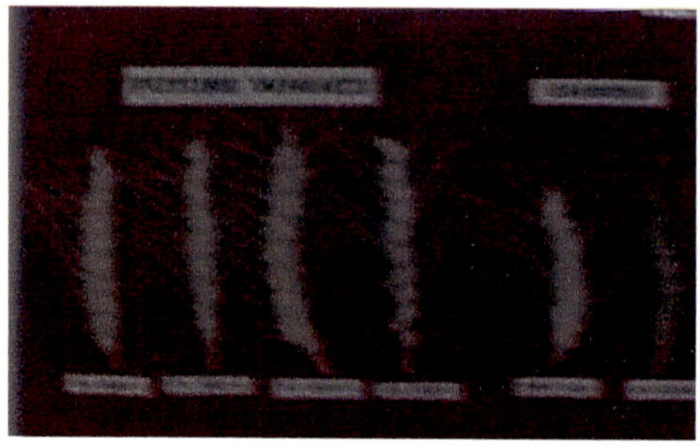

Figure 2.9. Future wheat

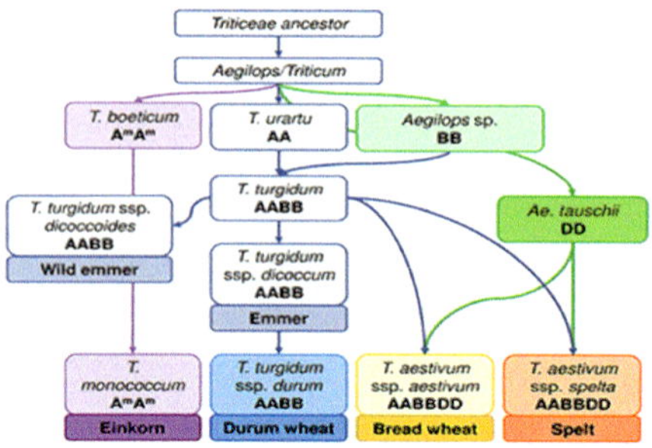

Figure 2.10. Evolution of the wheat species einkorn, durum wheat, spelt and common wheat

Figure 2.11. Wild ancestor; einkorn wheat, durum wheat; modern wheat, produced by crossing durum with goat grass; and a green variety with shorter stalks and larger seeds.

T. aestivum *T. durum* *T. dicoccum* *T. compactum* *T. sphaerococcum* *T. polonicum*

Figure 2.12a. Ear and spikelets of wheat and related species

T. timopheevii *T. carthalicum* *T. vavilovii* *Triticale* *Secale cereale* *T. monococcum*

Figure 2.12b. Ear and spikelets of wheat and related species

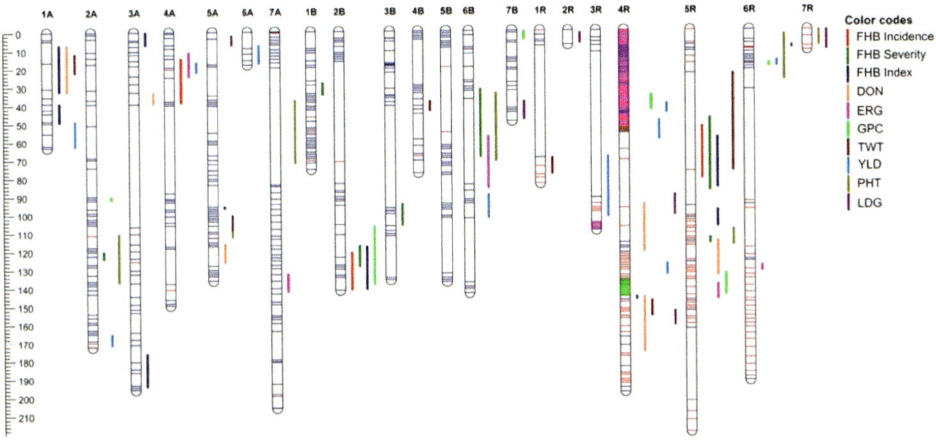

Figure 2.13. Ultima spring triticale genetic map. The twenty-one linkage groups (LGs; seven from each of three triticale genomes A, B and R) are labeled as 1A to 7R

Figure 2.14. Wild wheat has virtues scientists hope to tap

Figure 2.15. *Aegilops neglecta*, a wild wheat relative

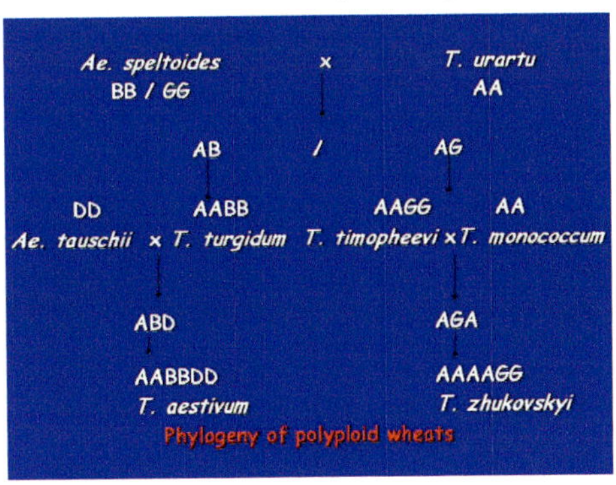

Figure 2.16. Phylogeny of polyploid wheats

List of species under primary gene pool

Species	Ploidy	Genome
Triticum urartu	2x	AA
T. monoccocum	2x	AA
T. boeoticum	2x	AA
Aegilops tauschii	2x	DD
T. dicoccoides	4x	AABB
T. dicoccum	4x	AABB
T. durum	4x	AABB
T. turgidum	4x	AABB
T. polonicum	4x	AABB
T. carthlicum	4x	AABB
T. abyssinicum	4x	AABB
T. turanicum	4x	AABB

Figure 2.17a. List of species under primary gene pool

List of species under primary gene pool		
Species	**Ploidy**	**Genome**
T. spelta	6x	AABBDD
T. compactum	6x	AABBDD
T. macha	6x	AABBDD
T. sphaerococcum	6x	AABBDD
T. vavilovi	6x	AABBDD
T. aestivum	6x	AABBDD

Figure 2.17b. List of species under primary gene pool

List of species under secondary gene pool		
Species	**Ploidy**	**Genome**
Ae. cylindrica	4x	CCDD
Ae. ventricosa	4x	DDMM
Ae. crassa	4x	DDMM
Ae. crassa	6x	DDDDMM
Ae. juvenalis	6x	UUDDMM
Ae. vavilovi	6x	DDMMSS

Figure 2.18a. List of species under secondary gene pool

List of species under secondary gene pool		
Species	**Ploidy**	**Genome**
Aegilops speltoides	2x	SS
Ae. longissima	2x	S^lS^l
Ae. bicornis	2x	S^bS^b
Ae. searsii	2x	$S^{sh}S^{sh}$
T. timopheevi	4x	AAGG
T. araraticum	4x	AAGG
T. militinae	4x	AAGG
T. zhukovkyi	6x	AAAAGG

Figure 2.18b. List of species under secondary gene pool

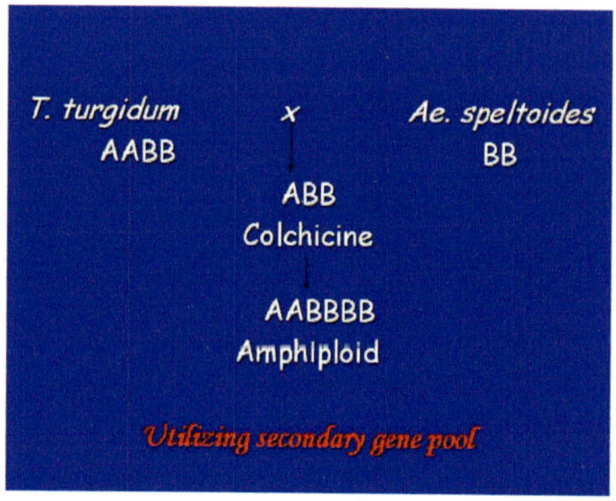

Figure 2.19. Utilizing secondary gene pool

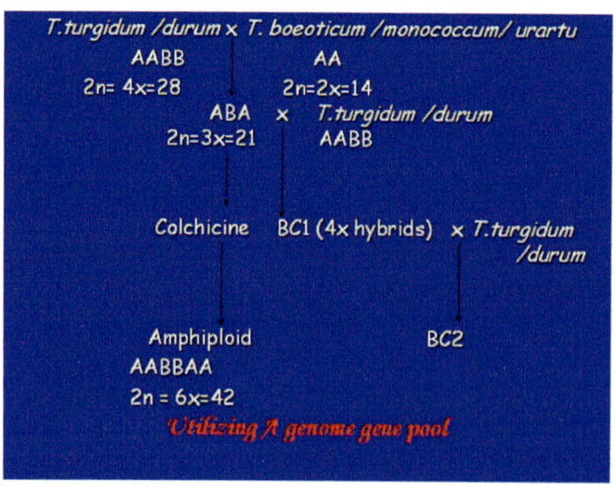

Figure 2.20. Utilizing A genome gene pool

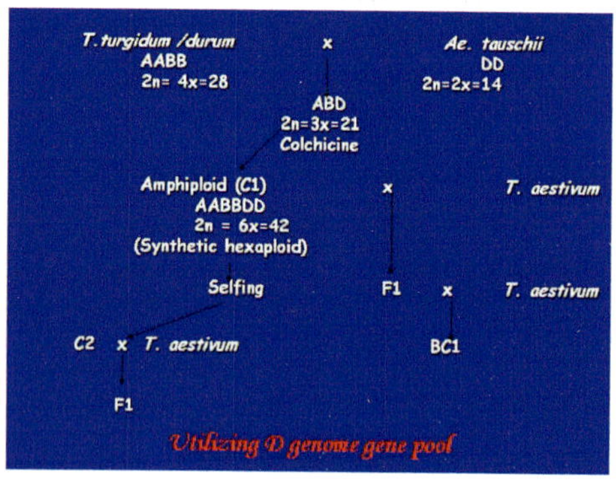

Figure 2.21. Utilizing D genome gene pool

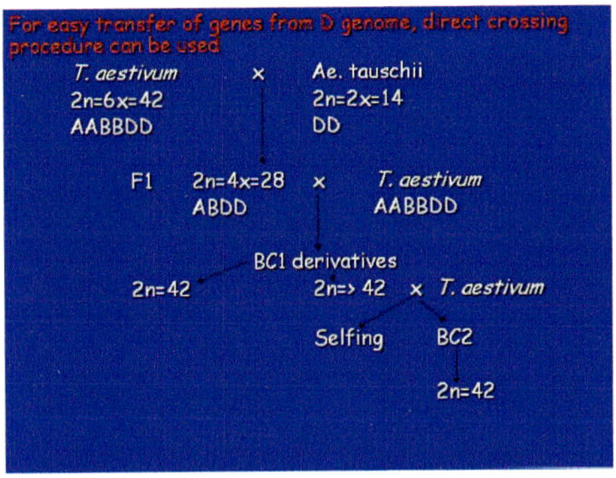

Figure 2.22. For easy transfer of genes from D genome, direct crossing procedures can be used.

List of species & genera under tertiary gene pool

Species	Ploidy	Genome
Aegilops caudata	2x	CC
Ae. umbellulata	2x	UU
Ae. peregrina	4x	UUSS
Ae. kotschyi	4x	UUSS
Ae. ovata	4x	UUMM
Ae. comosa	2x	MM
Agropyron spp	2x, 4x, 6x	PP
Thinopyrum junceum	2x, 4x, 6x	J^eJ^e, $J^bJ^bJ^bJ^b$, $J^bJ^bJ^bJ^bJ^eJ^e$
Th. bessarbicum	2x	J^eJ^e
Th. elongatum	2x	J^eJ^e
Elymus spp	4x, 6x	SSHH, SSYY
Secale cereale	2x	RR
Hordeum spp	2x, 4x	HH, HHHH

Figure 2.23. List of species genera under tertiary gene pool

Promising genetic stocks derived from wild crosses

Genetic stock	Wild species involved	Characteristics
WR180	*T. turgidum*	High 1000 kernel weight
WR201	*T. turgidum*	Early maturity
ISD8	*T. carthlicum* *T. turgidum*	High tillering, grain hardness
Pusa T3336	*Agropyron*	Leaf blight, 1000 kernel weight
Sel111	*Secale cereale*	1000 kernel weight, High protein
Sel 212	*Secale cereale*	Leaf and stem rust resistance
S2427	*Ae. speltoides*	Leaf & stem rust resistance, High protein

Figure 2.24. Promising genetic stocks derived from wild crosses

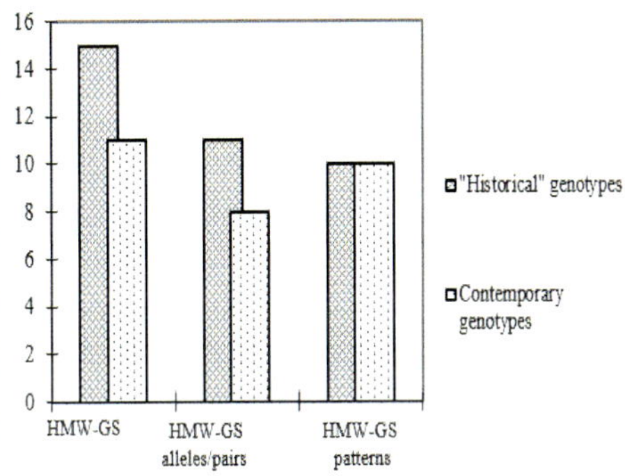

Figure 2.25. Variation in Glu-1 loci in old and modern wheats

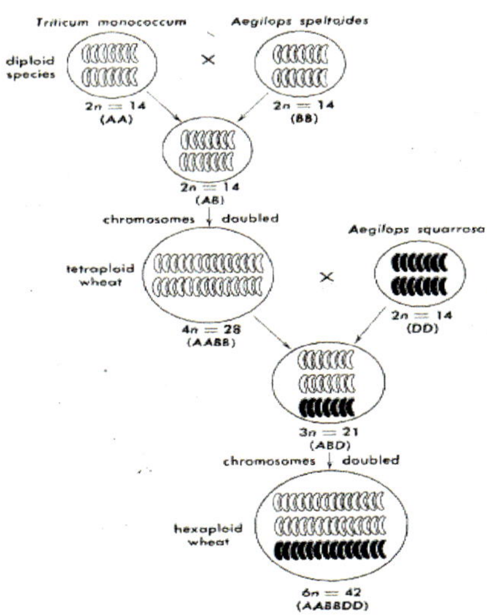

Figure 2.26. Origin of hexaploid wheat.

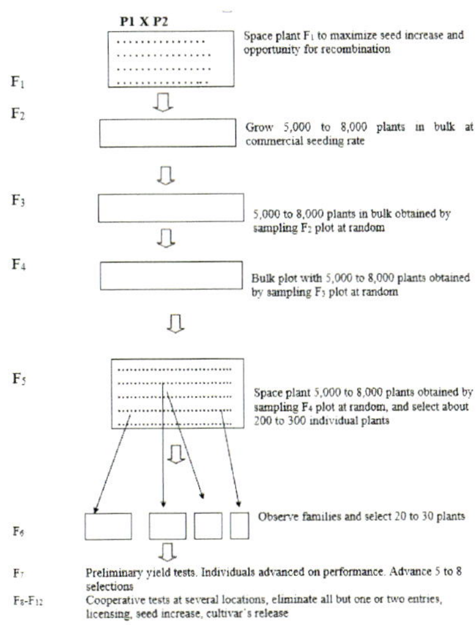

Figure 2.27. Selection Procedure for Segregating Materials

Figure 2.28. Dr D K Verma instructing Research Scholars in Crossing Block, 2019, IARI RS Indore, MP

Figure 2.29. A wheat spike whose anthers are ready to dehisce

Figure 2.30. A wheat spike being cut three-forth for emasculation

Figure 2.31. Pollen collection and extrusion of anthers

Figure 2.32. Pollen collection in parchment paper bag

Figure 2.33. A wheat spike being emasculated

Figure 2.34. Research Scholar Mr Ravikant Soni bagging emasculated wheat spike

Figure 2.35. Research Scholar Mr Pratik Patidar bagging emasculated wheat spike

Figure 2.36. Research Scholar Mr Lalit Chichkhede bagging emasculated wheat spike

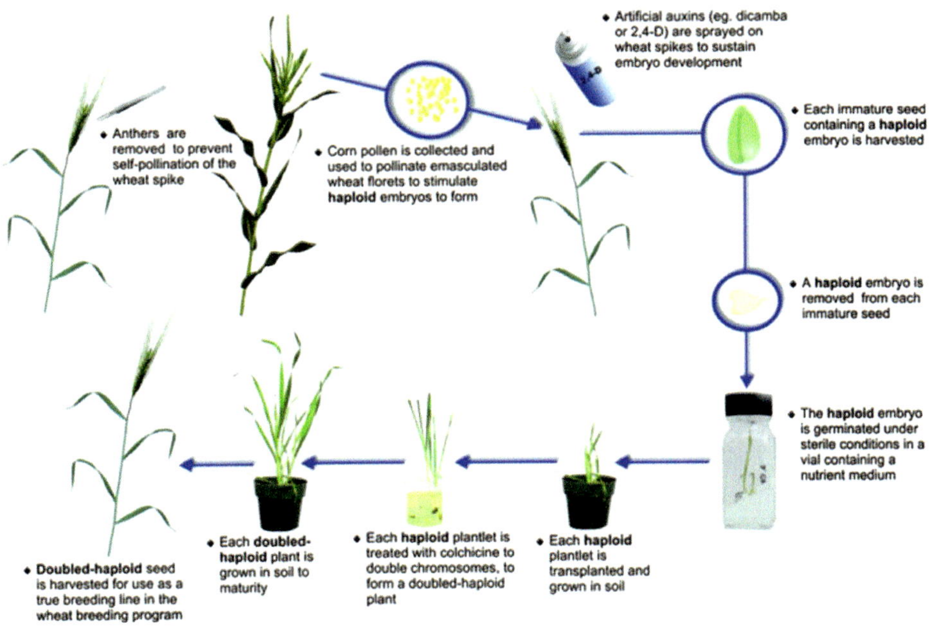

Figure 2.37. Doubled haploid plant production using the maize-pollen method

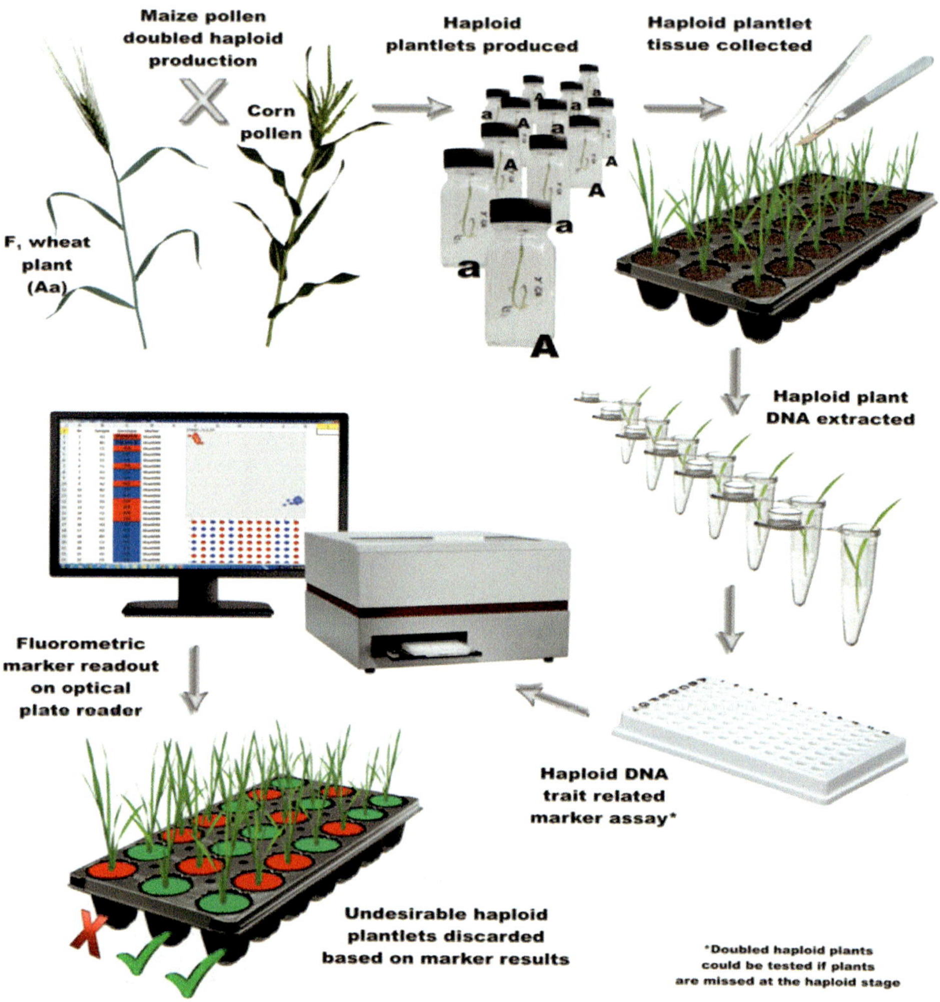

Figure 2.38. Merging of doubled haploid production and marker -assisted breeding

Figure 2.39. (a). Maize 1 week after seeding 25–30 seeds in a 10 cm pot. (b) One-week-old maize seedlings transplanted into pots

Figure 2.40. Wheat doubled haploid parental plants with segregating gametes (e.g. F1) grown in plastic containers

Figure 2.41. Spike preparation and emasculation. (a) Removal of center floret with forceps. (b) Removal of wheat anthers with forceps. (c) The yellow colour of the anthers with a slight green tinge indicating the appropriate time for emasculation

Figure 2.42a &b. Maize anthers ready for pollination, Spray application of dicamba

Figure 2.43. (a) Sterilization of caryopsis. (b) Wheat caryopsis with developing embryo

Figure 2.44. (a) Selfed wheat caryopsis with endosperm development compared to a caryopsis with a haploid embryo and no endosperm development. (b) Vial containing developing haploid plantlet on culture media

Figure 2.45. (a) Trimming leaves of haploids. (b) Trimming roots of haploids

Figure 2.46. Bagging after emasculation and pollination

Figure 2.47. Plants in container for colchicine treatment

characters cannot conveniently be used. Actual use of those morphologic characters best suited for implementing identification is exemplified. Section on description, anatomy and ontogeny are based on plants of one species only, *Triticum aestivum* (em Thell). It is estimated that to maintain self-sufficiency in the availability of wheat in India, annual production of wheat and rice together should increase by 2 mt every year. During the last 50 years, significant improvement in wheat production and productivity was achieved through exploitation of major genes for traits like dwarfness, photoperiod insensitivity and resistance to biotic stresses (Reynolds and Borlaug, 2006). In India, the productivity of wheat has improved, tremendously in the past three decades with the expansion of high yielding dwarf varieties and better use of inputs. However, a quantitative jump in wheat production and productivity is still needed to feed the fast expanding human population despite of drastic changes in world climatic conditions.

Table 2.1. Generalized wheat data sheet

Sl.No.	*Stage and particulars*	*Remarks*
1	Growth stage	
	1½ leaf stage	
	3½ leaf stage	Coincides with 3-4 leaf stage
	Early tillering stage	
	Fully tillering stage	
	Jointing Stage	Beginning of stem elongation
	Boot stage	Swelling at the uppermost node due to leaf sheath encircling the head which is not visible. Flag leaf fully developed
	Dough stage	
2	Germinating seedling	
3	Maximum tillering	
4	Flag leaf stage	
5	One leaf stage	
6	Two leaf stage	
7	Four leaf stage	
8	Jointing stage	
9	Late Jointing stage	
10	Booting stage	
11	Heading stage	Ear starts emerging from the leaf sheaths
	Flowering stage	Pollen is being shed and ovaries fertilized, some of the empty stamens are seen hanging outside the ears.
	Filling stage	Ovaries after fertilization start enlarging : milk → soft dough → hard dough
	Maturity	Colour of the glume changes
12	Wheat leaf	Dorsal
		Ventral
13	Auricle in wheat	

Sl.No.	Stage and particulars	Remarks	
14	Ligule in wheat		
15	Wheat rachis		
	Back rust of wheat		
	Brown rust of wheat		
	Yellow rust of wheat		
	Loose smut of wheat		
	Flag smut of wheat		
	Karnal bunt of wheat		
	Black point disease of wheat		
	Powdery mildew of wheat		
	Ear cockle of wheat		
	Tundu disease of wheat		
	Molya disease of wheat		
	Army worm		
	Mite		
	Rats		
	A typical burrow complex of the Indian mole rat		
	Test weight	31-49 grams/1000 kernel	
	Hectolitre weight	70-84 Kgs/hectolitre	
	Kernel size		
	Kernel shape		
	Kernel hardness and vitreousness	Measured in terms f Pearling Index (P.I.) varies between 35 to 55	
	P.I. = {(20 weight of pearls) x 100}/20	Less than 29	Hard wheat
		29-40	Semi hard wheat
		41-50	Soft wheat
		More than 50	Very soft wheat
	Colour	Red and white or amber	Depend on the colour of the bran
	Milling qualities	To ascertain the yield of straight grade flour or semolina (suji)	Related to commercial milling
	Protein content	6-20 per cent	
	= determined by Kjeldahl method = directly associated with water absorption	Flour-protein content (per cent)	End use suitability
		11.5-12.5	Bread
		9-11	Biscuits
		7-8.5	Cakes
	Gluten quantity and quality		

Sl.No.	*Stage and particulars*	*Remarks*	
	Pelshenke value (min)	Type of gluten	Recommended use
	30-60	Weak	Cakes, cookies
	60-100	Medium strong	Bread and chapatis
	Above 100	Strong	For mixing with weak flour
	Sedimentation value (ml)	Type of gluten	Recommended use
	Below 20	Weak	Cake, biscuits
	20-39	Medium strong	All purpose chapatis
	40-59	Strong	Bread making
	60 and above	Very strong	Belding with weak flour
	Yeild performance		
	Per plant yield		
	Tillering capacity		
	Plant height		
	Colour of stem		
	Colour of foliage		
	No. of days for 50 per cent heading		
	No. of days for maturity		
	Ear shape		
	Ear density		
	Awning		
	Awn colour		
	Glume colour		
	Fragidity of the rachis		
	Threshability		
	Seed colour		
	Seed texture		
	Seed size in gram		
	Recovery of flour in percentage		
	Flour colour		
	Chapati making quality		
	Bread making quality		

Nomenclature, Classification and Origin

Since 1753, when Linnaeus established the binomial system of naming species of plants and animals (Linnaeus, 1753), some 500 names of proposed wheat species have been published. Moreover, botanists working in different parts of the world sometimes assigned different names to the same species. The first name to be published of a valid species is given priority, and other names published for the same species are known as 'synonyms'. Another complication that has occurred a number of times is that, although a population of plants has been designated as a species of *Triticum*, subsequent research has indicated that it should be assigned to as another genus (*e.g. Agropyron* or *Aegilops*). Thus, the large number of proposed scientific names for species of the genus *Triticum* has been.

Scientific Classification

Kingdom: Plantae

Division: Magnoliophyta

Class: Liliopsida

Order: Poales

Family: Poaceae

Genus: *Triticum* L.

Species: *T. aestivum; T. aethiopicum; T. araraticum; T. boeoticum; T. carthlicum; T. compactum; T. dicoccoides; T. dicoccon; T. durum; T. ispahanicum; T. karamyschevii; T. macha; T. militinae; T. monococcum; T. polonicum; T. spelta; T. sphaerococcum; T. timopheevii; T. turanicum; T. turgidum; T. Urartu; T. vavilovii; T. zhukovskyi.*

Within a species, wheat cultivars are further classified by growing season, such as winter wheat vs. spring wheat, by gluten content, such as hard wheat (high protein content) vs. soft wheat (high starch content), or by grain color (red, white or amber). Crop management decisions require the knowledge of stage of development of the crop. For example, current recommendations often indicate the second application of nitrogen be done when the ear (not visible at this stage) is about 1 cm in size (Z31 on Zadoks scale). Knowledge of stages is also interesting to identify periods of higher risk, in terms of climate. For example, the meiosis stage is extremely susceptible to low temperatures (under 4°C) or high temperatures (over 25°C). Farmers also benefit from knowing when the flag leaf (last leaf) appears as this leaf represents about 75 per cent of photosynthesis reactions during the grain filling period and as such should be preserved from disease or insect attacks to ensure a good yield. Several systems exist to identify crop stages, with the Feekes and Zadoks scale being the most widely used. Each scale is a standard system which describes successive stages reached by the crop during the agricultural season.

Classification of Wheat

Common bread wheat (*Triticum vulgare*) – It is a typical wheat of alluvial soils of Indo Gangetic plains *i.e.*, Punjab, Uttar Pradesh, Bihar and parts of Rajasthan. Bulk of Indian crop consists of this type.

Macaroni wheat (*Triticum durum*) – It is grown in Italy, USA, Canada, and Russia. They are descended from emmer wheat. It is drought tolerant and cultivated in Punjab, Madhya Pradesh, Karnataka, Tamil Nadu, Gujarat, West Bengal and Himachal Pradesh. It is used for suji preparation.

Indian dwarf wheat (*Triticum spherococcum*) – This is found in limited areas of Madhya Pradesh, Uttar Pradesh of India. They are characterized by very short and compact heads having shorter grains. This belongs to the club wheat of western countries.

Bread Wheat (*Triticum aestivum*) – This is the type presently grown in India in almost all the wheat-growing zones. It is introduced in India by Dr. N.E. Borlaug of Mexico and called as Mexican dwarf wheat. It is the bread wheat.

Einkorn – (*T. monococcum*) A diploid species with wild and cultivated variants. One of the earliest cultivated, but rarely planted today.

Emmer wheat (*Triticum dicoccum*) – It is grown in Spain, Italy, Germany and Russia. It was developed from *T. diccoides* koru., a wild form. In India, it is grown in Maharashtra, Tamil Nadu and Karnataka.

Spelt – (*T. spelta*) Another hexaploid species cultivated in limited quantities.

QK-77 – (*T. polonicum* or *T. durum*) A trademarked tetraploid cultivar grown in small quantities that is extensively marketed.

Table 2.2. Ploidy level and genome of various species of wheat

Ploidy Level	*Species*	*Common Name*	*Genome*
Diploid (2n=14) 2 species	*T. boeticum* (*T. aegilopoides*)	Wild einkorn	AA
	T. monococum	Einkorn	AA
Tetraploid (2n=28) 7 species	*T. dicoccoides*	Wild Emmer	AABB
	T. dicoccum	Emmer	AABB
	T. durum	Macaroni wheat	AABB
	T. persicum	Persian wheat	AABB
	T. turgidum	Rivet wheat	AABB
	T. polonicum	Polish wheat	AABB
	T. timopheevi	-	AABB
Hexaploid (2n= 42) 5 species	*T. aestivum*	Common or bread wheat	AABBDD
	T. compactum	Club wheat	AABBDD
	T. sphaerococcum	Dwarf wheat	AABBDD
	T. spelta	Spelt wheat	AABBDD
	T. macha	Macha wheat	AABBDD

Classes of Wheat

Hard Red Spring – Hard, brownish, high protein wheat used for bread and hard baked goods. Bread Flour and high gluten flours are commonly made from hard red spring wheat.

Hard Red Winter – Hard, brownish, mellow high protein wheat used for bread, hard baked goods and as an adjunct in other flours to increase protein in pastry flour for pie crusts. Some brands of unbleached all-purpose flours are commonly made from hard red winter wheat alone.

Soft Red Winter – Soft, low protein wheat used for cakes, pie crusts, biscuits, and muffins. Cake Flour, for example, is made from soft red winter wheat.

Hard White – Hard, light colored, opaque, chalky, medium protein wheat planted in dry, temperate areas. Used for bread and brewing

Soft White – Soft, light colored, very low protein wheat grown in temperate moist areas. Used for pie crusts and pastry. Pastry flour, for example, is sometimes made from soft white winter wheat.

Hard wheats are harder to process and red wheats may need bleaching. Therefore, soft and white wheats usually command higher prices than hard and red wheats on the commodities market.

Durum and red durum – Very hard, translucent, light colored grain used to make semolina flour for pasta.

The wheat breeder, before he initiates a breeding program, must decide which class of wheat he is going to improve. All the varieties of hard red winter, hard red spring, soft red winter, and most of the varieties of white wheat belong in the species *Triticum vulgare* ($2n = 42$). Varieties of white club wheat comprise the species *T. compactum* ($2n = 42$). The durum and red durum wheat varieties make up the species *T. durum* ($2n = 28$). Wheat may be classed under two principal divisions. The first is composed of all the varieties of red wheat. The second division comprehends the whole varieties of white wheat, which again may be arranged under two distinct heads, namely, thick-chaffed and thin-chaffed. Thick-chaffed wheat varieties were the most widely used before 1799, as they generally make the best quality flour, and in dry seasons, equal the yields of thin-chaffed varieties. However, thick-chaffed varieties are particularly susceptible to mildew, while thin-chaffed varieties are quite hardy and in general are more resistant to mildew. Consequently, a widespread outbreak of mildew in 1799 began a gradual decline in the popularity of thick-chaffed varieties. Wheat varieties are generally designated by names established through publication and usage. The name may be a word-the name of a place, a man, or a descriptive term-a number, or a combination of words and numbers. Earlier the same name might be used for different varieties, or one variety grown under different names. But with the improved sources of information now available to the farmer, confusion in names is less common than previously. The identification of wheat varieties based on such distinguishing features of the wheat plant as presence of awns, glume pubescence, glume colour, kernel colour, and kernel shape. Wherever possible, origins and histories of the varieties are recorded, synonymous names are listed, and distributions of the varieties are mapped. New, improved varieties of wheat, as well as other important field crops, are registered. The wheat varieties grown are continually changing as a result of the development and distribution of new and improved strains by breeders. Information about the varieties of wheat grown is more complete than for any other major field crop.

Crop Growth Stages

There are 5 important stages. Adequate soil moisture is an important factor at these stages.

1. **CRI**: In wheat, the first node of crown is formed near the soil surface irrespective of the depth of sowing. Crown roots start developing at this node. Depending upon temperature, CRI takes place 15-21 DAS. Plants at his stage are very sensitive to soil moisture stress. Hence, there is a need for adequate moisture at this stage.
2. **Tillering**: The lateral buds at these nodes give rise to tillers. Tiller initiation takes place 4 weeks after sowing and tillers continue to emerge for another 2 or 3 weeks.
3. **Jointing stage**: It represents the node production and active growth period. Upto this stage, the seedlings consists of a whorl of leaves. At this jointing stage, the stem becomes visible with distinct nodes.
4. **Heading**: This is the stage where the internodes begin to elongate. Flag leaf and ear emerge towards the end of this stage. Adequate moisture should also be present in the soil at the heading stage.
5. **Grain development**: It is upto the end of soft dough stage of the grain. At this stage also, the plant requires adequate moisture at dough stage.

Table 2.3. Irrigation scheduling of long duration wheat varieties

Crown Root Initiation	21 – 25 DAS
Late tillering	25 – 60 DAS
Late jointing	60 – 70 DAS
Flowering stage	90 – 95 DAS
Milk stage	100-105 DAS
Dough stage	120-125 DAS

DAS: Days after sowing

Table 2.4. Genome Formulas for Several Species of *Triticum* and some of its Close Relatives

Species	*Chromosome Number (2n)*	*Genome Formula*	*Common Name*	*Use*
		Diploid species		
Triticum aegilopoides	14	AA	Wild einkorm	Wild
Triticum monococcum	14	AA	Einkorm	Cultivated
Aegilops speltoides	14	BB		Wild
Aegilops caudata	14	CC		Wild
Aegilops squarrosa	14	DD		Wild
Secale cereale	14	EE	Rye	Cultivated
		Tetraploid species		
Triticum dicoccoides	28	AABB	Wild emmer	Wild
Triticum dicoccum	28	AABB	Emmer	Cultivated
Triticum durum	28	AABB	Durum wheat	Cultivated

Species	Chromosome Number (2n)	Genome Formula	Common Name	Use
Triticum persicum	28	AABB	Persian wheat	Cultivated
Triticum polonicum	28	AABB	Polish wheat	Cultivated
Triticum turgidum	28	AABB	(solid stem) wheat	
Triticum timopheevi	28	AAGG	Timopheevi	Wild
Aegilops cylindrica	28	CCDD	Goat grass	Wild
Hexaploid species				
Triticum compactum	42	AABBDD	Club wheat	Cultivated
Triticum spelta	42	AABBDD	Spelt	Cultivated
Triticum vulgare	42	AABBDD	Common wheat	Cultivated

a: Adapted from Sears and Sarkar and Stebbins.

The polyploidy has played in the evolution of the wheat's became obvious in 1918 when it was established that the three natural groups of wheat had the chromosome number n=7,14,21, respectively (1938). Shortly thereafter, Kihara and Sax analyzed hybrids between the groups and showed that allopolyploidy was involved. Accordingly the genomic formulae AA, AABB and AABBDD could be given to the three groups. There then remained the major task of identifying the diploid species, which had contributed the two genomes of unknown ancestry, B and D. By 1926, it was reasonably certain that the S genome was present in *Triticum cylindricum* were usually found to have seven bivalents, whereas, *Triticum cylindricum* x *Triticum turgidum* (AABB) had little or no pairing. Since *Triticum cylindricum* is a tetraploid with a second genome in addition to D, it was itself unlikely to be the source of the D genome. Instead, it seemed that there must exist a diploid species DD that had been involved in the formation of both *Triticum cylindricum* and mmuneidy wheat. In 1956, Sarkar and Stebbins suggested that the source of the B genome was *Triticum speltoides),* which has the proper morphological characteristics to convert enkorn (AA) to emmer (AABB). Since the amphidiploid *Triticum speltoides* x *Triticum monococcum* was already in existence and had crossed with emmer, *Triticum speltoides* could not immediately be accepted unequivocally as the source of the B genome. The diploid hybrid of *Triticum speltoides* x *Triticum monococcum* has shown far more pairing than was to be excepted in a plant of the constitution AB, and the amphidiploid had multivalent pairing and was very poorly fertile. In 1958, however, it was discovered that chromosome 5B of wheat inhibits pairing of homologous (related) chromosomes. This discovery suggested that if a pairing-suppressing mutation, such as 5B presumably carries, had occurred in the einkorn- *speltoides* amphidiploid, this amphidiploid would have immediately become fertile and stable, with diploid like pairing. Furthermore, it has been found that, of all the conceivable donors of the B genome, only *Triticum speltoides* and *Triticum tripsacoides* (*Aegilops mutica*) have two satellited chromosomes which correspond to those of the B genome of wheat,chromosoimes 1B (I) and 6B(X). It has thus far been impossible to determine just how closely the chromosomes of the B genome correspond with those of *Triticum speltoides* and *Triticum tripsacoides*. The genomes of both *speltoides* and *tripsacoides* suppress the action of the 5B inhibitor, enabling the *speltoides* or *tripsacoides* chromosomes to pair with the chromosomes of the A genome as well as the B, and also allowing A to pair with B.

In 1798, Thomas Malthus stated in his Essay on the Principles of Population that 'Population, when unchecked, increases in a geometrical ratio. Subsistence increases only in an arithmetical ratio. Yields now account for most of the production increments in developing countries. The three major factors responsible for this dramatic jump in yield are the spread of high-yielding semidwarf wheat varieties, the expansion of irrigation, and the increase in use of chemical fertilizers. Expansion of the bread wheat area, and in particular the dramatic increase in yield, have resulted in an impressive rate of production growth of 4.9 per cent per year in the past decade, thereby easily surpassing annual population growth rate, estimated at 2 per cent. Although research on the yield gap and its implementation on the farms of developing countries is a high priority through the end of this century, this chapter is directed towards explaining and implementing conventional plant breeding principles and methods that may be helpful towards increasing food production. Cytogenetic and genetic principles and methods will play a greater role in furnishing new knowledge that will be applied in plant breeding to sustain food production. Mixing maternal and paternal genomes in embryos is not only responsible for the evolutionary success of sexual reproduction, but is also a cornerstone of plant breeding. However, once an interesting gene combination is obtained, further genetic mixing is problematic. To rapidly fix genetic information, doubled haploid plants can be produced: haploid embryos having solely the genetic information from one parent are allowed to develop, and chromosome doubling generates fully homozygous plants. A powerful path to the production of doubled haploids is based on haploid inducer lines. A simple cross between a haploid inducer line and the line with gene combinations to be fixed will trigger haploid embryo development. However, the exact mechanism behind in planta haploid induction remains an enduring mystery. These findings enabled translation of haploid induction capacity to other crops as well as the use of haploid inducer lines to deliver genome editing machinery into various crop varieties.

Genetics of some Morphological and Physiological Characters

The technology of plant breeding and production relies on basic knowledge of the genetic control of plant characteristics. It also seeks to produce varieties of crops suitable for both the environments and management systems where they will grow. Plant scientists based at our research facility, the Plant Breeding Institute (PBI), therefore conduct their research and develop technologies and products in the GxExM space: Genotype x Environment x Management. On the genetic aspects of GxExM, our plant breeding researchers have demonstrated expertise in the development and use of molecular genomics to uncover the genetics of plant characteristics, and in using the knowledge and tools to undertake breeding that is quicker and more certain to produce plants containing the desired genotype. We also develop genetic variation and source it from our global partners, providing the raw material for breeding. Both the biotic and abiotic aspects of the changing environments and management in GxExM are covered by plant production research.

Genetics of a scale for recording foliar blight disease caused by *Helminthosporium sativum* spp. Was developed. Due to the agro-climatic conditions prevailing in this part of the country, this place is considered to be the 'Hot Spot' for foliar blight. It was inferred that resistance to the disease is controlled by incomplete dominant major and minor genes in some Chinese and *Agropyron* transfer lines. Seedling resistance based on lesion size and type is controlled by dominant complementary genes. A new scale, based on genetic parameters was worked out in place of the Saari-Prescot scale. This new scale useful for genetical studies at seedling and adult stages of the plant has been universally accepted as the two digit scale. Increased seed rate along with reduced row spacing enhance wheat yield in this area. Under Furrow Irrigated Raised Bed System (FIRBS) of wheat planting, high tillering varieties like HD 2733, HD 2824 are more suited then shy tillering varieties like HP 1633. Overall yield of wheat-rice system as a whole was improved by incorporation of grain legume like urad in between rice and wheat crops of this system. An additional advantage of improvement of physical and chemical nature of soil was also recorded when legume incorporation was made in the system. Uses of bio-fertilizers like p-culture and *Azotobacter* in wheat crop were found to improve the available phosphorus in the soils of rice- wheat cropping system which compensates for 50 per cent of recommended nitrogenous and phosphatic fertilizers in wheat and shows residual effect also.

Height

The widespread adoption of semidwarf varieties has played a significant role in world wheat production. These varieties have short straw, which makes the plant less prone to lodging. The breeding and cultivation of semidwarf wheat varieties began when very little knowledge of the genetic control of plant height was available. In the last decade, considerable information has been revealed about genes for plant height and their function. The development of short-strawed varieties probably started in Japan in the 19th century. These varieties were used by breeders in Italy and the United States, followed by breeders in Mexico and certain other countries, to reduce plant height. Although the genetics of dwarfism has been extensively studied, relatively few genes have been genetically characterized. The general symbol for dwarfing genes (*i.e., tor* reduced height) is *Rht 'n'*, which has replaced the symbol used previously *(Sd 'n)* (Allan, 1970).

Gibberellic Acid-insensitive Dwarfing Genes

Rht1 and *Rht2* (the Norin 10 dwarfing genes) were incorporated into several important cultivars, through the initial cross that used the Japanese variety/, Norin 10 (Reitz and Salmon, 1968). *Rht1* is located in the alpha arm of chromosome 4A and is available in both *aestivum* and *durum* wheats. Similarly, *Rht2* is located in the short arm of chromosome 40. The two dwarfing genes appear to have very similar effects on plant height, but these can be modified by genotypic background. *Rht3* (the Tom Thumb dwarfing gene) is located in the alpha arm of chromosome 4A. *Rht3* was originally thought to be genetically independent from *Rht1* and *Rht2*; however, *Rht1* and *Rht3* were alternative alleles at the same locus of chromosome 4A. Because of the extreme dwarfism conferred by *Rht3*, its commercial use in wheat is doubtful. However, *Rht3* has potential for triticale breeding, in which its severe dwarfing effects are reduced by the rye genome. The genetic linkage between *Rht1*, *Rht2*, and *Rht3* and genes *tor* gibberellic acid (GA) insensivity (Gaf.genes) has been investigated by various researchers. Varieties with any of the three dwarfing genes fail to respond to the application of gibberellic acid. Although Hu and Konzak (1974) reported recombination between dwarfing genes and genes *tor* gibberellic acid insensitivity, Gale and Law (1976) consider *Rht1* and *Gai/1*, *RhtP*. And *Gai2*, and *Rht3* and *Gai3* to be pleiotropic genes. *Rht* 10 (the Ai-Bian 1 dwarfing gene) has been reported dominant GA insensitive dwarfing gene in Chinese variety Ai-Bian 1 (Izumi *et al.*, 1981). *Rht1 o* produces more severe dwarfism than *Rht3* and is located in the short arm of chromosome 40. The recombination values obtained by telocentric mapping indicate that the locus is not linked to the centromere and therefore non-allelic with *Rht2"*, however, the critical segregation, from a cross of Ai-Bian 1 with an *Rht2* genotype, has not been reported.

Gibberellic Acid-Sensitive Dwarfing Genes

Rht4 was induced by X-ray and fast neutron irradiation in the variety Burt and extracted in the line Burt ert 937. Hu (1980) reported that *Rht4* segregates as a single recessive factor and reduces plant height by about 4S per cent. *Rht5* was induced by ethyl methyl sulphonate (EMS) treatment of the variety Marted into the mutant line Marted Cl (Cl 13988) and is reported to be partly dominant. In the homozygous state, *Rht5* reduces plant height by SO per cent. Although the genetics is not clear, the symbol *Rht6* was assigned to account for the recessive height difference between the variety Burt and taller wheats. The recessive gene *Rht7was* induced by EMS treatment of the variety Bersee, located in chromosome 2A. This gene reduces plant height by 24 per cent. *RhtB* and *Rht9* are the Akakomugi dwarfing genes. The genetics of Akakomugi dwarfism is now becoming clear, more than 70 years after its first use in Italy. *RhtB* is located in chromosome 20 and *Rht9* in the short arm of chromosome 7B (Gale and Youssefian, 198S).

Awnedness

Awnedness can be inhibited by dominant alleles at three independent loci: *Hd*, *81*, and *82*. The first allele, *Hd*, causes the hooded awn phenotype and is located in the short arm of chromosome 4B. Alleles *81* and *82* cause awnless or tipped awned phenotype and are located in long arms of chromosomes SA and 6B, respectively. A bearded or fully awned phenotype occurs with genotype *hd b1 b2*. Watkins and Ellerton (1940) also noted the probability of a third allele, *b1a*, leading to a half-awned condition, and in their discussion, they consider the possibility of a similar third allele at the *82* locus. In view of more recent cytogenetic analyses, it seems that the half-awned condition could result from epistatic interactions between alleles *81* and/or *82* and various promoter genes. Although hooded, half-awned, tip-awned, and awnless variants occur among tetraploid wheats, these are relatively infrequent.

Pubescence

Pubescence in the glume, leaf, peduncle, and node is reported to be controlled by different dominant alleles. Gene *Hg* (hairy glume), located in the short arm of chromosome 1A controls the pubescence of the glume and is fairly widespread. Gene *HI* (hairy leaf) controls leaf pubescence and is located in chromosome 4A. Wheats possessing the translocation of rye chromosome SR can carry the gene *Hp* (hairy peduncle). The translocation of SR has been reported to have occurred with wheat chromosomes 4A-beta arm, 58-short arm and 60. Pubescent node phenotype is also known to be controlled by dominant allele *Hn* (hairy node), located in chromosome SAL and linked with the awn-inhibitor loci *Bt* (Gaines and Carstens, 1926).

Red Grain Colour

Red colour of the grain is controlled by three independent dominant alleles: *Rt, R2,* and *R3,* which are located in chromosomes 30, 3A, and 38. White grain is obtained when recessive alleles *rt, r2,* and *r3* are present together. Although alleles *Rt, R2,* and *R3* are capable of conferring red grain phenotype, they work in an additive manner; therefore, a variety with all three alleles will have very dark red grain, compared to a variety with only one of the alleles.

Anthocyanin Pigmentation

Simple inheritance is reported for the accumulation of anthocyanin pigmentation in various plant parts. Sharman (1944) reported that a single dominant factor is responsible for controlling purple anthers. Red auricles are reported to be controlled by allele *Ra,* located in chromosome 20 of variety Kenya S8. Three independent alleles, *Rct, Rc2, and Rc3,* located in chromosomes 7A, 7BS, and 7DS, respectively, are reported to control red/purple phenotype of the coleoptiles. Sutka (1977) has reported a fourth factor in chromosome 68 and found that the phenotype can be suppressed by alleles located in chromosomes 2A, 28, 20, 48, and 6A. His results are to be confirmed by other scientists. Purple (or red) culm (straw or stem) phenotype is controlled by a single dominant allele, *Pc,* located in chromosome 7BS. The genetics of purple grain or pericarp is not well understood. Genes for this phenotype have been transferred from tetraploid wheats to the mmuneidy level. Mcintosh and Baker (1967) reported that duplicate dominant genes could be involved.

Glume Colour

Two independent and dominant alleles, *Rgt* and *Rg2,* are reported to control the red (or brown/bronze) phenotype of the glumes. *Rgt* is known to be located in chromosome IBS (Mcintosh, 1983), whereas *Rg2.,* derived from *Aegilops squarrosa,* is located in chromosome IDL. The black colour of the glumes is reported to be controlled by a single factor, *Bg,* located in chromosome 1 A. This allele has been reported in diploid, tetraploid, and hexaploidy wheats. The *Bg* allele is epistatic to alleles *Rgt* and *Rg2.* PseuJo-black chaff phenotype got transferred from Yaroslav emmer (tetraploid) to Hope (mmuneidy) wheat at the same time that McFadden transferred the stem rust resistance gene, currently known as *Sr2.* A number of scientists have studied the pseudo-black chat phenotype and conflicting reports are available regarding its genetics. However, it is well established that this phenotype is controlled by a single recessive or partly dominant allele located in chromosome 38S (Singh, 1983) and is very closely linked with *Sr2* (Hare and Mcintosh, 1979).

Bhowal and Narkhede (1981) believe that the pseudo-black chaff phenotype can be suppressed by a number of alleles in chromosomes 18, 28, 3A, 30, 48, 40, 58, and 7A.

Hybrid Weakness

Hybrid weakness can be of two major types: hybrid necrosis (also described as progressive lethal necrosis and firing) and hybrid chlorosis.

Hybrid Necrosis

This phenotype is controlled by complementary dominant alleles at *Ne1* and *Ne2* loci in chromosomes 58L and 28S. Various alleles are reported at each locus and are described as *w* (weak), *m* (medium) and *s* (strong). The magnitude of necrosis therefore depends on the interaction of *Ne1w, Ne1m* or *Ne1s* with complementary counterparts *Ne2w, Me2, Me2ms,* or *Ne2s.* Hybrids with *Ne 1 s* and *Ne2s* are completely necrotic, whereas all other combinations show only partial necrosis. Furthermore, the phenotype is also influenced by the environment and genetic background of the variety. *Ne1* occurs very commonly in tetraploid wheats, whereas *Ne2* is rare or absent (Nishikawa, 1967).

Hybrid Chlorosis

As with hybrid necrosis, in hybrid chlorosis complementary dominant genes, *Ch 1* and *Ch2,* are again involved. *Ch1* is located in chromosome 2A and *Ch2* is located in chromosome 30-alpha. Allelic variation at the *Ch2* locus has been suggested (Mcintosh, 1983).

Grass Clump or Hybrid Dwarfs

The expression of grass-clump phenotype is attributable to the complementary gene action of at least four loci. Allele *01* is located in chromosome 20-alpha. *02* is in chromosome 2BL, *03* in chromosome 4BL, and *04* in chromosome 20-beta. Grass clump dwarfs are produced through the interaction of alleles in the following genotypic combination: *010203, 010304, 01020304, 010202, 010204,* and *010404.* Worland and Law (1980) indicates multiple allelism.

Photosensitivity

Photosensitivity is reported to be simply inherited and controlled by alleles at three independent loci of group 2 chromosomes. Alleles *ppd1, ppd2,* and *ppd3* are reported to be recessive (photoinsensitivity is thus the dominant phenotype) and are located in chromosomes 20, 2BS, and 2A, respectively (Law *et al.,* 1978). The degree of photosensitivity is reported to be influenced by the modifying genes (Halloran and Boydell, 1967).

Vernalization

Response to vernalization has been reported to be due to recessive alleles at five loci. Allele *Vm 1* is located in chromosome SAL and is insensitive to vernalization. *Vm 1* is also epistatic to other genes. The chromosome location of *Vm2* is not known but has been identified in many cultivars. *Vm3* is located in chromosome SDL and is also widely distributed. *Vm4* is reported to be present in Australian variety Gabo together with *Vm2*. *Vm5* is located in chromosome 7BS of variety Hope. Pugsley (1973) indicated that the winter wheats possibly possess all vernalization alleles, and that differences among cultivars with respect to vernalization requirements are possibly due to multiple recessive alleles.

Cytogenetics

Cytogenetic studies of interphase chromosomes have greatly contributed to a better understanding of gene regulation. Immuno-cytogenetic and molecular cytogenetic techniques have proven powerful tools to dissect the nuclear architecture and correlate chromosome substructures with covalent modifications of chromatin and with biochemical processes, such as transcription and replication. The advent of live-cell imaging using fluorescent protein technology and 3-D microscopy has provided cytogenetics with a new dimension because we are now able to monitor the dynamics of chromatin subdomains in real time. Along with novel in vivo microscopical approaches, such as photo bleaching and fluorescence recovery techniques, new insights into nuclear architecture and chromatin dynamics are likely in the future. Moreover, the ability to quantify fluorescence intensity in time and space provides us with a tool to measure dynamic properties of multiple chromatin proteins. In combination with computer modeling, this will allow us to establish a kinetic framework for the interaction between the molecular components of epigenetic mechanisms of gene regulation. The wheat varieties of IARI like HP-1761, HP- 1744, HP- 1731, HD- 2643, HD- 2733, HD- 2824, HW- 2045, NW- 2036, *etc.* are having a huge demand in Bihar, in particular, and NEPZ, in general. About 30-35 per cent area in Bihar is saturated by IARI varieties which have become very popular among the farming community due to high yield.

The aneuploid progenies, *i.e.* monosomics (2n=41); nullisomics (2n=40); trisomics (2n=43); tetrasomics (2n=44), *etc.*, dependent on their frequency and specific modification of plant morphology can be easily confused with other off-types resulted from outcrosses or mechanical impurifications. Moreover, some types of aneuploids, especially monosomics are not recognizable as natural genetic variants by the testing mmuneidy. The aneuploids for chromosome 5A with speltoid ear (monosomics or nullisomics) and compactoid ear (trisomics or tetrasomics) are exceptionally excluded from the evaluation of varietal impurity in some countries (Worland and Law, 1985). However, the aneuploid frequency in wheat cultivars may be higher than officially admitted levels of impurity and this could cause problems at certification. For example, in five traditional (tall) wheats, an average 1.08 per cent aneuplolid progeny, ranging from 1.60 per cent in Hybrid 46 to 0.37 per cent in Cappelle – Desprez was found; some values being above the standards (Riley and Kimber, 1961). The introduction of semidwarf varieties at the end of last century has led to a new reconsideration on the meaning of "atypical form". The majority of tall plants which usually occurred in these varieties are indeed aneuploid types – especially monosomics for the chromosomes 4B and 4D that carry height reducing gene RhtB1 and respectively RhtD1 (previously Rht2). The increased height of monosomic plants is the results of dosage effect of RhtB1/rhtB1 and RhtD1/rhtD1 loci. Such types of aneuploids cannot easily be distinguished from unacceptable deviants and add new question on the matter of seed certification.

In Basic seed of the English semi-dwarf variety Brigand with RhtD1 gene a rate of 1.53 aneuploids per thousand plants was found, this value exceeding the total level of off-types permissive in the population. Even higher frequencies of aneuploids, from 5.77 per cent to 10.42 per cent were registered in some Argentinian semi-dwarf varieties carrying RhtB1 gene. Outstanding differences of aneuploid frequencies between years of storage in the LasRosas-INTA cultivar were also noted (Suarez and Favret, 1986).

A preliminary evaluation on aneuploids presence in three Romanian wheats carrying RhtB1 gene revealed frequencies ranged from 0.08 per cent to 0.10 per cent reaching the standard limits of varietal purity. This paper presents new data concerning off-types frequencies in Romanian wheats and some morphological characteristics of monosomics that differentiate them from euploids. Plant material is represented by four semi-dwarf winter wheat varieties Fundulea 133, Fundulea 4, Dropia and Glosa. These varieties carry RhtB1 gene for reducing height. The translocation 1B/1R is present in Fundulea 4, uncertain in Fundulea 133 and absent in Dropia and Glosa. Samples of Breeder's seed and Prebasic I stocks were grown in different years in field plots as spaced plants. After anthesis off-type plants were easily identified by their taller distinctive phenotype. The plant chromosome complement was established by means of chromosome counts on root-tip squashes stained by Feulgen method. The acetocarmine anther smear technique was used to check the monosome 4B identity on the F_1 hybrids between monosomic plants and Chinese Spring ditelocentric 4BL tester line. An experimental field layout consisting of single plant randomization design was used to assess the variation in height, tiller number and other plant traits of monosomics and disomics (euploids). Field observation in seed multiplication plots of semi-dwarf Romanian cultivars showed the presence of atypical plants easily recognizable by their height after anthesis. All of the four cultivars produced tall off-types with various frequencies in different years and multiplication steps, too. Generally, an increased offtype plant number was found in the second and third year of the experiment, both in Breeder's seed and Prebasic I plots. Such tall plants could be resulted *de novo* as aneuploids for chromosome 4B by abnormal chromosome pairing due to the temperature variation during meiosis. Other factors could also contribute to the karyotypic instability and taller plants I like precocious ageing of seed due to improper storage condition. The size of the seeds after seed conditioned process may be another factor, because a higher percent of aneuploids was found in wrinkled ones and, not less important factor, the correctness of purification process in previous seed multiplication fields. Random samples of tall plants of Fundulea 133 and Fundulea 4 were grouped in two height classes and seeds of individual plants were then cytologically analysed. The majority of the analysed plants in taller classes were monosomics. Only three plants out of 20 were monosomics (2n=41) in the class of 91-96 cm while in the class of 97-104 cm, 23 plants out of 29 proved to be monosomics. A similar high proportion of monosomics (85.0 per cent) was found among taller plants of Fundulea 4.

Meiotic analysis of F_1 hybrids (monosomics x ditelocentric 4BL tester line) validated the 4B monosomic identity of tall analyzed aneuploids. The presence of euploid taller plants is difficult to explain. They could be a result of outcrosses during seed multiplication or have originated from accidental contamination. Another explanation is that the Fundulea 4 may contain a rare biotype with different photoperiod reaction and vernalization requirements reflected in an early spring development that could influence the final plant height. A misclassification was also possible since some plants classified as disomics could be monotrisomics or carry other combinations of 2n=42 chromosomes. On the other hand, a small part of chromosome 4B carrying the dwarfing RhtB1 gene might be lost. The cytologically selected monosomics of Fundulea 133 and Fundulea 4 together with normal disomic plant were studied in the field using the single plant randomization design. The results of morphological measurements revealed significant differences between euploid and monosomic plants regarding plant height, yield per plant and some spike characteristics. The presence of recognizable aneuploids for chromosome 4B in Breeder's seed and Prebasic I plant population of the analysed cultivars led us to the presumption that overall aneuploid frequency should be higher if other chromosomes were lost at a similar rate. In order to test this hypothesis a mitotic analysis was carried out on Breeder's seed of Glosa variety. Out of 451 analysed seeds 448 were euploids, one plant was monosomic (2n=41) one plant trisomic (2n=43) and another one had 63 chromosomes, probably resulted by pollination of a non-reduced meiocyte with a normal grain pollen. Aneuploid frequency of 0.44 per cent that we found in analysed seeds of Glosa Breeder's stock did not exceed the admissible standards of varietal purity at this seed category. The identity of monosomic and trisomic plants has not yet been established. However, it is worth to mention that the field plot was purified after anthesis and off-types eliminated before harvest time. Aneuploids occurrence in common wheat cultivar population should be considered a natural genetic variant and treated in an adequate manner. Among the tall off-type group of plants found in Romanian cultivars the majority were aneuploids for chromosome 4B which carry RhtB1 height reducing gene. The seed producers should take all necessary measures to keep the seed stocks as pure as possible during multiplication chain to prevent any troubles at the seed certification.

Cell identity is determined by the nuclear program and established by a complex molecular interplay between DNA sequence and proteins. The past few years have witnessed major breakthroughs in the elucidation of this intricate mechanism of epigenetic gene regulation. Covalent modifications in cytosine residues of DNA and in histone tails are revealed as hallmarks that establish the functional status of chromatin domains. In particular, our knowledge of the functional implication of histone tail modifications, known as the histone code, has rapidly increased. Cytogenetics has taken a prominent share in this expanding research field by linking nuclear structures with biochemical features such as gene activity, DNA replication, histone modification and DNA methylation. Cytogenetics have resulted in numerous karyotypes and laid the basis for the present knowledge and understanding of chromosome dynamics in relation to mitosis and meiosis. Moreover, cytogenetics has been applied to identify chromosomal rearrangements such as deletions and translocations and position these aberrations onto the linear chromosome relative to centromeres, telomeres and chromosome bands.

The resulting cytogenetic maps have provided important information about the physical distances between genetic markers on a linkage map. Fluorescence in situ hybridization (FISH) has enabled us to map molecular markers, ranging in size from large regions with repetitive elements to small single copy genes [Hans de Jong, 1999], and dramatically increased the accuracy of cytogenetic maps. A major spin-off was the high-resolution integration of genetic linkage maps with physical molecular maps, particularly when using FISH to pachytene chromosomes [Cheng, 2001]. These studies provide detailed information about the relation between molecular distances and recombination frequencies along the chromosome arm as well as the identity of DNA sequences in and at the boundaries of heterochromatin. Although cytogenetic studies of interphase nuclei have been carried out for more than a century, our understanding of interphase chromosomes is limited compared to their mitotic and meiotic counterparts. This is mainly because of their ostensibly unorganized appearance during interphase. Electron microscopic studies have monitored ultrastructural changes in interphase chromatin upon cell differentiation. However, apart from the nucleolus and heterochromatic domains, identifying subnuclear structures had been difficult. With the advent of molecular cytogenetic and immune-cytogenetic techniques, the molecular dissection of the interphase nucleus at the light microscopic level came within reach. The individual detection of protein complexes, RNAs and DNA sequences revealed many nuclear compartments and enabled detailed identification and localization of different nuclear structures [Spector, 2003]. The variation in nuclear architecture allows us to list cytological characteristics that reflect the relationship between nuclear organization and gene regulation. The list includes several quantitative and qualitative parameters, such as: (i) nuclear size and shape; (ii) relative content and distribution of heterochromatin; (iii) global patterns of chromatin modification (*e.g.* DNA methylation and histone tail modifications); (iv) presence of nuclear compartments (*e.g.* number and shape of nucleoli, chromocenters, chromosome territories, nuclear bodies); (v) organization of chromosomes (*e.g.* position and orientation of chromosomes and local differences in chromatin compaction). An attempt has been made to discuss progress and applications of cytogenetic techniques to study plant nuclear phenotypes in relation to gene expression. Because most of what we know about interphase chromosomes comes from studies with human cells, we regularly refer to mammalian studies in this review. Special attention is paid to the relationship between chromosome organization and gene activity and the influence of covalent chromatin modifications on nuclear structure and live cell imaging.

Bread wheat (*Triticum aestivum* L.) is an allohexaploid with 21 pairs of chromosomes and three genomes, A, B, and D. Based on chromosome pairing behaviour, it was believed for a long time that einkorn wheat *Triticum monococcum* contributed the A genome to both tetraploid durum (*Triticum turgidum*) and mmuneidy bread wheat. However, more recent molecular genetic evidences show that *T. immune* Tumanian ex Gandilyan, *Aegilops speltoides* Tausch, and *Aegilops tauschii* Coss. Are donors of A, B, and D genomes, respectively (Gill and Friebe, 2002). The mmuneidy wheat shows disomic inheritance and behaves like a diploid. This has been possible mainly due to the presence of two major pairing regulator genes, *Ph1* (Riley and Chapman, 1958) and *Ph2* (Mello-Sampayo, 1971). Wild relatives of cultivated wheat have been used in transferring new genetic diversity not known to occur in cultivated wheat germplasm. Although mutants of *Ph* genes

(Sears, 1977), which enhance pairing between wheat and chromosomes of related alien species, have been used successfully to induce translocations of chromosome segments onto wheat chromosomes, such translocations have also been achieved through radiation and tissue culture.

Relation of Interphase Chromosomal Position with the Domains of Transcriptional Activity

Chromosomes occupy nuclear territories, that vary in shape and position. Previous cytogenetic studies on human chromosomes reported the preferential position of genes at the surface of these territories [Kurz, 1996], suggesting that these are the sites of transcriptional activity. However, BrUTP labeling of nascent RNA revealed that transcription occurs inside the territory, almost exclusively at the surface of condensed chromatin domains. Moreover, a FISH approach with probes for constitutively expressed genes showed signals inside the chromosome territory. By contrast, genomic regions pertaining to the histocompatibility complex locus (MHC) appeared to localize outside the chromosome territory depending on cell type and on the activation state of their genes. By using a transcriptional activator, gene activation has been shown to be correlated with more-frequent repositioning of the genes away from the territory. Further analysis of the MHC locus and other gene-dense regions, suggests that both transcription activity and gene-density account for the extra-territorial location [Chubb and Bickmore, 2003]. Apparently, transcription sites in species with large chromosomes are not restricted to the more decondensed areas. However, transcription might not only come from coding regions. Considering the cytogenetic observations in humans and in plants, the correlation between chromosome arrangement and distribution of transcription sites might depend on chromosome size and on the linear organization of coding sequences along the chromosomes. Alternating gene-dense and repeat-rich segments might occupy separate nuclear domains. In this context, it is tempting to speculate about the extent to which clusters of co-regulated genes determine the functional organization of interphase chromosomes. These so-called RIDGEs (Regions of Increased Gene Expression) have been reported in the human genome. RIDGEs could globally regulate the expression levels of their embedded genes and govern chromatin compaction, nuclear positioning or local concentrations of transcription factors. In addition, the gene density in the euchromatic loops is tenfold higher. The decondensed loops in Arabidopsis might therefore be equivalent to RIDGEs.

Heterochromatin: Indefinable, Yet Measurable

Heterochromatin was originally described as highly condensed chromatin. In the past decade, several disciplines have revealed additional properties of heterochromatin, underscoring it as transcriptionally inert chromatin. The nuclear phenotype in differentiating cells often displays changes in heterochromatin. Although heterochromatin is difficult to define, it is possible to quantify the variation in heterochromatin between different cells, particularly in nuclei with discrete chromocenters. These distinct heterochromatic domains are present in several organisms, such as mouse, *Arabidopsis* and many other plant species. DAPI staining of nuclei reveals bright fluorescent, intensely stained chromocenters;

the area and the intensity of which are easily measured. To establish the relative heterochromatin fraction (RHF), the fluorescence intensity of all chromocenters is determined relative to the fluorescence intensity of the entire nucleus. RHF values in *Arabidopsis* nuclei are,10–15 per cent, which matches the percentage of repetitive sequences in the *Arabidopsis* genome. The method has been applied to establish quantitative differences in heterochromatin during development and between wild type and a gene silencing mutant affected in DNA methylation. For example, a DNA hypomethylation mutant has 30 per cent less heterochromatin compared with the wild type, whereas the F_1 (mutant and wild type) shows intermediate levels. However, a minor fraction of mutant cells display wild-type levels of heterochromatin. During leaf development, mesophyll cells differentiate and the nuclei become elliptic and larger. The cytogenetic analysis of the nuclear phenotype provides a powerful tool to quantify visible differences in heterochromatin and thus to examine subtle changes in chromatin.

DNA Methylation Directs Changes in Chromatin Structure

DNA methylation plays an important role in epigenetic gene repression. In plants, up to 80 per cent of cytosines at CpG sites are methylated. Indeed, immune-localization studies of 5-methylcytosine in *Arabidopsis* nuclei, revealed that chromocenters contain the majority of densely methylated sequences. One of the questions is whether DNA methylation is involved in the establishment of heterochromatin domains, or if it is merely a consequence of it. Valuable information came from cytogenetic studies of gene-silencing mutants, such as ddm1 and met1-1. Both mutants derepress transcriptional gene silencing and prompt transposon reactivation. They show strongly reduced DNA methylation levels, but the proteins have different functions. DDM1 is an SWI/SNF-like chromatin remodeling protein, whereas MET1 is a maintenance methyltransferase involved in CpG methylation. Cytogenetic examination of ddm1-2 and met1-1 revealed a striking nuclear phenotype. In contrast to the prominent heterochromatic chromocenters in wild-type plants, nuclei of the DNA methylation mutants show small chromocenters and 30 per cent less heterochromatin compared with the wild type, irrespective of the genetic background or the tissue type. Furthermore, mmune-localization showed that 5-methylcytosine is dispersed and no longer concentrated at chromocenters contrary to the wild type. FISH analysis revealed that the reduction in heterochromatin content is caused by re-localization of pericentromeric non-transposon sequences away from chromocenters. This implies that the localization of these sequences in heterochromatic domains is controlled by DNA methylation. By contrast, the major tandem repeats and the high-copy transposable elements remain in the heterochromatic environment of the reduced chromocenters. However, the ddm1-5 mutant, a functional null mutant with a stronger reduction of DNA methylation and a more severe phenotype than ddm1-2, has even less heterochromatin than ddm1-2 or met1-1 [Probst, 2003]. FISH analysis of ddm1-5 nuclei shows dispersed signals of the tandemly repetitive 180 bp repeat and the ribosomal genes, indicating decondensation of centromeric, pericentromeric and NOR regions, thus accounting for the further disintegration of the chromocenters. The change in chromatin upon demethylation is well illustrated with reactivated transposons and genes. Hypomethylation of the CAC_1 transposon in the ddm1-2 background leads to its reactivation and migration into euchromatin. Similarly, in the transgenic *Arabidopsis* line A, a region containing multiple copies of a silent hygromycine phosphotransferase (HPT) gene is strongly condensed and forms a small neo-chromocenter. In a ddm1-5 mutant, the transgene is reactivated and the locus is drastically decondensed. These cytogenetic data indicate that global demethylation of DNA is accompanied by a decrease in heterochromatin. Similar effects of DNA hypomethylation have been demonstrated in species with a large genome. The treatment of root tips with 5-azacytidine (5-Aza), a DNA methylation inhibitor, results in decondensation of chromatin fibers in *Vicia faba* and in wheat. In addition, FISH revealed that nuclear colocalization of allelic transgenes, as observed in untreated plants, is lost in 5-Aza-treated seedlings. The relocation was accompanied by increased transgene activity. However, the global distribution of transcription sites was not affected by 5-Aza-induced demethylation. Accordingly, the Rabl configuration of chromosomes remained unchanged upon 5-Aza treatment. These data underline the strong correlation between DNA methylation and the condensed chromatin. Although it is generally accepted that both cytosine methylation and condensed chromatin are involved in gene silencing, the causal relationship between DNA methylation and chromatin compaction remains unclear. A minor fraction of nuclei in DNA methylation mutants still display the wild-type nuclear phenotype. This indicates that in some cells, the role of DNA hypomethylation in chromatin condensation can be compensated by another, as yet unknown, mechanism [Abranches, 2000].

Modifications in Histone Tail: Gene Regulation through an Epigenetic Code

Covalent modifications of histone tails at specific residues are hallmarks of epigenetic gene regulation. Immuno-localization has proven indispensable for linking nuclear architecture with large-scale modification of histone acetylation, methylation and phosphorylation along mitotic chromosomes and in interphase nuclei. Histone acetylation is generally associated with transcription and DNA replication. In barley and *Vicia faba*, acetylation at histone H_4 lysine residues 5 and 16 (K_5 and K_{16}) is linked with DNA replication rather than with transcription, similar to the situation in mammals for K_5 and K_{12}. The histone acetylation of the same residue has different effects in different organisms, suggesting that the histone code might be species specific. The relationship between histone H_3 methylation at lysines K_4 (H_3meK_4) and K_9 (H_3meK_9) and gene activity was recently reported in several plant cytogenetic papers. The causal relationship between histone H_3K_9 methylation and DNA methylation has been a matter of discussion. DNA methylation at CpNpG was shown to be histone code dependent because loss of the histone H3 methyltransferase (KYP_2) resulted in loss of CpNpG methylation, comparable to the situation in *Neurospora crassa* and in mammalian pericentric heterochromatin. However, the inverse relationship appears to account for DNA methylation at CpG and histone methylation of H_3K_9 as reported in two recent *Arabidopsis* papers. Global H_3K_9 methylation is significantly reduced in the DNA methylation mutants ddm1 and met1. Similar results were obtained from immunolocalization and chromatin immunoprecipitation (ChIP) experiments with a null mutant of MET1. In addition, chromocenters in the kyp2 mutant show reduced histone H_3K_9 methylation, but global DNA methylation was strong in chromocenters, comparable to wild-type nuclei. These data indicate that CpG methylation controls histone methylation of H_3K_9 [Tariq, 2003]. By contrast, histone H_3K_9 methylation at centromeric minor satellites occurs independently of $SUV_{39}h$. Strikingly, the pericentromeric heterochromatin contains trimethylated H_3K_9, typical for $SUV_{39}h$ activity, whereas centromeric H_3K_9 is di-methylated. This implies the specific interactions between DNA methylation and di- or tri-methylation of histone H_3K_9 are dependent on the type of histone methyltransferase. The importance of target and product specificity of the histone modifying enzymes, adding another level of complexity to the histone code.

Chromatin Imaging in Living Cells

The use of immuno-labeling and FISH has provided us with invaluable data on nuclear organization. However, these techniques are applied to fixed cells and can therefore only provide static in situ data. In addition, the denaturing conditions of FISH treatment can result in reduced structural preservation. These problems are circumvented by the introduction of fluorescent proteins. For example, the fusion of green fluorescent protein (GFP) to chromatin proteins enables us to monitor chromatin in living mammalian cells while causing only limited phototoxicity damage. Waves of synchronous nuclear divisions, starting in the micropylar endosperm and moving to the opposite chalazal pole via the peripheral endosperm, have been reported. These regions represent three distinct mitotic domains during the development of the embryo. Movement of the fluorescent locus has been observed during interphase, albeit in confined volumes. This suggests that the locus is anchored to fixed sites within the nucleus that allow constrained movement. Endoreduplicated pavement cells showed a 6.6 times larger confinement volume than diploid guard cells, whereas the DNA-content-to-nuclear-volume ratio remained more-or-less constant. It is suggested that endoreduplicated nuclei contain less chromatin-interacting proteins and anchoring sites, therefore providing more free space for chromatin movements [Misteli, 2001].

Pollen Viability

Materials

Forceps, Glass slides, cover slip, dropper, Triphenyl Tetrazolium Chloride stain (1 per cent TTC in 5 per cent sucrose (Oberle and Watson, 1953), cellotape.

Method

Collect pollen grains in the morning time (8.00 AM-10 AM) depends on the crop. Dust pollen on the slides using a forceps or brush and mixed with a drop of the stain. Put cover slip on the specimen and seal it with cellotape. Insert the slide in the foldscope after 15 m of time and observe. Select a field with a minimum of 10 pollens. Record images using mobile camera. Calculate the percent viability by counting total pollens and viable pollens.

Pollen Germination

Materials

Forceps, cavity glass slides, cover slip, dropper, germination medium (depends on crop), cellotape.

Method

Collect pollen grains as the above procedure. Dust pollen on the cavity slides using a forceps or brush and mixed with a drop of the medium. Put cover slip on the specimen and seal it with cellotape. Incubate the slides in a humid chamber at 22-25°C. Insert the slide in the foldscope and observe. Select a field with a minimum of 10 pollens. Record images by using mobile camera. Calculate the percent germination by counting total pollens and germinated pollen.

Over the years, numerous cytogenetic stocks have been developed. Sears (1954) and Sears and Sears (1978) identified numerous wheat aneuploids, such as monosomics, nullisomics, and telosomics, for different wheat chromosomes or chromosome arms. Such stocks were very useful initially in determining the location of genes governing a specific trait on wheat chromosomes and later in determining the locations of molecular markers. More recently, development of deletion stocks (Endo and Gill, 1996) has highlighted the relationship between the chromosomes' physical maps with that of genetic maps. Moreover, these deletion stocks have assisted in targeting a specific region of the wheat chromosome that is useful for identifying closely linked markers with genes of interest, positional cloning of genes, and determining synteny with other cereal species, such as rice, which has had its whole genome sequenced (Mateos-Hernandez *et al.*, 2006). Although cytogenetic stocks are not used directly in wheat breeding, they have played a vital role in enhancing the understanding of wheat genetics, identification of molecular markers linked to genes of interest for marker-assisted selection, and perhaps more importantly in introducing new genetic diversity from alien species and genera. Identification of gene(s) for dwarfing and pigmentation in wheat using monosomic lines of Chinese Spring and C 591. Identification of gene(s) responsible for resistance to leaf and stem rusts. Common wheat, *Triticum aestivum* (2n = 6x = 42), is an allohexaploid comprising three distinct genomic complements, namely A, B, and D, with the genomic structure AABBDD. *Triticum monococcum* and *Aegilops squarrosa* L. [*T. tauschii* (Coss.) Schmal.] are recognized as the A and D genome donors, respectively, while the B genome donor is controversial [Kimber, 1974]. The diploid wheats have been assigned the genomic symbols AA or BB and the tetraploid or the emmer wheats have the genomic constitution AABB. On the basis of similarity between different nullisomics and a study of the nullisomic-tetrasomic combinations, Sears [1954] ascribed the 21 pairs of wheat chromosomes to seven homoeologous groups of three with similar genetic activities. Each genome thus has one chromosome pair in each homoeologous group, while each homoeologous group has one pair in each genome [Okamoto and Sears, 1962]. Wazuddin and Driscoll (1986) have proposed an alternative hypothesis on the evolution of polyploid wheats based upon their discovery of the absence of chromosome *d4* (chromosome

4 of diploid wheat) from hexaploid wheat. This involves the retention of two versions of chromosome 4B at the early tetraploid stage and a loss of *d4*, thereby implying that the A, B, and D genome progenitors contributed six, eight, and seven chromosomes, respectively, to hexaploid wheat. In southwestern Asia wheat was already an important crop when history was first recorded. It was cultivated in ancient Greece, Persia, and Egypt and throughout Europe in prehistoric times. Wheat was brought to the United States by the first colonists along our eastern shores. The genetic origin of wheat is of interest for it is a classical example of how closely related species may be combined in nature into a polyploid series. The species of *Triticum* and their close relatives may be divided into diploid, tetraploid, and hexaploid groups, with chromosome numbers of $2n = 14, 28$, and 42, respectively. Important species within each group. Species within the tetraploid group have apparently originated from combinations of two diploid species as indicated by the genomic formulas. The hexaploid species originated by the addition of a third genome to a tetraploid species. Present evidence indicates that the tetraploid emmers *(AABB)* evolved from amphiploids between *Triticum monococcum (AA)* and *Aegilops speltoides (BB)* or close relatives of these species, and that the hexaploid wheats originated as amphiploids between tetraploid emmers *(AABB)* and *Ae. squarrosa* (DD). A hexaploid wheat that closely resembles *T. spelta (AABBDD)* and forms fertile hybrids with it has been synthesized from a cross between *T. dicoccoides (AABB)* and *Ae. squarrosa* (DD). Inheritance studies in common wheat are often difficult to analyze owing to the polyploid nature of the crop. Many characters are dependent upon *two* or three genes, each gene having originated from a different genome. As a result, progress in developing linkage maps of common wheat by conventional genetic methods has been very slow. This has been facilitated by the development of *monosomics* (plants with one chromosome less than the normal) and *nullisomics* (plants with one chromosome pair less than the normal) in common wheat. By the use of these aberrant types the genetic analysis is simplified considerably, because a gene for a certain character can be identified with a specific chromosome. For example, if a chromosome pair carrying a rust-resistant gene is eliminated from a variety, the variety will no longer be resistant. Genes conditioning stem rust resistance, leaf rust resistance, solid stems, and other characters have been located by this procedure. The use of nullisomics also facilitates the substitution of particular chromosomes with desirable genes from other varieties or from closely related species. This technique may prove to be a valuable tool for the wheat breeder.

Wheat Pollination

Wheat is a self-pollinating crop. Blooming starts several days after the wheat spike emerges. The main culm flowers bloom first and the tillers later, in the order of their formation. Flowering begins in the upper part of the spike and proceeds in both directions. Flowering continues throughout the day, with two to three days required for a spike to finish blooming. The glumes normally open during the flowering process, the anthers protrude from the glumes, and part of the pollen is shed outside the flowers. Entry of foreign pollen while the flower is open may result in a small amount of crosspollination. Normally, cross-pollination is less than 1 per cent. If conditions are unfavourable for the opening of the glumes, the anthers may shed their pollen without being extruded. Wheat flowers are emasculated by clipping back the glumes and removing the anthers with fine-pointed tweezers. Pollinations are made one or two days later by breaking a ripe anther over the stigma.

Haploidy and Aneuploidy

Polyploidy in plants is usually associated with hybridization and sexual reproduction, a common event that forms a very significant aspect of evolution and speciation. Plants can adequately tolerate polyploidy as well as other variations (aneuploidy) from the normal euploid condition where the individual is considered a disomic. The nomenclature description for the aneuploids is reviewed by Khush [1973]. Different species have varied levels of tolerance to the monosomic and nullisomic condition, with the latter being tolerated adequately by the hexaploid status of *T. aestivum*. In *T. aestivum* all monosamics and nullisomics are viable. The morphology and vigour of the monosomics, in general, is comparable to the disomic plant. However, this expression is very varied in the nullisomics, many of them being weak and seed-sterile. These expressions distinctly vary in the tetraploids for monosomy, where nullisomy is not present. Monosomic analysis in some cases is used for locating gene(s) in chromosome(s) and chromosome arm(s), hemizygous ineffective genes, genes via chromosome substitutions, and genes by lack of expression in nullisomics, *etc.* For additional applications of monosomics, nullisomics, and a list of major genes for wheat characters located through their applications or via chromosome substitutions [Mcintosh, 1988].

Barrier to Introgression of Alien Genes into Wheat

Lack of crossability, ploidy differences, embryo abortion, hybrid sterility, useful and deleterious gene linkages, lack of synapsis, and crossing-over between wheat and alien chromosome are some factors that have limited alien germplasm exploitation. Besides the above, there are barriers to the production of wide hybrids that operate at various stages in the ontogeny of the hybrid, which include choice of parents, emasculation procedures, pre-pollination treatments, pollination, post-pollination treatments, and embryo rescue and culture [Mujeeb-Kazi and Kimber, 1985]. Several genotypes are often used to circumvent the non-crossability and sterility barriers. For instance, Riley and Chapman [1967a] reported the *kr1 kr1 kr2kr2* crossability genes in Chinese Spring wheat, which facilitate crossing to rye. Fedak and Jui [1982] identified a third crossability locus *(kr3kr3)*. The three genes are located on wheat chromosomes 58, SA, and 50 respectively and act in a complementary manner. To overcome ploidy level differences, genetic bridges and chromosome number doubling have been beneficial. Alien autotetraploids *(Agropyron cristatum, Psathyrostachys juncea)* have been of significant value in overcoming persistent wheat x diploid alien species crossability barriers. In general, it has been the practice, at least in the wheat group of species, to use the species with the higher ploidy as the female parent. There are at least two reasons for this practice. First, there would appear to be less imbalance between the chromosome numbers of the embryo and endosperm, and second, it is generally easier to emasculate hexaploid *T. aestivum* than the tightly invested florets of the related alien species. However, the early production of wheat-barley hybrids was accomplished with the lowerchromosome number barley as the female. Similarly, the hybrids between *T. tauschii* and *T. aestivum, Agropyron ciliare (Elymus ciliaris)* and *A. yezoense* (*E. yezoensis), A. trachycaulum* (*E. trachycaulus* and *T. aestivum, A. fibrosum (E. fibrosus)* and *T. aestivum*, and *E. canadensis* and *T. aestivum* all had the lowerchromosome-number parent as the female. So it would appear that other hybrids that have not been possible may be recovered if the reciprocal combination is attempted. Embryo rescue techniques have now become routine practice in interspecific and intergeneric hybridization, but additional media manipulations may facilitate recovery and differentiation of several complex combinations [Mujeeb-Kazi *et al.*, 1989]. While unfavourable linkages on alien chromosome segments are alterable with adequate backcrosses and large population size, special cytogenetic procedures may sometimes be necessary. In wheat, chromosome pairing Is restricted to homologues by a diploidizing mechanism located in the long arm of chromosome 5B, preventing normal synapsis between wheat and alien chromosomes. On the 3D-beta, 3A and 4D chromosomes, minor suppressors of allosyndesis have been located. The pairing promoting gene(s) exist on 3D, in the long arms of wheat chromosomes SA and SD and in the chromosome SD short arm. The regulatory influence over meiotic pairing of chromosomes SA and SD is related to temperature. Chromosome SD deficiencies lead to almost total failure of chromosome pairing at 15°C or lower temperatures. Nullisomic SA plants are normal at low temperatures and plants with the compensatory SA dosage but nullisomic for SD are normal. Thus, chromosome pairing in wheat is apparently a balance between genes that promote or those that suppress homoeologous pairing [Fedak and Armstrong, 1981]. The mode of action of the main pairing suppressor gene(s) on SB long arm has been controversial. Feldman [1966] hypothesized that the gene(s) affecting premeiotic association of chromosomes were a prerequisite for synapsis, whereas Driscoll *et al.* (1972) suggested that the genetic action by which associated chromosomes advance to synapsis operates after completion of homologous association. There occurs a formation of multivalents involving homeologues and synapsis of chromosomes from related genera in the absence of the chromosome SBL suppressor gene. The genetic variation among *T. aestivum* cultivars influences the level of chromosome pairing in intergeneric hybrids. Knowledge about the control of chromosome pairing in *Triticum* has assisted many workers in alien introgression research. It will be a key factor in success towards introducing alien chromosome segments into wheat via induced recombination, though researchers may still use ionizing radiation or callus culture.

Induced Recombination

In the effort to diversify the genetic base of *Triticum* species, production of intergeneric hybrids is only the first step. For the introduction of desirable alien variation into *T. aestivum,* complicated cytogenetic manipulations may often be necessary. Presumably, an excellent example of dealing with induced recombination is offered by the wheat/rye system. The activity of the *Ph* gene in chromosome 5BL of wheat is seemingly suppressed by genes in rye. More than two alleles were proposed in rye, that probably acted additively to promote homoeologous pairing in wheat. The short arm of rye chromosome 5R may carry pairing promoters. A probable influence of 5R on homoeologous pairing of wheat chromosomes, while the inhibition of a *Ph* allele dose by two doses of 5R was attributed as a cause of the high homoeologous pairing. The long arm of chromosome 5R lacks the pairing regulator gene(s) of 5BL. Furthermore, concluded a null effect on homoeologous pairing by 5R, when 5R was substituted for its three wheat homoeologues. There have been several reports of wheat/rye chromosome pairing at low frequencies when the *Ph* gene was absent or suppressed [Naranjo, 1982]. This is in contrast to earlier scepticism. Even though the activity of the *Ph* gene had been suppressed, and any wheat chromosome. Certain accessions of *Ae. speltoides* Tausch, *Ae. mutica* Boiss, and *Ae. longissima* Schweinf et Muschl inhibit the diploidization mechanism in common wheat [Riley and Kimber, 1966]. To promote pairing between chromosomes of wheat and those of an alien species, a hybrid between wheat and one of these *Aegilops* spp. could be crossed with the alien species. However, it is difficult to pick up the desired transfer chromosome because of the ability of the *Aegilops* chromosomes to pair with those of wheat and the alien species. Despite this limitation, Riley *et al.* (1968) used a modification of this method to transfer stripe rust resistance from *Aegilops comosa* to chromosome 2D of wheat to produce 'Compair'. Crossing wheat monosomic S8 or nullisomic S8, tetrasomic SA or SD to an alien species or to disomic alien addition or substitution lines produces some F_2 progeny that are nullisomic for S8 and hence devoid of the *Ph* gene. Alternatively, the mutant wheat line *phlbphlb* [Sears, 1977] could be used in place of monosomic S8 and nullisomic S8 terasomic SA or SO. These methods are expected to replace wheat chromatin. The low fertility of S8 and Phi-deficient hybrids restricts the opportunity for transfers. To reduce sterility problems, the use of less effective pairing mutants such as *phla* or *ph2* on 3DS (Sears, 1977). but added that the decreased frequency of allosyndesis may negate any advantage of increased fertility. For greater precision of alien chromosome segment transfer, it is better to eliminate or suppress *Phi* activity in a monosomic alien substitution line. Transfers are more likely to involve a particular, predetermined wheat chromosome. For instance, 17 of 20 transfers obtained from the offspring of a mono-3Ag, mono-30, nulli S8, tri-SO line involved chromosome 3D. Using a telosome of the critical arm of the alien chromosomes, instead of the full chromosome, facilitates the cytological determination of homoeologous pairing frequency and the identification of transfer chromosomes. Even when using induced allosyndesis, it is difficult to introduce only a short interstitial alien chromosome segment into wheat unless the critical gene is located close to the telomere. Two crossovers close to each other are required, and this is rare even between homoeologues. The crossing of two transfer chromosomes with exchanges proximal and distal to the critical gene, in order to produce a wheat chromosome with an intercalated alien segment. In the absence of exchanges on both sides of the critical gene, the transfer chromosome with its exchange closest to the gene may be combined with a homoeologous wheat chromosome in a Phi-deficient background. In a majority of the intergeneric hybrids, a codominant phenotype is expressed that Mujeeb-Kazi *et al.* (1989) associate with alien genetic penetration and expressivity that is not so evident in certain hybrids, *e.g.*, *T. aestivum* or *T. turgidum/H. vulgare*. Most of the modification of the wheat phenotype in their wheat-rye addition lines to interaction between wheat and rye genes. To be useful in wheat breeding, alien genes must be epistatic to those of wheat or interact with them to produce the desired effect (Riley, 1960). When resistance genes are introduced into a new background there is often a modification of expression of resistance to pests and diseases (Knott and Dvorak, 1976). Compared to the immunity of the rye parent, incomplete epistasis was the cause for the resistance observed in the wheat/rye amphiploid (Riley and Chapman, 1958b). The high resistance to yellow rust in *Ae. comosa* and *Ae. ovata* was not expressed in their amphiploids with wheat (Riley and Bell, 1958). The genes on chromosome 7DL of *T. aestivum* are reported (Kerber and Green, 1980) to have suppressed the stem rust resistance genes in tetraploid wheats. A similar decrease in the level of leaf rust resistance (as of the wheat parent) was expressed in some triticales even though the rye parents possessed strong resistance.

There are many examples of successful use of alien genes for wheat improvement. Genome analysis in hybrids gives objectivity to alien transfer programmes and the recent development of numerical methods of assessing genomic affinity has greatly facilitated gene transfer methodology. It allows for decisions to be made on a logical basis from a consideration of the values of X and C in the initial hybrids, where X = the relative affinity of the most closely related chromosome, and C is the mean arm-pairing frequency. C is derived from observation and X is that value that gives the closest fit between observed and expected chromosome pairing, that is, minimum sums of squares of differences. There are three working levels available: triploid (only one model), tetraploid (three models), and pentaploid (five models). Breeding strategies for hybrids can thus be formulated based upon chromosome pairing data analyses. The philosophy behind such categorization was associated with the earlier concept, relative to techniques of alien transfers. The derivation of numerical or relative affinity between the genomes of alien species and cultivated wheats provide the basis for the logical introduction of variation. When the mean arm-pairing frequency and the relative affinity have values approaching one, the transfer can be made by recombination. At intermediate values of mean arm-pairing frequency, increased homoeologous pairing can be induced by changes in the systems regulating chromosome pairing. At very low values approaching one, the transfer can be made by recombination. At intermediate values of the mean arm-pairing frequency, irrespective of the value of the relative affinity, irradiation or centric break-and-fusion in derived aneuploids is the optimum method. There is theoretical merit in attempting to enhance recombinations in the F_1 hybrids that involve. The *Ph* system. The procedures could involve as maternal parents the use of (1) nulli SB wheat stocks, (2) *PhPh* mutant, *etc.*, in the production of F_1 hybrids. Though such hybrids have exhibited a high meiotic chromosome pairing frequency, obtaining backcross derivatives from them has been problematic. Forster and Miller (1985) obtained normal and SB-deficient hybrids of *T. aestivum/A. junceum (Thinopyrum bessarabicum)*. The pairing in the normal hybrid was similar to that observed for this combination (Mujeeb-Kazi *et al.*, 1987), while that in one SB-deficient hybrid was 8.46 I + 1. 71 II rings 2.39 II rods + 2.S4 Ill + O.S8 IV+ 0.07 V, with VI and VII chromosomal exchanges observed in the second SB-deficient hybrid. From both the 5B-deficient hybrids, backcross derivatives have not been obtained, thus offsetting the demonstrated recombination advantages. They hybridized several *Aegilops* species with the Chinese Spring *PhPh* mutant stock and did indeed observe high pairing in the hybrids. Backcross derivatives, however, could not be obtained and the inference was made that the unreduced egg formation may have been adversely influenced by the *Ph* stock. The authors suggested additional studies for wider applications. On a limited scale, a few Chinese Spring *PhPh/Thinopyrum* species hybrids have resulted in BC, seed production that requires additional work with a larger combination of hybrids.

Translocations

The misdivision of wheat and alien chromosomes in the hemizygous condition associated with the fusion of non-homologous chromatids. Irradiation applications may also induce translocations between non-homologous chromosomes. This procedure led to a reciprocal translocation involving chromosome 6B and an *Ae. umbellulata* chromosome imparting leaf rust resistance to wheat leading to the release of the variety Transfer. The transfer of stem rust resistance from *A. elongatumto* wheat chromosome 6A and leaf rust resistance from *Agropyron* to chromosome 7D of wheat are other examples of irradiation applications. Even though translocations between homoeologous chromosomes predominate, induced translocations involve many deleterious changes. Successful translocations were detected by observing segregation for the desired characteristic in the progeny of the irradiated lines. The former researchers consequently produced a wheat-rye translocation stock Transec.'. This line carried both resistance genes for leaf rust and powdery mildew from rye. Despite the genetic segregation of translocation, cytological observations as confirmatory evidence are necessary. A major disadvantage is the large quantity of cytological analyses involved. In situations where the desired alien genes are located at the end of a chromosome, transfers occur readily, whereas diverse gene locations yield longer alien segments with accompanying deleterious effects. *Triticum aestivum* L. has evolved its genetic constitution over many years. Thus, gain of alien chromatin or any significant loss is likely to cause a genetic imbalance. The low fertility of rye addition lines exemplifies such an imbalance as a consequence of genetic interaction. This interaction has also been concluded as a cause of poor kernel characteristic in triticale for which heterochromatin is considered to be responsible. When homozygous, the yield limitations of induced translocation products influence commercial utilization and the quality levels may be depressed. There are, of course, variations that offset the quality depression when translocated lines are highly adapted, have high yields, and are disease resistant and where the nature of the consumed bread is remarkably different so that the quality constraint does not remain a negative factor (Peria *et al.*, 1990).

In intergeneric hybridization, the advance of the F_1 (wheat/alien species) hybrid classically occurs by first producing the amphiploid. Where wheat/alien chromosomal recombination events are a minimum, the next logical step is to pollinate the emasculated amphiploid by wheat to generate the backcross (BC_1) derivatives. Repeated backcrossing and cytological identification lead to eventual isolation of addition lines that *may* comprise intact alien chromosomes or may be traslocation products because of univalency in the BC progenies. F_1 hybrids do not generally lend themselves to fertile amphiploid production. The male-sterile hybrids, however, upon pollination by wheat set a low frequency of BC_1 seeds that are normal or aneuploid (Mujeeb-Kazi *et al.*, 1995). Such BC_1 production methodology provides the option of maximizing aneuploldy and structural change effects that may permit beneficial reorganization of otherwise separated genetic factors.

Alien Chromosome Addition and Substitution Lines

Doubling the chromosome number of wheat/alien species hybrids leads to fertile amphiploids. These amphiploids have poor fertility, expressing the undesirable genes of the wild species and a genetic antagonism between the parental genotypes. It is hence only the wheat/rye amphiploid that has established a place in agriculture on a commercial scale. The perennial wheat derived from *wheat/Agropyron* amphiploid is the only other partially doubled species to receive noticeable academic attention. It may be worthwhile to consider another amphiploid crop concept where a highly complex growth condition restricts wheat production. Salinity is one such stress that prohibits normal wheat cultivation. Presumably the genetics is quite complex and alien genetic transfers may not be able to overcome the constraint. It has been suggested that an amphiploid of wheat salt-tolerant alien species may be a reasonable crop on salt-stressed lands that prohibit a wheat crop. The *T. aestivum/A. junceum (Th. bessarabicum)* hybrid and its amphiploid produced have demonstrated tolerance under controlled liquid culture conditions. Its salt-saline site testing is forthcoming, and it is feasible to explore the concept despite the disease-susceptible wheat cultivar in the amphiploids pedigree. If successful, this may diversify our thinking of amphiploid-based crops and would also open up avenues for use of BC-selfed derivatives from partial autopolyploids or segmental autoallohexaploids (Mujeeb-Kazi and Miranda, 1984). O'Mara (1940) suggested a procedure aimed at reducing the undesirable characteristics of the amphiploids. This entailed the addition of single pairs of alien chromosomes to the allohexaploid AA88DD complement of wheat. This would facilitate identifying the gene information in an alien chromosome and its interaction with wheat chromosomes. Alien addition lines revert to the euploid through gametophytic selection, have low fertility, and possess poor agronomic characteristics. They are of use in irradiation applications and have been used to generate disomic substitutions of the alien chromosomes for some of the wheat chromosomes. Alien disomic substitution lines show more cytological stability than disomic additions, but due to the nullisomic wheat condition may be beset with severe chromatin loss problems and may also be non-viable. The imbalance could be reduced by substituting critical alien chromosome arms. Rye chromosomes do not compensate well for wheat chromosomes, as do those of *Agropyron* and *Aegilops* species. Homoeologous substitutions are particularly successful, but only a limited number have been of commercial use (Driscoll, 1983). Chromosome 1 of rye (1 R) with genes for mildew and stripe rust resistance has been spontaneously substituted for chromosome 18 of wheat (Zeller, 1973). Another cultivar, Weigue, has resistance to stem, leaf, and stripe rust. It has two sublines: a 1 R (18) substitution and a 18/1 R translocation. Spontaneous substitutions were obtained in advance derivatives of *T. aestivum/ Leymus racemosuslth T. aestivum* where wheat chromosomes SA, S8, and SD were replaced by two different *Leymus* chromosomes, tentatively identified as F (for 5B) and M (for 5A and 5D). Substitution lines may lead to translocation products and are useful for genetic analysis.

Hybrid Identification

Since characters of generic importance may be entirely suppressed in intergeneric hybrids, the morphological recognition of wide hybrids can be very unreliable. A convincing demonstration of hybridity can only be made by cytological investigations, and then it is possible to come to erroneous conclusions even if a detailed meiotic analysis is made. The difficulties hindering the positive cytological identification of hybrids tend to be of four types: first, technical problems in the collection and preparation of material; second, unreliability of the cytological technique chosen; third, somatic chromosome elimination or chimeras; and fourth, misinterpretation of meiotic data. A new technique (Mujeeb-Kazi and Miranda, 1985) can result in very clear preparations from which not only chromosome numbers may be determined but arm ratios, secondary constrictions, and intergeneric differences in chromatid thickness (*e.g. Thinopyrum/Secale)* can be identified. If both parents of a hybrid have the same chromosome number, somatic chromosome counts, at best, can only give an indication of hybridity if there are large and characteristic karyotypic differences between the parental species. The reliability of the chosen cytological technique must also be considered for hybrid identification. Simple counting of somatic chromosome number, if the parents differ in their chromosome number, can be reliable. If the somatic chromosome numbers are the same, it is still possible the parents differ in their ability to C- or N-band. The reliability of the recognition of hybridity on the basis of chromosome banding is directly proportional to the number, Intensity, and chromosomal distribution of the bands. If the arm ratio of the chromosomes of the parents is sufficiently different, this too may be used as an indication of hybridity. Again, the reliability of arm ratio as a pointer to hybridity depends on the number of easily recognized differences between the parents. The presence or absence of secondary constrictions is not a good method of recognizing hybrids. First, it requires that superior somatic preparations must be made, and second, no nucleolar organizer competition must take place. The length of somatic chromosomes generally provides a very poor method for the identification of hybridity. The general absence of large differences in relative length in the Triticeae and the inaccuracy involved in making measures of length (Kimber, 1970) both contribute to a low reliability. Even if an interspecific or an intergeneric hybrid zygote is recovered, it is possible that the seedling may not be a hybrid, due to chromosome elimination in the early zygotic divisions. The production of haploid barleys and wheats as a consequence of pollination by *Hordeum bulbosum* is an example of this type (Barclay, 1975). The spontaneous production of chimeras may also hinder the recognition of hybrids or make their use more difficult. Chimeras have been recognized. Kasha and Sadasivaiah (1971) recorded the expected chromosome number in only 40 per cent of the somatic cells of a diploid hybrid of *Hordeum vulgare* x *H. bulbosum.* Also, Kimber (unpublished) and Mujeeb-Kazi and Kimber (1985), recovered a haploid/diploid chimera instead of a hybrid following pollination of wheat by a related diploid. The origin of such plants is, of course, obscure because there is no proof of hybridity or the tissue from which the embryo actually developed. When such difficulties abound, any claim of hybridity must be accompanied by clear meiotic analyses. Since both the determination of species relationships and the choice of the most suitable method for the introduction of alien variation depend on the ability of the chromosome to pair, the interpretation of meiotic analyses is of some importance. In the earliest work the judgements were largely subjective, but essentially correct conclusions were reached. More recently, numerical methods for the analysis of meiosis in hybrids have provided some objectivity in determining genomic relationships.

Practicality and Prospects

As phylogenetic distance increases, so does the difficulty of the introduction of alien variation, but a rationale can be established on the basis of measurements of the relative affinity and mean arm-pairing frequency of the chromosomes involved. As the genomic relationships decrease, it first becomes necessary to induce recombination by manipulations of the chromosome-pairing regulator systems, and when the relationships are so remote that this is not possible, it is essential to resort to methods causing the breakage and reunion of chromosomes. There is a qualitative difference in the type of alien transfer that is required and it has considerable significance. In all cases involving disease reaction, mutations can influence the durability of the alien transfer und the system is considered 'dynamic'. Contrary to this is the 'static' system where the physiological traits influenced by alien transfers have the potential of being longer lasting and are capable of producing breakthroughs in commercial production. Most of the introduced loci have been initially placed in the cultivar Chinese Spring. This is a consequence of the use of this variety in the production of the essentially complete range of aneuploids (Sears and Sears, 1978) and also its ease of crossability. Its agronomic characteristics are, however, totally unacceptable. The phenotype of wide hybrids and their derived amphiploids mitigates against their commercial use; consequently, additional cytogenetic manipulations must be made before practical application becomes possible. In general, these manipulations will be directed at introducing the smallest piece of genetic material capable of controlling the desired phenotypes without affecting the other essential attributes of the recipient species (Mujeeb-Kazi and Kimber, 1985). The closeness of phylogenetic relationship, the ease of hybridization, and a high recombination frequency of the donor and recipient chromosomes are all manifestations of genetic commonality of the two species. Consequently, it is to be expected and has been observed that the majority of the introductions of alien variation in the Triticeae have been accomplished by recombination. Additionally, in alien genetic transfers so far, only rather simply inherited traits have been introgressed and major impact has been from species of the *Thinopyrum* group. Transferring polygenically controlled characters that are presumably recessive in nature, such as disease resistance *(Fusarium graminearum, Helminthosporuim sativum)* or stress tolerance (salt, drought, or heat tolerance), appears too difficult to achieve if short-term projections are made. Alien diploids would appear to have high priority in effecting gene transfers even if these would be for quantitative traits. The arsenal of a researcher ranges from inducing F_1 recombinations, if this does not occur normally, to developing alien addition lines, alien substitution lines, and translocations. Where genes are located in more than one alien chromosome, each gene can be manipulated in a wheat background prior to pyramiding these independently transferred genes into a single line or variety. More effective though slightly complex would be the exploitation of the *ph* loci for effecting alien transfers (Mujeeb-Kazi *et al.*, 1993).

Some tetraploid *Thinopyrum* species may be considered either strict autotetraploids or partial allopolyploids. In the former case, autosyndetic genome associations would be maximized, whereas structural differences of the partial allopolyploids seemingly prevent such close alien genome associations. Where maximum autosyndetic associations were expected in the F_1 these did not occur, presumably due to the active *T. aestivum* 58 Ph locus, *e.g., T. aestivumfrh. curvifolium*. Surprisingly, the level of pairing in *T. aestivum* hybrids with two partial allopolyploids *(Th. junceiforme* and *Th. scirpeum)* was considerably higher than that of the *T. aestivumfrh. curvifolium* partial autopolyploid hybrid. The type species of the *Thinopyrum* group is *T. junceum* and the basic or core genome is J-E, though other genomes may occur in certain complex polyploid species. Several species are autotetraploids (A. or *P. stipaefolium)*, segmental allotetraploids, or partial autopolyploids (A. or *Th. curvifolium, A.* or *Th. junceiforme,* A. or *Th. scirpeum)* or segmental allohexaploid (A. or *Th. junceum)* or segmental autoallohexaploids (A or *Th. podperae, A.* or *Th. varnense)* where there are two closely related genomes and the third genome is distinctly different from the other two. These relationships need to be observed to reach inferences about pairing in wheat/alien species hybrids since the advance of the F_1 hybrid for agricultural practicality is seemingly swifter where wheat-alien chromosome recombinations occur. Apart from F_1 wheat-alien species recombination events there is a high probability that the alien genome composition may provide adequate options for unique genetic manipulation. In the tetraploid partial autopolyploids, repeated selfing of the backcross self-fertile derivatives may lead to the formation of a synthetic genome (Mujeeb-Kazi *et al.,* 1984). In segmental autoallohexaploids, there is likelihood of the unrelated third genome to be eliminated and of fertile derivatives leading to the formation of stabilized synthetic genomes. With refinement of techniques, more difficult combinations will be produced. The polyploid derivation from sexual crosses of *T. aestivum* with *Zea mays, Sorghum bicolor* L. Moench, *Teosinte* (Ushiyama *et al.,* 1991), and *Tripsacum* are examples. Alien diploids have a decided advantage in intergeneric hybridization and, though they are difficult to hybridize, some success has been achieved. *Thinopyrum elongatum, Th. bessarabicum,* A. *cristatum,* and *H. vulgare* are some diploids hybridized with wheat. Recently, Mujeeb-Kazi and A.siedu (1990) successfully hybridized *T. aestivum/A. cristatum* and *T. aestivum!Ps. juncea.* The alien sources were autotetraploids and this may be a significant step towards the success of future hybridization attempts with complex diploids. The test with other auto- or induced autotetraploids are forthcoming. By backcrossing the F_1 mentioned above with wheat one can produce spontaneous amphiploids by fertilization of unreduced ABDPP or ABDNN eggs with the ABO pollen source. Such developments are encouraging as the demands for food production increase. It is difficult to predict the future genetic demands that may be placed on wheats and a stock of alien genetic material introduced from wild hybrids may prove to be of great value. Though the emphasis will remain on subtle genetic transfers, other techniques will also emerge. In this facet would be areas associated with callus culture, anther culture, radical distant hybridization, irradiation applications, somatic cell fusion, probes for diagnostic application, and presumably a greater emphasis on interspecific hybridization. The latter will optimistically facilitate A, B, and D genome recombinations, extraction of AB genomes, and development of synthetic wheats. It thus appears that wide hybridization will provide an expanding source of genetic variation for the plant breeders. Though futuristic, the gene(s) transfer(s) may presumably amount to quantum changes in either the production of wheat or in its distribution, or both.

Breeding for High Genetic Yield Potential, Stability, and Wide Adaptability

Wheat yield is a complex character dependent on three basic components: number of spikes, fertility of the spike, and weight of the grain. Yield is therefore the function of number of spikes x number of grains/spike x weight of grain, or number of grains/unit area x weight of grain in that unit area. There is an ideal balance for each of these three yield components for each agroecological niche. The balance depends on day length and heat units, which determine the flowering and maturity of the genotypes and is also highly influenced by mineral fertilizers and moisture supplies. Since the component balance is agroecologically stable, any attempt to change it may result in a phenotype unsuitable for some agricultural practices. A breeding programme must determine the best yield component balance before embarking on a selection programme for spike type, spike fertility, or grain size. Due to inherent negative correlation, simultaneous selection for increasing all components is not possible. Normally, the increase in one component simultaneously selects for a decrease in the other components. The yield components together form the yield potential (sink size) of a crop, which is very much dependent on agroclimatic conditions. Given the same agronomic practices and soil type, the response of a wheat variety is dependent on photosynthesis and heat units. There are optimal temperature regimes at which photosynthesis operate more efficiently. For best C02 assimilation, the optimum temperatures for a temperate crop are between 2° and 30°C; the rate decreases slightly to about 10°c, and approaches zero when the temperature is between 0 and 5°C (Monteith, 1977). For example, in the absence of disease stress, the variety Seri 82 is one of the highest-yielding varieties with different genetic yield potential at five different latitudes. The yield potential declines with decreasing latitudes, where heat units are above optimum levels. A wheat yield of 15 t/ha may be realized in an environment where the photoperiod is long and temperatures are optimum for photosynthesis, as in Western Europe or in the Pacific Northwest in the United States. In wheat- producing regions with short day lengths, such as Mexico or India, the sink size of the same variety shrinks and yield declines. There is clearly an optimum sink size (genetic yield potential) for each ecological niche, and breeders must be realistic about genotype and environment interaction and its impact on yield.

Manipulation of Harvest Index and Biomass

The first semidwarf spring wheats using the *Rhtt* and *Rht2* genes from Japanese wheat Norin 10 were produced in Mexico. Incorporation of these genes not only resulted in lodging resistance, but invariably showed higher harvest index and increased biomass. It was the higher rate of partitioning of assimilates into grain of these wheats that resulted in higher grain yield. The variety Yecora 70, a double dwarf, has a higher harvest index than the tall variety Yaqui 50 and, at comparable biomass production, produces a higher grain yield. The variety Genaro 81 (one-gene dwarf), even with a slightly lower harvest index than Yecora 70, produces a higher yield because of increased biomass. The manipulation of harvest index in the breeding programme was an indirect activity. Harvest index was manipulated through selection for dwarf genotypes. These dwarfs or semidwarf genotypes, when found to be higher yielding, invariably had an increased harvest index. Since the dwarf characteristic in phenotypes is easily identifiable, it is not necessary to measure the harvest index of each plant in the segregating population. However, not all short genotypes are expected to have a high harvest index. Disease susceptibility or physiological abnormalities may block the assimilates and limit grain filling or sink size.

Table 2.5. Dwarfing status, yield, harvest Index, and biomass of wheat varieties

Variety	*Dwarfing Status*	*Yield (t/ha)*	*Biomass (t/ha)*	*Harvest Index*
Yaqui50	Tall	5.33	14.9	32.8
Yecora 70	Double dwarf	7.11	13.7	45.7
Genaro 81	Single dwarf	7.82	16.5	41.9

Source: Waddington *et al.* (1986). Yields are significantly different at 0.05 probability.

A frequent question in wheat breeding is whether genetic yield potential has reached a plateau? The answer to this question is a resounding 'no'; genetic manipulation can still increase yield provided that a properly guided approach is employed in the breeding. It appears that further increase in genetic yield of wheat can be realized by increasing biomass higher than Genaro 81 and maintaining the Yecora 70 level harvest index. This can be done by restructuring plant type, which will enable higher tiller survival without upsetting other components such as grain number and grain weight. Preliminary results in the breeding programme indicate that this can be done. Higher tiller production with small, erect or semierect leaves in the canopy would facilitate light penetration in the early growth stages. This physiological characteristic may allow higher tiller survival because of the light penetration deep in the canopy. Provided the temperatures are right and nutrient and moisture supplies and photosynthesis are not limiting, this increased sink capacity can provide higher yield.

Manipulation of 1 B/1 R Translocation for Increased Yield

Circumstantial evidence suggests that the 1 B/1 R translocation (the small arm of the 1 R chromosome of rye trans located on the long arm of the 1 B chromosome of wheat) enhances genetic yield potential. This is inferred from the preliminary yield results of the isogenic lines of the variety Anza, when compared to the yield of Anza in the Yaqui Valley, Sonora, Mexico. The 1 B/1 R translocation line is best represented by the variety Veery, which has given higher yield response in many environments across world than other widely adapted varieties such as Anza,

Pavon 76, and Siete Cerros. It is not known exactly what physiological character is contributing to higher yield, though 1 B/1 R lines tend to 'stay green' and show delayed senescence.

Shuttle Breeding and Wide Adaptability

Shuttle breeding is defined as the growing of segregating generations in different environments. For example, in this scheme the F_2 is grown in one environment, and selected F3 progenies are planted in a second location, and so on. The underlying philosophy of shuttle breeding is to incorporate adaptation and resistance to all pathogens and stresses encountered in the different locations. Shuttle breeding methodology has been discussed in detail by Rajaram *et al.* (1984). CIMMYT uses two contrasting locations in Mexico, Ciudad Obregon (in the Yaqui Valley, Sonora) (27°N latitude, 39 masl) and Toluca (19°N latitude, 2640 masl), for generation advancement and selection. Obregon is characterized as a high-yielding environment. Toluca is a typical medium-yielding environment, where the disease spectrum is higher, represented by the rusts, fusarium and *septoria* diseases, barley yellow dwarf virus, and bacterial blight. At Obregon only stem rust, leaf rust, *Altemaria triticina,* and kamal bunt organisms are present. Furthermore, these two locations differ in soil type, photoperiod, temperature, and rainfall pattern. In addition to these locations, the breeding programme uses Huamantla (19°N latitude, 2500 masl) for drought resistance, El Batan (19°N latitude, 2249 masl) for leaf rust resistance, Tepatitlan (20°N latitude, 1700 masl) for drought resistance, Patzcuaro (18°N latitude, 2180 masl) for septoria resistance. Tampico (22°N latitude, 4 masl) for heat tolerance, and Poza Rica (21°N latitude, 60 masl) for heat tolerance and helminthosporium screening. This network of combined locations is used to breed and screen germplasm that can result in high-yielding varieties with wide adaptation and disease resistance. The underlying objective of shuttling segregating generation materials in several environments is to produce widely adapted varieties while maintaining high genetic yield potential. Some of the world varieties such as Siete Cerros, Anza, Nacozari 76, Pavon 76, and Veery were produced. It should be

Table 2.6. Countries in which Veery Selections have been Released

Country of Release	*Veery Release*	*Country of Release*	*Veery Release*
India	HUW 206	Mexico	Glennson 81, Genaro 81, Ures 81, Seri 82
Tanzania	Viri	Pakistan	Pak 81, Pirsabak 85
Kenya	R253	Peru	La Molina 82
Zambia	Serie	Zimbabwe	Rusape
Rwanda	Gicinya	Chile	MIllaleu Inla, Nobo Inia
Guatemala	Olintepeque	Ethiopia	Dashen
Turkey	Seri 82, Anadolu	Paraguay	Cordillera 3
Nepal	NI549	Portugal	Lima 1
Algeria	Numidie	South Africa	Gamtoos
Uruguay	LE. Cardenal	Zambia	Loerie
		Spain	Arganda, Cartaya

emphasized that while maintaining wide adaptation, no sacrifice has been made of genetic yield potential. It appears that the genetic systems governing yield and wide adaptation are different and hence can be combined.

Yield and Stability

In the breeding programme, due recognition is given not only to high yield and wide adaptability, but also to a third character known as stability. High stability is defined as ability of a genotype to produce stable and high yield over time in one environment or a variety which shows low genotype x environment interaction. There are four basic ways in which a variety may respond as production improves. Average yields of the progressively better sites are shown on the x-axis, and the variety yields are shown on the y-axix. The yield of a given hypothetical variety at each site is plotted and a regression line is drawn through the points that represent the performance of each genotype in the different environments. The slope of this regression line measures the responsiveness of the variety to better growing conditions. The regression line that represents the average response of all entries grown in each environment would have a slope of 1.00 (a 45° angle line). Thus, if the slope of the variety regression line is greater than 1.00 responsiveness to improving conditions is above average. The greater the slope, the higher the responsiveness of the variety. Variety 1 types; variety 2 types are rejected because they yield less than the average in poorer production environments. Variety 3 and 4 types do not respond well to improving conditions and are only promoted in stress environments. The key point here is that the regression of variety 1 types lies entirely above the 45° angle line, that is, variety 1 performs at least as well as the average under low input conditions and it responds better than average as production conditions improve. An analysis of the 15th International Spring Wheat Yield Nursery (ISWYN) shows that the yield of Veery 'S' is consistently better than the average yield of all entries in the 73 environments. The fact that the Veery regression line is above the average yield regression line indicates that Veery 'S' performs better than the average, in both poor and favourable environments. In other words, Veery 'S' uses available inputs efficiently in low-input environments and is able to respond favourably as production conditions improve. The impressive stability of the Veery 'S' lines stems from their enhanced disease resistance, tolerance to certain environmental stresses, and improved plant types.

Handling of Germplasm and Selection Methodology

The breeding methodology and selection criteria are described in detail by Rajaram *et al.* (1984), and the germplasm requirement and selection criteria presently in use in Mexico can only be briefly described here. Genetic resources for different characters are of utmost importance if different genetic variabilities are to be combined, and they should be the best currently available. A recombination programme depends on the breeding objective and number of characters sought in the variety to be developed. Simple crosses, backcrosses, three-way crosses, and double crosses are used to recombine genetic variability. The choice of the cross depends greatly on the complexity of the characters to be handled and the suitability of the parental stocks available. Three-way and double crosses produce

500-SOO cross seed for evaluation. Selection criteria and the types of spring and winter wheat germplasm used.

Spring and Winter Wheat Germplasm

Germplasm with a high genetic yield base, displaying variation for different yield components. Germplasm resistant to rusts, smuts, powdery mildew, *Septoria* spp., *Helminthosporium* spp., *Fusarium* spp., BYDV, bacterial blight, *etc.* Germplasm displaying genetic variation for agronomic characters such as height, lodging, flowering, and maturity. Germplasm displaying genetic variation for resistance to physiological stresses such as drought, heat, cold, *etc.* Germplasm displaying genetic variation for tolerance to mineral toxicities such as aluminium and salt, and deficiencies such as phosphorus. Germplasm that produces grain with good milling and baking quality characteristics.

Selection Procedure for Segregating Materials

1. F_1 Simple crosses are evaluated on the basis of disease resistance, agronomic characters, and hybrid vigour, and harvested in bulk for F_2.
2. F_1 Three-way, double cross, and backcross populations are handled on an individual plant basis. Materials are space-planted. Outstanding plants are identified, selected, and advanced to F_2.
3. F_2 Two to three thousand plants are space-grown in F_2 in well-watered, high fertility conditions. Individual plants are identified, on the basis of disease resistance and agronomic suitability (*i.e.*, tillering, lodging, maturity). Visual spike characters such as fertility must be satisfactory.
4. F_3 Progeny of selected F_2 plants are grown solid in 2 m three-row plots. Each plant is treated as a separate population. Selection is done by plot first and individual heads in the selected plot are bulked. The environment of the F_3 nursery depends on the conditions for which materials are being selected: irrigated or rainfed environments, acid soils, hot climate, *etc.*
5. F_4 Bulk F_4 is planted solid and selection criteria and the choice of environment are the same as for F_3.
6. F_5 Approximately 100 plants/plot selected in F_4 are space-planted in a favourable, high-yielding environment. Individual plants are selected on the basis of disease resistance, acceptable agronomic characters, and spike fertility.
7. F_6 Each plant selected in F_5 is grown individually as F_6 in plots of three rows, 2 m long. Agronomically outstanding lines acceptable in disease resistance are bulked and advanced to yield evaluations on the basis of choice in irrigated, rainfed, acid soil, and hot climate environments according to breeding objectives. The lines developed in purely irrigated condition of high fertility are also checked in rainfed condition of low fertility and *vice versa*.

Objectives in Wheat Breeding

The ultimate goal of the wheat breeder is to develop new varieties improved in some important features. This goal can be reached only by carefully planned selection and hybridization procedures which lead toward well-established and clearly defined objectives. The breeder needs to know what improvements will increase the productiveness and quality of wheat varieties and thus will be useful and profitable to both grower and processor. He must search out parent materials superior in these features and combine them into a superior variety. The objectives of wheat breeding are not always the same, for the environmental conditions that affect wheat production and adversities that limit wheat yields differ from one production area to another. However, certain broad objectives are important over wide production areas. These include *(a)* yield of grain, *(b)* maturity, *(c)* standing ability, *(d)* winter hardiness, *(e)* disease resistance, (f) insect resistance, and *(g)* quality.

Grain Yield

Yield measures the total returns to the wheat grower and is affected by all the environmental conditions influencing the growth of the wheat plant as well as the plant's heredity. The inherent capacity for yield may be expressed through such morphological features of the plant as tillering, length and density of the spike, number of grains per spikelet, or size of the grain. But no one of these physical components of yield can by itself be considered as an index to yield. *Vigo* is a wheat variety with only a moderate tillering capacity and long spikes. Pawnee has short spikes and a high tillering capacity. Both are high-yielding varieties. To represent the three dimensions of the box we may use *(a)* the number of heads per unit area, *(b)* the number of grains per head, and *(c)* the average weight per grain. The volume of the box, which will be the yield of the variety, is determined by the product of these components. An increase in anyone of the three components would result in an increase in total yield, provided there is no corresponding decrease in the other two components. With this representation of yield the problem of breeding for higher yield becomes that of finding the inherent combination of the three components that will result in the greatest volume of the box. In practice, as one component of yield is increased, the others tend to decline. As tillering capacity is increased, the heads tend to become shorter, or the size of the grains may be reduced. Thus selection cannot be made for one component without some consideration of the others. Yield of a variety is measured in terms of pounds (or bushels) of grain per acre. The ability to yield is expressed through the photosynthetic and metabolic processes within the plant. Perhaps we should say that the yield capacity of a variety is its inherent ability to synthesize starch, proteins, and other food materials, and to translocate and store them in the grain. If the breeder is to increase one component of yield of grain without reducing the other components by a corresponding amount, it will be necessary to increase the efficiency of many vital processes within the plant. This means of course that, since many complex physiological processes within the plant are influencing yield, many genes affecting the functioning of these processes must contribute to the final production of the grain. Since individual genes affecting complex yield processes cannot be identified, all of them are often lumped together by the breeder and referred to as "yield genes." Thus to breed for high yield it is necessary to combine into a variety a favourable combination of yield genes. All this is assuming that the wheat plant has a favourable environment in which to grow; that no factor such as cold, moisture, or disease will limit the final yield. To find such an environment would be rare indeed. So the ability of the plant to produce well in spite of an adverse environment will also contribute to the final production of grain. Thus we breed for cold resistance in winter wheat varieties in areas where non-hardy plants would be injured, or for resistance to drought, disease, or insect damage. A variety of wheat with resistance to stem rust might yield more than a susceptible variety in the presence of a heavy stem rust epidemic, even though the resistant variety was less productive in the absence of the rust. The ability of the wheat to stand until harvested without loss of grain from lodging or shattering will also affect the final yield, as well as its ability to mature within the limits of a favourable season. While constantly striving to improve the potential yielding ability by grouping into one strain of wheat the more favourable combination of yield genes, it is also necessary to stabilize production by breeding for resistance to the many adversities that may limit the final harvest.

Maturity

There has been a continuing trend toward the development of earlier maturing varieties. One of the features important in the early popularity of the Marquis variety of spring wheat was its early maturity, which enabled it in some seasons to escape stem rust. It has since been replaced by spring type varieties even earlier in maturity, although their inherent rust resistance was perhaps a major factor in this change. In the southern hard red winter wheat area, where late-maturing varieties may be injured by hot weather or drought, the search for earlier maturing wheats has led to the development of the popular Pawnee, Wichita, Triumph, and other varieties. The long adaptation of Purple straw in the south eastern states was largely a result of its earliness. Knox and Vermillion are proving popular in the southern Corn Belt, partly as a result of their early maturity. The advantages of early maturity are many. It enables the wheat to escape hot weather, drought, and rust. It permits the harvest on many occasions to proceed ahead of rainstorms, hail, or Rood damage. Early wheat often escapes damage from chinch bugs or grasshoppers. Early harvest enhances the growing of a legume seeded as a companion crop by removing the competitive effect of the wheat at an earlier date. Most early wheats have shorter straw and are less likely to lodge. There are also certain disadvantages of early maturity. Extremely early wheats may be lower in yield and less winter hardy. Early wheats tend to be lower in yield because the wheat plant has a shorter spring growth period in which to tiller, bloom, and to manufacture and store food materials in the wheat kernel. Many recently developed early varieties, such as those cited, have demonstrated, however, that *it is* possible to obtain a favourable combination of genes for both yield and earliness within the same variety. Early varieties generally start growth earlier in the spring than later maturing varieties. Sometimes the advanced growth of the early varieties causes them to be injured by late spring freezes. This is a factor that may limit the earliness to which one may go in some areas in the breeding of early-maturing winter varieties. The inheritance of earliness *is* complex and apparently depends upon the specific varieties being crossed. For example, in a cross between two spring varieties, multiple inheritance (three factors or more) was reported with earliness at least partially dominant. In a cross between Kawvale and Early premium varieties of winter wheat, lateness was reported to be dominant, with three factors involved in the inheritance of maturity.

Standing Ability

The ability of a wheat variety to stand in the field until harvest without loss of grain is important in obtaining a high yield. The use of the combine harvester and the application of higher rates of fertilizer amendments, especially nitrogen, have increased the need for the breeder to improve varieties in standing ability. The standing ability of the wheat plant involves its resistance to lodging and its resistance to shattering.

Lodging Resistance

Lodging in wheat occurs as a result of the bending or breaking of the wheat culm. Losses from lodging occur with any of the conditions, *viz.*, the wheat lodges before it is ripe and does not fill properly; the fallen wheat cannot be picked up in the harvesting operation and is left in the field; the lodged wheat provides a favourable environment for the development of rust, mildew, or other diseases. Rain, hail, and windstorms occurring after the wheat has headed, but before it ripens, are common causes of lodging. The wheat at this stage is green and heavy and is easily bent or broken over by the added weight of the rain or the force of the wind. If the plant is merely bent over, it may rise upright again after it has dried off, with little reduction of final yield. But if the culm is broken, the spike does not fill normally or may be lost entirely. Breakage of the culms at this stage of development may reduce the final yield by 20 to 30 per cent. Test weight and protein content are also reduced. Plants with culms which are inherently weak or plants which are very succulent as a result of excessive nitrogen fertilization or soil moisture will be more susceptible to lodging injury. Resistance to lodging may be improved by the development of varieties with *(a)* stiff, sturdy stems, *(b)* short straw, (c) a vigorous root system that will anchor the plant firmly in the soil, *(d)* more resilient straw that will not break in the wind, and *(e)* resistance to diseases and insects that weaken the straw or the root system. Lodging may also occur after the wheat is ripe and before it can be harvested. Delay in harvesting as a result of continued rainfall will result in this over-ripened wheat bending over. Eventually the culms may break. Development of varieties with short, stout stems will increase the length of time the wheat will stand without breaking. Some varieties are inherently stronger than others, even though the diameter of the culm may be similar. This is apparently associated with the structure of the cells and the deposition of lignin in the cell wall. The brittleness of the straw is also important since some apparently stiff-strawed varieties break under the pressure of strong winds. Disease and insect damage may result in lodging, as when the wheat straw deteriorates from a stem rust infection, or breaks over from damage by the Hessian fly or joint worm. Presence of root rots may also result in lodging of the wheat plant, for the anchorage of the plant will be weakened. Breeding for resistance to these diseases and insect pests will increase the ability of the wheat plant to stand in the field until harvest without lodging. Improvement in straw stiffness has received much attention in the breeding of new varieties. In the hard wheat area, newer varieties such as Pawnee and Triumph are much improved over the old Turkey, which was notoriously weak strawed. Thorne, Vigo, Knox, and Vermillion are considerably stiffer strawed. than soft wheat varieties grown earlier, such as Fulcaster, Mediterranean, or Red May. The stronger straw of Mida was an important factor in its replacing Rival, Pilot, Thatcher, and Regent in the hard-red spring wheat area. Some of the stiffest-strawed strains are found in the white wheats grown in the Pacific Coast area where the Federation and Rex varieties have set high standards in this respect. Short straw has the additional advantage that less straw will need to be handled by the combine in the harvesting operation. Excessive growth of straw slows down the speed at which the combine can be operated and increases the cost of harvesting. Shortness is often associated with earliness, but short varieties do not necessarily need to be early varieties. The popularity of Pawnee, Triumph, Knox, and Vermillion can be attributed partly to their short straw as well as to their earliness. Semi-dwarf types are being developed in the Pacific northwest from crosses of adapted varieties with Norin 10, a very short variety introduced from Japan. These and selections made elsewhere from crosses with Norin and other short strains promise to set new standards in the development of short wheat varieties in the United States.

Shattering Resistance

Losses from shattering usually occur when harvesting with the combine is delayed for some time after ripening, especially if the wheats ripen during a period of hot, dry weather. Varieties of wheat differ in their tendency to shatter. Most attention to the breeding of shatter-resistant wheat has been in the Pacific northwest, where in certain low summer-moisture areas, wheat ripens and stands until harvest without rainfall. There Rex, Ridit, Triplet, and the club variety Elmar are superior in this feature. The more shatter resistant spring wheats include Regent, Vesta, and the rust-resistant varieties Marquillo and Hope. Clarkan in the soft red winter area is resistant to shattering. Varieties especially susceptible to shattering include Fulcaster, Kawvale, Mercury, Rival, Garnet, and many others. The varieties with the greater shatter resistance have a larger amount of lignified tissue at the breaking point of the outer glume.

Winter Hardiness

The varieties of wheat are commonly divided on the basis of their *growth habit* into *spring* or *winter types,* although there is an intergrading from the spring into the winter growth habit. The spring type includes the hard-red spring, the durum, and a few of the white varieties. The hard and soft red winter varieties and most of the varieties of white wheat are winter types. Some spring type varieties of the white wheats are fall-planted in states where the winters are mild, such as the Baart and Hard Federation varieties. Winter varieties differ from spring varieties in *(a)* the development of a more or less prostrate type of seedling growth until subjected to cold weather, *(b)* the capacity for hardening when exposed to cold, and *(c)* the ability to withstand freezing temperatures after hardening. Within the winter wheats there are wide differences in the inherent hardiness of varieties. The varieties range from non-hardy, some of which can scarcely be distinguished from spring varieties, to the most winter resistant. Varieties grown in the southeast, such as Atlas and Coastal, or varieties grown along the Gulf Coast of Texas, such as Bowie, differ greatly in inherent hardiness from varieties like Turkey, Minturki, Yogo, and Marmin, which are grown along the northern limits of the winter wheat-producing region. The introduction of the Turkey hard wheat from Russia, with its inherent resistance to winter injury, made possible the safe production of winter wheat in the central Plains states. With the development of more hardy varieties, winter wheat production is gradually pushing northward into areas where only spring-seeded wheats could be grown before.

The most common causes of winter injury in wheat are *(a)* freezing as a result of low temperatures and/or insufficient soil moisture and *(b)* heaving. Freezing is in many respects a desiccation process and wheats resistant to winter injury are often resistant also to drought. Heaving results from the alternate freezing and thawing of heavy soils and results in the wheat plant being lifted and its roots sheared off or torn loose from the soil. Many environmental factors may affect the amount of winter injury. These include *(a)* moisture content of the soil, *(b)* snow cover, *(c)* hardening of the plant, *(d)* physical condition and fertility of the soil, *(e)* time and rate of planting, and (f) disease and insect injury. Winter injury as a result of low temperature and/or insufficient soil moisture is more common in the dry Plains area than in the humid eastern states where wheats are better protected by snow. The main cause of winter damage is usually insufficient soil moisture combined with low temperatures. In general, the more hardy varieties of soft wheat are fairly resistant to winter injury by heaving. But when the soft wheats were carried westward by the early settlers, they did not survive the combined effects of low temperature and inadequate soil moisture. The Turkey hard wheats survive better than the soft wheats under dry, cold conditions of the Plains, but they may be more severely injured than the soft wheats if exposed to heaving. This distinction is important in the adaptation of hard and soft winter wheats to the areas where each is now produced. The geographic areas in which winter injury by freezing and by heaving predominate are not sharply defined.

The comparative hardiness of wheat varieties can be measured by growing them over a large geographic area, where they will be subjected to winter injury in a wide range of climatic conditions. Uniform nurseries, cooperative between various agricultural experiment stations have been developed to provide for more extensive testing of varieties and new experimental strains. Another method of testing the hardiness of wheat varieties is to grow them in pots or flats in a greenhouse and, after suitable hardening procedures, subject them to low temperatures in specially designed freezing chambers. The latter technique measures only the resistance to cold. Measurements of injury by freezing tests have been found in many cases to correlate closely with field tests. The complex inheritance of winter hardiness is indicated by the results of a cross made between Poole and Minhardi varieties of wheat. The Poole parent variety averaged 24.1 and 23.3 per cent survival in artificial freezing tests, and the Minhardi variety averaged 53.0 and 51.5 per cent survival. The survival of the F_1 plants was 36.5 per cent, midway between the parents. The survival of the F_2 plants, estimated from the survival of the F_2 families, varied from 7.5 to 52.5 per cent with an average of 34.6 per cent. Non-winter-hardy transgressive segregates were recorded as well as lines as hardy as Minhardi. In this cross, no lines more hardy than the hardy parent were obtained. The results indicate that the inheritance of cold resistance is determined by many genes, which is characteristic of the inheritance of quantitative characters.

The International Maize and Wheat Improvement Center (CIMMYT) bread and durum wheat germplasm are developed that target the diverse wheat-production environments found in the developing world. The CIMMYT spring wheat breeding program is based in Mexico and shuttles germplasm between two contrasting environments, thereby achieving two generations a year. The lines developed through this process are then tested widely around the world and selected materials, based on international performance, are identified for continued crossing. In order to better facilitate the crossing program in Mexico and the deployment of appropriate germplasm to these diverse production environments, the concept of mega-environments (ME) was introduced in the early 1990s. MEs are geographical areas where wheat adaptation can be expected to be similar in terms of climatic, disease, or crop management constraints. This concept of ME has continued to be refined across the years as more sophisticated tools have become available and access to information has improved. DeLacy *et al.* (1994), in an analysis of historical yield data from the International Spring Wheat Yield Nursery (ISWYN), concluded that ME definitions could be improved by integrating genotype environment interaction (GEI) analyses. In recent years advances in GIS (global information systems) and the availability of spatial maps for factors as diverse as wheat production statistics, soil composition, rainfall, and annual temperature at critical growth stages have helped refocus ME based breeding objectives (Trethowan *et al.*, 2005a). Drawing on some specific examples from CIMMYT's wheat-breeding program, this chapter describes options available to wheat breeders to improve wheat yield, adaptation to diverse environments, and processing and nutritional quality.

High Yield

High yield depends on the number of heads/unit area, the number of grains/heads, the average weight of grain. While breeding for high yielding varieties all the above three components must be looked into. Omitting any one of them may not yield results. Further while breeding for high yield it is necessary to combine into a variety a favourable combination of genes influencing all yield process. *Breeding non- lodging varieties:* This is achieved after the identification of dwarfing gene in Japanese variety Norin 10. Most of our dwarf wheats are two gene dwarfs. *e.g.*, Sonara 63, Sonara 64, Kalyan Sona. Emphasis is now on triple gene dwarfs. *Breeding for disease resistance:* Rust is the major disease. Both stem rust and leaf rust are important ones. There are different races of rust. So, while breeding for rust resistance horizontal resistance is to be looked into. Back cross method of breeding and development of multi lines are the methods. *Breeding for insect resistance:* Hessian fly is the major pest. Resistance in most varieties is through antibiosis. *Breeding for quality:* Different wheat varieties vary greatly in their chemical composition which is considerably influenced by environment. The varieties of hard wheat or bread wheat which have higher gluten content. The soft wheat contain lesser gluten content which is suitable for cake making, pastries *etc.* The durum wheats are unsuited for either cakes or bread but they are suitable for making macaroni. So, depending upon the use the quality breeding objective is to be fixed.

Yield Potential, Yield Stability and Wide Adaptation

Yielding capacity of a cultivar in a particular environment in the absence of biotic and abiotic stresses can be defined as the yield potential of the cultivar in that environment. Factors that lead to annual fluctuations in yield potential at a site are often associated with variations in temperature and radiation (Lobell *et al.,* 2005). The yield potential of semidwarf wheat cultivars, regardless of their origin, has continued to increase at the rate of about 1 per cent annually. According to Rajaram *et al.* (2002), yield potential of bread wheat in north-western Mexico increased from 4.5 t/ha for the tall cultivar 'Yaqui 50' to 6.5 t/ha for the first semidwarf mega-cultivars 'Siete Cerros 66' and 'Sonalika,' which were associated with the widespread Green Revolution. This jump in yield potential is attributed mainly to the incorporation of dwarfing genes *Rht1* and *Rht2* that provided lodging tolerance, which allowed more fertilizers and irrigation to be applied to crops to increase production. In comparison with the first successful semidwarfs, yield potential of cultivar 'Baviacora 92,' released in 1992, was about 9 t/ha. Some of the new wheat genotypes developed in recent years have shown further increases in yield potential of 8–10 per cent over 'Baviacora 92' (Singh *et al.,* 2006). Yield stability and wide adaptation are important traits that must be present together with yield potential to ensure that a genotype maintains its superiority in a range of environments, management practices, and biotic and abiotic stresses. Wide adaptation of CIMMYT-derived semidwarf wheat cultivars has been attributed to their photo-insensitivity, which resulted from shuttle breeding at two contrasting sites in Mexico (Rajaram *et al.,* 2002). Breeding for specific adaptation has not been very successful because in most areas temperatures and rainfall patterns shift annually. Moreover, farmers are not able to maintain the same agronomic practices, and diseases and pest pressures vary from year to year. The wide adaptation and stable performance of CIMMYT-derived wheats are largely due to the shuttle breeding in Mexico where segregating populations were selected from two environments with differing disease and abiotic stresses; this was followed by international, multilocation testing of advanced lines. Using this approach, it is possible to identify the best stable performers in a single year of testing. Wide adaptation and stable yield performance of a new breeding line 'Weebill 1'. This genotype was yield-tested together with 49 other genotypes in each of the two international yield nurseries, 23rd ESWYT and 10th SAWYT; the first targeted for irrigated wheat environments, the second for semiarid. Regression of yield performance of this line over the best local checks used by co-operators shows that 'Weebill 1' not only had superior yield potential but also highly stable performance in low, intermediate, as well as high-yielding environments, regardless of irrigation or drought stress. Widely adapted genotypes are the best parents for targeting future crosses for further genetic improvements. Strategy of using widely adapted parents was used in designing crosses involving 'Weebill 1,' which gave rise to new lines with significantly higher yield potential.

Breeding Procedures

Introduction

Semi dwarf wheat from Mexico, Sonara 63, Sonara 64, Mayo 64, Lerma Roja 64. *Pure line selection:* Earlier varieties like P4, P6, P12 evolved at Pusa institute are result of pure line selection from local population. *Hybridisation and selection, a) Inter varietal:* A number of successful derivatives were developed at IARI New Delhi and Punjab. NP 809 - New pusa multiple cross derivative. However, all these varieties were lodging and poor yielder when compared to other countries. Hence the wheat hybridization programme was changed by Dr MS Swaminathan during 1963. Borloug was invited to our country and he suggested for introduction of semi dwarf varieties from Mexico. As a result, four commercial spring wheat varieties *viz.*, Sonara 63, Sonara 64 Mayo 64 and Lerma Roja 64 were introduced. However, they had red kernel hard wheats. These were utilized in our breeding programme and amber colour wheat varieties like Kalyan Sona, Safed Lerma, Sharbati Sonara were released, these are double gene dwarfs. *b) Inter specific crosses:* To get Hessian fly resistance. So also for rust resistance. *c) Back cross method of breeding:* Rust resistance in Chinese spring from Thatcher. *Hybrid wheat:* Male sterile lines were identified by crossing T.*timophevi x T. aestivum* Bison variety By repeated back crossing a male sterile line resembling Bison was evolved. *Mutation breeding:* Dr. MS Swaminathan did extensive work on this with gamma rays. *Sharbati Sonara* with increased protein content was evolved. *Development of multilines:* Borlaug developed multilines against rust. MLKS 15 was developed at IARI. Multiline is a mixture of pure lines which are phenotypically similar but genotypically dissimilar. Each line is produced by separate back cross method of breeding. Each line having resistance against a particular race of a disease. The semi dwarf varieties of wheat have been developed through the use of Japanese line Norin 10 as a source of dwarfing gene which led to "green revolution" in wheat production. The productivity of Semi dwarf varieties is about two and half times more than old tall growing varieties. More over these varieties are highly resistant to lodging and are highly responsive to fertilizer doses.

Wheat Introductions

The first introductions of wheat were made by the early, colonists who brought seed to this country. The emigrants from western Europe brought with them the soft wheats of that area. The exact origins of most of the early forms have been lost, but a few varieties widely grown in the eastern states until quite recently may be cited. One of these is Purple straw, which has been grown for over 150 years. Red May, known also by numerous other names, is believed to have descended from some of the earliest introductions into Virginia. The variety Mediterranean was introduced into America over 100 years ago from southern Europe. The hard red spring wheat had its beginning in this country from an introduction made by David Fife of Canada about 1842. This introduction, a winter type which originated in Poland, was planted in the spring. A spring type plant was found in it which produced seed and became the progenitor of the Red Fife variety. Red Fife later was used as a parent in producing the widely grown Marquis spring wheat variety. Introductions from the Kenya Colony in Africa are being used currently as sources of resistance to stem rust. The first introduction of hard red winter wheat was made by Mennonites who emigrated from Russia and settled in central Kansas in 1873. When the resistance of the Turkey Red wheat to cold and drought was learned, the variety spread rapidly in the Plains states and for many years was the most extensively grown variety in the United States. Turkey and later introductions of the Turkey type have since entered into the parentage of every important hard red winter variety grown in this country. Commercial culture of the durum wheats began with the introduction of the Arnautka variety into North Dakota in 1898, again by Russian emigrants. The Kubanka variety, introduced from Russia by the United States Department of Agriculture, was also a leading variety in the durum class for many years. The white wheats have apparently originated as introductions from several areas. Sonora, one of the oldest varieties in California and Arizona, is reported to have been brought into California by the Spanish over 150 years ago. Pacific Bluestem, widely grown in the Pacific Coast states for many years, was introduced from Australia about 1850. More recent introductions from Australia include Baart in 1900 and Federation ill 1914. The white club wheats were imported from Chile between 1850 and 1870. Little Club, which has entered into the parentage of many of our present Club varieties, and Big Club were among the early introductions.

Varieties Originating from Selection

Many of the older varieties were either impure when introduced or became impure after years of production through mixing, natural hybridization, or mutation. Selection was commonly practiced to purify these mixtures, and in many cases new varieties were developed from off-type plants found in the field by farmers and breeders. The methods by which true breeding lines may be isolated from these genetic mixtures were discussed in earlier chapters. Many varieties could be cited which were developed by this procedure, but only a few of the more important will be listed here. Fultz, for many years the leading soft wheat variety in the eastern states, was selected by Abraham Fultz in 1862 when he observed three spikes of awnless wheat in a field of bearded Lancaster (Mediterranean). Fulhio and Trumbull, varieties with resistance to the loose smut disease, originated as plant selections from Fultz. Trumbull later became widely grown in Ohio and surrounding states. Other selections were also made from Fultz, including a leaf rust-resistant strain which, along with Trumbull, were the parents of the Vigo variety. Other examples of varieties developed by selection are the Nittany variety of soft wheat, which originated as a plant selection from Fulcaster made in Pennsylvania in 1909; Red Rock, which had its origin as a single red kernel picked out of a white wheat called Plymouth Rock in Michigan in 1908; and Royal, a mosaic disease- resistant variety which was increased and distributed in Illinois in 1947 from a single plant of Illinois 2 that survived in a heavily mosaic-infested area. Important hard red winter wheats which originated as selections include Kanred, Cheyenne, and Blackhull. Kanred originated from one of 554 heads selected in 1906 from Crimean, which had been introduced by the United States Department of Agriculture from Russia. Kanred was resistant to some races of stem rust and leaf rust, although otherwise similar to Turkey. It was one of the first rust-resistant varieties to be developed in the United States. Cheyenne was selected from Crimean in 1922 in Nebraska. The Blackhull variety originated from three black heads found in a field of Turkey in 1912 by Earl G. Clark of Sedgwick, Kansas. The Kawvale variety, a semi hard wheat, was selected in Kansas from a soft wheat variety, Indiana Swamp in 1918. It is important because it later became a parent of Pawnee and other varieties. The hard red spring wheat, Haynes Bluestem, was developed from a selection made by LH Haynes in 1895. Mindum, a durum variety, was selected from a field of common durum. It proved to be rust resistant and later was widely grown.

Varieties Developed by Hybridization

Most varieties of wheat distributed before 1930 originated from introductions or by selection. Since about 1930, most improved varieties have been developed by hybridization. This sequence in obtaining new varieties is logical, for an intelligent hybridization program can be developed only after the parent materials have been sorted, tested, and the best strains have been isolated from them. Also, the large accumulation of knowledge in the field of genetics during the early part of this century has made possible a clearer understanding of the mechanics and principles involved in combining the desirable characteristics of parent varieties through hybridization. But we should not be misled into believing that the practice of hybridization awaited a clarification of these principles for, while not understood, hybridization was practiced much earlier, as some of the following examples will show. The Fulcaster variety of soft red winter wheat was produced in 1886 by a farmer-breeder, SM Schindel, from a cross between Fultz and Lancaster. This variety was widely grown in the eastern states until a decade or two ago. Other important soft red winter varieties more recently developed by hybridization include Thorne, from the cross Portage X Fulcaster made at the Ohio Agricultural Experiment Station in 1917; and Vigo, from the cross Trumbull X Fultz made at the Indiana Agricultural Experiment Station in 1932. Yorkwin, an important soft white wheat in the north-eastern states, was developed at the New York State Agricultural Experiment Station (Cornell) from a cross made in 1919 between Fulcaster X Goldcoin. In the hard-red winter wheat area, Pawnee, from the cross Kawvale X Tenmarq, made at the Kansas Agricultural Experiment Station, became the leading variety in acreage in the United States. The Tenmarq parent originated from a cross made in Kansas in 1917 between Marquis and P-1066. The latter is a sister selection of Kanred, made from Crimean. Many hard-red spring wheats have been developed by hybridization. Marquis, developed in Canada from the cross Hard Red Calcutta X Red Fife made in 1892, was the leading spring variety in acreage and the standard of baking quality in the spring wheat area for many years. Marquis has been a parent in many of the spring wheat varieties developed since. Hope was developed from a cross between Yaroslav emmer ($2n = 28$) and Marquis ($2n = 42$) made in 1915 in

South Dakota. Although seldom grown by the farmer, Hope ($2n = 42$) was the first hard red spring wheat to have stem rust resistance derived from an emmer, and it was used as a parent variety in the development of many important rust-resistant spring wheat varieties. Another hard red spring variety, Thatcher, developed in Minnesota, derived its stem rust resistance from Iumillo durum ($2n = 28$) in the double cross (Marquis X Iumillo) X (Marquis X Kanred). In Australia, William J. Farrer began an extensive program of wheat improvement in 1886. His most famous variety, Federation, produced in 1901, resulted from a three-way cross involving Fife, Etewah, and Purple straw. Fife was introduced from Canada to give quality. Etewah, an Indian variety, was used to provide earliness. Later, Purple straw was added to increase productiveness. Federation was introduced into the United States in 1914, and it is still grown on limited acreages. The white club variety, Hybrid 128, was developed from a cross between Jones Winter Fife and Little Club made in 1899 at the Washington Agricultural Experiment Station. Hybrid 128 in turn became a parent in crosses producing Alicel, Hymar, Albit, and other varieties.

Varieties Developed by the Backcross

The backcross method of breeding new varieties has been used extensively at the California Agricultural Experiment Station and has resulted in the improvement of several varieties of wheat by the systematic addition of bunt resistance, rust resistance, Hessian fly resistance, and other characters. Examples of improvements made in the Baart variety by backcrossing are cited. The number after the variety name designates the year in which the variety was released. For example, Baart 35 was distributed to farmers in 1935. The Carleton and Stewart varieties of durum were developed in North Dakota by crossing Mindum to Vernal emmer and backcrossing twice to Mindum.

Interspecific Crosses

Interspecific crosses have been used in wheat to transfer genes for stem rust resistance, leaf rust resistance, Hessian fly resistance, and other characters to common wheat. Stem rust resistance genes were transferred from Yaroslav emmer to the Hope variety, and from Vernal emmer to Carleton and Stewart varieties of durum wheat. Iumillo durum was used in the development of the stem rust-resistant variety of spring wheat, Thatcher. Hessian fly resistance from a durum imported from Portugal, P. 1. (Plant Introduction) 94587, is being used in the breeding of fly-resistant varieties. Wheat like selections from crosses of common wheat with *Agropyron elongatum* are resistant to wheat streak mosaic and to wheat joint worms. The transfer of a leaf rust-resistance gene from *Aegilops umbellulata* to common wheat was described in Chapter 3. Many other species, closely related to common wheat, have been used as sources of other desirable genes. Irradiation Breeding. Irradiation may be used to increase mutation frequencies in wheat as with other crops. To date no variety of wheat has been developed as a direct result of irradiation breeding. Irradiation with X-rays was instrumental in obtaining a crossover between a wheat chromosome and an alien chromosome from *Aegilops umbellulata*, which resulted in the transfer of a gene for leaf rust resistance from *Ae. umbellulata* to common wheat.

Table 2.7. Wheat mega-environments (ME) and main targeted traits for wheat improvement in current wheat improvement research

ME	*Moisture Regime*	*Temperature*	*Per cent Area**	*Main target traits*
ME1	Irrigated	Temperate	36.1	High yield potential, lodging tolerance Water and nutrient use efficiency Resistance to three rusts Large white grain with leavened and fl at bread quality
ME2	High rainfall (>500 mm)	Temperate	8.5	High yield potential and lodging tolerance Resistance to rusts, septoria tritici and fusarium head blight Large red grain and leavened bread quality
ME4	Low rainfall (<500 mm)	Temperate	14.6	Drought tolerance with responsiveness to water or hot availability Better root and emergence characteristics Adaptation to conservation agriculture Resistance to three rusts, septoria tritici, tan spot, root diseases Large white grain and leavened and fl at bread quality
ME5	Irrigated or High rainfall	Warmer	7.1	High yield with early maturity, lodging tolerance Heat tolerance Resistance to rusts and spot blotch for low rainfall areas Resistance to rusts and fusarium head blight for high rainfall areas Large white or red grain depending on low or high rainfall, leavened and fl at bread or noodle quality depending on country
ME6	Low rainfall	Cool	6.2	Drought tolerance, photosensitivity, tall stature Resistance to rusts Red grain with leavened bread quality

*Per cent of developing country wheat area.

Breeding Strategies to Increase Yield Potential

Increase in yield potential are mainly associated with increased biomass, kernel number, and harvest index (Singh *et al.*, 1998c). Yield components, such as grain size and number, or harvest index in more recent germplasm do not correlate to increased yield potential. Large kernel size continues to be an important trait in some local markets of developing countries and appears to be associated with better emergence under poor management. Some of the recent wheat germplasm developed at CIMMYT has not only shown increased grain yield potential but also kernel weight as high as 57 g/1,000 kernels in north-western Mexico compared with about 40 g for most of the wheat germplasm developed. Although the early increases in yield potential of semidwarf wheat cultivars came from the incorporation of dwarfing genes, subsequent progress can be attributed to quantitatively inherited additive genes. It is likely that intense breeding efforts during the last three decades in the post-green revolution era had already selected additive genes that greatly contributed to enhancing yield potential. If that is the case, then further progress is expected from selecting genes that have much smaller effects. Such progress would necessitate modification of commonly used traditional breeding schemes. Alternatively, introgression of new genetic diversity from unrelated wheat germplasm, including wide hybridization, can create a new genetic pool and bring in small-effect genes that may not be present in wheat germplasm commonly used in a breeding program. Wheat improvement or improvement of other self-pollinated crops often utilizes simple, three-way (top), four-way (double), or repeated backcrossing approaches. Various wheat breeders also commonly practice pedigree or bulk methods of selection. Wheat breeding at CIMMYT in the 1960s and 1970s relied on simple, top, and double crosses, followed by the pedigree method of selection. It was realized later that best advanced breeding lines were rarely derived from double crosses, possibly because the genetic variation generated by such crosses was too large and the chances of recovering plants with desirable combinations of genes were reduced due to insufficient population sizes. During the early 1980s, CIMMYT breeders relied upon simple and three-way crosses and occasionally single backcrosses, followed by a modified bulk selection scheme where individual plants were harvested in the F_2 generation to grow the F_3 generation, however, bulk selection was practiced in the F_3-F_5 generations. Individual plants or spikes were once again harvested in the F_5 or F_6 generation (Rajaram *et al.*, 2002).

Following the study by Singh *et al.* (1998c), which showed that selection schemes had little or no effect on the performance of progeny lines, but that it was the choice of parents that determined the progeny response, a selected bulk-breeding scheme was introduced in bread wheat improvement. Under this scheme, in all segregating generations until F_5 or F_6, one spike from each of the selected plants is harvested as bulk and a sample of seed is used to grow the next generation. Individual plants or spikes are harvested in the F_5 or F_6 generation. This scheme allows the retention of a larger sample of selected plants without increasing the cost and was found to be highly efficient in terms of operational costs. Moreover, retaining a large sample of plants in segregating populations increases the probability of identifying rare segregates that carry the most desired genes. Based on the view "germplasm is paramount to increasing yield potential" (Rasmusson, 1996), beginning were made to utilize a single-backcross crossing approach that was initially aimed at incorporating resistance to rust diseases based on multiple additive, minor genes (Singh and Huerta-Espino, 2004). However, soon it became apparent that the single-backcross approach also favoured selection of genotypes with higher yield potential. The reason why single backcross shifts the progeny mean toward the higher side of the curve is that it favours the retention of most of the desired additive genes from the backcross or recurrent parent while simultaneously allowing the incorporation and selection of additional useful small-effect genes from the donor parent. The shift in mean results in a higher frequency of new breeding lines that have superior yield potential than for the check cultivar. Repeated backcrossing is not desirable as it was devised to incorporate a single–or only a few–major gene(s) with the least disturbance to the genetic make-up of the recurrent parent. Crossing of spring wheat germplasm with winter wheat germplasm in the 1970s and 1980s not only increased the biomass of spring wheats, but it also increased yield potential, as observed in numerous cultivars released in many countries from *'Veery'* and *'Attila'* crosses. Although thousands of crosses were made in this effort, the above two crosses with winter wheat parents 'Kavkaz' and 'Nord Desperez,' respectively, can be considered the most successful because they led to the development of mega-cultivars that were subsequently grown on millions of hectares in the eastern African highlands, North Africa, the Middle East, Asia, and the Americas. These mega-cultivars showed 8– 10 per cent higher yield potential over previously developed cultivars. This yield increase has often been associated with the presence of the alien translocation commonly known as 1B.1R, where the short arm of chromosome 1B is replaced by the short arm of chromosome 1R of rye (*Secale cereale*). Alternatively, the presence of genes for resistance to major diseases, such as all three rusts and powdery mildew, on the rye chromosome arm could have provided superior disease resistance of Veery lines and some unknown small-effect genes from winter wheat germplasm could have provided increased yield potential (Singh *et al.*, 1998b).

Alien translocations were often selected for incorporating genes that confer resistance to diseases and pests or tolerance to abiotic stresses. However, often their effects on increasing grain yield potential remains undetermined. Alien translocation T7DS.7DL-7Ae#1L from *Thinopyrum elongatum* that carries leaf and stem rust resistance genes *Lr19* and *Sr25*, respectively, has been shown to increase yield potential ranging from almost non-significant levels to over 15 per cent (depending on genetic background and under irrigated conditions) through increased biomass production caused by increased photosynthetic rate (Reynolds *et al.*, 2001). Thus its widespread incorporation could lead to a quantum jump in yield potential. The initial translocation carried a gene that caused higher endosperm pigmentation that is an undesirable quality trait for bread wheat. However, Knott (1980) developed a non-yellow mutant and this non-yellow variant translocation maintains the yield-enhancing effect. Incorporation of rye translocation T4BS.4BL-2R#1L into three wheat genotypes showed that in a 'Weaver' background significant increases in yield potential, ranging between 5–15 per cent, occurred in all three experimental environments: irrigated normal sown, irrigated-late sown, and post-anthesis drought-stressed. However, this translocation had negative effects when incorporated into two other genetic backgrounds. These exemplify the importance of genotype translocation interactions, which are expected to occur when a portion of a wheat chromosome of a highly balanced genotype is replaced with an alien segment. Such imbalance could result in negative effects and needs to be rebalanced by compensating genes located in other homoeologous wheat chromosomes or the rest of the genome. Development of molecular markers for these translocations can aid their rapid identification and in corporation and could help enhance yield potential.

Molecular Markers in Wheat Breeding

Recent improvements in molecular-marker assay techniques, reduced cost, and projected improvements expected from single nucleotide polymorphism (SNP) technology, coupled with increasing numbers of markers for traits of importance, have resulted in increased use of marker-assisted selection (MAS) in applied wheat breeding. The SNPs and large-scale sequencing will enable the development and deployment of more gene-based markers, thereby improving the detection and application of markers in breeding. Molecular markers offer wheat breeders considerable scope to improve low heritability traits (bioassay or field test), traits that cannot be evaluated at the breeding site, or traits that require gene pyramiding, as is the case with some disease resistances. Most of the currently available wheat markers identify simply inherited Mendelian traits, although quantitative trait loci (QTL) analyses have identified and improved our understanding of some complex traits. William and Trethowan (2006) provide a description of many of the markers currently used in applied wheat breeding. An immediate application of molecular markers is the characterization of parents. Markers are used to create gene profiles of all parental material. All known markers for traits are applied at this stage and parents are selected for crossing on the basis of predicted segregation of different characters. If a trait has a high value and the alternative assays have low heritability, as is the case with many root-disease characters, then markers can be applied through the segregating phase. However, cost and time constraints will limit the number of assays that can be applied, and the wheat breeder must identify the most effective and efficient intervention points for MAS in the breeding process. If multiple markers for multiple traits are to be applied it is important that they be used as early as possible in the breeding process to improve the probability of identifying germplasm carrying not only the MAS target traits but also the suite of yield and quality attributes necessary for cultivar release. If markers are expected to segregate in a top cross, the F_1 top cross progeny can be screened for markers segregating in the original bi-parental cross; those in the top cross parent are not screened because they will all be positive and heterozygous. Following this period of allele enrichment, markers can be applied at an intermediate stage of development, such as F_4, or once the fixed-line progeny have been identified; in this instance all target markers are applied. In simple crosses in which several markers are segregating, MAS can be applied in the F_2 or F_3 generation. The number of plants selected for screening from each population will depend on the number of markers segregating, the cost of the assay, and the availability/effectiveness of alternative assays. The majority of the currently available wheat markers are dominant in nature, which makes it impossible to identify heterozygotes. Once fixed-line progeny are identified that are positive for all target alleles it will be necessary to screen individual plants within each line or conduct progeny tests to ensure the target gene is homozygous before crossing.

Doubled Haploid Breeding in Wheat

Anther Culture

Androgenesis is the production of haploids from anther culture or isolated microspore culture. Anther culture is well established in barley and is reportedly less technically demanding than isolated microspore culture. Successful anther culture benefits from the fact that a fraction of the pollen grains *in situ* are embryogenic. When cultured on artificial media under appropriate conditions, the embryogenic pollen grains can develop into calli or embryoids. Calli can be induced to give rise to embryoids with suitable culture media. Embryoids will develop into plantlets although regenerant fertility and numbers of green plants can impact the number of doubled haploid s lines produced. Careful attention to a number of procedural details will improve the response of anther culture. Anthers must be harvested from healthy, stress-free donor plants because plant stress significantly impacts androgenic response. Donor plants require optimal conditions of temperature, light intensity, photoperiod, nutrition and water. Pesticide application is discouraged two weeks before spike collection. Spikes of wheat or barley are selected that contain anthers with pollen at the mid- to late-uninucleate stage. The stage of pollen development is critical to allow the reprogramming of gametophytic development into pollen embryogenesis and / or embryogenic calli production in receptive pollen grains. Spikes or anthers are subjected to an induction-stress treatment to increase the efficiency of androgenic response. In barley, cold shock and carbohydrate starvation have been used successfully, and cold shock and auxin treatment have been reported to stimulate wheat androgenesis. Following induction stress, anthers are transferred to a culture medium for approximately one month although induction treatment times and culture media vary for the various published protocols and among genotypes used. Anther culture is concluded when small calli or embryoids are observed as white beads in the culture medium. When calli or embryoids are 1.5–2 mm in size, they are transferred to a solid regeneration medium where they will be cultured for 3–4 weeks. Improved root formation in developing barley and wheat regenerants has been reported when they are transferred to jars or vials containing regeneration medium without growth hormones for approximately one week before transplanting into soil. In wheat anther culture, regenerants are often treated with colchicine in order to improve the proportion of doubled haploid plants. After approximately 5–6 weeks, plantlets are carefully freed from the agar, rinsed with tap water and transferred to a soilless mix or potting compost. Plantlets should be covered for approximately 1–2 weeks to prevent desiccation. After this time, the plantlets can be transferred to larger pots to develop further. In the case of winter wheat, plants need to be vernalized to induce seed set which is completed before final transplanting. Many researchers have adopted anther culture because of the high numbers of pollen grains within each anther that can theoretically produce DH plants compared to one haploid plant per floret with intergeneric crossing approaches. Anther culture is reportedly more cost-effective than the intergeneric crossing methods for the large-scale production of doubled haploids, and in barley spontaneous doubling of haploids eliminates the need for the use of highly toxic chemical doubling agents such as colchicine. However, the anther culture method is highly genotype dependent and high frequencies of albino plants among regenerants is a significant impediment. Anther culture-derived plants can range from pale green to variegated (green and white sectors) to completely unpigmented. The unpredictability in wheat androgenic response is a major limitation to the adoption of the anther culture method in commercial wheat breeding programs. While anther culture has been widely used in barley DH breeding, the approach has not been widely adopted in wheat DH breeding programs.

Microspore Culture

Microspore culture involves harvesting immature microspores that are totipotent. Staging of the harvest is important to optimizing the efficiency of the microspore approach. Plants are grown in a chamber to provide optimal conditions for growth. Sporophyte induction from microspores requires a stress on the plants. Considering wheat, tillers are excised with spikes in the boot, and held in a refrigerator at 4°C for 21 days. The cool temperature provides conditions to induce sporophytes and reduce albinism. Staging is performed by microscopic examination of a representative floret from the spikes to ensure microspores are present at the late uninucleate stage. Immature spikes are harvested and homogenized in a Waring blender. The homogenate is processed to isolate the microspores for culture on an induction medium. Embryos that form are transferred to a regeneration medium where the embryos germinate and form roots and shoots. Subsequently the plantlets are transferred to a solid medium. The major disadvantage of microspore culture in certain species, including wheat, is the genotype dependency. Some genotypes are amenable to microspore culture, whereas other genotypes resist culture. Currently the genetic factors involved in controlling microspore culture amenability are unknown. Therefore, from a breeding perspective, the unpredictability of whether crosses will generate doubled haploid s or not, makes the method inefficient and too costly to be sustainable, particularly when compared to the wide-cross method which is less genotype dependent. Another disadvantage of the microspore approach is difficulty in adequately staging the microspores for culturing. Although experience will reduce the difficulty of staging, plants must be monitored for the appropriate stage by sampling florets and looking microscopically for microspores at the late uninucleate stage. Advantages to the microspore approach are mainly the spontaneous doubling of chromosomes to reconstitute the diploid condition and the high volume of progeny. The rate of spontaneous doubling ranges from 25 per cent to 70 per cent in wheat. Without spontaneous doubling of chromosomes, a great deal of labour is involved, along with the use of the hazardous chemical colchicine. Nevertheless, the approach has potential over the wide-cross method for greater productivity and efficiency. Moreover, the microspore approach is more amenable to applications such as gene editing.

Wheat-Maize-Pollen Method

The wide-cross approach in wheat involves crossing maize pollen onto emasculated wheat spikes. The major appeal to the approach is its consistency for producing haploids across a range of wheat genotypes. While some genotypes are more productive than others, haploid plants can be generated from all. This consistency is important as a breeding tool, because resources dedicated to generating doubled haploid s are not being wasted on unresponsive crosses. Maximizing productivity from the maize -pollen method depends on a plant production system that generates strong, healthy wheat plants for use as parents in DH production. It is also important to make sure maize plants are grown under optimized conditions for consistent, high quality pollen production. Embryo rescue is critical to the success of this approach. Therefore, excellent sterile techniques are necessary to achieve high rates of production, thereby minimizing contamination by fungi and bacteria that can impede embryo survival. Haploid plants and plants treated with colchicine for chromosome doubling are often weak, and require extra care and attention for their successful culture. As a starting point, we discuss a process used in wheat that meets the criteria for a productive doubled haploid system. Growth of spring wheat is optimized under cool temperature conditions with high light intensity. A long light cycle satisfies any requirement for day length in day-length sensitive genotypes and optimizes rate of development. Genotypes with a winter growth habit require a vernalisation treatment. Growing plants in a growth chamber allows the type of control necessary to optimize plant performance and development. A growth chamber set at 15°C during a dark cycle of 8 h and 18°C during a light cycle of 16 h is a good compromise between duration of development and optimizing the robustness of plants. Light quantity and quality are important, both for producing robust plants and optimizing the rate of phenological development. A traditional incandescent and fluorescent combination at a light intensity of 400 µmol $m^{-2}s^{-1}$ is effective for wheat development, but other lighting types may be suitable after testing on the species of interest. Humidity controlled at 70 per cent will provide an adequate balance of plant growth and minimize development of diseases such as powdery mildew. Throughput is important in breeding and can be optimized by growing plants in appropriately-sized pots and transplanting to larger pots as plants grow. Knox *et al.* (2000) suggested three pot changes for durum wheat. For wheat, two pot size changes, one at the three-leaf stage and another at tillering, after the initial planting in a small container has since been found to be sufficient. Sizing pots to plant development stage helps improve productivity in a high throughput doubled haploid system because it allows watering without a lot of time-consuming individual plant attention. A potting medium that provides good drainage but good nutrient holding capacity is important. Repotting plants has an additional advantage of providing fresh potting medium and diluting the build up of fertilizer salts and compounds that hinder plant growth. Pre-germinating F_1 or other seed used as DH parents should be considered if germination rate is thought to be a problem. To achieve full growth potential of the plants, it is important to fertilize regularly and provide adequate water.

In wheat there are two main methods of achieving anther removal. In the authors' experience, the method of clipping back the florets with scissors to expose the anthers reduces embryo production compared to the method of removing anthers through inserting forceps between an intact lemma and palea. The plant can tolerate some trimming such as the clipping of awns and the removal of tertiary florets from each spikelet. Placing a glassine bag on the spike is important to prevent stray pollen from pollenating florets. Embryo production is further optimized with the use of fresh maize pollen. We have found pollen is best when collected about 2 h after supplemental lighting has started the day cycle. The pollen is a vibrant yellow colour with a fine grain. Pollination on day 2 post-emasculation of a wheat spike, when stigmas are fluffy, is generally the most favourable pollen application time. Critical to optimizing haploid production from the wide-cross method is the application of auxin. We have found the artificial auxin, dicamba, to be superior in producing germinated haploid embryos. Timing of application and concentration need to be optimized for maximum embryo production. We have found a 100 mgL $^{-1}$ concentration of dicamba applied on both days 3 and 4 after emasculation using an aerosol spray bottle to be effective. Until now we have discussed factors relating to the efficiency of haploid embryo production, *i.e.* conditions that maximize the number of embryos produced per emasculated spike. Under optimized conditions, one can expect 60–80 per cent of florets on a wheat spike to produce embryos. Another area of attrition is the embryo-culture-to-germination stage. The two major factors that interfere with conversion of embryos to haploid plantlets is failure of embryos to germinate and contamination. Factors that affect germination include, as alluded to previously, the type of hormone treatments while the embryos are developing on the spike, but also the timing of embryo rescue. We have found that rescuing embryos 16–19 days post-emasculation is the best compromise between maximizing embryo growth while minimizing the necrosis that begins at about 19 days. The type of medium affects germination of the embryos and development of plantlets. Use of Gamborg's B5 nutrient medium with minimal organics has worked successfully in the authors' experience. A chilling stress of freshly-cultured embryos helps to promote germination. We use 5°C for 48–72 h. Following the cold treatment, we have found stratification with rapid temperature cycling to further improve germination of embryos. We use a 6 h cycle of 11°C dark and 18°C with lights on. Germinated embryos, while still in vials, are placed in a chamber at 12°C for 8 h darkness and 13°C with 16 h light (the difference in temperature being from heat generated by the lights).

Haploids developed using the wide-cross method must be subjected to a chromosome doubling treatment. We have found that colchicine continues to be the most effective treatment to achieve chromosome doubling and minimize attrition of plants post-treatment. We apply colchicine when the plants start to tiller, which coincides phenologically with a switch in development from the vegetative phase to the reproductive phase. The plants seem to be advanced enough to sustain the shock of the chromosome doubling treatment, while at the same time early enough to affect cells that give rise to eventual reproductive structures of florets, spikelets and sometimes whole spikes. The doubled plants are treated similar to the haploid plants to reintroduce them into the growth chamber. The plants regrow and meristematic cells that successfully experienced chromosome doubling give rise to fertile chimeras of florets, spikelets or spikes. Often F_1 DH parent plants (or whichever generation of parent) deriving from multiple crosses are handled in concurrent streams, particularly in the wide-cross system. Often there are peaks and troughs in productivity of a doubled haploid program for various reasons, such as a pest infestation, growth chamber failure and so on. To mitigate the impact on any one population in the system, it is better to plant F_1 DH parental plants derived from a number crosses at the same time and spread the planting over several days to a few weeks. That way each of a few F_1 plants from multiple crosses experiences the low production issues, rather than all the F_1 plants from one cross experiencing the issues which would be the case if only F_1 plants from one cross were planted at a time sequentially. The disadvantage to this approach is the extended production of doubled haploids as they come out the other end of the pipeline. Production duration of doubled plants of a population from a particular cross is further extended by the differential rate of embryo germination, recovery of plants doubled with colchicine and differences in plant maturity.

Critical to the high throughput required for doubled haploid production in a breeding program is the implementation of a tracking system. To be able to expend more energy on production of doubled haploid plants and less time documenting the plants at each transitional stage, a documentation system that can follow individual plants through the process is essential. As an example, we use a system whereby each cross is documented with a cross number in a computer file and each F_1 parental plant is assigned a unique number. As embryos are rescued from a parental plant, they are assigned a unique code. A combination of letters and numbers helps to make the plant identifier easy to read. For example a doubled haploid plant would indicate it was derived from an F_1 parental plant crossed, the cross being the first cross in a series of crosses, the F_1 plant being the first plant from which an embryo was rescued, and the plant being derived from the tenth rescued embryo of the AA F_1 plant. By printing the number with a barcode on a pot marker and attaching it to the vial in which an embryo was rescued, the pot marker can follow the embryo through the entire doubled haploid process without any transcribing of information. Thus the pot marker is transferred to the pot in which the haploid plantlet is transplanted, tied to the plant as it is treated with colchicine, transferred to the pot in which the colchicine treated plant is transplanted, and applied to the envelope into which the doubled haploid spikes are harvested. Another time-consuming process is the identification of spikes on F_1 parental plants being processed for the variety of manipulations including emasculation, maize pollination and multiple-hormone treatments. The process can be sped up by uniquely identifying each spike among a group of plants and multiple spikes on each plant. We identify the spikes with the cross and plant number on a coloured tag that is uniquely assigned to a particular day. Then, when it is time to pollinate at the same time all spikes emasculated on the same day across a range of plants, the plants with the same colour of tag can be assembled. Rather than writing treatments on tags, the tags are stamped or punched with a unique symbol.

Resources for Doubled Haploid Production

Doubled haploid production, whether through microspore culture or wide-cross approaches, is a sophisticated technology that requires appropriate segregating plant stocks, equipment, and appropriately-trained personnel for high throughput. Personnel must be trained in sterile techniques and plant husbandry. Knowledge of computer and basic software use are important for documenting and tracking large numbers of unique lines. Proper facilities and equipment are essential for optimizing the environment and procedures necessary for high throughput. The ideal facility for DH production will have controlled environment rooms of some combination of greenhouse and growth chambers. There needs to be a header house equipped for handling potting media and processing plants at the front end of the growth facilities. A laboratory equipped to handle toxic chemicals and materials under sterile conditions is essential. Haploid plants tend to be less vigorous than normal plants, therefore it is important to use materials that enhance their survival. For example, it may be necessary to pasteurize or sterilize potting media to prevent the media from being a source of pests and pathogens. Similarly it is important to periodically clean pots and equipment used in potting. The preferred facility will have an autoclave or pasteurizer and pot-washing facilities. As an alternative to pot washing, disposable containers can be used. The choice is a trade-off between investments in labour as opposed to supplies. The facility should be equipped with adequate benching for planting and transplanting and dust control for protection of workers. Air-borne pests are a persistent problem within growth rooms, therefore, the facility will also be used to handle pesticides. As a result, the header house facility should be equipped with a fume hood for mixing pesticides. The facility will need space to stock potting media, potting containers, fertilizers, pesticides and tools. There is no one system of handling plants, but some requirements are important to consider. The type of potting media should resist compaction and maintain porosity with good drainage and nutrient-supplying capability. As previously mentioned, sizing planting containers to plant size can save on expensive growth space and reduce labour by allowing general or automated watering and fertilizing. With pots optimized to plant size there are fewer issues with waterlogging or pots drying out too much, conditions that can set plants back or result in death.

Paramount to the production of a segregating DH population is segregating source materials. These materials are usually F_1 plants, but could be derived from any generation where there is a reasonable expectation of segregation. The DH process can be used to fix all loci in cultivars intended as parents in a genetic study to eliminate the potential confounding heterozygosity in the resulting population. The maximum genetic variation will be achieved by producing DH lines from F_1 plants. Single cross F_1 plants derived from inbred or DH parents are homogeneously heterozygous. However, a round of selection may be considered on F_2 plants with DH progeny produced either on the F_2 plant or on a sample of F_3 derivatives of the most favourable F_2 plants. In the former case, it would be impractical to perform selection in the field, which would limit the number of traits that could be selected, particularly if selection is performed prior to gamete sampling. In the latter case, however, the F_2 plants could be grown in the field and selected for a variety of traits including resistance to diseases and insect pests known to occur in the target environment. When creating doubled haploid s from top cross F_1, backcross F_1, or F_1 progeny from a four-way cross, caution must be taken to carry out sufficient sampling of plants to obtain the variation for traits targeted within the cross. If more than a very few loci are under consideration in breeding objectives for a cross, doubled haploidy may not be the best approach with other than simple crosses. In wheat, with the maize - pollen method, for example, large, healthy, vigorous plants are generated because they generally produce high rates of embryos. Using large plants with multiple spikes, many DH lines can be produced from only a few F_1 plants. However, using only a few F_1 plants runs contrary to requirements of sampling non-simple cross F_1 plants. Similarly with microspore culture, only a few spikes need be harvested to generate large numbers of haploid progeny; however, it is suggested that additional F_1 plants should be grown with a single spike used from each plant. Populations of segregating F_1 plants could be pre-selected using markers prior to use in DH production. The costs and benefits of different strategies will be discussed in more detail later.

Routine glassware and plastic ware are required, but also specialized containers may be needed. For example, specialized culture vessels can be used in microspore culture to improve productivity of the system. High volumes of glassware, particularly vials, require automated cleaning systems to reduce labour cost, or disposable products may be used if the cost offset can be justified in terms of reduced labour. As already mentioned, a high-throughput DH system tailored for breeding will require a tracking system. A computer with proper software and a printer that can directly print pot markers are required. Pot markers may be barcoded to facilitate tracking and documentation. It is important to minimize transcribing of information at various steps of production. Use of grid systems for positional identification, colour coding and multi-use labels can assist with tracking DH materials without the need to write out information. As previously mentioned, in the case of the maize - pollen method in wheat, we use computer printed pot markers with a unique identifier that are fixed to the vials in which embryos are rescued. When the embryo germinates, the pot marker is removed from the vial and placed in a potting tray cell with the plantlet. When the plantlet is ready for colchicine treatment, the pot marker is strung to the plantlet through the process of root washing, colchicine solution and rinse water root emersion. The marker is then placed in the greenhouse pot with the colchicine treated plant. The pot marker follows the plant through pot size changes and is placed in the harvest envelope.

Marker Assisted Breeding and Doubled Haploidy

Molecular markers are available for an ever-increasing number of traits in a wide range of crops. For example, in wheat there are DNA markers available for disease and insect resistance, as well as grain quality and agronomic traits. A doubled haploid production system can be integrated with molecular breeding to fix qualitative traits and improve gene frequencies of quantitative traits. The greatest efficiency of molecular breeding with doubled haploid production is achieved by applying markers to haploid plants. While marker testing can be performed at any stage, provided adequate lab personnel and resources are available; however, with delayed testing, resources are used to promote lines that could have been discarded at an earlier stage. The mechanics of applying marker-assisted breeding (MAB) to doubled haploid production requires good planning. Considering the maize pollen method in wheat, doubled haploid production and marker testing must be coordinated so that a small portion of leaf tissue can be sampled from haploid plantlets. This is a narrow window between when plantlets are transferred out of sterile media culture into potting media but prior to chromosome doubling. By sampling this stage, fewer plants need be treated with colchicine and replanted after the colchicine step. A reliable and efficient system of plant identification is crucial for effective marker assessment so that correct plants will be culled based on marker results. This approach is used within the wheat breeding program. Parents of crosses are evaluated with a panel of markers listed by Randhawa *et al.* (2013). Crosses that are segregating for traits to which markers are available are scrutinized for marker polymorphism. For example, a cross between the wheat stem sawfly (*Cephus cinctus*) resistant parent BW925 carrying *Sst1* and the *Fusarium* head blight (*Fusarium* spp.) resistant parent Carberry carrying *Fhb1* were considered for MAB. Markers for *Sst1* and *Fhb1* applied to the parents showed polymorphism. Crosses that show polymorphism are used to generate haploid plants that are sampled for DNA and tested for the molecular marker s. In the example, F_1 plants from the BW925/Carberry cross were used to generate haploid plantlets using the maize-pollen method. The haploid plantlets were grown in labelled multi-cell growth flats. Tissue was carefully sampled from each haploid plantlet to maintain the identity of each sample with each plant. The DNA was extracted and markers for stem solidness and FHB resistance were evaluated for each of the haploid plantlets. The haploid plantlets were culled based on marker data. A total of 394 haploid plantlets were evaluated and 48 per cent were retained while 52 per cent were discarded because they lacked the favourable molecular-marker variant for *Sst1*. Similarly, 57 per cent of the progeny expressed the favourable molecular variant for *Fhb1*, while 43 per cent did not. The *Fhb1* markers were not used for selection but rather for marker validation. The haploid plantlets remaining after selection with the *Sst1* marker were treated with colchicine to double the chromosomes and generate a doubled haploid population with an enriched allele frequency for *Sst1*.

Where possible, segregating F_1 plants generated from backcrosses or topcrosses (*i.e.* three-way crosses) should be pre-screened with markers to ensure the trait(s) of interest. Testing segregating F_1 plants with markers is an extremely powerful method of gene enrichment and substantially reduces the attrition of lines coming out of the other end of the breeding pipeline. Testing the segregating F_1 plants will eliminate one-half for each marker tested. If the cross is configured such that a double dose of the favourable allele is incorporated, undesirable heterozygotes are eliminated and none of the valuable DH lines will need to be discarded based on the locus. If the cross is configured such that the favourable allele is only in the heterozygous condition, half the F_1 plants will be eliminated with the marker and only one-half the DH lines will be eliminated instead of three quarters. Eliminating a few F_1 plants with a DNA marker is far more cost effective than eliminating many DH lines. The enriched population of DH lines resulting from testing F_1 lines will improve the success rate for identifying superior genotypes for other trait combinations when the DH lines are field evaluated. Nevertheless, marker-assisted selection requires considerably greater haploid production. If one marker is applied to haploids from a F_1 plant where the favourable allele is in the heterozygous condition, 50 per cent of haploids will not have the allele of interest and will be discarded; thus, in order to produce a population of, for example, 200 doubled haploids, approximately 400 haploids must be generated. To maximize stacking of genes for resistance to multiple pests Somers *et al.* (2007) used a complex crossing scheme in which they applied DNA markers to identify progeny carrying the desirable alleles. They then used the segregating progeny plants for the production of DH lines. The DH lines were then phenotyped in the field. Similarly, Wessels and Botes (2014) used markers on lines generated from crosses designed to recombine rust resistance genes to identify progeny possessing the accumulated favourable alleles. In addition to using markers on the segregating parental plants, DH plants were also subjected to marker evaluation.

Future Needs in Doubled Haploidy

The needs for DH methodology will depend in part on the crop type. For example, in wheat, a robust, genotype-independent anther or microspore culture DH protocol is still not available. Albinism can still be problematic in some protocols, such as barley anther culture. Methods that require regeneration or embryo generation from callus can suffer from unacceptable somaclonal variation. Thus, on-going development and improvement of methodology is required. Improved production efficiencies through the use of barcode and computerized labelling and the invention of methods that require even less handling of materials are desirable. The expanded commercial availability of DH production services *e.g.* Haplotech and Heartland Plant Innovations should ensure the on-going development and increased application of the technology in plant breeding. Future developments in DH technologies will require a closer integration of DH production with marker -assisted breeding to facilitate selection for major genes as well as quantitatively inherited traits. Due to the advantages that the DH method provides for the application of induced mutagenesis and transgene incorporation, these technologies will likely be deployed to a greater extent in the future; however, their use will be limited to crops where these technologies are accepted by producers, end users and the consumer.

How marker-assisted breeding can be applied to haploid plants has been demonstrated. This will become more sophisticated with the application of array or genotype by sequencing technologies in which the whole genome can be characterized. However, with these genotyping technologies the rate of attrition can be high. The high attrition rates in doubled haploid populations occur because only 50 per cent of the population is retained for each locus considered. For example only 1 plant would remain from 1000, if 10 loci were evaluated. This outcome is without considering the population size required to account for a reasonable probability of ensuring that 1 remaining plant will have all 10 desired loci. Microspore culture, with its greater potential to generate haploids and to bring the cost of doubled haploid production down, will be important to the success of genotype-based breeding of DH lines. As the cost of genotyping technologies drops, the feasibility is greatly increased of using such technology for full genome assessment of the genetic potential of each breeding line at an early generation for the purpose of discarding genotypes with poor genetic potential. Genotyping technologies will give the opportunity to measure the genetic potential of a cross between two parents, *i.e.* selecting at a genotypic level. A choice may be made to abandon the cross of particular parents for the DH stream based on assessment of the parental combining potential. Crosses can be designed based on genotypic information where the focus is on maintaining population size for selection for complex traits such as yield and protein and relegated to a classical inbreeding stream for germplasm development. Other crosses predicted to have a high combining ability would be entered in the DH stream for commercial cultivar development. Technology for single-gene editing has advanced in recent years to become an option as a breeding tool. Being able to edit a gene sequence in a haploid to modify traits greatly enhances this technology over working with a diploid. With a diploid, gene editing would likely result in a heterozygote or hemizygote depending on the nature of the editing. Such conditions can mask expression of the gene and additional work is required to obtain homozygous true breeding lines. However, by editing the haploid with chromosome doubling, the plant will be in the homozygous condition and expression of the gene will be immediately apparent. Similar to gene editing, transformation technologies are enhanced by working with a haploid compared to a diploid because the transformed locus will undergo doubling along with all other genes. Microspore culture has an advantage over the maize -pollen method in genetic modification systems. As single cells, microspores can be manipulated for transformation or gene editing and all cells in the plant derived from the modified micropsore will be genetically identical. In contrast, it would be technically more challenging to genetically modify the egg cell for subsequent haploid development with the wide-cross approach. The modification would have to be done on a multicellular embryo which would create genetic chimeras that would give rise to variable seeds. A subsequent selection stage would have to be applied not to the DH_1 (the first generation of doubled haploid) plant as with microspore culture, but to all DH 1 seed produced by each DH_1 plant.

Wheat-Maize-Pollen Doubled Haploid

Maize Pollen Donor Plants

Maize is planted each week to ensure a constant supply of pollen. Approximately 25–30 seeds are placed in a 10 cm paper towel lined plastic pot containing a 6: 1 mixture of soil and turface (a soil amendment similar to cat litter). From newly planted seeds, plants will take approximately 6–8 weeks before tasseling, so the planting of maize must be started before the planting of spring wheat parent plants. Once the seedlings are approximately 7–10 cm tall, they are transplanted individually into 15 cm clay pots containing the same soil mixture. The number of seedlings that are transplanted is dependent upon (1) the number of parental wheat plants, and (2) the number of spikes being emasculated. If 15 plants are transplanted per week, then at any given time there will be approximately 120 pots of maize at different stages of development. Once the maize is in clay pots the plants should be fertilized every week with a fertilizer solution of 20-20-20 (approx. 10 gL^{-1}) and 2 gL^{-1} of ammonium sulphate (21-0-0). The plants at all stages are grown at 22 °C light/20 °C dark with a photoperiod of 16 h. It is important to have the lights turn on around 7: 00 a.m. as pollen shed is affected by light and temperature. The maize pollen starts to shed between 1 and 2 h after the lights are illuminated. Before or shortly after the lights turn on, strip all the old anthers off the tassels. By removing all the old anthers, any new anthers that start to protrude are easily noticed and use of fresh pollen is ensured. As soon as the new anthers start to emerge, either shake the pollen from the tassels onto a large Petri dish or move the maize plants to the area where the pollination will be performed. Once the pollen starts to shed, the maize plants should not be moved as the pollen is easily lost from the anthers. When the maize plants start to silk, the period of pollen shed is nearly complete and the plants should be discarded although, some plants may be kept for seed production. In which case, tap the tassels of one plant onto the silks of the others. Repeat with several plants so that there is a high level of inter-crossing. Isolate the maize plants for seed production in a separate chamber because they tend to harbour mites and aphids which are difficult to control on maize and undesirable on pollen donor plants.

Wheat Parental Plants

Seed of wheat plants to be used to produce doubled haploid lines are started in 8 cm plastic pots containing Sunshine 5 (a soilless mix containing peat moss, perlite and vermiculite). Seed can be planted directly or germinated in dampened filter paper lined Petridishes. Transplanting from the Petri dish occurs when the shoot is approximately 3–5 cm long. Once the potted plants have reached the two leaf stage they are transplanted into a 25 cm plastic pot (8 L pots commonly used by nurseries) containing a 6: 1 mixture of soil and turface. At this stage, the plants are watered with 20-20-20 and ammonium sulphate (21-0-0). It is important to grow large, healthy parent plants with many tillers. The plants are grown at a 16 °C light/15 °C dark with 16 h of light. They are fertilized daily with a siphon mixer of a ratio of 16: 1 water to fertilizer (fertilizer 20-20-20, 200 g/20 L). Depending on the genotype of the wheat parent plant, 6–8 weeks is required for spike emergence. If aphids appear they should be spot controlled with a pyrethroid such as Raid especially in the later stages to minimize pesticide exposure of technical staff as the plants will be handled often. Other pesticides (*e.g.* Malathion) and fungicides (*e.g.* Tilt) do not appear to interfere with doubled haploid production, but there are handling and safety issues associated with their use.

Crossing

Emasculation

The timing of emasculation is critical when pollinating wheat with maize. A floret approximately 2/3 of the way up the spike should be checked to determine if the spike is ready. If the anthers are starting to change from a light green/yellow colour to a bright yellow colour and the stigma is fluffy, the spike should be emasculated, which is often the day before natural pollen shed would occur. If the spike has awns, they should be clipped fairly close to the glumes before starting to emasculate. Starting at the bottom of the spike, using fine tipped forceps, remove the central florets of each spikelet leaving only the two lateral florets. Remove the anthers from the lateral florets being careful not to damage the female structures. Continue this procedure up to the top of the spike. Remove the very top spikelet and if any of the spikelets near the top or bottom of the spike are too small to work with, they can also be removed. Finish emasculating the other side of the spike. Once the emasculation of the spike is complete, it should be bagged with a glassine bag and closed with a paper clip. Tag the spike with a crossing tag marked with the cross and the date of emasculation. Place the marked tag around the stem at the first internode. A second tag of a different colour is placed on the tiller at the second internode. This is an indicator tag and nothing is to be marked on it. It merely alerts the person crossing that something must be done to that particular spike, *i.e.* pollination or spraying with dicamba.

Pollination

Being careful not to disturb other anthers, select anthers from the maize plants when they are starting to emerge. The end of the anther should look like it is beginning to open. Remove several anthers from the maize plant. Remove the glassine bag from the emasculated wheat spike to be pollinated (the day after emasculation) and beginning with the bottom of the spike, carefully open the floret. Using tweezers, take a maize anther at its bottom end (opposite of the open end), tip the anther over the open floret and the pollen should flow easily from the anther to the stigma in the floret. One maize anther may be used for several florets. Alternatively, if maize pollen was collected in a Petri dish, use a soft brush to transfer the pollen from the dish to stigmas of the wheat floret. Once the entire spike is pollinated, mark the date of pollination on the crossing tag and replace the bag over the spike. The crossing bag may be left on the spike until the time of excision. The pollinated wheat plants are returned the growth cabinet. Ensure the lights in the growth cabinet are at a sufficient distance from the wheat spikes so the temperature is not elevated in the crossing bags.

Hormone Spray

The day following pollination, the spikes should be sprayed with a dicamba solution (100 mgL^{-1}). Remove the bag from the spike, and using a small hand-held sprayer, (a pump or a pressurized sprayer) apply the dicamba liberally to the spike, until a run off of liquid is seen (about 2 passes with the pressurized sprayer). Replace the crossing bag. Mark the tag to indicate the spike was treated. Remove the indicator tag as nothing else needs to be done to the spike until the time of excision. It is important to keep the parent plants with the crossed spikes under the same conditions as each other (*i.e.*, 15°C dark/16°C light, 16h light) as differing temperatures will affect the timing of excision. If warmer conditions were used the spike would be ready to excise at an earlier date. The crossing procedure and the hormone spray is a three-day process. In order to maximize production, lab work could be scheduled for 7 days a week. If this is not possible there are some modifications to the procedure to allow for a 5-day work week schedule. These include: Reduce the temperature of the cabinet to 9°C day and night on Friday. This will slow down the development of the plants over the weekend. Spikes which were emasculated on Friday are pollinated on Monday. Spikes that were pollinated on Friday receive the dicamba treatment on Monday. Return the cabinet to normal operating temperature on Monday morning. Slight adjustment of the timing of excision is required. Spikes that were pollinated on Monday and Tuesday are excised at 18 days post pollination. The timing of the application of dicamba is critical to this procedure. Dicamba treatment approximately 24 h after pollination produces the best results in terms of the number of embryos recovered per spike.

Excision and Embryo Rescue

Culturing of the embryos should be done under sterile conditions in a laminar flow hood. Equipment needed for excision in the laminar flow hood include: Dissecting microscope and light source; Sterilized distilled water (for rinsing after bleach/ethanol); 70 per cent ethanol solution (for sterilization); 50 per cent bleach solution (for sterilization); 95 per cent ethanol solution (for wiping down surfaces in the hood); 90 per cent alcohol for instrument sterilization; 3 vials for sterilization of caryopses (1 vial for sterilized water, 1 for 50 per cent bleach, 1 for 70 per cent ethanol solution) and 1 vial for 90 per cent ethanol solution to disinfect scalpel and forceps during excision; Beaker for waste from sterilizing caryopses; Pair of forceps; Scalpel; Cheesecloth for wiping down surfaces with alcohol; Sterilized containers of media and 10 mL syringes. At 16 days after pollination, cut off the appropriate spikes retaining approximately 15–20 cm of the stem with the spike and place the stems in a container of water until they are ready to be excised. Carefully spread the lemma and palea open with forceps and remove the caryopsis and place into the barrel of a 10 mL syringe. Repeat for all caryopses on the spike. Be careful not to squeeze the caryopsis as they are fluid filled and lack endosperm. Identify the cross in the syringe by affixing the crossing tag to the syringe with tape or by writing the corresponding number of the syringe onto the applicable crossing tag. If more than one spike from a plant has been harvested, they can be combined, however, be sure to keep different genotypes as well as those that have been treated differently, separate. Surface sterilize the caryopses by filling the syringe with the 70 per cent ethanol solution for 40s. Discard the ethanol and rinse the caryopses with sterilized distilled water. Next, draw in the 50 per cent bleach solution (Javex, approx. 6 per cent sodium hypochlorite), hold for 60 s, discard and rinse twice with sterilized distilled water. Place caryopses onto viewing field of dissecting microscope. Dip the forceps and the scalpel in the 90 per cent ethanol to sterilize and allow to dry momentarily. Using forceps and a scalpel, carefully open each caryopsis. This can be achieved in one of two ways: (1) At the embryo end, hold the caryopsis with the forceps, using the scalpel cut the brush end of the caryopsis. Tilt the open end of the caryopsis towards the microscope with the forceps; (2) Steady the caryopsis with the forceps or scalpel and use another set of forceps to grab and tear the seed coat about two-thirds of the way from the embryo end and pull back exposing the embryo. The embryo is a small white or somewhat translucent structure in the caryopsis. Because there is no solid endosperm present, the embryo is easy to see. Once the embryo is found, carefully touch the forceps to the embryo. The embryo will stick to the forceps and then it can be placed on the sterilized growth media. Open a vial of media in the flow hood, and touch the embryo to the media surface about half way up the slant. Keep the vials containing embryos from the different parental crosses separated. Only one embryo is placed in each vial. Once excision is complete, the embryos are given a 3 day cold and darkness treatment in a fridge at 4°C. After the cold treatment, the embryos are placed in the dark at room temperature for 2 days then placed under light banks for a 16-h light/8 h dark cycle at room temperature. A combination of fluorescent and incandescent light is used. The temperature under the lights is usually around 23–25°C. Each week, all the vials under the light banks are checked to see if the embryos are growing and are ready for transplanting.

Second Cold Treatment

If the embryos fail to grow after 2 weeks under the light banks, they are given a second cold treatment. The dormant 2-week-old embryos are placed back in the fridge for 2 weeks. At this time the vials are placed back under the light banks (a separate set of light banks are used to regulate the flow of vials from the first light treatment to the fridge and back under the lights after a second cold treatment). These vials are checked for growth periodically and returned to the main light banks when significant shoots and roots are visible. After 4 months, those embryos which have received a second cold treatment and were returned to under the light banks and still show no signs of growth are discarded. The medium constituents are agar, sucrose and Gamborg's B5 medium. From one liter of media, 120 one-oz Quorpak vials are filled with approximately 8 mL of medium. The medium is autoclaved in the vials and cooled on a slant.

Transplanting Haploid Plants

Once the embryos grow to a size of 2–4 cm, they are ready to be transplanted. Haploids are transplanted into 14 cm long plastic Containers (like those used in the forestry industry) containing Sunshine 5 soilless mix which has been loosely packed and watered. To remove the haploid from the vial, use a blunt spoonula and dig around the root, freeing it completely, and gently tip the vial over and withdraw the plantlet on the instrument. Dip the root into water to completely remove the agar from the plantlet. A hole is punched in the Sunshine 5 medium and the plantlet is placed so that the crown is covered. Firm the mix around the plantlet. The tray of Containers is subsequently set into a tray of water and covered with a plastic lid to provide a humid environment. The lid is left on for 3 days or until the haploids are established. These haploid plantlets are kept at 15 °C with a 16 h light/8 h dark cycle. It is important to keep the developing haploid plantlets cool so that tillering is encouraged.

Colchicine Treatment – Doubling

Plant Preparation

When the haploid plantlets are between the three- and five-leaf stage, they are ready for the colchicine treatment. In order to identify the plants when they are removed from the cones, the pot stake has a hole punched, and an elastic string (approximately 10 cm long) is fed through this hole and tied in a knot at the ends. The elastic is pulled around the base of the plant, the stake is then fed through the loop of the elastic pulled through to ensure a snug fit. This step can be done the day before the plants are to be colchicine treated. When preparing the plants, about 1/3 of the leaf tissue is trimmed off. Next, the plants are removed from the cones and the soil is stripped from the roots. The roots are cut back leaving about 5 mm and given a quick rinse with tap water. It is important to prepare the plants first thing in the morning because the colchicine treatment process requires steps throughout the day.

Colchicine Treatment and Rinse

Once the plant's leaves and roots have been trimmed they are placed in a beaker (a 250 ml beaker will hold about 100 plants) and are ready for the colchicine treatment. When working with colchicine, it must be handled very carefully as it is highly toxic. It is recommended to wear two pairs of nitrile gloves. Six drops (from an eye dropper) of DMSO (dimethyl sulfoxide) is added to 60 ml of colchicine (1 drop DMSO/10 ml of colchicine). This solution is transferred to the beaker containing the trimmed haploids (the crowns of the plants should be submerged). An aerator (a fish tank air pump will do) with a rubber tube leading from it, is placed in the beaker. The beaker is placed in a plastic pail and covered with a lid. A hole is made in the lid to allow the air tube to pass through. The covered pail is placed in a fume hood with the sash ajar to allow only for the air tube to run from the beaker to the air pump. This step ensures that the procedure is exposed to as little light as possible as well as containment of any chemical which may percolate from the beaker. The plants are treated for 3½ h at room temperature. Pour the used colchicine from the beaker into a labelled container for use in a subsequent day's treatment (may be used twice, no additional DMSO is required). Once the colchicine solution has been used twice it is discarded into a container designated for toxic waste disposal. To rinse residual colchicine from the plants, place the aeration tube back into the beaker (in sink) along with a rubber hose leading from the faucet (a backflow valve should be installed on the faucet). Have the water running at a flow that is strong enough to allow the water to trickle out of the beaker. Adjust the cold and hot water taps so that the temperature of the rinse water is between 15–18°C. Rinse for a minimum of 2 h.

RT-PCR Analysis

Total RNAs were extracted from leaves using ISOGEN and cDNAs were synthesized from the total RNAs with oligo dT primer in accordance with the protocol for the Ready-To-Go T-Primed First-Strand Kit (GE Healthcare, http: // www.gehealthcare.com). Leaves were sampled from the plants 2 h after the start of the light period. The RT-PCR analysis of einkorn wheat was performed using gene-specific primer sets for VRN1 (VRN1-BAC 81655L and VRN1-BAC 82017R), WCO1 (WCO1-L and WCO1-R) and FT (WFT-all-3L and WFT-all-3R). The Ubiquitin gene was used as the control with the primer set Ubi-1L and Ubi-1R. For the expression analysis of transgenic plants, total RNA was extracted from the first leaf of each plant using a Get Pure RNA Kit. Leaves were sampled from the plants 2 h after the start of the light period. RT-PCR was performed using the RNA PCR Kit (AMV) Version 2.1 (Takara Bio, http: //www.takara-bio.com) with AmpliTaq Gold DNA Polymerase. Gene-specific primer sets were used for FT (WFT-FW and WFT-RV), VRN1 (WAP1-556L and WAP1-982R) and VRN2 (VRN2-FW2 and VRN2-RV2). Wheat actin gene (Actin) was used as the control with the primer set ACT-FW and ACT-RV. The sequences of the primer sets and the annealing temperatures are shown in S1. In all experiments, the RT-PCR analyses were performed in the exponential range of amplification. To determine the sequence of RT-PCR products of VRN3 amplified from transgenic plants, RT-PCR products from positive transgenic lines 4 and 5 were directly sequenced.

In the diurnal expression study and mutant analysis, real-time PCR analyses were performed using a LightCycler 2.0 with gene-specific primer sets for GI (TaGI-3L and TaGI-3R), TaHd1 (TaHd1-2L and TaHd1-2R), FT (WFT-F4 and WFT-R4), VRN1 (WAP1-545L and WAP1-698R) and VRN2 (ZCCT1-1Lt and ZCCT1-1Rt). The PCR primers and annealing temperatures for WCO1 were the same as in the RT-PCR study. The quantity of transcripts was determined by the SYBR Green fluorescence of Actin using the primer set actin 361-L and actin 361-R. The Ubi-1 gene was also used as the internal control using the same primer set as in the RT-PCR study. In the transgenic study, real-time PCR was carried out using the Stratagene MX3000 Real-Time PCR system. The template cDNAs were amplified with Brilliant SYBR Green QPCR Master Mix using the same primer sets for WFT, WAP1 and VRN2 as in the RT-PCR experiment, at and annealing temperature of 60°C. The data were analyzed using Stratagene MXPRO ver. 3.0 software (Stratagene). The sequences of the primer sets and the annealing temperatures.

Generation of Transgenic Wheat Lines

Transgenic wheat plants were produced by a particle bombardment method using immature embryos. The expression plasmid 35S: VRN3-D was constructed and co-transformed with the plasmid pUBA (Toki *et al.*, 1992), which contains the bar gene as a selection marker under the control of the maize ubiquitin promoter, according to the method of Pellegrineschi *et al.* (2002). The integration of the transgene was confirmed by PCR of genomic DNA using primers designed from the 35S promoter sequence (35S-WFT-FW: 52 -CGCACAATCCCACTATCCTT-32), and from the WFT cDNA sequence (35S-WFT-RV: 52 -GAAGAGCACGAGCACGAAG-32). One transgenic plant, designated WF3, was selected from six independent transgenic T0 plants. Using 16 T1 plants derived from the WF3 T0 plant and two non-transgenic plants, RT-PCR experiments were carried out to examine the transcription levels of FT, VRN1 and VRN2. Grain protein content (GPC) and gluten quality are the most important factors determining the end-use quality of wheat for pasta-making. Both GPC and gluten quality are considered to be polygenic traits influenced by environmental factors and other agricultural practices. Two related $F_{8:9}$ recombinant inbred line (RIL) populations were generated to localise genetic factors controlling seven quality traits: GPC, wet gluten content (WGC), flour whiteness (FW), kernel hardness (KH), water absorption (Abs), dough development time (DDT) and dough stability time (DST). These lines were derived by crossing Weimai 8 and Jimai 20 (WJ) and by crossing Weimai 8 and Yannong 19 (WY). In total, WJ comprised 485 lines, while WY comprised 229 lines. Data on these seven quality traits were collected from each line in five different environments. Up to 85 putative QTLs for the seven traits were detected in WJ and 65 putative QTLs were detected in WY. Of these QTLs, 31 QTLs (36.47 per cent) were detected in at least two trials in WJ, while 24 QTLs (36.92 per cent) were detected in at least two trials in WY. Three QTLs from WJ and 25 from WY accounted for more than 10 per cent of the phenotypic variance. The total 150 QTLs were spread throughout all 21 wheat chromosomes. Of these, at least thirteen pairwise were common to both populations, accounting for 20.00 and 15.29 per cent of the total QTLs in WJ and WY, respectively. A major QTL for GPC, accounting for 53.04 per cent of the phenotypic variation, was detected on chromosome 5A. A major QTL for WGC also shared this interval, explained more than 36 per cent of the phenotypic variation, and was significant in two environments. Though co-located QTLs were common, every trait had its unique control mechanism, even for two closely related traits. Due to the different sizes of the two line populations, we also assessed the effects of population size on the efficiency and precision of QTL detection. In sum, this study will enhance our understanding of the genetic basis of these seven pivotal quality traits and facilitate the breeding of improved wheat varieties.

The aim is to compare of the copper (Cu), zinc (Zn), iron (Fe), calcium (Ca) and magnesium (Mg) contents in the grains of wild, ancient and cultivated wheat species. The results do not confirm the hypothesis that the ancient wheat species - einkorn, emmer and spelt - have higher mineral nutrient contents in their grains than those of common wheat. Otherwise, the results showed that the increase in grain weights correlated significantly with the decrease in Ca and Mg concentrations. Micronutrient malnutrition, particularly iron, iodine and vitamin A deficiencies, is a serious problem worldwide for the human population. Cereal-based foods have a great importance in the compensation for micronutrient deficiency, because 50 per cent of the foods produced worldwide (in terms of dry matter) is made up of cereal grains. Many authors suggested that ancient wheat species - especially the einkorn wheat - have a high nutritional value, because these species accumulate more mineral elements in their grains than the recently cultivated wheat varieties. Galbács *et al.* (1998) found that there is a significant correlation between ploidy levels of different wheat species and the mineral contents of their grains. Based on this theory, the aim of the present research was to study the existing variability in seed nutrient compositions of various *Triticum* L. and *Aegilops* L. species, and to verify the hypothesized correlation between the ploidy level and seed mineral nutrient concentrations. Our results showed that generally higher Cu, Zn, Ca and Mg contents were observed in the grains of *Aegilops* L. species, while higher Fe concentrations were found in the grains of *Triticum* L. accessions. The results do not confirm the hypothesis that the ancient wheat species - einkorn, emmer and spelt wheats - generally have higher mineral contents in their grains than common wheat. Additionally, no correlation was found between the ploidy level and genome types (A, AB, AG, ABD, S, D, UM, DMS) and the mineral content of the grains. Similar results were found using the hierarchical cluster analysis. As the 1000-kernel weight of wild species is generally lower than that of the cultivated species, the correlation between seed weight and mineral nutrient content was also studied. We found that the increase in seed weights correlates significantly with the decrease in Ca and Mg concentrations. The results showed that to improve the human nutritional quality of wheat varieties, the best sources of higher Cu, Zn, Ca and Mg contents could be found in the *Aegilops* L. genus, while genotypes with higher Fe contents could be found in the *Triticum* L. genus.

Table 2.8. 1000-kernel weight and mean concentration of mineral elements in grains of *Triticum* L. and *Aegilops* L. species harvested

Species	1000-Kernel Weight (g)	Mineral Element Concentration (mg/kg)				
		Cu	Zn	Fe	Ca	Mg
T. urartu var. *binartulutiro*	10.2	6.8	38.8	65.3	496	1550
T. monococcum ssp. *aegilopoides* var. *thaoudar*	13.5	4.5	37.8	52.3	477	1950
T. monococcum ssp. *monococcum* (landrace)	25.2	5.6	33.9	48.9	452	1139
T. monococcum ssp. *monococcum* (landrace)	19.5	6.3	34.5	22.3	690	1300
T. sinskajae var.*sinskajae*	35.1	7	41.8	46.8	404	1125
T. turgidum ssp. *dicoccoides*	23.0	6.9	21.1	33.2	614	1079
T. turgidum ssp. *dicoccon*	41.2	8.8	42.8	32.3	443	1100
T. turgidum ssp. *durum* var. *valenciae*	35.6	5.5	31.8	26.8	404	825
T. turgidum ssp. *polonicum*	32.6	6.8	26.8	33.9	413	1023
T. turgidum ssp. *turgidum*	20.3	6.8	25.5	31.3	611	1300
T. timopheevii ssp. *timopheevii*	20.2	6.3	37	33.3	532	1700
T. aestivum ssp. *spelta*	45.5	11.3	35.8	37.5	355	1325
T. aestivum ssp. *macha*	31.3	9	30.5	47.3	461	900
T. aestivum ssp. *compactum*	12.9	6.5	21.8	36.9	818	1250
T. aestivum ssp. *aestivum* cv. *Fertodi293*	41.6	8.8	36.8	25.8	567	1150
Ae. speltoides	12.8	10	37	37.3	813	1750
Ae. tauschii	9.8	8.5	27.5	24.8	707	1575
Ae. geniculata	13.3	13	41.8	40	691	2050
Ae. biuncialis	16.2	12.8	52.8	14.8	467	1950
Ae. vavilovii	13.9	11	49.8	36.5	876	2650
Mean	23.6	8.1	35.3	36.4	564.6	1434
Standard Deviation	11.6	2.4	8.4	11.6	155.5	458.9
Range	35.7	8.5	31.7	50.5	521	1825
Minimum	9.8	4.50	21.1	14.8	355	825
Maximum	45.5	13.00	52.8	65.3	876	2650
S.E. Mean	2.585	0.545	1.874	2.595	34.760	102.62

Polymorphism in seed storage proteins analyzed by gel electrophoresis is still one of the valuable tools for the evaluation of plant genetic resource collections in cereals, especially in wheat. Storage proteins of mature seed are used, mainly glutenin and gliadin. Protein analyses can be used effectively in all *Triticum* species and subspecies, respectively, regardless of their origin, habitus, ploidy level, *etc.* The aim of this contribution was to mention the applicability of seed storage protein analyses in four *Triticum* species and subspecies, respectively.

Protein profiles and the composition of alleles at the *Glu-1* locus of 34 *Triticum aestivum* L. subsp. *aestivum*, 21 *Triticum turgidum* L. subsp.*durum* (Desf.) Husn., 11 *Triticum turgidum* L. subsp.*turgidum*, 13 *Triticum aestivum* L. subsp.*spelta* Thell. accessions have been studied and determined. All wheat samples originated from the wheat collection maintained in the Genebank of the Slovak Republic, Research Institute of Plant Production, Pieš any. Glutenins were extracted and separated by sodiumdodecylsulphate polyacrylamide gel electrophoresis (SDS-PAGE), using the standard procedure. The set of analysed *Triticum aestivum* L. subsp.*aestivum* genotypes represents recent and former wheat cultivars extensively cultivated in Slovakia. There were detected HMW-GS null, 1, and 2* encoded by alleles at the *Glu-A1* locus, subunits 7+8, 6+8, and 7+9 at the *Glu-B1*, and 5+10 and 2+12 at the *Glu-D1* locus. The most frequent HMW-GS pattern was - null, 7+9, 5+10 possessed by 25.7 per cent of analyzed genotypes. Cultivars from this set were produced by Slovak and Czech breeders and assigned for specific agroecological conditions. Therefore, they are similar in morphological and agronomical parameters and also express a relatively low variation in HMW-GS. Protein homogeneity within cultivars is high, only in cultivar Barbara two protein lines (phenotypes) have been detected. Others were uniform in their protein patterns. Most of the analyzed cultivars possess HMW-GS with a high positive effect on the bread-making quality - subunits 7+8, 7+9, and 5+10, respectively. In the set of *Triticum turgidum* L. subsp. *turgidum* genotypes, the following HMW-GS were detected : null, 1, 2* (at the locus *Glu-A1*), and 6+8, 13+16 (*Glu-B1*). The most frequent, at relevant loci, were alleles 1 (81.8 per cent of genotypes) and 6+8 (90.9 per cent). *Triticum aestivum* L. subsp. *spelta* Thell. genotypes possessed only HMW-GS 1 at the *Glu-A1* locus, alleles 6+8, 20, 7+9 (*Glu-B1*), and 2+12, 5+10 (*Glu-D1*). The most frequent HMW-GS pattern was 1, 6+8, 2+12 (76.9 per cent of genotypes). The most frequent allele at the *Glu-A1* locus in *Triticum turgidum* L. subsp.*durum* (Desf.) Husn. genotypes was allele null, alleles 1 and 2* were also present. At the *Glu-B1* locus alleles 6+8, 7+8, 20, 14+15, and 13+16 have been observed. The combination null and 7+8 was the most frequent one (42.9 per cent of genotypes). Durum wheats are mostly used for pasta production. Their quality is referred to specific low molecular glutenin proteins. The composition of alleles at the *Glu-B3* locus in *Triticum turgidum* L. subsp. *durum* (Desf.) Husn. has not been studied by gliadin or glutenin electrophoresis and by the evaluation of either the presence or absence of specific gliadin and LMW-GS, but by polymerase chain reaction (PCR) and by the detection of specific alleles coding the allelic variant at the *Glu-B3* locus. Higher technological quality of durum wheat should be found in those with allele LMW-2. This study do not describe complete seed storage protein profiles analyzed by SDS-PAGE, but only the composition of glutenins with the highest molecular weight - HMW-GS. Analyzed protein profiles by SDS-PAGE are valuable tools for the identification and differentiation of wheat accessions maintained in wheat genetic resources collections. Based on glutenins, wheats with different ploidy levels can be differentiated very easily. Genotypes homogeneous and heterogeneous in protein profiles, respectively, can also be detected. Moreover, HMW-GS can be detected, their composition evaluated and included into the descriptive data set. On the basis of these data wheat breeders are able to select accessions and use them in breeding programmes for bread-making quality improvement. Seed storage proteins can be also used for molecular taxonomy studies and other purposes. Polymorphism in glutenins revealed by SDS-PAGE can be used also in other *Triticum* species.

The selection and use of different types of molecular markers depend on plant species to be characterized and evaluated. In our hands, protein fingerprinting is still one of the basic identifications, differentiation, and evaluation tools, especially in wheat and barley, due to its responsibility, simplicity, reliability, and low cost. In the collection of wheat genetic resources, the largest of our collections, we use the variation in glutenins and gliadins. We performed protein analysis and composition of alleles at the *Glu-1* loci, encoded high molecular weight subunits of glutenins (HMW-GS), we analyzed and described in several hundreds of diploid, tetraploid, and hexaploid wheats. Protein analyses were used also for heterogenity studies of more than 120 European wheat landraces and obsolete cultivars. Protein heterogenity was very high. Furthermore, sometimes other new allele encoding HMW-GS can be discovered as in this set of wheats - in one of the glutenin lines of the Russian obsolete cultivar Eritrospermum 917 new HMW-GS with the relative molecular weight of 98.6 kDa has been found. The above-mentioned protein approach has been used also for the genetic erosion study of wheat (*Triticum aestivum* L.) in Slovakia during the last decade. Based on these analyses we concluded that some of the HMW-GS (pairs 3+12, 17+18, and subunit 20), present in old historical"wheats, are missing in modern cultivars grown in Slovakia nowadays. There were observed together 15 different HMW-GS encoded by 11 alleles or allelic pairs in the old genotypes. A lower number of different HMW-GS and their alleles was observed in the set of modern wheat cultivars - 11 different HMW glutenin subunits encoded by 8 alleles or allelic pairs, but the same number of different HMW-GS patterns was revealed in both sets of wheats. From the point of view of genetic variability, it could be concluded that the long-term effort of breeders and decreasing the cultivation of landraces and old obsolete cultivars are associated with the loss of several HMW-GS alleles and decreasing the genetic variability of wheat. Molecular characterization confirmed a broader allelic variation of old wheat genotypes and landraces in comparison with modern cultivars. This study also showed that the maintenance of genetic resources in a gene bank could prevent losses of interesting genes.

The second step in the application of molecular markers in our Gene bank is the evaluation and characterization of genotypes and populations from the molecular point of view. This is probably their most promising utilization in the future gene banking and also in classical and molecular breeding. It is based on the evaluation of the presence or absence of important genes (alleles) or molecular markers linked with them. Based on the identified markers and genetic maps, collections of plant genetic resources can be screened to select the donors of important genes for prebreeding or breeding processes. At present we are cooperating with wheat and barley breeders. On the basis of the protein and DNA analyses of wheat and barley accessions, we can select and recommend donors of genes contributing to good breadmaking and technological quality, respectively, in bread and durum wheats, resistance to important phytopathogens - *Pseudocercosporella herpotrichoides, Puccinia recondita* f. sp. *tritici* in wheat, BaYMV and BaMMV in barley, *etc.* Other very important traits of the collection such as genetic diversity, the conditioned genetic variation in the progeny in breeding, the detection of genetic shifts in populations, the study of genetic erosion in specific territories, molecular taxonomy, *etc.*, may possibly be determined using molecular markers. Molecular markers describe plant genetic resources unambiguously. They uncover the genetic potential and enhance the efficiency of the collection, manipulation, maintenance, study, distribution, and utilization of plant genetic resources.

The effect of NaCl on the seed germination and seedling growth was analysed. The seed samples (10 seeds per sample) were placed on Petri dishes with filter paper moistened in Hoagland solution with 0 (control), 50, 100, 200, 300 and 400 mM NaCl. In each combination of NaCl, 100 seeds of every species were germinated. The Petri dishes with seeds were incubated in a thermostat at 20°C. After 2 days the energy of seed germination and after 5 days the capacity of seed germination, the number of seedling roots and the maximal length of roots and shoots were assessed. On the basis of shoot length, the tolerance index (TI) to salt treatment was calculated on the basis of the following formula:

TI= (mean shoot length of treated seedlings/mean shoot length of control) x 100 per cent.

Mean values of the measurements were compared using the Tukey test.

The analysis of seed germination revealed a strong influence of NaCl on the germination of *Aegilops* species and *T. aestivum*. NaCl decreased most strongly seed germination of Ae. *crassa* 4x and *Ae. crassa* 6x. The influence on the germination of *Ae. umbellulata, Ae. columnaris, Ae. biuncialis* and *Ae. ovata* was the lowest. The shoot length of 5-day seedlings declined with the increase in NaCl concentration. NaCl had a weaker influence on seedling root length and the number of roots. Kuiper *et al.* (1990) found that root growth in saline environments is generally less inhibited than leaf growth. They concluded that the root is an organ that can reduce the negative results of ion accumulation in the environment. Considering the effects at the concentrations of 300-400 mM NaCl, *Ae. comosa, Ae. squarrosa, Ae. biuncialis, Ae. columnaris, Ae. ovata, Ae. triuncialis* and *Ae. umbellulata* exhibited a higher salt tolerance than the other species. At 200 mM NaCl a higher tolerance was also shown by cv. 'Gama'. Gorham (1990) supposed that salt tolerance gene(s) are located in genomes U, M and D. The D genomes are possessed by 'Gama'. The donor of genomes D to hexaploid wheat was *Ae. squarrosa* (Kimber and Zhao 1983). Hence, it may result in a relatively higher salt tolerance of cv. 'Gama'. NaCl inhibited shoot elongation in *Ae. crassa* 4x, *Ae. crassa* 6x seedlings most strongly. The seedlings of *Ae. ventricosa, Ae. cylindrica, Ae. triaristata* 4x, *Ae. triaristata* 6x, *Ae. variabilis, Ae. variabilis* (*peregrina*), *Ae. uniaristata* and *T. aestivum* exhibited intermediate reaction to salt treatment. Farooq *et al.* (1990, 1992) identified several salt tolerant accessions of *Ae. ovata, Ae. cylindrica* and *Ae. squarrosa*. These accessions were used to obtain intergeneric hybrids with *Triticum aestivum* and *T. turgidum*. But not all the identified salt tolerant accessions of *Aegilops* species might be amenable to hybridization for the transfer of salt tolerance to cultivated wheat. Gorham (1990), on the basis of investigations on *Aegilops* species, noticed that the differences between them in salt tolerance also depend on a gene(s) which enhances the discrimination

Table 2.9. Effect of different concentrations of NaCl on the seeds germination and the shoot length of 5-day seedlings of *Aegilops* species and *Triticum aestivum* L. cv. Gama

Species	*Genome Composition*	*Germination Capacity (per cent)*						*Shoot Length (mm)*					
		NaCl (mM)						*NaCl (mM)*					
		0	*50*	*100*	*200*	*300*	*400*	*0*	*50*	*100*	*200*	*300*	*400*
Ae. comosa	M	65.6	70.5	54.8	45.0	26.5*	12.8*	48.25	33.05	29.58	9.08*	6.11*	1.83*
Ae. uniaristata	U_n	65.4	55.2	41.4	28.6*	25.0*	0.0	35.60	25.41	22.16	3.94*	2.60*	0.00
Ae. squarrosa	D	58.5	50.8	45.0	40.5	30.7	28.6*	37.91	28.18	26.78	4.25*	3.89*	1.66*
Ae. umbellulata	U	96.0	98.6	95.4	95.8	80.6	55.4	63.28	45.36	36.98	12.67*	4.98*	2.88*
Ae. columnaris	UM	93.4	92.8	91.5	90.6	92.6	70.5	74.03	55.99	46.98	18.16*	6.46*	3.29*
Ae. ventricosa	DU_n	64.6	58.6	50.4	32.8	25.4*	15.3*	52.16	21.44	15.63*	5.48*	2.36*	1.26*
Ae. crassa 4x	DM	56.5	37.8	24.6	9.8*	0.0	0.0	43.89	26.27	5.40*	1.22*	0.00	0.00
Ae. ovata	UM	96.6	88.4	80.3	70.5	71.6	35.5*	43.46	37.28	34.09	7.36*	3.65*	2.04*
Ae. triaristata 4x	UM	65.4	65.0	60.4	55.6	40.3	25.1*	69.10	43.36	26.37	8.30*	4.40*	1.80*
Ae. biuncialis	UM	85.6	70.5	65.5	54.8	45.2	40.5	57.73	54.41	39.69	26.00	5.22*	2.75*
Ae. variabilis	US	78.5	65.4	55.8	45.7	35.4*	15.0*	86.38	45.68	33.45*	9.27*	4.29*	2.67*
Ae. variabilis (peregrina)	US	75.4	70.7	66.5	32.6*	20.5*	10.4*	58.16	28.14	23.27	3.16*	2.46*	1.18*
Ae. triuncialis	UC	60.5	56.3	48.1	35.9	21.6*	12.8*	42.65	41.86	21.17	5.21*	3.86*	1.64*
Ae. cylindrica	CD	76.6	70.5	55.9	40.2	32.5	20.4*	58.72	51.89	34.74	4.26*	2.35*	2.12*
Ae. crassa 6x	DDM	64.5	45.0	30.5	15.5*	0.0	0.0	35.50	28.96	24.51	2.68*	0.00	0.00
Ae. triaristata 6x	UMU_n	85.6	85.0	76.7	60.5	44.2*	30.5*	59.17	42.57	26.67	6.34*	3.18*	1.76*
T. aestivum Gama	ABD	100.0	100.0	91.6	70.4	35.5*	20.8*	82.85	56.74	41.24*	20.28*	2.64*	2.18*

*: Means significantly different from the control (0 mM NaCl) at $P<0.05$.

between Na and K. The obtained results corroborated the Gorham's suggestions that genes responsible for K/Na discrimination are present in U, M and D genomes. According to Gorham, the S genome diploid *Aegilops* species all lacked the enhanced K/Na discrimination, as C genome species *Ae. caudata* and the U_n genome species *Ae. uniaristata*. The obtained results seem to confirm a higher salt tolerance of *Aegilops* with U, M and D genomes. High sensitivity to NaCl in *Ae. crassa* 4x and *Ae. crassa* 6x species perhaps resulted from an essential modification in their D and M genomes (Kimber and Feldman 1987).

Gene resources of *Triticum durum* accumulated in our collection are routinely assessed in a three-year cycle of field experiments. The grains of valorized genotypes are successively delivered to the air-conditioned store to keep their viability, as well as preserve the material for Polish centers and exchange them with foreign gene banks. Resources collected till now include over 1500 varieties and populations originating from all the regions of the species cultivation around the world. Durum wheat was cultivated in one replication on $2m^2$ plots. Plant spacing in a row was about 1.5-2.0 cm, with inter-spaces of 20 cm (each plot was sown with 630 grains). For every genotype, the length of the period from emergence to heading and from emergence to full maturity was measured. Plant height was calculated on the basis of plants' measurements at three random points of the plot. Spike rachis length, the number of spikelets per spike and spike density were calculated from the measurements of 10 spikes. The number and weight of grains per spike, as well as 1000-grain weight were assessed on the basis of combined threshing of 50 spikes randomly selected from each plot. Total protein content in the grain was estimated by applying the coefficient of 5.7 for recalculating nitrogen into protein. The accepted limits do not make it possible to present the full results. The range of genotype variation for spring population of *T. durum* under analysis (on the basis of multi-year mean values calculated for every genotype). Great diversity among the collected material in relation to all the analyzed trait levels should be emphasized, notably for such field-forming elements as the number of grains per spike and spikelet and the weight of grains per spike. The lowest variability coefficient was obtained for the length of all vegetation periods. A relatively low variability coefficient was also found for the length of the "emergence-heading" period, the number of spikelets per spike, and total protein content in the grain. Although there were considerable differences between minimum and maximum levels of the above parameters, the majority of genotypes generally had values similar to the average for all the population, so that the variability coefficient was low. Extreme values were represented by few genotypes.

A significant influence of years and genotype year interactions on such an important yielding element as the weight of grains per spike, was found. Generally, a low level of the weight of grain per spike was obtained in 1997 and 1999. However, it is worth mentioning that in the examined material there were genotypes with quite high and stable levels of the weight of grains per spike. Varieties and lines possessing relatively high values of the weight of grains per spike in all the cultivation years (which was always above the mean value for the whole analyzed population) and the regression coefficient (b) below 1.0. They were Polish lines and varieties of European origin. Only one genotype from Ethiopia was found in this group. In general, these genotypes have short straw, resistant to lodging, dense spikes and good spikelet fertility. Other genotypes of foreign origin showed low levels of the weight of grains per spike, or its value varied strongly. Thus, it can be concluded that the materials of foreign origin should be critically assessed before their application in Polish weather conditions.

Table 2.10. Mean values and range of variation for the studied traits in *Triticum durum* Desf. accessions

Traits	*Average*	*Min. - Max.*	*CV (per cent)*
Number of days: emergence-heading	47.6	41.7- 55.7	5.7
Number of days: emergence-maturity	99.3	90.0 - 111.3	4.8
Plant height [cm]	82.8	49.6 - 112.2	15.4
Spike rachis length [cm]	5.4	3.4 - 8.1	17.9
Number of spikelets per spike	14.8	11.4 - 17.3	6.5
Spike density [D]	26.5	17.1 - 40.4	19.1
Number of grains per spike	27.2	14.9 - 39.4	22.0
Spikelet fertility	1.8	1.1 - 2.6	19.4
Grain weight per spike [g]	1.02	0.40 - 1.76	31.7
1000 grain weight [g]	36.6	24.4 - 48.2	14.9
Total protein content (per cent)	16.2	11.0 - 21.8	11.1

Table 2.11. Mean values for studied traits in selected varieties and lines of *Triticum durum* Desf. (means over 3 years).

Cultivar Line	*Days to Heading*	*Plant Height [cm]*	*Spike Density [D]*	*Spikelet Fertility*	*1000 Grain Weight [g]*	*Total Protein Content (per cent)*
LGR 896/60/a/b	49.3	86.3	33.0	2.4	46.3	14.7
LGR 897/3/a	53.3	90.6	28.7	2.5	45.2	15.5
LGR 1360/10	50.7	83.2	25.2	2.3	45.7	15.0
Svetlana	46.3	104.0	25.9	2.2	48.2	13.3
LGR 1360/22	48.7	84.2	27.9	2.4	46.6	15.1
ICDW 13340	52.7	110.5	29.5	2.3	46.4	14.4
Clairdoc	53.3	80.2	24.6	2.3	45.8	11.0
LGR 898/4	54.0	84.7	26.5	2.6	44.6	15.0
Florentino	47.7	75.4	30.2	2.4	43.8	13.8
Topdur	45.0	86.9	23.8	2.4	44.5	15.7
Elan	49.3	109.9	27.5	2.4	43.1	14.0

Cultivars and strains tested in the working collection may serve as donors of useful agronomic traits for breeding purposes, especially if plant height, resistance to lodging, leaf and stem rusts, protein content and the sedimentation value are concerned. The forms combining several useful traits are the most valuable material, especially those which have good yield parameters, resistance to diseases and quality traits simultaneously. There are only few genetic sources of resistance to certain diseases and yield parameters, so there is an urgent need for searching or creating new genotypes with high grain weights per ear and good resistance to leaf and ear septorias.

Engineering Nuclear Male Sterility

The production of hybrids is an essential component of crop breeding programmes but till date, hybrid wheat has remained elusive! The development of a suitable hybridisation system for wheat requires a high degree of male sterility in all parts of the female parent to avoid self-fertilization. De Block and coworkers (De Block *et al.*, 1997) have developed a nuclear male sterile system in wheat by introducing the barnase gene under the control of a tapetum specific promoter, the expression of which prevents normal pollen development at specific stages of anther development. This system employed the ribonuclease-inhibitor barstar gene to restore the fertility of male sterile plants. To avoid complicated gene integration patterns, the target tissues were incubated on niacinamide containing medium before bombardment. The authors suggest that the enzyme poly (ADP-robose) polymerase (PARP), which plays a key role in the processes of cell division and recombination, is inhibited by niacinamide, thereby resulting in simple integration pattern of the transgene (De Block *et al.*, 1997). Expressing the *barnase* gene at specific stages of anther development destroys the tapetum, thereby preventing normal pollen development and causes pollen sterility.

Transgene Silencing in Wheat

Success at developing improved wheat cultivars through genetic engineering depends on stable and predictable expression of the inserted gene. However, gene silencing is a common phenomenon in the production of transgenic plants and needs to be effectively controlled for the desired results. This is all the more important because gene silencing is an important phenomenon involved during natural plant defense. Gene silencing is a complicated phenomenon as it includes both transcriptional gene inactivation and post transcriptional gene inactivation. The complex and the large genome size of wheat is expected to be prone to silencing by introduced transgenes. With the development of transformation methodologies for wheat, more information has started to accumulate regarding the inheritance and expression of the introduced transgenes. Information regarding the long term stability of transgenes is thus of immense significance for the use of genetic manipulation as a tool in wheat crop improvement and is necessary for the introduction and subsequent expression of desirable agronomic trait. DNA methylation plays a significant role in establishing and maintaining an inactive state of the gene by rendering the chromatin structure inaccessible to the transcription machinery. Methylation of DNA is expected to result in reduced gene expression. The variability of transgene expression in clonal cell lines of wheat and found a negative correlation between the degree of methylation and marker gene expression. PEG-mediated approach was used to transform protoplasts isolated from suspension cultures of *T. aestivum*. The integration and expression of the selectable marker gene *npt*II fused with different promoters, namely, alcohol dehydrogenase, *shrunken* from maize, and *actin1* from rice, was confirmed by Southern analysis and enzyme activity test. Transgenic cell lines under selection were 'protoplasted' and clonal callus lines were cultivated from genetically identical single cells without selection pressure. A reduction/loss of marker gene expression was observed due to a reduction in the *npt*II transcript level and was seen to be associated with hypermethylation of the integrated DNA. The silencing effect was reversed by a 4-week culture phase on media supplemented with demethylation agent, 5-azacytidine, thus confirming the role of methylation in transgene silencing.

The inheritance and stability of an *Act1*D-*uid*A: *npt*II expression cassette in T_4 and T_5 transgenic plants. based on the histochemical localization of GUS activity, this study demonstrated the lack of any cytoplasmic effect on the inheritance of the transgene. The transgenes maintained a multiple integration pattern similar to that observed in the T_1 generation. The transgenic plants which produced low *gus* and *npt*II activity in seeds had an intact expression cassette. Southern blot analysis of genomic DNA performed with the methylation sensitive enzyme, *Hpa*II, showed the transgene in *GUS* negative plants to be highly methylated relative to the transgene in *GUS* positive plants. Some of the studies on gene silencing also indicate that multiple integration pattern and copy number is also associated with DNA methylation. Cannell and coworkers studied the inheritance of *gus* and *bar* marker genes over three generations. The integration, inheritance and expression of these marker genes in the population studied were stable and predictable with a few exceptions. The inheritance of integration patterns was stable, and transmission/inheritance of the transgenes followed Mendelian ratios in a majority of lines. From this study, the authors claim that the transformation procedure, transgene integration, and marker gene expression had little effect on the transmission of transgenes to the subsequent progenies. However, over the three generations studied, a uniform, 'progressive' transgene silencing, specific to the *gus* gene, was observed. This gradual loss of the *gus* gene expression was not accompanied by a reduction in *bar* expression. Since in this case, the *gus* and the *bar* genes are located on the same plasmid, observations indicate a post-transcriptional gene silencing. The silencing of HMW glutenins was also observed in transgenic wheat expressing extra HMW subunits. Characterization of six independent events involving the transformation of wheat HMW-GS genes, 1Ax1 and 1Dx5, in a cultivar expressing five subunits by particle bombardment resulted in partial or complete gene silencing. Silencing of all the HMW glutenin subunits was observed in two different events of transgenic wheat expressing the 1Ax1 subunit transgene and overexpressing the 1Dx5 subunit. Control of gene silencing thus remains a challenge for the immediate future. The problem of gene silencing in wheat can be minimized by optimising methods for simple integration patterns, use of promoters and gene sequences isolated from cereals, use of matrix associated regions (MARs) or scaffold attachment regions (SARs), which insulate transgenes from surrounding chromatin, might help in reducing the gene silencing problem.

Transposon Tagging in Wheat

Transposon mutagenesis has been widely exploited in various organisms to isolate genes that encode unidentified products. The maize activator (*Ac*) and Dissociation (*Ds*) elements are the best-studied transposable elements in heterologous host plants. Initial studies have reported the introduction and activation of the *Ac/Ds* elements into cultured wheat cells by using a wheat dwarf virus by particle bombardment. A transposon tagging system in wheat by introducing the *Ac* transposase gene under the CaMV 35S promoter into cultured wheat embryos by particle bombardment. For the development of a transposon tagging system, embryos isolated from a stable *Ac* line were bombarded with a plasmid containing the maize dissociator (*Ds*) element located between the rice *Act1* and the *gus* genes. The transient expression of the *gus* gene was observed after the excision of *Ds* elements. Southern and northern analysis of the T_0 and T_1 plants has exhibited the stable expression and inheritance of the *Ac* transposase gene. These results have demonstrated the precise processing of the maize *Ac* transposase gene and the synthesis of the transposase protein in transgenic *Ac* lines. The *Ac* transposase gene causes the transactivation and excision of the *Ds* in the *Ac* transgenic lines transformed with the maize *Ds* elements. Thus in near future we can expect traits of commercial importance to be tagged for a wider utility in important crop plants.

Marker Assisted Selection in Wheat

Conventionally, plant breeding depends upon morphological/phenotypic markers for the identification of agronomic traits. With the development of methodologies for the analysis of plant gene structure and function, molecular markers have been utilized for identification of traits. Molecular markers act as DNA signposts to locate the gene(s) for a trait of interest on a plant chromosome, and are widely used to study the organization of plant genomes and for the construction of genetic linkage maps. Molecular markers are independent from environmental variables and can be scored at any stage in the life cycle of a plant. Over the last several years, there has thus been marked increase in the application of molecular markers in the breeding programmes of various crop plants. Molecular markers not only facilitate the development of new varieties by reducing the time required for the detection of specific traits in progeny plants, but also fasten the identification of resistance genes and their corresponding molecular markers, thus accelerating efficient breeding of resistance traits into wheat cultivars by marker assisted selection (MAS). The availability of back-cross derived near isogenic lines (NILs) have also facilitated the analysis of various lines by using different marker systems. The introduction of alien genetic variation into wheat is a valuable and proven technique for wheat improvement. Wild relatives of wheat provide an enormous resource of new genes for wheat improvement, particularly disease and stress tolerance. These genes can be of use if recombined into lines adapted to the conditions of a particular region. Initial studies on the application of molecular markers in wheat relied on the hybridisation based restriction fragment length polymorphism (RFLP) system. RFLP maps provided a more direct method for selecting desirable genes via their linkage to easily detectable markers thereby expediting the movement of desirable genes among varieties. The factors that had been instrumental for the use of RFLP in wheat was the limited number of polymorphisms observed among wheat lines and more significantly the availability of aneuploid stocks for the determination of chromosomal location of genes. In wheat, RFLP's have been used to map seed storage protein loci, loci associated with flour colour, cultivar identification, vernalization (*Vrn*1) and frost resistance gene on chromosome 5A, intrachromosomal mapping of genes for dwarfing (*Rht*12) and vernalization (*Vrn*1), resistance to preharvest sprouting, quantitative trait loci (QTL's) controlling tissue culture response (*tcr*), nematode resistance, milling yield, resistance to chlorosis induction by *Pyrenophora tritici-repentis*. RFLP markers are also useful in selection programs for resistance against pests and pathogens, which is otherwise labor and time consuming, and to detect homozygous individuals and have been used for resistance to barley yellow dwarf virus, resistance to wheat spindle streak mosaic virus, resistance against powdery mildew, resistance against leaf rust, resistance against cereal cyst nematode. The use of RFLP analysis in wheat has, however, been of limited use in the intervarietal analysis due to low level of polymorphism and the high cost for screening in breeding situations.

With the development of polymerase chain reaction (PCR) methodologies, Random Amplified Polymorphic DNA (RAPD) emerged as a convenient and effective technique for tracing alien chromosome segments in translocation lines. RAPD markers provide a useful alternative to RFLP analysis for screening markers linked to a single trait within near isogenic lines and bulked segregants. The development of a DNA polymorphism detection method by combining RAPD with DGGE (denaturing gradient gel elecrophoresis) for pedegree analysis and fingerprinting of wheat cultivars. RAPD markers can be converted to more user-friendly Sequence Characterized Amplified Region (SCAR) markers, that display a less complex banding pattern. SCAR markers linked to resistance genes against fungal pathogens have been characterized in combination with RAPD and RFLP. In recent years, RAPD and other PCR based markers like Sequence Characterized Amplified Regions (SCAR), Sequence Tagged Sites (STS) and Differential Display Reverse Transcriptase PCR (DDRT-PCR) are increasingly being used for identification of desirable traits in wheat and related genera. These markers have been used in particular for disease resistance against viral and fungal pathogens and also for insect and nematode pests and have the potential of pyramiding of resistance genes for effective breeding programs. PCR based markers have been extensively characterized for genes of resistance against common bunt, *Tilletia tritici*,; powdery mildew, *Erysiphe graminis*, leaf rust, *Puccinia recondita*, resistance against Hessian fly, *Mayetiola destructor* and Russian wheat aphid, *Diuraphis noxia*. Simple sequence repeats or microsatellites are more promising molecular markers for the identification and differentiation of genotypes within a species. The high level of polymorphism and easy handling has made microsatellites extremely useful for different applications in wheat breeding. Microsatellites have also been used to identify resistance genes like *Pm6* from *Triticum timopheevii* and *Yr15* from breadwheat. In near future, molecular markers can provide simultaneous and sequential selection of agronomically important genes in wheat breeding programs allowing screening for several agronomically important traits at early stages and effectively replace time consuming bioassays in early generation screens.

Genomics and Future

Genomics is the discovery and study of many genes simultaneously on a genome-wide scale. In the pregenomic era, gene discovery was the limiting factor for the improvement of crops by genetic engineering. For the upcoming postgenomic era, an exponential number of genes will be discovered through genome sequencing efforts on *Arabidopsis* and rice. Wheat genomes are large in comparison to the current plant models and are ideal targets for partial genome sequencing. However, due to a high level of genetic collinearity at a broader genetic level and also at the gene level, comparative genome analysis studies can serve as efficient tools for transferring information and resources from well studied genomes to the related genera. DNA mapping and sequence information generated in the rice genome project would be highly useful for the characterization of larger genomes like wheat. Molecular isolation of agronomically important genes and an understanding of their action will contribute immensely to breeding of cultivars. With the availability of detailed information regarding the location and function of gene(s) encoding for useful traits, scientists in future will be well equipped for efficiently creating varieties with exact combinations of desirable traits. However, genetic transformation will remain a significantly important tool for understanding gene functions and for testing the utility of new sequences. In near future, crop varieties could be tailor-made to meet both local consumer preferences and the demands of particular niche or environment. The new tools of biotechnology not only have the potential for increasing the effectiveness and efficiency of wheat breeding programs, and but will also provide insights into the genetic control of key traits to be used for genetic manipulation.

References

Abranches, R (2000) Widely separated multiple transgene integration sites in wheat chromosomes are brought together at interphase. *Plant J.* **24**: 713–723.

Agache, S; Bachelier, J and de Buyser, J (1989) Genetic analysis of anther-culture response in wheat using aneuploid, chromosome substitution and translocation lines. *Theor Appl Genet* **77**: 7–11.

Alonso, LC and Kimber, G (1981) The analysis of meiosis in hybrids. II Triploid hybrids. *Can Genet. J. Cytol.* **23**: 221-234.

Altpeter, F; Vasil, V; Srivastava, V; Stöger, E and Vasil, IK (1996a) Accelerated production of transgenic wheat (*Triticum aestivum* L.) plants. *Plant Cell Reports* **16**: 12-17.

Amoah, BK; Wu, H; Sparks, C and Jones, HD (2001) Factors influencing *Agrobacterium*-mediated transient expression of uidA in wheat inflorescence tissue. *Journal of Experimental Botany*. **52**: 1135-1142.

Anderson, JA; Sorrells, ME and Tanksley, SD (1993) RFLP analysis of genomic regions associated with resistance to preharvest sprouting in wheat. *Crop Science* **33**: 453-459.

Arain, SM; Sial, MA; Jamali, KD and Laghari, KA (2018) Grain yield performance, correlation and cluster analysis in elite bread wheat (*Triticum aestivum* L.) lines. *Polish Botanical Society Acta Agrobotanica*. **71**(4): 1747.

Asif, M; Eudes, F and Randhawa, H (2014) Induction medium osmolality improves microspore embryogenesis in wheat and triticale. *In Vitro Cell Dev Biol Plant* **50**: 121–126.

Ausemus, ER; Harrington, JB; Reitz, LP and Worzella, WW (1946) A summary of genetic studies in hexaploid and tetraploid wheats. *J. Amer.* Soc. *Agron.* **38**: 1082-1099.

Badiyal, A; Chaudhary, HK and Jamwal, NS (2014) Interactive genotypic influence of triticale and wheat on their crossability and haploid induction under varied agroclimatic regimes. *Cereal Res Commun* 1(1): 1–10.

Bajaj, YPS (1990). Biotechnology in wheat breeding. In: Bajaj YPS ed. Biotechnology in agriculture and forestry Wheat, Vol.3. springer-Verlag, Berlin, Heidelberg, New York, USA. pp. 3-23.

Barkley, A and Chumley, FG (2012) A DH laboratory for Kansas wheat breeding: an economic analysis of biotechnology adoption. *Int Food Agribus Rev* **15**(2): 99–120.

Barnabas, B (2003) Protocol for producing doubled haploid plants from anther culture of wheat (*Triticum aestivum* L.). In: Maluszynski M, Kasha K, Forster BP, Szarejko I (eds) Doubled haploid production in crop plants. A manual. Kluwer Academic Publishers, Norwell, pp. 65–70.

Barro, F; Cannell, ME; Lazzeri, PA and Barcelo, P (1998) The influence of auxins on transformation of wheat and tritordeum and analysis of transgene integration patterns in transformants. *Theoretical and Applied Genetics* **97**: 684-695.

Barro, F; Rooke, L; Bekes, F; Gras, P; Tatham, AS; Fido, R; Lazzeri, PA; Shewry, PR and Barcelo, P (1997) Transformation of wheat with high molecular weight subunit genes results in improved functional properties. *Nature Biotechnology* **15**: 1295-1299.

Beales, J; Turner, A; Griffiths, S; Snape, JW and Laurie, DA (2007) A Pseudo-Response Regulator is misexpressed in the photoperiod insensitive Ppd-D1a mutant of wheat (*Triticum aestivm* L.) *Theor. Appl. Genet.* **115**: 721–733.

Ben Amer, IM; Korzun, V; Worland, AJ and Börner, A (1997) Genetic mapping of QTL controlling tissue-culture response on chromosome 2B of wheat (*Triticum aestivum* L.) in relation to major genes and RFLP markers. *Theoretical and Applied Genetics* **94**: 1047-1052.

Bieri, S; Potrykus, I and Futterer, J (2000) Expression of active barley seed ribosome-inactivating protein in transgenic wheat. *Theoretical and Applied Genetics* **100**: 755-763.

Blummel, M; Upadhyay, SR; Gautam, N; Barma, NCD; Hakim, MA; Hussain, M; Mujahid, MY; Chatrath, R; Sohu, VS; Mavi, G; Mishra, VK; Kalappanavar, IK; Naik, R; Biradar, S; Prasad, SVS; Singh, RP and Joshi. AK (2019) Comparative assessment of food-fodder traits in a wide range of wheat germplasm for diverse biophysical target domains in South Asia. *Field Crops Res* **236:** 68-74.

Bommineni, VR; Jauhar, pp. and Peterson, TS (1997) Transgenic durum wheat by microprojectile bombardment of isolated scutella. *Journal of Heredity* **88**: 301-313.

Bonjean, AP and Angus, WJ (editors) (2001) The World Wheat Book: a history of wheat breeding. Lavoisier Publ., Paris. 1131 pp.

Borlaug, NE (1968) Wheat breeding and its impact on world food supply. In *Proc. 3d Int.* Wheat Genet. *Symp.*, Australian Academy of Science, Canberra, Australia, pp. 1-36.

Braun, HJ; Rajaram, S and Van Ginkel, M (1996) CIMMYT's approach to breeding for wide adaptation. *Euphytica* **92**: 175–183.

Brinch-Pederson, H; Oleson, A; Rasmussen, SK and Holm, PB (2000) Generation of transgenic wheat (*Triticum aestivum* L.) for constitutive accumulation of an *Aspergillus* phytase. *Molecular Breeding* **6**: 195-206.

Bryan, GJ; Collins, AJ; Stephenson, P; Orry, A; Smith, JB and Gale MD (1997) Isolation and characterization of microsatellites from hexaploid bread wheat. *Theoretical and Applied Genetics* **94**: 557-563.

Cannell, ME; Doherty, A; Lazzeri, PA and Barcelo, P (1999) A population of wheat and *tritordeum* transformants showing a high degree of marker gene stability and heritability. *Theoretical and Applied Genetics* **99**: 772-784.

Castillo, AM; Cistué, L and Vallés, MP (2009) Chromosome doubling in monocots. In: Touraev A, Forster BP, Jain SM (eds) Advances in haploid production in higher plants. Springer, Berlin, pp. 329–338.

Cauvain, SP and Cauvain, PC (2003) *Bread Making*. CRC Press. p. 540.

Chaudhary, HK; Sethi, GS; Singh, S; Pratap, A and Sharma, S (2007) Efficient haploid induction in wheat by using pollen of *Imperata cylindrica, Plant Breeding* **124** (1): 96-98.

Chawla, HS; Cass, LA and Simmonds, JA (1999a) Expression of anthocyanin pigmentation in wheat tissues transformed with anthocyanin regulatory genes. *Current Science* **76**: 1365-1370.

Chawla, HS; Cass, LA and Simmonds, JA (1999b) Developmental and environmental regulation of anthocyanin pigmentation in wheat tissues transformed with anthocyanin regulatory genes. *In Vitro Cell and Developmental Biology* **35**: 403-408.

Chen, WP; Gu, X; Liang, GH; Muthukrishnan, S; Chen, PD; Liu, DJ and Gill, BS (1998) Introduction and constitutive expression of a rice chitinase gene in bread wheat using biolistic bombardment and the *bar* gene as a selectable marker. *Theoretical and Applied Genetics* **97**: 1296-1306.

Chibbar, RN; Kartha, KK; Leung, N; Qureshi, J and Caswell, K (1991) Transient expression of marker genes in immature zygotic embryos of spring wheat (*Triticum aestivum* L.) through microprojectile bombardment. *Genome* **34**: 453-460.

Cho, SW; Ishii, T; Matsumoto, N; Tanaka, H; Eltayeb, AE and Tsujimoto, H (2011a) Effects of the cytidine analogue zebularine on wheat mitotic chromosomes. *Chromosome Sci* **14**: 23–28.

Cho, SW; Moritama, Y; Ishii, T; Kishii, M; Tanaka, H; Eltayeb, AE and Tsujimoto, H (2011b) Homology of two alien chromosomes during meiosis in wheat. *Chromosome Sci* **14**: 45–52.

Chong, K; Bao, S; Xu, T; Tan, K; Liang, T; Zeng, J; Huang, H; Xu, J and Xu, J (1998) Functional analysis of the *ver* gene using transgenic wheat. *Physiologia Plantarum* 102: 87-92.

Choudhry, SL and Bains, SS (1966) Growing wheat - the intensive way. *Indian Farming* 16(7): 4-7.

Cockram, J; Jones, H; Leigh, FJ; O'Sullivan, D; Powell, W; Laurie, DA and Greenland, AJ (2007) Control of flowering time in temperate cereals: genes, domestication, and sustainable productivity. *J. Exp. Bot.* **58**: 1231–1244.

Cook, JP; Wichman, DM and Martin, JM (2004). Identification of microsatellite markers associated with a stem solidness locus in wheat. *Crop Sci* **44**: 1397–1402.

Copp, LGL (1965) Purple grain in hexaplold wheat. *Wheat Inform. Serv.* 19-20: 18.

Cox, TS; Hatchett, JH; BS; Ranpp, WJ and Sears, RG (1990). Agronomic performance of hexaploid wheat lines derived from direct cross between wheat and *Aegilops squarrosa*. *Pl. Breeding*, **105**: 271.

Crasta, OR; Francki, MG; Bucholtz, DB; Sharma, HC; Zhang, J; Wang, RC; Ohm, HW and Anderson, JM (2000) Identification and characterization of wheat-wheatgrass translocation lines and localization of barley yellow dwarf virus resistance. *Genome* **43**: 698-706.

Daniell, H; Krishnan, M and McFadden, BF (1991) Transient expression of β -glucuronidase in different cellular compartments following biolistics delivery of foreign DNA into wheat leaves and calli. *Plant Cell Reports* **9**: 615-619.

Danyluk, J; Kane, NA; Breton, G; Limin, AE; Fowler, DB and Sarhan, F (2003) TaVRT-1, a putative transcription factor associated with vegetative to reproductive transition in cereals. *Plant Physiol.***132**: 1849–1860.

Darwin, CR (1876) The effects of cross and self-fertilisation in the vegetable kingdom. John Murray, London.

De Block, M; Debrouwer, D and Moens, T (1997) The development of a nuclear male sterility system in wheat. Expression of the *barnase* gene under the control of tapetum specific promoters. *Theoretical and Applied Genetics* **95**: 125-131.

Dear, PH (2001) Genome mapping. Encyclopedia of life science. McMillan Pub Ltd, Nature Publishing Group.

Demeke, T; Hucl, P; Baga, M; Caswell, K; Leung, N and Chibbar, RN (1999) Transgene inheritance and silencing in hexaploid spring wheat. *Theoretical and Applied Genetics* **99**: 947-953.

DePauw, RM; Knox, RE and Clarke, FR (2009) Stettler hard red spring wheat. *Can J Plant Sci* **89**: 945–951.

Devaux, P; Kilian, A and Klienhofs, A (1995) Comparative mapping of the barley genome with male and female recombination-derived, doubled haploid populations. *Mol Gen Genet* **249**: 600–608.

Devkota RN (1994) Wheat Breeding objectives in Nepal: The national testing system and recent progress. In: Saunders DA, Hettel GP (eds), Wheat in Heat-Stressed Environments: Irrigated, Dry Areas and Rice-Wheat Farming Systems. CIMMYT, Mexico, pp. 216-223.

Devos, KM and Gale, MD (1992) The use of random amplified polymorphic DNA markers in wheat. *Theoretical and Applied Genetics* **84**: 564-577.

Devos, KM and Gale, MD (1997) Comparative genetics in the grasses. *Plant Molecular Biology* **35**: 3-15.

Devos, KM; Bryan, GJ; Collins, AJ; Stephenson, P and Gale, MD (1995) Application of two microsatellite sequences in wheat storage proteins as molecular markers. *Theoretical and Applied Genetics* **90**: 247-252.

Dhir, SK; Pajeau, ME; Fromm, ME and Fry, JE (1994) Anthocyanin genes as a visual marker for wheat transformation. In: Henry, RJ and Ronalds, JA, eds. Improvement of Cereal Quality by Genetic Engineering, Plenum Press, New York, USA. pp. 71-75.

Dirks, R; van Dun, K; de Snoo, CB; Vandenberg, M; Lelivelt, CLC; Voermans, W; Woudenberg, L; de Wit, JPC; Reinink, K; Schut, JW; van der Zeeuw, E; Vogelaar, A; Freymark, G; Gutteling, EW; Keppe, MN; van Drongelen, P; Kieny, M; Ellu, P; Touraev, A; Ma, H; de Jong, H and Wijnker, E (2009) Reverse breeding: a novel breeding approach based on engineered meiosis. *Plant Biotechnol. J.* **7**: 837–845.

Ditsch, DC and Grove, JH (1991) Influence of tillage on plant populations, disease incidence, and grain yield of two soft red winter wheat cultivars. *Journal of Production Agriculture* **4**: 360–365.

Dreccer, FM; Borgognone, MG; Ogbonnaya, FC; Trethowan, RM and Winter, B (2006) CIMMYT selected synthetic bread wheats for rainfed environments: Yield evaluation in Mexico and Australia. *Australian Journal of Agricultural Research.*

Driscoll, CJ (1983) Third compendium of wheat-alien chromosome lines. In: *Suppl. 6th Int. Wheat Genet. Symp.* 37 pp.

Dubcovsky, J; Echaide, M; Giancola, S; Rousset, M; Luo, MC; Joppa, LR and Dvorak, J (1997) Seed-storage-protein loci in RFLP maps of diploid, tetraploid and hexaploid wheat. *Theoretical and Applied Genetics* **95**: 1169-1180.

Dubcovsky, J; Loukoianov, A; Fu, D; Valarik, M; Sanchez, A and Yan, L (2006) Effect of photoperiod on the regulation of wheat vernalization genes VRN1 and VRN2. *Plant Mol. Biol.***60**: 469–480.

Dvorak, J and Knott, DR (1977) Homoeologous chromatin exchange in a radiation-induced gene transfer. *Can. J. Genet. Cytol.* **19**: 125-131.

Dweikat, I; Ohm, H; Mackenzie, S; Patterson, F; Cambron, S and Ratcliffe, R (1994) Association of a DNA marker with Hessian fly resistance gene *H9* in wheat. *Theoretical and Applied Genetics* 89: 964-968.

Eagles, HA; Hollamby, GL; Eastwood, RF and Podlich, DW (2001) Use of markers for monogenically and polygenically inherited traits. In *Proc. 10th Australian Wheat Breeding Assembly*, pp. 35–37. Mildura, Australia.

Eastwood, R.F.; Lagudah, E.S. and Appels, R. (1994). A directed search for DNA sequence tightly linked to cereal cyst nematode resistence genes in *Triticum tauschii*. *Genome* 37: 311-319.

El-Hennawy, MA; Abdalla, AF and Shafey, SA (2011) Production of doubled haploid wheat lines (*Triticum aestivum* L.) using anther culture technique. *Ann Agric Sci* 56: 63–72.

Endo, TR and Gill, BS (1996) The deletion stocks of common wheat. *J. Hered.* **87**: 295–307.

Fedak, G and Jui, PY (1982) Chromosomes of Chinese Spring wheat carrying genes for crossability with Betzes barley. *Can. J. Genet. Cytol.* **24**: 227-233.

Fehr, WR (1987) Principles of cultivar development. Vol.1. McMillan Publishing Co. NY.

Feldman, M (2001) Origin of cultivated wheat. In: Bonjean AP, Angus WJ, editors. The World Wheat Book, A History of Wheat Breeding. Paris: Lavoisier Publishing. pp. 3–56.

Fowler, DB (2014) CDC chase hard red winter wheat. *Can J Plant Sci* **94**: 183–186.

Fox, SL; Wise, IL and Smith, MA (2013) Shaw hard red spring wheat. *Can J Plant Sci* **93**: 299–305.

Frame, BR; Drayton, PR; Bagnall, SV; Lewnau, CJ; Bullock, WP; Wilson, HM; Dunwell. JM; Thompson, JA and Wang, K (1994) Production of fertile transgenic maize plants by silicon carbide whisker-mediated transformation. *Plant Journals* **6**: 941-948.

Frey, KJ (1971) Improving crop yields through plant breeding. *Amer Soc. Agron.* (Spl) **20**: 15-58.

Friedt, W; Breun, S and Zucher, S (1986) Comparative value of andro-genetic doubled haploid and conventionally selected barley lines. *Plant Breed* 97: 56–63.

Gao, M and Chibbar, RN (2000) Isolation. characterization and expression analysis of starch synthase IIa cDNA from wheat. *Genome* **43**: 768-775.

Gill, BS and Friebe, B (2002) Cytogenetics, phylogeny and evolution of cultivated wheats. In *Bread Wheat Improvement and Production*, edited by B.C. Curtis, S. Rajaram and H. Gomez Macpherson, pp. 71–88. FAO, Rome.

Gill, KS and Gill, BS (1994) Mapping in the realm of polyploidy: The wheat model. *BioEssays* **16**: 841-846.

Gill, KS; GS Bhullar and A Bhatia (1979) Analysis of generation means for metric traits in durum wheat. *Crop Improv.* **6**: 103-109.

Gonzalez, JM and Cubero, JI (1993) Selection strategies and choice of breeding methods in pp. 281-314. M.D. Hayward (ed) Plant Breeding Principles and prospects. Chapman and Hall, NY.

Gororo, NN; Eagles, HA; Eastwood, RF; Nicolas, ME and Flood, RG (2002) Use of *Triticum tauschii* to improve yield of wheat in low-yielding environments. *Euphytica* **123**: 241–254.

Graf, RJ; Thomas, JB and Beres, BL (2012) Flourish hard red winter wheat. *Can J Plant Sci* **92**: 183–189.

Grauda, D; Lepse, N and Strazdiòa, V (2010) Obtaining of doubled haploid lines by anther culture method for the Latvian wheat breeding. *Agron Res* **8**: 545–552.

Guang-Min, XIA and Zhong-Ti, LI (1999) Transgenic plant regeneration from wheat (*Triticum aestivum* L.) mediated by *Agrobacterium tumefaciens*. *Acta Phytophysicologica Sinica* **25**: 22-28.

Gupta, PK; Kumar, J; Mir, RR and Kumar, A (2010) Marker-assisted selection as a component of conventional plant breeding. *Plant Breed. Rev.* **33**: 145–217.

Gupta, PK; Langridge, P and Mir, RR (2010b) Marker-assisted wheat breeding: present status and future possibilities. *Mol Breed* **26**: 145–161.

Gupta, PK; Rustgi, S and Mir, RR (2013) Array-based high-throughput DNA markers and genotyping platforms for cereal genetics and genomics. In: Gupta PK, Varshney RK (eds) Cereal genomics II. Springer, The Netherlands, pp. 11-55.

Gupta, PK; Varshney, R; Sharma, P and Ramesh, B (1999) Molecular markers and their applications in wheat breeding. *Plant Breeding*, **118**: 369-390.

Gupta, VP and Indu (1988) Tissue specific variation for physiological and genetic parameters. In 'Recent Advances in Plant Tissue Culture' Ed. JB Chowdhry and DR Sharma, ICAR, New Delhi. pp. 53-56.

Guzy, WJ; Labocha, PA and Kwasniewski, M (2007) Different recombination frequencies in wheat doubled haploid populations obtained through maize pollination and anther culture. *Euphytica* **156**(1): 173–183.

Harrington, JB and Waywell, CG (1950) Testing Resistance to Shattering and Lodging in Cereals. *Scientific Agriculture* **30**: 51-60.

Harvey, A; Moisan, L; Lindup, S and Lonsdale, D (1999) Wheat regenerated from scutellum callus as a source of material for transformation. *Plant Cell Tissue and Organ Culture* **57**: 153-156.

Haugerud, A and Collinson, MP (1990) Plants, genes and people: improving the relevance of plant breeding in Africa. *Experimental Agriculture* **26**: 341–362.

Hauptmann, RM; Vasil, V; Ozia-Akins, P; Tabaeizadeh, Z; Rogers, SG; Fraley, RT; Horsch, RB and Vasil, IK (1988) Evaluation of selectable markers for obtaining stable transformants in the Gramineae. *Plant Physiology* **86**: 602-606.

Hawkesford, MJ; Araus, JL; Park, R; Calderini, D; Miralles, D; Shen, T; Zhang, J and Parry, MAJ (2013) Prospects of doubling global wheat yields. *Food and Energy Secur* **2**: 34–48.

Hayward, MD; Bosemark, NO and Romagosa, I(Eds.) (1993) Plant Breeding Principles and prospects. Chapman and Hall, London, NY.

He, DG; Mouradev, A; Yang, YM; Mouradeva, E and Scott, KJ (1994) Transformation of wheat (*Triticum aestivum* L.) through electroporation of protoplasts. *Plant Cell Reports* **14**: 192-196.

He, GY and Lazzeri, PA (1998) Analysis and optimization of DNA delivery into wheat scutellum and tritordeum inflorescence explants by tissue elctroporation. *Plant Cell Reports* **18**: 64-70.

He, S; Ohm, H and Mackenzie, S (1992) Detection of DNA sequence polymorphisms among wheat varieties. *Theoretical and Applied Genetics* **84**: 573-578.

Henikoff, S (2000) Reaching for new heitz. *Genetica* **109**: 7–8.

Hermsen, JG Th and Waninge, J (1972) Attempts to localize the gene *Ch1* for hybrid chlorosis in wheat. *Euphytica* **21**: 204-208.

Hess, D; Dressler, K and Nimmrichter, R (1990) Transformation experiments by pipetting *Agrobacterium* into the spikelets of wheat (*Triticum aestivum* L.). *Plant Science* **72**: 233-244.

Hobbs, SLA; Warkentin, TD and DeLong, MO (1993) Transgene copy number can be positively or negatively associated with transgene expression. *Plant Molecular Biology* **21**: 17-26.

Houshmand, S; Knox, RE and Clarke, FR (2007) Microsatellite markers flanking a stem solidness gene on chromosome 3BL in durum wheat. *Mol Breed* **20**: 261–270.

Hu, ML (1980) A study of the gamma-ray induced semidwarfing gene in wheat (*Trilicum aestivum* L. em Thell.). *J. Agric. Assoc. of China* 109: 5-16. Cited in *Plant Breed. Abstr.* 52: 332.

Huang, XQ; Cloutier, S and Lycar, L (2006) Molecular detection of QTLs for agronomic and quality traits in a doubled haploid population derived from two Canadian wheats (*Triticum aestivum* L.). *Theor Appl Genet* **113**(4): 753–766.

Humphreys, DG; Townley, STF and Czarnecki, E (2013) Snowstar hard white spring wheat. *Can J Plant Sci* **93**(1): 143–148.

Iglesias, VA; Gisel, A; Bilang, R; Leduc, N; Potrykus, I and Sautter, C (1994) Transient expression of visible marker genes in meristem cells of wheat after biollistic micro-targeting. *Planta* **192**: 84-91.

Ingram, HM; Power, JB; Lowe, KC and Davey, MR (1999) Optimisation of procedures for microprojectile bombardment of microspore-derived embryos in wheat. *Plant Cell Tissue Organ Culture* **57**: 207-210.

Jaccard, P (1908) Nouvelles recherches sur la distribution florale. Bull. Soc. Vaud. Sci. Nat. **44**: 223–270.

Jain, KBL (1989) Breeding procedures for self-pollinated crops. In: Plant breeding Theory and Practice (Ed. Chopra, VL). Oxford and IBH Pub Co. Pvt. Ltd., New Delhi, pp. 471.

Jensen, NF (1970) A diallel selective mating system for cereal breeding. *Crop Sci.* **10**: 629-635.

Joppa, LR; Du, C and Hart, GE (1997) Mapping gene(s) for grain protein in tetraploid wheat (Triticum *turgidum* L.) using a population of recombinant inbred chromosome lines. *Crop Sci* **37**: 1586–1589.

Jordan, MC (2000) Green fluorescent protein as a visual marker for wheat transformation. *Plant Cell Reports* **19**: 1069-1075.

Joshi, AK; Mishra, B; Chatrath, R; Ortiz Ferrara, G and Singh, RP (2007) Wheat improvement in India: present status, emerging challenges and future prospects. *Euphytica* **157**: 431–446.

Juhasz, A; Gardonyi, M; Tamias, L and Bedo, Z (2003) Characterization of the promoter region of *Glu-1Bx7* gene from over expressing lines of an old Hungarian wheat variety. *Proc. 10th Int. Wheat Genetics. Symp.*, edited by N.E. Pogna, M. Romano, E.A. Pogna and G. Galterio, Pasteum, Italy, pp. 1448–1350.

Kane, NA; Agharbaoui, Z; Diallo, AO; Adam, H; Tominaga, Y; Ouellet, F and Sarhan, F (2007) TaVRT2 represses transcription of the wheat vernlization gene TaVRT1. *Plant J.***51**: 670–680.

Karunaratne, S; Sohn, A; Mouradev, A; Janet, S; Steinbib, HH and Scott, KJ (1996) Transformation of wheat with the gene encoding the coat protein of barley yellow mosaic virus. *Australian Journal of Plant Physiology* **23**: 429-435.

Khurana, P; Sehgal, A and Chugh, A (1998) Transgene interactions in transgenics. In: Srivastava, PS, ed. Plant Tissue Culture and Molecular Biology-Applications and Prospects. Narosa Publishing House, New Delhi, India. pp. 728-750.

Khush, GS (1999) Green Revolution: Preparing for 21st century. *Genome* **42**: 646–655.

Kim, KM and Baenziger, PS (2005) A simple wheat haploid and doubled haploid production system using anther culture. *In Vitro Cell Dev Biol Plant* **41**: 22–27.

Klein, TM and Jones, TJ (1999) Methods of genetic transformation: The gene gun. In: Vasil, I.K. ed. Molecular Improvement of Cereal Crops. Kluwer Academic Publishers, Dordrecht, The Netherlands. pp. 21-42.

Kloti, A; Iglesias, VA; Wijnn, J; Burkhardt, PK; Datta, SK and Potrykus, I (1993) Gene transfer by electroporation into intact scutellum cells of wheat embryos. *Plant Cell Reports* **12**: 671-675.

Knox, R; Clarke, JM and DePauw, RM (2000) Dicamba and growth condition effects on DH production in durum wheat crossed with maize. *Plant Breed* **119**: 289–298.

Knox, RE; DePauw, RM and Clarke, FR (2008) Alvena hard red spring wheat. *Can J Plant Sci* **88**: 513–518.

Koba, T and Tsunewaki, K (1978) Mapping of the sand *Cff2* genes on chromosome 3D of common wheat. *Wheat Inf.* Serv. **45-46**: 18-20.

Kobayashi, Y; Kaya, H; Goto, K; Iwabuchi, M and Araki, T (1999) A pair of related genes with antagonistic roles in mediating flowering signals. *Science.* **286**: 1960–1962.

Konzak, CF (1976) A review of semidwarfing gene sources and a description of new mutants useful for breeding short-stature wheats. In *Induced Mutations in Cross-breeding*. IAEA, Vienna, Austria, pp. 79-93.

Korzun, V; Röder, M; Worland, AJ and Borner, A (1997) Intrachromosomal mapping of genes for dwarfing (*Rht12*) and vernalization response (*Vrn1*) in wheat by using RFLP and microsatellite markers. *Plant Breeding* **116**: 227-232.

Kuchel, H; Fox, R; Reinheimer, J; Mosionek, L; Willey, N; Bariana, H and Jefferies, S (2007) The successful application of a marker assisted wheat breeding strategy. *Molecular Breeding*, **20:** 295-308.

Kumar, J; Mir, RR; Kumar, N; Kumar, A; Mohan, A; Prabhu, KV; Balyan, HS and Gupta, PK (2010) Marker-assisted selection for pre-harvest sprouting tolerance and leaf rust resistance in bread wheat. *Plant Breed*. **33**: 145–217.

Kumar, J; Pratap, A and Solanki, RK (2009b) Basics in molecular genetic mapping and QTL analysis in plants. In: P.Bhowmik, S. K.Basu, and A.Goyal (eds), Advances in Biotechnology, 35–52. Bentham Sciences Publishers, IL, USA.

Kumar, L (1991) Association analysis in late sown wheat (*Triticum aestivum* L (Em) Thell. M.Sc (Ag). Thesis, Rajendra Agricultural University, Bihar, Pusa (Samastipur) – 848 125, India pp. 94.

Kuspira, J and Unrau, J (1960) Determination of gene-chromosome associations and establishment of chromosome markers by aneuploid analysis in common wheat. I. F_2 analysis of glume pubescence, spike density and culm colour. *Can. J. Genet. Cytol*. **2**: 301-310.

Lage, J and Trethowan, RM (2006) Synthetic hexaploid wheat improves bread wheat adaptation to rainfed environments globally. *Field Crops Res*.

Lagudah, ES; Moullet, O and Appels, R (1997) Map-based cloning of a gene sequence encoding a nucleotide-binding domain and a leucine-rich region at the *Cre*3 nematode resistance locus of wheat. *Genome* **40**: 659-665.

Langridge, P; Lagudah, ES; Holton, TA; Appels, R; Sharp, PJ and Chalmers, KJ (2001) Trends in genetic and genomic analyses in wheat: A Review. *Australian J Agric Res* **52**: 1043-1078.

Laufs, J; Wirtz, U; Kammann, M; Matzeit, V; Schafer, S; Schell, J; Czernilofsly, AP; Baker, B and Gronenborn, BH (1990) Wheat dwarf virus Ac/Ds vectors: Expression and excision of transposable elements introduced into various cereals by a viral replicon. *Procs of the National Academy of Sciences of the United States of America* **87**: 7752-7756.

Laurie, D and Bennett, MD (1986) Wheat x maize hybridization. *Can. J. Genet. Cytol.* **28**: 313-316.

Laurie, DA (1989) The frequency of fertilization in wheat pearl millet crosses. *Genome* **32**: 1063–1067.

Law, CN and Worland, AJ (1986) An effect of temperature on the fertility of wheats containing gibberellic acid insensitive dwarfing genes. *EWAC Newsletter* 4 -5.

Law, CN; Sutka, J and Worland, AJ (1978) A genetic study of day-length response in wheat. *Heredity* **41**: 1~191.

Lee, B; Murdoch, K; Topping, J; Kreis, M and Jones, MGK (1989) Transient gene expression in aleurone protoplasts isolated from developing caryopses of barley and wheat. *Plant Molecular Biology* **13**: 21-29.

Lentner, M and Bishop, T (1993) Experimental Design and Analysis (Second Ed.). Valley Book Co, CT: Blacksburg.

Liu, PY; Jun, Z and Lu, Y (2006) Impacts of QTL environment interactions on genetic response to marker-assisted selection. *Acta Genet. Sin.* **33**: 63–71.

Liu, S; Rudd, JC and Bai, G (2014) Molecular markers linked to genes important for hard winter wheat production and marketing in the U.S. Great Plains. *Crop Sci* **54**: 1304–1321.

Lobell, DB; Ortiz-Monasterio, I; Asnar, GP; Matson, PA; Naylor, RL and Falcon, WP (2005) Analysis of wheat yield and climatic trends in Mexico. *Field Crops Res.* **94**: 250–256.

Loeb, TA and Reynolds, TL (1994) Transient expression of the *uidA* gene in pollen embryoids of wheat following microprojectile bombardment. *Plant Science* **104**: 81-91.

Longin, CFH and Würschum, T (2016) Back to the future–Tapping into ancient grains for food diversity. *Trends Plant Sci.* **21**: 731.

Lonsdale, DM; Lindup, S; Moisan, LJ and Harvey AJ (1998) Using firefly luciferase to identify the transition from transient to stable expression in bombarded wheat scutellar tissue. *Physiologia Plantarum* **102**: 447-453.

Lorz, H; Baker, B and Schell, J (1985) Gene transfer to cereal cells mediated by protoplast transformation. *Molecular and General Genetics* **199**: 178-192.

Loukoianov, A; Yan, L; Blechl, A; Sanchez, A and Dubcovsky, J (2005) Regulation of VRN-1 vernalization genes in normal and transgenic polyploid wheat. *Plant Physiol.***138**: 2364–2373.

Lynch, M and Ritland, K (1999) Estimation of pair wise relatedness with molecular markers. Genetics, **152:** 1753–1766.

Ma, ZQ; Sorrells, ME and Tanksley, SD (1994) RFLP markers linked to powdery mildew resistance genes *Pm1*, *Pm2*, *Pm3* and *Pm4* in wheat. *Genome* **37**: 871-875.

Mahalakshmi, A and Khurana, P (1995) *Agrobacterium*-mediated gene delivery in various tissues and genotypes of wheat (*Triticum aestivum* L.). *Journal of Plant Biochemistry and Biotechnology* **4**: 55-59.

Mahalakshmi, A; Chugh, A and Khurana, P (2000) Exogenous DNA uptake via cellular perneabilization and expression of foreign gene in wheat zygotic embryos. *Plant Biotechnology* **17**: 235-240.

Mahalakshmi, A; Maheshwari, SC and Khurana, P (1993) High frequency divisions in leaf base protoplasts of wheat (*Triticum aestivum* L.). *Journal of Plant Biochemistry and Biotechnology* **2**: 61-65.

Maheshwari, N; Rajyalakshmi, K; Baweja, K; Dhir, SK; Chowdhury, CN and Maheshwari SC (1995) *In vitro* culture of wheat and genetic transformation: Retrospect and Prospect. *Critical Reviews in Plant Science* **14**: 149-178.

Maluszynski, M; Kasha, K and Forster, BP (eds) (2003) Doubled haploid production in crop plants. A manual. Kluwer Academic Publishers, Norwell.

Marks, MS; Kemp, JM; Woolston, CJ and Dale, PJ (1989) Agroinfection of wheat: A comparison of *Agrobacterium* strains. *Plant Science* **63**: 247-256.

Masood, MA; Qamar, M and Raza, I (2018) Comparative efficiency of alpha lattice design versus randomized complete block design in wheat field trials. *International Journal of Scientific and Engineering Research*. **9**(11): 646-650.

McCormac, AC; Wu, H; Bao, M; Wang, Y; Xu, R; Elliot, MC and Chen, DF (1998) The use of visual marker genes as cell-specific reporters of *Agrobacterium*-mediated T-DNA delivery to wheat (*Triticum aestivum* L.) and barley (*Hordeum vulgare* L.). *Euphytica* **99**: 17-25.

Mi, QL; Velazhahan, R; Muthukrishnan, S and Liang, GH (2000) Wheat transformation mediated by *Agrobacterium tumefaciens*. *Annual Wheat Newsletter* 45.

Miflin, B (2000) Crop improvement in the 21st Century. *J. Exp. Bot.* **51**: 1-8.

Mir, RR; Kumar, N; Jaiswal, V; Girdharwal, N; Prasad, M; Balyan, HS and Gupta, PK (2012) Genetic dissection of grain weight in bread wheat through quantitative trait locus interval and association mapping. *Mol Breed* **29**: 963–972.

Misteli, T (2001) Protein dynamics: implications for nuclear architecture and gene expression. *Science* **291**: 843–847.

Mizuno, T and Nakamichi, N (2005) Pseudo-response regulators (PRRs) or True oscillator components (TOCs) *Plant Cell Physiol*.**46**: 677–685.

Mochida, K and Tsujimoto, H (2001) Production of wheat doubled haploids by pollination with Job's tears (*Coix lachryma-jobi* L.). *J Hered* **92**(1): 81–83.

Moehring, AJ (2011) Heterozygosity and its unexpected correlations with hybrid sterility. *Evolution* 65(9): 2621–2630.

Mohler, V and Singrun, C (2004) General considerations: marker assisted selection. In: H.Lorz, and G.Wenzel (ed.), Biotechnology in Agriculture and Forestry: Molecular Marker Systems in Plant Breeding and Crop Improvement, Springer, Berlin,pp305–318.

Mooney, PA and Goodwin, PB (1991) Adherence of *Agrobacterium tumefaciens* to the cells of immature wheat embryos. *Plant Cell Tissue Organ Culture* **25**: 199-208.

Mooney, PA; Goodwin, PB; Dennis, ES and Llewelleyn, DJ (1991) *Agrobacterium tumefaciens*-gene transfer into wheat tissues. *Plant Cell Tissue Organ Culture* **25**: 209-218.

Mu, HM; Liu, SJ; Zhou, WJ; Wen, YX; Zhang, WJ and Wei, RX (1999) Transformation of wheat with insecticide gene of arrowhead proteinase inhibitors by pollen tube pathway and analysis of transgenic plants. *I Chuan Hsueh Pao (Chinese)* **26**: 634-642.

Mujeeb-Kazi, A (2003) Wheat improvement facilitated by novel genetic diversity and *in vitro* technology. *Plant Tissue Cult* **13**(2): 179–210.

Muller, E; Lorz, H and Lutticke, S (1996) Variability of transgene expression in clonal cell lines of wheat. *Plant Science* **114**: 71-82.

Murai, K; Miyamae, M; Kato, H; Takumi, S and Ogihara, Y (2003) WAP1, a wheat APETALA1 homolog, plays a central role in the phase transition from vegetative to reproductive growth. *Plant Cell Physiol.***44**: 1255–1265.

Myburg, AA; Cawood, M; Wingfield, BD and Botha, AM (1998) Development of RAPD and SCAR markers linked to the Russian wheat aphid resistance gene *Dn*2 in wheat. *Theoretical and Applied Genetics* **96**: 1162-1169.

Nagarajan, S (2005) Can India produce enough wheat even by 2020? *Curr Sci* **89**: 1467–1471.

Naranjo, T (1982) Preferential occurrence of wheat-rye meiotic pairing between chromosomes of homoeologous group 1. *Theor. Appl. Genet.* **63**: 219-225.

Nehra, NS; Chibbar, RN; Leung, N; Caswell, K; Mallard, C; Steinhauer, L; Baga, M and Kartha, KK (1994) Self-fertile transgenic wheat plants regenerated from isolated scutellar tissues from microprojectile bombardment with two distinct gene constructs. *Plant Journal* **5**: 285-297.

Nevo, E; Korol, AB; Beiles, A and Fahima, T (2002) Evolution of Wild Emmer and Wheat Improvement: Population Genetics, Genetic Resources, and Genome. Springer. p8.

Niu, Z; Jiang, A and Hammad, A (2014) Review of doubled haploid production in durum and common wheat through wheat maize hybridization. *Plant Breed* **113**: 313–320.

Oard, JH; Paige, D and Dvorak, J (1989) Chimeric gene expression using maize intron in cultured cells of bread wheat. *Plant Cell Reports* **8**: 156-160.

Oard, JH; Paige, DV; Simmonds, JA and Gradziel, TM (1990) Transient gene expression in maize, rice and wheat cells using an air gun apparatus. *Plant Physiology* **92**: 334-339.

Ogbonnaya, F; Dreccer, F; Ye, G; Trethowan, R; Lush, D; Shepperd, J and Van Ginkel, M (2006) Yield of synthetic backcross-derived lines. In *Extended Abstracts of the International Symposium on Wheat Yield Potential: Challenges to International Wheat Breeding*, March 20–24th, 2006, edited by M.P. Reynolds and D. Godinez, p. 12. Ciudad Obregón, Mexico.

Okamoto, M and Sears, ER (1962) Chromosomes involved in translocation obtained from haploids of common wheat. *can. J. Genet. Cytol.* **4:** 24-30.

O'mara, JG (1940) Cytogenetic studies on Triticinae: I. A method for determining the effects of individual *Secale* chromosomes on *Triticum. Genetics*, **25**: 401-409.

Ortiz, JPA; Reggiardo, MI; Ravizzini, RA; Altabe, SG; Cervigni, GDL; Spitteler, MA; Morata, MM; Elias, FE and Vallejos, RH (1996) Hygromycin resistance as an efficient selectable marker for wheat stable transformation. *Plant Cell Reports* **15**: 877-881.

Pandey, AK (1984) G x E studies in wheat (*Triticum aestivum* L) Ph D. Thesis, Rajendra Agricultural University, Bihar, Pusa (Samastipur) – 848 125, India.

Parker, GD; Chalmers, KJ; Rathjen, AJ and Langridge, P (1998) Mapping loci associated with flour colour in wheat (*Triticum aestivum* L.). *Theoretical and Applied Genetics* **97**: 238-245.

Parker, GD; Chalmers, KJ; Rathjen, AJ and Langridge, P (1999) Mapping loci associated with milling yield in wheat (*Triticum aestivum* L.). *Molecular Breeding* **5**: 561-568.

Patidar, P (2012) Study of the genetic network controlling root and shoot architecture to yield in segregating populations of bread wheat in Central India. BU/Ag/11/3110 dt 13.10.2011. M.Sc (Plant Breeding), Bundelkhand University, Jhansi, India.

Patnaik, D and Khurana, P (2001) Wheat biotechnology: A minireview, *Electron. J. Biotechnol.* 4(2).

Patterson, HD and Hunter, EA (1983) The efficiency of incomplete block designs in national list and recommended list of cereal variety trials. *The Journal of Agricultural Science*. **101**(2): 427-433.

Patterson, HD and Williams, ER (1976) A new class of resolvable incomplete block designs. *Biometrika*, **63**: 83-90.

Pellegrineschi, A; Noguera, LM; Skovmand, B; Brito, RM; Velazquez, L; Salgado, MM; Hernandez, R; Warburton, M and Hoisington, D (2002) Identification of highly transformable wheat genotypes for mass production of fertile transgenic plants. *Genome*.**45**: 421–430.

Perl, A; Kless, H; Blumenthal, A; Galili, G and Galun, E (1992) Improvement of plant regeneration and GUS expression in scutellar wheat calli by optimization of culture conditions and DNA-microprojectile delivery procedures. *Molecular and General Genetics* **235**: 279-284.

Pingali, PL and Rajaram, S (1999) Global wheat research in a changing world: Options for sustaining growth in wheat productivity. In: Cimmyt. World Wheat Facts and Trends. D.F., Mexico. pp. 1-18.

Powell, WM; Morgante, M; Andre, C; Hanafey, M; Vogel, J; Tingey, S and Rafalski, A (1996) The comparison of RFLP, RAPD, AFLP and SSR (microsatellite) markers for germplasm analysis. Mol. Breed. **2**: 225–238.

Pratap, A; Sethi, GS and Chaudhary, HK (2005) Relative efficiency of different Gramineae genera for haploid induction in triticale and triticale x wheat hybrids through the chromosome elimination technique, *Plant Breeding* **124** (2): 147-153.

Preston JC and Kellogg EA. (2008) Discrete developmental roles for temperate cereal grass vernalization1/fruitful-like genes in flowering competency and the transition to flowering. *Plant Physiol.***146**: 265–276.

Probst, AV (2003) Two means of transcriptional reactivation within heterochromatin. *Plant J.* **33**: 743–749.

Qi, L; Cao, M; Chen, P; Li, W and Liu, D (1996) Identification, mapping and application of polymorphic DNA associated with resistance gene *Pm*21 of wheat. *Genome* **39**: 191-197.

Quisenberry, KS (1938) Survival of Wheat Varieties in the Great Plains Winter Hardiness Nursery, 1930-1937. *Journal of the American Society of Agronomy* 30: 399-405.

Rajaram, S (1988) Breeding and testing strategies to develop wheat for rice-wheat rotation areas. In: Klatt AR (ed), Wheat Production Constraints in Tropical Environments. CIMMYT, Mexico, pp. 187-196.

Rajaram, S; Borlaug, NE and Van Ginkel, M (2002) CIMMYT international wheat breeding. In *Bread Wheat Improvement and Production*, edited by B. C. Curtis, S. Rajaram and H. Gomez-Macpherson, pp. 103–117. FAO, Rome.

Randhawa, HS; Asif, M and Pozniak, C (2013) Application of molecular markers to wheat breeding in Canada. *Plant Breed* **132**: 458–471.

Rasco-Gaunt, S; Riley, A; Barcelo, P and Lazzeri, PA (1999) Analysis of particle bombardment parameters to optimise DNA delivery into wheat tissues. *Plant Cell Reports* **19**: 118-127.

Rasul, G; Humphreys, DG and Brûlé, BAL (2009) Mapping QTLs for pre-harvest sprouting traits in the spring wheat cross 'RL4452/AC Domain'. *Euphytica* **168**(3): 363–378.

Razin, A (1998) CpG methylation, chromatin structure and gene silencing- the three-way interaction. *EMBO Journal* **7**: 4905-4908.

Razzaque, MA and Hossain, ABS (1991) The wheat development program in Bangladesh. In: Saunders DA (ed), Wheat for the Nontraditional Warm Areas. CIMMYT, Mexico, pp. 44-54.

Redha, A and Talaat, A (2008) Improvement of green plant regeneration by manipulation of anther culture induction medium of hexaploid wheat. *Plant Cell Tiss Org Cult* **92**: 141–146.

Reeves, TG; Pingali, PL; Rajaram, S and Cassaday, K (2001) Crop and natural resource management strategies to foster sustainable wheat production in developing countries. In Wheat in a Global Environment, edited by Z. Bedo and L. Lang, Kluwer Academic Publishers, the Netherlands, pp. 23–36.

Reynolds, MP and Borlaug, NE (2006) International collaborative wheat improvement: impacts and future prospects. *J Agric Sci* **144**: 3–17.

Reynolds, MP; Pask, AJD; Hoppitt, WJE; Sonder, K; Sukumaran, S; Molero, G; Pierre, CS; Payne, T; Singh, RP; Braun, HJ; Gonzalez, FG; Terrile, II; Barma, NCD; Hakim, A; He, Z; Fan, Z; Novoselovic, D; Maghraby, M; Gad, KIM; Galal, EHG; Hagras, A; Mohamed, MM; Morad, AFA; Kumar, U; Singh, GP; Naik, R; Kalappanavar, IK; Biradar, S; Sai Prasad, SV; Chatrath, R; Sharma, I; Panchabhai, K; Sohu, VS and Mavi, GS (2017) Strategic crossing of biomass and harvest index - source and sink - achieves genetic gains in wheat. *Euphytica* **213** (11): 257.

Ribaut, JM and Hoisington, D (1998) Marker-assisted selection: New tools and strategies. *Trends Plant Science*,**3**: 236-239.

Riley, R (1965) Cytogenetics and plant breeding Genetics today, Proc 11th *Int. Gen. Congr* **3**: 681 - 688.

Risch, N and Merikangas, K (1996) The future of genetic studies of complex human diseases. Science **273**: 1516–1517.

Roder, MS; Plaschke, J; Konig, SU; Borner, A; Sorrells, ME; Tanksley, SD and Ganal, MW (1995) Abundance, variability and chromosomal location of microsatellites in wheat. *Molecular and General Genetics* **246**: 327-333.

Rooke, L; Bekes, F; Fido, R; Barro, F; Gras, P; Tatham, AS; Barcelo, P; Lazzeri, P and Shewry, (1999) Overexpression of a gluten protein in transgenic wheat results in greatly increased dough strength. *Journal of Cereal Science* **30**: 115-120.

Roy, JK; Prasad, M; Varshney, RK; Balyan, HS; Blake, TK; Dhaliwal, HS; Singh, H; Edwards, KJ and Gupta, PK (1999) Identification of a microsatellite on chromosomes 6B and a STS on 7D of bread wheat showing an association with preharvest sprouting tolerance. *Theoretical and Applied Genetics* **99**: 336-340.

Sadasivaiah, RS; Perkovic, SM and Pearson, DC (2004) Registration of 'AC Andrew' wheat. *Crop Sci* **44**: 696–697.

Sautter, C; Waldner, H; Neuhaus-Url, G; Galli, A; Neuhaus, G and Potrykus, I (1991) Micro-targeting: high efficiency gene transfer using a novel approach for the acceleration of micro-projectiles. *Biotechnology* **9**: 1080-1085.

Sayre, KD; Rajaram, S and Fischer, RA (1997) Yield potential progress in short bread wheats in Northwest Mexico. **37**: 36–42.

Schachermayr, GM; Messmer, MM; Feuillet, C; Winzeler, H; Winzeler, M and Keller, B (1995) Identification of molecular markers linked to the *Agropyron elongatum*–derived leaf rust resistance gene *Lr24* in wheat. *Theoretical and Applied Genetics* **90**: 982-990.

Schmidt, R (2000) Synteny: recent advances and future prospects. *Current Opinion in Plant Biology* **3:** 97-102.

Searle, I and Coupland, G (2004) Induction of flowering by seasonal changes in photoperiod. *EMBO J*.**23**: 1217–1222.

Sears, ER (1977) An induced mutant with homoeologous pairing in wheat. *Can. J.* Genet. *Cytol.* **19**: 585-593.

Sears, ER and Sears, LMS (1978) The telocentric chromosomes of common wheat. In *Proc. 5th Int. Wheat Genet. Symp.* New Delhi, pp. 389-407.

Serik, O; Ainur, I; Murat, K; Tetsuo, M and Masaki, I (1996) Silicon carbide fiber-mediated DNA delivery into cells of wheat (*Triticum aestivum* L.) mature embryos. *Plant Cell Reports* **16**: 133-136.

Sharp, PJ; Johnston, S; Brown, G; McIntosh, RA; Pallotta, M; Carter, M; Bariana, HS; Khatkar, S; Lagudah, ES; Singh, RP; Khairallah, M; Potter, R and Jones, MGK

(2001) Validation of molecular markers for wheat breeding. *Aust. J. Agric. Res.* **52**: 1357–1366.

Shewry, PR and Jones, HD (2005) Transgenic wheat: Where do we stand after the first 12 years? *Annals of Applied Biology* **147**: 1–14.

Shewry, PR and Lazzeri, P (1997) Molecular approaches to wheat quality improvement. *Chemistry and Industry* **21**: 559-562.

Shimada, S; Ogawa, T; Kitagawa, S; Suzuki, T; Ikari, C; Shitsukawa, N; Abe, T; Kawahigashi, H; Kikuchi, R; Handa, H and Murai, K (2009) A genetic network of flowering-time genes in wheat leaves, in which an APETALA1/FRUITFULL-like gene, VRN1, is upstream of FLOWERING LOCUS T *Plant J.* **58**(4): 668–681.

Shimada, T; Seki, M and Morikawa, H (1991). Transient expression of glucuronidase (Gus) gene in wheat pollen embryos via microprojectile bombardment. *Wheat Information Service* **72**: 106-108.

Shimoni, Y; Blechl, AE; Anderson, OD and Galili, G (1997) A recombinant protein of two high molecular weight glutenins alters gluten polymer formation in transgenic wheat. *Journal of Biological Chemistry* **272**: 15488-15495.

Simmonds, J; Stewart, P and Simmonds, D (1992) Regeneration of *Triticum aestivum* apical explants after microinjection of germ line progenitor cells with DNA. *Physiologia Plantarum* **85**: 197-206.

Singh, AK (1984) Study on variability and associations among quantitative traits of some promising strains of wheat (*Triticum aestivum* L. (Em) Thell under rainfed condition. Ph D Thesis, Rajendra Agricultural University, Bihar, Pusa (Samastipur) – 848 125, India.

Singh, PN (1990) Combing ability and heritability studies in wheat (*Triticum aestivum* L) M.Sc (Ag). Thesis, Rajendra Agricultural University, Bihar, Pusa (Samastipur) – 848 125, India, pp. 100.

Singh, SS (1993) Crop production - It's scope, limitation and strategies. In: Crop Management under Irrigated and Rainfed Conditions. pp. 102.

Singh, SS and Singh, BB (2001) IARI wheats for evergreen revolution, IARI, New Delhi, pp.92.

Sivamani, E; Bahieldin, A; Wraith, JM; Al-Niemi, T; Dyer, WE; Ho, TDH and Qu, R (2000) Improved biomass productivity and water use efficiency under water deficit conditions in transgenic wheat constitutively expressing the barley *HVA1* gene. *Plant Science* **155**: 1-9.

Slattery, CJ; Kavakli, IH; Okita, TW (2000) Engineering starch for increased quantity and quality. *Trends in Plant Science* **5**: 291-298.

Snape, JW; Simpson, E and Parker, BB (1986) Criteria for the selection and used doubled haploid systems in cereal breeding programmes. In: Horn W, Jensen CJ, Odenbach W, Shieder O (eds) Genetic manipulation in plant breeding. Walter de Gruyter, Berlin, pp. 217–229.

Somers, DJ and Humphreys, G (2009) State of QTL detection and marker-assisted selection in wheat improvement. In: Carver B (ed) Wheat: science and trade. Blackwell Publishing Inc, USA, pp. 309–326.

Sorokin, AP; Ke, XY; Chen, DF and Elliott, MC (2000) Production of fertile transgenic wheat plants via tissue electroporation. *Plant Science* **156**: 227-233.

Spector, DL (2003) The dynamics of chromosome organization and gene regulation. *Annu. Rev. Biochem*. **72:** 573–608.

Srivastava, V; Anderson, OD and Ow, DW (1999) Single-copy transgenic wheat generated through the resolution of complex integration patterns. *Procs in National Academy of Science of the United States of America* **96**: 11117-11121.

Srivastava, V; Vasil, V and Vasil, IK (1996) Molecular characterization of the fate of transgenes in transformed wheat (*Triticum aestivum* L.). *Theoretical and Applied Genetics* **92**: 1031-1037.

Stachel, M; Lelley, T; Grausgruber H and Vollman, J (2000) Application of microsatellites in wheat (*Triticum aestivum* L.) for studying genetic differentiation caused by selection for adaptation and use. *Theoretical and Applied Genetics* **100**: 242-248.

Stoger, E; Vaquero, C; Torres, E; Sack, M; Nicholson, L; Drossard, J; Williams, S; Keen, D; Perrin, Y; Christou, P and Fischer, R (2000) Cereal crops as viable production and storage systems for pharmaceutical scFv antibodies. *Plant Molecular Biology* **42**: 583-590.

Stroup, WW; Baenziger, PS and Multize, DK (1994) Removing spatial variation from wheat yield trials: a comparison of methods. *Crop Science*. **86**: 62-66.

Sundaresan, V (1996) Horizontal spread of transposon mutagensis: New uses of old elements. *Trends in Plant Sciences* **1**: 184-190.

Sung, S and Amasino, RA (2006) Molecular genetic studies of the memory of winter. *J. Exp. Bot*.**57**: 3369–3377.

Syvanen, AC (2005) Toward genome-wide SNP genotyping. Nat. Genet. **37:** S5–S10.

Szarejko I (2003b) Doubled haploid mutant production. In: Maluszynski M, Kasha K, Forster BP, Szarejko I (eds) Doubled haploid production in crop plants. A manual. Kluwer Academic Publishers, Norwell, pp. 351–361.

Tadesse, W; Inagaki, M and Tawkaz, S (2012) Recent advances and application of doubled haploids in wheat breeding. *Afr J Biotechnol* **11**(89): 15484–15492.

Takumi, S and Shimada, T (1996) Production of trangenic wheat through particle bombardment of scutellar tissue: frequency is influenced by culture duration. *Journal of Plant Physiology* **149**: 418-423.

Takumi, S; Otani, M and Shimada, T (1994) Effect of six promoter-intron combinations on transient reporter gene expression in einkorn, emmer and common wheat cells by particle bombardment. *Plant Science* **103**: 161-166.

Tanksley, SD; Young, ND; Paterson, AH and Bonierbale, MW (1989) RFLP mapping in plant breeding: New tools for an old science. *Biotechnology* **7**: 257-264.

Tao, WJ; Liu, JY; Liu, DJ and Chen, PD (1999) Identification of molecular markers linked to the *Pm6* gene conferring powdery mildew resistance in wheat. I *Chuan Hsueh Pao (Chinese)* **26**: 649-656.

Tariq, M (2003) Erasure of CpG methylation in *Arabidopsis* alters patterns of histone H3 methylation in heterochromatin. *Proc. Natl. Acad. Sci. U. S. A.* **100**: 8823–8827.

Thomas, WTB; Forster, BP and Gertsson, B (2003) Doubled haploids in breeding. In: Maluszynski M *et al.* (eds) Doubled haploid production in crop plants. A manual. Kluwer Academic Publishers, Norwell, pp. 337–350.

Töpfer, R; Gronenborn, B; Schell, J and Steinbiss, HH (1989) Uptake and transient expression of chimeric genes in seed derived embryos. *Plant Cell* **1**: 133-139.

Townley-Smith, TF; Humphreys, DG and Czarnecki, E (2010) Superb hard red spring wheat. *Can J Plant Sci* **90**: 347–352.

Tranquilli G and Dubcovsky J. (2000) Epistatic interaction between vernalization genes Vrn-Am1 and Vrn-Am2 in diploid wheat. *J. Heredity*.**91**: 304–306.

Trethowan, RM and Hodson, D; Braun, HJ and Pfeiffer, WH (2005a) Wheat breeding environments. In *Impacts of International Wheat Breeding Research in the Developing World: 1988–2002*, edited by J. Dubin, M.A. Lantican and M.L. Morris, pp. 4–11. Mexico, D.F.: CIMMYT.

Trevaskis, B; Hemming, MN; Dennis, ES and Peacock, WJ (2007) The molecular basis of vernalization-induced flowering in cereals. *Trends Plant Sci*.**12**: 352–357.

Ushiyama, T; Shimizu, T and Kuwabara, T (1991) High frequency of haploid production of wheat through intergenetic cross with teosinte. *Japan J Breed* **41**: 353–357.

Uze, M; Potrykus, I and Sautter, C (1999) Single-stranded DNA in the genetic transformation of wheat (*Triticum aestivum* L.): Transformation frequency and integration pattern. *Theoretical and Applied Genetics* **99**: 487-495.

Vaccino, P; Accerbi, M and Corbellini, M (1993) Cultivar identification in *T. aestivum* using highly polymorphic RFLP probes. *Theoretical and Applied Genetics* **86**: 833-836.

van de Wouw, M; van Hintum, T; Kik, C; van Treuren, R and Visser, B (2010) Genetic diversity trends in twentieth century crop cultivars: a meta analysis. *Theor. Appl. Genet.* **120**: 1241–1252.

Van Ginkel, M; Ortiz-Monasterio, I; Trethowan, RM and Hernandez, E (2001) Methodology for selecting segregating populations for improved N-use efficiency in bread wheat. *Euphytica* **119**: 223–230.

Varshney, RK; Graner, A and Sorrells, ME (2005) Genic microsatellite markers in plants: features and applications. Trends Biotechnol. **23**: 48–55.

Vasil, IK and Anderson, OD (1997) Genetic engineering of wheat gluten. *Trends in Plant Sciences* **2**: 292-297.

Vasil, IK and Vasil, V (1999) Transgenic cereals: *Triticum aestivum* (wheat). In: Vasil, I.K., ed. Molecular Improvement of Cereal Crops. Kluwer Academic Publishers Dordrecht, The Netherlands. pp. 137-147.

Vasil, V; Brown, SM; Re, D; Fromm, ME and Vasil, IK (1991) Stably transformed callus lines from microprojectile bombardment of cell suspension cultures of wheat. *Biotechnology* **9**: 743-747.

Vasil, V; Castillo, AM; Fromm, ME and Vasil, IK (1992) Herbicide resistant fertile transgenic wheat plants obtained by microprojectile bombardment of regenerable embryogenic callus. *Biotechnology* **10**: 667-674.

Vasil, V; Srivastava, V; Castillo, AM; Fromm, ME and Vasil, IK (1993) Rapid production of transgenic wheat plants by direct bombardment of cultured immature embryos. *Biotechnology* **11**: 1553-1558.

Vepa, SS (2003) Impact of globalisation on food consumption of urban India. In: FAO Technical Workshop on Globalization of Food Systems: Impacts on Food Security and Nutrition. 8–10 October 2003, Rome, Italy.

Verma, DK (1991) Morphological characterization and genetical evaluation of local landraces of rice (*Oryza sativa* L) M.Sc (Ag). Thesis, Rajendra Agricultural University, Bihar, Pusa (Samastipur) – 848 125, India.

Wang, YC; Klein, TM; Fromm, M; Cao, J; Sanford, JC and Wu, R (1988) Transient expression of foreign genes in rice, wheat and soybean cells following particle bombardment. *Plant Molecular Biology* **11**: 433-439.

Wazuddin, M and Driscoll, CJ (1986) Chromosome constitution of polyploid wheats: Introduction of diploid wheat chromosome 4. *Proc. Nat. Acad. Sci.* USA **83**: 4870-4874.

Weeks, JT; Anderson, OD and Blechl, AE (1993) Rapid production of multiple independent lines of fertile transgenic wheat (*Triticum aestivum*). *Plant Physiology* **102**: 1077-1084.

Weeks, T (2000) Transformation of wheat with the cyanamide hydratase gene (09/518,988). Patent. U.S. Department of Agriculture. Agricultural Research Service.

Williams, JGK; Kubelik, AR; Livak, KL; Rafalsks, JA and Tinget, SL (1990) DNA polymorphism amplified by arbitrary primers are useful as genetic markers. *Nucleic Acids Research* **18**: 6531-6535.

Williams, KJ; Fisher, JM and Langridge, P (1996) Development of a PCR-based allele-specific assay from an RFLP probe linked to resistance to cereal cyst nematode in wheat. *Genome* **39**: 798-801.

Witrzens, B; Brettell, RIS; Murray, FR; McElroy, D; Li, Z and Dennis, ES (1998) Comparison of three selectable marker genes for transformation of wheat by microprojectile bombardment. *Australian Journal of Plant Physiology* **25**: 39-44.

Woolston, CJ; Barker, R; Gunn, H; Boulton, MI and Mullineaux, PM (1988) Agroinfection and nucleotide sequence of cloned wheat dwarf virus DNA. *Plant Molecular Biology* **11**: 35-43.

Worland, AJ; Borner A; Korzun, V; Li, WM; Petrovic, S and Sayers, EJ (1997) The influence of photoperiod genes on the adaptability of European winter wheats. In: Braun HJ (ed) Wheat: prospects for global improvement. Dordrecht, Netherlands, Kluwer Academic Press, pp. 517-526.

Worland, T and Snape, JW (2001) Genetic basis of worldwide wheat varietal improvement. In: Bonjean AP, Angus WJ, editors. The World Wheat Book, A History of Wheat Breeding. Paris: Lavoisier Publishing. pp. 59–100.

Xing, L; Munoz, P; Inostroza, L and Acuna, H (2014) Comparing accuracy of Alpha-lattice, row-column and randomized complete block design methods to estimate genetic parameters. *Crop Breeding and Genetics. Poster Number* 546.

Yan, L; Fu, D; Li, C; Blechl, A; Tranquilli, G; Bonafede, M; Sanchez, A; Valarik, M; Yasuda, S and Dubcovsky, J (2006) The wheat and barley vernalization gene VRN3 is an orthologue of FT. *Proc. Natl Acad. Sci. USA.***103**: 19581–19586.

Yan, L; Loukoianov, A; Blechl, A; Tranquilli, G; Ramakrishna, W; San, MP; Bennetzen, JL; Echenique, V and Dubcovsky, J. (2004) The wheat VRN2 gene is a flowering repressor down-regulated by vernalization. *Science.***303**: 1640–1644.

Yan, L; Loukoianov, A; Tranquilli, G; Helguera, M; Fahima, T and Dubcovsky, J (2003) Positional cloning of the wheat vernalization gene VRN1. *Proc. Natl Acad. Sci. USA.***100**: 6263–6268.

Yang, Y; Shah, J and Klessig, DF (1997) Signal perception and transduction in plant defense responses. *Genes and Development* **11**: 1621-1639.

Yau, SK (1997) Efficiency of alpha-lattice designs in international variety yield trials of barley and wheat. *The Journal of Agricultural Science*. **128**: 5-9.

Ye, G and Smith, KF (2008a) Marker-assisted gene pyramiding for inbred line development: basic principles and practical guidelines. *Int J Plant Breed* **2**: 1–10.

Ye, G and Smith, KF (2008b) Marker-assisted gene pyramiding for inbred line development: practical applications. *Int J Plant Breed* **2**: 11–22.

Ye, G and Smith, KF (2009) Marker-assisted gene pyramiding for cultivar development. *Plant Breed. Rev*. **33**: 219–256.

Ye, G and Smith, KF (2010) Marker-assisted gene pyramiding for cultivar development. In: Janick J, editor. Plant breeding reviews. Hoboken: Wiley; pp. 219–256.

Young, ND (1999) A cautiously optimistic vision for marker assisted selection. *Mol Breed*. **5**: 505-510.

Yu, J; Holland, JB; McMullen, MD and Buckler, ES (2008) Genetic design and statistical power of nested association mapping in maize. *Genetics* **178**: 539–551.

Yu, J; Pressoir, G; Briggs, WH; Bi, IV; Yamasaki, M; Doebley, JF; McMullen, MD; Gaut, BS; Nielsen, DM; Holland, JB; Kresovich, S and Buckler, ES (2006) A unified mixed-model method for association mapping that accounts for multiple levels of relatedness. *Nat. Genet*. **38**: 203–208.

Zadoks, JC; Chang, TT and Konzak, CF (1974) A decimal code for the growth stages of cereals. *Weed Research* **14**: 415–421.

Zaghmout, OM and Trolinder, NL (1993) Simple and efficient method for directly electroporating *Agrobacterium* plasmid DNA into wheat callus cells. *Nucleic Acids Research* **21**: 1048.

Zaghmout, OMF (1994) Transformation of protoplasts and intact cells from slowly growing embryogenic callus of wheat (*Triticum aestivum* L.). *Theoretical and Applied Genetics* **89**: 577-582.

Zhang, H; Tan, E and Suzuki, Y (2014) Dramatic improvement in genome assembly achieved using doubled-haploid genomes. *Sci Rep* **4**: 6780.

Zhang, K; Tian, J and Zhao, L (2008) Mapping QTLs with epistatic effects and QTL environment interactions for plant height using a doubled haploid population in cultivated wheat. *J Genet Genomics* **35**(2): 119–127.

Zhang, L; Rybczynski, JJ; Langenberg, WG; Mitra, A and French, R (2000) An efficient wheat transformation procedure: transformed calli with long-term morphogenic potential for plant regeneration. *Plant Cell Reports* **19**: 241-250.

Zhou, H; Stiff, CM and Konzak, CF (1993) Stably transformed callus of wheat by electroporation-induced direct gene transfer. *Plant Cell Reports* **12**: 612-616.

Zhu, C; Gore, M; Buckler, ES and Yu, J (2008) Status and prospects of association mapping in plants. *Plant Genome* **1**: 5–20.

Zhu, Z; Sun, B; Liu, C; Xiao, G and Li, X (1993) Transformation of wheat protoplasts mediated by cationic liposome and regeneration of transgenic plantlets. *Chinese Journal of Biotechnology* **9**: 257-261.

Chapter 3

Bread Wheat Breeding at IARI RS Pusa, Bihar: Past, Present (2006-2010) and Future

Outlines

1. Wheat Research at IARI Regional Station, Pusa (Bihar)
   a. Early introductions
   b. Post-Independence Period
2. Coordinated Approach for Wheat Research in India
3. All India Co-ordinated Wheat Improvement Project (AICWIP) at IARI RS Pusa, Bihar
4. Enriching Wheat Germplasm through Materials Generation Programme and Conduct of Various Trials
5. Selection Criteria and Constitution of Different Trials
6. Hybrid Wheat

a. Hybrid Wheat Production
b. Male Sterility
c. Genic Male Sterility
d. Problem of Hybrid Wheat
e. Development of Wheat Hybrid through CMS System

7. Dormancy in Wheat Seeds
8. Biotechnology Application to Wheat Improvement
   a. Wheat Genetic Resources

b. Wheat Molecular Genetics and Molecular Linkage Maps
c. Mapping of Single or Major Genes
d. Quantitative Trait Mapping

9. Marker Assisted Selection
10. Wheat Molecular Genetics and Wheat Genetic Engineering
11. Candidate Genes for Wheat Genetic Engineering
    a. Resistance to Abiotic Stresses
    b. Induced Fungal Resistance
    c. Quality Traits
    d. Other Possible Targets
    e. Gene Pyramiding for Multiple Disease Resistance
    f. Marker Assisted Selection Strategy for Wheat Improvement
12. Production and Utilization of Synthetic Hexaploid Wheat
    a. Diversity of Ae. tauschii and SHW
    b. Interspecific Crosses of *Ae. tauschii* and *T. turgidum*
    c. Direct Cross of Diploid *Ae. tauschii* to Common Wheat
    d. Direct Cross of Tetraploid Wheat to Common Wheat
    e. Direct Versus Synthetic Approaches
    f. Genetic Resources in Synthetic Hexaploid Wheat
    g. Using Synthetic Hexaploid Wheat in Breeding
    h. Expression of *Ae. tauschii* and *T. turgidum* Genes in Synthetic Hexaploid Wheat
    i. Synthetic Hexaploid in Wheat Improvement
    j. Disease and Pests Resistance
    k. Enhancing Yield Productivity
    l. Environmental Stress Tolerance
       i. Improving Salinity Tolerance
       ii. Mineral/Nutrient Efficiency
       iii. Boron Toxicity
    m. Quality Improvement
       i. Bread Making Quality
       ii. Preharvest Sprouting
    n. Future Prospects
13. First-Generation Sequencing Resources for Yield Improvement in Wheat
    a. Restriction Fragment Length Polymorphism (RFLP) Markers
    b. Random Amplified Polymorphic DNA (RAPD) Markers
    c. Amplified Fragment Length Polymorphism (AFLP) Markers

d. Simple Sequence Repeat (SSR) and Inter Simple Sequence Repeat (ISSR) Markers

14. Second-Generation Sequencing Resources for Yield Improvement in Wheat
15. Translational Genomics
16. HI 1563 (Pusa Prachi) identified for high fertility, irrigated, late sown conditions of North Eastern Plains Zone

Wheat Research at IARI Regional Station, Pusa, Bihar

IARI Regional Station, Pusa, Bihar is situated in Samastipur district of Bihar in the 'U' shaped southwestern lap of river Burhi Gandak at 52m above msl, 25.98°N latitude and 85.67°E longitude. On July 5, 1784 East India Company acquired the land at Pusa for Horse Breeding Stud Farm, but in 1874, the Stud Farm was closed due to epidemic gland disease. In 1902, a plan was mooted for animal husbandry, dairy development, educational institution and Agricultural Institute at Pusa. On 26 December 1903, the Government of India took over all the land from the Bengal Government and with the donation of 30,000 pounds by Mr. Henry Phipps, a US citizen, the Agricultural Research Institute (ARI) was established at Pusa. On 1st April 1904, the Agricultural Research Institute was founded by Lord Curzon, the then Viceroy and Governor General of India. The main building was popularly called as Phipps Laboratory or Naulakha. In 1905, five departments (Agriculture, Cattle Breeding, Chemistry, Economic Botany and Mycology) were started at the ARI. In 1907, Bacteriology unit was added to ARI. The name of ARI was changed to the Imperial Institute of Agricultural Research in 1911, and in 1919, the Institute was renamed as Imperial Agricultural Research Institute. On 15th January 1934 at 2.00 P.M., devastating earthquake damaged the main building of the institute *i.e.* Phipps Laboratory, and on 29th July 1936, the institute was shifted to Delhi. On 7 November 1936 new institute's building was inaugurated at Delhi. A small unit called Botanical Sub Station was left over at Pusa, Bihar, and in 1969, the Botanical Sub-Station was renamed as the Regional Research Station, and in 1975, the Regional Research Station was upgraded as the IARI Regional Station, Pusa, Bihar. Development of wheat varieties for irrigated and rainfed conditions in North Eastern and Eastern Zones, in general, and Bihar, in particular; Breeding of high yielding rice varieties for low lying areas of North Bihar, pigeon pea for East Indian conditions; high yielding and powdery mildew resistant varieties of field pea; suitable varieties of rapeseed and mustard for North Eastern Plain Zone; improvement of papaya through breeding; Management of oil seed and pulses in cropping systems in North Bihar; pathogenesis and epidemiology of foliar disease.

The station has released many high yielding varieties namely HP 1633, HP 1731, HP 1744 and HP 1761 playing a significant role in sustaining the wheat production in Bihar. Other recent varieties being a part of the station are HD 2733, HD 2824, HD 2643 and HW 2045. Station judiciously manages the breeder and IARI seed production programme of many varieties. Moreover, station regularly utilizes the off-Season Screening Nursery facilities at Lahaul Spiti and Wellington. Approximately 4500 breeding lines have been generated under varietal development programme, HP 1883, 14 promising cultures such as HP 1893, 1894, 1895, 1896, 1897,

Figure 3.1. Every bit of food on your plate is a result of someone's undue hard work, blood and sweat

Figure 3.2. If you ate today, thank a farmer

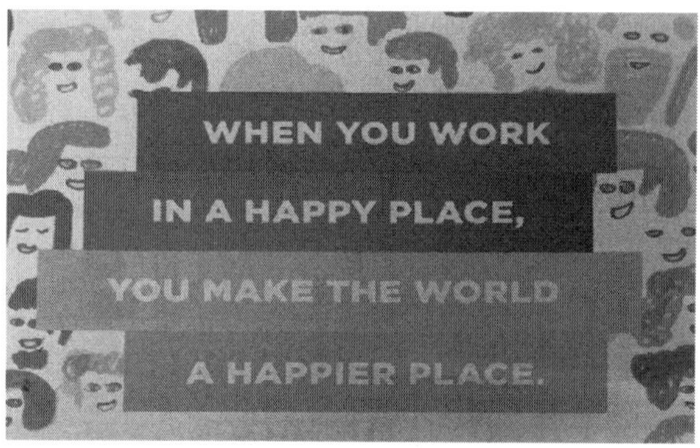

Figure 3.3. When you work....

Figure 3.4. Dr D K Verma with trainee colleagues at IIWBR, Karnal, Haryana (erstwhile DWR, Karnal, Haryana)

Figure 3.5. Dr D K Verma, delivering lecture on Integrated Disease Management of Wheat at LN Mithila University, Darbhanga, Bihar, 2019

Figure 3.6. Dr D K Verma, delivering lecture on Integrated Disease Management of Wheat at TM Bhagalpur, University, Bhagalpur, Bihar

Figure 3.7. VC, Purnea Univ and LN Mithila Univ, appraising Dr D K Verma, for lecture on Integrated Disease Management of Wheat

Figure 3.8. Dr D K Verma, during 'Breeding of Bread Wheat' Assignment during 2009-2013

Figure 3.9. Litti (stuffed wheat ball) +chokha (smash)

Figure 3.10. Panchkosi mela, Buxar, Bihar, where litti mela is being celebrated on grand scale

Figure 3.11. Wheat improvement for high productive environment of Northern India

Figure 3.12. Standardization of Application for wheat under zero tillage under NEPZ, cultivar HD 2733

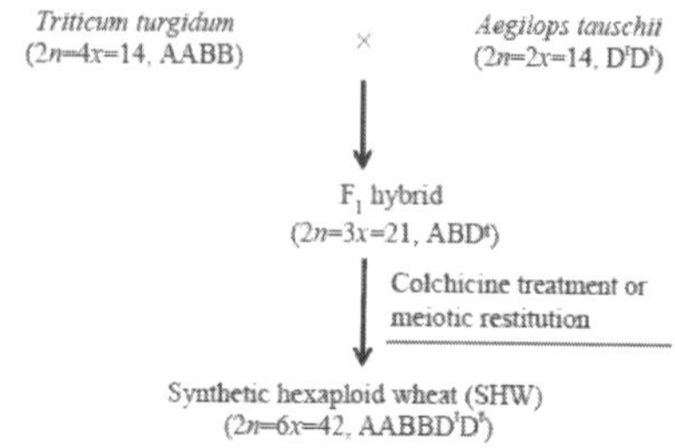

Figure 3.13. The production of synthetic hexaploid wheat (SHW)

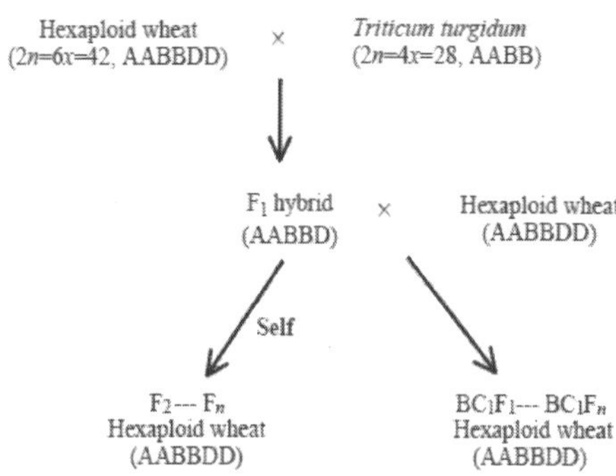

Figure 3.14. Direct cross for genetic transfer from tetraploid wheat into hexaploid wheat

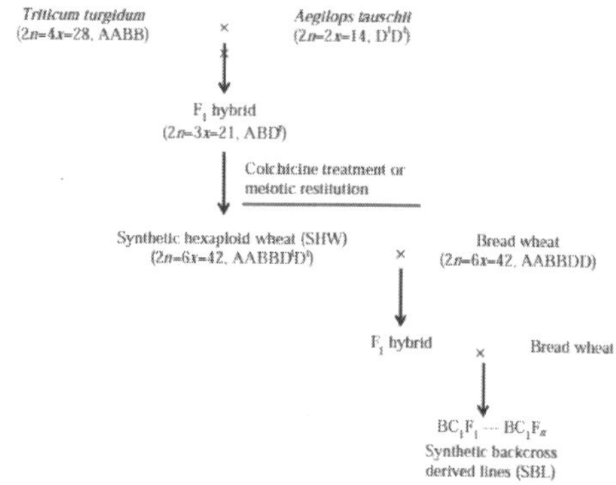

Figure 3.15. Procedure for the transfer of desirable traits from synthetic hexaploid wheat into elite cultivars (synthetic backcross derived lines, SBLs)

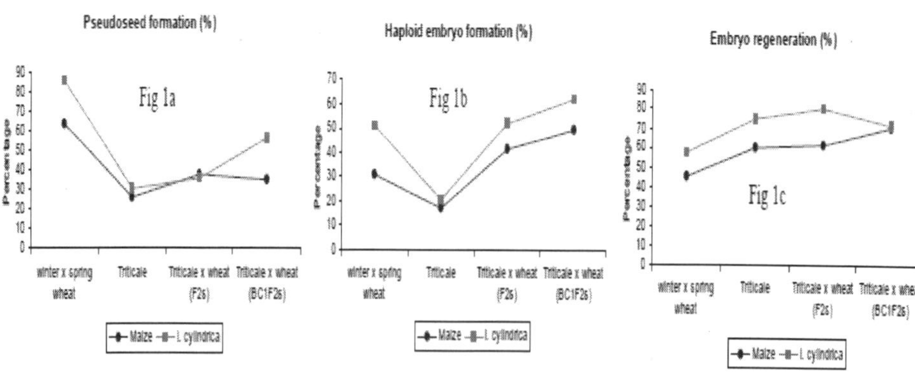

Figure 3.16. Relative frequencies of pseudo seed formation, haploid embryo formation and their regeneration obtained from winter x spring wheat F1s, triticale, triticale x wheat F2s and BC1 F2s after hybridization with maize and I. cylindrica. Data pooled over the five spikes each of five genotypes of each group

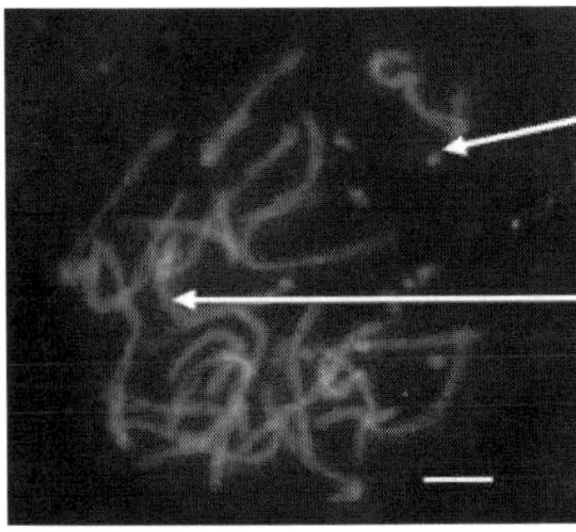

Figure 3.17. GISH analysis of first zygotic prometaphase of wheat x I . cylindrica hybrid. Bar represent 10 µm

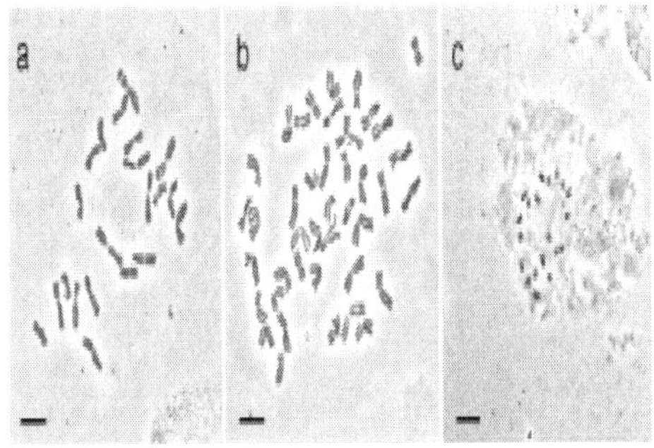

Figure 3.18. Metaphase chromosomes: (a) wheat haploid (n=21) generated through wheat x *Imperata cylindrica* hybridization (b) normal wheat (2n=42) and (c) *Imperata cylindrica* (2n=20).

Figure 3.19. Anthesis in *I. cylindrica* spikes

Figure 3.20. Wildly growing *I. cylindrica* plants in the surroundings of wheat fields in North-West Himalayan regions

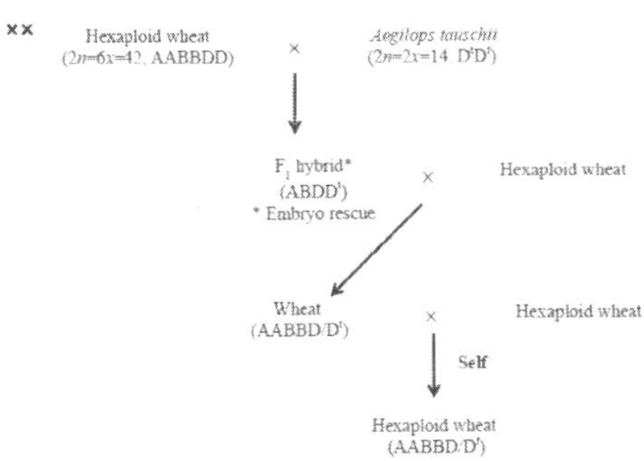

Figure 3.21. Direct cross for genetic transfer from diploid Ae. tauschii into hexaploid wheat

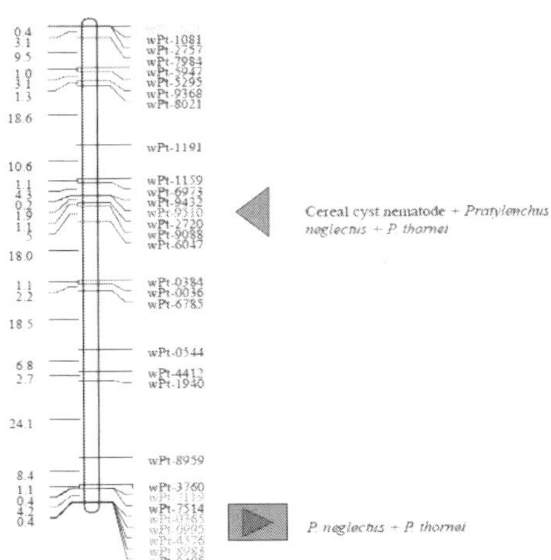

Figure 3.22. Loci on chromosome 3B that confer multiple root pathogen resistance in synthetic hexaploid

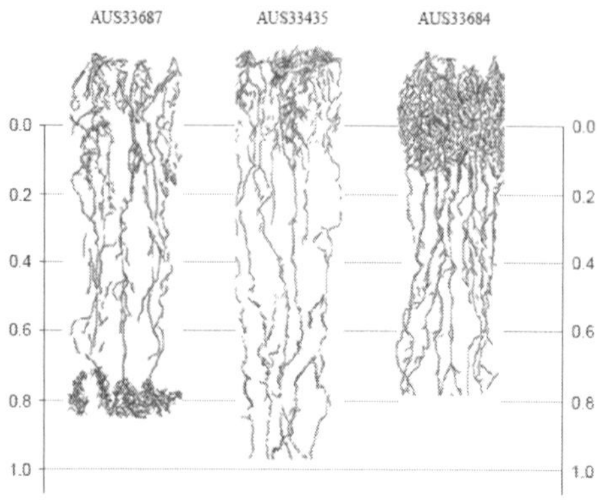

Figure 3.23. Rooting patterns of three synthetic backcross derived lines illustrating desirable traits for root distribution at depth (AUS33687), depth of rooting (AUS33435), and larger root system (AUS33684)

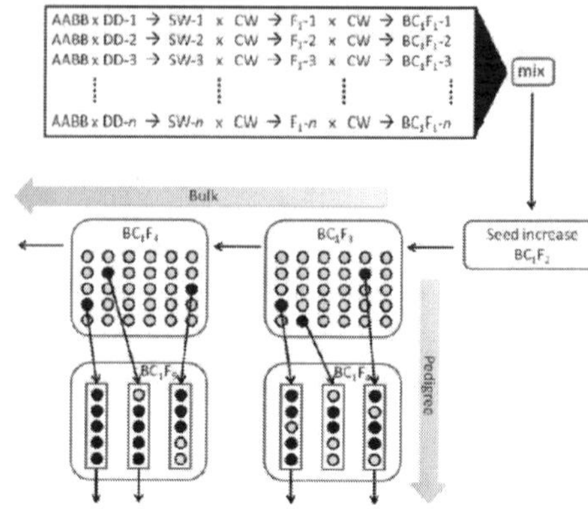

Figure 3.24. Multiple Synthetic Derivatives (MSD) wheat population with variation from many accessions of *Aegilops tauschii*. AABB, durum wheat cultivar; DD-1 to DD-n, *Ae. tauschii* accessions

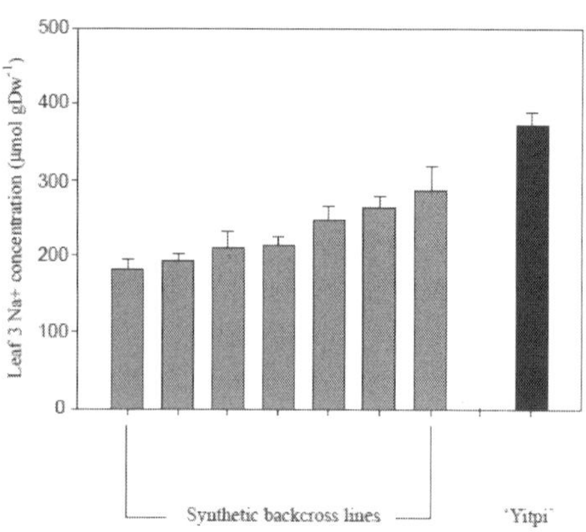

Figure 3.25. Synthetic backcross derived lines with lower Naþ exclusion than the recurrent elite Australian wheat cultivar, 'Yitpi'

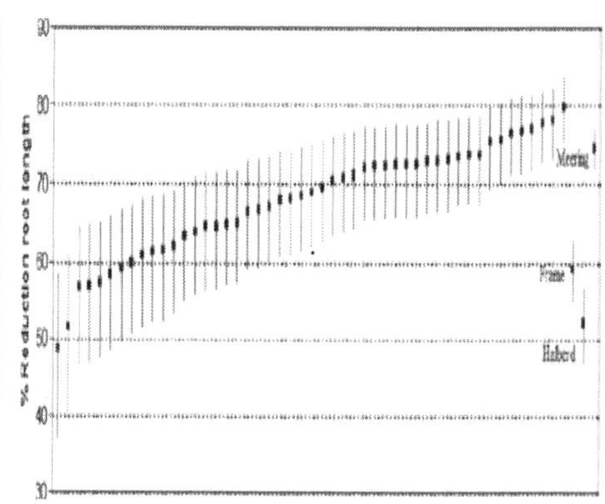

Figure 3.26. Genetic variation for boron tolerance in synthetic hexaploid wheat (SHW)

1898, 1899, 1900, 1901, 1902, 1903, 1904, 1905 and HP 1906 have been tested under NIVT testing. Efforts have been taken to introgress potent alien gene(s) such as Lr 9, Lr 24+Sr 24, Lr 28, Lr 19+Sr 25, Lr 32 and Lr 37 at the background of agronomically superior cultivars either singly or in association with other *Secale cereale* gene. Many leaf blight resistant/tolerant sources have been exploited in breeding programme besides Chirya-3 and Chinese sources. Synthetic wheat such as PHR 1001 and PHR

1026 and other buitre lines such as PHR 1004 and PHR 1007 have been exploited. For terminal heat tolerance, the sources such as PBW 510, WH 730, WH 775, HD 2808, Raj 3777 *etc.* were utilized in breeding programme. Under pre-breeding approach, many cultures were selected at NBPGR, New Delhi and brought to our gene pool after thorough screening at Lahaul Spiti and Wellington under shuttle breeding programme. Promising materials selected from the National and International Nurseries conducted at IARI, RS Pusa were also amalgamated to gene pool.

Early Introductions

In 1905, Sir Albert Howard, with his wife GLC Howard assisted by Abdur Rahman Khan initiated wheat improvement work with collection of wheat sorts from the mixture of landraces belonging to *Triticum aestiviium, T. durum, T. sphaerococcum* and *T. Trigdum* grown by the farmers. One of the selections, Pusa 4 isolated from the local type, Mundia was outstanding with regard to maturity, yield, adaptability and quality. It won first prize for quality in several exhibitions in Australia and America. Other important varieties were Pusa 6 and Pusa 12. In the year 1911 a prized wheat variety P.4 (now NP.4) was produced which spread not only in the Indian Sub-continent but also cultivated in Australia and other foreign countries for its quality. In 1913 Butler's classical booklet on "**Wheat and Wheat cultivation**" was published. "**Fungal Diseases of Plants**" the classical book written by EJ Butler was published in 1918. This was followed by another classical publication "**List of Indian Crop Pests**" by TB Fletcher. The development of famous wheat variety P.52 (NP 52) in 1925 was the result of systematic hybridization work on wheat by Sir Albert Howard followed by P.80-5 (NP 80-5) and P 165 (NP 165). IARI Pusa was the first institution to systematically assess richness of germplasm. In 1934 it was identified that dwarf varieties of barley had the potential of becoming high yielding. Several popular varieties of wheat such as Sharbati, Dara, Safed Pissi, Chandausi, Karachi Choice White, Hard Red Calcutta, Lal Kanak, Lal Pissi, Bansi Kathia, Khandwa, Malvi *etc.* had their origin at Pusa. One of the selections NP 4, isolated from the local type Mundia was outstanding with regard to maturity, yield, adaptability and quality. Among these, Pusa 4 was extensively used as a parent in wheat improvement programme around the world. In 1925, a famous wheat variety, Pusa 52 (NP 52) was developed as a result of the systematic hybridization, followed by Pusa 80-5 (NP 80-5) and Pusa 165 (NP 165). Several popular varieties of wheat such as Sharbati, Dara, Safed Pissi, Chandausi, Karachi Choice White, Hard Red Calcutta, Lal Kanak, Lal Pissi, Bansi Kathia, Khandwa, Malvi, *etc.* had their origin at Pusa. IARI, Pusa is one of the important centres in NEP zone which has developed many new high yielding varieties such as HP1102, HP1731 and HP1761 for timely sown irrigated condition while HP1209, HP1633, HP1744 for irrigated late sown condition of NEP zone. Due to introduction of HP1102 a variety developed from the cross 8156 (Blanko)/Nadadores63 occupied more than 60 percent area in Bihar. There were long queue to receive the seed of HP1102 a dwarf variety of that period. HP1209 is a variety for late sown cultivation and was released in 1976. It became very popular as more than 70 percent area of Bihar comes under late sown cultivation. Therefore, demand of the seed of HP1209 was very high and still it persists. HP1633 (Sonali) is grown under late sown condition. This is substitute of popular variety Sonalika, which is the first *Lr9* alien gene introduced variety in the country shows very good tolerance to foliar blight in addition to resistance against brown rust. HP1744 is an outstanding variety for chapatti quality and is most popular among farmers.

This Station has to its credit two major contributions in this category in recent times. Genetics of a scale for recording foliar blight disease caused by *Helminthosporium sativum* spp. Foliar blight has been one of the major diseases. Due to agro-climatic conditions of this part of the country is considered to be the '**Hot Spot**' for foliar blight. It was inferred that resistance to the disease is controlled by incomplete dominant major gene + minor genes in some Chinese and *Agropyron* transfer lines. Seedling resistance based on lesion size and type is controlled by dominant complementary genes. A new scale based on genetic parameters was worked out in place of the Saari-Prescot scale. This new scale useful for genetical studies at seedling and adult stages of the plant has been universally accepted as the two-digit scale. Increased seed rate along with reduced row spacing enhance wheat yield in this area.

Post-Independence Period

In 1950-51, work on wheat breeding was restarted after creation of new posts under **Wheat Rust Control Scheme**. During the sixties, three famous varieties of wheat, namely NP 835, NP 852 and NP 884 were released which are still grown in certain pockets under rainfed condition. In 1969, a new post was created under All India Coordinated Project on Wheat and Barley and **PL 480 Scheme** for identification of new sources of resistance to wheat rusts. In 1978, a high yielding variety of wheat, HP 1102, containing very high resistance to rusts was identified for North Eastern Plain Zone for timely sown conditions. This was quickly followed by another high yielding variety, HP 1209 for late sowings. Subsequently, other wheat varieties, such as HP 1493, HP 1633 (Sonali), HP 1731 (Raj Laxmi), HP 1744 (Rajeshwari) and HP 1761 (Jagdish) were released and these varieties are very popular among the cultivators. The second phase of wheat breeding may be stated to have commenced with the work of Dr B P Pal and Dr Ram Dhan Singh who added the dimensions of disease resistance and quality (particularly chapati making quality) in the breeding programme. In between 1950-51, three promising wheat varieties namely NP-835, NP-852 and NP-884 were developed which are excellent in chapati quality. It is chapati quality of NP-852 and 884, which attracted tremendously the farmers of Bihar and grew the temptation of wheat cultivation in each family. The more recent phase of wheat programme initiated in 1963 with the introduction of Norin-10 dwarfing gene with the generous help of Dr NE Borlaug. Dr Borlaug made the first prediction at his Kanpur visit that India can make his wheat production double or triple if high yielding varieties were used with improved technology. This prediction was proved true under the dynamic leadership of IARI when the country imported 18,000 tonnes of dwarf wheat from Mexico in 1966. This seed revolutionizes the wheat production of the country from 10.4 million tonnes to 11.4 million tonnes in

1966-67, 16.5 million tonnes in 1967-68, 18.6 million tonnes in 1968-69 and so on. At this juncture, a railway wagon loaded with Mexican dwarf wheat-seed could not be accepted by Bihar government and it was returned back from Mughalsarai. However, cultivation of dwarf wheat was popularized by IARI, Pusa to its contact farmers. IARI developed its first dwarf variety HP 1102 in 1976 for NEP zone from the crosses of 8156 (Blanco) and Nadadoses 63. It revolutionized the production of Bihar. No doubt, Sonalika and Kayansone were in cultivation practice but HP 1102 was proved superior to these two cultivars. In 1977, HP 1209 was introduced for cultivation under late sown irrigated condition with an average yield potential upto 33.3 q/ha. This variety is still popular in eastern states and in Nepal as Lumbini. In due course of time a good number of varieties were produced by IARI such as HP 1493, HP 1633, HP 1731, HP 1761, HP 1744, HD 2643, HW 2045, HW 2034, HDR 77 and WR 544. From IB/IR translocation lines, HD 2733 and HD 2824 were developed which are the most recent varieties with high yielding ability upto 55q/ha.

Coordinated Approach for Wheat Research in India

Wheat is the staple food and main source of energy in the Indian diet and has been under cultivation in the Indian subcontinent from ancient times. Though wheat research in India started long back during 20th century, the major jump in yield potential was realized after 1960 with the initiation of All India Coordinated Wheat Improvement Project (AICWIP) in 1961 by the Indian Council of Agricultural Research (ICAR). The project was launched formally in 1965 with the financial support by the centre to strengthen wheat research activities in the country. The project was elevated to the status of Directorate of Wheat Research in 1978. The Directorate was moved to the present location at Karnal in 1990, and with addition of Barley Network it was renamed as All India Coordinated Wheat and Barley Improvement Project (AICW and BIP). The establishment of the project was an important milestone in the history of wheat improvement programme in India to meet the demand of increasing population. This led to the development of dwarf wheat cultivars using the material brought by Dr. N.E. Borlaug, which were highly responsive to inputs, non-lodging types and possessed improved levels of disease resistance. The dwarf wheat varieties improved under the AICRIP (ICAR) could yield exceptionally high under better management production conditions and led to green revolution in India.

All India Co-ordinated Wheat Improvement Project (AICWIP) at IARI RS Pusa, Bihar

The mandates are to evolve, organize, coordinate and monitor multi-locational research for developing and identifying superior high yielding varieties tolerant to biotic and abiotic stresses under varied agro-climatic zones; collect, evaluate, catalogue, maintain and share working collections with a focus on identifying suitable donors for different stresses and quality attributes; develop appropriate crop production procedure and technology for wheat based cropping systems to optimize resource utilization in a sustainable manner; monitor the rust pathogen dynamics and develop technology to mitigate crop losses due to pests; establish national and international linkages for strengthening wheat and barley improvement programme; provide off-season nursery facility for rapid generation advancement and seed multiplication; serve as a core facility for proper data analysis and information management and coordinate and organize breeder seed production and technology transfer activities. The mandate of the Institute is to conduct basic and strategic research with a view to understanding the processes, in all their complexity, and to undertake need-based research, that lead to crop improvement and sustained agricultural productivity in harmony with the environment. The Regional Station, Pusa work as a service center and center of excellence for increasing production and productivity bridging the gap in farmer's field and seed production. The thrust in the areas of Crop improvement are to organize and monitor trials/nurseries for different production conditions; combine short duration, adaptation, blight resistance, heat tolerance for late planting, pre breeding aided with related species for yield components, quality traits, biotic and abiotic stresses; organize and produce breeder seed of recommended varieties. Improved barley and durum for domestic and industrial usage; augment, maintain and evaluate germplasm to identify and catalogue genetic resources; marker aided selection for crop improvement; develop technology to exploit heterosis and generate new variability through transgenics in addition to Resource management, Crop protection and health; Quality and basic sciences; Statistics and social sciences and the Service Functions. It would enable the trainees to correctly record disease reactions, various ancillary data and their reporting in data booklets. The broad areas covered were Constitution and conduction of wheat varietal trials; Recording of ancillary data under field conditions; Zonal monitoring of coordinated trials; Constitution, conduction and recording data in national nurseries; Conduction and recording of data in International trials/nurseries; Generation and interpretation of molecular data in AVT entries; Physiological trials; Germplasm evaluation: methodology and data recording; Breeder and nucleus seed production; Plant pathological nurseries and scoring disease reactions; Agronomy trials: conduction and data recording; Quality parameters: data recording and interpretation.

Enriching Wheat Germplasm through Materials Generation Programme and Conduct of Various Trials

Each year we are conducting various trials and are actively engaged in material generation programme. The list of trials being conducted at IARI RS Pusa, Bihar are being given in table 2007-2008. A total of two hundred ninety-nine cross combinations were being made among the different wheat breeding lines whose pedigree number ranges from 8563 to 8861. These crossed materials will be further evaluated during 2008-2009 rabi season. A system of allotment of HP (Hybrid Pusa) No has been adopted at IARI RS Pusa, Bihar to PS (Pusa Selection) No by earlier breeder at this station, which again was continued with more precision by **Initial Plant Pathological Screening Nursery (IPPSN).** One-hundred-gram seed of each entry for testing in different CVTs along with their parentage for testing in IPPSN. A **Preliminary Disease Screening Nursery** (PDSN) is being constituted and grown in Delhi, Karnal, Pusa, Dhaula Kuan, Wellington, Indore and Mahabaleshwar.

Table 3.1. Constitution of Crossing Block at IARI RS Pusa, Bihar

Sarangpur local	PDW 215	HI 8627
Motia	PDW 233	HI 8629
Trinakaria	PDW 245	HI 8651
Kathia 25	PDW 274	HI 8652
NP 404	Raj 1555	HI 8653
A 206	Raj 6562	HI 8654
A 9-30-1	Raj 6566	HI 8638
Guji 'S'	IWP 5007	HI 8639
Meghdoot	IWP 5061	HI 8645
Bij. Yellow	IWP 5070	HI 8647
Bij. Red	NP 200	HI 8661
NIDW 15	RS 749	HI 8662
Jairaj	CPAN 6236	HI 8663
B 662	CPAN 6201	HI 8664
IWP 5019	HD 4672	HI 8665
HG 110	HD 4676	HI 8666
Line 1172	HD 4692	HI 8667
P 6046	HD 4693	HI 8668
AKDW 3347	HD 4694	ID 312
ED 2398 A	HD 4695	ID 313
B 146	HD 4696	DBP 01-9
B 206	HD 4703	DBP 01-11
B 414	HI 7747	UAS 403
Marquis	HI 8381	UAS 404
EC 463473	HI 8498	Altar 84 (35th IDYN DS-3)
MACS 1967	HI 8550	HG 1250
MACS 2846	HI 8591	HG 1251
MACS 3125	HI 8596	HG 1261
MPO 215	HI 8620	HG 1282
WH 896	HI 8623	HG 1298
HG 1300	HG 1393	RS 847
HG 1300A	HG 1374 (WA)	AKDW 4151
HG 1313	HG 1374 (BA)	AKDW 4240
HG 1327	RD 935	D 18
HG 1353	RD 932	D 52
D 61	D 658	A 739
D 71	D 665	NIDW 9
D 83	Yuk	CPAN 6200
D 282	N 59	

Table 3.2. Crosses performed during 2006-07 crop season at ICAR-IARI RS Pusa, Bihar

Cross No.	*Pedigree*	*Cross No*	*Pedigree*
8487	HP 1761 x Raj 4101	8525	NW 3069 x HP 1761
8488	HP 1761 x NW 3069	8526	NW 3069 x Raj 4101
8489	PBW 533 x HP 1883	8527	NW 3069 x UP 2680
8490	PBW 533 x PBW 587	8528	NW 3069 x CBW 12
8491	PBW 533 x PBW 579	8529	NW 3069 x DBW 17
8492	PBW 533 x Raj 4101	8530	NW 3083 x HP 1883
8493	PBW 533 x DBW 17	8531	NW 3083 x PBW 579
8494	PBW 533 x NW 3083	8532	CBW 12 x Raj 4101
8495	PBW 587 x UP 2679	8533	CBW 12 x Raj 4120
8496	PBW 587 x NW 3069	8534	CBW 12 x NW 3069
8497	PBW 579 x Raj 4101	8535	DBW 14 x UP 2679
8498	PBW 579 x UP 2680	8536	DBW 14 x WH 1022
8499	PBW 579 x DBW 17	8537	DBW 22 x HP 1883
8500	PBW 579 x NW 3069	8538	DBW 22 x Raj 4101
8501	HD 2329 x Raj 4101	8539	DBW 22 x NW 3083
8502	HD 2329 x NW 3069	8540	DBW 22 x UP 2679
8503	HD 2329 x CBW 12	8541	DBW 22 x UP 2680
8504	HD 2643 x WH 1022	8542	DBW 22 x RNB 45
8505	HD 2643 x UP 2679	8543	HW 2045 x PBW 579
8506	HD 2733 x PBW 533	8544	HW 2045 x PBW 599
8507	HD 2733 x HP 1883	8545	HW 2045 x HD 2932
8508	HD 2733 x WH 1022	8546	HW 2045 x Raj 4101
8509	HD 2733 x HI 1539	8547	WH 1022 x PBW 533
8510	HD 2733 x DBW 17	8548	WH 1022 x UP 2679
8511	HD 2733 x NW 3083	8549	Raj 4101 x DBW 17
8512	HD 2839 x DBW 14	8550	Raj 4120 x Raj 4101
8513	HD 2932 x UP 2683	8551	Raj 4120 x HD 2643
8514	UP 2679 x Raj 4101	8552	Raj 4120 x K 306
8515	UP 2679 x Raj 4120	8553	Raj 4120 x UP 2679
8516	UP 2679 x NW 3069	8554	Raj 4120 x NW 3069
8517	UP 2679 x NW 3079	8555	C 7971 x PBW 587
8518	UP 2679 x CBW 12	8556	C 7971 x Raj 4101
8519	UP 2680 x Raj 4101	8557	NIG-I-S/NO 18 x HD 2733
8520	UP 2680 x PBW 579	8558	NIG-I-S/NO 18 x HD 2824
8521	UP 2680 x CBW 12	8559	NIG-I-S/NO 18 x PBW 533
8522	UP 2680 x CBW 17	8560	NIG-I-Fi-S/NO 6 x UP 2683
8523	UP 2683 x Raj 4101	8561	C 7973 x PBW 587
8524	UP 2683 x NW 3069	8562	C 7973 x Raj 4101

Table 3.3. Three way crosses performed during 2006-07 crop season at ICAR-IARI RS Pusa, Bihar

Cross No.	Pedigree	Cross No.	Pedigree
8763	UP 2679 x Raj 4101 x MP. 3224	8779	HW 2045 x PBW 579 x HUW. 612
8764	UP 2679 x Raj 4101 x HD. 2824	8780	HW 2045 x PBW 579 x PBW. 373
8765	UP 2679 x Raj 4101xMP. 3224	8781	HW 2045 x Raj 4101 x DBW. 52
8766	UP 2680 x Raj 4101 x DBW. 52	8782	HW 2045 x Raj 4101 x PBW. 373
8767	UP 2680 x Raj 4101 x HD. 2824	8783	WH 1022 x PBW 533 x CBW. 38
8768	UP 2680 x Raj 4101 x MP. 3224	8784	WH 1022 x PBW 533 x HUW. 612
8769	UP 2683 x Raj 4101 x WH. 1069	8785	Raj 4101 x DBW 17 x DBW. 52
8770	UP 2683 x Raj 4101 x HUW. 612	8786	Raj 4101 x DBW 17 x PBW. 373
8771	UP 2683 x Raj 4101 x CBW. 38	8787	Raj 4120 x Raj 4101 x CBW. 38
8772	UP 2683 x NW 3069 x HUW. 612	8788	Raj 4120 x Raj 4101 x PBW. 373
8773	UP 2683 x NW 3069 x CBW. 38	8789	Raj 4120 x Raj 4101 x HUW. 612
8774	UP 2683 x NW 3069 x WH. 1069	8790	Raj 4120 x HD 2643 x HUW. 612
8775	NW 3069 x Raj 4101 x HUW. 612	8791	Raj 4120 x HD 2643 x CBW. 38
8776	NW 3069 x Raj 4101 x WH. 1069	8792	Raj 4120 x NW 3069 x PBW. 373
8777	NW 3069 x Raj 4101 x CBW. 38	8793	Raj 4120 x NW 3069 x DBW. 52
8778	CBW 12 x Raj 4120 x HD. 2824	8794	Raj 4120 x NW 3069 x HUW. 612

Table 3.4. Crosses performed during 2007-08 crop season at ICAR-IARI RS Pusa, Bihar

Cross No.	Variety	Cross No.	Variety
8565	DBW 52 x PBW 574	8598	K.701 x HI. 1558
8566	DBW 52 x HI. 1558	8599	K.701 x PBW. 533
8567	DBW 52 x RAJ 4120	8600	K.701 x PBW. 574
8568	DBW 52 x HD. 2733	8601	K.701 x RAJ. 4120
8569	DBW 52 x HUW. 612	8602	DBW. 28 x DBW. 46
8570	DBW 52 x CBW. 38	8603	DBW. 28 x NW. 4060
8571	DBW 52 x HD. 2824	8604	DBW. 28 x HI. 1558
8572	CBW.38 x PBW. 533	8605	DBW. 28 x HUW. 612
8573	CBW.38 x NW. 4060	8606	DBW. 28 x CBW. 38
8574	CBW.38 x HP. 1761	8607	DBW. 28 x DBW. 52
8575	CBW.38 x HI. 1558	8608	DBW. 28 x RSP. 529
8576	CBW.38 x HD.2733	8609	DBW. 28 x RAJ. 4120
8577	CBW.38 x RAJ. 4120	8610	DBW. 39 x DBW. 52
8578	CBW.38 x DBW. 46	8611	DBW. 39 x HD. 2733
8579	CBW.38 x PBW. 574	8612	DBW. 39 x CBW. 38
8580	CBW.38 x K. 702	8613	DBW. 39 x UAS. 307
8581	FLW. 8 x RAJ. 4120	8614	DBW. 39 x RAJ. 4166
8582	FLW. 8 x UAS. 307	8615	GW. 404 x NW. 4046

Cross No.	*Variety*	*Cross No.*	*Variety*
8583	FLW. 8 x HI. 1558	8616	GW. 404 x K. 704
8584	CBW.17 x HUW. 612	8617	GW. 404 x RSP. 505
8585	CBW.17 x HI. 1558	8618	GW. 404 x UAS. 307
8586	CBW.17 x PBW. 373	8619	RSP. 505 x RAJ. 4170
8587	CBW.17 x DBW. 52	8620	RSP. 505 x RAJ. 4166
8588	RAJ. 4120 x PBW. 533	8621	RSP. 505 x HI. 1558
8589	RAJ. 4120 x PBW. 574	8622	PHS. 623 x HI. 1558
8590	RAJ. 4120 x HI. 1558	8623	PHS. 623 x UAS. 307
8591	DBW. 38 x NW. 4060	8624	PHS. 623 x RAJ. 4120
8592	DBW. 38 x HUW. 612	8625	DBW.31xRAJ. 4166
8593	DBW. 38 x HI. 1558	8626	UAS. 307xWH. 1046
8594	DBW. 38 x RAJ. 4120	8627	UAS. 307xDBW.31
8595	DBW. 38 x CBW. 38	8628	RSP. 529 x UAS. 307
8596	DBW. 38 x RSP. 529	8629	RSP. 529 x RAJ. 4166
8597	DBW. 38 x DBW. 52	8630	RSP. 529 x PBW. 574
8631	RSP. 529 x PBW. 533	8664	PHR. 203 x DBW. 46
8632	RSP. 529 x RAJ. 4170	8665	PHR. 203 x CBW. 38
8633	RSP. 529 x HI. 1558	8666	PHR. 203 x DBW. 52
8634	HD. 2733 x CBW. 38	8667	PHR. 1005 x DBW. 31
8635	HD. 2733 x HI.1558	8668	PHR. 1005 x WH. 1046
8636	RSP. 561 x DBW. 31	8669	PHR. 1005 x RAJ. 4125
8637	RSP. 561 x DBW. 52	8670	PHR. 1006 x HUW. 612
8638	RSP. 561 x RAJ. 4125	8671	PHR. 1006 x HD. 2824
8639	RSP. 561 x DBW. 38	8672	PHR. 1006 x PBW. 533
8640	RSP. 561 x DBW. 18	8673	PHR. 1006 x HD. 2733
8641	RSP. 561 x WH. 1046	8674	PHR. 1006 x CBW. 38
8642	HD. 2967 x CBW. 38	8675	PHR. 1009 x PBW. 533
8643	HD. 2967 x DBW. 46	8676	PHR. 1009 x HD. 2967
8644	HD. 2967 x DBW. 52	8677	PHR. 1009 x HI. 1558
8645	HUW. 612 x PBW. 373	8678	PHR. 1009 x CBW. 38
8646	HUW. 612 x CBW.38	8679	PHR. 1009 x DBW. 52
8647	HUW. 612 x PHS. 623	8680	PHR. 1009 x WH. 1067
8648	HUW. 612 x HD. 2824	8681	PHR. 1014 x RAJ. 4166
8649	HUW. 612 x DBW. 52	8682	PHR. 1014 x RAJ. 4170
8650	HUW. 612 x RAJ. 4166	8683	PHR. 1014 x PBW. 574
8651	UP. 2736 x RAJ. 4120	8684	PHR. 1014 x PBW. 533
8652	UP. 2736 x UAS. 307	8685	PHR. 1014 x PBW. 612
8653	UP. 2736 x HI. 1558	8686	PHR. 1014 x HI. 1558
8654	HD. 3005 x PBW.533	8687	PHR. 1015 x HD. 2824
8655	HD. 3005 x HI. 1558	8688	PHR. 1015 x HUW. 612

Cross No.	Variety	Cross No.	Variety
8656	HD. 3005 x PBW. 607	8689	PHR. 1015 x CBW. 38
8657	HUW. 624 x RAJ. 4120	8690	PHR. 1015 x HD. 2733
8658	HUW. 624 x HI. 1558	8691	PHR. 1015 x PBW. 533
8659	HUW. 624 x PBW. 607	8692	PHR. 1022 x DBW. 52
8660	HD. 3000 x UAS. 307	8693	PHR. 1022 x PBW. 533
8661	HD. 3000 x RAJ. 4120	8694	PHR. 1022 x CBW. 38
8662	HD. 3000 x PBW. 607	8695	PHR. 1022 x HD. 2733
8663	HD. 3000 x HI. 1558	8696	PHR. 1022 x HUW. 612

Table 3.5. Three way crosses performed during 2007-08 crop season at ICAR-IARI RS Pusa, Bihar

Cross No.	Variety	Cross No.	Variety
8697	PHR.1022 x DBW. 46	8730	PHR. 1015 x HD. 2733 x PBW. 373
8698	PHR.1022 x HD. 2824	8731	PHR. 1015 x HD. 2733 x DBW. 52
8699	PHR.1025 x K. 704	8732	PHR. 1015 x HD. 2733 x K. 701
8700	PHR.1025 x K. 710	8733	PBW 533 x PBW 579 x DBW. 52
8701	PHR.1025 x RSP. 505	8734	PBW 533 x PBW 579 x PBW. 607
8702	PHR.1025 x NW. 4046	8735	PBW 533 x PBW 579 x WH. 1069
8703	PHR.1025 x GW. 404	8736	PBW 533 x Raj 4101 x WH.1069
8704	PHR.1025 x K. 707	8737	PBW 533 x Raj 4101 x PBW. 607
8705	PHR.1031 x DBW. 52	8738	PBW 533 x Raj 4101 x DBW. 52
8706	PHR.1031 x DBW. 38	8739	PBW 533 x NW 3083 x DBW. 52
8707	PHR.1032 x HUW.612	8740	PBW 533 x NW 3083 x PBW. 373
8708	PHR.1032 x DBW. 52	8741	PBW 533 x NW 3083 x K. 701
8709	PHR.1032 x DBW. 38	8742	PBW 587 x NW 3069 x PBW. 373
8710	PHR.1032 x HD. 2967	8743	PBW 587 x NW 3069 x DBW. 52
8711	PHR.1032 x PBW. 373	8744	PBW 587 x NW 3069 x K. 701
8712	PHR.1032 x HI. 1558	8745	PBW 579 x Raj 4101 x HUW. 612
8713	MIYCSN. 16 x PHS.623	8746	PBW 579 x Raj 4101 x K. 701
8714	MIYCSN. 16 x PHR.1032	8747	PBW 579 x Raj 4101 x WH. 1069
8715	MIYCSN. 17 x PHS. 623	8748	PBW 579 x UP 2680 x HD. 2824
8716	MIYCSN. 17 x PHR. 1032	8749	PBW 579 x UP 2680 x PBW. 373
8717	MIYCSN. 17 x PHR. 1009	8750	PBW 579 x UP 2680 x HD. 3009
8718	MIYCSN.19 x PHR.1032	8751	PBW 579 x DBW 17 x PBW. 373
8719	MIYCSN.19 x PHR. 1009	8752	PBW 579 x DBW 17 x HD. 3009
8720	MIYCSN.19 x PHS. 623	8753	PBW 579 x DBW 17 x HD. 2824
8721	MIYCSN.31 x PHR.1009	8754	HD 2329 x Raj 4101 x PBW. 373
8722	MIYCSN.31 x PHR. 1032	8755	HD 2329 x NW 3069 x HD. 2824
8723	MIYCSN.31 x PHS. 623	8756	HD 2329 x NW 3069 x PBW. 373

Cross No.	Variety	Cross No.	Variety
8724	NI. 5439 x GW. 404	8757	HD 2329 x NW 3069 x HD. 3009
8725	NI. 5439 x NW. 4046	8758	HD 2839 x DBW 14 x PHS. 622
8726	NI. 5439 x K. 704	8759	HD 2839 x DBW 14 x MP. 3224
8727	HP 1761 x Raj 4101 x WH. 1069	8760	HD 2839 x DBW 14 x HD. 2824
8728	HP 1761 x Raj 4101 x PBW. 607	8761	HD 2932 x UP 2683 x DBW. 52
8729	HP 1761 x Raj 4101 x DBW. 52	8762	HD 2932 x UP 2683 x HD. 2824

Table 3.6. F_1's developed and evaluated during 2006-07 at ICAR-IARI RS Pusa, Bihar

Cross	H. No.	Cross	H. No.
DBP 01-9 x PDW 233	H 9518	ID 35 x Raj 6566	H 9570
DBP 01-9 x Raj 6566	H 9519	ID 36 x HD 4693	H 9571
DBP 01-11 x HI 8498	H 9520	ID 36 x PDW 274	H 9572
DBP 01-11 x Raj 6566	H 9521	ID 37 x HD 4676	H 9573
HD 4676 x HI 8634	H 9522	ID 37 x HD 4693	H 9574
HD 4676 x PDW 233	H 9523	ID 38 x HI 8498	H 9575
HD 4685 x HI 8634	H 9524	ID 38 x Raj 1555	H 9576
HD 4685 x WH 896	H 9525	ID 39 x C 44	H 9577
HD 4694 x WH 896	H 9526	ID 39 x V 21-4	H 9578
HI 8550 x HI 8634	H 9527	ID 310 x HI 8498	H 9579
HI 8550 x MACS 3125	H 9528	ID 310 x V 21-4	H 9580
HI 8591 x HD 4676	H 9529	ID 311 x HI 8498	H 9581
HI 8591 x HD 4694	H 9530	ID 311 x Raj 1555	H 9582
HI 8596 x HD 4676	H 9531	ID 312 x HI 8381	H 9583
HI 8596 x HD 4694	H 9532	ID 312 x HI 8498	H 9584
HI 8620 x HI 8381	H 9533	ID 313 x HI 8498	H 9585
HI 8620 x HI 8498	H 9534	ID 313 x Raj 1555	H 9586
HI 8620 x HI 8591	H 9535	ID 314 x C 44	H 9587
HI 8620 x HD 4708	H 9536	ID 314 x HI 8381	H 9588
HI 8622 x PDW 274	H 9637	ID 315 x HI 8381	H 9589
HI 8622 x Raj 6566	H 9538	ID 315 x HI 8498	H 9590
HI 8636 x HI 8498	H 9539	ID 315 x Raj 1555	H 9591
HI 8636 x HD 4672	H 9540	ID 316 x HI 8498	H 9592
HI 8636 x V 21-4	H 9541	ID 316 x Raj 1555	H 9593
HI 8637 x HD 4672	H 9542	ID 317 x HG 110	H 9594
HI 8637 x A 9-30-1	H 9543	ID 318 x HI 8498	H 9595
HI 8639 x A 9-30-1	H 9544	ID 318 x Raj 1555	H 9596
HI 8639 x C 44	H 9545	ID 319 x HI 8498	H 9597
HI 8639 x HD 4672	H 9546	ID 319 x Raj 1555	H 9598
HI 8640 x HI 8627	H 9547	ID 323 x HD 4672	H 9599

Cross	*H. No.*	*Cross*	*H. No.*
HI 8640 x HD 4672	H 9548	ID 324 x A 9-30-1	H 9600
HI 8641 x A 9-30-1	H 9549	ID 324 x HD 4672	H 9601
HI 8641 x C 44	H 9550	ID 329 x HI 8627	H 9602
HI 8641 x HI 8627	H 9551	ID 329 x HD 4672	H 9603
HI 8652 x Trinakaria	H 9552	ID 330 x A 9-30-1	H 9604
HI 8653 x HG 110	H 9553	ID 330 x HD 4672	H 9605
HI 8653 x HI 8591	H 9554	ID 333 x HI 8627	H 9606
HI 8653 x P 6046	H 9555	ID 333 x HD 4672	H 9607
HI 8653 x Trinakaria	H 9556	Kathia 25 x HI 8627	H 9608
HI 8654 x PDW 245	H 9557	Kathia 25 x HI 8638	H 9609
HI 8656 (Buttah) x HI 8627	H 9558	MACS 1967 x A 9-30-1	H 9610
HI 8656 x MACS 1967	H 9559	MACS 1967 x HI 8627	H 9611
ID 31 x HI 8498	H 9560	MACS 1967 x HD 4672	H 9612
ID 31 x Raj 1555	H 9561	MACS 1967 x ID 327	H 9613
ID 32 x HI 8381	H 9562	MACS 1967 x RS 886	H 9614
ID 32 x HI 8498	H 9563	Malavraj x P 6046	H 9615
ID 32 x Raj 1555	H 9564	NP 404 x HI 8638	H 9616
ID 33 x HI 8620	H 9565	PDW 233 x HI 8591	H 9617
ID 33 x PDW 233	H 9566	PDW 245 x WH 896	H 9618
ID 34 x HD 4693	H 9567	PDW 245 x DBP 01-11	H 9619
ID 34 x PDW 274	H 9568	PDW 274 x B 662	H 9620
ID 35 x PDW 233	H 9569	PDW 274 x HD 4692	H 9621

Table 3.7. Station Breeding Programme, Screening Nurseries; Coordinated Trials at Institute Level; Coordinated Trials at Directorate Level being executed during the crop season (2006-2007)

Station Breeding Programme			
1	Hybridization programme	102	Hybridization programme was performed to create the new genetic variability for further use in the breeding programme
1a	Hybrid F_1 seeds	500 [x 1 Rep]	Breeding material generation advanced testing
1b	Hybrid F_2 seeds	1500 [x 1 Rep]	Breeding material generation advanced testing
1c	Hybrid F_3 seeds	1210 [x 1 Rep]	Breeding material generation advanced testing
1d	Hybrid F_4 seeds	1108 [x 1 Rep]	Breeding material generation advanced testing
2	Seed multiplication		To supply the different testing trials conducted all over India
3	Crossing block	273 [x 1 Rep]	Maintenance and purposeful utilization for crossing
4	STHF-I	36 [x 3 Rep]	Station Trial High Fertility
5	STHF-II	36 [x 3 Rep]	Station Trial High Fertility
6	STHF-III	36 [x 3 Rep]	Station Trial High Fertility
7	STHF-IV	36 [x 3 Rep]	Station Trial High Fertility

Screening Nurseries			
8	IPPSN	1520	Initial Plant Pathological Screening Nursery
9	LBSN	118 [x 1 Rep]	Leaf Blight Screening Nursery
10	NGSN	117 [x 1 Rep]	National Genetic Stock Nursery
11	PDSN	632	Plant Disease Screening Nursery
12	TPN	20[x 1 Rep]	Trap Plot Nursery
13	SASCN	13	Salinity Alkalinity Screening Nursery
14	WSNTSAC	150 [x 1 Rep]	Wheat Screening Nursery Tolerant to Salinity/Alkalinity condition
15	YCSN	94 [x 1 Rep]	Yield Component Screening Nursery
16	11th EGPSN	150	11th Eastern Gangetic Plains Screening Nursery
17	39th IBWSN	1001-1174	39th International Bread Wheat Screening Nursery
18	15th SAWYT	80 [x 2 Rep]	15th Semi-Arid Wheat Yield Trial (Biofortification Trial)
19	EIGN	164 [x 1 Rep]	Elite International Genetic Nursery
Coordinated Trials at Institute Level			
20	CVT-1A	36 [x 3 Rep]	Common Varietal Trials
21	CVT-1B	36 [x 3 Rep]	Common Varietal Trials
22	CVT-3	49 [x 3 Rep]	Common Varietal Trials
23	CVT-5A(Res-IR)	36 [x 3 Rep]	Common Varietal Trials (Restricted Irrigation)
24	CVT-5A(RF)	36 [x 3 Rep]	Common Varietal Trials (Rainfed)
Coordinated Trials at Directorate Level			
25	NIVT-1A	49 [x 2 Rep]	National Initial Varietal Trials
26	NIVT-1B	49 [x 2 Rep]	National Initial Varietal Trials
27	NIVT-2	49 [x 2 Rep]	
28	NIVT-3	49 [x 2 Rep]	National Initial Varietal Trials
29	NIVT-4	49 [x 2 Rep]	National Initial Varietal Trials
30	NIVT-5A	49 [x 2 Rep]	National Initial Varietal Trials
31	NIVT-5B	49 [x 2 Rep]	National Initial Varietal Trials
32	NIVT-6	49 [x 2 Rep]	National Initial Varietal Trials
33	NIVT-7	49 [x 2 Rep]	National Initial Varietal Trials
34	AVT(LS-IR)	9 [x 4 Rep]	Advanced Varietal Trials (Late Sown- Irrigated)
35	AVT(TS-IR)	13 [x 4 Rep]	Advanced Varietal Trials (Timely Sown- Irrigated)
36	AVT(TS-RF)	11 [x 4 Rep]	Advanced Varietal Trials (Timely Sown- Rainfed)

Table 3.8. Entries promoted during crop season 2006-2007 at ICAR-IARI RS Pusa, Bihar

	PS	*HP No.*	*Pedigree*
NIVT-1A	PS-834	HP -1915	K9330/UP2338//1843
	PS-837	HP -1916	Raj4004/HD2652
NIVT-1B	PS-841	HP -1917	UP2532/DL7882//CPAN3048
	PS-842	HP -1918	UP2338/HW3034//PBW532

	PS	*HP No.*	*Pedigree*
NIVT-3 NWPZ	PS-862	HP -1919	PBW524//HUW516/Raj3814
NEPZ	PS-855	HP-1920	HP1744/BL2053
IVT-SHZ	PS-870	HP-1921	RL6043/Raj3765
	PS-859	HP-1924	PBW483/WR544

Table 3.9. Presence of Sr, Lr and Yr genes in the following wheat varieties

Variety	*Sr Genes*	*Lr Genes*	*Yr Genes*
PBW 343	2+5+31+	26+	9+27+
PBW 502	2+5+31+	26+	9+
HD 2687	2+31+	26+23+	9+
DBW17	31+	26+23+	9+
PBW 550	31+	26+34	9+18+
WH 542	31+	26+34+	9+18+
Raj 3765	2+	10+13+	2KS+
HD 2967	11+	13+	-
HS 295	2+8b+	23+34+	2+18+
HS 490	2+9b+	23+	A+
HW 2044	2+24+	24+	2+
VL 804	5+31+	26+34+	9+18+
VL 892	-	13+10+	-
WH 896	9e+	-	-
HI 1500	2+24+	24+	A+
HD 2733	31+	26+34+	9+18+
HD 2781	2+	-	-
HD 4672	11+	23+	-
Hw 1085	2+31+	26+	9+

Source: AICWBIP Progress Report, Crop Protection for 2011-12, 2009-10, 2003-04, 2002-03 and 2000-01. DWR, Karnal.

Table 3.10. Cultivar being included in Crossing Block for Material Generation

Aestivum Wheat	*Durum Wheat*
CPAN 1235, CPAN 1696, CPAN 1842, GW 322, GW 326, HD 2135, HD 2380, HD 2402, HD 2501, HD 2733, HI 784, HI 1418, HI 1434, HI 1479, HP 1731, HP 1761, HS 295, HS 365, K 7827, K 9107, K 9465, NIAW 34, PBW 498, Raj 3765, Raj 3777, UP 301, UP 2003, UP 2390, UP 2473, VL 804, WH 542, Pissi Local, Sonalika, DWR 241, HD 2668, HD 2669, HI 1404, HI 1459, HP 1744, HP 1748, HP 1749, HP 1807, HP 1831, HPW 82, HPW 147,	AKDW 3347, AKDW 4256, Bijaga Red, Castel Porziano, DBP 02-08, ED 2398-A, Guji 'S', GW 1114, GW 1170, GW 1209, Local Red, HD 4502, HD 4530, HD 4672, HD 4676, HD 4696, HI 8381, HI 8498, HI 8592, HI 8653, IWP 5019, MACS 3061, MACS 3063, MPO 615, NIDW 9, NIDW 295, PDW 289, Raj 6516, Raj 6562, RD 930, Local Red, RS 749, Trinakaria, UAS 2021, UP 45, VD 97-15, WG 7143, Yuk

Aestivum Wheat	*Durum Wheat*
HPW 157, HPW 160, HS 364, HUW 482, HUW 507, HUW 524, JWJ 3016, K 9223, K 9527, K 9533, NW 1067, PBW 426, PBW 442, PBW 452, Raj 3897, Raj 3900, Raj 3965, Raj 3974, UP 2418, VL 775, VL 791, VL 801, VL 809, HW 3021, HW 3024, HW 3025, HW 3063, HW 3067, HW 3068, HW 3070, HW 3072, HW 3080, HW 3081, HW 4011, HW 4028, J 98-20, HS 365, HW 3014, GW 295, HS 365, HD 2705, HD 2716, HPW 162, Raj 3777, HPW 162, HS 396, PDSN 98, PDSN 983, PDSN 919, PDSN 909, PDSN 979, PDSN 918, PDSN 951, PDSN 1024, PDSN 956, PDSN 945, PDSN 905, PDSN 908, PDSN 961, PDSN 921, Raj 4055, NIAW 907, PATH 17, JWS 132, WR 678, UP 36, BW 1765, Frontana, HUW 468	

Table 3.11. Entries being tested during 2007-2008 at ICAR-IARI RS Pusa, Bihar

CVT-1A		*CVT-1B*		*CVT -3*	
Entries	*Pedigree*	*Entries*	*Pedigree*	PS-899	CBW 14/K 2008
PS-877	GW 273/RW 3482//HD 2733	PS-888	PBW 343/PBW 506	PS-900	HP 1761/WH 595//GW 373
PS-878	HD 2733/CDWR 9522	PS-889	HW 2045/GW 173	PS-901	HD 2733/HUW 234// HP 1744
PS-879	HP 1731/UP 2532	PS-890	HD 2733/HW 3029	PS-902	HW 3047/UP 2382// GW 373
PS-880	PBW 343/HD 2687	PS-891	PBW 343/HD 2687	PS-903	HW 2039/DBW14//WH 423
PS-881	PHR 1006/UP 2592	PS-892	HD 2733/CBW 17	PS-904	UP 2584/VEE/LIRA// BOW/3/BEN/4/KAUZ
PS-882	HD 2733/GW 273//Sonata	PS-893	HUW 468RAJ 3765	PS-905	RW 3482/PBW 502// SONALI
PS-883	PBW 343/K 9107// PBW 532	PS-894	PBW 226//HD 2687	PS-906	CBW 13/PBW 503
PS-884	PBW 343/DL-153-2//HW 2045	PS-895	PBW 343/HD 2687	PS-907	HP 1856/WH 765
PS-885	HD 2733/HP 1744//HUW 234	PS-896	HD 2733/WH 736//HD 2285	PS-908	RW 3482/UP 2592// PBW 343
PS-886	PBW 343/HW 2062	PS-897	HW 2045/ SITTU//PGO/ SEVI//DBW 14	PS-909	HD 2733/WH 736// PBW 373
PS-887	HW 3039/HD 2826	PS-898	Sonalika/HW 2008//HD 2795	PS-910	HW 2045/PBW 502

Table 3.12. Entries being evaluated in CVT 5A RF and RI, 2008-09 in three replications

Entries					
RS-1028	IND-230	RS-1030	IND-229	P-1	C-306(C)
WBM 2213	P-2	IND-222	P-3	RS-1025	P-5
IND 221	RS-1024	P-4	RS-1031	RS-1029	WBM-2118
IND 225	DW-1456	PBW-175(C)	RS-1026	IND-223	P-7
IND 228	IND-224	P-6	IND-220	RS-1027	DL-1162
RS-1023	RS-1022	IND-226	IND-227	K-8027(C)	NI-5439(C)

Table 3.13. Entries being tested during Advanced Varietal Trial (Timely Sown- Irrigated)

Sl.No.	*Variety*	*Sl.No.*	*Variety*	*Sl.No.*	*Variety*	*Sl.No.*	*Variety*
1	DBW 39	23	PBW 618	45	PBW 617	67	CBW 38 (I)
2	RSP 561	24	HD 2997	46	RW 3673	68	RAJ 4120 (I)
3	RAJ 4120 (I)	25	HD 2967	47	CBW 38 (I)	69	DBW 39
4	DBW 46	26	PBW 617	48	K 0707	70	K 0307 (C)
5	K 0708	27	K 0708	49	HD 2998	71	K 0607
6	PBW 343 (C)	28	RW 3673	50	HUW 612	72	RW 3673
7	CBW 38 (I)	29	K 0707	51	K 0607	73	PBW 618
8	HUW 612	30	DBW 39	52	HD 3003	74	HD 2967
9	HUW 468 (C)	31	DBW 46	53	HD 3002	75	K 0707
10	HD 2997	32	HUW 468 (C)	54	RAJ 4120 (I)	76	HD 2997
11	K 0607	33	HD 3003	55	HD 2997	77	PBW 343 (C)
12	HUW 616	34	PBW 343 (C)	56	DBW 46	78	HUW 620
13	PBW 618	35	HUW 612	57	HD 2967	79	DBW 46
14	K 0307 (C)	36	HD 3002	58	PBW 618	80	HUW 616
15	HD 2967	37	HUW 616	59	HUW 468 (C)	81	RSP 561
16	HD 3003	38	HUW 620	60	HUW 620	82	HUW 612
17	K 0707	39	RAJ 4120 (I)	61	RSP 561	83	HUW 468 (C)
18	HUW 620	40	HD 2998	62	HUW 616	84	HD 3002
19	HD 3002	41	RSP 561	63	K 0307 (C)	85	HD 3003
20	RW 3673	42	K 0607	64	K 0708	86	K 0708
21	HD 2998	43	CBW 38 (I)	65	DBW 39	87	PBW 617
22	PBW 617	44	K 0307 (C)	66	PBW 343 (C)	88	HD 2998

Table 3.14. Entries being tested during Advanced Varietal Trial (Late Sown- Irrigated)

Sl.No.	*Variety*	*Sl.No.*	*Variety*	*Sl.No.*	*Variety*	*Sl.No.*	*Variety*
Sl.No.	*Variety*	*Sl.No.*	*Variety*	*Sl.No.*	*Variety*	*Sl.No.*	*Variety*
1	DBW 14 (C)	18	HD 2982	35	HUW 234 (C)	52	NW 4035
2	K 0710	19	DBW 52	36	DBW 51	53	K 0711
3	NW 2036 (C)	20	DBW 51	37	MP 3224	54	K 0710
4	DBW 51	21	RAJ 4176	38	K 0710	55	HUW 234 (C)
5	WH 1078	22	HI 1563	39	K 0711	56	WH 1078
6	WH 1076	23	PBW 625	40	HI 1563	57	RAJ 4176
7	HD 2982	24	WH 1076	41	NW 2036 (C)	58	DBW 52
8	K 0711	25	WH 1078	42	PBW 625	59	DBW 51
9	RAJ 4176	26	HUW 234 (C)	43	HD 2985	60	HD 2982
10	PBW 625	27	K 0710	44	HD 2983	61	HD 2985
11	HD 2985	28	MP 3224	45	RAJ 4176	62	DBW 14 (C)
12	MP 3224	29	HD 2985	46	HD 2982	63	PBW 625
13	HI 1563	30	HD 2983	47	DBW 52	64	HD 2983
14	DBW 52	31	NW 2036 (C)	48	DBW 14 (C)	65	HI 1563
15	NW 4035	32	NW 4035	49	NW 4035	66	MP 3224
16	HUW 234 (C)	33	K 0711	50	WH 1076	67	NW 2036 (C)
17	HD 2983	34	DBW 14 (C)	51	WH 1078	68	WH 1076

Table 3.15. Entries being tested during Advanced Varietal Trial (Timely Sown-Rainfed)

Variety	*Variety*	*Variety*	*Variety*
HD 3016	PBW 612*	C 306 (C)	K 0616*
HD 2888 (C)	C 306 (C)	HUW 629	C 306 (C)
K 0716	K 0616*	HD 2888 (C)	K 0617*
PBW 612*	HUW 629	HD 3016	HD 2888 (C)
K 8027 (C)	K 0617*	K 8027 (C)	HUW 629
HUW 629	K 8027 (C)	K 0617*	K 8027 (C)
K 0617*	HD 3016	PBW 612*	HD 3016
C 306 (C)	HD 2888 (C)	K 0716	K 0716
K 0616*	K 0716	K 0616*	PBW 612*

Selection Criteria and Constitution of Different Trials

A set of selected nurseries during the onset of cropping season for evaluation and screening are being planted each year. Several rounds of selections at various crop growth stages are made by a multidisciplinary team comprising of breeders and pathologists. The elite material thus selected is being provided to the active wheat breeding centers across the country in the form of **Elite International Germplasm**

Nurseries (EIGN-I for bread wheat and EIGN-II for durum wheat). The disease resistance level of these selected lines was confirmed in the summer nurseries at Dalang Maidan (Lahaul and Spiti) and Wellington, Nilgiri Hills (TN) before their inclusion in EIGN. The lines in EIGN-I and BIGN-II had a good blend of adaptation, disease resistance and yield contributing traits. The distribution of these nurseries are made in such a way that about 4-5 centres are included from each wheat growing zone,getting exposed to the maximum diverse agroclimatic conditions. In addition, the durum wheat (EIGN-II) was distributed to only such centers where durum improvement work is in progress. The material supplied through BIGN-I and II was evaluated at respective centres in an augmented design having two row plot of 2.5 m length spaced at 23 cm apart by raising them under recommended agronomic practices. The crop was grown under natural conditions and no artificial inoculation was made. The data were recorded for days to heading, plant height, maturity, grains 1spike, 1000-grain weight and grain yield. In addition, it was also evaluated against rusts and leaf blight under natural conditions and the necessary observations on disease were at various crop growth stages. The feedback reports suggested that most of the active centres have utilized elite materials for various purposes *viz.*, hybridization, direct testing and as genetic stock for future use.

Table 3.16. Entries being tested during Station Trial (Rainfed) during 2008-2009

Variety	*Variety*	*Variety*
PS- 899	3303/5901	3654/5343
PS- 900	3636/5293	PS 899
PS- 907	2445/5763	2331/5631
PS- 908	PS- 907	1605-6/6416
PS- 910	K. 8027	2376/5694
PS- 911	PS- 908	PS-900
HP- 1911	HP- 1915	PS-964
HP- 1915	2388/5712	HP-1915
4403/4729	4403/4729	PS-907
3636/5293	3654/5343	4403/4729
3654/5343	PS- 899	K-8027
2331/5681	2376/5694	3636/5293
2376/5694	PS- 911	PS-908
2388/5712	4472/5714	C-306
4472/5714	PS- 910	2388/5712
2445/5763	1605-6/6416	PS-910
3303/5901	PS- 964	4472/5714
1605-6/6416	PS- 900	P-911
PS- 964	C- 306	2445/5763
K. 8027	PS- 911	HP-1911
C. 306	2331/5681	3303/5901

Table 3.17. Elite crosses advanced for advanced generations at IARI RS Pusa, Bihar and IARI RS Indore, M P

Crosses Performed at IARI RS Pusa, Bihar	
HP 1894: PBW 343/HP 1761//HW 2045	HP 1896: HD 2733/HP 1761
HP 1903: PBW 343/Raj 3765//HW 2045	
HP 1907: HUW 516/HD 2643	HP 1908: PBW 343/HD 2733
HP 1909: Site/Mo/4/Nac./Th.Ac//3*Pvn/3/Mirlo/ Bue	HP 1910: Turaco/2*Borl. 95
HP 1911: PBW 343/Raj 3777//WH 736	HP 1912: HUW 516/HD 2643
HP 1913: Raj 3765/GW 322	HP 1914: HD 2733/GW 273//Sonata
HP 1915: HD 2733/RI 6043	

Table 3.18. HP No allotted to PS series from Pusa, Bihar (2008-2009)

PS No	*HP No*	*PS No*	*HP No*	*PS No*	*HP No*
PS930	HP 1929	PS941	HP 1938	PS953	HP 1947
PS933	HP 1930	PS942	HP 1939	PS956	HP 1948
PS934	HP 1931	PS943	HP 1940	PS958	HP 1949
PS935	HP 1932	PS944	HP 1941	PS960	HP 1950
PS936	HP 1933	PS946	HP 1942	PS972	HP 1951
PS937	HP 1934	PS947	HP 1943	PS974	HP 1952
PS938	HP 1935	PS948	HP 1944	PS975	HP 1953
PS939	HP 1936	PS949	HP 1945		
PS940	HP 1937	PS950	HP 1946		

Table 3.19. Entries promoted during 2006-2007

	PS	*HP No*	*Pedigree*
NIVT-1A	PS-834	HP -1915	K9330/UP2338//1843
	PS-837	HP -1916	Raj4004/HD2652
NIVT-1B	PS-841	HP -1917	UP2532/DL7882//CPAN3048
	PS-842	HP -1918	UP2338/HW3034//PBW532
NIVT-3 NWPZ	PS-862	HP -1919	PBW524//HUW516/Raj3814
NEPZ	PS-855	HP-1920	HP1744/BL2053
IVT-SHZ	PS-870 PS-859	HP-1921 HP-1924	RL6043/Raj3765 PBW483/WR544

Table 3.20. Entries being tested/evaluated/promoted during rabi 2008-2009 at IARI RS Pusa, Bihar

NIVT-1B	HP 1925	PBW 226/Lr9//HD 2687
NIVT-3	HP 1926	CBW 13/PBW 503
NIVT-5A	HP 1928	HD 2865//HI5439//PBW 175
	HP 1927	C306//HD-2685//UP2691
IR-LS-TAS-DOS-PZ trial	HP 1913	Raj -3765 x GW-322
CVT-2	PS 969	DBW 14 x PBW 531
	PS 970	PS 668 x 10TH HTWYT.Acc.No.2
	PS 971	SUNC0/2*PASTOR
	PS 972	RW 3482/20TH ESWYT-22/HW 2043
CVT-6	PS 973	SUNC0/2*PASTOR
CVT-7	PS 974	HP 1731/UP2529/HD 2687
	PS 975	DBW 14 x PBW 531
QCSN	HP 1913	Raj -3765 x GW-322
	HP 1925	PBW 226/Lr9//HD 2687
	HP 1926	CBW 13/PBW 503
	HP 1928	HD 2865//HI5439//PBW 175
	HP 1927	C306//HD-2685//UP2691
	HP 1913	Raj -3765 x GW-322
CVT-1A	PS 929	SUNCO/2*PASTOR
	PS 930	JNR B 5/PIFED
	PS 931	QT6581/4/PASTOR/SITE/MO/3/CHEN/.
	PS 932	Attila/HP 1761//HW 2045
	PS 933	HP 1744/Lok 1/PBW 343
	PS 934	HD 2733/HP 1761
	PS 935	HD 2733/10TH WAWSN-61
	PS 936	Hahn/2*Weaver//SW895193/3/JKW 4
	PS 937	Narpatganj/GW 273/HW 2045
	PS 938	CBW 14/PBW 506
	PS 939	PBW 343/HW 3026
CVT-1B	PS 940	HD 2865/35TH IBWSN. Acc. No.405
	PS 941	Kauz/5/PA710/ALD/PAT 72300/3/PUN/4/BOW/6/Raj 1482/ UP 2338//PBW 302
	PS 942	HP 1731/UP 2529/HD 687
	PS 943	HD 2733/10TH WAWSN76/HS 418
	PS 944	DBW 14/PBW 531
	PS 945	DL 457/HD 2733
	PS 946	PBW 226 Lr 9/WR 544/WH 770
	PS 947	HD 2733/10TH WAWSN76/PHR 1015
	PS 948	HP 1870/PBW 522

	PS 949	CBW17/K103
	PS 950	Attila/HP 1761//HW 2045
CVT-5A	PS 963	UP 2532/DL 788-2//CPAN 3038
	PS 964	CPAN1990/Kauz//K 8027
	PS 965	SUNCO/2*PASTOR
	PS 966	HP 1493/HW 2004//K 8027
	PS 967	ALTAR 84/*Ae squarrosa* (TROS)/OPATA/3)
	PS 968	HP 1744/Lok 1/PBW 343
CVT-3	PS 951	HP 1731/HUW 541/HW 2045
	PS 952	PBW 343/NW 1012
	PS 953	CBW 12/PBW 483
	PS 954	HD 2733/UP 2532
	PS 955	Attila/HP 1761
	PS 956	HD 2733/NW 1014//HD 2733
	PS 957	K 9107/HUW 541
	PS 958	CBW 13/PBW 503
	PS 959	HD 2733/CDWR 9522
	PS 960	HD 2733/CBW12//HD 2733
	PS 961	PBW 343/PBW 503/PHR 1006
	PS 962	CHEN/Ae squarrosa//2*0PATA/3/BABAX/4/JARU
CVT-1A	PS 929	SUNCO/2*PASTOR
	PS 930	JNR B 5/PIFED
	PS 931	QT6581/4/PASTOR/SITE/MO/3/CHEN/.
	PS 932	Attila/HP 1761//HW 2045
	PS 933	HP 1744/Lok 1/PBW 343
	PS 934	HD 2733/HP 1761
	PS 935	HD 2733/10TH WAWSN-61
	PS 936	Hahn/2*Weaver//SW895193/3/JKW 4
	PS 937	Narpatganj/GW 273/HW 2045
	PS 938	CBW 14/PBW 506
	PS 939	PBW 343/HW 3026
CVT-1B	PS 940	HD 2865/35TH IBWSN.Acc. No.405
	PS 941	Kauz/5/PA710/ALD/PAT 72300/3/PUN/4/BOW/6/Raj 1482/ UP 2338//PBW 302
	PS 942	HP 1731/UP 2529/HD 687
	PS 943	HD 2733/10TH WAWSN76/HS 418
	PS 944	DBW 14/PBW 531
	PS 945	DL 457/HD 2733
	PS 946	PBW 226 Lr 9/WR 544/WH 770
	PS 947	HD 2733/10TH WAWSN76/PHR 1015
	PS 948	HP 1870/PBW 522

	PS 949	CBW17/K103
	PS 950	Attila/HP 1761//HW 2045
CVT-5A	PS 963	UP 2532/DL 788-2//CPAN 3038
	PS 964	CPAN1990/Kauz//K 8027
	PS 965	SUNCO/2*PASTOR
	PS 966	HP 1493/HW 2004//K 8027
	PS 967	ALTAR 84/*Ae squarrosa* (TROS)/OPATA/3)
	PS 968	HP 1744/Lok 1/PBW 343
CVT-3	PS 951	HP 1731/HUW 541/HW 2045
	PS 952	PBW 343/NW 1012
	PS 953	CBW 12/PBW 483
	PS 954	HD 2733/UP 2532
	PS 955	Attila/HP 1761
	PS 956	HD 2733/NW 1014//HD 2733
	PS 957	K 9107/HUW 541
	PS 958	CBW 13/PBW 503
	PS 959	HD 2733/CDWR 9522
	PS 960	HD 2733/CBW12//HD 2733
	PS 961	PBW 343/PBW 503/PHR 1006
	PS 962	CHEN/Ae squarrosa//2*0PATA/3/BABAX/4/JARU

Table 3.21. Entries being tested/evalutated in IPPSN during *rabi* 2008-2009 at IARI RS Pusa, Bihar

Entries	*Pedigree*
PS 929	SUNCO/2*PASTOR
PS 930	JNR B 5/PIFED
PS 931	QT6581/4/PASTOR/SITE/MO/3/CHEN/.
PS 932	Attila/HP 1761//HW 2045
PS 933	HP 1744/Lok 1/PBW 343
PS 934	HD 2733/HP 1761
PS 935	HD 2733/10TH WAWSN-61
PS 936	Hahn/2*Weaver//SW895193/3/JKW 4
PS 937	Narpatganj/GW 273/HW 2045
PS 938	CBW 14/PBW 506
PS 939	PBW 343/HW 3026
PS 940	HD 2865/35TH IBWSN.Acc. No.405
PS 941	Kauz/5/PA710/ALD/PAT 72300/3/PUN/4/BOW/6/Raj 1482/UP 2338//PBW 302
PS 942	HP 1731/UP 2529/HD 687
PS 943	HD 2733/10TH WAWSN76/HS 418
PS 944	DBW 14/PBW 531

Entries	*Pedigree*
PS 945	DL 457/HD 2733
PS 946	PBW 226 Lr 9/WR 544/WH 770
PS 947	HD 2733/10TH WAWSN76/PHR 1015
PS 948	HP 1870/PBW 522
PS 949	CBW17/K103
PS 950	Attila/HP 1761//HW 2045
PS 951	HP 1731/HUW 541/HW 2045
PS 952	PBW 343/NW 1012
PS 953	CBW 12/PBW 483
PS 954	HD 2733/UP 2532
PS 955	Attila/HP 1761
PS 956	HD 2733/NW 1014//HD 2733
PS 957	K 9107/HUW 541
PS 958	CBW 13/PBW 503
PS 959	HD 2733/CDWR 9522
PS 960	HD 2733/CBW12//HD 2733
PS 961	PBW 343/PBW 503/PHR 1006
PS 962	CHEN/Ae squarrosa//2*0PATA/3/BABAX/4/JARU
PS 963	UP 2532/DL 788-2//CPAN 3038
PS 964	CPAN1990/Kauz//K 8027
PS 965	SUNCO/2*PASTOR
PS 966	HP 1493/HW 2004//K 8027
PS 967	ALTAR 84/*Ae squarrosa* (TROS)/OPATA/3)
PS 968	HP 1744/Lok 1/PBW 343

Table 3.22. Entries being tested/evalutated in Quality Screening Nursery (QCSN) during *rabi* 2008-2009 at IARI RS Pusa, Bihar

Entries	*Pedigree*	*Entries*	*Pedigree*
HP-1900	HD-2733/CP-305.RK.45	HP-1912	HUW-516/HD-2643
HP-1911	PBW-343/RAJ-3777//WH-736	HP-1913	RAJ-3765/GW-322

Breaking the Yield Barriers

Development of semi dwarf, medium to early maturing, lodging resistance (having strong Rht genes to sustain grain weight by its thick culms) with tolerance/resistance to rusts and foliar blight and with good grains are needed. The varieties required shoul withstand high fertility dose of 175: 75: 60 kg NPK. Introduction of high yielding varieties responsive to high dose of fertilizers and increased number of irrigations will attract diseases therefore, efforts are under way to incorporate foliar blight resistance genes which cause yield loss as well as deterioration in grain quality due to associated diseases namely black point of seeds. Resistance sources will further be explored besides Chinese, Yugoslavian and alien genes from *Agropyron* derivatives. Incorporation of thermo-insensitive genes to save the crop from forced ripening on account of early dry westerly wind. Since a considerable area is affected by salinity/alkalinity, tolerant varieties will be developed which can perform well under these conditions. Regarding alien gene(s), all effective alien gene(s) such as Lr 9, Lr 24+Sr 24, Lr 28, Lr 19+Sr 25, Lr 32, Lr 37 *etc.* were exploited in breeding Programme. Due to knocking down of Sr 31 gene against the virulent race *Ug99* in Uganda and Kenya, a preventive measure was taken this year and Sr 25 gene was extensively utilized in breeding programme through it's genetic sources such as FLW-8, HW 1085, HW 2044 and HW5030. A good deal of heterosis with low inbreeding depression was detected in many crosses. Efforts have been taken to introgress potent alien gene(s) such as Lr 9, Lr 24+Sr 24, Lr 28, Lr 19+Sr 25, Lr 32 and Lr 37 at the background of agronomically superior cultivars either singly or in association with other *Secale cereale* gene. Many leaf blight resistant/tolerant sources have been exploited in breeding programme besides Chirya-3 and Chinese sources. Synthetic wheat such as PHR 1001 and PHR 1026 and other buitre lines such as PHR 1004 and PHR 1007 have been exploited in breeding programme. For terminal heat tolerance, the reckoned sources such as PBW 510, WH 730, WH 775, HD 2808, Raj 3777 *etc.* were utilized in breeding programme. Approximately 4500 breeding lines have been generated under varietal development programme, HP 1883 is at the stage of anvil. Efforts have been taken to introgress potent alien gene(s) such as Lr 9, Lr 24+Sr 24, Lr 28, Lr 19+Sr 25, Lr 32 and Lr 37 at the background of agronomically superior cultivars either singly or in association with other *Secale cereale* gene. Many leaf blight resistant/tolerant sources have been exploited in breeding programme besides Chirya-3 and Chinese sources. Synthetic wheat such as PHR 1001 and PHR 1026 and other buitre lines such as PHR 1004 and PHR 1007 have been exploited in breeding programme. For terminal heat tolerance, the reckoned sources such as PBW 510, WH 730, WH 775, HD 2808, Raj 3777 *etc.* were utilized in breeding programme.

Hybrid Wheat

Hybrid varieties are the F_1 generation from crosses between two inbreds. Commercially hybrids are either single crosses or double crosses. In the first concrete heterosis wheat breeding trials Lelley (1953b) attempted to produce the necessary male sterility first by mechanical, then by chemical intervention. These attempts, though failed to achieve concrete results revealed some aspects of the hybrid effect and difficulties of *Hybrid wheat* production. Efforts have ever since been made to induce male sterility by chemical means, so far without any considerable success. Meyar (1967) obtained some results in spring wheat by using omnidel as gametecide but the effect of the chemical depended on the weather. Jos and Singh (1967) studied the effect of Na-2, 3 dichloro-iso-butyrate on pollen sterility. According to Borghi *et al.* (1973) spraying a 2000 ppm solution of 2-chloroethylphosphoric acid (etaphon) at the late and early stage of 'boot' as well as in the initial phase of heading caused complete male sterility with considerable allogamic seeds setting. In their opinion this result is promising. Kihara in 1951 and Fukasawa in 1963 reported on plasmic male sterility in *Aegilops caudata* and *Aegilops ovata* respectively. Wilson and Ross

(1962) substituted *Triticum eastivum* nucleus into the plasm of *Triticum timopheevi* and obtained male sterile derivatives. As it was found later, male sterility caused by the plasm of *Triticum timopheevi* was the most suitable for practical use.

Hybrid Wheat Production

The actual measure and stability of heterosis effect observed in wheat had to be determined, Adequate methods for establishing and forecasting combining ability should be found out. Utilization of genic male sterility had to be studied. Reliable and full restoration of fertility had to be solved. It was necessary to reasses the biological conditions of wheat flowering and fertilization from allogamic point of view, and, Possible obstacle to the practical production of hybrid seed-grain were to be indicated. Heterosis effect, of which the excess height of culm is an undesirable feature. Quality is mostly intermediate. In productive tillering mostly an outstanding heterosis effect is felt. The extent of hybrid effect changes mostly from year to year, hybrids must therefore be studied over several years. Reciprocal combination does not always give the same heterosis effect. It is worth elucidating a heterosis effect is still felt in the F_2 generation. The prospective measure of heterosis effect can be assessed by the analysis of combining ability. This establishes the value of varieties or lines as heterosis partners. The combining ability is indicated by cyclic or diallel crossing.

Male Sterility

On the course of investigating the resources including plasmic male sterility, in addition to the plasms of *Aegilops ovata, Aegilops caudata* and *Triticum timopheevi,* a number of other plasmic sources, have been found to cause male sterility. The plasms of *Aegilops caudata* have the disadvantage that besides male sterility it may also cause pistilloidy. The undesirable effect of *Aegilops ovata* plasm is felt in the prolongation of flowering from. Both phenomena result in deficient seed setting (Porter *et al.,* 1965). Further, male sterile sources are *Triticum zhukovskii* and the octoploid *Triticum timonovum* Heslot. *Triticum timonovum* also gives the most viable male sterile progeny with hexaploid wheats, while *Triticum timopheevi* with *Triticum durum. Triticum monococcum* as well as *Triticum arraticum* had also been used as sources of male sterility. Whether it was from *Triticum aestivum* or *Triticum durum* that nuclei were substituted into the plasms of *Aegilops bicornis, Aegilops umbellata, Aegilops sparonensis, Aegilops longissimum, Aegilops heldereichi* and *Aegilops variuabilis,* makes no difference, male fertility and vitality are equally reduced. According to Kihara (1973) hexaploid nucleus in *Aegilops squarrosa* plasm does not cause sterility, while tetraploid nucleus does. Accordingly, genome D obviously contains a gene that restores male sterility, in *Aegilops squarrosa.* The biological and genetic explanation of male sterility originating from alien cytoplasm is not quite sufficient or yet.

Genic Male Sterility

Suneson (1964) described genic male sterility. Athwal *et al.* (1967) gives account of ploy-factorial male sterility, transmitted in a recessive form. Gill and Anand (170) described a recessive male sterile from determined by three genes where the extent of seed setting depend on the number of genes. In the case of three genes present male sterility was 95 per cent while two genes caused male sterility was 65 per cent and one gene caused male sterility as 35 per cent. Briggle (1970) described a male sterile combination where the segregation ratio of normal to male sterile was 3: 1 without intergrades. Trupp (1971) likewise gave account of a male sterile form of simple genic heredity. From a practical part of view only the recessive monogenic male sterility is of significance which can be completely restored by a dominant fertility zone. Homozygous recessive male sterile lines cannot, however, be maintained by autogamic fertilization only through allogamy, and even then dominant restorer gene must not be contained in the pollen of the partner. Driscoll (1973) elaborated a method of solving the problem of male sterility and restoration and the maintenance of male sterile lines on the basis of genic transmission. In his opinion, the gene pair causing recessive male sterility (ms). The dominant "Ms" restorer gene-pair has to be incorporated in a chromosome pair homologous with that carrying the 'ms' gene pair. This should be additioned to the genome of the male sterile variety (Driscoll, 1972b). The system requires three Lines: X,Y,Z. All three are homologous for gene "ms", each contains a male sterile gene-pair. Further, line X contains two additioned chromosomes with a "Ms" gene pair, and line Y are additioned chromosomes with the same gene. The X plants produce only hyperploid gametes with 22 chromosomes. The Y lines produce gametes with 21 and 22 chromosomes of which the former shows a higher vitality than the hyperploids. The Z plants do not contain alien chromosomes are male sterile, but only remain so when fertilized by pollen containing (ms) gene. Such pollen is, however, produced only by the Y plants. With the Y plants removed after flowering and the Z plants separately harvested homozygous ms ms sterile seed-grain is obtained. In practice the operation can be carried out by the following scheme:

	Male Block		*Female Block*		*Seed Grain*
First step	X	x	Z	=	Y
Second step	Y	x	Z	=	Z

After this Z seed is the recipient partner of the F_1 seed grain mass production, the pollen donor and restorer is a normal variety, and so the hybrid F_1 seed grain Ms ms heterozygous for the property in question, it is, fertile and does not contain alien chromosomes. According to Driscoll (1973) the chromosome mapping confirmed that chromosome 4A,4B,5A and 5D carry "Ms" gene or genes which restore male sterility. The chromosomes carry the "Ms" gene on one of the arms only. The deletion of this arm may cause male sterility. Hybridization with monosomic F_1 plants may result such deletions. Recessive male sterile "Z" forms can thus be produced. The 4E chromosomes of *Aegilops elongatum* contain an Ms gene and a factor determining the blue color of aleurone. There is an Ms gene on chromosome. 5R of *Secale* too. The C chromosome of *Aegilops umbellata* likewise carries an Ms gene. Thus, the incorporation of an Ms gene carried by an alien chromosome in the X line can be carried out as well (Bielig and Driscoll,1973; Driscoll,1973). This form of utilization of genic male sterility is much more complicated than plasmic male sterility induced by nucleus substitution. The restoration of gene male sterility means no problem, but, it is all the more difficult to restore plasmic male sterility. Restoration of plasmic male sterility would be easier if we, knew the mechanism of male sterility. Unfortunately, this point has not been fully clarified yet. According to Joppa *et al.* (1966) the pollen of plants with plasmic male sterility contains only a little. Start the filament shows an irregular growth, its vascular system is abnormal. Srivastava *et al.* (1969) pointed out a decreasing tendency for the biological activity of mitochondria in the sterile plants. A total of 6Rf genes restoring plasmic male sterility have been identified so far, which are uptake to restore the *Triticum timopheevi* type of male sterility. According to McIntosh (1973) gene Rf_1 is in chromosome 1A and is found in the crossing *Triticum timopheevi* from *Marquis;* gene Rf_2 is in chromosome 7D and is found in the crossing *Triticum timopheevi* from *Marquis;* gene Rf_3 is in chromosome 1D chromosome. It is identified in *Triticum spelta* var *duhamelianum;* gene Rf_4 is in 6D chromosome; gene Rf_5 is in 6D chromosome. It is identified in *Triticum timopheevi,* X *Aegilops squarrosa* x *Ca nothatch;* gene $Rf_6$is in *Triticum zhukovskii.* Modifiers influencing the Rf gene have been identified in chromosomes 1B,2A,6A and 6B (Robertson and Curtis,1967). Deficient cross pollination 1B,2A,6A and 6B cross pollination caused by the autogamic fertilization of wheat is one of the main obstacles to the introduction of hybrid wheat in commercial production (Briggle,1964). Cross pollination has been found to have relatively wide possibilities depending on meteorological conditions and variety (Medvedeva,1968). Cross pollination is unfavorably influenced by the fact that the wheat pollen is heavy and quickly, drops. It is therefore, better if the pollen donor is higher in growth (Lelley,1966). In some wheat variety 80 per cent of the pollen is carried away still it is not enough.

Problem of Hybrid Wheat

Qualitative deterioration, Full restoration of male sterility, Cross pollination, Lack of seed setting, High seed requirement. For wheat, as a plant of exclusive generative production. *Hybrid wheat* has been known so far to be the only possibility of fixing the F_1 state. But till today *Hybrid wheat* had not become popular in India, but using the hybridization program had developed several varieties.

Table 3.25. Important varieties of wheat evolved by hybridization

NP- 65	NP- 785	NP- 798	C- 281
NP- 710	NP- 786	NP- 799	C- 285
NP- 718	NP- 789	NP- 809	RS-31-1
NP- 783	NP- 790	NP- 825	NP-201
NP- 784	NP- 797	NP- 828	Hyb-3B
			NP-412

Results of such demonstrations has shown that our promising variety HD 2733 gave an average yield of 47.5 q/ha yield at farmers field and other varieties like HD 2643, HP 1731 also performed better. Varietal development of wheat is a continuous process for the breeder. After joining at Regional Station, Pusa attempt was made to develop genotypes combining desirable maturity with yield and resistance to disease particularly for foliar blight, which is serious problem under Rice-Wheat cropping system. Following advance lines have been developed and these are under evaluation in All India coordinated various trials. HP 1882, 1883, 1884, 1885, 1886, 1887, 1888, 1889, 1890, 1891, 1892, 1893, 1894, 1895, 1896, 1897, 1898, 1899, 1900, 1901, 1902, 1903. As a result of gene Pyramiding, the culture HP 1879 consistently showed superiority both in Yield and resistance against three rusts and promoted to 2nd year AVT (Restricted irrigation) in Southern Hill Zone. A good number of promising cultures with diverse genetic background have qualified for different NIVTS and shows very good promise. They can be enlisted as HP 1882, 1883 in NIVT – IA, HP 1884,1885,1886,1887,1892 in NIVT 1B, HP 1889, 1890 in NIVT-3 and HP 1891 in NIVT-7. Introgression of FBR/FBT (Foliar blight resistant/tolerant) gene from chirya-3 and *Agropyron curnfolium* derivative W 76 at agronomically superior genotypes are found very promising and occupied place in CVTs such as PS 752 (Chirya-3 HP-1731 in CVT-1, PS 768 (chirya-3, HP-1731) in CVT-3 and PS 7629 (W-76 HD-2329 Lr 24+Sr 24) in CVT-3. Under prebreeding approach, many cultures selected but Ps 754(Prn//Lar 422/Ana.), PS747 and PS777 (Attila/3Prl Oasis//Kauz) were found very promising and occupied place in CVTs. For rainfed breeding, parents with high adoptability such as C306, MACS 6145 and K 8027 were intercrossed with new genotype AKAW 3717 exhibiting better stand and adoptability over different mega environments. Bridging necrotic gene: - HD 2733 is a high yielding released cultivar. But it inherits strong necrotic gene Ne^s. Some desirable genotypes such as HW 2045, PHR 1010, CBW- 17, GW 173 and WR544 carrying Ne_g^s produce lethal hybrids with HD 2733. Therefore, their necrotic gene was firstly bridged by DL 450, HP 1731, JKW-1, HD 2824 and DL457 respectively and then developed some recombinants with HD 2733. Some additional sources carrying strong necrotic gene (Ne_g^s) were identified in gene pool grown at R.S. Pusa *Viz.* WH 423, HI1525, WR 901, RWP 9912, HPW 185, KBRL- 10, and NIAW 1076 *etc.* The proven high temperature tolerant gene from PBW509, RAJ 4037, WH 730 and HD 2808 were tried to introgress in the genetic background of PBW 524, PBW 569 and GW373 to have some novel recombinants.

Table 3.24. Gene combinations produced

Variety	*Gene Combinations*	*Comment*
HD 2932	Lr 19+Yr 10+Sr 26	Homozygous
	Lr 19+Yr 10	Homozygous
	Lr 19+Sr 26	Homozygous
	Yr 10+Sr 26	Homozygous
HD 2733	Lr 19+Yr 15+Sr 24	Heterozygous
	Lr 19+Yr 15	Homozygous
	Lr 19+Sr 24	Homozygous
	Yr 15+Sr 24	Homozygous
HD 2967	Lr 19+Yr 10	Heterozygous

Development of Wheat Hybrid through CMS System

It was need of the hour to decide to re-address hybrid wheat development through CMS and CHA approach. Through CMS Approach, cytoplasmic male sterile lines were developed by using *Triticum timopheevi, T. araraticum, Ae. caudata and Ae. speltoides* as source parents. The results indicated more than 80 per cent fertility restoration by the exotic stocks PWR 4099 and PWR 4101, in *T. timopheevi* and *T. araraticum* based CMS lines. Few hybrids developed through CMS approach showed heterosis for yield components *viz.* spikelet number, spike length and tillers/plant. Through CHA approach, chemicals were identified that induced male sterility when sprayed at 10-15mm of spike length at 50-60 days after germination in most of the genotypes. Over 1500 single cross hybrids developed through CHA were evaluated at DWR. Of these, 6 hybrids were identified depicting standard heterosis over the best check and superior quality parameters. In addition, a large number of hybrids through CHA (and a few by CMS system) were evaluated at co-operative centers namely, PAU, CCSHAU and IARI. Based on the results and experiences gained from earlier work done on hybrid wheat, the following conclusions can be drawn. Out of two pollination control systems (CHA and CMS), the cytoplasmic genetic male sterility system, though cumbersome, provides useful and eco-friendly approach for hybrid seed production. During 6-7 years of study, the CHA approach indicated certain inherent problems beyond control and, therefore, not found practically feasible to produce hybrid seed. The CHA system was effective to attempt large number of hybrid combinations and preliminary testing of these hybrids on multi locations. The multi locations and multiyear data revealed that only few hybrids performed well. The parents of these superior hybrids might be used as basic material for CMS line development and restorer conversion. The *timopheevii* based CMS system was found most appropriate since the fertility restoration was highly satisfactory.

In addition, male sterile cytoplasm and restorer gene(s) have to be incorporated in agronomically superior heterotic parents. Male sterile cytoplasm can be transferred easily by backcrossing but for transfer of fertility restorer (*Rf*) genes test cross to identify plants carrying fertility restorer gene(s) is required. If proper molecular markers are developed and *Rf* genes are tagged, it will facilitate easy identification and incorporation of fertility restoration gene(s) in diverse genetic/ heterotic backgrounds. Development of molecular markers for restorer gene(s) will help in marker assisted selection for *Rf* genes with normal cytoplasmic background, screening of plants with restorer gene(s) at early seedling stage in backcross programme, pyramiding of restorer genes for enhancing restoration of fertility and making selection of *Rf* genes independent of environmental influence. Keeping in mind the above facts, the present project is proposed for basic and strategic research on various aspects of development of commercially feasible wheat hybrids including floral biology studies to enhance out-crossing, identification of parental lines that exhibit better heterosis, production and maintenance of CMS system, fertility restoration system and to attempt large number of crosses to identify promising hybrids with high commercial heterosis.

Keeping the aforesaid points the broad objectives should be to investigate parental diversity through establishment of genetic diversity based on morphological, geographical, pedigree relationship; to identify heterotic gene pool through combining ability analysis and estimation of heterosis; to investigate variability in the floral characteristics of diverse bread wheat genotypes; development of CMS lines in different genetic backgrounds; development of restorer germplasm using diverse genetic sources; molecular tagging of fertility restorer genes; evaluation of experimental hybrids for yield and component traits, resistance to biotic and abiotic factors and quality attributes; standardizing hybrid seed production technology in terms of M: F ratio, planting time and geometry of commercial hybrids. This will help in better understanding about the feasibility of hybrid wheat through CMS approach and following information will be generated. The diversity pattern for identification of diverse parents for heterosis. Extent of variability and association among floral traits that promotes out-crossing. Extent of CMS based male sterility and fertility restoration. Development of promising and stable CMS lines and better restorers. Magnitude of standard heterosis through available germplasm. Seed production techniques for hybrid such as planting ratio, time, seed rate including suitable tillage options, *etc.* Development of hybrids which will be commercially suitable at farmer's field.

Year 1

Study of parental diversity for morphological and yield components and using molecular markers, crossing programme using elite germplasm for identification of heterotic gene pool, study of floral traits that promote out-crossing and identification of donor parents, collection of CMS, maintainer and restorer lines and their maintenance, conversion of better agronomic bases into new CMS lines by hybridization, deployment of fertility restoration in better agronomic bases, identification of restorer lines through crossing elite germplasms with CMS line, parental polymorphism survey using molecular markers (RAPD, AFLP and SSR) for molecular tagging of fertility restorer gene(s).

Year 2

Steps 1-8 as in 1st year, combining ability analysis and estimation of heterosis for identification of heterotic parents, Evaluation of performance of CMS lines for floral traits, stability and heterosis, Evaluation of performance of restorer lines for floral traits, level of fertility restoration and heterosis, Selection of F_2 mapping population and Bulked Segregant Analysis (BSA) for identifying molecular marker(s) linked to fertility restorer gene(s). Attempting large number of crosses using CMS converted lines and restorers and their evaluation in order to identify heterotic pool/combinations.

Year 3

Steps 9-14 as in 2nd year, Markers linked to fertility restorer gene(s) identified based on BSA will be used for mapping of Rf gene(s) using F_2 population. Evaluation of experimental hybrids in order to identify heterotic combinations.

Year 4

Steps 15-17 as in 3rd year, Evaluation of selected experimental hybrids on plot basis for yield and component traits, resistance to diseases (rust, blight) and insects (aphids) and abiotic stresses. Evaluation of hybrid seed production using different row ratios, planting time, *etc.*

The durum wheat varieties '**Pusa Gaurav**' (HI 8691) and '**Malav Kranti**' (HI 8638) Both these varieties have proven yield potential, adequate level of rust resistance and desired quality traits as highlighted in our representation. There is an urgent need to increase the area and production of durum wheat, as it has several advantages to offer. Durum wheat has rust resistance spectrum different from that of the bread wheat, with high levels of resistance to currently prevalent and virulent pathotypes of leaf rust race 77-group. The durum wheat has better field tolerance to loose smut and Karnal bunt diseases, compared to bread wheat. Durum wheat is rich in protein, β carotene (precursor of Vitamin A) and essential micro-nutrients like iron and zinc, and hence, can help in overcoming the problem of malnutrition. Having excellent pasta making quality, durum wheat can provide quality raw material for the rapidly growing fast-food industry in the country, generating additional employment and economic growth. However, a number of high yielding and rust resistant durum wheat varieties for different cultivation conditions have recently been developed. Providing a premium price for durum wheat will contribute significantly to the expansion in its area and increased production, leading to the benefits.

Dormancy in Wheat Seeds

Resistance of mature seed to sprout for an extended period of time is termed as dormancy. It is not a permanent phenomenon but one which disappears slowly with time. Unfavorable wet and humid climatic conditions at the time of harvest or during post-harvest storage of wheat seeds may cause considerable damage both for its consumption as food and market value as was as a good quality seed. Wheat seeds often sprout in the ear head when weather during harvest is wet and humid. In autumn 1952, wheat crop in New Zealand had to be resown because of poor germination. On closer examination, it was found that the seed used had been carried over from the wet harvest. The wheat seeds had not sprouted but had been so affected by the wet conditions that a year later its germination was much reduced. Some lines by then, had less than 0 of live seeds, some had over 90, but all when tested were slow to germination hence easily killed by fungi in the field before they are through the ground (Anonymous, 1953). It was Nilsson-Ehle (1914) who first draw attention to the tendency of several white wheat types to sprout readily in the field under wet and humid conditions whereas red wheats showed marked resistance under similar circumstances. Subsequent investigations (Voss, 1934; Hutchinson *et al.*, 1948) confirmed these findings but also noted that the red wheat varieties have a variation in dormancy, whereas the white wheat varieties have little or no dormancy or range in dormancy. The newly harvested wheat kernels which do not germinate after at least 8 days are classified as dormant and these which germinate in three days are classified as nondormant. Everson and Hart (1961) tested about 300 wheat selections for post –harvest dormancy of threshed and unthreshed seeds. The sixteen varieties with red seed coat and five with white seed coat indicated that red varieties had post-harvest dormancies ranging from the inhibition of germination for 5 or more days during a period upto 3 weeks after maturity, to germination in less than three days after maturity, with the exception of variety (Brevor) which possessed a weak inhibitory system.

Table 3.25. Important crosses advanced for advanced generations at ICAR-IARI RS Indore, Madhya Pradesh

Crosses Performed at IARI RS Indore, M P	
Single Cross	*Three Way and Double Crosses*
GW 322/EGYPT 11	Ag Local/Raj 3974/HI 1544
GW 322/ESWYT 141	CPAN 1676/GW366/DL 788-2
GW 322/HI 1415	CPAN 1676/HI 1544/GW 322
GW 366/HHW 3026	CPAN 1766/GW 366/HW 2004
GW 366/LBPY 06-13	CPAN 1990/HI 1479/MP 1203
HI 1531/EGPYT 11	CPAN 3034/HD 2987/GW 322
HI 1544/PHS 725	GW 32/GW 326/HI 1406
HW 4067/HRWYT 217	GW 366/HI 1535/CPAN 3013
HD 2932/HW 2640	HD 2189/GW 322/HI 1405
HI 1500/HW 4032	HD 2705/NIAW 907//CPAN 1676/HD 2987
HI 1418/WR 2010	HD 2705/PBW 442//CPAN 3050/GW 366
Crosses attempted involving superior bread wheat varieties/advanced generation stocks, and at least one line having Ug99 resistance	
DL 788-2*/4TH SAMNYT 401	HI 1500*/EGPYT 11
DL 788-2/EGPYT 5	HI 1500/ESWYT 130
DL 788-2/EGPYT 17	HI 1500/ESWT 141
DL 788-2/EGPYT 141	HI 1500/GW 2008-127
DL 788-2/HW 4013	HI 1500/HI 1405
DL 788-2/LBPY 07-7	HI 1500/HW 4032
DL 788-2/LBPY 07-16	HI 1500/LBPY 06-12
DL 788-2/PHS 729	HI 1500/07-13
DL 788-2/PHS 823	HW 2043/MP 4010*
DL 788-2/PHS 829	HW 3063/MP 4010*
HW 3063/DL 788-2	
DL 788-2/HI 1535//GW 322	
DL 788-2/HI 1535//HW 2004	
CPAN 1676/GW366//DL 788-2	
CPAN 1676/HW 2004//DL 788-2	
HW 2040/Lok 1*	
HW 2045/Lok 1	*Varieties resistant against Ug 99

Doubled haploidy breeding serves as a useful tool in wheat improvement by providing instant homozygosity, which leads to fixation of desirable characters of recombinants as well as reduction in the number of generations required to achieve success. Of the various techniques for haploid production in wheat, chromosome elimination has been studied and investigated by researchers to a great extent and nowadays is being utilized for large-scale practical applications. The technique was initiated in wheat with the investigations of Barclay (1975), who recovered wheat haploids in crosses between the wheat variety Chinese Spring and *Hordeum bulbosum*. The technique was, however, genotype-specific due to the presence of the dominant crossability inhibitor genes *Krl* and *Kr2* in wheat, which are expressed in many wheat varieties. The wheat x maize for haploid production (Laurie and Bennett, 1987) was genotype non-specific because of the insensitivity of maize pollen to the action of *Kr1* and *Kr2* genes thereby rendering the chromosome elimination technique more efficient and of practical value. Although the wheat x maize system of haploid production is quite successful yet the maize crop needs to be grown in greenhouse so as to coincide flowering with wheat thereby increasing the cost of haploid production. Hence there was a dire need to search for another species which may serve, as an efficient pollen source for haploid production in wheat and so dispense with the need to grow maize under greenhouse conditions during the winter season. Keeping in view the necessity to search for winter season growing plant species as a pollen parent having equal or higher potential of haploid induction in wheat than that of maize, the present investigation was planned. Four diverse groups of wheat and triticale encompassing winter x spring wheat F_1s, triticales, triticale x wheat F_2s and triticale x wheat BC_1F_2s were hybridized with maize (*Zea mays*) and cogon grass (*Imperata cylindrica*) as pollen sources and the relative performance of both the pollen parents was compared in respect of various haploid induction parameters on the basis of overall results. The protocol for developing the haploids was used as per Laurie and Bennett (1987) and Chaudhary *et al.* (2002)3 with certain modifications for hormonal applications especially in case of triticale and triticale x wheat derivatives. The roots of the haploid regenerants obtained after embryo rescue in wheat were randomly screened for cytological investigations. Hybrid origin of the haploids through *Imperata cylindrica* – mediated approach was confirmed following genomic *in situ* hybridization (GISH) as per the protocol of Mukai and Gill (1991).

The results obtained after crossing from each group of wheat and triticale with maize (2n = 20) and *I. cylindrica* (2n = 20) were compiled and average frequency of the haploid induction parameters *viz.*, pseudo seed formation, haploid embryo formation and embryo regeneration was worked out and compared. The *I. cylindrica* is performing better than the maize in all the groups of wheat and triticale in respect of all the haploid induction parameters and hence identified as an efficient pollen source. Amongst different groups, the performance of winter x spring wheat crosses was observed significantly better than the triticale and the crosses of the triticale with wheat. It is expected that the D genome might also be playing significant role in determining the success of the haploid production. Genomic *in situ* hybridization analysis of the zygotes of wheat x *I. cylindrica* was exercised and 21 chromosomes of wheat and 10 of *I. cylindrica* were seen in the same prometaphase cell. It was also observed that the chromosomes of *I. cylindrica* are eliminated faster than that of maize and there is no endosperm formation in the former case. Metaphase preparations from the roots of haploid regenerants obtained in wheat after hybridization with *I. cylindrica* were investigated and haploid status (n = 21) of the plants was confirmed.

Confirmation for such status of the wheat plants obtained through maize mediated system has already been exercised by Laurie and Bennett (1988).

Overall, wheat x *I. cylindrica* system of doubled haploidy (DH) breeding in wheat following chromosome elimination approach has following advantages over the existing wheat x maize system:

Coincidence of Flowering

In wheat x maize system, maize has to be raised under artificial conditions in the green houses/polyhouses so as to coincide the flowering with that of wheat or triticale whereas such exercise is not required in case of wheat x *I. cylindrica* system because *I. cylindrica* is a winter season plant and coincides well for flowering with that of wheat and triticale under natural conditions.

Raising of Pollen Source

Maize has to be raised every season whereas *I. cylindrica* being a wild weedy perennial grass doesn't require its repeated sowings. *I. cylindrica* is available under natural conditions in almost all parts of the world wherever wheat is cultivated.

Genotype Specificity

Both the plant species *viz.* maize and *I. cylindrica* are genotype non-specific for hybridization with any variety of wheat, triticale or their derivatives.

Availability of Pollen

Likewise, maize, pollen of *I. cylindrica* is readily available in abundance during the wheat hybridization period.

Haploid Induction Performance

I. cylindrica performs significantly better than maize for all the haploid induction parameters in wheat and triticale and their derivatives and can produce more DH populations in comparison to the maize-mediated system in the same time period. Hence, the *I. cylindrica* mediated system of DH breeding has been proved a very efficient and economically viable approach for the acceleration of wheat breeding endeavours with enhanced precision and efficiency. This innovation has globally opened new vistas for the commercial utilization of the new system for large scale production of doubled haploids of wheat to be used directly as improved varieties or gene mapping populations.

Incorporation of genetic diversity into elite wheat (*Triticum aestivum* L., 2n=6x=42, AABBDD) cultivars has long been recognized as a means of improving wheat productivity and securing global wheat supply. Synthetic hexaploidy wheat (SHW) genotypes recreated from its two progenitor species, the tetraploid, *Triticum turgidum* (2n=4x=28, AABB) and its diploid wild relative, *Aegilops tauschii* (2n=2x=14, DD) are a useful resource of new genes for hexaploidy wheat improvement. These include many productivity traits such as abiotic (drought, heat, salinity/sodicity, and waterlogging) and biotic (rusts, septoria, barley yellow dwarf virus (BYDV), crown rot, tan spot, spot blotch, nematodes, powdery mildew, and fusarium head blight) stress resistance/tolerances as well as novel grain quality traits. Numerous

SHWs have been produced globally by various institutions. This examines the varied aspects in the utilization of synthetics for wheat improvement including the traits and genes identified, mapped, and transferred to common wheat. It has also been demonstrated that synthetic backcross- derived lines (SBLs, *i.e.*, when SHW is crossed to adapted local bread varieties) show significant yield increases and thus, enhanced yield performance across a diverse range of environments, demonstrating their potential for improving wheat productivity worldwide. This is particularly evident in moisture limited environments. The use of SBLs, advanced backcross QTL analysis, chromosome introgression lines, and whole genome association mapping is contributing to the elucidation of the genetic architecture of some of the traits. The contribution of transgressive segregation to enhanced phenotypes and the mechanisms including its genetic and physiological basis are yet to be elucidated. Understanding these would further enhance the utility of SBLs. Considerable progress has been made in the identification of useful quantitative trait loci (QTL) and genes, however the transfer of such rich genetic diversity into elite wheat cultivars is still quite limited. Gaps still exist in data cataloging; and access to such information could serve as an important community resource. Future production of new SHW should extend to under-exploited AB genome tetraploids such as *T. turgidum* ssp. *carthlicum, T. turgidum* ssp. *dicoccum*, and *T. turgidum* ssp. *dicoccoides* and identifying gaps in the *Ae. tauschii* germplasm used for existing SHW. Identifying geographical areas where the progenitor species of the existing SHW were collected would assist in guiding future collection missions. The recent advances in molecular technologies with whole genome sequencing becoming affordable will provide researchers with opportunities for more detailed analysis of traits and the deployment of more efficient strategies in the use of the unique exotic alleles derived from SHW for common wheat improvement. Thus, the contribution of SHW and the derived SBLs to wheat cropping systems worldwide is likely to grow in significance. However, these potential benefits are only realizable if phenotyping is equally extensive and effective.

Biotechnological Application to Wheat Improvement

Today, the world's population is increasing at the most rapid rate ever. Two hundred people are being added to the planet every minute. It is forecast that by the year 2050, the world's population will double to nearly 12 billion people. To feed this population, these people will require a staggering increase in food production. In fact, it has been estimated that the world will need to produce more than twice as much food during the next 50 years as was produced since the beginning of agriculture 10 000 years ago. The focus on current approaches for the use of modern molecular-based technologies to develop improved varieties and discusses areas for future applications. Biotechnology can be defined in many different ways, all areas that use molecular approaches to understand and manipulate a plant genome will be considered. However, the techniques are divided between those that make use of molecular markers for studying the genetic material already present within the wheat plant and genetic engineering aimed at the introduction of novel genetic material.

Wheat Genetic Resources

Harlan and de Wet (1971) proposed the concept of three gene pools, primary, secondary, and tertiary, based on the evolutionary distance between the species and success rate of hybridisation among species. The primary gene pool consists of biological species and crossing within this pool is easy. Common wheat arose recently (6,000–8,000 years ago) from hybridization of tetraploid *T. turgidum* L. (2n=4x=28, AABB) and diploid, *Ae. tauschii* Coss. (2n=2x=14, DD) and these two species constitute the primary gene pool. Gene transfers from these species can be made by homologous recombination either by direct crosses of these species with common wheat or by the production of synthetic wheat. No special cytogenetic manipulation, except embryo rescue, is necessary to produce the F_1 hybrids. Gene transfers from *Triticum urartu* or *Triticum monococcum* as well as *Aegilops speltoides* is difficult because of massive hybridization barriers including endosperm abortion, death of hybrids seedlings, female sterility of F_1 hybrids, reduced recombination, and the lack of recombination between 4A of diploid wheat with 4A of polyploid wheat. Thus, these species are described as secondary gene pool. Wild relatives of wheat have long been recognized as a source of useful genes for cultivated wheat improvement. Amongst wild species in the primary pool, considerable genetic variability in both *Ae. tauschii* and *T. turgidum* potentially adaptive traits, such as resistance to biotic and abiotic stresses are present. One of the strategies in exploiting the wild relatives of wheat has been through synthetic hexaploid wheat (SHW). The SHW, often designated as primary synthetic (PS) wheat, is usually created by artificially hybridizing durum wheat (T. turgidum ssp. durum; 2n=4x=28, AABB) with accessions of *Ae. tauschii* (syn *Ae. squarrosa, T. tauschii*; 2n=2x=14, DD), the donor of the D genome of hexaploid bread wheat (*T. aestivum*; 2n=6x=42, AABBDD), in a process analogous to the natural hybridization event that produced hexaploid wheat. Some SHWs have been produced by crossing different tetraploid wheat ancestors, cultivated, and wild emmer, *Triticum dicoccum* or *T. dicoccoides* or *T. carthlicum*; 2n=4x=28 AABB with *Ae. tauschii*. The preference for *Ae. tauschii* as the D genome donor in SHW is its close homology with the D genome of hexaploid wheat. It has been demonstrated that wheat and *Ae. tauschii* could be crossed directly and that the D genome chromosomes of the two species pair in meiosis. Recombination between the D genome of *T. aestivum* and *Ae. tauschii* has been reported to occur at a frequency similar to that of an intraspecific cross. The close pairing affinity of *Ae. tauschii* D genome chromosomes with that of *T. aestivum* indicates that only one or two rounds of backcrossing to elite wheat are needed to incorporate both qualitative and quantitative characters from *Ae. tauschii* (Gororo, 1999). Trethowan and Mujeeb-Kazi (2008) documented the contributions of the unique Triticeae gene pool diversity to common wheat improvement. The wider adaptation provided by increasing the genomic diversity of bread wheat via SHW provides a means to sustain productivity gains in the face of increasing disease epidemics and worsening climate change scenarios. The current review attempts to address several important issues: (i) increasing genetic diversity of traits and genes in SHW contributing significantly to overcoming the challenges; (ii) novelty and diversity of genes in SHW; (iii) the genetic basis of biotic and abiotic stress tolerance in SHW; (iv) the best strategies for exploiting novel diversity in wheat breeding programs; (v) potential advances in genetic technologies to mine new and diverse alleles in SHW for hexaploidy wheat improvement; and (vi) the current gaps in knowledge.

Wheat Molecular Genetics and Molecular Linkage Maps

Molecular genetics, or the use of molecular techniques for detecting differences in the DNA of individual plants, has many applications of value to crop improvement. The differences are called molecular markers because they are often associated with specific genes and act as '**signposts**' to those genes. Such markers, when very tightly linked to genes of interest, can be used to select indirectly for the desirable allele, and this represents the simplest form of marker-selection (MAS), whether used to accelerate the back-crossing of such an allele or in pyramiding several desirable alleles. Markers can also be used for dissecting polygenic traits into their Mendelian components or quantitative trait loci (QTL), thus increasing understanding of the inheritance and gene action for such traits and allowing the use of MAS as a complement to conventional selection procedures. Molecular markers are also used to probe the level of genetic diversity among different cultivars, within populations, among related species, *etc.* The applications of such evaluations are many, including varietal fingerprinting for identification and protection, understanding relationships among the units under study, efficiently managing genetic resources, facilitating introgression of chromosomal segments from alien species, and even tagging of specific genes. In add ition, markers and comparative mapping of various species have been very valuable for improving the understanding of genome structure and function and have allowed the isolation of genes of interest via map-based cloning. Several molecular marker types are available and they each have their advantages and disadvantages. The characteristics and usefulness of the most commonly used marker systems are shown. Restriction fragment length polymorphisms (RFLPs) were the first to be developed (some 15 years ago) and have been widely and successfully used to construct linkage maps of various species, including wheat. With the development of the polymerase chain reaction (PCR) technology, several marker types emerged. The first one was random amplified polymorphic DNA (RAPD), which quickly gained popularity over RFLPs due to the simplicity and decreased costs of the assay. However, most researchers now realize the weaknesses of RAPDs and use them with much less frequency. Microsatellite markers, or simple sequence repeats (SSRs), combine the power of RFLPs (codominant markers, reliable, specific genome location) with the ease of RAPDs and have the advantage of detecting higher levels of polymorphism. The amplified fragment length polymorphism (AFLP) approach takes advantage of the PCR technique to selectively amplify DNA fragments previously digested with one or two restriction enzymes. Playing with the number of selective bases of the primers and considering the number of amplification products per primer pair (Hoisington *et al.*, 1998).

Table 3.26. Characteristics and usefulness of molecular marker types for wheat molecular genetics

Use[b]	*RFLPs*	*RAPDs*	*SSRs*	*AFLPs*
Varietal fingerprinting	++	-/+	+++	+++
Genetic diversity	++	-	++	+
Qualitative gene tagging	++	++	++	+++
QTL mapping	++	-/+	++	++
MAS	+	-	++	+/++
Comparative mapping	++	-	-	-
Principle	Endonuclease restriction; Southern blot; hybridization	DNA amplification with random primers	Amplification of simple sequence repeats using specific primers	Endonuclease restriction; use of adapters and specific primers
Types of probe/ primers	gDNA, cDNA	Random 9- or 10-mer oligo-nucleotides	Specific 16- to 30-mer primers	Specific adapters and selective primers
Type of polymorphism	Single base changes; insertions/ deletions	Single base changes; insertions/ deletions	Changes in length of repeats	Single base changes; insertions/ deletions
Genomic abundance	High	Very high	Medium	Very high
Level of polymorphism	Medium	Medium	High	High
Inheritance	Co-dominant	Dominant	Co-dominant	Dominant
Number of loci detected	3-9	1-10	1-3	70-140
Need for sequence information	No	No	Yes	No
Technical difficulty	Medium	Low	Low	Medium/High
Reliability	High	Intermediate	High	Medium/High
Quantity of DNA required	10-15 mg	10-50 ng	50-100 ng	100-1 000 ng
Use of radioisotopes	Yes/No	No	Yes/No	Yes/No
Start-up costs	Medium	Low	Medium	High
Development costs	Medium	Low	High	Medium/High

a RFLP = restriction fragment length polymorphism; RAPD = random amplified polymorphic DNA; SSR = simple sequence repeat; AFLP = amplified fragment length polymorphism; - = not useful; +/- = somewhat useful; ++ = useful; +++ = very useful. bQTL = quantitative trait locus; MAS = marker-assisted selection.
Source: Adapted from Rafalski and Tingey, 1993.

The developments in molecular genetics in wheat have been relatively slower than maize, rice, due to wheat's ploidy level, the size and complexity of its genome, the very high percentage of repetitive sequences and the low level of polymorphism. Much fewer maps exist in wheat and far fewer QTL studies have been reported when compared to other grass species. However, due to the large number of disease and pest resistances controlled by major genes, the mapping of such genes has dominated the research activities in wheat molecular genetics. On the other hand, the hexaploid nature of wheat and its amenity to cytogenetic manipulation have offered unique tools for molecular geneticists of wheat. These include the use of various aneuploid stocks, such as nullitetrasomic and ditelosomic lines, to assign molecular markers to specific chromosome arms (Plaschke *et al.*, 1996), of chromosomal deletion stocks (Endo and Gill, 1996) for the physical mapping of markers (Röder *et al.*, 1998a) and of single chromosome substitution lines to map genes of known chromosomal location (de la Peña *et al.*, 1997).

Table 3.27. Characteristics of the bread wheat genome that explain the slow progress in mapping as compared to a diploid, highly polymorphic species such as maize

Characteristic[a]	*Wheat*	*Maize*
Ploidy level	6x	2x
Number of chromosomes	21	10
Genome size *(number of base pairs x 10^6)*	16 000	4 500
Polymorphism level	Low	High
· RFLPs: probe x enzyme combinations (per cent)	20-30	80-85
· SSRs: primer pairs (per cent)	40-50	50-60
Repetitive sequences (per cent)	>80	60
To construct linkage maps of same density (15 markers/chromosome):		
Number of loci needed	315	150
Number of RFLP probes needed	1000-1500	200-250
Number of SSR primer pairs needed	700-800	250-300

a RFLP = restriction fragment length polymorphism; SSR = simple sequence repeat.

The establishment of genetic linkage maps provides the basis for mapping the gene(s) responsible for the expression of traits of interest. In wheat, such maps have also corroborated cytological evidence of major chromosome rearrangements (Nelson *et al.*, 1995a) and have allowed the comparative mapping among related species (Börner *et al.*, 1998). The first RFLP maps wee reported by Chao *et al.* (1989) for the group 7 homoeologous chromosomes. Using mapping populations developed at the John Innes Centre, Norwich, England, Devos *et al.* (1992) published the group 3 maps. These were followed by group 2 (Devos *et al.*, 1993), group 5 (Xie *et al.*, 1993), groups 4, 5 and 7 (Devos *et al.*, 1995) and group 6 (Jia *et al.*, 1996) maps. The Norwich wheat RFLP linkage map has also been published altogether (Gale *et al.*, 1995) and now contains over 500 loci. Another important mapping population was developed by crossing a synthetic (amphihexaploid) wheat (*Aegilops tauschii* [syn. *Triticum tauschii*] x Altar 84 durum) to a spring bread wheat cultivar Opata 85 and was genotyped. The use of such a non-intervarietal cross resulted in a very dense map (about 1 000 RFLP loci) due to the higher polymorphism level. Maps of group 1, group 2, group 3, groups 4, 5 and 7 and group 6 have been published. Recently, Röder *et al.* (1998b) placed 279 SSR loci on the map also referred to as the ITMI (International Triticeae Mapping Initiative) map. In addition to two other linkage maps in wheat, a number of RFLP physical maps have been constructed using Chinese Spring deletion lines. These deletion lines were also used to construct group two SSR physical maps (Röder *et al.*, 1998a). In general, genetic maps have revealed a lower level of polymorphism in the D genome. Furthermore, a large number of RFLP loci and a fair number of microsatellite loci (Bryan *et al.*, 1997) have been assigned to chromosome arm locations using nullisomic-tetrasomic and ditelosomic lines.

Mapping of Single or Major Genes

In the last five years, a large number of genes of various functions have been mapped to specific wheat chromosomal regions. A vast majority of those genes controlling disease and pest resistance, stress tolerance, quality and other traits. Several mapping/tagging strategies using mostly RFLPs and RAPDs have lead to these results. Several RFLP and RAPD linked markers were then converted to PCR-based, more robust markers, such as sequence tagged sites (STSs), sequence characterized amplified regions (SCARs) or allele specific amplicons (ASAs). The existence of numerous sets of wheat near-isogenic lines (NILs) differing in the presence or absence of a resistance allele has facilitated the mapping of genes for which such lines exist (*Pm1*, *Pm2* and *Pm3*; *Lr9* and *Lr24*; *Bt-10*; *Yr15*). Dweikat *et al.* (1997) screened a series of NILs in Newton for Hessian fly resistance alleles using 1 600 random 10-mer primers. One to three RAPD markers were identified for each of the 11 genes being tagged, and linkage determined by screening F_2 populations segregating for each individual gene. On the other hand, Feuillet *et al.* (1995) screened Thatcher NILs for *Lr1* (on 5DL) with 37 RFLP probes mapping to group 5 chromosomes and found three to be linked to the gene after testing on F_2 populations between Thatcher and *Lr1*/Thatcher. The same approach was used by Williams *et al.* (1994) who found two RFLP markers flanking the *Cre* (*Cre1*) gene on the long arm of 2B. When the chromosomal location of a particular gene is known from previous genetic studies but no NILs are available, one can still utilize the markers mapped to that chromosome to score the parental lines for polymorphisms, construct a single chromosome map and determine which marker is closest to the gene of interest. This strategy was followed by Dubcovsky *et al.* (1996) to tag the *Kna1* locus in wheat responsible for higher potassium/sodium (K+/Na+) accumulation in leaves, a trait correlated with higher salt tolerance. Single chromosome maps and markers for genes on these chromosomes have also been developed using single-chromosome recombinant lines (*Vrn1* and *Fr1*, *Pch2*; *Rht8*). These mapping populations were derived according to Law (1966) by crossing lines of the same background but differing for a single chromosome, back-crossing to a monosomic line for the chromosome under study, identifying the monosomic plants with a hemizygous recombinant chromosome, selfing those and detecting disomic recombinants. Despite the difficulties of producing such mapping populations, the main advantage they offer is that they allow the scoring of the phenotypic effect of the gene of interest without the confounding effects of other genes (on other chromosomes) involved in the expression of the same trait. Bulk segregant analysis (BSA), developed by Michelmore *et al.* (1991) to tag disease resistance genes in lettuce, has been successfully applied in wheat. This approach has been mostly used with RAPDs (*Pm1* and *Pm2*; Hu *et al.*, 1997 for *Pm1*) although it is now being used with AFLPs. Either marker technique is used to screen two bulks of DNA samples from individuals identified in the two opposite tails of a segregating population for a target trait. For a major gene, all loci in the genome should appear to be in linkage equilibrium, except in the region of the genome linked to the target gene. To overcome the problems of limited repeatability of RAPDs, and the fact that repetitive sequences are often amplified (Devos and Gale, 1992), Eastwood *et al.* (1994) and William *et al.* (1997) used BSA and RAPDs on DNA enriched for low-copy sequences. In both cases, there was a noted increase in repeatability and levels of polymorphism detected compared with non-enriched DNA. The AFLP technology offers the advantage of the high number of DNA fragments amplified with one primer combination, and the problem of highly repetitive DNA is overcome by using methylation sensitive endonucleases, such as *Pst*I and *Sse*I. The fact that several of the resistance genes mapped in wheat have been introgressed from alien species explains the success of tagging them since a higher level of polymorphism is detected compared to segments where no alien DNA is transferred.

Quantitative Trait Mapping

The low number of quantitative traits dissected into their QTL in wheat is a reflection of the focus given to simply inherited traits and the difficulty of building comprehensive genetic linkage maps. In addition, more work is involved to generate good quality, reliable phenotypic data from replicated field (or greenhouse) evaluations of the trait under study.

Table 3.28. Published markers for important genes in wheat

Trait	*Locus*	*Source*	*Marker*	*Chromosome*
Disease resistance				
Leaf rust	*Lr1*	*Triticum aestivum*	RFLP/STS	5DL
	Lr3	*T. aestivum*	RFLP	6BL
	Lr9	*Aegilops umbellulata*	RAPD/STS	6BL
			RFLP	
	Lr10	*T. aestivum*	RFLP/STS	1 AS
	Lr13	*T. aestivum*	RFLP	2 BS
	Lr18	*T. timopheevii*	N-band	5BL
	Lr19	*Thinopyrum*	RFLP	7DL
			Isozyme	
	Lr20	*T. aestivum*	RFLP	5 AL
	Lr23	*T. turgidum*	RFLP	2BS
	Lr24	*Agropyron elongatum*	RFLP	3DL
			RAPD/STS	
			RAPD/SCAR	
	Lr25	*Secale cereale*	RAPD	4BL
	Lr27	*T. aestivum*	RFLP	3BS
	Lr29	*Ag. elongatum*	RAPD	7DS

Trait	*Locus*	*Source*	*Marker*	*Chromosome*
	Lr31	-	RFLP	4BL
	Lr32	*Ae. tauschii*	RFLP	3DS
	Lr34	*T. aestivum*	RFLP	7DS
	QTL	*T. aestivum*	RAPD/RFLP	7BL, 1BS, 1DS
Suppressor	*SuLr23*	-	RFLP	2DS
Stem rust	*Sr2*	*T. turgidum*	RFLP/STS	3BS
	Sr5	*T. aestivum*	RFLP	6DS
	Sr9e	*T. aestivum*	RFLP	2BL
	Sr22	*T. monococcum*	RFLP	7AL
	Sr36	*T. timopheevii*	RFLP	2BS
Stripe rust	*Yr15*	*T. dicoccoides*	RFLP/RAPD	1BS
Powdery mildew	*Pm1*	-	RFLP	7AS
			RFLP	
			RAPD-STS	
	Pm2	-	RFLP	5D
			RFLP, STS	
	Pm3	-	RFLP	1A
			RFLP	
	Pm4a	-	RAPD	.
	Pm4b	-	AFLP	-
	Pm12	*Ae. speltoides*	RFLP	6B/6S
	Pm18	-	RFLP	7AL
	Pm21	*Haynaldia villosa*	RAPD	6VS, 6AL
	Pm25	*T. monococcum*	RAPD	1A
Suppressor	SuPm8	-	Storage protein	1AS
Wheat streak mosaic virus	*Wsm1*	*Ag. elongatum*	STS	-
Common bunt	*Bt-10*	-	RAPD	
Loose smut	Ut-X (T10)	-	RFLP/RAPD-	-
			STS	
	T19	-	Antibody	6A
Eyespot	*Pch1*	-	RFLP/Isozyme	7DL
	Pch2	*T. aestivum*	RFLP	7AL
Tan spot	QTL	-	RFLP	1AS, 4AL, 2DL
Fusarium	QTL	*T. aestivum*	AFLP/RFLP	3BS, 2AL,
scab				6BS, 4BL
Karnal bunt	QTL	*T. turgidum*	RFLP	3BS, 5AL
Pest resistance				
Hessian fly	*H3,5,6,9,*	-	RAPD	1A, 5A
	10, 11, 12,			

Trait	*Locus*	*Source*	*Marker*	*Chromosome*
	13, 14, 16, 17			
	H9	-	RAPD	-
	H21	S. *cereale*	RAPD	2RL
	H23, H24	*Ae. tauschii*	RFLP	6D, 3DL
	H27	*Ae. ventricosa*	Isozyme	$4M^{V}$
Cereal cyst nematode	*Cre1*	*T. aestivum*	RFLP-STS	2BL
	Cre2	-	RFLP	6BL
	Cre3	*Ae. tauschii*	RAPD	2DL
	(Ccn-D1)			
Stres tolerance				
Cadmium uptake	-	-	RAPD	-
Aluminium	*Alt2*	-	RFLP	4D
tolerance			RFLP	4DL
Drought	-	-	RFLP	5A
induced ABA				
Na^+/K^+	*Kna1*	*T. aestivum*	RFLP	4D
discrimination			RFLP	4DL
Quality traits				
Kernel	*Ha*		RFLP	5D
hardness	*Hn*, QTL	.	RFLP	5DS, 2A, 2D,
				5B, 6D
Grain protein	QTL	*T. turgidum*	RFLP	4BS, 5AL, 6AS,
				6BS, 7BS
High protein	-	*T. dicoccoides*	ASA	6B
LMW glutenin	-	*T. turgidum*	-	1B
HMW glutenin	*Glu-D1-1*	*T. aestivum*	ASA	1DL
Flour colour	-	-	RFLP/AFLP	7A
Other trait				
Pre-harvest sprouting	QTL	*T. aestivum*	RFLP	-
Vernalization	*Vrn1*	.	RFLP	5AS
	Vrn3	-	RFLP	5DS
Photoperiod	*Ppd1*	*T. aestivum*	RFLP	2DS
	Ppd2	*T. aestivum*	RFLP	2BS
Dwarfing	*Rht8*	-	SSR	2DS
	Rht 12	-	SSR	5AL
Fertility	*Rf1, Rf3*	-	RFLP	6BS, 1BS
restoration	*Rf4*			
Meiotic	*ph1b*	-	RFLP/STS	5BL, 5BL
pairing	Deletion	-	AFLP/STS	

Source: Modified from Langridge and Chalmers, 1998.

The ITMI map is one of the densest available, and the population from which it was developed is segregating for a number of traits. It has therefore been used to map some important traits in addition to several major genes. Known genes include vernalization (*Vrn1* and *Vrn3*), red-coleoptile (*Rc1*), kernel hardness (*Ha*) and powdery mildew (*Pm1* and *Pm2*) genes, as well as genes conferring and suppressing leaf rust resistance (Nelson *et al.*, 1997). QTL mapped for kernel hardness, resistance to tan spot and Karnal bunt resistance. Aside from the ITMI population, Anderson *et al.* (1993) reported on QTL for pre-harvest sprouting (PHS) after using around 40 RFLP markers on two segregating populations of recombinant inbred lines (RILs), which were evaluated for PHS in up to seven environments. Using a combination of RFLP markers on most of a RIL population and selective genotyping with AFLPs on a sub-set of the population, Anderson *et al.* (1998) identified five putative QTL associated with Fusarium scab resistance. Efforts in breeding for disease resistance in general, and leaf rust resistance in particular, have focused on the use of durable resistance. Such resistance is controlled by a number of minor genes also referred to as adult plant resistance (APR) genes. In order to determine the number and location of these genes, and find tightly linked markers that will enhance the breeding efforts for such resistance, CIM-MYT has been involved in mapping APR loci in the leaf rust resistant cultivars Parula and Frontana. William *et al.* (1997), using BSA on RILs from a cross between Parula and Siete Cerros, identified three RAPD markers associated with two leaf rust resistance loci. Nullisomic-tetrasomic analysis showed that these are located on 7BL and 1BS or 1DS. A genetic linkage map using RFLP, SSR and AFLP markers in a segregating population of Frontana x INIA66 in order to map primarily durable leaf rust resistance but also other important traits that are segregating in the same population. Although the map now includes about 450 marker loci, some gaps still exist and efforts are focused on filling those with the SSR markers that are becoming available. With the current map and using composite interval mapping, CIMMYT has identified five and seven QTL for leaf rust resistance and barley yellow dwarf virus (BYDV) tolerance, respectively (Khairallah *et al.*, 1998).

Marker Assisted Selection

Three factors are required for the effective implementation of molecular markers in breeding programmes: (i) the availability of 'user-friendly' markers (cheap, easy and reliable); (ii) the validation of markers across different genetic backgrounds; and (iii) the possibility of implementing them within a breeding programme. MAS is a breakthrough technology that changes the process of variety cultivation from traditional field-based format to a laboratory format. It is the use of molecular markers to track the location of genes of interest in a breeding programme. MAS is a form of indirect selection and most widely used application of DNA markers. Once traits are mapped a closely linked marker may be used to screen large number of samples for rapid identification of progeny that carry desirable characteristics. MAS is one of the most widely used applications of molecular marker technologies and one that plant breeders have been quick to embrace. Biotechnology have provided additional tools that do not require the use of transgenic crops to revolutionize plant breeding progress in molecular genetics has resulted in the development of DNA tags and marker assisted selection strategies for cultivar development.

Several molecular marker types are available and they each have their advantages and disadvantages. With the development of the polymerase chain reaction (PCR) technology, several marker types emerged. The first of those were random amplified polymorphic DNA (RAPD), which quickly gained popularity over RFLPs due to the simplicity and decreased costs of the assay. However, most researchers now realize the weaknesses of RAPDs and use them with much less frequency. Microsatellite markers or simple sequence repeats (SSRs) combine the power of RFLPs (codominant markers, reliable, specific genome location) with the ease of RAPDs and have the advantage of detecting higher levels of polymorphism. The amplified fragment length polymorphism (AFLP) approach takes advantage of the PCR technique to selectively amplify DNA fragments previously digested with one or two restriction enzymes (Hosington *et al.*). Later, microsatellite markers or SSRs (Simple Sequence Repeats) were developed, which took advantage over RAPD and RFLP. Playing with the number of selective bases of the primers and considering the number of amplification products per primer pair, this approach is certainly the most powerful in terms of polymorphisms identified per reaction. The essential requirements for MAS in a plant-breeding program are as follows (Mohan *et al.*, 1997): Marker(s) should co-segregate or be closely linked (1 cM or less) with the desired trait; An efficient means of screening large populations for the molecular markers should be available. A relatively easy analysis based on PCR technology is the best option and the screening technology should have high reproducibility across laboratories, be economical to use and user friendly. RFLPs, RAPDs and AFLPs do not fit the first requirement. However, techniques are available to turn them into user-friendly markers. RFLP clones can be sequenced and primers designed to amplify the DNA fragments are shown by hybridization to be polymorphic. However, the resulting STS or SCAR does not always turn out to be polymorphic and further manipulations are needed if this is the case. The amplified fragment is usually digested with one or two restriction endonucleases to detect small length differences, or the fragment from two or more cultivars is cloned and sequenced again to create ASAs. ASAs are usually based on single nucleotide differences. RAPD and AFLP fragments can be isolated from the gel, cloned and sequenced to generate STSs or SCARs. Attempts to generate such markers for wheat are neither always successful nor easily achieved. SSRs, on the other hand, if tightly linked to genes of interest are probably the most attractive markers since no further manipulations are needed for implementation. Despite the large number of markers for wheat genes in few of those markers are close enough to the genes of interest to be useful in breeding applications.

Breeders used molecular markers to increase the precision of selection for best trial combinations. Variety developed by MAS are not considered genetically modified organisms (GMOs) and accepted by local and international market. Molecular marker aided selection methods have resulted in significant improvement in breeding efficiency by reducing trial and error aspect of breeding process and by allowing for time and cost savings. Molecular marker systems will benefit from the constant increase in the integration of biotechnological production of segregating populations such as homozygous double haploids in wheat breeding cycles since, the major requirement of being co-dominant for molecular markers will disappear. Marker assisted selection (MAS) offers an opportunity to select desirable lines based on genotype rather than phenotype. Marker assisted selection is an invaluable tool for gene pyramiding (Bringing genes from different individuals together in one individual) and has been fairly successful for combining single gene traits. Marker assisted selection (MAS) is based on the identification and use of markers, which are linked to the gene(s) controlling the trait of interest. By virtue of linkage, selection may be applied to the marker itself. The advantage consists in the opportunity of speeding up the application of the selection procedure. For instance, a character which is expressed only at the mature plant stage may be selected at the plantlet stage, if selection is applied to a molecular marker. Selection may be applied simultaneously to more than one character. Selection for a resistance gene may be carried out without needing to expose the plant to the pest, pathogen or deleterious agent. If linkage exists between a molecular marker and a quantitative trait locus (QTL), selection may become more efficient and rapid. The construction of detailed molecular and genetic maps of the genome of the species of interest is a prerequisite for most forms of MAS. However, the current cost of the application of these techniques is significant, and the choice of one cost technique rather than others may be dictated by factors. There are few examples of crop varieties in farmer's fields, which have developed through MAS.

Genetic resistance in wheat against diseases like leaf rust, stripe rust, stem rust, spot blotch, hill bunt *etc.*, is generally governed by one, two or three genes and these genes can be tagged with any of the above DNA markers, specially which are based on PCR technology. In wheat, RFLPs have been used to map seed storage protein loci, loci associated with protein flour colour, cultivar identification, vernalization and frost resistance gene, intrachromosomal mapping of genes for dwarfing and vernalization, resistance to preharvest sprouting, quantitative trait loci (QTL) controlling tissue culture response, nematode resistance and milling yield. PCR based markers have been useful for characterization of genes for resistance against common bunt, powdery mildew, leaf rust resistance against hessian fly and Russsian wheat aphid. RFLP, DNA sequencing, and a number of PCR-based markers are being used extensively for reconstructing phylogenies of various species. The techniques are speculated to provide path breaking information regarding the fine time scale on which closely related species have diverged and what sort of genetic variations are associated with species formation. Efforts are being made for studying the genetic variation in plants to understand their evolution from wild progenitors and to classify them into appropriate groups (Jeffrey, 1995). RFLP markers have proved their importance as markers for gene tagging locating and manipulating quantitative trait loci (QTL), in evolutionary studies for deducing the relationship between the hexaploid genome of bread wheat and its ancestors (Gill 1991). Specific markers like STMS (Sequence-tagged microsatellite markers) ALPs (Amplicon length polymorphisms) or STS markers have proved to be extremely valuable in the analysis of gene pool variation of crops during the process of cultivar development and classification of germplasm.

Wheat biotechnological research have been relatively slow, due to its ploidy level, the size and complexity of its genome, the very high percentage of repetitive sequences and low level of polymorphism. Lack of genetic polymorphism in crops like wheat and soybeans and the consequent problems to identify molecular markers have been a major limitation to the impact of marker assisted selection (MAS) in wheat breeding. However, the identification of a high number of polymorphisms in Single Sequence Repeats (SSR) should therefore, greatly enhance the potential to find molecular markers in wheat. RAPDs emerged as a convenient and effective technique for tracing alien chromosome segments in translocation lines. RAPD markers provide a useful alternative to RFLP analysis for screening markers linked to a single trait within near isogenic lines and bulked segregants. He *et al.* (1992) reported the development of a DNA polymorphism detection method by combining RAPD with DGGE (denaturing gradient gel elecrophoresis) for pedigree analysis and fingerprinting of wheat cultivars. RAPD markers can be converted to more user-friendly Sequence Characterized Amplified Region (SCAR) markers, which display a less complex banding pattern. SCAR markers linked to resistance genes against fungal pathogens have been characterized in combination with RAPD and RFLP (Liu *et al.,* 1999). In recent years, RAPD and other PCR based markers like Sequence Characterized Amplified Regions (SCAR), Sequence Tagged Sites (STS) and Differential Display Reverse Transcriptase PCR (DDRT-PCR) are increasingly being used for identification of desirable traits in wheat and related genera. These markers have been used in particular for disease resistance against viral and fungal pathogens and also for insect and nematode pests and have the potential of pyramiding of resistance genes for effective breeding programs. PCR based markers have been extensively characterized for genes of resistance against common bunt, *Tilletia tritici*, powdery mildew, *Erysiphe graminis*, leaf rust, *Puccinia recondita*, resistance against Hessian fly, *Mayetiola destructor* and Russian wheat aphid, *Diuraphis noxia*. SSRs or Microsatellites are more promising molecular markers for the identification and differentiation of genotypes within a species. The high level of polymorphism and easy handling has made microsatellites extremely useful for different applications in wheat breeding. Microsatellites have also been used to identify resistance genes like *Pm6* from *Triticum timopheevii* and *Yr15* from breadwheat (Chague *et al.,* 1999).

Some markers have been tested across a number of cultivars as a first step towards marker validation (*e.g. Lr1; Pm1* and *Pm2; Bt-10;* Hessian fly resistance genes; *Cre1* and *Cre3*). Although, in general, not a large enough number of varieties are tested, some markers seem to be promising, such as *Lr9, Lr 10* and *Lr19*. Another factor contributes, though to a lesser extent, to the scarcity of markers used in breeding programmes. Often the scientists developing the markers are not directly connected with breeding activities and/or their laboratories are not set up to handle the numbers that would come out of a breeding programme. The authors are aware of very few me presently using markers as additional selection tools. Examples of those include the use of ASAs or SCARs for cadmium uptake, high protein (HP) content and the 1B/1R translocation. Ogbonnaya *et al.* (1998) report on very robust markers for both *Cre1* and *Cre3* and will be able to use those to pyramid the two resistance alleles. In the process of validating a number of markers for genes of interest (*e.g. Ph1, Sr2, Lr1, 9, 10, 24* and HP) across germplasm and of designing a marker service facility for the easy and reliable implementation of validated markers in the breeding programmes. The facility is being designed to implement only PCR-based markers and will be equipped with a DNA sequencer and an automated pipetting station in order to be able to handle the large number of samples coming from the breeding activities.

Wheat Molecular Genetics and Wheat Genetic Engineering

Loss of genetic diversity has become a problem not only of the natural plant and animal population but also agriculturally important species. Ancient cultivars or landraces and wild relatives of domesticated species are being lost as modern varieties become adopted by farmers. This has led to calls for genetic conservation of crop germplasm. Microsatellites are commonly used to study genetic relationships among genotypes within species because of their high level of polymorphism. Microsatellites markers are currently used to identify genotypes Quantitative trait loci (QTLs) and genetic diversity. Knowledge of genetic diversity of wheat varieties is a prerequisite for the successful management of conservation programs. The first microsatellite in wheat possessed 279 microsatellites by now a total of 1235 microsatellite loci were developed and mapped. Recently 101 microsatellite markers derived from expressed sequence tags EST-SSRs were added into a set of microsatellites (Gao *et al.*, 2004). There is little doubt that wheat has been a difficult species for the application of molecular genetics. The low level of polymorphism between elite varieties coupled with the hexaploid nature of the crop provide significant challenges for those attempting to develop molecular markers and to use then in genetic studies. With the development of AFLP and microsatellite marker systems, renewed studies are underway to analyse the genetic basis of many important traits in wheat.

While always difficult to predict, there are some significant developments in marker technologies and functional genomics. While the PCR-based marker systems have allowed more effective and efficient genotyping, DNA-array technology offers to increase substantially the number of genes that can be analysed. Currently, the cost of the arrayer (to develop the chips containing the desired genes), the array reader (to detect the presence of each gene) and a set of gene sequences (to develop primers to be arrayed) have limited the application of this new technology to wheat. Both the arrayer and reader are decreasing in price and this will make this technology available to many laboratories in the near future. Efforts are also underway to develop complete expressed sequence tag (EST) databases for wheat and related species. If this data can remain in the public sector (such databases for wheat are currently available in the private sector), chips containing a significant number of wheat genes will be produced and used in the not too distant future. Perhaps the next challenge facing wheat researchers will be gene isolation. Examples of transposon-based cloning and even map-based cloning are available in many species. A few researchers have had success in wheat, although the approaches possible are by no means routine. What is promising is the availability of large DNA libraries (yeast artificial chromosomes and bacterial artificial chromosomes) for *Ae. tauschii* (D genome) and *T. monoccocum* (A genome). These represent valuable resources for the identification and isolation of genes from wheat. The use of degenerate primers and probes from other species can readily provide candidate sequences. The only components lacking, or at least limiting, are a reverse-genetic system for wheat and a reliable and efficient genetic engineering system. The latter is becoming more effective, while the former will require significant work to develop. The insertion of Ac/Ds into rice indicates that it is feasible, but the hexaploid nature of bread wheat will make it more complex. It may be more practical to develop the system in a diploid species, such as *Ae. tauschii or T. monoccocum*. Wheat has been and will continue to be a difficult species to investigate at the molecular level; however, recent innovations in technology have opened the door for renewed efforts to use wheat for molecular genetic investigations. While *Arabidopsis* and rice may provide interesting model systems, each plant species will require a certain level of study, hopefully utilizing what is known in other species. One can predict that researchers will have in their hands all the genes from most of the major crop species, including wheat, in the near future. The challenge then will be to determine the function of each and how to use this information to develop improved wheat varieties for feeding the world's growing population.

Cereals, including wheat, have been prime targets for genetic manipulation since the first reports of successful production of transgenic plants. However, the progress towards efficient cereal transformation has been slow, mainly due to difficulties encountered in the establishment of embryogenic cell culture methods and a lack of efficient DNA delivery systems. Several different methods have been attempted for transforming wheat. Direct transfer of DNA into protoplasts mediated by polyethylene glycol and electroporation have proven to be ineffective. A more recent and versatile method for cereal transformation is microprojectile bombardment or biolistics. In this method, the DNA transfer process is genotype and tissue independent, although the regeneration of transformed cells still requires competent cells, which do demonstrate genotype differences. The biolistics method involves shooting cells with micro-particles coated with the desired DNA. By a still somewhat unknown process, the DNA is removed from the particles and ultimately inserts itself into the cell's (usually nuclear) genome. Usually, the insertion events are random and characterized by multiple copies and a certain degree of rearrangements. The most recent cereal transformation method involves the use of a naturally occurring bacterium, *Agrobacterium tumefaciens*. For several years, cereals were classified as non-hosts for *Agrobacterium*, as they were not infected *in vivo* or *in vitro*. Recent investigations have shown that *Agrobacterium* can attach to cereal cells, that these cells produce factors that induce *Agrobacterium* virulence genes and that the bacterium can transfer its T-DNA into the cell (Tinland, 1996).

The first reports of successful *Agrobacterium*-mediated transformation of wheat were those of Hess *et al.* (1990), which involved pipetting *Agrobacterium* onto the spikelets of wheat. Mooney *et al.* (1991) reported the first transformed cells from wheat embryos co-cultivated with *Agrobacterium tumefaciens*. These reports were considered promising, but *Agrobacterium*-mediated transformation of wheat was not considered practical until the recent reports by Cheng *et al.* (1997). Whatever the method used, an effective selection regime is required for the isolation of transformed cells. Often, cereal tissue culture cells show a high level of natural resistance to the antibiotics or herbicides commonly used for selection. Currently, the most common selectable markers systems in use are PPT (phosphinothrycin) and bialaphos for the *bar* gene, G418 and paromomycin for the *npt*II gene, and hygromycin for the *aph*III-IV (or *hpt*) gene. None of these are as selective as required, and usually large numbers of regenerants must be screened to identify the few that are transformed. Even so, fertile transgenic wheat plants presenting herbicide resistance have been produced in several laboratories. Hygro-mycin-resistant wheat plants have also been produced (Ortiz *et al.*, 1996). Chamberlain *et al.* (1994) reported a new promoter (Emu) which drives very high levels of gene expression in wheat. Attention is also given to promoters that can regulate the spatial and/or temporal expression of a gene. The modification of various characters, such as grain quality and disease resistance, will depend also on the availability of promoters to regulate gene expression in specific tissues. Recent reports have demonstrated that the use of a native wheat glutenin promoter effectively controls the expression of an introduced high molecular weight wheat glutenin gene. Expression of the *barnase* gene under the control of a tapetum specific promoter resulted in male sterile wheat plants (De Block *et al.*, 1997). For practical application of genetic engineering technology, it is essential that transformed plants have continued expression and stable inheritance of the inserted transgene(s). There are a number of reports describing non-Mendelian inheritance and inactivation of the foreign genes. Such non-Mendelian inheritance and loss of expression appear to be independent of the cereal species transformed and the nature of the introduced genes. Current evidence indicates that the copy number and insertion position in the genome influence the level of stability and expression. A better understanding of transgene inactivation is needed to improve the efficiency of the transformation process and the stability of the transgenes under field conditions. In addition, the development of site-specific recombination and transposon-mediated delivery systems may provide for improved transgene stability (Bretell and Murray, 1995).

Candidate Genes for Wheat Genetic Engineering

Genetic engineering has opened up new opportunities for plant breeders by enabling them to incorporate genes isolated from organisms outside the gene pools to which they usually have access. This broadens the possibilities they have for overcoming a number of biotic and abiotic stresses. This describes some of the gene strategies that are being considered to provide useful products for breeders in a relatively short period of time.

Resistance to Abiotic Stresses

Traditional approaches at transferring resistance to crop plants are limited by the complexity of stress tolerance traits, as most of these are quantitatively linked traits (QTLs). The direct introduction of a small number of genes offers convenient alternative and a rapid approach for the improvement of stress tolerance. Although, present engineering strategies rely on the transfer of one or several genes that encode either biochemical pathways or endpoints of signalling pathways, these gene products provide some protection either directly or indirectly against environmental stresses. Drought is a major abiotic factor that limits crop productivity, thereby causing enormous loss. Low temperature (LT) tolerance is a complex quantitative character that is expressed in anticipation of and during exposure of plants to temperatures that approach freezing. This environmentally reduced character is determined by a highly integrated system of structural and developmental genes that are regulated by environmentally responsive complex pathways. The superior LT-tolerance genes have been tagged using molecular markers that allow plant breeders to select hardy genotypes without having to wait for a test frost in the field (Fowler and Limin, 2007).

Table 3.29. Candidate genes and targeted traits for wheat genetic engineering

Target Trait	*Candidate Gene(s)*[a]	*Effects*
	Quality	
Bread-making quality	HMW-GS 1Ax1 HMW-GS Dx5B HMW-GS Dy10A	Increased levels of HWM-GS proteins in the endosperm of those varieties lacking these alleles; an increase in dough elasticity and strength
Nutritional quality	a, b and g zeins	Increased levels of proteins in the endosperm providing enhanced nutritional quality
	Biotic stresses	
Fungal diseases	b1,3-glucanase, chitinase	Degradation of fungal cell walls
	Osmotin	Disruption of fungal membranes
	Ribosome-inhibiting protein	Disruption of fungal protein synthesis
Insects	Lectins	Induction of plant defence responses; enhanced resistance to certain grain weevils and aphids
	a-amylase inhibitors	Prevention of weevil growth and development
Nematodes	Cysteine proteinase inhibitors	Resistance to cereal cyst and root knot nematodes
Viruses	Coat protein genes	Prevention of disassembly and movement of viruses
	Abiotic stresses	
Drought, heat, cold Aluminium	Peroxidases citrate synthase	Protection against oxidative stress binding to aluminium leading to prevention of aluminium from entering the roots

a HMW-GS = high molecular weight glutenin subunits.

Induced Fungal Resistance

The fungal infection process is a complex mechanism that usually includes three stages: (i) prior-to-entry relationships; (ii) penetration; and (iii) establishment of the pathogen in the host. The plant resistance mechanisms involve the interaction of several factors, such as the environment, morphological peculiarities of the host and, in particular, biochemical defence genes. The later factors can be improved by the introduction of genes that modify the reaction of the plant metabolism to the infection. Examples are the introduction of the â1,3-glucanase gene that may stop the penetration of the fungus or the introduction of chitinase and osmotin genes that interact with the development of the fungal haustorium by changing the chitin structure and the osmolarity of the membrane. Since single genes encode many of the active anti-microbial/anti-fungal factors, these defence systems are amenable to manipulation by gene transfer. An interesting observation is that the combination of different anti-fungal proteins can lead to synergistic protection against a broad range of phytopathogenic fungi. Zhu's study demonstrated that coexpression of a rice chitinase and a â1,3-glucanase derived from alfalfa gave substantially higher protection against the pathogen *C. nicotianae* than either transgene alone. It is likely that this battery of inducible defences represents a series of complementary mechanisms for protection against both the initial attack and possibly secondary, opportunistic infections.

Quality Traits

Other than reporter genes, perhaps the most targetted trait for genetic engineering in wheat is quality. Seed storage proteins (SSP) are contained in the seed of higher plants. These proteins have been classified as albumins, globulins and glutenins on the basis of their solubility in solvents. The high molecular weight glutenin subunits (HMW-GS) genes in wheat are located on the long arm of the homeologous chromosomes 1A, 1B and 1D. Bread-making properties are particularly associated with variation at the *Glu-D1* and *Glu-A1* loci. The HMW-GS 1Ax1, 1Ax2, 1Dx5 and 1Dx10 have been shown to be associated with stronger dough, better elasticity and, hence, improved bread-making quality. Many elite wheat varieties lack the desired studies have demonstrated that the introduction of one or two HMW-GS genes results in a stepwise increase in dough elasticity. The transgenic lines produced so far have also demonstrated a very high level of expression and stability over several generations. This may imply that native genes are more tolerated by a plant genome. subunits and, thus, many research groups are attempting to introduce these via genetic engineering. In addition to increasing the bread-making quality, altered amino acid composition of the SSP is feasible and could result in improved nutritional properties. For example, the insertion of genes for proteins, such as zeins or albumins, could lead to an increase in the desired amino acid. Other approaches are also being considered, such as reducing the level of anti-nutritional factors and modifying starch and oil composition and content.

Other Possible Targets

Many other targets exist for improving wheat via genetic engineering. Several strategies are available for engineering resistance to insects, for example, the use of lectins for resistance to aphids and lepidopteran pests and á-amylases for resistance to various coleopteran pests. In addition, proteinase inhibitors have been used to engineer both insect and nematode resistance. Viral coat protein genes for enhanced resistance to viruses (Grumet, 1994), citrate synthase for tolerance to aluminium (de la Fuente *et al.*, 1997) and even various options for improving the tolerance of wheat to drought, heat salinity and waterlogging are also being investigated. Often the most advanced studies are in the private sector and, thus, current information regarding their success or failure remains confidential. Hopefully, the reports, or at least the products, will be made available for use in developing new approaches for the multitude of stresses that wheat is subjected to in its growing environment. The bread wheat genotypes Attila, Kauz, Baviacora and Bobwhite and the durum wheat genotypes Minimus, Ariza, Altar and Bajio were found to be highly regenerable and excellent candidates for transformation efforts. Efforts are now focused on the production of transgenic plants via biolistics and *Agrobacterium*-mediated methods. Current target traits are enhanced fungal resistance via various pathogen-related proteins, including chitinase, glucanase and ribosome-inhibiting proteins, and enhanced quality via HMW-GS genes. The first putative transgenic plants are being produced and investigated in a biosafety greenhouse. As these are confirmed, progeny are produced and analysed for the inheritance of each transgene. Ultimately, the best events will be taken to the field under appropriate biosafety regulations for proper evaluation of the expression of the inserted transgene and for effects on other agronomic characteristics.

Gene Pyramiding for Multiple Disease Resistance

Development of resistant varieties for single disease is not enough to save plant product and to feed growing population especially in developing countries. In India, there are number of constraints in the successful production of wheat like leaf and yellow rust, powdery mildew and spot blotch. These all diseases together toll heavy yield losses and put us in the situation to redesign our experiments to develop multiple pest resistant varieties in wheat. The grown varieties on the market do carry some known disease resistance genes against powdery mildew and rusts and low levels of quantitative resistance against leaf spot diseases. However, most of the resistance genes against rusts and powdery mildew are already broken down. In the sustainable agriculture, which is economical both for the farmer and nature, multiple disease resistance is an essential tool against pathogens attack beside cultural practices like crop rotation. With the biotrophic fungi like rusts and powdery mildew, the only solution is the durable disease resistance. In late 1980's Fusarium head blight (FHB) become a world wide problem. Earlier, it was sporadic and localized to China, South America and a few European countries. Varieties resistant to FHB such as Wuhan 1, 2 and 3 and Shangai 7 and 8 and Suzhoe wheats were the product of generous collaboration between various Chinese institutions. Fusarium head blight is a serious disease of wheat (*Triticum aestivum* L.) in humid and semi humid areas of the world. Evaluation of *Fusarium* head blight resistance is time consuming, laborious and costly as the inheritance of resistance is complex phenomenon. The most recent biotic threat to global wheat production is *Ug99*, a virulent race of stem rust which has emerged from Uganda and has been confirmed at widely distributed testing locations in Kenya and Ethiopia. Yield losses of up to 71 percent have been recorded under experimental conditions (Dixon, 2003).

Marker Assisted Selection Strategy for Wheat Improvement

Wheat is most widely grown crop in the world, best adapted to temperate region and is a staple food of about 35 per cent of the world population. Molecular markers have been introduced over last two decades, which has revolutionized the entire scenario of biological sciences. Traditionally, breeders have relied on visible traits to select improved varieties however; MAS rely on identifying marker DNA sequences that are inherited alongside a desired trait during the first few generations. Molecular markers are also considered as useful tools for pyramiding of different resistance genes and developing multi-line cultivars targeting for durable resistance to the disease. With the development of methodologies for the analysis of plant gene structure and function, molecular markers have been utilized for identification of traits to locate the gene(s) for a trait of interest on a plant chromosome and are widely used to study the organization of plant genomes and for the construction of genetic linkage maps. Breeders used molecular markers to increase the precision of selection for best trial combinations. With the development of AFLP and microsatellite marker systems, renewed studies are underway to analyze the genetic basis of many important traits in wheat. In light of the fresh challenges CIMMYT is giving emphasis on molecular breeding, functional genomics, deployment of transgenes for abiotic stresses *etc.*, should get priority to maintain pace with time and growth. In near future, molecular markers can provide simultaneous and sequential selection of agronomically important genes in wheat breeding programs allowing screening for several agronomically important traits at early stages and effectively replace time consuming bioassays in early generation screens.

Production and Utilization of Synthetic Hexaploid Wheat

Since then numerous trait transfers aimed at testing expression levels in particular D genome variants at the allohexaploid level have been performed.

Diversity of *Ae. tauschii* and SHW

Ae. tauschii Coss. is adapted to a variety of environments such as desert margins, steppe regions, stony hills, wastelands, roadsides, sandy shores, and even humid temperate forests, in the edges of wheat fields in eastern Turkey, Iraq, Iran, Pakistan, India (Kashmir), China (the Himalaya), Afghanistan, most of central Asia, Transcaucasia, and the Caucasus region. The diversity of the D genome of this species is much larger than that of bread wheat. Based on morphology, taxonomists have divided *Ae. tauschii* into two subspecies: ssp. *tauschii* and ssp. *strangulata* (Eig) Tzvel. Ssp. strangulata is regarded to be the direct donor of D genome to common wheat. The diversity of *Ae. tauschii* has been studied using molecular tools such as chloroplast DNA variation, AFLP, SSR, isozymes, and random amplified polymorphic DNA (RAPD) markers. Recently, Sohail *et al.* (2012) analyzed the diversity using 4,449 polymorphic DArT markers and found the diversity of ssp. *strangulata*, that is the origin of D genome of bread wheat, contains only a limited part of whole diversity of *Ae. tauschii.* Thus, SHWs produced by crosses between tetraploid wheat and any subspecies of *Ae. tauschii* include untapped amount of genetic variation in which useful genes for bread wheat breeding must be present.

Interspecific Crosses of *Ae. tauschii* and *T. turgidum*

The crossing of *T. turgidum* (AABB) with *Ae. tauschii* (DtDt) followed by the induction of chromosome doubling of the hybrid, via colchicine treatment, produces an amphiploid referred to as a synthetic hexaploid (AABB DtDt). When SHW is crossed with cultivated hexaploid wheat, typically there are 21 chromosome pairs at meiotic metaphase I that are associated primarily as ring bivalents indicating complete chromosome homology and full fertility. In this approach, SHWs are often spontaneously produced from partially fertile hybrid plants with a high frequency of unreduced gametes arising from meiotic restitution [Xu and Joppa, 2000].

Direct Cross of Diploid *Ae. tauschii* to Common Wheat

The second approach uses direct crosses between *Ae. tauschii* (DtDt) and common wheat (AABBDD) to produce a hybrid with the genome constitution ABDDt. The embryo of the hybrid (F_1) seed must be excised within 12–18 days postpollination depending on the prevailing environment and grown on an artificial culture medium to prevent seed abortion. The hybrid is then backcrossed to the common wheat parent and selection made in the progeny for 42-chromosome plants, which have 21 pairs of bivalents at meiosis. This eliminates the confounding effects of segregation in the A and B genomes. This method also has the advantage that no colchicine treatment is necessary in the cross between wheat and the *Ae. tauschii* hybrid. However, lines derived from direct-crossed hybrids have the disadvantage of segregating for the D genome, and exhibit instability because of aneuploidy, potentially making genetic analysis more difficult [Mujeeb-Kazi 1993].

Direct Cross of Tetraploid Wheat to Common Wheat

Following the determination of the chromosome numbers of bread and durum wheat by Sakamura and Sax in 1918, the first pentaploid products of crosses between hexaploid and tetraploid wheats were produced and cytologically examined. Meiotic behaviour of the pentaploid (2n=5x=35, AABBD) was subsequently described with parental (28 or 42 chromosome) euploid progeny being recovered. Similar to the crosses of *Ae. tauschii* to common wheat, direct crosses of tetraploid to hexaploid wheat have been used to transfer useful genes from various tetraploid subspecies into common wheat. The pioneering work of McFadden [1930] in transferring resistance to stem and leaf rust from cultivated emmer wheat (*T. turgidum ssp. dicoccum*) into common wheat heralded the use of this approach in interspecific crosses. Through this procedure, McFadden (1930) successfully introduced Sr2 for adult plant stem rust resistance from emmer wheat into common wheat and named it 'Hope'. Sr2 has been deployed in many cultivars developed in the regions where wheat production is vulnerable to stem rust [McIntosh *et al.*, 1995] and continues to confer effective rust resistance more than 80 years later. Both Sr2 and Sr14 are presently an important source of resistance to *Ug99* lineage races of stem rust. Several other genes, including Lr23 for leaf rust resistance from durum wheat, Yr15 for resistance to stripe rust, Pm30 and pm42 for resistance to powdery mildew and gene(s) for high grain protein content Gpc-B1 and stripe rust resistance Yr36 from wild emmer, and Hdic for resistance to Hessian fly from cultivated emmer, were transferred into common wheat through direct tetraploid– hexaploid crosses. Recently, Xie and Nevo (2008) provided an overview of genes transferred to common wheat from *T. dicoccoides*. Compared with SHW approach, a common problem in direct cross of tetraploid to hexaploid is the low fertility of the F_1 pentaploid (AABBD) plants. The pentaploid interspecific hybrids between the wild emmer accession G-25 and common wheat cultivars were highly sterile (fertility of l per cent –2 per cent). They overcame the sterility by exposing the F_1 progeny to open mass pollination with the common wheat cultivars or using a bridging cross between the progeny of an emmer-durum cross and common wheat cultivars. Presently, the sterility of the pentaploid hybrids is usually overcome by further crossing or backcrossing the hybrids with common wheat cultivars. Direct cross of tetraploid to hexaploid followed by backcross would hasten the recovery of euploid progeny (2n=42) with introgressed genes.

Direct Versus Synthetic Approaches

The use of restitution nuclei and the restoration of the parental wheat phenotype by limited backcrossing. The direct hybridization between common wheat and *Ae. tauschii,* followed by two backcrosses to wheat, permitted rapid recovery of the desired genes from *Ae. tauschii* in euploid backgrounds. They postulated that two backcrosses involving a diploid donor and hexaploid recurrent parent result in approximately 96 per cent restoration of the total recurrent parent genome, a proportion almost equivalent to that resulting from four backcrosses involving the same ploidy. This suggests that the time scale for germplasm development possessing agronomically acceptable progenies with *Ae. tauschii* as a donor parent is shorter than that usually associated with interspecific crossing. However, aneuploidy is generated in such a progeny and it is necessary to either make further crosses using the direct cross genotype as the male parent and rely on gametic selection or cytologically select euploid plants. Cox (1997) reported that F_1 plants between wheat and *Ae. tauschii* set approximately one seed per spike, of which about 50 per cent are germinable. Chromosome numbers of BC_1 plants range from 35 to more than 50, of which 25 per cent –50 per cent arise from fertilization of an unreduced female gamete. The backcrossing BC_1 plants as males would screen out aneuploid pollen and produce primarily 42-chromosome BC_2 progeny. However, Cox (1997) reported that among all BC_2 plants produced from their studies, only 60 per cent had 42 chromosomes. The rates of transmission of alleles from the diploid parent to BC_2 progenies and recombination among loci are generally at the rates expected under the assumptions of non-preferential segregation and complete chromosome homology. Lagudah *et al.* (1993) argued that in the direct cross approach, an a priori assumption is made that the introgressed trait is sufficiently expressed to allow detection and selection in the background of the recurrent wheat parent while in a SHW, the expression of the *Ae. tauschii* trait can be determined. Cox (1997) suggested that several considerations influenced the use as direct backcrossing approach to improve winter wheat.

Some of those considerations were:

1. Threshability of the Grain

In direct-backcross populations, only one gene conditioning non-threshability, Tg from *Ae. tauschii* segregates whereas the additional gene q from the tetraploid parent segregates in crosses between wheat and SHW. Thus, single gene segregation results in a higher frequency of threshable progeny.

2. Adaptation of the Durum Parents

The SHW have been more useful as donor parents in northern Mexico and Australia, where durum and spring wheat are both well adapted. In contrast, durum wheat is not as productive, where winter wheat is produced. Thus, introgression of genes from the non-adapted durum parent of a synthetic into the A and B genomes of hard winter wheat, at the same time genes are being introduced into the D genome, disrupts adaptation to a much greater extent than do direct crosses between hard winter wheat and *Ae. tauschii.*

3. Rapid Recovery of Recurrent Parent Genotype

Genes may be backcrossed directly from *Ae. tauschii* into almost any elite winter wheat cultivar or breeding line, and progeny similar to the winter wheat parent can be selected from BC_2 populations. Recently, the direct hybridization approach has been reexamined where wheat pollen was applied to the female *Ae. tauschii* parent yielding a high 35 per cent embryo formation frequency. However, the F_1 hybrid plant recovery frequency was only 6.8 per cent (Sehgal *et al.*, 2011). During the production of SHW via reciprocal crosses extremely high embryo excision frequencies were observed (between 65 per cent and 80 per cent) against the reciprocal range of 3 per cent –28 per cent. Different durum wheat parents were used in producing SHW and a few of these parental lines appear to exhibit very high levels of crossability with *Ae. tauschii.* The durum cvs 'LARU', 'Altar 84', 'Aconchi', 'Croc-1', and a line with the pedigree of 'Cando/R143//Erite/Mexi' were better combiners and later additions included cultivars 'Chen', 'Croc', 'Doy', 'Dvergand', and a line 'D67.2'.

Apart from precision transfers into the D genome only, the speed with which advanced lines of good agronomic phenotype are recovered warrants more attention to this strategy. It is evident from the international nursery that only a few *Ae. tauschii* accessions are represented in their pedigrees. Hence, there is a need to widen the *Ae. tauschii* accessions sources by including collections from areas of particular stresses, for example, drought and thus broaden the variability for such stress tolerance in addition to choose of *Ae. tauschii* on the basis of genotype diversity. These accessions could enter future direct crossing programs that could target breeding goals more effectively. This range in the future could be widened not only for bridge crosses but also for the well-targeted traits.

Genetic Resources in Synthetic Hexaploid Wheat

Since the first SHW was produced in the 1940s (McFadden and Sears 1944), over 1,500 SHW have been developed globally. The production of SHW on a large scale was conducted from 1988 to 2010 with an output of approximately 1,300 SHW using approximately 900 *Ae. tauschii* accessions. The amount of genetic diversity in the *Ae. tauschii* populations were considerably higher than the D genome of bread wheat for many disease and insect resistance, isozymes, and seed storage protein. Additionally, in the 1980s, LR Joppa developed a number of spontaneous SHW from partially fertile hybrids between 'Langdon' durum and different *Ae tauschii* accessions. Among these Langdon-derived SHW, 46 were recently characterized for resistance to various diseases and high molecular weight glutenin subunits. Amongst the two elite sets of 128 SHW were selected based on agronomic performance and disease resistance. Both Langdon- derived SHW and elite SHW are widely distributed and are used as base germplasm for genetic study and breeding. A limited number of winter habit SHW were also produced in and are maintained. These are not yet in wide distribution but have the potential to aid winter wheat breeding programs. The winter SHW, like the spring habit SHW, also exhibit a wide range in diversity for major phenological traits and allelic diversity profile. On the basis of percentage of polymorphic amplified fragment length polymorphism (AFLP) markers, SHW contained a higher level of molecular diversity than observed in cultivated hexaploid wheat. The significantly higher number of alleles per locus and Nei's gene diversity within SHW than in common bread wheat.

Using Synthetic Hexaploid Wheat in Breeding

There are a number of approaches to SHW breeding.

1. Direct Phenotyping

In this approach, SHW are evaluated for genetic variation for disease or insect resistance and tolerance to a range of biotic and abiotic stresses. This is the traditional and most widely reported procedure in the literature.

2. Genetic Analysis via Crosses and Backcrossing

Promising SHWs that carry desired traits are crossed with adapted elite wheat cultivars to fast track the development of elite breeder-friendly parental stocks. In some cases, genetic and quantitative trait analyses are undertaken concurrently. This traditional approach involves, first, the screening and identification of SHW with the desired characteristic followed by backcrossing to elite cultivars to generate "synthetic derived backcrossed bread wheat lines '(SBLs)', introducing the targeted trait from the SHW donor into adapted germplasm and generating novel recombinant genotypes to widen the existing primary gene pool of common wheat. Such novel genotypes result from either new Dt or AdBd (from durum, *T. turgidum ssp. dicoccum* or *T. turgidum ssp. dicoccoides*) derived alleles or interactions between these new variants and the hexaploid wheat genome. Thus, the impact of SBLs is not exclusive to the introgression of the D genome variants. The additional backcrossing results in considerable time and resources invested in prebreeding efforts before the genetic stock with desirable traits can be available for use in breeding depending on the tetraploid used. For SHW produced using cultivated durum, one BC may be enough. Regardless, population size is probably more important and warrants further investigation to determine the appropriate population size needed to recover desirable traits in an adapted genetic background.

3. Advanced Backcross-Quantitative Trait Loci (AB-QTL) Analysis

Tanksley and Nelson (1996) proposed the advanced backcross AB-QTL approach, which has some advantages over classical QTL analysis. In this method, the detection of QTL and the development of improved breeding lines are conducted simultaneously. Unadapted germplasm from wild species is used to detect exotic alleles superior to the elite alleles, which simultaneously broadens the genetic variation of modern cultivars. Since the AB-QTL strategy is based on BC2 or BC3 generations, the portion of the exotic donor genome, and thus linkage drag, is considerably reduced. A number of AB-QTL populations have been developed using synthetics. Disadvantages include smaller classes that reduce power to detect

QTL and a large investment in time and labor to develop the populations and considerable investment to phenotype them.

4. Development and Use of Introgression Lines (ILs) or Substitutions Lines

This technique offers the potential for further gains in the use of SHW. Development of specific D genome introgressed segments starting from *Ae. tauschii* single chromosome substitutions or ILs in common wheat has received limited attention. An IL set represents the complete or nearly complete genome of a wild species in the uniform background of an elite cultivar, where each single line solely contains a single marker-defined segment of the exotic parent. Advantages of using ILs include evaluating the effects of introgressed segments in different target recipient backgrounds, the ability to dissect the contribution of the diploid D genome from the other genomes, and the potential to reveal cryptic variation. However, ILs preclude the assessment of epistasis (known to be important for some traits) and cannot dissect the effect of loci on the same segment. One of the impediments in using the IL as a tool in breeding is the time taken to develop them.

There is no documented evidence on the number of backcrosses that are best when making crosses between SHW and adapted common wheat cultivars. Generally, one to two backcrosses appear necessary, and it is advisable to target larger than normal BC_1F_1 and BC_1F_2 populations. To estimate the relative size and genetic structure of BC_1 and BC_2 progeny lines derived from synthetic wheat, recently conducted molecular marker analysis of crosses between spring wheat Paragon and four SHW lines. Tracing DArT haplotypes of progeny composing individual streams from BC_1 into BC_2 indicated that, on average, over the whole genome, a population size of approximately 300 individuals would capture the whole synthetic genome at BC_1, but an average population of more than 800 individuals would be needed to capture its entirety at BC_2. However, conducting the same analysis on a genome-by-genome basis suggested that the number of lines needed to capture each genome differed markedly with estimated population sizes of 450, 600, and 1500 for the A, B, and D genomes, respectively. In a global evaluation of SBL for yield, the SBL possessing a smaller portion of the SHW genome tended to perform better, indicating that at least one backcross is required to improve agronomic type sufficiently to exploit underlying variation for yield and adaptation. The application of appropriate selection pressure against tenacious glumes, hard threshing and non-adapted plant types at the early segregating generations would result in the identification of good agronomic types with the desired traits. One or two backcrosses are sufficient to introgress desirable traits from SHW into adapted wheat with minimal linkage drag. They suggested that after one or two backcrosses the coefficient of parental of SHW in derived lines should stabilize between 12.5 per cent and 25 per cent. Zhang *et al.* (2005) used microsatellite (SSR) markers to quantify introgression patterns of the SHW into locally adapted cultivars and found that only a small segment of the SHW remained after crossing and selection. This indicates that through back- and/ or top-crossing and the use of large F_2-populations, it is possible to select progeny with favorable additive alleles from the SHW, thus avoiding undesirable linkage drag. One of the impediments in the use of SHW is the relatively high frequency of F_1 hybrid necrosis when crossed with adapted common wheat cultivars. Hybrid necrosis is the gradual premature death of leaves or plants in certain F1 hybrids of wheat, and it is caused by the interaction of two dominant complementary genes Ne1 and Ne2 located on chromosome arms 5BL and 2BS, respectively. When the two dominant alleles for both loci were present either in homozygous or heterozygous condition (*i.e.*, Ne1Ne1Ne2Ne2, Ne1ne1Nc2ne2, Ne1Ne1Ne2ne2, and Ne1ne1-Ne2Ne2), necrosis occurs and seedlings plants die early in plant development at the one or two leaf stage. Hexaploid and durum wheat differ in the frequency of alleles at these two loci; hence, crosses between SHW carrying the durum complement of necrosis genes and hexaploidy wheat frequently result in hybrid necrosis. The frequency of necrosis is even higher in crosses with *T. turgidum ssp. dicoccum* or *T. turgidum ssp. dicoccoides*. Consequently, hybrid necrosis quite often limits successful transfer of desirable traits from *Ae. tauschii* into common wheat. The difficulties in using some of the SHWs in a wheat breeding program was further demonstrated in their attempt to transfer *Septoria tritici* blotch (STB) resistance identified in *Ae. tauschii* to common wheat. The F_1 hybrids of six wheat varieties (Egret, Owlet, Vulcan, Suneca, Sunfield, and Hartog) were severely necrotic and hybrids with Millewa segregated for the presence or absence of hybrid necrosis. Seven F_1 hybrids (*Grebe, Janz, Kite, M1696, Rosella, Summit, and Teal*) were not necrotic. Similarly, synthetic wheat lines were crossed as male parents to elite winter wheat cultivars (*Robigus, Alchemy, Cordiale, Gallant, Humber, Oakley, Q Plus, Scout, Shamrock, Timber, and Viscount*). Severe hybrid necrosis occurred in all progeny except those produced from crosses with Robigus. Therefore, in attempts to exploit SHW for common wheat improvement it is necessary to use different wheat cultivars to minimize the incidence of necrosis, the failure of the cross, and maximize the chance of finding healthy combinations. It is essential to keep records on necrosis in crosses between common wheat and SHW so that its incidence is minimized by avoiding the use of wheat parents more likely to possess the complementary necrosis genes. This is important to consider for a wider use of diploid and tetraploid species in making new SHW. Chu *et al.* (2006) mapped Ne1 and Ne2 using microsatellite markers and the closely linked markers identified in that study may be used to genotype SHW and wheat parents for Ne1 and Ne2.

Expression of *Ae. tauschii* and *T. turgidum* Genes in Synthetic Hexaploid Wheat

Evaluation of SHW provides a means of determining if traits from progenitor species are expressed at a higher ploidy level. Expression of genes for traits of both simple and complex inheritance is reported to change when transferred from the diploid relative to a higher ploidy. The background of the tetraploid genome modulates the expressions of some traits. Sohail *et al.* (2011) used 33 *Ae. tauschii* accessions along with their corresponding SHWs produced by crosses with common cultivar of durum wheat, Langdon and compared the expression of drought tolerance related traits (the net photosynthetic rate, stomatal conductance, leaf water potential, *etc.*) under well-watered and drought conditions. Although there was large variation in both *Ae. tauschii* and SHWs, the traits of *Ae. tauschii* were not significantly correlated with their corresponding SHW lines, indicating that these quantitative traits are expressed as a result of complex interaction of many genes in hexaploids. Tolerance of abiotic stresses such as salinity, heat, and drought cannot be directly defined by the performance of their respective *Ae. tauschii* parents. They proffered the following explanation (i) the homoeologous genes A and B in background of SHW leading to epistatic interactions and differential gene expression; (ii) a modified gene expression caused by genomic changes during artificial hybridization; and (iii) a high genetic diversity within *Ae. tauschii* accessions, which can result in phenotypic variation in the SHW, especially if unintentional selection or drift changes the frequencies of alleles present in the SHW compared with the *Ae. tauschii* parents. There have been many studies on the expression of disease resistance genes from *Ae. tauschii* in the hexaploid background. Early researchers suggested that the transfer of resistance to leaf rust (*Puccinia recondita f. sp. tritici*) and stem rust (*Puccinia graminis f. sp. tritici*) from *Ae. tauschii* into common wheat has been difficult because it is either partially or completely suppressed in SHW. This raises major implications in the use of such sources of resistance for wheat improvement. The suppressor gene(s) on chromosome 7D of wheat that inhibits expression of stem rust resistance in the A and/or B genome. They indicated that the mechanisms underlying this phenomenon are yet not fully understood. The identification of alternative alleles in *Ae. tauschii* that lack the suppressor effect, and crossing them with common wheat (substitution for the corresponding homoallele in common wheat), may allow expression of some rust resistance genes present in the A or B genome, which otherwise would have been suppressed. Alternatively, suppressor alleles may be screened out using molecular markers. Deletion or inactivation through mutation of the suppressor gene might allow expression of genes for resistance already present in hexaploid wheat or in related donor species (Williams *et al.*, 1992).

Bai and Knott (1992) indicated that suppressors of disease resistance were in the D genome of common wheat. Additionally, there was evidence that suppressors were disease specific. In the first study, 10 rust resistant wild tetraploid wheats were crossed with both durum and common wheat. Resistance to leaf rust and stem rust was expressed in the hybrids with durum wheats but suppressed in the hybrids with common wheats suggesting that the suppressor(s) may be localized in the D genome of the common wheat. To test this possibility, 12 pentaploid F_1 hybrids between durum and common wheat cultivars were tested with leaf rust race 15. Leaf rust resistance from the durum parents was suppressed in all cases. Further, to localize the genome position of these suppressors, a set of 14 D-genome disomic chromosome substitution lines in 'Langdon' durum were used. The result demonstrated that chromosomes 1B, 2B, and 7B carried genes for resistance to stem rust while chromosomes 2B and 4B possessed genes for resistance to leaf rust and 1D and 3D carried suppressors. Crosses between seven D-genome monosomics of '**Chinese Spring**' (CS) with three *T. turgidum* ssp. *dicoccoides* accessions, showed that 'CS' carries suppressors on chromosomes 1D, 2D, and 4D which suppressed the stem rust resistance of all three *T. turgidum* ssp. *dicoccoides* accessions. Therefore, in the absence of any one of the three 'CS' chromosomes carrying suppressors, resistance was expressed indicating that some form of complementary gene interaction is involved. Evidence for specificity was supported by results from a later study in which 'CS' chromosome 3D suppressed resistance to leaf rust race 15 in all three *dicoccoides* accessions, but had no effect on resistance to the other strains. Similarly, a suppressor on CS chromosome 1D suppressed resistance to leaf rust in *dicoccoides* accession PI276000 only. Similar results were obtained, that chromosome 7D of cv. Canthatch, carried a dominant effect gene that suppressed resistance to all stem rust strains tested except for race C64. In another study, gene(s) on the A and B genome of cv. Tetra Canthatch suppressed resistance from *Ae. tauschii* accession RL5459 to four out of nine leaf rust races in SHW (Kema *et al.*, 1995). The expression of seedling resistance to both stripe and leaf rust in SHW was also determined by suppressors present in the A or B and D genomes. The presence of suppressors to stripe rust (*Puccinia striiformis*) resistance in A or B and D genomes of *T. turgidum* ssp. *durum* and *Ae. tauschii.* Of 13-hexaploid-wheat lines synthetized by crossing resistant T. turgidum ssp. durum with susceptible/intermediate *Ae. tauschii,* only six were resistant while others were either intermediate or susceptible. This suggests that resistance was fully expressed, partially expressed, or suppressed depending on the presence of suppressor gene(s) in the *Ae. tauschii* genotypes, again, implicating suppressors within the D genome. Similarly, of 62 SHW derived from crosses of susceptible/intermediate *T. turgidum ssp. durum* with resistant *Ae. tauschii,* only 10 showed a level of resistance similar to the *Ae. tauschii* parents while 52 displayed intermediate to susceptible reactions. The high frequency of partial to complete suppression suggests that suppressors of stripe rust resistance are also in A or B genomes. Knott (2000) also found suppressors to stem rust to be present on either A or B genome chromosomes of durum near-isogenic line 'Medea'.

The mechanism of suppression of stem and leaf rust resistance in wheat is genetically complex. Resistance genes located on one genome are inhibited by a suppressor(s) on another. Suppressors of rust resistance are on the D-genome chromosome of wheat and suppressed genes on the A- and B-genome chromosomes of tetraploid wheats and vice versa. The genes also appear to suppress resistance to specific parental genotypes and pathogen isolates. Badebo *et al.* (1997) investigated the genetics of suppressors of resistance to stripe rust in SHW derived from *T. turgidum subsp. dicoccoides* and *Ae. squarrosa*. They reported that resistance in wild emmer accession, 342-2M was suppressed to race 7E134b in three SHWs due to the D genome in *Ae squarrosa*. However, they also observed reciprocal suppression consistent. In addition to specific suppressor loci, genome-wide processes appear to be active. In this regard, it should be noted that the process of creating new SHW, that is, the process of allopolyploidization, generates two "shocks." One is hybridity, by which two diverged genomes are joined together to form one nucleus and the other is polyploidy, resulting in duplicated genomes. Gene expression, for example, gene silencing and gene activation, was changed in the newly synthesized allopolyploid. Gene silencing was a general response in polyploidization and mainly resulted from DNA sequence elimination or methylation. Analysis of genome-wide gene expression (cDNA-AFLP, RT-PCR, and reverse-Northern analysis) of SHW suggested that gene expression changes are an intrinsic part of polyploidy and that gene loss and, or silencing may be partly or wholly responsible for variation in trait expression. Analysis of SHW developed have undergone numerous rounds of meiosis, suggested that a large proportion (up to 7.7 per cent) of parental cDNA-AFLP fragments were altered in the SHW lines. In addition, *Ae. tauschii* genes were reported to suffer a significantly higher level of suppression than *T. turgidum* genes; a similar situation was found in the common wheat Chinese Spring. Moreover, transcript analysis during resynthesis indicates that gene loss and or silencing occurs early in synthetic allohexaploid formation, possibly during the development of F_1 intergeneric hybrids or just after chromosome doubling (Kashkush *et al.*, 2002). The aforementioned studies suggest that two mechanisms may control gene expression during artificial allohexaploidization in SHW, "suppressor genes" on the alternate parent or silencing of some genes upon polyploidization. Further research is needed to differentiate between these two mechanisms as the former may be overcome by selecting an alternative parental partner whilst the latter may be more stochastic and overcome merely by repeating the cross. Taken together, one is tempted to suggest that trait expression in *Ae. tauschii* is down regulated to a greater degree than that of tetraploid parents. Therefore, a fuller understanding of trait expression in SHW is, therefore, critical to efforts to increase and efficiently utilize the genetic diversity in SHW for common wheat improvement.

Synthetic Hexaploid in Wheat Improvement

Within the last two decades, numerous SHW lines and their derivatives have been produced and evaluated and useful genes/traits have been identified. These include disease resistance genes and traits that contribute to increased yield both under rainfed-drought or moisture limiting conditions and optimal yield expression and the underlying yield components such as large kernel size, high biomass, high above ground biomass, and a large root system that captures water from deep soil zones, tolerance to salinity, waterlogging, and sodicity. Additional traits include improved quality and micronutrient contents such as Fe and Zn. It was reported that by the year 2003–2004, 26 per cent of all new advanced lines made available through international screening nurseries to cooperators for either irrigated or semiarid conditions were synthetic derivatives. However, in the most recent survey carried out as part of this review using data from 7 years, on average 17 per cent of all entries were synthetic derived. However, the percentage varied among nurseries with 9 per cent for Elite Spring Wheat Yield Trial (ESWYT), 11 per cent for International Bread Wheat Screening Nursery (IBWSN), 13 per cent for Semiarid Screening Nursery (SAWSN), and 35 per cent for Semiarid Wheat Yield Trial (SAWYT). Data from ICARDA's Spring Bread Wheat Observation Nursery, "SBW-ON," 2007–2012, revealed that an average of 17 per cent of all entries distributed were synthetic derivatives and the percentage between nurseries ranged from 5 per cent in SBW-ON-2007 to 24 per cent in SBW-ON-2012. Lage and Trethowan (2008) in an earlier study using SAWYT 5-15, examined the proportion of SHW derived lines in each SAWYT, as well as their coefficient of parentage (COP). They reported that the first SHW-derived lines appeared in the 5th SAWYT in 1997. Eight percent of the lines were SHW-derived increasing to 46 per cent by the 15th SAWYT. The average of COP of SHW in the SHW derived lines in the 5th SAWYT was as high as 75 per cent but decreased steadily to an average of 19 per cent by the 15th SAWYT. Several groups around the world have or are currently producing publically available synthetic wheat germplasm. Recently, a population of 215 doubled haploid and recombinant inbred line populations consisting of 1,700 F_6 derived lines from the cross Synthetic

W7984_Opata M85 were produced (Sorrells *et al.*, 2011). In recent years, a number of initiatives were established to more systematically explore the utilization of SHW to increase wheat productivity. For example, the Australian Grain Research and Development Corporation (GRDC) funded the Synthetic Evaluation Project from 2001 to 2007–a multi-institutional project to improve the productivity and sustainability of rainfed wheat production in the Mediterranean environments. The project aimed to identify sources of disease resistance and abiotic stress tolerance in SHW and to incorporate them into Australian adapted germplasm for yield and drought. More than 5,000 BC_1 derived F_5 lines in two elite spring (Paragon) and facultative (Xi-19) wheat varieties have been produced. The approach operated on two levels, (1) production of non-selected backcross lines for mapping and gene discovery, (2) development of precommercial germplasm through field selection for introduction into elite UK wheat germplasm by commercial plant breeding companies.

The first documented release of a wheat cultivar derived from SBL by backcrossing SHW with locally adapted Chinese cultivar was reported in 2006. The cultivar, "**Chuanmai 42**" which now grows in more than 100,000 ha reportedly out-yielded all commercial varieties over 2 years in Sichuan provincial yield trials, including the commercial check cultivar Chuanmai 107 by 22.7 per cent (Yang *et al.*, 2006, 2009). Similarly, two lines have been released in Uruguay by INIA named 'Genesis 2354' and 'Genesis 2359'. A third variety designated '**NOGAL**' was reported to be released by a private company. In the United States, one of the most successful uses of SHW in wheat breeding was the deployment of Gb3 for resistance to greenbug (*Schizaphis graminum* Rondani) in wheat cultivars. Gb3 is derived from *Ae. tauschii* chromosome 7D and was originally identified in Langdon-derived SHW line Largo. It has been deployed in wheat cultivars '**TAM 110**' and '**TAM 112**' (**PI 643143**) and provides effective protection from greenbug (Lu *et al.*, 2010).

Disease and Pests Resistance

Significant genetic diversity for a wide range of biotic stresses has been reported in SHW, direct crosses involving *Ae. tauschii* and tetraploid and their derived wheat lines. These include resistance to Hessian fly, greenbug, Karnal bunt, powdery mildew, stem rust, stripe rust, leaf rust, *Septoria nodorum* blotch, *Septoria tritici* leaf blotch (STB), spot blotch, *Fusarium* head blight, tan spot (yellow leaf spot), cereal cyst nematode (CCN) (*Heterodera avenae* Woll.) and root lesion nematodes (RLN) (*Pratylenchus thornei* and *Pratylenchus neglectus*). The identification of SHWs with varying levels of resistance to *P. thornei*. Of the resistant SHW, with both durum and *Ae. tauschii* parents resistant, six SHW were based on *Ae. tauschii* accessions from subsepecies typica, seven from meyeri, five from *strangulata* and one from an intermediate meyeri/typica accession. There have also been reports of SHW possessing resistance to both P. neglectus and P. thornei. Recently, SHWs possessing resistance to several different diseases were identified (Zwart *et al.*, 2010). Following the identification of CCN resistance genes in *Ae. tauschii* including the Cre3 gene, the resistant *Ae. tauschii* accessions were used to produce SHW and subsequently backcrossed with Australian cultivars. The CCN resistant lines carrying the Cre3 gene from *Ae. tauschii* were developed and advanced to the final stages of evaluation

Table 3.30. Genes identified and transferred from various sources into bread wheat and germplasm produced including traits whose genes are not yet characterized and localized

Germplasm	*Pedigree*	*Description*	*Gene(s)*
		From *Ae tauschii*	
KS85WGRC01	TA1644 (*Ae tauschii*)/Newton//Wichita	Hessian fly-resistant, hard red winter wheat germplasm, also resistant to soil borne mosaic virus.	*H22* 1DL
KS80WGRC03	TA1042 (*Ae tauschiii*)/2'Wichita		*H23* 6DS
KS89WGRC04	TA1695 (*Ae tauschii*)//3"Wichita	Hessian fly-resistant, hard red, also resistant to soil borne winter wheat germplasm mosaic virus.	*H22* 1DL
KS89WGRC06	TA2452 (*Ae tauschiii*)/TA1642 (*Ae tauschiii*)//2'Wichita/3/Newton	Hessian fly-resistant, hard red, also resistant to soil borne winter wheat germplasm	*H24* 3DL
KS89WGRC07	Wichita//TA1649 (*Ae tauschiii*)/'Z' Wichita 3/Newton	Leaf rust-resistant, hard red, also resistant to soil borne winter wheat germplasm	*Lr21* 1DS
KS90WGRC10	TAM107'3/TA2460 (*Ae tauschiii*)	Leaf rust-resistant, hard red, also resistant to soil borne winter wheat germplasm	*Lr39* 2DS
KS91WGRC11	Century'3/TA2450 (*Ae tauschiii*)	Leaf rust-resistant, hard red, winter wheat germplasm	*Lr42* 2DS
KS91WGRC12	Century'3/TA2541 (*Ae tauschiii*)	Adult leaf rust-resistant, hard red, winter wheat germplasm. Also segregating for resistance to wheat soil borne mosaic and wheat spindle streak mosaic viruses	
KS92WGRC15	TAM200/KS86WGRC02//Karl	Leaf rust-resistant, hard red, winter wheat germplasm. Molecular analysis indicates that this line was derived from TA 1640, similar to KS89WGRC07	*Lr40*1DS
KS92WGRC16	Triumph 64/3/KS8010-71/TA2470(*Ae tauschiii*)//TAM 200	Leaf rust-resistant, hard red, winter wheat germplasm.	*Lr21.Lr39* 1DS.2DS
KS92WGRC21	TAM 200'3/TA2570 (*Ae tauschiii*)	Hard red, winter wheat germplasm with resistance to powdery mildew and wheat soil borne and spindle streak mosaic viruses.	
KS92WGRC22	Century'3/TA2567 (*Ae tauschiii*)	Hard red, winter wheat germplasm with resistance to powdery mildew and wheat soil borne and spindle streak mosaic viruses.	
KS93WGRC25	Karl'3/TA273 (*Ae tauschiii*)	Hard red, winter wheat germplasm with resistance to powdery mildew and wheat soil borne and spindle streak mosaic viruses.	*H26* 3DL
KS95WGRC33	KS93U69 (sister line of KS90WGRC10)'3/TA2397 (*Ae tauschiii*)	Hard red, winter wheat germplasm with resistance to powdery mildew and wheat soil borne and spindle streak mosaic viruses.	*Lr41* 1DS

Germplasm	*Pedigree*	*Description*	*Gene(s)*
KS96WGRC39	TAM107'3/TA2460 (*Ae tauschiii*)	Tan spot-resistant, hard red, winter wheat germplasm	
KS96WGRC40	TAM107'3/TA2460 (*Ae tauschiii*)// TA2397/3/3TAM107-3TA2460	Hard red, winter wheat germplasm with resistance to soil borne cereal mosaic virus	*SBWMV5*DL
KS00WGRC44	TAM107'3/TA1715 (*Ae tauschiii*)	Leaf rust-resistant hard red, winter wheat germplasm. Resistance due a single dominant gene	
KS04WGRC49	Karl 92'3/TA2473 (*Ae tauschiii*)	Hard red, winter wheat germplasm with unique high-molecular-weight glutenin and gliadin subunits	
NC97BGTD7	Saluda '3/*Ae tauschiii* TA2492	Hard red, winter wheat germplasm with unique high-molecular-weight glutenin and gliadin subunits	*Pm34* 5DL
NC96BGTD3	Saluda '3/*Ae tauschiii* TA2377	Wheat germplasm with resistance to powdery mildew	*Pm35* 5DL
RL5713	Thatcher'7//RL5713 (*Ae tauschiii*) MarquisK	Wheat germplasm with resistance to leaf rust	*Lr32* 3DS
RL5404		Wheat germplasm with resistance to leaf rust	*Lr22a*
XX194	Moroccos183/AE457-78 (*Ae tauschiii*)	Wheat germplasm with resistance to powdery mildew	*Pm2*
Largo	Langdon/*Ae tauschiii PI268210*	Hexaploid synthetic wheat with resistance to green bug	*Gb3*7DL
W7084	Altar84/TA1651 (*Ae tauschiii*)	Hexaploid synthetic wheat with resistance to green bug	G*b7 7DL*
L-18913	Langdon/*Ae tauschiii* AUS18193	Hexaploid synthetic wheat with resistance to green bug	*Gb7*7DL
L-18911	Altar84/*Ae tauschiii* AUS 18911	Hexaploid synthetic wheat with high molecular weight glutenin subunit	*GluDx5Dy10 1DL*
L-18913	Langdon/*Ae tauschiii* AUS 18913	Hexaploid synthetic wheat with high molecular weight glutenin subunit	*GluDx2DyT10 1DL: Sr4g, 1DL, 2DS*
L-18964	Langdon/*Ae tauschiii* AUS 18964	Hexaploid synthetic wheat with high molecular weight glutenin subunit	*GluDx5Dy12 1DL*
L-21712	Langdon/*Ae tauschiii* AUS 21712	Hexaploid synthetic wheat with high molecular weight glutenin subunit	*GluDx2.1Dy10 1DL*
RL54056	Tetra Canthatch/RL5288 (*Ae tauschiii*)	Hexaploid synthetic wheat with resistance to stem rust	*Sr331DS*

Germplasm	*Pedigree*	*Description*	*Gene(s)*
SW8	Langdon/*Ae tauschiii* Clae 26	Hexaploid synthetic wheat with resistance to Hessian fly	*H25* 3DL
SW39	Langdon/*Ae tauschiii* RL5561	Hexaploid synthetic wheat with resistance to Hessian fly	
Molly	Newton'7/3/KU221-10(T PERSICUM ku138/Ae *tauschiii* KU2076) Eagle// KS806	Isogenic line for Hessian fly resistance gene H13 in winter wheat cultivar Newton	*H13* 6DL
CIGM86.940		Hexaploid synthetic wheat with resistance to stripe rust	*Yr28* 4DS
CIG87M6-2-1	RL5289 (Ae *tauschiii)*	Hexaploid synthetic wheat with resistance to stem rust	*Sr45* 1DS
Sear's synthetics		Hexaploid synthetic wheat with resistance to *Septoria tritici blotch T. dicoccoides/37-1 [Ae tauschii]*	*Stb5 7DS*
Sear's synthetics	Produra/Ae tauschii G3489	Hexaploid synthetic wheat with resistance to *root knot nematodes*	*Rkn*
	SQ24 (*Ae tauschii*)/TD65 (*T turgidum*)	Hexaploid synthetic wheat with resistance to Russian wheat aphid	*Dn3*
TAM 110	TAM 105*4/Amigo'5//Largo (Langdon)/*Ae tauschii* PI 268210	Hard red winter wheat cultivar with resistance to green bug	*Gb3 7DL*
TXGBE272	TAM 105*4/Amigo'5//Largo (Langdon)/*Ae tauschii* PI 268210	Hard red winter wheat cultivar with resistance to green bug	*Gb3 7DL*
TXGBE273	TAM 105*4/Amigo'5//Largo (Langdon)/*Ae tauschii* PI 268210	Winter wheat germplasm with resistance to green bug biotype E	*Gb3 7DL*
TXGBE285	TAM 105*4/Amigo'5//Largo (Langdon)/*Ae tauschii* PI 268210	Winter wheat germplasm with resistance to green bug biotype E	*Gb3 7DL*
TXGBE292	TAM 105*4/Amigo'5//Largo (Langdon)/*Ae tauschii* PI 268210	Winter wheat germplasm with resistance to green bug biotype E	*Gb3 7DL*
TXGH10563B	TAM 105*4/Amigo'5//Largo (Langdon)/*Ae tauschii* PI 268210	Winter wheat germplasm with resistance to green bug biotype B, C and E	*Gb3 7DL*
TXGH10989	TAM 105*4/Amigo'5//Largo (Langdon)/*Ae tauschii* PI 268210	Winter wheat germplasm with resistance to green bug biotype B, C and E	*Gb3 7DL*
TXGH13622	TX71562-6*4/Amigo'5//Largo (Langdon)/*Ae tauschii* PI 268210	Winter wheat germplasm with resistance to green bug biotype B, C and E	*Gb3 7DL*
TAM 112	TU1254-7-9-2-1 [TAM 200/*Ae tauschii* TA2460)/TXGH10440 [TAM105'4/ Amigo'5//Largo)	Hard red winter wheat germplasm with resistance to green bug biotype and leaf rust resistance from *Ae tauschii*	

Germplasm	*Pedigree*	*Description*	*Gene(s)*
XX186	Durum 'Santa Marta'/*Ae tauschii* TA2460)//BGRC 1458	Hexaploid synthetic wheat germplasm with resistance to powdery mildew	*Pm197D*
WX-SYN.B-92-52	CIGM87.2768-1B-0Y-0M-0Y[SH12]; (Alter84/T tauschii (Acc 198)]	Karnal bunt resistant (0 per cent infection VS check with 65 per cent susceptibility), 6.1t/ha, 61.3, 1000 grain weight	
WX-SYN.B-92-81	CIGM86.953-1B-0Y-0M-0Y[SH46]; (Duergand/*T tauschii* (Acc 221)]	Karnal bunt resistant 4.2t/ha, 55.7, 1000 grain weight	
WX-SYN.B-92-91	CIGM86.953-1B-0Y-0M-0Y[SH46]; (Duergand/*T tauschii* (Acc 221)]	Karnal bunt resistant 4.2t/ha, 55.7, 1000 grain weight	

in the Horsham, Victoria based wheat breeding program. The Cre3 gene is now widely dispersed in wheat germplasm. Adiagnostic PCR based SCAR marker and an SSR marker for selecting Cre3, were developed (Martin *et al.*, 2004). Wheat germplasm with an apparent yield advantage over the recurrent parent were developed from the CCN resistant *Ae. tauschii* material. This yield advantage was expressed mostly in environments with high terminal moisture stress and low yield levels. The SBL germplasm continues to be utilized in the Australian wheat breeding programs although no varieties have yet been released from these lines (Eastwood *et al.*, 2006). The percentage of SHW that displayed resistant reactions varied from 1 per cent for *S. nodorum* glume blotch to 82 per cent for STB. The resistance in SHW for crown rot was significantly better than what exists in current common wheat cultivars. In the work recently carried out, 914 SHWs were evaluated for resistance to *Ug99* stem rust in Ethiopia and to Russian Wheat aphid, Hessian fly and Sunn pest. A total of 10 per cent of the SHW had adult plant resistance to stem rust indicating the presence of potentially new race-nonspecific genes to be exploited for improving stem rust resistance in wheat. A total of1 per cent, 2 per cent, and 15 per cent were resistant to Russian wheat aphid, Hessian fly, and Sunn pest, respectively (El Bouhssini *et al.*, 2012). However, the key questions to be answered, are these genes different from those previously reported in cultivated wheat and what is the genetic basis of their resistance?

The durum wheat parents for resistance to Hessian fly and identified 52 SHW with high or moderate resistance. Because all the durum parents were susceptible, the resistance genes in these SHW should be contributed by *Ae. tauschii* D genome. They haplotyped the 52 SHW using eight PCR-based markers closely-linked to five resistance genes (H13, H22, H23, H26, and H32) previously identified in *Ae. tauschii.* The marker data showed that 32 SHW had the same haplotypes as the wheat lines containing the five known resistance genes, but 19 SHW had different haplotypes, suggesting that these lines may contain new genes for resistance to Hessian fly. Thus, haplotype analysis on the uncharacterized SHW using the markers closely linked to the known genes might provide preliminary information on the resistance genes in SHW. As in previous studies, some of the SHW possessed multiple disease and insect pest resistances and thus, provide a valuable resource for identification of resistance genes using SHW-derived mapping populations. The identification of the genetic and molecular basis of major resistance determinants to different pathotypes will aid in the selection of favorable alleles during cultivar development. In parallel with the use of specific pathogen virulence races, SHWs contribute to a diversified wheat gene pool to assist in studies on mechanisms for multiple host pathogen-interactions and ultimately to obtain durable resistance. Recently, association mapping was used to investigate the genetic architecture of resistances to CCN, and root lesion nematodes–*P. neglectus* (PN) and *P. thornei* (PT). Clusters of disease resistance QTL were identified at multiple genomic locations. There were coincident QTL for CCN and PN on chromosome 1A, 1D, 2D, 3D, 5D, 6B, and 7B; CCN and PT on chromosomes 4B and 6A; PN and PT on chromosomes 3B, 6B, and 7A. Of significance is the identification of novel QTL that are involved in the control of root diseases in wheat (Mulki *et al.*, 2012). The loci with combined resistance to multiple diseases such as CCN-PN, CCN-PT, or PN-PT, constitute valuable resources for the transfer of multiple disease resistance into locally adapted wheat cultivars using marker-assisted selection (MAS).

Due to their wider genetic variation relative to common wheat, SHW are excellent parental materials for developing mapping populations such as recombinant inbred lines and doubled haploids. The "**International Triticeae Mapping Initiative**" (ITMI) mapping population was developed from the cross between the CIMMYT SHW line W-7984 and the hard-red spring wheat (HRSW) 'Opata 85' and it has been extensively used for mapping studies (Xu *et al.*, 2004). A consensus genetic map consisting of over 2400 molecular markers has been constructed based on the ITMI population. A number of genes or QTLs for resistance to leaf rust, stripe rust, Karnal bunt, Hessian fly, and greenbug have been mapped using the ITMI population. A doubled haploid population from the cross between CIMMYT SHW line TA4152-60 (Scoop 1/*Ae. tauschii* WPI 358) with resistance to multiple diseases and a hard red spring wheat line ND495 and used the population to construct a linkage map consisting of 632 markers. Several novel QTLs for resistance to tan spot, leaf rust, and *Stagonospora nodorum* blotch have been identified using this population and the linkage map. The population is currently being used for mapping the genes and QTLs for resistance to Hessian fly, stripe rust, and Fusarium head blight. Zwart *et al.* (2010) documented the location of coincident QTLs for both species of RLN on chromosomes 6DS and 2BS in a cross between SHW line CPI133872 and the bread wheat cultivar, 'Janz'. The QTL analysis revealed six putative QTLs for seedling resistance and seven for adult plant resistance. For field resistance, two loci had stable main effects across environments and five loci exhibited marker by environment interaction effects. The strongest effects were detected at marker locus Xbarc149 on chromosome 1D, at which the exotic allele decreased seedling symptoms by 46.3 per cent and field symptoms by 43.6 per cent. Some of the detected QTLs colocalized with known resistance genes, while others appear to be novel resistance loci indicating, that the exotic SHW, Syn 022L may be useful for the improvement of leaf rust resistance in cultivated wheat. Similarly, a synthetic backcross F1-derived doubled haploid population between SHW line, CPI133872 and the common wheat cultivar, Janz, was phenotyped for seven disease resistance traits, namely: STB, yellow leaf spot (YLS) also known as tan spot (*Pyrenophora triticirepentis*), stripe rust, leaf rust, stem rust, and two RLN species in multiple years and/or locations. A multiple-QTL model to identify a tightly linked cluster of foliar disease resistance QTL on chromosome 3DL. Major QTL for resistance to STB and YLS were contributed by the SHW parent CPI133872 and linked in repulsion with the coincident *Lr24*/*Sr24* locus carried by parent Janz. Additional QTL for YLS were detected in 5AS and 5BL. Consistent QTL for stripe rust resistance were identified on chromosomes 1BL, 4BL and 7DS, with the QTL in 7DS corresponding to the Yr18/Lr34 region. Three major QTL for *P thornei* resistance (2BS, 6DS, and 6DL) and two for *P neglectus* resistance (2BS, 6DS) were detected. They also reported the identification of recombinants combining resistance to STB, YLS, rust diseases and RLN from parents CPI133872 and Janz that constitute valuable germplasm for the transfer of multiple disease resistances into new wheat cultivars.

Enhancing Yield Productivity

Wheat is grown under irrigated and rain-fed conditions worldwide and increasing yield is a major thrust of most wheat improvement programs. Arguably, increases in grain yields of common wheat over the last several decades came in part from the use of gibberellic acid sensitive dwarfing genes (Rht1 and Rht2), which were widely distributed during the "**green revolution**". In addition to reducing plant height, these genes contributed to the repartitioning of photosynthates from vegetative to reproductive tissues. The increased grain yield results from the production of more grains per square meter. Given the importance of wheat as a food crop, continued genetic gains in yield are necessary, given that there will be a need to provide food for an approximately double the current world population by 2050, with less available resources such as land, oil and fertilizer. To achieve this, the rate of genetic gain in breeding programs can increase by either extending the amount or nature of variation available for selection, or by accelerating the selection process to produce varieties more rapidly (Langridge and Fleury 2011). The yield increases in some of the progenies were a result of changes in yield components and in particular, increases in 1,000-grain weight relative to the recurrent parent. Similar observations were made in SBL where higher yielding lines were associated with increased rates of grain filling and higher 1,000-grain weight, especially in lower yielding environments with drought stress (Gororo *et al.*, 2002).

Similarly, yield advantages of SBLs over elite cultivars of bread wheat have been reported to be as high as 30 per cent. In the high yielding environments, grain yields of the SBLs were similar to those of their Australian recurrent common wheat parent, while in the lower yielding environments, the SBLs produced significantly higher grain yields, up to 149 per cent of their Australian recurrent wheat parent. Fourteen of these families, each derived from a different primary synthetic wheat backcrossed to the Australian bread wheat cultivar Yitpi, yielded significantly ($p<0.05$) more grain on average than their recurrent parent. In higher-yielding environments, grain yield advantages were mostly associated with increases in grain weight, while in lower-yielding environments; grain yield advantages were mostly associated with increases in the number of grains per square meter. For example, in the lowest yielding environments, increases in the number of grains per $320m^2$ were attributable to superior grain yields, whereas in the highest yielding environments, improvements to grain weight (up to 112 per cent of that of Yitpi) were responsible." Christopher *et al.* (2006) examined the physiological attributes associated with SBL, that exhibited high yield in trials when grown in a trial. The SBL exhibited a stay-green phenotype and higher yields compared to standard northern cultivars. The yield of SBL examined was 55 per cent, 31 per cent, and 15 per cent higher than the local adapted cultivars–Hartog, Banks, and Baxter, respectively. The yield was also 12 per cent higher than that of the stay-green line Seri-M82. The performance of a limited number of SBLs against that of their parents in trials at CIMMYT, Mexico and Horsham, Victoria, Australia. In an extensive study of the utility of SBL, SBL appear to offer greatest benefits for improving grain size and the maintenance of grain size across environments, even where grain yield is low. An economic benefit to farmers from larger grains is achieved through lower levels of grain screenings (small grains), especially in environments prone to high screenings levels. The synthetic derived wheat germplasm, **'Vorobey'** was a top performing line the world (trial means ranged from 1 to 8 t/ha).

The release of four cultivars derived from SHW backcrossed to locally adapted cultivars in China. Of the four cultivars, Chuanmai 42 had large kernels and resistance to stripe rust, and produced the highest average yield (>6 t/ha) among all elite cultivars over 2 years in Sichuan provincial yield trials, out-yielding the commercial check cultivar Chuanmai 107 by 22.7 per cent. They reported that Chuanmai 42 increased grain yield by 0.45–0.75 t/ha. Average yields on the fertile Chengdu plain increased to 7.5 t/ha in 2005 and 2006. Based on this level of superiority, Chuanmai 42 also became an important parent in other breeding programs in China. The conclusions that can be drawn from these comparisons between the SBLs and elite adapted cultivars under a range of environmental conditions from low rainfall to favorable high rainfall conditions are that: The grain yields of the top SBLs are higher than their recurrent parents and the best locally adapted check cultivars under moisture stress, non-moisture limiting environments and diverse agroecological zones. These results indicate that SBLs are also responsive to more favorable, higher yield conditions and could be used to improve elite wheat cultivars, for example, for irrigated high-input environments. The significantly enhanced performance of SBLs is not solely associated with mechanisms related to escape from moisture stress. Other characters must play a role including root architecture. One of the unexplored areas of research is the root systems of SHW and their potential to contribute to improved productivity. This is partly because of lack of efficient phenotyping for root traits. The root systems of three SBLs, AUS33687, AUS33435, and AUS33684 measured at ear emergence (Z59 in Zadoks' growth scale for cereals; Zadoks *et al.*, 1974) on a deep sandy soil. The root system of AUS33684 neither had seminal roots as deep as those of AUS33435 nor as dense a root at depth as AUS33687, but its root system had a profuse branching in the top 0.35m of the soil profile. The root proliferation in AUS33684 they posited led to a higher total root length density (in the 0.0–1.2m soil layers) than in AUS33687 and AUS33435. They argued that total root length density is considered to be directly related to the amount of water uptake and to indicate the size of the root system. However, there was no indication of the yield performance of the three SBL genotypes and therefore the potential benefit of the diversity in root systems to yield improvement in their studies. Nevertheless, they argued that vigorous or large root system contributes to adaptation in dry environments and dry seasons where crop growth depends on seasonal rainfall. However, a large root system may be of less value in environments where crop growth is dependent on stored soil water where access to more soil water runs the risk of exhausting soil water before completing grain filling. However, the exemption would be when water can be accessed from deeper in the profile. The improved performance of SBL compared to recurrent parents was in part due to its increased root biomass at depth and better water extraction capacity–an important attribute on soils where wheat is grown on stored moisture. Lopez and Reynolds (2011) reported that SBL showed on average a 26 per cent yield increase when compared with the parental hexaploid wheat under terminal drought. Among the traits associated with improved performance was earliness to flowering, greater root mass at depth, greater water extraction capacity and increased water use efficiency at anthesis. They reported some degree of independence between these traits indicating the potential for the pyramiding of favorable alleles. In a hydroponic based study carried out under controlled glasshouse, 20 out of 100 SHW had higher initial aerial and root biomass than current Australian check cultivars used. The SHW exhibited significantly thicker root systems than elite common wheat cultivars with some possessing longer root systems. Deeper rooting depth impacts water use and could be beneficial in exploiting water at depth under drought conditions.

Environmental Stress Tolerance

Improving Salinity Tolerance

Wheat production areas globally are divided into irrigated and rainfed sectors. The former is beset with a variable salinity level constraint that penalizes yield outputs. It is estimated that 20 per cent of the irrigated land in the world is presently affected by salinity excluding the regions classified as arid and desert lands. In rainfed production systems where transient salinity occurs, yields can be well below theoretical for the rainfall received, when subsoil salinity is present, and unused water at harvest is one of its symptoms. The cost of transient salinity and associated constraints in sodic soils for the farming economy has been estimated at ca. AU$1.5 billion per year (Rengasamy, 2002). Conventional diversity for salinity tolerance has been limiting with just one land race **Kharchia** 65 playing a major role in salt tolerant varietal development in India where the cultivars KRL1-4 and later KRL 19 emerged. An encouraging transfer strategy emerged when the D genome of wheat was shown to be of significance for salinity tolerance (Shah *et al.*, 1987) paving the way for follow-up experiments and exploitation of the D genome synthetics. Evaluation of an elite set of SHW resulted in several SHW being categorized as tolerant (Mujeeb-Kazi 2003b). Independent studies indicated and confirmed that SHW possessed considerable variation for salinity tolerance based on Na^+ exclusion. In their study, some SHW exhibited superior differential response to salt compared to the common wheat check in a hydroponic system at a concentration of 100mM NaCl (Dreccer *et al.*, 2004). In a further confirmation of these results, more than 300 SHW were evaluated over 4 years. The SHW displayed significant genetic variation for sodium tolerance, often exhibiting much lower Na^+ concentrations and higher K^+/Na^+ ratios in leaves than most Australian commercial common wheat cultivars. However, these studies did not evaluate the two mechanisms that could contribute to it: (i) the ability of transporters in the leaf sheath to retain Na^+ there and (ii) the size of the sheath versus the blade results in the exclusion of Na^+ from blade. In addition to Na^+ exclusion, SHW with significantly superior tissue tolerance than elite wheat cultivars after exposure to 250mM NaCl were identified. Earlier studies established that relative to durum wheat, salinity tolerance of common wheat is associated with a relatively high ability to exclude Na^+ from the leaf blades resulting in an overall increase in the K^v/Na^+ ratio, in some cases associated with an increase in K^+ uptake. The successful transfer of salinity tolerance measured as Na^+ exclusion in SHW into an elite Australian common wheat cultivar, Yitpi with some of the SBLs showing significantly enhanced Na^+ exclusion compared to either the SHW or the recurrent common wheat cultivar. A major locus, Kna1, controlling K^+/Na^+ uptake has been shown to be present on chromosome 4DL. QTL analysis of SBL using two of the SHW indicates that different mechanisms contribute to enhanced salinity tolerance of SBL. The results suggest that in addition to the locus on 4D, other loci on chromosomes 2B, 3B, 4B, 6D, 7A, and 7D influence salinity tolerance, either as Na^+, K^+/Na^+ ratio or as tissue tolerance. There were coincident QTL for traits such as predicted mean tissue necrosis (tissue tolerance) and, K^+/Na^+ on chromosomes 3B and 7A. For predicted mean necrosis, a group of markers on 4D consistently explained most of the variation at 21 per cent. A similar trend was observed for the predicted mean Na^+ with 8 per cent –18 per cent phenotypic variance and 6 per cent –14 per cent of phenotypic variance for K^+/Na^+ ratio on chromosome 4B. The identification of such loci independent of Kna1 suggests that the salt tolerance in SHW could be deployed to mitigate the impact of salinity and help stabilize and improve wheat production in saline soils.

Mineral/Nutrient Efficiency

The search for genetic resources that accumulate high levels of iron (Fe) and zinc (Zn) led to in-depth evaluation of wheat landraces, tetraploid and diploid progenitors of hexaploid wheat for their enhanced micronutrient status. *Ae. tauschii, T. turgidum* ssp. *dicoccoides, T. monococcum,* and *T. boeoticum* were among the most promising sources of high Fe and Zn levels in the grain. Some of these genotypes showed values as high as 142 mg/g of Zn; however, in some locations manure had been applied in the past cropping cycles that probably contributed to the high Zn values. Cakmak *et al.* (1999) concluded that SHWs represent invaluable sources of useful genes for improvement of mineral efficiency, and these genes can be readily transferred to wheat. A recent study evaluated a set of high yielding lines under field conditions. The Zn values generally ranged between 15 and 35mg/g but increased to 43 mg/g in some genotypes, while the Fe concentrations ranged from 20 to 60 mg/g. Taking into account the bioavailability, the daily intake and the estimated average requirements, some general estimations were made by "Harvest Plus" to set a breeding target for wheat. For example, in Pakistan and northern India the target is to increase Fe and Zn levels by 25 and 10 mg/g, respectively, above the mean of genotypes grown in the region. This translates into total Zn and Fe levels in the grain of 45 and 60mg/g, respectively. From these results, whilst there is sufficient unexplored genetic variability in SHW to develop wheat varieties with increased mineral levels in the grain, it is worthy of note that Zn and Fe are concentrated in the aleurone layer of the grain, it is therefore likely that low yield and high mineral content may be coupled together. This is because the aleurone/endosperm ratio changes with increasing yield with implication in for breeding. However, it should also be kept in mind that mineral bioavailability is limited by the presence of phytic acid (PA) in the aleurone layer, which forms insoluble complexes with dietary cations, thus hindering their intestinal absorption (Cheryan and Rackis 1980). PA breakdown strongly depends on the phytase activity in the flour. Consequently, both mineral and phytase concentrations should be taken into account in breeding programs. Recently, Ram *et al.* (2010) reported higher genetic variability of phytase in SHW compared to Indian cultivars. Since there was a larger genotypic effect on phytase levels and also larger variation in activity as compared to phytate levels, there is greater scope for manipulating phytase levels than phytate content through breeding in wheat. The release of cultivars with high mineral concentrations complemented with high intrinsic phytase activity could greatly improve the nutritional value of common wheat, provided that whole-wheat flour is utilized to preserve the source of the minerals. CIMMYT nearly a decade ago had screened some wheat progenitor genetic resources and identified accessions of *T. turgidum ssp. dicoccum* with elevated levels of iron and zinc. These tetraploids were used to develop SHW in the wide crossing unit and those stocks (*T. turgidum ssp. dicoccum / Ae. tauschii)* entered the wheat-breeding program.

Boron Toxicity

Boron toxicity limits wheat yield in some environments by limiting root growth and affecting water uptake. Dreccer *et al.* (2003) evaluated the variability for boron tolerance in entries of the primary SHW set and identified boron tolerance in many entries that was similar to the tolerant conventional check **'Frame'**. Further, 21 out of 87 SHW performed as well as **Halberd**, the most boron tolerant cultivar in Australia, and were significantly different from elite check cultivars. However, the genes controlling boron tolerance are whether different from those identified earlier in the primary wheat gene pool are not known.

Quality Improvement

The SHWs possessed significantly more genetic variation for quality traits including those associated with colour and colour stability of Asian noodles such as near zero extremes for polyphenol oxidase and lipoxygenase, than currently available in common wheat (Mares and Mrva, 2008).

Bread Making Quality

The major determinants of end use quality in wheat include grain hardness or endosperm texture, protein content and protein quality. Functional differences in grain and flour quality between hexaploid and tetraploid (*T. turgidum* ssp. *durum*) wheats have been attributed to the influence of the D genome. With the growing interest during the last decade in exploitation of the D genome SHWs and expected varietal outputs it is crucial that quality is considered in the breeding scheme particularly when an alien resource is utilized. The unique and beneficial quality glutenin subunits exist. Desirable sub-units at the Glu-Dt1 locus are 5x10, 1.5x10, 2.1x10, 1.5x12, and 3x10. In the sets of SHW evaluated, 50 per cent –60 per cent of these subunits were observed in Elite-1 and Elite-II subsets of SHW. The superiority of two *Ae. tauschii* encoded subunits 1Dx1.5 and 1Dx2.1 for grain quality improvement prompted the development of an allele-specific molecular marker to facilitate their early selection in segregating generations when SHW were targeted for breeding. The synthetic derivatives carrying excellent bread-making quality can be bred if the common wheat parent(s) in the cross have good baking quality. Variation at the Ha locus on chromosome 5D, a major determinant of milling yield, has revealed complete deletions in the Pina and Pinb genes present in the A and B genomes of common wheat. Phenotypic differences in grain hardness/

softness are largely due to the locus on the D genome resulting from deletions in puroindoline a (Pina) and/or mutations in puroindoline b (Pinb). *Ae. tauschii* is generally soft-grained and so are the SHWs. Characterization of the *Ae. tauschii* and SHWs identified eight different Pina alleles and six unique Pinb alleles that are all associated with a soft endosperm (Lillemo *et al.*, 2006). To further understand the mechanism of grain hardness formation in *Ae. tauschii* and SHW, a total of 57 new SHWs derived from crosses of *T. turgidum ssp. dicoccoides* and *Ae. tauschii* were analyzed to compare puroindoline alleles to grain hardness using gene-specific PCR primers and the single kernel characterization system (SKCS) of Perten 4100. Four allelic variations of Pina and Pinb were identified, that is, Pina-D1a, Pina-D1c, Pina-D1h, and Pina-D1j. Eight (14 per cent) of the lines contained Pina-D1a/Pina-D1j while the remaining 86 per cent of the lines, 49, carried Pina-D1c/Pina-D1h. Kernel texture differed significantly among the SHWs; SKCS hardness ranged from 10.5 to 42.6, with significant differences between the parental types, *T. turgidum* ssp. *dicoccoides* or *Ae. tauschii.* However, variation in kernel texture did not strictly correspond to puroindoline genotypes Pina-D1a/Pina-D1j and Pina-D1c/Pina-D1h, indicating genes besides those found at the Ha locus affected kernel texture.

Narasimhamoorthy *et al.* (2006) using an AB-QTL population of hard winter wheat by synthetic cross identified a major hardness QTL which coincided with the well-known Ha and Gsp puroindoline genes on 5DS, in addition to a novel QTL on chromosome arm 3BL that have not been previously reported for grain hardness. The overall visco-elastic properties that confer bread making characteristics of hexaploid wheat in contrast to the pasta making characteristics of durum wheat reflects the impact of the D genome. While various aspects of grain, flour, and dough quality have been studied to further partition components of the D genome that exert major influence, most of the studies have revealed the significant role of gluten and puroindoline proteins in flour functionality. Although variation at the Glu-Dt1 locus that controls HMW glutenins in *Ae. tauschii* is much greater than the corresponding locus in *T. aestivum*, none of the novel variants transferred into common wheat from *Ae. tauschii* has so far been found to improve bread-making characteristics compared to what already exists in common wheat varieties (Bennett 1994). The predominant influence of the D-genome in determining the hardness/ or softness of common wheat is of some concern to breeders who appear hesitant in using SHW for improvement of common wheat due to fear of compromising the quality of their elite germplasm. In response to these concerns, Garg *et al.* (2010) investigated the end use quality, in particular bread-making quality by dough strength and storage protein profile, of SBL derived from backcrossing SHW with elite commercial Australian varieties. The SBLs investigated were predominantly from the following lines: Altar 84/*Ae. squarrosa* (211)//Janz/3/Westonia, Croc-1/Ae. squarrosa (224)//Kulin/3/Westonia and Cpi/Gediz/3/Goo//Jo69/Cra/4/Ae. squarrosa (208)/5/2_Westonia. The results from the study indicated that all the SBLs were primehard like parental cultivars or even harder. This suggests that backcrossing using appropriate parents obviates the concerns of breeders in the use of SHW because the results demonstrate the full recovery of hard-grain textured SBLs due largely to the use of hard grain parents and a favorable selection of hard genotypes.

The evaluated SHW, SBLs with bread wheat and durum wheat cultivar for traits that determine colour and colour stability in Asian noodles and the frequency of a genetic defect know as late maturity a-amylase (LMA). They reported that the SHW harboured substantial genetic variation for quality traits associated with colour and colour stability of Asian noodles including near zero extremes for polyphenol oxidase and lipoxygenase. They indicated that these extremes represent a significant advantage compared with current bread wheat cultivars and are similar to the best durum wheat. While alternative strategies for reducing polyphenol oxidase and lipoxygenase are available, the SHW provide a useful resource for wheat breeders attempting to develop improved wheat cultivars for the Asian noodle market. However, they found that most SHW were prone to late maturity a-amylase and mature grain contained unacceptably high levels of a-amylase. Elimination of this genetic defect, or selection within breeding populations for low or non-LMA, is both time consuming and labour intensive and presents a significant obstacle to exploitation of variation for other traits. Nonetheless, as proof of concept of the potential of exploiting SHW, they recovered near-zero polyphenol oxidase (PPO) SBL, free from LMA, from backcross populations involving a high LMA SHW. In perhaps the most comprehensive evaluation of the quality of SBL, Nelson *et al.* (2006) used QTL analysis to study the milling and baking quality of 114 recombinant inbred lines of the ITMI SHW (W7985) _common wheat (Opata, 85) crosses. The traits investigated included kernel-texture, protein concentration and quality, dough strength, and mixing traits. They reported that some of the RILs had consistently superior quality traits than their parents, indicating the potential of a new combination of alleles from the diploid and tetraploid parents, especially alleles of known storage proteins for improvement of quality traits in wheat cultivars. For example, mixogram traits were influenced most by chromosomal regions containing gliadin or low-molecular-weight glutenin loci on chromosome arms 1AS, 1BS and 6DS, with the SHW contributing the favorable alleles. The advanced backcross QTL (AB-QTL) strategy to locate QTLs for baking quality traits in two BC2F3 populations of winter wheat. The BC2F3 were from SHW produced from *T. turgidum* spp. *dicoccoides* and *Ae. tauschii.* The exotic allele improved trait performance at 14 QTLs (36.8 per cent), while the elite genotype contributed a favorable effect at 24 QTLs (63.2 per cent). The favorable exotic alleles were mainly associated with grain protein content, though the greatest improvement of trait performance due to the exotic alleles was achieved for the traits falling number and sedimentation volume. At the QTL on chromosome 4B, the exotic allele increased the falling number by 19.6 per cent; and, at the QTL on chromosome 6D, the exotic allele increased sedimentation volume by 21.7 per cent, indicating that the SBL derived from wild emmer *Ae. tauschii* carried favorable QTL alleles for baking quality traits.

Preharvest Sprouting

Most cultivated hexaploid wheat varieties are characterized by low embryo dormancy and are sensitive to preharvest sprouting (PHS). Thus, PHS is one of the most important quality constraints in wheat production in environments characterized by summer rainfall and high humidity. PHS resistance is inherited as a quantitative trait controlled by a large number of genes, and is significantly influenced by genetic background, environmental conditions and their interactions. The extent of genetic diversity of 28 wheat lines, including durum and SHW, reputed to possess PHS resistance. Across most current varieties, the alleles at some of the QTL regions associated with PHS tolerance were largely the same, depending on the gene pool used, while the natural variation present in *Ae. tauschii* provided a novel source of allelic diversity. This knowledge of the different haplotypes can be used to design optimum strategies for the incorporation of PHS resistance into cultivated varieties. Consistent with previous studies, their results indicated clear differences in the level of dormancy between red grained and white-grained genotypes. However, the white-grained genotypes were more variable, with and a few genotypes displayed levels of dormancy comparable to red-grained genotypes. *Ae. tauschii* accessions and the SHW derived from them expressed high levels of seed dormancy. In an attempt to incorporate PHS tolerance into wheat, a SBL population was generated from a cross between an Australian white common wheat cultivar, '**Janz**,' and a SHW, '**Syn36**'. The resultant SBLs were grown in field trials at Horsham in 2004/2005 and evaluated for grain dormancy (GI), sprouting index (SI), and visibly sprouted seeds (VI), the latter two following artificial weathering. There were significant (P<0.01) genotypic differences among the SBLs across the three indices studied using a weighted germination index of 0–1, where 1 represents extreme susceptibility. The range in value in the BC_1F_6 population for GI value after 14 days (GI-14) were 0.13–0.85 compared to '**Syn36**' the resistant parent with a GI-14 value of 0.30, and '**Janz**', the susceptible parent with GI-14 value of 0.86, indicative of transgressive segregants towards high dormancy. A similar trend was observed for all measures of PHS tolerance. Some of the SBLs were identified with significantly superior preharvest sprouting tolerance across all three measures (GI, SI, and VI), compared to the current Australian PHS tolerant standard SUN325 with a GI-14 of 0.46. A similar result was obtained in a SBL involving SHW, 'Syn37' and the cultivar, '**Janz**' (Imtiaz *et al.*, 2008). In a further elucidation of genetic architecture of preharvest sprouting resistance (PHSR), Imtiaz *et al.* (2008) identified seven QTL controlling resistance to PHS on chromosomes 1D, 2D, 3D, 6D, and 4A. The SHW parent, 'Syn37' contributed resistance on 1D, 2D, 3D, and 4A, while the susceptible cultivar, '**Janz**' contributed alleles on chromosome 6D, which indicated the presence of complementary alleles in the two parents. A large component of the phenotypic variation was explained by red grain color (RGC) depending on the indices used. They mapped RGC or R-D1b locus at a distance of 5cM from the marker wms1200 located on chromosome 3D, with the highest level of variation explained by GI (43 per cent), VI (17 per cent), and SI 8 per cent. On aggregate, the red-seeded genotypes overall displayed higher levels of dormancy than the white seeded genotypes. Similarly, in a cross between a PHS tolerant SHW and common wheat cultivar, '88-1643', The identification of a major QTL associated with PHS tolerance designated Qphs.sau-2D within the marker intervals between Xgwm261 and Xgwm484 in two consecutive years which accounted for 26 per cent –28 per cent of the phenotypic variation for PHS tolerance. Some white-grain synthetic-derived wheat lines that are highly resistant to preharvest sprouting were developed using SHW produced with highly dormant *Ae. tauschii* accessions (Imtiaz *et al.*, 2008).

Future Prospects

Wheat breeding targets are linked with the increasing population trends and with 2030 being very close, about 730 million tons of wheat will be required to feed a population of about nine billion people. With genetic diversity being paramount for improving productivity through breeding, looking at resources beyond bread wheat is imperative and SHWs and the SBLs provide a rich source of genetic diversity for improving wheat production. The potential benefit of SHW gained prominence in the mid-1980s with the simultaneous use of direct crossing of *Ae. tauschii* with wheat and the random production of SHW aimed at capturing the value of the three different genomes for the improvement of common wheat. Two decades later, the exploitation of genetic variability in SHW is being translated into the successful identification, mapping, characterization, and transfer of several agronomically important traits including disease and pest resistance into common wheat. Recently, new sources of resistance were identified in *Ae. tauschii* against the stem rust race, *Ug99*, (Rouse *et al.*, 2011). This was also evident in the reports of the successful development of high yielding *Ug99* resistant SBLs. The identification of both major and minor gene resistance in SHWs against the stripe rust Yr27 virulent race that devastated wheat crops in Ethiopia and Syria in 2010 thus providing genetic resource that could be useful in broadening the genetic bases of stripe resistance in wheat.

In the short-term this trend is expected to continue by deploying populations derived from SHW using SSR, DArT, and SNP markers for genetic mapping, and trait identification. However, while the approach that first involves screening for targeted biotic and abiotic stresses would continue, emphasis must now include the use of genome wide approaches that involves low-cost sequencing technologies. With about 1,500 SHWs catalogued globally, the validation of existing genes and the discovery of new genes for resistance and tolerance to abiotic and biotic stresses in SHW can be enhanced by employing genotyping by sequencing of the SHW and undertaking of genome wide association studies. The release of the 5_ sequence coverage of the wheat genome will facilitate efforts towards full sequencing of the bread wheat genome which, if accomplished, will accelerate gene discovery, allow even more targeted mining of novel genes/alleles in SHW and will enable their subsequent introgression into bread wheat with impact on breeding efficiency. However, these potential outcomes are equally depended on precision phenotyping. One of the key features of SHW is the availability of lines with multiple disease resistance (MDR). However, this potential has yet to be exploited. It would be necessary to assemble a set with MDR and conduct allelism tests with previously described genes. The understanding of the mechanisms involved in conferring MDR is likely to have implications for resistance spectra and durability associated with specific MDR loci in wheat improvement. As pathogens mutate or new stresses become threats this unique diversity in SHW will gain further significance. Where major genes are involved, it is quite likely that a pathogen shift will defeat single gene resistance particularly for rust diseases. The availability of adult plant resistance QTL in the SHW, if proven to be race nonspecific, provides the opportunity to use gene pyramiding as a means of improving durability. Ultimately, the cloning of the genes underlying MDR will lead to more precise and targeted incorporation of MDR genes into wheat varieties with negligible linkage drag. It has also been demonstrated that SHW and their derived SBLs (*i.e.*, when crossed to adapted local varieties) show significant yield increases and thus enhanced yield performance across a diverse range of environments, demonstrating their potential for use in improving wheat productivity worldwide. This is particularly evident in moisture limited environments. It is crucial that this potential be realized at the farm level. These should be used to accelerate the development of improved varieties that can provide high yields and improved livelihoods, and at the same time meet the challenges of marginal environments and the threat of climate change in rainfed wheat production systems. However, the genetic and physiological bases of enhanced performance are little understood, thus key questions associated with understanding and unraveling the basis of enhanced yield performance remain to be answered. The precise contribution of genes or genomic segments derived from either the AdBd, Dt or specific interactions with the ABD genomes of hexaploid wheat remain largely undefined. In the absence of any a priori basis for selecting SHW that are predictive of the outcomes of significant yield boost in SBLs there is the need to analyze and define genomic regions with consistent contributions to yield increases/yield potential in introgression lines from different recipient wheat backgrounds across different environments. While there are now many reports of potentially beneficial QTLs identified in SBLs, most are yet to be validated in diverse genetic background, followed by fine mapping and ultimately cloning.

Future production of new SHWs should extend to evaluating underexploited tetraploids A and B genome such as *T. turgidum ssp. carthlicum, T. turgidum ssp. dicoccum*, and *T. turgidum ssp. dicoccoides* and identifying gaps in the existing *Ae. tauschii* germplasm used for existing SHWs. SSRs to explore the relationship between the D-genomes of *Ae. tauschii*, SHWs and a 190 elite diverse cultivars of *T. aestivum*. At both K=3 and K=7 placed bread wheat (coloured purple) into a single exclusive sub-population. The D-genome donors of the SHWs and *Ae. tauschii* accessions were partitioned between the remaining groups. Clade A is a mixture of ssp. *tauschii* and *ssp. strangulata* encompassing *Ae. tauschii* accessions with no apparent geographical split observed for sub-clades 1 and 2. Bread wheat is mostly associated with clade A supporting the hypothesis that it evolved around the shores of the Caspian Sea. From this analysis, the available SHWs mostly exploit clade A and, therefore, by inference mostly capture D-genome diversity from *Ae. tauschii* accessions more closely related to bread wheat. Clade B encompasses accessions to the north (sub-clade B1) and east (sub-clade B2) of the Caspian Sea which is apparently under-exploited within current CIMMYT SHWs germplasm. Identifying geographical areas where the progenitor species of the existing SHW where collected would assist in guiding future collection missions. Coupled with eco-environmental and disease prevalence data, such efforts would guide future SHW and wide crossing programs toward a more targeted approach to exploiting the rich vein of genetic diversity provided by the primary, secondary and tertiary gene pools. Such a targeted approach is encapsulated in the focused identification of germplasm strategy (FIGS).

FIGS develop environmental profiles, using statistical approaches and GIS technologies, to predict where selection pressures are likely to occur for specific traits. Germplasm collection sites that fall within the appropriate environmental profile are identified, and the resulting high probability trait-specific subset of germplasm is advanced for screening. Thus, implementation of FIGS could result in the production of new SHWs focused on maximizing allelic diversity for key abiotic and biotic stress tolerances that continue to impede wheat productivity. Targeted traits include drought, heat, salinity, and multiple disease resistance, all of which can contribute to sustainable wheat production in vulnerable dry-land communities that in turn can contribute to poverty reduction and increased food insecurity. On the other hand, many studies have indicated inconsistency of the traits in diploid (*Ae. tauschii)* and allohexaploid (SHW) levels, suggesting difficulty of selection of useful traits in diploid level. Furthermore, the expected traits of SHWs may not always appear in the SBLs produced by backcrosses with elite wheat cultivars. This is because the genetic background of SHWs are largely different from that of elite cultivars. For the enhancement of the genetic diversity of bread wheat through SHWs, the genes of *Ae. tauschii* should be evaluated in the genetic background of elite wheat cultivars. For this purpose, populations including large diversity of *Ae. tauschii* in genetic background of elite cultivars are necessary. Tsujimoto (2010) proposed to produce multiple synthetic derivative (MSD) populations by backcrossing elite cultivar and SHW lines produced using variety of *Ae. tauschii* accessions. Once MSD population is produced, it can be the source of genes for breeding with any breeding objectives. Arguably, the potential of SHW and SBLs described here demonstrate that such germplasm can provide an unquestionable benefit to wheat improvement. Nonetheless, currently only a limited number of varieties have been released from SBLs. Major limitations include some of the perceived undesirable characters such as red seed color in some countries or excessive plant height as well as the fact that it usually takes between 10 and 15 years to develop and release cultivars with heightened interest in the use of SHW and SBL being relatively recent. Such undesirable characters could be addressed via multiple back-crossing to parents with desirable traits. Results presented above indicate that positive transgression for quantitative traits associated with abiotic stress tolerance is common and that such novel genotypes result from either new Dt or AdBd derived alleles or interactions between these new variants and the common wheat genome. Thus, a measure of the impact of SBLs is not exclusive to the either the introgression of the Dt or the AdBd variants, but rather from the interactions including epistatic interactions between the genomes.

Although the potential of biotechnology has often been exaggerated, a high level of optimism is clearly justified for its use in the improvement of wheat. Undoubtedly, functional genomics, as it is now termed, will revolutionize the way in which plant breeding is undertaken in the future. Basic research is leading to an improved understanding of the genetic mechanisms operating within a plant in response to the diverse stresses that it is exposed to, as well as the overall production of biomass and grain. New methodology, such as automated marker systems and array technology, will allow the large number of samples to be processed that are encountered in a typical breeding programme. In fact, perhaps the biggest challenge facing scientists is how to store, process and access the vast amounts of information generated using these new marker systems. This knowledge base offers promise for making germplasm improvement faster, cheaper and more effective. Emerging genetic engineering techniques are providing breeders with the never-be-fore-seen capability to create novel plants by combining genetic material from a wide array of sources. Although not without controversy, the options seem limitless and, with the proper oversight and understanding, should provide extremely powerful options to develop durable and highly productive plant varieties for almost any production environment. The challenge for developing countries is to tap as much of this emerging technology as possible. This does not necessarily mean that countries must establish inhouse capabilities. What is required is that nations recognize the importance of the new approaches and ensure that appropriate legislation and regulations are enacted to allow the country to acquire, evaluate and most importantly deploy the new plant varieties produced via biotechnology. All available tools to ensure an adequate supply of food for the world must be employed.

This addresses the development and use of molecular markers for yield enhancement in wheat. Since their key goal for breeding is to maximize yield, extensive efforts have been made toward the improvement of yield. Agronomic traits related to yield, yield-related, disease resistance, and abiotic stresses are considered to be quantitative traits (QTLs), also known as complex traits, because they are controlled by numerous genes and are affected by environmental factors. Researchers have been studying such traits in the past decades for the development of molecular markers which can be used in various wheat breeding studies mainly involving restriction fragment length polymorphism (RFLP), simple sequence repeat (SSR), single nucleotide polymorphism (SNP), random amplified polymorphic DNA (RAPD), and amplified fragment length polymorphism (AFLP). Furthermore, the advent of next-generation sequencing (NGS) has accelerated the discovery of agronomically important genes. All of the technologies have enabled great advances for increasing the productivity of wheat. Here, the past history of first-generation sequencing, present status of second-generation sequencing, and future potential of translational genomics linked to the yield will be discussed. As the world's population is projected to reach approximately 9 billion by 2050, grain production of major staple crops needs to double to meet global food needs. Together with rice and maize, wheat is one of the major staple crops widely grown in many countries, providing one-fifth of the calories and the protein for the world's population. In addition, bioethanol is made primarily from wheat in Europe. Wheat starch is a major component for the production of bread, porridge, cakes, biscuits, and cereals which is a highly versatile crop for the human diet. There are two distinct types of wheat, spring wheat and winter wheat, cultivated in many countries based on growing seasons, of which spring wheat is planted in most countries except in the United States and Northern Europe where the predominant crop is winter wheat. The global consumption of wheat has increased at a much faster rate than all other crops, because of the scale-up cultivation in developing countries. Therefore, there has been an extensive effort over the past decades to increase wheat production through the application of molecular techniques which are powerful tools for enhancing effectiveness in breeding. The most common or bread wheat species, *Triticum aestivum*, is an annual grass cultivated in temperate zones worldwide that belongs to the genus *Triticum* of the tribe *Triticeae*, and the family Poaceae.

The most economically important cereals in the Poaceae family are maize, wheat, rice, barley, and millet. In the genus *Triticum*, there are approximately 25 species including wild and domesticated consisting of a series of diploid, tetraploid, and hexaploid forms, including the diploid einkorn wheat, *Triticum monococcum* (AA genome), the diploid wild wheat, *Triticum Urartu* (AA genome), the allotetraploid emmer wheat, *Triticum turgidum* var. *durum* (AABB genome), the allohexaploid common wheat, *T. aestivum* L. (AABBDD genome), and the autoallohexploid *Triticum zhukovskyi* (AAAAGG genome). Of these, *T. aestivum* has 42 chromosome pairs that are derived from two rounds of polyploidization events. Its genome size is approximately 17 Gb, composed of A, B, and D genomes created from the hybridization of three different species. The first hybridization between *T. urartu* (2n = 2x = 14) and a B genome species that has not yet been identified occurred 0.20–1.3 million years ago (MYA) to form the tetraploid *Triticum dicoccoides* (2n = 4x = 48). The B genome is still unclear and may be extinct, but cytological evidence suggests that the S genome of *Aegilops speltoides* is a closely related species or an ancestral progenitor to the B genome of wheat. The second hybridization event resulted in the complete form of the hexaploid genome *T. aestivum* (2n = 6x = 42), which occurred between the domesticated *Triticum dicoccum* or *T. durum*, a wild goatgrass, and *Aegilops tauschii* about 8000–10,000 years ago. Like most allopolyploid plants, wheat also has the diploid-like chromosome pairing behavior during meiosis, preventing multivalent formation created by multiple homologous or homoeologous chromosomes [Brenchley, 2012].

In wheat breeding, a strong emphasis has been put toward the improvement of grain yield as it the most important goal in wheat breeding. It has been detected that 37 per cent of wheat areas have experienced the yield stagnation. If the breeder develops an improved wheat variety, having a superior of trait, but produces low yields, producers unlikely will grow it because the yield is necessary for economic feasibility. The grain yield is a complex character with low heritability which is influenced not only by genes but also by the effects of the environment. In wheat, it has been documented that the higher yield is inversely related to the protein content and can also delay maturity. Furthermore, abiotic stress factors including drought, salinity, extreme temperatures, and acidity contribute the most to yield loss, ranging between 60 and 82 per cent [Cakmak, 2005]. As a consequence, extensive efforts have been made to identify the QTLs associated with the yield and its related traits which can be deployed by breeders through marker-assisted selection (MAS). The first report of the genome-wide assessment of molecular marker-based map in the nuclear genome of wheat began in the 1989 with the use of restriction fragment length polymorphism (RFLP). Subsequent analyses have been performed for construction of genetic maps to improve the efficiency of conventional breeding based on amplified-fragment-length polymorphism (AFLP), single-nucleotide polymorphism (SNP), diversity arrays technology (DArT), simple sequence repeat (SSR) or microsatellite, random DNA marker (RDM), gene targeted marker (GTM), and functional marker (FM). Based on these markers, there are 180 genetic maps extrapolated in wheat, most of which are developed by SSR markers. For example, molecular markers such as SSR and DArT have been used to detect QTLs for fusarium head blight (FHB) resistance, which can further be implemented in breeding. Since 2007, a number of research studies have been taken to identify QTLs related to yield based on different mapping populations such as kernel length, kernel width, spike length, spike number, the grain number of spikes, sterile spikelet number per spike, fertile spikelet number per spike, and thousand kernel weight [Liu, *et al.*, 2014]. However, the development of molecular markers and their applications in breeding have been relatively difficult in wheat because of its three closely related sub-genomes and a large genome size consisted of high amounts (80 per cent) of repetitive sequences. In 2012, the availability of wheat whole genome sequences has provided a framework for understanding of polyploidization, and domestication by comparing its sequences with ancestral and progenitor genomes, enabling us to understand the genetic diversity of wheat, which may help accelerate breeding programs. The chapter addresses molecular areas of research for yield improvement in wheat, focusing on finding QTLs for traits that affect yield. There are three objectives of this chapter as follows: (1) to explore the use of major molecular markers that have been used to identify yield and its-related QTLs in the past; (2) the current progress of molecular markers for linkage map construction; (3) to assess genomic studies of wheat; and (4) to discuss the potential of translational genomics in wheat using well-studied grasses such as rice and barley.

First-Generation Sequencing Resources for Yield Improvement in Wheat

Restriction Fragment Length Polymorphism (RFLP) Markers

Earlier molecular studies in wheat have shown that the RFLP markers are the common tools used as the oldest method of molecular markers for the construction of genetic maps. RFLP are typically inherited as simple Mendelian codominant markers, and are not influence by the environment, which could serve as highly heritable genetic markers for the study of inheritance of a trait. Chao *et al.*, created the first linkage maps of the homoeologous group 7 chromosomes on the wheat genome using 18 cDNA clones across six mapping populations from ten varieties. They mapped 31 RFLP loci on chromosomes 7A, 7B, and 7D. In 1991, detailed linkage maps were constructed with a total size of 1800 cm consisting of 197 RFLP loci. RFLP markers associated with two Hessian fly resistance genes from *Triticum taushii*, and preharvest sprouting genes from two recombinant inbred populations of white wheat (*T. aestivum* L. em. Thell.). Based on comparative mapping among cereal species such as wheat, rye, and barley, the development of detailed RFLP maps of the homoeologous group-2 chromosomes has established and showed that they had collinear relationships, indicating the high degree of conservation in gene orders. The RFLP markers flanking the resistance to cereal cyst nematode were identified to be *Xglk605* and *Xcdo588*, which were mapped at 7.3 and 8.4 cm from the *Cre* locus. Later, RFLP-based genetic maps belonging to different homoeologous groups were generated including homoeologous group 1, group 2, group 3, and group 5, and group 6. During this period, the powdery mildew genes were detected by RFLP probes. To increase the frequency of polymorphism detection in wheat, a non-intervarietal cross (W7984 X Opata85) developed was used creating a dense map which consisted of about 1000 RFLP loci on group 1, group 2, group 3, and group 6. Other disease-resistant genes of *Lr9* responsible for leaf rust and *Sr22* responsible for stem rust were characterized and mapped using RFLP markers. Furthermore, the feasibility of the 37 RFLP probes on group 5 chromosomes from Thatcher near isogenic lines (NILs) for leaf rust resistance gene *Lr1* was conducted by Feuillet *et al.*, of which three were investigated to be linked the gene. Galiba *et al.*, analyzed the cross between a frost-sensitive, vernalization-insensitive substitution line, *Triticum spelta* 5A and a frost-tolerant, vernalization-sensitive line, Cheynne 5A and showed co-segregation with *Vrn1* and *Xpsr426* RFLP marker. There was also a linkage detected between *Vrn1* and *Xwg644*. The mapping of a single locus controlling the aluminum tolerance gene *Alt2* was completed on the Chinese Spring chromosome arm 4DL derived from homoeologous recombination between *T. aestivum* cv. Chinese Spring chromosome 4D and *Triticum turgidum* cv. Cappelli chromosome 4B. Starting from 1998, QTLs for yield and yield-related traits were investigated. The dwarfing genes, *Rht-B1* and *Rht-D1*, associated with plant height in wheat were firstly mapped on the short arms of chromosomes 4BS (*Xfba1-4B*) and 4DS (*Xfba211-4D*). A year later, thirteen RFLP probes and one morphological marker locus, *Eps*, were used to develop a genetic linkage map and identified that an RFLP marker *Xcdo549* on the short arm of chromosome 3A was associated with plant height, kernel number spike-1, and 1000-kernel weight. As wheat lodging can result in yield losses, nine QTLs for lodging resistance were detected with the genetic distance between the flanking RFLP markers, of which seven coincided with QTLs for morphological traits. Since the year 2000, there have been many studies published for loci associated with grain yield, heading date, disease, and spike morphology. Kato *et al.*, conducted mapping of QTLs for grain yield and its components on chromosome 5A and confirmed that the grain-yield QTLs were closely linked to QTLs for yield components. They found RFLP markers associated with grain yield, tiller number/plant, ear grain weight, 50-grain weight, and spikelet number/ear based on a homozygous population of single-chromosome recombinant lines. Genetic mapping of QTLs conferring resistance to stripe (yellow) rust and powdery mildew was performed. The resistance gene, *Yr28*, derived from *Ae. tauschii and* the adult-plant resistance (APR) gene *Yr18* were mapped on chromosome arm 4DS and 7DS, respectively. They found that *Yr18* was closely correlated with leaf-rust gene Lr34. In addition, the RFLP *xbcd135* and *xbcd266* loci mapped at a genetic distance of 1.6 and 4.8 cm were identified to be closely linked to *Pm6*, a gene conferring resistance to the powdery mildew. Later, two of the powdery resistant genes, *Pm26* and *Pm29*, were mapped on the RFLP linkage map. Several RFLP markers within six major QTL regions have been identified to be tightly linked traits related to compactness such as spike length and number of spikelets. Based on a set of 114 recombinant lines (RILs) of the '**International Triticeae Mapping Initiative**' mapping population, morphological, agronomical, and disease resistance traits have been studied. This gene was located between the RFLP *Xcdo393* and *Xwg241* on the log arm of chromosome 1A, which showed a large effect on heading date with a phenotypic variance of 0.47. Despite RFLP markers are highly reproducible and used as the primary way for most genetic work in wheat, it has been difficult to use RFLP markers due to the low levels of polymorphism detected in wheat.

Random Amplified Polymorphic DNA (RAPD) Markers

The applications of RAPD markers have been beneficial to improve breeding programs in wheat because they are simple and fast PCR based, require no prior knowledge of target DNA sequence, and are analyzed either by the presence or absence of an amplicon via agarose gel electrophoresis. In wheat, RAPD has been used since 1990. Devos and Gale confirmed that a degree of polymorphism detected by six RAPD primers was comparable with RFLP markers. They identified four RAPD markers with bread wheat cultivar 'Chinese Spring.' The application of both bulk segregant analysis (BSA) and RAPD has started in 1994. Using BSA on DNA enriched for low-copy sequences by RAPD markers, the *Cre3* gene resistance to cereal cyst nematode (CCN) in *Triticum tauschii*, was mapped on the long arm of chromosome 2D (Ccn-D2). More efforts have been made to distinguish the wheat varieties resistant to cadmium stress, powdery mildew, and common bunt [58]. The RAPD marker *OPC20* was closely correlated with a gene controlling cadmium uptake in durum wheat (*T. turgidum* L. var. *durum*). In addition, a RAPD marker *OPH17-1900* located on the chromosome arm 6VS, was detected that could be used for the detection of a powdery mildew resistance gene, *Pm21*, in breeding. For common bunt, also known as stinking smut and covered smut, the polymorphic RAPD marker, *BW553*, was identified between resistant and susceptible NILs. Furthermore, identification of RAPD markers linked to the *Yr15* gene controlling stripe rust resistance was conducted using 340 RAPD primers, six of which were detected to be polymorphic. The *OPB13* RAPD marker was the only one that produced polymorphism in 123 F2 individuals and showed that it was linked to *Yr15* through screening a series of NILs each consisted of a different gene for Hessian fly resistance using 1600 random 10-mer primers. Another RAPD marker, *OPE-13*, showed the complete co-segregation with the H21 Hessian fly resistance gene in wheat. Dweikat *et al.*, developed RAPD markers linked to 11 different Hessian fly resistance loci that could be used for determining the presence or absence of specific Hessian fly resistance genes. More RAPD markers including *OPX061050*, *OPAG04950*, and *OPAI14600*, was observed to be linked to the new powdery mildew resistance *Pm25* gene, where the linkage distance between them was 12.8, 17.2, and 21.6 cm, respectively. The application of the BSA approach on DNA enriched for low-copy sequences, generating an increased level of polymorphism and in repeatability. Two RAPD markers, *UBC320420* and *UBC638550*, that co-segregated with *Pm1a* and one RAPD marker, *OPF12650*, tightly linked to the resistance gene. In these studies, it has been observed that RAPD markers also have been difficult to use in wheat like RFLP markers due to the very low level of polymorphism. Furthermore, the reproducibility of this RAPD bands have been found to be questionable. Despite this, continuous efforts have been made to develop RAPD genetic markers for the discrimination of species/cultivars from each other [Liu, *et al.*, 1999].

Amplified Fragment Length Polymorphism (AFLP) Markers

For assessment of large numbers of polymorphic loci, the AFLP technology has been implemented as a powerful tool because of its advantage of having good levels of reproducibility, insensitivity, fast, and no need of sequence information required for primer design. In 1995, a novel PCR-based assay for plant DNA fingerprinting using AFLP markers has resulted in high levels of DNA polymorphism. In fact, the AFLP technique has observed to be more efficient and less expensive and less labor intensive compared to the RFLP technique in wheat. Earlier AFLP-based marker studies have been found to be informative in assessment of genetic diversity in wheat varieties started in 1998. Regarding the investigation of traits associated with yield, initiated the AFLP technique to develop an AFLP marker associated with resistance to *Septoria tritici* blotch and powdery mildew, respectively. Several AFLP markers closely linked to the *Pm1c* and *Pm4a*. Out of 92 AFLP primer combinations, 31 polymorphic fragments detected between the resistance and susceptible lines, of which eight were found to be the most reliable polymorphic markers. One of the AFLP markers, *18M2*, was detected as being highly specific for the *Pm1c* gene, while the 4aM1 AFLP marker was identified at 3.5 cm from *Pm4a*. A couple of years later, an AFLP analysis was conducted using NILs of the strip rust resistance gene *Yr10* and designed AFLP primers that can be useful in detecting the *Yr10* gene. They found the two markers, *PM14(301)* and *PM42(249)*, of which *PM14(301)* was converted to PCR marker that could be a useful tool for MAS. In 2006, nine AFLP markers that showed polymorphism between the Argentinian wheat cultivar Sinvalocho MA and the rust leaf susceptible cultivar Gamma 6 were used to construct a linkage map of the *Lr3* gene for leaf rust resistance on chromosome 6BL of wheat. They detected seven markers linked to *Lr19* resistance to wheat leaf rust using *Mse I* and *Pst I* based on Thatcher, 23 NILs and F2 generation of Tc*Lr19* x Thatcher. Dhillon *et al.*, detected putative AFLP markers linked to leaf rust resistance genes *Lr9*, *Lr19*, and *KLM4-3B* using NILs of wheat. More likely, the AFLP technique has been approached for developing polymorphic markers underlying a trait.

Simple Sequence Repeat (SSR) and Inter Simple Sequence Repeat (ISSR) Markers

Among all available markers, SSR (or microsatellite) markers have become the best suited tool in plant breeding programs, because they are practical, convenient, easy to use, and inexpensive. Moreover, SSRs with tandem repeats of a motif of <6 bp are the most polymorphic, codominant, easy for scoring banding patterns, and have wide genomic distribution, high reproducibility, and a multiallelic nature. SSR analysis has been conducted in most of the QTL studies for mapping various traits. Up to 2015, there are more than 4000 SSR markers that have been developed and used for the construction of wheat genetic maps. The high level of variability and Mendelian inheritance of SSR DNA markers have been first reported by Devos *et al.*, [1995]. For example, SSR markers have been implemented for tagging and mapping important yield-related genes such as the dwarfing genes *Rh8*, *Rht12*, and the vernalization gene *Vrn1*, and a gene for preharvest sprouting tolerance. SSR primers also have been used to detect some of the wheat resistance genes. Peng *et al.*, found nine microsatellite loci found to be linked to *YrH52* with recombination frequencies of 0.2–0.35, and LOD scores of 3.56–54.22. The identification chromosomes with QTLs underlying 1000 grain weight (GW) has conducted by Varshney *et al.*, [2000]. They found the SSR marker Xwmc333 as being linked to GW and the major QTL for GW (*QGw1. ccsu-1A*) with a R2 value of 0.1509. Moreover, the SSR markers of *Xgwm210*, *Xgwm296*, and *Xgwm455*, were detected to be polymorphic and linked to the leaf rust resistance gene *Lr39*, of which *Xgwm210* was the closest marker mapped 10.7 cm from *Lr39*. These markers showed a signification association with gene effects of 84.662.92, and 2.61 per cent, contributing a total of 31 per cent of the phenotypic variance in seed size. Additional SSR markers have been reported for strip rust resistance genes, including the *Xgwm501* marker linked to *Yr5*, and the *Xpsp3000* marker linked to *Yr10*. A microsatellite linkage map of the powdery mildew resistance gene *Pm5e* on chromosome 7B has constructed with 20 microsatellite loci, consisting of two codominant markers *Xgwm783* and *Wgwm1267* located close to *Pm5e* with a linkage distance of 11.0 cm and 6.6 cm, respectively. The detection of QTL linked to FHB resistance has found that two microsatellite loci, *Xgwm533* and *Xgwm*, were significantly associated with QTL for FHB. Furthermore, five QTLs, *QYld.crc-1A*, *QYld.crc-2D*, *QYld.crc-3B*, *QYld.crc-5A.1*, and *QYld.crc-5A.2*, controlling grain yield have been identified on chromosomes 1A, 2D, 3B, and 5A. Other QTLs associated with 1000 grain weight, spikes meter–2, seed number spike–1, average seed weight spike–1, harvest index, days to heading, days to maturity, and grain filling time have also been detected in the Superb/BW278 mapping population. Liu *et al.*, [98] performed association mapping by genotyping wheat germplasm accession from China using SSR markers and EST-SSR markers, detecting 10 SSR markers on chromosome 4A associated with plant height, spike length, spikelets per spike, spikelet density, grains per spike and thousand-kernel weight. Even though studies involving RFLP, RAPD, and AFLP markers have only been used for identification and mapping of QTLs and genes, they are less likely to be applied into breeding programs because the application of these markers is likely to be less efficient for MAS [Akhunov *et al.*, 2009].

Second-Generation Sequencing Resources for Yield Improvement in Wheat

Researchers have focused on discovering SNPs to be used as genetic markers, because they have many advantages. SNPs act as codominant, single-locus, biallelic markers offering a lower error rate, and higher accuracy than SSR markers. However, the nature of polyploidy and having similar sequences among the A, B, and D genomes makes it difficult to identify SNPs. Therefore, progress in SNP detection has been limited, especially related to yield and its related traits. With the advancement in next-generation sequencing (NGS), sequencing has become increasingly popular due to the rapid development of NGS technologies, including SOLiD/Ion Torrent PGM from Life Science, Genome Analyzer/HiSeq 2000/MiSeq from Illumina, and 454 FLX Titanium/GS Junior. These technologies have led to the implementation of high-density SNP genotyping in wheat [Li, *et al.*, 2015]. An important milestone in wheat genomic research was accomplished in 2012 with the completion of *de novo* sequencing of bread wheat, the variety Chinese Spring (CS42), facilitating advances in genomic research into the genus *Triticum* and providing insights into the polyploidization and domestication of wheat. Brenchley *et al.*, sequenced the wheat genome using Roche 454 pyro-sequencing technology (GS FLX Titanium and GS FLX+ Platforms. To identify the three component genomes (A, B, and D) of hexaploid wheat, the following technologies have been applied: Illumina sequence assemblies of *Triticum monococcum* (related to the A-genome donor), *Ae. speltoides* complementary DNA assemblies, and 454 sequences from *Ae. tauschii* (the D-genome donor). Sequence analysis revealed that the A, B, and D genomes have estimated to consist of approximately 28,000, 38,000, and 36,000 genes, respectively. As a consequence, they identified the number of genes in wheat to be between 94,000 and 96,000. Wheat also has experienced a recent whole genome duplication (WGD) at 0.5 MYA and about 3.5 MYA. Brenchley *et al.*, also examined gene conservation between wheat and its most closely related species *Brachypodium distachyon* and detected a high degree of conservation between the two species. A large set of SNPs (132,000 SNPs) in A, B, and D genes will enhance future studies aimed at identifying QTLs and discovering associations of traits. Since the completion of the draft sequence of wheat, extensive efforts have put into the identification of various molecular markers influencing yield to increase MAS efficiency. Based on the genotype by sequencing (GBS) approach, the linkage map of wheat comprised of markers including 538 GBS Bin, 258 AFLPs, 175 SSRs, and an EST has been constructed in 2014. They identified five QTL regions linked to thylakoid membrane damage (TMD), SPAD chlorophyll content (SCC), and plasma membrane damage (PMD), known as indicatives of high temperature tolerance, on chromosomes 6A, 7A, 1B, 2B and 1D and also detected some of the SSR markers associated with these traits such as the SSR marker *Xbarc121* and *Xbarc49* for all three traits and *gwm18* and *Xbarc113* for SCC. More SNP and SSR markers have been investigated using the 9 K Infinium iSelect Beadchips. The SNP distribution between cultivars and landraces has provided impacts on our understanding of crop improvement on the structure of genetic diversity and insight into signatures of selection. Liu *et al.*, reported six SNPs from two genes, *wsnp_CAP11_c209_198467* and *wsnp_JD_c4438_5568170*, showed significant association with soil-borne wheat mosaic virus (SBWMV) resistant which can be used in MAS to improve SBWMV resistance in wheat breeding. Using the high-density Illumina iSelect 90K SNP assay, a linkage map spanning 3609.4 cm was constructed based on 5636 polymorphic SNP markers, with an average length of 171.9 cm per chromosome and marker density of 0.64 cm. Association of agronomic traits with 1,366 SNP markers in durum wheat has been performed. By genotyping 150 accessions of durum wheat germplasm based on the Illumina Bead Array platform and Golden Gate Assay, a large amount of SNP markers were detected associated with key yield-related traits including plant height, number of effective spikes, length of main spike, number of spikelets per plant, panicle neck length of main spike, grain number per plant, grain weight per plant, and 1000-grain weight. These SNP markers have enhanced the previous QTL analyses and can be utilized for MAS to improve yield in wheat.

Translational Genomics

The wheat genome is very complex as it is a polyploid species consisted of three diploid progenitor genomes. Recent advances in genome sequencing technologies have accelerated efforts to complete genomes of many crop species, which has opened the door for discovery and knowledge of the genetic basis of a number of important agronomic traits, with the final aim of crop improvement and production. Moreover, the completion of sequences has enabled assessment of translational genomics which is an effective way for researchers and breeders

Table 3.31. Comparative analysis of yield-related genes in wheat, barley, and maize

Species	*Barley*		*Rice*		*Wheat*	
Trait	*Locus*	*Identity*	*Locus*	*Identity*	*Locus*	*Identity*
Early flowering	KR706151	42.43	XM_015770121	32.86	KX161741.1	44.78
Late flowering	EU331897.1	44.13	XM_015787003	38.04	KJ711539.1	43.16
Dwarf/small	AY750996.1	42	AB630963.1	43.05	AY747606.1	43.1
Growth defective					AY244509.2	42.97
Stems Waxy	AK366020.1	58.91	FJ487950.1	58.98	KU376264.1	58.75
Stem Fasciation	AJ567377.2	60.29	FJ501983.1	59.64	EU981913.1	59.82
Stem short petiole	AK360706.1	36.39	AY585350.1	54.03	M11336.1	55.81
Stem twisted petiole	A14G25		EF190873.1	58.4	EU189003.1	54.23
Leaves, abnormal shape			Dc828333.1	58.37	AY831792.1	56.52
Leaves, rough surface						
Flowers, carpel	AK368772.1	CT831121.1	AY887064.1		NM00112060.1	
Flowers, stamen	AB085818.1		AK334664.1		DQ343238.1	
Flowers, petal						
Flowers, sepal						
Caryopsis, short	EU333963.1		AP014958.1		PJ373211.1	
Caryopsis, abnormal shape						

Table 3.32. Promising materials selected from the National and International Nurseries conducted at IARI, RS Pusa, Bihar

NP852	For late sown irrigated condition
NP884	For normal sowing irrigated condition
BW 11 (Purbali)	$(Lr_1 + Lr_3 + Lr_{10})$, $(Sr_2 + Sr_8 + Sr_{9b} + Sr_{11}+)$, Yr_2 + (ks), for Eastern Zone (West Bengal, Assam, Eastern states) for non-traditional wheat growing area; released by virtue of its yield potentiality and adoptability for far eastern zone under timely and irrigated cultivation.
HD 2045	The variety was recommended for its cultivation in 2002 for NEPZ under irrigated late sown condition; the plant height is about 100 cm and matures in 115 days; highly resistant to brown rest and tolerant to foliar blight.
HD 2285	This is also very popular high yielding short duration thermo tolerant variety, released for late and very late irrigated condition of sowing in Peninsular, North Western and Northern India; double dwarf of early maturity; with high yield potential; resistant to rusts and Karnal bunt; possesses minimal heat tolerance and wider adaptability; grains are amber, hard, medium bold which makes good chapattis.
HD 2307	A double and early maturity variety; resistant to rusts; grains are amber, bold and hard; possesses terminal heat tolerance.
HD 2329	It was released for irrigated timely sown for North West and North East Zone, since 1984, this variety is still in farmers field in different areas of the wheat zones.
HD 2380	$(Lr_{23} + Lr_{13})$, is widely adoptable wheat variety and released for Peninsular Zone as well as for Northern hills; very bold grain with resistant to rusts; grown in some pockets of Nepal.
HD 2402	$(Lr_{23}+)$, $(Sr_8 + Sr_{9b} + Sr_{11}+)$, Yr_2 + (ks), initially released for timely irrigated sown condition for far Eastern Zone but by virtue of its thermo tolerant again it was released for very late sowing condition; double dwarf variety of medium early maturity; tolerant to rusts; possesses wider adoptability and terminal heat tolerance; grains are amber, hard and medium bold.
HD 2501	$(Lr_{23} + Lr_{34}+)$, $(Sr_2 + Sr_{11}+)$, $Yr_{18}+$, released for Peninsular Zone under late sown irrigated cultivation; as good as HD 2189 in this zone; even today it is popular in this zone.
HD 2610	developed for similar agro climatic condition; not much popular.
HD 2643 (Ganga)	developed for North Eastern Plain Zone for late sown cultivation; even today, Ganga is very popular in eastern U.P., Bihar, West Bengal and North Eastern parts of the country; a double dwarf variety with medium maturity; excellent plant type with profuse tillering; resistant to all the 3 rusts; grains are excellent which make good chapattis.
HD 2687 (Shrestha)	released for North Western plain zone (Punjab, Haryana, Western U.P., Parts of Rajasthan *etc.*) under timely and irrigated high fertility condition; very popular high yielding variety in this zone; due to high yield this variety occupied larger areas in eastern U.P. and Bihar.
HD 2733 (VSM)	Another breakthrough in production and productivity has been in North Eastern Zone after releasing HD 2733 for timely irrigated cultivation; long duration high yielding resistant; a high yielding double dwarf variety developed by IARI, New Delhi, and is recommended for cultivation under timely sown irrigated conditions of North Eastern plain zone; plant height about 85cm. tall; at ear emergence was is present on its flag leaf, leaf sheath, spike and peduncle; has erect and small green leaves with slightly non- synchronous tillering; grains are very bold, hard and amber; is resistant to leaf rust and leaf blight.

HD 2824 (Poorva)	(Lr_{23} + Lr_1 +) (Sr_{7a} + Sr_{11} +), Yr_2 + (ks), is the high yielding dwarf wheat variety released for North Western Plain under irrigated high fertility condition; non-lodging, medium, early maturity with high yield. It was very popular in ninties among farmers; has profuse tillering and dwarf stature (85-90 cm); dark green leaves and can produce 50 q/ha under well fertilized irrigated conditions; wide adaptability, high degree of resistance to rusts (brown and yellow) and tolerance to leaf blight; grains are amber, hard, lustrous and attractive with a 1000 grain weight of 41gm and suitable for excellent chapatti making and other products.
HD 2888	released for this zone for rainfed condition; serve the hope of rainfed wheat growers in Bihar and Northern States; developed for cultivation in 2006 for NEPZ under timely grown rain-fed situation; shows high resistance against brown rust and tolerance to leaf blight; average height of the plant is 95 cm and matures within 130 days; protein content in this verity is 10.31 per cent.
HD 4502 (Malavika)	the first dwarf durum wheat in India released in 1974 for irrigated timely sown condition in peninsular zone; heat tolerant, resistant to black and brown rusts with high yield; served the durum wheat zone for more than 20 years.
HDR 77	It was developed by IARI New Delhi and is a very promising wheat variety; recommended for cultivation in 1990 for rain-fed late sown condition of NEPZ; short duration maturing (100 days) variety and shown tolerance to pre-harvest sprouting; average height of the plant is 75 cm and is tolerant to lodging; grains are bold lustrous and 1000 grain weight is near about 36 to 38 gm.
HP1102	8156 (Blanco) x Nad.63; for timely sown irrigated condition of NEPZ; plant height 85 cm, matures in 110 days, grains bold, amber in color and mottled
HP1209	E 4871/Pengamo 62; for late sown irrigated condition of NEPZ; plant height 85 cm, matures in 106 days, Grains amber, hard and bold.
HP1493	E 8841/NP830; for rain fed timely sowing in Bihar, plant height 135 cm, matures in 135 days, Grains very bold and lustrous, resistant to brown rust and tolerant to foliar blight.
HP1633 (Sonali)	RL 6010/5* Sonalika; for late sown irrigated condition of NEPZ; plant height 90 cm, matures in 115 days, grains very bold, amber, lustrous, highly resistant against leaf rust, karnal bunt and tolerant to leaf blight.
HP1731 (Rajlaxmi)	LIRA/PARULA//TONICHI; for timely sown irrigated condition of NEPZ, plant height 85 cm, matures in 120 days, grains amber, hard with moderate protein (11.4 per cent) and good chapatti making quality, resistant to all the three rusts (leaf, stem and stripe), tolerant to foliar blight.
HP1744 (Rajeshwari)	Ciano/Parula//Chilero/Garuda; for late sown irrigated condition of NEPZ; plant height 85 cm, matures in 110 days, grains hard, bold, amber and very good for chapatti quality, tolerant to leaf blight and highly resistant to brown and yellow rust.
HP1761 (Jagdish)	R 6010/6* Inia 66//3* Kauz; for timely sown irrigated condition of NEPZ, plant height 84 cm. matures within 123 days, grains small, hard and amber in colour, moderately resistant to leaf blight, heat tolerant.
HW 2034 (MACS 6145)	This variety was developed by IARI New Delhi and its regional station Wellington and was recommended for its cultivation in NEPZ under rain-fed condition; tall variety (110 cm) and highly resistant to brown rust.
HW 2045	(Sr_{25} + $Lr_{19)}$, this is the improved line of the variety HD 2402 (Lr 19 is incorporated) released for late sown condition for North Eastern Zone; gradually it is becoming popular in this zone.

to transfer knowledge of genetic and genomic information among related species, such as rice and wheat. The translational genomics tool is known as 'model to crop' translation that can be contributed to the implementation of genetic and genomics in crop species. There are three well-characterized model grass species rice, *Brachypodium*, and barley, which can be used to accelerate the application of translational genomics among the Poaceae family. The large amount of insufficient knowledge such as biological processes that are controlled by genes can be filled from well-studied species through translational biology approach. In translational genomics, comparative genomics studies can take advantage from available genomes and can provide information on the extrapolation of knowledge of gene functions among species. Genome sequences are currently available for wheat, but information QTL mapping as compared to other model grass species have been less reported. Comparative analysis based on the candidate gene approach (CGA) is known as the most powerful tool for exchange information among species. We conducted a brief comparative analysis of yield-related genes in Arabidopsis, barley, rice, wheat, and maize. The 37 Arabidopsis genes controlling flowering, growth, stems, leaves, flowers, and fruits, have been used to investigate homologous sequences in the barley, rice, wheat, and maize genomes using BLAST. Using the detected homologs among the grass species, we found a number of homologous sequences with the sequence identity values ranging between 36.39 and 60.39 based on MUSCLE v3.51. The release of genomic sequence of wheat, barley, and maize, provides a new opportunity for translational genomics. Since comparative genomics focuses on comparing genomes among plant species looking for similarities and differences of DNA sequence, protein sequence, and gene orders, information from well-studied and analyzed species can be applied for less studied crops to improve a specific target trait, which can be implemented in crop breeding and improvement. For example, genome-wide comparative analysis of flowering related genes in *Arabidopsis*, wheat, and barley has revealed that there are 900 and 275 putative orthologs in wheat and barley, respectively. In addition, they showed many orthologous genes having similar expression profiles in different tissues of wheat and barley based on their *in-silico* expression analyses. Such a work will help researchers to investigate candidate genes controlling the time of flowering in rice and barley which can be incorporated into molecular breeding for early flowering in wheat and barley in short-season cropping region [Peng, *et al.*, 2015].

HI 1563 (Pusa Prachi) identified for High Fertility, Irrigated, Late Sown Conditions of North Eastern Plains Zone

Bread wheat variety HI 1563, developed at IARI Regional Station, Indore, Madhya Pradesh, was identified for high fertility, irrigated, late sown conditions of North Eastern Plains Zone (NEPZ) at the 49th All India Wheat and Barley Research Workers' Meet at Punjab Agricultural University, Ludhiana during August, 27-30th 2010. HI 1563 is a genotype characterized by high and stable yield, good quality and strong resistance to stripe, stem and leaf rusts. It has given 11.1 per cent higher yield over HUW 234, 9.0 per cent over NW 2036 and 6.7 per cent over DBW 14, and can serve as a good source of improving the genetic diversity in North Eastern Plains Zone for wheat cultivation. In addition to very high field resistance to leaf, stripe

Table 3.33. Comparative distinguishing features of HP 1731 (Raj Laxmi), HP 1744 (Rajeshwari) and HP 1761 (Jagdish)

Variety	*HP-1731 (Raj Lakshmi) [Lira//Parula/Tonichi]*	*HP-1744 (Rajeshwari) [Ciano/Parula//Chilero/ Garuda]*	*HP-1761 (Jagdish) [RL 6010/6*Inia//3*KAUZ]*
Year of Release	1995	1997	1997
Average Yield (q/ha)	41	36.40	42.47
Distinguishing Features	Plant height: 85 cm Growth habit: Semi-spreading in early stage. Anthocyanin Pigmentation of – Coleoptiles: Absent. Leaf sheath: Absent Auricle (at flag leaf stage): Absent Foliage colour (Boot stage): Dark green Leaf width (Boot stage): Narrow Average days to heading: 74-76 days. Average days to maturity: 119 days. Waxiness (at ear emergence) Flag leaf: Present Leaf sheath: Present Ear: Present Peduncle: Present Ear colour at maturity: White Ear shape: Parallel Ear density: Intermediate Awn length: Normal Awn colour at maturity: White Glume shoulder: Oblique Glume beak: Medium Glume pubescence: Absent Grain Colour: Amber Texture: Semi hard to hard	Plant height: 85.5 cm Growth habit: Semi-spreading in early stage and erect after tillering initiation. Anthocyanin Pigmentation of – Coleoptiles: Absent. Leaf sheath: Absent Auricle (at flag leaf stage): Absent Foliage colour (Boot stage): Dark green Leaf width (Boot stage): Narrow Average days to heading: 63 days. Average days to maturity: 103 days. Waxiness (at ear emergence) Flag leaf: Present Leaf sheath: present Ear: Present Peduncle: Present Ear colour at maturity: White Ear shape: Parallel Ear density: Intermediate Awn length: Normal Awn colour at maturity: White Glume shoulder: Oblique Glume beak: Medium Glume pubescence: Absent Grain Colour: Amber	Plant heinght: 84 cm Growth habit: Semi-spreading in early stage and erect after tillering initiation. Anthocyanin Pigmentation of – Coleoptiles: Absent. Leaf sheath: Absent Auricle (at flag leaf stage): absent Foliage colour (Boot stage): dark green Leaf width (Boot stage): Narrow Average days to heading: 75-88 days. Average days to maturity: 108-133 days. Waxiness (at ear emergence) Flag leaf: Present Leaf sheath: present Ear: Present Peduncle: Present Ear colour at maturity: White Ear shape: Parallel Ear density: Intermediate Awn length: Normal Awn colour at maturity: white Glume shoulder: Oblique Glume beak: medium Glume pubescence: Absent Grain Colour: amber

Variety	*HP-1731 (Raj Lakshmi) [Lira//Parula/Tonichi]*	*HP-1744 (Rajeshwari) [Ciano/Parula//Chilero/ Garuda]*	*HP-1761 (Jagdish) [RL 6010/6*Inia//3*KAUZ]*
	Cheeks: Angular Crease width: Narrow Brush end hair: Not prominent Shape: Ovoid to Oblong Avg 1000 grain weight (g): 35 Flowering: 65-87 days Maturity: 118-137 days Medium early maturity Agronomic feature- Resistance to lodging: Straw strength – 4 in 1-9 scale Resistance to shattering, leaf rust and leaf blight	Texture: Semi-hard to hard Cheeks: Angular Crease width: Narrow Brush end hair: Not prominent Shape: Ovoid to oblong Avg 1000 grain weight (g): 39 Maturity group: Early Reactions to diseases: Resistant to leaf rust and leaf blight Yield (q/ha): 36.4 q/ha.	Texture: hard Cheeks: Angular Crease width: Narrow Brush end hair: Not prominent Shape: Oblong Avg 1000 grain weight (g): 41 Maturity group: Medium Reactions to diseases: resistant to leaf rust and leaf blight Yield (q/ha): 42.47q/ha.
Suitability	Irrigated Timely Sown condition of NEPZ. However, it shows flexibility both for early as well as late sown conditions.	Irrigated Late Sown condition of NEPZ.	Irrigated Timely Sown condition of NEPZ. However, it shows flexibility both for early as well as late sown conditions.

and stem rusts, HI 1563 has seedling resistance to almost all the pathotypes of the three rusts; and adult plant resistance to virulent pathotypes 46S119 and 78S84 of stripe rust; pathotypes 40A and 117-6 of stem rust; and pathotypes 77-5 and 104-2 of leaf rust. It has good levels of resistance to leaf blight, powdery mildew and foot rot. Hence, release of HI 1563 will enrich genetic diversity under wheat cultivation in NEPZ. HI 1563 is a good quality wheat genotype with high hectoliter weight, good grain appearance, high sedimentation value, high extraction rate, good bread, chapati and biscuit quality with good levels of micronutrients like iron, zinc and copper, and therefore, will ensure better market price, thereby making wheat cultivation in that area more profitable. In North Eastern Plains Zone, normally due to the late harvesting of paddy, wheat is mainly sown in late November or early December. Therefore, this Zone urgently needs high yielding, widely adaptable, rust resistant bread wheat varieties under late sown conditions for which HI 1563 holds good promise.

References

Ahn, S; Anderson, JE; Sorrells, ME and Tanksley, SD (1993) Homoeologous relationships of rice, wheat and maize *improvement*188 *Bread wheat: improvement and production189*189 chromosomes. *Mol. Gen. Genet.*, **241**: 483-490.

Akhunov, E; Nicolet, C and Dvorak, J (2009) Single nucleotide polymorphism genotyping in polyploid wheat with the Illumina Golden Gate assay. *Theor Appl Genet.***119**: 507–517.

Altschul, SF; Gish, W; Miller, W; Myers, EW and Lipman, DJ (1990) Basic local alignment search tool. *J. Mol. Biol.* **215**(3): 403–410.

Anderson, JA; Sorrells, ME and Tanksley, SD (1993). RFLP analysis of genomic regions associated with resistance to preharvest sprouting in wheat. *Crop Science* **33**: 453-459.

Anonymous (2003) Consolidated Progress Report, Increasing Wheat Production and Building up of Research Capabilities in Warmer Areas and Eastern India, (NATP MM/III-58), IARI Regional Station, Pusa, District-Samastipur, Bihar-848125.

Athwal, DS; Phul, PS; Minocha, JL (1967) Genetic male sterility in wheat. *Euphytica* **16**: 54-360.

Auriau, P (1973) ***Hybrid wheat***: Results obtained in France. *Proc Fourth Intl Wheat Gene Symp* (Missouri): 321.

Bahl, PN and Maan, SS (1973) Chromosomal location of male fertility restoring genes in six lines of common wheat. *Crop Sci* **13**: 317-320.

Barclay, IR (1975) High frequencies of haploid production in wheat (*Triticum aestivum*) by chromosome elimination. *Nature* **256**: 410-411.

Barro, F; Rooke, L; Bekes, F; Gras, P; Tatham, ST; Fido, R; Lazzari, PA; Shewry, PR and Barcelo, P (1997) Transformation of wheat with high molecular weight genes results in improved functional properties. *Nature Biotech* **15**: 1295-1299.

Becker, D; Brettschneider, R and Lorz, H (1994) Fertile transgenic wheat from microprojectile bombardment of scutellar tissue. *Plant J.*, **5**: 299-307.

Bérard, A; Le Paslier MC; Dardevet, M; Exbrayat-Vinson, F; Bonnin, I and Cenci, A (2009) High-throughput single nucleotide polymorphism genotyping in wheat (*Triticum* spp.). *Plant Biotechnol J.* **7**(4): 364–374.

Bieling, LM and Driscoll, CJ (1973) Release of series of MAs lines. *Proc Fourth Intl Wheat Gene Symp* (Missouri): 147-153.

Bohorova, NE; van Ginkel, M; Rajaram, S and Hoisington, DA (1995) Tissue culture response of CIMMYT elite bread wheat varieties and evaluation of regenerated plants. *Cer. Res. Commun.*, **23**: 243-249.

Bordbar, F; Rahiminejad, MR; Saeidi, H and Blattner, FR (2011) Phylogeny and genetic diversity of d-genome species of *Aegilops* and *Triticum* (Triticeae, Poaceae) from Iran based on microsatellites, ITS, and trnL-F. *Pl. Syst. Evol;***291**: 117–131.

Börner, A; Korzun, V and Worland, AJ (1998) Comparative genetic mapping of loci affecting plant height and development in cereals. *Euphytica*, **100**: 245-248.

Brenchley, R (2012) Analysis of the bread wheat genome using whole-genome shotgun sequencing. *Nature.* **491**(7426): 705–710.

Brettel, RI and Murray, FR (1995) DNA transfer and gene expression in transgenic cereals. *Biotech. Genet. Eng. Rev*, **13**: 315-334.

Briggle, LW (1963) Heterosis in wheat- a review. *Crop Sci* **3**: 407-412.

Briggle, LW (1964) Hybrid wheat. What it means to wheat production. *Cereal Sci. Today* **9**: 59-65.

Briggle, LW (1970) A recessive gene for male sterility in hexaploid wheat. *Crop Sci* **10**: 693-696.

Broglie, K; Chet, I; Holliday, M; Cressman, R; Biddle, P; Knowlton, C; Mauvais, CJ and Broglie, R (1991) Transgenic plants with enhanced resistance to the fungal pathogen *Rhizoctonia solani*. *Science*, **254**: 1194-1197.

Bryan, GJ; Collins, AJ; Stephenson, P; Orry, A; Smith, JB and Gale, MD (1997) Isolation and characterisation of microsatellites from hexaploid bread wheat. *Theor Appl Genet* **94**: 557-563.

Cadalen, T; Boeuf, C; Bernard, S and Bernard, M (1997) An intervarietal molecular marker map in *Triticum aestivum* L. Em. Thell. and comparison with a map from a wide cross. *Theor. Appl. Genet.*, 94: 367-377.

Cassman, KG (1999) Ecological intensification of cereal production systems: Yield potential, soil quality, and precision agriculture. *Proc. Natl. Acad. Sci. (USA)* **96**: 5952–5959.

Chamberlain, DA; Brettel, RIS; Last, DI; Witrzens, B; McElroy, D; Dolferus, R and Dennis, ES (1994) The use of the Emu promoter with antibiotic and herbicide resistance genes for the selection of transgenic wheat callus and rice plants. *Aust. J. Plant Physiol.*, **21**: 95-112.

Chao, S; Sharp, PJ; Worland, AJ; Warham, EJ; Koebner, RM and Gale, MD (1989) RFLP-based genetic maps of wheat homoeologous group 7 chromosomes. *Theor Appl Genet.***78**(4): 495–504.

Chaudhary, HK; Singh, S and Sethi, GS (2002) Interactive influence of wheat and maize genotypes on haploid induction in winter x spring wheat hybrids. *Journal of Genetics and Breeding* **56**: 259-266.

Chen, F; Xia, XC; Wang, DS; Lillemo, M and He, ZH (2006) Detection of allelic variation for Puroindoline alleles in CIMMYT germplasm developed from synthetic wheat crossing with common wheats (in Chinese with an English abstract). *Sci. Entia Agr. Sinica* **39**: 440–447.

Cheng, M; Fry, J; Pang, Sh; Zhou, H; Hironaka, CM; Duncan, DR; Conner, TW and Wan, Y (1997) Genetic transformation of wheat mediated by *Agrobacterium tumefaciens*. *Plant Physiol.*, **115**: 971-980.

Christensen, A and Quail, P (1996) Ubiquitin promoter-based vectors for high-level expression of selectable and / or screenable marker genes in monocotyledonous plants. *Trans. Res.*, **5**: 213-218.

Chu, CG; Faris, JD; Friesen, TL and Xu, SS (2006) Molecular mapping of hybrid necrosis genes Ne1 and Ne2 in hexaploid wheat using microsatellite markers. *Theor. Appl. Genet.* **112**: 1373–1381.

Cox, TS (1997) Deepening the wheat gene pool. *J. Crop Prod.* **1**: 1–25.

Cox, TS; Sears, RG and Bequette, RK (1995) Use of winter wheat x *Triticum tauschii* backcross populations for germplasm evaluation. *Theor. Appl. Genet.* **90**: 571–577.

Curtis, BC (2002) 'Wheat in the World' in B.C. [Internet].

De Block, M; Debrouwer, D and Omens, T (1997) The development of a nuclear sterility system in wheat. Expression of the barnase gene under the control of tapetum specific promoters. *Theor. Appl. Genet.*, **95**: 125-131.

de la Pena, RC; Murray, TD and Jones, SS (1997) Identification of an RFLP interval containing *Pch2* on chromosome 7AL in wheat. *Genome* **40**: 249-252.

De Vries,A.P (1971) Flowering biology of wheat, particularly in view of hybrid seed production. *Euphytica* **20**: 152-170.

Devey, ME and Hart, GE (1993) Chromosomal localization of intergenomic RFLP loci in hexaploid wheat. *Genome,* **36**: 913-918.

Devos, KM and Gale, MD (1993) Extended genetic maps of the homoeologous group-3 chromosomes of wheat, rye and barley. *Theor Appl Genet.* **85**: 649–652.

Devos, KM; Dole el, J and Feuillet, C (2009) Genome organization and comparative genomics. *Wheat Science and Trade*. 327–367.

Devos, KM; Millan, T and Gale, MD (1993) Comparative RFLP maps of the homoeologous group 2 chromosomes of wheat, rye and barley. *Theor. Appl. Genet.,* **85**: 784-792.

Directorate of Wheat Research (1999) Elite International Germplasm Nurseries. In "Genetic Resources Identified from National and International Nurseries. GRU, DWR Kamal. p29.

Directorate of Wheat Research (2000) Elite International Germplasm Nurseries. In "Germplasm evaluation and enhancement" eds (D. Mohan and Jag Shoran) GRU, DWR Kamal. p53.

Dixon, J (2007) The Economics of Wheat. Developments in Plant Breeding **12**: 9-22.

Dreccer, MF; Ogbonnaya, FC and Borgognone. G (2004) Sodium exclusion in primary synthetic wheats. p. 118–121. In: Procs of 11th Wheat Breeding Assembly, Symposium on Seeding the Future Conference, Sept. 21–24, Canberra, Australia.

Driscoll, CJ (1972) XYZ systems of producing Hybrid wheat. *Crop Sci* **12**: 516-517.

Driscoll, CJ (1973) A chromosomal male sterility system of producing *Hybrid wheat. Proc Fourth Intl Wheat Gene Symp* (Missouri): 669-674.

Dvorak, J and Akhunov, ED (2005) Tempos of gene locus deletions and duplications and their relationship to recombination rate during diploid and polyploid evolution in the *Aegilops-Triticum* alliance. *Genetics*. **171**(1): 323–332.

Edgar, RC (2004) MUSCLE: multiple sequence alignment with high accuracy and high throughput. *Nucleic Acids Res.* **32**(5): 1792–1797.

Endo, TR and Gill, BS (1996) The deletion stocks of common wheat. *J. Hered.*, **87**: 295-307.

Englendow, FL and Pal, BP (1935) Hybrid vigour in wheat. *Indian Journal of Agricultural Science.* **V** (6): 693-704.

Frankel, OH and Bennet, E (1970) Genetic resources in plants-their exploration and conservation. Davis FA. Philadelphia, USA.

Fritz, AK; Cox, TX; Gill, BS and Sears, RG (1995) Molecular marker-facilitated analysis of introgression in winter wheat, Triticum tauschii populations. *Crop Sci.* **35**: 1691–1695.

Gale, MD; Atkinson, MD; Chinoy, CN; Harcourt, RL; Jia, J; Li, QY and Devos, KM (1995) Genetic maps of hexaploid wheat. *In* Z.S. Li and Z.I. Xin, eds. *Proc. 8th Int. Wheat Genetics Symp.*, Beijing, 20-25 Jul. 1993, Beijing, China Agricultural Scientech pp. 29-40.

Gao, LF; Jing, RL; Huo, NX; Li, Y; Li, XP; Zhou, RH; Chang, XP; Tang, JF; Ma, ZY and Jia, JZ (2004) One hundred and one new microsatellite loci derived from ESTs (EST-SSRs) in bread wheat. *Theor Appl Genet* **108**: 1392-1400.

Gill, KS and Anand, SC (1970) Genetic male sterility for hybrid seed production in wheat. *Crop Sci* **10**: 385-386.

Gill, KS and Gill, BS (1996) A PCR-based screening assay of *Ph1*, the chromosome pairing regulator gene of wheat. *Crop Sci* **36**: 719-722.

Gill, KS; Gill, BS; Endo, TR and Boyko, EV (1996) Identification and high-density mapping of gene rich regions in chromosome group 5 of wheat. *Genetics*, **143**: 1001-1012.

Godfray, HCJ; Beddington, JR; Crute, IR; Haddad, L; Lawrence, D; Muir, JF; Pretty, J; Robinson, S; Thomas, SM and Toulmin, C (2010) Food security: the challenge of feeding 9 billion people. *Science* **327**: 812–818.

Gororo, NN (1999) Growth, development and yield of wheat lines bearing *Triticum tauschii* germplasm. Ph.D. thesis. Univ. Melbourne, Australia.

Hauptmann, RM; Vasil, V; Ozias-Atkins, PM; Tabaeizadeh, Z; Rogers, SG; Fraley, RT; Horsch, RB and Vasil, IK (1988) Evaluation of selectable markers for obtaining stable transformants in the Gramineae. *Plant Phys.*, **86**: 602-606.

He, S; Ohm, H and Mackenzie, S (1992). Detection of DNA sequence polymorphisms among wheat varieties. *Theor Appl Genet* **84**: 573-578.

Hess, D; Dressler, K and Nummrichter, R (1990) Transformation experiments by pipetting *Agrobacterium* into the spikelets of wheat (*Triticum aestivum* L.). *Plant Sci.*, **72**: 233-244.

Hohmann, U; Endo, TR; Herrmann, RG and Gill, BS (1994) Characterization of deletions in common wheat induced by an *Aegilops cylindrica* chromosome:

detection of multiple chromosome re-arrangements. *Theor. Appl. Genet.*, **91**: 611-617.

Hu, X; Ren, J; Ren, X; Huang, S; Sabiel, SA and Luo, M (2015) Association of agronomic traits with SNP markers in Durum Wheat (*Triticum turgidum* L. durum (Desf.). *PLoS One.* **10**(6): e0130854.

Humphreys, DG; Procunier, JD; Mauthe, W; Howes, NK; Brown, PD and Mac-Kenzie, RIH (1998) Marker-assisted selection for high protein concentration in wheat. In: Fowler DB, Ged-des WE, Johnston AM, Preston KR (eds.) Wheat Protein, Production and Marketing. Proc Wheat Protein Symp, Saskatoon, Saskatchewan, Canada, 9-10 Mar. (1998), pp. 255-258.

Imtiaz, M; Ogbonnaya, FC; Oman, J and.van Ginkel, M (2008) Characterization of quantitative trait loci controlling genetic variation for preharvest sprouting in synthetic backcross-derived wheat lines. *Genetics* **178**: 1725–1736.

Jag Shoran, Hariprasad, AS; Lakshmi Kant; Chauhan, VS and Sinha, MK (1997) Contribution of different characters towards seed yield in wheat under environments. *Ibid.*

Jag Shoran, Mohan, D; Tyagi, BS and Kumar, J (1998) Yield Trial/Nurseries: Information and benefits at what cost? pp116-117 Ibid.

Jenes, B; Moore, H; Cao, J; Zhang, W and Wu, R (1993) Techniques for gene transfer. *In* S. Kung and R. Wu, (Eds) *Transgenic plants*, vol. 1, *Engineering and utilization*, p. 125-126. New York, NY, USA, Academic Press.

Jha, MP; Sinha, BP; Singh, KMP and Sinha, AK (1990) Breeding for high yield potential in wheat. Annual Wheat Newsletter.**36**: Coarado State University.

Joppa, LRF; Neal, HM and Welsh, IR (1966) Pollen and anther development in cytoplasmic male sterile wheat (*Triticum aestivum* L). *Crop Sci.***6**: 296-297.

Kashkush, K; Feldman, M and Levy, AA (2002) Gene loss, silencing, and activation in a newly synthesized wheat allotetraploid. *Genetics* **160**: 1651–1659.

Kato, K; Miura, H; Akiyama, M; Kuroshima, M and Sawada, S (1998) RFLP mapping of the three major genes, *Vrn1*, *Q* and *B1*, on the long arm of chromosome 5A of wheat. *Euphytica* **101**: 91-95.

Kearney, J (2010) Food consumption trends and drivers. *Phil. Trans. R. Soc. B.* **365**: 2793–2807.

Korzun, V; Röder, MS; Ganal, MW; Worland, AJ and Law, CN (1998) Genetic analysis of the dwarfing gene (*Rht8*) in wheat. Part I. Molecular mapping of *Rht8* on the short arm of chromosome 2D of bread wheat (*Triticum aestivum* L.). *Theor. Appl. Genet.*, **96**: 1104-1109.

Kota, RS; Gill, KS; Gill, BS and Endo, TR (1993) A cytogenetically based physical map of chromosome 1B in common wheat. *Genome*, **36**: 548-554.

Kundu, S and Nagarajan, S (1998) Distinguishing characters of Indian Wheat Varieties. Research Bulletin No. 4. Directorate of Wheat Research, Karnal. pp-160.

Lage, J and Trethowan, RM (2008) CIMMYT's use of synthetic hexaploid wheat in breeding for adaptation to rainfed environments globally. *Aust. J. Agr. Res.* **59**: 461–469.

Lagudah ES; Appels, R; McNeil, D and Schachtman, DP (1993) Exploiting the diploid D genome chromatin for wheat improvement. In: JP Gustafson, R Appels and P Raven (eds) Gene conservation and exploitation. Plenum Press, New York, pp. 87–107.

Langridge, P and Chalmers, K. (1998) Techniques for marker development *In* A.E. Slinkard, ed. *Proc. 9th Int. Wheat Genetics Symp.*, Saskatoon, Saskatchewan, Canada, 2-7 Aug. 1998, vol. 1, p. 107-117.

Langridge, P and Fleury, D (2011) Making the most of 'omics' for crop breeding. *Trends Biotechnol.* **29**: 33–40.

Larkin, PJ; Taylor, BH; Gersmann, M and Brettel, RIS (1990) Direct gene transfer to protoplasts. *Aust. J. Plant Physiol.*, **17**: 291-302.

Laurie, DA and Bennett, MD (1988) Cytological evidence for fertilization in hexaploid wheat x sorghum crosses. *Plant Breeding* **100**: 73-82.

Law, CN (1966) The location of genetic factors controlling a number of quantitative characters in wheat. *Genetics*, **56**: 445-461.

Liu, G; Jia, L; Lu, L; Qin, D; Zhang, J and Guan, P (2014) Mapping QTLs of yield-related traits using RIL population derived from common wheat and Tibetan semi-wild wheat. *Theor Appl Genet.* **127**: 2415–2432.

Liu, L; Wang, L; Yao, J; Zheng, Y and Zhao, C (2010) Association mapping of six agronomic traits on chromosome 4A of wheat (*Triticum aestivum* L.). *Mol Plant Breed.* **1**(5): 1–10.

Liu, YG (1991) Tsunewaki K. Restriction fragment length polymorphism (RFLP) analysis in wheat. II. Linkage maps of the RFLP sites in common wheat. *Jpn J Genet.* **66**(5): 617–633.

Liu, ZQ; Pei, Y and Pu, ZJ (1999) Relationship between hybrid performance and genetic diversity based on RAPD markers in wheat. *Triticum aestivum* L. *Plant Breed.* **118**(2): 119–123.

Ma, ZQ and Sorrells, ME (1995) Genetic analysis of fertility restoration in wheat using restriction fragment length polymorphisms. *Crop Sci.*, **35**: 1137-1143.

Mahajan, V and Nagarajan, S (1997) Indian initiative for the development of hybrid wheat. The Genetic and Exploitation of Heteresis in Crop. Aug., 7-22,1997, Mexico City, Mexico.

Mahajan, V; Singh, K and Kelkar, RG (1997) Possibility of using CHA route in developing wheat hybrids. In Proceed. In Proceed. International Group Meeting on "Wheat Research Needs Beyond 2000 AD" Aug. 12-14. 1997 D.W.R., Karnal. pp. 101-106.

Marino, CL; Tuleen, NA; Hart, GE; Nelson, JC; Sorrells, ME; Lu, YH (1996) Molecular genetic maps of the group 6 chromosomes of hexaploid wheat (*Triticum aestivum* L. em. Thell.). *Genome*.**39**(2): 359–366.

Mass, C and Werr, W (1989) Mechanism and optimised conditions for PIG mediated DNA transformation transfection into plant protoplasts. *Plant Cell Rep.*, **8**: 148-151.

McElroy, D; Zang, W; Xu, D; Witrzens, B; Gubler, F; Jacobsen, J; Wu, R; Brettel, RIS and Dennis, ES (1994) Development of promoter systems for the expression of foreign genes in transgenic cereals. *In* R. Henry and J. Ronalds, eds. *Improvement of cereal quality by genetic engineering*, New York, NY, USA, Plenum Press, pp55-59.

McFadden, ES (1930) A successful transfer of emmer characters to vulgare wheat. *J. Am. Soc. Agr.* **22**: 1020–1034.

Medini, M; Hamza, S; Rebai A and Baum, M (2005) Analysis of genetic diversity in Tunisian durum wheat cultivars and related wild species by SSR and AFLP markers. *Genet Resour Crop Evol* **52**: 21-31.

Mihalev, IM (1973) A new fertility restorer source in common wheat. *Proc Fourth Intl Wheat Gene Symp* (Missouri): 375-379.

Mitra, M (1934) A leaf spot disease of wheat caused by *Helminthosporium tritici*-Repentis Died. *Indian Journal of Agricultural Science*. **IV** (4): 692-700.

Mohan, M; Nair, S; Bhagwat, A; Krishna, TG; Yano, M; Bhatia, CR and Sasaki, T (1997) Genome mapping, molecular markers and marker assisted selection in crop plants. *Molecular Breeding* **3:** 87-103.

Mooney, P; Goodvin, PB; Dennis, ES and Llewellyn, DJ (1991) *Agrobacterium tumefaciens* - gene transfer into wheat tissue. *Plant Cell Tissue Organ Cult.*, **25**: 209-218.

Mujeeb-Kazi, A (2003b) Wheat improvement facilitated by novel genetic diversity and *in vitro* technology. *Plant Cell Tissue Organ Cult.* **13**: 179–210.

Myburg, AA; Cawood, M; Wingfield, BD and Botha, AM (1998) Development of RAPD and SCAR markers linked to the Russian wheat aphid resistance gene *Dn2* in wheat. *Theor Appl Genet* **96**: 1162-1169.

Nagarajan, S (1999) Large yield gap to fill. Survey of Indian Agriculture 1999. "The Hindu" Chennai. pp49-53.

Nagarajan, S and Singh, G (1998) Sustaining Wheat (*Triticum aestivum*) revolution-retrospect and prospect. *Ind. J. Agri Sci.* 68 (8, special issue): 396-404.

Nagarajan, S; Tyagi, BS and Singh G (1998) Extent of genetic diversity in Indian Wheat: The inadequacy to Break: yield barrier. In Genetics and Biotechnology in Crop Improvement Eds. PK Gupta *et al.* pp. 276-382.

Nanda, GS; Rao, VSP and Mohan, D (1998) Genetic Resources utilized and underutilized. In Proceed. International Group Meeting on "Wheat Research Needs Beyond 2000 AD" Aug. 12-14. 1997 DWR, Karnal. pp. 51-58.

Nehra, SN; Chibar, R; Leung, N; Caswell, K; Mallard, C; Steinhauer, L; Baga, M and Kartha, K (1994) Self-fertile transgenic wheat plants regenerated from isolated scutellar tissues following microprojectile bombardment with two distinct gene constructs. *Plant J.*, **52**: 285-297.

Niwa, K; Aihara, H; Yamada, A and Motohashi. T (2010) Chromosome number variations in newly synthesized hexaploid wheats spontaneously derived from self-fertilization of *Triticum carthlicum* Nevski/*Aegilops tauschii* Coss.F_1 hybrids. *Cereal Res.Commu.*38: 449–458.

Ortiz, JPA; Reggiardo, MI; Ravizzini, RA; Altabe, SG; Cervigni, GDL; Spitteler, MS; Morata, MM; Elias, FE and Vallejos, RH (1996) Hygromycin resistance as an efficient selectable marker for wheat stable transformation. *Plant Cell Rep.*, **15**: 877-881.

Pal, BP (1966) Wheat. Indian Council of Agricultural Research, New Delhi, pp370.

Palve, SM and Raghavaiah (2002) Genetic evaluation of interspecific derivation of wheat. *Ind J Genet* **62** (2): 107-12.

Pang, S; DeBoer, DL; Wan, Y; Ye, G; Layton, LG; Neher, MK; Armstrong, CL; Fry, JE; Hinchee, MAW and Fromm, ME (1996) An improved green fluorescent protein as a vital marker in plants. *Plant Physiology* **112**: 893-900.

Pathak, D (2001) Introgression of quantitative trait loci from tetraploid *Triticum* species into *T. durum*. Ph.D. Dissertation, PAU Ludhiana.

Patterson, AH; Lander, ES; Hewitt, JD; Peterson, S; Lincoln, SE and Tanksley, SD (1988) Resolution of qualitative traits into Mendelian factors by using a complete RFLP linkage map. *Nature* **335**: 721-726.

Paull, JG; Chalmers, KJ; Karakousis, A; Kretschmer, JM; Manning, S and Langridge, P (1998) Genetic diversity in Australian wheat varieties and breeding material based on RFLP data. *Theor Appl Genet* **96**: 435-446.

Peng, FY; Hu, Z and Yang, RC (2015) Genome-wide comparative analysis of flowering-related genes in *Arabidopsis*, wheat, and barley. *Int J Plant Genomics*. (874361).

Pestsova, EG; Beorner, A and Reoder, MS (2001) Development of a set of *Triticum aestivum–Aegilops tauschii* introgression lines. *Hereditas* **135**: 139–143.

Plaschke, J; Börner, A; Wendehake, K; Ganal, MW and Röder, MS (1996) The use of wheat aneuploids for the chromosomal assignment of microsatellite loci. *Euphytica*, **89**: 33-40.

Porter, KB; Lahri, KA; Atkins, IM (1965) Cross pollination of male sterile winter wheat (*Triticum aestivum* L) having *Aegilops caudata* L and *Aegilops ovata* L cytoplasm. *Crop Sci* **5**: 161-163.

Potrykus, I (1990) Gene transfer to cereals: an assessment. *Bio/Technology*, **8**: 535-542.

Qu, LJ; Foote, TN; Roberts, MA; Money, TA; Aragón-Alcaide, L; Snape, JW and Moore, G (1998) A simple PCR-based method for scoring the *ph1b* deletion in wheat. *Theor. Appl. Genet.*, **96**: 371-375.

Rafalski, JA and Tingey, SV (1993) Genetic diagnostics in plant breeding: RAPDs, microsatellites and machines. *Trends Genet.*, **9**: 275-280.

Rajenkar, PK; Dhaliwal, HS; Rao, VS and Tiwari, R (1997) Molecular biology approaches to wheat Improvement in India- Present status and future strategy. pp. 145-160 Ibid.

Ray, DK; Ramankutty, N; Mueller, ND; West, PC and Foley, JA (2012) Recent patterns of crop yield growth and stagnation. *Nat Commun*. **3**: 1293.

Robertson, LD and Curtis, BS (1967) Monosomic analysis of sterility restoration in common wheat (*Triticum aestivum* L). *Crop Sci* (7): 493-495.

Roder MS, Korzun V, Wendehake K, Plaschke J, Tixier MH, Leroy P, Ganal MW (1998) A microsatellite map of wheat. *Genetics* **149**: 2007-2023.

Röder, MS; Korzun, V; Gill, BS and Ganal, MW (1998a) The physical mapping of microsatellite markers in wheat. *Genome*, **41**: 278-283.

Röder, MS; Korzun, V; Wendehake, K; Plaschke, J; Tixier, MH; Leroy, P and Ganal, MW (1998b) A microsatellite map of wheat. *Genetics*, **149**: 1-17.

Roy, JK; Prasad, M; Varshney, RK; Balyan, HS; Blake, TK; Dhaliwal, HS; Singh, H; Edwards, KJ and Gupta, PK (1999) Identification of a microsatellite on chromosomes 6B and a STS on 7D of bread wheat showing an association with preharvest sprouting tolerance. *Theor Appl Genet* **99**: 336-340.

Sautter, C; Waldner, H; Neuhaus-Uri, G; Galli, A; Neuhaus, G and Potrikys, I (1991) Microtargeting: high efficiency gene transfer using a novel approach for the acceleration of micro-projectiles. *Bio/Technology*, **9**: 1080-1085.

Sax, K (1922a) Sterility in wheat hybrids III. Endosperm development and F_2 sterility. *Genetics* **7**: 554–559.

Sax, K (1922b) Sterility in wheat hybrids. II. Chromosome behaviour in partially sterile hybrids. *Genetics* **7**: 513–552.

Schena, M; Shalon, D; Davis, RW and Brown, PO (1995) Quantitative monitoring of gene expression patterns with a complementary DNA microarray. *Science*, **270**: 467-470.

Shalon, D; Smith, SJ and Brown, PO (1996) A DNA microarray system for analyzing complex DNA samples using two-color fluorescent probe hybridization. *Genome Meth.*, **6**: 639-645.

Sharma, AK; Singh, GP; Rane, J and Nagarajan, S (1998) Alternate selection approach for targeted cropping sequences. pp. 227-237 *Ibid.*

Shaw, FJF (1931) Milling and baking tests with some Indian wheat grown at Pusa and Mirpurkhas (Sind). *Indian Journal of Agricultural Science*. **I** (3): 395-414.

Sinha, AK (2003) Combining ability analysis for quantitative traits in bread wheat. *RAU J.Res.***13** (1 and 2): 57-60.

Sinha, AK; Jha, MP and Sinha, BP (1999) Correlation studies in long ear wheat. *Ann. Agric.Res.***20** (2): 148-150.

Sinha, AK; Kumari, S; Jha, MP; Sinha, BP and Singh, KMP (1992) Additional sources of necrotic genes identified. *Annual wheat news letter*.**39**: Colorado State University.

Smale, M; Reynolds, MP; Warburton, M; Skovmand, B; Trethowan, R; Singh, RP; Ortiz- Monasterio, I and Crossa. J (2002) Dimensions of diversity in modern spring bread wheat in developing countries from 1965. *Crop Sci.* **42**: 1766–1779.

Sohail, Q; Shehzad, T; Kilian, A; Eltayeb, AE; Tanaka, H; and Tsujimoto, H (2012) Development of diversity array technology (DArT) markers for assessment of population structure and diversity in *Aegilops tauschii. Breed. Sci.* 38–45.

Somers, DJ; Issac, P and Edwards, K (2004) A high density microsatellite consensus map for bread wheat (*Triticum aestivum* L.) *Theor Appl Genet* **109**: 1105-1114.

Sorrells, ME; Gustafson, JP; Somers, D; Chao, S; Benscher, D; Guedira-Brown, G; Huttner, E; Kilian, A; McGuire, PE; Ross, K; Tanaka, J; Wenzl, P; Williams, K and Qualset, CO (2011) Reconstruction of the synthetic W9784_Opata M85 wheat reference population. *Genome* **54**: 875–882.

Srivastava, HK; Sarkissan, IV and Shanda, JL (1967) Mitochondrial complementation and cytoplasmic male sterility in wheat. *Genetics* **63**: 611-618.

Suneson, CS and Cox, EL (1964) Promiscuity in barley and wheat. *Crop Sci* **4**: 551.

Takumi, S and Shimada, T (1996) Production of transgenic wheat through particle bombardment of scutellar tissues: frequency is influenced by culture duration. *J. Plant Physiol.*, **149**: 418-423.

Tanksley, SD and Nelson, JC (1996) Advanced backcross QTL analysis: a method for the simultaneous discovery and transfer of valuable QTLs from unadapted germplasm into elite breeding lines. *Theor. Appl. Genet.* **92**: 191–203.

Taylor, MG; Vasil, V and Vasil, IK (1993) Enhanced GUS gene expression in cereal grass cell suspensions and immature embryos using the maize ubiquitin-based plasmid pAHC25. *Plant Cell Rep.*, **12**: 491-495.

Tinland, B (1996) The integration of T-DNA into plant genomes. *Trends Plant Sci.*, **1**: 178-184.

Trupp, CR (1971) Genetic Male Sterility as a means of developing maintainer lines and restorer parents of Hybrid wheat. *Crop Sci* **11**: 460-461.

Varshney, R; Prasad, M; Roy, J; Kumar, N; Singh,H and Dhaliwal, HS (2000) Identification of eight chromosomes and a microsatellite marker on 1AS associated with QTL for grain weight in bread wheat. *Theor Appl Genet*, **100**(8): 1290–-1294.

Vasil, V; Brown, SM; Re, D; Fromm, ME and Vasil, IK (1991) Stably transformed callus lines from microprojectile bombardment of cell suspension cultures of wheat. *Bio/Technology*, **9**: 743-747.

Vasil, V; Castillo, A; Fromm, M and Vasil, I (1992) Herbicide resistant fertile transgenic wheat plants obtained by micro-projectile bombardment of regenerable embryogenic callus. *BioTechnology*, **10**: 667-675.

Venter, E and Botha, AM (2000) Development of markers linked to *Diuraphis noxia* resistance in wheat using a novel PCR-RFLP approach. *Theor Appl Genet* **100**: 965-970.

Walton, PD (1971) Heterosis in Spring wheat. *Crop Sci* **11**: 422-424.

Weeks, J; Anderson, O and Blechl, A (1993) Rapid production of multiple independent lines of fertile transgenic wheat. *Plant Physiol.*, **102**: 1077-1084.

Wells, DG and Lay, CL (1970) Hybrid vigor in hard red spring wheat crosses. *Crop Sci* **10**: 22-23.

Worland, AJ; Borner A; Korzun, V; Li, WM; Petrovic, S and Sayers, EJ (1997) The influence of photoperiod genes on the adaptability of European winter wheats. In: Braun HJ (ed) Wheat: prospects for global improvement. Dordrecht, Netherlands, Kluwer Academic Press, pp. 517-526.

Xie, DX; Devos, KM; Moore, G and Gale, MD (1993) RFLP-based genetic maps of the homoeologous group 5 chromosomes of bread wheat (*Triticum aestivum* L.). *Theor. Appl. Genet.*, **87**: 70-74.

Xie, WL and Nevo, E (2008) Wild emmer: genetic resources, gene mapping and potential for wheat improvement. *Euphytica* **164**: 603–614.

Young, ND (1999) A cautiously optimistic vision for marker assisted selection. *Mol Breed*. **5**: 505-510.

Zadoks, JC; Chang, TT and Konzak, CF (1974) A decimal code for the growth stages of cereals. *Weed Research* **14**: 415–421.

Zeven, SC (1972) Plant density effect on expression of heterosis for yield and its component in wheat and F_1 versus F_3 yield. *Euphytica*. **21**: 468-488.

Zhou, H; Arrowsmith, JW; Fromm, ME; Hironaka, CM; Taylor, ML; Rodrigues, D; Pajeau, ME; Brown, SM; Santino, CG and Fry, JE (1995) Glyphosate-tolerant CP4 and GOX genes as a selectable marker in wheat transformation. *Plant Cell Rep.*, **15**: 159-163.

Chapter 4

Bread Wheat and Durum Wheat Breeding at IARI RS Indore, Madhya Pradesh: Past, Present (2010-2013; 2018-2020) and Future

Outlines

1. Constraints in the Growth of Wheat Agriculture in Madhya Pradesh
2. Wheat Research at IARI Regional Station, Indore, MP
3. All India Coordinated Wheat Improvement Programme at IARI Regional Station, Indore, MP
4. Application of Molecular (DNA) Markers in DUS Testing
   a. Hybridization Based Marker System
   b. PCR-Based Markers Systems
   c. Simple Sequence Repeats
   d. Amplified Fragment Length Polymorphism
   e. Inter Simple Sequence Repeats
   f. Random Amplified Polymorphic DNA
   g. Single Nucleotide Polymorphic Markers
   h. Molecular markers and DNA Fingerprinting in Wheat
   i. Generation of DNA Fingerprint Database
   j. Variety Characterization Based on PCR-Markers Technology

5. Initiatives by UPOV
   a. Variety Identification using Bar Code
   b. Grain Characterization for Export
   6. Other Uses of DNA Fingerprinting in Wheat Germplasm Management
   a. Diversity Analysis
   b. Phylogenetic Studies
   c. Gene Bank Management
   d. Resolving IPR Disputes
7. Future of Wheat Variety Testing
8. Material Generation and other Programmes
   a. Alleviating malnutrition through popularization of cultivation and consumption of durum wheat
   b. Differential patterns of resistance of durum and bread wheats to important rust pathotypes
   c. Development of protocols for evaluation of rust resistance in durum and bread wheats
   d. Identification of sources of combined field resistance to stem and leaf rusts in durum wheat
   e. Generation of promising material for crop improvement
   f. Identification of diverse genes for resistance to leaf rust in durum wheat
   g. Narrow genetic base for leaf rust resistance in bread wheat
   h. Wheat Disease Survey and Surveillance
9. Evolving Improved Varieties of Durum Wheat
10. Bread Wheat Genotypes under Pipeline of Evaluation
    a. Identification of hard threshing (non shattering) genotypes of bread wheat:
    b. Breeding for Heat Management
    c. Breeding for Terminal Heat
    d. Breeding for High Yielding, Early Maturing, Drought Tolerant Genotypes
    e. Breeding for Limited Tillage after Soybean Harvest
    f. Technology Developed and Commercialized
11. Wheat Disease Management in Central India
12. Dynamics of Doubled Haploidy in Bread Wheat Improvement
    a. Protocols for haploidy breeding in wheat
    b. Wheat x Maize Hybridization Technique
    c. Advantages of Doubled Haploidy Breeding
    d. Protocol for Hardening and Chromosome Doubling

   e. Heterobeltiosis for grain yield and its components in Triticum durum
   f. Hierarchical classification of allelic variation of gliadins in durum wheat
   g. Genetics Stocks
13. Malavshakti (HI 8498) – A High Yielding Durum Wheat Cultivar for Central India
   a. Wider Adaptability
   b. Disease Resistance
   c. Grain Quality
14. Future Challenges for Wheat Cultivation in Madhya Pradesh

Wheat is the first in production among all cereal crops in the world and is the staple food of a larger segment of population. Among different wheat species, durum wheat, *Triticum durum,* is next in importance to bread/common wheat, *Triticum aestivum.* In India, it is generally cultivated in Central (Madhya Pradesh, Chattisgarh, Gujarat, Southern parts of Rajasthan, and Bundelkhand region of Uttar Pradesh) and Peninsular (Maharasthra and Karnataka) India. It occupies about 1.5-million-hectare area in the states of Karnataka, Maharashtra, Madhya Pradesh, Gujarat, Punjab and Uttar Pradesh (Bundelkhand region) with an estimated production of 8 to 10 per cent of the total wheat production. Normally, durum wheat is most popular to prepare various eating dishes in regions of Malwa of Madhya Pradesh, therefore, it is locally called as **Malvi** or **Kathia** wheat. Durum wheat is highly priced for making semolina (suji) and for preparing Indian counterparts of sewayan known as vermicelli. Durum wheat consists of high protein, high yellow pigment (Vitamin A) and higher micronutrients compared to bread wheat, which makes it a nutritious wheat. The best quality durum grains are yellowish, translucent, very hard and difficult to grind. Grains are vitreous with excellent appearance. Durum wheat is less elastic but more nutritious and easily digestible than bread wheat. When produced under favourable conditions, it has more protein and mineral matter and less starch content as compared to bread wheat. As a result of all these characteristic features, durum wheat fetches as premium price of 20-25 per cent over bread wheat. With the spread of semi-dwarf bread wheat varieties under cultivation, indigenous durum wheat went out of cultivation from Central India by 1970 due to its poor yield and high susceptibility to rusts. But with the release of Raj 1555, a rust resistant durum wheat variety had given hope for the farmers of Gujarat and Madhya Pradesh, that the durum cultivation can be restarted for the household preparations like Bati, Bafla, Dalia, Suji, and related macaroni products. The climate of central India had a lot of advantages for production of quality wheat *i.e.,* less vegetative period, more grain filling period, slow increase of temperatures during the crop season with frost and fog free conditions, *etc.,*

Since olden days, the rainfed durum wheat of central India is very popular with old varieties like **A 9-30-1, Bijaga Yellow** and **Bijaga Red** *etc.,* but with the recent release of durum wheat varieties for various growing conditions like early sown rainfed and limited irrigation; and timely sown and high fertility conditions,

Figure 4.1. Dr D K Verma during IRC-1, 2019 at ICAR-IARI RS Indore, Madhya Pradesh

Figure 4.2. Indore being cleanest city of India is also in forefront in wheat research in India

Figure 4.3. Trap Plot Nursery at ICAR-NRCS, Indore, MP

Figure 4.4. Dr HJ Barun in RI and RF field

Figure 4.5a. Dr DK Verma, being honoured for Wheat Research during Mithila Mahotsav, 2019.

Figure 4.5b. Dr DK Verma, being honoured for Wheat Research during Mithila Mahotsav, 2019.

Figure 4.6. ICAR-IARI Regional Station Indore, Madhya Pradesh team

8 COMMODITIES

Madhya Pradesh pips Punjab to emerge as the second largest wheat producer

High-yielding wheat varieties developed by IARI-Indore help improve output

Advance estimates

Area, output in Madhya Pradesh

Different varieties

Figure 4.7. Newspaper excerpts: Madhya Pradesh pips Punjab to emerge as the second largest wheat producer

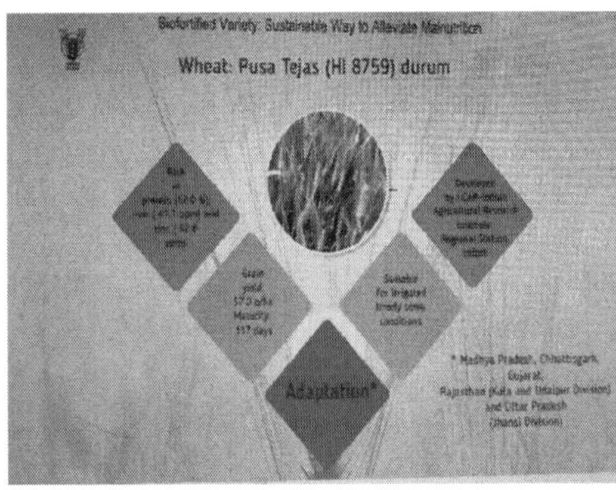

Figure 4.8. Biofortified Variety: Sustainable way to achieve malnutrition

Figure 4.9. Dr KV Prabhu as guiding force behind wheat research at IARI Regional Station, Indore, MP during 2012

Figure 4.10. Ornamental exposition of durum wheat at ICAR-IARI RS Indore, MP

Figure 4.11. The Prophet (SAW) said: whoever guides someone to goodness will have a similar reward

Figure 4.12. Observation being recorded at CIAE, Bhopal, MP

Figure 4.13. Ms Lalita Glande and Mrs Mandakini Deoras, the women of success at Wheat Breeding Programme at IARI RS Indore, MP

Figure 4.14. Preparation of durum wheat trails at Durum Laboratory, ICAR-IARI RS Indore, MP

Figure 4.15a. Wheat seed sorting at ICAR-IARI RS Indore, MP

Figure 4.15b. Wheat seed sorting at ICAR-IARI RS Indore, MP

Figure 4.16. Display of wheat cultivars live specimens at ICAR-IARI RS Indore, MP

Figure 4.17. SRFs who had devoted themselves in bread wheat and durum wheat breeding at ICAR-IARI Indore

Figure 4.18. Sowing lay out of Durum trials at NRCS, Indore (erstwhile DSR).

Figure 4.19. Field view of AVT (Late Sown) at NRCS, Indore, MP

Figure 4.20. Field view of CIMCOG at NRCS, Indore, MP

Figure 4.21. Field view of AVT (Timely Sown) at NRCS, Indore, MP

Figure 4.22a. Field view of Crossing Block at NRCS, Indore, MP

Figure 4.22b. Field view of Crossing Block at NRCS, Indore, MP

Figure 4.22c. Field view of Crossing Block at NRCS, Indore, MP

Figure 4.22d. Field view of Crossing Block at NRCS, Indore, MP

Figure 4.22e. Field view of Crossing Block at NRCS, Indore, MP

Figure 4.22f. Field view of Crossing Block at NRCS, Indore, MP

Date x Variety under different date of sowing

Period 2012-13 to 2014-15 (mean of 3 seasons grain yield in q/ha)

Variety	5th Oct.	20 Oct.	5th Nov.	Mean
HI 8663	44.1	46.9	44.3	45.1
HI 8713	44.9	48.2	44.6	45.9
HI 8627	43.9	42.3	40.9	42.4
HI 1531	37.5	40.8	34.4	37.6
HD 2987	34.7	45.2	49.4	43.1
GW 273	36.7	45.1	42.0	41.3
MP 3288	36.9	47.9	41.9	42.2
Lok 1	29.9	42.6	50.0	40.8
GW 366	27.9	44.1	44.7	38.9
HI 1544	35.7	47.1	51.6	44.8
HDCSW 18	44.8	41.5	39.5	41.9
Mean	37.8	44.2	44.3	

Despite application of more number of irrigations under extra early sown conditions (5th October), poor productivity of wheat suggest that the sowing of wheat can only be done after 20th October in vertisols of Central India.

Figure 4.23. Date x Variety under different date of sowing

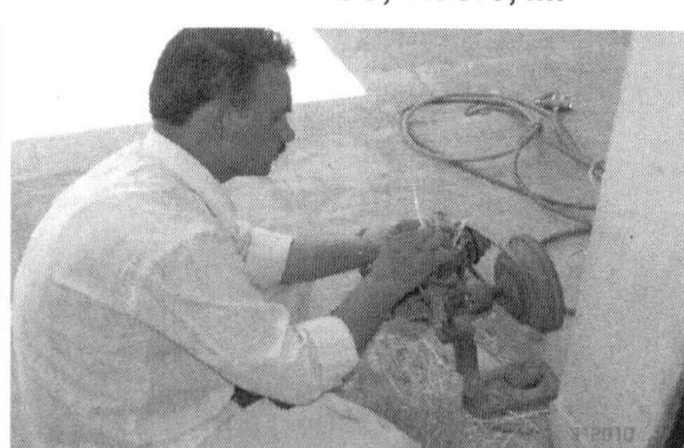

Figure 4.24. Wheat straw sample preparation at ICAR-IARI RS Indore

Figure 4.25. Field and grain view of HI 1544 (Purna)

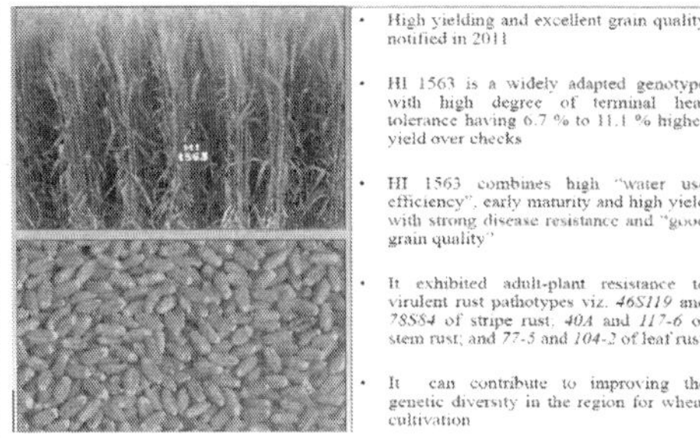

Figure 4.26. Field and grain view of HI 1563 (Pusa Prachi)

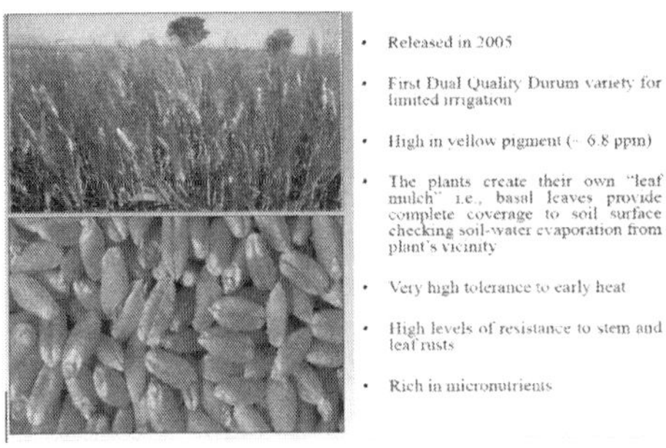

Figure 4.27. Field and grain view of HI 8627 (Malav Kirti)

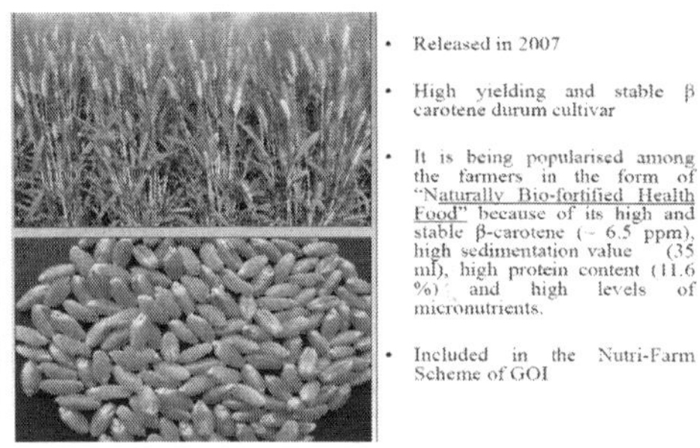

Figure 4.28. Field and grain view of HI 8663 (Poshan)

Figure 4.29. Field and grain view of HI 8713 (Pusa Mangal)

Figure 4.30. Field and grain view of HI 8737 (Pusa Anmol)

Figure 4.31. Tagging in crossing block at ICAR-IARI RS Indore, 2019

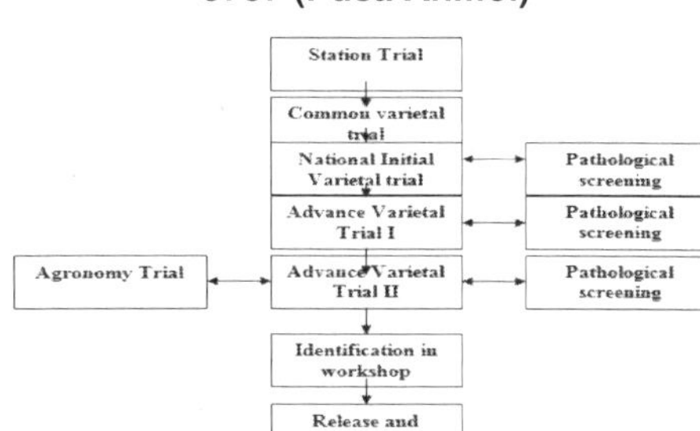

Figure 4.32. Stages of evaluation of improved strains before identification and release of variety

Figure 4.33. Dr D K Verma with Dr G P Singh and others during pre-workshop meeting at ICAR-IARI RS Indore

Figure 4.34. Rainfed wheat trial at ICAR-IARI RS Indore, MP

Figure 4.35. Germination percentage being observed

Figure 4.36. After dry sowing, irrigation channels are being created

Figure 4.37a. Layout for sowing of trials at ICAR-IARI RS Indore, MP

Figure 4.37b. Layout for sowing of trials at ICAR-IARI RS Indore, MP

Figure 4.38. Display of IARI Regional Station, Indore, Varietal Trial at CIAE, Bhopal, 2012.

Figure 4.39. CTD data recording MLHT trial at ICAR-IARI RS Indore, 2019

Figure 4.40. CAST, 2019 ICAR-IARI RS Indore sowing

Figure 4.41. Harvesting of Wheat from Conservation Agriculture Station Trial, 2019-2020

Figure 4.42. Inoculation of wheat trials for rust resistance breeding

Figure 4.43. 58th AIWBR Workers Meet, 2019 was engraved with Rangoli at Ravindra Natya Griha, Indore, 2019

Figure 4.44. Dr D K Verma chairing the session during examination of Ph D Scholar at BRAUSS, Mhow, Indore, MP

Figure 4.45. Dr D K Verma being honoured at Radiance Hotel by USI (Rotarian), Indore for his contribution to Wheat Research in Central India

Figure 4.48. Field view of HI 8691 (Pusa Gaurav) [(HG 509 / HI 8381) // (Boomer-33/Plata 8)]

Figure 4.49. Grains of HI 8691

Figure 4.50. HI 1612: A field view

Figure 4.51. HI 8777: A field view

the area under durum is on an increasing trend. Due to its distinct resistance to virulences of rusts, *durum* cultivation is essential to minimize the build-up of rusts in Central India, which acts as secondary focus for "Northern Wheat Belt", thereby keeping entire country free from the epidemics of leaf (brown) and stem (black) rusts. Durum being highly resistant to Karnal bunt and loose smut, will ensure the availability of produce free from seed borne diseases. Similarly, its high yield potential and high tolerance to drought and heat will ensure more production with less irrigation, making wheat cultivation "highly profitable" in Central India. With intensive popularization of newly evolved *durum* varieties HI 8381 (Malavshree), HI 8498 (Malavshakti), HD 4672 (Malav Ratna) and HI 8627 (Malav Kirti) and their "low-cost cultivation technology", wheat productivity in Madhya Pradesh.

Constraints in the Growth of Wheat Agriculture in Madhya Pradesh

Large turn off and soil erosion in most parts of the state resulting into (a) water congestion due to impeded drainage in early parts of the monsoon season and (b) inadequate moisture in latter part when needed the most. Nearly 72 per cent of cultivated area is subject to rain fed agriculture. Low cropping intensity (135 per cent) due to practice of keeping land fallow in kharif and taking only one crop on residual/conserved moisture in rabi some part of the state. Large proportion of cultivable waste and fallow land arising out of rain fed agriculture. Inadequate technology development suitable to specific regions of the state, particularly varieties suitable to match rainfall patterns. Large tribal population, marginal and small farmers having low investment capacity for adoption of high technology. High proportion of in-descript animals and fodder shortage in spite of large proportion of fallows and wastelands. Vagaries of monsoon and frequent natural calamities.

Wheat Research at IARI Regional Station, Indore (Madhya Pradesh)

IARI – Regional Station, Indore catering to the needs of Central India *i.e.*, Madhya Pradesh, Chhattisgarh, Rajasthan (Kota and Udaipur divisions), Gujarat and Bundelkhand region of Uttar Pradesh worked systematically on various basic and applied aspects of heat and drought tolerance in heavy to light heavy black cotton soils of Central India, a region that suffers from water shortage, short winters, very high October temperatures and cracking soils. After the cessation of monsoon rains in the month of October, there is a period of almost 40 days, which could not be utilized for wheat sowings because of detrimental effects of high October temperatures on germination and high seedling mortality, in spite of having a rich underground stored water. A number of morpho-physiologically different genetic stocks including the existing wheat varieties in bread and durum wheats were evaluated in target environment and their behaviour studied in detail. Performance of different varieties in early October sown conditions in two different years, clearly indicate the problem and the emerging solutions of early heat. This resulted in a number of revelations that helped in the development of "Appropriate Technology" for the management of early heat. Incorporating diverse rust resistance genes in newly generated material for evolving improved cultivars. Evolving genotypes tolerant to high temperature, moisture stress conditions and hard threshing. Evolving wheat genotypes for good grain quality and for other quality parameters like high protein content and chapati, bread and biscuit making purposes. Evolving early maturing genotypes combining high yield.

Table 4.1. List of Station Breeding Programme, Screening Nurseries; Coordinated Trials at Institute Level; Coordinated Trials at Directorate Level being executed during the crop season (2011-2012)

Station Breeding Programme			
1	Hybridization programme		Hybridization programme was performed by creating 220 fresh single cross and 24 fresh three way combinations
2	Breeding materials		
2a	Hybrid F_1 seeds	148 [x 1 Rep]	H-15351- H 15498
2b	Hybrid F_2 seeds	195 [x 1 Rep]	H 15156 - H 15350
2c	Hybrid F_3 seeds	58 [x 1 Rep]	H 14956 - H 15013
2d	Hybrid F_4 seeds	1208 [x 1 Rep]	H 14418 - H 14625
2e	Hybrid F_6 seeds	46 [x 1 Rep]	H 14330 - H 14375
2f	Hybrid F_7 seeds	626 [x 1 Rep]	H 13700 - H 14325
2g	Hybrid F_8 seeds	44 [x 1 Rep]	H 14054 - H 14097
3	Seed multiplication	Elite materials	To supply the different testing trials conducted all over India
4	Germplasm Rejuvenation Trial	758 aestivum + 127 durum	
5	Crossing block	478 [x1 Rep]	Maintenance and purposeful utilization for crossing
6	STRF-I	28 [x 3 Rep]	Small Scale Trial (Rainfed) -I
7	STRF-II	28 [x 3 Rep]	Small Scale Trial (Rainfed) –II
8	STRF-III	28 [x 3 Rep]	Small Scale Trial (Rainfed) -III

9	STRF-IV	28 [x 3 Rep]	
10	STMF-I	28 [x 3 Rep]	Small Scale Trial (Med Fert) -I
11	STMF-II	33 [x 3 Rep]	Small Scale Trial (Med Fert) –II
12	STMF-III	28 [x 3 Rep]	
13	STMF-IV	28 [x 3 Rep]	
14	STHF-I	33 [x 3 Rep]	Small Scale Trial (High Fert) -I
15	STHF-II	33 [x 3 Rep]	Small Scale Trial (High Fert) –II
16	STHF-III	33 [x 3 Rep]	Small Scale Trial (High Fert) –III
17	STHF-IV	33 [x 3 Rep]	Small Scale Trial (High Fert) –IV
18	STHF-V	33 [x 3 Rep]	Small Scale Trial (High Fert) –V
19	STHF-VI (Bhopal)	33 [x 3 Rep]	Station Trial High Fertility
Screening Nurseries			
20	IPPSN	1520	Initial Plant Pathological Screening Nursery
21	41st IDYN	50[x2 Rep]	41st International Durum Yield Nursery
22	30th ESWYT	50[x2 Rep]	30th Elite Spring Wheat Yield Trial
23	17th SAWYT	50[x2 Rep]	17th South Asia Wheat Yield Trial
24	NGSN	90[x1 Rep]	National Genetic Stock Nursery
25	EIGN-I	83[x1 Rep]	Elite International Germplasm Nursery
26	EIGN-II	65[x1 Rep]	Elite International Germplasm Nursery
27	DHTSN	42[x2 Rep]	Drought and Heat Tolerance Screening Nursery
28	QCSN	67[x1 Rep]	Quality Component Screening Nursery
29	PDSN	632	Plant Disease Screening Nursery
30	TPN	20[x 1 Rep]	Trap Plot Nursery
31	BARC, Mumbai (derivatives of C 306)	26[x 1 Rep]	Screening for shattering resistance; Replacement of Lok-1 with HI 1544 having hard threshing ability like Lok-1
32	BHU, Varanasi	90	Nursery
33	6th HTWSN	167	
34	29th SAWSN	145	Salinity Alkalinity Screening Nursery
35	YCSN	100[x1Rep]	Yield Component Screening Nursery
36	43rd IBWSN	236 [x1 Rep]	43rd International Bread Wheat Screening Nursery
37	42nd ITYN	30 [x 2Rep]	42nd International Triticum Yield Nursery
38	CSISA		Cereal Systems Initiative for South Asia
39	DUS-Testing	349	
40	5th EBWYT	30 [x 3Rep]	5th Elite Bread Wheat Yield Trial
41	31st ESWYT	50[x2 Rep]	31st Elite Spring Wheat Yield Trial
42	17th SAWYT-repeat	50[x2 Rep]	17th South Asia Wheat Yield Trial
43	18th SAWYT-repeat	50[x2 Rep]	18th South Asia Wheat Yield Trial
44	1st CRFPL	56[x1 Rep]	1st CIMMYT Rainfed Parental Lines

Coordinated Trials at Institute Level			
45	CVT-1A	36 [x 2 Rep]	Common Varietal Trials
46	CVT-1B	36 [x 2 Rep]	Common Varietal Trials
47	CVT-2 (Indore)	36[x2 Rep]	Common Varietal Trials
48	CVT-2 (Bhopal)	36[x2 Rep]	Common Varietal Trials
49	CVT-2 (Pune)	36[x2 Rep]	
50	CVT-3 IR-LS	49[x2 Rep]	Common Varietal Trials
51	CVT-4 (Pune)	36[x2 Rep]	Common Varietal Trial- 4
52	CVT-4 (Bhopal)	36[x2 Rep]	Common Varietal Trial- 4
53	CVT-4 (Indore)	36[x2 Rep]	Common Varietal Trial- 4
54	CVT-5A(Res-IR)	36[x2 Rep]	Common Varietal Trial- 5A-RI
55	CVT-5A(RF)	36[x2 Rep]	Common Varietal Trial- 5A-RF
56	CVT-5B-RF	36[x2 Rep]	Common Varietal Trial- 5B-RF
57	CVT-5B-RI	36[x2 Rep]	Common Varietal Trial- 5B-RI
Coordinated Trials at Directorate Level			
58	NIVT- 2-TS	36[x2 Rep]	National Initial Varietal Trial- 2-TS
59	NIVT- 3-TS	49[x2 Rep]	National Initial Varietal Trial- 3-TS
60	NIVT- 4-TS	36[x2 Rep]	National Initial Varietal Trial- 4-TS
61	NIVT- 5A-RF	36[x2 Rep]	National Initial Varietal Trial- 5A-RF
62	NIVT- 5A-RI	36[x2 Rep]	National Initial Varietal Trial- 5A-RI
63	NIVT- 5B-RF	25[x2 Rep]	National Initial Varietal Trial- 5B-RF
64	NIVT- 5B-RI	25[x2 Rep]	National Initial Varietal Trial- 5B-RI
65	AVT-RF	15[x4 Rep]	Advanced Varietal Trial-RF
66	AVT-RI	15[x4 Rep]	Advanced Varietal Trial-RI
67	AVT-TS-IR-HF	21[x4 Rep]	Advanced Varietal Trial-TS-IR-HF
68	AVT-TS-IR-HF (Bhopal)	21[x4 Rep]	Advanced Varietal Trial-TS-IR-HF
69	AVT-LS-IR	7[x4 Rep]	Advanced Varietal Trial-LS-IR

Table 4.2. Entries promoted from CVT to NIVT for further evaluation

Trial	*IND No.*	*HI Nos.*	*Pedigree*	*Promoted To*
CVT-1	IND 319	HI - 1591	HW 2107/HW 524	NIVT-1
CVT-2 (5)	IND 322	HI - 1592	HW 2045/HI 1479//DVK 283	NIVT-2
	IND 323	HI - 1593	HI 1531/GW 373	
	IND 324	HI - 1594	HW 2045/HI 1479	
	IND 327	HI - 1595	HI 1563/Lok-1	
	IND 328	HI - 1596	HUW234+Lr34/PRINIA*2//YANAC	

Table 4.3. Selection of genotypes from Station Trials

Trial Condition	*Trial Number*	*Number of Genotypes*	*Selections Made*	*Name of Genotypes*
RF	I	30	6	HAS-700, 730, 731, 742, 729, 749
RF	II	30	8	HAS -755, 758, 813, 826, 827, 828, 829, 853
RF	III	30	7	HAS - 930, 933, 942, 944, 945, 946, 947
RF	IV	20	5	HAS - 982, 984, 991, 999, 1012
MF	I	30	4	HAS - 692, 707, 730, 743
MF	II	36	9	HAS - 750, 751, 759, 829, 833, 834, 854, 873, 874
MF	III	30	9	HAS - 931, 933, 951, 955, 962, 963, 964, 982, 983
MF	IV	36	7	HAS - 686, 687, 691, 694, 696, 737, 1059
HF	I	36	1	HAS - 777
HF	II	36	4	HAS - 785, 799, 858, 862
HF	III	36	1	HAS - 935
HF	IV	36	1	HAS - 1078
HF	V	36	3	HAS - 716, 720, 790
HF	VI (Bhopal)	36	4	HAS - 686, 687, 691, 694, 696, 737, 1059

Table 4.4. Parents used in hybridization

Parents used in hybridization (Female = 69 + 27F_1s)	**Parents used in hybridization (male = 49)**
Baaj, Chewink, CPAN 1796, DL 153-2, DL 788-2, Francolin, GW 366, GW 273, GW 295, GW 322, HD 2307, HD 2402, HD 2135, HD 2177, HD 2329, HD 2501, HD 2888, HD 2380, HD 2987, HI 1123, HI 1418, HI 1454, HI 1459, HI 1544, HI 1563, HI 1077, HI 1500, HI 1531, HP 1744, HPW 160, HS 365, HS 295, HUW 468, HW 2004, HW 3063, JW 3269, JW 3211, K 7827, K 9533, Kingbird, Lok 1, Lok Bold, MP 3288, MP 4010, Mukta, Narmada 4, Narmada 112, NI 5439, NIAW 34, NL 297, NW 1067, PHS 1108, PHS 1109, QUAIV, Raj 3777, Raj 4037, Raj 4268, Raj 3765, Raj 3900, Raj 3974, Raj 4055, Raj 4083, Raj 4215, Raj 4264, Super 152, Super 172, UP 2390, UP 2418 and WH 1078 + F1 as female: H-15351, H-15352, H-15353, H-15354, H-15358, H-15362, H-15376, H-15377, H-15386, H-, 15390, H-15396, H-15398, H-15399, H-15400, H-15402, H-15403, H-15419, H-15426, h-15427, H-15430, H-15431, H-15432, H-15435, H-15441, H-15445, H-15447 and H-15479	Baaj, Chewink, CPAN 1796, DL 788-2, FRANCOLIN, GW 295, GW 366, GW 496, HD 2307, HD 2329, HD 2380, HD 2987, HI 1123, HI 1418,HI 1454, HI 1479,HI 1500, HI 1531, HI 1544, HI 1563, HI 4010, HP 1744, HP 1860, HPW 157, HPW 160, HS 295, HS 365, HW 2004, K 7827, K 9533, King bird, LOK BOLD, LOK-1, MP 3288, MP 4010, QUAIV, Qwisec, Raj 3765, Raj 3900, Raj 3974, Raj 4032,Raj 4037, Raj 4055, Raj 4083, Raj 9533, Super 152, Super 172, UP 2418 and UP 2425
Ug 99 resistant parents used in hybridization as female: DL 153-2, DL 788-2, GW 322, HD 2987, HI 1531, HS 295, Lok 1, MP 3288	**Ug 99 resistant parents used in hybridization as male:** DL 788-2, HD 2987, HI 1531, HS 295, LOK-1, MP 3288

Table 4.5. Breeding materials (2011-2012)

H Number	*Pedigree*	*No.of SPLs*	*Generation*
H-14956	Agra Local x DL 788-2	4	F3
H-14957	Agra Local x GW 366	2	F3
H-14958	Agra Local x HD 2705	2	F3
H-14959	Agra Local x HD 2987	5	F3
H-14960	Agra Local x HI 1535	6	F3
H-14963	Agra Local x MP 4010	5	F3
H-14968	CPAN 1676 x GW 366	8	F3
H-14969	CPAN 1676 x HD 2987	5	F3
H-14970	CPAN 1676 x HI 1479	4	F3
H-14971	CPAN 1676 x HI 1544	11	F3
H-14972	CPAN 1776 x GW 366	6	F3
H-14973	CPAN 1776 x HD 2987	7	F3
H-14974	CPAN 1776 x HI 1479	3	F3
H-14975	CPAN 1842 x CPAN 4148	5	F3
H-14981	CPAN 1990 x HD 2987	6	F3
H-14983	CPAN 1990 x HI 1544	14	F3
H-14984	CPAN 2063 x HD 2987	11	F3
H-14985	CPAN 2063 x Raj-3777	12	F3
H-14987	CPAN 2063 x GW- 366	7	F3
H-14988	CPAN 3031 x GW- 366	11	F3
H-14989	CPAN 3031 x HD 2987	10	F3
H-14990	CPAN 3031 x HI 1544	6	F3
H-14991	CPAN 3031 x Raj-3777	5	F3
H-14992	CPAN 3050 x GW- 366	8	F3
H-14993	CPAN 3050 x HD 2987	9	F3
H-14994	CPAN 3050 x HI 1544	16	F3
H-14995	CPAN 3050 x Raj-3777	6	F3
H-14996	CPAN 3065 x GW 366	11	F3
H-14997	CPAN 3065 x HD 2987	11	F3
H-14998	CPAN 3065 x HI 1544	14	F3
H-14999	CPAN 3065 x Raj-3777	5	F3
H-15000	DL-788-2 X GW 366	10	F3
H-15001	DL-788-2 X HD 2987	19	F3
H-15002	DL-788-2 X HI-1535	11	F3
H-15008	GW 322 X GW 326	4	F3
H-15021	Giant 3 X PHR 1012	3	F3
H-15023	H 13560-1 X Giant 3	3	F3
H-15025	H 13560-2 X Giant 3	2	F3
H-15028	H 13560-2 X PHR 0623	6	F3

H Number	Pedigree	No.of SPLs	Generation
H-15030	Giant 3 X H 14292	7	F3
H-15031	Giant 3 x H 14290	4	F3
H-15033	H 13533(6)-4-2 x HI 1544	7	F3
H-15035	H13533(6)-4-2 x Ind 221	2	F3
H-15037	H 13560-7 x PHR 1012	7	F3
H-15038	HD 2705 x HI 1459	7	F3
H-15039	HD 2705 x NIAW 907	9	F3
H-15040	HD 2705 x PBW 442	13	F3
H-15041	HD 2705 x Raj 3974	12	F3
H-15042	HD 2189 x GW 322	3	F3
H-15044	HD 2189 x H 135060-1	5	F3
H-15048	HD 2380 X HI 1544	13	F3
H-15049	HI 1136 x HI 1531	2	F3
H-15050	HI 1385 x HW 4076	9	F3
H-15051	HI 1418 x GW 322	8	F3
H-15052	HI 1418 x GW 326	10	F3
H-15053	HI 1418 x HI 1479	8	F3
H-15054	HI 1418 x HI 1539	1-0	F3
H-15055	HI 1479 x GW 322	3	F3
H-15056	HI 1479 x GW 326	7	F3
H-15057	HI 1479 x HI 1539	5	F3
H-15058	HI 1459 x HD 2705	10	F3
H-15059	HI 1459 x NIAW 907	10	F3
H-15060	HI 1459 x PBW 442	4	F3
H-15061	HI 1459 x Raj 3974	10	F3
H-15063	HI 1531 x HI 1544	7	F3
H-15064	HI 1535 x HD 2987	7	F3
H-15065	HI 1539 x GW 322	18	F3
H-15066	HI 1539 x GW 326	8	F3
H-15070	HI 1544 x HD 2987	25	F3
H-15071	HI 1544 x HI 1500	11	F3
H-15072	HI 1544 x HI 1535	29	F3
H-15073	HI 1544 x Raj 3777	14	F3
H-15074	HUW 541 x HW 4076	9	F3
H-15075	HW 888 x HW 3005	6	F3
H-15079	HW 3005 x HW 888	2	F3
H-15080	HW 3008 x HW 4076	9	F3
H-15081	HW 3008 x VL 830	11	F3
H-15082	HW 4076 x GW 366	24	F3
H-15083	HW 4076 x VL 830	11	F3

H Number	Pedigree	No.of SPLs	Generation
H-15091	MACS 2496 x GW 366	6	F3
H-15092	MACS 2496 x HD 2987	12	F3
H-15094	MACS 2496 x HI 1531	4	F3
H-15095	MACS 2496 x GW 366	6	F3
H-15096	MACS 2496 x HW 2004	3	F3
H-15098	PBW 442 x NIAW 907	5	F3
H-15099	Raj 3974 x HI 1459	8	F3
H-15100	Raj 3974 x NIAW 907	11	F3
H-15101	Raj 3974 x PBW 442	5	F3
H-15102	VL 738 x Ind 231	6	F3
H-15105	DL-788-2 x H 14952	7	F3
H-15109	H 14585 x HD 2987	9	F3
H-15110	H 14601 x HD 2987	8	F3
H-15111	H 14646 x H 14293 (4)-6	6	F3
H-15115	H 14678 x HI 1544	5	F3
H-15116	H 14682 x HD 2189	5	F3
H-15118	H 14739 x HW 3035 x HD 2380	12	F3
H-15120	H 14836 x HI 1544	5	F3
H-15123	H 14951 x Giant 3	5	F3
H-15125	H 14952 x HI 1544	8	F3
H-15142	H 14845 x H 14955	2	F3
H-15143	H 14846 x H 14700	5	F3
H-15144	H 14953 x H 14762	1	F3
H-15149	[HI 1400 x HD 2380]x HD 2987	8	F3
H-15155	VL WW 55 x GW 366 x HI 1531	8	F3
PKV (3)-1	C 126/HI 1544	7	F3
H-15013	GW 366 x HI 1500	1	F3
H-14418	DL 230-1 x HW 2010	10	F4
H-14419	DL 230-7 x HW 2001	10	F4
H-14420	DL 230 x HW 2005	19	F4
H-14421	DL 377 x GW 322	28	F4
H-14422	DL 377-1 x HI 1531	9	F4
H-14423	DL 377-1 x HI 1500	6	F4
H-14425	DL 765-1 x Ind 165	4	F4
H-14426	DL 765-1 x 15th IBWSN-2	11	F4
H-14427	HW 502 x HW 5021	5	F4
H-14433	Giant 3 x HI 1424	14	F4
H-14434	Giant 3 x HI 1395	17	F4
H-14435	Giant 3 x HI 1434	3	F4
H-14436	Giant 3 x HI 1476	7	F4

H Number	Pedigree	No.of SPLs	Generation
H-14440	Giant 3 x HW 2045	17+11	F4
H-14441	Giant 3 x HW 3006	10	F4
H-14442	Giant 3 x HP 117	Not selected	F4
H-14443	Giant 3 x Lok 1	10	F4
H-14447	HI 1289 x GW 366	5	F4
H-14448	HI 1345 x GW 366	20+17+11	F4
H-14449	HI 1345 x H 1479	7	F4
H-14451	HI 1358 x GW 322	5	F4
H-14453	HI 1368 x HW 4032	6	F4
H-14454	HI 1568 x HW 2013	4	F4
H-14455	HI 1369 x GW 366	7	F4
H-14457	HI 1372 x GW 366	9	F4
H-14460	HI 1384 x 2nd IBWSN	4	F4
H-14463	HI 1394 x Ind 147	11	F4
H-14547	AKAW 4627 x HI 1418	6	F4
H-14548	GW 03-2 x PBW 223	9	F4
H-14550	GW 03-2 x HD 2380	5	F4
H-14551	GW 03-2 x HW 4011	5	F4
H-14554	GW 03-4 x HW 4028	13	F4
H-14555	GW 03-4 x PBW 498	6	F4
H-14557	GW 173 x PBW 1496	9	F4
H-14559	GW 173 x PBW 1496	7	F4
H-14564	GW 173 x K 9107	12	F4
H-14565	GW 173 x HI 1531	2	F4
H-14566	GW 173 x HW 4028	2	F4
H-14568	GW 322 x HW 4028	6	F4
H-14571	GW 322 x K 9107	4	F4
H-14574	GW 322 x PBW 498	3	F4
H-14579	GW 295 x PBW 498	8	F4
H-14580	GW 295 x K 9107	3	F4
H-14584	GW 295 x HW 4083	5	F4
H-14585	GW 295 x HW 4028	5	F4
H-14587	GW 366 x HW 4028	11	F4
H-14588	GW 366 x PBW 498	14	F4
H-14592	GW 366 x HW 3083	3	F4
H-14596	GW 366 x BW 1765	2	F4
H-14598	HD 2380 x HW 2004	5	F4
H-14600	HD 2932 x HI1418	5	F4
H-14601	HD 2932 x GW 173	6	F4
H-14611	HI 1390 x PBW 498	2	F4

H Number	Pedigree	No.of SPLs	Generation
H-14612	HI 1390 x Raj 3779	7	F4
H-14614	HI 1390 x HP 1761	5	F4
H-14615	HI 1390 x HW 2028	9	F4
H-14617	HI 1390 x HW 4011	9	F4
H-14618	HI 1390 x K 9107	8	F4
H-14619	HI 1390 x HD 2380	11	F4
H-14620	HI 1390 x HI 1531	12	F4
H-14623	HI 1400 x K 9107	5	F4
H-14625	HI 1400 x HW 4028	8	F4
H-13793	GW 273 x HW 2045	3+4	F7
Totals		1373	

Table 4.6. F_1 list 2012-13

H.No	Pedigree	No.of Seed
H-15499	Baaj x HI- 1563	51
H- 15500	Baaj x HI- 1479	37
H-15501	Baaj x HI- 1418	16
H-15502	Baaj x HI- 1544	46
H-15503	Baaj x HI- 1454	51
H-15504	Chewink x HW-2004	9
H-15505	CPAN-1796 x FRANCOLIN	26
H- 15506	CPAN-1796 x LOK-1	15
H-15507	DL-153-2 x HI-1418	61
H- 15508	DL-153-2 x HI-1454	21
H- 15509	DL-788-2 X HI-1544	9
H- 15510	DL-788-2 X HI-1563	7
H- 15511	DL-788-2 X HI-1123	96
H- 15512	FRANCOLIN X HI-1563	40
H- 15513	FRANCOLIN X HI-1479	35
H- 15514	FRANCOLIN X HI-1418	48
H- 15515	FRANCOLIN X HI-1531	38
H- 15516	FRANCOLIN XHW-2004	55
H- 15517	FRANCOLIN X HI-1500	57
H- 15518	FRANCOLIN X HI-1454	28
H- 15519	FRANCOLIN X HI-1544	42
H- 15520	GW273 X HP-1744	23
H- 15521	GW273 X HP-1860	11
H- 15522	GW273 X K-9533	26
H- 15523	GW-295 X HI-1418	29

H.No	Pedigree	No.of Seed
H- 15524	GW295 X HI-1563	14
H- 15525	GW295 X HD-2987	5
H- 15526	GW-322 X FRANCOLIN	14
H- 15527	GW-322 X SUPER-172	46
H- 15528	GW-322 X Baaj	34
H- 15529	GW-322 X K-9533	34
H- 15530	GW-322 X Raj-4083	69
H- 15531	GW-322 X Raj-3765	30
H- 15532	GW- 366 X Qwisec	70
H- 15533	GW- 366 X Super- 152	39
H- 15534	GW- 366 X Chewink	53
H- 15535	HUW-468 X Baaj	45
H- 15536	HUW-468 x Super- 152	15
H- 15537	HPW-160 X Raj-4037	14
H- 15538	HPW-160 X Raj-4037	7
H- 15539	HPW-160 X HD-2380	18
H- 15540	HP-1744 X FRANCOLIN	41
H- 15541	HP-1744 X K-9533	62
H- 15542	HP-1744 X MP-4010	29
H- 15543	HS-295 X Baaj	34
H- 15544	HS-295 X FRANCOLIN	17
H- 15545	HS-295 X 4037	19
H- 15546	HS - 365 X Baaj	11
H- 15547	HS- 365 X FRANCOLIN	10
H- 15548	HS - 365 X Raj - 4037	10
H- 15549	HS- 365 X Super - 172	60
H- 15550	HD - 2307 X HI-1418	5
H- 15551	HD - 2307 X HD- 2987	5
H- 15552	HD- 2307 X HI - 1563	6
H- 15553	HD- 2135 X Super- 152	49
H- 15554	HD- 2135 X Baaj	88
H- 15555	HD- 2135 X Chewink	38
H- 15556	HD- 2177 X Baaj	42
H- 15557	HD- 2177 X FRANCOLIN	28
H- 15558	HD- 2177 X Super-152	55
H- 15559	HD- 2329 X HI-1418	5
H- 15560	HD- 2329 X HI-1544	7
H- 15561	HD-2380 X Baaj	29
H- 15562	HD - 2380 X UP- 2418	28
H- 15563	HD - 2380 X Raj-4055	65

H.No	Pedigree	No.of Seed
H- 15564	HD - 2402 X Super - 152	61
H- 15565	HD- 2402 X Baaj	41
H- 15566	HD- 2402 X FRANCOLIN	21
H- 15567	HD- 2501 X Super-152	16
H- 15568	HD - 2501 X King bird	73
H- 15569	HD- 2501 X Super-172	43
H- 15570	HD- 2888 X Super- 172	39
H- 15571	HD- 2888 X FRANCOLIN	43
H- 15572	HD-2888 X Baaj	41
H- 15573	HD-2987 X Raj- 4032	18
H- 15574	HD- 2987 X LOK BOLD	24
H- 15575	HD- 2987 X GW-295	18
H- 15576	HI-1077 X Super-172	41
H- 15577	HI-1077 X LOK-1	52
H- 15578	HI-1077 X GW-496	56
H- 15579	HI- 1123 X Super-172	18
H- 15580	HI- 1123 X LOK-1	33
H- 15581	HI-1123 X GW- 496	36
H- 15582	HI- 1418 X GW-295	12
H- 15583	HI - 1418 X Super-152	22
H- 15584	HI - 1418 X Chewink	65
H- 15585	HI- 1418 X DL-788-2	6
H- 15586	HI- 1454 X Baaj	26
H- 15587	HI -1454 X FRANCOLIN	41
H- 15588	HI - 1454 X Super - 172	44
H- 15589	UP-2390 X HI-1454	11
H- 15590	HI-1500 X Super-152	77
H- 15591	HI- 1500 X Baaj	51
H- 15592	HI- 1500 X FRANCOLIN	16
H- 15593	HI-1531 X Baaj	19
H- 15594	HI-1531 X FRANCOLIN	17
H- 15595	HI - 1531 X Super-172	18
H- 15596	HI- 1544 X FRANCOLIN	56
H- 15597	HI- 1544 X Super-152	39
H- 15598	HI- 1544 Xbaaj	52
H- 15599	HI- 1459 X CPAN- 1796	13
H- 15600	HI- 1459 X DL-788-2	26
H- 15601	HI- 1563 X Super-172	50
H- 15602	HI - 1563 X K-7827	6
H- 15603	HI- 1563 X Raj-4037	17

H.No	*Pedigree*	*No.of Seed*
H- 15604	HI - 1563 X Raj-3900	7
H- 15605	HW-2004 X Baaj	47
H- 15606	HW-2004 X Super-152	31
H- 15607	HW-2004 X FRANCOLIN	70
H- 15608	HW-2004 X Super-172	118
H- 15609	HW-3063 X DL-788-2	40
H- 15610	HW-3063 X UP-2425	32
H- 15611	HW-3063 X HD-2307	32
H- 15612	HW-3063 X UP-2425	21
H- 15613	PHS-1108 X HD-2987	5
H- 15614	PHS-1108 X Raj-4037	7
H- 15615	PHS-1108 XMP-4010	12
H- 15616	PHS-1109 X HI- 1454	7
H- 15617	PHS-1109 X HI-4010	5
H- 15618	PHS-1109 X Raj- 4037	12
H- 15619	Raj-3765 X HD-2987	7
H- 15620	Raj-3765 X HI-1418	25
H- 15621	Raj-3765 X Baaj	5
H- 15622	Raj- 3777 x Baaj	18
H- 15623	Raj-3777 X HD-2329	28
H- 15624	Raj-3900 x HI- 1563	57
H- 15625	Raj-3900 X HD-2987	23
H- 15626	Raj-3900 X HI - 1454	27
H- 15627	Raj-3900 X HI - 1418	32
H- 15628	Raj-3974 X HI-1563	26
H- 15629	Raj - 3974 X HI-1418	9
H- 15630	Raj- 4037 X HI 1563	34
H- 15631	Raj- 4037 X HI- 1418	23
H- 15632	Raj- 4037 X 1454	48
H- 15633	Raj-4055 X HD-2987	22
H- 15634	Raj-4055 XMP-4010	44
H- 15635	Raj-4055 X HI-1544	79
H- 15636	Raj-4215 X HS-295	33
H- 15637	Raj-4215 X HS-365	18
H- 15638	Raj-4264 X FRANCOLIN	12
H- 15639	Raj-4264 X Baaj	25
H- 15640	Raj-4264 X HD- 2987	5
H- 15641	Raj- 4268 X Baaj	54
H- 15642	Raj- 4268 X HD-2987	39
H- 15643	Raj-4083 X HI- 1418	23

H.No	*Pedigree*	*No.of Seed*
H- 15644	Raj- 4083 X HD-2987	9
H- 15645	Raj- 4083 XHI- 1454	41
H- 15646	WH-1078 X Raj-4037	24
H- 15647	WH-1078 X MP-4010	54
H- 15648	WH-1078 X Raj-4083	37
H- 15649	JW-3211 X FRANCOLIN	34
H- 15650	JW-3211 X Baaj	33
H- 15651	JW-3211 X Super-152	34
H- 15652	JW- 3269 X Raj-3900	27
H- 15653	JW- 3269 Xraj-4037	39
H- 15654	JW- 3269 X GW-295	45
H- 15655	K-7827 X UP-2425	10
H- 15656	K-7827 X HD-2307	5
H- 15657	K-7827 X HI- 1454	7
H- 15658	K-9533 X DL-788-2	15
H- 15659	K-9533 X Raj-4037	31
H- 15660	LOK-BOLD X Chewink	10
H- 15661	Lok-Bold X Baaj	16
H- 15662	LOK-BOLD X QUAIV	25
H- 15663	LOK-1 X Chewink	46
H- 15664	LOK-1 X Baaj	42
H- 15665	LOK-1 X Super-152	152
H- 15666	LOK-1 X QUAIV	57
H- 15667	LOK-1 X FRANCOLIN	111
H- 15668	MP- 3288 X HPW-157	7
H- 15669	MP-3288 X K-9533	6
H- 15670	MP-4010 X Chewink	21
H- 15671	MP-4010 x FRANCOLIN	56
H- 15672	MP-4010 X QUAIVE	50
H- 15673	MP-4010 X HD-2987	42
H- 15674	MP-4010 X HI- 1563	32
H- 15675	MP-4010 X HI - 1418	49
H- 15676	MUKTA X Baaj	17
H- 15677	MUKTA x FRANCOLIN	16
H- 15678	MUKTA X Super-172	8
H- 15679	NIAW- 34 X K- 7827	8
H- 15680	NIAW-34 X Baaj	76
H- 15681	NIAW- 34 X Raj-3974	22
H- 15682	NI-5439 X Super-172	50
H- 15683	NI-5439 X FRANCOLIN	39

H.No	Pedigree	No.of Seed
H- 15684	NI-5439 X Baaj	29
H- 15685	NARMADA -4 X FRANCOLIN	8
H- 15686	NARMADA -4 X Super-172	19
H- 15687	NARMADA-4 X Baaj	31
H- 15688	NARMADA-112 X FRANCOLIN	29
H- 15689	NARMADA -112 X Super-172	26
H- 15690	NARMADA-112 X Baaj	6
H- 15691	NL -297 X Raj-3765	30
H- 15692	NL -297 X HI-1454	12
H- 15693	NL-297 X HI - 1418	30
H- 15694	NW-1067 X HPW-160	5
H- 15695	NW-1067 X Raj 9533	53
H- 15696	NW-1067 X HPW-157	50
H- 15697	QUAIV X MP-3288	27
H- 15698	QUAIV X GW-366	8
H- 15699	SUPER-152 X HI 1544	50
H- 15700	SUPER-152 X HI 1500	23
H- 15701	SUPER-152 X HI-1479	58
H- 15702	SUPER-152 X HI-1563	47
H- 15703	SUPER-152 X HI -1531	25
H- 15704	SUPER-152 X HW-2004	21
H- 15705	SUPER-152X HI1418	49
H- 15706	SUPER-152 X HI-1454	49
H- 15707	SUPER-172 X HI 1454	45
H- 15708	SUPER-172 X HI 1418	48
H- 15709	SUPER-172 X GW - 496	84
H- 15710	SUPER-172 X LOK-1	50
H- 15711	UP-2418 X HI-1563	22
H- 15712	UP-2418 X HD-2987	14
H- 15713	UP-2418 X HI -1418	16
H- 15714	H-15351 X FRANCOLIN	121
H- 15715	H-15352 X GW -496	88
H- 15716	H-15353 X CHWINK	93
H- 15717	H-15354 X SUPER-152	112
H- 15718	H-15358 X HD-2987	29
H- 15719	H-15362 X MP-4010 [CPAN 1796 x HI 1522]	33
H- 15720	H-15376 X Baaj [GW 322 x UP 2473]	115
H- 15721	H-15377 x chwink [GW 322 x VL 804]	100
H- 15722	H-15386 X HI -1454 [HD 2380 x AKW 381]	13
H- 15723	H-15390 X HI 1500 [HD 2428 x DL 788-2]	15

H.No	Pedigree	No.of Seed
H- 15724	H-15396 X Super -152 [HD 2781 x CPAN 1676]	78
H- 15725	H-15398 X Baaj [HD 2781 x HP 1761]	68
H- 15726	H-15399 x Super 152 [HD 2781 x K 9107]	79
H- 15727	H-15400 X Quaiv [HD 2781 x Raj 3765]	81
H- 15728	H-15402 X Super-152 [HP 1744 x CPAN 1676]	13
H- 15729	H-15403 X Kingword [HP 1744 x CPAN 1842]	20
H- 15730	H-15419 X Super-152 [HW 2015 x GW 366]	72
H- 15731	H-15426 X Baaj [HW 2061 x HD 2932]	21
H- 15732	H- 15427 x Baaj [HW 2061 x HD 2962]	95
H- 15733	H-15430 X Francolin [HW 4042 x HI 1462]	11
H- 15734	H-15431 X Baaj [HW 4042 x HI 1567]	61
H- 15735	H-15432 X Super-152 [HW 4055 x HI 1462]	20
H- 15736	H-15435 X Francolin [HW 4059 x HD 2932]	100
H- 15737	H-15441 X Baaj [K 9006 x HI 784]	92
H- 15738	H- 15445 x Baaj [K 9107 x HS 295]	12
H- 15739	H-15447 X Francolin [K 9465 x HI 1484]	34
H- 15740	H-15479 X Super-152 [GW 322 x LBPY 06-13) X NIAW 907]x Super-152	35
H- 15741	Kingword X HI-1544	13
H- 15742	Kingword X HD-2380	32

Table 4.7. Seed multiplication of elite material

Trial	Name of Genotypes
CVT-1	IND – 318, 319, 320
CVT-2	IND – 321, 322, 323, 324, 325, 326, 327, 328, 329
CVT-3	IND – 330, 331
CVT-5A	IND – 332, 333, 334, 335, 336, 337
NIVT -2	HI – 1588, HI – 1589, HI – 1590
AVT 1st year	HI – 1584
AVT 2nd year	HI – 1579

Table 4.8. Entries promoted from CVT to NIVT for further evaluation

Trial	IND Nos.	HI Nos.	Pedigree	Promoted to Trial
CVT-1	IND 319	HI - 1591	HW 2107/HW 524	NIVT-1
CVT-2	IND 322	HI - 1592	HW 2045/HI 1479//DVK 283	NIVT-2
	IND 323	HI - 1593	HI 1531/GW 373	
	IND 324	HI - 1594	HW 2045/HI 1479	
	IND 327	HI - 1595	HI 1563/Lok-1	
	IND 328	HI - 1596	HUW234+Lr34/PRINIA*2//YANAC	

Table 4.9. Bread Wheat (*Aestivum*) Germplasm [2009-2010 and 2010-2011]

A-85, A-115, A-206, A-240, A-375, A-376, A-469, AC-214, AC-301, AC-319, AC-350, AO-88, AD-168, AM-206 [Dicoccum], Ariana, Arjun, Arjun (HI-2009), Atocava, Atlas-82, Abbysinia, Australia, Australia-153, Australian Line, Alipura-6-1, Agra Local, BR-3142, BW-11, BAU-2177, BAU-2187, Bundi Local, Bundi Local-41, Bundi Local (Pissi Type), Bundi Local (Chanousi Type), Bundi-112B, Bansi Local, Bansi R_3H, Bansi-R-3-4, Baxi-3-12 (D), Buckey-13, Buckey-57, Boult Shark, Breeze, Bonanza, BW-17, C-13, C-281, C-285, C-286, C-306, C-518, CA-82, CC-44, CC-256, CC-336, CC-352, CC-358, CC-401, CC-404, CC-443, CC-444, CC-464, CC-466, CC-476, CC-527, CC-591-1, CC-1351, CP-14-1, CP-274, CP-275, CP-276, CMA-5948 AT, COLL-16A, COLL-34, COLL-41B, Crazia, Charter, Ciano-27829, CPAN-1676 (Rohini), E-3931, E-3933, E-3942, E-3954, E-3967, E-3975, E-3982, E-3983, E-3985, E-3986, E-3987, E-4002, E-4042, E-4057B, E-4077, E-4078, E-4079B, E-4081A, E-4115B, E-4118, E-4138, E-4150, E-4192, E-4221, E-4224, E-4245A, E-4251, E-4282,. E-4307, E-4356B, E-4405, E-4470B, E-4476, E-4521, E-4529, E-4543, E-4545A, E-4576, E-4590, E-4601, E-4603, E-4604, E-4606, E-4608, E-4665, E-4671, E-4752, E-4782,	CPAN-1796, CPAN-1817, CPAN-1823, CPAN-1909, CPAN-1922, CPAN-1990, CPAN-1992, CPAN-1996, CPAN-1998, CPAN-2016, CPAN-2045, CPAN-2051, CPAN-2055, CPAN-2063, CPAN-2069, CPAN-2074, CPAN-2091, CPAN-2092, CPAN-2094, CPAN-2099, CPAN-3004, CPAN-3006, CPAN-3013, CPAN-3016, CPAN-3017, CPAN-3022, CPAN-3027, CPAN-3030, CPAN-3034, CPAN-3044, CPAN-3045, CPAN-3046, CPAN-3050, CPAN-3051, CPAN-3056, CPAN-3057, CPAN-3058, CPAN-3065, CPAN-3066, CPAN-3067, CPAN-3068, CPAN-3069, CPAN-3070, CPAN-3071, CPAN-3072, CPAN-3073, CPAN-3074, CPAN-3075, CPAN-3076, CPAN-3077, CPAN-3078, CPAN-3079, CPAN-3080, CPAN-3081, CPAN-3082, CPAN-3083, CPAN-3084, CPAN-6117, D-305, DG-977, DL-153-2, DL-153-12, DL-177-2, DL-205-1, DL-209, DL-215-1, DL-230-7, DL-330-7, DWR-39, DWR-118 [Diococcum], Gwalior-351, Grevner A, H-3-1, H-16, H-18A, H-48A, H-50, H-125, H-125A, H-126A, H-129A, H-138, H-140, H-164, H-177A, H-217, H-263-1A, H-281, H-298A, H-301-1A, H-339-2-5A, H-391-2-4A, H-430-1, H-599, H-704A, H-710A, HB-88, HB-102-102A, HB-116A, HB-117, HB-134A, HB-135, HB-155, HB-158, HB-187, HB-192, HB-208, HB-229, HB-244, HB-317, HB-340, HB-501, HB-602, HB-608, HD-168, HD-193,	DWL-5023, Durum, Durum-1, Dia Den, Dhar Local, E-150, E-153, E-219, E-221B, E-240, E-260, E-312, E-340, E-406, E-474, E-479, E-483, E-492B, E-497B, E-547, E-557, E-579, E-601, E-674, E-688, E-706, E-708, E-714, E-720A, E-743, E-750, E-802, E-811, E-816A, E-830, E-875A, E-877, E-894, E-898, E-909, E-921A, E-934, E-938A, E-957, E-958, E-993, E-996, E-189B(D), E-469B, E-1013, E-1052, E-1131, E-1134, E-1617B, E-1769, E-1774, E-1784, E-1812, E-1815, E-1818A, E-1834, E-1834A, E-1850A, E-1851A, E-1864A, E-1869A, E-1887B (A), E-1889, E-1893, E-1906(A), E-1913, E-1914A, E-1915A, E-1934, E-1934A, E-1948, E-1951, E-1952, E-1959, E-1983, E-1994, E-1995, E-261-B, E-262B, E-494A, E-2008, E-2011, E-2024B, E-2026, E-2028, E-2055, E-2056, E-2059, E-2067B, E-2098B, E-2115, E-2125, E-2125A[D], E-2125B, E-2127, E-2131, E-2148, E-2154, E-2160, HI-690, HI-694, HI-695, HI-703, HI-707, HI-712, HI-714, HI-715, HI-716, HI-720, HI-723, HI-728, HI-730, HI-732, HI-739, HI-724, HI-744, HI-752, HI-753, HI-757, HI-761, HI-766, HI-784, HI-800, HI-872, HI-895, HI-914 [Durum], HI-930, HI-935, HI-955, HI-977, HI-1077, HI-1114, HI-1399, HI-1467, HI-2130 (Two samples), HI-3638, NI-5439, NI-6778,	E-2175, E-2178, E-2179, E-2190, E-2217, E-2230, E-2231, E-2248, E-2255, E-2259, E-2281, E-2285, E-2287, E-2297, E-2298, E-2299, E-2300, E-2302, E-2312, E-2315, E-2317, E-2319, E-2326, E-2329, E-2337, E-2338, E-2348, E-2360, E-2374, E-2379, E-2407A, E-2465, E-2540, E2565, E-2570, E-2770, E-2772, E-2781, E-2831, E-2843B, E-2893, E-2905, E2905A, E303B, E-3007, E-3084, E-3088, E-3089, E-3090, E-3096, E-3097b, E-3123a, E-3131A, E-3152A, E-3155, E-3217, E-3243, E-3280, E-3318, E-3321, E-3324, E-3327, E-3339, E-3340, E-3343, E-3345, E-3347, E-3353, E-3356, E-3409, E-3409, E-3426, E-3427, E-3470, E-3500, E-3507, E-3517, E-3527, E-3528, E-3535, E-3558, E-3563, E-3571, E-3572, E-3575A, E-3578A, E-3580, E-3591, E-3638, E-3644, E-3672, E-3674, E-3679, E-3682, E-3691, E-3698, E-3830B, E-3852, E-3857, E-3895, E-3909, E-3918B, E-3925, E-3927, E-3928, J-340, J-373, J-393, J-394, J-405, J-M-1-4, JU-12[Durum], JW-10, JWJ-36-37, JWJ-76-42, JWJ-78-1, JWJ-175, JWJ-176, JWJ-278, Janak, JNK-6W-124, JNK-6W-189, JNK-6W-190, JNK-6W-193, JBRR, JBRR-20, K-33, K-46, K-47, K-52, K-65, K-68, K-73, K-98, K-103, K-129, K-309, K-338, K-449, K-830, K-1229,

E-4795, E-4812B, E-4825B, E-4845, E-4851B, E-4870, E-4950, E-4958, E-4982, E-4987, E-5776, E-5988, E-5690, E-5895, E-629B, E-6006A, E-6950, E-6968, E-6983, E-6984, E-759B, E-715, E-7149, E-7174, E-7222, E-7223, E-7236, E-7284, E-7295, E-7305, E-7317, E-7322, E-7323, E-7412, E-7443, E-7480, E-7549B, E-7709, E-7753, E-7775, E-7794, E-7796B, E-8161, E-8671, E-8672, E-8678, E-8684, E-8685, E-8686, E-8779, E-8834B, E-9126, E-9132, E-9139, E-9152, E-9233B, E-9259, E-9311, E-9313, E-9596, E-9745, E-9801, E-9990, E-10048, E-10187, E-10278, E-10303, E-10435, E-10702, E-10713, E-10862, E-10869B, E-10872, E-10903, E-10916, E-10917, E-10924, E-10943, E-10947, E-10962, E-11021, E-11026, E-11036, E-11061, E-11073, E-11086, E-11089, E-11090, E-11169, E-11173, E-11185, E-11191, E-11228, E-11252, E-11255B, E-11258, E-11260, E-11515, E-11535, E-11536, E-11537, E-11538, E-11635, E-11812, E-11813, E-11814,. E-11851, Frex A, Flether-A, GL-1, GL-2, Lok-18, GL-4, GL-5, GL-6, GL-7, GL-8, GL-9, GL-10, GL-11, GL-13, GL-18, GL-20, GL-21, GL-22, GL-23, GL-24, GL-25, GL-26, GL-27, GL-28, GL-29, GL-30, GL-31, GL-32, GL-34, GL-35, GL-36, GL-37, GL-38, GL-39, GL-40, GL-42, GL-44, GL-45, GL-46, G-49805A,	HD-211, HD-219, HD-462, HD-501, HD-1160B, HD-1275, HD-1459A, HD-1464, HD-1788, HD-1918, HD-1922, HD-1941, HD-1944, HD-1949, HD-1973, HD-1982 (Janak), HD-1993, HD-1999, HD-273-1, HD-2001, HD-2004-1, HD-2012, HD-2028, HD-2037, HD-2074, HD-2108, HD-2114, HD-2122, HD-2134, HD-2137, HD-2150, HD-2157, HD-2160, HD-2179, HD-2189, HD-2197, HD-2204, HD-2212, HD-2232, HD-2236, HD-2258, HD-2260, HD-2270, HD-2275, HD-2278, HD-2280, HD-2281, HD-2285, HD-2298, HD-2236, HD-2307,HD-2327, HD-2328, HD-2329, HD-2329B, HD-2332, HD-2336, HD-2363, HD-2380, HD-2385, HD-2402, HD-2427, HD-2428, HD-2501, HD-2502, HD-2504, HD-2518, HD-2599, HD-4502, HD-4512, HD-4530, HD-4578, HD-4579, HDR-2, HDR-64, HDR-70, HDR-77, HI-76, HI-146, HI-171, HI-260, HI-281, HI-282, HI-334-7-19, HI-346, HI-360, HI-363, HI-366, HI-367, HI-368, HI-388 (Mukta), HI-385-2, HI-386, HI-390-3,HI-392, HI-393, HI-394, HI-395, HI-402, HI-405, HI-407, HI-412, HI-418, HI-423A, HI-426, HI-430, HI-431, HI-443, HI-471, HI-473, HI-475-2, HI-476, HI-487-1, HI-511, HI-527, HI-533, HI-545, HI-555, HI-559, HI-563, HI-563-1-2, HI-574, HI-579, HI-600, HI-602, HI-611, HI-612, HI-615, HI-618, HI-624, HI-625, HI-629, HI-640, HI-646, HI-651, HI-652, HI-657,	HI-588, HY-65, HY-656, HP-82, HP-720, HP-833, HP-1034, HP-1102, HP-1179, HP-1209, HP-1267, HP-1327, HP-1379, HP-1385, HP-1388, HP-1410, HP-1604, HP-1628, HP-1640, HP-2107, HP-2182,HPW-30, HPW(DL)30, HPW-39, HPS-73-3, HS-13, HS-27, HS-43, HS-45, HS-51, HS-65, HS-81, HS-86[Durum], HS-100, HS-115-3-1, HS-131, HS-135, HS-144, HS-146, HS-147, HS-139, HS-207, HS-217, HS-223, HS-225, HS-240, HS-276, HS-277, HS-295, HS-296, HS-409, HS-528-537, HS-656, HS-679/1, HS-729, HS-1090, HS-1098, HS-1139/1, HS-1140-3-1, HUW-234, HUW-266, HUW-55,HUW-105, HUW-107, HUW-206, HUW-217, HUW-234, HUW-258, HUW-261, HUW-269, HUW-280, HUW-284, HUW-319, HUW-323, HUW-325, HUW-326, HUW-327, HUW-328, HUW-329, HUW-330, HUW-334, HUW-350, HUW-2001, HW-36, HW-105, HW-112, HW-113, HW-120, HW-121, HW-123, HW-135, HW-151, HW-153, HW-155, HW-166, HW-167, HW-169, HW-195, HW-205, HW-305, HW-517, HW-544, HW-550, HW-558, HW-560, HW-571, HW-657, HW-741, HW-1045, HW-1095, HW-1211, HW-2001, HW-2001A, HW-2002, HW-2002A, HW-2003, HY-11, HY-18, HY-65, HY-105, HY-152, HY-228, HY-469-1, HY-862-3, HYB-12, Hindi-62, IWP-129, I-1361 [Durum],	K-2041A, K-3010, K-4500-4, K-4970-1, K-720-3, K-7244, K-7284, K-7310, K-7404, K-7410, K-7416, K-7417, K-7419, K-7420, K-7425, K-7482, K-7504, K-7520, K-7525, K-7538, K-7539, K-7611, K-7827, K-8027, K-8020, K-8029, K-8230, K-8565, K-8660, KA-9001, KA-9003, KA-9004, KA-9005, KA-9006, KA-9007, KA-9010, KA-9012, KA-9013, KA-9015, KA-9017, KA-9018, KA-9019, KA-9020, KA-9021, KA-9022, KA-9023, KA-9024, KA-9025, KA-9026, KA-9027, KA-9028, KA-9029, KA-9031, KA-9032, KA-9033, KA-9034, KA-9035, KA-9036, KA-9037, KA-9038, KA-9039, KA-9040, KA-9041, KA-9042, KA-9043, KA-9044, KA-9045, KA-9046, KA-9047, KA-9048, KA-9049, KA-9050, KA-9051, KA-9052, KA-9053, KA-9054, KA-9055, KA-9056, KA-9058, KA-9060, KA-9061, KA-9063, KA-9065, KA-9066, KA-9067, KA-9069, KA-9070, KA-9071, KA-9072, KA-9073, KA-9074, KA-9077, KA-9079, KA-9081, KA-9082, KA-9083, KA-9085, KA-9087, KA-9088, KA-9089, KA-9090, KA-9091, KA-9092, KA-9094, KA-9095, KA-9096, KA-9100, KA-9015, KA-90103, KA-90104, KA-90105, KA-90107, KA-90108, KA-90110, KA-90111, KA-90112, KA-90113, KA-90114, KA-90116, KA-90117, KA-90118, KA-90120, KA-90121, KA-90122, KA-90123, KA-90124, KA-90125,

GS-2292, GL-7313A, GW-116A, GW-1004, Gaza-115.HRA, Gwalior, Gwalior-5, Gwalior-101, Gwalior-101A, Gwalior-331, KA-90147, KA-90148, KA-90149, KA-90150, KA-90151, KA-90152, KA-90153, KA-90154, KA-90155, KA-90156, KA-90158, KA-90159, KA-90160, KA-90161, KA-90162, KA-90163 [Durum], KA-90164, KA-90165, KA-90166, KA-90167, KA-90168, KA-90171, KA-90172, KA-90173, KA-90174, KA-90175, KA-90176, KA-90178, KA-90180, KA-90181, KA-90182, KA-90183, KA-90186, KA-90187, KA-90188, KA-90189, KA-90190, KA-90191, KA-90192, KNB-3B, KN-3, KRL-1-4, Kolaz-27, Kolaz-37, Kenya-276, Kanphad, Kenphad Khapli, Kanphad-21, Kanphad-25, Kanphad-30, Kalyan Sona, Koper Gaon, Kharchia, Kharchia-65, Karalic-1, Kirac-66, Kiran, Kubanka, Kathia Red, Kunda Local, Kunden (DL)-1532, Local Red Karnatka [Durum], Lok-1, Lal Bahadur, LGM-15, LGM-17, LGM-18, LGM-19, LGM-20, LGM-29, NP-202[Dicoccum], NP-710, NP-715, NP-718, NP-720, NP-737, NP-751, NP-758, NP-770, NP-787, NP-824, NP-829, NP-830, NP-832, NP-834, NP-839, NP-846, NP-852, NP-876, NP-890, NP-896, NP-8395, NI-747-19, NI-749-19, NI-817B, NI-5439, NI-5643, NI-7178, NI-7861, NI-8188, NI-8771, NI-8796, NDW-14, NDW-87, ND-422A, NG-14-8-2-3, Narmada-4,	HI-660, HI-662, HI-667, HI-671, HI-672, HI-674, HI-677, HI-682, HI-687, LGM-31, LGM-33, LGM-34, LGM-35, LGM-45, LGM-51, LGM-55, LGM-56, LGM-59, LGM-66, LGM-67, LGM-70, LGM-72, LGM-77, LGM-79, LGM-81, LGM-82, LGM-83, LGM-86, LGM-87, LGM-89, LGM-92, LGM-94, LGM-98, LGM-103, LGM-105, LGM-109, LGM-110, LGM-111, LGM-112, LGM-113, LGM-114, LGM-115, LGM-116, LGM-117, LGM-118, LGM-123, LGM-124, LGM-126, LGM-130, LGM-132, LGM-133, LGM-134, LGM-135, LGM-136, LGM-137, LGM-138, LGM-139, LGM-140, LGM-142, LGM-143, LGM-145, LGM-147, LGM-148, LGM-151, LGM-153, LGM-159, LGM-160, LGM-162, LGM-164, LGM-165, LGM-167, LGM-172, LGM-174, LGM-180, LGM-182, LGM-185, LGM-188, LGM-191, LGM-193, LGM-194, LGM-197, LGM-198, PBW-258, PBW-215[Durum], PBW-216[Durum], PBW-218[Durum], PLC-28-1, PLC-217, PLC-253, PLC-273, PLC-281-1, PLC-4082, PL-85, PL-90, PL-456-1, PKD-56, PLA-8A, PKV[Durum], PN-306-1, PV-18, PW-3, Raj-711-2, Raj-861[Durum], Raj-977, Raj-1052, Raj-1152, Raj-1260, Raj-1337, Raj-1342, Raj-1389, Raj-1418, Raj-1450, Raj-1457, Raj-1459, Raj-1482, Raj-1455, Raj-1649, Raj-1656, Raj-1700, Raj-1720, Raj-1774, Raj-1777, Raj-1865, Raj-1945, Raj-1948, Raj-1953, Raj-1972, Raj-1975, Raj-1981, Raj-2184, Raj-2535,	I-2047[Durum], IWP-72, IC-8572 [Dicoccum], J-9, J-12, J-13, J-17, J-24, J-25, J-41, J-56, J-62, J-122, J-123, J-124, J-134, J-139, J-160, J-280, J-304 LGM-199, LGM-201, LGM-205, LGM-206, LGM-207, LGM-213, LGM-214, LGM-220, LGM-225, LGM-226, LGM-227, LGM-229, LGM-237, LGM-238, LGM-239, LGM-241, LGM-242, LGM-243, LGM-244, LGM-245, LGM-246, LGM-248, LGM-249, LGM-250, LGM-251, LGM-252, LGM-257, LGM-259, MP-10, MP-13, MP-17, MP-20, MP-23, MP-24-1, MP-25, MP-27, MP-28, MP-29, MP-30, MP-31, MP-35, MP-37, MP-39, MP-53, MP-56, MP-57, MP-58, MP-60, MP-61, MP-62, MP-65, MP-69, MP-70, MP-76, MP-78, MP-84, MP-85, MP-90, MP-112, MP-114, MP-115, MP-118, MP-119, MP-124, MP-125, MP-134A, MP-137, MP-138, MP-143, MP-152, MP-153, MP-154, MP-156, MP-157, MP-160, MP-162, MP-164, MP-165, MP-166, MP-167, MP-168, MP-169, MP-170, MP-171, MP-176, MP-177, MP-178, MP-179, RCB-850,. RCB-855, RCB-Lib-857, RCB-895, RCB-1094, RS-1-4, RS-30-1, RS-31-1, RS-64-1, RTMP-83, Ratlam, River, Redley, Ratha Katha, Rence, RR-25, RWS-3223, RWS-3225, SV-7, SV-27, SV-28, SV-63-1-4399, SV-66A, SV-73, SV-84, SV-93, SV-63-11-4398, SV-183, SV-404, SV-634[Durum], SV-637, SV-638, SV-639, SV-642, SV-644, SV-660, SV-665, SV-671, SV-673, SV-679,	KA-90126, KA-90127, KA-90128, KA-90129, KA-90130, KA-90131, KA-90132, KA-90133, KA-90134, KA-90135, KA-90137, KA-90138, KA-90139, KA-90141, KA-90143, KA-90145, KA-90146 MP-181, MP-182, MP-192, MP-194, MP-195, MP-196, MP-211, MP-227, MP-233, MP-235, MP-238, MP-239, MP-254, MP-259, MP-261, MP-267, MP-305, MP-409, MP-411, MP-412, MP-415, MP-427, MP-428, MP-502[Durum], MP-505, MP-510, MP-529, MP-539, MP-541, MP-586, MP-607, MP-608, MP-645, MP-788, MP-825, MP-843, MP-845, MP-883A, MP-847, MP-2037, MG-14-4-1, MG-513-2, MPO-501, MPO-503, Magura Local-1, Magura Local-2, Magura Local-3, Magura Local-4, Magura Local-5, Magura Local-6, Magura Local-7, Magura Local-8, Magura Local-9, Magura Local-10, Magura Local-11, Mandya, Mandya-3-2, Mandya-3-2-5, Malvika, Mukta, MLKS-11, Mandawal Red, Mandya, Metro, Mandsaur Local [Durum], MACS-2496 [Durum], N-1, N-602, N-761, N-7707, NP-4, NP-52, NP-123[Dicoccum], NP-163[Dicoccum], NP-165, NP-200[Dicoccum], SV-855, SV-864, SV-894, SV-900A, SV-900B, SV-901, SV-904, SV-905, SV-908A, SV-908B, SV-910, SV-914, SV-915, SV-916, SV-917, SV-918, SV-920, SV-933, SV-936A, SV-938, SV-939A, SV-945, SV-960, SV-966, SV-973, SV-975,

Narmada-112, P-226, Pissi Local, PL-228, PL-4219 [Durum], P Athion, Poona-71, Poona-74, Poona-82, Poona-84, Poona-85, Poona-86, Poona-880, Power Kheda, Power Kheda White, Pathkunda-10, PA/2NP63/2807176-1-3-40, PBW-34 [Durum], PBW-36, PBW-54, PBW-65, PBW-120, PBW-138, PBW-159[Durum], PBW-175, PBW-213[Durum], PBW-222, PBW-225, PBW-226, PBW-230, PBW-255, SV-6808, SV-7958, SV-7988, SV-10183, S-301-1, S-305, S-314, S-411, S-700, S-903, Son-63, SPM, Sujata Lyke, Sonalika, Sagwara Local, Sagarika, Swati, Shera (HD-1925), Super, Suv Soka, Sujata Type, Spherococcum, TP-552, TPS-5, Triveni, Tu-167, Tol-8156, Toreen-23, Tw-11, Tr-29, T_2-68, T_{14}-15-1, UP-225, UP-262, UP-281, UP-289, UP-301, UP-368,	Raj-3038, Raj-3039, Raj-3063, Raj-3077, Raj-3232, Raj-3392, Raj-3395, Raj-3496, Raj-3520, Raj-6351, RCB-725, RCB-816, RCB-817, RCB-821, RCB-Lib-840, RCB-Lib-843, RCB-845, RCB-Lib-847, RCB-848, UP-1109, UP-2003, UP-2104, UP-2121, UP-2042, UP-2255, Ujjain Local, V-975, V-980, V-1026, V-1030, V-120, VL-401, VL-404, VL-482, VL-527, VL-616, VL-641, VL-651, VL-678, VL-682, VL-687, VL-1788, VL-689, VL-690, VL-691, VL-692, VL-693, VL-694,VL-695, VL-696, VL-697-1, VL-698, VL-699, VL-700, VL-1131, W-128, W-471, WL-261, WL-381, WL-382, WL-383,	SV-683B, SV-686, SV-687, SV-690, SV-692, SV-693, SV-697, SV-698, SV-701, SV-707, SV-711, SV-715, SV-717, SV-718, SV-722, SV-728, SV-740, SV-756, SV-758A, SV-759, SV-760, SV-761, SV-774, SV-776, SV-776B, SV-777, SV-779, SV-789, SV-791, SV-792, SV-793B, SV-801, SV-805, SV-807, SV-809, SV-813, SV-814, SV-815, SV-819, SV-820, SV-825, SV-826A, SV-826B,SV-829, SV-830, WL-388, WL-389, WL-391, WL-392, WL-393, WL-400, WL-404, WL-405, WL-406, WL-410, WL-411, WL-412, WL-414, WL-415, WL-416, WL-418, WL-419, WL-421, WL-422, WL-423, WL-424, WL-425, WL-430, WL-432, WL-438, WL-442, WL-443, WL-444, WL-445, WL-446, WL-448, WL-449, WL-445, WL-450, WL-467, WL-453,	SV-976, SV-978, SV-980, SV-980B, SV-981, SV-982, SV-988, SV-998, SV-999, SV-999A, SV-1001, SV-1003, SV-1006, SV-1008, SV-1018A, SV-1020, SV-1023, SV-1024, SV-1029, SV-1030, SV-1035, SV-1036, SV-1037, SV-1038, SV-1040, SV-1050, SV-1051, SV-1062, SV-1069, SV-1075, SV-1078, SV-1082, SV-1085, SV-1086, SV-1097, SV-1098, SV-1099, SV-1107, SV-1110, SV-1117, SV-1157, SV-1168, SV-1183, SV-1211, SV-1217, SV-1222, SV-1225, SV-1227, SV-1228, SV-1229B, SV-1230, SV-1233, SV-1247, SV-3133, WL-487, WL-500, WL-528, WL-598, WL-553, WL-950, WL-991AL, WL-998, WL-1001, WL-1006 [Diococcum], WL-1073A, WL-1073B, WG-1809, WG-2009, WG-2009-1, WG-2009-2, WH-106, WH-111, WH-147, WH-157, WH-261, WH-283, WH-291, WH-331, WH-416, WH-494, WH-802, WH-869, WL-1, WL-2, WL-410, WL-711, WL-1562, WL-2265

All India Coordinated Wheat Improvement Programme at IARI Regional Station, Indore, MP

The ultimate aim of any breeding programme is to develop varieties superior to the existing ones in yielding ability, disease resistance, quality and other characteristics. For this reason, the strains are extensively evaluated for their performance, disease resistance, quality *etc.*, in multilocation trials. The multilocation trials of wheat are conducted under the All India Coordinated Wheat Improvement Programme, which play a key role in testing, identification and release of new varieties. Before the new strains are included in the multilocation tests, the breeder evaluates their performance at his own station to make sure that the new strain is superior to the existing varieties. The various activities and operations in the release of a strain as a variety may be grouped into the following three classes for convenience in description: 1. Evaluation, 2. Identification and 3. Release and Notification. The Notification of a variety is a prerequisite for seed certification. Superior strains are developed in wheat like any other crop through different breeding strategies like introduction, hybridization and selection in the segregating generations over years. **Evaluation** of a strain as a variety consists of various trials and tests to determine its superiority over the best existing variety in terms of yield and other agronomic traits, and its suitability for consumption. It includes station trials – conducted by a breeder who has developed the new strain and tested for one or more years at that location to evaluate its superiority in performance compared to the best available variety for the region. Presently, common varietal trials (CVT) are being conducted at Institute or University level to evaluate the best performing lines of different stations working in co-ordination for one year to select strains showing superiority in performance, compared to the best available variety for the region before it is included in the trials of the concerned All India Coordinated Programme. The station trial may be repeated to make sure that the new strains proposed for Initial Evaluation Trial are superior to the existing varieties. National Initial Varietal Trial (NIVT) – The strains included in the initial evaluation trial for various sowing conditions *i.e.*, Timely sown, Late sown, Rainfed, Restricted Irrigation *etc.*, are evaluated over the zones for evaluating their performance over the zones like North Hills Zone, North Western Plains Zone, North Eastern Plains Zone, Central Zone, Peninsular Zone and South Hill Zone *etc.*, Based on their performance, strains are promoted to Advance Varietal Trials of the respective zones. The pattern of zones for a crop is based on several considerations, including the distribution pattern of the crop and the area under it in the various regions. Advance Varietal Trials (AVT) will be conducted for two years. Every year, based on the performance in AVT, entries poor in yield and/or disease resistance are rejected. The data from evaluation for two years under Advance Varietal Trials and one year of Agronomical trial (simultaneously conducted for strains promoted to AVT II for one year) is essential for the identification and release of a wheat entry as a variety. The data from agronomical trials, and disease scores are important considerations before identifying any variety for release and must be included in the release proposal. Thus, wheat entries are evaluated for one year in National Initial Varietal Trial and two years under Advance Varietal Trial before their **identification** by the workshop of the project for the prerelease multiplication and for adoptive trials; identified strains are proposed to the Central Sub-committee on Crop Standards, Notification and Release of Varieties for final **release and notification**. Only after the notification of the variety by the Central Sub-committee on Crop Standards, Notification and Release of Varieties, the variety enters into the seed chain for multiplication and distribution to the farmers. During *rabi*, a number of lines were evaluated along with check varieties PBW 343, Raj 1555, HI 8498 and Raj 3765. Synthetics with more than 55 grains/spike are SYN 02-1, SYN 02-6, SYN 02-7, SYN 02-9, SYN 02-16, SYN 02-18 and SYN 02-19 and yield/plot more than 300 g were SYN 02-8, SYN 02-15, SYN 02-16, SYN 02-17, SYN 02-18, SYN 02-19, SYN 02-24 and SYN 02-38.

Table 4.23. AICRP Trials and performance of genotypes under different Production Condition and Zone

AICRP Trial	*Zone*	*Production Condition*	*Name of Genotypes*
AVT – II	NWPZ	TS, RI	HI 1620
AVT – I	NWPZ	TS, RI	HI 1628
AVT – I	NEPZ	TS, RI	HI 1628
AVT – I	PZ	TS, IR	HI 1624, HI 1625, HI 8800 (d)
AVT – I	PZ	TS, RI	HI 8802 (d), HI 8805 (d)
SPL Trial–Ist	NWPZ/NEPZ	VLS, IR	HI 1621
NIVT – 2	CZ/PZ	TS, IR	HI 1629, HI 1630, HI 1631, HI 1632
NIVT – 3B	CZ/PZ	LS, IR	HI 1633, HI 1634, HI 8807 (d), HI 8808 (d)
NIVT – 4	CZ/PZ	TS, IR	HI 8809 (d), HI 8810 (d), HI 8811 (d), HI 8812 (d), HI 8813 (d)
NIVT – 5B	CZ/PZ	TS, RI	HI 1635, HI 8814 (d), HI 8815 (d)

Application of Molecular (DNA) Markers in DUS Testing

Wheat (*Triticum aestivum*) is the world's leading cereal grain and most important food crop grown worldwide. The species of *Triticum* are grouped into diploids (2n=2x=14), tetraploids (2n=4x=28) and hexaploids (2n=6x=42). Only two species of *Triticum* are commercially important: the hexaploid species, T. aestivum (the bread wheat and the principal wheat in commerce), and the tetraploid species, *T. turgidum*. The total numbers of accessions in international and local gene banks around the world are estimated to be in excess of 400,000, although many accessions may probably be duplicated in the different collections. Traditionally, the assessment of the genetic composition of crop germplasm has been conducted on the basis of morphological and phenotypic characters, which frequently lack the resolving power needed to identify individual genotypes. FAO has developed a set of descriptor for wheat to facilitate uniformity in describing wheat accessions in different gene banks and to assist breeders in locating genes specific breeding programs (Poelhman and Sleper 1995). These are also the type of descriptors used for registration of varieties and for awarding Plant Breeders' Rights (PBR) following distinctness, uniformity and stability (DUS) testing. Even though these descriptors are useful, they are limited in number, multigenic, quantitative or continuous and may be affected by environmental factors. These characters often require the varieties to be grown to full maturity. Estimation from biochemical markers *viz.* isozyme and storage proteins may also be biased by the general consideration that only a minor portion of the genome is represented by these markers. Another limiting factor for these descriptors is the increasing number of registered varieties over the time so it has become difficult to compare efficiently each newly submitted variety against all existing varieties. Molecular markers are a useful complement to morphological and physiological characterization of cultivars. These are also useful to evaluate potential genetic erosion, defined here as a reduction of genetic diversity in time. In the last decade, molecular markers such as RFLP, RAPD, SCAR, AFLP *etc.* have been used to characterize and assess genetic variation at the DNA level, allowing an estimation of degree of relatedness between individuals (Tanksley 1993). The advantages in using molecular markers for variety testing are as follows: These markers are tissue, growth and season independent; They are highly polymorphic so can efficiently discriminate between two genotypes; They have excellent genome coverage; They are independent of environmental variations; They have ability to distinguish homozygotes from heterozygotes (SSRs); Only small tissue sample is required for molecular genotyping. DNA markers are alleles of loci where sequence variation or polymorphism is present in given genotypes. These are broadly classified into two categories. The thrust of today's researchers is that if you can identify the variety genetically, you know almost everything about it. Therefore, main purpose of developing molecular profiles of varieties these days is to use them as an alternative to protein fingerprinting and to complement the visual grading system. Current techniques used for variety identification in wheat include gel electrophoresis and High-Performance Liquid Chromatography (HPLC). These methods require a significant level of expertise to perform and are not well suited to routine, high throughput screening operations. The advent of DNA markers and extensive physical and genetic mapping of traits in wheat has recently presented the opportunity to revolutionize the way in which new wheat breeding material is developed. DNA markers have been described for a wide range of quality traits, including starch quality, dough strength and grain.

Hybridization Based Marker System

Restriction Fragment Length Polymorphism (RFLP) markers detect the inherited mutational changes in DNA between the genotypes that result in loss or gain of restriction sites. The unique restriction profiles are thus detected by hybridization with labeled probes followed by autoradiography. RFLP (Bostein *et al.*, 1980) was one of the first DNA marker techniques used to characterize wheat cultivars and assess genetic diversity (Kim and Ward,2000). However, the relatively low level of polymorphism observed among elite wheat cultivars and the complexity and cost of the technique, limit the use of RFLP for routine cultivar identification.

PCR-Based Markers Systems

These are arbitrary or specific DNA primers used to amplify specific nucleotide sequences of genomic DNA. These bands are resolved on agarose/PAGE gels to score polymorphism between individuals. *e.g.* Randomly Amplified Polymorphic DNA (RAPD), Amplified Fragment Length Polymorphism (AFLP), Inter Simple Sequence Repeats (ISSR), Simple Sequence Repeat (SSR), Single nucleotides (SNPs) *etc.*

Simple Sequence Repeats

These loci are amplified by PCR using primers (18–25 bp long) specific for sequences flanking hypervariable regions of tandem repeats of 2 to 4 base pairs. The variation in the number of repeats present in these loci determines differences in length of the amplified fragments. The SSR technique gained rapid acceptability because of its codominant nature, reproducibility, and high information content. This methodology is useful in identifying genotypes in self-pollinated species with low levels of genetic variability such as and wheat (Dograr *et al.*, 2000)

Amplified Fragment Length Polymorphism

The AFLP technique combines the RFLP reliability with the power of PCR to amplify simultaneously many restriction fragments. This technique operates as a RAPD, but a primer consists of a longer (app.15bp) fixed portion and a short (2-4bp) random primer. This marker system produces many loci so proves useful in differentiating genotypes. This technique was used successfully to evaluate genetic diversity and genetic relationships in wheat (Domini *et al.*, 2000) and barley (Qi and Lindhout, 1997).

Inter Simple Sequence Repeats

The ISSRs (inter-simple sequence repeats) are PCR products obtained with primers based on dinucleotide, trinucleotide, tetranucleotide and pentanucleotide repeats (Zietkiewicz *et al.*, 1994). These primers are generated by microsatellite repeat anchored primers amplifying regions between adjacent SSR loci and one of the widely used markers for genotypic analysis. Thus, ISSR DNA has been proposed as a new source of genetic marker that overcomes many of the technical limitations of RFLP and RAPD analyses. ISSR analyses are easier than SSR analyses in some steps, as there is no need of any prior knowledge of the target sequences of the SSR. These markers have been used for cultivar identification in wheat (Nagaoka and Ogihara, 1997).

Random Amplified Polymorphic DNA

Random-amplified polymorphic DNA (RAPD) analysis is based on the amplification of DNA fragments with the polymerase chain reaction (PCR) starting from primers with arbitrary sequences of typically 10-mer random primers. This technique is considerably faster and simpler than some other molecular techniques. RAPD markers have been used to examine both interspecific and intraspecific variation in a number of plant species RAPDs have been used for measuring genetic diversity in several plant species, *i.e.* wheat (Teshale *et al.*, 2003) and Triticum (Cao *et al.*, 1999).

Single Nucleotide Polymorphic Markers

SNPs (single nucleotide polymorphisms) belong to the last-generation molecular markers and occur at high frequencies in both animal and plant genomes. These markers detect difference in one base at marker site therefore; allow automatizing and enhancing tenfolds the effectiveness of genotype analysis. A bioinformatics strategy was developed to discover SNPs within a large wheat EST database and to demonstrate the utility of SNPs in genetic mapping and genetic diversity applications (**Salina *et al.*, 2003**). The Wheat SNP Database includes conserved and genome-specific PCR primers for amplification of STSs from genomic DNA of wheat and its diploid and tetraploid ancestors, DNA sequences, sequence annotations, intron/exon predictions, electropherograms, haplotypes, SNPs, and positions of SNP markers on wheat linkage maps. As of May 2006, the database contained 17,174 primers (forward and reverse), 1,102 wheat polymorphic loci, and 2,224 polymorphic sequence tagged sites in diploid ancestors of polyploid wheat (Franky You, 2006).

Molecular Markers and DNA Fingerprinting in Wheat

In the 1970s, researchers began using a technique called protein fingerprinting to discern varieties of wheat that could not distinguish by the naked eye. Today, protein fingerprinting is routinely used to support the visual grading system. However, a new technology called DNA fingerprinting is on the horizon. This is highly specialized work with rapid, automated, portable and cost-effective technology to characterize varieties. A DNA fingerprint is a genetic photograph or the blue print of a genotype which in an absolute measure of the genetic makeup of that genotype. It must be unique to that genotype in order to distinguish it from other genotypes. DNA fingerprinting often targets the highly variable tandemly repeated DNA sequences, minisatellites (9-100bp repeats) and microsatellites (1-6bp repeats) to develop database. Choice of DNA fingerprinting technique depends upon the infrastructure, technical expertise and operational funds available as well as requirements of the experiments. However, the two most important considerations in the use of molecular techniques for genotyping are the discrimination power and reliability test of the marker system. SSR is regarded as the method of choice in cereals especially wheat because of the abundance, high polymorphism and codominant inheritance of microsatellites.

Table 4.24. Comparison of PCR-based DNA fingerprinting techniques

Feature	*RAPD*	*ISSR*	*SSR*	*AFLP*
Development cost	Low	Low	High	Low
Level of polymorphism	Medium	Medium	High	Medium
Reliability	Low	Medium	High	Medium
Level of skill required	Low	Low	Medium	Medium
Automation cost	Medium	Medium	High	High
Samples/day	50	50	50	50

Generation of DNA Fingerprint Database

The characterization of germplasm at molecular level has become necessary these days to meet the demands of international markets for export to display authenticity and quality of the agricultural produce. This molecular polygraph is also important in assessing genetic variability that is important because a decrease in genetic variability might result in a reduction of the plasticity of the crops to respond to changes in climate, pathogen populations, or agricultural practices. During the genotyping of desired germplasm at DNA level the fingerprint database is developed that includes the following steps (for wheat). **DNA isolation:** Usually CTAB-method (Saghai - Maroof) is used to isolate wheat DNA from 1-2 weeks old plants. Isolated DNA is purified by RNAse treatment and quantified on 0.8 per cent agarose gel/by fluorimeter. **PCR-amplification:** is done with the desired primers at required annealing temperature in thermocycler. **Gel-Electrophoresis:** is carried out depending on the type of marker system used. RAPD and ISSR amplified products can be resolved on 1.5 per cent agarose gel. For AFLP and SSR marker systems normally 4-6 per cent Polyacrylamide gel is used as the bands resolved in these marker systems is small and cannot be resolved to reveal polymorphism on agarose gels. The automated sequencer is becoming common in advanced labs for developing amplification profiles PCR based molecular markers. In this system, amplified band is obtained in the form of peak and then database is developed. Although this technology is more reliable and user friendly than PAGE but the running cost of sequencer is very high. **Allele scoring and Data analysis:** Scoring is done on the basis of presence and absence of bands/peaks for all types of markers. The DNA fingerprint database developed is then used in software program for varietal characterization on the basis of differences in banding pattern of various genotypes. This database developed is further analyzed by suitable software program like NT-SYS, PCA, Winboot or Mapmaker for similarity index to differentiate various genotypes and establish genetic relatedness in the germplasm.

Variety Characterization Based on PCR-Markers Technology

The development of more than 400 SSR loci in wheat initiated the use of these markers for variety characterization in wheat. The first SSR markers available were used to characterize eight European cultivars (Devos *et al.*, 1995) of wheat bread. The DNA Fingerprint Database of European wheat germplasm has been developed recently (Roder *et al.*, 2002). 105 Argentine bread wheat (*Triticum aestivum* L.) cultivars released between 1932 and 1995 were characterized by simple sequence repeat (SSR) and amplified fragment length polymorphism (AFLP) markers. A selected subset of 10 highly informative SSR was used to construct an Identification Matrix that allowed the discrimination of the 105 cultivars. Data obtained from SSR markers were complemented by information derived from AFLPs. Molecular data were used to quantify genetic diversity across Argentine wheat breeding programs and to determine if modern wheat cultivars have a lower genetic diversity than earlier cultivars (Manifesto *et al.*, 2001). The DNA profiling was done for high-value monovarietal pasta to test its authenticity. AFLP profiles obtained from eight monovarietal pasta samples suggested the presence of unknown variety in monovarietal pasta by referring to a durum wheat molecular markers database (Terzi *et al.*, 2004). In SNP system, SNaPshot primer extension technique can distinguish different genome specific PCR products although it can not exactly identify the A, B or D genomes. These SNaPshot fingerprinting patterns can be used to distinguish the wheat cultivars from one another. Sequence scanning will allow to identify a large number of single base variations and develop variety specific SNP markers, which can be utilized for variety identification.

Initiatives by UPOV

In the countries that are members of the UPOV, new variety must undergo DUS testing that constitutes the basis of intellectual property protection in the form of PBR. Furthermore, with the increasing number of UPOV members, variety reference collections are expanding making discrimination difficult. In this framework various authors have demonstrated in principle that D (distinctness) can be established in collections of wheat varieties by the use of DNA microsatellites analyzed as SSRs or STSs (sequence tagged sites). Efforts at international level are underway to find the suitability of microsatellites across different laboratories. UPOV has constituted a working group on **Biochemical and Molecular Techniques** (BMT) and DNA profiling in particular to study the utility of molecular markers in the variety registration. The BMT in its Seventh Session (2001, BMT/7/2) recommended that the current greatest need for the development of molecular techniques are in "**pre-screening**" in the process of examining distinctness.

Variety Identification using Bar Code

The variety bar code, **CSIRO ID System,** tests leaf or grain samples using a panel of DNA markers in Australia. Each marker gives a 'yes' or 'no' result. The pattern can be used to develop an individual 'bar code' for each variety. DNA markers are used to develop a variety 'bar code'. For example, testing using a panel of 14 markers gave a characteristic bar code for each of the 36 major commercial varieties for western region in Australia. Some closely related varieties can return the same bar code. This is not surprising in view of the extensive backcrossing that is employed to develop new varieties in many breeding programs. To overcome this problem, CSIRO researchers augmented the test by using a DNA fingerprinting technique that shows up even the smallest genetic difference between varieties. Results from the fingerprinting are then used to add detail to the original bar code, allowing even very closely related varieties to be successfully differentiated. Aside from its accuracy, the variety ID system was also designed to be practical. Results are displayed using a colour-based test. These results can be read by machine and software has been developed to interpret the data. This enables this system to efficiently run the high volume of tests required by the grains industry. CSIRO's wheat DNA variety testing system has been developed through Graingene (a joint venture between CSIRO Plant Industry, the Grains Research and Development Corporation, AWB Ltd and Syngenta) and licensed by Agri-food Technology. The barley variety identification system was developed in collaboration with ABB Grain Ltd.

Grain Characterization for Export

The grain industry has started feeling heat because of IPR and trips demands imposed through WTO and quality assurance pressed by importers. Therefore, efforts has been initiated to look for more reliable methods for characterization other than phenotypic descriptors for agricultural produces. So DNA fingerprinting is now being looked upon as an attractive alternative and investments in research is happening in the world to obtain an appropriate set of diagnostic DNA fingerprints technology. Few examples in Canada, **Canada Grain Commission (CGC)** is answering new and challenging quality assurance needs for exports. DNA tests are now in the research and development stage in Canada supported by a partnership of government, industry, producer groups, and universities. Very soon they will be supporting Identity Preservation (IP) systems by automated, portable tests using DNA fingerprinting so that during export, shipments will be able to be certified for a certain desired variety, for the GMO status of grain, or for a variety with a new, valuable trait. For variety identification however, there is an added complexity because of the need to quantitatively determine variety composition of commingled grain shipments typically containing ten or more varieties. Accurate detection levels as low as 1 per cent may be required. But DNA fingerprinting has even greater advantages. Regions of plant DNA are analyzed to identify differences in DNA coding sequences that uniquely distinguish one variety from another. These sequences can then serve as probes to identify the DNA of varieties commingled in grain shipments.

Other Uses of DNA Fingerprinting in Wheat Germplasm Management

Diversity Analysis

Knowledge of genetic diversity and relationship among a set of germplasm and the potential merit of the genetic diversity would be beneficial to all phases of crop improvement. Assessments of genetic diversity of the elite germplasm have been sought and used by plant breeders for numerous reasons *e.g.* genetic relationships, parent selection, germplasm management and protection among others. Bread wheats have been extensively selected by traditional breeding techniques for over a hundred years. This process has yielded modern varieties with important agronomic and quality attributes, including reduced plant height, grain hardness, good dough forming properties and resistance to a wide variety of diseases and pests. Producers in different geographical areas use distinct wheat varieties to suit their region and market; more than two hundred varieties have been released so far in India that provides an important assurance of food for the nation. But this profit-oriented breeding has led to narrowing of genetic variability in germplasm that is otherwise required to find new genes, study phylogeny and introgress useful genes from diverse accessions. Molecular markers like SSRs (Dograr *et al.*, 2000), AFLPs (Domini *et al.*, 2000) and RAPDs (Teshale *et al.*, 2003) have been extensively used to assess genetic diversity in various types of wheat like hexaploids, tetraploids and diploids along with their wild progenitor.

Phylogenetic Studies

DNA markers have been proved useful in detailed information about the genetic structure of natural populations in wheat. RAPDs were used to reclassify misclassified accessions of Triticum (Cao *et al., 1999*). Phylogenetic relationship among argentine wheat has been studied using SSR and AFLP markers (Manifesto *et al.,* 2001).

Gene Bank Management

The objective of PGR conservation is to preserve, as broad sample of the genetic variability as is scientifically and economically feasible, including currently recognized genes, traits and genotypes. Molecular genetics has an important role to play in conservation including resource acquit ion (sampling strategies and defining core sets), maintenance (monitoring trueness to type and changes in allele frequencies in populations), characterization (of newly acquired accessions when little or no data available) and utilization (aiding in pre-breeding and marker assisted selection). However, the application of molecular markers for the resolution of problems of genetic resource conservation is at infancy and requires extensive collaboration among conservation workers and molecular biologists.

Resolving IPR Disputes

DNA fingerprinting can be used to provide proof or defense against allegations of breach of IPRs. Proof of infringement in all the above cases would require a rapid and unambiguous method of identification for which the molecular methods are highly recommendable (Basmati rice quality control)

Future of Wheat Variety Testing

It is believed that the limitations of current protein-based variety identification methods can be overcome by a DNA-based approach. Every variety has a unique DNA fingerprint because of the differences in the genetic makeup of each variety. The differences are reflected in the coding sequences of the DNA. Simple sequence repeats (SSR), or microsatellites, can be used for wheat variety identification. Other DNA sequence patterns that can be used included are sequence tagged sites (STS) and Single Nucleotide Polymorphisms (SNP). SNPs have the best potential for rapid variety identification and database is now available for SNPs for wheat variety identification. However, SNP databases must be optimized for use with rapid diagnostic technology. DNA fingerprinting requires specialized equipment, chemicals and skilled technicians. Currently, DNA analysis is more expensive and takes longer than protein analysis. Variety identification involving DNA extraction and analysis can take at least one day. This will require additional dedicated resources. Therefore, to fully harness the potential of DNA markers technology, it is required to establish a network of molecular biology labs using uniform set of standardized protocols and should be harmonized with the conventional DUS testing system to identify wheat germplasm. Development of high yielding, rust resistant, good quality *durum* genotypes with wide adaptability; To assist in development of Quality lab and the analysis of all the samples of *durum* and *aestivum* material. To involve in the characterization of the released varieties under DUS testing. Collection and evaluation of durum wheat accessions for high β-carotene (10 ppm) content under the Project "Biofortification of wheat for micronutrients through conventional and molecular breeding approaches". Performance of durum entries in Co-ordinated Wheat Trial: HI 8663 from AVT I (HF-TS) was promoted for 2nd year of testing for Central and Peninsular Zone. HI 8671 from NIVT 4 was promoted to AVT I (HF) for Central and Peninsular zone and HI 8672 and HI 8673 from NIVT 4 was promoted to AVT I (HF) for Central zone and may be noted that HI 8663 and HI 8671 had wide adaptability as it has performed well across two agroclimatic zones. HI 8678 was promoted from NIVT 5B to AVT-(RF and RI) in Central Zone.

Material Generation and other Programmes

Durum Entries in Common Varietal Trials

From CVT –4, ID 514 (HI 8679), ID 51 (HI 8680), ID 516 (HI 8681), ID 55 (HI 8682), ID 56 (HI 8683) and ID 515 (HI 8684) showed better performance compared to checks over the locations and were promoted to NIVT 4. From CVT – 5B (Rainfed), ID 526 (HI 8685), ID 522 (HI 8686), ID 537 (HI 8687), ID 525 (HI 8688) and ID 521 (HI 8689) showed better performance compared to checks and were promoted to NIVT 5B. **Generation of new material:** About 235 **durum genotypes** in crossing block were used for making crosses by giving emphasis for rust resistance, grain quality, drought tolerance and yield potential. Emphasis is given while selection on characters like yield, earliness and medium early, E1 and E2 height groups, long ears, bold grains and good lustre along with good tillering and erect leaves. **National/International nurseries and trials:** Three nurseries (563 lines) were grown and analysed for yield and other parameters. 62 entries were selected based on their superiority in grain yield and other characteristics. **Performance of durum genotypes under late sown conditions:** Out of 42 durum wheat genotypes planted in December, Twenty two (22) showed significant superiority in grain yield, compared to *aestivum* check DL 788-2. Genotypes ID 58, ID 519 and ID 51 had performed better than all the other genotypes. It was also observed that durum wheat genotypes are yielding better than the *aestivum* varieties under late sown conditions with 2 or 3 irrigations (overall mean yield of 33.6 q/ha in durums, compared to 29.0 q/ha in case of bread wheat varieties). Under late sowing, the grain weight got reduced in both aestivum and durums, whereas, β–carotene and protein content showed an increasing trend due to cool conditions prevailing during grain filling. **Behaviour of quality traits in durum wheat :** Durums are having good sedimentation value and high β–carotene and protein content, compared to *aestivums*.

Grain yield showed positive relationship with grain weight and protein; negative with SDS value; and independent association with β–carotene. However, positive relationship of grain weight with SDS value, protein and β–carotene content showed that 1000 grain weight can be used as dependable parameter for selecting high yielding good quality lines in durums.

Alleviating Malnutrition through Popularization of Cultivation and Consumption of Durum Wheat

India being predominantly a rice eating country, its population suffers from malnutrition. In M.P., malnutrition is 10 per cent higher than the average malnutrition of the country. Total nutrition of wheat is much higher than that of rice; and it compares with pulses for vitamins and minerals. Of the two major wheat species under cultivation in India, ***durum*** **is nutritionally superior to** ***aestivum*** **as it contains higher protein, β-carotene (Vitamin-A) and minerals.** Hence, **an extensive campaign for popularizing the consumption of durum wheat** through sale of durum products *e.g. dalia* (porridge), semolina and flour, **launched in 2004, was further intensified during 2006, creating wide awareness for inclusion of durum preparations in daily diet.**

Differential Patterns of Resistance of Durum and Bread Wheats to Important Rust Pathotypes

While all 59 durum entries included in CVT 2006-07 were seedling resistant to leaf rust pathotype 77-5, around 27 per cent (61/229) of the *aestivum* entries showed susceptibility to this currently most prevalent leaf rust pathotype in India. In the field tests, out of the 21 durum entries (other than the checks') tested in the APR Nursery, while 20 were resistant to the stem rust pathotype 40A, currently the most prevalent stem rust pathotype in India, only 6 were resistant to the stem rust pathotype 117-6. In contrast, only one of the 144 *aestivum* entries (other than the checks') was susceptible to the pathotype 117-6, but 15 lines showed susceptibility to the pathotype 40A. These findings confirmed the earlier observations on the differential patterns of resistance to stem and leaf rust pathotypes among durum and bread wheats, and which can be utilized for protecting the entire wheat crop from both the rusts through simultaneous cultivation of the two wheat species.

Development of Protocols for Evaluation of Rust Resistance in Durum and Bread Wheats

In continuation of the work done during 2003-06, the 120 selected representative genotypes each of durum and bread wheats were seedling evaluated with additional 2 pathotypes *viz.*, 117 of stem rust and 12-5 of leaf rust. Both the pathotypes were more virulent on durums, compared to bread wheats. In fact, 12-5 has proved to be the most virulent one on durum wheat among 37 leaf rust pathotypes tested so far. These studies revealed that **leaf rust pathotypes 12-5, 12-2, 104-2, 11, and 106,** and **stem rust pathotypes 117-6, 117A, 117-1, 117-3 and 117A-1** should be used **for evaluation of rust resistance in durum wheat**; while **leaf rust pathotypes 77-6, 77-5, 77-7, 77-2 and 77-1,** and **stem rust pathotypes 40A, 40-1, 295, 11 and 117-4** need to be included **for evaluating resistance in bread wheat,** based on the relatively high virulence frequencies of these pathotypes on the two wheat species.

Identification of Sources of Combined Field Resistance to Stem and Leaf Rusts in Durum Wheat

A total of 108 durum wheat genotypes (selected from over 1100 lines evaluated at Indore during 2002-03) including released varieties, land races and indigenous as well as exotic genetic stocks, maintained their field resistance to both stem and leaf rusts (terminal rust severity up to 10S) at Indore and Mahabaleshwar during 2006-07 as well. Hence, having shown continued resistance since 2002-03, these genotypes **hold promise as stable sources of combined resistance to stem and leaf rusts for use in durum wheat improvement**.

Generation of Promising Material for Crop Improvement

A large F_2 population derived from crosses involving five rust resistant durum wheat genotypes *viz.*, AKDW 4339, B 276, CPAN 6118, Guji S', and VD 2001-14 was field planted and evaluated for rust resistance. A large number of plants from this population were selected by the durum wheat breeder combining rust resistance with desirable agronomic traits for use in crop improvement.

Identification of Diverse Genes for Resistance to Leaf Rust in Durum Wheat

Seedling tests involving F_2 and F_3 populations derived from 'S x R' crosses showed that resistance to leaf rust pathotype 12-2 was conditioned by a single dominant gene each in HG 110 and Line 1172, and by a pair of dominant genes in B 662, ED 2398-A and IWP 5019. Resistance to leaf rust pathotype 104-2 was controlled by a single dominant gene each in B 662 and IWP 5019, while two dominant genes-controlled resistance to this pathotype in ED 2398-A. The other two genotypes, HG 110 and Line 1172, being seedling susceptible to this pathotype, were not included in the study. **Allelic tests involving 'R x R' crosses showed that these genes were different from each other.** Thus, **at least eight diverse dominant genes were identified for leaf rust resistance** among five resistant durum genotypes-'B 662', 'ED 2398-A', 'HG 110', 'IWP 5019' and 'Line 1172'. These genes are **different from *Lr23*, the leaf rust resistance gene commonly postulated in Indian durum wheat genotypes**, as *Lr23* is ineffective against the pathotypes 12-2 and 104-2.

Narrow Genetic Base for Leaf Rust Resistance in Bread Wheat

Inheritance studies showed the presence of a single dominant gene in Amrita (HI 1500) and Harshita (HI 1531) for seedling resistance to leaf rust pathotype 77-5. Absence of susceptible segregants in F_2 populations of Amrita/RL 6064 (NIL carrying *Lr24*), Harshita/RL 6064, and Amrita/Harshita crosses confirmed the presence of *Lr24* in both Amrita and Harshita. It may be noted that 'Amar' (HW 2004), 'Abha' (HI 1454) and 'Vidisha' (DL 788-2), released for cultivation in central India, also carry *Lr24*. Thus, there is an urgent need to diversify the leaf rust resistance base of bread wheat through incorporation of additional *Lr* genes and popularization of durum wheat cultivation in the region, since central India serves as secondary focus of rust infection for northern plains.

Wheat Disease Survey and Surveillance

Leaf rust infection was observed in varying degrees in farmers' fields as well as in the trap nurseries. Trace infection of stem rust occurred on Agra Local and Pissi Local in the trap nursery. Rust samples were mailed to Flowerdale and

Mahabaleshwar for virulence analysis. No incidence of Karnal bunt was noticed in any of the 130 wheat grain samples collected from Indore grain market. These samples were mailed to DWR-Karnal for confirmation, and for the quality analysis toward 'classification of Indian wheat'.

Table 4.25. Pedigree of some important Single crosses, Three way crosses performed at IARI RS Indore, MP

Single Crosses	*Resistant Sources*	*Three Way Crosses*
DL 788-2/4th SAMNYT 401	CPAN-1235	DL 788-2/HI 1535//HW 2004
DL 788-2/EGPYT 5	CPAN-1676	CPAN 1676/GW 366//DL 788-2
DL 788-2/LBPY 07-7	CPAN-1796	DL 788-2/HI 1535//GW 322
DL 788-2/PHS 823	CPAN-1842	CPAN 1676/HW 2004//DL 788-2
DL 788-2/EGPYT 17	DL-788-2	
DL 788-2/HW 4013	GW-322	Varieties
HI 1500*/EGPYT 11	GW-326	NS732/HER/3/PRL/SARA//TSI/VEE#5/4/ FRET2
HI 1500*/GW 2008-127	GW-366	SOKOLL/3/PASTOR//HXL7573/2*BAU
HI 1500*/LBPY 06-12	HD-2135	WBLL4//OAX 93.24.35/WBLL1
HW 3063/MP 4010*	HD-2380	CHRZ//BOW/CROW/3/WBLL1/4/CROCK_1/ *Ae.squarrosa* (213)//PGC
HW 2045/LOK 1	HD-2402	
HW 2043/MP 4010*	HD-2501	
DL 788-2/ESWYT 141	HD-2733	
DL 788-2/LBPY 07-16	HI-784	
DL 788-2/PHS 829	HI-1434	
DL 788-2/HW 4032	HI-1479	
DL 788-2/PHS 729	HP-1761	
HW 2040/LOK 1	HS-295	
HI 1500*/ESWYT 130	HS-365	
HI 1500*/HI 1405	K-7827	
HI 1500*/LBPY 07-13	K-9107	
HI 1500*/ESWYT 141	K-9465	
HI 1500*/HW 4032	NIAW-34	
HW 3063/DL 788-2	PBW-498	
	Raj-3765	
	Raj-3777	
	UP-2473	
	VL-804	

* Varieties previously reported resistant against Ug 99.

It is a high yielding variety as is evident from its average 8 per cent (*i.e.* 4q/ ha) higher yield over HI 8498, the most widely cultivated durum variety, and an average 20 per cent (*i.e.* 9q/ha) higher yield over Lok-1, the most popular wheat

Table 4.26. Performance of BARC special collaborative trail during 2011-12

Sl. No.	*Entry*	*No. of Plants*	*Days to Heading*	*Plant Height*	*Tillers/ Plant*	*Spike Length (cm)*	*Grains per Spike*	*1000 Grain wt.*	*Yield per Plot (g)*	*Yield/Plant (g)*
1A	C 306	12	87	97.5	11.3	12	56.6	46.1	615	51.3
1	BARC-1	8	96	96	12.0	11	51.0	46.5	301	37.6
2	BARC-2	18	90	91	11.7	10	48.2	45.4	568	31.6
3	BARC-3	25	85	90	13.0	9	65.0	38.1	517	20.7
4	*C 306*	12	89	94	9.3	10	47.8	46.5	454	37.8
5	BARC-5	18	76	87	16.0	8	70.4	36.5	489	27.2
6	BARC-6	24	74	96.5	13.0	8	53.2	39.0	500	20.8
7	BARC-7	15	72	103.5	12.3	12	67.0	44.9	456	30.4
8	BARC-8	20	67	96.5	13.7	9	51.4	42.8	444	22.2
9	BARC-9	22	66	100	17.3	10	61.0	31.0	495	22.5
10	BARC-10	18	71	102.5	13.3	11	53.2	44.3	529	29.4
11	BARC-11	18	73	98	16.7	12	68.8	43.6	503	27.9
12	BARC-12	12	68	96.5	17.3	10	60.2	45.5	336	28.0
13	BARC-13	13	69	99	15.0	13	68.6	39.5	473	36.4
13A	*C 306*	15	83	101	9.7	10	49.0	45.4	558	37.2
14	BARC-14	16	66	87	17.0	10	58.8	40.3	652	40.8
15	BARC-15	10	73	93	13.0	11	60.4	35.7	521	52.1
16	BARC-16	12	73	80	13.3	12	70.4	38.6	340	28.3
17	BARC-17	18	68	77.5	16.3	10	65.0	40.1	556	30.9
18	BARC-18	20	66	92	14.0	9	56.2	40.4	612	30.6
19	BARC-19	18	71	83.5	10.0	12	64.6	43.0	598	33.2
20	BARC-20	19	80	80.5	19.0	13	55.6	47.2	541	28.5
21	BARC-21	27	71	86	19.0	13	51.2	38.4	470	17.4

Sl. No.	Entry	No. of Plants	Days to Heading	Plant Height	Tillers/ Plant	Spike Length (cm)	Grains per Spike	1000 Grain wt.	Yield per Plot (g)	Yield/Plant (g)
22	BARC-22	24	67	95	12.3	10	52.4	39.5	498	20.8
23	BARC-23	18	72	86.5	10.0	11	48.0	47.9	372	20.7
24	BARC-24	22	80	94	14.3	9	49.4	41.9	585	26.6
25	BARC-25	15	82	93	14.7	8	45.0	44.6	378	25.2
26	BARC-26	20	85	80.5	10.7	11	49.6	51.9	516	25.8
27	BARC-27	22	87	86.5	12.7	12	54.0	49.4	539	24.5
27A	*C 306*	18	89	93.5	11.7	10	54.2	47.2	447	24.8

cultivar in central India in the coordinated trials conducted during 2007-2010 at 17 location in M P. It showed resistance to all the currently prevalent and virulent pathotypes of stem and leaf rusts. It exhibited rust resistance spectrum, different from that of HI-8498, and hence, will broaden the rust resistance base, ensuring protection from rust resistance diseases to the wheat cultivation in central India. In addition, it displayed good levels of field resistance to loose smut and foot rot diseases. Its high hectolitre weight, high protein content (> 12.0 per cent), and high hardness index (82) will ensure high semolina recovery, while less cooking time and less gruel solid loss will provide good quality to pasta products. In addition, its high 1000 grain weight, good grain appearance, and very low mottling (in fact, least among all the test durum varieties) will fetch premium price for the farmers, making wheat cultivation in the state of Madhya Pradesh more profitable.

Evolving Improved Varieties of Durum Wheat

Evolving high yielding, widely adapted, "dual-quality" durum genotypes rich in protein and yellow pigment to meet the demands of consumer preference, "Nutritional Security", pasta industries and export. Incorporating diverse genes for resistance to stem and leaf rusts in newly generated material for evolving improved cultivars. Evolving early maturing genotypes combining high yield and limited irrigation requirements. Evolving durum cultivars having high tolerance to early heat and moisture stress conditions. Screening international nurseries/material under biotic and abiotic stress conditions for their further utilization. Maintenance Breeding: Production of breeder seed and nucleus seed of IARI – durum varieties required for central zone. Crossing programme for evolving high yielding, widely adapted, "dual-quality" durum genotypes rich in protein and yellow pigment will be continued. Diverse sources for resistance to stem and leaf rusts will be given importance in the crossing programme to diversify the genetic base of the durum wheat genotypes. Selection will be practiced in the breeding material to develop the improved genotypes in durum wheat. Screening station trials, IARI common varietal trials and co-ordinated trials and international nurseries/material under biotic and abiotic stress conditions for their further utilization. Production of breeder seed and nucleus seed of IARI – durum varieties released from the station. Durum wheat variety **HI 8704** completed three years of testing under Co-ordinated trials and proposed for identification during 2011 Wheat Workshop for cultivation under timely sown, high fertility, irrigated conditions of Central Zone. Durum wheat variety **HI 8713** performed superior for grain yield over checks, and was promoted to final year of testing (AVT II) for timely sown, high fertility, irrigated conditions of Central Zone and will be proposed for identification during 2012 Wheat Workshop for cultivation under timely sown, high fertility, irrigated conditions of Central Zone. **HI 8724 (NWPZ, CZ and PZ), HI 8725 (CZ and PZ), HI 8726 (CZ), HI 8727 (CZ and NWPZ), and HI 8728 (NWPZ, CZ and PZ)**, was promoted for AVT (HF) in those zones. It is evident from this that the durum wheat genotypes HI 8724, HI 8725, HI 8727, and HI 8728 showed wide adaptability, as these were promoted for testing under more than one agro-climatic Zone. **HI 8730 and HI 8731 for Central Zone**, having resistance to black and brown rusts and was promoted for AVT (RF/RI). A total of 2721 quintals of wheat breeder seed including 1103 quintals of six durum wheat varieties was produced. ***Durum* wheat variety HI 8704** was proposed for identification during the 50th All India Wheat and Barley Research Workers' Meet, New Delhi in August, 2011 for irrigated, timely sown conditions of Central Zone. It is a widely adapted genotype with average yield of 4.9 t/ha, being 1.4 per cent to 7.0 per cent higher over checks *viz.*, HI 8498, MPO 1215, Lok-1 and HI 1544. It showed high levels of seedling and field resistance to stem and leaf rusts. It exhibited resistance to the African stem rust race '*Ug99*' and its variants also during screening in Kenya. **Durum wheat genotypes under Pipeline of Evaluation:** The under listed genotypes performed superior for grain yield over checks, and were promoted to various All India Co-ordinated Advanced Trials.

Durum wheat variety **HI 8713** performed superior for grain yield over checks (approximate 6.8 per cent superiority over HI 8498 for 2 years), and was promoted to final year of testing (AVT II) for timely sown, high fertility, irrigated conditions of Central Zone and will be proposed for identification during 2012 Wheat Workshop for cultivation under timely sown, high fertility, irrigated conditions of Central Zone. **HI 8724 (NWPZ, CZ and PZ), HI 8725 (CZ and PZ), HI 8726 (CZ), HI 8727 (CZ and NWPZ), and HI 8728 (NWPZ, CZ and PZ)**, which are occupied 1st non-significant group in NIVT 4 during 2010-11 and having resistance to black and brown rusts and was promoted for AVT (HF)-I in those zones. It is evident from this that the durum wheat genotypes HI 8724, HI 8725, HI 8727, and HI 8728 showed wide adaptability, as these were promoted for testing under more than one agro-climatic Zone. **HI 8730 and HI 8731 for Central Zone**, which are occupied 1st non-significant group in NIVT 5B during 2010-11 and having resistance to black and brown rusts and was promoted for AVT (RF/RI)-I.

HI 8734 showed good performance compared to check varieties in IARI common varietal trials **(CVT 3)** during 2010-11 and promoted to NIVT 3. **HI 8735, HI 8736, HI 8737, HI 8738, HI 8739 and HI 8740** showed superiority over high yielding check variety HI 8498 in the IARI – Common Varietal Trials **(CVT-4)** during 2010-11 and promoted to NIVT 4. **HI 8741, HI 8742, HI 8743, HI 8744 and HI 8745** showed good performance compared to check varieties in IARI common varietal trials **(CVT 5B)** during 2010-11 and promoted to NIVT 5B. 81 selections were made out of 466 genotypes in the station trials which had shown superiority over the checks. Of which, 30 genotypes were promoted to IARI – Common Varietal Trials under different growing conditions. A total of 231 fresh crosses including simple and three-way combinations were attempted among different durum wheat breeding lines, which are under further evaluation. **Wheat variety proposed for identification :** High yielding durum wheat genotype HI 8704 (parentage : **HG 822/ HI 8498)** having good resistance to rusts and distinctively different from HI 8498 with good (>50 g) was proposed during 50th Wheat and Barley Research Worker's Meet, 2011.

Wheat Variety in the Final Year of Testing

Genotype **HI 8713** (parentage : **HD 4672/PDW 233) for h**igh fertility, timely sown conditions under Central Zone, will be tested for the third year, based on its superior performance in AVT I, compared to high yielding checks. 256 durum wheat genetic stocks were rejuvenated for future use. 341 crosses were made involving various donors of agronomic traits, quality, rust resistance and abiotic stress. Single plant selections (48 plants) were made from crosses which involve bread wheat for incorporation of tillering character into durum wheat with an aim to develop drought tolerant varieties in durum wheat.

Genetics and Maintenance Breeding

Maintenance and evaluation of genetic stocks with particular reference to heat tolerance, water-use efficiency, rust resistance, and hard-threshability in bread wheat, and total yellow pigment in bread wheat. Nucleus and breeder seeds production of durum and bread wheat varieties released during the last five years. **The programme is designed to provide additional support to the ongoing work on maintenance breeding at the station.** Maintenance and evaluation of genetic stocks with particular reference to heat tolerance, water-use efficiency, rust resistance, and hard-threshability in bread wheat, and total yellow pigment in durum wheat. **Maintenance Breeding**: Production of Nucleus and Breeder Seeds of latest varieties like HI 1531, HI 1544, HI 8627 (durum), HI 8663 (durum), HI 2987, HI 2932 and HI 1563. **Genetic stocks:** 256 durum wheat genetic stocks were planted in the season for rejuvenation. These genotypes are being used in crossing programme for incorporation of various yield component characters in enriching the genetic variability in bread wheat and durum breeding programme. Genotype, HI 8730 and HI 8731were found to have higher early and terminal heat tolerance, and promoted to further testing to confirm its stability for high grain yield under high temperatures. **Wheat variety proposed for identification :** High yielding durum wheat genotype HI 8704 (parentage : **HG 822/HI 8498)** having good resistance to rusts and distinctively different from HI 8498 with good (>50 g) was proposed during 50th Wheat and Barley Research Worker's Meet, 2011.

Table 4.27. Genotypes carrying the genes for specific traits

Traits	*Genotypes Carrying the Genes for Specific Traits*
Rust resistance	B 662, IWP 5019, B 146, HG 110, ED 2398 A, Line 1172, P 6046, GW 1139, HI 8591, HI 8627, HI 8663, PDW 245, CPAN 1235 and CPAN 1676
Grain quality	PDW 233, WH 896, HD 4696, HI 8596, Trinakaria, HI 8638, HI 8663, DBP 01-9, DBP01-11, V21-79, HI 8627
Earliness	DD 64, GW 1139, HI 8651, IWP 5061, HI 8498, HI 8639, HI 8691, HI 8704, HI 8703, HI 8713 and HI 8715
Drought tolerance	HD 4694, HD 4672, GS 27, HI 8627, HI 8638, A 9-30-1, Bijaga Yellow, HI 8666, HI 8722, HI 8730 and HI 8731
Heat tolerance	Raj 3777 and HD 2135
Water use efficiency	Raj 3765, HD 2962
Yield potential	DL 153-2, GW 322, HI 8591, HD 4687 (HI 8596), HD 4692, HI 8653, HD 4693, HI 8498, HD 4672, HI 8381, HI 8652, Raj 1555, MACS 3125, PDW 245, HI 8627, HI 8663, NIDW 295, HI 8691, HI 8671, HI 8703, HI 8704, HI 8713, HI 8715, HI 8724, HI 8728, HI 8730 and HI 8638

Bread Wheat Genotypes under Pipeline of Evaluation

The genotypes under listed performed superior for grain yield over checks, and were promoted to various All India Co-ordinated trials. 2103 entries were evaluated in 48 trials. HI 1572 was proposed for identification and HI-1579 was promoted to final year in NWPZ and HI 1584 in CZ. Six bread genotypes *viz.*, **HI 1586, HI 1587 and HI 1588 for NIVT 2** are being promoted from NIVTs. From altogether 357 entries from different SSTs 89 selections were made for different characters in the station trials based on field and table screening, which had shown superiority over the checks. From breeding material 2440 genotypes, from F_1 148 genotypes, from F_2 195 genotypes and from crossing block 485 entries were being maintained, evaluated and fresh crosses were being made. The important crosses were H-15351 (AKW 1071/CPAN 1235), H-15361 (CPAN 1796/HI 1484), H-15371 (DL 784-3/HD 2501), H-15381 (HS 295/GW 322), H-15391 (HD 2428/HI 1479), H-15401 (HPW 42/HS 295), H-15411 (HI 1434/DL 788-2), H-15451 (Lok-1/HI 1434) and H-15598 (H 15250/GW 326). The parents used in bread wheat hybridization programme were AKAW-1071, CPAN-1235, CPAN-1796, DL-788-2, DL 784-3, GW-326, GW-322, GW-366, HD-2428, HD-2501, HI-1434, HI-1567, HI-784, HI-1434, HI-1479, HI-1484, HPW-42, HS-295, HW-2031, HW-4042, K-9006, Lok–1, Raj-4161 and many more. A total of 148 fresh crosses including simple and three-way combinations were attempted among different wheat breeding lines, which are under further evaluation.

Identification of Hard Threshing (Non-shattering) Genotypes of Bread Wheat

Hard-threshing trait contributing to the non-shattering phenotype of the widely cultivated bread wheat variety Lok 1 has been responsible for its popularity in central and peninsular India. Hence, emphasis is being laid on the incorporation of this trait in our bread wheat breeding programme. Out of 758 germplasm accessions evaluated during 2010-11, 3 bread wheat genotypes *viz.*, A-1292, A-1306 and A-1305 exhibited hard threshing trait, and are being utilised in crop improvement.

Breeding for Heat Management

Genetic variation exists in seed germination under high temperatures. Most of the early varieties germinate early. Durum varieties take one to two days more than bread wheat for germination. Good germination under high temperatures can be achieved by ensuring an optimum supply of moisture around seed by incorporating organic manure in the soil. Gibberellic acid can induce 100 per cent germination under very high temperatures (35°C – 40°C), but make the seedlings tender through activation of phasic development. Dry sowing and irrigation will save water and ensures 100 per cent germination under high temperatures, rather than relying on *paleva* (pre-sown irrigation). Total biomass has a positive relationship with yield. Selected medium late, tall rainfed varieties like Amar (HW 2004) give good biomass under high temperatures in early October sowing condition yielding moderate quantities of good quality grain restricted only by an inherent low harvest index of tall genotypes. Medium late and semi dwarf durum variety Malav Kirti (HI 8627) combines a good biomass, harvest index and grain yield and can be very successfully grown in early October yielding 50 q/ha of quality grain. Early maturing wheat varieties like Lok-1 and HI 1418 and even durum variety HI 8498 are not suitable for early sowings because of their activated phasic development, resulting in 8 – 10 days early flowering, low biomass and low grain yield. Wheat develops two seemingly opposite strategies to complete its life cycle under abiotic stress conditions. Escaping: Most of the early maturing wheat varieties like Lok-1, HI 1418, HI 1454 and HI 8498 have the potential of modifying their life cycles according to the intensity of water/heat stress. These are the types suitable for late sown conditions circumventing "Terminal Heat". True heat tolerant genotypes: These medium late to late genotypes, prolong their vegetative phase under high temperatures to meet their vernalization requirements. Bread wheat varieties – HW 2004, HI 1531 and durum wheat variety – HI 8627 belong to this class. Grain weight: Maximum grain weight was found in early October sown conditions in all the wheat varieties, irrespective of their duration, indicating clearly that early heat has no deleterious effects whatsoever on grain development. Early sowings facilitated more number of cold days during the grain development in the month of January and February. Roots: Roots, both seminal and nodal, play an important role in meeting the challenges of abiotic stresses. Root hairs play a very important role in utilizing water vapours in the rhizosphere. Adverse effects of high temperature are clearly evident even in seedlings raised from Gibberellic acid treated seeds. Very high temperatures inhibit the growth of root hair impairing the normal growth and physiology.

Breeding for Terminal Heat

Late sown and very late sown wheat crop, usually sown after potato crop, suffers considerably from high temperatures during grain development. Only, high tillering, early maturing varieties with good biomass and grain weight yield better under these conditions. A number of early maturing rust resistant varieties like DL 788-2, HI 1454, HI 1418 and GW 173 have been released for late sown conditions. Terminal heat is a major problem in North Western Plains Zone and North Eastern Plains Zones of India, where light soils are more vulnerable to moisture stress, which aggravates the intensity of damage done by heat waves in the later half of the wheat crop. Heavy soils in larger parts of Central India, maintain a better flow and utilization of underground water by the crop. Medium late or late genotypes in both durum and *aestivum* wheat, suffered heavily under terminal heat as reflected by a heavy loss of test weight and grain quality. Thermodynamic energy flow in wheat, like any other living near equilibrium system exchanging energy and matter with the environment, works in a specific range of minimax temperature gradient. Anything beyond these limits, affect the entire physiology of the plant. Averages give only a superficial idea, and air temperatures are always much lower than the surface soil temperatures of black cotton soils, which may reach as high as 46°C in day time, affecting the growth and functioning of the nodal roots, water use efficiency and ultimately yield. It is ironical that high temperatures can largely be managed only by timely application of proper irrigation water, a resource already in limited supply. Irrigation plays a crucial role in mitigating the effects of high temperatures both in early and terminal heat conditions. IARI – Regional Station,

Indore has averted a serious crisis of wheat production by developing suitable heat tolerant wheat varieties like HI 8627, HD 4672, HI 8663, HW 2004, HI 1531 for early October sown conditions and high yielding heat escaping early maturing wheat varieties like HI 1454, DL 788-2, HI 1418 and HI 8498 for late sown conditions along with the entire package of technology. Two entirely different genotypes are required to be tailored for the management of early heat and terminal heat, although genes imparting resistance against heat shocks are required in both the types. We can achieve 50 – 60 q/ha of good quality grain even under high temperatures by sowing in early October utilizing the stored moisture to its best and supplemental irrigation depending on the variety.

Table 4.28. Terminal Heat tolerance breeding of wheat genotypes

Wheat	*Triticum aestivum* (L.)	KS89WGRC9[e]	Heat/drought stress tolerance	Sears *et al.*, 1992
		Jinghua 1	High yield	Doafen, 1986
		Florin	Doubled haploid	De Buyser *et al.*, 1987
		Anther culture 28	Powdery mildew resistance	Zhao *et al.*, 1990

Breeding for High Yielding, Early Maturing, Drought Tolerant Genotypes

With the evolution of wheat varieties HI 8498 and HI 1479, it is evident that early maturity and high yield can be combined. Such genotypes with high tolerance to drought will need limited irrigation to produce high yield, thereby saving irrigation water for other useful crops or to meet drinking water supply. Early maturity will add to water saving in addition to saving the crop from terminal drought under limited irrigation availability. Because of high degrees of tolerance to drought and high temperatures, durum wheats need less irrigation; and hence, in Central and Peninsular India, they should be promoted under cultivation on priority to save irrigation water. *Durums* need 3 to 4 irrigations to yield 50 q/ha., while, *aestivums* need 5 to 6 irrigations to give similar yield. Central India acts as secondary focus of rust infection for northern wheat belt. Because of high resistance of *durums* to leaf rust and of *aestivums* to stem rust, diversity created by the two wheat species

Table 4.29. Water use efficient wheat varieties

Bread Wheat (Sharbati)	*No. of Irrigations Required*	*Durum Wheat (Malvi/Kathia)*	*No. of Irrigations Required*
HW 2004 (Amar)	1-2	HI 8498 (Malavshakti)	3-4
HI 1500 (Amrita)	1-2	HI 8381 (Malavshree)	3-4
HI 1531 (Harshita)	1-2	HI 8663 (Poshan)	3-4
HI 1418 (Naveen Chandausi)	3-4	HD 4672 (Malav Ratna)	1-2
HI 1544 (Purna)	3-4	HI 8627 (Malav Kirti)	1–2
DL 788-2 (Vidisha)	3-4	HI 8638 (Malav Kranti)	1-2

together in Central India, will cut down the multiplication of both the rusts due to their complementary resistance, eliminating the chances of rust epidemics in the country.

Breeding for Limited Tillage after Soybean Harvest

After soybean harvest in September-October, soil is in easily cultivable condition for a week. Hence, one cross ploughing followed by planking within a week of soybean harvest is sufficient to conserve the residual moisture and make the field ready for wheat sowing. Crop should be sown at appropriate time according to availability of irrigation. Compared to talls, semidwarfs show better water use efficiency even under limited water supply. Semi-dwarf and medium early maturing HD 4672 and HI 8627 (durum wheats) and HD 2781 (aestivum wheat) are better yielders than their respective tall and medium late check varieties, which proved that dwarfness and early maturity can be combined along with drought tolerance. These semi-dwarf genotypes are more desirable under low input conditions because of their better water use efficiency. Similarly, HI 1500 being bearly 10 days early in heading and maturity to "Sujata" and "HW 2004", had outyielded both the check varieties in production and productivity, proving that early maturity and drought tolerance can be easily combined. Such genotypes will escape terminal drought at maturity and hence, will yield better because of their good grain filling. So, these varieties must be popularised in the areas of Tropical India, where limited irrigation facilities are available to improve the productivity and profitability of the wheat crop. Dry shallow sowing followed by irrigation instead of giving *paleva* (pre-sowing irrigation) ensures good germination, uniform crop stand and saves one irrigation, energy of repeated ploughing and 10 to 15 days of time. Use of balanced fertilizers (N: P: K in 4: 2: 1 ratio) by drilling three inches deep in the soil before sowing in wheat ensures fertilizer use efficiency. Applying 5- 10 tons/ ha FYM or 2.5 tons/ha poultry manure or green manuring and Zinc Sulphate application at least once in every three years along with recommended doses helps in maintaining soil health and sustain productivity in wheat-based crop rotation. In warmer areas, where two irrigations are available, use 1st as post-sowing to ensure good germination and 2nd after 40 to 45 days of 1st irrigation. Irrigating fields by making beds and channels, rather than long strips increases water use efficiency. Studies conducted at IARI – Indore indicated that by providing adequate nutrition to wheat crop, a good harvest of both wheat and soybean crops can be ensured without giving any additional nutrition to soybean crop, which will reduce the input cost in wheat-soybean system, improving profitability of both the crops. Dry sowing and clean cultivation methods will eliminate the use of weedicides too. Studies indicated that durums have high resistance to leaf rusts while aestivums to stem rusts for ex., pathotype 77-5 of leaf rust, being most virulent on *aestivum*, is the weakest pathotype on *durum* and similarly, pathotype 117-6 of stem rust, being most virulent on durums, is the weakest pathotype on *aestivum*. So, *aestivum* and *durum* varieties complement the weaknesses of each other for the control of stem and leaf rusts. Hence, equal emphasis must be given on the popularization of both *durum* and *aestivum* varieties in tropical area to minimize the chances of leaf and stem rusts epidemics in the country. Thus, popularization of durum wheat and early maturing high yielding varieties requiring less irrigation will improve **"water use efficiency"** and **"per day productivity"**. In addition, growing drought and heat tolerant varieties, eliminating pesticide use, following integrated cultivation methods in any wheat based cropping system and practicing appropriate low-cost water use efficient cultivation methods will help in improving wheat productivity and profitability in these tropical areas of the country.

Technology Developed and Commercialized

Limited input *durum* variety HD 4672 (*Malav Ratna*) and *aestivum* varieties like HW 2004 (Amar) and HI 1500 (Amrita) exhibited very high early heat tolerance and hence, can yield > 30 q/ha with just one supportive irrigation. They maintained normal vegetative phase, even if planted in 1st half of October. Such genotypes performed well under limited irrigation as well. Hence, to evolve early heat tolerant and limited irrigation varieties, evaluation of breeding material under rainfed condition appears to be essential. Another limited input *durum* variety HI 8627 (Malav Kirti), possessing very high early heat tolerance and having about 7 ppm β-carotene has been identified. Popularization of this genotype will provide "household nutritional security" to eradicate malnutrition, and help in controlling night blindness, for reducing "vitamin-A" deficiency and to promote good vision and for maintaining healthy skin, mucous membranes, teeth, skeleton and soft tissues. Malav Kirti exhibited very wide adaptability and hence, is suitable for planting under all types of cultivation conditions.

Wheat Disease Management in Central India

Economically important diseases of wheat crop in central India include leaf and stem rusts, loose smut, and Karnal bunt. **Leaf rust:** It is most common of the three wheat rust diseases as it develops under moderate temperatures (15-25°C), and can appear during December-February. Also known as brown rust, it is caused by the fungus *Puccinia triticina*. The disease is characterized by circular or slightly elliptical orange to orange-brown pustules (uredia) occurring on upper surface of the leaves and leaf sheaths. These remain usually scattered. **Stem rust:** It is favoured by warmer temperatures (25-35°C), and hence appears late in the season (February-March). Also known as black rust, it is caused by the fungus *Puccinia graminis tritici*. Presence of elongated dark reddish-brown pustules (uredia) is the characteristic symptom of the disease, which may occur on leaf sheaths, both sides of the leaves, stems and even spikes. With light infections, these pustules are usually separate and scattered, but with heavy infections, these may coalesce. **Wheat stem rust symptoms:** In case of wheat rust diseases, the uredia produce uredospores which contribute to spread of the disease by infecting healthy plants. Later, the uredia cease to produce uredospores, turn dark in colour and become telia (the resting stage of the fungus). Recurrence of rust diseases in central India is through the winter rain showers carrying uredospores blown by wind currents from Nilgiri and Pulney hills in south (focus of infection for both leaf and stem rusts), and Himalayan ranges in the north (focus of infection for leaf rust).

Table 4.30a. Yield potential and other characteristics of the superior Synthetics

Genotypes	*Yield/ Plot (g)*	*Grains/S pike*	*1000 Grain Weight (g)*	*Days to Flowering*	*Days to Maturity*
Synthetics					
Syn 02-8	314	42	37.5	76	109
Syn 02-15	318	46	33.7	80	112
Syn 02-16	324	64	37.5	70	105
Syn 02-17	310	54	37.7	70	105
Syn 02-18	301	61	36.1	71	106
Syn 02-19	316	58	35.9	71	107
Syn 02-24	305	46	39.6	70	103
Syn 02-38	306	48	38.3	72	104
Checks					
HI 8498 ©	**277**	**45**	**54.0**	**70**	**107**
Raj 3765 ©	**313**	**47**	**44.6**	**65**	**105**

During *rabi*, thirty-seven lines were evaluated along with check varieties PBW 343, Raj 1555, HI 8498 and Raj 3765. The lines showing superior performance in grain yield per plot, compared to the best aestivum check, Raj 3765 are J 2001-38, J 96-6, VA 2002-21, ABV 2002-62, ABV 2002-113, 20th SAWSN 44, 20th SAWSN –71 and 20th SAWSN –73. Lines superior for 1000 grain weight (> 45 g) are J 2001-38, VA 2002-21 and 20th SAWSN –79. During *rabi*, a number of lines were evaluated along with check varieties HI 8498, Lok-1 and WH 147. Derivatives with more than 48 grains per spike were PHT – 18, WR-HT-10, DL –335, DL – 339 and VA 2000-31; and for yield/plot more than 350 g were WR-HT-6, AKAW – 3294, AKW -65-1, AKDW -4151, AKAW -4221, AKAW – 4228, DL – 264, DL – 246, VA 99-62, VA 2002-36, VA 2002-35, VA 2000-31, VA 2000-35, PHT –17 and PHT –18. For 1000 grain weight (> 45 g), superior lines are WR-HT-8, AKDW 4151, VA 2000-35, DL –192, AKW 381, AKDW 4240. DL –246 and VA 2002-35. During *rabi*, a set of lines selected for their superior performance during *rabi* 2002-03, were evaluated along with four check varieties PBW 343, Lok-1, HI 8498 and Raj 3765. Six varieties *i.e.*, DL 337, DL 349, DL 246, VA 2000-13, VA 2000-31 and VA 2002-35 performed on par in grain yield/plot, compared to check varieties. For 1000 grain weight (> 45 g), superior lines are WR-HT-8, AKAW 3294, AKAW 4228, DL 246, VA 2002-36, VA 2002-41, VA 2002-13, VA 2002-35, VA 2000-31, VA 2000-41 and VA 2000-13. Three varieties *i.e.*, DL 246, VA 2002-35 and VA 2000-31 showed superiority over the years for 1000 grain weight and yield and hence, will be evaluated in the *aestivum* station trials for their further utilisation.

A limited input *durum* variety HI 8627 (Malav Kirti) having about 7 ppm β-carotene has been identified by the 43rd Wheat Workshop in August, 2004 for cultivation under both rainfed and limited irrigation conditions in Central India. Popularization of this genotype will provide "household nutritional security" to eradicate malnutrition, and help in controlling night blindness, for reducing "vitamin-A" deficiency and to promote good vision and for maintaining healthy skin, mucous membranes, teeth, skeleton and soft tissues. Three genotypes *i.e.*, HI 8638 (*durum*), HI 1531 (*aestivum*) and HI 8645 (*durum*) evolved for rainfed and limited irrigation conditions are in 3rd year of testing in All India Co-ordinated Wheat Trials. To check the **"Rural Migration"** and to improve **"Nutritional Security"**, early heat tolerant varieties of durum and aestivum wheats grown under low-cost technology were introduced in tribal villages of Bagali (Dewas District) for the first time in the history of "**Crop Cultivation**".

Table 4.30b. Yield potential and other characteristics of the superior genotypes

Genotypes	*Yield/ Plot (g)*	*Grains/ Spike*	*1000 Grain Weight (g)*	*Tillers/ Metre*	*Days to Flowering*	*Days to Maturity*	*Plant Height (cm)*
J 2001-38	766	62	46.3	45	77	124	99
J 96-6	695	61	37.0	55	72	121	92
VA 2002-21	850	68	46.0	58	71	122	91
ABV 2002-62	708	53	38.0	61	75	117	95
ABV 2002-113	682	48	44.1	64	74	122	94
20th SAWSN –44	750	54	37.0	68	76	120	119
20th SAWSN –71	692	49	42.2	73	72	113	95
20th SAWSN –73	699	49	44.2	72	80	124	103
Raj 3765 ©	**680**	**46**	**42.6**	**71**	**71**	**118**	**94**

Table 4.30c. Yield potential and other characteristics of the superior genotypes

Genotypes	*Yield/Plot (g)*	*Grains/ Spike*	*1000 Grain Weight (g)*	*Days to Flowering*	*Days to Maturity*
WR-HT-6	397	40	40.2	69	103
AKAW 3294	372	35	42.7	61	103
AKW 65-1	388	42	37.1	72	108
AKDW 4151	421	37	50.1	78	111
AKAW 4221	377	48	38.9	61	97
AKAW 4228	381	40	43.6	64	102
DL 264	407	46	40.7	73	109
DL 246	400	38	46.8	64	103
VA 99-62	353	37	43.4	62	102
VA 2002-36	355	41	42.4	66	104
VA 2002-35	353	41	46.4	62	103
VA 2000-31	376	49	41.9	60	104
VA 2000-35	383	42	49.9	62	101
PHT –17	378	37	34.4	79	115
PHT –18	384	56	38.5	70	101
HI 8498 ©	**303**	**41**	**53.8**	**69**	**107**
Lok-1 ©	**392**	**42**	**50.6**	**63**	**102**
WH 147 ©	**443**	45	42.4	66	**102**

Table 4.30d. Yield potential and other characteristics of the superior genotypes.

Genotypes	*Yield/Plot (g)*	*Tillers/ Metre*	*1000 Grain Weight (g)*	*Days to Flowering*	*Days to Maturity*	*Plant Height (cm)*
DL 337	1096	88	40.0	74	121	89
DL 349	1075	108	43.4	70	123	95
DL 246	1097	96	48.1	66	123	88
VA 2000-13	1131	104	45.6	69	118	91
VA 2000-31	1102	94	40.0	75	118	91
VA 2002-35	1072	84	48.7	68	117	87
Raj 3765 ©	1094	105	45.9	68	109	**98**

Table 4.31. Introduction of early heat tolerant varieties of durum and aestivum wheats in tribal villages of Bagali, Dewas, MP

Village	*Population*	*1st Year*		*2nd Year*		*3rd Year*	
		Area (ha)	*Production (q)*	*Area (ha)*	*Production (q)*	*Area (ha)*	*Production (q)*
Dangadungi	165	0	0	8	80	80	210
Seetapuri	583	0	0	8	80	12	150
Khuntkhal	572	0	0	4	40	7	80
Sivanpani	616	7	70	15	225	70	1400
Shampura	636	8	80	27	300	48	600

Table 4.32. Crosses involving durum line

Cross between Nulli and Ditelo Cytostocks and Durum Line HI 8663	F_1 were selfed or backcrossed with durum parent HI 8663
N1ATB x HI 8663	HI 8663 x N1ATB x HI 8663
N6AT68 x HI 8663	HI 8663 x N6AT68 x HI 8663
Ditl 1AL x HI 8663	HI 8663 x Ditl 1AL x HI 8663
Ditl 6AL x HI 8663	HI 8663 x Ditl 6AL x HI 8663

Table 4.33. Appropriate water use efficient wheat varieties developed for Central India

Bread Wheat Varieties	*Durum Wheat Varieties*
HW 2004 (Amar)	HD 4672 (Malav Ratna)
HI 1500 (Amrita)	HI 8381 (Malavshree)
HI 1531 (Harshita)	HI 8498 (Malavshakti)
HI 1418 (Naveen Chandousi)	HI 8627 (Malav Kirti)
HI 1479 (Swarna)	
DL 788-2 (Vidisha)	

Table 4.34. Yield potential of HI 1500 under rainfed conditions

Varieties	*Maturity*	*Average Yield (q/ha)*	*Per Day Productivity (kg/ha)*
HI 1500	ME	16.0	12.9
Checks			
Sujata	ML	14.3	11.0
HW 2004	ML	15.2	11.9

Dynamics of Doubled Haploidy in Bread Wheat Improvement

Doubled haploidy breeding as a biotechnological tool has been a favourite choice for many plant breeders because it provides with completely homozygous lines in a single generation, thereby saving considerable time, space and expenses in the breeding of new lines which can be directly released as cultivars. Besides, doubled haploidy may be extensively used for developing mapping populations required for the study of linkage maps through the use of molecular markers. These DH lines also serve as efficient breeding material for the application of marker-assisted selection. As a result of continued efforts over years, a range of doubled haploid lines derived from winter x spring wheat and triticale x wheat F_1s have been developed, which are under various stages of evaluation for productivity, nutritional grain quality and resistance to various abiotic and biotic stresses at Palampur. More potential maize genotypes which yield higher haploid embryo formation and regeneration frequency in crosses with wheat has been identified, particularly Early Composite. Also, *Imperata cylinderica* has been identified to be a potential alternative to maize as the pollen source foe haploid production. However, to make doubled haploidy breeding more successful, identification of still more efficient/alternative pollen source(s) for wide hybridization (chromosome elimination technique), would be desirable and made use of for the purpose. Increasing the efficiency of colchicine treatment for the doubling of chromosomes in haploids will further enhance the efficiency of the technique. Marker-assisted selection using RFLP- linked QTLs in the potential F_2 populations for achieving stress tolerance can be effectively utilized by fixing the selected material in the form of DH lines.

This has been possible with the introduction of the dwarf wheat genotypes carrying the dwarfing genes *viz., Rht-B*1b and *Rht-D1b* from Norin 10-Brevor 14 background (Borner *et al.,* 1996) and the genes for photoinsensitivity from the Mexican spring wheats. This led to an overall genetic upgradation of wheat, leading the nation to reach the status of self-sufficiency. However, at present, the wheat production has almost reached a plateau and another breakthrough is required in order to meet the ever-increasing food demand of the country. Wheat is generally grown under diverse conditions in different regions of the country. Drought, cold and edaphic stresses and susceptibility to various diseases are the major constraints, which drastically reduce the production of wheat in several areas. Therefore, the breeding objectives must essentially comprise the development of high-yielding varieties with superior nutritional grain quality (high protein, lysine and β-carotene content) and resistance to abiotic (drought, cold, soil salinity and acidity) and biotic (yellow rust, brown rust, loose smut, powdery mildew, Karnal bunt and hill bunt) stresses. For this purpose, sources within and outside the species carrying genes for desirable quality, physiology and tolerance/resistance to the various stresses have to be utilized. The success of breeding programmes depends mainly upon careful selection of parents in order to isolate potential recombinants from the segregating populations. The time, space and labour requirements for raising the segregating generations of crosses render such important breeding programmes very costly. However, a breeding procedure is available in wheat for the development of completely homozygous lines from the F_1 generation itself in a single step and is commonly referred to as 'doubled haploidy breeding'. The procedure comprises use of *in vitro* haploid production techniques such as anther culture (Wang *et al.*, 1973) or intergeneric hybridization with *Hordeum bulbosum* (effective in wheat genotypes which are non-carrier of *Kr* genes – *Bulbosum* technique), *Zea mays* L. (Laurie and Bennett, 1987) and *Imperata cylinderica* (Pratap *et al.*, 2005) for the development of haploid plantlets (plants with gametic chromosome number) and, subsequently, doubling the chromosome number of haploids through colchicine treatment to obtain completely homozygous diploid plants.

Protocols for Haploidy Breeding in Wheat

The procedure followed in the anther culture experiments is described as under:

Determination of Pollen Development Stage

The developmental stage of microspores in the anthers is determined by squashing the anther in the drop of acetocarmine stain and analysing under microscope. The growth stage of anther donor spikes correlated with the development stage of the pollen, which should be at mid-to late uninucleate stage of development.

Pretreatment of Anther Donor Spikes

Prior to anther culture, the spikes are subjected to chilling treatment at 4°C for 48 hours, so as to stimulate the microspores to undergo somatic division.

Callus Induction Medium

Potato II medium (Chuang *et al.*, 1978) is the standard anther culture medium which is to be supplemented with 0.5 mg/l glutamine (de Buyser and Henry, 1979) and 6 g/l agarose gelling agent.

Anther Isolation and Culture

After chilling treatment, the spikes are sterilized by cutting off the boot leaf and dipping the spike along with the leaf sheath in 70 per cent alcohol for 10 seconds. Thereafter, the outer leaf sheath is removed using sterile forceps under aseptic (laminar air flow) conditions. One-third of the upper and lower parts of the spike are cut off and the middle part is sterilized with 0.1 per cent mercuric chloride solution for 2-3 minutes followed by three washings in sterile distilled water. The anthers are excised from spikes by opening the florets with the help of two sterile forceps and detaching the anthers individually. The anthers are cultured on Potato II callus induction medium.

Incubation Conditions

The cultured anthers are incubated in plant growth chamber at 28±1°C temperature at 80 per cent relative humidity in dark for the interval of 8 days. Then, they are transferred to plant growth room where a temperature regime of 25±1°C, 80 per cent relative humidity and alternate photoperiod of 10h light/14 h dark are maintained.

Regeneration

The developed calli are to be sub-cultured on the plant regeneration medium. Regeneration medium supplemented with 0.5 mg/l of NAA (Chuang *et al.*, 1978) is used for achieving callus differentiation.

Incubation Conditions for Regeneration

The calli cultured on the regeneration medium are kept at 25±1°C temperature, 80 per cent relative humidity and alternate photoperiod of 10h light/14h dark. Shoot and root initiation are observed in the totipotent calli, 15-20 days of culture. Some of the calli which fail to regenerate develop chlorophyll pigmentation or rhizogenesis, while still others turn necrotic. There is also incidence of albinism among the regenerating calli.

Rooting of the Regenerants

For proper rooting the differentiated calli are transferred on the liquid rooting medium comprising half the strength of MS salts (Murashige and Skoog, 1962), 1 mg/l each of NAA (Naphthalene acetic acid) and IBA (Indole butyric acid) and devoid of sucrose and agar. The sub-cultures are maintained at the same incubation conditions as required for regeneration.

Wheat x Maize Hybridization Technique

The generalized procedure of wide hybridization experiments in wheat F_1s is described as under:

Synchronization of Flowering Wheat F_1s and Maize

Different maize genotypes are raised under polyhouse conditions during *rabi* season on different dates so as to coincide their flowering with that of wheat F_1s during the month of March and April.

Wide Hybridization Procedure

Emasculation is done three days prior to anthesis by removing the anthers manually with the help of forceps without cutting lemma and palea in the wheat florets. Fresh pollen from the maize plant is collected in petri plates and applied gently onto the feathery stigma of the F_1 hybrids with a fine camel hair brush. The pollinated spikes are immediately covered with butter paper bags and tagged.

In vivo Hormone Application

To stimulate seed and embryo formation in wheat x maize crosses, 2,4-D solution of 100 ppm concentration (Matzk and Mahn, 1994) is injected (after 24 hours of pollination) at the base of uppermost internode of the pollinated spikes.

Embryo Rescue Medium

The Murashige and Skoog (1962) medium is used for the rescue of haploid embryos. This is supplemented with kinetin, glutamine, certain essential amino acids such as arginine, cysteine and leusine and sodified with agar (Bains *et al.*, 1998).

Embryo Rescue Procedure and Incubation Conditions

The crossed spikes are harvested from the tiller base after 18-20 days of pollination. The embryo-carrying seeds (lacking endosperm tissue) are identified by placing the seeds in a petriplate under a light source and viewed from below (Bains *et al.*, 1998). The embryos settle at the bottom and become visible. The embryos are excised under strict aseptic conditions and cultured on the nutrient medium. The cultures are given cold treatment at 4°C temperature in dark for the initial 24 hours. Thereafter, they are incubated in dark 20±2°C for about a week, till roots and shoots start developing. The regenerants are shifted to 10h day length regime at the same temperature for further expansion.

Transfer of Haploids to Rooting Medium

The haploid regenerants are transferred to liquid rooting medium (½ strength MS salts, 1mg/l each of NAA and IBA, devoid of sucrose and agar). The sub-cultures are maintained at the same incubation conditions as required for regeneration.

Advantages of Doubled Haploidy Breeding

Doubled haploidy breeding offers the following unique advantages to breeders with respect to savings in time, space and labour by development of 100 per cent homozygous genotypes in a single step; no need to grow segregating generations; efficient selection possible on the basis of gametic frequency, instead of zygotic frequency, thereby requiring smaller population size; easy *in vitro* screening for the complex traits like drought tolerance during the culture process; creation of gametoclonal variation and use in genetic and mutation experiments as the hemizygous constitution of gametes allows direct expression of dominant and recessive mutations among haploid regenerants, while the deleterious mutations are not recovered. In order to harness quick and desirable results from winter x spring wheat and triticale x wheat hybridization programmes, as well as utilize the aforesaid advantages for wheat improvement, the potential haploidy breeding approach is underway following androgenesis (anther/microspore culture) and/or wheat x maize system(s) so as to isolate potential and completely homozygous recombinants with increased selection efficiency and precision.

Protocol for Hardening and Chromosome Doubling

Hardening of Haploid Plantlets

Well pulverized soil and FYM are thoroughly mixed in equal proportions and double sterilized. The haploids are transferred from the liquid rooting medium to the soil in small pots (10cm diameter). High soil moisture as well as relative humidity conditions are maintained for 10-15 days to allow the hardening of the plants.

Chromosome Doubling

The haploid plantlets are treated with colchicine at 3- to 5-tiller stage according to the method given by Inagaki (1985) with slight modifications. The roots of the haploids are cut off leaving a zone of 2-3cm, the crown of the haploid is submerged in colchicine solution supplemented with dimethyl sulphoxide at 20°C for 5 hours, the treated plants are transferred in soil and maintained up to maturity.

Genetics of Anther Culture and Wheat x Maize System

The inheritance studies revealed preponderance of non-additive gene action for the component traits of anther culture (calli induction, regeneration and green haploid plantlet formation) (Chaudhary *et al.*, 2003) as well as wheat x maize system (seed formation, embryo formation and regeneration). Further, in case of wheat x maize system, the maize genotypes exhibited a significant influence on embryo formation and regeneration of haploids (Chaudhary *et al.*, 2002), for embryo formation frequency in wheat x maize system, high heritability estimates (broad sense) have been obtained. Of the maize genotypes evaluated through a line (wheat genotypes) x tester (maize genotypes) analysis, Early Composite gave much higher GCA effect for embryo formation following wheat x maize hybridization. Bitsch *et al.* (1998), Verma *et al.* (1999) and Zhang *et al.* (1996) also reported differential effects of maize genotypes studied.

Potential Maize Genotypes

Of the maize genotypes tested at Palampur, Early Composite, Mudgram Local and L118 have been identified to be more efficient ones for haploid embryo formation and regeneration frequencies in crosses with bread wheat.

Relative Efficiency of Anther Culture and Wheat x Maize System for Haploid Induction

A comparison of the efficiency of the anther culture technique and the wheat x maize system at Palampur and elsewhere revealed high efficiency of wheat x maize system with respect to embryo formation, regeneration and doubled haploid production and the anther culture technique was recorded to be less efficient due to constraints like low calli induction, high genotypic specificity and incidence of albinism (Sadasivaiah *et al.*, 1999 Singh *et al.*, 2001, Chaudhary, 2003, and Singh *et al.*, 2004). Also, in case of the triticale x wheat hybrids, doubled haploid lines have been successfully obtained through androgenesis (Sethi, 2000, Sethi and Chaudhary, 2001) as well as maize-mediated system. At the same time, the limitations/constraints of doubled haploidy breeding, mainly the following ones, cannot be overlooked *viz.*, technological constraints and inadequate DH population.

Heterobeltiosis for Grain Yield and its Components in *Triticum durum*

Triticum durum is the second most important wheat species in India, mainly grown in Central India and Punjab. In a self-pollinated crop like wheat, the scope for utilization of heterosis depends mainly upon the direction and magnitude of heterosis. Estimation and exploitation of heterosis over better parent (heterobeltiosis) may be useful in identifying true heterotic cross combinations, but these crosses can be of immense practical value if they show superiority over the standard or the best varieties of the area. The present study aims at estimating the magnitude of heterosis over released standard varieties of durum wheat apart from that over better parent in grain yield and nine other characters. Ten genotypes of durum wheat *viz.*, HD 4694, HI 8498, HI 8381, HI 8596, HD 4672, HD 4692, HI 8591, HD 4695, HI 8651 and HI 8653 as lines were crossed with four male parents *viz.*, Bijaga Red, HG 110, IWP 5019 and MPO 215, which were resistant to black rust to develop forty F_1 crosses using line x tester mating design during rabi 2000 – 01. The fourteen parents and 40 F_1's were grown in a randomized block design with three replications under normal sown irrigated conditions during rabi 2001-02 at Indian Agricultural Research Station – Regional Station, Indore. Plot size comprised of a single row of 2m length spaced 30 cm apart. Seeds were dibbled 10 cm apart within a row. The recommended package of practices was followed to raise a good crop. Observations were recorded on five randomly selected competitive plants in each entry and in each replication for days to 50 per cent flowering, plant height, number of tillers/plant, grain number/spike, grain yield/plant, 1000 grain weight, protein content (per cent), sedimentation value (ml), β-carotene (ppm) and harvest index. Heterosis over better parent (BP) and standard varieties (SV) were calculated as per the standard procedures. HI 8498 (Malavshakti), HI 8381 (Malavshree) and HD 4672 (Malav Ratna) were kept as standard varieties of durum wheat for the present investigation. The analysis of variance showed significant estimates of mean squares for all the ten characters indicating sufficient diversity among the genotypes. The range of mean performance and heterosis, best parent and hybrids, identified on the basis of per se performance and heterobeltiosis and number of crosses showing heterobeltiosis were presented. A lot of range for per se performance of parent, hybrids and heterosis per cent over better parent is noticed for all the characters. Similar observations were observed by Sharma and Maloo (1986) and Khan and Sher (1999). The cross HI 8591 x HG 110 showed higher potential for yield based on both per se performance and heterobeltiosis. Out of 40 crosses studied, sixteen most promising combinations were identified and their heterotic expression for different characters analysed. The crosses showing heterosis for grain yield per plant were not heterotic for all the characters. This implied that heterosis in a complex character like yield can be registered by single or several characters (Gautam and Jain, 1985). Grain number per spike and harvest index is the two most important characters contributing to yield heterosis. It was also observed that high yielding purelines did not produce high heterotic response for grain yield. Similar results were also reported by Darell and Charles (1970).

Of the sixteen promising crosses, five exhibited positive heterosis for grain number/spike and harvest index along with grain yield. The cross HD 4694 x Bijaga

Red showed the highest heterobeltiosis for early maturity, but HI 8651 x HG 110 manifested the highest better parent heterosis for number of tillers per plant. HI 8591 x HG 110 showed highest better parent heterosis for two characters like grain number per spike and harvest index along with grain yield per plant. Majority of the high heterobeltiosis crosses involve high x low or low x high performing (gca) parents and a few high x high and low x low combinations. Whereas, majority of the high heterotic crosses over standard varieties involved high x high or low x high cross combinations. Similar results were reported by Gupta and Ahmed (1999) in macaroni wheat. Out of forty crosses investigated, four crosses showed superiority over HD 4672 (Malav Ratna), nine over HI 8381 (Malavshree) and eighteen over HI 8498 (Malavshakti). In case of high x high crosses as observed for HI 8591 x HG 110 and HI 8596 x HG 110, there are possibilities of complementary epistatic effects acting in the direction of additive effects of the good performer that can be exploited for developing high yielding purelines through progeny selection. The crosses of high x low or low x high group with the expression of positive heterotic effects as observed for HI 8591 x MPO 215, HI 8596 x MPO 215, HI 8381 x HG 110, HI 8653 x Bijaga Red *etc.*, may be due to the dominant x recessive interaction showing considerable additive genetic variance, expected to produce desirable segregates. In case of low x low crosses with high heterosis as recorded by HI 8498 x MPO 215, HI 8381 x MPO 215, non-additive type of variation can be exploited by multiple crosses followed by intermating among desired segregants.

The five top ranking cross combinations selected on the basis of *per se* performance and heterosis for yield and quality characters for durum wheat *i.e.*, HI 8591 x HG 110, HD 4694 x Bijaga Red, HI 8591 x MPO 215, HI 8596 x HG 110 and HD 4692 x Bijaga Red for grain yield/plant, HI 8651 x MPO 215, HI 8381 x HG 110, HI 8651 x Bijaga Red, HD 4672 x IWP 5019 and HI 8696 x IWP 5019 for 1000 grain weight, HD 4694 x IWP 5019, HD 4694 x HG 110, HI 8596 x Bijaga Red, HI 8596 x IWP 5019 and HD 4695 x HG 110 for protein content, HI 8653 x IWP 5019, HD 4692 x IWP 5019, HD 4692 x HG 110, HI 8596 x IWP 5019 and HI 8381 x IWP 5019 for sedimentation value, HD 4694 x IWP 5019, HD 4695 x MPO 215, HD 4672 x MPO 215, HI 8596 x MPO 215 and HD 4692 x MPO 215 for β-carotene, which can be exploited to select desired segregants for those traits in later generations. Standard heterosis for grain yield over the released standard varieties *i.e.*, HD 4672, HI 8381 and HI 8498 in the present study was not of appreciable magnitude being 18.75, 24.81 and 37.90 percent respectively, for the cross-combination HI 8591 x HG 110. It appears, therefore, that production of hybrid wheat on commercial basis is limited by the lack of appreciable magnitude of standard heterosis. Similar findings were observed by Atale and Vitkara (1990). Although the ranking of cross combinations based on per se performance and value of heterosis varied, the best hybrid combinations for yield were HI 8591 x HG 110, HI 8591 x MPO 215, HI 8596 x HG 110 and HD 4692 x Bijaga Red which showed superiority based on per se performance, heterotic over the better parent and standard varieties. Exploitation of these cross combinations would be highly rewarding for selecting good segregants in the subsequent generations to develop high yielding pure lines through progeny selection.

Table 4.35. Range, heterosis, best parent and hybrids for grain yield and its components in durum wheat

Characters	*Range*		*Heterosis Per cent Pver Better Parent*	*Best Parent*	*Best Hybrids Based on*	
	Parents	*Hybrids*			*Performance*	*BP Heterosis*
Days to 50 per cent flowering	61 - 83	62 - 82	-4.17 to 25.27	IWP 5019	HD 4672 x IWP 5019	HD 4672 x Bijaga Red
Plant height (cm)	73.6 - 107.6	79.8 - 110.6	-0.41 to 31.81	IWP 5019	HI 8653 x IWP 5019	-
No. of tillers/plant	4.1 - 7.9	4.5 - 6.5	-42.37 to 12.0	HD 4694	HI 8596 x Bijaga Red	HI 8651 x HG 110
Grain number/spike	35.7 - 55.1	33.3 - 57.7	-39.66 to 18.37	MPO 215	HI 8591 x MPO 215	HI 8591 x HG 110
Grain yield/plant (g)	7.8 - 16.2	8.0 - 17.1	-34.88 to 33.28	HD 4692	HI 8591 x HG 110	HI 8591 x HG 110
1000 grain weight (g)	44.4 - 59.4	48.2 - 60.4	-8.37 to 23.67	HG 110	HI 8651 x MPO 215	HI 8653 x Bijaga Red
Protein content (per cent)	12.9 - 14.2	12.5 - 14.4	-11.38 to 3.47	IWP 5019 HI 8653	HD 4694 x IWP 5019 HD 4694 x HG 110 HI 8596 x Bijaga Red	HD 4692 x HG 110
Sedimentation value (ml)	19.3 - 49.0	20.7 - 42.3	-36.05 to 9.47	HI 8653	HI 8653 x IWP 5019	HI 8498 x HG 110
β-carotene (ppm)	2.32 - 6.93	2.65 - 6.04	-44.01 to 14.60	HD 4694	HD 4694 x IWP 5019	HD 4692 x MPO 215
Harvest index (per cent)	0.35 - 0.48	0.27 - 0.51	-36.43 to 21.05	HI 8596	HD 4692 x IWP 5019	HI 8591 x Bijaga Red

Table 4.36. ***Per se*** **performance and heterobeltosis of promising crosses for grain yield in** ***Triticum durum***

	Hybrids	*Grain Yield/Plant (g)*		*1000-Grain Wt (g)*		*Protein (Per cent)*		*Sedimentation Value (ml)*		*β-carotene (ppm)*		*Harvest Index (Per cent)*	
		P	*H*	*P*	*H*	*P*	*H*	*P*	*H*	*P*	*H*	*P*	*H*
LxH	HD 4694 x BIJ red	15.7	8.39**	54.0	16.73**	13.4	-4.1**	23.3	-6.67**	4.70	-32.18**	0.48	20.17**
HxL	HI 8591 x MPO 215	15.4	19.64**	51.8	-4.92**	13.3	-2.9**	24.7	-10.84**	4.14	-11.16**	0.45	-15.52**
HxH	HI 8591 x BIJ red	13.2	2.49*	57.9	6.22**	13.3	-4.7**	22.3	-10.67**	3.30	-29.26**	0.46	21.05**
HxH	HJ 8591 x HG 110	17.1	33.28**	58.2	-2.00	13.2	0.28	23.7	-21.98**	3.94	-15.38**	0.45	10.00**
LxH	HI 8653 x BIJ red	14.5	30.96**	56.1	23.67**	13.6	-3.7**	31.3	-36.05	2.73	-51.77**	0.47	10.24**
HxH	HI 8596 x HG 110	15.2	2.85*	57.8	-2.72	13.9	-2.6**	30.0	-1.10	4.34	-26.25**	0.46	-3.47**
HxH	HD 4672 x HG 110	14.8	2.94	57.5	-3.26	13.6	2.7**	32.0	5.49**	4.05	-19.17**	0.44	-1.48**
LxL	HI 8498 x MPO 215	13.1	5.50**	55.7	-4.85**	13.4	-2.2**	29.3	-7.37**	4.83	2.91**	0.43	-3.76**
LxH	HI 8498 x BIJ red	13.6	10.08**	58.7	0.23	13.5	-3.3**	33.0	4.21**	2.88	-38.52**	0.49	9.77**
LxH	HI 8498 x HG 110	13.4	8.06**	58.1	-2.09	13.4	0.63**	34.7	9.47**	4.56	-2.77**	0.43	-4.48**
LxL	HI 8381 x MPO 215	14.2	3.25*	53.1	4.80**	13.2	-3.4**	29.7	7.23**	3.50	-30.26**	0.43	2.96**
LxH	HJ 8381 x HG 110	14.7	7.21**	59.5	0.13	13.1	-1.0**	27.7	-8.79**	2.90	-42.34**	0.47	4.44**
LxH	HD 4695 x BIJ red	13.4	20.63**	50.4	-1.02	13.9	-0.29	34.3	-14.88**	4.44	-21.68**	0.45	5.43**
LxH	HD 4695 x HG 110	13.2	20.32**	57.9	-2.47	13.9	1.96**	36.3	-9.92**	4.94	-12.98**	0.44	-2.24**
LxH	HI 8651 x BIJ red	13.1	7.40**	59.4	4.87**	13.9	0.07	28.3	-27.35**	3.23	-22.40**	0.41	1.65**
LxH	HI 8651 x HG 110	14.0	15.18**	57.3	-3.59	13.4	0.05	3.33	-14.53**	3.77	-9.60**	0.42	-5.22**
	SEm (h) ±		1.31		1.87		0.18		0.70		0.26		0.039

Table 4.37. Crosses showed superiority for grain yield per plant over the standard released varieties

Standard Heterosis Over the Released Standard Checks								
Superior Over HI 8498 (Malavshakti)			*Superior Over HI 8381 (Malavshree)*			*Superior Over HD 4672 (Malav Ratna)*		
Cross		*Per cent Superior*	*Cross*		*Per cent Superior*	*Cross*		*Per cent Superior*
HI 8591 x HG 110	H x H	37.90	HI 8591 x HG 110	H x H	18.75	HI 8591 x HG 110	H x H	24.81
HI 8591 x MPO 215	H x L	24.20	HI 8591 x MPO 215	H x L	6.94	HI 8591 x MPO 215	H x L	12.41
HI 8596 x HG 110	H x H	22.60	HI 8596 x HG 110	H x H	5.55	HI 8596 x HG 110	H x H	10.95
HD 4692 x Bijaga Red	H x H	20.96	HD 4692 x Bijaga Red	H x H	4.16	HD 4692 x Bijaga Red	H x H	9.49
HD 4672 x HG 110	H x H	19.35				HD 4672 x HG 110	H x H	8.03
HI 8381 x HG 110	L x H	18.55				HI 8381 x HG 110	L x H	7.30
HI 8596 x Bijaga Red	H x H	17.74				HI 8596 x Bijaga Red	H x H	6.57
HI 8653 x Bijaga Red	L x H	16.94				HI 8653 x Bijaga Red	L x H	5.84
HI 8596 x MPO 215	H x L	16.94				HI 8596 x MPO 215	H x L	5.84
HI 8381 x MPO 215	L x L	14.52						
HI 8651 x HG 110	L x H	12.90						
HD 4692 x HG 110	H x H	9.68						
HI 8498 x Bijaga Red	L x H	9.68						
HD 4672 x Bijaga Red	H x H	8.87						
HD 4692 x MPO 215	H x L	8.06						
HI 8498 x HG 110	L x H	8.06						
HD 4695 x Bijaga Red	L x H	8.06						
HI 8591 x Bijaga Red	H x H	6.45						

Table 4.38. Five top ranking cross combinations selected in basis of per se performance and heterosis for yield and quality traits of wheat (*Triticum durum*)

Characters		*Crosses*	*F_1 Mean*	*Heterosis*
Grain yield per plant (g)	1	HI 8591 x HG 110	17.1	33.28**
	2	HD 4694 x BIJ red	15.7	8.39**
	3	HI 8591 x MPO 215	15.4	19.64**
	4	HI 8596 x HG 110	15.2	2.85*
	5	HD 4692 x BIJ red	15.0	7.62**
1000-grain weight (g)	1	HI 8651 x MPO 215	60.4	6.53**
	2	HI 8381 x HG 110	59.5	0.13
	3	HI 8651 x BIJ red	59.4	4.87**
	4	HD 4672 x IWP 5019	59.2	11.58**
	5	HI 8596 x IWP 5019	59.2	14.73**
Protein content (per cent)	1	HD 4694 x IWP 5019	14.4	1.77**
	2	HD 4694 x HG 110	14.4	3.22**
	3	HI 8596 x BIJ red	14.4	1.36**
	4	HI 8596 x IWP 5019	14.2	0.09
	5	HD 4695 x HG 110	13.9	1.96**
Sedimentation value (ml)	1	HI 8653 x IWP 5019	42.3	13.61**
	2	HD 4692 x IWP 5019	40.0	6.25**
	3	HD 4692 x HG 110	39.7	7.03**
	4	HI 8596 x IWP 5019	38.7	7.41**
	5	HI 8381 x IWP 5019	38.0	5.56**
Beta carotene (ppm)	1	HD 4694 x IWP 5019	6.04	12.89**
	2	HD 4695 x MPO 215	5.70	0.47*
	3	HD 4672 x MPO 215	5.35	6.86**
	4	HI 8596 x MPO 215	5.33	9.30**
	5	HD 4692 x MPO 215	4.87	3.03**

Hierarchical Classification of Allelic Variation of Gliadins in Durum Wheat

Durum wheat genotypes were analysed for protein (gliadin) polymorphism to find out the existing genetic diversity, and to assess its utility for improvement in grain yield along with quality traits. Six different *Gli-B*$_1$ alleles were found in land races, rust resistance sources and old released varieties, while two in recently released and advance lines. Most of the recently released varieties and advance lines showed γ-45/*Gli-1* alleles, which is the best type for pasta making quality, remaining showed γ-42/*Gli-1* allele, which are not good for pasta making. It is advisable to select for γ-45/*Gli-1* as a bio-chemical marker in the future breeding programmes. The rust resistance sources do not possess γ-45/*Gli-1* alleles, so these lines can be used as donors to introduce disease resistance in the good quality recently released varieties, which are containing γ-45/*Gli-1* alleles. From hierarchical analysis, it was found that landraces, released varieties and rust resistance sources are genetically distinct. The presence of new γ-gliadin patterns are interesting in rust resistance sources and need to be investigated for their role in pasta making as well as overall technological quality of durum wheat. Wheat is one of the world's most important food crops, which provides over 20 per cent of the calories and protein in human nutrition. In India, about five per cent of the total wheat area is under durum wheat (*Triticum turgidum* var. *durum*) cultivation and approximately 2.5 million tons of durum wheat is produced annually. Durum wheat is unique in that it is generally considered the hardest of all wheats and has high protein content and especially suited for production of pasta products with high yellow pigment content required for attractive appearance of pasta in dry and cooked form. With changing consumer demands, improvement in processing quality of durum wheat, (mainly gliadin and gluten profiles) is receiving attention. The gluten complex is composed of two main groups of proteins, gliadins and glutenins constituting about 80-85 per cent of the total grain proteins. The gliadins are responsible for the dough's cohesiveness and along with strong

Table 4.39. Materials being tested at Wellington/Mahabaleshwar/ Indore, 2010 for Leaf rust and Stem rust

Bulk No: HI 1418, HI 1479, HI 1500, HI 1531, HI 1554, HW 2004, HI 1454, HI 8663, Bulk 33, Bulk 39, Bulk 42, Bulk 46, Bulk 49, Bulk 55, Bulk 62, Bulk 63, Bulk 65, Bulk 76, Bulk 77, Bulk 78, Bulk 79, Bulk 83, Bulk 84, Bulk 89, Bulk 97, Bulk 99, Bulk 101, Bulk 112, Bulk 127, Bulk 141, Bulk 142, Bulk 143, Bulk 145, Bulk 149, Bulk 170, Bulk 172, Bulk 177, Bulk 184, Bulk 186, Bulk 192, Bulk 193, Bulk 202, Bulk 206, Bulk 207, Bulk 213, Bulk 214, Bulk 215, Bulk 220, Bulk 226, Bulk 228, Bulk 229, Bulk 230, Bulk 233, Bulk 235, Bulk 237, Bulk 238, Bulk 240, Bulk 242, Bulk 243, Bulk 245, Bulk 257, Bulk 259, Bulk 272, Bulk 294, Bulk 296, Bulk 301, Bulk 305, Bulk 314, Bulk 322, Bulk 324, Bulk 340, Bulk 341, Bulk 342, Bulk 343, Bulk 350, Bulk 361, Bulk 366, Bulk 375, Bulk 376, Bulk 378, Bulk 386, Bulk 388, Bulk 398, Bulk 412, Bulk 415, Bulk 419, Bulk 426, Bulk 431, Bulk 433, Bulk 458, Bulk 462, Bulk 463, Bulk 466, Bulk 467, Bulk 469, Bulk 475, Bulk 476, Bulk 488, Bulk 489, Bulk 491, Bulk 493, Bulk 496, Bulk 502, Bulk 504, Bulk 505, Bulk 506, Bulk 507, Bulk 510, Bulk 511, Bulk 516, Bulk 517, Bulk 518, Bulk 520, Bulk 521, Bulk 522, Bulk 523, Bulk 527, Bulk 542, Bulk 544, Bulk 545, Bulk 546, Bulk 548, Bulk 552, Bulk 553, Bulk 558, Bulk 559, Bulk 564, Bulk 565, Bulk 573, Bulk 574, Bulk 578, Bulk 579, Bulk 583, Bulk 584, Bulk 587, Bulk 602, Bulk 608, Bulk 609, Bulk 613, Bulk 615, Bulk 630, Bulk 632, Bulk 634, Bulk 635, Bulk 641, Bulk 644, Bulk 646, Bulk 647, Bulk 649, Bulk 651, Bulk 656, Bulk 670, Bulk 672, Bulk 674, Bulk 677, Bulk 681, Bulk 686, Bulk 691, Bulk 694, Bulk 695, Bulk 699, Bulk 705, Bulk 706, Bulk 716, Bulk 717, Bulk 720, Bulk 725, Bulk 726, Bulk 729, Bulk 733, Bulk 734, Bulk 739, Bulk 741, Bulk 742, Bulk 743, Bulk 745, Bulk 747, Bulk 749, Bulk 759, Bulk 760, Bulk 763, Bulk 764, Bulk 767, Bulk 771, Bulk 778, Bulk 783, Bulk 790, Bulk 791, Bulk 792, Bulk 795, Bulk 796, Bulk 798, Bulk 803, Bulk 808, Bulk 810, Bulk 812, Bulk 815, Bulk 820, Bulk 822, Bulk 824, Bulk 827, Bulk 833, Bulk 834, Bulk 836, Bulk 838, Bulk 839, Bulk 841, Bulk 846, Bulk 847, Bulk 849, Bulk 851, Bulk 852, Bulk 853, Bulk 855, Bulk 859, Bulk 860, Bulk 862, Bulk 865, Bulk 868, Bulk 871, Bulk 873, Bulk 877, Bulk 878, Bulk 885, Bulk 887, Bulk 889, Bulk 890, Bulk 895, Bulk 908, Bulk 909, Bulk 917, Bulk 920, Bulk 928, Bulk 936, Bulk 937, Bulk 938, Bulk 939, Bulk 940, Bulk 947, Bulk 949, Bulk 950, Bulk 951, Bulk 952, Bulk 955, Bulk 956, Bulk 960, Bulk 962, Bulk 966, Bulk 968, Bulk 970, Bulk 972, Bulk 975, Bulk 983, Bulk 984, Bulk 985, Bulk 986, Bulk 989, Bulk 990, Bulk 994, Bulk 995, Bulk 1011, Bulk 1012, Bulk 1017, Bulk 1020, Bulk 1021, Bulk 1062, Bulk 1063, Bulk 1073, Bulk 1074, Bulk 1078, Bulk 1080, Bulk 1087, Bulk 1089, Bulk 1099, Bulk 1112, Bulk 1136, Bulk 1141, Bulk 1147, Bulk 1150, Bulk 1163, Bulk 1166, Bulk 1175, Bulk 1177, Bulk 1197, Bulk 1201, Bulk 1205, Bulk 1215, Bulk 1222, Bulk 1233, Bulk 1244, Bulk 1303, Bulk 1319, Bulk 1321, Bulk 1340, Bulk 1393, Bulk 1407, Bulk 1418, Bulk 1423, Bulk 1430, Bulk 1447, Bulk 1458, Bulk 1464, Bulk 1470, Bulk 1485, Bulk 1489.

Genetics Stocks

HI 8751 (IC0623451; INGR 17032) and **HI 8765** (IC0624495; INGR17038) **were registered as durum wheat germplasm with multiple disease resistance:** Durum wheat production is affected by several biotic factors like rusts, leaf blight, karnal bunt, flag smut, loose smut *etc.*, and it is often very difficult to get multiple disease resistant high yielding genetic stocks. ICAR-IARI Regional station has developed two durum genetic stocks *viz.*, **HI 8751 (**HD 4685/HI 8634**)** and **HI 8765 (**HI 8504/ CPAN 6206//HI 8627**)**, which were registered at ICAR-NBPGR, New Delhi as genetic stocks with multiple disease resistance. These durum wheat genotypes were identified to be resistant to stem, leaf and stripe rusts, Karnal bunt and flag smut in multi-location testing. They showed high levels of adult-plant resistance to most prevalent and virulent pathotypes like 77-5 and 104-2 of leaf rust, 40A and 117-6 of stem rust; and 46S119 and 78S84 of stripe rust in isolated nurseries. They also showed resistance to other diseases like flag smut and Karnal bunt. Hence, they can be utilized in durum wheat improvement as potential resistance donors to breed new varieties.

Malavshakti (HI 8498) – A High Yielding Durum Wheat Cultivar for Central India

'Malavshakti' is the highest yielding genotype, compared to all popular cultivars of durum and aestivum wheats in Central India. Its high resistance to leaf rust will add to the genetic diversity in wheat cultivation, ensuring greater protection against "leaf rust epidemics" in Central India and also in northern wheat belt by cutting down inoculum supply along the "*Puccinia* path". High grain quality of HI 8498 will fetch premium to the farmers in the national market and will boost durum export from the country. Thus, HI 8498 meets a long felt need of a highly rust resistant, high yielding, quality durum cultivar to replace old varieties under irrigated conditions. Its spread will help in stabilizing wheat production and in raising productivity and profitability of wheat cultivation in Central India. Central India is known for its durum, kathia or macaroni wheats, possessing excellent grain appearance, golden amber colour and bold lustrous grain. *'Malavshakti'*, in addition to meeting these unique grain traits, has high yield potential of 6-7 t/ha, combined with high rust resistance. Due to its high-water use efficiency, under water stress conditions of Malwa region, it can yield 4-5 t/ha with 3-4 irrigations only. Because of its wider adaptability to diverse agroclimatic conditions, early maturity and exceptionally high yield potential, it has been termed as a historical variety at 38th All India Wheat Worker's Workshop held at Ludhiana during August, 1999. "Malavshakti" (HI 8498), was developed from the cross-involving Cr 'S' – Gs" S"/A-9-30-1//Raj 911 at IARI, Regional Wheat Research Station, Indore and released in June, 1999 for the timely sown, irrigated, high fertility conditions of the Central zone comprising Madhya Pradesh, Gujrat, southern parts of Rajasthan and Bundelkhand region of Uttar Pradesh. During last five years of multi-locational experiments at different research centers and on farmer's fields, it was observed that Malavshakti meets the requirements of both the resourceful and marginal farmers. Due to its diverse adaptability, it can be planted from timely sown (November) to late (December) and very late (January) sown conditions. Central Zone has low average productivity (about 1.8 t/ha) and the existing popular bread wheat variety Lok-1 has become susceptible to leaf and stem rusts. The newly evolved durum wheat genotype 'Malavshakti' combines high yield potential, adequate disease resistance and excellent grain quality.

Wider Adaptability

"Malavshakti" has been released for the Central zone, but its performance in NIVT-4 as a check was equally encouraging in the North Western plains and the adjoining Peninsular zone. During 98-99, it was tested in Advance Varietal trial under late sown conditions of central zone and out yielded all the genotypes under test. The versatility of the genotype to adjust even under rainfed situations enabled it to rank at par to A-9-30-1 and Sujata, the most popular rainfed *durum* and *aestivum cultivar, respectively* with an average yield of 1.8 to 2.1 t/ha in station trials during 1997-98. In on-farm Front Line Demonstration during 97-98, it outyielded all the bread wheat cultivars by a margin of 8 to 10 q/ha. Because of its uniform synchronous tillering, early maturity and erect leaves combined with lustrous, attractive, bold amber grains, it ranked most preferred variety among farmers.

Disease Resistance

Race 77 group of leaf rust pathotypes have knocked down almost all the present bread-wheat varieties in the country, creating an alarming situation. The two durum cultivars, Raj 1555 and HI 8381 are also showing signs of increasing susceptibility to dominant rust races including Race 77. High resistance of HI 8498 to these races, particularly to Races 77-2, 77-5 and 104-2 will enrich the genetic diversity among wheat cultivars, ensuring more safety against leaf–rust epidemic in central India and also in northern wheat belt by cutting down the inoculum supply along the "*Puccinia* path". In Central India, resistance to stem rust is important. HI 8498, is highly resistant to this rust and is postulated to carry *Sr* 2, *Sr* 11 and additional genes for resistance to stem rust (PPSN, 95-96) (*Sr*-2 is known to have contributed to durable resistance to this rust throughout the world). This variety is resistant to dominant stem-rust races 40-A, 40-1 and 122. *Sr* 2, a gene for adult plant resistance, expressed from the flag leaf stage onwards, in conjunction with other genes present in HI 8498 will provide adequate protection to stem-rust in central India where this-rust incidence occurs at the post heading stages of wheat crop. HI 8498 is highly resistant to stripe rust whereas, check varieties showed moderate (Raj 1555) to high susceptibility (Lok-1 and GW 190) to this rust. For other foliar and seed borne diseases, HI 8498 proved to be more tolerant under natural and artificial epiphytotic conditions in Central India.

Grain Quality

"Malavshakti" (HI 8498) scored best grain appearance because of its very bold attractive kernels. Its bold and well-filled grains have high 1000-grain weight and hectolitre weight a pre requisite for high semolina recovery. Similarly, it has higher protein content, SDS - sedimentation value and β-carotene content, compared to Raj

1555 and HI 8381, which are an essential requirement for export and for producing high quality pasta-products.

Table 4.40. Comparative quality-characteristics of Malavshakti (HI 8498) and other check varieties

Characters	*Checks*			
	HI 8498	*Raj 1555*	*HI 8381*	*Lok-1*
Grain appearance (Max. score 10)	5.8	5.6	5.6	5.7
1000 grain weight (g)	50.5	48.5	48.5	46.0
Hectolitre weight (kg)	82.1	80.2	80.2	77.7
Protein content (per cent)	12.4	12.0	12.3	11.4
Sedimentation value (ml)	35	34	28	57
β-carotene (ppm)	4.3	3.7	4.1	-

Source: DWR, Karnal.

Future Challenges for Wheat Cultivation in Madhya Pradesh

To make the wheat cultivation more productive and profitable, it is essential to cut down the cost of production, to compete more effectively in global market. With increasing pressure on ground water, irrigation is becoming a major limiting factor in improving wheat production. Hence, evolution of highly water use efficient genotypes and water use efficient cultivation technology is essential to minimise the irrigation requirements of wheat crop. With decreasing per capita land availability, it is essential to increase cropping intensity to improve vertical production for ensuring "Food Security" to exploding population of the country. Survey conducted in Dewas, Ujjain, Dhar and Indore districts of the state indicated that farmers are harvesting 30-32 qtls. of wheat per hectare, using 4 to 6 irrigations. Same is the case in command areas having assured 4-6 irrigations during the crop season. Under proper management, comparable yields can be realised with just 2 to 3 irrigations, using appropriate wheat varieties. Thus, there is large gap in technology available and the technology being implemented by the farmers, demanding immediate attention on this front. To improve the wheat productivity and to reduce the cost of cultivation, the following breeding strategy needs to be adopted and the low-cost technologies need to be popularized with dedicated efforts under soybean-wheat cropping system.

An increasing global demand, value addition potential, resistance to diseases, better market price are some of the key factors which make Indian durum wheat an export commodity. It has an export potential similar to basmati rice. The potential markets for Indian durum can possibly be countries of Middle East, Mediterranean region and Africa. Even within the country, the preference for low calorie food is increasing. Affluent people and office-goers have found these instant foods as a convenient source of their dietary meal. Demand for baby food and pasta products as well as backed food is on the rise in cities. With the availability of high-yielding

Table 4.41. Prominent wheat varieties developed for Rainfed and Limited irrigation conditions

Malav Kranti (HI 8638)	It is highly suitable for sowing early in October under rainfed or with one or two irrigations, released in 2009 for Madhya Pradesh state. It is a lodging resistant, semi-tall variety with good quality characteristics like high protein and yellow pigment. It is highly suitable for preparing chapatti along with suji and dalia. It yields about 25 – 30 quintals per hectare. It is a pubescent and black awn variety, which helps for early identification of the variety.
Malav Ratna (HD 4672)	It is highly suitable for sowing last fortnight in October under limited irrigation (1 or 2) with semi-dwarf nature. It is a glabrous, and black awn variety with non-synchronous flowering, which yields to about 30 – 45 quintals per hectare.
Malav Kirti (HI 8627)	It is highly suitable for sowing last fortnight in October under limited irrigation (1 or 2) with semi-dwarf nature. It is a glabrous, and white awn variety with non-synchronous flowering, which yields to about 30 – 45 quintals per hectare. It is the first highly nutritious durum wheat variety released for cultivation with high levels of protein, micronutrients (50 per cent higher) and vitamin A (nearly 2.5 times), compared to most popular bread wheat variety Lok-1, which makes this variety as a naturally bio-fortified.

Table 4.42a. Prominent wheat varieties developed for Timely sown and highly irrigated

Malav Shree (HI 8381)	It is highly suitable for sowing under timely sown i.e, November with 4 or 5 irrigations. It is having high tillering, pubescence with black awns with synchronous flowering, which yields to about 50 – 60 quintals per hectare. It is suitable for preparation of pasta products.
Malav Shakti (HI 8498)	It is the most popular variety of durum wheat highly suitable for sowing under timely sown i.e, November with 3 or 4 irrigations. It is an early maturing variety having high tillering, glabrous with white awns with synchronous flowering, which yields to about 50 – 60 q/ha under timely sown; and 40 – 45 q/ha under late sown conditions. It is suitable for preparation of pasta products and qualified for all specified characteristics for export.
Poshan (HI 8663)	It is a recently released durum wheat variety highly suitable for sowing under timely sown i.e, November with 3 or 4 irrigations. It is a medium-early maturing variety having high tillering, glabrous with white awns with synchronous flowering, which yields to about 55 – 60 q/ha under timely sown. It is first officially released as a natural bi-fortified variety with a stable β-carotene of 6.5 ppm and high protein and micronutrients; and is suitable for preparation of fine pasta products.
MPO 1215	It is a durum wheat variety released in 2009, which is highly suitable for sowing under timely sown i.e, November with 3 or 4 irrigations. It is a medium-early maturing variety having high tillering, glabrous with white awns with synchronous flowering, which yields to about 55 – 60 q/ha under timely sown.

superior quality varieties of durum wheat for various zones, a part of the area under bread wheat can be diverted to durum wheat production, in addition to increasing its area in the central zone to fetch higher price to increase the profit margin of farmers. Thus, it is an ideal wheat to be grown in Central and Peninsular parts of the country for "ensuring food and nutritional security" and "production stability", increasing employment opportunities through fast food industry and sustainability. An awareness campaign is, therefore, urgently required for the growers, traders and consumers about the importance of durum wheat as "Health Food".

Table 4.42b. Prominent wheat varieties recently developed

HI 1612	It is a bread wheat genotype released for timely sown, restricted irrigation conditions of North Eastern Plains Zone (NEPZ). This is the first bread wheat variety released for restricted irrigation conditions of NEPZ. It is a significantly high yielding (37.6 q/ha) genotype, compared to check varieties K 8027 (33.6 q/ha), C 306 (31.0 q/ha) and HD 2888 (30.3 q/ha) under timely sown, restricted irrigation (RI) conditions. It has significant yield advantage of 24.1 per cent, 21.3 and 11.9 per cent over the checks HD 2888, C 306 and K 8027, respectively. It has a potential yield of 50.5 q/ha with high levels of field resistance to stripe and leaf rusts. It also showed high levels of adult plant resistance to prevalent and virulent stripe rust pathotypes 46S119, 78S84 and 110S119; and leaf rust pathotypes 77-5, 77-9 and 104-2. It showed good levels of resistance to leaf blight, Karnal bunt, Fusarium head blight and loose smut. It is a good quality bread wheat genotype with excellent chapati quality (7.64), biscuit quality (7.39), high protein content (11.5 per cent) and high sedimentation value (58 ml). It has high levels of essential micronutrients like iron (41.5 ppm) and zinc (35.5 ppm) content.
HI 8777	It is a durum wheat genotype released for rainfed, timely sown conditions of Peninsular Zone. It is a widely adapted and high yielding durum wheat genotype (18.5 q/ha), compared to durum wheat check varieties AKDW 2997-16 (16.2 q/ha) and UAS 446 (16.2 q/ha) under rainfed, timely sown conditions. It has significant yield advantage of 14.2 per cent over the checks AKDW 2997-16 and UAS 446. It has a potential yield of 28.8 q/ha with high 1000 grain weight (43.27 g). It showed significantly superior level of leaf rust resistance. In isolated nurseries, HI 8777 showed high levels of adult plant resistance to prevalent and virulent stem rust pathotypes 40A and 117-6; and leaf rust pathotypes 77-5, 104-2 and 77-9. It showed better resistance to Karnal bunt, loose smut, flag smut and foot rot than the checks. It has high levels of essential micronutrients *viz.*, Iron (48.7 ppm) (9.4 per cent > AKDW 2997-16, 9.4 per cent > UAS 446 and 5.6 per cent > MACS 4028) and Zinc (43.6 ppm) (22.5 per cent > AKDW 2997-16, 6.6 per cent > UAS 446 and 8.2 per cent > MACS 4028), and high overall acceptability (7.0), compared to durum checks. It showed 0 per cent yellow berry incidence over 2 years.

Table 4.43. Comparative quality and mineral contents of Lok-1 and HI 8627

Variety	*Hectolitre Weight (kg)*	*Protein Content (per cent)*	*β-carotene (ppm)*	*Iron Content (ppm)*	*Zinc Content (ppm)*	*Copper Content (ppm)*	*Manganese Content (ppm)*
Lok-1 (*aestivum*)	80.6	10.6	2.3	35.5	27.2	4.5	33.5
HI 8627 (*durum*)	82.3	11.0	5.7	49.6	42.1	6.0	27.4

References

Atale, SB and Vitkare, DG (1990) Heterotic expression for yield and yield components in 15 x 15 diallel in bread wheat. *Indian J Genet*. **50**: 153-156.

Bains, NS; Mangat, GS; Singh, K and Nanda, GS (1998) A simple technique for the identification of embryo carrying seeds from wheat x maize crosses prior to dissection. *Plant Breeding* **117**: 191-192.

Bhalla, GS; Hazell, P and Kerr, J (1999) Prospects for India's cereal supply and demand to 2020. International Food Policy Research Institute, USA.

Bitsch, C; Groger, S and Lelley, T (1998) Effect of parental genotypes on haploid embryo and plantlet formation in wheat x maize crosses. *Euphytica* **103**: 319-323.

Borner A; Plaschke, J; Korzun, V and Worland, AJ (1996) The relationships between the dwarfing genes of wheat and rye. *Euphytica* **89**: 69-75.

Botstein, D; White, RL; Skolnick, M and Davis, RW (1980) Construction of a genetic linkage map in man using restriction fragment length polymorphisms. *Am. Human Genet.* **32**: 314-333.

Cao, W; Scoles, G; Hud, P and Chibbar, RN (1999) The use of RAPD analysis to classify *Triticum* accessions. *Theor. Appl. Genet.* **98**: 602-607.

Chatrath, R; Mishra, B and Shoran, J (2006) Yield potential survey–India. In: Reynolds MP, Godinez D (eds) Extended abstracts of the international symposium on wheat yield potential "Challenges to International Wheat Breeding", March 20–24, 2006, Cd. Obregon, Mexico. CIMMYT, Mexico, D.F., p 53.

Chaudhary, HK (2003) Genetic amelioration of winter and spring ecotypes of wheat through androgenesis and wheat x maize systems. Final progress report of the ICAR ad-hoc project.

Chaudhary, HK; Dhaliwal, I; Singh, S and Sethi, GS (2003) Genetics of androgenesis in winter and spring wheat genotypes. *Euphytica* **132**: 311-319.

Chaudhary, HK; Sethi, GS; Singh, S; Pratap, A and Sharma, S (2005) Efficient haploid production in wheat by using pollen of *Imperata cylindrica*. *Plant Breeding* **124**: 96-98.

Chaudhary, HK; Singh, S and Sethi, GS (2002) Interactive influence of wheat and maize genotypes on haploid induction in winter x spring wheat hybrids. *Journal of Genetics and Breeding* **56**: 259-266.

Chowdhury, S; Kumbhare, NV; Sinha, AK and Prasad, Y (2002) Increasing wheat production and Building up of Research Capabilities in the warmer Areas and Eastern India. Sub Project, National Agricultural Technology Project (NATP) IARI Regional Station, Pusa (Bihar).

Chu, CC; Wang, CC; Sun, CS; Chun, NP; Yin, KC and Hsu, C (1973) Investigation on the induction and morphogenesis of wheat (*Triticum vulgare*) pollen plants. *Acta Botanica Sinica* **15**: 1-11.

Chuang, CC; Ouyang, J; Chia, H; Chou, SM and Ching, CK (1978) A set of potato media for wheat anther culture. *In*: Procs China-Australia Plant Tissue Culture Symposium. Peking, 1978, pp. 51-66.

Consolidated Progress Report (2003) "Increasing Wheat Production and Building up of Research Capabilities in Warmer Areas and Eastern India" (NATP MM/ III-58), IARI Regional Station, Pusa, District-Samastipur, Bihar-848125.

Darrell, GW and Charles, LL (1970) Hybrid vigour in hard red spring wheat crosses. *Crop Science*. **10**: 220-223.

de Buyrer J and Henry, Y (1979) Androgenese sur des Bles tenders en cours de selection. I. L.' obtention des plants *in vitro Z. pflanzenzucht* **83**: 49-56.

Devos, KM and Gale, MD (1992) The use of random amplified polymorphic DNA markers in wheat. *Theor. Appl. Genet.* **84**: 567-577.

Domini, P and Cooke, RJ (2000) Temporal trends in the diversity of UK wheat. *Theor. Appl. Genet.* **100**: 912–917.

Dorgrar, N; Akin-Yalin S and Akkaya, MM (2000) Discriminating durum wheat cultivars using highly polymorphic simple sequence repeat DNA markers. *Pl. Breed.* **119**: 360-362.

Gautam, PL and Jain, KBL (1985) Heterosis for various characters in durum wheat. *Indian J Genet.* **45**: 159-165.

Gupta, PK; Balyan, HS; Kumar, J; Kulwal, PL; Kumar, N; Mir, RR; Kumar, A and Prabhu, KV (2008). QTL analysis and marker assisted selection in bread wheat. Procs of 11th International wheat genetics symposium (IWGS), August 24-29, 2008, Brisbane Australia.

Gupta, RR and Ahmad, Z (1995) Line x tester analysis for some metric traits in macaroni wheat. *Crop Science*. **22** (2): 244-249.

Inagaki, MN (1985) Chromosome doubling of the wheat haploids obtained from cross with *Hordeum bulbosum* L. *Japanese Journal of Breeding* **35**: 193-195.

Joshi AK, Mishra B, Chatrath R, Ortiz Ferrara G, Singh RP (2007) Wheat improvement in India: present status, emerging challenges and future prospects. *Euphytica* **157**: 431–446.

Kantwa, SR; Sai Prasad, SV and Parihar, CM (2006) Agriculture and its impact on ecological balance. *Agriculture Today*. October: pp. 48-49.

Khan, AS and Sher, A (1999) Estimation of heterotic effect for yield and its component in bread wheat (*Triticum aestivum* L.). *Pakisthan J. of Bio. Sci.* **2** (3): 928-930.

Kim, HS and Ward, RW (2000) Patterns of RFLP-based genetic diversity in germplasm pools of common wheat with different geographical or breeding program origins. *Euphytica*. **115**: 197-208.

Manifesto, MM; Schlatter, AR; Hopp, HE; Suarez, EY and Dubcovsky, J (2001) Quantitative evaluation of genetic diversity in wheat germplasm using molecular markers. *Crop Sci.* **41**: 682-670.

Matzk, F and Mahn, A (1994) Improved techniques for haploid production in wheat using chromosome elimination. *Plant Breeding* **113**: 125-129.

Mishra, AN; Kaushal, K and Pandey, HN (2001) Appropriate pathotypes of stem rust and leaf rust for evaluating resistance in durum wheat and bread wheat. *Wheat Information Service* **93**: 38-39.

Mishra, AN; Kaushal, K; Shirsekar, GS; Yadav, SR and Pandey, HN (2005) Diverse sources of stem rust resistance identified in durum wheat. *Indian Phytopathology* **58**: 335-338.

Mishra, AN; SaiPrasad, SV; Verma, DK; Singh, AK and Saxena, DC (2012) Gehun ki labhdayak kheti Kasm lagat: adhik utpadan. Gehun Vistar Bulletin No 5.

Shatabdi Sanskaran. In Hindi. IARI-Regional Station, Indore: Diamond Jubilee (1951-2011). pp. 1-6.

Mishra, AN; Sharma, KC; SaiPrasad, SV; Verma, DK; Singh, AK and Saxena, DC (2012) *Madhya bharat mein mavi gehun ki kheti: Avashyakta aur upyogita. Gehun Vistar Bulletin* No 3. Hirak Jayanti Sanskaran. In Hindi. IARI-Regional Station, Indore: Diamond Jubilee (1951-2011). pp. 1-6.

Murashige, T and Skoog, F (1962) A revised medium for rapid growth and bioassays with tobacco tissue cultures. *Physiology Plantarum* **15**: 473-497.

Nagaoka, T and Ogihara, Y (1997) Application of ISSR polymorphism in wheat cultivars using RAPD fingerprinting. *Theor Appl Genet* **94** (5): 597-602.

Nagarajan, S (2005) Can India produce enough wheat even by 2020. *Curr Sci* **89**: 1467–1471.

Ouyang, J; Hu, H; Chuang, CC and Tseng, CC (1973) Induction of pollen plants from anthers of *Triticum aestivum* L. cultured *in vitro. Scientica Sinica* **16**: 79-95.

Pandey, HN (2004) Low-Cost "Water Use Efficient" Wheat Cultivation Technology for Tropical India Presented at *National Symposium on Wheat Improvement for Tropical Areas.* September 23-25, 2004, jointly organized by IARI – Wellington and TNAU, Coimbatore pp. 100 -101.

Pandey, HN (2004) Strategies to improve wheat productivity and export opportunities in Central and Peninsular India. *"Wheat Technologies for Warmer Areas" Ed. Rao et al.*, pp. 40-49.

Poelham, JM and Sleper, DA (1995) Breeding field crops, 4th edition. Panama Publishing Corporation, New Delhi.

Porwal Y, Sai Prasad, SV and Billore, M (2006) Association and path analysis for grain yield, its components and quality traits in bread wheat. *National Journal of Plant Improvement* **8**(1): 24-27.

Prabhu KV (2006) Pyramiding disease resistance genes using molecular markers. In Rai M, Singh M, Kumar S, Kumar R, Rai N (eds), National Symposium on Molecular Breeding in Crop Plants, March 20-21, IIVR, Vanarasi, India.

Pratap, A; Sethi, GS and Chaudhary, HK (2005) Relative efficiency of different Gramineae genera for haploid induction in triticale and triticale x wheat hybrids through the chromosome elimination technique. *Plant Breeding* **124**: 147-53.

Roder, MS; Wendehake, K; Korzum, V; Bredemeijer, G; Laborie, D; Bertand, L; Issac, P; Rendell, S; Jackson, J; Cooke, RJ; Vosman, B and Ganal, MW (2002) Contruction and analysis of a microsatellite-based database of European wheat varieties. *Theor Apppl. Genet.* **106**: 67-73.

Roeder, MS; Huang, XQ and Borner, A (2007) Fine mapping of the region on wheat chromosome 7D. *Funct Integr Genomics* **8**: 79–86.

Roy, SK and Salahuddin, ABM (1994) Effect of spikelet thinning on individual seed weight and seed yield of wheat under two sowing dates. *Jour of Appl Seed Production*. **12**: 83-85.

Sadasivaiah, RS; Orshinsky, BR and Kozub, GC (1999) Production of wheat haploids using anther culture and wheat x maize hybridization techniques. *Cereal Research Communications* **27**: 33-40.

Sai Prasad, SV; Rao, VS and Pandey, HN (2006) Hierarchical classification of allelic variation of gliadins in durum wheat.*Cereal Research Communications* **34**(2-3): 1153-1159.

Salina, EA and Khalestkina, EK (2006) SNP markers: Methods of analysis, ways of development, and comparison on an example of common wheat. *Russian Journal of Genetics*. 585-594.

Sethi GS (2000) Androgenesis mediated transfer of rye chromatin into wheat through triticale x wheat hybrids. Final project report submitted to DST, GoI, New Delhi.

Sethi GS and Chaudhary, HK (2001) Androgenesis- a potential system for the transfer of rye chromatin into wheat. Procs Diamond Jubilee Symposium – Hundred Years of Post-Mendelian Genetics and Plant Breeding – Retrospect and Prospects, Nov. 6-9, 2001, New Delhi. pp: 379.

Sharma, SN and Maloo, SR (1986) Combining ability studies in durum wheat. *Madras Agric. J.* **73** (7): 370-374.

Singh RP, Huerta-Espino J, Sharma R, Joshi AK, Trethowan R (2007) High Yielding Spring Bread Wheat Germplasm for Global Irrigated and Rainfed Production Systems. *Euphytica* (In this issue).

Singh, AK; Jain, R; Singh, U; SaiPrasad, SV; Verma, DK; Sharma, KC; Kantwa, SR and Mishra, AN (2012) Transfer of wheat production technology in Central India: Approaches and Strategies. In Extension activities and Institutional Linkages. In 60 years of wheat research IARI-Regional Station, Indore: Diamond Jubilee (1951-2011). pp. 59-69.

Singh, N; Behl, RK and Punia, MS (2001) Production of doubled haploids via maize population in wheat. *Cereal Research Communications* **29**: 289-296.

Singh, S; Sethi, GS and Chaudhary, HK (2004) Relative efficiency of androgenesis and maize-mediated production frequencies of polyhaploids in winter x spring wheat and triticale x wheat hybrids. *Acta Agronomica Hungarica* **52**: 205-209.

Tanksley, SD (1993) Mapping polygenes. *Annual Review of Genetics*. **27**: 205-233.

Terzi, V and Morcia, C (2004) DNA-based analysis for authenticity assessment of monovarietal pasta. *European Food Research and Technology*. **219**(4): 428-431.

Teshale, ET; Bansal, S; Mishra, A and Khanna, VK (2003) DNA fingerprinting of wheat genotypes by RAPD markers. *WIS* **96**: 23-27.

United Nations (1995) World urbanization prospects, the 1994 revision. United Nations, New York.

Varma, PK (2004) A new wheat ideoytype for warmer areas: A participatory model. *"Wheat Technologies for Warmer Areas" Ed. Rao et al.*, pp. 36 - 39.

Varma, PK and Pandey HN (2002) Wheat in Madhya Pradesh: Challenges ahead. Presented at *National Symposium on Agriculture in Changing Global Scenario.*

February 21-23, 2002, Indian Agricultural Research Institute, New Delhi 110 012 pp. 266 - 267.

Verma, DK; Samdur, MY; Kalra, SC; SaiPrasad, SV; Bakode, S and Mishra, AN (2012) Bread Wheat Improvement. In Crop Improvement. In 60 years of wheat research IARI-Regional Station, Indore: Diamond Jubilee (1951-2011). pp. 25-32.

Verma, V; Bains, NS; Mangat, GS; Nanda, GS; Gosal, SS and Singh, K (1999) Maize genotypes show striking differences for induction and regeneration of haploid wheat embryos in the wheat x maize system. *Crop Science*, **39**: 1722-1727.

Wang, CC; Chu, CC; Sun, CS; Wu, SH; Yin, KC and Hsu, C (1973) The androgenesis in wheat anthers of *Triticum aestivum* cultured *in vitro*. *Scientica Sinica* **16**: 218-222.

William HM, Trethowan RM, Crosby-Galvan EM (2007) Wheat breeding assisted by markers: CIMMYT's experience. *Euphytica* **157**: 307–319.

Ziethkiewicz, E; Rafalski, A and Labuda, D (1994) Genome fingerprinting by simple sequence repeat (SSR) - anchored polymerase chain reaction amplification. *Genomics*. **20**: 176-183.

Chapter 5

Leymus racemosus: Potential Species of Gene Pool Enrichment for Wheat Improvement

Outlines

1. Botany of *Leymus racemosus*
2. Crossability of *Leymus racemosus* with Wheat and Chromosome Introgression Lines
3. Use of Genes from *Leymus racemosus* for Wheat Improvement
4. Impact of *Leymus* Genes on Improvement of Wheat Traits
    a. Scab Resistance
    b. Biological Nitrification Inhibition
    c. Aluminum Tolerance
    d. Heat Stress
5. Genetic Maps and Molecular Markers
6. Future Prospects

Leymus racemosus is a wild species belonging to tribe Triticeae (Poaceae), which includes important cereal crops such as bread wheat and barley. This perennial species grows along the coast and in dry lands, and is reportedly tolerant to various biotic and abiotic stresses. Although *L. racemosus* is evolutionarily distant from wheat (*Triticum spp.*) within the tribe, it has been successfully hybridized with wheat, and several wheat–*L. racemosus* chromosome introgression lines have been selected from among the backcrossed progenies of the hybrid. *L. racemosus* is, therefore, a promising wild species for wheat improvement. The production of wheat, as one of

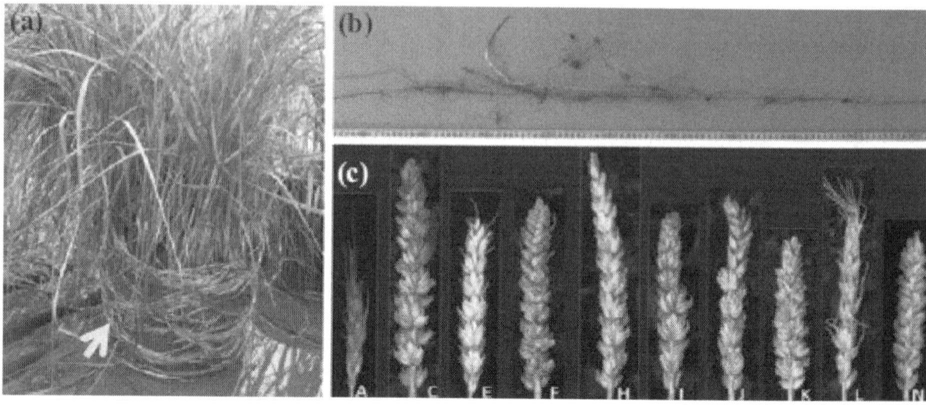

Figure 5.1. Morphology of ***Leymus racemosus*** and its addition lines. a ***L. racemosus maintained*** in a pot of 50 cm diameter. This plant is perennial, stress-tolerant and forms long rhizomes as indicated by the arrow. b The long rhizome of ***L. mollis***, the species most closely resembling ***L. racemosus***. c Spikes of wheat–***L. racemosus*** chromosome introgression lines (CILs). The morphology of the CILs is similar to common wheat

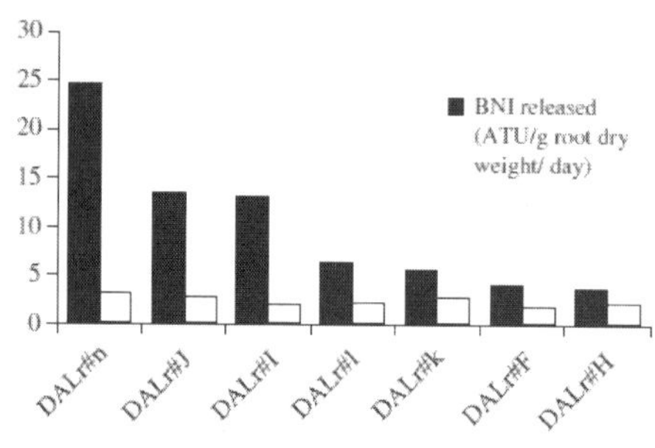

Figure 5.2. BNI release and dry matter production of some wheat–***L. racemosus*** disomic addition lines

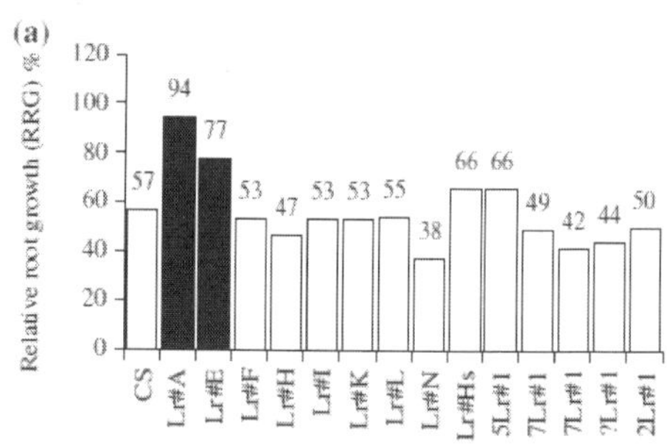

Figure 5.3a. Performance of common wheat cultivar Chinese Spring (CS) and ***Leymus racemosus*** chromosome addition lines

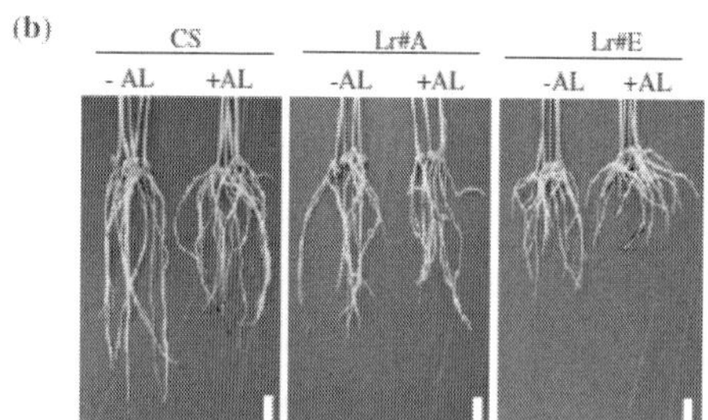

Figure 5.3b. Performance of common wheat cultivar Chinese Spring (CS) and ***Leymus racemosus*** chromosome addition lines

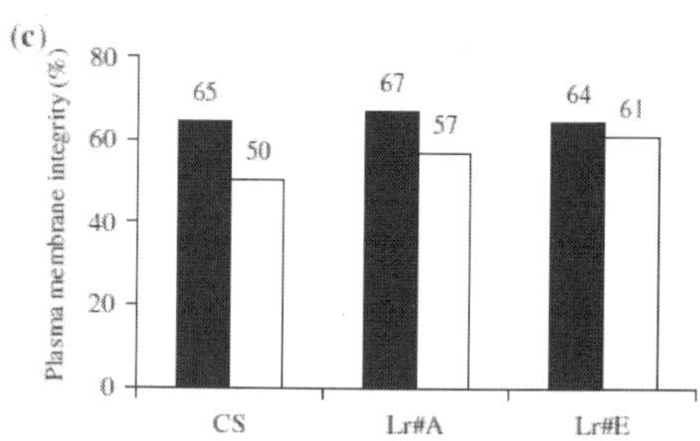

Figure 5.3c. Performance of common wheat cultivar Chinese Spring (CS) and ***Leymus racemosus*** chromosome addition lines

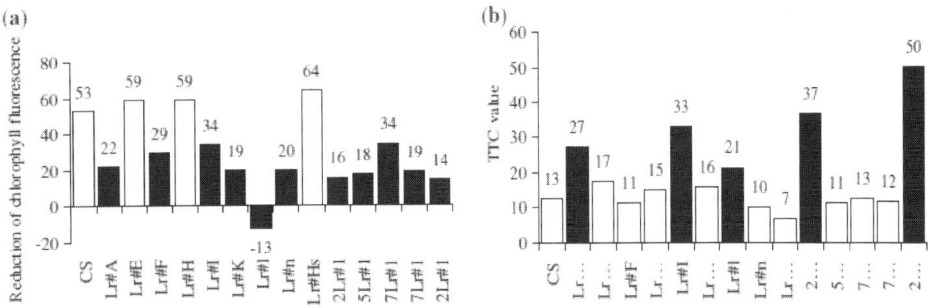

Figure 5.4. Reduction in chlorophyll fluorescence (a) and triphenyl tetrazolium chloride (TTC) reduction assay values (b) in CS and *L. racemosus* addition lines grown in a growth chamber under normal and heat-stressed conditions; dark bars indicate lines with significant differences from CS

the world's most important staple cereals, must increase to provide food for growing population, but wheat is threatened by the effects of climate change, especially increased drought and heat. In addition, depletion of natural resources makes the likelihood of success for increasing future productivity unclear. Maximizing yield to help achieve food security under such challenging conditions will not be easy. Wheat productivity could be improved by enhancing tolerance to biotic and abiotic stress, and tolerance to soil macro- and micronutrient deficiencies or toxicities. However, the genetic base of variation available for most of these traits is very narrow in the available elite germplasm. Wild relatives of wheat, including *L. racemosus,* are an important source of wheat genetic variation and have contributed much to the improvement of wheat productivity. This review describes the potential of this wild wheat relative as a germplasm for wheat improvement, and the characteristics that affect its efficient use in breeding, including its chromosomes, traits, and some genes that have been identified and transferred from this species to common wheat.

Wheat (*Triticum* spp.) is the world's most widely grown grain and provides nearly 20 per cent of the calories humans require. The world's population is growing exponentially and will exceed nine billion by 2050. According to recent FAO estimates, projected demand will require global agricultural production to increase by about 60 per cent relative to yields obtained between 2005 and 2007. Between 2010 and 2012, nearly one in eight people in the world (870 million) had less than the minimum amount of food required. Of these, 852 million comprised up to 15 per cent of the total population of developing countries. Further, impact of climate change is expected to reduce wheat gain yield, which will further complicate food production in the future. The requirement for higher productivity under more limiting conditions, whether by increasing yield or acreage, is an urgent objective that must be met. However, the enhancement of yield is limited by biotic and abiotic stresses and limited availability of arable land. The modern bread wheat (*Triticum aestivum*), which accounts for 95 per cent of the production of all wheat species, has evolved only 8500 years ago from a rare interspecific hybridization event between tetraploid and diploid wheat. Because this event is thought to have occurred only once, all modern wheat is thought to have descended from that single cross. Consequently, the genetic variability of hexaploid bread wheat germplasm has a rather narrow base, and after domestication and several decades of extensive selection and breeding, this already limited genetic base has become even narrower. With such a narrow base, the required increases in wheat yields are hard to imagine. Therefore, there is now growing interest in the great genetic diversity present in wild relatives of wheat. Certain traits in some of these wild relatives have already been characterized and successfully used to improve the tolerance of wheat to abiotic (such as heat, salinity, and drought) and biotic stresses (*e.g.*, diseases, insects, and kit-nematodes), and to improve nutrient use efficiency, grain yield, and bread making quality (Liu *et al.*, 2013), via molecular biology and cytogenetic techniques.

Botany of *Leymus racemosus*

Among the wild relatives of wheat, the genus Leymus (wildrye) is a distantly related source of many potentially adaptive alien genes that could be useful for the breeding of common wheat. Leymus is an allopolyploid genus with chromosomes numbers ranging from 2n = 4 = 28 to 2n = 12 = 84 in the tribe Triticeae, which comprises about 50 species and subspecies worldwide. The genome of tetraploid *L. racemosus* is comprised of two genomes of different origins (2n = 28, NsNsXmXm); the Ns genome originated from *Psathyrostachys*, while the origin of the Xm genome remains unknown. *L. racemosus* (Lam.) Tzvelev, is a perennial grass that is also known by the common name 'mammoth wildrye' as well as other names including *Elymus arenarius* var. *giganteus*, *E. giganteus*, *E. racemosus*, and *L. giganteus*. The species *L. racemosus* distributes in a range of dry coastal or inland sites from Central Asia to Eastern Europe. It is also found growing in a variety of environments including saline, alkaline, dry, or semi-dry lands and forests. Different from the other species, in Leymus only *L. racemosus* and L. mollis have exceptionally large spikes, strong rhizomes, and vigorous growth. *L. racemosus* has been used for ecological purposes such as stabilizing and protecting sand dunes from wind erosion, revegetating mine tailings, and making firebreaks and green strips. It is also used to create wildlife habitat for birds. Monsen *et al.* (2004) reported that *L. racemosus* is generally unsuitable and unpalatable for grazing animal. *L. racemosus* is quite salt- and drought-tolerant and is also resistant to various diseases including scab. The ability of this species to tolerate heat stress; plants grew vigorously and flowered from May to June, and grain-filling, which is particularly a heat-sensitive stage, succeeded from July through August during high summer temperatures and humidity.

Crossability of *Leymus racemosus* with Wheat and Chromosome Introgression Lines

Several *Leymus* species, including *L. racemosus*, have been successfully hybridized and backcrossed with wheat. The first pentaploid hybrid between wheat and *L. racemosus*, which included the genomes ABDNsXm was developed. The second attempt of hybridization was made in China, aimed at introducing the genes for scab resistance from *L. racemosus* into the common wheat. The cytogenetics of the F_1 hybrid between wheat and *L. racemosus* and its BC_1 and BC_2 derivatives. Subsequently, seven disomic addition lines and two di-telosomic addition lines were developed and screened for scab resistance. One disomic addition line and one double-disomic addition–substitution line via anther culture. Another five disomic addition lines and several additional double-disomic additions, di-telosomic addition, and disomic substitution lines were characterized in 23 lines derived from the bread wheat cultivar 'Chinese Spring' (CS). Kishii *et al.* (2004) produced ten chromosome addition lines using a *L. racemosus* accession collected along the Bulgarian Black Sea coast.

Use of Genes from *Leymus racemosus* for Wheat Improvement

There are two possible strategies currently available to use genes from *L. racemosus*, or other wild relatives of wheat, for wheat improvement. In the first approach, the desired phenotype of wild species is identified by direct examination of the species. This is followed by crossing wheat and the wild species to produce a sterile F1 hybrid by embryo rescue. Then, the F_1 plants are repeatedly backcrossed to wheat to generate CILs (including addition, substitution, and translocation lines) that harbor entire or partial chromosomes from the wild species. Finally, these CILs are screened to identify the line that has the desired phenotype. The advantage of this method is reliability; and because useful traits can be easily followed from the wild species to common wheat and it may be particularly efficient for qualitative traits. However, this strategy is time-consuming, and expected traits from the wild species may not exhibit sufficient expression in the genetic background of common wheat. This happened especially in case of quantitative characters. In addition, wild relatives such as *L. racemosus* are often perennials with substantial

Table 5.1 Wheat–*Leymus* chromosome introgression lines produced or maintained at Tottori Alien Chromosome Bank

Strain ID	*Strain Name and Chromosome Name*	*Homoeologous Group*	*Number of Chromosomes*
TACBOW0001a	Leymus racemosus A addition	2	44
TACBOW0003	*L. racemosus* E addition	NDb	44
TACBOW0004	*L. racemosus* F addition	4	44
TACBOW0005	*L. racemosus* H addition	3	44
TACBOW0006	*L. racemosus* I addition	5	44
TACBOW0008	*L. racemosus* k addition	6	66
TACBOW0009	*L. racemosus* l addition	2	44
TACBOW0010	*L. racemosus* n addition	3,7	44
TACBOW0011	*L. racemosus* H substitution	3	42
TACBOW0012 (TA#7643)a	*L. racemosus* 2Lr#1 addition	2	44
TACBOW0013 (TA#7646)	*L. racemosus* 5Lr#1 addition	5	44
TACBOW0014 (TA#7648-1)	*L. racemosus* 7Lr#1 addition	6	44
TACBOW0015 (TA#7648-3)	*L. racemosus* 7Lr#1 addition	3,7	44
TACBOW0016 (TA#7652)	*L. racemosus* ? Lr#1 addition	NAc	44
TACBOW0017 (TA#7644)	*L. racemosus* 2Lr#1 substitution	2	42

a TA#7643: Number from the Kansas State University gene bank; b ND: not determined; c NA: not available.

vegetative biomass, small seeds, and poor agronomic traits. This nature hinders proper evaluation of their traits and response to different environmental conditions. Thus, it is not an easy task to hunt for the useful genes or traits from wild species.

The second strategy for using genes from *L. racemosus* or other wild relatives of wheat depends on phenotyping the CILs and not the wild species itself. The unique property of these CILs is that they are phenotypically similar to wheat, which eases evaluation for any useful genes they might carry. The advantage of this method is that the presence of these genes in the wheat genetic background allows evaluation of their positive or negative impact results from interactions with the wheat genome, and the influence of the wheat genome on the expression of the alien genes. Once a CIL with desired phenotype is identified, translocation lines could be developed for a target chromosome that bears the desired phenotype. Translocation lines could then be screened to identify lines with the desired phenotype. The utility of CILs is not limited to wheat breeding; they can also be used for other applications. For example, Kikuchi *et al.* (2009) and Cho *et al.* (2011a) used wheat–*L. racemosus* CILs to investigate the effects of heavy-ion beams and the DNA methylation inhibitor zebularine on chromosome structure and rearrangement. As those CILs harbor a pair of alien chromosomes that serve as markers that can be easily detected by genomic *in situ* hybridization (GISH), Cho *et al.* (2011b) used double monosomic addition lines carrying *L. racemosus* and *L. mollis* chromosomes to study meiotic chromosome pairing. The line having two homologous chromosomes painted in different colors by GISH was quite useful for investigating factors affecting meiotic chromosome pairing and recombination.

Impact of *Leymus* Genes on Improvement of Wheat Traits

Within the last few decades several wheat–*L. racemosus* CILs have been developed and evaluated, and useful chromosomes and traits have been determined, and analyzed (Mohammed *et al.*, 2014). These lines include useful genes for disease resistance, biological nitrification inhibition, aluminum tolerance, and heat stress tolerance.

Scab Resistance

L. racemosus is highly resistant to scab, also known as Fusarium head blight. A wheat breeding program was later initiated in China to transfer this scab tolerance to common wheat. Three scab-resistant wheat–*L. racemosus* disomic addition lines (DALr.2, DALr.7, and DALr.14) were identified. The additional chromosomes in DALr.2 and DALr.14 belong to homoeologous group 7 and 5, respectively, so they were designated 7Lr#1 and 5Lr#1. From these disomic addition lines, new wheat–*L. racemosus* translocation lines were then developed by inducing chromosome breakage with irradiation and gametocidal genes. In that study, nine scab-resistant wheat–*L. racemosus* translocation lines were identified. In one of these lines, NAU614, the long arm of 5Lr#1 had been translocated to wheat chromosome 6B. In four other lines, NAU601, NAU615, NAU617, and NAU635, part of the short arm of 7Lr#1 had been transferred to various wheat chromosomes. Another four lines, NAU611, NAU634, NAU633, and NAU618 contained translocations involving *Leymus* chromosome Lr.7 and various wheat chromosomes. The translocation lines containing a single alien chromosome segment were more resistant to scab than was the susceptible wheat parent CS but were less resistant than the resistant parent *L. racemosus*. At least three resistance genes were identified in *L. racemosus*, including one located on chromosome Lr.7, another that could be assigned to the long arm of 5Lr#1, and a third that could be assigned to the short arm of 7Lr#1. One novel resistance gene designated Fhb3 was also mapped to the distal region of the short arm of chromosome 7Lr#1, and specific PCR-based markers were also developed to facilitate marker-assisted selection in scab resistance breeding programs (Qi *et al.*, 2008).

Biological Nitrification Inhibition

Biological nitrification is a serious problem in agriculture that results in costly nitrogen loss. The first example of biological nitrification inhibition (BNI) identified among the wild relatives of cereals in *L. racemosus*, then introduced and successfully expressed this BNI trait in common wheat. *L. racemosus* releases 20-fold higher amounts of BNI compounds than does cultivated wheat. In their study, root exudates of cultivated wheat were unable to inhibit biological nitrification in the soil, but the root exudates of *L. racemosus* could suppress NO3– formation and maintain over 90 per cent of the inorganic nitrogen in the soil as NH4+ for 60 days. Chromosomes Lr#n, Lr#J, and Lr#I were associated with the high BNI capacity of *L. racemosus*. The BNI trait from the Lr#n chromosome was introduced into wheat by inducing a Robertsonian translocation between wheat and the Lr#n chromosome (Kishii *et al.*, 2008). Recently, a program was established at CIMMYT to identify and map the genes responsible for BNI and introduce them into all commercial wheat cultivars in order to reduce production costs and minimize the nitrogen pollution generated by wheat production systems.

Aluminum Tolerance

Aluminum toxicity is a key factor that limits wheat production in the acidic soil with pH below 5.5. Mohammed *et al.* (2013) screened 15 wheat–*L. racemosus* CILs for aluminum (Al) toxicity tolerance. *L. racemosus* chromosomes Lr#A and Lr#E were found to significantly enhance relative root growth under Al stress and thereby increase the Al tolerance of wheat. A line containing Lr#E chromosome was more aluminum-tolerant than a line containing Lr#A chromosome at the highest Al concentration tested (200 µM). The introgressed chromosomes did not affect the aluminum uptake or expression of the wheat ALMT1 gene that controls the exclusion of Al from root tips, but a gene or genes on chromosome Lr#E did improve wheat cell membrane integrity of wheat under Al stress conditions. The results suggest that there are Al tolerance mechanisms other than exclusion of Al from the root tips. Eventually, chromosome engineering using these two lines could result in wheat cultivars with enhanced Al tolerance.

Heat Stress

The response of 14 wheat–*L. racemosus* CILs to high temperature to determine whether they have potential for breeding improved wheat cultivars. These introgression lines and their parent CS were evaluated either in a growth chamber at the seedling stage and in the field at the reproductive stage in two heat-stressed environments in Sudan. To be sure that the plants would be exposed to heat stress at the reproductive stage; optimum and late planting strategies were used. For seedlings under growth chamber conditions, the presence of chromosomes Lr#A, Lr#I, Lr#l, and 2Lr#1 were found to improve heat stress tolerance as measured by improved photosynthesis and mitochondrial electron transport. Under field conditions, the presence of chromosomes Lr#I and Lr#n enhanced grain number per spike and heat tolerance. Their results also indicated that chromosome 7Lr#1 might improve yield potential as it has higher grain yield than the wheat parent CS under non-stressed conditions. Based on a heat susceptibility index (HSI) calculated from grain yield obtained from optimum and late sowings, *L. racemosus* chromosomes Lr#I and 2Lr#1 were classified as highly heat tolerant, whereas Lr#A and 5Lr#1 were classified as moderately heat tolerant. This study clearly indicated that several *Leymus* chromosomes have good potential for improving wheat adaptation and tolerance to heat stress (Mohammed *et al.*, 2014). In addition to scab resistance, BNI, aluminum tolerance, and heat stress tolerance, lines carrying the Lr#J and Lr#l chromosomes also expressed gene(s) for resistance to leaf or stem rust, and the Lr#k line possessed a gene for a novel seed storage protein. Further, the Lr#H chromosome addition line exhibited early heading (Kishii *et al.*, 2004).

Table 5.2. Kernel number per spike, grain yield, and heat susceptibility index (HSI) of some addition lines and their parent cultivar 'Chinese Spring' (CS) during evaluation for heat stress tolerance at Gezira (2012/2013), Sudan

Line	*Kernel Number/Spike*				*Grain Yield*				*HSI*
	Shambat		*Gezira*		*Shambat (g/plant)*		*Gezira (g/m²)*		
	OP	*LP*	*OP*	*LP*	*OP*	*LP*	*OP*	*LP*	
Lr#A	53b	50b	49a	32a	1.9a	1.7a	122a	56a	0.59
Lr#I	68a	58a	47a	29b	2.3a	1.7a	132a	78a	0.24
Lr#n	59a	58a	45b	26b	2.1a	1.9a	133a	50a	0.71
2Lr#1	54b	49b	36b	27b	2.1a	1.7a	121a	74a	0.25
5Lr#1	43c	43c	32b	14c	1.9a	1.8a	110a	56a	0.59
7Lr#1	49c	57b	47a	36a	2.1a	1.9a	265a	29a	1.12
CS	55b	51b	37b	26b	1.5b	1.3b	51b	14b	1.40

a, b Significant difference from CS; c Compatible to CS.

Genetic Maps and Molecular Markers

Genetic maps, genomic information, linkage mapping are extensively to identify genes or QTLs in plant genomes that control important traits. It relies on random recombination at meiosis and segregation of the recombinant chromosomes among offsprings. For genes introduced on alien chromosomes, it is difficult to perform genetic analysis and construct genetic maps using CILs because no recombination occurs at meiosis due to absence of homology between the alien chromosome and the wheat chromosomes. Therefore, deletion mapping is the best way to identify and map the genes located on the alien chromosome to allow their transfer into wheat. To create an efficient and informative deletion map, specific molecular markers for the target chromosome are essential. No molecular markers specific for *L. racemosus* have been developed to date. However, expressed sequence tags (EST) markers from barley have been used to develop molecular markers that are polymorphic between wheat and some important wild species. About 182 markers polymorphic between wheat and *L. racemosus* were identified in their study. Although genetic maps are not yet available for *L. racemosus*, several genetic maps have been constructed for other Leymus species. Molecular genetic maps were constructed for two full-sib families (TTC1 and TTC2) derived from L. cinereus and L. triticoides. The genetic maps of the TTC1 and TTC2 families included a combined total of 1583 AFLP markers and 67 heterologous anchor markers (from other grass species) in 14 linkage groups. Two sets of seven homoeologous linkage groups were tentatively identified based on synteny of the heterologous anchor markers between wheat, barley, and cereal rye. These populations were found to segregate for genes and QTLs controlling growth habit, forage quality, and seed shattering (Larson and Kellog 2009). From a collection of 28786 ESTs derived from subterranean rhizome and tiller buds of *L. triticoides L. cinereus* hybrids, 11281 unigene contigs were assembled. Of these, 9389 had at least one match with the rice genome. In that study, 1798 SSR primers were designed of which 1575 were polymorphic. Further, Leymus EST-SSR primers that aligned to rice chromosome 2 were tested for synteny in Leymus. Using the same mapping populations, an integrated consensus map including 375 AFLP, 350 SSR, and 48 heterologous markers that cover 2381 cM. A total of 146 of these markers were tested in 17 wheat–*L. racemosus* CILs and were found to be polymorphic between wheat and *L. racemosus.* All CILs tested positively for between 8 and 24 Leymus EST-SSR marker loci representing at least one homoeologous group; most of these matched the previous chromosome identification. Using suppressive-subtractive hybridization, 112 osmotic-stress-responsive genes (ESTs) from L. mollis that are differentially expressed under osmotic stress. All these types of genetic information that are available could be used to generate molecular markers for *L. racemosus* to facilitate the introduction of its useful adaptive genes in wheat breeding.

Future Prospects

The objectives of wheat breeding programs are linked with the demands of cumulative population growth and the necessity to increase food production. Due to the narrow genetic base available in modern bread wheat germplasm, achieving this objective will be the challenge of the century. Therefore, finding broader sources of adaptive genes in the wheat gene pool will be essential for introducing traits that could help increase yields in the warmer, drier climate predicted for most of the world's breadbasket. *L. racemosus* represents a potentially valuable source of genetic diversity for the improvement of stress tolerance specifically, and wheat production generally. In this review, we have described *L. racemosus* genes that will be potentially useful for improving the tolerance of wheat to biotic and abiotic stresses and improving its nitrogen use efficiency. However, for the most efficient exploitation of *L. racemosus* genes in wheat breeding, the targeted screening of wheat–*L. racemosus* cytogenetic stocks for resistance to biotic and abiotic stresses should be continued. The wealth of molecular genetic information that is becoming available should be efficiently exploited for breeding purposes, and the shortage should be replenished through genome wide approaches.

References

Bushman, BS; Larson, SR; Mott, IW; Cliften, PF; Wang, RRC; Chatterton, NJ; Hernandez, AG; Ali, S; Kim, RW; Thimmapuram, J; Gong, G; Liu, L and Mikel, MA (2008) Development and annotation of perennial Triticeae ESTs and SSR markers. *Genome* **51**: 779–788.

Cho, SW; Ishii, T; Matsumoto, N; Tanaka, H; Eltayeb, AE and Tsujimoto, H (2011a) Effects of the cytidine analogue zebularine on wheat mitotic chromosomes. *Chromosome Sci* **14**: 23–28.

Cho, SW; Moritama, Y; Ishii, T; Kishii, M; Tanaka, H; Eltayeb, AE and Tsujimoto, H (2011b) Homology of two alien chromosomes during meiosis in wheat. *Chromosome Sci* **14**: 45–52.

Curtis, T and Halford, NG (2014) Food security: the challenge of increasing wheat yield and the importance of not compromising food safety. *Ann Appl Biol* **164**: 354–372.

Dear, PH (2001) Genome mapping. Encyclopedia of life science. Macmillan Publishers Ltd, Nature Publishing Group.

Fan, X; Sha, LN; Yang, RW; Zhang, HQ; Kang, HY; Ding CB, Zhang L, Zheng YL and Zhou YH (2009) Phylogeny and evolutionary history of *Leymus* (Triticeae; Poaceae) based on a single copy nuclear gene encoding plastid acetyl-CoA carboxylase. *BMC Evol Biol* **9**: 247.

Habora, MEE; Eltayeb, AE; Tsujimoto, H and Tanaka, K (2012) Identification of drought stress responsive genes from *Leymus mollis*, a wild relative of wheat (*Triticum aestivum* L.) *Breed Sci* **62**: 78–86.

Hagras, AA; Kishii, M; Sato, K; Tanaka, H and Tsujimoto, H (2005) Extended application of barley EST markers for analysis of alien chromosomes added to wheat genetic background. *Breed Sci* **55**: 335–341.

Kaur, P; Larson, SR; Bushman, BS; Wang, RRC; Mott, IW; Hole, D; Thimmapuram, J; Gong, G and Liu, L (2008) Genes controlling plant growth habit in Leymus (Triticeae): maize barren stalk1 (ba1), rice lax panicle, and wheat tiller inhibition (tin3) genes as possible candidates. *Funct Integr Genomics* **8**: 375–386.

Kikuchi, S; Saito, Y; Ryuto, H; Fukunishi, N; Abe, T; Tanaka, H and Tsujimoto, H (2009) Effects of heavy-ion beams on chromosomes of common wheat, *Triticum aestivum*. *Mutat Res/Fund Mol Mech Mutagen* **669**: 63–66.

Kishii, M (2011) Production of wheat-*Leymus racemosus* translocation lines. *Wheat Inf Serv* **111**: 11–13.

Kishii, M; Yamada, T; Sasakuma, T and Tsujimoto, H (2004) Production of wheat-*Leymus racemosus* chromosome addition lines. *Theor Appl Genet* **109**: 255–260.

Larson, SR and Kellog, EA (2009) Genetic dissection of seed production traits and identification of a major-effect seed retention QTL in hybrid *Leymus* (Triticeae) wild ryes. *Crop Sci* **49**: 29–40.

Larson, SR and Mayland, HF (2007) Comparative mapping of fiber, protein, and mineral content QTLs in two interspecific Leymus wildrye full-sib families. *Mol Breed* **20**: 331–347.

Larson, SR; Kishii, M; Tsujimoto, H; Qi, LL; Chen, PD; Lazo, GR; Jensen, KB and Wang, RRC (2012) *Leymus* EST linkage maps identify 4NsL-5NsL reciprocal translocation, wheat–*Leymus* chromosome introgressions, and functionally important gene loci. *Theor Appl Genet* **124**: 189–206.

Larson, SR; Wu, XL; Jones, TA; Jensen, KB; Chatterton, NJ; Waldron, BL; Robins, JG; Bushman, BS and Palazzo, AJ (2006) Comparative mapping of growth habit, plant height, and flowering QTLs in two interspecific families of *Leymus*. *Crop Sci* **46**: 2526–2539.

Liu, J; Chang, Z; Zhang, X; Yang, Z; Li, X; Jia, J; Zhan, H; Guo, H and Wang, J (2013) Putative *Thinopyrum intermedium*-derived stripe rust resistance gene Yr50 maps on wheat chromosome arm 4BL. *Theor Appl Genet* **126**: 265–274.

Lobell, DB; Schlenker, W and Cost-Roberts, J (2011) Climate trends and global crop production since 1980. *Science* **333**: 616–620.

Löve, A (1984) Conspectus of the Triticeae. *Feddes Repert* **95**: 425–521.

Lu, RJ; Chen, PD and Liu, DJ (1995) Transfer of useful germplasm from *Leymus racemosus* Lam to common wheat. IV. Development of double alien disomic addition and substitution lines by anther culture. *J Nanjing Agric Univ* **18**: 1–6.

McGuire, PE and Dvorak, J (1981) High salt-tolerance potential in wheat grasses. *Crop Sci* **21**: 702–705.

Mohammed, YSA; Tahir, ISA; Kamal, NM; Eltayeb, AE; Ali, AM and Tsujimoto, H (2014) Impact of wheat-*Leymus racemosus* added chromosomes on wheat adaptation and tolerance to heat stress. *Breed Sci* **63**: 450–460.

Monsen, S; Stevens, R and Shaw, N (2004) Grasses. USDA forest service Gen. Technical report. *RMRS-GTR*-136, pp. 199–294.

Mujeeb-Kazi, A and Rodriguez, R (1981) An intergeneric hybrid of *Triticum aestivum* L. X *Elymus giganteus*. *J Hered* **72**: 253–256.

Mujeeb-Kazi, A; Bernard, M; Bekele, GT and Minard, JL (1983) Incorporation of alien genetic information from *Elymus giganteus* into *Triticum aestivum*. In: Sakamoto S (ed) Procs of the 6th International wheat genetic symposium, Kyoto, pp. 223–231.

Pérez-de-Castro, AM; Vilanova, S; Cañizares, J; Pascual, L; Blanca, JM; Díez, MJ; Prohens, J and Picó, B (2012) Application of genomic tools in plant breeding. *Curr Genomics* **13**: 179–195.

Qi, LL; Friebe, B; Zhang, P and Gill, BS (2007) Homoeologous recombination, chromosome engineering and crop improvement. *Chromosome Res* **15**: 3–19.

Qi, LL; Wang, S; Chen, PD; Liu, DJ; Friebe, B and Gill, BS (1997) Molecular cytogenetic analysis of *Leymus racemosus* chromosomes added to wheat. *Theor Appl Genet* **95**: 1084–1109.

Reif, JC; Zhang, P; Dreisigacker, S; Warburton, ML; van Ginkel, M; Hoisington, D; Bohn, M and Melchinger, AE (2005) Wheat genetic diversity trends during domestication and breeding. *Theor Appl Genet* **110**: 859–864.

Sun, WX; Chen, PD and Liu, DJ (1998) Transfer of useful germplasm from *Leymus racemosus* Lam. To common wheat. VI. Development and identification of T. aestivum-*L. racemosus* telosomic addition lines. *Acta Genet Sinica* **25**: 259–264 (in Chinese).

Wang, RRC; Zhang, JY; Lee, B; Jensen, KB; Kishii, M and Tsujimoto, H (2006) Variations in abundance of two repetitive sequences in Leymus and *Psathyrostachys* species. *Genome* **49**: 511–519.

Wang, S; Yin, L; Tanaka, H; Tanaka, K and Tsujimoto, H (2010) Identification of wheat alien chromosome addition lines for breeding wheat with high phosphorus efficiency. *Breed Sci* **60**: 372–379.

Wang, YN; Chen, PD; Wang, ZT and Liu, DA (1991) Transfer of useful germplasm from *E. giganteus* to common wheat. II. Cytogenetics and scab resistance of backcross derivatives. *J Nanjing Agric Univ* **14**: 1–5.

Chapter 6

Breeding Wheat for Abiotic Stresses (Heat, Drought, Waterlogged, Cold) Resistance

Outlines

1. Abiotically Stressed Environments
   a. Resistance to Abiotic Stress
   b. Challenges of Heat in Wheat Production
   c. Screening Wheat Germplasm for Heat Tolerance at Terminal Growth Stage
   d. *In vitro* Screening for Abiotic Stresses
   e. Identification of the Components Involved in Drought Tolerance Mechanism
   f. Enhancement of Water Stress Tolerance in Crop Plants Employing Genetic Engineering Tools
   g. Engineering Genes involved in Plant's Stress Surveillance
   h. Engineering Genes for Osmotic Regulation and Ionic Balance
   i. Engineering Genes Involved in Transcriptional Regulation
   j. Drought Stress Induced Genes
2. Transition of Drought Tolerance Capability in Crop Plants from Laboratory to Farmer's Field
3. Abiotic Stress Resistance
   a. Breeding for Lodging Resistance
   b. Breeding for Salt Tolerance

4. Breeding for Drought Tolerance
   a. Concept of Breeding for Semi-arid Environments
   b. Drought-Tolerance Breeding Techniques
   c. Mechanisms of Drought Resistance
   d. Breeding for Mineral Stress/Ion stress Conditions
   e. Breeding for Cold Tolerance
   f. Breeding for Freezing Resistance
5. Phenotyping for Abiotic Stress Tolerance in Crop Plants
   a. Phenotyping under Controlled Condition
   b. Phenotyping under Field Condition
   c. Phenotyping Sites for Different Abiotic Stress (Drought stress)
   d. Precise Phenotyping for Drought Tolerance and Related Dynamic Traits
   e. Infrared Thermography (IRT)/Canopy Spectral Reflectance (SR) Temperature Trait for Phenotyping
6. Phenotyping Sites for Different Abiotic Stress
7. Precise Phenotyping for Drought Tolerance and Related Dynamic Traits
8. IR Thermography (Canopy Temperature) Trait for Phenotyping
9. Chlorophyll Fluorescence Measurements and Use in Plant Phenotyping
   a. Fluorescence Parameters
   b. Instrumental Setting
   c. Infrared thermography (IRT)/Canopy Spectral Reflectance (SR) Temperature Trait for Phenotyping
   d. Magnetic Resonance Imaging (MRI) and Positron Emission Tomography (PET)
   e. Nuclear Magnetic Resonance (NMR)
   f. Photosynthetic System (PSII) of Spikes
   g. Photosynthetic System (PSII) Sensitivity to Temperature
   h. Photosystem Sensitivity to Desiccation
10. Biotechnological Approaches for Climate-Smart Crops
   a. Crop Genetic Transformation
   b. Integrated Genomics, Physiology and Breeding Approaches for Improving Drought Tolerance
   c. Mini-Rhizotron Technique for In-Situ Root Study
   d. Equipment and *in-situ* Root Monitoring
   e. Mobilisation of Pre-Anthesis Reserves During Grain-Filling
   f. Opportunities of Molecular Plant Breeding
   g. Estimation of Heat Stress

   h. Strategic Trait-Based Crossing
   i. QTL Discovery for Drought Tolerance-Related Traits
   j. Association Mapping in Wheat (Phenotyping for Heat Tolerance)
   k. Identification of Genes Associated with Drought Tolerance
   l. Modern Breeding Approaches for Developing Superior Germplasm for Drought Tolerance

11. Future Outlook
   a. Breeding for Tolerance to Abiotic Stresses
   b. Breeding for Input-Use Efficiency

Prime Minister Narendra Modi's aim of doubling farmer's income by the year 2022, has evoked strong responses from various analysts, experts and the media. The goal has been dubbed as impossible and unrealistic. It is the responsibility of all agriculture stakeholders to make it possible so that India will produce sufficient food to achieve food and nutritional security. For this we have to work in scientific and holistic manner so that we may increase the agriculture production besides increasing the input use efficiency in eco-friendly manner. Indian agriculture is passing through difficult times due to two consecutive drought situations in several parts of the country, thereby resulting into wide spread distress among farmers. The rural areas in these parts are facing food and livelihood crisis, more specifically the shortage of fodder and drinking water. Despite the earlier history of famines and continuing patterns of droughts, remarkable successes have been achieved towards food security at the national level. Yet the country continues to face challenges in terms of availability, access, balanced diet, diversity and equity associated with consumption of nutritious food at sufficient levels. For meeting requirement of the burgeoning population, the country will need an estimated 400 million tonnes of food grains by 2050, from a current level of about 260 million tonnes. Current population is 1.21 billion, it is projected to reach 1.6 billion by 2050.

"Vulnerability of the wheat crop to climate change, particularly the threat of rising temperatures due to an 'earlier-than-normal' onset of summer (terminal heat) affecting the crop at the grain filling/maturity stages (March/April), is a major concern among researchers," the attaché added. Of 30 million hectares under wheat cultivation, about 10 million to 12 million are reported to be prone to terminal heat stress. The attaché cited market sources as reporting that 55 per cent to 60 per cent of wheat is marketed by farmers, with the rest saved for personal food, feed and seed use. More than half of the wheat marketed is procured by the government under the minimum selling price program for distribution through the public distribution system and other food programs for sale on the open market to private millers. "The balance of marketed wheat is procured directly by the private sector for milling, processing and other uses," the report said. The attaché forecast that in the near future, India is likely to continue to be an erratic participant in the international wheat market, importing wheat when domestic supplies are depressed or of particularly low quality and exporting wheat when international prices are favorable.

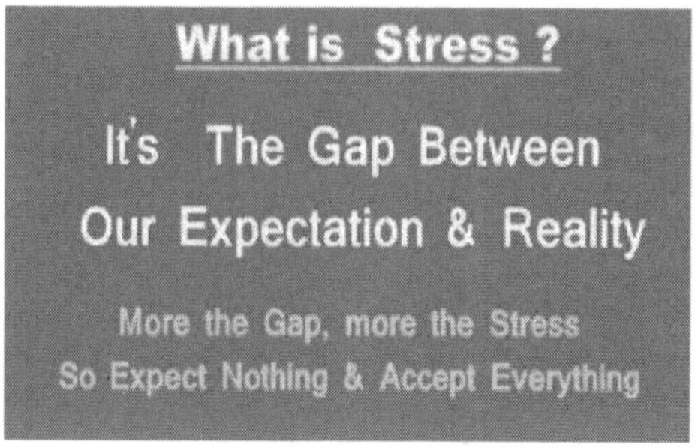

Figure 6.1. What is Stress?

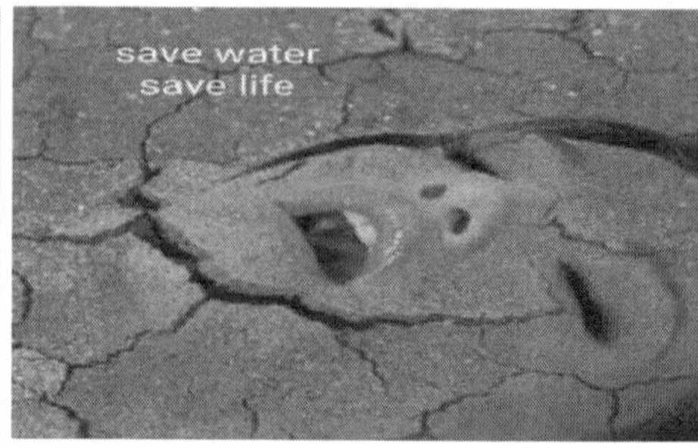

Figure 6.4. Depiction of drought situation by mother earth

Figure 6.2. Dr D K Verma in lab during COVID-19 period at IARI RS Indore, MP

Figure 6.3. Women symbol of water conservation

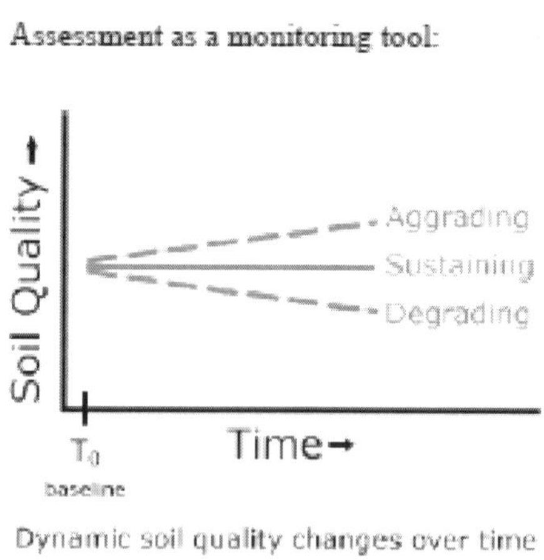

Figure 6.5. Assessment as a monitoring tool

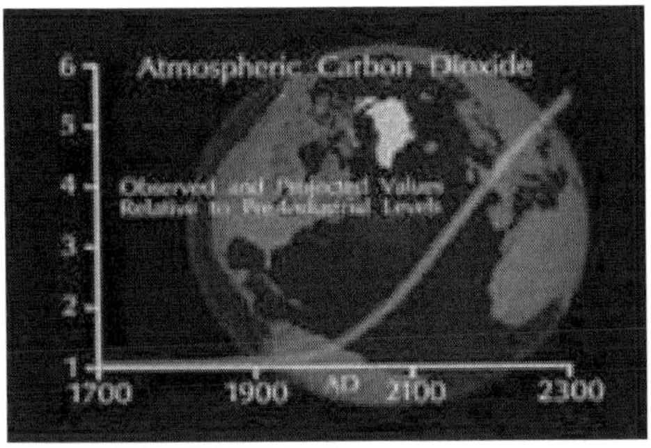

Figure 6.6. Global Warming: Global Threat

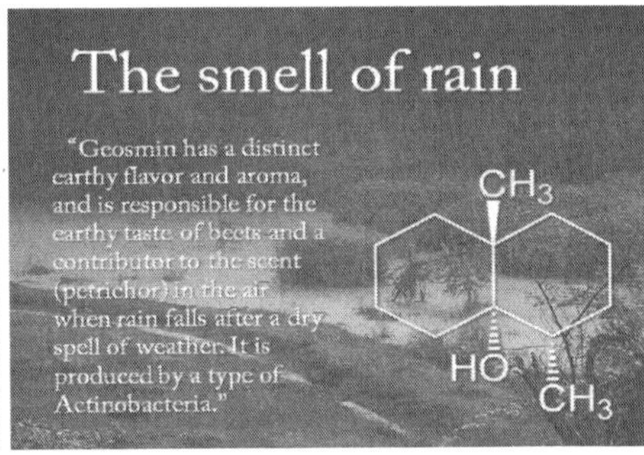

Figure 6.7. The smell of rain

Figure 6.8. No rain in the eye of Cartoonist

Figure 6.9. Dry and dusty field, where even sowing is impossible

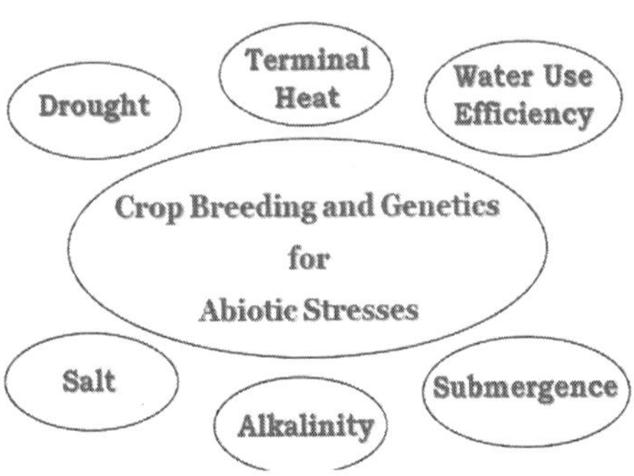

Figure 6.10a. Pictorial presentation of Crop Breeding and Genetics for Abiotic Stresses

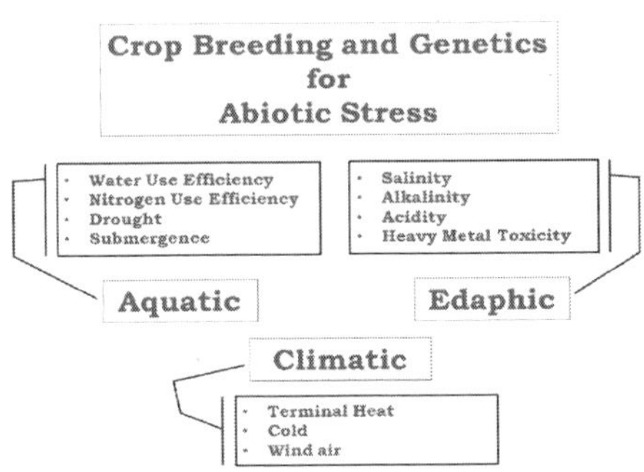

Figure 6.10b. Pictorial presentation of Crop Breeding and Genetics for Abiotic Stresses

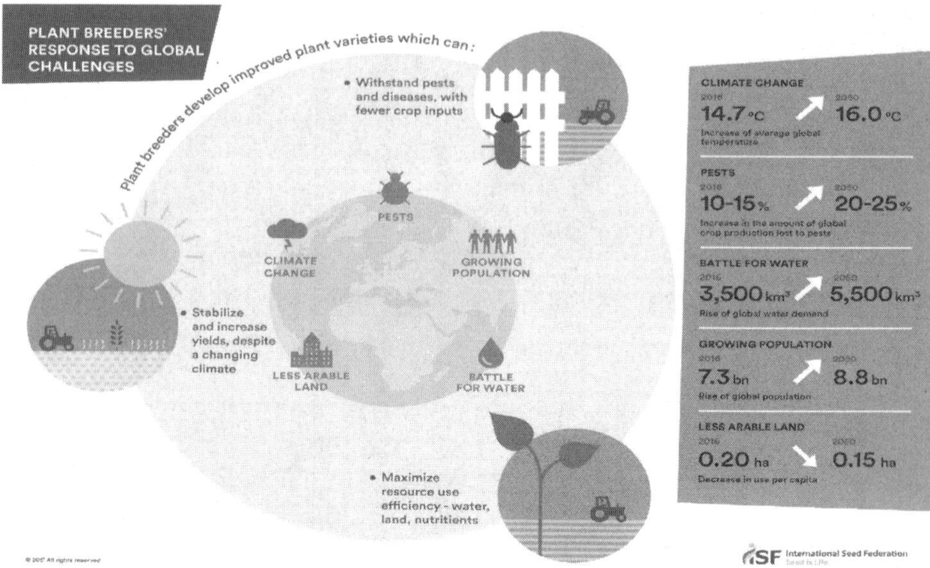

Figure 6.11. Plant Breeders Response to Global Challenges

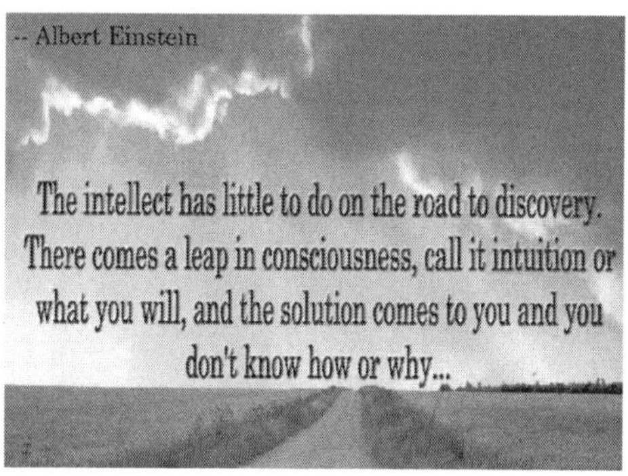

Figure 6.12. Albert Einstein – The intellect has little to do....

Figure 6.13. Global warming will reduce the world`s crop production by up to two percent every decade and wreak $1.45 trillion of economic damage

Figure 6.14. Harvesting of DTSN, 2020 at ICAR-IARI RS Indore

Figure 6.15. Wheat beat the terminal heat in Bihar, BISA, Pusa, Samastipur, Bihar

Figure 6.16. Display of abiotic stress resistant varieties of wheat

Figure 6.17. Pathetic condition of a farmer, whose wheat crop got lodged

Figure 6.18. Lodged wheat crop

Figure 6.19. Utilization of atmospheric moisture by hanging crown roots in wheat

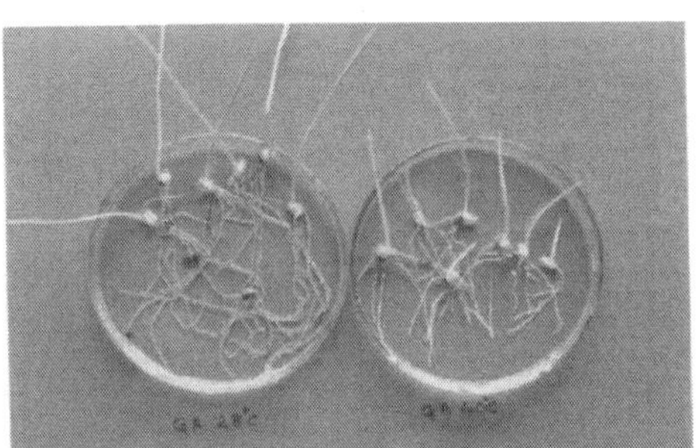

Figure 6.20. Root hair formation at normal and high temperature in GA treated seeds / seedlings

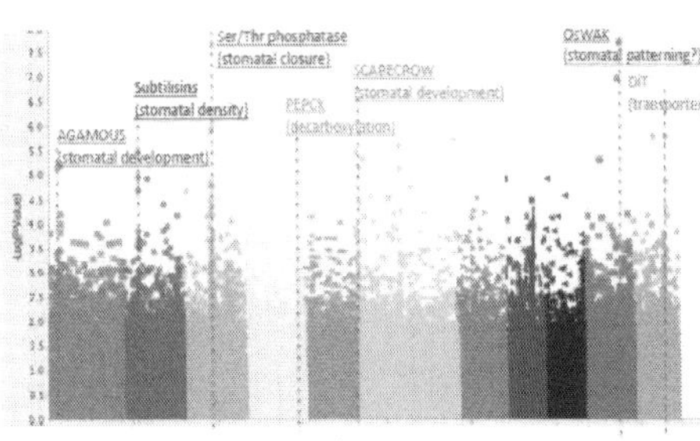

Figure 6.21. Intercellular CO_2 concentration

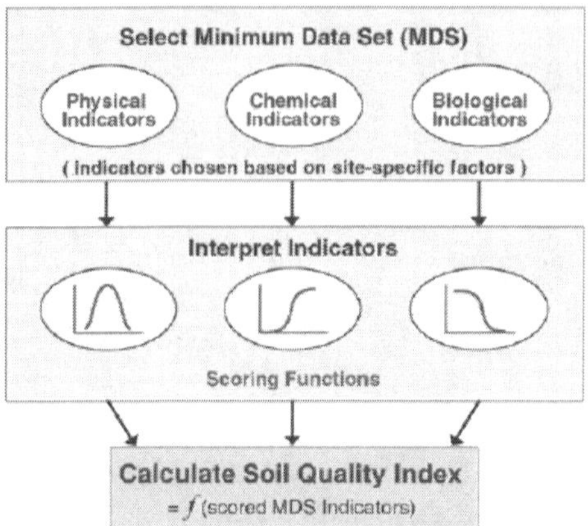

Figure 6.22. Calculation of Soil Quality Index

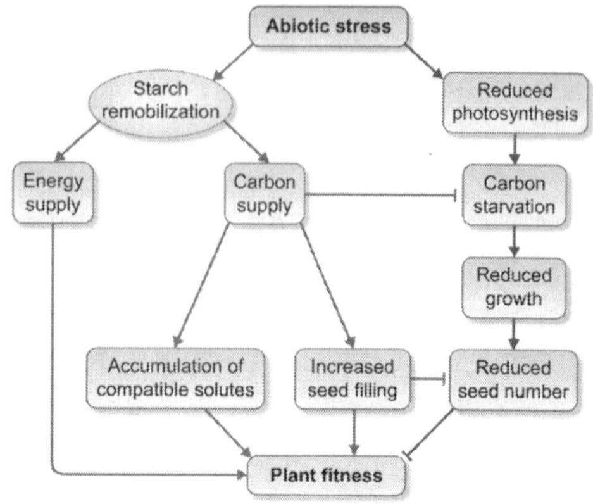

Figure 6.23. Abiotic stresses and Plant fitness

Figure 6.24a. Wheat trials at KVK Zanskar, Jammu & Kashmir for abiotic stress, 2018

Figure 6.24b. Wheat trials at KVK Zanskar, Jammu & Kashmir for abiotic stress, 2018

Figure 6.25. Heat tolerant and susceptible lines inside the plastic sheet tunnel

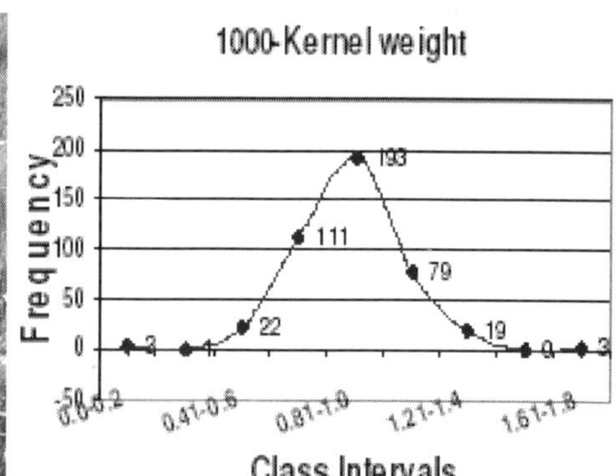

Figure 6.26. Frequency Distribution of number of 1000-grain weight under heat stress

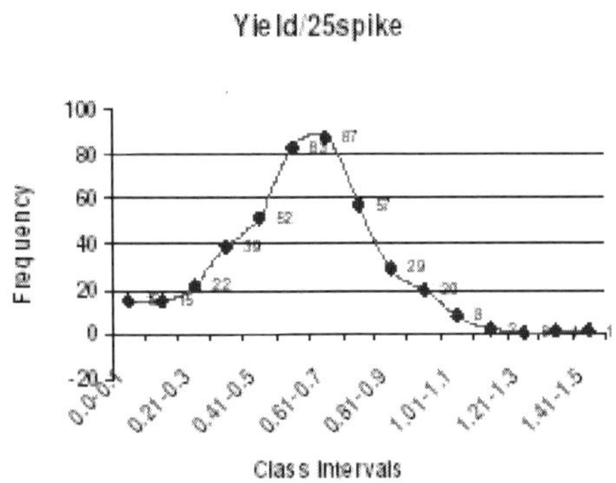

Figure 6.27. Frequency Distribution of No. of Yield/25 spike under heat stress

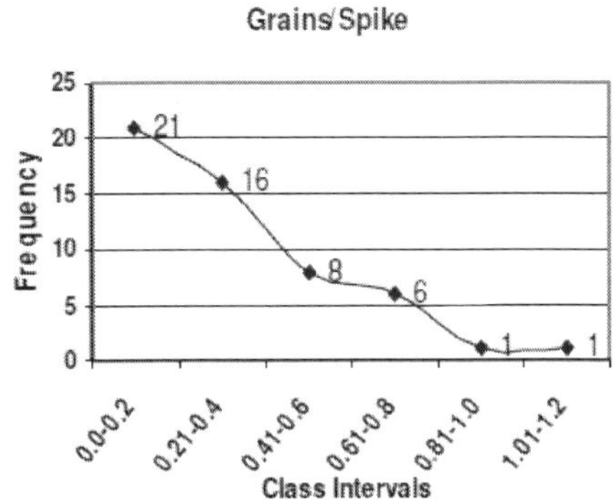

Figure 6.28. Frequency Distribution of No. of grains/spike under heat stress

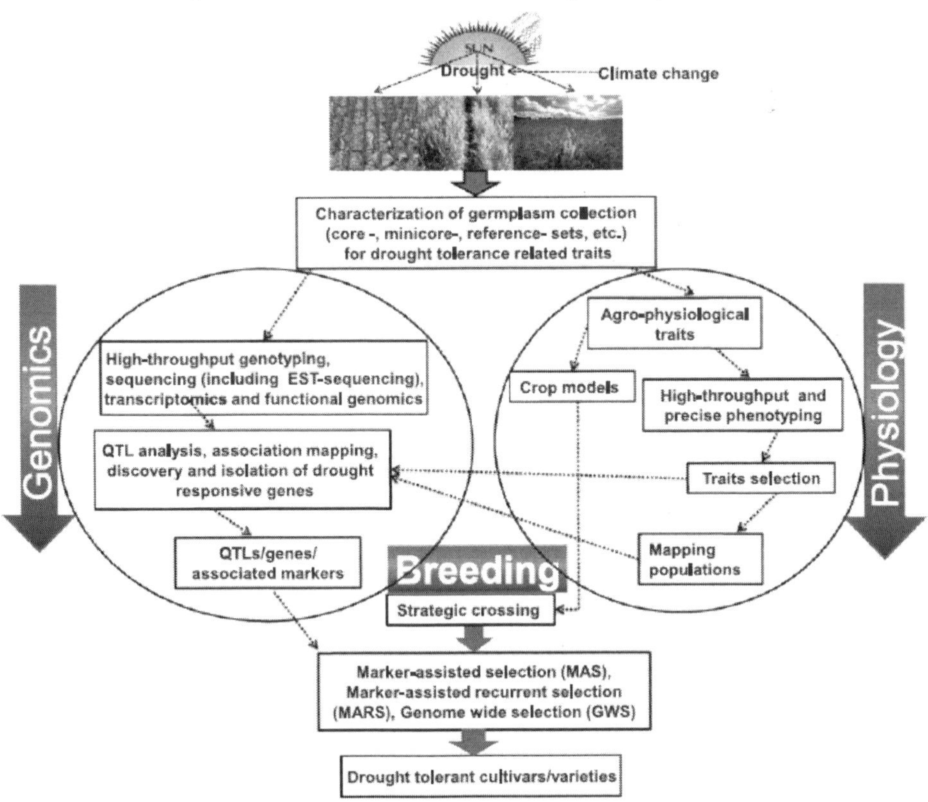

Figure 6.29. A holistic approach for integrating genomics, physiology and breeding approaches for developing the superior varieties with enhanced drought tolerance

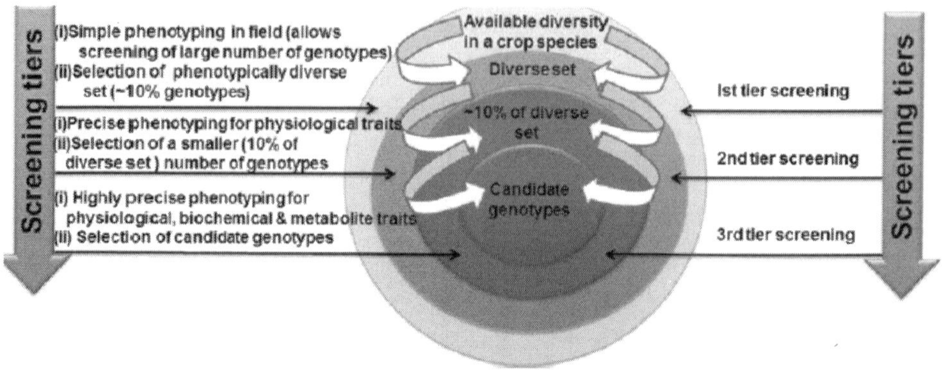

Figure 6.30. An overview of three-tier screening of germplasm collection for traits related to drought tolerance. Three screening tiers are shown on the right side and the procedure of selection of germplasm followed in each tier of screening is provided on the left side

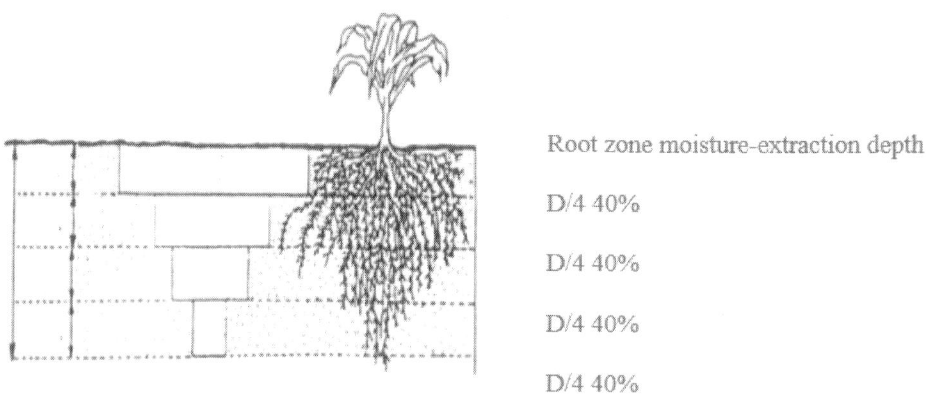

Figure 6.31. Moisture extraction pattern

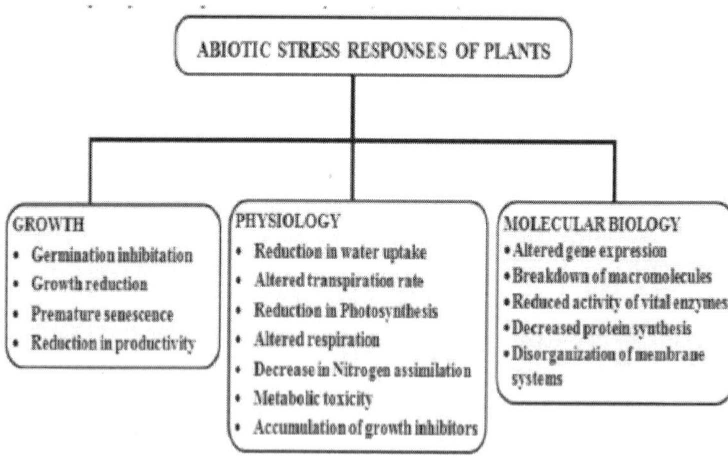

Figure 6.32. Abiotic stress response of plants

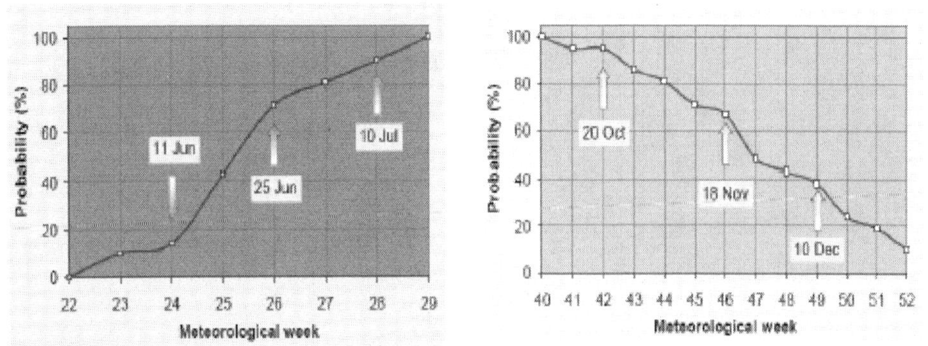

Figure 6.33. Variation in beginning and end of LGP at a place

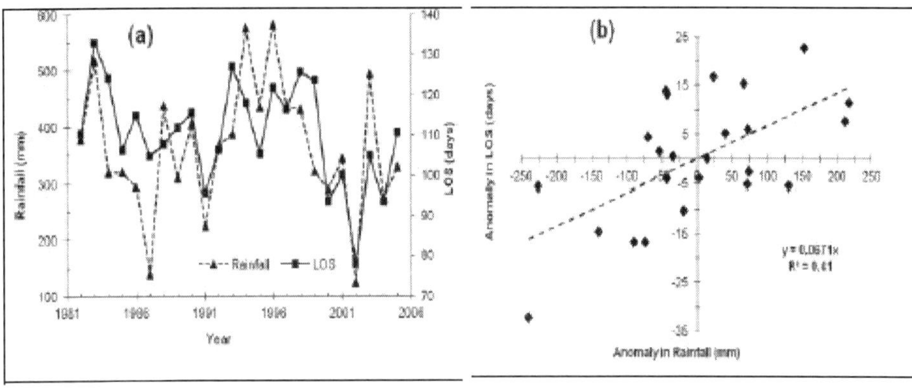

Figure 6.34. Relationship between length of season (LOS) and rainfall as (a) temporal pattern and (b) scatter plot of their anomalies with regression line

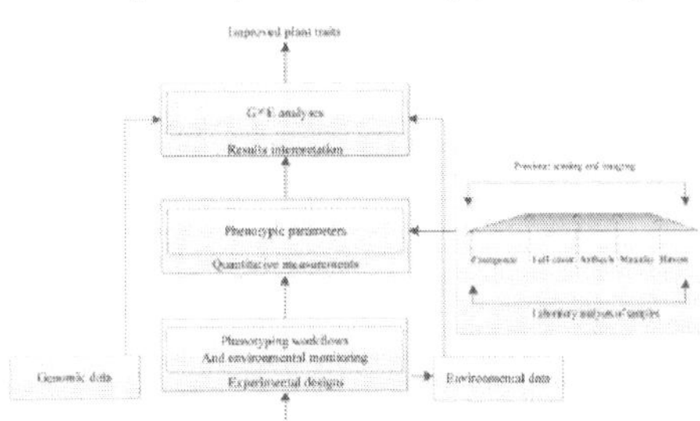

Figure 6.35a. Scheme for plant phenotyping

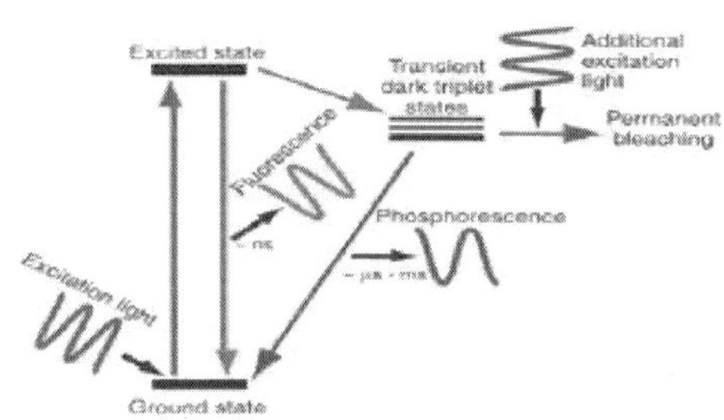

Figure 6.35b. Scheme for plant phenotyping

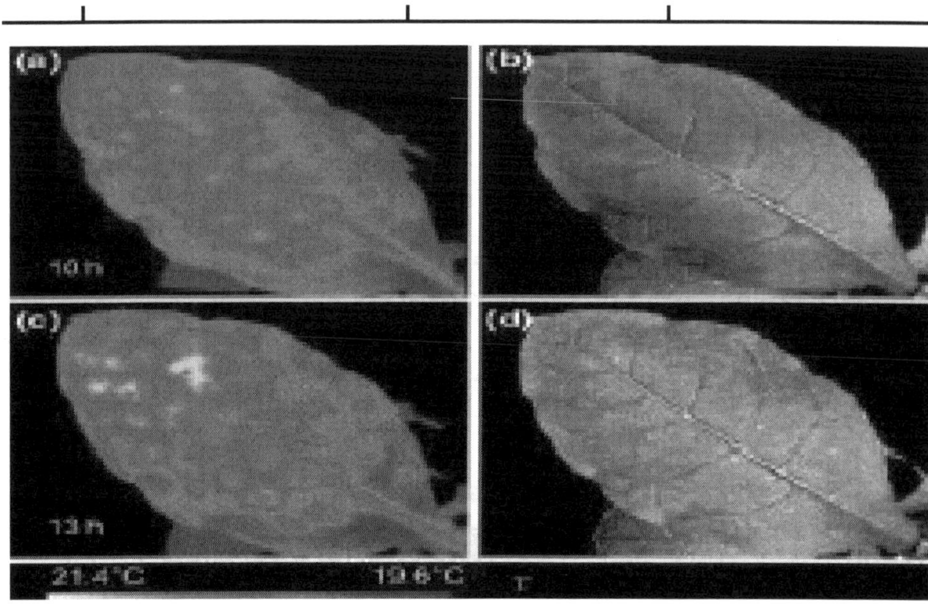

Figure 6.36a, b, c & d. Plant phenotyping

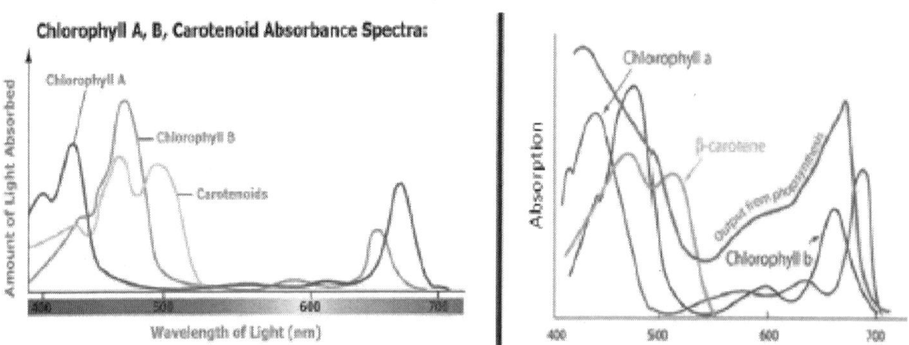

Figure 6.37a & b. Chlorophyll A, B, Carotenoid absorbance spectra

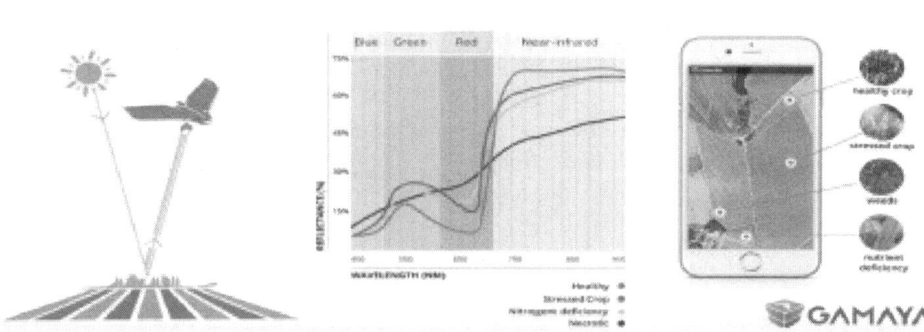

Figure 6.38a, b & c. Measure reflectance of crop making proprietary hyperspectral imaging camera, b. Analyse spectrum of reflected light and correlate with crop and characteristics, c. Identify perceived problems of farmland

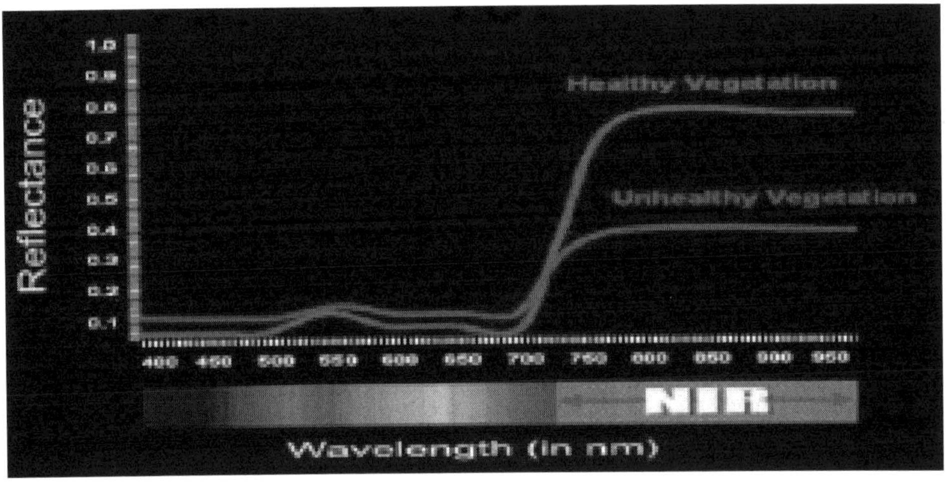

Figure 6.39. Graph between Reflectance and Wavelength

Figure 6.40. Prototypes to screen genotypes in (a) field, (b) pot culture (c) tools for rapid analysis of images

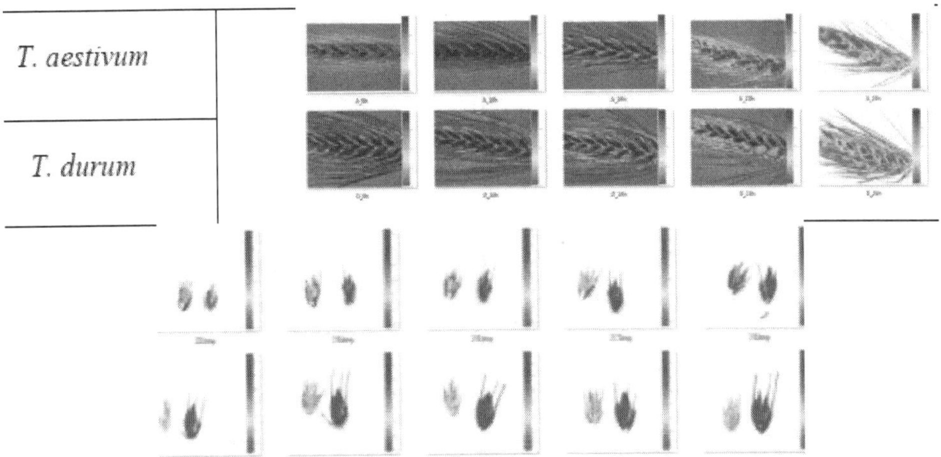

Figure 6.41. Chlorophyll fluorescence imaging to assess desiccation tolerance

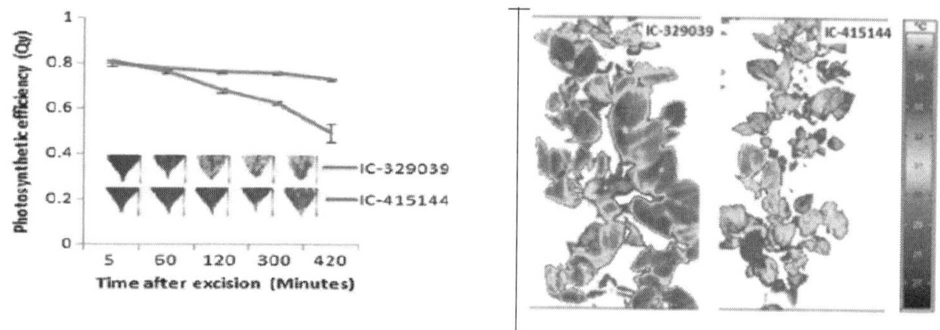

(a) (b)

Figure 6.42. (a) Fluorescence images depicting change in chlorophyll fluorescence of excised mungbean leaves of high and low LWL genotypes over a period of indicated time points (b) Thermal images of the high and low LWL genotypes exposed to drought

Figure 6.43. Mini-Rhizotron for *in-situ* root image acquisition

Figure 6.44a. *In-situ* root image acquisition by using CI-600 root scanner

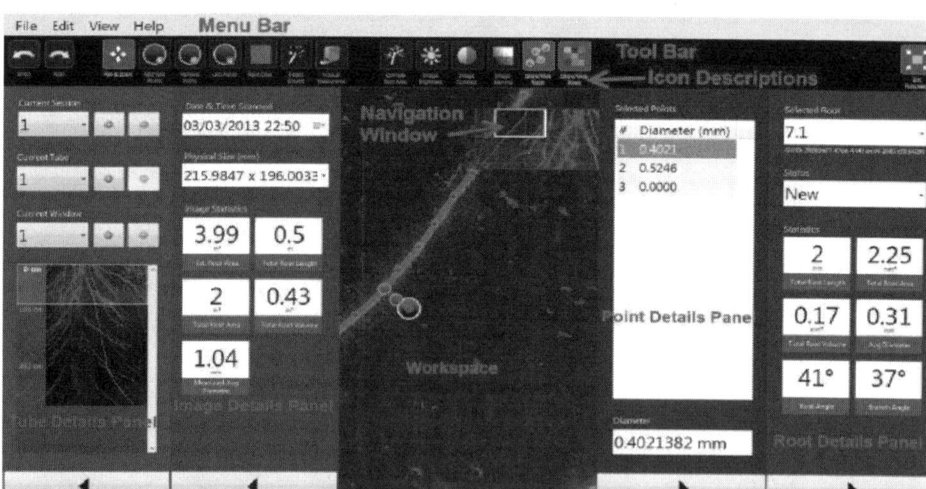

Figure 6.44b. *In-situ* root image acquisition by using CI-600 root scanner

Breeding to Enhance Crop Productivity in Scarce Water Conditions

To enhance crop yield per unit of scarce water requires both improved cultivars and better management. The challenge is to manage the crop or progress its genetic makeup to: capture more of the water supply for use in transpiration; exchange transpired water for CO_2 more effectively in producing biomass; and convert more of the biomass into grain harvest. The upper limit of water productivity of well-managed disease-free water-limited cereal crops is typically 20 kg $ha^{-1}mm^{-1}$. When water is the predominant limitation, there is a little scope for improving overall water productivity by better matching the crop development to the pattern of water supply, thereby reducing evaporative and other losses and fostering a good balance of water-use before and after flowering, which is needed to give a high harvest index. Marker-assisted selection can help in controlling some root diseases that limit water uptake, and in maintaining fertility in water-stress. The ideas of drought resistance and drought tolerance are giving way to the idea of water productivity ("**more crop per drop**"). Thus water can be quantified, with crop yield per unit volume of water. While it is true that a crop plant that does not survive severe water deficits will not produce any yield. "**Water productivity**" can also mean different things to different people (Kijne *et al.*, 2003, Pereira *et al.*, 2002). An attempt has been made primarily on improving water productivity.

Water as a Limiting Resource

This implies two questions *viz.*, whether the water is mainly limiting crop yield and when water is the main limiting factor how it can be overcomed. The limiting water resource can bee dissected into three components (Araus *et al.*, 2002), namely, how can one manage a crop or improve its genetic makeup to: transpire more of the limiting water supply, exchange transpired water for CO_2 more effectively in producing biomass, and, convert more of the biomass into grain or other harvestable product. There is a wide range of biochemical, physiological, agronomic and ecological processes that may affect water productivity and that variously influence these components.

Water as the Determinant to Yield

There is a lot of available information on winter-cereal crop yields in relation to rainfed water supply. provided that estimates of available water in the soil at sowing are good and taken into account when assessing the seasonal water supply. The rough estimate of the amount of water lost by direct evaporation from the soil (Hanks *et al.*, 1969) commonly varies by a factor of two, in specific cases, around the 110 mm average. The reason for such deviations is that the distribution of rain during the growing season can be unfavourable; for example, a preponderance of small falls of rain can lead to larger losses by direct evaporation from the soil (Sadras, 2003); or water deficits at flowering can lead to infertility even though the crop may have had good water supply at other times. (Hammer *et al.*, 1993). Subsequent to the range of years there has been a large increase in yields such that many farmers are now achieving yields close to the upper bound (Passioura 2002). In the earlier period, root diseases were rife, and thus exposed the crops to unexpectedly large water deficits because of the inability of the roots to effectively exploit water in the subsoil. Such data are rare, though Musick *et al.* (1994) provide a good example. The more common way of expressing water productivity is as the ratio of yield to water supply or total evapotranspiration. The empirical evidence accords with knowledge of the processes governing the net exchange by leaves between water and CO_2 leading to the production of biomass (transpiration efficiency (TE), and by the fraction of that biomass that can be converted into harvest index (HI), a dimensionless quantity usually defined as the ratio of grain yield to above-ground biomass. Transpiration efficiency depends on photosynthetic type (C_3, C_4, CAM) and on the evaporative demand of the environment. While the transpiration efficiencies of C_4 plants are famously much larger than those of C_3 plants at the level of gas exchange of leaves, the differences between the two in water productivity may be much smaller in the field, partly because of constraints that arise when scaling up (Gifford, 1974) and partly because C_4 crops are generally grown in hotter climates with larger evaporative demands (Fischer and Turner, 1978). As a general rule, well-managed cereal crops attain a maximum water productivity of about 2.0 kg of grain per cubic meter of transpired water (20 kg ha^{-1} mm^{-1}). Comparable data for oilseeds and grain legumes are scarce, but generally these crops have smaller maximum water productivities, ranging from about 8 to 15 kg ha^{-1} mm^{-1} (Johnston *et al.*, 2002). Because of the much greater energy cost of producing oil compared with carbohydrate, oilseeds have a maximum water productivity of about two thirds that of wheat (Hocking *et al.*, 1997). Yields markedly less than these maxima are worth exploring for maladies other than low water supply.

Capturing more Water

A rainfed crop's water supply comprises available water in the soil at the time of sowing plus rainfall during the growing season. The main losses are by direct evaporation from the soil surface and vertical drainage of water beyond the reach of the roots. Run-off from the soil surface may be substantial during heavy rain, but much of that run-off may become run-on in lower parts of a field, with little net loss from the field as a whole (Batchelor *et al.*, 2002) unless infiltration rate is poor or rainfall is intense. Runoff during intense rainfall can be greatly reduced with a good (>50 per cent) trash cover on the soil (Silburn and Glanville 2002). For crops that rely on growing-season rainfall, much water can be lost by direct evaporation from the soil, especially if there are many small falls of rain (Sadras, 2003). Crops that rely largely on water stored in the soil at the time of sowing lose much less in this way while they are growing, though evaporative losses before sowing are typically large (Hatfield *et al.*, 2001). Cabangon *et al.* (2002) report especially large evaporative and other losses in rice fields in the few weeks it takes to prepare them for planting. The rate of development of leaf area by the young crop strongly affects this loss, for the larger the leaf area the greater is the proportion of evapotranspiration that passes through the leaves. Indeed, the marginal increase in evapotranspiration induced by additional leaf area can approach zero if the distribution of rain is such that the soil surface remains wet during much of the crop's vegetative phase, for the evaporative demand by the environment may be met no matter what the ratio of transpiration to direct evaporation (Shepherd *et al.*, 1987).

Canopy temperature depression (CTD): CTD will be measured at two stages *i.e.* at anthesis and one week after anthesis. For measurement of CTD, canopy temperature and the ambient air temperatures will be measured using handheld infrared thermometer and the CTD will be calculated as follows:

CTD = Ambient air temperature-Canopy temperature

Chlorophyll content: As measured by chlorophyll meter (SPAD/CCM) at two stages *i.e.* at anthesis and one week after anthesis.

No. of tillers per meter length: The number of tillers in per meter length of two central rows will be counted and averaged.

No. of grains/spike: Ten spikes will be randomly selected from the plot and the number of grains will be counted. Average no. of grains/spike will be calculated.

Grain filling duration (GFD): The GFD will be calculated as the number of days required from anthesis to physiological maturity (when 75 per cent glumes turn yellow).

Green flag leaf duration (GFLD): GFLD will be calculated as the number of days required from anthesis to physiological maturity (when 75 per cent glumes turn yellow).

1000-grain weight (g): Ten spikes will be randomly selected from two central rows from each plot, hand threshed and 1000 grain weight estimated.

Grain yield/plot: The grain yield of the entire plot will be determined in grams.

Total biomass production: The plot will be harvested and total biomass production will be determined.

Harvest index: The harvest index (Grain yield per plot/Total biomass produced per plot) will be calculated

Abiotically Stressed Environments

These are the most fragile systems with biophysical constraints of Soil Edaphic Constraints: shallowness, coarseness, stony, low fertility, acidity, alkalinity, salinity, pollutants, hypoxia *etc.* Water Constraints: deficits: access and quantities, droughts, saline water, excess/waterlogging. Landscape constraints: steep terrain, boulders, undulations, sand dunes. Unfavourable climatic conditions: droughts, cyclones, hailstorms, heat/cold waves *etc.* Socio-economic constraints: small and fragmented holdings, low access to markets, poor infrastructure, low literacy, price instability *etc.*

Resistance to Abiotic Stress

Traditional approaches at transferring resistance to crop plants are limited by the complexity of stress tolerance traits, as most of these are quantitatively linked traits (QTLs). The direct introduction of a small number of genes offers convenient alternative and a rapid approach for the improvement of stress tolerance. Although, present engineering strategies rely on the transfer of one or several genes that encode either biochemical pathways or endpoints of signalling pathways, these gene products provide some protection either directly or indirectly against environmental stresses. Drought is a major abiotic factor that limits crop productivity, thereby causing enormous loss. Low temperature (LT) tolerance is a complex quantitative character that is expressed in anticipation of and during exposure of plants to temperatures that approach freezing. This environmentally reduced character is determined by a highly integrated system of structural and developmental genes that are regulated by environmentally responsive complex pathways. The superior LT-tolerance genes have been tagged using molecular markers that allow plant breeders to select hardy genotypes without having to wait for a test frost in the field (Fowler and Limin 2007). The impact of hardly earned green revolution is fading, primarily due to biotic and abiotic stresses, limiting the growth and productivity of almost all crop plants. The biotic stresses include the stress caused to plants vis a vis reduced productivity due to insect, pest and diseases, while the abiotic stresses include salt, drought, temperature and ion stresses *etc.* Hence evolving the biotic/abiotic stress resistant/tolerant plants is the only alternative to harness the fruits of green revolution. Conventional breeding is a time consuming and cumbersome approach. Moreover, it is difficult to screen and handle larger populations. As an attractive alternative, the tissue cultures technique of generation and *in vitro* screening of mutants for various stresses have shown a promise in a recent past, in several crop plants. *In vitro* screening for biotic and abiotic stresses holds a significant importance, as millions of cells can be screened in a single Petri dish, which reduces both time and space. Several other advantages of *in vitro* screening for disease resistance over conventional breeding include: relatively high mutation frequency, less chances of chimera formation, strictly controlled nutritional and physical environments, easy administration of selective agents. Moreover, the *in vitro* screening can be carried out throughout the year, without any seasonal constraints. Enormous variability is accumulated by plant cell proliferating in a culture (Bajaj, 1996). The major reasons for occurrence of the genetic variability include karyotypic changes, cryptic changes, associated with chromosome rearrangements, transposable elements, gene amplification and somatic crossing over. The genetic variation occurring in somatic cells is known as somaclonal variation. Treating the cell cultures with various physical and chemical mutagens can further enhance this variation. The prerequisite for *in vitro* screening is the callus induction, which can be established from almost every plant part. However, callus induction from tender/young explants such as from in vitro seedlings, leaves, embryos and inflorescence is generally more proliferative and exhibit efficient plant regeneration. Murashige and Skoog (MS) 1962 medium supplemented with higher concentrations (2-10 mg/l) of auxins *viz.*, 2, 4-D (2, 4 dichlorophenoxy acetic acid), NAA (Naphthalene acetic acid) and low concentrations (0.1-1 mg/l) of cytokinins *viz.*, kinetin (6 furfuryl amino purine) are used for callus induction. Cell suspension cultures can further be obtained by transferring actively growing calli to liquid medium and incubating the culture on shaker, throughout. For *in vitro* screening millions of cells are placed in the petridish on the medium containing elevated levels of biotic and abiotic stress factors. The susceptible cells get automatically eliminated and only resistant ones survive which divide and redivide and form colonies. Thus, obtained calli are regenerated into whole plant, which are finally assessed for a particular stress.

Challenges of Heat in Wheat Production

Heat poses a real serious threat to wheat cultivation under the fast-changing scenario of 21st century. Adverse effects of long exposure to moderately high temperatures and short exposure to very high temperatures on grain growth explains the complexity of high temperatures and vulnerability of wheat crop. In spite of having a large number of genes from harsh climates of the world in its present genomic composition, wheat continues to be a cold loving cereal. High temperatures affect all the vital stages of life cycle in bread wheat and durums. Wheat seeds sense the threshold temperatures very effectively and enter into a secondary dormancy at very high temperatures. A substantial reduction in wheat yields has been reported under heat stress due to acceleration in phasic development. Mean maximum temperatures during grain development between 25°C to 35°C and between 35°C to 40°C are considered moderately high and very high, respectively. Heat stress has been reported to reduce the grain weight by reducing grain growth period. Range of mini. – max. temperatures from October to April at Indore, representing a Malwa location, show the severity of high October temperatures and a big diurnal variation in temperatures. The hexaploid wheat species showed highest heat tolerance in pre-anthesis biomass accumulation rate when compared to that of diploid and tetraploid wheat species. Results of heat studies done in a particular agro-climatic zone under different experimental conditions from isolated cells and single potted plants in a lab to wheat crop in the field can hardly be replicated *in toto* in any one condition and hence need to be synthesized afresh in a specific target environment. Temperature tolerant cell lines/plant have been obtained for both low and high temperature. It has been found that mannitol and ABA increase proline, when added to the medium, which resulted in increased chilling tolerance. It suggested that proline accumulation can be used as selection criteria for cold resistance. Likewise, the insensitivity of cultured cells to ABA, has been used to select for heat tolerance in wheat.

Screening Wheat Germplasm for Heat Tolerance at Terminal Growth Stage

Wheat is a winter season crop grown in the tropics and subtropics despite the relatively high temperature that occur during the growth cycle. Heat stress is an important constraint to wheat productivity affecting different growth stages specially anthesis and grain filling. It has already been established that heat stress can be a significant factor in reducing the yield and quality of wheat (Stone and Nicolas 1995). In Pakistan terminal heat stress is a major reason of yield decline in wheat due to delayed planting. Similarly heat stress is a major challenge to wheat productivity in India (Joshi *et al.*, 2007). Late planted wheat suffers drastic yield losses which may exceed to 40-50 per cent. Therefore, there is a dire need to develop genotypes that are either tolerant to terminal heat stress or that mature early without yield losses and thus escape the stress. Terminal heat stress is a common abiotic factor for reducing the yield in certain areas of West Asia and North Africa (Ferrera *et al.*, 1993). Heat tolerance thus should be essential characteristic of wheat cultivars to be developed. Stay green is a trait that has been used to indicate heat tolerance in hot environment (Kohli *et al.*, 1991). Photosynthetic rate is maximum at 20-22°C and decreases abruptly at 30-32°C (Murata and Lyama, 1963). Heat stress injuries of the photosynthetic apparatus during reproductive growth of wheat diminish source activity and sink capacity which results in reduced productivity (Harding *et al.*, 1990). Source activity is damaged by heat because both leaf area (Slovacek and Hind 1981) and photosynthesis is reduced (Kuroyanagi and Paulsen 1985). Heat injury limits sink growth potential particularly when stress is imposed during early sink developmental stages (Nicolas *et al.*, 1984). Grain yield was negatively related to the thermal time accumulated above the base temperature of 310C (Mian *et al.*, 2007). High temperature above 32°C has been reported reducing grain yield and grain weight (Wardlaw *et al.*, 2002). This study was conducted to find out sources of terminal heat tolerance in bread wheat germplasm for utilization in the breeding program. varieties/lines were sown in a plastic sheet tunnel. One set of the material was sown in the open adjacent to the tunnel. Each entry was sown in a single meter row by maintaining 30cm and 7.5cm inter and intra row spacing. Three seeds per hill were placed at the time of planting and then at two leaf stage, thinned to single seedling. NPK fertilizer was applied @ 100-85-0 kg/hectare. During the 1st year the experiment was sown on 1st December while in the next year, the crop was planted on 10th November, the crop was exposed to heat stress by covering the tunnel with plastic sheet for two weeks (11th to 25th March). However, the tunnel was covered for four weeks (25th February to 24th March) exposé the material to heat stress. Daily temperature inside and outside the tunnel was recorded and maintained above 32°C inside the tunnel. At maturity twenty five spikes from each entry were randomly selected for data recording and data for the following characters were recorded: 1) Grains/spike 2) 1000-grain weight 3) yield/spike The heat effects for each character were expressed as ratio of stressed/nonstressed (relative ratio) and for each variety/line as relative value (mean of ratios of all entries for the characters studied). Data for plant survival (ability to stay green) was also recorded after 21 days exposure to heat stress. The capability to set seed/survival under heat stress for a longer period was regarded as heat tolerance. The germplasm comprising of 442 wheat varieties/lines was sown in one meter long row in a plastic sheet tunnel to screen the material for heat tolerance. A set of the material was sown in the open adjacent to the tunnel. The material was exposed to heat shock (>32°C) by covering the tunnel with plastic sheet during grain formation for two weeks in 2004-05 and for four weeks in 2005-06. Data was recorded from 25 randomly selected heads from each row for 1000 grain weight, grains per spike and yield per spike during both the years. The data regarding survival (ability to stay green under heat stress) was also recorded. Heat effect was expressed as ratio of stressed/non stressed plants. The effects of heat stress were lesser in shorter period exposure and more drastic in prolonged exposure of the genotypes to heat. The ability of lines to stay green for longer period in heat shock had no direct relationship with seed setting. Three entries CB-367 (BB#2/PT//CC/INIA/3/ALD'S') CB-333 (WL 711/3/KAL/BB// ALD 'S') and CB-335 (WL711/CROW 'S'//ALD#1/CMH77A. 917/3/HI 666/PVN 'S') showed maximum grain development and survival. This study revealed that these genotypes can be utilized in breeding programs for development of wheat varieties having heat tolerance at terminal growth stage.

Table 6.1. Selected heat tolerant lines and their relative ratios for the characters studied in heat shock for two weeks.

Parentage	1000 Grain Weight	Yield/ Spikes	Relative Value
PBW 343	0.9	1.03	1.02
Inqilab/Fink's	2.36	1.44	1.47
Kauz*2/Yaco//Kauz	0.9	0.92	0.89
PVN//CAR422/ANA/3/Kauz*2/TRAP//Kauz	0.91	1.08	1.0
PVN/YACO/3/Kauz*2/TRAP//Kauz	0.95	1.06	1.04
-DO-	0.88	0.9	0.93
ATTILA/3/HUI/CARC//CHEM/CHTO/4/ATTILA	0.93	0.98	0.99
FRET-	0.88	0.91	0.95
CNO'S'/PJ62//ON1160.147/3/BB/GALLO	1.25	1.06	1.05
SKA/PRL'S'//YDING	0.91	0.98	1.06
SW 89.3064*2/BORL 95	1.02	0.92	0.95
LU26/3/KWZ/JB 216.67//SIS'S	1.03	0.96	0.9
HD 2359/3/OVA=/AZ/MUS'S'	0.86	0.9	0.91
MAYA/PVN	0.96	0.9	0.93
KVZ/3/TOB/CTFN//BB/4/BLO'S'/5/VEE#5/6/BOW'S'/3/ YDING'S'//BB/CHA	1.07	0.94	0.98
WL711/3/KAL/BB//ALD	0.93	1.17	1.12
WL711/CROW'S'//ALD#1/CMH77A.917/3/HI666/PVN'S	0.86	0.92	0.95
ZHONG 69/PRL'S	0.88	0.91	0.95
WL711//ALD'S'/PVN'S'	0.78	1.38	1.27
CHIS/ALGER 86//HD2204/3/KVZ/TRM//PTM/INIA	1.03	1.02	1.02
HI 666/PVN'S'	0.82	0.9	0.94
BCN/4/RABI//GS/CRA/3/AENTRYSQ (190)	0.83	0.91	0.94
NEW CHENAB//HD2204/JUN'S'	0.9	0.9	0.93
CHEN/AENTRYSQ (205)//3* KUUZ'"S'	0.92	0.98	0.99
BOW/MOR//BAU	0.86	1.04	1.04
BAV 92//SAP/MOM	0.83	0.92	0.95
TRM//MAYA 74S/MON'S'/3/INQ 91	1.3	0.94	0.99
PND 88/BB'S'/T 0B-66	0.93	0.96	0.98

As far as population behavior is concerned when the material was exposed to heat stress for two weeks, the relative ratio for the characters studied *i.e.*, grains per spike, 1000 grain weight and yield per 25 spikes showed normal curves. Maximum number of entries (135) showed 0.61-0.80 relative ratio for grains per spike which means that they were able to maintain 61-80 per cent of the grains per spike under heat stress as compared to the normal while the 153 entries produced even less grains per spike as their relative ratio was less than 61 per cent and even there were some entries which had shown no seed formation under heat stress. There were 99 entries which showed 0.80-1.00 relative ratio. Some entries gave higher than 1.00 relative ratios which might be due to experimental error. When these entries were exposed to heat stress for four weeks after anthesis in 2005-06, the nature of the curve was changed and different type of curve was obtained. Most of the entries had shown no seed formation Only 57 entries were able to set seed. Out of which 21 entries showed relative ratio 0.0-0.2, 16 entries showed ratio 0.21-0.4, eight entries showed 0.41-0.60, six entries showed 0.61-0.80 and one entry showed greater than 0.80 relative ratio for number of grains per spike under heat stress.

Table 6.2. Selected heat tolerant lines and their relative ratios for the characters studied under prolonged heat shock.

Parentage	*Grains/ Spike*	*1000 Grain Weight*	*Yield/ Spikes*	*Relative Value*
CHAM 4	0.48	0.54	0.26	**0.43**
MEXIPAK 65	0.31	0.66	0.2	**0.39**
SHALIMAR 88	0.51	0.63	**0.32**	**0.49**
BAW 898	**1.06**	**0.41**	**0.43**	**0.63**
CMH 76A.912/CMH 76A.769	**0.06**	**2.5**	**0.15**	**0.90**
CHAT'S'	**0.71**	**0.39**	**0.28**	**0.46**
SW 89.3064 *2/BORL 95	**0.29**	**0.44**	**0.13**	**0.29**
CROW'S'/NAC//SA75/3/F3.71/TRM	0.46	0.51	0.24	0.40
MAYA/PVN	0.12	0.55	0.06	0.24
WL711/3/KAL/BB//ALD	0.32	0.57	0.18	0.36
WL711/CROW'S'//ALD#1/CMH77A.917/3/HI 666/ PVN'S'	0.12	0.67	0.08	0.29
BB#2/PT//CC/INIA/3/ALD'S'	0.65	0.7	0.45	0.60
NEW CHENAB//HD 2204/JUN'S'	0.61	0.49	0.3	0.47
TRM//MAYA 74'S'/MONS/3/INQ 91	0.53	0.54	0.29	0.45
ALTAR84/AENTRYSQ(219)//OPATA/3/3*INQ 91	0.5	0.53	0.27	0.43
91B 2061/FSD 83	0.94	0.49	0.46	0.63
7C/FANG.60	0.63	0.44	0.27	0.45
CHENAB//HD 2204/JUN'S'	0.7	0.58	0.4	0.56

During 1st year of study when the heat shock was given for shorter period 26 entries showed less than 0.60 relative ratio for 1000 grain weight whereas in 111 entries this character was slightly affected/not affected as the relative ratio was 0.61-1.00 whereas many entries showed higher grain weight which might be due to lesser grain number per spike as compared to normal. None of the entries had shown relative ratio near to the unity for all the traits when the heat stress was applied for four weeks. Similarly, grain yield was lesser affected under shorter heat stress period and drastically affected under longer heat shock. During shorter exposure 29 entries showed relative ratio near to unity. Whereas under longer exposure only 51 entries were able to give some yield. Only three entries were able to give 40-50 per cent yield as their relative ratio was between 0.40-0.50 followed by 3 other which realized 0.30-0.40.

Staying Green Ability

It was observed that greenness in some lines was affected and some lines maintained this trait during two weeks exposure during 1st year. So, during the second year the data regarding ability to stay green under heat stress was recorded. After three weeks of stress period, seventy two entries were green and after four weeks, thirty seven entries were green while rest of the entries was dried out due to heat shock. Under heat shock some lines showed tolerance as their relative values were near to unity. Among these the especial behavior observed for the line; Entry numbers 84, 198,432 was that in spite of the effect on number of grain per spike. They recovered after exposure and improved their grain weight to recovered level that their yields were near to unity but when exposed to shock for 4 weeks they were not able to recover and hence had shown very poor performance. Four entries have shown relative value and relative ratio for each character near to unity in shorter exposure and when exposed to heat shock for longer period. Entry numbers 367 and 436 have shown better relative ratio for yield and over all relative value by maintaining grain no and grain weight to certain extent. However, the three entries (Entry No. 333,335 and 367) were able to stay green for three weeks under heat stress but entry numbers 436 was not able to remain green under heat shock and thus was forced to maturity as it is evident from the study that it maintained its grain number.

Grains/Spike

However, its grain weight was reduced more than 50 per cent under prolonged heat shock. This indicated that prolonged exposure to heat shock at grain development phase have severe effects on grain weight (Wardlaw *et al.*, 2002). Heat shock affected all the three characters when heat stress was applied for shorter period, the effects were less pronounced. When the length of application was extended during the 2nd year of study, only few entries survived and none of these showed normal grain weight. One entry showed exceptionally normal grain weight which might be due to escape. Similarly none of them have normal grain weight under heat stress for longer period. Whereas reasonably good number of entries showed normal grain weight under heat stress for shorter period while some of the entries showed higher grain weight under stress which might be due to component compensation effect (Khan, 2003). However, yield was also affected under shorter exposure whereas much affected under heat shock for longer period. During the 1st year, some of the entries have shown better performance under short heat stress period because of heat shock recovery or escape phenomenon. Therefore, during the second year, heat stress period was prolonged and there was no chance of recovery or escape. So, the yield in case of all the entries was drastically affected. Similarly, negative effects of heat stress on yield were reported by Mian *et al.* (2007).

1000-kernel Weight

Best performing entries under heat stress for two weeks. These entries showed relative ratio/value near to unity. Relative value for each entry was calculated on the basis of its performance for the three characters. The relative ratio for each character was calculated and then was averaged to get relative value.

Table 6.3. Lines remained green during heat stress imposed from 25.2.2007 to 24.3.2007 at ICAR-IARI RS Pusa, Bihar

Varity/line
Chenab79
CHILERO=CHIL'S
FAISALABAD 85
IQBAL2000
KANCHAN
KASYON/BOW'S'
MILAN
MILAN/HD 832
PUNJAB 76
SA 42
WATAN/2*ERA
NG8319//SHA4/LIRA
SKAUZ*2//PRLII/CM65531
RABE/6/WRM/4/FN/3*TH//K58/2*N/3/AUS 6869/5/PELOTAS-ARTHUR/7/2*RABE/8/LAJ 3302
RABE/6/WRM/4/FN/3*TH//K58/2*N/3/AUS-6869/5/PELOTAS-ARTHUR/7/2*RABE/8/IRENA
KAUZ*2/YACO//KAUZ
PFAU/WEAVER
COMARA/TEG//WEAVER/3/LAJ 3302
KAUZ/TSC//2*SKAUZ
ACONCHI-89
MRB589-5
WH542
ATRIS-2
HAAMA-11
MON'S'/ALD'S'
CNO'S'/PJ 62//ON.1160.147/3/BB/GALLO
CONDOR'S'/ANA75//CONDOR'S'/MUS'S'
BOBWHITE'S'
BUTEO'S'
KVZ/3/TOB/CFTN//BB/4/BLO'S'/5/SNB'S'
PRL'S'/PEW'S'
SUNBIRD'S
UP 262
MEISE
SKA/PRL'S'//YDING

LU26/3/KVZ/JB216-67//SIS'S'
BURGUS/SORT-12-13//KAL/BB/3/PAK.81
HD 2359/3/GOV/AZ//MUS'S'
NAPHAL/CONDOR
CH.70/6/KN.83//CH.70/ALD/5/CH.70/4/INIA/CNO//CAL/3/LR/SON
HAHN'S'
K338/EOCH//KUDIAT 17/KTY/3/IMREN/4/6CGJ
PGO/SERI
BOW'S'//CMH75A.142/3/CMH74A.487/HD 2172
MAYA/PVN
SKA/PANILA//YDING
KAUZ//ALTAR84/AOS
VEE/KOEL//HEI/3*CNO79/3/SKAUZ
4777//FKN/GB/3/VEE'S'/4/BUC'S'/PVN'S'
BUC'S'/FLK'S'//MYNA'S'/VUL'S'
PRL'S'/VEE#6//MYNA'S'/VUL'S'
VS73.600/MRL'S'/3/BOW'S'//YR/TRF'S'
AMD/HN4/3/GTO/7C//BB/CNO67/5/PVN'S'/4/BB/CNO'S'//JAR/3/ORZ'S'/6/ TAN'S'/SNB'S'
CYUS'S'/MESCORIA/5/D68.11/3/GLL*2//T. DIC.V.VERN/4/DWL5023
KVZ/3/TOB/CTFN//BB/4/BLO'S'/5/VEE#5/6/BOW'S'/3/YDING'S'//BB/CHA
VEE'S'/SNB'S'

Table 6.4. Selected heat tolerant lines that stayed green and produced seeds

Parentage	*Grains/ Spike*	*1000 Grain Weight*	*Yield*	*Relative Value*
WL 711/3/KAL/BB//ALD = V-85054	0.79	0.74	0.67	**0.73**
WL 711/CROW "S"//ALD #1/CMH 77A.917/3/HI 666/PVN 'S'	0.59	0.77	0.50	**0.62**
BB # 2/PT//CC/INIA/3/ALD 'S'	0.78	0.87	0.62	**0.76**
MAYA/PVN	0.53	0.75	0.48	0.59

The lines showed relative ratios near to unity (1.00±0.1) and consequently their relative value near to unity. Some lines have shown unique behavior. Entry numbers 84,198,432 showed lesser number of grains per spike. However, when the heat shock was off, they improved their grain weight due to component compensation effect and ultimately their yield was normal under heat shock treatment. When the heat shock was applied for four weeks, none of the entries showed more than 50 per cent of their yield potential. The relative ratio for yield was less than 0.50 for each entry. Three entries (Entry No. 367, 436) have shown better relative value 0.60-0.63 and relative ratio for yield *i.e.* 0.43-0.46. This was due to their ability to maintain grain number and grain weight under heat shock. Entry numbers 333 and 335 are not worth mentioning as they produced nominal yield after two and four week's exposure. Stay green had been used as a selection criterion in hot areas (Kohli *et al.*, 1991). Many researchers (Herzog *et al.*, 1982) had reported negative affect of heat on plant. As far as relationship between the ability of the lines to stay green and their relative value is concerned, none of the lines that stayed green for four weeks were able to set seed in reasonable amount. However, some entries (Entry No. 333, 336 and 367) which were able to set some seed also stayed green for three weeks. Many of the entries that produced good yield under heat stress were not able to stay green for three weeks. This might be due to the fact that these lines were forced to maturity after setting as much seed as possible. Therefore, the ability of a line to stay green has no direct relationship with seed setting. To develop heat tolerant lines, selection should be made on the basis of yield, grain weight and grain number per spike. High temperature effects on yield and grain weight had also been reported (Mian *et al.* (2007). This study concluded that most of the wheat genotypes were more affected when exposed to heat stress for longer period of time and vise versa. However, the genotypes vary in their ability to tolerate heat stress. Some genotypes had the capability to stay green for longer period under heat stress. There was no direct relationship of the ability of lines to stay green for longer period in heat shock and seed setting. Three entries CB-367 (BB#2/PT//CC/INIA/3/ALD'S') CB-333 (WL 711/3/KAL/BB//ALD 'S') and CB-335 (WL711/CROW 'S'//ALD#1/CMH 77A. 917/3/HI 666/PVN 'S') displayed the uppermost grain development and survival under the stress. This study revealed that these genotypes can be utilized in breeding programs for development of wheat varieties having heat tolerance at terminal growth stage.

In vitro Screening for Abiotic Stresses

The abiotic stresses include high salt levels leading to salt stress, low water availability leading to drought stress, excess water leading to submergence and anoxia stress, sub-optimal temperature leading to low temperature stress and supra optimal temperature leading to high temperature stress. The ion stress due to heavy metals in soil is also important. The basic factor that decides the choice of crops in a given region is the *Length of Growing period (LGP)*. The LGP, however, depends not only on the rainfall distribution but also on the soil type, depth and water holding capacity. The depth, therefore, is expressed as the sum of the length of the rainy season and the period for which the stores soil moisture meets the crop water need after the cessation of the rainy season. As per FAO (1996), LGP starts when P > 0.5 PET and ends with utilization of assumed quantum of stored soil moisture (100mm), after P falls below PET. A long-term data analysis of LGP data by Kesava Rao *et al.*, 2013 at Chhattisgarh, India LGP generally begins by 20th June and ends by 25 Nov (LGP, 160 days). In 3 out of 4 years, the season begins by 25 Jun and ends by 18 Nov. In 9 out of 19 years, it begins by 10 Jul and ends by 20 Oct. Therefore the assured length of season (LGP) is 21 weeks (145 days) beginning from 25 Jun and ending by 18 Nov. The LGP of various crops given by FAO (1996) is given in Table 6.2.

Table 6.5. Length of Growing Period requirement of various crops

Crop	*LGP (days)*
Wheat/Barley/Oats	120-150
Rice	90-150
Maize, sweet	80-110
Sorghum	120-130
Millet	105-140

Dhakar *et al.* (2013) studied the relationship between average LOS and average seasonal rainfall and the scatter plot between their anomalies (from their normal). A close correspondence is seen between LOS and rainfall in most of the years. The LOS decreased during drought years when less rainfall was received. The Figure 3b clearly shows that anomaly in LOS is directly related to anomaly in seasonal rainfall. The nature of relation between the two is linear with a *R*2 value of 0.41 which is significant at 99 per cent confidence level. The relation implies that, on an average the length of season decreases by about 7 days for each 100 mm seasonal rainfall deficit.

Identification of the Components Involved in Drought Tolerance Mechanism

The genetic model of model plant *Arabidopsis* has allowed the identification of numerous pathways important to growth under limiting water (Hirayama and Shinozaki 2010], and these pathways tend to be conserved among agricultural crops (Nakashima *et al.*, 2009). In the last decade, researchers made promising breakthroughs in understanding ABA biosynthesis, ABA receptors, and other components of the ABA signal transduction pathway (Kline *et al.*, 2010). This valuable new mechanistic understanding of the complex ABA signalling pathway will help in expediting innovations around managing plant responses to drought. Biotechnological approaches to improve drought stress tolerance in plants may involve over-expression of genes involved in particular aspects of cellular homeostasis such as osmotic adjustment, chaperones, or antioxidants (Yang *et al.*, 2010). Alternatively, ectopic expression or suppression of regulatory genes could potentially activate multiple mechanisms of stress tolerance simultaneously (Century *et al.*, 2008). Genes encoding members of the AP2/ERF transcription factor family including the Dehydration Responsive Element Binding Proteins (Xu *et al.*, 2011), ABA Response Element Binding Proteins, and NAC transcription factors (Nakashima *et al.*, 2009) have all shown promise, as well as genes encoding proteins involved in other aspects of signal transduction, such as kinases and protein modification enzymes (Yang *et al.*, 2010). In addition, progress in identification of plant microRNAs, including those with expression altered by drought stress (Wang *et al.*, 2011), provides exciting new targets for controlling drought response pathways. Demonstration of drought stress tolerance in crops in controlled environments is proceeding at an encouraging rate (Yang *et al.*, 2010). Many of these recent discoveries have been in rice, which is both an excellent model species for basic research, and one of the world's most important crops. Study of rice mutants with altered stress tolerance led to the identification of genes in three pathways that can be manipulated to improve stress tolerance (Zhang *et al.*, 2009). Several genes that can provide drought stress tolerance were identified by altered expression of genes shown to be induced by drought stress in rice (Liu *et al.*, 2009). Findings from Arabidopsis continue to be a rich source of drought leads (Gao *et al.*, 2011). Some transgenes were derived from extremely stress-tolerant species such as *Thellungiella halophila* (Lv *et al.*, 2009), a salt-tolerant relative of Arabidopsis, and Atriplex hortensis (Wang *et al.*, 2010), although direct comparison of alleles from less tolerant species is needed to validate this approach. Overexpression of some regulatory proteins has led to dwarf phenotypes with reduced yields, but use of drought-inducible (Xu *et al.*, 2011) or tissue-specific (Jeong *et al.*, 2010) promoters may overcome this issue. The magnitude and consistency of gene effects may be improved by co-expression of two or more transgenes that each provide drought efficacy, ideally through different mechanisms (Wei *et al.*, 2011).

Enhancement of Water Stress Tolerance in Crop Plants Employing Genetic Engineering Tools

Abiotic stress, mainly drought, reduces average yields for most major crop plants by more than 60 per cent. Water stress or water deficit is one of the major environmental factor, which prevents plants from realizing their full genetic potential and has a major impact on crop yield (Salekdeh *et al.*, 2009). Abiotic stress leads to a series of morphological, physiological, biochemical and molecular changes that adversely affect plant growth and productivity (Wang *et al.*, 2001). To counter this adverse environmental factor, plants try to avoid the drought condition through various ways. They try to escape the season of drought by early flowering. They may decrease the leaf area (LA), increase the efficiency of roots to acquire water or may decrease activity of stomata. Slowing growth, osmotic adjustments and synthesis of antioxidants are some other mechanisms adapted by plants to combat adverse environmental conditions. These adaptations help plants to adapt to drought conditions. Some plants are able to adapt to water deficiency by shortening their growth cycle or they avoid drought stress by augmenting root growth to increase their water uptake (Molnar *et al.*, 2004). Unfortunately, the mechanisms by which crops maintain yield under drought stresses are poorly understood since drought can occur at different stages of the plant's development, with different effects on plant function, and thus requires distinct mechanisms for tolerance. In addition, a variety of abiotic stresses commonly occur during drought, such as high temperatures, high concentrations of salt and other toxic solutes and low availabilities of nutrients (Fleury *et al.*, 2010), and these vary by location and time. Signaling pathways are induced in response to environmental stress and recent molecular and genetic studies have revealed that these pathways involve many components. The multiplicity of information embedded in abiotic stress signals underlies one aspect of the complexity of stress signaling (Chinnusamy *et al.*, 2004). Nevertheless, most studies on water stress signaling have focused on primarily salt stress because plant responses to salt and drought are closely related and the mechanisms overlap (Zhu 2002). Responses to stress are not linear pathways, but are complicated integrated circuits involving multiple pathways and specific cellular compartments, tissues, and the interaction of additional cofactors and/or signaling molecules to coordinate a specified response to a given stimulus. Plants respond to these stresses at molecular and cellular levels as well as physiological level. Expression of a variety of genes has been demonstrated to be induced by these stresses. The products of these genes are thought to function not only in stress tolerance but also in the regulation of gene expression and signal transduction in stress response (Shinozaki *et al.*, 2003). In addition, the sensitivity of many crops to a particular abiotic stress varies depending on their developmental stage. For example, rice is sensitive to salt stress at the young seedling stage, but much less at the reproductive stage (Lutts *et al.*, 1995). It is suggested that stress tolerance mechanisms in a plants are controlled by a variety of genes, which are expressed at different times during the life of the plant (Fleury *et al.*, 2010). Plant adaptations to most abiotic stresses involve a range of traits which combine to contribute plant tolerance. Individual genes have been reported to improve the stress tolerance in some crops, for instance, the transcription factor ZmNF-YB2 has been reported to improve drought tolerance in maize (Nelson *et al.*, 2007). While in majority of cases, it is not a simple matter of identifying the single gene that will provide resistance to a particular abiotic stress. This review describes recent advances in improvement of crop plants for water stress tolerance using biotechnological tools. This review deals with transgenic approach for improvement of crop plants for water stress tolerance. Many plants have multiple physiological, biochemical and molecular mechanisms that enable them to tolerate environmental stresses. Understanding the mechanisms by which plants perceive and transduce the stress signals to initiate adaptive responses and their engineering using molecular biology and genomic approaches are essential for improving abiotic stress tolerance in crop plants. Several efforts in this direction have been carried out in many laboratories targeting manipulation of genes belonging to diverse categories. Genetic engineering strategies rely on the transfer of one or several genes that are either involved in signalling and regulatory pathways, or that encode enzymes present in pathways leading to the synthesis of functional and structural protectants, or that encode stress tolerance-conferring proteins. Attempts have been made to confer drought resistance to plants through biotechnological approaches and drought tolerant varieties of crops such as wheat, soya bean and rice have been produced.

Genes involved in many of the essential steps of the stress response have been identified and characterized. In particular, the recent discovery of ABA receptors, progress in understanding the transcriptional and post-transcriptional regulation of stress responsive gene expression, and studies on hormone interactions under stress have facilitated addressing the molecular basis of how plant cells respond to abiotic stress. Genetic approaches with more complex genomes are becoming increasingly tractable as genomic information in non-model crops increases and even whole crop genomes can be re-sequenced. Thus, genetic approaches to elucidating the molecular basis to abiotic stress tolerance in crops are becoming more easily achievable. This knowledge can be delivered to breeders through marker-assisted selection or genetic modification technologies. The mechanism underlying the environmental stress response in plants is probably more advanced and prominent than in animals. Moreover, the question of how plant cells react to

Table 6.6: Transgenic crops for water stress tolerance

Pathway Targeted	*Gene Family*	*Trans Gene*	*Transgenic Expression*	*Crop*	*Reference*
Osmoregulation	H+-PPase	AVP1	CaMV35S	Cotton	Pasapula *et al.*, 2011
Osmoregulation + glycinebetaine biosynthesis	H+-Ppase + choline dehydrogenase	BetA and TsVP	Zm Ubiquitin	Maize	Wei *et al.*, 2011
ABA biosynthesis	LOS5/ABA3	LOS5	OsHVA22P (stress inducible) and OsActin1	Rice	Xiao *et al.*, 2009
ABA sensing; farnesyltransferase	Farnesyltransferase	BnFTA	RNAi with AtHPR1 promoter (drought induced in shoot)	Canola	Wang *et al.*, 2009
Stress response	AP2/ERF	AP37	Os.Cc1 (constitutive)	Rice	Oh *et al.*, 2009
		CBF3	OsHVA22P (stress inducible) and OsActin1 (constitutive)	Rice	Xiao *et al.*, 2009
		HARDY	CaMV35S	Berseem	Abogadallah *et al.*, 2011
	NAC	OsNAC10	RCc3 (root) (constitutive expression not efficacious)	Rice	Jeong *et al.*, 2010
	C2H2-EAR zinc finger protein	ZAT1	OsHVA22P (stress inducible) and OsActin1	Rice	Xiao *et al.*, 2009, Mittler *et al.*, 2006]
	MAP kinase	NPK1	OsHVA22P (stress inducible) and OsActin1	Rice	Xiao *et al.*, 2009
Ion transport	Na+/H+ antiporter	NHX1	Actin1	Rice	Xiao *et al.*, 2009
	Ser/Thr kinase	SOS2	OsHVA22P (stress inducible)	Rice	Xiao *et al.*, 2009, Batelli *et al.*, 2007

various environmental stresses is one of the most attractive topics not only to plant biologists but also to agronomists, because abiotic stress is a particular threat to crop productivity. It is estimated that abiotic stress such as drought, salinity and extreme temperatures, which usually cause primary crop losses worldwide, lead to an average yield loss of >50 per cent for most major crop plants (Boyer 1982). Furthermore, world food production needs to be doubled by the year 2050 to meet the ever-growing demands of the population (Tilman *et al.*, 2002). For these reasons, understanding the mechanisms underlying plant abiotic stress responses and the generation of stress tolerant plants has received much attention in recent years. However, because of the complexity of stress tolerance traits, conventional approaches are less effective at directly connecting tolerance traits to the determinant genes that play key roles in the stress response. Owing to recent progress in functional genomics, genes involved in many of the essential steps of the stress response have been identified and characterized. In particular, the discovery of ABA receptors, progress in understanding the transcriptional and post-transcriptional regulation of stress-responsive gene expression, and studies on hormone interactions under stress have facilitated addressing the molecular basis of how plant cells respond to abiotic stress. Importantly, the physiological functions of a number of genes have been investigated in transgenic model plants and some crops, and an approach for utilizing useful genes for crop genetic improvement by gene transfer has been proposed. This review describes recent progress towards understanding plant abiotic stress responses, primarily focusing on ABA receptor identification, stress-responsive gene expression regulation by transcription factors (TFs), signal transduction mediated by protein modification and the roles of phytohormones in plant stress responses and development. The roles of small RNAs and RNA-directed DNA modifications in this regard have been extensively reviewed elsewhere (Chinnusamy and Zhu 2009).

Engineering Genes involved in Plant's Stress Surveillance

Sensors initiate a signalling cascade to transmit the signal and activate nuclear transcription factors to induce the expression of specific sets of genes. Genes involved in stress signal sensing and a cascade of stress-signalling in model plant such as *Arabidopsis thaliana* has been of recent research interest (Shinozaki and Yamaguchi-Shinozaki 1999). Components of the signal transduction pathway may also be shared by various stress factors such as drought, salt and cold (Shinozaki and Yamaguchi-Shinozaki 1999). Although there are multiple pathways of signal transduction systems operating at the cellular level for gene regulation, ABA is known component acting in one of the signal transduction pathways, while others act independently of ABA. The early response genes have been known to encode transcription factors that activate downstream delayed response genes (Zhu 2002). Although, specific branches and components exist (Lee *et al.*, 2001), the signalling pathways for salt, drought, and cold stresses all interact with ABA, and even converge at multiple steps (Xiong *et al.*, 1999). Abiotic stress signalling in plants involves receptor coupled phosphorelay, phosphoionositol-induced Ca2+ changes, mitogen activated protein kinase (MAPK) cascade, and transcriptional activation of stress responsive genes (Xiong and Zhu 2001). A number of signaling components are associated with the plant response to high temperature, freezing, drought and anaerobic stresses (Grover *et al.*, 2001). One of the merits for the manipulation of signalling factors is that they can control a broad range of downstream events that can result in superior tolerance for multiple aspects (Umezawa *et al.*, 2006). Alteration of these signal transduction components is an approach to reduce the sensitivity of cells to stress conditions, or such that a low level of constitutive expression of stress genes is induced (Grover *et al.*, 1999). Transgenic tobacco plants produced by altering stress signalling through functional reconstitution of activated yeast calcineurin not only opened-up new routes for study of stress signalling, but also for engineering transgenic crops with enhanced stress tolerance (Grover *et al.*, 1999). Overexpression of an osmotic-stress-activated protein kinase, SRK2C resulted in a higher drought tolerance in *A. thaliana*, which coincided with the upregulation of stress responsive genes (Umezawa *et al.*, 2004). Similarly, a truncated tobacco mitogens-activated protein kinase kinase kinase (MAPKKK), NPK1, activated an oxidative signal cascade resulting in cold, heat, salinity and drought tolerance in transgenic plants (Shou *et al.*, 2004). However, suppression of signalling factors could also effectively enhance tolerance to abiotic stress (Wang *et al.*, 2005).

Engineering Genes for Osmotic Regulation and Ionic Balance

Early attempts to develop transgenic plants resistant to water stress focused on single action genes responsible for the modification of a single metabolite or protein that would confer increased tolerance to drought stress. Recent reviews {Bhatnagar-Mathur *et al.*, 2008) document progress in this area. Osmoregulation is one of the most effective ways evolved by stress-tolerant plants to combat abiotic stress, but most crop plants lack the ability to synthesize the osmoprotectants naturally produced by stress-tolerant plants. Therefore genes concerned with the synthesis of osmoprotectants have been incorporated into transgenic plants to confer stress-tolerance (reviewed in (Bhatnagar-Mathur *et al.*, 2008). Overproduction of compatible solute osmoprotectants such as amino acids (*e.g.* proline), quaternary and other amines (*e.g.* glycine betaine and polyamines), and sugars and sugar alcohols (*e.g.* mannitol, trehalose and galactinol) has been achieved in various target plants. Glycinebetaine in particular has been extensively studied as a compatible solute, both by genetically engineering its biosynthesis in agriculturally important species and by its exogenous application (Chen and Murata 2008).When maize plants were transformed with the beta gene from Escherichia coli that encodes choline dehydrogenase, they accumulated glycinebetaine in tissues and were more tolerant to drought stress than wild-type plants at different developmental stages. Most importantly their grain yield was 10–23 per cent higher than that of wild-type plants after three weeks of drought stress (Quan *et al.*, 2004). In some cases the accumulation of compatible solutes also protects plants against damage by reactive oxygen species (ROS) (Bohnert and Shen 1999); in other cases the solutes have chaperone-like activities that protect other proteins maintaining their structure and function (Diamant *et al.*, 2001]. Genes coding for heat-shock proteins, molecular chaperones and LEA proteins (reviewed in [Bhatnagar Mathur *et al.*, 2008, Umezawa *et al.*, 2006) have been extensively used to improve drought responses in plants. An interesting recent example is the use of RNA chaperones of bacterial origin by Castiglioni *et al.*, to confer abiotic stress tolerance in several species, and improved grain yield in maize under water-limiting conditions (Castiglioni *et al.*, 2008). These authors demonstrated that constitutive expression of two cold shock proteins – CspA from E. coli and CspB from Bacillus subtilus (both RNA chaperones) – conferred abiotic stress tolerance to transgenic. They obtained a greater than 20 per cent increase in maize grain yield under water-limiting conditions in field trials, without observing pleiotropic effects on plant development. The improvement in drought response was observed in the late vegetative/flowering period as well as the grain-fill period: during these periods, three consecutive days of wilting can reduce grain yield by 30–50 per cent (Claasen and Shaw 1970). Stress tolerance conferred by manipulation of cold shock proteins is not only novel, but also appears as a highly promising approach to improving plant productivity in suboptimal growth conditions.

Engineering Genes Involved in Transcriptional Regulation

Transcription factors (TFs) are small molecules that attach to specific sites on a DNA molecule in order to activate or deactivate the expression of certain genes. A single gene encoding a specific stress protein does not always result in sufficient expression to produce useful tolerance, because multiple and complex pathways are involved in controlling plant drought responses (Bohnert *et al.*, 1995) and because modification of a single enzyme in a biochemical pathway is usually contrasted by a tendency of plant cells to restore homeostasis Konstantinova *et al.* (2002)]. Targeting multiple steps in a pathway may often modify metabolite fluxes in a more predictable manner. Another promising approach is therefore to engineer the overexpression of genes encoding stress inducible transcription factors. Transcription factors typically regulate several genes and are likely to be used extensively in the next generation of genetically modified crops (Yamaguchi-Shinozaki and Shinozaki 1994, Chinnusamy *et al.*, 2005). Numerous transcriptional regulators are known to be involved in plant responses to drought stress (Yamaguchi-Shinozaki and Shinozaki 2006); most fall into one of the large transcription factor families (AP2/ERF, bZIP, NAC, MYB, MYC, Cys2His2 zincfinger, NFY and WRKY); and some cis-elements, bound by these transcription factors, have been identified (Yamaguchi-Shinozaki and Shinozaki 2006). For example abscisic acid responsive elements (ABRE) (Mundy *et al.*, 1990) are 50 upstream regions of abscisic acid-responsive genes that are bound by AREB/ABF transcription factors belonging to the basic leucine zipper family. These mediate at least one of the abscisic acid-dependent pathways involved in responses to drought stress. Another cis-element is the dehydration responsive element/C-repeat (DRE/CRT) which is involved in one of the abscisic acid independent pathways (Yamaguchi-Shinozaki and Shinozaki, 1994). Various DRE/CRT-binding proteins, coding for ERF/AP2 transcription factors, are induced by desiccation, salt treatment, and cold in some plant species (Yamaguchi-Shinozaki and Shinozaki 2006). The first examples of transcription factor engineering to improve abiotic stress tolerance were overexpression of the ERF/AP2 factors CBF1, DREB1A and CBF4. Overexpression of these factors resulted in cold, drought and salt tolerance in Arabidopsis [Jaglo-Ottosen *et al.*, 1998, Kasuga *et al.*, 1999, Haake *et al.*, 2002) and it was later shown the similar tolerance could be induced in many crop plants by overexpression of these factors [Agarwal *et al.*, 2006). Numerous transgenic

Arabidopsis varieties with improved drought tolerance due to overexpression of various stress-regulated transcription factors have been reported, but similar results have also been obtained in crop plants (Saibo *et al.*, 2009). Typically a gene coding for a transcription factor in Arabidopsis is isolated, characterized and shown to improve drought response when overexpressed. The gene is then transferred to a crop plant where it often confers the same drought-tolerant phenotype. The HRD gene, coding for an AP2/ERF-like transcription factor (Karaba *et al.*, 2007) exemplifies this approach. *Arabidopsis* plants with a gain-of-function mutation in the HRD gene (hrd-D mutants) are droughtresistant, salt-tolerant, and overexpress abiotic stress marker genes. Overexpression of the same gene in rice significantly improves water use efficiency both under well-watered conditions (50–100 per cent increase) and under drought (50 per cent increase). These plants also show enhanced photosynthetic assimilation and reduced transpiration (Karaba *et al.*, 2007). HRD gene overexpression conserves drought tolerance in both dicots and monocots.

Drought Stress Induced Genes

It is primarily due to water deficit in the environment surrounding the plant, leading to growth inhibition, starvation and accumulation of proteins and toxins, enzyme inactivation and ion leakage. It can also induce secondary stresses such as heat and nutrient deficiency. Biochemical changes include decrease in carbon-di-oxide fixation, decrease in synthesis of protein and nucleic acids and increase in proline content. Drought tolerance has been achieved *in vitro* in two ways. Increased salt tolerance enhances drought tolerance or addition of high molecular weight chemicals like polyethylene glycol a non-penetrating osmoticum, or various dextrans which compete with cells for available water in similar way as in dry soil. Heyser and Nabors (1979) were first to report the selection of cell lines resistant to stress induced by PEG, which was followed by several other reports in crop plants. The complex plant response to abiotic stress involves many genes and biochemical-molecular mechanisms. Various genes respond to drought-stress in various species, and functions of their gene products have been predicted from sequence homology with known proteins. Many drought-inducible genes are also induced by salt stress and low temperature, which suggests the existence of similar mechanisms of stress responses. Genes induced during drought-stress conditions are thought to function not only in protecting cells from water deficit by the production of important metabolic proteins but also in the regulation of genes for signal transduction in the drought stress response (Shinozaki *et al.*, 2003). Thus, these gene products are classified into three major groups. (1) those that encode products that directly protect plant cells against stresses such as heat stress proteins (HSPs) or chaperones, LEA proteins, osmoprotectants, antifreeze proteins, detoxification enzymes and free-radical scavengers [Bray *et al.*2000]. (2) those that are involved in signalling cascades and in transcriptional control, such as MAPK, CDPK [Ludwig *et al.*, 2004] and SOS kinase [Zhu *et al.*, 2001], phospholipases [Frank *et al.*, 2000], and transcriptional factors [Choi *et al.*, 2000], those that are involved in water and ion uptake and transport such as aquaporins and ion transporters (Blumwald 2000). Stress inducible genes have been used to improve the stress tolerance of plants by gene transfer. It is important to analyze the functions of stress-inducible genes not only to understand the molecular mechanisms of stress tolerance and the responses of higher plants, but also to improve the stress tolerance of crops by gene manipulation. Hundreds of genes are thought to be involved in abiotic stress responses (Seki *et al.*, 2003).

Transition of Drought Tolerance Capability in Crop Plants from Laboratory to Farmer's Field

Many genes have been identified that can improve traits contributing drought tolerance. Demonstration of drought efficacy in the field is a critical step for showing commercially relevant drought tolerance, but resources for this testing are limited for many researchers, and governmental regulation of transgenic crops is often a barrier to field testing (Fgedoroff *et al.*, 2010). Benefit from several transgenes has been demonstrated in field trials. One example is the Cold Shock Protein B (CspB) RNA chaperone from Bacillus subtilis. CspB plays a role in adaptation of bacteria to low temperatures, and its overexpression was shown to provide stress tolerance to Arabidopsis, rice and maize (Castiglioni *et al.*, 2008). Results from field testing at multiple locations with controlled irrigation showed that maize lines expressing the CspB gene had higher yield under water-limiting conditions than controls, and also had yields equivalent to controls under optimal growing conditions. While this transgene provided significant yield improvements, it is expected that the addition of transgenes with different modes of action can complement the performance of this gene, and may expand the geographic regions and growing conditions under which benefit may be obtained. Several other recent examples in which transgenic lines have demonstrated improved drought tolerance in field testing. In one experiment, several transgenes with ability to improve stress tolerance in model species were tested in transgenic rice in field trials over two years (Xiao *et al.*, 2009). Each gene was tested with two promoters, one constitutively expressed and the other drought-responsive. Efficacy in promoting drought tolerance was demonstrated for six of these genes with one or both promoters. This experiment provides an excellent example of fairly rapid movement of transgenes with known efficacy from models into crops, and it is hoped that some of these genes will ultimately have commercial utility. An example that demonstrates the importance of testing the translation of greenhouse experiments to field performance was reported for the AP37 and AP59 genes in rice (Oh *et al.*, 2009). Over-expression of these genes in transgenic rice showed that either gene improved drought tolerance phenotypes in the growth chamber, but only AP37 showed yield improvement under drought in the field. One phenotype that may lead to a difference in greenhouse and field results is reduced plant size. Smaller plants use less water and thus have more water available compared to larger control plants in identical pots. However, in the field this mode-of-action may not give benefit, and may even produce yield drag (Blum 2011). Testing drought tolerance in field trials is difficult, even if controlled irrigation is available, because of the unpredictable variability of weather, soil, rain, and pests or diseases. Furthermore, some transgenes may function in pathways that interact with environmental parameters, leading to variable results. The conceptually simplest way to deal with these issues is to test at many locations over multiple years. However, this kind of testing is expensive and time consuming. More thorough characterization of transgenic lines may improve the ability to predict which lines are likely to show benefit in field conditions. This characterization may be enhanced by use of high-throughput phenotyping methods, which are often based on non-destructive imaging techniques to quantify biomass, shoot architecture, photosynthesis, pigmentation, water content, transpiration rate, and other traits (Reuzea *et al.*, 2010). High-throughput methods for imaging root architecture have been developed (Zhu *et al.*, 2011), creating opportunities to generate a more complete phenotypic profile. Field performance data from transgenic plants can be combined with thorough phenotypic data obtained in a greenhouse, using different stresses and taken at a variety of developmental stages, to develop models for predicting field performance based on greenhouse results (Tardieu and Tuberosa 2010). Such modelling could improve the success rate of greenhouse to field translation. Based on results obtained, it may be desirable to modify screening protocols. Screen modifications could involve the level of drought stress used, and also the developmental stage. Most drought research has been conducted by screening and testing under severe drought conditions. The types of mechanisms that can protect against this level of stress such as reducing plant size or decreasing stomatal conductance may be accompanied by reduced productivity under well-watered conditions. Identification of new gene leads by screening under moderate rather than extreme drought may identify genes that provide a mode-of-action more suitable for typical agricultural environments (Skirycz *et al.*, 2011). Most non-field screens for drought tolerance have focused on vegetative stages, because of the relative ease and speed of obtaining data, despite the knowledge that water limitation at the time of flowering is the most damaging to crop productivity. Therefore, it may be productive to conduct screening and follow-up testing using stress applied around flowering.

Abiotic Stress Resistance

Salt and drought resistance, freezing and chilling resistance, Enhanced photosynthesis, Nutrient uptake for Enhanced nutritional efficiency. To summarise, Abiotic stress management as a long term approach involves in understanding physiological mechanism and genetic improvement. There is a great scope for research and development for use of biostimulants as plant Abiotic stress management in India to avoid losses caused due to climate change and help horticultural farming community to improve their crop yield and quality for better returns of their farm income. This will be of great path for fulfilling for research scientists to accept challenges for better opportunities in climate change Agriculture for higher horticulture production in India.

Water Logging and Wetland Abiotic Stresses

In 1925, Kochler *et al.*, recorded plants leaning 300 from the vertical as lodged and discarded those leaning less. 1928, Hayes and McClelland placed plants in 4 classes, those leaning *viz.*, 00,200,400 and 600 and calculated LI (Lodging Index).

$$LI = \frac{\text{No of plants in each class x Angle of lodging}}{\text{Total no of plants}}$$

1948, Fellows

Defined lodging as the condition in which the culms in large areas of wheat bend at or near the surface of the ground, the bending varying in degree from slight deviation from the perpendicular to a nearly prostrate condition, though commonly the term is applied only if the deviation is more than slight. He differentiated three types of lodging: - (i) lodging proper, (ii) Buckling, (iii) Breaking. Lodging should be regarded as an "**A bundane disease**" which restricts the exploitation of otherwise yield promoting factors. Losses due to lodging may often amount to upto 30 per cent of the grain yield. There are two types of lodging *viz.*, stem lodging and root lodging. Stem lodging follows bending or breaking of the lower culm internodes whereas root lodging refers to straight and intact culms leaning from the crown, involving a certain disturbance of the root system. Mechanical aspects of lodging. Recovery from lodging. Lodging caused by foot rot and root-rot diseases. Lodging of insect attacked culms. Lodging has twofold effects both on crop development and yield, which can be described as methods of investigation; comparison between samples from lodged and from standing areas within the same plots; comparison of lodging plants in untreated plots with erect plants in CCC-treated plots; artificially prevented lodging; artificially induced lodging; effects on grain yield; degree of lodging and growth stage at which it occurs; effect on grain yields components; effect on grain quality: - Reduction in test weight. Reduce milling quality of wheat. Negligible effect on baking quality. Adversely affects the quality of barley. Leads to sprouting in the heads; effect on culm developments and tillering: - Reduces straw yield (Moulder, 1954). Promote the development of late tillers; physiological effects of lodging: - Interferes with carbohydrate assimilation (Moulder, 1954). Interferes with the translocation of carbohydrates and of minerals (Hashimoto, 1959); impact of lodging on grain harvest: - Harvest capacity is reduced (Baumgartner, 1969). Increases moisture content of lodged grain.). and incidence of diseases in lodging crops.

Plant Characters Associated with Lodging

Plant characters associated with lodging are any however, a few have been described as (a) Culm characters:

1. Length- Culm-length, which comprises the lever of the lodging inducing torque, is obviously associated with lodging. With regard to lodging at heading, the length of the 3 or 4 lowest internodes is of greater effect than that of the two uppermost internodes, which although comprising about two-thirds of the final length, have not yet completed their elongation at this stage.
2. Basal internodes: Varietal differences in lodging resistance were indeed found to be significantly positively associated with the diameter and wall thickness of the basal internode- primarily the second one (Sechler, 1961).

   Culm density, *i.e.* dry weight per unit length of culm measured at the base of the plant, is, of course, dependent on the diameter wall thickness. This variety affect stem lodging against root lodging.

3. Anatomical structure: - The changes in the lignification throughout the growing period (Heyland, 1956) presumably contribute to the inconsistent relationships between lodging and culm anatomy.
4. Chemical composition: High content of cellulose, hemicellulose and lignin in the culms of lodging resistant strains (Gaikovskaya and Baltaga, 1970).

Root and crown characters: No. of coronal roots/plant, coronal roots dry matter, tensile strength and length of the root crown induces lodging resistance.

Mechanical properties: Straw stiffness, straw length, breaking strength and root pulling resistance.

Other characters: Head density, head shape, nodding angle, flag-leaf shape, profuse tillering.

Breeding for Lodging Resistance

Resistance to lodging, unlike resistance to disease or pests, is similarly to grain yield, a character of the plant population rather than of the single plant. Moreover, it is strongly affected by environmental conditions. Therefore, its evaluation is rather complicated and plant breeders have made use of various lodging indices as selection criteria. Selection according to plant character associated with lodging resistance has been applied, especially in the evaluation of initial breeding material from rather comprehensive collections (Basistov, 1970). Evaluating lodging resistance should be based on Direct assessment in the field; Auxiliary methods *viz.*, breaking strength of straw (Helneick, 1915), culm density (Atkin, 1948) and pulling resistance (Hobert and Koechler, 1925); Lodging indices: - 0 to 4 scale. Inheritance of lodging resistance and associated characters. Achievements and prospects of breeding.As per Levitt (1980 b) flooding (*i.e.* water logging) is the presence of water in soil excess of field capacity. It leads to deficiency of O_2 and build up of CO_2, Ethylene and other toxic gases and this leads to reduction in aerobic respiration. Once soil becomes water logged, air space in soil is displaced with water, the O_2 in the soil in dissolved in water. *i.e.* O_2 decreases; CO_2 ethylene and other toxic gases increases, O_2 replacement in the soil is very inefficient. Diffusion of atmospheric O_2 into the water-logged soils is very inefficient (because of the slow diffusion of atmospheric O_2 to water logged soil); root systems are suddenly plunged into an anaerobic condition. This switching from aerobic to anaerobic respiration disrupts root metabolism; Carbohydrates level get depleted it is due to dissipation of metabolism; High water temperature and Low light. Characteristics of plants in response to water logging stress : Reduced growth/ elongation; Chlorosis, senescence and abscission of lower leaves; Wilting and leaf curling; Hypertrophy (increase in size of organ due to increase in cell size); Epinasty (downward growth of petioles). Floodplains are flat lands bordering rivers, which are subjected to periodic flooding from rivers, which tend to be most expansive along the lower reaches of rivers. Floodplain wetlands are associated with rivers and the streams. They are connected with the river or receive water from the backflow of rivers during monsoon floods. They are common in the Ganga and Brahmaputra river systems. Floodplain wetland forms an important source of fishery in the states of Assam, West Bengal, Bihar, Uttar Pradesh and Manipur (Das, 2017).

Mechanisms of Tolerance

Adventitious root formation on lower part of stem (close to surface so that O2 tension is quickly restored after transient water logging); Lenticel (*i.e.* raised pores in the stem of plants) formation; Aerenchyma formation (soft plant tissue continues air spaces found in aquatic plants) in the cortex that provide canal parallel to the axis of the root through which gases can diffuse longitudinally; Elongation capacity (In rice – best elongation response give 100 per cent recovery from submergence and poorest elongation gives upto 49 per cent recovery). Scoring for elongation can be done between booting and flowering stage after flooding the crop to varying depths. In sugarcane, *S. spontaneum* has more tolerance to flooding. Some canes gave upto 70 per cent of their production potential when in continuous flood for 5 months.

Ideotype for Flooded Areas

The postulated ideotype for flooded areas should have the following characteristics are Capacity to carry out functional activity at low O2 concentration (*i.e.* High cytochrome activity); Ability for photosynthesis under low light intensity; Capacity to synthesis food rapidly; Regeneration capacity of shoots when damaged by flood; Ability to withstand drought at later growth stage; Deep root system and Narrow, medium long and dark green leaves with high sugar and protein content.

Table 6.7. Genes encoding enzymes that synthesize osmotic and other protectants

Gene	*Gene action*	*Species*	*Phenotype*	*References*
Mt1D	Mannitol-1-phosphate dehydrogenase (mannitol synthesis)	Wheat	Drought and salinity tolerance of calli and plants	Abebe *et al.*, 2003
beta	Choline dehydrogenase (glycine betaine synthesis)	Maize	Drought resistance at seedling stage and high yield after drought	Ruidang *et al.*, 2004

Breeding for Salt Tolerance

Excess salt usually sodium chloride is the most widespread, inhibiting plant growth in nature. In India about 7 million ha of land is salt affected and another 15-20 million ha is on the verge of conversion. Growing salt tolerant plants can reclaim such lands. The usual method for selecting salt tolerant plants through cell culture is to incorporate lethal levels of salt into growing medium and isolate resistant cells visually. During salt stress the growth inhibition result mainly from increase of the osmotic pressure, while discoloration and necrosis is due to the toxicity of salt. The advantage of tissue culture technique for selecting salt resistant mutants was discovered and was followed and demonstrated in several crop plants (Bajaj, 1996) in subsequent years. Salt Tolerance refers to the ability of plants to prevent, reduce or overcome injurious effects of soluble salts present in their root zone. It is a global problem as saline and alkali soils are fond in almost all the countries of the world, more in Semi-Arid Tropic (SAT) of world. Problem of salinity can be overcome by two ways: Soil reclamation: costly, time consuming and short lived; Resistant varieties: less costly, more effective, long lasting require longer period to develop. Salt stress probably causes more damage to plants in term of economic losses than any other crop production stress factor. The primary reason for work on breeding plants for stress environments is that most expansion of crop production in the future must occur in marginal land that is now not suited to agricultural production. There are two reasons for the increased interest in breeding crop plants for salt tolerance. (1) Water quality is deteriorating throughout the world. Greater salinity results from increased urban and agricultural demands on water resources, and from over-irrigation and the excessive application of fertilizer. (2) The depletion of energy sources has increased the cost of such salinity management techniques as land reclamation, drainage, and irrigation management.

Salinity and Sodicity

A soil become saline when the concentration of soluble salts increases in the soil profile whereas a soil is considered sodic when the concentration of sodium ions is increased in comparison to calcium and magnesium ions on the soil colloids. Saline soils have good soil aggregation but poor plant establishment due to physiological dryness created by high osmotic pressure due to presence of soluble salts whereas soil is completely dispersed in case of sodic soils due to presence of high concentration of sodium ions which causes dispersion of soil aggregates. There are various factors contributing salinity which includes use of saline water for irrigation purpose, presence of soluble salts in parent material, use of saline soil amendments *etc.* Sodicity may be caused using poor quality irrigation water and poor water management that cause water logging on the soil surface. Anthropogenic sodicity is called secondary sodification. The three major problems induced by sodicity are poor soil structure, low water infiltration and surface crusting. Salinity causes disturbance in nutrient uptake resulting loss in crop yield and quality. Soil quality cannot be measured directly because it is a broad, integrative, context-dependent concept. Instead, we analyze a variety of proxy measurements that together provide clues about how the soil is functioning as viewed from one or more soil-use perspectives. These measurements are called soil quality indicators. A set of low-cost readily measured indicators that accurately predict soil functions of interest is called an efficient indicator set. Indicators of soil quality may include characteristics of soil solids, soil solutions, soil atmospheres, vegetation, and other soil biota, and possibly even economic analyses of land-uses or ecosystem services. Although the quantity and quality of data may differ, the process of soil quality evaluation follows the same basic steps regardless of the method used: identification of soil use issues followed by indicator selection and interpretation. More specifically, in order to select appropriate indicators, one must first determine the land-use objectives, and then indicators must be proposed, measured and assessed across a representative set of lands and management practices. An efficient indicator set should be used to inform land management decisions at specific sites and then be used to monitor trends in soil function after changing practices and over time. Soil quality is evaluated to learn about the effects of management practices on soil function. Reasons for evaluating soil quality fall into three categories: Awareness and education; Evaluation of practice effects and trouble-shooting and Evaluation of alternative practices. The soil quality concept emphasizes an ecological approach to land management. Management actions don't have simple, single effects in complex systems, such as soil. Management has multiple effects, both direct and indirect. For example, tillage is used to loosen surface soil, prepare the seedbed, and control weeds and pests. But tillage can also break up soil structure, speed the decomposition and loss of organic matter, increase the threat of erosion, destroy the habitat of helpful organisms, and cause compaction. Understanding the tradeoffs that exist for the range of management options is a first step towards improved land management and public policy. Assessment as an Educational Tool includes one-on-one and field day use of in-field testing tools. Evaluation of practice effects and trouble-shooting Soil quality is often referred to as "Soil Health" because of objectives similar to the monitoring and maintenance of human health. Doctors monitor health indicators and watch for irregularities or declines in status. The set of health indicators measured during a checkup is familiar to all of us: temperature, pulse, blood pressure, heartbeat, urine samples, *etc.* Monitoring of these indicators may reveal potential problems even before painful symptoms occur; the earlier problems are observed, the easier they are to treat. Likewise, soil indicators that appear irregular or decline over time provide a signal that some aspect of the management should be reconsidered. Although soil fertility testing already serves this role in regard to plant nutrition, soil quality assessment expands this to include the wider range of soil functions and environmental outcomes.

Screening Techniques

Common methods are Sand culture by using nutrient solution in sand and irrigation with saline water; Solution culture by using solution culture tanks (Hydroponic culture); Micro plot techniques by using small micro plots, Micro plot Techniques: By using small micro plots of size 6 x 3 x 1 m (CSSRI, Karnal, Haryana) at central soil salinity Research Institute. Then Multilocation Trial (MLT) conducted over seasons to get more reliable results. Genotypes which survive better under salinity are considered tolerant and tested further.

Selection Criteria

Germination (per cent) is saline medium; Dry matter accumulation (seeding/ plant dry wt.)/Early vigour; Leaf senescence or death – Estimated by total dead leaf area or No. of dead leaves; Leaf necrosis; Leaf ion content; Osmoregulation (Determined as maintenance of turgor under stress) Measured as proline or CHO accumulation or accumulation of glycine, betaine *etc.* and Economic yield.

Breeding for Drought Tolerance

Although the average yield per unit area has risen many times in high-moisture environments, increases in production under dry conditions have been less impressive. The genetics of drought-related characters is complicated and current research has not been able to describe it adequately. Crop plants occupy a great variety of environments, making it difficult for breeders to adapt crops to arid conditions. Understanding drought tolerance mechanisms, along with quantifying and clearly explaining the relationship between plant characters and drought tolerance, would reduce the dependence on empiricism and help in the development of cultivars with high stable yield under conditions of moisture stress. The primary climatic factors affecting crop production in dry areas are yearly precipitation, the seasonal precipitation distribution pattern, relative humidity, the average temperature, and soil type and depth (Quizenberry, 1981). The term 'drought' describes a condition in which available soil moisture is reduced to the point where the plant cannot absorb it rapidly enough to compensate for transpiration. Drought occurs as a result of low precipitation, high temperatures, or wind. It is not a uniform phenomenon, and a plant's reaction will depend on the stage of development at which drought occurs. Some plant processes are relatively insensitive to water stress, whereas others are distinctly affected. Drought tolerance is the plant's ability to obtain and retain water, as well as to continue its metabolic functions during a period of low water potential in its tissue. It is the result of many, often independent, morphological and physiological characters whose interaction has not yet been clearly explained. Arnon (1972) defined drought tolerance as the ability to survive drought conditions without injury. Levit (1972) divided drought resistance into 'drought avoidance' and 'drought tolerance'. He stated that drought-avoiding plants maintain a high internal water potential, in spite of low environmental water availability. Kozlowsky (1968) classified dry region plants into four groups: drought escaping, drought evading, drought enduring, and those that resist drought by storing a water supply to be used when soil moisture is scarce. For plant breeders, drought tolerance is the ability of one genotype to be more productive with a given amount of soil water than another (Quizenberry, 1981). The drought-tolerance mechanisms of most interest to the agronomist concern productivity. Yield is the integrated result of a number of interrelated physiological processes. Moisture stress can affect photosynthetic processes, growth (which provides photosynthetic tissue), and seed formation (which provides a reservoir for photosynthates). Many morphological changes in leaves and other green organs appear to improve tolerance through higher water use efficiency (WUE) and photosynthate accumulation. A stable metabolic system could be involved in the drought tolerance of plants exposed to a high level of stress. Nitrate reductase activity and the accumulation of proline and abscisic acid (ABA) are also reported to be associated with drought tolerance. There are three stages of physiological research concerning drought. The first relates plant characteristics to stress survival. The second is the much more challenging investigation of differences within a species related to drought tolerance. The third is the development of simple, rapid methods for measuring the characteristics identified in the two previous research stages.

Plant breeders engaged in breeding improved varieties for drought conditions try to increase the yield of crops grown under less-than optimum conditions. This can be achieved either through improvement of cultural practices, or by breeding varieties that are more productive under stress, the latter strategy being the one dealt with in this paper. Although the need for crop varieties better adapted to semi-arid conditions is widely recognized, plant breeders disagree on how to satisfy this need. Some think drought tolerance can be obtained simply by selecting for high yield. In this commonly used approach, it is assumed that some varieties with superior yields under optimum conditions will also yield relatively well in suboptimum conditions (Rajaram and Nelson, 1984). In such varieties, drought tolerance may be present as an unidentified component of performance stability. During the breeding process, yield and performance stability are handled as one complex. The second major breeding approach maintains that yield potential is irrelevant, and that a superior variety for semi-arid growing conditions must be selected, developed, and tested in a semi-arid environment. Varieties developed in such programmes often show a narrow adaptation to the relevant environment and are said to have high genotype-environment interaction (Blum, 1979).

Concept of Breeding for Semi-arid Environments

The proposed modifications to the drought-breeding concept presented by Rajaram and Nelson (1984) as long-term objective is to develop germplasm with high yield stability, disease resistance, and high yield potential under all conditions. The result of this approach has been the development of high-yielding, input-efficient, and input-responsive varieties that have gained unusually rapid worldwide acceptance. Worrall *et al.* (1980) analysed data from 14 years of the international Spring Wheat Yield Nursery (ISWYN). Their analysis showed that wheats and their derivatives perform well and fit into nearly every wheat-growing environment, including sem-arid zones. Based on ISWYN data they concluded that breeding programme has been effective in developing high-yielding spring wheat germplasm and that significant progress Is being made in breeding for stability over a wide range of environments. The key parts of a drought-tolerance breeding programme are tolerance mechanisms, the stages at which a crop faces water stress, and the severity of the stress. Lewis and Christiansen (1981) suggest that stress environments be selected at a level that differentiates between stress-susceptible and stress-tolerant genotypes. The following methods are used to identify germplasm for drought tolerance.

Table 6.8. Yield performance of six wheat varieties and advanced lines under two irrigation regimes In Cd. Obregon. Sonora, 1982

Variety	*Five Irrigations*		*Two Irrigations*	
	Yield (kg/ha)	*Rank*	*Yield (kg/ha)*	*Rank*
Genaro 81	7461	1	4745	2
Lira 'S'	6920	2	4747	1
Veery 'S'	6819	3	4378	5
Neelkant	6758	4	4585	4
Junco 'S'	6346	5	4681	3
Tanager 'S'	5202	6	3637	6
LSD (5 per cent)	639		711	

Gradient Irrigation (Line source) to Identify Cultivars with Water Use Efficiency

Hanks *et al.* (1976) described line sprinkler systems that can be used to create a continuous water gradient for studying genotypes for drought response. If used when the air is calm, this system allows plants to be subjected to a wide range of water-stress levels. At Ciudad Obregon, a sprinkler system was installed to apply water in a liner fashion with the amount decreasing as the distance from the irrigation line increased. To evaluate the system's usefulness in screening for moisture-stress performance, 25 genotypes were sown perpendicular to the irrigation line. Five plots were harvested at 3 m intervals to obtain information on the relative performance of each genotype at differing levels of applied water. Average yield was positively correlated with the amount of water applied. The genotype response was evaluated using regression analysis. The variety found to be best adapted to water stress conditions was Galvez 87 bread wheat.

Multisite International Trials for Identifying Drought Tolerance

Multilocation testing has been of great value in wheat breeding programmes for identifying cultivars that perform well over a range of environments, including moisture-stress situations. The ISWYN was the first systematically organized international yield trial. Begun in 1964, it was designed to disseminate the most recently released spring wheat varieties throughout the world and to supply information on yield, disease resistance, stability, and wide adaptation to national wheat breeding programmes. CIMMYT also distributes the Drought Screening Nursery (DSN) to provide genetic material for national programmes in semi-arid areas around the world. Through these programmes, input-efficient, input-responsive, and stable varieties, such as **Anza, Siete Cerros 66, Pavon 76, Nacozari 76, Glennson 81, Seri 82, Ures 81**, and **Genaro 81**, have been identified and used as parents for building future varieties. The programmes have produced promising results not only in irrigated areas of the world, but also in semi-arid regions (Rajaram *et al.*, 1984).

Drought-Tolerance Breeding Techniques

The approach to drought-tolerance breeding is to test the breeding material alternately in highly favourable environments where they can realize their yield potential and in moisture stress conditions, on the assumption that input efficiency and input responsiveness can be combined. The analyses of Worrall *et al.* (1980) and Pfeiffer (1984) support this assumption. Ciudad Obregon has dry weather from October to May, and long-term seasonal average rainfall is 48.2 mm. This is a suitable environment for drought trials if the number of irrigation applications is regulated. In an optimum breeding programme, five to six irrigations are applied to produce a good crop of wheat. In our drought experiments, plants receive only one irrigation to facilitate the selection of lines. High fertility and optimum moisture are applied in the F_2, F_5, and F_6 generations. The F_3 and F_4 generations are grown under low fertility and reduced moisture. Selection during the irrigated and well-fertilized F_2, and F_5, and F_6 generations is conducted normally. Rust epidemics are created and susceptible plants are discarded. Good tillering capacity, head development, leaf retention capability, and grain plumpness are the characters evaluated in the individual plant and line selections. In F_3 and F_4 materials grown under limited moisture and low fertility, the characters that are important for drought-stress situations and considered in the selection process are: (1) delayed leaf senescence (longer leaf retention, (2) tiller viability, (3) reduced spike sterility,

(4) grain plumpness, (5) relative higher test weight, and (6) relative higher yield. Advanced lines are simultaneously subjected to yield trials under optimum and stress conditions. Lines that perform well under both conditions are included in the DSN and distributed to co-operators in moisture-stress regions.

Mechanisms of Drought Resistance

There are three mechanisms of drought resistance. *Drought Escapes*: It is due to ability of a genotype to mature early, before occurrence of drought. Drought escape is most common is plants grown is desert region. *E.g.* Early maturing varieties of sorghum, maize, bajra, wheat, rice *etc.*; give more yield than late maturing under drought. *Drought Avoidance (Dehydration avoidance)*: It is due to the ability of plants to maintain favourable water balance even under stress. The plants which avoid drought retain high moisture content is their tissues and lose less water. This is possible either because of increased water uptake (due to increase in root development) plants are called water spenders (or), reduced water loss (due to reduction in growth of aerial parts are called water savers (*i.e.* to avoid transpiration). Dehydration avoidance is interpreted as the ability of genotypes to maintain high leaf water potential when grown under soil moisture stress. Several traits contribute to dehydration avoidance Such as: Leaf rolling, folding and reflectance narrow leaves, increased pubescence on aerial organs, presence of awns, osmotic adjustment of stomata, cuticular wax, increased water uptake; Reduced Transpiration: Increase is concentration of Abscisic Acid (ABA), closure of stomata, ABA plays role in reduction of leaf expansion, Promotion of root growth *etc. Drought Tolerance (Dehydration tolerance)*: Ability of plants to produce higher yield even under 'low water potential'. In cereals drought tolerance generally occur during reproductive phase. Tolerant cultivars exhibit better germination, seedling growth and photosynthesis. Drought tolerance may be because of high proline accumulation; maintenance of membrane integrity and Drought Resistance: is the sum total of avoidance and Tolerance. It refers to the genetic ability of plants to give good yield under moisture stress conditions.

Various morphological, physiological and biochemical features/parameters associated with drought resistance:

1. **Morphological**: Earliness, Reduced tillering, Leaf characters : Leaf rolling, Leaf folding, Leaf shedding, Leaf reflectance, Reduced leaf area: Narrow leaf, Change in leaf angle, Hairiness (presence of hairs on leaf and other parts, lowers leaf temperature and reduce transpiration), Colour of leaves, Wax content, Awns (*e.g.*, wheat and barley), Root system (rooting. depth and intensity)
2. **Physiological**: Photosynthesis (efficient system like C_4) under stress, photosynthetic efficiency is reduced due to chloroplast damage, Reduced Transpiration and reduced respiration losses, Stomatal behavior (closure of stomata, also change in size and number of stomata), Osmotic adjustment, Leaf enlargement (increase in thickness), Leaf cuticle wax (increases)

3. **Biochemical**: Accumulation of proline and betaine, Increase in Abscisic acid (barley) and Ethylene (maize and wheat), Protein synthesis (increases under stress), Nitrate – reductase activity.

Breeding for Mineral Stress/Ion Stress Conditions

The growth and productivity of plants are influenced by their genetic capability and the environment in which they are grown. Plant *x* environment Interaction, as well as the ability to convert water, CO_2, and mineral nutrients into tissues, determine the level to which plants grow and produce. It is due to presence of heavy metal ions in soil such as Mg, Pb, Cu, Zn, Al and cell lines resistant to these metals have been recovered in several crop plants. It has been observed that, the Al tolerant cell lines secreted citric acid and malic acid into the medium (Parrot and Button, 1990). It was thought that citric acid chelates aluminium which thus prevents uptake. Induction/screening of mutants using biotechnological approaches has contributed a lot towards the development of novel biotic and abiotic stress tolerant plants. However, in the recent past genetic engineering has shown a potential (Aggarwal *et al.*, 1999, Grover *et al.*, 1998), by which the specific gene for a particular stress can be directly transferred to plants. This could be the technique of future, compared to the in vitro screening method for biotic and abiotic stress tolerance. This process of conversion is photosynthesis. The rate of plant productivity (or photosynthesis) can be greatly inhibited by the physical environment. Genetic variability in the plant kingdom is important in providing plants that are adapted to specific environments. Plants that survive under natural ecological conditions must be tolerant to these conditions. Cross-pollination and mutation load to new genetic combinations, and natural selection leads to a concentration of genetic factors in certain environments. For example, plants differ in their ability to use or tolerate specific mineral elements in soils. Iron deficiency, boron toxicity, and salinity are common problems on alkaline soils, and aluminium and manganese toxicity and phosphorus deficiency are common problems in acid soils. An extensive review on toxicity and deficiency has been presented. The soils of the world can be divided roughly into three categories: acid, alkaline, and neutral. Each category occupies about one-third of the global land area. Mineral stress, drought stress, shallow soils, excess water, salinity, and rock debris are some of the adverse factors found in soils (Dudai, 1976).

Genetic Solutions to Mineral Stress

Genetic variation in plant response to mineral deficiency and toxicity has been reported in many crops. One major exploitation of these genes has been the breeding of wheat cultivars tolerant to high aluminium (Al) availability in acid soils. Although liming of soils can improve the aluminium toxicity problem, it only corrects the pH in the upper 15 to 20 cm of soils. Breeding has offered a more flexible solution to this problem. For example, the aluminium-tolerant wheat cultivar Maringa has permitted wheat production on several million hectares of high-aluminium soils. The mechanism of aluminium tolerance is the ability of plant roots to exude substances that raise the rhizosphere pH, which in turn precipitates aluminium from the soil solution.

Wheat Breeding Programme for Tolerance to Al Toxicity

As stated earlier, wheat varieties developed were highly tolerant to aluminium toxicity. This endeavour involved 50 years of research in Brazil to classify wheat varieties for aluminium toxicity (crestamento) and use them to breed cultivars for production. Despite these efforts, the yield potential of these varieties was very low. On the contrary, the wheats were very susceptible but highly productive in non-acid soils. A hybridization programme was undertaken using the gene pools to combine resistance of wheats with high yield potential of wheats. The F_2 germplasm was generally screened in Mexico for plant type, and resulting progenies were tested as F_3 for adaptation. Selected progenies in Brazil were grown in Mexico as F_4. Screen is also made for aluminium-tolerant seedlings in laboratory conditions. Using this system, it has been possible to combine resistance with higher yield potential.

Breeding for Al Tolerance Combined with Phosphorus Efficiency

Phosphorus is an essential element not available to plants in acid soils because it is fixed. As stated earlier, some wheat varieties use this element more efficiently than do others. One of the major breakthroughs in the use of the P-efficiency gene has been the transfer of a small piece of chromosome of rye in wheat. This was achieved through the 1 B/1 A translocation in Alondra from the variety Weique Red Mace. Attempts are being made to combine the 1 B/1 A translocation with conventional aluminium tolerance. This combination with proper plant type and resistance to other disease may produce higher biomass and yield in aluminium-toxic soils.

Table 6.9. Steps involved during *in vitro* screening

Selection of source material	*In vitro* seedlings, glasshouse/fields grown young/mature plants.
Explant selection	Seeds, epicotyl, cotyledons, hypocotyl, leaves, embryos *etc.*
Surface sterilization of explants	Surface sterilize using 0.1 per cent mercuric chloride for 2-8 min. depending upon explant used followed by rinsing in sterile water 2-3 times to remove excess sterilant.
Inoculation on callus induction medium	Inoculate on MS+ 2,4-D + kin medium using aseptic conditions.
Raising of cell suspension cultures	Raise cell suspensions in a flask kept on gyratory shaker for 80-120 rpm.
Cell plating	Plating of cells on a medium containing toxins/stress factors such as sodium chloride for salt tolerance, PEG for drought tolerance, exposure to low or high temperature for low and high temperature tolerance and relative ions *e.g.* Al for aluminium tolerance.
Selection of tolerant cells	Susceptible cells turn brown in colour and die while tolerant cells will survive and continue to proliferate on selective medium.
Plant regeneration	Regeneration of plants from tolerant cells.
Analysis	Genetic analysis of regenerants.

Breeding for Cold Tolerance

When temperatures remain above-freezing *i.e.* >0°C to <10-15°C it is called chilling. When temperature. remain below freezing *i.e.*<0°C it is called Freezing.

Chilling Resistance: Chilling sensitive plants are typically tropical plants. Temperate plants are generally tolerant to chilling injury.

Effects of Chilling Stress on Plants

Reduced germination; Poor seedling establishment; Stunted growth; Wilting; Chlorosis; Necrosis; Pollen sterility; Poor fruit set/seed formation; Reduced root growth; Locked open stomata and ABA accumulation. At sub cellular level reduces membrane stability; poor chlorophyll synthesis (affected); reduced photosynthesis and respiration; toxicity due to H_2O_2 formation. Chilling Tolerance is the ability of some genotypes to survive/perform better under chilling stress than other genotypes is called chilling tolerance. It is because of chilling hardening, *i.e.* an earlier exposure to a near chilling temperature for a specified period as a result of which chilling tolerance of the concerned plants increases.

Mechanisms of Chilling Tolerance

Membrane lipid un-saturation; Reduced sensitivity of photosynthesis; Increased chlorophyll accumulation; Improved germination; Improved fruit/seed set; Pollen fertility. Sources of chilling Tolerance are Late adopted breeding populations *e.g.* maize; Germplasm (*e.g.* collected from high altitude, low temperature geographic regions); Induced mutants for cold tolerance; Cold tolerant somaclonal variants and Related wild species.

Selection criteria should be based on Germination test; Growth under stress (measured as plant dry matter accumulation); Chlorophyll Loss under chilling stress *e.g.* rice, cucumber, tomato (measured as seedling colour); Membrane stability : (Assayed in terms of solute leakage from tissues); Photosynthesis: Chilling injury to photosynthesis is assayed as variable chlorophyll fluorescence at 685 nm; Seedling mortality; Seed/Fruit set and Pollen fertility (apply during injury at PMC).

Breeding for Freezing Resistance

Freezing injury/Frost injury/cryo injury/Freezing Stress: Dormant state is conducive to freezing resistance, while resistance in actively growing tissue is rare: Thus Freezing resistance largely involves surviving freezing stress in such a manner as to enable subsequent regrowth when the temperature rises. As water in plants cools below 00C, it may either 1. Freeze *i.e.* form ice or 2. Super cool without forming ice. Effects of freezing stress, Ice formation: Two types Intercellular ice formation and Intracellular ice formation. Intercellular Ice formation: Initiation of ice formation on plant surface is sufficient to induce freezing of the internal (intercellular and xylem vessels *etc.*) water is most plant species. Intracellular ice formation: It is more lethal may be due to physical disruption of subcellular structure by ice crystal. Intracellular ice formation is the major and terminal freezing stress. Extracellular ice formation in a cases the concentrations of extra cellular solutes, the re by water is withdrawn from the cells during extra cellular ice formation. This creates water stress in the frozen tissue/plant. Membrane disruptions: Freezing causes disruptions is and/or alter the semi permeable properties of plasma membrane; Loss of solutes from the cells occur; Cells remain plasmolyzed even after thawing which is often called as frost plasmolysis; Cells may become highly turgid due to uptake of excess water.

Suspercooling

Cooling of water below 0°C without ice crystal formation is called super cooling, In plants water may cool down to -1 to-15°C is herbaceous sps and to -40 to -45°C in hardy trees. This becomes possible apparently because internal ice-nucleates are absence in such cases. This is regarded as an important. Mechanism of freezing avoidance. Stress due to external factors: Consequent to freezing, Ice sheet formation below and above the ground causes reserve depletion anoxia *etc.* in plants. Tissues killed during freeze-thaw are highly prone to pathogen attacks and Auto toxicity may occur.

Mechanism of Freezing Resistance: The ability of a genotype to survive freezing stress and to recover and re grow after thawing is known as freezing resistance. Freezing resistance is a complex trait involving physiological, chemical and physical processes at the tissue and cell level.

Mechanism of Freezing Resistance

Freezing avoidance: The ability of plant tissues/or genes (but the whole plants) to avoid ice formation at sub zero temperature is called freezing avoidance. Super cooling is a mechanism of freezing avoidance it is controlled by Lack of ice-nucleates 2. Small cell size, Little or no intercellular space 4. Low moisture content, Barriers against external nucleates, Presence of antinucleators; Freezing Tolerance: Ability of plants to survive the stresses generated by extra cellular ice formation and to recover and regrow after thawing is known as freezing tolerance. The various components of freezing tolerance are Osmotic adjustment; Amount of bound water; Plasma membrane stability; Cell wall components properties and Cold-responsive proteins *E.g.* ABA.

Problems in Breeding for Freezing Tolerance

Freezing Tolerance is a complex trait and involves several components. So, it is not ready measurable under field conditions. Breeding work under field conditions is highly influenced by other environ factors and biotic stresses. Due to large G X E for the trait field survival shows poor heritability. Freezing tolerance also shows a large GXE interaction which limits progress under selection. Laboratory tests are yet to be developed to screen large breeding population.

Phenotyping for Abiotic Stress Tolerance in Crop Plants

The poor and depleted soil fertility remains a primary constraint to agricultural productivity in most of tropical and sub-tropical regions. The elevated temperatures, changing precipitation patterns and extreme weather events also greatly affected on agriculture production (IPCC, 2007). The average rate of crop production increase by only 1.3 per cent per year, but it cannot keep pace with population growth. By connecting the genotype to the phenotype, high yielding, stress-tolerant plants can be selected far more rapidly and efficiently than is currently possible. However, the lack of access to phenotyping capabilities limits our ability to dissect the genetics of quantitative traits related to growth, yield and adaptation to stress. Now days, plant phenotyping greatly helps the genetic analysis of abiotic stress tolerance to further elucidate the stress tolerance mechanisms. However, conventional methods of plant phenotyping are laborious and destructive as compared to the recently developed high-throughput, non-destructive imaging technologies (Yang *et al.*, 2013). The recent phenotyping techniques, being non-destructive, enable acquiring quantitative data on plant growth, health, and water use under abiotic stress by taking multiple images of the same plant at different time points and at different wavelengths (Jones *et al.*, 2009). Therefore, these technologies are being routinely applied to quantify traits related to stress tolerance in a number of crop plants (White *et al.*, 2012).

Phenotyping under Controlled Condition

Although filed phenotyping is the best option to select genotypes of our interest in the target environment for yield and its component, the phenotyping in controlled environment facilities is advantageous for imposing abiotic stresses uniformly, which is not possible in field conditions. The studies on influence of abiotic stress factors like excess or limited moisture stress, high temperature and salinity are conducted under controlled conditions. The controlled condition under which the plants are grown should be relevant to the conditions prevailing in the field (Izanloo *et al.*, 2008). Evaluation under controlled conditions is advantageous in terms of collecting data at a particular stage when genotypes being tested differ in durations to attain certain phenological stage. Growing plants in pots allows for strict control of water stress imposed on test genotypes and the homogeneity of stress severity; such control is seldom achieved under field conditions, particularly when genotypes under test differ in phenology and biomass.

Phenotyping under Field Condition

Ultimately, evaluation of crop plants for yield performance under particular abiotic stress needs to be done under field conditions. Field phenotyping helps to identify tolerance traits in the ultimate target environment and helps in evaluating many genotypes at a time. Unlike controlled growth condition, in field evaluations, there are certain factors which impact the quality of the phenotypic data to be collected (Tuberosa, 2011) listed the following factors to be evaluated carefully to ensure the collection of meaningful phenotypic data in field experiments under water limiting conditions. The factors are the experimental design, heterogeneity of experimental condition between and within experimental unit, size of the experimental unit and number of replicates, number of sampled plants with in each experimental unit and genotype-by-environment-by management interaction. Though the field evaluations are conducted on the ultimate target environmental conditions or crop management during the experimentation might influence the plant's phenotype. Thus the variability caused by these factors must be kept to the minimum so as to collect quality phenotyping information. In field evaluation, techniques like measuring canopy spectral reflectance (Gutierrez *et al.*, 2010) and screening under high temperature stress (Hazra *et al.*, 2009) and drought stress (Ashraf *et al.*, 2005) are employed. The phenotyping methodologies like line source irrigation, withholding irrigation to impose water stress (Rao and Bhatt,

Table 6.10. Different imaging techniques in plant phenotyping used to detect abiotic stresses

Imaging Techniques	*Sensor*	*Resolution*	*Phenotype Parameters*	*Examples*	*Environment Conditions*
Fluorescence imaging	Fluorescence cameras	Whole shoot or leaf tissue, time series	Photosynthetic status (variable fluorescence), quantum yield, non-photochemical quenching, leaf health status, shoot architecture	Wheat (Bürling *et al.*, 2010), Tomato (Mishra *et al.*, 2012)	Controlled environment, field
Thermal imaging	Near infrared cameras	Pixel-based map of Surface temperature in the infrared region	Canopy or leaf temperature	Wheat (Manickavasagan *et al.*, 2008)	Controlled environment, field
Visible light imaging	Visible spectral range	whole organs or organ parts, time series	Projected area, Growth dynamics, Shoot biomass, Yield traits, Panicle traits, Root architecture, Imbibition and germination rates, Early embryonic axis growth, Height, Size morphology, Flowering time	*Arabidopsis thaliana* (Joosen *et al.*, 2012), Rice (Clark *et al.*, 2011)	Controlled environment, field
Hyperspectral imaging	Near-infrared instruments, spectrometers, hyper spectral camera	Crop vegetation cycles, indoor time series experiments	Leaf and canopy water status; Leaf and canopy health status; panicle health status; leaf growth; Coverage density	Wheat (Moshou *et al.*, 2005)	Controlled environment, field
Near Infrared imaging	Near-infrared cameras, multispectral line scanning cameras, active thermography	Continuous or discrete spectra for each pixel in the near-infrared region	water content composition parameters for seeds, leaf area index	Soybean (Bolon *et al.*, 2011)	Controlled environment, field

1992), imposition of salinity stress and conducting evaluation trails during high-temperature periods in the hotspot areas are a few techniques that are followed under field condition.

Phenotyping Sites for Different Abiotic Stress (Drought Stress)

Long-term daily climate data and soil data are required to ensure a site allows drought stress to be applied at the required growth stage, with minimum variation in soil properties. Drought phenotyping is often conducted during the off (dry) season to control the timing, intensity and duration of the period of water stress and avoid the climatic uncertainty associated with conducting drought trials during the main season. Rainout shelters in the main season can be used as an alternative to screening in the dry season but cost and limited space are important considerations.

The quick development of germplasm and their tolerance to several complex polygenic inherited abiotic and biotic stresses combined is critical to the resilience of cropping systems in the face of climate change. Plant phenomics is a simply plant physiology in 'new clothes', but it promises to bring physiology up to speed with genomics by introducing the incredible recent advances made in computing, robotics and image analysis to the wider field of plant biology. Phenomics provides the opportunity to study previously unexplored areas of plant science, and it provides the opportunity to bring together genetics and physiology to reveal the molecular genetic basis of a wide range of previously intractable plant processes. The future challenges of characterizing crop plant for desirable traits require the advances we have seen in information technology, and there is a need to build on these advances for global food security. The better knowledge of the physiological, biochemical, molecular and genetic basis of the mechanisms promoting tolerance to abiotic stress will enhance the capacity to improve crop yield under hostile environments.

Precise Phenotyping for Drought Tolerance and Related Dynamic Traits

After establishing the most suitable target trait for selecting grain yield under drought stress, the next step is to establish a high-throughput precision phenotyping platform for pinning down the source trait most tightly connected to yield (Tuberosa 2010). The precise phenotyping of drought-related PTs often requires the utilization of sophisticated and expensive techniques herein listed:

Near-Infrared (NIR) Spectroscopy on Agricultural Harvesters

This method provides spectral information corresponding to the field plot in a single near-infrared spectrum, where physical and chemical characteristics of the harvested seed material are captured. By using calibration models (*i.e.*, mathematical and computational operations that relate the spectral information with phenotypic values), several traits can be determined on the basis of a single spectrum (dry matter, protein, nitrogen, starch and oil content, grain texture and grain weight, *etc.* The use of NIR spectroscopy on agricultural harvesters provides indexing of grain characteristics. In contrast to conventional sample-based methods, NIR spectroscopy on agricultural harvesters secures a good distribution of measurements within plots and covers substantially larger amounts of plot material (Welle *et al.*, 2003), thus reducing sampling error and providing more representative measurements of the plot material in terms of homogeneity.

Infrared Thermography (IRT)/Canopy Spectral Reflectance (SR) Temperature Trait for Phenotyping

The surface temperature of the canopy is related to the amount of transpiration resulting in evaporative cooling. IR based thermometer/camera allows canopy temperature (CT) to be directly and easily measured remotely and without interfering with the crop. It is well documented that CT is correlated with many physiological factors like plant water status, stomatal conductance, transpiration rate, crop yield *etc.* CT used routinely, particularly for stress diagnostic and breeding selection of stress adapted genotypes: (i) under drought conditions it is related to the capacity to extract water from deeper soil profiles and/or agronomic water use efficiency (WUE); (ii) under irrigated conditions it may indicate photosynthetic capacity, sink strength and/or vascular capacity –depending on the genetic background, environment and developmental stage; and (iii) under heat stress conditions is related to vascular capacity, cooling mechanism and heat adaptation. CT is an integrative measurement (*i.e.*, scoring the entire canopy of many plants within a plot), and so has advantages over other methods used for stress detection, such as stomatal conductance and water potential, because it integrates a larger area of plant/crop measurement, is non-destructive, does not interfere with stomata (which are sensitive), and is faster and not laborious. However, trait expression shows interaction with both developmental phase and time of day (*e.g.*, pre-heading and/or morning readings are usually lower due to lower incident solar radiation and air temperature), which can be used to relate different canopy traits and stress tolerances. Infra-red thermometry or IR thermography measures temperature of the target by measuring the radiant thermal energy emitted by the target. Infrared is a type of electromagnetic radiation, which is emitted, to greater or lesser degree, by all objects that have temperature. IR spectral region of 8 to 13 µm is typically used for thermal remote sensing.

Factors Influencing the Canopy Temperature

Stomatal features: shape, size and structure of stomata. **Leaf Characters:** Number and orientation of leaves, presence of cuticle and waxiness on leaf lamina and stem. **Plant water status:** Water content of plant/leaf in relation to that required for optimal growth. **Crop Yield:** high photosynthesis, Emissivity of objects. **Time of day:** Measurements typically from 11: 00h to 14: 00h, Avoid cloudy and rainy days. **Environment condition:** Measurements must be in clear sky and there is little or no wind. The plant surfaces are dry and not wet from dew, irrigation or rain. **Plant phenological stage:** Stage should be objective based and interval should be roughly 5-7 days between each measurement– to give a reasonably heritable estimate of trait expression. Always take measurements of the part of the plot which is most exposed to the sun, and ensure to avoid the shadow of the operator and/or shadows from the neighbouring plots.

Image Capturing Process in Variocam HR Inspect 575

Press the button AL. The thermographic system automatically focuses and the temperature scaling of the false-colour image is automatically optimised according to the current scene. Or Adjust the view manually to focus the object. In the live mode, joystick movements '!, "! change the selected temperature level and joystick movements ?, → change the selected temperature range. Pressing the Enter button switches between the live mode and the focus mode. In the focus mode, joystick movements '!, "!focus over larger or shorter distances to the object. For Storage of the thermal image press the S button. The live image freezes, *i.e.*, camera goes into stop mode. Pressing the S button again saves the thermal image on the SD card. Pressing the C button interrupts the saving process. The camera will return to the live mode after the saving process. For switch off press button CL, the dialogue for switching off is selected and confirmed by pressing Enter.

Image Processing

From menu "File", select submenu "Open file" and open the desired thermograms (*.irb files). Select the desired colour palette via the "Scale", which is located on the right-hand of the thermogram. Via menu "View", you can display further image elements, measurement data, annotations and parameters in addition to the thermogram. By pressing the right mouse key on the colour scale, the dialogue "Level/Range", where the temperature level and range can be adjusted as desired by moving the scroll bar. The adjustment is also adopted for subsequent thermal images. With the help of the respective functional buttons on the symbol bar, points of measurement, areas, *etc.* as well as the display of temperature maximum and minimum can be activated. For inserting the analysed thermal images into the reports, select the menu "Report". Alternatively the images can be stored in common image formats suc as.jpg,.bmp *etc.*

Prototype of Tools for Image-based Phenotyping

Taking into consideration the need to accelerate phenotyping in field efforts have been made to develop phenotyping tools. A hand operated track mounted trolley was designed for imaging purpose which hosts a camera and a Lap Top PC. The system acquires images of each plot in the experimental field after recognizing the barcode. Images are stored with plot name. Tools have also been developed to rapidly analyze these images. Promising results have been obtained with image acquisition and analysis tool. This field based, semiautomated platforms potentially allow high-throughput phenotyping at a low cost.

IR Thermometer Vs Thermal Camera

A IR thermometer gives number whereas, thermal imaging cameras generate an image. A IR thermometer reads the temperature of one single spot whereas, a thermal imaging camera gives you temperature readings for each pixel of the entire thermal image. Because of advanced optics, thermal imaging cameras can also resolve temperatures from a longer distance. This allows you to quickly inspect large areas and hance, simultaneous response recording for large set of genotypes.

Limitations of Thermal Imaging

Thermal cameras are more expensive and It is greatly Influenced by being around any object and environment hence, necessitating the use of reference. Imaging sensor calibration and atmospheric correction are often required for high efficiency

Chlorophyll Fluorescence Measurements and Use in Plant Phenotyping

Chlorophyll fluorescence is one of the most highly informative, rapid and non-destructive diagnostic methods for the detection and quantification of damage in the photosynthetic apparatus caused by environmental stress. No investigation into the photosynthetic performance of plants under field conditions seems complete without some fluorescence data. Light energy absorbed by chlorophyll molecules in a leaf can undergo one of three fates: it can be used to drive photosynthesis (photo-chemistry), excess energy can be dissipated as heat or it can be re-emitted as light–chlorophyll fluorescence. These three processes occur in competition, such that any increase in the efficiency of one will result in a decrease in the yield of the other two. Hence, by measuring the yield of chlorophyll fluorescence, information about changes in the efficiency of photochemistry and heat dissipation can be gained. Fluorescence occurs mostly from chlorophyll a of PSII in the red region of the spectrum (685nm) and therefore it is emitted as red light. More than 90 per cent of absorbed light is utilized by photosynthesis. Only about 1 to 2 per cent light is utilized by the fluorescence process (Maxwell and Johnson, 2000). The spectrum of fluorescence is different to that of absorbed light, with the peak of fluorescence emission being of longer wavelength than that of absorption. Therefore, fluorescence yield can be quantified by exposing a leaf to light of defined wave-length and measuring the amount of light re-emitted at longer wavelengths. Several research documents are available in support of chlorophyll a fluorescence to assess PSII status under light (Luttge 2000), cold (Koscielniak and Biesaga-Koscielniak 1999), heat (Bukhov and Carpentier 2000), and water stress (Goltsev *et al.*, 2012).

Fluorescence Parameters

Chlorophyll fluorescence can be used as a non-intrusive method of monitoring photosynthetic events. There have been very many fluorescence parameters defined in the literature. The aim is to provide information on the parameters that can be usefully used in crop improvement programmes to identify differences in plant performance non-destructively and rapidly. Consequently, the focus will be on the fluorescence parameters associated with the induction of fluorescence on exposure of dark-adapted leaves to light and the operation of photosynthesis under growth and other light conditions. It gives a good measure of the photochemistry and electron transport rate and can be related to photosynthetic efficiency. For example, environmental stresses affect the PSII efficiency and there is decrease in Fv/FM. The fluorescence parameters are very useful in monitoring and judging the physiological state of the plants under environmental stresses such as water deficit, temperature, nutrient deficiency, polluting agents, attack by pathogens. There are two types of chlorophyll fluorescence meters – time resolving (Continuous Light) fluorimeters and pulse modulated fluorimeters (Hall *et al.* 1993). Time resolving fluorimeters give the fluorescence parameters (F and Fo) for dark adapted leaves and only record Kautsky curves. However, the fluorescence measurements of the light adapted leaves require pulse modulated fluorimeters. Commercially available fluorimeters as well as integrated photosynthesis systems with fluorimeter are available for monitoring the fluorescence parameters. The fluorimeters models are PEA (Hansatech), PAM-2000 (Walz) *etc.* The models of integrated photosynthesis systems with fluorimeter include LI-6400 (LICor), CIRAS-2 (PP-Systems), HCM-1000 (Walz), LCpro+ (ADC).

Table 6.11. Fluorescence parameters and their physiological relevance (Baker, 2008)

Parameter(s)/Definition	*Physiological Relevance*
F= Fluorescence emission from dark adapted leaf	Provides little information on photosynthetic performance because these parameters are influenced by many factors. F' is sometimes referred to as Fs' when at steady state
F'=Fluorescence emission from light adapted leaf	
Fo=Minimal fluorescence from dark adapted leaf	Level of fluorescence when QA is maximally oxidized (PSII centers open)
Fo'=Minimal fluorescence from light adapted leaf	
Fm=Maximal fluorescence from dark adapted leaf	Level of fluorescence when QA is maximally reduced (PS II centers closed)
Fm'=Minimal fluorescence from light adapted leaf	
Fv=Variable fluorescence from dark-adapted leaves	Demonstrates the ability of PS II to perform photochemistry (QA reduction)
Fv'=Variable fluorescence from light adapted leaves	
Fq'= Difference in fluorescence between Fm' and F'	Photochemical quenching of fluorescence by open PS II centers.
Fv/Fm = Maximum quantum efficiency of PSII photochemistry	Maximum efficiency at which light absorbed by PSII is used for reduction of QA.
Fq'/Fm'=PS II operating efficiency	Estimates the efficiency at which light absorbed by PS II is used for QA reduction. At a given photosynthetically active photon flux density (PPFD) this parameter provides an estimate of the quantum yield of linear electron flux through PS II. This parameter has previously been termed $\Delta F/Fm'$ and φ PS II in the literature.
Fv'/Fm'=PS II maximum efficiency	Provides an estimate of the maximum efficiency of PS II photochemistry at a given PPFD, which is the PS II operating efficiency if all the PS II centers were _open‘ (QA oxidized).
Fq'/Fv'=PS II efficiency factor	Relates the PS II maximum efficiency to the PS II operating efficiency. Nonlinearly related to the proportion of PSII centers that are _open‘ (QA oxidized). Mathematically identical to the coefficient of photochemical quenching, *qP.*
NPQ=Non photochemical quenching	The non photochemical quenching from Fm to Fm'. Monitors the apparent rate constant for heat loss from PS II. Calculated from (Fm/Fm')–1.

Parameter(s)/Definition	*Physiological Relevance*
qE=Energy-dependent quenching	Associated with light-induced proton transport into the thylakoid lumen. Regulates the rate of excitation of PS II reaction centers.
qL=Fraction of PS II centers that are 'open'	Estimates the fraction of _open' PSII centers (with QA oxidized) on the basis of a lake model for the PSII photosynthetic apparatus. Given by (Fq'/Fv') (Fo'/F')
φF=Quantum yield of fluorescence	Number of fluorescent events for each photon absorbed

Instrumental Setting

Focus and Zoom

Place a testing plant under the FC. The testing plant should be similar to the plants that will be used in. Increase El. Shutter and Sensitivity to see a nice image. Change False- color scale (right mouse click on the scale right to the image) to Black and White. Zoom and Focus the objective to see sharp image. FOCUS and ZOOM should be kept during the whole experiment otherwise the level of absolute fluorescence signal could be changed. Then arrange the plant to final position, leaves of interest perpendicular to camera.

Camera Settings

Let measuring flashes switched on and adjust El. Shutter and Sensitivity in LIVE WINDOW. Change False- color scale to Extended spectrum or Extended spectrum 3_0_3 (the most sensitive color scales for human eye). Keep El. Shutter as low as possible (low resolution CCD between 0-1, high resolution CCD between 1-2), otherwise measuring pulses would be too strong causing actinic effect. Adjust Sensitivity by trucking the bar to get a signal in the range of 200-500 digital units (dark blue or blue color).

Light Settings

ACTINIC LIGHT: Choose intensity of Actinic light (Act1 or Act2): (a) either desired absolute light intensity can be chosen with respect to cultivation conditions, or (b) it can be adjusted according to the fluorescence transient. (a) Place a light meter under FC to the position and distance

Protocol

Click on the magic hat pictogram in top panel – Protocol and Menu Wizard and choose measuring protocol. There are some predefined protocols on the left side. User alone can define own protocols by using Wizards on the right side. The predefined protocol Fv/Fm Page 2 is a simple protocol determining Fo, Fm and Fv/Fm. This can be used to check if Saturating pulse is strong enough before running any other more complicated protocol.

Importing Settings

Click button Use in the bottom of LIVE WINDOW to import camera and light settings to the protocol.

Measure

Click red flash icon, Start Experiment, in the top panel to run the measurement. Unopened or partially opened leaf clips. Clip may shift in smooth leaves while attaching measuring head. Stray light may enter the leaf clip if leaf is not flat.

Table 6.12. Methods to achieve dark adaptation

Method	*Limitations*
A leaf can be put into a leaf clip shielding it from ambient light.	If the ambient light intensity is high, and the leaf is not entirely flat, there is a chance that some stray light reaches the shielded area
Detached leaves can be kept for a while between wet filter paper	Consequences for the physiological state of the leaf
Measurement in dim light under lab conditions	Leaves can still absorb and use most of the green light for photosynthesis
Measurements directly in the field at nigh	Measurements differ from measurements following a relatively short dark-adaptation

Infrared Thermography (IRT)/Canopy Spectral Reflectance (SR) Temperature Trait for Phenotyping

Spectral reflectance of plant canopy is a non-invasive phenotyping technique that enables several dynamic complex traits, such as biomass accumulation, to be monitored with high temporal resolution. It has many advantages including easy and quick measurements, integration at the canopy level and additional parameters can also be measured simultaneously via a series of diverse spectral indices like photosynthetic capacity, leaf area index, intercepted radiation and chlorophyll content. Therefore, canopy reflectance is considered as one of the valuable tools for high-throughput phenotyping. The canopy reflectance indices have been already used with great promise for measuring the effects of increasing atmospheric CO_2 and O_3 on soybean canopies. Investigations at the individual plant level under well controlled environmental conditions showed that spectral reflectance could be used to: (1) estimate the effects of environmental perturbations, such as changing atmospheric composition, on canopy structure and function, (2) monitor plant photosynthetic pigment composition and (3) assess plant water status and detect abiotic or biotic plant stresses. Plant water status as determined by plant water content or water potential integrates the effects of several drought-adaptive traits. Several methods are used to determine crop water content, including leaf water potential, leaf stomatal conductance and canopy temperature (CT), which is the relative measure of water flow associated with water absorption from the soil under water deficit. In addition to the above, one of the most commonly used indirect techniques for measurement of these variables is thermal infrared imaging, or infrared thermography, which involves the measurement of leaf or canopy temperature. Plant canopy temperature is a widely measured variable that is closely related to canopy conductance at the vegetative stage and therefore provides insight into plant water status. In any given environment, stomatal variation is the dominant cause of changes in canopy temperature. Although thermal imaging does not directly measure stomatal conductance, it has become a high-throughput tool for estimating differences in stomatal conductance (Merlot *et al.*, 2002). Thermal infrared imaging for estimating conductance can be used at the whole plant or canopy level over time.

Magnetic Resonance Imaging (MRI) and Positron Emission Tomography (PET)

These two methods are used at the Ju¨lich Plant Phenotyping Centre, Germany (Heike Schneider, personal communication) to investigate root/shoot systems growing in sand or soil which allow to assess structure, transport routes and the translocation dynamics of recently fixed photoassimilates labelled with short-lived radioactive carbon isotope ($\delta^{11}C$). Quantitative MRI and PET data not only help to study differences between species, but also provide a phenotype within a species, the growth pattern, water relations and/or translocation properties of assimilates. Therefore, the MRI–PET combination can provide new insights into structure–function relationships of intact plants. It also allows monitoring of dynamic changes in plant properties, which have previously not been possible to assess systematically, thus improving our understanding of plant performance (such as resource use efficiency or biomass production).

Nuclear Magnetic Resonance (NMR)

The short half-life of $\delta^{11}C$ (only 20 min) limits the utility of MRI to study source–sink relationships in minute structures such as developing seeds. NMR provides an alternative in vivo detection platform using 1H NMR and utilizes the signal emitted by protons associated with carbon nuclei, thereby sucrose and water movement may be imaged and quantified. Therefore, NMR technology is employed with $^{13}C/^{1}H$ double-resonant high-resolution coil to achieve better resolution for monitoring the structure of tissues like seeds through non-invasive visualization, mapping water movements and monitoring of sucrose allocation using ^{13}C-labelled sucrose (Melkus *et al.*, 2011).

Photosynthetic System (PSII) of Spikes

Chlorophyll fluorescence based photosynthetic efficiency was measured in spikes of two *T. aestivum* and three *T. durum* wheat cultivars which were developed in central zone of India. It was observed that the *T. durum* wheat had high photosynthetic efficiency than *T. aestivum* as indicated by chlorophyll fluorescence parameter (Fv/Fm) at similar phenological stage. In addition, the rate of decline in photosynthetic efficiency with increase in desiccation was high in *T. aestivum* than in *T. durum*. Similar trend was observed in each of the spikelets except terminal ones. Durum wheat had relatively less moisture than the *aestivum* throughout the measurements suggesting that better photosynthetic efficiency in the former than in the later was intrinsic. The results also indicated that chlorophyll fluorescence of spikes could be employed for phenotyping responses of wheat germplasm for drought tolerance.

Photosynthetic System (PSII) Sensitivity to Temperature

Chlorophyll fluorescence technique is used to identify crops tolerant to desiccation. Photosynthetic efficiency in terms of Fv/Fm was measured in five leaves of each of the above crops. Results revealed that photosystem were less sensitive to desiccation when compared with the same in other crops under this experiment. High sensitivity to desiccation was conspicuous as revealed by rapid decline in Fv/Fm values which indicate sensitivity of plants to stress. The rate of decrease in quantum efficiency with moisture stress was in the order indicating that was more tolerant than others.

Photosystem Sensitivity to Desiccation

Chlorophyll fluorescence imaging to study sensitivity of different crops to temperature. Chlorophyll fluorescence imaging technique was preferred for phenotyping for photosynthetic efficiency of plants based on photosystem performance. We conducted experiments with leaves of different fruit crops mentioned above clearly revealed that chlorophyll fluorescence ratio (Fv/Fm) decreases as temperature increased but with varying degree of sensitivity among the crops. Studies revealed that sensitivity of photosystem to temperature was less than that of other crops. The rate of decrease in quantum efficiency with rise in temperature was in the order indicating that was more tolerant to temperature than others.

Biotechnological Approaches for Climate-Smart Crops

The agricultural production and cultivation practices of all major crops is affected by climate change, the level of which depends on geographical areas. The climate variables such as increased mean temperatures, increase in frequency and magnitude of extreme weather events like rainfall, high temperatures, hailstorm *etc.* directly affect agricultural production. Perennial fruits crops are mainly selected for improved yield and quality. The adverse conditions are impacting productivity by affecting overall plant performance as well as influencing phenology of perennial crops. Influence on phenology leading to extended growth season has adverse effect not only on productivity but also on quality of the produce. Moreover, changing pattern of climate would result in increased attack of pathogens and pests and change in disease and pest epidemiology. However, not all changes in climate have adverse effect, for example increased atmospheric CO_2 and extended growing season could be favorable for improving productivity of some species. Medlyn (2011) reported 21 per cent decrease in stomatal conductance and transpiration in woody plants in parallel to rise in CO_2 concentration, resulting in improved water use efficiency and subsequent improvement in plant performance under water stress conditions. However, a concomitant increase of ambient temperature will increase evapotranspiration and the potential gain in yield caused by the higher CO_2 levels would be compensated or result in net losses due to the negative effect of high temperature (Vara Prasad *et al.*, 2005). It is also predicted that the elevated

CO_2 levels will reduce the nutritional quality of many crops. The severity of the effects of increased average temperature on crops will depend on the phenological state. In winter, increased temperature will result early occurrence of phenological events such as flower bud induction and during the fruit development phase, higher temperatures could affect final yield and quality. Besides high temperature, other associated stresses like moisture stress, salinity, radiation *etc.* are predicted to affect the plants adversely and warrant immediate attention to develop technologies to manage and temperatures are predicted to have effect on precipitation. Biotechnology has potential to mitigate effects of climate change. Genetically modified herbicide resistant crops require zero tilling thus significantly reduce fuel use, carbon sequestration and emission of CO_2 besides reducing soil erosion. Similarly, insect resistant crops require fewer pesticide sprays, thus resulting in saving of fuel and reduced CO_2 emission. According to an estimate, biotech crops resulted in reduction of 37 million kg of active ingredients, decreased rate of herbicide and insecticide sprays and ploughing reduced CO_2 emission by 23.1 billion kg of CO_2. Attempts are also being made to develop plants that can store more carbon dioxide in their roots. Another application of biotechnology is the development of varieties with resistance to biotic and abiotic stresses either through transgenic or marker assisted breeding. While success has been achieved for annual crop, little or no head way is achieved in crop. Life cycle, genetic heterozygosity, self-incompatibility, large body size and lack of knowledge of genetics of most traits are major constraints faced by the researchers involved in genetic improvement of crops. Moreover, understanding of fruit development physiology and metabolic pathways as well as identification of key genes for important traits are prerequisite for developing climate resilient varieties. In crops, so far, only the functions of a few genes are known, thus hampering the progress to large extent. Understanding of plant response to stress is important to develop strategies for cultivar improvement as well as for stress management. Big strides have been made in the development of techniques and instrumentations for big data generation and analysis of genomic, transcriptomic, proteomic and metabolomics data. The collective term used for the data obtained by global analysis of genome, RNA, protein, metabolites *etc.*, is Omics. Integration of omics data with phenotype and genotype data enables identification of genes involved in different traits and also allows understanding of effect of different environmental factors on biological process.

Genomics plays an important role to improve the efficiency of plant breeding for sustained production under changing environmental conditions by allowing efficient exploitation of adaptation traits present in germplasm accessions and by expanding the gene pool. Availability of genome sequence facilitate the identification of candidate genes for tolerance. On the other hand, Next generation sequencing techniques allows identification of SNPs which could be used for identification of genomic regions for traits either through QTL mapping or marker trait analysis and subsequently closely linked markers for traits related to tolerance to different stresses. In non-model crops where genome sequence is not yet available, transcriptome analysis is considered to be a viable approach to understand the gene function and molecular basis of many cellular responses of plants in response to abiotic stress. For non-model plants, *De novo* assembly of short RNA sequences is used to reconstruct the entire transcriptome and gene expression profile at the temporal and spatial basis to identify the key genes. Proteomics approach is frequently used to identify global changes in structure and abundance of proteins in response to environmental signals. Proteomics has the advantage of its capacity to reveal posttranslational modifications, which is important to estimate the impact of modification on plant productivity. Systems biology approaches to integrate genomics studies with omics data related to a particular phenotype pave the way to interpret the complex quantitative traits. As a case study, identification of genes for salt, moisture and heat stresses, which are important manifestations of climate change, in grapes using omics approaches will be discussed. Understanding mechanism of plant response to salinity is important to develop strategies not only for cultivar improvement but also for stress management. In different studies, whole transcriptome analysis of grapevine leaves in response to salt, heat and moisture stresses identified a large number of differentially expressed genes (DEGs). The placement of these DEGs on grape chromosomes indicated clustering of genes on chromosomes suggested putative QTLs for salt stress tolerance. Besides, pathways such as metabolic pathways, biosynthesis of secondary metabolites, membrane transport development related pathways were the major pathways affected by salinity stress. Microsatellite markers were also identified in selected DEGs and will have direct application in the development of varieties for improved tolerance.

Crop Genetic Transformation

Apart from herbicide-resistant crops, which enable excellent control of weeds, there are no immediate prospects for producing GM crops that could greatly improve water productivity. There are hundreds of patents that claim inventions that may improve "drought tolerance". Almost all are concerned with metabolic or stress-induced genes having doubtful functional significance at the level of a field-grown crop whose production is limited by water. Desiccation tolerance, the ability of plants to survive severe water deficits, has been a popular target. The problem is that droughts that are severe enough to kill crop plants are not commonly relieved by good rains during the particular growing season. Using a well-tuned crop simulation model driven by a long run of weather data would soon reveal how often genuine desiccation tolerance would be useful. While there are genes available that confer desiccation tolerance in transformed plants (Haake *et al.*, 2002) there remains doubt about whether this "tolerance" is simply due to the transformed plant growing more slowly or having smaller stomatal conductance than the wild type, thereby using water more slowly, and thereby not experiencing as severe water-deficits, despite having received no water for the same length of time. Evidently, assays need to be done more carefully, with the involvement of people experienced in exploring plant water relations. Indeed, progress in this general field is likely to remain slow unless teams are created which have the collective expertise not only to scale up from gene expression to the performance of plants in the field, but also to identify real problems in field-grown plants that may be amenable to effective genetic manipulation. The great prize of C_4 photosynthesis in rice, while feasible, and although the requirements are well understood at biochemical, structural and physiological levels, is still a long way off (Mitchell and Sheehy 2000). There are three areas that may prove useful before long. Carbohydrate metabolism in water-stressed maize at flowering, as noted earlier, can strongly affect embryo development. The work of Helentjaris *et al.* (2002) on invertase activity may prove to be effective in preventing severe embryo abortion during periods of water stress at flowering; it is soundly based on much physiological and biochemical analysis of floral development in water-stressed maize. Another example is the discovery of the CBF family of transcription factors (Thomashow *et al.*, 1999) which markedly improve freezing tolerance in Arabidopsis. Greater tolerance of freezing would enable breeders of crops with a winter-spring growing season to aim for earlier flowering, which would then give the crops a longer period of grain-filling in mild conditions before the heat and aridity of late spring and summer arrives. The third example is that of aluminium tolerance, now that a gene has been cloned that can enable roots of sensitive plants to grow in acid soils that commonly contain high levels of soluble aluminium (Sasaki *et al.*, 2004).

Improvements in water productivity will most likely come, approximately equally, from better agronomy and better cultivars, with improvements in one stimulating improvements in the other. Developing expectations about what water-limited yield might reasonably be, given the growing season's weather and other conditions, has proved to be an important stimulus to the way farmers' think about how they manage their crops. Analysing major discrepancies between actual and expected water-limited yields can reveal other major limitations such as inadequate nutrition, hitherto unrecognised root diseases, inadequate rooting depth (compaction, inhospitable subsoil), inappropriate choice of cultivar, poor establishment, or inadequate infiltration. One of the reasons for there being large differences in many regions between potential and actual yields is that farmers' management of their crops is often constrained by their perception and handling of risk. While crop breeders' historical success in improving yield potential has at the same time improved water productivity except perhaps when water is extremely scarce, there are also many specific opportunities to improve the water economy of crops by better tuning the development of a crop, both vegetative and floral, to particular environments – to be able to take tactical advantage of variable weather to sow at the right time to capitalise on opportunities for good establishment. As well there are reasonably well understood biochemical and physiological behaviours that may be amenable to selection, possibly even transformation, for ensuring that water-stressed crops do not lose more fertility than necessary or be unable to transport stored assimilate to the grain. The best immediate prospects come from well-understood single gene traits that affect the competence of roots, and therefore their ability to take up water, for example tolerance to the high levels of soluble aluminium that usually accompany soils of low pH.

Integrated Genomics, Physiology and Breeding Approaches for Improving Drought Tolerance

Drought is the most devastating abiotic stress affecting crop productivity, which is caused by insufficient rainfall and/or altered precipitation patterns. The seriousness of drought stress depends on its timing, duration and intensity. The impact of drought on crop production has been evidenced as early as the beginning of the seventeenth century, known as "***Sahel drought***", caused due to human intervention effects of deforestation, overgrazing and industrialization. Increase in greenhouse emissions has resulted in altered precipitation, increase in arid land, desertification and finally reduction in crop productivity. Moreover, it has been causing global warming, which in turn is responsible for raising the earth's surface temperature and sea water level. As of today, climate–yield predictions are well captured in several important major crop species through simulations. These important crops are in need of adaptation investments to avoid catastrophic yield losses and to meet the food demand of a fast-increasing population. Drought is often accompanied by relatively high temperatures, which promote evapotranspiration and affects photosynthetic kinetics, thus intensifying the effects of drought and further reducing crop yields. It is anticipated that the occurrence of drought in many food-producing regions will increase significantly in response to climate change (Reynolds and Ortiz, 2010). Tolerance to drought is a complex quantitative trait controlled by several small effect genes or QTLs and is often confounded by differences in plants phenology. To address the complexity of plant responses to drought, it is vital to understand the physiological and genetic basis of this response. Failure to understand the molecular mechanisms of seed yield stability has hampered both traditional breeding and the use of modern genetics in the improvement of drought tolerance of crop plants (Sinclair, 2011). Recent advances in crop physiology, systematic plant phenotyping and genomics have led to new insights in drought tolerance, thus providing crop breeders with greater knowledge of the gene networks and providing new tools for plant improvement to increase crop yield. While plant physiology improves our understanding of the complex network of drought tolerance- related traits thus improving selection efficiency, molecular biology and genomics approaches identify the candidate genes and quantitative trait loci (QTLs) associated with these traits. While QTLs can be deployed in crop improvement through molecular breeding, candidate genes are the prime targets for generating transgenics using genetic engineering. Identification of the "most appropriate" candidate genes along with selection of "most suitable promoters" and generation of a large number of events are critical for the development of desirable transgenics with enhanced drought tolerance using know-how knowledge (Varshney *et al.*, 2011). However, the expensive regulatory process and negative public perceptions of biosafety limit the application of genetic engineering approach, while there is a wider acceptance of products generated through molecular breeding and **Targetted Induced Local Lesions in Genome (TILLING)** (Barkley and Wang 2008). In the last decade, several important reviews of plant drought response and tolerance have been published. The importance of multifaceted strategies including genetic engineering, physiological approaches and genomics approaches have been described in several crop species. Also, the descriptions of molecular-physiological mechanisms of drought tolerance were outlined by several reviews. The importance of drought tolerance is to be highlighted, especially in a variable climate and discuss the recent progress made in the area of crop physiology for precise phenotyping and genomic approaches, such as identification and cloning of QTLs and identification of candidate genes associated with drought tolerance. In addition, new molecular breeding strategies such as marker-assisted recurrent selection (MARS) and genomic selection (GS) or genome-wide selection (GWS) are discussed as options to be integrated in crop improvement programmes for developing the next generation of drought-tolerant crops.

Physiological Traits in Breeding Programmes

The use of physiological traits (PTs) in a breeding programme, either by direct selection or through a surrogate such as molecular markers, depends on their relative genetic correlation with yield, extent of genetic variation, heritability and genotype 9 environment interactions. For instance, in drought environments, osmotic adjustment, accumulation and remobilization of stem reserves, superior photosynthesis, heat- and desiccation-tolerant enzymes, *etc.* are important PTs. However, it is important to establish their heritability and genetic correlation with yield in target environments. Identification of drought-adaptive PTs and mechanisms is time consuming and costly; however, if successful, the benefits are likely to be substantial. The information on important PTs can be collected on potential parental lines involving screening of entire crossing block, or a set of commonly used parents, thus producing a catalogue of useful PTs. This information can be used strategically in designing crosses, thereby increasing the likelihood of transgressive segregation events, which bring together desirable traits. However, if enough resources are available, screening for PTs could be applied to segregating generations in yield trials, or any intermediate stage, depending on when genetic gains from selection are optimal. It is important to note that using specific traits, breeding strategies are effective only when these traits are properly defined in terms of the stage of crop development at which they are relevant, the specific attributes of the target environment for which they are adaptive and their potential contribution to yield. The early escape from progressively intensifying moisture stress, through the manipulation of plant phenology, is the most commonly exploited genetic strategy used to ensure relatively stable yields under terminal drought conditions (Richards 1991). When significant genetic diversity for a physiological trait in a germplasm collection for the given species is established, it is imperative that the relevance of the trait as a selection criterion be determined.

Mini-Rhizotron Technique for *In-situ* Root Study

Monitoring of root architectural developments holds potential scope for the exploitation and desirable manipulation of root characteristics to enhance resource-use efficiency and crop productivity. Much emphasis has been to quantify the behavior of above ground development of the plants, and below ground developments of roots have been neglected in most of the field experiments conducted so far. This discrepancy might be ascribed to their hidden nature at below ground and of variable nature, both of which enormously complicate observation and conducting the research trials and not easily instrumented or observed due to relatively high cost and technical difficulties in sampling, data collection and analysis. Though, advance techniques and tools are required to record the complex and vital process of root architectural development for a desired period or at a particular critical crop growth stage. The growth and behavior of roots can be monitored for an entire crop growing season through successive measurements on the same site and same depth by mini-rhizotron technique. The in-situ root images are analyzed with an available root analysis software package and data thus collected from the mini-rhizotrons can be used to draw a dynamic picture of root system in terms of root count, length, diameter, surface area and volume. It is a non-destructive but expensive method for root studies.

Equipment and *In-situ* Root Monitoring

Image Acquisition Software

This system (CI-600 and CI-601 Root scanner) for automating image capture for minirhizotrons has recently been developed by CID, Inc, WA, USA. With the introduction of the CI-601, image acquisition can be performed automatically and remotely.

Mini-Rhizotrons

The clear acrylic tubes having 6.4 cm inner diameter and 7.0 cm outer diameter and 105 cm standard length are used for installation in the field. However, tubes with greater length (180 cm) are also can be utilized.

In-situ Root Image Acquisition using the CI-600

The CI-600 root scanner connected with a laptop computer through USB as shown in Figure 2 is inserted into already installed rhizo-tubes to monitor root growth at a desired depth. When the plant begins to build a network of roots, images of the structure and behavior of the roots can be recorded. The CI-600 scanner head rotates within the tube to scan roots and it provides nearly 3600 high-resolution images of soil and roots of 21.59 x 19.56 cm size. The connected images of different depths can be captured by moving the camera along the tube and these images can be saved in computer for further analysis.

Image Analysis System

It provide an opportunity to facilitate analyzing procedure. They offer a rapid assessment of root characteristics like length and surface area, diameter and tips, root branching patterns *etc.* The image analysis software CI-690 RootSnap! has been developed by CID, Inc, WA, USA. It allows users to measure root growth and turnover dynamics, disease, and behavior over time by analyzing scanned images collected with the CI-600. The analysed images are stored in rsp formats and the software supports exporting data to Microsoft Excel for further statistical analysis. RootSnap! is a faster and more reliable method for analyzing root images. It includes a revolutionary user interface that employs a combination of advanced image analysis and a multi-touch LCD screen, which allows users to quickly and easily trace roots using their fingers.

Analysis of *in-situ* Root Images with CI-690 RootSnap! Software

After the RootSnap! software is installed on a touchscreen computer, the application can be accessed from the Start Menu or by double-clicking on the RootSnap! icon on the desktop. To begin using RootSnap!, an image of a root needs to be imported into the program or an already saved project or session can be opened.

Table 6.13. Features displayed at the Menu Bar

Edit Menu	*View Menu*	*Help Menu*
1. Undo and Redo	1. Navigation window	
2. Delete current root	2. Zoom	
3. Lock current root	3. Panels	
4. Migrate roots	4. Layers	
5. Tube angle	5. Toolbar	
6. Default root status	6. Icon descriptions	
	7. Large icons	

Beginning to Map Roots

Select a window image to start mapping. Follow the steps as given below for mapping the roots: Import image; Rotate/flip the image; Verify correct physical size in Image Details Panel; Window alignment (if lower in tube then Window 1); Adjust image Brightness, Contrast and Gamma; Estimate root percent; Zoom in on root to map until it is at least as wide as your finger; Place the first point in the center of the root. The first point is critical; it must be on the root! Detect growth; Check automatically mapped points for accuracy and diameter. Move/place points past color change to keep automatically detecting growth. Detect growth again. Start mapping branches: Map a few points and detect growth, Dock branch to parent root. Detect growth - Move points to end of branches. Continue for rest of root system.

Exporting Data

Data from RootSnap! projects, tubes, windows or sessions can be exported to be opened as a spreadsheet. Exported data is saved as.csv (comma separated value) files which can be opened using Microsoft Excel or similar programs and saved. Root data is displayed at the top of the file including the root id, length, average diameter, area, volume, mean angle, branch count, branch ids and point count. Option is also available to save the root image while exporting the data.

Performance of New Wheat Genotypes Under Irrigated Saline and Alkali Conditions

Two new genotypes NW(S) 2-4 and KRL 119 were tested against established salinity tolerant varieties Kh 65 and KRL 19 in saline patch of this location under different nitrogen level application. Results of the trial revealed that at lower level of nitrogenous fertilizer application these new genotypes could not perform better than established check KRL 19 but contrary to this at higher level of nitrogenous test application KRL 119 proved to be better yielder genotype than established check.

Mobilisation of Pre-Anthesis Reserves During Grain-Filling

Crops that suffer water deficits during grain filling may produce a large biomass but be unable to match that with a good harvest index. Excessive vegetative growth, especially if induced by an oversupply of mineral nitrogen, can make the effects of water deficits worse by using too much water before flowering. The result is that the crop senesces prematurely and its yield falls with excessive nitrogen supply. Van Herwaarden *et al.* (1998) have argued that an oversupply of nitrogen worsens this imbalance in water use by reducing the amount of storage carbohydrates available for retranslocation to the grain. There is a negative correlation between nitrogen level and storage carbohydrate, not necessarily related to water deficits (Batten *et al.*, 1993), with the implication that excess nitrogen results in the investment of photosynthate into structural rather than storage carbohydrate when stimulating excessive vegetative growth.

Opportunities of Molecular Plant Breeding

The foregoing discussion describes how breeding and agronomy are closely intertwined in improving water productivity of rainfed crops. Marker-assisted selection (MAS) is becoming increasingly useful. Eagles *et al.* (2001) list about 20 markers in use in Australian wheat breeding programs. Most concern leaf diseases and grain quality. Apart from the leaf diseases, the control of which improves water productivity in both water-deficient and water-sufficient environments, several concern root diseases, which more directly affect water productivity. Table 3 lists several markers discussed by Eagles *et al.* (2001) and adds some more recent ones that are being used in breeding programs. Of these markers, the most important one so far is that for CCN resistance in wheat, which has contributed to the selection of several new cultivars. In maize, markers for ASI have proved useful in improving fertility in water-stressed crops.

Estimation of Heat Stress

Data on daily temperature (maximum and minimum), relative humidity, sunshine hours and rainfall will be collected. The heat stress will be estimated in terms of difference in maximum and minimum temperature during the crop season, heat degree days and high temperature incidence. The heat stress will be estimated separately for pre and post heading periods at all the locations. Mean minimum and maximum temperatures before and after heading will be calculated by taking into consideration the minimum days to heading and maximum days to maturity at respective locations. Heat degree days (HDD) will be computed with 0°C as base temperature. High temperature incidence (HTI) will be computed on the basis of daily minimum and maximum temperatures when the crop was exposed to greater than 30°C.

Estimation of Response to Heat Stress

The percent reduction (per cent reduction) in all the traits under stress conditions will be calculated as follows:

Per cent reduction in X = (X control – X heat stress)/X control 100

The heat susceptibility index (HSI) for GFD, TGW and GY/P will be calculated as follows:

HSI of X= [(1- Xheat stress/Xcontrol)/D]

where,

GFD, TGW and GY/P each are represented by X.

D (stress intensity) = (1- Yheat stress/Ycontrol)

Yheat stress= Mean of Xheat stress of all genotypes

Ycontrol= Mean of Xcontrol of all genotypes

Table 6.14. Phenotyping for yield component and heat tolerance traits

Year 1	Collection of 2000-3000 wheat accessions (comprising of released varieties, AVT entries, breeding material developed by various centers, synthetic hexaploids and exotic germplasm). Identifying about 1000 accessions by analyzing the data on yield components, pedigree, response to stress conditions and adaptability to stress environments. Genotyping the ~1000 accessions with at least two SSRs per arm of 21 wheat chromosomes. Assembling the association panel with about 300 diverse wheat genotypes based on the data on phenotypic traits, pedigree, SSR genotype and performing structure analysis. Seed multiplication of the Association Mapping Panel and distribution of the adequate quantity to the participating institutes to conduct multi-location yield component and heat tolerance trials
Year 2	Performing replicated yield component and heat tolerance trials at eight locations in alpha-lattice design with three replications. Phenotyping the association panel at each location. Genotyping the association panel either with a large number of SNPs or by 'genotyping-by-sequencing' technology.
Year 3	Multi-location trials and phenotyping to continue
Year 4	Multi-location trials and phenotyping to continue
Year 5	Analyzing the multi-location and multi-year data of the yield component and heat tolerance traits of the association panel. Performing marker-trait association mapping using suitable software tools to identify the QTLs for the five yield component traits

Table 6.15. Early heat and drought tolerant varieties

Bread wheat	
HI 1500 (Amrita)	Amrita' was released for cultivation in Central Zone by the **CVRC**, for cultivation under rainfed/limited irrigation conditions of Central Zone. High yield potential and early maturity of HI 1500 is bound to improve the wheat productivity under limited input conditions in central belt of the country.
HI 1531 (Harshita)	HI 1531 (Harshita) is the 1st early maturing, high yielding, semi-dwarf bread wheat genotype ever evolved in the history of drought tolerance breeding in Central India and released by **CVRC** for rainfed as well as limited irrigation cultivation in Central Zone. HI 1531 being nearly one week early in heading and maturity, compared to HW 2004, will escape frosting and terminal drought, ensuring stability in wheat production in Central Zone.
Durum wheat	
HI 8627 (Malav Kirti)	HI 8627 (`Malav Kirti'), the 1st high yielding durum wheat genotype which performed equally well under both rainfed and restricted irrigation conditions, was released for cultivation in Central Zone by the **CVRC**. Its "dual quality" (suitable for chapati as well as macaroni preparations), rich combination of protein, β-carotene, iron, zinc and copper in HI 8627 will ensure "Nutritional security" at household level, providing "Health Food" to rural India.

HI 8638 (Malav Kranti)	'Malav Kranti', the 1st "Tall-dwarf" high yielding durum wheat genotype which performed equally well under both rainfed and restricted irrigation conditions, was released for cultivation in Madhya Pradesh by the **SVRC**. Due to its high β-carotene, hectolitre weight, SDS value and protein content, it can serve as a "dual purpose" durum variety for making nutritious chapati and semolina (suji) as well to enrich "Nutritional Security" in rural areas of the state.

Table 6.16. "Water Use Efficient" varieties for optimum conditions

	Bread wheat
HI 1479 (Swarna)	Swarna' (HI 1479) was released for timely sown, high fertility conditions of Madhya Pradesh (India) by **SVRC**. It combines early maturity and high yield with rust resistance and good chapatti making quality. Swarna is one week early in maturity to Lok 1, the dominant wheat cultivar in central India, and has given 14 per cent higher yield over it.
HI 1544 (Purna)	Purna' is "complete" in all respects as it combines high "water use efficiency", early maturity and high yield with strong disease resistance and excellent grain quality and released by **CVRC** for high fertility and timely sown conditions in Central Zone, ensuring stability in wheat production in Central Zone.
	Durum wheat
HI 8663 (Poshan)	'Poshan' has high nutritional value. It can serve as a "naturally bio-fortified health food" and can be used for "dual purpose", *i.e.*, both for "nutritive chapati" and for fast food preparations for improving "nutritional security", as it contains high and stable β-carotene, high sedimentation value, high protein content and high levels of micronutrients, particularly iron and zinc.

Strategic Trait-Based Crossing

The conceptual models of drought-adaptive traits have been found useful for accumulating complementary PTs in selected progeny. The key steps in this type of PT breeding include: (1) characterization of crossing block lines for stress-adaptive mechanisms, (2) strategic crossing among parents with different but potentially complementary PT expression, thus ensuring cumulative gene action in selected progeny, and (3) early generation selection (EGS) of bulks for canopy temperature (CT). This type of physiological characterization is used to assess variation and thereby increases the rates of genetic gains. The main objective of strategic trait-based crossing is to accumulate traits that will be complementary for a given target environment. Under water-limited situations, traits that improve water uptake, water use efficiency and partitioning to yield, respectively, are likely to work synergistically to maximize productivity in the target environment. This has resulted in the distribution of advanced lines to rain-fed environments worldwide by the International Maize and Wheat Improvement Center (CIMMYT) and it has been confirmed that PT crossing results in cumulative gene action in selected progeny, resulting in increased yield under drought environments. Characterization of candidate parents for better targeted crossing should have the highest priority in terms of physiological interventions in breeding for a number of reasons including: (1) since a large investment is needed in trait measurement and the information obtained can be used for many generations of crossing once the initial characterization has been made, (2) the number of lines in a crossing block are relatively small (*100/target environment) and the detailed characterization is even possible for traits which are relatively time-consuming, *e.g.* for traits like soil moisture depletion or stem carbohydrates. The genetic gains in yield can be accelerated by incorporating complex PTs deterministically in modern plant breeding in addition to simply agronomic inherited traits like plant height, flowering time, resistance to prevalent diseases, quality parameters and yield based on multilocation trials. PT based breeding approaches have been already implemented successfully in Australian breeding programmes as well as by CIMMYT, leading to international distribution of a new generation of elite drought adapted lines (Reynolds *et al.*, 2009b).

Molecular Dissection of Drought Tolerance

In several genetic studies, drought tolerance has been found to be a complex quantitative trait controlled by a large number of minor genes/QTLs. Recent advances in genome mapping and functional genomics technologies have provided powerful new tools for molecular dissection of drought tolerance (Worch *et al.*, 2011). The molecular markers and/or candidate genes identified provide a better understanding of the molecular basis of drought tolerance and, once validated, can be used in molecular breeding.

QTL Discovery for Drought Tolerance-Related Traits

Traditional QTL mapping involves: (1) development of mapping populations segregating for drought tolerance related traits, (2) identification of polymorphic markers, (3) genotyping of the mapping populations with polymorphic markers, (4) construction of genetic maps, (5) precise phenotyping for drought tolerance-related traits, as mentioned above, and (6) QTL mapping using both genotypic and phenotypic data. This process is commonly called linkage mapping/linkage analysis-based QTL mapping. During the past decade, a large number of studies involving linkage mapping have been conducted in several crops to identify QTLs linked to drought tolerance. However, linkage mapping-based QTL mapping does not provide precise information on QTLs because of inherent limitations associated with each mapping population. Some of these limitations are summarized by Myles *et al.* (2009) and include: (1) insufficient time for recombination to occur and shuffle the genome into small fragments, and as a result the QTLs identified are generally localized to large genomic regions/chromosomal segments, (2) insufficient phenotypic variation for the trait present in the mapping population and (3) segregation of different QTLs for the same trait in different mapping populations. To overcome some of above constraints, linkage disequilibrium (LD)-based association mapping, initially used in human genetics, has been suggested as an alternative approach for QTL mapping in crop species. The association mapping (AM) approach involves: (1) selection of a diverse association panel/group of individuals from a natural population/germplasm collection, (2) precise recording of phenotypic data on the panel, (3) candidate gene sequencing or high density marker genotyping of the panel, (4) study of population structure (the level of genetic differentiation among groups within the selected population) and kinship (coefficient of relatedness between pairs of each individual within the population) and (5) association analysis based on information gained through population structure, kinship, and correlation of phenotypic and genotypic/haplotypic data. AM offers several advantages over bi-parental linkage mapping and these include: (1) exploitation of all the recombination events that took place during the evolutionary history of a crop species, resulting in much higher mapping resolution, (2) less time required in mapping QTL as there is no need to develop a specialized mapping population, rather a natural germplasm collection of a crop species is sufficient, (3) cost-effectiveness because the same AM panel and genotyping data can be used for mapping of different traits, (4) populations can be structured to avoid randomly generated lines (recombinant inbred lines; RILs), many of which express substandard agronomic type and (5) a higher number of alleles can be sampled compared to linkage mapping where only two alleles are usually surveyed (in apple or potato where heterozygous parental lines are used, more than two alleles/locus can be present). Markers linked to drought tolerance traits, identified using AM, have been reported in wheat (Varshney *et al.*, 2012). However, obtaining a clean set of reproducible phenotypic data of drought tolerance from a larger germplasm collection for AM studies remains an open challenge even in the era of phenomics driven technology.

Table 6.19. QTLs identified for drought tolerance-related traits in wheat crop

Traits Studied	*Number of QTLs*	*Chromosome/ Linkage Group*	*Phenotypic Variation Explained (PVE per cent)*
Agronomic, phonological and physiological traits	104	1A, 1B, 1D, 2B, 3A, 3B, 4A, 4B, 4D, 5A, 5B, 6A, 6B, 7A, 7B, UA-b	11.2–33.5
Morpho-physiological traits	110	All 14 chromosomes	0.8–42.4
Grain yield and growth traits	42	1A, 1B, 2A, 2B, 3A,3B, 4A, 4B, 5A, 6B, 7A and 7B	3.4–53.9
Water-soluble carbohydrates and associated traits	48	All chromosomes except 2B, 3D, 4D, 5D, and 6D	1.1–7.6
Grain yield and yield components under drought	1	4AL	12.0–41.0
Yield and growth traits	16	1B, 1D, 2B, 3A, 4A, 4B, 4D, 5A, 5B, 6A, 6B, 6D, 7A, 7B	-
Stem reserves mobilization	3	2D, 5D, 7D	21.1–42.3

QTLs for drought tolerance have been identified for several major and important crop species like rice, maize, wheat, barley, sorghum, pearl millet, soybean and chickpea. These QTLs were identified for a variety of important traits including: (1) yield and yield contributing traits under water-deficit conditions (in the case of wheat, maize, rice, soybean and pearl millet), (2) physiological responses including water-soluble carbohydrates, carbon isotope ratio, osmotic potential, chlorophyll content, flag leaf rolling index, grain carbon isotope discrimination, relative water content, leaf osmotic potential, osmotic adjustment, chlorophyll and chlorophyll fluorescence parameters to drought stress (in the case of wheat, maize and rice), (3) flowering time including anthesis to silking interval (in maize), (4) root traits (rice, maize, wheat, soybean and chickpea), (5) stay green (sorghum) and (6) nitrogen fixation (soybean). However, so far QTL studies on the impact of drought on grain quality have not been documented. While some key QTL studies in some crop species, an updated compilation of mapped QTL and major genes associated with abiotic stress tolerance including drought tolerance in crop plants is available at PLANTSTRESS site. Most of the identified QTLs for drought traits explain a relatively small portion of total phenotypic variation. As a result, their direct deployment in breeding programmes through marker-assisted selection (MAS) may not be very effective.

Association Mapping in Wheat (Phenotyping for Heat Tolerance)

Screening and identification of Indian wheat genotypes possessing faster seedling growth, early vigor, cooler canopy which is able to utilize more soil moisture and tolerance from early and terminal heat stress. To study various morpho-phenological traits for identifying water use efficient genotypes under heat stress conditions. Development of new germplasm/mapping populations/ varieties with better yield, crop establishment and early growth under heat prone conditions. Available wheat varieties/genotypes (including drought and heat adaptive cultivars, advanced lines, elite germplasm *etc.*) from various centres will be used for this investigation. Based on morpho-physiological traits phenotyping, parental combinations will be identified and crosses will made involving tolerant donors and well adapted agronomic bases lacking for heat tolerance Susceptible x tolerant crosses will be attempted for developing mapping populations. The material generated will be advanced at off season nursery for speeding up the development of heat tolerant genotypes. Advance lines developed through marker assisted breeding will be evaluated in station, common and coordinated trials for release of varieties with heat tolerance.

Table 6.20. Identified early heat tolerant and terminal heat tolerant varieties

Early Heat Tolerant		*Terminal Heat Tolerant*	
Bread Wheat	*Durum Wheat*	*Bread Wheat*	*Durum Wheat*
HI 1500, HI 1531	HD 4672, HI 8627	DL 788-2, HD 2932	HI 8627, HD 8498

Observations Recorded

Early Vigour, Days to 50 per cent flowering, Days to anthesis, Days to maturity, Plant Height (cm), No. of tillers per meter row length, Spike length, Grain number per spike, Spike weight, 1000 Grain Weight, Grain yield per plot, Biomass per plot, Leaf area, Leaf area index, Canopy temperature measurement at vegetative stage and 7 day's interval (at 12h, 14 h and 16 h), Stay green Trait (0-9 scale), Leaf area under greenness, Grain filling duration

Wheat x Maize Technique

Doubled haploid production is a technique which has the potential of accelerating the wheat breeding cycle. Using this technique homozygous wheat lines can be derived directly from F_1 as compared to the conventional approach where sufficiently homozygous material becomes available generally in the F_6 to

F_8 generations. In wheat, several methods are available for haploid production, but the wheat x maize system has emerged with distinct advantages. Main advantages include the absence of genotypic specificity, low gametoclonal variation and freedom from albinism. A protocol specific to local conditions was generated as a result of series of studies over years. The optimizations for each step were made one by one; a simple technique for the identification of embryo carrying seeds from wheat x maize crosses prior to dissection was developed, identification of superior maize genotype for pollination showing striking differences for induction and regeneration of haploid wheat embryos in wheat x maize system by Verma *et al.* (1999), rapid crossing techniques in wheat x maize crosses was developed by Sandhu *et al.* (2003). Later, two studies standardized chromosome-doubling methodology. Dhawan *et al.* (2003) studied wide range of colchicine doses applied prior to regeneration and identified non-toxic treatments in terms of embryo formation and plant regeneration. Sood *et al.* (2003) investigated novel chromosome doubling technique involving post-pollination colchicine injection (0.5 to 1 per cent) along with 2,4-D (100ppm) to the tiller 24 hours after pollination. When colchicine is applied at pre-regeneration stage it can also affect the frequency of embryo formation and plant regeneration. Therefore, a thorough examination of the effect of these colchicine treatments and their time of application on embryo formation and plant regeneration frequency remains to be made. It has been found that tiller culture greatly affects the plant regeneration frequency increasing it by four folds. The tiller culture technique however needs to be standardized with regard to tuning of culture also along with tiller culture medium composition so as to further enhance the embryo formation and plant regeneration frequency.

Durum Wheat x Maize System

With the success of bread wheat x maize system the method was adopted to produce haploids in durum wheat as well (O' Donougue and Bennett 1994). Almouslem *et al.* (1998) reported durum wheat x maize cross in several cultivars including Langdon substitution lines. They reported that substitution of chromosome 5D for 5B confers in durum higher ability to produce haploids. Battlisteros *et al.* (2003) reported tiller culture in durum wheat x maize crosses. They observed the effect of humidity on haploid formation. At PAU, studies on durum wheat x maize has not been initiated fully yet preliminary investigations reveal that the protocol developed for wheat x maize for bread wheat may not be equally successful for durum wheat.

Crossing Techniques

The crossing work is performed once the ear start emerging. Wheat spikes, partially emerged from the boot leaf should be selected and central (secondary) florets removed with a forceps. The primary florets are emasculated, (emasculations could be escaped but the choice of spike at appropriate stage should be ensured by selecting the spikes where anthers are green while the feathery stigma had opened up, indicating receptivity). Maize pollen should be collected in the forenoon (9-11 a.m.) on the butter paper by gently tapping the tassels. To ensure collection of fresh pollen the previously extruding anthers should be shed by shaking the tassels 1 hr prior to pollen collection. Pollinations are be performed in the forenoon by opening up the wheat florets with a pointed forceps and dispensing maize pollen on the stigma with a fine paint brush (Camel, number zero). Pollinations are generally done from 11 am to 1 pm. After pollination spikes were covered with glassine bags and tagged appropriately. The pollen should not be used for more than half hour or till the color remains light yellow once the color changes to dark yellow or pollen shows granulation it should be discarded and fresh pollen should be collected.

Hormone and Colchicine Treatment

In order to sustain the development of caryopsis, 2,4-D has to be administered. For injecting prick a hole in the upper part of the internode, with the help of injection needle and then inject the solution into the base of the uppermost internode till drops came out from the hole pricked in the upper part. Approximately 1-2 ml of solution could be injected.

Table 6.21. injections administered and time after pollination

Injection Composition	*Time after Pollination*
2,4-D (100 ppm) + Colchicine (1 per cent) +DMSO	24 hrs
2,4-D (100 ppm) + Colchicine (1 per cent) +DMSO	48 hrs
2,4-D (100 ppm)	72 hrs

Preparation of Injecting Solution

First prepare solution of 2,4-D by mixing 100 mg of 2,4-D in 1 l of water (2,4-D dissolves in alcohol) and from this solution take 98ml and to this add 100mg of colchicine and 2ml of DMSO (dimethyl sulphoxide). The remaining solution of 2,4-D can be used for third injection and spray.

Tiller Culture and Embryo Rescue

The following stock solutions were prepared for use in preparation of tissue culture medium. MS salts: Readymade MS Salt mix (Duchefa, Haarlem, Netherlands) for 10 liters dissolved in distilled water and volume made to 1 liter. This solution is used at rate of 100 ml per liter of MS medium prepared.

Hormones. Cytokinins (Kinetin, BAP) – stock of concentration 1mg/ml is made by dissolving 20 mg cytokinin in few drops of IN HCl and then making the volume to 20 ml with distilled water. Stocks are used at rate of 0.5-1 mg/l MS medium or as required. Auxins (2, 4-D, NAA, IAA, IBA) – stock of concentration 1mg/ml made by dissolving 20 mg auxin in few drops of absolute alcohol and making the volume to 20 ml with distilled water (preferably warm water). Stocks are used at rate of 0.5-1 mg/l MS medium or as required.

Solid medium (for embryo culture), Gelrite (Duchefa, Harlem, The Netherlands) was added at the rate of 2gm/l, as a solidifying agent. The medium prepared is autoclaved at 121?C for 15 minutes at 15 psi. For sub culturing the medium was changed to ½ MS + sucrose (10 gm/l) + NAA (1 mg/l) + IBA (1 mg/l).

Table 6.22. Composition of MS (1962) medium stock solution

Strength	*Constituent Salts*	*Quantity per 2 Liters*	*Use ml/l*	*Actual Amount in the Culture Medium (mg/l)*
X20	Ammonium nitrate (NH_4NO_3)	66.0g	50	1650
	Potassium nitrate (KNO_3)	76.0g	-	1900
	Potassium dihydrogen phosphate (KH_2PO_4)	6.800g	-	170
	Boric acid (H3BO$_3$)	0.248g	-	6.2
	Manganese sulphate ($MnSO_4.7H_2O$)	0.892g	-	22.3
	Zinc sulphate ($ZnSO_4.7H_2O$)	0.344g	-	8.6
	Potassium iodide (KI)	0.033g	-	0.825
	Cobalt chloride ($CoCl_2.6H_2O$)	1.0ml	-	0.25
	Copper sulphate ($CuSO_4.5H_2O$)[a]	1.0ml	-	0.025
	Sodium molybdate ($Na_2MoO_4.2H_2O$)[b]	1.0ml	-	0.25

Table 6.23. One ml of conc. HCl was added to this stock solution to avoid precipitation.

X50	Calcium chloride ($CaCl_2.2H_2O$)	11.0g	20	440
X50	Magnesium sulphate ($MgSO_4.7H_2O$)	9.250g	20	370
X50	Ferrous sulphate ($FeSO_4$. $7H_2O$)[c]	0.695g	20	27.8
	Na_2EDTA	0.932		37.28

Table 6.24. Vitamins and Amino acids g/2 litre

X100	Thiamine HCl	0.020	10	0.1
	Nicotinic Acid	0.100		0.5
	Pyroxidine HCl	0.100		0.5
	Glycine	0.400		2.0
	Myo-inositol[d]			100
	Sucrose[d]			30000

a: Dissolved 100 mg $CuSO_4.5H_2O$ in 100 ml distilled water; b: Dissolved 100 mg $Na_2MnO_4.2H_2O$ in 100 ml distilled water; c: Dissolved 0.695 g $FeSO_4$. $2H_2O$ and 0.932 g Na_2 EDTA in distilled water and heat slightly to get yellow solution; d: Myo-inositol and sucrose were added in solid form while preparing the final medium.

Tiller Culture

The pollinated tillers are to be detached from the plant by giving a cut about 2″ below the second internode (counting from top). The detached tillers are transferred to a beaker, the cut ends being dipped in water and ears covered with a polythene bag to avoid desiccation. The stems should be thoroughly washed in running water followed by sterilization with $HgCl_2$ (0.2 per cent) for 10 mins in the laminar air flow chamber. After this treatment the wheat stems are to be swabbed with 70 per cent ethanol and given two washings in autoclaved distilled water. This is followed by giving a final slanted cut to the tiller, about 1" below the second internode, keeping the part to be cut under sterile water in a pertridish. The sterilized tillers were cultured in liquid media, The basal part of the stem remained in the test tube whereas the spikes remained exposed on the top. After culture the tillers were given hormone and colchicine injections as already described. The tiller cultures should be maintained at 25°C, with 12 h light/12 h dark regimes and high humidity. A mist sprayer was used intermittently to increase humidity at the level of the spikes. Tiller culture medium: The regular MS used for embryo rescue can be used except for following modifications: Add 10mg/l 2,4-D; Gelrite (or agar) is not used; sucrose be used is 4 per cent.

Embryo Rescue

The caryopses are to be carefully removed from the spikes, 13 days after tiller culture. The caryopsis should be washed under running tap water and transferred to sterilized conditions under laminar air flow for surface sterilization with $HgCl_2$ (0.1 per cent) for 8 mins. Caryopses are then transferred to ethanol (70 per cent) for 30 secs., followed by two washings with autoclaved distilled water. Caryopsis are taken out in a sterile petri dish and dissected. Embryos are cultured on regeneration medium having MS basal salts + sucrose (30g/l) + Inositol (100mg/l) + BAP (0.5mg/l) + Kinetin (0.5mg/l) + Gelrite (2g/l) and incubated at 25°C.

QTL Cloning for Drought Tolerance-Related Traits

In general, QTLs identified through linkage mapping-based approaches have low resolution and have been located in 10–20 cM intervals. The support interval of the QTL may also span several hundreds of genes and identifying the right candidate gene(s) with causal effect on the trait is like finding a 'needle' in the 'genomic haystack'. Therefore, to identify the causal gene(s), positional cloning of QTLs have been undertaken in several crop species. QTL cloning, in general, involves the following steps: (1) delimiting the QTL region by using a large mapping population ([1,500 plants) derived from a cross between two NILs for the target QTL, (2) identifying the contig covering the QTL region by screening the closely linked molecular markers with a large insert library like BAC (bacterial artificial chromosome) library, (3) sequencing the contig and candidate gene identification based on sequence data and (4) validating the effect of candidate gene(s) on phenotype. Although many reports are available on cloning of QTLs associated with different traits, there are few reports addressing QTL cloning for drought tolerance traits. For instance, a major flowering time QTL "Vgt1" associated with drought tolerance has been cloned in maize. Recently, the gene encoding ATP-binding cassette (ABC) subfamily G (HvABCG31) full transporter was cloned from eibi1 mutation responsible for leaf water conservation in wild barley and rice (Chen *et al.*, 2011). Newer genomics approaches like association mapping and next-generation sequencing (NGS) hold great promise for accelerating QTL cloning of drought tolerance-related traits. The cloning of drought tolerance QTLs provides an opportunity to validate candidate genes that can be used to develop transgenic plants, not only in the original crop species but also in other crop species.

Identification of Genes Associated with Drought Tolerance

The significant advances made in the model plant systems of major crop species provide an opportunity to identify candidate genes associated with drought tolerance. Some approaches are discussed in the following section:

Candidate Genes (CGs) from Model Plant Species

Genome sequences have recently become available for several model and major plant species. Genome annotation, molecular physiological as well as functional genomics studies undertaken in model and/or major crop species provide evidence of the candidate genes (CGs) involved in conferring drought tolerance. The CGs can be: (a) genes involved in cell protection under drought stress (*e.g.* proteins involved in osmotic adjustment, degradation, repairs, detoxification and structural adaptations) and/or (b) genes involved in regulation of other genes involved in the drought response (protein kinases and transcription factors such as DREB, bZIP, MYB, *etc.*). Knowledge of the CGs responsible for drought tolerance is useful for understanding the functional basis of drought tolerance and assists in their subsequent use, once they are validated, in molecular breeding through MAS. For instance, a set of nearly 30 important candidate genes associated with drought tolerance have been compiled. Validation of the CGs, an important and essential step before they can be deployed, can be undertaken using several approaches including integration with QTL maps, association mapping, expression analysis using qRT-PCR, allele mining and TILLING. Several of these approaches in relation to breeding programme applications are discussed by Varshney *et al.* (2005). However, candidate genes have not delivered as much as anticipated for crop breeding, especially for drought tolerance.

Transcriptomics and Functional Genomics

Transcriptomics and functional genomics have been used extensively in recent years to better understand the stress responsive mechanisms in crop plants. The candidate genes associated with drought tolerance mechanisms have been identified, characterized and assessed for their transcriptome responses using whole-genome sequencing or through micro-array technologies. The generation of ESTs from either normalized or non-normalized cDNA libraries from drought-challenged tissues of drought-responsive genotypes is one of the most common approaches for isolation of drought-responsive candidate genes. A large number of drought-responsive genes have been generated in several crop species. This study was further extended by the National Research Centre on Plant Biotechnology (NRCPB) in India and 5,494 high quality drought-responsive ESTs were isolated using suppression subtraction hybridization (SSH) of drought challenged root and shoot tissues. Such studies provide an important resource for marker development and also act as resource for the identification and selection of candidate genes (both up- and down-regulated) associated with drought tolerance. Although bioinformatics analysis (*e.g.* BLASTX) of such ESTs can help to identify the most promising EST/gene(s), it is essential to prove the function of the most promising genes using wet laboratory experiments such as qRT-PCR. Another approach to identify candidate genes is transcript profiling that involves analysis of differential gene expression in the given tissue at different time points after exposure of the plant to drought stress or between drought tolerant and susceptible genotypes. However, it is important to target the right tissue and the precise stage of tissue in addition to the dynamics (*i.e.* timing and intensity) of the stress treatment imposed to mimic drought conditions for isolation of RNA for use in transcriptomics studies. Instead of using genotypes with different genetic background, near-isogenic lines (NILs), which differ only in the target trait, are the ideal genetic material that ensures differentially expressed genes are linked to the trait and not to the genetic background. More recently, it was demonstrated that miRNAs are also involved in drought stress response/tolerance in crop plants and their validation revealed their possible involvement in drought tolerance (Kulcheski *et al.*, 2011).

Several platforms have become available for transcript profiling: (a) PCR-based differential display PCR (DDRTPCR) analysis, (b) cDNA–Amplified Fragment Length Polymorphism (cDNA–AFLP) analysis, (c) cDNA and oligonucleotide microarrays and (d) digital expression analysis based on counts of ESTs. SuperSAGE, an improved version of the serial analysis of gene expression (SAGE) technique, has been also successfully applied in several crop plants including chickpea for expression analysis of *80,000 transcripts from unstressed and drought-stressed roots. However, with the advent of NGS technology, the sequence-based transcriptome analysis is in many ways considered superior to microarrays in orphan crops where genome sequence information is lacking, since the sequencing-based method is real time, digital and highly accurate. Therefore, it is anticipated that microarrays may soon be replaced by sequencing-based digital gene expression analysis. The application of NGS technologies to gene expression analysis has catalysed the development of techniques like Digital Gene Expression TAG (DGE-TAG), DeepSAGE and RNA-Seq. RNA-seq based on NGS technologies has several advantages for examining transcriptome fine structure including detection of allele-specific expression and splice junctions and may allow direct high-throughput sequencing of RNA from the stress (*e.g.* drought)-challenged tissues of different genotypes. Such transcript profiling (including RNA-seq) based on drought tolerant and drought-sensitive genotypes can identify candidate genes associated with drought tolerance that can be used as genic molecular markers (GMMs) and integrated into genetic/QTL maps. It is possible that some candidate genes identified as above may be associated with QTLs for drought tolerance traits. In such cases, a genetical genomics approach that involves quantitative analysis of transcript profiling of the candidate genes can provide the expression QTLs for drought tolerance-related traits. In case eQTLs are found in the cis-condition, then the candidate gene-based molecular markers should act as the functional and diagnostic markers for the respective traits (Potokina *et al.*, 2008). It is anticipated that NGS-based transcript profiling should be routinely used for major crop species in the identification of candidate genes for drought tolerance and for subsequent use in genetical genomics or molecular breeding.

Modern Breeding Approaches for Developing Superior Germplasm for Drought Tolerance

Once the candidate genes or markers associated with QTLs for drought tolerance are identified, the next step is their deployment in breeding practices. Some of these approaches are discussed below.

Marker-assisted Backcrossing (MABC)

When the QTLs identified for drought tolerance traits contribute higher phenotypic variation, they are considered major QTLs. These QTLs, after validation in desired germplasm, can be used for introgressing drought tolerance from the donor genotypes (generally used for identification of the QTL for the trait) into elite, less drought-tolerant cultivars or breeding lines (recipient parents) without transfer of undesirable or deleterious genes from the donors (linkage drag). The process is commonly referred to as marker-assisted backcrossing (MABC). Superior lines or cultivars are developed that contain only the major gene/QTL from the donor parent, while retaining the whole genome of the recurrent parent. Although MABC has been used extensively for introgressing resistance to biotic stresses, only a few reports are available on the use of MABC to develop the superior lines/varieties for drought tolerance. The relatively low success of MABC for improving drought tolerance can be attributed to the complex nature of drought. In many instances, the expression of drought tolerance is controlled by minor main-effect QTLs or epistatic QTLs. These studies highlight the need to transfer several QTLs/genes to achieve a significant impact, assuming additive variance, and this may require unmanageable population sizes (Ribaut *et al.*, 2010).

Marker-assisted Recurrent Selection (MARS)

To overcome the limitations of MABC, particularly when multiple QTLs control the expression of a complex trait, the MARS approach, which involves intermating selected individuals in each selection cycle, has been recommended. It generally involves the use of an F_2 base population, and can be used in self-pollinated crops like wheat, barley and chickpea for developing pure lines with superior per se performance. MARS has the additional advantage of overcoming the limitation of inadequate improvement in the frequency of superior alleles in F_2 enrichment, since MAS is practised in each cycle following intermating to improve the frequency of favourable alleles. In case of wheat, MARS for water use efficiency is being exercised under an Indo-Australian project involving partners from DWR, Karnal, PAU Ludhiana, IARI, New Delhi and Australia. Generation Challenge Programme (GCP) also launched a challenge initiative to improve heat/drought tolerance in wheat through MARS approach involving the Indian Agricultural Research Institute (IARI), New Delhi, India, Chinese Academy of Agricultural Sciences (CAAS), China, and partners from Australia.

Genome-Wide Selection (GWS)

Genome-wide selection (GWS) or genomic selection (GS) is another important approach to develop superior germplasm lines with overall excellent performance in a target environment. Genome-wide marker genotyping is used for GWS rather than selected markers showing significant associations (as in case of MARS) with the traits of interest. In summary, individuals in a phenotyped population (generally referred to as the 'training population') are genotyped using genome-wide markers and breeding values of alternative alleles of all the markers are fitted as random effects in a linear model. Individuals in subsequent recurrent selection generations are then selected based purely on the sum of those breeding values [genomic estimated breeding value (GEBV)]. Therefore, GWS reduces the frequency of phenotyping and similarly also increases annual gains from selection by reducing cycle time. Several groups have recently started exploring the GWS approach in both self- and cross-pollinated crops with some modifications for both types of crops. The success of the GWS approach is dependent on the availability of a diverse and representative training population. Furthermore, the phenotyping of the training population is crucial and additional lines should be integrated over time to increase the effectiveness and relevance of the gene effect estimates. This approach has been recently used to improve durable stem rust resistance in wheat (Rutkoski *et al.*, 2010) and eventually could be systematically explored to bring different components of multigenic drought tolerance using the GWS approach.

Future Outlook

It is evident that precise phenotyping is essential to screen larger core collection/ mapping population for identifying the most appropriate QTL and candidate genes for use in plant breeding. A number of phenotyping approaches are available and this area of research is currently referred to as 'phenomics'. Nevertheless, re-integration of the pieces of the 'phenomics' puzzle into a comprehensive and relevant crop improvement framework of 'seed yield stability' will involve crop modelling. The combination of phenomics and modelling offers great potential to rapidly assess the value of certain traits on plant performance. The use of models to understand gene-to-phenotype relationships provides an efficient platform for a new and creative interaction between genetics– genomics and crop physiology. To meet the real-world challenge of increased crop production, the information available from functional genomics and systems biology needs to be integrated at the crop level; thereby, crop physiology will have a fundamental role in achieving this goal. A new generation of crop models combined with systems biology studies should enable us to significantly narrow the gap between genes and complex phenotypes by predicting the field performance of crop genotypes. Crop models will significantly contribute to higher level of integration by directly linking physiological processes to complex crop phenotypes within the scope of source–sink relationships. Similarly, recent advances in genomics make it possible to not only conduct large-scale and high-throughput marker genotyping, but also sequence or re-sequence the genomes of germplasm collections, thus facilitating the identification of QTLs and candidate genes associated with drought tolerance. While commonly used MABC has not been very effective in developing superior lines for drought tolerance, modern breeding approaches such as MARS and GWS are powerful tools for pyramiding multiple QTLs for drought tolerance or introgressing multiple complex traits such as heat tolerance in addition to drought tolerance. It is essential to integrate crop physiology, genomics and breeding approaches to dissect complex drought tolerance traits, understand the molecular basis of drought tolerance and develop the next-generation crops for our changing climate. Although work is ongoing in some major crops, it is anticipated that integrated physiology, genomics and breeding approaches will be initiated/accelerated in the so-called orphan crops that are important for food security in many developing countries.

Breeding for Tolerance to Abiotic Stresses

Wheat is increasingly being grown on marginal lands and in farming systems where inputs are limited. Depressed international and local prices have caused farmers to use their best land for higher-value crops, and in southern and central Asia, where wheat is grown under irrigation, water availability for irrigation is declining as demand for water from urban centers increases. In addition, wheat production in low-latitude areas is likely to be most affected by climate change; some estimates predict wheat yield reductions of 12 per cent if average global temperatures rise by 2°C. Wheat breeders can mitigate the effects of changing production environments to some extent. Farmers have been selecting wheat for drought tolerance for thousands of years, gradually accumulating those genes that confer adaptation to limited moisture, so variability already exists within the wheat gene pool for drought tolerance. Our understanding of the genetic basis of drought tolerance is poor. Nevertheless, considerable progress has been made in yield improvement under drought in recent decades using the wheat gene pool and selection under drought stress (Trethowan *et al.*, 2002). The opportunity exists to improve tolerance further if new genetic variability can be combined with existing variability and if the underlying genetic control of tolerance can be better understood. Synthetic hexaploid wheat, developed by crossing modern durum wheat (*Triticum durum*) with *Aegilops tauschii*, the probable donor of the D-genome in hexaploid wheat, has introduced new genetic variation into the wheat gene pool for many characters. Not surprisingly, synthetic wheat has also been a source of variation for drought and heat tolerance. The original crosses that gave rise to hexaploid wheat occurred some 8,000 years ago and probably involved only a few accessions of *A. tauschii*. Hundreds of accessions of *Aegilops tauschii* have been collected from some of the harshest environments and are maintained by various gene banks, which represent considerable genetic variation. Once the synthetic hexaploid is made it can be easily crossed with cultivated wheat and target traits transferred. Although synthetic wheats can also introduce undesirable traits, some synthetic-derived advanced materials have improved wheat adaptation around the world, especially in drought-stressed environments. The performance of a synthetic hexaploid derivative 'Vorobey' sown in 33 different environments, with mean yield ranging from less than 1 t/ha to more than 9 t/ha compared to the best locally adapted cultivar at each location. Clearly the synthetic derivative out-yields each of 33 locally adapted check cultivars across the range of environments.

Recent research has shown that considerable new genetic variation also exists among the wild tetraploid progenitors, *T. dicoccum*, of cultivated durum wheat; the resulting synthetic hexaploids when recombined with adapted wheat through limited backcrossing have greatly improved yield under drought conditions. Remember that the observed drought tolerance of the synthetic derived materials is not just due to the genes introgressed from synthetic wheat. Many original primary synthetics do not have exceptional drought tolerance, but when recombined with existing genetic variation in adapted wheat, they can express high levels of tolerance; it is likely that new additive genes from synthetic wheat contribute to additive and additive additive variation. Drought tolerance is a complex character that is greatly influenced by GEI. However, water-use efficiency (WUE) can be increased by improving root resistance to soil-borne diseases as well as micronutrient imbalances, because healthier roots will use available soil moisture more efficiently. Fortunately, many of these characters are simply inherited, and molecular markers are available to assist selection (William *et al.*, 2003a). To improve breeding efficiency for drought tolerance, strategy is to ensure that drought-tolerant germplasm maintain responsiveness if more moisture becomes available in a season. High yield potential and drought-stress tolerance are not mutually exclusive; they can be bred simultaneously by selecting segregating populations grown under favourable environments and drought-stressed environments in alternate generations–a practice used. Moreover, to generate a more precise drought stress at different growth stages. This system allows the application of exact amounts of water at chosen growth stages to generate different drought scenarios representing different parts of the world. The genetic control of heat tolerance, like drought tolerance, is poorly understood. Nevertheless, significant variation for heat tolerance exists in the wheat gene pool. In many environments, late planting can expose crop and selection nurseries to high temperatures from flowering onward, giving wheat breeders the opportunity to select lines with high levels of heat tolerance. Lines are selected during the segregating phase for adaptation to heat by late planting. A gravity is used to separate bulk populations into those that can maintain seed weight under high temperature; the derived lines are then tested under heat stress in yield trials. The yield of synthetic-derived lines tested under late heat stress with the highest yielding local check cultivar; the derived lines are clearly higher yielding than the locally adapted cultivar. Physiological tools, such as the infrared thermometer that measures canopy temperature depression (CTD), are also available to assist the plant breeder in discriminating among progenies. The CTD is easy to measure, improves the heritability of selection, and can be used to select in segregating generations (Trethowan, 2006b). Heat avoidance or early maturity is an extremely important trait for wheat to have in order to circumvent the effects of high temperature at grain filling. All currently grown popular cultivars are earlier maturing than cultivars popular in the north-western Gangetic Plains. A simultaneous improvement in heat tolerance and yield potential for earlier maturing germplasm is the only option to increase production in heat-stressed environments. The performance of two new high-yielding, early maturing lines and the best check, popular cultivar 'HUW234,' in farmers' fields under zero-tillage in six villages in a heat-stressed ME5 environment. The two lines performed better than the check at all sites and showed an average advantage of about 18 per cent over 'HUW234.' Selecting segregating populations for improved heat tolerance in the north-eastern Gangetic Plains may provide future wheat germplasm for the north-western Gangetic Plains and other parts of the world where heat is expected to cause major stress in coming years due to global warming. There is also significant variation in the wheat gene pool for tolerance to soil toxicities and deficiencies, such as salinity, aluminium, boron, and zinc. Nevertheless, new genetic variation will need to be found or created and introgressed if adaptation to some extreme environments is to be improved.

Breeding for Input-Use Efficiency

Wheat with better N- and P-use efficiency will reduce the amount of these essential elements that need to be applied by farmers, thereby increasing the profitability of the farming system and reducing leaching to ground water, river systems, and the sea. There is significant variation in the wheat gene pool for both N- (Ortiz-Monasterio *et al.*, 1997) and P-use efficiency (Jones *et al.*, 1992). To illustrate how these traits can be improved genetically, the case of N-use efficiency (NUE) will be discussed citing a study conducted by Van Ginkel *et al.* (2001). The NUE is comprised of two components, N-uptake efficiency (UPE) and N-utilization efficiency (UTE). In crosses between lines with high UPE and high UTE they found that the selection environment was critical in identifying genotypes with improved NUE. They tested four environments for the selection of segregating populations: (1) always sown under low nitrogen, (2) always sown under medium nitrogen, (3) always sown under high nitrogen and, (4) alternating low nitrogen with high nitrogen. The alternating treatment identified lines that performed well at intermediate and high nitrogen levels. No differences were found among selection regimes when the resulting lines were evaluated at low nitrogen. Generally, if nitrogen is limiting, the contribution of UPE to overall NUE is greater than UTE (Ortiz-Monasterio *et al.*, 1997).

References

Abebe, T; Guenzi, AC; Martin, B and Cushman, JC (2003) Tolerance of Mannitol-Accumulating Transgenic Wheat to Water Stress and Salinity. *Plant Physiol.*, **131**: 1748-1755.

Acevedo, E; Nachit, M and Ferrara, GO (1991) Effects of heat stress on wheat and possible selection tools for use in breeding for tolerant, D.A. pp. 401-421.

Agarwal, SK; Agarwal, M and Grover, A (1999) Emerging trends in agriculture biotechnology research: use of abiotic stress-induced promoter to drive expression of a stress resistance gene in the transgenic system leads to high level stress tolerance associated with minimal negative effects on growth. *Current Science* **77**: 1577-1579.

Agarwall, RR; Yadav, JSP and Gupta, RN (1982) Saline alkali soils of India. ICAR., New Delhi.

Ahmad, M and Sorrells, ME (2002) Distribution of microsatellite alleles lined to Rht8 dwarfing gene in wheat. *Euphytica* **123**: 235-240.

Al-Ashkar, I (2013) Anther culture response and salt tolerance in some wheat genotypes. *Ann Agric Sci* **58**(2): 139–145.

Al-Khatib, K and Paulsen, GM (1984) Mode of high temperature injury to wheat during grain development. *Physiol Plant* **61**: 363-368.

Al-Khatib, K and Paulsen, GM (1999) High temperature effects on photosynthetic processes in template and tropical cereals. *Crop Sci* **39**: 119-125.

Allen, GJ; Jones, RGW and Leigh, RA (1995) Sodium transport measured in plasma membrane vesicles isolated from wheat genotypes with differing K+/Na+ discrimination traits. *Plant Cell Environ* **18**: 105-115.

Almouslem, AB; Jauhar, PP; Peterson, TS; Bommineni, VR and Rao, MB (1998) Haploid durum wheat production via hybridization with maize. *Crop Sci* **38**: 1080-87.

Alvim, CFC; Cascardo, JCM; Nunes, CC; Martinez, CA; Otoni, WC and Fontes, EPB (2001) Enhanced accumulation of BiP in transgenic plants confers tolerance to water stress. *Plant Physiol*, **126**: 1042-1054.

Alwine, JC; Kemp, DJ and Stark, GR (1977) Method for detection of specific RNAs in agarose gels by transfer to diazobenzyloxy- methyl-paper and hybridization with DNA probes. *Proc. Natl. Acad. Sci. USA* **74**: 5350-5354.

Amrani, N; Sarrafi, A and Alibert, G (1993) Genetic variability for haploid production in crosses between tetraploid and hexaploid wheats with maize. *Plant Breed* **110:** 123-28.

Anderson, JA; Sorrells, ME and Tanksley, SD (1993) RFLP analysis of genomic regions associated with resistance to preharvest sprouting in wheat. *Crop Science* **33**: 453-459.

Anderson, WK; Heinrich, A and Abbotts, R (1996) Long-season wheats extend sowing opportunities in the central wheat belt of Western Australia. *Australian Journal of Experimental Agriculture* **36**: 203-208.

Andrews, SS and Carroll, CR (2001) Designing a decision tool for sustainable agroecosystem management: Soil quality assessment of a poultry litter management case study. *Ecol. Applic.* **11** (6): 1573-1585.

Angus, JF and van Herwaarden, AF (2001) Increasing water use and water use efficiency in dryland wheat. *Agronomy Journal*, **93**: 290-298.

Angus, JF; Gault, RR; Peoples, MB; Stapper, M and van Herwaarden, AF (2001) Soil water extraction by dryland crops, annual pastures, and lucerne in south-eastern Australia. *Australian Journal of Agricultural Research* **52**: 183-192.

Aniol, A and Gustafson, JP (1984) Chromosome location of genes controlling aluminum tolerance in wheat, rye and triticale. *Can. J. Genet. Cytol.* **26**: 701–705.

Anonymus (2010) National Disaster Management Guidelines: Management of Drought. A publication of the National Disaster Management Authority, Government of India. ISBN 978-93-80440-08-8, September 2010, New Delhi.

Araus, JL; Bort, J; Steduto, P; Villegas, D and Royo, C (2003) Breeding cereals for Mediterranean conditions: Ecophysiology clues for biotechnology application. *Ann. Appl. Biol.*, **142**: 129-141.

Araus, JL; Slafer, GA; Reynolds, MP and Royo, C (2002) Plant breeding and drought in C_3 cereals: What should we breed for? *Annals of Botany* **89**: 925-940.

Arnon, I (1972) Crop production in dry regions, vol-1, Leonard Hill, London.

Ashraf, M (2010) Inducing drought tolerance in plants: recent advances. *Biotechnol Adv* **28**: 169–183.

Babu, CR; Nguyen, BD and Chamarerk, V (2003) Genetic analysis of drought resistance in rice by molecular markers: association between secondary traits and field performance. *Crop Sci* **43**: 1457–1469.

Babu, R; Nair, SK; Prasanna, BM and Gupta, HS (2004) Integrating marker-assisted selection in crop breeding-prospects and challenges. *Curr Sci* **87**(5): 607-619.

Bachem, CWB; van der Hoeven, RS; De Bruijn, SM; Vreugdenhil, D; Zabeau, M and Visser, RGF (1996) Visualization of differential gene expression using a novel method of RNA fingerprinting based on AFLP: analysis of gene expression during potato tuber development. *Plant J* **9**: 745–753.

Badaruddin, M; Saunders, DA; Siddique, AB; Hossain, MA; Ahmed, MU; Rahman, MM and Parveen, S (1994) Determining yield constraints for wheat production in Bangladesh. In: Saunders DA, Hettel GP (eds), Wheat in Heat stressed Environments: Irrigated, Dry Areas and Rice-Wheat Farming Systems. CIMMYT, Mexico, pp. 265-271.

Bahieldin, A; Mahfouz, HT; Eissa, HF; Saleh, OM; Ramadan, AM; Ahmed, IA; Dyer, WE; El-Itriby, HA and Madkour, MA (2005) Field evaluation of transgenic wheat plants stably expressing the HVA1 gene for drought tolerance. *Physiol. Plant*, **123**: 421-427.

Bains, NS; Mangat, GS; Singh, K and Nanda, GS (1998) A simple technique for the identification of embryo carrying seeds from wheat x maize crosses prior to dissection. *Plant Breed* **117:** 191-92.

Bajaj, YPS (1996) Biotechnology in Agriculture and forestry series. Somaclonal variation in crop improvement. Springer-Verlag, Heidelberg, New York.

Bal, SK and Minhas, PS (2017) Atmospheric Stressors: Challenges and Coping Strategies. In: P.S. Minhas *et al.* (eds) Abiotic Stress Management for Resilient Agriculture. Springer Nature Singapore Pte. Ltd., pp. 9-50.

Balla K; Karsai, I; Bónis, P; Kiss, T; Berki, Z and Horváth, Á (2019) Heat stress responses in a large set of winter wheat cultivars (*Triticum aestivum*) depend on the timing and duration of stress. *PLoS ONE* **14**(9): e0222639.

Ballesteros, J; Llamas, GC; Ramirez, MC and Martin A (2003) Low relative humidity increases haploid production in durum wheat x maize crosses. *Plant Breed* **122** (3): 276-78.

Balyan, HS; Gupta, PK; Kulwal, PL and Kumar, N (2006) QTL analyses for three grain quality traits in bread wheat using intervarietal mapping populations. *Czech J Genet Plant Breed* **41**: 281-283.

Balyan, HS; Sreenivasulu, N; Riera-Lizarazu, O; Azhaguvel, P and Kianian, F (2008) Mutagenesis and high-throughput functional genomics in cereal crops: Current status. *Advances in Agronomy* **98**: 357-417.

Barkley, NA and Wang, ML (2008) Application of TILLING and EcoTILLING as reverse genetic approaches to elucidate the function of genes in plants and animals. *Curr Genomics* **9**: 212–226.

Barnabas, B; Jager, K and Feher, A (2008) The effect of drought and heat stress on reproductive processes in cereals. *Plant Cell Environ* **31**: 11–38.

Bartels, D and Sunkar, R (2005) Drought and salt tolerance in plants. *Crit Rev Plant Sci* **24**: 23–58.

Batchelor, WD; Basso, B and Paz, JO (2002) Examples of strategies to analyze spatial and temporal yield variability using crop models. *European Journal of Agronomy* **18**: 141-158.

Batelli, G; Verslues, PE; Agius, F; Qiu, Q; Fujii, H; Pan, S; Schumaker, KS; Grillo, S and Zhu, JK (2007) SOS2 promotes salt tolerance in part by interacting with the vacuolar H+-ATPase and upregulating its transport activity. *Molecular and Cellular Biology*, **27**: 7781-7790.

Batten, GD; Blakeney, AB; McGrath, VB; Ciavarella, S and Barrow, NJ (1993) Non-structural carbohydrate: analysis by near infrared reflectance spectroscopy and its importance as an indicator of plant growth. *Plant and Soil*, **155/156**: 243-246.

Beltrano, J; Ronco, MG and Montaldi, ER (1999) Drought stress syndrome in wheat is provoked by ethylene evolution imbalance and reversed by rewatering, amino ethoxyvinyl glycine, or sodium benzoate. *J. Plant Growth Regul.* **18**: 59–64.

Berger, B; Parent, B and Tester, M (2010) High-throughput shoot imaging to study drought responses. *Journal of Experimental Botany*, **61**: 3519-3528.

Bernardo, R (2010) Genome wide selection with minimal crossing in self-pollinated crops. *Crop Sci* **50**: 624–627.

Bhatia, S; Mamrutha, HM; Tiwari, R; Rane, J and Tiwari, V (2012) Molecular marker assisted detection of vernalization genes in Wheat. International conference on plant biotechnology for food security: New Frontiers. New Delhi.

Bhatnagar-Mathur, P; Vadez, V and Sharma, KK (2008) Transgenic approaches for abiotic stress tolerance in plants: retrospect and prospects. *Plant Cell Rep* **27**: 411–424.

Black AL, Brown PL, Halvorson AD and Siddoway FH (1981) Dryland cropping strategies for efficient water-use to control saline seeps in the Northern Great Plains, U.S.A. *Agricultural Water Management* **4**: 295-311.

Blanco, A; Degiovanni, C; Laddomada, B; Sciancalepore, A; Simeone, R; Devos, KM and Gale, MD (1996) Quantitative trait loci influencing grain protein content in tetraploid wheats. *Plant Breed* **115**: 310-316.

Bluementhal, C; Bekes, F; Gras, PW; Barlow, EWR and Wrigley, CW (1995) Influence of wheat genotypes tolerant to the effects of heat stress on grain quality. *Cereal Chem* **72**: 539-544.

Blum, A (1979) Genetic improvement of drought resistance in crop plants: a case for sorghum. In *Stress Physiology In Crop Plants.* H. Mussello and R.C. Staples, eds. New York, Wiley lnterscience, pp. 429-44S.

Blum, A (1988) Plant Breeding for Stress Environments. CRC Press.

Blum, A (1998) Improving wheat grain filling under stress by stem reserve mobilisation. *Euphytica* **100**: 77–83.

Blum, A (2009) Effective use of water (EUW) and not water use efficiency (WUE) is the target of crop yield improvement under drought stress. *Field Crop Res* **112**: 119–123.

Blum, A (2011) Drought resistance – is it really a complex trait? *Functional Plant Biology*, **38**: 753-757.

Bohnert, HJ and Shen, B (1999) Transformation and compatible solutes. *Sci. Hortic.*, **78**: 237-260.

Bohnert, HJ; Nelson, DF and Jenson, RG (1995) Adaptation to environmental stresses. *Plant Cell*, **7**: 1099-1111.

Bolaños, J and Edmeades, GO (1993) Eight cycles of selection for drought tolerance in lowland tropical maize. II. Responses in reproductive behavior. *Field Crops Research* **31**: 253-268.

Borrell, A; Hammer, G and van Oosterom, E (2001) Stay-green: A consequence of the balance between supply and demand for nitrogen during grain filling? *Annals of Applied Biology*, **138**: 91-95.

Boyer JS (1982) Plant productivity and environment. *Science*, **218**: 443-448.

Boyle, MG; Boyer, JS and Morgan, PW (1991) Stem infusion of liquid culture medium prevents reproductive failure of maize at low water potential. *Crop Science* **31**: 1246-1252.

Brauman, KA; Siebert, S and Foley, JA (2013) Improvements in crop water productivity increase water sustainability and food security - a global analysis. *Environment Research Letters* **8**: 8024030 (Open access).

Braun, HJ; Atlin, G and Payne, T (2010) Multi-location testing as a tool to identify plant response to global climate change. In: Reynolds MP (ed) Climate change and crop production. CABI Publishers, Wallingford, pp. 115–138.

Brennan, JP and Morris, ML (2001) Economic issues in assessing the role of physiology in wheat breeding programs. In *Application of Physiology in Wheat Breeding* (Eds Reynolds, MP., Ortis-Monasterio, JI. and McNab, A), Mexico: CIMMYT. pp. 78–86.

Bressan, R; Bohnert, H and Zhu, JK (2009) Abiotic stress tolerance: from gene discovery in model organisms to crop improvement. *Mol Plant* **2**: 1–2.

Bulavas, J (1936) Hybrids of wheat and rye. *Zemes. Ukio. Akad. Metrastis.* **10**: 204–215.

Cabangon, R; Tuong, TP and Abdullah, NB (2002) Comparing water input and water productivity of transplanted and direct-seeded rice production systems. *Agricultural Water Management* **57**: 11-31.

Cattivelli, L; Rizza, F; Badeck, FW; Mazzucotelli, E; Mastrangelo, AM; Francia, E; Mare, C; Tondelli, A and Stanca, AM (2008) Drought tolerance improvement in crop plants: an integrated view from breeding to genomics. *Field Crops Res* **105**: 1–14.

Ceccarelli, S; Grando, S; Maatougui, M; Michael, M; Slash, M; Haghparast, R; Rahmanian, M; Taheri, A; Al-Yassin, A; Benbelkacem, A; Labdi, M; Mimoun, H and Nachit, M (2010) Plant breeding and climate changes. *Journal of Agricultural Science, Cambridge* **148**: 627–637.

Ceccarelli, S (1994) Specific adaptation and breeding for marginal conditions. *Euphytica* **77**: 205–219.

Chaerle, L and van der Straeten, D (2000) Imaging techniques and the early detection of plant stress. *Trends Plant Sci* **5**: 495–501.

Chapman, SC (2008) Use of crop models to understand genotype by environment interactions for drought in real-world and simulated plant breeding trials. *Euphytica* **161**: 195–208.

Charron, AJ and Quatrano, RS (2009) Between a rock and a dry place: the water-stressed moss. *Mol Plant* **2**: 478–486.

Chauhan, H; Desai, SA and Khurana, P (2007) Comparative analysis of the differential regeneration response of various genotypes of *Triticum aestivum, Triticum durum* and *Triticum dicoccum*. *Plant Cell Tiss Organ Cult*. **91**(3): 191-199.

Chinnusamy V, Jagendorf A, Zhu JK (2005) Understanding and improving salt tolerance in plants. *Crop Sci*, **45**: 437-448.

Chinnusamy V, Schumaker K, Zhu JK (2004) Molecular genetic perspectives on crosstalk and specificity in abiotic stress signalling in plants. *J. Experimental Botany*; **55**: 225-236.

Christiansen, MN and Lewis, CF (1982) Breeding Plants for Less Favourable Environments. Wiley International.

Chuang, CC; Ouyang, J; Chia, H; Chou, SM and Ching, CK (1978) A set of potato media for wheat anther culture. *In*: Procs China-Australia Plant Tissue Culture Symposium. Peking, pp. 51-66.

CI-600 manual, CID, Inc, WA, USA, https: //cid-inc.com/plant-science-tools/ rootmeasurement-with-minirhizotron/ci-600-in-situ-root-imager/.

CI-690 Rootsanp manual, https: //www.cid-inc.com/plant-science-tools/root-measurementwith-minirhizotron/ci-600-in-situ-root-imager/.

Clark, RT; MacCurdy, RB; Jung, JK; Shaff, JE; McCouch, SR; Aneshansley, DJ and Kochian, LV (2011) Three-Dimensional root phenotyping with a novel imaging and software platform. *Plant Physiol.*, **156**: 455–465.

Collins, NC; Tardieu, F and Tuberosa, R (2008) Quantitative trait loci and crop performance under abiotic stress: Where do we stand? *Plant Physiol* **147**: 469–486.

Condon, AG; Richards, RA; Rebetzke, GJ and Farquhar, GD (2002) Improving intrinsic water-use efficiency. *Crop Science* **42:** 122-131.

Cooper, PJM; Keatinge, JDH and Hughes, G (1983) Crop evapotranspiration – a technique for calculation of its components by field measurements. *Field Crops Research* **7**: 299-312.

Covarrubias, AA and Reyes, JL (2010). Post-transcriptional gene regulation of salinity and drought responses by plant microRNAs. *Plant, Cell and Environment.*, **33**: 481-489.

Crook, MJ and Ennos, AR (1994) Stem and root characteristics associated with lodging resistance in four winter wheat cultivars. *Journal of Agricultural Science.* **123(**2): 167-174.

Cushman, JC and Bohnert, HJ (2000) Genomic approaches to plant stress tolerance. *Current Opinion in Plant Biology* **3**: 117-124.

D'Ovidio, R and Anderson, OD (1994) PCR analysis to distinguish between alleles of a member of a multigene family correlated with wheat bread-making quality. *Theor Appl Genet* **88**: 759-763.

Dadshani, S; Sharma, RC; Baum, M; Ogbonnaya, FC; Léon, J and Ballvora, A (2019) Multi-dimensional evaluation of response to salt stress in wheat. *PLoS ONE* **14**(9): e0222659.

Dastane, NG (1972) A practical manual for water use research in agriculture, Navabharat Prakashan, Pune.

Davies, WJ and Zhang, JH (1991) Root signals and the regulation of growth and development of plants in drying soil. *Annual Review of Plant Physiology and Plant Molecular Biology*, **42**: 55-76.

de la Fuente, JM; Ramírez-Rodriguez, V; Cabrera-Ponce, JL and Herrera-Estrella, L (1997) Aluminium tolerance in transgenic plants by alteration of citrate synthase. *Science*, **276**: 1566-1568.

de Sousa, CNA (1998) Classification of Brazilian wheat cultivars for aluminium toxicity in acid soils. *Plant Breeding* **117**: 217–221.

Desai, SA; Lohithaswa, HC; Hanchinal, RR; Kalappanavar, IK and Math, KK (2005) Combining ability analysis for quantitative traits in bread wheat (*Triticum aestivum* L.). *Indian J. Genet.*, **65** (4): 311-312.

Deyholos, M; Wang, H and Galbraith, D (2001) Microarrays for gene discovery and metabolic pathway analysis in plants. *Life Science* **2**(1) 2-4.

Dhawan, R; Singh, K and Bains, NS (2003) Effect of pre-regeneration colchicines treatments on the efficiency of wheat x maize system of haploid production in wheat. *Indian J. Genet.* **63** (3): 247- 248.

Diamant, S; Eliahu, N; Rosenthal, D and Goloubinoff, P (2001) Chemical chaperones regulate molecular chaperones *in vitro* and in cells under combined salt and heat stresses. *J. Biol. Chem.*, **276**: 39586–39591.

Doorenbos, I and Fruitt, WO (1977) Crop Water Requirement, FAO Irrigation and Drainage Paper. Vol-24.

Dreccer, MF; Ogbonnaya, FC; Borgognone, G and Wilson, J (2003) Boron tolerance is present in primary synthetic wheats. p. 1130–1132. In: N. Pogna (ed.), Procs of 10th Intl. Wheat Genetics Symposium, Sept. 1–6, 2003, Paestum, Italy.

Driscoll, CJ (1966) Gene-centromere distances in wheat by aneuploid F_2 observations. *Genetics* **54**: 131–135.

Driscoll, CJ and Jensen, NF (1964) Chromosomes associated with waxlessness, awnedness and time of maturity of common wheat. *Can. J. Genet. Cytol.* **6**: 324–333.

Dubcovsky, J; Santa Maria, G; Epstein, E; Luo, MC and Dvorak, J (1996) Mapping of the K+/Na+ discrimination locus *Kna1* in wheat. *Theor Appl Genet* **92**: 448-454.

Eagles HA, Bariana HS, Ogbonnaya FC, Rebetzke GJ, Hollamby GJ, Henry RJ, Henschke PH, and Carte M 2001 Implementation of markers in Australian wheat breeding *Australian Journal of Agricultural Research*, **52:** 1349–1356.

Eathington, SR; Crosbie, TM; Edwards, MD; Reiter, RS and Bull, JK (2007) Molecular markers in a commercial breeding program. *Crop Sci* **47**: S154–S163.

Edmeades, GO; McMaster, GS and White, JW (2004) Genomics and the physiologist: bridging the gap between genes and crop response. *Field Crops Res* **90**: 5–18.

Edwards, M and Johnson, L (1994) RFLPs for rapid recurrent selection. In: Joint plant breeding symposium series of CSSA and ASHA, 5–6 Aug, Corvallis, Alexandria, VA.

Enclona, EA; Thenkabail, PS; Celis, D and Diekmann. J (2004) Within-field wheat yield prediction from IKONOS data: A new matrix approach. *Int. J. Remote Sens.* **25**: 377–388.

English, MJ; Solomon, KH and Hoffman, GJ (2002) A paradigm shift in irrigation management. *Journal of Irrigation and Drainage Engineering-ASCE* **128**: 267-277.

Farooq, S; Iqbal, N; Asghar, M and Shah, TM (1992) Intergeneric hybridization for wheat improvement – IV. Expression of salt tolerance gene (s) of *Aegilops cylindrica* in hybrids with hexaploid wheat. *Cer. Res. Comm.* **20** (1-2): 111-118.

Farquhar GD and Richards RA 1984 Isotopic composition of plant carbon correlates with water-use efficiency of wheat genotypes. *Australian Journal of Plant Physiology*, **11**: 539-552.

Farre, G; Ramessar, K; Twyman, RM; Capell, T and Christou, P (2010) The humanitarian impact of plant biotechnology: recent breakthroughs vs bottlenecks for adoption. *Curr Opin Plant Biol* **13**: 219–225.

Fayez, KA and Bazaid, SA (2014) Improving drought and salinity tolerance in barley by application of salicylic acid and potassium nitrate. *J. Saudi Soc. Agric. Sci.* **13**: 45–55.

Federizzi, LC; Fantini-AC; Carvalho-FIF-de and De-Carvalho-FIF (1994) Effect of artificial lodging on wheat genotypes of tall and dwarf stature. *Ciencia-Rural.* **24(**3): 465-469.

Ferrera, OG; Rajaram, S and Mosaad, MG (1993) Breeding strategies for improving wheat in heat stressed environments. *Proc. Int. Conf. Wad Medani, Sudan and Dinajpur, Bangladesh*. pp. 24-32.

Ferris, R; Ellis, RH; Wheeler, PR and Hadley, P (1998) Effect of high temperature stress at anthesis on grain yield and biomass of field grown crops of wheat. *Annl Bot* **82**: 631-639.

Feuillet, C; Leach, JE; Rogers, J; Schnable, PS and Eversole, K (2010) Crop genome sequencing: lessons and rationales. *Trends Plant Sci* **16**: 77–88.

Filella, I and Penuelas, J (1994) The red edge position and shape as indicators of plant chlorophyll content, biomass and hydric status. *Int. J. Remote Sens.* **15**: 1459–1470.

Fischer RA and Turner NC (1978) Plant Productivity in the Arid and Semiarid Zones. *Annual Review of Plant Physiology* **29**: 271-317.

Fischer, RA (1979) Growth and water limitation to dryland wheat yield in Australia: a physiological framework. *Journal of the Australian Institute of Agricultural Science* **45**: 83-94.

Fleury, D; Jefferies, S; Kuchel, H and Langridge, P (2010) Genetic and genomic tools to improve drought tolerance in wheat. *J Exp Bot* **61**: 3211–3222.

Fleury, D; Jefferies, S; Kuchel, H and Langridge, P (2010) Genetic and genomic tools to improve drought tolerance in wheat. *Journal of Experimental Botany*, **61**: 3211-3222.

Fodor, SPA; Read, JL; Pirrung, MC; Stryer, L; Lu, A and Solas, D (1991) Light directed, spatially addressable parallel chemical synthesis. *Science* **251**: 767-773.

Fowler, DB and Limin, AE (2007) Progress in Breeding wheat with tolerance to low temperature in different phenological developmental stages. *Development in Plant Breeding* **12**: 301-312.

Gadekar, DA; Rasal, PN; Gavhane, VN and Gorve, S.R (2006) Elimination of shattering habit in wheat through irradiation. Poster presented at BARC Golden Jubilee and DAE-BRNS Life sciences symposium 2006 on, " Trends in Research and Technologies in Agriculture and Food Sciences. pp. 106.

Gadekar, DA; Rasal. PN; Borkar, SG and Gavhane, VN (2004) Capturing male sterility in wheat. Wheat In Tropical Areas Edit. Proc. of the National Symposium on Wheat Improvement for the Tropical areas, held at Tamilnadu Agricultural University Coimbtore and IARI, Wellington, Tamil Nadu, India.pp.217-219.

Galiba, G; Quarrie, SA; Sutka, J; Morgounov, A and Snape, JW (1995) RFLP mapping of the vernalization (*Vrn1*) and frost resistance (*Fr1*) genes on chromosome 5A of wheat. *Theoretical and Applied Genetics* **90**: 1174-1179.

Garg, D; Pandey, GC; Dalal, S; Tiwari, R; Singh, R and Sareen, S (2011) Morph-physiological and molecular response of wheat genotypes under heat stress. 3rd International Group Meeting on Wheat Productivity Enhancement under changing climate from Feb. 9-12, 2011 at UAS, Dharwad. pp. 72.

Garg, D; Sareen, S; Dalal, S; Tiwari, R and Singh, R (2012) Grain Filling duration and temperature pattern influence the performance of wheat genotypes under late planting. *Cereal Research Communications*.

Garg, D; Sareen, S; Dalal, S; Tiwari, R and Singh, R (2012) Heat shock protein-based SNP marker for terminal heat stress in wheat (*Triticum aestivum* L.) *Aust. Jour. Crop Sc.*

Gibson, LR and Paulsen, GM (1999) Yield component of yield grown under high temperature stress during reproductive growth. *Crop Sci* **39**: 1841-1846.

Gill, BS; Friebe, B and Endo, TR (1991) Standard karyotype and nomenclature system for description of chromosome bands and structure observations in wheat (*Triticum aestivum*). *Genome*, **34**: 830-39.

Girma, K; Freeman, KW; Teal, RK; Arnall, DB; Tubana, B; Holtz, S and Raun, WR (2007) Analysis of yield variability in winter wheat due to temporal variability, and nitrogen and phosphorus fertilization. *Arch. Agron. Soil Sci.* **53**: 435–442.

Gorshkova, VA and Selezneva,TG (1987) Use of diallel analysis in barley breeding to identify sources of productivity and lodging resistance. *Novoe-v-selektsii-i-semenovodstve-sel'skokhozyai\br\stvennykh-kul'tur*. 26-32.

Goyal, A; Bandopadhyay, R; Sourdille, P; Endo, TR; Balyan, HS and Gupta, PK (2005) Physical molecular maps of wheat chromosomes. *Funct. and Integr. Genomics* **5**: 260-263.

Graham, RD (1975) Male sterility in wheat plants deficient in copper. *Nature* **254**: 514-515.

Grover, A; Kapoor, A; Satya Lakshmi, O; Agrawal, S; Sahi, C; Katiyar-Agarwal S, Agarwal M, Dubey H (2001) Understanding molecular alphabets of the plant abiotic stress responses. *Curr. Sci.*, **80**: 206-216.

Grover, A; Sahi, C; Sanan, N and Grover, A (1999) Taming abiotic stresses in plants through genetic engineering: Current strategies and perspective. *Plant Sci.*, **143**: 101-111.

Grover, A; Sanan, N and Sahi, C (1998) Genetic engineering for high level tolerance to abiotic stresses through over-expression of transcription factor genes: The next frontier. *Current Science* **75**: 178-179.

Gupta, PK (2008) Single-molecule DNA sequencing technologies for future genomics research. *Trends Biotechnol.* **26**: 602-611.

Gupta, PK and Rustgi, S (2004) Molecular markers from the expressed region of the genome in higher plants. *Funct. and Integr. Genomics* **4**: 139-162.

Gupta, PK; Balyan, HS; Edwards, KJ; Isaac, P; Korzun, V; Roder, M; Joudrier, P; Schlatter, AR; Dubcovsky, J; de la Pena, RC; Khairallah, M; Hayden, M; Keller, B; Wang, R; Hardouin, JP; Jack, P and Leroy, P (2002) Genetic mapping of sixty six new SSR loci in bread wheat (*Triticum aestivum*). *Theor. Appl. Genet.* **105**: 413-422.

Gupta, PK; Balyan, HS; Kulwal, PL and Gahlaut, V (2012) Phenotyping, genetic dissection, and breeding for drought and heat tolerance in common wheat: Status and prospects. *Pl Breed Rev.*

Gupta, PK; Balyan, HS; Kulwal, PL; Kumar, N; Kumar, A; Mir, RR; Mohan, A and Kumar, J (2007) QTL analysis for some quantitative traits in bread wheat. *J Zhejiang University Sci B* **8**: 807-814.

Gupta, PK; Kulwal, PL and Rustgi, S (2005) Wheat cytogenetics in the genomics era and its relevance to breeding. In: Puertas MJ, Naranjo T (eds) *Cytogenetic Genome Research*, **109**: 315-327.

Gupta, PK; Kumar, J; Mir, RR and Kumar, A (2010) Marker assisted selection as a component of conventional plant breeding. *Plant Breed Rev* **33**: 145–217.

Gupta, PK; Langridge, P and Mir, RR (2010) Marker-assisted wheat breeding: present status and future possibilities. *Mol Breed*.

Gupta, PK; Rustgi, S and Kulwal, PL (2005) Linkage disequilibrium and association studies in higher plants: Present status and future prospects. *Plant Mol Biol* **57**: 461-485.

Gustafson, JP and Ross, K (1990) Control of alien gene expression for aluminum tolerance in wheat. *Genome* **33**: 9–12.

Gutierrez, M; Reynolds, MP and Klatt, AR (2010) Association of water spectral indices with plant and soil water relations in contrasting wheat genotypes. *J Exp Bot* **61**: 3291–3303.

Gutierrez, M; Reynolds, MP; Raun, WR; Stone, ML and Klatt, AR (2010) Spectral water indices for assessing yield in elite bread wheat genotype grown under well irrigated, water deficit stress, and high temperature conditions. *Crop Sci.* **50**: 197-214.

Haesaert, G; Baets, A; De Baets, A; Bockstaele, E van (ed.) and Heursel, J (1994) Triticale breeding: problems and possibilities. Plant breeding for mankind, Agribex symposium, Brussels, Belgium, 7-8 February 1994. *Acta Horticulturae*. No. 355, 121-134.

Hal, NLW and Vorst, O (2000) The application of DNA microarrays in gene expression analysis. *J. Biotech.* **78**: 271-80.

Hammer, GL; McCown, RL and Freebairn, DM (1993) APSIM: the agricultural production system simulator – its role and structure. Procs of the 7th Australian Agronomy Conference, Adelaide pp. 232-235.

Hanks, RJ; Gardner, HR and Florian, RL (1969) Plant growth-evapotranspiration relations for several crops in the central great plains. *Agronomy Journal*. **61**: 30-34.

Harding, SA; Guikema, GA and Paulsen, GM (1990) Photosynthetic decline from high temperature stress during maturation of wheat. II Interaction with source and sink processes. *Plant Physiol* **92**: 654-658.

Hazen, SP; Wu, Y and Kreps, JA (2003) Gene expression profiling of plant responses to abiotic stress. *Funct Integr Genomic* **3**: 105–111.

Held, IM; Delworth, TL; Lu, J; Findell, KL and Knutson, TR (2005) Simulation of Sahel drought in the 20th and 21st centuries. *Proc Natl Acad Sci USA* **102**: 17891–17896.

Herzog, H (1982) Relation of source and sink during the grain filling period in wheat and some aspects of its regulation. *Physio Plant* **56**: 155-166.

Hideg, NT; Oberschall; A; Dudits, D and Vass, I (2003) Detoxification function of aldose/aldehyde reductase during drought and ultraviolet-B (280-320 nm) stresses. *Plant Cell Environ.*, **26**: 513-522.

Hirayama, T and Shinozaki, K (2010) Research on plant abiotic stress responses in the postgenome era: past, present and future. *The Plant Journal*, **61**: 1041-1052.

Hobbs, PR and Gupta, RK (2003) Rice-wheat cropping systems in the Indo-Gangetic plains: issues of water productivity in relation to new resource-conserving technologies. Chapter 15 in Kijne *et al.* (2003).

Hospital, F (2003) Marker-assisted breeding. In: Newbury HJ (ed) Plant molecular breeding. Blackwell Publishing, Carlton, pp. 30–56.

Huerta-Espino, J and Singh, RP (2005) Effects of 7DL.7Ag (mutated for reduced flour yellowness) and 4BL.2R translocations on grain yield of wheat under irrigated and stressed environments. In *Abstr. of 7th International Wheat Conference: Wheat Production in Stressed Environments,* Nov 27–Dec. 2, p. 339. Mar de Plata, Argentina.

Intergovernmental Panel on Climate Change (IPCC) (2007) In: Pachauri RK, Reisinger A (eds) Climate change 2007 synthesis report. IPCC, Geneva.

IPCC (Intergovermental Panel on Climate Change) (2014) Climate Change 2014: impacts, adaptation, and vulnerability. part A: global and sectoral aspects. In: Field, CB; Barros, VR; Dokken, DJ; Mach, KJ; Mastrandrea, MD; Bilir, TE; Chatterjee, M; Ebi, KL; Estrada, YO; Genova, RC; Girma, B; Kissel, ES; Levy, AN; MacCracken, S; Mastrandrea, P.R; White, LL (Eds.) Contribution of Working Group II to the Fifth Assessment Report of the Intergovernmental Panel on Climate Change. Cambridge University Press, Cambridge, United Kingdom and New York, NY, USA, p. 1132.

Ishii, T and Yonezawa, K (2007b) Optimization of the marker-based procedures for pyramiding genes from multiple donor lines: II. Strategies for selecting the objective homozygous plant. *Crop Sci* **47**: 1878–1886.

Ishii, T; Hayashi, T and Yonezawa, K (2008) Optimization of the marker-based procedures for pyramiding genes from multiple donor lines: III. Multiple-gene assemblage using background marker selection. *Crop Sci* **48**: 2123–2131.

Izanloo, A; Condon, AG; Langridge, P; Tester, M and Schnurbusch, T (2008) Different mechanism of adaptation to cyclic water stress in two south Australian bread wheat cultivars. *J. Exp. Bot.* **59**: 3327-3346.

Jadhav, SS; Desai, SA; Biradar, SS; Kalappanavar, IK; Naik, R; Patil, BN; Arun Kumar, GS; Pradeep, PE; Yadawad, A and Sridevi, BM (2011) Assessment of morphological and molecular diversity in bread wheat germplasm lines. Abstract presented in 3rd International Group Meeting on "Wheat productivity enhancement under changing climate" from Feb. 9-12, 2011. pp: 129.

Jahnke, S; Menzel, MI; van Dusschoten, D; Roeb, GW; Bu¨hler, J; Minwuyelet, S; Blu¨mler, P; Temperton, VM; Hombach, T; Streun, M; Beer, S; Khodaverdi, M; Ziemons, K; Coenen, HH and Schurr, U (2009) Combined MRI–PET dissects dynamic changes in plant structures and function. *Plant J* **59**: 634–644.

Jaiswal, SK; Sweta, Prasad, LC; Sharma, S; Kumar, S; Prasad, R; Pandey, SP; Chand, R and Joshi, AK (2007) Identification of molecular marker and aggressiveness for different groups of *Bipolaris sorokiniana* isolates causing spot blotch disease in wheat (*Triticum aestivum* L.). *Current Microbiology* **20**: 1-8.

Jefferies, SP; Pallotta, MA; Paull, JG; Karakousis, A; Kretschmer, JM; Manning, S; Islam, AKMR; Langridge, P and Chalmers, KJ (2000) Mapping and validation of chromosome regions conferring boron toxicity tolerance in wheat (*Triticum aestivum*). *Theor. Appl. Genet.* **101**: 767–777.

Johnson, GV and Raun, WR (2003) Nitrogen response index as a guide to fertilizer management *J. Plant Nutr.* **26**: 249–262.

Jones, HG (2007) Monitoring plant and soil water status: established and novel methods revisited and their relevance to studies of drought tolerance. *J Exp Bot* **58**: 119–130.

Jones, HG; Serraj, R; Loveys, BR; Xiong, L; Wheaton, A and Price, AH (2009) Thermal infrared imaging of crop canopies for the remote diagnosis and quantification of plant responses to water stress in the field. *Funct Plant Bio* **36**: 978–989.

Jordan, BR (1998) Large scale expression measurement by hybridization methods: from high density membranes to DNA chips. *J. Biochem.* **124**: 251-58.

Joshi, AK; Mishra, B; Chatrath, R; Ferrara, GO and Singh, RP (2007) Wheat improvement in India: Present status, emerging challenges and future prospects. *Euphytica* **157**: 431-446.

Joshi-Saha, A; Valon, C and Leung, J (2011) Abscisic acid signal off the STARTing block. *Molecular Plant*, **4**: 562-580.

Kalagouda, BG; Shenoy, RV; Vadavi, RS; Patil, MS; Patil, SA; Hanchinal, RR; Desai, SA and Lohithaswa, H (2006) Lanthanide (III) and Yttrium (III) Complexes of Benzimidazole-2-Acetic Acid: Synthesis, Characterisation and Effect of La(III) Complex on Germination of Wheat. *Bioinorganic Chemistry and Applications*, 1–8.

Kalappanavar, IK; Desai, SA; Ramya, G; Pradeep, PE; Das, BK and Bhagwat (2010) Breeding free-threshable, dwarf emmer wheat (*Triticum dicoccum* (Schrank.) Schulb.) genotypes through mutagenesis. Paper presented in 8th International Conference on Wheat held at St.Petersburg, Russia from June 1-4, 2010.

Kale, SM; Pardeshi, VC; Kadoo, NY; Ghorpade, PB; Jana, MM and Gupta, VS (2011) Development of genomic simple sequence repeat markers for linseed using next generation sequencing technology. *Molecular Breeding*.

Kanke, Y; Raun, W; Solie, J; Stone, M and Taylor, R (2011) Red edge as a potential index for detecting differences in plant nitrogen status in winter wheat. *J. Plant Nutr.* (in press).

Karim, S; Aronsson, H; Ericson, H; Pirhonen, M; Leyman, B; Welin, B; Palva, ET; Patrick, VD and Kjell-Ove, H (2007) Improved drought tolerance without undesired side effects in transgenic plants producing trehalose. *Plant Mol. Biol.*, **64**: 371-386.

Kasuga, M; Qiang, L; Miura, S; Shinozaki, KY; Shinozaki, K and Liu, Q (1999) Improving plant drought, salt, and freezing tolerance by gene transfer of a single stress-inducible transcription factor. *Nature Biotechnol.*, **17**: 287-291.

Kathiresana, A; Lafittea, HR; Chena, J; Mansuetoa, L; Bruskiewicha, R and Bennetta, J (2006) Gene expression microarrays and their application in drought stress research. *Field Crops Res* **97**: 101–110.

Kato, K; Miura, H; Akiyama, M; Kuroshima, M and Sawada, S (1998) RFLP mapping of the three major genes, *Vrn1*, *Q* and *B1*, on the long arm of chromosome 5A of wheat. *Euphytica* **101**: 91-95.

Katterman, G (1935) Genetic results with wheat-rye hybrids upto F_4. I. Pubescence of the stems and the relationship of this character to length of straw and set of grains. *Pflanzenbau* **12**: 131–149.

Khan, IMR; Fatma, M; Per, TS; Anjum, NA and Khan, NA (2015) Salicylic acid–induced abiotic stress tolerance and underlying mechanisms in plants. *Front. Plant Sci.* **6**: 462.

Khurana, P; Chauhan, H and Desai, SA (2007) Wheat. In Transgenic cereals and grasses. Compendium of Transgenic crop plants. Edited by Chittaranjan Kole and Timothy C. Hall, Wiley-Blackwell, USA Vol I: 83-100.

Kijne, JW; Barker, R and Molden, D (2003) Water Productivity in Agriculture: Limits and Opportunities for Improvement. CABI Publishing, Wallingford.

Kirigwi, FM; Ginkel, MV; Brown-Guedira, G; Gill, BS; Paulsen, GM and Fritz, AK (2007) Markers associated with a QTL for grain yield in wheat under drought. *Mol Breed* **20**: 401–413.

Kohli, MM; Mann, C and Rajaram, S (1991) Global status and recent progress in breeding wheat for the warmer areas. pp. 225-241, DA Saunders, ed. Wheat for the nontraditional warmer areas. Mexico, D.F: CIMMYT.

Kovtun, Y; Chiu, WL; Tena, G and Sheen, J (2000) Functional analysis of oxidative stress activated mitogen-activated protein kinase cascade in plants. *Proc Natl Acad Sci.*

Kudoyarova, GR; Melentiev, AI; Martynenko, EV; Timergalina, LN; Arkhipova, TN; Shendel, GV; Kuz'mina, LY; Dodd, IC and Veselov, SY (2014) Cytokinin producing bacteria stimulate amino acid deposition by wheat roots. *Plant Physiol Biochem.* **83**: 285-91.

Kuiper, D; Suit, J and Kuiper, PJC (1990) Actual cytokinin concentrations in plant tissue as an indicator for salt resistance in cereals. *Plant and Soil* **123**: 243-250.

Kulkarni, P; Desai, SA; Lohithashwa, HC and Hanchinal, RR (2008) Breeding for free threshability im emmer wheat [*Triticum dicoccum* (Schrank.) Schubl.] through induced mutagenesis. *Indian J. Genet.*, **68**(4): 380-386.

Kulkarni, P; Desai, SA; Lohithashwa, HC; Kalappanavar, IK and Hanchinal, RR (2008) Hybridization followed by mutagenesis is a novel method of isolation of rare free-threshable mutants of dicoccum wheat (*Triticum dicoccum* (schrank.) Schubl.). *Paper presented in (Joint FAO/IAEA Programme) International Symposium on Induced Mutations in Plants (ISIM). 12-15 August, 2008* Vienna, Austria (IAEA-CN-167-287P).

Kulwal, P; Ishikawa, G; Benscher, D; Feng, Z; Yu, LX; Jadhav, A; Mehetre, S and Sorrells, ME (2012) Association Mapping for Pre-harvest Sprouting Resistance in White Winter Wheat. *Theor Appl Genet.*

Kulwal, P; Kumar, N; Kumar, A; Gupta, RK; Balyan, HS and Gupta, PK (2005). Gene networks in hexaploid wheat: interacting quantitative trait loci for grain protein content. *Funct. and Integr. Genomics* **5**: 254-259.

Kulwal, PL; Kumar, N; Gaur, A; Khurana, P; Khurana, J; Tyagi, AK; Balyan, HS and Gupta, PK (2005) Mapping of a major QTL for pre-harvest sprouting tolerance on chromosome 3A in bread wheat. *Theor. Appl Genet.* **111**: 1052-1059.

Kulwal, PL; Mahendar, T and Varshney, RK (2012) Genomics interventions in crop breeding for sustainable agriculture. In: Robert A. Meyers [Ed.] *Encyclopedia of Sustainability Science and Technology.*

Kulwal, PL; Mir, RR; Kumar, S and Gupta, PK (2010) QTL analysis and molecular breeding for seed dormancy and pre-harvest sprouting tolerance in bread wheat. *J Plant Biol* **37**: 59-74.

Kulwal, PL; Roy, JK; Balyan, HS and Gupta, PK (2003) QTL analysis for growth and leaf characters in bread wheat. *Plant Sci* **164**: 267-277.

Kulwal, PL; Singh, R; Balyan, HS and Gupta, PK (2004) Genetic basis of pre-harvest sprouting tolerance using single-locus and two-locus QTL analyses in bread wheat. *Funct Integr Genomics* **4**: 94-101.

Kulwal, PL; Thudi, M and Varshney, RK (2011) Genomics interventions in crop breeding for sustainable agriculture. In: Meyers RA (ed) Encyclopedia of sustainability science and technology, Springer, New York.

Kumar, A; Kumar, J; Singh, R; Garg, T; Chhuneja, P; Balyan, HS and Gupta, PK (2009) QTL Analysis for grain colour and pre-harvest sprouting in bread wheat. *Plant Sci* **177**: 114-122.

Kumar, J; Mir, RR; Kumar, N; Kumar, A; Mohan, A; Prabhu, KV; Balyan, HS and Gupta, PK (2010) Marker-assisted selection for pre-harvest sprouting tolerance and leaf rust resistance in bread wheat. *Plant Breed.*

Kumar, M; Meena, KK and Rane, J (2018) Manual of Climate smart agricultural technologies for resource conservation and increasing farmer's income. National Institute of Abiotic Stress Management pp. 196.

Kumar, N; Kulwal, PL; Balyan, HS and Gupta, PK (2007) QTL mapping for yield and yield contributing traits in two mapping populations of bread wheat. *Mol. Breeding* **19**: 163–177.

Kumar, N; Kulwal, PL; Gaur, A; Tyagi, AK; Khurana, JP; Khurana, P; Balyan, HS and Gupta, PK (2006) QTL analysis for grain weight in bread wheat. *Euphytica* **151**: 135-144.

Kumar, N; Roy, JK; Kulwal, PL; Balyan, HS and Gupta, PK (2010) QTL analysis for growth related traits in intervarietal mapping population of common wheat. *J Genet Breed* **64**.

Kumar, R; Kumari, S and Chaudhary, A (2011) Biotechnological Interventions to enhance thermal tolerance in wheat (*Triticum aestivum* L.) enhance thermal tolerance in wheat *(Triticum aestivum L.)* A Review. Souvenir and Abstracts of National Symposium on Biotechnological Perspective of Plants, Microbes and their Interaction, Muzaffarpur, Bihar 2011, 139-140.

Kumar, R; Sharma, VK and Kumari, S (2011) Assessment of genetic variability in wheat (*Triticum aestivum* L.) for heat tolerance. Book of Abstracts – National Seminar on Contemporary Approaches to Crop Improvement, UAS, Bangalore, 2011, 296.

Kumar, S; Rana, B; Yadav, MK; Singh, AK; Singh, R and Joshi, AK (2011) Impact of DREB2A transcription factor and associated QTLs on drought tolerance in Indian wheat. In Proc: 3rd International Group Meeting on "Wheat Productivity Enhancement under Changing Climate", 9-12 February, 2011 UAS, Dharwad, Karnataka, India.

Kumar, S; Sehgal, SK; Kumar, U; Prasad, PVV; Joshi, AK and Gill, BS (2012) Genomic characterization of drought related traits in spring wheat. *Euphytica.*

Kumar, S; Sehgal, SK; Vara Prasad, PV; Bai, G; Joshi, AK and Gill, BS (2009) QTL Mapping For Traits Associated With Drought Tolerance in Spring Wheat. Plant and Animal Genome XVII conference, 10-14 January, 2009 San Diego, CA (USA).

Kumar, S; Singh, AK and Singh, R (2010) Global wheat program: Feeding a hungry world: National conference on traditional conservation of plant biodiversity and energy resources vis a vis biotechnological tools for sustainability, CCS University, Merrut, UP.

Kumar, S; Singh, AK; Kumari, J and Singh, R (2011) Diversity analysis of wheat genotypes based on the morphophysiological and RAPD markers. *Vegetos* **24**(2): 217-224.

Kumar, U; Joshi, AK; Kumar, S; Chand, R and Röder, MS (2010) Quantitative trait loci for resistance to spot blotch caused by *Bipolaris sorokiniana* in wheat (*T. aestivum* L.) lines 'Ning 8201' and 'Chirya 3'. *Mol Breeding* **26**: 477–491.

Kumar, U; Joshi, AK; Kumari, M; Paliwal, R; Kumar, S; Röder, MS (2010) Identification of QTLs for stay green trait in wheat (*Triticum aestivum* L.) in the 'Chirya 3' x 'Sonalika' population. *Euphytica* **174**: 437-445.

Kumar, U; Kumar, S; Tyagi, K; Chand, R and Joshi, AK (2005) Microsatellite markers for resistance to spot blotch in spring wheat. 57th International Symposium on Crop Protection, 10th May, 2005, Gent, Belgium pp. 19.

Kundu, S; Sareen, S; Malik, R; Kumar, S and Jag Shoran (2009) Cutting cost for germplasm conservation. *Gene Conserve* **31**: 698 – 707.

Kuroyanagi, T and Paulsen, GM (1985) Mode of high temperature injury to wheat. II. Comparison of wheat and rice with and without inflorescences. *Physio Plant* **65**: 203-208.

Laurie, DA and Bennett, MD (1986) Wheat x maize hybridization. *Can J Genet Cytol* **28**: 313-316.

Lbova, MI (1990) Length of the stem and some of its elements in a tall late mutant of spring wheat under different growing conditions. *Tsitogenetika-zernovykh-kul'tur*. 56-61.

Lekes, J (1988) Development of important traits of spring barley from the beginning of selection and the detection of suitable genetic resources - part II. *Scientia Agriculturae Bohemoslovaca*. **20(**2): 91-100.

Lemerle, D; Gill, GS; Murphy, CE; Walker, SR; Cousens, RD; Mokhtari, S; Peltzer, SJ; Coleman, R and Luckett, DJ (2001) Genetic improvement and agronomy for enhanced wheat competitiveness with weeds. *Australian Journal of Agricultural Research* **52**: 527-548.

Levitt, J (1980) Response of plant to environmental stress, John Wiley and Sons. Inc New york pp. 347.

Lewis, CF and Christiansen, MN (1981) Breeding plants for stress environments. In *Plant Breeding II*. KJ Frey, ed. Proc. Plant Breeding Symposium 112, 16 March 1979. Iowa State University Press, Ames, Iowa, pp. 151-177.

Li, FR; Zhao, SL and Geballe, GT (2000) Water use patterns and agronomic performance for some cropping systems with and without fallow crops in a semi-arid environment of Northwest China. *Agriculture Ecosystems and Environment* **79**: 129-142.

Liang, P and Pardee, AB (1992) Differential display of eukaryotic messenger RNA by means of the polymerase chain reaction. *Science* **257**: 967–971.

Liu, L; Wang, J; Song, X; Li, C; Huang, W and Zhao, C (2004b) Study on winter wheat yield estimation model based on NDVI and seedtime. p. 4045–4047. In Proc. 07-27, 2018), ICAR-NIASM, Baramati Geoscience and Remote Sensing Symp., Anchorage, AK. 20–24 Sept. 1004. IEEE, New York.

Liu, W; Ming, YZ; Enrique, AP and Calvin, FK (2002) Highly efficient doubled haploid production in wheat (*T. aestivum* L.) via induced microspore embryogenesis. *Crop Sci* **42**: 686-97.

Lizarazu, OR and Mujeeb-Kazi, A (1990) Maize (*Zea mays* L.) mediated wheat (*Triticum aestivum* L.) polyploid production using various crossing methods. *Cereal Res Commun* **18:** 339-43.

Lobell, DB; Schlenker, W and Costa-Roberts, J (2011) Climate trends and global crop production since 1980. *Science* **333**: 616–620.

Lohithaswa, HC; Desai, SA and Hanchinal, RR (2010) Genotype x environment interaction for grain quality traits in emmer wheat (*Triticum dicoccum* (schrank.) Schubl.). *Indian J. Genet.*, **70**(4): 370-372.

Lohithaswa, HC; Giriyappanavar, SA; Desai, SA and Hanchinal, RR (2011) Deciphering traits association pattern in imparting drought tolerance in durum wheat. 3rd International Group Meeting on Wheat Productivity Enhancement under Changing Climate, University of Agricultural Sciences, Dharwad, pp. 40.

Long, SP and Hillgren, JE (1993) Measurement of CO_2 assimilation by plants in the field and the laboratory. *In*: Hall DO, Scurlock JMO, Bolhar-Nordenkampf HR, Leegood RC, Long SP, eds. Photosynthesis and production in a changing environment. London: Chapman and Hall, pp.129-167.

Long, SP; Farage, PK and Garcia, RL (1996) Measurement of leaf and canopy photosynthetic CO_2 exchange in the field. *Journal of Experimental Botany*, **47** (304): 1629-1642.

Luo, MC and Dvorak, J (1996) Molecular mapping of an aluminium tolerance locus on chromosome 4D of Chinese Spring wheat. *Euphytica* **91**: 31-35.

Maccaferri, M; Sanguineti, MC; Corneti, S; Ortega, JL; Salem, MB; Bort, J; DeAmbrogio, E; del Moral, LF; Demontis, A; El-Ahmed, A; Maalouf, F; Machlab, H; Martos, V; Moragues, M; Motawaj, J; Nachit, M; Nserallah, N; Ouabbou, H; Royo, C; Slama, A and Tuberosa, R (2008) Quantitative trait loci for grain yield and adaptation of durum wheat (*Triticum durum* Desf) across a wide range of water availability. *Genetics* **178**: 489–511.

Maccaferri, M; Sanguineti, MC; Demontis, A; El-Ahmed, A; del Moral, LG; Maalouf, F; Nachit, M; Nserallah, N; Ouabbou, H; Rhouma, S; Royo, C; Villegas, D and Tuberosa, R (2011) Association mapping in durum wheat grown across a broad range of water regimes. *J Exp Bot* **62**: 409–438.

Mahantashivayogayya, K; Hanchinal, RR; Desai, SA; Salimath, PM and Patil, BN (2011) Genetic investigations for improving heat-tolerance and productivity in *Dicoccum* wheat (*Triticum dicuccum*). 3rd International Group Meeting on Wheat Productivity Enhancement under Changing Climate, University of Agricultural Sciences, Dharwad, Pp. 48.

Malik, R; Kundu, S; Sareen, S; Kumar, R; Jagshoran and Mishra, B (2008) SSR and ISSR Markers for Assessing DNA Polymorphism and Genetic Diversity among Bread Wheat Varieties of India *Proc. 11th International Wheat Genetic Symposium.*

Malik, R; Sareen, S; Kundu, S; Kumar, R and Shoran, J (2010) The Use of SSR and ISSR markers for assessing DNA Polymorphism and genetic diversity among Indian Bread wheat cultivars. *Environment and Ecology* **28** (2B): 133-1337.

Malone, JH and Oliver, B (2011) Microarrays, deep sequencing and the true measure of the transcriptome. *BMC Biol* **9**: 34.

Manickavasagan, A; Jayas, D and White, N (2008) Thermal imaging to detect infestation by cryptolestes ferrugineus inside wheat kernels. *J. Stored Prod. Res.* **44**: 186–192.

Marigoudra, RM;, Desai, SA; Lohithaswa, HC; Kalappanavar, IK and Hanchinal, RR (2010) New sources of resistance to spot blotch in emmer wheat developed through mutagenesis. *Indian J. Genet.*, 70(3): 288-291.

Marioni, JC; Mason, CE; Mane, SM; Stephens, M and Gilad, Y (2008) RNA-seq: an assessment of technical reproducibility and comparison with gene expression arrays. *Genome Res* **18**: 1509–1517.

Mathew, I; Shimelis, H; Shayanowako, AIT; Laing, M and Chaplot, V (2019) Genome-wide association study of drought tolerance and biomass allocation in wheat. *PLoS ONE* **14**(12): e0225383.

Mathews, KL; Malosetti, M; Chapman, S; McIntyre, L; Reynolds, M; Shorter, R and van Eeuwijk, F (2008) Multi-environment QTL mixed models for drought stress adaptation in wheat. *Theor Appl Genet* **117**: 1077–1109.

Matsumura, H; Yoshida, K; Luo, S; Kimura, E; Fujibe, T; Albertyn, Z; Barrero, RA; Kru¨ger, DH; Kahl, G; Schroth, GP and Terauchi, R (2010) High-throughput SuperSAGE for digital gene expression analysis of multiple samples using next generation sequencing. *PLoS ONE* 5: e12010.

McNeil, SD; Nuccio, ML and Hanson, AD (1999) Betaines and related osmoprotectants. Targets for metabolic engineering of stress resistance. *Plant Physiol.*, **120**: 945-949.

Medlyn BE (2011) Comment on Drought-Induced Reduction in Global Terrestrial Net Primary Production from 2000 Through 2009 . *Science* **333**: 1093-1093.

Meena, KK; Sorty, AM; Bitla, UM; Choudhary, K; Gupta, P and Pareek, A (2017) Abiotic Stress Responses and Microbe-Mediated Mitigation in Plants: The Omics Strategies. *Front. Plant Sci.* **8**: 172.

Melkus, G; Rolletschek, H; Fuchs, J; Radchuk, V; Grafahrend, BE; Sreenivasulu, N; Rutten, T; Weier, D; Heinzel, N; Schreiber, F; Altmann, T; Jakob, PM and Borisjuk, L (2011) Dynamic (13) C/(1) H NMR imaging uncovers sugar allocation in the living seed. *Plant Biotechnol J* **9**: 1022–1037.

Meuwissen, T; Hayes, BJ and Goddard, ME (2001) Prediction of total genetic value using genome-wide dense marker maps. *Genetics* **157**: 1819–1829.

Mian, MA; Mahmood, A; Ihsan, M and Cheema, NM (2007) Response of different wheat genotypes to post anthesis temperature stress. *J Agric Res* **45**: 269-276.

Mittler, R and Blumwald, E (2010) Genetic engineering for modern agriculture: challenges and perspectives. *Ann Rev Plant Biol* **61**: 443–462.

Mogali, S; Basavaraj, M; Athani, V; Biradar, NK; Patil, L; Naik, K and Desai, SA (2011) Molecular characterization of wheat genotypes using ISSR markers. 3rd International Group Meeting on Wheat Productivity Enhancement under Changing Climate, University of Agricultural Sciences, Dharwad, Pp. 125.

Mogali, SC; Basavaraj, M; Athani, V; Naik, LK; Nadaf, HL and Desai, S (2011) STMS-based DNA fingerprints of the wheat varieties. 3rd International Group Meeting on Wheat Productivity Enhancement under Changing Climate, University of Agricultural Sciences, Dharwad, pp. 126.

Mohan, A; Kulwal, PL; Singh, S; Kumar, V; Mir, RR; Kumar, J; Prasad, M; Balyan, HS and Gupta, PK (2009) Genome-wide QTL analysis for pre-harvest sprouting tolerance in bread wheat. *Euphytica* **168**: 319–329.

Monasterio, I and Graham, RD (2000) Breeding for trace minerals in wheat. *Food and Nutrition Bulletin* **21**: 393–396.

Montes, JM; Melchinger, AE and Reif, JC (2007) Novel throughput phenotyping platforms in plant genetic studies. *Trends Plant Sci* **12**: 433–436.

Morgan, JM and King, RW 1984 Association between loss of leaf turgor, abscisic acid levels and seed set in two wheat cultivars. *Australian Journal of Plant Physiology* **11**: 143-150.

Morran, S; Eini, O; Pyvovarenko, T; Parent, B; Singh, R; Ismagul, A; Eliby, S; Shirley, N; Langridge, P and Lopato, S (2011) Improvement of stress tolerance of wheat and barley by modulation of expression of DREB/CBF factors. *Plant Biotechnology Journal*, **9**: 230-249.

Munns, R (2002) Comparative physiology of salt and water stress. *Plant, Cell and Environment*, **25**: 239-250.

Murata, Y and Lyamam J (1963) Studies on photosynthesis of forage crops II, Influence of air temperature upon the photosynthesis of forage and grain crops. *Proc Crop Sci Society of Japan*, **31**: 315-321.

Musick, JT; Jones, OR; Stewart, BA and Dusek, DA (1994) Water-yield relationships for irrigated and dryland wheat in the U.S. Southern Plains. *Agronomy Journal* **86**: 980-986.

Myles, S; Peiffer, J; Brown, PJ; Ersoz, ES; Zhang, Z; Costich, DE and Buckler, ES (2009) Association mapping: critical considerations shift from genotyping to experimental design. *Plant Cell* **21**: 2194–2202.

Nagalakshmi, U; Wang, Z; Waern, K; Shou, C; Raha, D; Gerstein, M and Snyder, M (2008) The transcriptional landscape of the yeast genome defined by RNA sequencing. *Science* **320**: 1344–1349.

Nagarajan, S and Rane, J (1997) Techniques for quantifying Heat tolerance in wheat. In Proceed. International Group Meeting on "Wheat Research Needs Beyond 2000 AD" Aug. 12-14. 1997 D.W.R., Karnal. pp. 209-226.

Naik, R; Desai, SA; Hanchinal, RR; Biradar, SS; Yadwad, A; Shridevi, BM; Kalappanavar, IK; Patil, BN; Jadhav, S and Kumar, A (2011) Adaptation of wheat genotypes for drought prone areas of peninsular India. 3rd International Group Meeting on Wheat Productivity Enhancement under Changing Climate, University of Agricultural Sciences, Dharwad, pp. 208.

Nataraja, KN; Mamrutha, HM and Rama, N (2008) Lead talk on Plant cuticular waxes and their role in water conservation and drought tolerance. Golden jubilee conference of Indian society for plant physiology, November 12-14th,2008, IARI, New Delhi, India.

Nelson, DE; Repetti, PP; Adams, TR; Creelman, RA; Wu, J; Warner, DC; Anstrom, DC; Bensen, RJ; Castiglioni, pp. and Donnarummo, MG (2007) Plant nuclear factor Y(NF-Y) B subunits confer drought tolerance and lead to improved corn yields on water-limited acres. In *Proceedings of the National Academy of Sciences,* **104**: 16450-16455.

Neuberger, T; Sreenivasulu, N; Rokitta, M; Rolletschek, H; Go¨bel, C; Rutten, T; Radchuk, V; Feussner, I; Wobus, U; Jakob, P; Webb, A and Borisjuk, L (2008) Quantitative imaging of oil storage in developing crop seeds. *Plant Biotechnol J* **6**: 31–45.

Nicolas, ME; Gleadon, RM and Dalling, MJ (1984) Effects of drought and high temperature on grain growth in wheat. *Aust J Plant Physiol* **11**: 553-566.

Nidumolu, U; Sadras, V; Hayman, P and Crimp, S (2008) Comparison of NDVI seasonal trajectories and modeled crop growth dynamics. In M.J. Unkovich (ed.) Global issues, paddock action: Proc. Aust. Agron. Conf., 14th, Adelaide, SA, Australia. September 2008.

Nielsen, KL; Høgh, AL and Emmersen, J (2006) Deep SAGE–digital transcriptomics with high sensitivity, simple experimental protocol and multiplexing of samples. *Nucleic Acids Res* **34**: e133.

Nielsen, KL; Petersen, AH and Emmersen, J (2008) Deep SAGE: Tag based transcriptome analysis beyond microarrays. In: Janitz M (ed) Next-generation genome sequencing: towards personalized medicine. WILEYVCH Verlag GmbH and Co. KGaA, Weinheim, pp. 229–244.

Nilsson, HE (1995) Remote sensing and image analysis in plant pathology. *Annu. Rev. Phytopathol.* **33**: 489–527.

Noh, H; Zhang, Q; Han, S; Shin, B and Reum, D (2005) Dynamic calibration and image segmentation methods for multispectral imaging crop nitrogen deficiency sensors. *Trans. ASAE* **48**: 393–401.

O'Donoughue, LS and Bennett, MD (1994) Durum wheat haploid production using maize wide-crossing. *Theor Appl Genet* **89:** 559-66.

Oehler, E (1931) Researches on the grain-setting, morphology and fertility in wheat-rye hybrids. *Z. Züchtung* **36**: 357–393.

Oenema, O and Pietrzak, S (2002) Nutrient management in food production: Achieving agronomic and environmental targets *Ambio* **31**: 159-168.

Oettler, G (1983) Crossability and embryo development in wheat-rye hybrids. *Euphytica* **32**: 583–600.

Ogbonnaya, FC; Subrahmanyam, NC; Moullet, O; deMajnik, J; Eagles. HA; Brown, JS; Eastwood, RF; Kollmorgen, J; Appels, R and Lagudah, ES (2001) Diagnostic DNA markers for cereal cyst nematode resistance in bread wheat. *Australian Journal of Agricultural Research* **52**: 1367-1374.

Pandey, GC; Sareen, S; Siwach, P; Garg, D; Singh, R and Tiwari, R (2011) Putative marker associated with yield trait in wheat under terminal heat stress. 3rd

International Group Meeting on Wheat Productivity Enhancement under changing climate from Feb. 9-12, 2011 at UAS, Dharwad. pp. 226.

Passioura JB (2004) Water-use efficiency in farmers' fields. In "Water-Use Efficiency in Plant Biology" (Ed. M Bacon) Blackwell, Oxford pp.302-321.

Passioura, JB (1977) Grain yield, harvest index, and water use of wheat. *Journal of the Australian Institute of Agricultural Science* **43**: 117-120.

Passioura, JB (2002) Environmental biology and crop improvement. *Functional Plant Biology*, **29**: 537-546.

Passioura, JB (2010) Scaling up: the essence of effective agricultural research. *Funct Plant Biol* **37**: 585–591.

Paull, JG; Rathjen, AJ and Cartwright, B (1991) Major gene control of tolerance of bread wheat (*Triticum aestivum* L.) to high concentrations of soil boron. *Euphytica*, **55**: 217-228.

Peleg, Z; Fahima, T; Krugman, T; Abbo, S; Yakir, D; Korol, AB and Saranga, Y (2009) Genomic dissection of drought resistance in durum wheat 9 wild emmer wheat recombinant inbreed line population. *Plant Cell Environ* **32**: 758–779.

Penner, GA; Clarke, J; Bezte, LJ and Leisle, D (1995) Identification of RAPD markers linked to a gene governing cadmium uptake in durum wheat. *Genome*, **38**: 543-547.

Pfeiffer, WH; Trethowan, RM; van Ginkel, M; Ortiz-Monasterio, I and Rajaram, S (2005a) Breeding for stress tolerance in wheat. In M Ashraf and PJC Harris (eds.) Abiotic Stresses: Plant Resistance through Breeding and Molecular Approaches. New York: The Haworth Press, Inc. pp. 401–489.

Pfeiffer, WH; Trethowan, RM; van Ginkel, M; Ortiz-Monasterio, I and Rajaram, S (2005) Breeding for abiotic stress tolerance in wheat. In Abiotic Stresses: Plant Resistance through Breeding and Molecular Approaches, edited by M. Ashraf and P.J.C. Harris, pp. 401–489. The Haworth Press, Inc. N.Y.

Pierret, A; Moran, CJ and Pankhurst, CE (1999) Differentiation of soil properties related to the spatial association of wheat roots and soil macropores. *Plant and Soil*, **211**: 51-58.

Pinto, RS; Reynolds, MP; Mathews, KL; McIntyre, CL; Olivares-Villegas, JJ and Chapman, SC (2010) Heat and drought adaptive QTL in a wheat population designed to minimize confounding agronomic effects. *Theor Appl Genet* **121**: 1001–1021.

Porter, JR and Gawith, M (1999) Temperatures and the growth and development of wheat: A review. *Eur. J. Agron*. **10**: 23–36.

Prasad, M; Kumar, N; Kulwal, PL; Röder, MS; Balyan, HS; Dhaliwal, HS and Gupta, PK (2003). QTL analysis for grain protein content using SSR markers and validation studies using NILs in bread wheat. *Theor Appl Genet* **106**: 659-667.

Prasad, M; Varshney, RK; Kumar, A; Balyan, HS; Sharma, PC; Edwards, KJ; Singh, H; Dhaliwal, HS; Roy, JK and Gupta, PK (1999) A microsatellite marker associated with a QTL for grain protein content on chromosome arm 2DL of bread wheat. *Theor. Appl. Genet.* **99**: 341-345.

Prasad, M; Varshney, RK; Roy, JK; Balyan, HS and Gupta, PK (2000) The use of microsatellites for detecting DNA polymorphism, genotype identification and genetic diversity in wheat. *Theor. Appl. Genet.* **100**: 584-592.

Quarrie, SA; Gulli, M; Calestani, C; Steed, A and Marmiroli, N (1994) Location of a gene regulating drought-induced abscisic acid production on the long arm of chromosome 5A of wheat. *Theor Appl Genet* **89**: 794-800.

Rafalski, JA (2010) Association genetics in crop improvement. *Curr Opin Plant Biol* **13**: 174–180.

Rama, N; Nethra, P; Mamrutha, HM and Nataraja, KN (2006) Validation of CER1 gene function by post transcriptional gene silencing. International conference on biotechnology approaches for alleviating malnutrition and Human health,9-10th January 2006, University of Agriculture sciences, GKVK, Bangalore.

Ramsay, G (1998) DNA chips: State of the art. *Nature Biotechnology*. **16**: 40-44.

Ramya, P; Chaubal, A; Kulkarni, K; Gupta, L; Kadoo, N; Dhaliwal, HS; Chhuneja, P; Lagu, M and Gupta, V (2010) QTL mapping of 1000-kernel weight, kernel length, and kernel width in bread wheat (*Triticum aestivum* L.). *Journal of Applied Genetics* **51**(4): 421-429.

Ramya, P; Kadoo, N; Desale, C; Kore, P; Dhaliwal, HS; Chhuneja, P and Gupta, V (2012) Kernel morphometric traits in hexaploid wheat (*Triticum aestivum* L.) are modulated by intricate QTL x QTL and genotype environment interactions. *Journal of Cereal Science*.

Ramya, P; Kulkarni, KP; Dholakia, BB; Lagu, MD; Kadoo, NY; Dhaliwal, HS and Gupta, VS (2011) Molecular marker analysis reveals Yellow berry tolerance loci on chromosomes 2D and 5D in bread wheat (*Triticum aestivum* L.). *Journal of Wheat Research* **3**(1): 33-36.

Rasal, PN; Gadekar, DA; Bhoite, KD and Gavhane, VN (2007) NIAW 917-New variety of bread wheat. *Indian J. Genet.*, **67**(2): 217-218.

Rasal, PN; Gavhane, VN; Gadekar, DA; Kamble, MS; Kashid, NV and Shirpurkar, GN (2005) NIDW 15 (Panchavati)- A nobel variety of wheat for rainfed conditions of Maharashtra. *Res. on Crops* **6** (3): 421-423.

Rasal, PN; Gavhane, VN; Kusalkar, DV and Gadekar, DA (2007) Variation in heat susceptibility index and thermal requirement of bread wheat (*Triticum aestivum* L.) genotypes under high temperature stress. *Res. on Crops* **8**(2): 398-400.

Raun, WR; Johnson, GV; Stone, ML; Solie, JB; Lukina, EV; Thomason, WE and Schepers, JS (2001) In-season prediction of potential grain yield in winter wheat using canopy reflectance. *Agron. J.* **93**: 131–138.

Raun, WR; Solie, JB and Stone, ML (2011) Independence of yield potential and crop nitrogen response. *Prec. Agric.* **12**: 508–518.

Raun, WR; Solie, JB; Stone, ML; Martin, KL; Freeman, KW; Mullen, RW; Zhang, H; Schepers, JS and Johnson, GV (2005) Optical sensor based algorithm for crop nitrogen fertilization. *Commun. Soil Sci. Plant Anal.* **36**: 2759–2781.

Raun, WR; Solie, JB; Taylor, RK; Arnall, DB; Mack, CJ and Edmonds, DE (2008) Ramp calibration strip technology for determining mid-season N rates in corn and wheat. *Agron. J.* **100**: 1088–1093.

Rebetzke, GJ; Chapman, SC; Mcintyre, L; Richards, RA; Condon, AG; Watt, M and Van Herwaarden, A (2009) Grain yield improvement in water-limited environments. In: Carve BF (ed) Wheat: science and trade. Wiley–Blackwell, Ames, IA, pp. 215–249.

Rebetzke, GJ; Richards, RA; Fischer, VM and Mickelson, BJ (1999) Breeding long coleoptile, reduced height wheats. *Euphytica,* **106**: 159-168.

Reddy, DMR; Aspatwar, A; Dholakia, BB and Gupta, VS (2008) Evolutionary analysis of WD40 super family proteins involved in spindle checkpoint and RNA export: Molecular evolution of spindle checkpoint. *Bioinformation* **2**: 461- 468.

Rengasamy P (2002) Transient salinity and subsoil constraints to dryland farming in Australian sodic soils: an overview. *Australian Journal of Experimental Agriculture* **42**: 351-361.

Reynolds, MP (2002) Physiological approaches to wheat breeding. In: Curtis BC, Rajaram S, Macpherson HG (eds) Bread wheat. FAO Plant production and protection series-30.

Reynolds, MP and Ortiz, R (2010) Adapting crops to climate change: a summary. In: Reynolds MP (eds) Climate change and crop production, CAB international, pp. 1–8.

Reynolds, MP and Trethowan, RM (2007) Physiological interventions in breeding for adaptation to abiotic stress. In: Spiertz JHJ, Struik PC, van Laar HH (eds) Scale and complexity in plant systems research, gene–plant–crop relations. Wageningen UR frontis series, vol 21, pp. 129–146.

Reynolds, MP and Tuberosa, R (2008) Translational research impacting on crop productivity in drought-prone environments. *Curr Opin Plant Biol* **11**: 171–179.

Reynolds, MP; Bonnett, D; Chapman, SC; Furbank, RT; Mane's, Y; Mather, DE and Parry, MAJ (2011) Raising yield potential of wheat. I. Overview of a consortium approach and breeding strategies. *J Exp Bot* **62**: 439–452.

Reynolds, MP; Dreccer, F and Trethowan, R (2007a) Drought-adaptive traits derived from wheat wild relatives and landraces. *J Exp Bot* **58**: 177–186.

Reynolds, MP; Foulkes, JM; Slafer, GA; Berry, P; Parry, MAJ; Snape, J and Angus, WJ (2009b) Raising yield potential in wheat. *J Exp Bot* **60**: 1899–1918.

Reynolds, MP; Manes, Y; Izanloo, A and Langridge, P (2009a) Phenotyping approaches for physiological breeding and gene discovery in wheat. *Ann Appl Biol* **155**: 309–320.

Reynolds, MP; Mujeeb-Kazi, A and Sawkins, M (2005) Prospects for utilising plant-adaptive mechanisms to improve wheat and other crops in drought- and salinity-prone environments. *Ann Appl Biol* **146**: 239–259.

Reynolds, MP; Saint Pierre, C; Saad Abu, SI; Vargas, M and Condon, AG (2007b) Evaluating potential genetic gains in wheat associated with stress-adaptive trait expression in elite genetic resources under drought and heat stress. *Crop Sci* **47**: S-172–S-189.

Reynolds, MP; Singh, RP; Ibrahim, A; Ageeb, OA; Larqué-Saavedra, A and Quick, J (1998) Evaluating physiological traits to compliment empirical selection for wheat in warm environments. *Euphytica* **100**: 85–94.

Reynolds, MP; Skovmand, B; Trethowan, R and Pfeiffer, W (2000) Evaluating a conceptual model for drought tolerance. In: Ribaut J-M, Poland D (eds) Molecular approaches for the genetic improvement of cereals for stable production in water-limited environments: a strategic planning workshop held at CIMMYT, El Batan, Mexico, 21–25 June 1999. CIMMYT, Mexico, pp. 49–53.

Rhoads, DM; Samuel, JW; You, Z; Mrinalini, M and Elthon, TE (2005) Altered gene expression in plants with constitutive expression of a mitochondrial small heat shock protein suggests the involvement of retrograde regulation in the heat stress response. *Physiol. Plant*, **123**: 435-444.

Ribaut, JM; Bänziger, M; Setter, T; Edmeades, G and Hoisington, D (2004) Genetic Dissection of Drought Tolerance in Maize: A Case Study. In H. Nguyen and A. Blum (eds.) Physiology and Biotechnology Integration for Plant Breeding. Marcel Dekker, Inc. New York.

Ribaut, JM; de Vicente, MC and Delannay, X (2010) Molecular breeding in developing countries: challenges and perspectives. *Curr Opin Plant Biol* **13**: 213–218.

Richards, RA (1991) Crop improvement for temperate Australia: Future opportunities. *Field Crops Research* **26**: 141–169.

Richards, RA; Rebetzke, GJ; Condon, AG and van Herwaarden, AF (2002) Breeding opportunities for increasing the efficiency of water use and crop yield in temperate cereals. *Crop Science*, **42**: 111-121.

Riede, CR and Anderson, JA (1996) Linkage of RFLP markers to an aluminium tolerance gene in wheat. *Crop Sci* **36**: 905-909.

Risi, JLD; Iyer, VR and Brown, PO (1997) Exploring the metabolic and genetic control of gene expression on a genomic scale. *Science* **278**: 680-686.

Rockström, J and Falkenmark, M (2000) Semiarid crop production from a hydrological perspective: gap between potential and actual yields. *Critical Reviews in Plant Sciences* **19**: 319-346.

Rossel, JB; Philippa, BW; Luke, H; Soon, CW; Andrew, P; Philip, MM and Barry, JP (2006) A Mutation Affecting Ascorbate Peroxidase 2 Gene expression reveals a link between responses to high light and drought tolerance. *Plant Cell Environ.*, **29**: 269-281.

Roy, JK; Prasad, M; Varshney, RK; Balyan, HS; Blake, TK; Dhaliwal, HS; Singh, H; Edwards, KJ and Gupta, PK (1999) Identification of a microsatellite on chromosome 6B and a STS on 7D of bread wheat showing association with preharvest sprouting tolerance. *Theor. Appl. Genet.* **99**: 336-340.

Roy, SJ; Tucker, EJ and Tester, M (2011) Genetic analysis of abiotic stress tolerance in crops. *Current Opinion in Plant Biology* **14**(3): 232–239.

Russel, GE (1978) Plant Breeding for Pest and Disease Resistance. Butterworths.

Sadras, V; Roget, D and Krause, M (2003) Dynamic cropping strategies for risk management in dry-land farming systems. *Agricultural Systems* **76**: 929-948.

Sadras, VO (2003) Influence of size of rainfall events on water-driven processes - I. Water budget of wheat crops in south-eastern Australia. *Australian Journal of Agricultural Research* **54**: 341-351.

Saini, HS and Westgate, ME (2000) Reproductive development in grain crops during drought. *Advances in Agronomy*, **68**: 59-96.

Salam, A; Hollington, PA; Gorham, J; Jones, RGW and Gliddon, C (1999) Physiological genetics of salt tolerance in wheat (*Triticum aestivum* L.): Performance of wheat varieties, inbred lines and reciprocal F-1 hybrids under saline conditions. *Journal of Agronomy and Crop Science* **183**: 145–156.

Salekdeh, GH; Reynolds, M; Bennett, J and Boyer, J (2009) Conceptual framework for drought phenotyping during molecular breeding. *Trends in Plant Science*, **14**: 488-496.

Salekdeh, GH; Reynolds, MP; Bennett, J and Boyer, J (2009) Conceptual framework for drought phenotyping during molecular breeding. *Trends Plant Sci* **14**: 488–496.

Salem, KFM; Roder, MS and Borner, A (2007) Identification and mapping quantitative trait loci for stem reserve mobilisation in wheat (*Triticum aestivum* L.). *Cereal Res Commun* **35**: 1367–1374.

Salimath PM and Biradar, S (2002) Genetics of drought and breeding for drought tolerance. Book chapter In: Physiological approaches for enhancing productivity potential under drought condition. Eds: M. B. Chetti *et al.*, pp. 117-134 Publ. by UAS, Dharwad.

Salvi, S and Tuberosa, R (2005) To clone or not to clone plant QTLs: present and future challenges. *Trends Plant Sci* **10**: 297–304.

Salvi, S and Tuberosa, R (2007) Cloning QTLs in plants. In: Varshney RK, Tuberosa R (eds) Genomics-assisted crop improvement, vol 1. Springer, New York, pp. 207–226.

Sandhu, APS; Dhawan, R; Gill, MS and Bains, NS (2003) Evaluation of rapid crossing techniques for haploid production in wheat x maize crosses. *Indian J. Genet* **63** (2): 155-156.

Sanguineti, MC; Li, S; Maccaferri, M; Corneti, S; Rotondo, F; Chiari, T and Tuberosa, R (2007) Genetic dissection of seminal root architecture in elite durum wheat germplasm. *Ann Appl Biol* **151**: 291–305.

Sardans, J; Pen^uelas, J; Lope-Piedrafita, S (2010) Changes in water content and distribution in Quercus ilex leaves during progressive drought assessed by in vivo 1H magnetic resonance imaging. *BMC Plant Biol* **10**: 188.

Sareen, S; Kundu, S; Malik, R; Kumar, S and Shoran, J (2008) Durability of packing systems for germplasm conservation under cold dry conditions in western Himalayas. *Ind. Jour. Plant Genetic Resources.* **20**(3): 204-207.

Sareen, S; Meena, BK; Dhillon, OP and Singh, SS (2011) Exploring indigenous wheat germplasm for terminal heat tolerance. 3rd International Group Meeting on Wheat Productivity Enhancement under changing climate from Feb. 9-12, 2011 at UAS, Dharwad. pp. 216.

Sareen, S; Munjal, R; Singh, NB; Singh, BN; Verma, RS; Meena, BK; Shoran, J; Sarial, AK and Singh, SS (2012) Genotype x Environment Interaction and AMMI Analysis for Heat Tolerance in Wheat. *Cereal Research Communications* **40**(2): 267–276.

Sarial, AK; Haile, JK; Assefa, S and Sareen, S (2008) AMMI analysis for stability and location effect on grain yield and its components of durum wheat genotypes. *Proc. 11th International Wheat Genetic Symposium http: //hdl.handle.net/2123/3238.*

Sasaki, T; Yamamoto, Y; Ezaki, B; Katsuhara, M; Ahn, SJ; Ryan, PR; Delhaize, E and Matsumoto, H (2004) A wheat gene encoding an aluminium-activated malate transporter. *Plant Journal* **37**: 645-653.

Saulescu, NN; Ittu, G; Balota, M; Ittu, M; Mustatea, P; Braun, HJ (ed.); Altay, F (ed.); Kronstad, WE (ed.); Beniwal, SPS (ed.) and McNab, A (1997) Breeding wheat for lodging resistance, earliness and tolerance to abiotic stresses. Wheat: prospects for global improvement. Procs of the 5th International Wheat Conference, Ankara, Turkey, 10-14 June 1996. 181-188.

Saunders, (Undated) Wheat for the nontraditional warmer areas. Mexico, D.F: CIMMYT.

Saurmel' kh IA (1988) Breeding bread wheat for resistance to lodging. *Puti-intensifikatsii-vozdelyvaniya-sel'skokhozyaistvennykh kul'tur v Kustanaiskoi oblasti.* 102-107.

Schena, M and Heller, RA (1998) Microarrays: biotechnology's discovery platform for functional genomics. *TIBTECH* **16**: 301-306.

Schena, M and Shalon, D (1995) Quantitative monitoring of gene expression patterns with a complementary *DNA Microarray Science* **270**: 467-470.

Schilling, KE and Libra, RD 2000. The relationships of nitrate concentrations in streams to row crop land use in Iowa. *Journal of Environmental Quality* **29**: 1846-1851.

Schillinger, WF; Cook, RJ and Papendick, RI (1999) Increased dryland cropping intensity with no-till barley. *Agronomy Journal* **91**: 744-752.

Schütz, L; Gattinger, A and Meier, M (2018) Improving Crop Yield and Nutrient Use Efficiency via Bio fertilization–A Global Meta-analysis. *Front Plant Sci* **8**: 2204.

Sehgal, D and Yadav, R (2010) Molecular markers-based approaches for drought tolerance. In: Jain SM, Brar DS (eds) Molecular techniques in crop improvement. Springer, New York, pp. 207–230.

Shankar, U; Prasad, U; Kumari, S; Chaudhary, A and Kumar, R (2011) A Miniprep method modified for rapid and economic isolation of genomic DNA from local wheat varieties without using liquid Nitrogen, Souvenir and Abstracts of National Symposium on Biotechnological Perspective of Plants, Microbes and their Interaction, Muzaffarpur, Bihar, 114-115.

Sharma, S; Balyan, HS; Kulwal, PL; Kumar, N; Varshney, RK; Prasad, M and Gupta, PK (2002) Study of interspecific SSR polymorphism among 14 species from *Triticum-Aegilops* group. *Wheat Inform Services* **95**: 23-28.

Shendure, J (2008) The beginning of the end for microarrays? *Nat Methods* **5**: 585–587.

Shinozaki, K and Yamaguchi-Shinozaki, K (1999) Molecular responses to drought stress. In: Shinozaki K, Yamaguchi-Shinozaki K (eds) Molecular responses to cold, drought, heat and salt stress in higher plants. R.G. Landes Co., Austin, pp. 11-28.

Shinozaki, K; Yamaguchi, SK and Seki, M (2003) Regulatory network of gene expression in the drought and cold stress responses. *Curr Opin Plant Biol* **6**: 410–417.

Shinozaki, K; Yamaguchi-Shinozaki, K and Seki, M (2003) Regulatory network of gene expression in the drought and cold stress responses. *Current Opinion in Plant Biology*, **6**: 410-17.

Sinclair, TR (2011) Challenges in breeding for yield increase for drought. *Trends Plant Sci* **16**: 289–293.

Singh, R; Tyagi, BS; Pahwa, G; Sareen, S and Tiwari, R (2012) Molecular marker-based detection of *Ph1b* mutation to increase homoeologous pairing in wheat (*Triticum aestivum*)* *Indian Journal of Agricultural Sciences* **82** (4): 363–365.

Singh, Y; Kumar, M and Singh, NP (2018) Training Manual of ICAR Sponsored Summer School on Climate Change and Abiotic Stress Management Strategies for Doubling Farmers Income. ICAR-National Institute of Abiotic Stress Management, Publication No. 26.

Slovacek, RE and Hind, G (1981) Correlation between the photosynthesis and the transthylakoid proton gradient. *Biochem Biophys Acta* **35**: 393-404.

Sood, S; Dhawan, R; Singh, K and Bains, NS (2003) Development of novel chromosome doubling strategies for wheat x maize system of wheat haploid production. *Plant Breed* **122** (6): 493-96.

Spiertz, JHJ (1974) Grain growth and distribution of dry matter in the wheat plant as influence by temperature, light energy and seed size. *Neth. J Agri Sci* **22**: 207-220.

Sreenivasulu, N; Sopory, SK and Kishor, PBK (2007) Deciphering the regulatory mechanisms of abiotic stress tolerance in plants by genomic approaches. *Gene* **388**: 1–13.

Sreenivasulu, N; Sunkar, R; Wobus, U and Strickert, M (2010) Array platforms and bioinformatics tools for the analysis of plant transcriptome in response to abiotic stress. *Methods Mol Biol* **639**: 71–93.

Sreenivasulu, N; Varshney, RK; Kavi Kishor, PB and Weschke, W (2004) Functional genomics for tolerance to abiotic stress in cereals. In: Gupta PK, Varshney RK (eds) Cereal genomics, Kluwer Academic Publishers, New York, pp. 483–514.

Srinivasa Rao, Ch; Prabhakar, M; Maheswari, M; Srinivasa Rao, M; Sharma, KL; Srinivas, K; Prasad, JVNS; Rama Rao, CA; Vanaja, M; Ramana, DBV; Gopinath, KA; Subba Rao, AVM; Rejani, R; Bhaskar, S; Sikka, AK and Alagusundaram, K (2016) National Innovations in Climate Resilient Agriculture (NICRA), Research Highlights 2015-16. Central Research Institute for Dryland Agriculture, Hyderabad. 112 p.

Stapper, M and Fischer, RA (1990) Genotype, sowing date and plant spacing influence on high-yielding irrigated wheat in southern New South Wales. III. Potential yields and optimum flowering dates. *Australian Journal of Agricultural Research*. **41**(6): 1043-1056.

Stone, PJ and Nicolas, ME (1995) Effect of heat stress during grain filling on two wheat varieties differing in heat tolerance grain growth. *Aust J Plant Physiol* **22**: 927-934.

Sud, S; Bains, NS; Nanda, GS; Singh, K and Arya, L (2010) Genetic diversity and Heterosis in wheat. *SABRAO Journal of Breeding and Genetics* **42**(2): 96-105.

Suenaga, K and Nakajima, K (1989) Efficient production of haploid wheat (*Triticum aestivum*) through crosses between Japanese wheat and maize (*Zea mays*) *Plant Cell Rep* **8**: 263-66.

Tahsildhar, N; Hanchinal, RR; Vijayakumar, AG; Desai, SA; Lohithaswa HC and Nadaf, HL (2007) Quantifying diversity in wheat using D^2 technique and molecular profiling. *Indian J. Genet.*, **67** (3): 287-289.

Talame', V; Ozturk, NZ; Bohnert, HJ and Tuberosa, R (2007) Barley transcript profiles under dehydration shock and drought stress treatments: a comparative analysis. *J Exp Bot* **58**: 229–240.

Tanner, CB and Sinclair, TR (1983) Efficient water use in crop production: research or re-search? In Limitations to Efficient Water Use in Crop Production (eds. HM Taylor, WR Jordan and TR Sinclair) American Society of Agronomy, Wisconsin, USA, pp. 1-27.

Tardieu, F and Tuberosa, R (2010) Dissection and modelling of abiotic stress tolerance in plants. *Current Opinion in Plant Biology*, **13**: 206-212.

Thomashow, MF; Stockinger, EJ; Jaglo-Ottosen, K; Zarka, D and Gilmour, SJ (1999) Method for regulating cold and dehydration regulatory genes in a plant. Michigan State University, East Lansing, MI US Patent 5891859.

Tottman, DR and Makepeace, RJ (1979) An explanation of the decimal code for the growth stages of cereals with illustrations. *Ann. Appl. Biol.* **93**: 221-234.

Trethowan, RM and Mujeeb-Kazi, A (2008) Novel germplasm resources for improving environmental stress tolerance of hexaploid wheat. *Crop Sci.* **48**: 1255–1265.

Tuberosa, R (2010) Phenotyping drought-stressed crops: key concepts, issues and approaches. In: Monneveaux P, Ribaut JM (eds) Drought phenotyping in crops: from theory to practice. Generation Challenge Programme, c/o CIMMYT, Mexico, pp. 3–35.

Tuberosa, R (2011) Phenotyping drought-stressed crops: key concepts, issues and approaches. In: Monneeveux Philippe, Jean MarcelRibaut (eds) Drought phenotyping in crops: from theory to practice, CGIAR Generation Challenge Programme, Texcoco.

Tuberosa, R and Salvi, S (2006) Genomics approaches to improve drought tolerance in crops. *Trends Plant Sci* **11**: 405–412.

Umezawa, T; Fujita, M; Fujita, Y; Yamaguchi-Shinozaki, K and Shinozaki, K (2006a) Engineering drought tolerance in plants: discovering and tailoring genes to unlock the future. *Current Opinion in Biotechnology*, **17**: 113-122.

UNEP (2009) Water: in the transition to a green economy: a UNEP brief available online at www.unep.ch/etb/ebulletin/pdf/GE per cent 20and per cent 20Water per cent 20Brief.pdf.

Van der Plank, JE (1982) Host-Pathogen Interactions in Plant Disease. Academic Press.

van Herwaarden, AF and Passioura, JB (2001) Improving estimates of water-use efficiency in wheat. *Australian Grain* **11**(4): 3-5.

van Herwaarden, AF; Angus, JF; Richards, RA and Farquhar, GD (1998) "Haying-off", the negative grain yield response of dryland wheat to nitrogen fertiliser. II. Carbohydrate and protein dynamics. *Australian Journal of Agricultural Research* **49**: 1083-1093.

Varshney, RK; Bansal, KC; Aggarwal, PK; Datta, SK and Craufurd, PQ (2011) Agricultural biotechnology for crop improvement in a variable climate: hope or hype? *Trends Plant Sci* **16**: 363–371.

Varshney, RK; Graner, A, and Sorrells, ME (2005) Genomics-assisted breeding for crop improvement. *Trends Plant Sci* **10**: 621–630.

Varshney, RK; Nayak, SN; May, GD and Jackson, SA (2009b) Next generation sequencing technologies and their implications for crop genetics and breeding. *Trends Biotechnol* **27**: 522–530.

Velculescu, VK; Zhang, L; Vogelstein, B and Kinzler, KW (1995) Serial analysis of gene expression. *Science* **270**: 484-487.

Verma, D K (2008) Cytogeneticist's insight about molecular markers and their role in crop improvement. *The Indian Journal of Crop Science*. Communicated on 17.03.2008.

Verma, DK (2008) Breeding to enhance crop productivity in scarce water conditions. *SAARC Journal of Agriculture*. Communicated on 17.03.2008.

Verma, V; Bains, NS; Mangat, GS; Nanda, GS; Gosal, SS and Singh, K (1999) Maize genotypes show striking differences for induction and regeneration of haploid wheat embryos in wheat x maize system. *Crop Sci* **39**: 1722-1727.

Vogel, B (2009) Marker-assisted selection: a non-invasive biotechnology alternative to genetic engineering of plant varieties. In: Erwood S, Truchi N (eds) Smart breeding. Greenpeace International, The Netherlands, pp. 4–25.

von Korff, M; Grando, S; Del Greco, A; This, D; Baum, M and Ceccarelli, S (2008) Quantitative trait loci associated with adaptation to Mediterranean dry land conditions in barley. *Theor Appl Genet* **117**: 653–669.

Wakchaure, GC; Minhas, PS; Ratnakumar, P and Choudhary, RL (2016a) Optimising supplemental irrigation for wheat (*Triticum aestivum* L.) and the impact of plant bioregulators in a semi-arid region of Deccan Plateau in India. *Agric. Water Manag*. **172**: 9–17.

Wang, BK; Jin, ZY; Zhao, MZ; Wang, BK; Jin, ZY and Zhao, MZ (1995) Effects of TM on lodging resistance and yield components of barley. *Acta-Agriculturae-Nucleatae-Sinica*. **9**(3): 200-202.

Wang, F; Wang, X and Sayre, K (2004) Comparison of conventional, flood irrigated, flat planting with furrow irrigated, raised bed planting for winter wheat in China. *Field Crops Research* **87**: 35-42.

Wang, GP; Hui, Z; Li, F; Zhao, MR; Zhang, J and Wang, W (2010). Improvement of heat and drought photosynthetic tolerance in wheat by overaccumulation of glycinebetaine. *Plant Biotechnology Reports*, **4**: 213-222.

Wang, W; Vinocur, B and Altman, A (2003) Plant responses to drought, salinity and extreme temperatures: towards genetic engineering for stress tolerance. *Planta*, 218: 1-14.

Wang, WX; Vinocur, B; Shoseyov, and, Altman, A (2001) Biotechnology of plant osmotic stress tolerance: physiological and molecular considerations. *Acta Hort*; **560**: 285-292.

Wang, Y; Ying, J; Kuzma, M; Chalifoux, M; Sample, A; McArthur, C; Uchacz, T; Sarvas, C; Wan, J and Dennis, DT (2005) Molecular tailoring of farnesylation for plant drought tolerance and yield protection. *Plant J.*, 43: 413-424.

Wanjura, DF and Hatfield, JL (1987) Sensitivity of spectral vegetative indices to crop biomass. Trans. *ASAE* 30: 810–816.

Wardlaw, IF; Blumenthal, C; Larroque, O and Wrigley, CW (2002) Contrasting effects of chronic heat stress and heat shock on grain weight and flour quality in wheat. *Functional Plant Biol.* **29**: 25-34.

Watkins, AE and Ellerton, S (1940) Variation and genetics of the awn in *Triticum*. *J. Genet.* **40**: 243–270.

Wegulo, SN and Carlson, MP (2011) Ergot of small grain cereals and grasses and its health effects on humans and livestock. University of Nebraska-Lincoln Extension. EC 1880.

Westgate, ME; Passioura, JB and Munns, R (1996) Water status and ABA content of floral organs in drought-stressed wheat. *Australian Journal of Plant Physiology* **23**: 763-772.

White, JW; Andrade-Sanchez, P; Gore, MA; Bronsona, KF; Coffelt, TA; Conley, MM; Feldmann, KA; French, AN and Heun, JT (2012) Field-based phenomics for plant genetics research. *Field Crops Res* **133**: 101–112.

Winicov, I and Bastola, DR (1997) Salt tolerance in crop plants: New approaches through tissue culture and gene regulation. *Acta Physiol. Plant,* **19**: 435-449.

Witcombe, JR; Hollington, PA; Howarth, CJ; Reader, S and Steele, KA (2008) Breeding for abiotic stresses for sustainable agriculture. *Philosophical Transactions of the Royal Society B: Biological Sciences,* **363**: 703-716.

Yamada, M; Morishita, H; Urano, K; Shiozaki, N; Kazuko, YS; Shinozaki, K and Yoshiba, Y (2005). Effects of free proline accumulation in petunias under drought stress. *J. Exp. Bot.,* **56**: 1975-1981.

Yamaguchi-Shinozaki, K; Kasuga, M; Liu, Q; Nakashima, K; Sakuma, Y; Abe, H; Shinwari, ZK; Seki, M and Shinozaki, K (2002) Biological mechanisms of drought stress response. *JIRCAS Working Rep.* **45**: 1–8.

Yamamori, M (1994) An N-band marker for gene *Lr18* for resistance to leaf rust in wheat. *Theor Appl Genet* **89**: 643-646.

Yang, DL; Jing, RL; Chang, XP and Li, W (2007) Identification of quantitative trait loci and environmental interactions for accumulation and remobilization of water-soluble carbohydrates in wheat (*Triticum aestivum L.*) stems. *Genetics* **176**: 571–584.

Yang, J and Zhang, J (2006) Grain filling of cereals under soil drying. *New Phytol* **169**: 223–236.

Yang, S; Vanderbeld, B; Wan, J and Huang, Y (2010) Narrowing down the targets: towards successful genetic engineering of drought tolerant crops. *Mol Plant* **3**: 469–490.

Yang, S; Vanderbeld, B; Wan, J and Huang, Y (2010) Narrowing down the targets: towards successful genetic engineering of drought tolerant crops. *Molecular Plant,* **3**: 469-490.

Yin, X; Struik, PC and Kropff, MJ (2004) Role of crop physiology in predicting gene-to-phenotype relationships. *Trends Plant Sci* **9**: 426–432.

Zahour, A (1989) Breeding for moderate rainfall areas. Cereal-Improvement-Program-annual-report1988 ICARDA; Aleppo; Syria. 18-25.

Zeven, AC (1985) The genetics of auricle colour in wheat (*Triticum aestivum* L.): A review. *Euphytica* **34**: 233–236.

Zhang, J; Dell, B; Conocono, E; Waters, I; Setter, T and Appels, R (2009) Water deficits in wheat: fructan exohydrolase (1-FEH) mRNA expression and relationship to soluble carbohydrate concentrations in two varieties. *New Phytol* **181**: 843–850.

Zhang, J; Oweis, TY; Garabet, S and Pala, M (1998) Water-use efficiency of wheat under rain-fed conditions and supplemental irrigation in a Mediterranean environment. *Plant and Soil* **201**: 295-305.

Zhu, J; Ingram, PA; Benfey, PN and Elich, T (2011) From lab to field, new approaches to phenotyping root system architecture. *Current Opinion in Plant Biology*, **14**: 310-317.

Zhu, JK (2002) Salt and drought stress signal transduction in plants. *Annu. Rev. Plant Biol.*, **53**: 247-273.

Zhu, T (2003) Global analysis of gene expression using gene chip microarrays. *Current Opinion in Plant Biology* **6**: 418-425.

Zivy, M and Vienne, DD (2000) Proteomics: a link between genomics, genetics and physiology. *Plant Molecular Biology*. **44**: 575-580.

Chapter 7

Breeding Wheat for Biotic Stresses (Disease, Insect Pests and Nematodes) Resistance

Outlines

1. Agro-Ecological Zoning of Wheat in India
   a. Rust Epidemics
   b. Evolution of New Pathotypes of Different Rust
   c. Leaf Rust
   d. Stem Rust
   e. Stripe Rust
   f. Gene Postulation
   g. Identification, Distribution and Prevalence of Rust Races
   h. Resistance/Susceptibility
   i. Durability of Resistance
   j. Gene Deployment
   k. 1BL.1RS Translocation
   l. Genetics of Pest Resistance
   m. Resistance to Pathogenic Diseases
2. Disease Resistance
   a. Rust Resistance Breeding in India
   b. Gene for Gene Hypothesis

    c. Concept of Vertical, Horizontal and Durable Resistance
    d. Breeding for Disease Resistance
    e. Methods in Resistance Breeding
    f. Combination of Multiple Hypersensitive Resistance Genes
    g. Scoring Backcross Population for Leaf Rust Resistance
    h. *In vitro* DNA amplification through Polymerase Chain Reaction
    i. *In vitro* Screening for Biotic Stresses
    j. Exploration of Strategies for Development of Safe Wheat
    k. Data Analysis/Interpretations
    l. Strategies to Control Rust Diseases
3. Identification and Creation of Epiphytotic Conditions for Wheat Diseases
    a. Yellow or Stripe Rust (*Puccinia striiformis*)
    b. Stem Rust (*Puccinia graminis tritici*)
    c. Leaf Rust (*Puccinia rubigovera tritici*)
    d. Loose Smut (*Ustilago tritici*)
    e. Karnal Bunt (*Neovossia indica*)
    f. Flag Smut (*Urocystis agropyri*)
    g. Foliar Blights (*Drechslera, Alternaria, Septoria*)
    h. Powdery Mildew (*Erysiphe graminis tritici*)
    i. Hill Bunt or Stinking Smut (*Tilletia caries* and *Tilletia foetida*) and Dwarf Bunt (*Tilletia controversa*)
    j. Foot Rot/Root Rot/Seedling Blight (*Sclerotinia rolfsii, Drechslera sorokiniana*)
    k. Wheat Blast (*Magnaporthe oryzae*)
4. Gene Cataloguing for Disease Resistance Stocks
    a. Germplasm Collection
    b. Purification and Maintenance
    c. Race Specific and Race Non-Specific Resistance
    d. Characterization of Novel Genes
    e. Linkage with Markers and Known Genes
    f. Gene Location of Recessive Resistance Gene
    g. Cytological Markers
    h. Molecular Technique
    i. Catalogue of Symbols
    j. Characterization of Genetic Stocks of Leaf Rust Resistance Genes Using Molecular Markers in Wheat
    k. Screening for Leaf Rust Reaction
    l. *Lr9, Lr19, Lr28, Lr28* and *Lr32, Lr19* and *Lr24*

   m. Validation of the SCAR and SSR markers linked to genes *Lr9*, *Lr24*, Lr 28 and Lr 19
   n. Influence of Resistance Genes in Cultivars in Different Areas
   o. DNA Isolation
   p. *Lr28* and *Lr32*
   q. Lr34, Lr46 and other Minor Genes for Durable Resistance to Leaf Rust
   r. *Lr19* and *Lr24*
   s. Screening of Backcross Population for Foreground Selection of Sr36 Gene
   t. Recurrent Parent Genome Recovery (RPG) (per cent) through Marker Assisted Background Selection (MABS)
   u. Development of Backcross Populations
   v. Yr18, Yr29 and other Minor Genes for Durable Resistance to Stripe Rust
   w. Recent Advances in Gene Expression Analysis
   x. Serial Analysis of Gene Expression (SAGE)
   y. DNA Microarray Technology
   z. Hybridization and Scanning
   aa. Screening Methodology and Selection Environment
   bb. Current Status of Resistance Breeding
   cc. Preparedness to Stem Rust Race, *Ug99*
5. Identification, Screening and Control of Seed Borne Disease in Wheat
   a. Methods for Detection and Identification of Seed Borne Pathogens
   b. Detection of Fungal Organisms
   c. Examination after Softening or Soaking Seeds
   d. Incubation Methods
   e. Staining Methods
   f. Seedling Symptom Tests
   g. Infectivity or Indicator-Inoculation Test
   h. Immunodiagnostic and Nucleic Acid Methods
   i. Agglutination, Precipitin and Immunodiffusion Methods
   j. Enzyme-Linked Immunosorbent Assay (ELISA)
   k. Polymerase Chain Reaction (PCR) Technology
6. Conceptual Framework of Wheat Rust Management
   a. Syringe Inoculation in Wheat
   b. Enhanced Susceptibility to Stem Rust
   c. Validity of the Phenotypic Markers in Wheat

   d. Development of Locally Adapted Host Differentials for Indian Pathotypes
   e. Evaluation of Wheat Genotypes Against Stem and Leaf Rust Resistance at Indore
7. Gene Cloning
   a. Pedigree-Based Genome Mapping for Marker-Assisted Selection and Recurrent Parent Recovery in Wheat
   b. Pedigree Based Marker Assisted Selection System) including Graphical Genotype with IBD (PBMASS)
   c. New Donors of Stem Rust Resistance in Durum Wheat
   d. Diverse Sources of Adult-Plant Resistance to Leaf Rust in Bread Wheat
   e. Sources of Rust Resistance in Wheat and Triticale
   f. Combining Classical and Novel Resources
   g. Gene Pyramiding for Multiple Disease Resistance
   h. Gene Yr9, *Lr24/Sr24*
   i. Breeding for Durable Resistance to Rust Diseases
   j. Breeding for Durable Resistance to Powdery Mildew
   k. Breeding for Resistance to Fusarium Head Blight (Scab)
   l. Breeding for Resistance to Septoria tritici Blotch
   m. Breeding for Resistance to Spot Blotch and Tan Spot
   n. Preparedness to Stem Rust Race, *Ug99*
   o. Development of Wheat Genotypes with Resistance to Black Rust Race *Ug99*
8. Identification and Creation of Epiphytotic Conditions for Resistance to Insect Pests in Wheat
   a. Termites (*Microtermes obesi* Holmgren; *Odontotermes obesus* Rambur)
   b. Aphids (*Sitobion miscanthi* (Takahashi) *S. avenae* Fabricius, *Rhopalosiphum maidis* Fitch, *R. padi*)
   c. Shoot fly (*Atheriqona naqvii, A. oryzae*)
   d. Brown Mite (*Petrobia latens* Muller)
   e. Army Worm (*Mythimna separata* Walker)
   f. American Pod Borer (*Helicoverepa (Heliothis) armigera* Hub)
   g. Pink Stem Borer (*Sesamia inferens* Walke)
   h. Cut Worm (*Cirphis loreyi* Dup., *Euxoa spinifera* Hub. and *Agrotis ypsilon* Rott)
   i. Gujhia Weevil (*Tanymecus indicus*)
   j. Sugarcane Leaf Hopper (*Pyrilla perpusilla*)
   k. Flea Beetle (*Chetocnema basilis* Baly)
   l. Surface Grass Hopper (*Chrotogonus trachypterus*)

m. Leaf Hopper (*Typhlocylbea maculifrons* Motsch and *Zygnidia manaliensis* (Singh)
n. Hessian Fly (*Phytophaga destructor*)
o. Wheat Stem Sawfly (*Cephus cinctus*)
p. Greenbug (*Toxoptera graminum*)
q. Other Insects

9. Breeding for Biotic Stress (Insect) Resistance
   a. Mechanism of Insect Resistance
   b. Nature of Insect Resistance/Factors for Insect -Resistance
   c. Genetics of Insect Resistance
   d. Breeding Methods for Insect Resistance
   e. Problems in Breeding for Insect Resistance
10. Breeding for Biotic Stress (Nematode) Resistance
   a. Ear Cockle Mamni and Yellow Ear Rot or Tundu (*Anguina tritci* and *Clavibacter tritici*)
   b. Molya Disease (*Heterodera avenae*)
   c. Nematode Pathotypes/Physiological Races/Biological Races/Biotypes/Trophotypes
   d. Pot Screening
   e. Nematode Pathotypes
   f. Nematode Physiological Races
   g. Nematode Biotypes

Agro-Ecological Zoning of Wheat in India

The evaluation of germplasm and pre-breeding lines for rust resistance using most virulent pathotypes and pathogenicity surveys is the major strength of the system. In the beginning the gene deployment schemes with aim to prevent large-scale buildup of wheat rusts were considered. After entering the genotypes in initial variety trials, these are tested against different diseases including the rusts consecutively for three years to ensure their resistance against prevailing pathotypes in the particular zone. Cultivars of semi-dwarf stature carrying dwarfing genes are grown in this zone. Resistance to leaf blight or foliar blights (spot blotch) like *Alternaria*, *Helminthosporium*, [*Cochliobolus sativus* (anamorph, *Bipolaris sorokiniana* Sacc. (Shoem)] and leaf rust are the major thrust areas. Spot blotch is a serious problem of this zone, which affects the leaves, stem and the spike, it can result in shrivelled grain and black point, a dark staining of the embryo end of the seed. There is no host immunity to spot blotch and even resistant wheat genotypes can suffer from a reduction of grain yield. Therefore, an integrated approach is needed to reduce losses due to spot blotch. The prominent varieties grown in this zone are: HUW206, K8027, HUW234, HP1633 (Sonali), DL784-3 (Vaishali), K9107, DBW39, NW2036, HD2733, DBW14, MACS6145 (HW2034), HD2888 and K307. Recently,

Figure 7.1. Healthy wheat crop

Figure 7.2. Screening plot for wheat healthy crop and rust infected crop

Figure 7.3. Prevalence of yellow rust on wheat at KD Farm, Srinagar, J&K

Figure 7.4a & b. Wheat leaf/brown rust and Stem/black rust

Figure 7.5a & b. Wheat loose smut

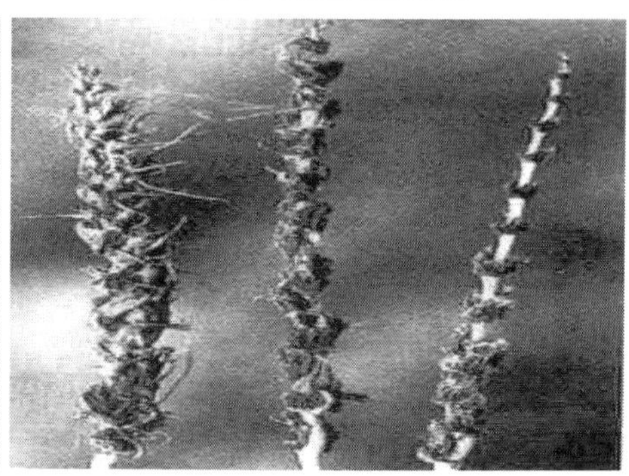

Figure 7.6. Loose smut of wheat is caused by *Ustilago segatum* and occurs throughout the India.

Figure 7.7. Karnal bunt of wheat is caused *Neovossia indica*.

Figure 7.8. Leaf spots / Foliar blight caused by *Helminthosporium aestivum* and *Alternaria triticina*

Figure 7.9. Wheat crop affected by three different rusts, viz., yellow, brown and black in NEPZ and CZ

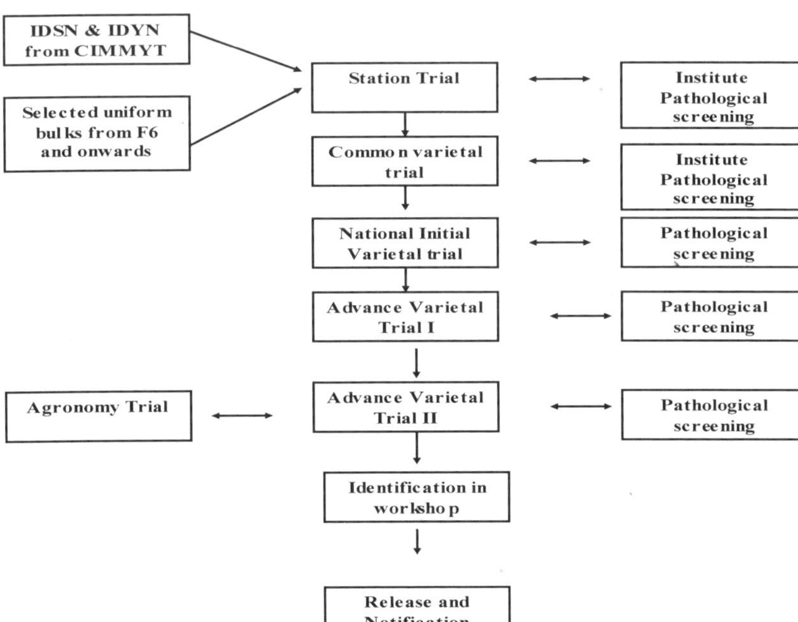

Figure 7.10. Stages of evaluation of improved strains before identification and release of variety.

Figure 7.11. Dr Kamini Kaushal inoculating with rust pathogens to wheat seedling

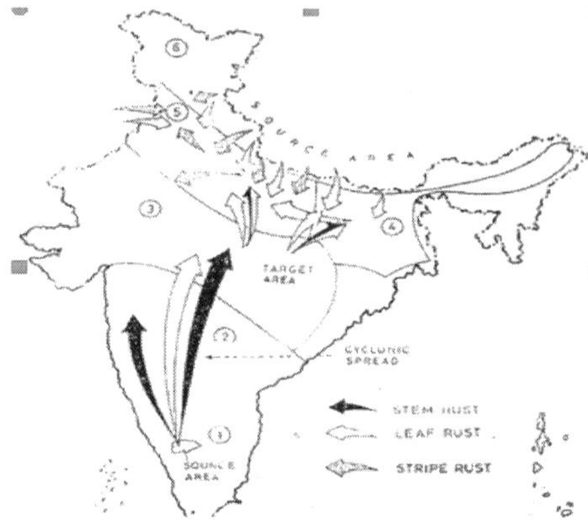

Figure 7.12. *Puccinia* pathway

Figure 7.13a. Syringe inoculation of wheat rust pathogen into the trial for creation of epiphytotic

Figure 7.13b. Syringe inoculation of wheat rust pathogen into the trial for creation of epiphytotic

Figure 7.14. Spray of suspension of wheat rust pathogen into the trial for creation of epiphytotic

Figure 7.15. Seedling resistance study at ICAR-IARI RS Indore, MP

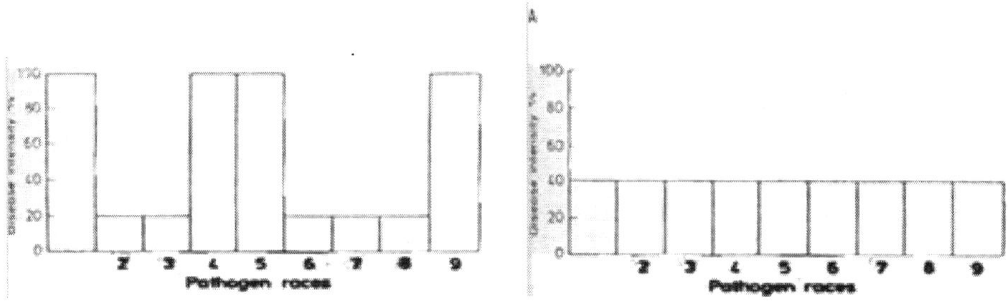

Figure 7.16. Graphical representation of Vertical & Horizontal resistance

Figure 7.17. Transfer of yellow rust resistance gene (Yr15) to HD 3059, BC_2 and BC_1F_3, seed were generated after identification of resistant BC_2F_1 and BC_1F_2 individuals phenotypically against yellow rust race 110S119.

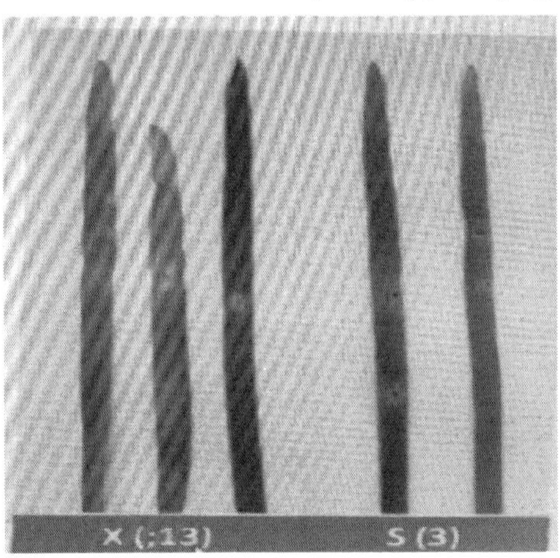

Figure 7.18. Phenotyping for leaf rust resistance gene LrS2427.

Figure 7.19. Dr D K Verma, delivering lecture on Integrated Disease Management of Wheat at TM University, Bhagalpur, Bihar, 2019

Figure 7.20. Multiplication of wheat rust inoculum in glass house. Multiple races of I;eaf rust were revived, multiplied and use for multipathotype testing to understand the spectrum of resistance in targeted genotypes to identify novel gene(s). Specific leaf rust races viz., 77-7,77-8 and 77-10 virulent against Lr9, Lr119 and Lr 28 were multiplied through single spore culture on Thatcher NIL for respective gene to obtain maximum purity of inoculum. Mixed inoculum for yellow was also multiplied and used for screening of genotypes.

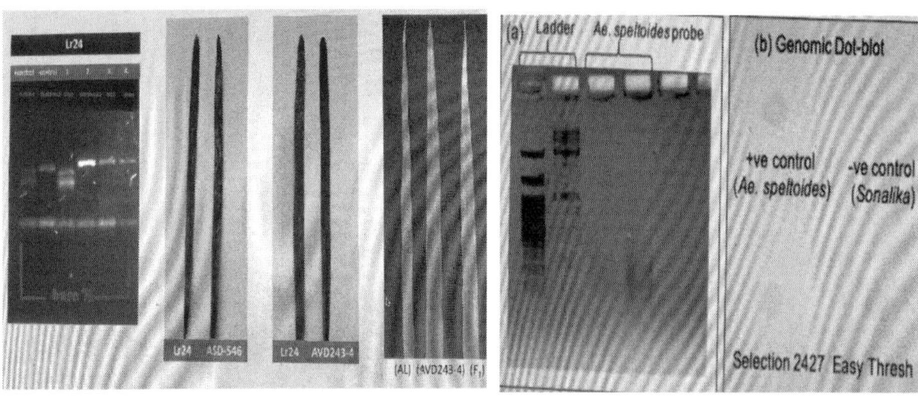

Figure 7.21 a & b. Dot-blot hybridization assay in *Ae speltoides* derived lines viz., Sel 2427 and Essay using digoxigenin-labelled *Ae speltoids* DNA as probe

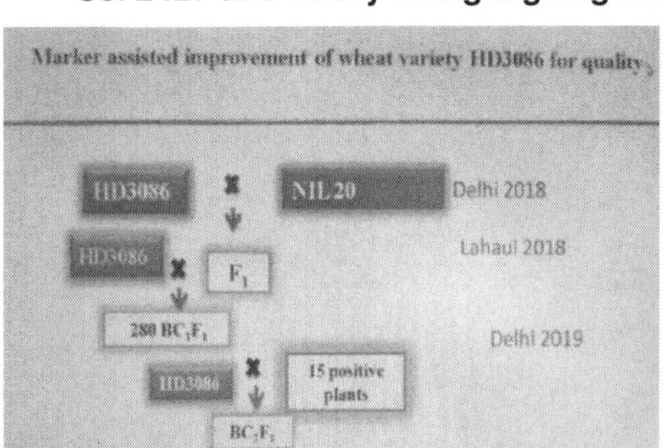

Figure 7.22. Marker assisted improvement of wheat variety HD 3086

Figure 7.23. Wall display of aestivum and durum varieties at ICAR-IARI RS Indore

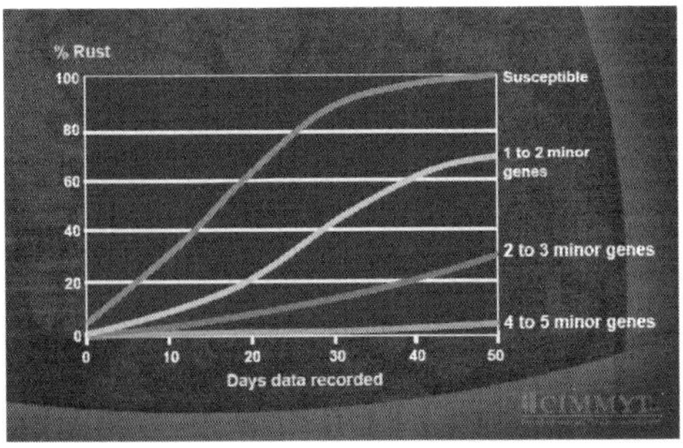

Figure 7.24. Durum seed and products display at ICAR-IARI RS Indore

Figure 7.25. Pyramiding Polygenic Resistance to Leaf and Stripe Rusts in Wheat

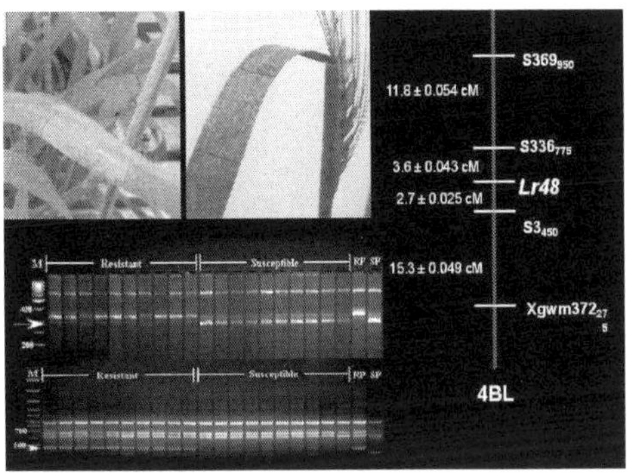

Figure 7.26. Screening of APR and susceptible individuals of the cross [Agra Local x CSP44(lr48)] with SSR marker Xgwm372$_{275}$ (top) and RAPD S3$_{450}$

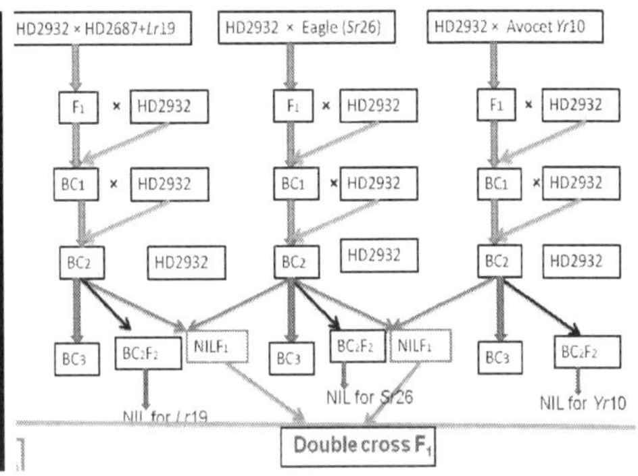

Figure 7.27. Marker Assisted Breeding employed in HD2932

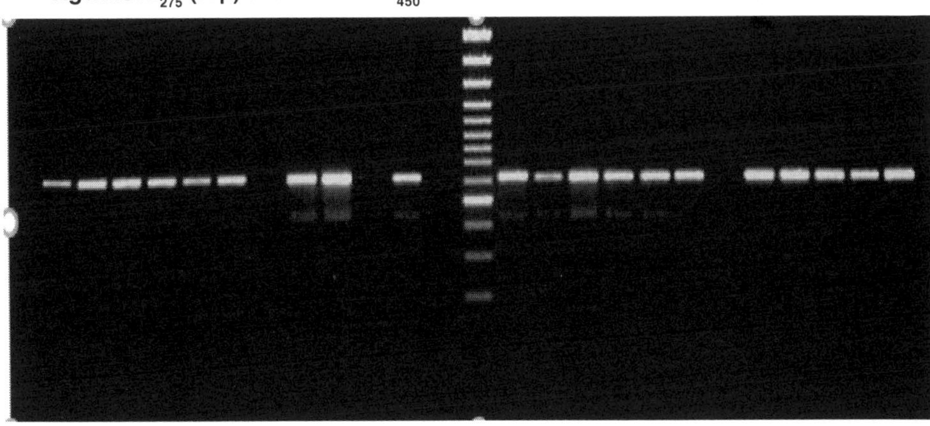

Figure 7.28. Expression of *Lr28* SCAR marker S421$_{640}$ for leaf rust resistance in 11 wheat NIL pairs carrying genes *Lr32* and *Lr28* Lane 1 & 13: C306, 2 & 14 : HD 2285, 3 & 15: HD2329, 4 & 16: HUW 234; Lane 5 & 17: Kalyansona, 6 & 18 :Lok 1, 7& 19 : NI 5439 8 & 20 : PBW 226; Lane 9 & 21: Sonalika, 10 & 22 : WH 147, 11& 23 : WH 542 12 : Lr32 donor , Lane 24 : Lr 28 donor

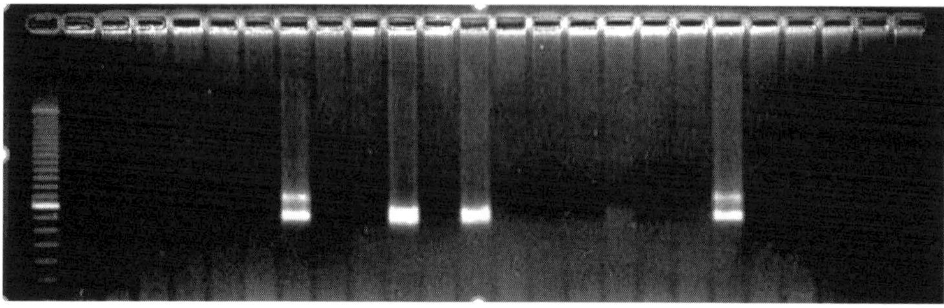

Figure 7.29. Expression of *Lr28 linked* SSR marker Xgwn160 for leaf rust resistance in 11 NIL pairs carrying genes *Lr32* and *Lr28* Lane 1 & 13: C306, 2 & 14 : HD 2285, 3 & 15: HD2329, 4 & 16: HUW 234; Lane 5 & 17: Kalyansona, 6 & 18 :Lok 1, 7& 19 : NI 5439 8 & 20 : PBW 226; Lane 9 & 21: Sonalika, 10 & 22 : WH 147, 11& 23 : WH 542 12 : Lr32 donor , Lane 24 : Lr 28 donor

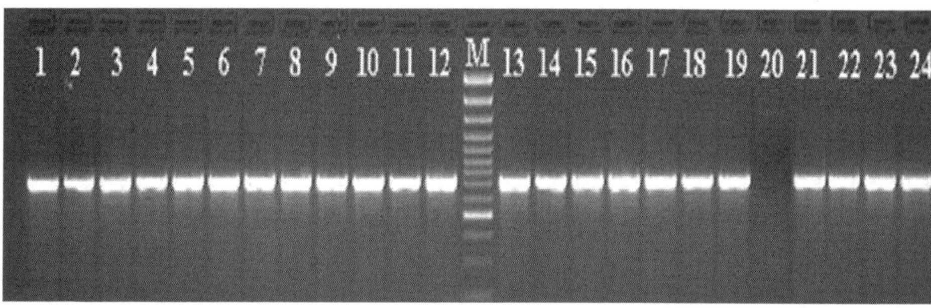

Figure 7.30. Lr19 SCAR SCS73$_{719}$ on Lr19 and Lr24 carrying NILs in different wheat backgrounds. Lr 19 NILs. Lane 1. HW2049 (HD 2285), 2. HW2046(HD 2329), 3. HW2080(HS 240), 4. HW2043(HUW 234), 5. H W2081 (Kalyansona), 6. HW2041(Lok 1), 7. HW2044(PBW 226), 8. HW2050 (Sonalika), 9. HW2042(WH 147), 10. HW2082 (NI 5439), 11. HW2083 (C 306), 12. Lr19 Donor (Sunstar*6/C80-1); Lr24 NILs: Lane 13. HW2008(HD 2285), 14. HW2007 (HD 2329), 15. HW2020(HS 240), 16. HW2015(HUW 234), 17. HW2002 (Kalyansona), 18. HW2006 (Lok 1), 19. HW2016 (PBW 226), 20.HW2001 (Sonalika), 21. HW2005(WH 147), 22. HW2003(NI 5439), 23. HW2004(C 306), 24. Lr24 Donor(TR 380-14*7/3Ag#14)

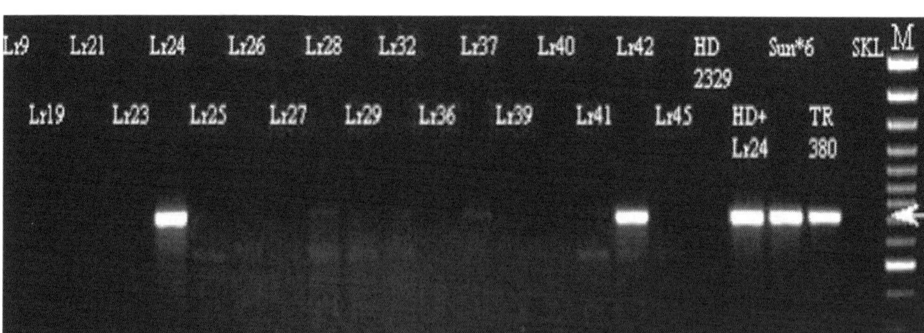

Figure 7.31. Expression of *Lr19* SCAR marker SCS73$_{719}$ for leaf rust resistance in alien Lr gene NILs of Thatcher

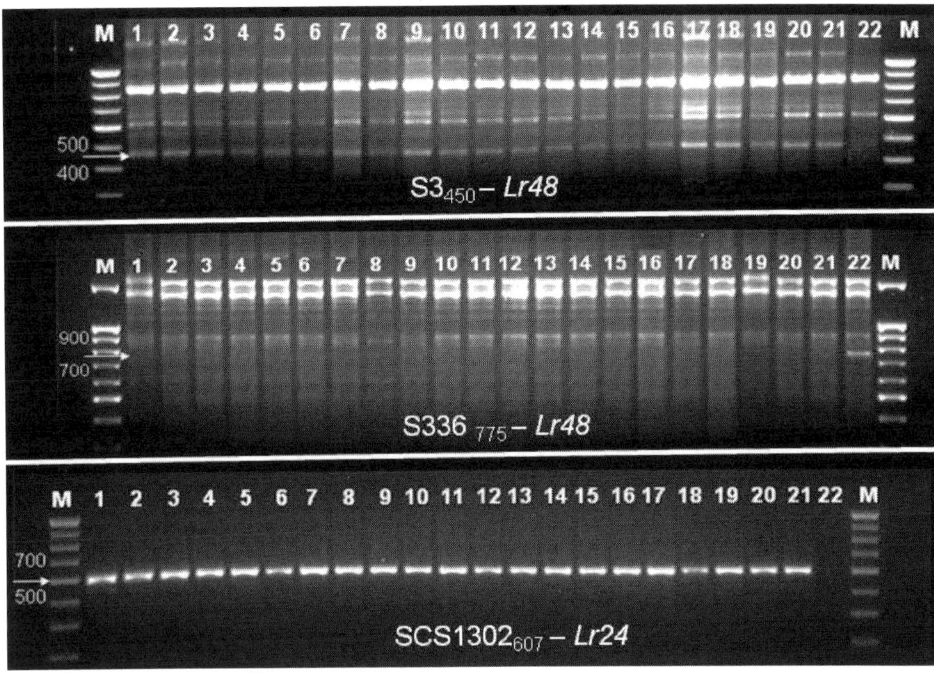

Figure 7.32. Confirmation of the presence of pyramided gene combination of Lr 24 + Lr 48 in the F_4 generation

Figure 7.33. Phenotyping of F_3 families for leaf rust resistance, (a) F_3 families being phenotyped under controlled environment

Figure 7.34. Phenotyping of F_3 families for leaf rust resistance, (b) Typical resistance at adult plant stage exhibited by a family due to the presence of both Lr24 and Lr 48

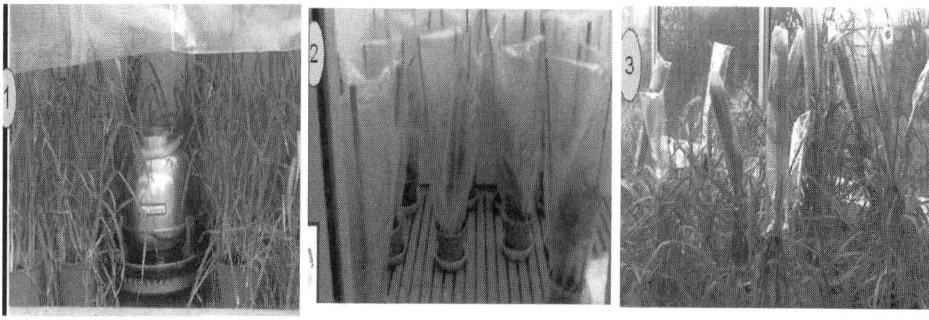

Figure 7.35a,b and c. Standardization of methodology for APR for leaf rust resistance under controlled condition

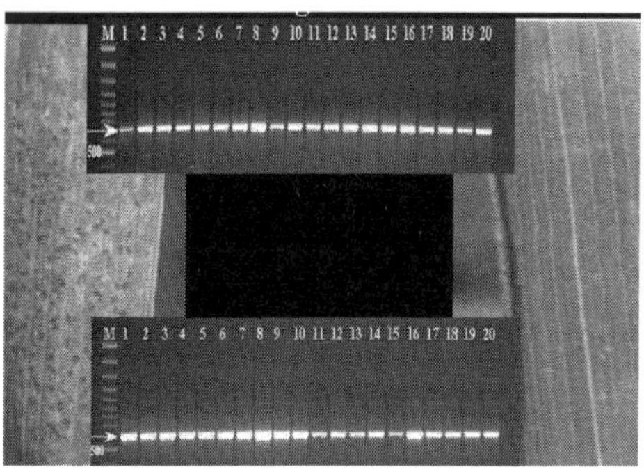

Figure 7.36. MAS for Adult Plant Resistance (APR), Lr 24 and Lr 28 are fixed in HD 2329 and PBW 343 background

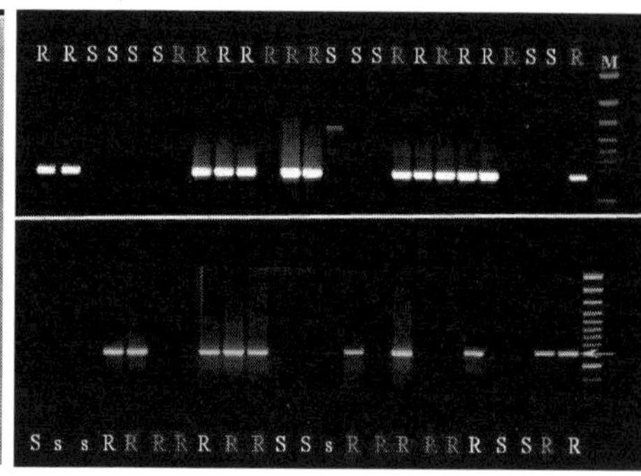

Figure 7.37. Selection in BC_2F_1 generation for Lr24 + Lr28 pyramids using the SCAR marker $SCS421_{640}$ and SCAR marker $scs73_{719}$ in two sub-families PBW 343 background

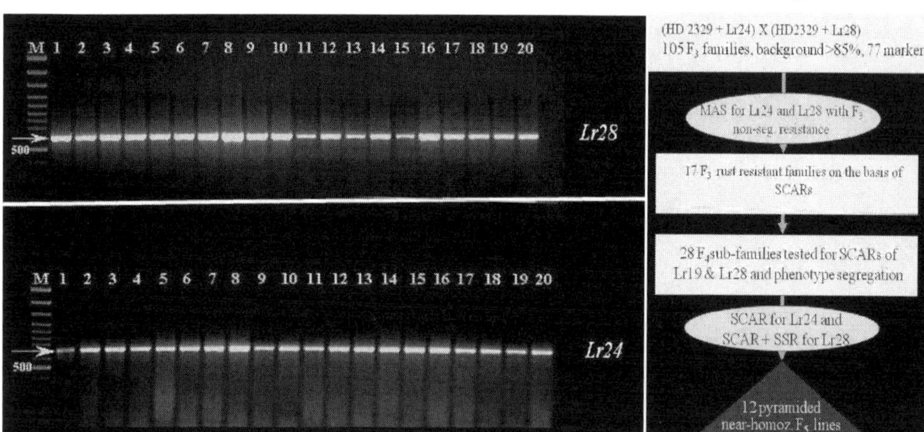

Figure 7.38. Selection in F_5 generation for Lr24 + Lr28 pyramids using the SCAR marker $SCS421_{640}$ and SCAR marker $scs73_{719}$ in two sub-families of the cross HW 2037 x HW 2046

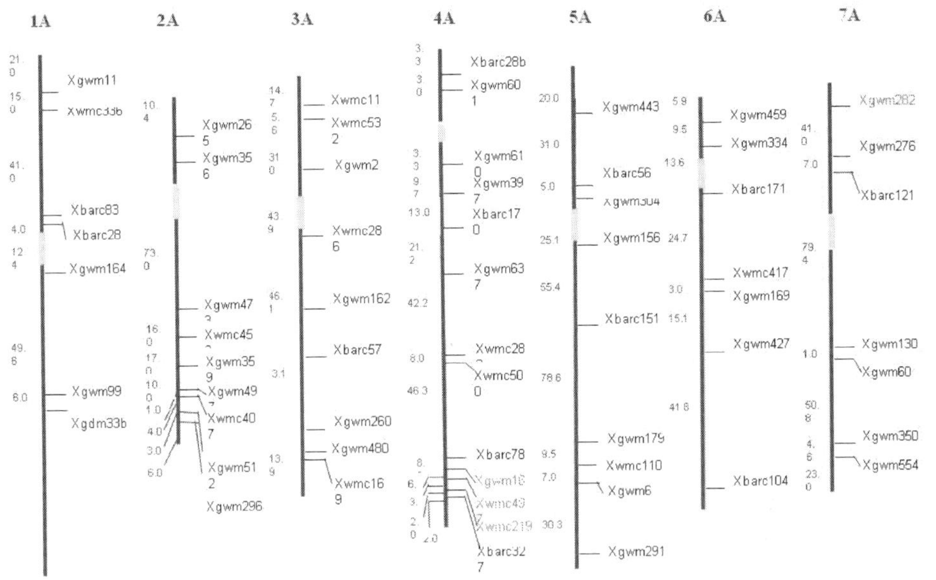

Figure 7.39. Co-dominant informative SSR marker suite for polymorphism framework in Indian wheat - A genome

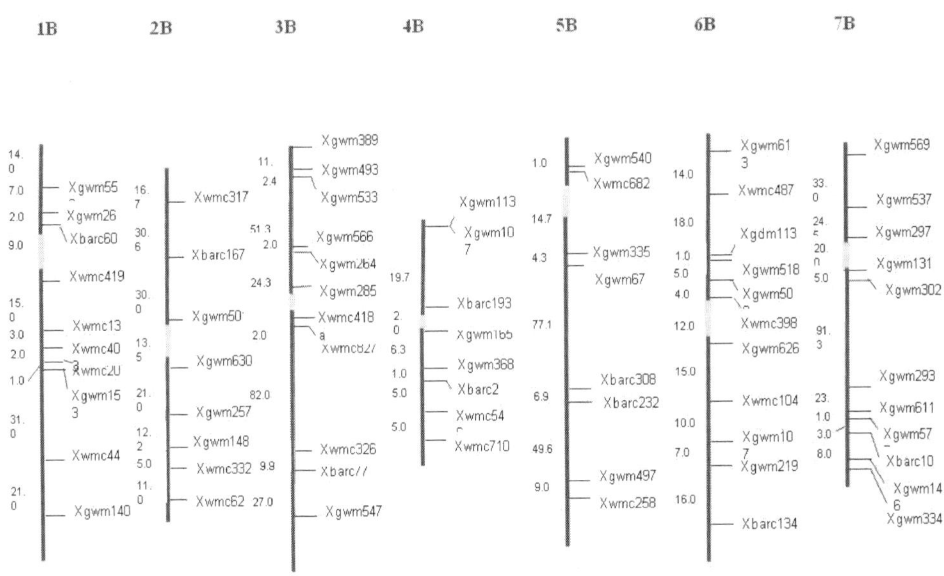

Figure 7.40. Co-dominant informative SSR marker suite for polymorphism framework in Indian wheat - B genome

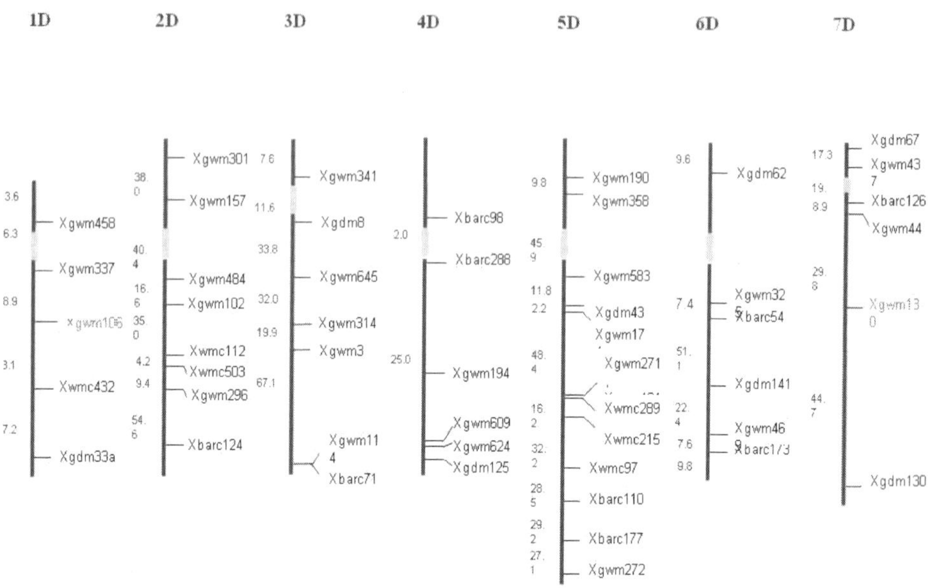

Figure 7.41. Co-dominant informative SSR marker suite for polymorphism framework in Indian wheat - D genome

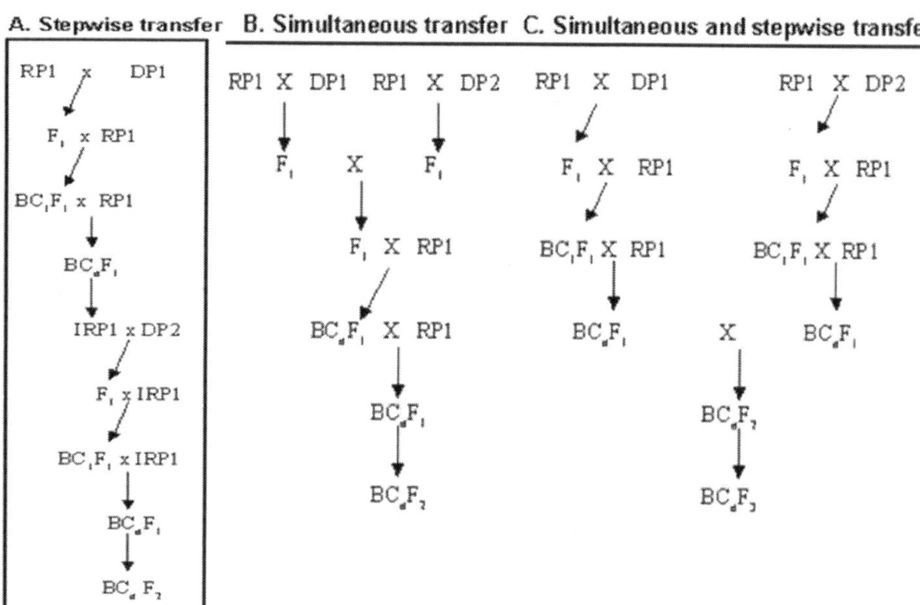

Figure 7.42. Different schemes of backcrossing, [Abbreviations: RP-Recurrent parent, DP – Donor parent; BC – Backcross; IRP: Improved recurrent parent]

No. of genes present	Varieties
Lr 19 +, Sr 25	HW 2045
Lr 26 + Sr 31 +, Yr 9 +	HD 2733
Lr 26 + Lr 23 + Lr 1, Yr 9 + Sr 31 +	HD 2643
Lr 23 + i Sr 9b + Sr 11 + and Yr 2 +	HD 2285
Lr 1 + Lr 3 + Lr 10, Sr 2 + Sr 8+ Sr 9b + Sr 11+, Yr 2 + (ks)	BW 11 Pubali
Lr 13 + Lr 10 + Lr 34, Sr 8+ Sr 9b + Sr 11+, Yr 2 + Yr 18 +	HD 2329
Lr 23 + Lr 1, Sr 7a + Sr 11+, Yr 2 + (ks)	HD 2428
Lr 23 + Lr 34, Sr 2+ Sr 11+, Yr 18 +	HD 2501
Lr 23 + Lr 34 +, Sr 2 + Sr 8+ Sr 11, Yr 18 + Sr 2+ Sr 31; Lr 26, Lr 23, Lr 1, Lr 34, Yr 9 + Pm 9	HD 2687
Lr 19 +, Sr 25	HW 2045
Lr 26 + Sr 31 +, Yr 9 +	HD 2733
Lr 26 + Lr 23 + Lr 1, Yr 9 + Sr 31 +	HD 2643

Figure 7.43. Gene deployment in bread wheat varieties

	PS	HP No	Pedigree
NIVT-1A	PS-834	HP -1915	K9330/UP2338//1843
	PS-837	HP -1916	Raj4004/HD2652
NIVT-1B	PS-841	HP -1917	UP2532/DL7882//CPAN3048
	PS-842	HP -1918	UP2338/HW3034//PBW532
NIVT-3			
NWPZ	PS-862	HP -1919	PBW524//HUW516/Raj3814
NEPZ	PS-855	HP-1920	HP1744/BL2053
SHZ	PS-870	HP-1921	RL6043/Raj3765
	Ps-859	Hp-1924	PBW483/WR544

Figure 7.44. Entries promoted (2006-2007)

Trial Name	Entries	D/s	D/h	Rem	Data
CVT 1B	36	24.11.06	15.04.07	Good	Sent to
CVT 3	49	15.12.06	14.05.07	Good	DWR
39th IBWSN	1001-1174	27.11.06	18.04.07	Good	Karnal
NGSN	117	29.11.06	-	Good	
IPPSN	1520	18.12.06	-	Good	
PDSN	632	18.12.06	-	Good	
TPN	20	04.12.06	-	Good	
LBSN	118	11.12.06	-	Good	

Figure 7.45a. Coordinated trials: 2006-2007

Trial Name	Entries	D/s	D/h	Rem	Data
AVT (RF)	8	10.11.06	09.04.07	Good	Sent to
AVT TS-Irr	20	25.11.06	16.04.07	Good	DWR
AVT LS-Irr	23	14.12.06	11.05.07	Good	Karnal
NIVT-1A	49	20.11.06	13.04.07	Good	
NIVT-1B	49	20.11.06	13.04.07	Good	
NIVT-3	49	14.12.06	14.05.07	Good	
CVT5A(RF)	36	10.11.06	10.04.07	Good	
CVT5A(RI)	36	10.11.06	10.04.07	Good	
CVT1A	36	24.11.06	15.04.07	Good	

Figure 7.45b. Coordinated trials: 2006-2007

Figure 7.46. Analysis of Host-Pathogen Interaction in Leaf Rust Infection of Wheat: A transcriptomic approach, Seedling Resistance (*Lr28*); APR (*Lr48*)

Figure 7.47. Figure Selection for minor genes, D 2329 + *Lr 28* and HD 2329; CSP-44 + *Lr 48*

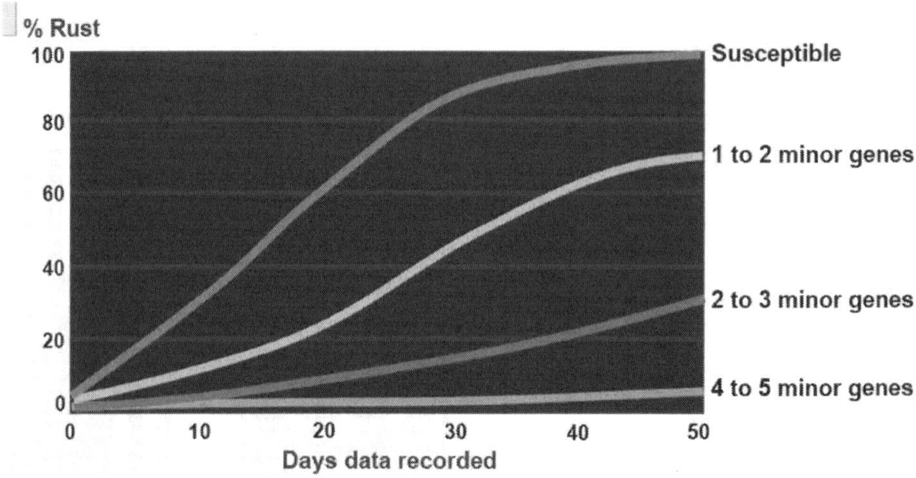

Figure 7.48. Genetic basis of durable resistance to rust diseases of wheat

Variety	Severity	Village	District	Date
Lok 1	tS	Kotri	Sehore	4/3/11
Local Mix	10S	Bhauransa	Dewas	4/3/11
Local Mix	60S	Fanda	Bhopal	4/3/11
Local Mix	tS	Kolukhedi	Bhopal	26/2/11
Local Mix	10S	Agera	Dewas	26/2/11
Local Mix	10S	Arniya	Dewas	23/2/11
Local Mix	tS	Singar Chauri	Sehore	17/2/11
		STEM RUST		
Local Mix	tS (with leaf rust)	Fanda	Bhopal	4/3/11

Figure 7.49a. Leaf rust infection in Farmers field

Variety	Severity	Village	District	Date
Local Mix	60-100S	Sonkatch	Deaws	29/3/11
Lok 1	tS-5S	Hon PirPipliya	Dewas	17/3/11
Local Mix	40S	Hon PirPipliya	Dewas	17/3/11
Sujata	tS-30S	Kotri	Sehore	17/3/11
Lok 1 (OC)	30S	Kotri	Sehore	17/3/11
Local Mix	100S	Kotri	Sehore	4/3/11
Local Mix	5S	Semlibari	Sehore	4/3/11
Lok 1	tS	Semlibari	Sehore	4/3/11
Sujata	tS	Kotri	Sehore	4/3/11

Figure 7.49b. Leaf rust infection in Farmers field

AVT

T. aestivum: HI 1571*, HI 1572*, HI 1579, and HI 1580

T. durum: HI 8703*, HI 8704*, HI 8713, HI 8714, HI 8715, and HI 8722

NIVT

T. aestivum: HI 1582, HI 1583, HI 1584, HI 1585, HI 1586, and HI 1587

T. durum: HI 8723, HI 8724, HI 8725, HI 8726, HI 8727, HI 8728, HI 8729, HI 8730, HI 8731, and HI 8732

IPPSN

T. aestivum: IND Nos. 298-303, 305-309, 311-312, 313 (NG), 315-316 (IND 298-317)

T. durum: ID Nos. 1001, 08, 09, 12-14, 18, 20, 22-25, 27-30 (ID 1001-1030)
Only ID 1006 and ID 1020 susceptible to leaf rust in multilocation data

Elite PPSN

HI 1563, HI 1567, HI 1568, HI 1569, HI 8702, HI 8708, and HI 8709

MDSN

HI 1560, HI 8680 – HI 8682, HI 8690-HI 8693, HI 8696, and HI 8699

APR Resistant to both stem rust pathotypes *40A* and *117-6*

T. aestivum: All AVT entries

T. durum: All AVT entries

Figure 7.50. Indore entries resistant to both stem and leaf rusts

Wild relatives and their desirable traits

Species	Ploidy	Genome	Characteristics/ traits
Ae. speltoides	2x	SS	Suppress Ph locus, resistant to leaf and stem rusts, powdery mildew
Ae. tauschii	2x	DD	Tolerance to leaf, stem and yellow rusts, powdery mildew, abiotic stress
Ae. cylindrica	4x	CCDD	Spike length and spikelet number
Ae. peregrina	4x	UUSS	Resistant to water logging conditions, heat
Ae. squarrosa	2x	DD	Leaf and stripe rust resistance
Ae. geniculata	4x	UUMM	Environmental stresses tolerance
Agropyron elongatum	2x	J^eJ^e	Leaf rust resistance, high β carotene
A. intermedium			Leaf rust resistance
Secale cereale	2x	RR	Resistance to leaf, stem and yellow rusts, powdery mildew, winter hardiness

Figure 7.51. Wild relatives and their desirable traits

Wild relatives and their desirable traits

Species	Ploidy	Genome	Characteristics/ traits
Triticum armenicum			Leaf rust resistance
T. araraticum	4x	AAGG	Stem rust resistance
T. baeoticum	2x	AA	Disease resistance and spikelet number
T. comosum	2x	MM	Leaf and yellow rust resistance
T.dicoccoides	4x	AABB	Resistance to leaf rust, powdery mildew, Septoria sp., drought tolerant and higher yield potential
T. kotschyi	4x	UUSS	Drought, heat and salt tolerant
T. longissimum	2x	SS	Resistance to leaf and stem rusts, powdery mildew, heat, drought
T. monococcum	2X	AA	Stem rust resistance
T. spelta	6x	AABBDD	Resistance to leaf rust and powdery mildew, frost tolerance
T. sphaerococcum	6x	AABBDD	Drought tolerant

Figure 7.52a. Wild relatives and their desirable traits

Wild relatives and their desirable traits

Species	*Ploidy*	*Genome*	*Characteristics/ traits*
T. turgidum ssp turgidum	4x	AABB	Grain number, and weight, earliness, resistance to leaf and yellow rust
T. turgidum ssp carthalicum	4x	AABB	High tillering and grain weight
T. turgidum ssp polonicum	4x	AABB	Drought tolerance, grain weight
T. turgidum ssp dicoccoides	4x	AABB	High biomass and grain weight
T. timopheevii ssp timopheevii	4x	AAGG	Source of cytoplasmic male sterility, resistance to leaf and stem resistance and powdery mildew
T. urartu	2x	AA	Spike and spikelet size, abiotic stress tolerance
Thinopyrum bessarabicum			Salt tolerance

Figure 7.52b. Wild relatives and their desirable traits

Genes transferred from species belonging to secondary & tertiary gene pool

Trait	*Genes*	*Source*
Stem rust resistance	Sr32, Sr39, Sr34,Sr31, Sr26,Sr25, Sr37, Sr38, Sr40, Sr27,	*Ae. speltoides*, *Ae. comosa*, *S. cereale*, *A. elongatum*, *Ae. ventricosa*, *T. timopheevi*
Leaf rust resistance	Lr28, Lr35, Lr36, Lr9, Lr29, Lr24, Lr18, Lr37, Lr25,	*Ae. speltoides*, *Ae. umbellulata*, *S. cereale*, *A. elongatum*, *A. intermedium*, *Ae. ventricosa*, *T. timopheevi*

Figure 7.53a. Genes transferred from species belonging to secondary & tertiary gene pool

Genes transferred from species belonging to secondary & tertiary gene pool

Trait	*Genes*	*Source*
Powdery mildew resistance	Pm12, Pm13, Pm8, Pm7	*Ae. speltoides*
		Ae. longissima
		S. cereale
Wheat streak mosaic		*A. elongatum*
BYDV		*A. intermedium*
Yellow rust	Yr8,Yr9,Yr17	*Ae. comosa*
		S. cereale
		Ae. ventricosa

Figure 7.53b. Genes transferred from species belonging to secondary & tertiary gene pool

Promising genetic stocks derived from wild crosses

Genetic stock	*Wild species involved*	*Characteristics*
WR740	*T. sphaerococcum*	Drought, heat tolerance, high tillering, round grains
ISD 215	*T. turgidum*, *T. carthlicum*	High protein content, stem rust resistance
WR251	*T. turgidum*	Early maturity, Heat tolerance, stem rust resistance
WR196	*T. turgidum*, *T. carthlicum*	Early maturity, high 1000 grain weight
WR95	*T. turgidum*, *T. carthlicum*	High grain no./spike/unit area
WR180	*T. turgidum*	High 1000 kernel weight

Figure 7.54. Promising genetic stocks derived from wild crosses

Genes transferred from species belonging to primary gene pool

Trait	*Genes*	*Source*
Stem rust resistance	Sr21, Sr22,Sr2, Sr9e,	*T. monococcum*, *T. dicoccum*
	Sr13, Sr14, Sr, 17, Sr33	*Ae. tauschii*
Leaf rust resistance	Lr21, Lr22a, Lr39, Lr40,	*Ae. tauschii*
	Lr41, Lr41, Lr42, Lr43, Lr44	*T. spelta*
Powdery mildew resistance	Pm4a, Pm5, Pm4b, Pm11	*T. dicoccum*
		T. carthlicum, *T. spelta*
Stripe rust resistance	Yr15, Yr5, Yr28	*T. dicoccoides*, *T. spelta*
		Ae. tauschii
Curl mite colonization	Cmc1	*Ae. tauschii*

Figure 7.55. Genes transferred from species belonging to primary gene pool

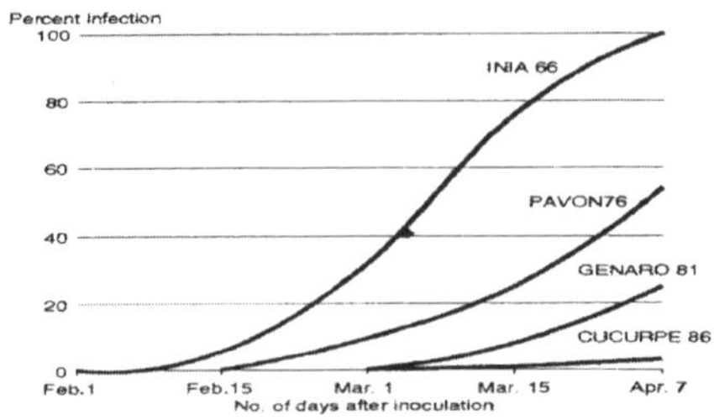

Figure 7.56. Slow rusting resistance to leaf rust (Cd. Obregon)

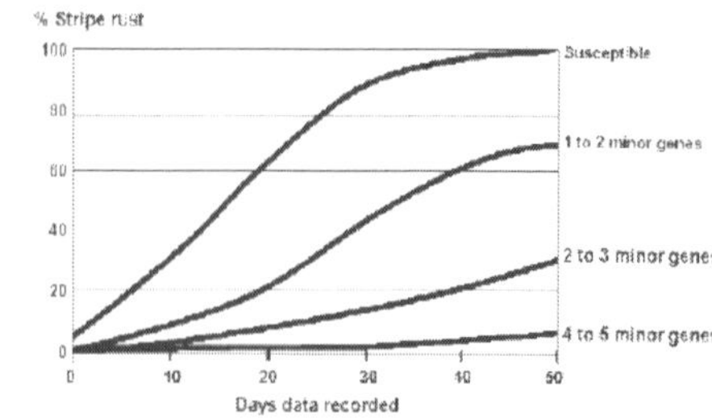

Figure 7.57. Relationship between the progress of stripe/leaf rust and the number of minor genes

Figure 7.58. Wheat blast training, BWMRI, Bangladesh

Figure 7.59. Hand-on experience at the wheat blast training, BWMRI, Bangladesh

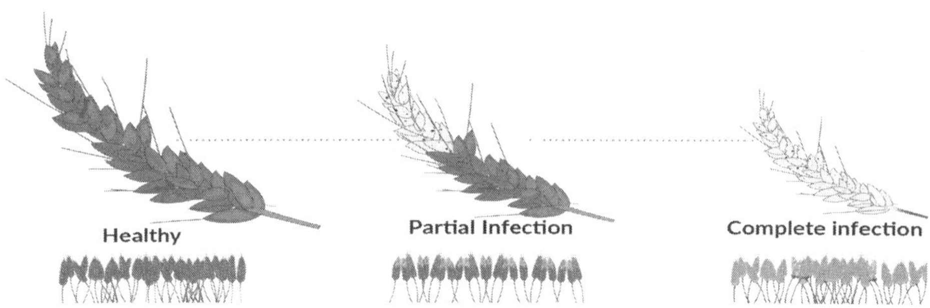

Figure 7.60. Wheat blast at different developmental stages

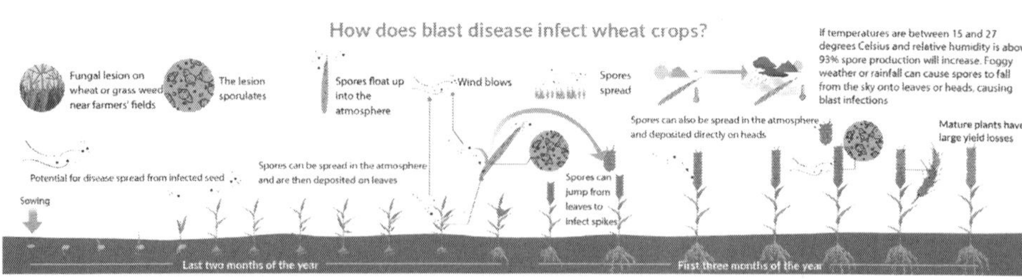

Figure 7.61. How does blast disease infect wheat crops?

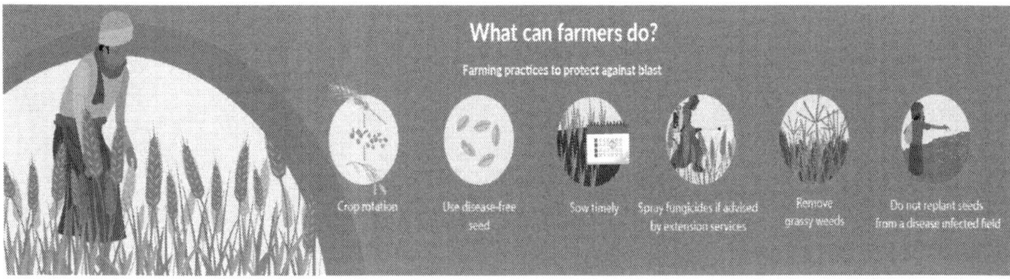

Figure 7.62. What can farmers do?

Figure 7.63. How can farmers prevent and manage wheat blast?

Figure 7.64. Wheat spikes damaged by blast

Figure 7.65. The grain in this blast-blighted wheat head has been turned to chaff

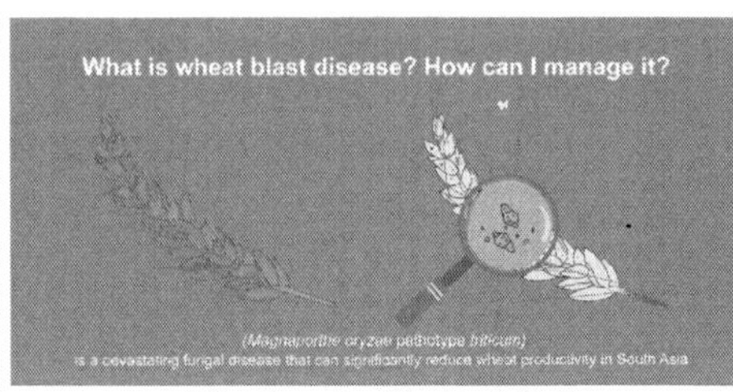

What does wheat blast look like?

Figure 7.66. Management of blast disease

Figure 7.67. Ear cockle disease of wheat is caused by a nematode, *Anguina triticina,* the first plant parasitic nematode recorded. It Is also known as seed gall, purples, pepper corns and hard smut

varieties *viz.*, CBW38, HD2967, HD2985, K307 and Raj4120 are being cultivated. Only a few races of the existing leaf rust, stem rust race flora are present in this Zone. Leaf rust pathotypes, 104-2, 104-3 and 77-5 are prominent, whereas stem rust is sporadically identified from this zone. This zone is also not affected by yellow rust but recently 46S119 race of stripe rust has been identified with insignificant impact on wheat crop. The genes postulated in the varieties of this zone are: *Lr10, Lr13, Lr23* and *Lr26*, which are providing effective resistance to leaf rust, whereas for stem rust the genes *Sr2, Sr7a, Sr11, Sr5* and *Sr31* have been identified after the introduction of Veery derivatives in this zone during the year 1983 (HUW206), at present many varieties carry 1BL.1RS translocation. Prior to introduction of *Secale cereale*-derived gene *Lr26*, the gene *Lr9* was deployed in 1992 through the release of a variety Sonali (HP1633), provides effective resistance for a brief period till a virulence 77-7 (biotype of race 77) of leaf rust was identified. Wheat is attacked by a number of diseases. Major diseases of wheat crop are the rusts, smuts, bunts, powdery mildew, leaf spots and ear cockle. For a person associated with the wheat seed and grain production programme it is must to know the symptoms of these diseases so that they can identify these diseases in the field or lab. They must also know the ways to manage these diseases in the seed crop. Symptoms, certification standards and control methods of these important diseases of wheat are summarized.

Rust Epidemics

Rust diseases are a potential threat to destabilize wheat production in the countries growing wheat. India has witnessed several rust epidemics in past resulting into heavy yield losses in 1827 and 1839 and the damage was immense that the harvest may even fell short of the quantity of seed sown (Barclay 1892). The first stem rust epidemic record goes back to 1786 A.D. in central India. It was usually the black rust that does the greatest damage in the Indian plains and huge losses were estimated (Vasudeva and Prasada, 1948). About 20 per cent production was destroyed in central India during 1946-47 epidemic. Mehta (1950) estimated the loss about 200 million rupees due to rusts damage for wheat grain every year. States like Bihar and eastern Uttar Pradesh recorded severe damage. Widespread occurrence of leaf rust was observed affecting a popular cultivar Kalyansona in Punjab, Haryana, Delhi and western Uttar Pradesh. Higher hills of North India, though harbor stem rust infection, but due to very low temperature, have not been suggested to play any significant role in epidemics of north India. However, stray examples of outbreak of local epidemic of stem rust experienced at Sanchore in Rajasthan can not be explained without considering the inoculums dissemination from northern hills. To avoid such epidemics, the breeders have recommended short duration wheat varieties which fit well in wheat-paddy crop rotation in the Indo-Gangetic plains. The spread of stem rust is more common from hills of Karnataka than it appears. In general, the spread of stem rust inoculum flowing continuously to neighbouring plains from Nilgiri hills and travels as far as the plains in central India. However, stem rust is able to survive on susceptible local durums occupying a very small area in the hills of Karnataka and the infection is almost endemic in this area. Therefore, we have to keep in view that the occurrence of stem rust in this particular area needs consideration for judicious gene deployment in future. Nayar *et al.* (1997) reported that both leaf rust and stripe rust occurred each year from 1967 to 1974 but the losses were estimated only twice. A severe leaf rust epidemic was noticed in an approximately 4 million hectares (Nagarajan, 2008) in north western plain zone affecting widely grown varieties HD2285 and HD2329. Sporadic high incidences of stripe rust are recorded in some parts of Punjab. Recently stripe rust appeared in significant proportion in north-western areas (Gangwar *et al.*, 2013). Stem rust and leaf rust cause heavy yield losses in the countries which have warm congenial temperatures (Kolmer, 2002). However, after the introduction of dwarf wheats, the stem rust is under control because most of the cultivars were resistant at the time of their release. Leaf rust is also widely distributed and occurs regularly in other countries of the world; severe leaf rust can cause losses up to 70 per cent (Huerta-Espino *et al.*, 2011, Singh and Rajaram 2002). Singh *et al.* (2004) reported that both leaf and stripe rusts could affect yield approximately by 63 and 43 per cent respectively, if the susceptible cultivars are grown. The knowledge of resistance genes before their transfer into high yielding genetic background has proved beneficial in controlling rust epidemics. Evaluation of advanced lines in seedling stage and multi-location testing against widely prevailing rust pathotypes, monitoring of rust variation and detection of new rust variants has supported the release of varieties in India and thus deployment of such genotypes has avoided rust epidemics to a large extent. Protecting from rust diseases by cultivation of resistant cultivars, India is now comfortably placed with high production of more than 90 mt. Regarding producing adequate wheat by 2030 some distress about the rust epidemics has been expressed in the world (Singh *et al.*, 2011).

Evolution of New Pathotypes of Different Rust

The rust perpetuates throughout the year in the southern and northern hills, neighbouring country Nepal, from where they disseminate in peninsular and central India as well as Indo-Gangetic plains (Nagarajan *et al.*, 1997). A longer and continuous association of host and pathogen becomes available for the proliferation of the pathogen. Further, self-sown wheat plants are also available in the area for this obligate parasite and hence rust perpetuate throughout the year due to favourable ecological conditions needed for their adaptation. The population of rust pathogen thus mostly remains in a dynamic state. Shift in race flora may be due to either changes in the frequency of existing components in the population or due to evolution of new races caused by the introduction of new gene(s) into the host. Disease monitoring system was started after the introduction of Mexican wheats in the mid 1960s. Therefore, systematic surveys for disease prevalence, severity and related information are being conducted periodically since then.

Leaf Rust

The resistance of Democrat and Mediterranean (both carrying *Lr3*) was overcome by a new virulent pathotype 77, when their resistance was exploited in resistance breeding programme. After the introduction of dwarf wheats from Mexico, the new virulences were picked up for the genes *Lr1, Lr10* and *Lr23*. Matching virulence(s) in the races 12, 77 and 104 for *Lr23* were evolved when the resistance of Gabo was utilized. Later, when *Lr26* was utilized, pathotypes with matching virulence appeared. Further, both *Lr23* and *Lr26* had combined virulence overcoming the resistance of these genes. Changes in the population of *Puccinia recodita* (now *P. triticina*) have also been recorded. The elegant example being the evolution of new races, 12-1 and 77-1, which overcame the resistance of *Secale cereale*-derived gene *Lr26*. Prior to this, race 77 was identified on Sonalika, 77A-1 on WL711 and HD2009 (Arjun), 77-2 and 77-4 on HD2285, 77-5 and 104-2 on UP2338 (*Lr26*). Durability of resistance seemed fleeting because the extensive deployment of gene combination *Lr23+Lr26*. More than 40 per cent of the leaf rust isolates identified possessed the combined virulence for these two genes. The evolution of a virulence 77-7 for the genes *Lr9* and 77-10 for *Lr28* were also virulent on *Lr23* and *Lr26*, which were identified in 1974 and 1983, respectively. Why these virulence(s) selected the genes already present in the cultivated varieties being grown for several years is to be understood. The genes were not introduced in the place of the origin of these races, where only a few plants of the stocks carrying both these genes were grown. This indicated that a new resistant host to create selection pressure in the pathogen is needed for the evolution of a new virulence. The mutation is spontaneous and a random event, which continues to happen without the deployment of a particular resistance gene. It is likely that it gets selected on a differential host present at the place congenial for pathogen proliferation. A typical example of *Yr8* in United Kingdom can be cited as a new race virulent on *Yr8* was discovered before it was introgressed into common wheat from *Aegilops comosa*. To a great extent the pathotypic variation is coming from across the border as the evolution of new races has not yet been demonstrated in India.

Stem Rust

In case of stem rust, new pathotypes evolved and the genes Sr2, Sr11 and Sr9b were rendered ineffective. When the cultivars Safed Lerma (S307), Chhoti Lerma (S331) both carrying *Sr11*, HD2135, HW517 and HW741 were grown in cyclic shifts in the Nilgiri hills, within 3-4 years of cultivation, Safed Lerma became susceptible to races 42B and 122 of stem rust. By 1974, these cultivars lost their resistance to race 40A of stem rust. Subsequently, varieties HD2135 carrying *Sr2, Sr8a* and *Sr30*, HW517 (1978) and HW741 (1982) having *Sr2, Sr8b*, and *Sr11* became susceptible to race 40A. A virulent pathotype, 40-1 was isolated from the Nilgiris in 1990, which overcame the gene *Sr24* before the cultivars, Vaishali (DL784-3) in 1993 (NEPZ) and Vidisha (DL788-2) in 1996 (CZ) were commercially deployed. A new race in group 117 was detected showing virulence for *Sr37*. Virulence 58G13-3 collected from Karnataka state in 2006 overcame the resistance of Sr25 (Jain *et al.*, 2009).

Stripe Rust

When Kalyansona released in 1967 showed resistance to all the prevailing pathotypes of stripe rust, but soon in 1970 it became susceptible to three pathotypes *viz.*, 14A, 20A and 38A. Presumably all these pathotypes had migrated from across the border. The wide spread cultivation of Sonalika (*Yr*2Sk) and Kalyansona (*Yr*2Ks) in India exposed the resistance against stripe rust pathogen I (Sharma *et al.*, 1973). Subsequently, races K and N evolved and thus the new races were either evolved or migrated. Later the gene present in Hybrid 46 was also overcome by a new virulence

(Kumar *et al.*1994) and a virulence for Yr9 was picked up from Gurdaspur, Punjab. The testing of materials against physiological races of all the rusts in seedling stage, gene identification and incorporation of rust resistance genes led to the development of resistant cultivars.

Gene Postulation

The utility of leaf rust (*Lr*) resistance genes was first evaluated by Sawhney *et al.* (1977) and subsequently various identification procedures were suggested. Leaf rust races in India were identified based on the response of 13 hosts with different *Lr* genes. However, during the early phase of resistance breeding, before the introduction of Mexican dwarfs, stem rust pathotypes virulent on Vernal, Khapli and Kota were detected. These isolates were virulent on *Sr9d* and *Sr9g*. Another allele *9e* of the gene *Sr9* showed differential behaviour. Stem rust resistant Australian varieties Gabo and Ridley were identified to carry *Sr11* and the same resistance was introgressed into Indian genotypes. Timstein was also used in Indian breeding programme. A pathotype 7G35 was detected on Khapli in 1947, which later became virulent on *Sr11*. Following the introgression, variation within the pathotype(s) was observed and several virulent pathotypes evolved. The gene *Sr30* present in Mentana and Festiguay was overcome by new races. Mentana also carries *Sr8a*. The evolution of new races continued overcoming the resistances of *Sr5, Sr8b, Sr11* and *Sr30*.

Table 7.1. Percent virulence frequencies

Pathotype	*Durum Wheat*	*Bread Wheat*
77	2	24
77-1	6	48
77-2	4	52
77-3	9	44
77-4	5	38
77-5	7	78
77-6	12	92
77-7	11	57
77-8	5	32
77A	2	19
77A-1	7	31

The release of the cultivars Sonalika and Kalyansona (both carrying *Lrl3)* and subsequently Arjun led to a decline of the area sown to Sonalika and Kalyansona by 1976. Arjun, released in 1970, was identified to carry *Lr13* (Sawhney, 1987) provided a high degree of adult plant resistance. Saini *et al.* (1991) identified additional factor for resistance in Arjun, however, the identity of the additional gene(s) could not be established. A variety Kundan, released in 1985, maintained its level of resistance to leaf rust for a very long period even though several new pathotypes have evolved since then. This variety has been postulated to carry *Lr34* and *Lr23* (Sawhney *et al.*,

1992). Evidently, both these genes provide durable resistance. Sawhney and Sharma (1990) reported that *Lr34* mainly exhibited an infection type 3 to many pathotypes of leaf rust in the seedling, but showed moderate susceptibility (30-60MS) at adult stage (Tomar and Menon, 2001). The adult plant resistance gene *Lr34* has been found to be associated with a stripe rust resistance gene *Yr18* (McIntosh, 1992b). No systematic work to incorporate race specific resistance to stripe rust has been started in India. An attempt was made to introduce durable resistance from European winter/ semi winter wheats into two Indian genotypes, Kalyansona and WL711. Several derivatives were identified to possess stripe rust resistance other than that controlled by specific genes. Historically, the introduction of dwarf varieties *viz.*, Kalyansona and Sonalika carrying *Yr2+A* genes provided resistance to the then prevalent races of stripe rust. The genes, *Yr2* and *Yr6* have been postulated in a number of Indian varieties. However, these genes including *Yr7* have been reported ineffective in north as well in south India. The gene *Yr7* floating in Indian cultivars prior to 1990 was reported to be ineffective against many stripe rust pathotypes previously. *Triticum spelta album* derived gene *Yr5* has been effective in India, however, an isolate of race 13 identical has been found virulent on *Yr5*. The *Yr5* has not been incorporated in Indian germplasm. *Triticum spelta album* and Moro (Prabhu *et al.*, 1991) both carrying *Yr5* and Heines VII carry additional resistance (*Yr25*) to stripe rust races (Singh and Johnson, 1988). The gene *Yr18* associated with leaf rust resistance gene *Lr34* has been identified in a number of genotypes. This gene combination probably has been introduced in Indian wheat varieties during 1960s.

Table 7.2. Leaf rust infection in Trap Nurseries

Genotype	*Severity*
C 306	5S
DL 803-3	20S
Lal Bahadur	80S
Morocco	80S
Pissi Local	40S
Tc+Lr19	tS
Tc+Lr24	tMS
WH147	10S

Table 7.3. Developing near-isogenic *Lr* lines in NP 4 background and genetics of rust resistance in wheat

Final: Lr1,2a,2c,3a,9,10	Durum wheat: Seedling studies on resistance to stem and leaf rusts in GW1114, HD 4672, MPO615 Raj 6562 and RS 749
BC_6F_2: Lr18,19,20 BC_6F_3: Lr15,17	Bread wheat: Seedling studies on resistance to stem and leaf rusts in HD2705, HI1459, NIAW907. PBW 442 and Raj 3974
BC_6F_1: Lr13,23,26 BC_3: Lr24 BC_2: Lr14a	Bread wheat: Seedling studies on resistance to stem and leaf rusts in GW322, GW366, HI1418, HI1479 and HI1539

The introduction of dwarf wheat's, though indicated further variation but there was no major change in the pattern of rust races. Only a few of these changes could be related to hosts. Variety HD2329 released in 1982 became most popular variety in the NWPZ occupying more than 5 million ha of area once upon a time. The variety was postulated to carry *Lr13, Lr10* and *Lr34* and some additional genes for leaf rust resistance, *Sr8b, Sr9b* and *Sr11* for stem rust and *Yr2* and *Yr18* providing resistance against stripe rust. The durability to leaf rust may be ascribed to the presence of *Lr13* and *Lr34* together. Several workers have reported that *Lr34* interacted favourably with *Lr13*. Another variety HD2285 released in subsequent year 1983 was also equally popular under late sown conditions, which carried a combination of *Lr23, Sr9b, Sr11* and *Yr2* genes. The popularity of these two genotypes indicated that the genes they possess were protective against rust epidemic. A highly adaptive to the conditions prevailing in peninsular zone, genotype HD2189 carrying the combination of *Lr13, Lr34, Sr2, Sr8a, Sr11, Yr2* and *Yr18* is still in cultivation occupying a large area since its release in 1979 (Nayar *et al.*, 2001). Prior to the release of varieties carrying 1BL.1RS translocation, a few most popular Indian wheat varieties, namely, C306 (*Lr34, Yr2KS, Yr18*), NI5439 (*Lr34, Sr11, Yr18*), Kalayansona (*Lr13, Lr34, Sr11, Yr2KS, Yr18*), WL711, HUW234 (*Lr14a, Sr9b, Sr11 Yr2Ks*), Sonalika (*Lr13, Sr2, Sr11, Yr2Sk*), HD2285 (*Lr23, Sr5, Sr9b, Sr11* and *Yr2*), HD2329 (*Lr13, Lr10, Lr34, Sr8b, Sr9b, Sr1, Yr2* and *Yr18*), Lok 1 (*Lr13, Sr2, Sr9b, Sr11, Yr2Ks*) and WH147 (*Lr13, Lr34, Sr7, Sr11, Yr2Ks, Yr18*) occupied major area across the wheat zones in the country. These cultivars provided durable resistance against the prevailing rust races, which is ascribed to the presence of diverse resistance genes carried by them. These genotypes might have carried additional resistance genes in their genetic analysis. Two major and relevant changes in stem rust have been the detection of *Sr24* and *Sr25* virulences. A pathotype 40-1 virulent on *Sr24* was detected in 1990, while *Sr25* became ineffective to a pathotype, 58G13-3 collected from Karnataka state in 2006 (Jain *et al.*, 2009). Similarly, the changes in pattern of leaf rust races has also appeared as, virulence(s) against *Lr9, Lr19* and *Lr28* have been detected, but these pathotypes occurred in very low frequency and confined to limited area. The postulation of genes present in the varieties prior to their release was initiated in the beginning of 1980s. The postulation of genes in Indian wheat was started with seedling tests carried out with a range of Australian pathotypes of stem and leaf rusts, which led to the postulation of several genes for stem rust *viz.*, *Sr2, Sr5, Sr6, Sr9b, Sr11* and *Sr12* and leaf rust, namely, *Lr1, Lr3, Lr10, Lr13, Lr14a, Lr23* and *Lr26*. From 1980 onward the seedling tests are regularly conducted at Wheat Research laboratory, Flowerdale, Shimla to postulate genes in pre-released Indian wheats. The work on identification of probable rust resistance genes in a set of 53 Indian wheat genotypes started by selecting and inoculating appropriate pathotypes differing in their avirulence/virulence genes and by comparison of the host-pathogen interaction (HPI) matrix. The pedigree analysis indicated further evidence for the presence of the postulated genes. Genes for leaf rust resistance, namely, *Lr1, Lr3, Lr10, Lr14a* or *Lr22, Lr15* and *Lr26* were detected either independently or in combination. The presence of stripe rust resistance genes, *Yr2, Yr7,* and *Yr9*; stem rust, *Sr5, Sr7a, Sr8, Sr11* and *Sr31* in various combinations were also postulated. In a genetic analysis conducted on a large number of Indian wheats, near-isogenic lines each with *Lr14b, Lr14ab, Lr30* and certain international cultivars. Sawhney and Joshi, (1996) identified diverse sources of adult plant resistance (APR) to leaf rust. Prolonged leaf rust resistance in some of the Indian cultivars was attributed to the likely presence of *Lr34* either alone or in combination with other APR components. Widely grown cultivar Arjun was postulated to carry a diverse gene (*Lr13*) for conferring adult plant resistance. Gene postulation in the advanced materials and the released varieties is a regular practice to help the breeders. In some of the cultivars carrying 1BL.1RS translocation, gene *Sr5* and *Sr2* have also been postulated along with *Sr31*. Since the gene *Sr31* is highly effective in seedling stage against all the prevailing races of stem rust it obscures the seedling resistance exhibited by *Sr2* and *Sr5*. The *Sr2* could be postulated in any genotype because of its linkage with black chaff but it is difficult to postulate *Sr5*. The postulation of this gene could authentically be done through either through test of allelism or F_2 analysis in a cross involving *Sr31* and *Sr5*. The molecular analysis is the best option for correct identification of genes.

The leaf rust resistance genes postulated in Indian wheat cultivars from are Lr1, Lr3, Lr10, Lr13, Lr14a, Lr23, *Lr24*, Lr34 with very low frequency of *Lr9* and *Lr28*. The most frequently identified genes floating among the cultivars are Lr13 in bread wheat and Lr23 in durum genotypes. Lr13 in combination provides durable resistance to leaf rust at adult plant stage. However, recently *Lr26* and Lr34 have been identified as most frequent and valuable. The significance of Lr34 in Indian wheat breeding programme has been elaborated earlier. Many cultivars carry two and three gene combinations and a very few possess four genes. The information generated over a period of time reveals that the genes listed here have the potential to provide durable resistance in combination. Sawhney *et al.* (1989) suggested that the genes Lr10 and Lr23, when added to the Chinese Spring background provide a good source for durable resistance. The suitable combinations *viz.*, *Lr26*+Lr1, *Lr26*+Lr23, *Lr26*+Lr34, *Lr26*+Lr34+Lr1 have been the major protective resistance to Indian cultivars. Lr34 is an important gene providing resistance to multi-pathogen. Lillemo *et al.* (2008) reported that that Lr34 has been shown to provide resistance to powdery mildew and spot blotch (Lillemo *et al.*, 2013) caused by *Bipolaris sorokiniana*. Earlier Singh *et al.* (1992) identified that Lr34 is associated with leaf tip necrosis (Ltn1), morphological marker. The cultivars carrying *Lr26* occupy a major area under cultivation in India. A single variety PBW343 (*Lr26*) occupied about 7 million hectare area in NWPZ and NEPZ. Before it came into cultivation cultivar HD2329 had occupied about 4-5 million ha of area. That means that the resistance exhibited by *Lr26* and Sr31 is one of the most commonly deployed in modern plant breeding. Among more than 50 designated genes for stem rust resistance, the commonly identified genes are, Sr2, Sr11, Sr31 and *Sr24* followed by Sr2, Sr7b, Sr9e, Sr11 and Sr12 among durum cultivars (Bhardwaj, 2011) while of stripe rust only a few such as Yr2, Yr9 and Yr18 have been the most frequent. Recently some of the genotypes carrying Yr27 (*e.g.*, HS490) have also been identified.

Identification, Distribution and Prevalence of Rust Races

Physiologic specialization in rusts of wheat has been investigated extensively by many workers and the presence of different races varying in pathogenicity has also been studied. A set of differential varieties for identification of races was proposed for each rusts of wheat. However, there was no consideration of the involvement of specific genes in the parasite, although genes for resistance to leaf rust were recognized. The specific gene interaction was better understood and the gene-for-gene hypothesis proposed by Flor (1956) for flax and flax rust was accepted, which was later elaborated by Person (1959) into a general concept of host-parasite interactions. Subsequently a number of procedures of screening resistance against rusts were suggested based on near isogenic lines. Now a considerable progress in the genetic aspects of host-parasite interactions has been made. Gene postulations are done based on the theory of gene-for-gene hypothesis (Flor, 1956), however, the interactions between host gene and corresponding gene in parasite influences the correct postulation. Hence from the genetics point of view the pattern of pathogenic reaction may sometimes lead to biased results. Since the identification of two stem rust races, 15 and 27 by MN Levine from wheat samples collected from Pusa, Bihar in 1923 (Mehta, 1940) and subsequently during the period 1931 to 1935 races 21, 40, 42 of stem rust and 10, 63, 106, 107 and 108 of leaf rust were identified, a large number of races of different rusts have been reported in India. Out of 126 races reported so far, 32 of stem rust, 49 of leaf rust and 28 of stripe rust are being maintained at DWR, RS Flowerdale (Gangwar *et al.*, 2013).

The distribution of rusts races is given zone wise in preceding discussion. The frequent arrival of the new pathotype renders the cultivars susceptible and each wheat zone in India has different pattern of pathotypes showing a definite distribution. The distribution of pathotypes and their frequency has been worked out periodically. However, the races of all the rusts commonly distributed across the country during the last 15 years are: 12-4, 12-5, 12-7, 77-1, 77-2, 77-5, 77-9, 104A, 104-2 and 104-3, 162A, 162-1 and 162-2 of *Puccinia triticina* (earlier known as *Puccinia recondita*) whereas stripe rust races, namely, I, L,T, U, 47S102, 47S103, 46S102, 78S84, 46S119 and 67S64 are commonly occurring across the zones prone to *Puccinia striiformis*. The stem rust is a problem of warmer areas and for last 30 years the races 21-1, 40A, 40-1, 40-2, 117-A, 117-1, 117-2, 117-3 117-4 117-5 and 117-6 have been identified. At different times, distribution pattern of races has been different in the zones with some races occurred with >60 of frequency. Varying degree of race prevalence has been reported in different years (Singh *et al.*, 1979; Bhardwaj *et al.*, 1994, 1996; Kumar, 2013. Among them races 12, 77, 104B, 77A-1, 77-5 of leaf rust, 40A, 40-1 and 117-6 of stem rust and 47S102, 46S102 and Yr9 virulences, 78S84, 46S119 and 67S64. Generally, when a new pathotype is picked up, it takes few years for its population do become predominant. Nayar *et al.* (1987) reported that pathotypes, 77A and 77A-1 identified in 1974 and 1976, respectively became predominant during late 1980s and about 38 per cent of samples in 1990 contained the above races. Pathotypes 77-5 and 104-2 of leaf rust identified in 1991 still have very high frequency. In a given time there are not more than 2-3 races of any wheat rust prevailing across the zones. Sinha *et al.* (1979) reported 96 per cent of samples carried 40A and 40-1 pathotypes of stem rust in southern hills zone. The moderate to high incidence of particular rust is sporadic and localized. Also most of the time the rusts appear at the terminal stage of crop growth and hence does not threat the crop of any consequence. Thus, there is no real threat of rusts in India. The low incidence or no incidence of rust is mainly ascribed to existing diversity of resistance genes present in the cultivars grown in a mosaic pattern in each wheat zone.

Resistance/Susceptibility

Resistance in a genotype can be described as the inherent capacity of a plant to restrict the establishment and subsequent activities of potential pathogen, while the susceptibility is its opposite condition where a plant shows its inability to resist the effect of a pathogen. There are various type of resistances described in literature. Van der Plank (1968) put forwarded vertical and horizontal resistance. Hypersensitive type of resistance is probably the most common race-specific resistance that interacts differentially with pathogen races. A large number of resistance genes (>50 of stem rust,> 65 of leaf rust, >40 of stripe rust and powdery mildew) have been designated (McIntosh, *et al.*, 2012). Most of them are classified as race specific resistance genes associated with hypersensitive reaction in the host. This type of resistance has been commonly utilized in breeding programmes. It is observed that this type of resistance is short lived due to evolution of new races (McDonald and Linde, 2002) Robinson (1976) observed lesions on leaves after the pathogen infection but no colonization of pustules on host could be recorded. Flor (1955) described that race-specific resistance is conditioned by the interaction of specific genes in the host with those of the pathogens. Stakman *et al.* (1962) described immunity as the absence of visible lesions on the host plant is considered to be immune response. It is generally agreed that plants immune to diseases are immune to infection by the pathogen. Gene *Lr28* showed immunity in India over a very long period, even flecks were not observed on the host leaves. Many of the genes for hypersensitive reaction can produce an immune reaction to some rust races and visible fleck infection types to other cultures. Some of the genes exhibit moderate or intermediate type of resistance, when the seedling produces 1 to 3 type of reaction. Such genes minimize the intensity and sporulation percent at adult plant stage. A *Triticum timopheevi* derived temperature sensitive gene *Lr18* produce intermediate type of infection. Usually genes that give moderate resistance in the seedling stage give the same type of resistance in the adult stage. Sometimes two kind of resistance may be conditioned by the same gene. Many researchers have observed that a host gene may give hypersensitive resistance to some isolates and a rate reducing resistance to others. Gireesh *et al.*(2014) reported that a durum variety A-9-30-1 exhibited ';1' infection type against 12-A, 12-2, 12-4, 12-3, 106, 108, 162-2 and 162 pathotypes of leaf rust, while it produced IT '3' against 12-5, 104B, 104A and 77-5 and similarly a common wheat variety NI5439 gave infection type '3' to stem rust races, 11, 40A, 295 and 117-6, whereas it produced ';' reaction against 11A, 15-1 and 21A-1 at seedling stage. It is field resistance with which the farmer is directly concerned. Resistance expressed at adult plant stage termed as horizontal (Robinson 1976) is effective against a wide spectrum of rust races. Parlevliet (1985) described it as partial resistance or slow rusting (slow development of rust) as it is associated with race non-specificity and is often referred as adult plant resistance (APR). Slowing effect on disease development may be attributed to resistance controlled by minor genes. A combination of several genes may confer durability of resistance on the premise that when components such as increased latent period, smaller pustule size, production of reduced number of uredia per unit area, humidity *etc.* may contribute towards slow development of rust. Chopra and Kulkarni (1980) observed slow rusting among many Indian wheat genotypes. While studying stripe rust of wheat, Johnson (1981) introduced a term durable resistance, which refers to resistance that has remained effective in cultivars for a very long period. In other terms, this resistance remains effective during the prolonged and widespread use in an environment favourable to disease. Bahadur (1997) studied 37 Indian wheat genotypes for adult plant resistance and reported that HP1102, HP1209, HW741, HW971 carrying Lr13 in combination of Lr34 or Lr23 produced moderate infection type against 77-1, 77-2 and 104 races of leaf rust. HD2189 showing adult plant resistance for more than 30 years in peninsular India carries race-specific genes, Lr1, Lr13, Lr17 and Lr34. It has also been observed that genes governing resistance can also be suppressed by non-allelic genes *e.g.*, the resistance of Lr23 is inhibited by a gene present in Thatcher and powdery mildew resistance gene Pm8 is suppressed by Su-Pm8 located in 1A chromosome of wheat (Ren *et al.*, 1997).

Durability of Resistance

The rust in India has been controlled effectively using race-specific genes for the last 5-6 decades, because the utilization of hypersensitive type of resistance which is easy to incorporate into susceptible cultivars. As an effective strategy, the major genes that can be assayed in seedling plants has played a significant role in controlling rusts, although they may be responsible for the boom and bust pattern. It is also well known that virulent mutants arise and make the resistance ineffective. In such conditions these major genes can by either combined (pyramids) with another major gene(s) or with the genes which have minor effect. However, some alien genes have played a great role in stabilizing wheat production by protecting from heavy rust infection. It is Breeder's endeavour to replace the ineffective genes and therefore, new sources for resistance have to be searched from wild species. Genetic diversity and durability are the two most important features of the resistance. Proof for durability comes only after the deployment of a cultivar over a large area. It is known that a few wheat cultivars maintain their resistance to rust diseases for a long period of time. Since genetic diversity serves as an insurance against vulnerability, several genes have been placed in agronomically superior genetic backgrounds for their judicious use in pre-breeding for introducing novel genetic diversity for resistance in wheat. Leaf rust resistance genes provide resistance at seedling stage and some of them are effective at adult plant stage and called APR (adult plant resistance). Most of the time single genes become ineffective due to newly evolved virulent forms of rust within a few seasons of cultivation due to heavy selection pressure exerted by the cultivation of resistant cultivars. Therefore, to provide durability in resistance the judicious usage of genes may help in building up a strategy to control rust. The systematic deployment of cultivars in India has averted, barring a few instances, any possible outbreak of rust epiphytotics. Tomar *et al.* (1998) identified some cultivars expressing slow development of rust at adult stage. Besides many stocks showing resistance at adult plant stage, an Australian cultivar CSP44 and Indian genotype VL404 exhibited 40MSS and 30S low reactions to leaf rust in adult stage respectively. Later on Saini *et al.* (2002) identified that CSP44 and VL404 carry

APR genes *Lr48* and *Lr49*, respectively. Bansal *et al.* (2008) confirmed the presence of durable leaf rust resistance gene *Lr34* in CSP44 and VL404 using tightly linked marker csLV34. A recessive gene *Lr48* has displayed adult plant resistance against the leaf rust pathotypes of sub-continent. Looking at its utility in wheat breeding, Samsounpour *et al.* (2009) pyramided *Lr48* with *Lr24* in the background of popular cultivar PBW343. Saritha *et al.* (2005) studied wheat x *T. timopheevi* derivatives and some Indian wheat cultivars for leaf rust resistance at adult stage. Test of allelism demonstrated that cv. WR544 and a *T. timopheevi* derivative Sel. T3073 possessed APR genes different from that of *Lr13* and *Lr34* present in Kundan and Thatcher. Genes like Lr34, Lr46, Lr67, Lr68, and Sr2 may indeed be highly durable and useful. The slow rusting is observed in other than Lr34 genotypes such as Pavon76, which carries Lr46. This gene also functions similar to Lr34. Nayar *et al.* (2001) analysed that about 20 per cent genotypes from India possesses Lr34 validated by marker developed by Lagudah *et al.* (2006). A limited number of hypersensitive seedling resistance genes (*Lr24*, *Lr28*, Lr35), there are few other durable ones Lr34, Lr46 and Lr67) effective in India.

Gene Deployment

Breeding for disease resistance is the only eco-friendly option available to minimize loss due to rusts. The cultivation of resistant varieties in a given agro-ecological zone is followed by the changes in pathogen population enabling the pathogen to overcome the resistance gene(s). An additional strategy for using host resistance to control diseases is called 'gene deployment'. This refers to using different sets of resistance genes in the various subareas of an epidemiological subunit. This strategy interferes with the annual development of an epidemic and the formation of complex races. The deployment of multilines makes it necessary to divide an ecological area into epidemiological subunits and to allot a different gene or genes to each area. This operation requires the cooperation of breeders to provide genetic stock with designated genes and of pathologists to police the gene allotment. Many national programmes now have coordinated wheat improvement programmes and this strategy can best be applied through them. The distribution of pathotypes in different agro-ecological zones permits the deployment of cultivars with different genetic constitution. It is expected that by increasing genetic diversity in the field the losses due to rusts are likely to be curtailed. To study the pathogenic variability and its monitoring, a proper and well-defined virulence identification system to exploit host resistance genes should be followed. This necessitates the continuous search for effective genes and their efficient deployment in the area, where a variety is released for cultivation. With the pathogens displaying high degree of pathogenic variability, it is important to evaluate the breeding materials against these potential pathotypes. While detailed knowledge of the genetic basis is not essential for an effective resistance breeding (Simmonds, 1985), much can be gained if the genetics of resistance is known. However, considerable care should be taken when the knowledge gained in one geographical area is applied to another, as previously unknown genes may provide resistance in a new region. The rust diseases of wheat cause much serious threat to the crop and hence, a proper disease management strategy to keep these problems at bay, thus avoiding the epidemics.

To reduce host vulnerability to pathogenic variation, generally diverse source of resistance (variation) must be used. Browning *et al.* (1969) advocated the use of vertical resistance in different wheat zones along with the *Puccinia* path in United States. McIntosh (1985) advocated the use of overlapping oligogenic resistance or the combinations of oligogenes which exhibited low coefficient of rust infection. Gene deployment is a promising and effective strategy to curtail the rust epidemics. Like the other countries the existence of a *Puccinia* path in the Indian subcontinent was established. Therefore, strategically devised gene deployment can be followed in central and peninsular India against stem rust of wheat. Similar possibilities exist in the Indo-Gangetic Plain for managing leaf rust also.

To achieve it, some of the important points need consideration for effective gene deployment. First of all, a good source of resistance, which do not impair grain yield must be available. For all practical purposes, harmony and understanding between the breeders leading a program, farmers and governmental agencies in a zone where gene(s) are to be deployed is needed. The inoculum source of the area, exogenous or indigenous where disease control is targeted must be ascertained. Adequate/sufficient quantities of seed should be ensured to cover the target area. The strategy for gene deployment along the "*Puccinia path*" for managing stem and leaf rust of wheat has been proposed by Nagarajan *et al.* (1984) and Bahadur and Nagarajan (1985) following the effectiveness of resistance genes and the distribution of pathotypes. Bahadur *et al.* (1994) advocated the use of resistance genes, namely, *Sr24*, Sr31 and Sr36 for stem rust and *Lr9*, *Lr19*, *Lr24* and Lr34 for the management of leaf rust in different agro-ecological zones in combination of other desirable genes. Since the major gene Yr9, provided resistance for a very long period in India has become now ineffective to newly evolved virulences the deployment of genes such as Yr5, Yr10 and Yr15, which are individually as well as collectively effective against the prevailing races would be an effective strategy to check the losses due to stripe rust. There are several approaches, such as gene pyramiding, gene deployment, gene cycling and multilines which have been suggested to use the host resistance effectively. Use of resistance genes over a large area to control the unforeseen threats of rust epidemic can be described as "gene deployment". Fundamentally the deployment of rust resistance genes is to create genetic diversity of effective genes which may act as barriers to rust spread within a region prone to rust inoculum multiplication and its dissemination. India has successfully achieved the restriction of rust infection build ups by creating the mosaic of varieties carrying diverse resistance genes. Both, race specific and non-race specific (adult plant resistance) resistance genes are being used against wheat rusts in India. India has been deploying rust resistant cultivars from the very beginning. Before 1962, cultivars with tall stature, namely, NP4, NP12, NP789, NP824, C591, C218, K68, PKD4, PV18 and C306 exhibiting rust resistance at adult plant stage against prevailing rust pathotypes were cultivated. Rust resistant Mexican wheats Lerma Rojo and Sonora 64 were introduced after 1962 and the cultivars, Sonalika and Kalayansona selected from segregating materials were grown on large scale shaping the Green Revolution. The susceptible cultivars were replaced rapidly by new releases. A successful, though unintentional, deployment for stem rust resistance is the large scale cultivation of HD2189 in Peninsular India. Presently, this cultivar is resistant to Indian stem rust pathogen. Another cultivar DWR162 carrying Sr31 was released in 1993 is being cultivated in Karnataka falling in this zone is also resistant. So the inoculum which spreads from Karnataka or Nilgiris is not able to multiply as it lands on the resistant cultivars. Consequently, three popular cultivars of Central Zone, namely, Lok-1, Sujata and WH147, though susceptible but are protected against any transported inoculum. In addition, resistant cultivars, HW2004, HI1500 (both rainfed wheats sown early) and DL788-2 (timely sown) carrying Lr 24 are being cultivated on large scale in the central zone. The prevalence of pathotypes in specific region is usually influenced by host and prevailing weather conditions. This knowledge can also facilitate the process of gene deployment. The wide spread cultivation of rust resistant varieties possessing different genes and their combinations trigger the **Boom and Burst** cycle. The breakdown of resistance genes due to the evolution of new races reduced the life span of cultivars. Most of the time it was unplanned use of genes, because the breeders always wanted to release a rust resistant varieties carrying either race specific or race non specific genes. Many seedling genes deployed thus rendered ineffective to matching virulences (Nagarajan *et al.*, 1986).

The genetic uniformity of the genotypes leads to the evolution of new races. The gene Lr1, present in some of the varieties then, thus reduced in effectiveness. Only 5 pathotypes of *Puccinia triticina* were virulent on Lr1 before 1972, however, at present more than 20 pathotypes are virulent on Lr1. Similarly, about 24 pathotypes of leaf rust are virulent on Lr3 (Democrat), whereas only one pathotype was known to be virulent on this gene in 1955. Before 1974 the gene Lr13 was ineffective to race 77 but at present there are more than 10 pathotypes. A matching virulence for Lr23 was detected in 1974 and about 10 pathotypes are known to be virulent on this gene. The *Secale cereale*- derived gene *Lr26* was rendered ineffective in 1983 and since then about 12 pathotypes with different virulence combination have developed. Interestingly only the races 12, 77 and 104 of leaf rust have undergone a rapid evolution of new biotypes.

1BL.1RS Translocation

Spontaneous-centric fusion translocation between wheat and rye was identified by Zeller (1973) conferring spectacular resistance to multiple diseases such as leaf rust, stem rust, stripe rust and powdery mildew. Although in Europe the gene *Lr26* derived from *Secale cereale* L. was an important source of resistance, but virulent strains of leaf rust appeared soon after cultivars with this gene were released. The genetic stocks, namely, Benno, Kavkaz, Avrora, Clement, Skorospelka 35 carrying *Secale cereale*-derived rust and powdery mildew resistance (1BL.1RS translocation carrying linked genes *Sr31 Lr26 Yr9 Pm8*) were received in late 1970s and their utilization in breeding programme begun. These genes were observed to be very effective against the virulences of respective rust and powdery mildew pathogens. Since they were late in flowering (winter wheats) and produced red kernel, their use was restricted. Changes in the population of *Puccinia triticina* have been recorded regularly in India. The deployment of *Lr26*+ combination of genes a virulence matching it was observed at Wellington (India) in the beginning of 1980s and confirmed later in 1987. Since then, a number of pathotypes with different avirulence/virulence combinations have evolved. All these lines had shown differential reaction to leaf rust, which may be due to the presence of additional resistance genes in them. Benno gave low coefficient of infection at multi-location, the low level of resistance was perhaps due to the presence of a new pathotype in the natural population that was identified later as 77-1. Subsequently, four new pathotypes *viz.*, 12-1, 12-3, 77-3 and 107-1 virulent on *Lr26* were identified. In spite of virulences identified for *Lr26* the utility of this translocation was not lowered by its ineffectiveness. The elegant example being the evolution of the new races, 12-1 and 77-1 which overcame the reistance of resistance gene *Lr26* in the year 1983 and 1985. The wheat-rye 1BL.1RS translocation has been most extensively exploited for the development of cultivars in India and other parts of the world. The first variety HUW206 carrying these genes was released for general cultivation in 1983 in the North Eastern Plains Zone, where both stem rust and stripe rust were not a problem. A cultivar DWR162 carrying *Lr26* was released in 1993 for peninsular zone was resistant to leaf rust for many years. This is not unique but perhaps may be the first example when an alien gene deployed in 1993 on a large scale was rendered ineffective in India prior to its adaptation. No doubt that the gene *Lr26* has played a very crucial role in providing durable resistnace and protecting wheat from any epidemic threat to stable wheat production. Although the resistamce was overcome shortly after its deployment, but the gene still provided effective resistance in combination with different genes *viz., Lr23, Lr1, Lr3, Lr10* and *Lr34.* Out of 60 varieties released so far, 19 varieties carry *Lr26+Lr23* combination. This spontaneous beneficial translocation exhibited heterotic effect and threfore, increased yield, stability and wider adaptability not only in India but elswhere too. The incorporation of these genes through breeding was done on the basis of resistance during selection and not that these genes were deliberately pyramided. But these genes were deployed because these seem to be associated (particularly, 1RS) with increased yield. There after a series of varieties have been released in all the six wheat zones and are being cultivated on a large scale under different sowing conditions. A total of about 60 varieties have been released till 2013. One of the cultivar PBW343 occupied more than 6 mh area for a number of years. Although the resistance of *Lr26* was overcome early in 1987 but the resistnace available in lines originating from 1BL.1RS translocation continued to be utilized in India with the combination of other race specific genes *viz., Lr1, Lr23* and *Lr34* (Nayar *et al.*, 1994, 2001). In recent years many cultivars carrying resistance gene combinations of *Lr23+Lr34, Lr26+Lr23, Lr26+Lr23+Lr1, Lr26+Lr34+Lr1, Lr26+Lr23+Lr34+Lr1* have been regularly identified by Directorate of wheat Research, Karnal. The gene *Lr23* has been commonly used with *Lr26*. The cmmercial use of this gene is perhaps the most valuable for Indian farmers and provided economic security. According to Van der Plank (1982) mutations arise in the population of a plant pathogen on susceptible host and may be more frequent in nature than revealed experimentally. There are evdences to show that mutations are are an important source of variability and the rates of mutations vary greatly from locus to locus. Differential rates of mutations in *P. graminis tritici* to stem rust resistance genes and in *P. triticina* to leaf rust resistance genes have been observed. The variation in pathogen may be caused by other mechanisms too. However, in Indian sub-continent the sexual phase does not occur in the absence of *Berberis* and

Thalictrum, therefore the assumption that only the spontaneous mutations are the major sources for origin of new biotypes of rust pathogens.

Genetics of Pest Resistance

Numerous interaction studies involving various pathogenic diseases and insects of wheat have indicated that the gene-for-gene hypothesis was followed. This hypothesis was developed by Flor (1956) from concurrent studies of the genetics of resistance in flax, *Unum usitatissimum* L., and of inheritance of pathogenicity in its rust pathogen, *Melampsora lini* (Ehrenb.) Lev. The hypothesis stated that 'for each gene conditioning rust reaction in the host there is a corresponding gene conditioning pathogenicity in the parasite.' The host genes and relevant parasite genes are known as corresponding genes. Incompatibility (low infection type or resistance) occurs only when a host resistance allele (an allele for low reaction) and a corresponding pathogen avirulence allele (an allele for low pathogenicity) participate in the same interaction. The substitution of the resistance allele with the susceptibility allele, or of the avirulence allele with the virulence allele, results in compatibility (*i.e.*, high infection type or susceptibility). The 'genetic background' and the environmental conditions significantly affect host response, even though resistance may be based on a single gene. One example of this was clearly demonstrated by Luig and Rajaram (1972), who reported the expression of stem rust resistance genes *Sr5* and *Sr9b* with increasing temperatures, with heterozygosity, and with progressive backcrossing to a wheat genotype considered to have a highly susceptible genetic background. Resistance to pests in wheat may be expressed at various growth stages. Seedling resistance is often monogenic and is usually operative in mature plants when the same avirulent pathotype (or race) of the pathogen is used. Resistance expressed only in adult plants is not uncommon in wheat, and inheritance studies have shown that this trait may also be simply inherited and specific in nature. However, there are also some reports of non-specific resistance controlled by quantitative genes. When pathogen isolates vary in pathogenicity with respect to a particular host genotype, the host genotype is said to possess specific resistance, *i.e.*, the expression of resistance is specifically dependent upon the presence of the corresponding avirulence allele in the pathogen. Specific resistance is described in various terms, including race-specific, physiologic, seedling, hypersensitive, major gene, non-uniform, differential, monogenic, oligogenic, racial, and vertical resistance (Black, 1970). By definition, non-specific resistance is effective against all pathotypes of the pathogen and should, therefore, apply no selection pressure on the pathogen. Consequently, races of the pathogen and the resistance should remain stable. The terminology in the literature is confusing. Various terms have been used to describe the concept of non-specific resistance, *e.g.*, race-non-specific, general, field, adult-plant, partial, horizontal, durable, slow rusting, non-hypersensitive, generalized, multigenic, polygenic, minor gene, multiple allele, and quantitatively inherited. Genetics of resistance to certain major pathogenic diseases and pests are briefly reviewed. All known genes for resistance (low reaction) are designated with a capital letter, even though they may act as recessive alleles (Mcintosh, 1983). In other instances, the dominance of individual alleles may vary with the environment, the genetic background and the particular culture of the pathogen.

Resistance to Pathogenic Diseases

Stem rust. Stem rust, caused by the fungus *Puccinia graminis tritici,* is also known as black rust. A number of specific resistance genes have been identified and at certain loci multiple allelism is reported. Some resistance genes listed in table were transferred from related species to hexaploid *Triticum aestivum (Sr21, Sr22,* and *Sr35* from *T. monococcum; Sr2, Sr9e, Sr9g, Sr12, and Sr13,* from tetraploid wheats; *Sr32* and *Sr39* from *T. speltoides; Sr33* from *T. tauschii; Sr34* from *T. comosa; Sr38* from *T. ventricosa, Sr36* and *Sr37* from *T. timopheevii; Sr40* from *T. araraticum; Sr27* and *Sr31* from *Secale cereale;* and *Sr24, Sr25,* and St2.6 from *Agropyron elongatum).* These resistance genes can be easily combined in the same variety in the absence of allelism and close genetic linkage. Most current cultivars in areas prone to stem rust usually possess these specific resistance genes in various combinations. Resistance to *P. graminis tritici* controlled by multiple genes having small effects or non-specific genes has also been reported. This kind of resistance appears to occur, but the genetics is not well understood because of the complexity of analysis. Complication in analysis may also be the result of the residual effects of specific genes overcome by the pathogen. Durable resistance to stem rust has been attributed to the '*Sr2* gene complex' (Roelfs, 1988).

Table 7.4. Specific genes for resistance to *Puccinia graminis tritici, Puccinia recondita tritici*

Resistance Gene	*Chromosome Location*	*Resistance Gene*	*Chromosome Location*	*Resistance Gene*	*Chromosome Location*	*Resistance Gene*	*Chromosome Location*
Sr2	3BS	*Lr1*	5DL	*Yr1*	2A	*Pm1*	7AL
Sr5	6DS	*Lr2a*	2D- α	*Yr2*	7B	*Pm2*	5DS
Sr6	2D-α	*Lr2b*	2D- α	*Yr3a*	-	*Pm3a*	1AS
Sr 7a	4BL	*Lr2c*	2D- α	*Yr3b*	-	*Pm3b*	1AS
Sr 7b	4BL	*Lr3*	6BL	Yr3c	-	*Pm3c*	1AS
Sr 8a	6A- α	*Lr3ka*	6BL	*Yr4a*	-	*Pm4a*	2AL
Sr8b	6A- α	*Lr3bg*	6BL	*Yr4b*	-	*Pm4b*	2AL
Sr9a	2BL	*Lr9*	6BL	*Yr5*	2BL	*Pm5*	7BL
Sr9b	2BL	*Lr10*	1AS	*Yr6*	7BS	*Pm6*	2B
Sr9d	2BL	*Lr11*	2A	*Yr7*	2BL	*Pm7*	4A-β
Sr9e	2BL	*Lr12*	4A	*Yr8*	2D	*Pm8*	1BL
Sr9f	2BL	*Lr13*	2BS	*Yr9*	1B	*Pm9*	7A
Sr9g	2BL	*Lr14a*	7BL	*Yr10*	1BS	*Pm10*	1D
Sr 10	-	*Lr14b*	7BL	Yr11	-		
SR11	6BL	*Lr15*	2D- α	*Yr12*	-		
Sr 12	3BS	*Lr16*	2BS	*Yr13*	-		
Sr13	6A-β	*Lr17*	2AS	*Yr14*	-		
Sr14	1BL	*Lr18*	5BL	*Yr15*	1B		
Sr15	7AL	*Lr19*	7DL	*Yr16*	2DS		

Resis-tance Gene	Chromo-some Location	Resis-tance Gene	Chromo-some Location	Resis-tance Gene	Chromo-some Location	Resis-tance Gene	Chromo-some Location
Sr 16	2BL	*Lr20*	7AL	Yr17	2AS		
Sr17	7BL	*Lr21*	1DL	*Yr18*	7D		
Sr18	1DL	*Lr22a*	2D- α				
Sr19	2BS	Lr22b	2D- α				
Sr20	2BL	*Lr23*	2BS				
Sr21	2AL	*Lr24*	3DL				
Sr22	7AL	*Lr25*	4A-β				
Sr23	2BS	*Lr26*	1BL				
Sr24	3DL	*Lr27*	3BS				
Sr25	7DL	*Lr28*	4BL				
Sr26	6A-β	*Lr29*	70S				
Sr27	3A	*Lr30*	4BL				
Sr28	2BL	*Lr31*	4A-β				
Sr29	6DL	*Lr32*	3D				
Sr30	5DL	*Lr33*	1BL				
Sr31	1BL	*Lr34*	7D				
Sr32	2A,2B, 2D	*Lr35*	2B				
Sr33	1DL	*Lr36*	6BS				
Sr34	2A,2D	*Lr37*	2AS				
Sr35	3AL	*Lr38*	2AL				
Sr36	2BS	*Lr39*	2DS				
Sr37	4A-β	*Lr40*	1D				
Sr38	2AS	*Lr41*	1D				
Sr39	2B						
Sr40	2BS						

Disease Resistance

Cereal plants have naturally evolved genes to restrict or slow down diseases. Plant breeders have used these traits to develop high levels of resistance in cereal plants. Descriptions and genetic maps of these cultivars are large-scale with hundreds of candidates for resistance. Cloning and in-depth study at the molecular level has only taken place in recent years to identify specific genetic codes for disease resistance. Plant pathologists have witnessed changes in pathogen populations at least hundreds of times over the nearly hundred years. The most dramatic changes were those that accompanied the breakdown of major resistance genes. **Boom and Bust cycles** have been documented most thoroughly with regards to cereal rusts (Johnson, 1992) and powdery mildews of cereals (Brown *et al.*, 1993, Brown *et al.*, 1997) that exhibit a gene-for-gene interaction with their hosts. In the majority of these cases, a single resistance gene with large effect became widely distributed over a large geographical area because it was effective against a large fraction of the pathogen population (the **'boom'**). The pathogen population adapted to this new environment (the presence of a major resistance gene) by evolving a new population that could overcome this resistance gene (the **'bust'**). In these cases, the 'breakdown' of genetic resistance was not due to a mutation in the resistance gene of the host. Rather, it was due to the evolution of the local pathogen population as a result of selection for mutants, recombinants, or immigrants that were better adapted to the resistant variety. The singlegene resistance's have both advantages and disadvantages. The advantages are complete protection against the parasite in question, and compatibility with breeding for wide climatic adaptation. These characteristics are attractive to large, centralized breeding institutes since these institutes target large areas and thus broad adaptability in their breeding material. The main disadvantage of vertical resistance is its temporary nature, since it breaks down to new strains of the parasite. Other disadvantages include a loss of horizontal resistance while breeding for vertical resistance, and the fact that single genes for resistance cannot always be found. This is termed as **'Vertifolia effect'** by Vanderplank (1968). Thus, it has appeared impossible to breed for vertical resistance to some species of crop parasites, including many of the insect pests of crops. On the other hand, scientific acceptance of horizontal resistance began slowly but many scientists still doubt the value of horizontal resistance, and some even its very existence. The reasons are that horizontal resistance is often difficult to measure, is dependent on environmental effects, and is often partial. Genes in the nucleus generally control both horizontal and vertical resistance. Vanderplank also pointed out that both types of resistance can, and often do, coexist. The use of disease resistant or tolerant cultivars is the most economical and efficient way of controlling disease. Conventional methods of breeding, tissue culture and biotechnological tools are used for production of disease-free seed. The rust diseases of wheat have historically been one of the major biotic production constraints. The criteria to choose cultivars is mainly based upon the potential of these genotypes, to restrict epidemics occurring due to respective races/pathotypes of rust pathogens prevalent. The last few decades representing post green revolution era supported " boom and bust cycle" leading to emergence/build up of several new races of rust pathogens as a result of wide spread use of narrow genetic bases in our wheat cultivars. Leaf rust and stripe rust could affect production on approximately 60 (63 per cent) and 43 (46 per cent) million hectares, respectively in Asia, if susceptible cultivars were grown there. Development of genetic resistance to rusts is the most efficient, cost effective and ecofriendly approach to overcome the losses caused by rust epidemics. The use of single vertical gene resistance exerts pressure on the pathogen and because of selection pressure and evolution, new and virulent races of the fungus appear, which increase the need to develop durable resistance. The use of gene combinations, irrespective of whether they are major or minor, has been suggested as the best genetic control of leaf rust (Roelfs, 1988).

Table 7.5. Molecular markers employed for MAS for leaf, stem and stripe rust resistance genes

Gene	*Marker*	*Marker Type*
Lr24	SCS1302	SCAR
Lr34	CSLV34	STS
Lr57/Yr40	XLr57/Yr40-MAS-CAPS 16	PCR-CAPs
Lr28	SCS421	SCAR
Lr9	SCS5	TPSCAR
Lr28	SCS421	SCAR
Lr9	SCS5	TPSCAR
Lr58	PS16	SSR
Sr25/Lr19	Xwmc221	SSR
Sr26	43BE518379	SCAR
Yr10	Xpsp3000	SSR
Yr15	Xgwm273	SSR
Lr48	Xwmc332	SSR
Lr25	Xgwm251	SSR

Rust Resistance Breeding in India

Breeding for disease resistance plays a key role to protect yield potential of wheat varieties. Systematic work on wheat improvement in India started only during the first decade of 20th century when the Agricultural Research Institute was established in 1905 at Pusa Bihar. Howard (1910) had developed some improved varieties. Subsequently, the research on wheat had started at Niphad, Shimla, Gurdaspur, Durgapura, Kalyani and Indore. During the early phase, the varieties were developed through pure line selection from landraces. In this effort, the varieties Pusa 4, Pusa 12, Type 9, Type 8a, C13, C46, AO13, Ekdania 69, Motia, Gulab, Arnej 206 *etc.* were developed. A breeding programme to develop rust resistant varieties was initiated in 1934 by B.P. Pal, the control of wheat rusts by breeding resistant varieties in the country was reported by Pal (1948).

Table 7.6. Confirmed sources of *aestivum* wheat for resistance to wheat rusts

Aestivum Wheat		*Durum Wheat*	
Variety	*Parentage*	*Variety*	*Parentage*
DBW 17	CMH79A.95/3*CN079//RAJ 3777	AKDW 3347	Immer `S'(Jacana*2/RU//Celta/ Yav 1)
DL 788-2	K 7537/HD 2160 MUT//HD 2278/ DL 896-2	B 662	PBW 34*2/Chuanmai # 18
GW 273	TW 275/7/7/10/lok-1	DWR 1006	LULA/CREX//AAZ
GW 322	PBW 173/GW 196	ED 2398A	Ethiopian local variety from the germplasm collection at IARI-Indore
GW 366	DL 802 -3/GW 232	GW 1139	MACS 2340/IWP 5070

Aestivum Wheat		*Durum Wheat*	
Variety	*Parentage*	*Variety*	*Parentage*
HD 2009	LR 64A/NAI 60	GW 1189	SU-CULLUGU-7/VD 90-12
HD 2135	H 41-3 (HD 1962* (E 4870*K65)	HD 4672	BRED/PBW 34//ALTAR 84
HD 2189	HD 1963/HD 1931	HD 4696	AAZ/MORUS -1
HD 2380	HD 2255/HD 2257	HG 110	Sarangpur Local/HI 8185
HD 2402	HD 2267/HD 2236	HI 8381	JO 'S'/AA 'S'//FGO 'S'
HD 2643	VEE'S'//HD 2407//HD 2329	HI 8498	RAJ 6070/RAJ 911
HD 2864	DL 509 -2/DL 377-8	HI 8591	HI 8144/NI 8625
HD 2932	KAUZ/STAR//HD 2643	HI 8627	HD 4672 X PDW 233
HI 784	NAPO/TOB 'S'/3/8156/KAL/BB	HI 8638	SELECTION FROM BUTTAH
HI 1418	HI 999/HI 601	HI 8663	HI 8177/HI 8185
HI 1454	HI 1076/CC 505/HI 1136	IWP 5019	HD 4519'2/NP 200
HI 1479	HD 2195/HD 2160//WH 283	LINE 1172	MACS 9'2/*T. militinae*
HI 1500	HW 2002*2/STREMPALLI/PNC 5	MACS 2846	CPAN 6079/MACS 2340
HI 1531	HI 1182/CPAN 1990	NIDW 15	DOM 50
HI 1535	BOW/BUC//BUL//CLRP6/HW3034	NIDW 295	BOOMER 33/PLATA 8
HI 1544	HD 2402/HW 3007	P 6046	CPAN 1438/3* HD 4519//CBD 274
HP 1744	CIANO/PARULA//CHILERO/ GARUDA	WH 921	INIA 66/AA/INIA 66/3/SEN 81
HP 1761	RL 6010/6* INIA 66//3*Kauz		
HS 295	CQT/AZ//IAS 55/ALD 'S'/3/ALD 'S'/ NAFN/4/PIN 'S'/PEZSI – 127		
HS 365	HS 207/SKA		
HPW 42	VEE'S'/4/PUN 'S'/CBB//CNO 'S'/#/ JAR/ORZ 'S'		
HUW 468	CPAN 1962/TON//LIRA 'S'/PRL 'S'		
HW 1085	C 306*7/TR 380-14 *7/3AG 14		
HW 2044	PBW 226 *5/SUNSTAR *6/C801		
K 9107	K 8101/K68		
K 9465	B 1153/CB 85 = HD 2402/CPAN 1830//VEE 'S'		
NIAW 34	CIANO 79/PRL 'S'		
NW 1012	PARANA#2/1 JUP/BJY 'S'/3/VEE#5		
PBW 426	KT BAGE/HP 2449//RAJ 3219/ WL 711		
PBW 502	WH 485/PBW 343//RAJ 1482		
PBW 506	KAUZ/PBW 232		
RAJ 1972	HD 2195/HD 2160		
RAJ 3765	HD 2402/VL 639		
RAJ 3777	RAJ 3160/HD 2449		
VL 804	CPAN 3018/CPAN 3004//PBW 65		

Table 7.7. *Lr* markers validated and used in wheat breeding

Gene	*Marker*	*Cycle Conditions*
Lr9	$SCS5_{550}$	94°C-2min; 30 cycles (94°C-1min;60°C-1min; 72°C-1min); 72°C-7min
Lr19	SCS265512	94°C-5min; 35 cycles (94°C-1min;65°C-1min; 72°C-1min); 72°C-7min
	$SCS253_{736}$	94°C-5min; 35 cycles (94°C-1min;63°C-1min; 72°C-1min); 72°C-7min
Lr24	$SCS5_{719}$	94°C-2min; 35 cycles (94°C-1min;55°C-1min; 72°C-1min); 72°C-7min
Lr28	$SCS421_{570}$	94°C-2min; 35 cycles (94°C-1min;60°C-1min; 72°C-1min); 72°C-7min
Lr26	P6M12-P	94°C-3min; 40 cycles (94°C-0.5min;57°C-1min; 72°C-0.5min); 72°C-10min

Gene for Gene Hypothesis

The concept of gene for hypothesis was first developed by Flor in 1956 based on his studies of host pathogen interaction in flax rust caused by *Melampsora lini*. The gene for gene hypothesis states that for each gene controlling resistance in the host, there is a corresponding gene controlling pathogenicity in the pathogen. The resistance of host is governed by dominant genes and virulence of pathogen by recessive genes. The genotype of host and pathogen determine the disease reaction. When genes in host and pathogen match for all the loci, then only the host will show susceptible reaction. If some gene loci remain unmatched, the host will show resistant reaction. Now gene-for -gene relationship has been reported in several other crops like potato, Sorghum, wheat *etc.* The gene for gene hypothesis is known as "**Flor Hypothesis**".

Table 7.8. A simple scheme to explain gene for gene relationship hypothesis (Fehr, 1987)

Varieties	*Host Genotype*	*Pathogen Genotypes*	*Disease Reaction*
One gene pair	AA	aa	Susceptible
	Aa		
	BB	bb	Susceptible
	Bb		
Two gene pair	AA CC	aa	Resistant
	Aa CC	cc	Resistant
	Aa Cc	aacc	Susceptible
Three gene pair	AA BBCC	Aabb	Resistant
	AABBCC	aacc	Resistant
	Aa BbCc	aabbcc	Susceptible

Note: Dominant genes in the host are responsible for resistance and recessive genes in the pathogen for virulence.

Vertifolia Effect

Vander plank introduced the term Vertifolia effect and refers to epidemic development in a variety carrying vertical resistance genes (oligogenes) leading to heavy economic losses. Total failure of vertical resistance leading to a disease epidemic is known as Vertifolia effect. This failure occurs because of two reasons, *viz.*, the level of horizontal resistance in varieties carrying oligogenes is usually low and the pathogen is able to evolve new virulent pathotypes.

Polygenic Inheritance

In this type the disease resistance is governed by many genes with small effects and a continuous variation for disease reaction is produced. The genes show additive and non-additive effects and the environmental effect is also observed. The polygenic resistance does not show pathotype-specificity as against the oligogenic resistance. It is almost same as horizontal resistance. In some cases the polygenic inheritance may have a oligogenic component, the oligogenes acting in an additive manner *e.g.* bacterial blight resistance in cotton.

Cytoplasmic Inheritance

Resistance in some cases is determined by cytoplasmic genes or plasma gene(s). *E.g.* The T-male sterility cytoplasm (CMS-T) in maize is extremely susceptible to *Helminthosporium* leaf blight, while the non-T cytoplasm is resistant to this disease.

Table 7.9. Vertical and Horizontal Resistance (Vander plank)

Feature	*Vertical Resistance*	*Horizontal Resistance*
Pathotype – specificity	Specific	Non-specific
Nature of gene action	Oligogenic	Polygenic; rarely oligogenic
Response to pathogen	Usually, hypersensitive	Resistant response
Phenotypic expression	Qualitative	Quantitative
Stage of expression	Seedling to maturity	Expression increases as plant matures
Selection and evaluation	Relatively easy	Difficult
Host pathogen interaction	Present	Absent
Commonly used, synonyms	Major: gene, race –specific seedling, monogenic, pathotype specific resistance	Polygenic, race nonspecific, pathotype-nonspecific, mature plant, adult plant, field uniform resistance
Efficiency	Highly efficient against specific races	Variable, but operates against all races

Concept of Vertical, Horizontal and Durable Resistance

Before 1900, the science of genetics consisted of biometrics. Biometricians look at inherited characteristics as expressed on a quantitative basis. For example, a susceptible plant crossed with a resistant plant would produce progeny with every degree of susceptibility, ranging from the one extreme of minimum resistance, to the other extreme of maximum resistance. However, the recognition of *Mendel's* laws of inheritance in 1900 drastically changed the science of genetics. Genetic characteristics are either present or absent, with no intermediates. There was conflict between the Mendelians and the biometricians on whether genetically inherited characteristics are based on one single gene or on polygenes. All the characteristics of interest to plant breeders appeared to be quantitative in their inheritance, not qualitative. However, in 1905 a British scientist, *RH Biffen,* discovered that resistance to the stripe rust disease wheat was controlled by single gene. As a result, resistance of many plant diseases was shown to be controlled by single genes. Until this period concept of races was not clear. Following two devastating stem rust epidemics in North America in 1904 and 1916, Stakman and his coworkers were able to show a biological form (race) of *P. graminis* (Stakman and Piemeisel, 1917). During 1925-50, physiological races, pathogen life cycle and epidemiology was studied by many scientists like, EC Stakman, KS Chester, IA Watson, KC Mehta, WL Waterhouse, JH Craigie, KC Mehta studied rust epidemiology in India and found that the alternate host present in hill did not play significant role in the life cycles of stem and leaf rusts. He also determined that rust inoculum survive in foothills of Himalayas in North in summers and the Nilgiris and Pulney Hills in south and was blown to different wheat growing areas in the plains (Mehta, 1929). Flor was the first person to initiate simultaneous inheritance studies involving the pathogenicity in Flax rust and resistance in its host Flax (Flor 1942). These studies lead to the formulation of the gene-for-gene relationship concept (Flor 1956) which states that, "For each gene conditioning rust reaction in the host there is a specific corresponding gene conditioning pathogenicity in the parasite".

Table 7.10. Disease reaction in the common gene-for-gene pattern of interaction between resistance in the host and virulence in the pathogen.

Parasite	*Host*	
	RR	*rr*
AA	Resistant	Susceptible
aa	Susceptible	Susceptible

Vanderplank (1963, 1968) classified resistance into two major types, namely, vertical resistance and horizontal resistance.

Vertical Resistance

In this type of resistance host shows resistance to one or some specific races of pathogen while it completely fails to check the establishment of other races so that it displays completely resistant reaction for some and susceptible for other races. (Specific, qualitative, differential). Characteristics are usually controlled by one or a few genes (monogenic or oligogenic); these genes apparently control a major step in the interaction of the pathogen with the host plant and therefore play a major role in the expression of vertical resistance; responds like an incompatible, hypersensitive reaction, complete resistance; limits initial inoculum.

Horizontal Resistance

A lack of differential behavior of different races of pathogen on a number of varieties of host gives a characteristic type of disease reaction where no variety is completely resistant to any of the races. Similarly, no race of pathogen shows completely susceptible reaction on any of the varieties. Varieties can be ranked for intensity of disease development irrespective of the race of pathogen but no one is completely resistant. Such a type of disease reaction of host varieties is called Horizontal resistance (HR). This resistance is also called nonspecific, general, quantitative, adult-plant, field, or durable resistance. Many genes (multigene resistance) control horizontal resistance. These genes appear to exert their influence by controlling the numerous steps of the physiological processes in; the plant that provide the materials and structures that make up the defense mechanisms of the plant. *e.g.* cutin depth, plant architecture. Characteristics are affected by environment; does not protect against infection but slows down disease development. A comparison of some characteristic attributes of VR and HR makes their difference apparent especially with respect to genetic control of resistance and mode of expression.

Table 7.11. Comparative factors of HR and VR

Factor	*Vertical Resistance*	*Horizontal Resistance*
Synonyms	Complete resistance Race-specific resistance Monogenic Major gene resistance Qualitative Non-durable	Partial resistance Non-race specific Polygenic Minor gene resistance Quantitative Durable, Field resistance Rate-reducing resistance, Tolerance
Genetic control	One or few genes	Few to many genes
Expression by host	Seedling and adult plants	Increases with plant development; lower expression by seedlings
Mechanism	Limits initial infection; none to low symptom development	Slows progress of infection, colonization, lesion development and pathogen reproduction – resulting in slowed progress of epidemics
Efficiency	Host reacts differentially to races of pathogen	Similar reaction to all races of pathogen.
Primary inoculum	Primary inoculum has little effect	Effectiveness declines as primary inoculum increases
Secondary inoculum	None to low amounts produced	Slowed pathogen reproduction; plus fewer successful infections
Environment	Usually stable across range of environments	Effectiveness declines as environment favors pathogen or predisposes host
Pathogen virulence	Exerts strong selection pressure on pathogen population; emergence of new races	Exerts weak selection pressure on pathogen population; slows emergence of new races
Relative durability	Low	High
Supplementary control tactics	Minor effect	Enhance effectiveness

Durable Resistance

Durable resistance is that which remains effective in a cultivar during its widespread cultivation for a long sequence of generations or period of time in an environment favorable to a disease or pest. Johnson (1979) proposed the concept of durable resistance associated with horizontal or non-specific resistance having non-hypersensitive response to infection. Both partial resistance and slow rusting resistance are believed to be durable. Working with leaf rust of wheat, Caldwell (1968) emphasized that slow rusting is a form of general or horizontal resistance with intermediate infection type. He stated that such resistance to rusts involves exclusion of the fungus, limitation of the pustule size without hypersensitivity or possibly slow growth and development of the fungus. Parlevliet (1975) used the term partial resistance to characterize slow rusting to leaf rust in barley. According to him, partial resistance is a form of incomplete resistance characterized by a reduced rate of epidemic development despite a high or susceptible infection type. Partial resistance to leaf rust involves interaction of minor genes and that the components of resistance may be under pleiotropic genetic control. These types of resistances involve host pathogen combinations in which the rust develops slowly and never reaches a high degree of severity. The South American cultivar Frontana is considered as one of the best sources of durable resistance. The most durable resistance to leaf rust is associated with a few gene combinations. *Lr12* and *Lr13* in combination with Lr34 are the basis of most of durable resistance according to Roelfs (1988). Gupta and Saini (1987) gave the concept of non-hypersensitive resistance. According to them cultivars showing such resistance show low coefficient of infection despite their high (compatible) seedling and adult plant reaction against same race(s). Such resistance has also been considered to be durable. Sawhney (1989) suggested that durable resistance to leaf rust might be obtained by combining seedling specific genes with adult plant resistance genes. The ability of a cultivar to remain resistant can be influenced by several factors which include: the inherent durability of resistance; the inherent variability in pathogenicity of the pathogen; the life cycle of the host and the pathogen; the manner and extent of deployment of the cultivar and of other resistant and susceptible host cultivars or host species; epidemiological factors affecting the size of the pathogen population. These include climate and weather; and the level of susceptibility of the cultivar and other hosts.

Durable Resistance to Black Stem Rust of Wheat

Sr 2 is a single recessive gene for adult plant resistance that has provided durable resistance in many wheats, in Australia, the USA and in CIMMYT wheats. It was derived from an 'Emmer' (tetraploid) wheat, via the cultivar 'Hope'. This gene has been successfully transferred to many new wheat cultivars by classical breeding, assisted by an unusual linked character known as pseudo black chaff (Hare and McIntosh, 1979).

Resistance to the Eyespot Disease of Wheat

'Cappelle Desprez', French bread wheat, was recognized to possess a measure of resistance to eyespot caused by *Pseudosporella herpotrichoides.* The resistance was durable and remains effective today, despite some variation in pathogenicity in the eyespot pathogen, particularly in relation to wheat, rye and other grasses (Scott *et al.*, 1978).

Resistance to Yellow and Brown Rust of Wheat

The linked genes *Lr34* and *Yr18* control incomplete or partial resistance and durable resistance to brown and yellow rust, respectively. The linked character of leaf-tip necrosis has been used to assist in their selection. European wheats with durable resistance to yellow rust have a range of levels of disease resistance, and it appears to be under rather complex genetic control, not including *Yr18*, except in 'Besostaja 1' wheat. Another gene that may provide durable resistance to leaf rust has been identified as *Lr46* as one of two genes present in the cultivar 'Pavon 76' which is linked with stripe rust resistance gene *Yr29* that had displayed durable resistance. Combining *Lr34* and *Lr46* might be a useful strategy. Molecular markers for these genes would be very useful. The plant breeding aims to introduce into cultivars disease resistance that will give useful disease control for long enough to ensure that the commercial life of a cultivar is not curtailed by a break down in resistance.

Breeding for Disease Resistance

Historically, plant breeders have introduced a considerable number of resistance genes into different crops, but the pathogen often has remarkable ability to adapt to the gene-for-gene type of resistance. Nevertheless, this simple type of resistance has, in some cases, fasted for many decades. A prerequisite to breeding for durable resistance is a full understanding of the host-parasite relationship. Once this knowledge has been acquired, the breeder can devise methods of controlling epidemics by means of general or non-specific resistance. Although long-fasting resistance can probably never be attained in all crops, a system of Integrated pest control whose central core Is general resistance will probably provide the most successful approach to stabilize yield near its full genetic potential.

Types of Resistance

Resistance may be classified in two groups, based on the differences in the mechanism involved: 'avoidance' and 'true resistance'. The mechanisms of the first group reduce the chances of contact between the host and pathogen, whereas the mechanisms of the second operate from the moment of contact between host tissue and parasites (Parlevliet, 1979a).

Avoidance

Avoidance involves a heterogeneous group of mechanisms and often depends on the structure of the entire plant, of certain plant parts, or of certain plant tissue. A few examples are: (1) the resistance of barley varieties *(Hordeum vulgare)* with closed flowering to loose smut, *Ustilago nuda* (Pederson, 1960), and (2) the resistance of tall wheat cultivars to *Septoria nodorum* blotch *(Septoria nodorum)*, in which the height of the plant impedes the spread of the inoculum to the upper plant parts (Bronnimann *et al.*, 1973).

True Resistance

True resistance reduces the rate of colonization or growth of the parasite from the first moment of contact between the host tissue and the parasite. It implies an intimate contact between host and parasite and it involves their mutual interaction. The two typical examples are: (1) resistance of the hypersensitive type is frequently used in resistance breeding and is generally controlled by major genes. The resistance genes appear to operate in a gene-for-gene relationship with the virulence genes in the parasite (Flor, 1971; Day, 1974). This is called vertical resistance (VA) by Van der Plank (1963), and, (2) partial resistance-expressed as a reduction in the rate at which an epidemic builds up, occurring in spite of a susceptible reaction to infection. Such a reduction may be the result of a lower infection frequency, a longer latent period,.and/or reduced spore production. Van der Plank (1963) named this as horizontal resistance (HR).

Hypotheses about Horizontal Resistance

Van der Plank (1963) introduced the concept of vertical and horizontal resistance and Robinson (1976) assumed that genes involved in HR do not operate on a gene-for-gene basis with the genes in the pathogen. Horizontal resistance is supposedly stable, usually controlled by several additive genes. Some authors have suggested that HR represents the accumulated 'ghost' or 'residual' effects of resistance genes that have succumbed to corresponding virulence genes, with these residual effects being race-non-specific. Martin and Ellinboe (1976) observed that a wheat line carrying the resistance gene *Pm4,* which was susceptible to mildew races carrying the corresponding virulence genes, permitted less abundant pathogen growth than a near-isogenic line that possessed the allele for susceptibility, *pm4.* They concluded that the ghost effect was race-specific, but that it did contribute to the HR phenomenon. The view of Nelson (1975), in which VR and HR genes are assumed to be the same, is similar to the 'ghost gene' hypothesis. Eenink (1976) equated HR with stable resistance and assumed that the stability of resistance was determined by genetics of pathogenicity. Parlevliet and Zadoks (1977) concluded that all resistance genes in the host population and virulence genes in the pathogen population form an integrated system. Major or minor genes in the host interact on a gene-for-gene basis with major or minor virulence genes in the pathogen. When only a few genes with major effects operate, differential interaction can be detected easily and this is called VR. When several to many genes with small effects determine the resistance/ pathogenicity pattern collectively, it is difficult to separate the small differential interaction from experimental errors. This resistance behaves as HR and appears to be race-non-specific. These authors concluded that polygenic true resistance is stable owing to genetic homeostasis in the parasite population. In order to overcome such polygenic resistance, the parasite would have to change its genetic constitution at several or many loci, a change that could scarcely be achieved without a serious loss in overall fitness. HR can arise in two ways: (1) When the host genes do not operate in a gene-for-gene way with the pathogen genes, no differential interactions are possible. (2) When several to many host genes with small effects operate on a gene-for-gene basis with an equivalent number of genes in the pathogen population, differential effects are so small as to be undetectable, and the results appear to be

HR. Assuming that gene-for-gene relationships are common in host-specialized pathogens, VR and HR are extremes of continuum. When a few genes having large effects are operating, differential interactions are easily discernible and the result is a mixture of VA and HR. With many genes of small effects, differential interactions cannot be recognized and resistance is predominantly HR (Parlevliet, 1981).

Methods in Resistance Breeding

Irrespective of the controversy ensuing over the classification of VR and HR, plant breeders and pathologists generally recognize that certain resistances are more durable than other. Classic examples are 'Hope' and 'Kite' for stem rust resistance, 'Frontana' for leaf rust resistance, and 'little Joss' for stripe rust resistance, which have not been completely overcome by the pathogen. Some kinds of partial resistance against *Phytophthora infestans* in potatoes and leaf rust in wheat have remained fairly stable for a long time. In breeding cultivars for disease resistance, various choices are available in regard to resistance gene or gene combination. The breeding methodology would depend on the type of resistance gene chosen. These types of genes can be catagorised as: (1) Hypersensitive resistance genes used one or two at a time, (2) a combination of multiple hypersensitive resistance genes, (3) polygenic resistance genes each having minor effects (partial resistance), (4) a combination of a hypersensitive resistance gene and genes with partial resistance, and (5) major and/or minor genes in a multiline or varietal mixture.

Resistance Genes Used Singly

This strategy is poor for disease control, as the resistance gene loses effectiveness in a very short time. Single gene mutation for virulence is very common. This strategy predominated in wheat breeding in the 1950s and 1960s, but recently has focussed on the transfer of hypersensitive genes to wheat from related species and genera of wheat. Such genes as *Lr9, Lr19,* and *Lr24* for leaf rust resistance and *Sr24, Sr26,* and *Sr31* for stem rust resistance are classic examples of this procedure. However, some of these genes have already lost their effectiveness, and others lose effectiveness when exposed to disease in large areas. Single-gene protection, whether derived from the same gene pool or from related species, therefore has limited effectiveness. In this strategy, resistance genes are released one at a time. Whenever the pathogen overcomes the resistance, the second gene is released in production. Virulence surveys must be an integral part of this strategy as it is essential to stay ahead of the development of new virulence. The breeding methodology is simple. These genes can be transferred through backcrossing or through simple crossing and recovered in progeny, as their phenotypic effects are great. In the authors' opinion, use of single-gene resistance in breeding for agricultural protection should be discouraged, as this approach is frequently disappointing.

Combination of Multiple Hypersensitive Resistance Genes

The use of multiple resistance genes with major effects has given stability to stem rust pathogen in Australia and North America. This strategy was adapted just as it became clear that single-gene protection was ephemeral. Multiple gene strategy is based on the concept that the pathogen population has a barrier of several resistances presented to it simultaneously. As stated before, the system has worked in the case of stem rust. It has failed to provide long-lasting protection in Mexico against leaf rust fungus. The variety Anahuac 75, which possesses three effective genes, succumbed to a leaf rust epidemic in Mexico. It also appears that the relative merit of each gene in the combination is important. The variety Mendos, with *Sr11, Sr36,* and *Sr17* became susceptible to stem rust shortly after its release. According to Parlevliet (1981) this strategy can operate satisfactorily only when breeding is centrally coordinated and when the production area is isolated from other areas where the system is not applied. A complex breeding system such as three-way crosses or double crosses can be employed to breed multiple gene combinations. The right choice of race spectrum in selecting the gene combination is very important. Genetic analysis is important to discern the right combination in the variety before releasing it as an agricultural commodity.

Use of Partial Resistance (Polygenic resistance with minor gene effect)

Partial resistant is generally polygenic, having minor gene effects. Whenever a large number of minor genes are involved, their accumulation and combination with other agronomic and yield characters is difficult but not impossible to achieve. This character is also more vulnerable to environmental effects. In a severe epidemic, partial resistance seems to be less effective, and hence selecting for this character by creating artificial epidemics is a doubtful procedure. However, accumulating polygenes for partial resistance seems to be quite effective in many host-pathogen systems (Parlevliet and van Ommeron, 1975). In Mexico, through conventional breeding, partial resistance to leaf rust has been combined in varieties such as Pavon 76, Genaro 81, and Cucurpe 86. Compared to INIA 66, these varieties rust slowly; however, high temperatures may hasten the development of leaf rust on the first two varieties. The best breeding strategy to combine polygenes for partial resistance. is recurrent selection in which improvement. of resistance is a continuous process. It is possible to make resistance more durable if the right combination of polygenes for partial resistance is achieved. Although resistance is never complete, the host-pathogen interaction is more balanced and stable.

Use of Hypersensitive Genes and Partial Resistance

In certain agricultural systems, the release and promotion of partly resistant varieties is a difficult task. Farmers and research agencies may discard a partly resistant variety because it is partly susceptible. The breeding strategy for combining these two distinct types of resistance has two steps. Since hypersensitive resistance is epistatic to partial resistance, it is not possible to make the cross and guarantee that both types of resistance will be present. Hence, in the first attempt, partial resistance should be combined with acceptable agronomic type and high yield potential. Once this is achieved, hypersensitive genes can be combined through backcrossing to this partly resistant high-yielding type using the latter variety as recurrent parent. This process would ensure that partial resistance is not lost.

Use of Major or Minor Genes in Multilines and Varietal Mixture

The multiline concept was first proposed by Jensen (1952) and Borlaug (1958) to combat the disease problem. Since then, commercially grown multilines have been promoted in oats and wheat. A multiline is a mixture of isogenic or near-isogenic lines with different resistance genes. Varietal mixtures are a more heterogeneous population, in which the resistance gene or genes is represented by each mixture component. Van der Plank (1968) assumed that stabilizing selection would prevent the development of complex pathogen races in a multiline. Marshall and Pryor (1978) showed that the strength of stabilizing selection needed per virulence gene to prevent the development of complex races was reduced when the number of component genotypes in the cultivar or blend increased. When the multiline is grown extensively in the area where the epidemic develops, and when pure line cultivars carrying resistance genes also present in the multiline are grown at the same time, a rapid evolution of complex races may occur (Parlevliet, 1979b).

Multiline breeding is a very conservative approach in regard to yield, because newer cultivars may supersede it. But to maintain yield stability, especially in high disease hazard areas, the multiline approach may potentially be the best. Agricultural losses due to major diseases are great throughout the world. The evolution of a virulent pathogen in high disease hazard areas may create an epidemic of devastating proportion, as happened with leaf rust in 1977. A multiline cultivar would have prevented that situation. Breeding methodology is simple, straight backcrossing to transfer the resistance genes. A highly productive recurrent parent is chosen as agronomic and yield base for the area where multilineal variety is to be promoted. Recurrent parent with partial resistance as agronomic and yield base would be even better.

Scoring Backcross Population for Leaf Rust Resistance

DNA Extraction

The small scale DNA extraction method is used for isolating DNA from large number of plants (two persons can isolate DNA from 100 plants in one day).

Midi-prep CTAB DNA Extraction from Wheat Leaves (modified)

Collect young leaf samples from each plant (one leaf/plant) in a small polythene bag (10x10cm) and place in ice (these leaves can be used immediately for DNA extraction or these can be stored in –80°C freezer until needed). Grinding the leaf in liquid nitrogen in a 3″ mortar and transfer the ground samples in 2-ml tubes (place tubes on ice or at –70°C if the number of samples is more). Add **900µl** of extraction buffer. Vortex to suspend ground samples. Incubate for 40-50 minutes at 65°C. Stir after regular intervals (10min. interval) to homogenize the mixture. Add **800 µl** chloroform, mix well for 15-20 minutes and spin for 7 min at 13 000 rpm in a microcentrifuge. Transfer the supernatant into a clean tube and add **600µl** of isopropanol. Mix well by slowly inverting the tubes until DNA precipitates. Fish out precipitated DNA or spin down by centrifugation. Rinse the DNA with 70 per cent ethanol and air dry by inverting tubes on a clean filter paper. Add about **50 - 100µl** of 1X TE (Note: This procedure yields up to 50µg DNA).

TaExtraction Buffer

Table 7.12. 2X CTAB Extraction Buffer (100ml)

Component	*Quantity*	*Final Concentration*
1M Tris, pH 8.0	20.0ml	100 mM
5M NaCl	28.0ml	1.4 M
0.5M EDTA	4.0ml	20 mM
10 per cent CTAB	20.0ml	2.0 per cent
Sod. bisulphite	0.5g	0.5 per cent
merceptoethanol	1.0g	1.0 per cent
dds H_2O	28.0	

Tris – Trizma base (Tris [hydroxymethyl] amino methane); Mol. Wt. = 121.1.

NaCl - Sodium chloride; Mol. Wt. = 58.43.

EDTA – Ethylenediamine tetraacetic acid ($Na_2EDTA.2H_2O$); Mol. Wt. = 372.

2X CTAB Buffer

Mix the components (1-4) of table in that order. This can be stored at room temperature. Sodium bisulphite and Merceptoethanol are added just prior to using the buffer.

10 per cent CTAB (100ml)

Dissolve 4.1g NaCl in 80ml H_2O and slowly add 10g CTAB while heating and stirring. If necessary, heat to 65 °C to dissolve. Adjust final volume to 100ml

1M Tris (pH 8.0) (100ml)

Add 12.11g Tizma base in about 80ml H_2O. Stir vigorously, using a magnetic stirrer. Adjust pH to 8.0 by adding ~ 4ml conc. HCl. Cool the solution to room temperature and adjust volume of the solution to 100ml with ddH_2O. Sterilize the solution by autoclaving. (*Note: if solution turns yellow, discard it*)

0.5M EDTA (pH 8.0) (100ml)

Add 18.61g $Na_2EDTA.2H_2O$ to ~ 80ml H_2O. Stir vigorously, using a magnetic stirrer. Add about 2g NaCl pellets (*EDTA will not dissolve without adding NaCl*). Adjust volume of the solution to 100ml with ddH_2O. Sterilize the solution by autoclaving.

Table 7.13. 1X TE Buffer (Tris, pH 8.0 = 0.01M, EDTA, disodium, pH 8.0 = 0.001M)

Component	*Quantity*	*Final Concentration*
1M Tris, pH 8.0	1.0 ml (1000µl)	0.01M or 10mM
0.5M EDTA, pH 8.0	0.2 ml (200µl)	0.001M or 1mM
dds H_2O	98.8 ml	

DNA quality and Quantity Check can be estimated/checked by two methods, *viz.*, Spectophotometer method and Biophotometer Method.

Spectophotometer Method

Measurement of the amount of ultraviolet radiation absorbed by nucleic acid bases offers an accurate means of quantifying DNA concentration in a sample. This method, however, is not accurate if the DNA is contaminated with proteins, polysaccharides and other nucleic acids. Since the equipment is expensive, the second cheaper method can be used easily and reliably for quality and quantity analysis of DNA.

Biophotometer Method

This method is used for rapid, simple and convenient measurement of DNA and protein concentration. It measures the absorbance at various wavelengths of 230,280 and 320 nm. For pure samples the absorbance at 320 nm is approximately zero. The absorbance of impure samples can be calculated by applying the correction factor.

Gel Electrophoresis

The procedure involves loading about 2µl of DNA sample (2µl DNA + 6.5µl H2O + 1.5µl 6X loading dye) in 0.8 per cent agarose gel. A control with known DNA concentration is loaded on both sides of the gel. Run the gel for 1h, stain with ethidium bromide and visualize under UV light. Photograph the gel or acquire the image in a photo documentation system. A single high molecular weight (MW) band indicates good quality DNA and a smear will indicate the degraded DNA sample. Compare the band intensity of the sample DNA with the control DNA and infer the quantity of DNA of the samples. For PCR, dilute the sample DNA uniformly to a concentration of 20ng/µl in ddH_2O.

In vitro DNA Amplification through Polymerase Chain Reaction

In this activity one will learn how to set up a polymerase chain reaction (PCR) and amplify the target DNA. The PCR is a rapid procedure for *in vitro* enzymatic amplification of specific segments of DNA using specific primer sequences. The applications of PCR include, analysis of sequence variation, monitoring of specific genes in a population, genetic fingerprinting, cloning of genomic DNA and cDNA, in vitro mutagenesis, *etc.* Objective is to set up a PCR reaction mixture and to amplify DNA and detect bands from the amplified PCR products. Prepare a bulk reaction mix containing all the components listed below except DNA or primers depending on whether you are preparing several reactions using the same primers for different DNA samples or different primer pairs for the same DNA sample. We routinely use the 20µl reaction but if you need to save on primers and/or nucleotides, the 15µl reaction works as well. Table below gives concentrations of various components in a reaction mix.

Table 7.14. Quantity and concentration of various components of a PCR reaction (for 25µl reaction mixture)

Stock Solution	*Quantity Required*	*Final Concentration*
DNA (40-50ng/ul)	2.00µl	80-100ng
PCR buffer (10X) (without $MgCl_2$)	2.0µl	**1X**
$MgCl_2$ (15mM)	2.0µl	1.5mM
dNTPs (1mM)	2.0µl	100µM
Primer, F (5µM)	1.0µl	0.25µM
Primer, R (5µM)	1.0µl	0.25µM
Taq polymerase (5U/µl)	0.2µl	1U
ddH_2O	9.8µl	-
Total	20.0µl	

PCR Buffer (10X)

100mM Tris-HCl (pH 8.0)

500mM KCl

15mM $MgCl_2$

0.1 per cent Gelatin

Primers

Primers are available commercially as 200µM concentration. Dissolve the stock in sterile water to prepare working solution of 5µM

dNTPs

These are commercially available as 100mM concentration. Dissolve 25µl of each dNTP to prepare 100µl of 25mM stock solution. From this prepare 1mM working solution.

Procedure

Dilute the DNA to the desired concentration (40-50ng/µl). Aliquot 2µl DNA in 0.5ml eppendorf tubes. Prepare a cocktail of buffer, dNTPs, $MgCl_2$, primers, enzyme and water. For example for PCR amplification of 10 samples, mix together, 20µl PCR buffer, 20µl dNTPs, 10µl F primer, 10µl R primer, 2µl Taq polymerase, and 90.8µl ddH_2O. (Note: while preparing cocktail always prepare 5-10 per cent more to accommodate wastage due to pippeting). Add 18µl of cocktail to the genomic DNA. Give a brief spin and overlay the mixture with a drop of mineral oil.

Table 7.15. Temperature profiles (35 cycles) in a thermal cycler

Step I	94°C	5 minutes	Initial denaturation (one period)
Step II	94°C	1minute	Denaturation
Step III	72°C	2 minutes	Annealing
Step IV	60°C	1 minute	Primer extension (DNA synthesis)
Repeat steps II to IV		35 times	
Step V	72°C	5 minutes	completion of synthesis (one period)

Add 3µl of 6X loading dye to each tube. For STSs, load 12µl of each sample in a 1.5 per cent agarose gel prepared with 1X TBE gel buffer. Electrophorese in 1X TBE at 60 V, constant voltage, until the blue dye has migrated as required. Stain gel in 1µg/ml ethidium bromide (100µl of 10mg/ml ethidium bromide in 1000ml dH_2O) for 15 minutes with gentle shaking. Rinse gel in dH_2O for 20 min, slide gel onto a UV transilluminator and photograph. Score the data for both genes individually.

In vitro Screening for Biotic Stresses

Among biotic stresses, the disease resistance holds special significance, due to huge losses caused by different diseases in the plants. There are various biotechnological approaches for disease resistance in crop plants. The screening for disease resistance using *in vitro* tools has been successfully demonstrated in plants. Using toxins, toxin analogues and culture filtrates in the culture medium, the resistant plants have been developed to several diseases in wide range of crop plants. However, disease resistance through *in vitro* screening technique has some limitations *viz.*, it is only applicable to those diseases where the pathogen toxins are involved; toxin response to single cell deviate from the whole plant reaction; desired level of resistance is dependent upon the mode of action of toxin involved; efficient regeneration system from cell to plant is mandatory and it is not always certain to induce desired changes. Thus keeping in view the advantages and disadvantages, the tissue culture approach of *in vitro* screening for disease resistance, should be selectively and carefully used. It should be used, only if the disease resistance through conventional approach is cumbersome or not possible. In other words, tissue culture would be of great value in situations, where either there is no source of resistance to diseases or available source of resistance is not satisfactory and resistance is linked with undesirable characters.

Table 7.16. Forward MAS in wheat

Plant height	*Rht-B1, Rht-D1*
Vernalization	*Vrn-A1, Vrn-D1, Vrn-B3*
Photoperiod sensitivity	*Ppd-A1, Ppd-B1, Ppd-D1*
Yellow rust	*Yr5, Yr7, Yr 15, Lr37/Yr17/Sr38, Yr57, Yr 78*
Leaf rust	*Lr34, Lr48, Lr21, Lr16, Lr67, Lr68, Lr19, Lr47*
Stem rust	*Sr2, Sr22, SrCAD*
Quality	*Gpc-B1, Pina-D1, Pinb-D1, Dlu-A1, Glu-D1*
Drought tolerance	*Dreb, 1-FEH-6B*
Soil borne diseases	*Cre8*
TGW	*TaGW2-6A*
FHB	*Fhb1, QTL-FHB-2DLc*

Exploration of Strategies for Development of Safe Wheat

Development of nullisomic cytogenetic stocks for chromosome 1 (N1A) na 6 (N6A) in durum wheat. Development of ditelosomic cytogenetic stocks for chromosome 1AL and 6AL in durum wheat. Pyramiding of ditelosomic 1AL and

6AL chromosomes in durum wheat and/or Pyramiding of nullisomic 1A and 6A chromosomes in durum wheat. Evaluation of the non-immunogenic potential of cytogenetic stocks through immunoassay

Data Analysis/Interpretations

DNA amplification of an F_2 population segregating for *Lr24* and *Lr28*. SCAR markers linked to *Lr24* and *Lr28* were used for DNA amplification. Disease reaction of individual F_2 plant was recorded at seedling stage against leaf rust race 77-5, to which both the genes confer resistance individually. The plants having either *Lr24* or *Lr28* or both showed the amplification and the ones that does not showed amplification have neither *Lr24* nor gene. Scoring of SCAR markers is given. Since both markers are dominant in nature so we can't distinguish between homozygous and heterozygous plants. Plant number 5, 6, 8 have both genes whereas plant no. 2, 4, 7, 10 and 15 have only *Lr24* allele and plant no. 13, 14, 18, 19 and 20 have only *Lr28* allele. As with resistance to stem rust, multiple allelism is known to occur at certain loci. Several genes are also transferred from related species to wheat. Genes *Lr27* and *Lr31* do not confer resistance on an individual basis, but rather interact in a partly dominant complementary manner to give resistance. The resistance conferred by the *Lr27/Lr31* combination is also specific in nature. Interaction between known genes for enhanced resistance also occurs. Studies of the genetics of the slow rusting type of resistance (slow rusting may be defined as slow development of rust on a variety, compared to the rate with which the disease develops on a susceptible check) indicated that such resistance could also be simply inherited and could be influenced by various environmental factors. Durable resistance to leaf rust in bread wheats appears to be due to the additive interaction of *Lr34* with other minor genes (Singh and Rajaram, 1992).

Strategies to Control Rust Diseases

The three rust diseases–stem (or black), leaf (or brown), and stripe (or yellow)–are caused by fungi *Puccinia graminis* f.sp. *tritici, P. triticina,* and *P. striiformis* f.sp. *tritici.* They continue to cause losses–often major–in various parts of the world and hence receive more attention in breeding. The wheat crop can be protected from rust, or at least the occurrence of epidemics can be reduced, by emphasizing the following: (1) regional cooperation in monitoring the evolution and migration of new races of rust fungi; (2) enhanced information on the genetic basis of resistance in important wheat cultivars for their deployment; and (3) a shift toward breeding and deploying wheat cultivars with durable resistance. The phenomenon of the erosion of race-specific resistance genes, or their combinations, has led scientists to look for alternative approaches to resistance management. Van der Plank (1963) was the first epidemiologist to clearly define the theoretical basis of the concepts of resistance. In the late 1960s and 1970s, there was a revival of the concept of general (race-nonspecific) resistance and its application in crop improvement. This concept was widely used for breeding leaf-rust resistance by Caldwell (1968), stem-rust resistance in wheat by Borlaug (1972), and yellow-rust resistance by Johnson (1988). The wide application of such a concept in breeding for leaf-rust resistance, commonly known as slow rusting, has dominated bread wheat improvement program for almost 30 years, with major impact. Today, we better understand the genetic basis of durable resistance to rust diseases, and this knowledge is being applied in breeding. The genetic basis of durable resistance to three rusts is described below. We firmly believe that development and deployment of such resistance will provide a long-term genetic solution to rust control.

Identification and Creation of Epiphytotic Conditions for Wheat Diseases

Yellow or Stripe Rust (*Puccinia striiformis*)

Yellow pustules on leaves forming stripes. If attack is severe uredopustules appear on glumes and surface of the grains. ***Brown or leaf rust* (*Puccinia recondita*):** Round, orange pustules, irregularly arranged or in clusters on leaves, less common on leaf sheath and stalk. ***Black or stem rust* (*Puccinia graminis tritici*):** Dark brown to black, elongated and large pustules are severe on stalk but also appear on leaf sheaths, leaves and ears. Inoculum of the commonly prevalent races of yellow and brown rusts is maintained at Research Station (ICAR) Shimla and supplied to different research centers for screening of wheat material in the Rabi season. The rust inoculum is established and multipled on 10 15 days old seedlings of rust susceptible wheat varieties grown in glass house/alkathane chambers under temperature and humidity-controlled conditions. The small pots with multiplied inoculum are transferred in the field in infector rows. The infector rows are sown in between the breeding material and it constitutes of a mixture of highly susceptible varieties/lines to yellow and brown rusts. Attempt is made to establish several focii of infection in the infector row by placing the rust infected plants/directly inoculating the plants maintaining sufficient relative humidity using alkathane chambers. Inoculum of yellow rust is multiplied in the infector rows in the month of December and of brown rust in and February. The optimum temperature for spore germination of yellow rust is 9 13°C and for brown rust it is 19°C. Once the uredopustules are formed at the primary infection foci, the uredospores are released for secondary spread. Rusts are compound interest diseases which spread with every crop of uredospores released form the uredopustules. Once the disease gets established in the infector rows. the uredospores are collected with the rust collectors which have suction device. Suspension of the spores, thus collected, is made in soapy water which acts as spreader and avoids coagglutination of the spores. The spore suspension is sprayed over the breeding material in the evening. For the establishment of the rust in the breeding material, irrigation is given after rust inoculum has been sprayed. Several other methods of creating rust epiphytotics have been adopted at different research institutes. Amongst them syringe inoculation method is quite common being practised for large scale screening of breeding material. Observations of rust development on the breeding material is recorded visually on the basis of leaf area infected (severity) and type of disease reaction (response) according to modified cobb's scale. As severity is determined by observation, readings cannot be absolutely accurate. Therefore, different intervals have been used by different workers. A less than 5 per cent severity is referred to as a trace severity(denoted "t"). Beyond that, the intervals used most commonly for recording severity are 5 per cent, 10 per cent,

20 per cent, 40 per cent, 60 per cent, 80 per cent and 100 per cent. The field response of a variety or line refers to the type of disease reaction and is recorded by using the following letters:

Response

O: No visible infection on plants

R: Resistant, Necrotic areas with or without minute uredia

MR: Moderately resistant. Small uredia surrounded by necrotic areas

X: Intermediate Variable sized uredia, some with necrosis or chlorosis

MS: Moderately susceptible Medium uredia with no necrosis possibly some distinct chlorosis

S: Susceptible Large uredia with no necrosis and little or no chlorosis.

Combining Severity and Response Readings

Severity and field response readings are recorded at the same time with severity first and are combined in the following way:

tR: Trace severity of a resistant type infection

5MR: 5 per cent severity of a moderately resistant type infection

60S: 60 per cent severity of a susceptible type infection

Usually, a single severity and response gives an adequate picture of the reaction of a line or variety; occasionally there is obvious variability in reaction within a line. The variability may appear in several forms *viz.*, Clear cut separation of plants into 2 or even 3 classes; A range of reaction from plants without clear cut separation into classes and A range of reaction in each plant. It is usually impractical to try to determine what causes the variability. However, it is quite simple to record whether the variability is represented by a clear-cut separation of plants into classes or by a range in the reaction. A coma separating two severity and response readings indicate that the plants fall into clear cut classes with readings as given. For example; 5R, 40S means that there were two classes of plants in the row with respect to rust reaction; one group 5R and the other 40S. A dash separating two readings indicate a range in severity and response of the plants in the row *e.g.* 10R 5S means that there was a range of severity and response to rust from 10R to 5S. When using these combinations, the first reading is understood to represent the predominant class. If it seems desirable to give information on the number in each group, it may be done in following manner: 30P 5R, 4P 5S meaning that 30 plants had a severity and response reading of 10R and 4 plants scored 5S. Coefficient of infection (C.I.) for rusts of a particular variety is calculated when multilocation testing against rusts is carried out. Numerical values given to response are; t = 2, S=1, MR=0.4, MS=0.8, x = 0.6, tS = 2 x 1 = 2 and tR = 2 x 0.2 = 0.4. Multiple allelism at certain loci is known to occur and some genes were also derived from related species. The presence of a number of genes in wheat that modify the effect of 'major' genes toward a higher, hypersensitivity type of stripe rust resistance. In the absence of major genes, these 'minor' genes condition a high type of hypersensitivity resistance. Johnson (1988) described various cultivars that carry durable resistance. Singh (1992) indicated that *Yr18* has conferred partial resistance of durable nature in wheats.

Stem Rust (*Puccinia graminis tritici*)

Stem rust is one of the most destructive of plant diseases. The fungus inciting stem rust spends part of *its* life cycle on the common barberry plant. It also overwinters on growing wheat in Mexico and *is* spread by wind-borne spores that are carried northward in the spring. Damage from stem rust has been greatest in the spring wheat area, although occasional epiphytotics also occur in the winter wheat areas. Most of the breeding work for resistance has been with the spring wheats. In their improvement stem rust resistance is a major objective. In the central Plains states early-maturing varieties often escape stem rust damage. One of the earliest stem rust-resistant wheats was *Kanred,* a hard red winter variety selected from Crimean and distributed in Kansas in 1917. A spring variety with stem rust resistance, Ceres, from the cross Marquis X Kota, was distributed in North Dakota in 1926 and was widely grown during the following decade. In 1935, 1937, and 1938, Ceres was severely damaged by stem rust. Ceres was soon replaced by new varieties with resistance derived from durum and emmer. Thatcher from the double cross (Marquis X Iumillo durum, $2n = 28$) X (Marquis X Kanred) was distributed in Minnesota in 1934. Thatcher was resistant in the rust epidemics that attacked Ceres and so replaced large acreages of the latter variety. A selection from the Marquis X IumilJo cross, Marquillo, has been used also as a source of rust resistance in other crosses. Hope was developed from a cross made in South Dakota in 1915 between Yaroslav emmer *(*$2n = 28$) and Marquis. Hope never became important commercially, but varieties with the Hope resistance to stem rust, or that of a sister selection H-44, became leading varieties in the hard red spring area. Two stem rust-resistant varieties of durum, Carleton and Stewart, derived their resistance from Vernal emmer. In 1950, a heavy stem rust epiphytotic occurred in the spring wheat area, and this time even the varieties with resistance derived from durum and emmer were attacked. Why were these "resistant" varieties of wheat later damaged by the *stem* rust epiphytotics? The answer lies in the fact that the organism inciting stem rust *(Puccinia graminis tritici)* has numerous physiologic strains or races. Each race of the stem rust organism differs from other races, just as varieties of wheat are inherently different. One of the races of rust may infect certain varieties of wheat, but not others. This is another way of saying that a variety of wheat may be resistant to certain races of rust, but not others. The races of rust are morphologically similar, but pathologically different. The races are identified only by the response or reaction of the varieties of wheat to infection. Approximately 300 distinct races of stem rust have been identified by the differential reaction of a few varieties. New races of rust originate by hybridization between present races or by mutation, just as new varieties of wheat are developed. A new variety of wheat may be resistant to those races of stem rust that are prevalent in nature at the time of its distribution, but new races, to which the wheat variety is susceptible, may arise and become widespread. Then the "resistant" variety will be attacked. The Ceres variety was damaged by a new race, 56, which was widespread in 1935, 1937, and 1938. It was a new race of stem rust, 15B, that in 1950 attacked the varieties with Hope and Thatcher resistance derived from emmer and durum.

Breeding stem rust-resistant varieties is a continuing process because the rust race complex is always changing. As new races of rust appear which infect commercial varieties, genes for resistance to the new races must be searched out and incorporated into adapted strains. It is necessary for the breeder to know which races of stem rust are widespread, to learn about the new races that are increasing in nature, and to find varieties resistant to them. To aid in obtaining this information a "rust laboratory" has been established. Hundreds of collections of stem rust are examined each year at this laboratory and the physiologic race or races of each collection are identified. Thirteen varieties of wheat are used to differentiate the races of stem rust. The reactions of these "differential" varieties to four races-56, 17, 11, and 15B-are given. Whether a variety is resistant or susceptible to a specific race may be determined by the intensity and nature of the infection that the rust pathogen incites in the wheat plant. The severity of the infection is also affected by the temperature of the atmosphere and the age of the plant. Some varieties are resistant at low temperatures but not at high temperatures. Some varieties are resistant both in the seedling stage and as mature plants, while others are resistant only after maturity. Many of the races of stem rust are actually mixtures of genetic types and could be further subdivided into sub-races or biotypes if tested on additional differential varieties. The resistance of a variety of wheat to a particular race of stem rust can be determined only by growing the variety under conditions where it will be exposed to an attack of that race of the rust organism and observing the number, size, and character of the rust pustules that develop. The severity of the disease is generally expressed as a percentage of the maximum possible infection. Since rust does not occur in the field every year, artificial rust epiphytotics are frequently established in a disease nursery where the varieties and strains are growing. Wheat varieties may also be tested for their rust reaction in the greenhouse. Inoculations in the field are usually made by injecting a suspension of rust spores into the leaf whorl of a susceptible "spreader variety" early in the season. The rust spreads from the spreader variety into adjacent rows of the selections or varieties being tested. In the greenhouse, rust spores are usually rubbed or dusted on wheat plants, which are growing under temperature and humidity conditions favourable for germination of the spores and the development of the rust. By use of artificial inoculations, a variety may be tested with known races of rust, and thus information about the reaction of the variety to specific races of the rust organism may be obtained.

Interspecific crosses have been used extensively to transfer genes for resistance to stem rust to common wheat. Principal sources of stem rust resistance genes used in the past were Yaroslav emmer (used in Hope variety), Vernal emmer (used in Carleton and Stewart durum), and Iumillo durum (used in Thatcher)., Hope and Thatcher, or other derivatives from these crosses, have been used in the development of many resistant spring wheat varieties. After race 15B, which infects the above varieties with resistance derived from emmer and durum, became widespread, new sources of resistance were needed again. Certain wheats, which originated in the Kenya Colony of Africa, are resistant to race 15B, but some of the Kenya varieties are susceptible to other races of stem rust. No variety is known to be resistant to all races of stem rust. Khapli emmer is resistant to all races of stem rust collected, but it is susceptible to certain races identified from South America. It is usually necessary to combine several genes for stem rust resistance from different sources into a single variety in order to give protection to a large number of races. The inheritance of resistance varies with the variety being studied and the specific race of rust with which the variety is inoculated. Resistance in Hope was reported to be due to one or two dominant genes, although other workers indicated additional genes were involved. Kenya 58 has been shown to have a gene governing resistance to race 15B. Gabo, Lee, and Timstein have two complementary genes governing resistance to race 56. Other modes of inheritance have been reported with other parent varieties.

Table 7.17. Reactions to Four Physiologic Races of *Puccinia graminis tritici* of the Thirteen Wheat Varieties Used to Differentiate Races of Stem Rust

Variety	*Species*	*Reaction to Physiologic Race Number*			
		56	*17*	*11*	*158*
little Club	*Triticum compactum*	S	S	S	S
Marquis	*Triticum vulgare*	S	S	S	S
Reliance	*Triticum vulgare*	S	R	S	S
Kota	*Triticum vulgare*	S	S	S	S
Arnautka	*Triticum durum*	R	S	S	S
Mindum	*Triticum durum*	R	S	S	S
Spelmar	*Triticum durum*	R	S	S	S
Kubanka	*Triticum durum*	S	S	S	S
Acme	*Triticum durum*	S	S	S	S
Einkorn	*Triticum monococcum*	R	S	S	S
Vernal	*Triticum dicoccum*	R	R	R	S
Khapli	*Triticum dicoccum*	R	R	R	R
Lee	*Triticum vulgare*	R	R	R	S

S: Susceptible; R: Resistant.

Leaf Rust (*Puccinia rubigovera tritici*)

Leaf rust is found almost everywhere that wheat is grown. Loss in yield due to leaf rust is generally not so great as the loss caused by a severe stem rust epiphytotic. Yield reductions of 5 to 10 per cent are most common, but severe epiphytotics during early stages of growth may sometimes kill seedling plants. In the eastern and southeastern states damage from leaf rust occurs with more regularity than in the hard red winter and spring wheat areas, although damage in the latter areas has been increasing. The organism inciting leaf rust, like that inciting stem rust, has many physiologic races. One hundred sixty-three races had been identified by 1955. The prevalence and distribution of the races are constantly changing as new virulent races develop and become widespread. Many races of leaf rust, like those of stem rust, could be subdivided into biotypes or sub-races if additional varieties were to be used as differentials. The breeder must know the races present in his area when selecting a source of resistance. Many varieties have resistance to a small number of physiologic races of leaf rust and have at one time or another been reported resistant in the field when only races to which they are resistant were present. Examples are the hard wheat varieties Kanred, Tenmarq, and Kawvale, and the soft wheat varieties Mediterranean, Denton, Wabash, Vigo, and Blackhawk. The spring wheat varieties derived from Hope were considered resistant until new races of leaf rust to which they were susceptible became widespread in the spring wheat area. Certain varieties of durum and emmer have shown considerable resistance to leaf rust through many years. Varieties of common wheat currently used as sources of leaf rust resistance include Exchange, Frontana, Rio Negro, Surpreza, and Timstein. Genes for resistance to leaf rust have been transferred to common wheat from *Triticum durum* ($2n$ = 28), *T. timopheevi* ($2n$ = 28), *Aegilops umbellulata* ($2n$ = 14), and other species. The mode of inheritance of resistance to leaf rust may vary from one to many genes and depends upon the parent variety. As with stem rust, inheritance studies are influenced by the races of rust present, the stage of plant growth, and other factors. Artificial epiphytotics of leaf rust may be established in the field or in the greenhouse. The method is similar to that used for establishing artificial epiphytotics of stem rust.

Loose Smut (*Ustilago tritici*)

Loose smut is a common disease in most wheat-growing areas. The fungus inciting loose smut is borne inside the seed and the hot water or soak methods of seed treatment are required for its control. Since these methods are difficult to manipulate and seldom are used by the farmer, the use of resistant varieties offers the best means of preventing loss from this disease. Nineteen physiologic races of loose smut have been described. Kawvale was resistant to all these races. Thorne, Trumbull, and Leap were resistant to many of them. Derivatives of Kawvale, such as Pawnee and Ponca, also have excellent resistance. Hope and many varieties with Hope parentage are resistant to loose smut. Resistant strains are identified by their freedom from smutted plants after being subjected to natural infection or to artificial inoculation. The artificial inoculations are usually made by injecting dry spores or a spore suspension into the florets at the time of flowering. The smut reaction is usually expressed either as the percentage of smutted heads or as the percentage of smutted plants. Inheritance of the Trumbull resistance has been reported to be monogenic. Three factors for resistance have been identified in Hope and two factors in Kawvale. The fungus destroys the ear completely, turning them into a black loose powdery mass consisting of spores and leaving behind the rachis. The disease is created artificially at the flowering stage. Screening against loose smut disease gets completed in two Years. In one crop season artificial inoculations are made following "Go Go" method and in the second-year symptoms are observed by growing the infected seeds. The wheat material to be screened is sown in one-meter row in the field. At mid anthesis stage the central florets are removed with the help of forceps and the rest are cut with the scissors in such a way that the ovaries get exposed. Such ears are covered over with the crossing bags open from both ends. The lower end is stapled tightly at the stalk of the enclosed ear. Loose smut spores from the freshly collected smutted ear hands are dusted over the enclosed ears from the open end of the crossing bags. After inoculations the open end of the crossing bag is then stapled. The pathogen invades the embryo through stigma. Two three ears of each strain are inoculated and the seeds collected from the inoculated ears are again sown in one-meter row in the following season for appearance of the disease. The disease is systemic in nature and the internally seed borne pathogen moves along with the growing plant. The infected plant will bear the loose smut infected ears instead of normal ears. Several other methods of screening, such as drenching of ears in the spore suspension, spray inoculation methods *etc.* have also been used by different workers but "Go Go" method is generally followed for artificial screening of wheat lines against loose smut.

Visual percentage of the smutted heads within the host population is recorded.

F: Free of loose smut infection

TS: Varieties with trace infection (below 5 per cent of plants showing infection)

S: Susceptible (5 25 per cent plants showing infection)

HS: Highly susceptible (above 25 percent plants showing infection)

Karnal Bunt (*Neovossia indica*)

In an infected plant some of the tillers bear infected ears. In the ear only a few grains are infected and most of the grains get partially infected. The infected grains on pressing give out black powder of spore mass. This powder gives peculiar stinking smell. Screening against Karnal bunt is done using syringe method of inoculation. Sporidial cultures of the pathogen are made in PDA from the teliospores. The fungus is kept in sporulating condition by repeated subculturing on potato dextrose agar medium. For inoculations, the sporidial suspension is prepared in water from 7 10 days old sporidial cultures. Wheat strains to be screened are sown in one-meter row in the field and inoculations are done with sporidial suspension at just awn emergence stage of wheat plant growth. To maintain sufficient relative humidity in the field, mist sprayer is run at different time intervals depending upon prevailing weather conditions. Five to ten ears of each strain are inoculated and bunted grains are counted after harvesting to calculate percent infection. Besides syringe method of inoculation, several other methods of screening, such as spraying of inoculum, dipping of ears in sporidial suspension. have also been tried but syringe method of inoculation followed by sprinkling of water with mist sprayer in the field is widely adopted. The strains which score upto 5 per cent infection are rated resistant. As the pathogen affects the grains to varying extents, severity of infection in the grains is also taken into account for calculating coefficient of infection. No. of grains showing incipient infection, 1/2,3/4 and whole grain infected are multiplied with the numerical values 0.25, 0.5, 0.75 and 1.00. Finally, the added values, thus, obtained is divided by total grains and multiplied by hundred.

Flag Smut (*Urocystis agropyri*)

Long narrow lead grey or black streaks or stripes running parallel to veins are formed on the leaves. The stripes eventually ruptures and expose black sooty mass of spores. The reduction in crop yield is due to complete loss of productivity of infected plants. Screening against flag smut is carried out in the sick plots. Infected plants collected from different fields are buried in the field during off season. The infected plant debris are taken out from the field before sowing and crushed. The crushed infected plant material is mixed in the soil and also smeared over seeds before sowing. Every year infected plant debris are mixed in the soil in the same field for making sick plot. The disease scoring is based on the percentage of infected plants in test rows of 2 meter length accommodating more than 60 plants. The entries showing 0 10 per cent disease are rated as resistant. Four specific genes, *Ut1, Ut2,* (Nielsen, 1977), *Ut3,* and *Ut4* (Nielsen, 1982), have been found to confer resistance to loose smut, which is caused by the fungus *Ustilago tritici.* Chromosomal locations of genes are not known.

Foliar Blights (*Drechslera, Alternaria, Septoria*)

Many pathogens are associated with leaf spots and blights. *Helminthosporium (Drechslera)* and *Alternaria* are the most important pathogens responsible for producing spots, blights and bloches. Elongated straw coloured spots are common. Speckled leaf blotch (*Septoria tritici*) also occurs in Punjab. Black tip of seeds, black spot of glumes are also associated with these fungi which renders the grains black tipped and shrivelled. The grains loose weight and cause reduction in yield. Leaves from the blighted crop are dried under shade conditions. In the next season, cut pieces of the blighted leaves are broadcast in infector rows sown in the material to be screened. Frequent watering is done to maintain sufficient relative humidity in the field so as to provide congenial conditions for the pathogen to grow and multiply. Pathogens isolated, maintained and multiplied over potato dextrose agar medium have also been used for spraying of infector rows after making spore suspension. Recently at Directorate of Wheat Research Karnal, a new technique has been developed wherein sorghum seed medium is used for multiplying the pathogen (soaked sorghum seeds in water for 24 hrs is sterilized). Screening material is planted in temperature and humidity regulated polythene house as there is no irrigation constraint. Repeated transfer of pathogenic cultures on sorghum seed medium do not harm their pathogenicity. On the basis of infection on the leaves from lower leaf to flag leaf and percentage of leaf surface covered by lesion, numeric score was given. The varieties with no infection and probably representing an escape were rated O and OE, respectively. The varieties showing upto rating 3 were rated resistant,4 and 5 moderately resistant/susceptible and 6 to 9 as susceptible to highly susceptible, respectively. More recently double-digit system in 0 9 scale has been used. The data were recorded on flag leaf and one leaf below flag at 59 64 stage in Zodok's scale. Score 48 means 4 scale on flag and 8 on leaf below flag leaf. Scale used for categorizing reaction is as follows:

Score	*Category*
00 13	Highly resistant (HR)
14 35	Resistant(R)
36 57	Moderately Resistant (MR)
58-78	Susceptible (HS)
>78	Highly susceptible (HS)

Powdery Mildew (*Erysiphe graminis tritici*)

The fungus develops numerous superficial white colonies on all the above ground parts of the plant. The white colour of the colonies changes from grey to reddish brown when cleistothecia develop. Infected plants become stunted due to reduction in the size and number of leaves. Powdery mildew specimens are maintained and multiplied on 7 10 days old seedlings on wheat variety Agra Local or any known susceptible variety in 6" pots. The potted seedlings are covered with chimneys and the upper side covered with four layers of muslin cloth. The temperature of 20 22°C, relative humidity 70 80 per cent and 20,000 25,000 lux light intensity with 14 hrs day length is maintained by cool fluorescent tubes of 40W. For evaluation of germplasm at the seedling stage, the seeds of test variety are sown in 7 rows in trays\plastic bread boxes containing a mixture of sandy loam soil and Farm Yard Manure (3: 1). One row of highly susceptible variety was always sown in each tray for secondary spread of the pathogen. The leaves having profuse growth of powdery mildew conidia are gently tapped so as to retain younger generation of conidia. The seedlings to be inoculated are dusted with conidia of the pathogen. After inoculations, the boxes/trays are shifted to the growth chamber maintaining temperature, relative humidity and light. The disease reaction is recorded, 10 12 days after inoculations. The lines are rated resistant/susceptible on the basis of type of infection, development of mycelium and conidial production. The overall reaction is 'R' if there is no infection slight flecking, moderately developed mycelium and slight spore production. For field screening genotype may be sown in 1 2 meter row 25 cms apart. Susceptible variety may be planted after every 20 rows. The pots with powdery mildew inoculum are introduced in the field in the first week of February and on each alternate day, thereafter. Frequent watering is done for the rapid increase and progress of powdery mildew. Disease data is recorded on the plant on 0 9 scale as mentioned under foliar blight. The specific genes for resistance to powdery mildew disease caused by fungus *Erysiphe graminis* are listed in table. In hexaploid wheats, *Pm4a* and *Pm5* were derived from *T. dicoccum, Pm4b* from *T. carthlicum, Pm6* from *T. timopheevi,* and *Pml* and *PmB* from *Secale cerea/e.* Resistance due to non-specific genes of quantitative nature has also been described (Bennett, 1984).

Hill Bunt or Stinking Smut (*Tilletia caries* and *Tilletia foetida*) and Dwarf Bunt (*Tilletia controversa*)

Bunt or stinking smut is found throughout the wheat-growing areas. It is most severe in the Pacific northwest and the intermountain regions. Common bunt is both seed borne and soil borne. Seed treatments have been used extensively for many years to control the seed-borne spores; but until recently chemicals were unknown which would control the soil-borne spores. In the Northwest where the soil is often contaminated with smut spores, resistant varieties are desirable along with seed and soil treatments for control of common bunt. Dwarf bunt is not amenable to control by seed treatment, and resistant varieties, along with soil treatment, are the means for its control.

Varieties of wheat that are used as major sources of resistance to common bunt are:

1. *Hard red winter.* Turkey selections (including Rio and Oro), Hussar, and Ridit (from cross Turkey X Florence).
2. *White.* Florence (introduction from Australia), Martin, and Rex (from cross White Odessa X Hard Federation).
3. *Hard red spring.* Hope (from cross Yaroslav emmer X Marquis).

Various commercial varieties have been developed with resistance from one or more of these sources. Seventeen physiologic races of *Tilletia caries* and fifteen races of T. *foetida* which incite common bunt have been identified. Five major genes for resistance to specific races and two weak genes have been identified as follows: Martin MM and M_2M_2, Hussar *HH,* Rio *RR,* Turky *TT,* and the weak genes XX and *YY.* Five of these genes, *MHRTX,* are associated in the same linkage group. The Rio gene *(RR)* and the Turkey gene *(TT)* are closely linked and react similarly to the different races. Varieties carrying the Martin-Hussar gene complex are highly resistant to dwarf bunt. The Martin gene and either the Turkey or Rio gene together give resistance to twenty-five races of *T. caries* and *T. foetida* and to several races of *T. controversa.* The Hope variety and several strains with Hope parentage were resistant when tested with twenty-six races and spring planted, but the resistance derived from Hope has not proved satisfactory in fall-planted wheats. Wheat may be tested for resistance to bunt by dusting the seed before planting with the smut chlamydosp6res. Infections are reported as percentage of smutted plants. The ears of infected plants remain erect, dark or olive green and with open glumes within which the bunt balls are seen. Bearded varieties have deformed awns or may be completely absent. The disease is seed borne. The growth of the pathogen hyphae keeps pace with plant development and at maturity bunt balls are formed in place of wheat kernels. The disease is created artificially by sowing the seeds thoroughly mixed with chlamydosporic inoculum mixture of *T.caries* and *T. foetida*. The disease incidence is recorded by counting infected tillers in a row.The data is represented in the form of percentage of infected tillers in a test row of minimum of 60 plants. 'F' represents varieties free of infection, 1-10 percent resistant and above 10 percent susceptible. Ten specific genes are known to confer resistance to bunts caused by the fungi *Tilletia caries, T. foetida,* and *T. controversa.*

Table 7.18. Specific genes for resistance to *Tilletia caries, T. foetida* and *T. controversa;* Specific genes for resistance to Hessian fty [*Mayetiola destructor (Phytophaga destructor)]*

Resistance Gene	*Chromosome Location*	*Resistance Gene*	*Chromosome Location*
Bt1	2B	H1	-
Bt2	-	H2	-
Bt3	-	H3	A
Bt4	1B	H4	-
Bt5	1B	*H5*	-
Bt6	1B	*H6*	5A
Bt7	2D	*H7*	-
Bt8	-	*H8*	-
Bt9	-	*H9*	5A
Bt10	-	*H10*	5A
		H11	-
		H12	-
		H13	-

Foot Rot/Root Rot/Seedling Blight (*Sclerotinia rolfsii, Drechslera sorokiniana*)

The root and collar show browning when the disease is associated with *Drechslera sorokiniana.* Later, elongated brown spots appear on leaves, glumes and spikelets. Ears are poorly filled, grains shrivelled and black tipped. The disease adds to yield losses on establishment of epidemics due to some other major pathogens than *Sclerotinia rolfsii* or may itself acquire the potential to inflict harms above economic threshold. The sclerotial foot rot disease is created artificially by growing entries in pots having sterile soil thoroughly mixed with mycelium and sclerotia of the pathogen multiplied on corn meal sand medium. The disease incidence is noted as percent necrotic seedlings after 30 days of sowing

Wheat Blast

A fearsome fungal disease first reported in Bangladesh in 2016, is a huge threat to food safety and security in South America and South Asia. Directly striking the wheat spike, wheat blast can shrivel and deform the grain in less than a week from the first symptoms, leaving farmers no time to act. To mitigate the threat of wheat blast, a Precision Phenotyping Platform (PPP) was established in Jashore in 2018 to screen wheat germplasm for the disease. The platform has been a reliable source of screening for wheat lines around the world and has proven a great success towards taming the wheat blast disease. Currently around 5,000 wheat lines from Bangladesh, China, India, Japan, Mexico, Nepal and Pakistan, are being tested under natural and artificial inoculated conditions. "Bangladesh is our neighboring country and wheat blast is a very concerning issue for us. A wheat pathologist, is leading independent wheat blast research activities including mutation breeding and the use of nano technology and probiotic bacteria in controlling wheat blast. The pathogen can infect all aerial wheat plant parts, but maximum damage is done when it infects the wheat ear. It can shrivel and deform the grain in less than a week from first symptoms, leaving farmers no time to act. First officially identified in Brazil in 1985, affecting as much as 3 million hectares in the early 1990s. It continues to seriously threaten the potential for wheat cropping in the region. In 2016, wheat blast spread to Bangladesh, which suffered a severe outbreak. It has impacted around 15,000 hectares of land in eight districts, reducing yield on average by as much as 51 per cent in the affected fields. Wheat blast spreads through infected seeds, crop residues as well as by spores that can travel long distances in the air. Blast appears sporadically on wheat and grows well on numerous other plants and crops, so rotations do not control it. The irregular frequency of outbreaks also makes it hard to understand or predict the precise conditions for disease development, or to methodically select resistant wheat lines. At present blast requires concurrent heat and humidity to develop and is confined to areas with those conditions. However, crop fungi are known to mutate and adapt to new conditions, which should be considered in management efforts. There are no widely available resistant varieties, and fungicides are expensive and provide only a partial defense. They are also often hard to obtain or use in the regions where blast occurs, and must be applied well before any symptoms appear - a prohibitive expense for many farmers. The *Magnaporthe oryzae* fungus is physiologically and genetically complex, so even after more than three decades, scientists do not fully understand how it interacts with wheat or which genes in wheat confer durable resistance. It is a fast-acting and devastating fungal disease that threatens wheat production and food security in South America and South Asia. The disease, which originated in South America and first appeared in Bangladesh in 2016, can by dispersed by wind across large distances and spores can be seed borne. There is deep concern among scientists that the disease could spread further across South Asia. A *ex-ante* analysis found that in Bangladesh, India and Pakistan wheat blast could potentially cause losses of 0.89 – 1.77 million tons each year, with 7 million hectares of growing area at risk. The grain in this blast-blighted wheat head has been turned to chaff. In the last three decades, there has been a remarkable progress in breeding for rust resistant varieties, which in turn prevented the calamity of rust epidemic in different epidemiological sub-zones of the country and reduced the competition among pathogens (Bahadur *et al.*, 1994). However, very less progress has been made for other diseases such as foliar blight (Singh *et al.*, 1986). Spot blotch is a disease of importance mainly in warm, humid wheat growing areas where the mean temperature of the coolest month is higher than 17.5ÚC (Dubin *et al.*, 1998). It causes serious yield losses to wheat in India (Joshi *et al.*, 2002). More recently, spot blotch has also expanded into the cooler, non-traditional irrigated rice-wheat production areas (ME-1) (Van Ginkel and Rajaram 1998).

Sprague (1950) reported that *Bipolaris sorokiniana* has a large host range and almost all the plants belonging to family *Poaceae* come under its host range. Apart from wheat, it infects oat, barley, rye, *Phylaris, Agropyron, Pennisetumm, Lollium, Poae, Secale, Setaria etc.* (Bakonyi *et al.*, 1998). Some reports indicate its existence in dicot crops such as alfalfa, red and yellow clover (Gourley 1968) and buckwheat

(Renfro 1963). The host range of the pathogen is reported to differ from isolate to isolate; some isolates had broader, while other had a narrow range. Foliar blight is reported to be a complex of many pathogenic fungi occurring together or simultaneously at different growth stages of wheat crop (Singh 1993). Collectively the disease is called *Helminthosporium* Leaf Blight (HBL). The three important leaf blight diseases reported in South Asia are Spot blotch (*B. sorokiniana*) (Sacc. In sorok) shoem, Tan spot (*Pyrenophora tritici* var. *repentis* (Died.) Drechs.) and Alternaria blight (*Alterania tritici*) (Pras and Prab). However, spot blotch (*B. sorokiniana*) is of increasing concern in South East Asia (Kumar *et al.*, 2010). In addition, it occurs in warm humid wheat growing environments of Latin America, Africa and China. The symptoms of spot blotch appear as small, light brown lesions which are scattered throughout the leaves and increase in size with stage advancement. Later, these lesions coalesce and change to large spots after a week of infection (oval to oblong and measuring 0.5 to 10 mm long and 3 to 5 mm wide) (Chand *et al.*, 2002).

B. sorokiniana

Variabilities in the isolates of *B. sorokiniana* have been morphologically (Maraite *et al.*, 1998) and pathologically (Maraite *et al.*, 1998) reported. However, very little information is available on the aggressive pattern of the *B. sorokiniana* isolates. Pathogenic variability is of crucial significance in disease management where host resistance is the major component. The spot blotch resistance gene(s) in wheat are not known to interact in gene for gene manner, but resistant genotypes are known to show significant reduction in disease development as compared to the susceptible cultivars (Joshi *et al.*, 2002). Hetzler *et al.* (1991) found differences in disease causing ability of different strains on a set of wheat genotypes, which was further confirmed by Maraite *et al.* (1998). Recently, Chand *et al.* (2003) reported that the isolates of *B. sorokiniana* could be grouped in to five categories based on their morphological and pathological traits. In addition to morphological traits, *B. sorokiniana* varies in its pathogenicity on gramineous hosts. However, little information is available about aggressiveness of the *B. sorokiniana*. Sexual reproduction in *B. sorokiniana* is rare and reported only from Zambia (Raemaekers, 1987). As far as variability in asexual population is concerned, it is due to para-sexual recombination (Tinline 1962). However, these results were not verified and other means of variability were not explored. Hetzler *et al.* (1991) found differences in disease causing ability of different strain on a set of wheat genotypes, which was further confirmed by Maraite *et al.* (1998). Other workers also showed that isolates of *B. sorokiniana* possessed pathological variability (Nelson 1960; Hetzler *et al.*, 1991; Maraite *et al.*, 1998). Ten isolates of *B. sorokiniana* (*Cochliobolus sativus*) from different geographical regions of Brazil were analyzed for their virulence on wheat cultivars, morphological characteristics, and growth rate on PDA. Variability in cultural characteristics was observed in the morphology and growth rate between the original isolates and the re-isolates. However, no relationship between morphological variability and virulence was observed among the ten-origional isolates (Oliveira *et al.*, 1998). Chand *et al.* (2003) demonstrated that the five groups of the isolates of *B. sorokiniana* differed for their morphological appearance. Traditional methods used to study variability in pathogens were based on morphological, pathogenicity or biochemical tests. These methods distinguish pathogen isolates on the basis of their physiological characters *i.e.*, pathogenicity and growth behaviors. However, these are highly influenced by the host age, inoculum quality and environmental conditions. DNA markers are not influenced by any such conditions and therefore provide more reliable results. Kumar (2003) for the first time established variability among the isolates of *B. sorokiniana* by using RAPD markers. Jaiswal *et al.* (2007) identified molecular markers to assist the aggressiveness for different groups of *B. sorokiniana* isolates.

Yield losses due to leaf blight are variable, but are important in fields with low inputs and under late sown conditions. The average yield losses due to leaf blight for South Asia have been estimated to be 19.6 per cent (Saari 1998). Yield losses between 20-80 per cent have been reported by Duveiller *et al.* (1998). The losses due to foliar blight may be to the tune of 100 per cent under most severe conditions of infection (Mehta 1994). The pathogen not only reduces yield, but also reduces germination, seedling emergence and decrease the intensity of roots in the subsequent crops (Joshi 1986). Further, ear length has been observed considerably reduced, which in turn, affected the number of grains, which neither were shriveled nor was the test weight much affected. However, when the disease occurred only at flag leaf, the losses of grain/ear were up to 24.2 per cent. Sources for spot blotch resistance have been identified over the years and broadly their origin fall into three categories: Latin America, China and Wild relatives of the wheat or alien species (Van Ginkel and Rajaram 1998). The Latin American sources are mainly derived from Brazil and may trace back to Italian ancestry. Older resistant Brazilian commercial varieties are BH-1146 and CNT-11 (Mehta 1985). The Chinese sources are mostly materials from the Yangtze River Basin. Early Chinese sources of resistance used at CIMMYT include Sanghai#4, Suzhoe#8 and Yangmai#6 (Van Ginkel and Rajaram 1998). *Thinophyrum curvifolium* is used as alien resistance against spot blotch (Mujeeb-Kazi *et al.*, 1996). Using these alien sources in combination with Chinese resistance sources, outstanding lines such as Mayoor and the Chirya series were developed. More than 14000 entries of wheat and related wild species, representing 13 genera and 136 species have been studied at PAU, Ludhiana, Punjab (India) where, *Aegilops triuncialis, Ae. speltoides, Ae. squarrosa, Ae. triaristata, Ae. cylindrica, Triticum dicoccoides,* and *T. boeoticum* were found promising (Singh and Dhaliwal 1993). Chaurasia *et al.* (1999) evaluated 1387 spring wheat germplasm for tolerance to foliar blight disease and observed that none of the genotype was immune to the disease. However, CIMMYT lines showed more tolerance than Indian lines. Joshi and Chand (2002) evaluated 1407 spring wheat lines for their leaf angle and resistance to spot blotch in F_3, F_4 and F_5 generations and revealed that lines homozygous for erect leaf posture showed significantly lower mean AUDPC than those with drooping leaves. Joshi *et al.* (2002) observed a wide range of plant height and days to maturity when evaluated for resistance to spot blotch in two crosses of wheat, and indicated that resistance to spot blotch severity was independent of plant height and days to maturity.

Wheat genotypes have been investigated by wheat workers from time to time and some sources of resistance have been established. Chaurasia *et al.* (1999) evaluated spring wheat lines belonging to Indian and CIMMYT gene pool and found 43 lines as resistant; CIMMYT lines showed better tolerance than Indian lines.

Sources of spot blotch resistance that have been identified over the years broadly falls into three categories; Latin American, Chinese and wild relatives of wheat or alien species (Van Ginkel and Rajaram 1998). The Latin American sources are mainly derived from Brazil and may trace back to Italian ancestry. Older resistant Brazilian commercial varieties are BH-1146 and CNT-1. Chinese source of resistance used at CIMMYT include Shanghai#4, Suzhoe#8 and Yangmai#6. *Thinophyrum curvifolium* has also been used as alien resistance sources at CIMMYT. Inheritance studies on resistance to spot blotch are limited and nature of inheritance is still debatable. Reports indicate presence of both monogenic (Adlakha *et al.*, 1984) and polygenic (Velazquez Cruz 1994) types of resistance. However, the experience of wheat workers to achieve partial resistance in breeding populations (Dubin and Rajaram 1996) suggested polygenic type of resistance.

Genetics of Resistance

Little is known about the inheritance of resistance to spot blotch and is still debatable but seems to be polygenic with additive effect (Velazquez Cruz 1994). Indian resistance sources have shown one or two genes to be involved as previously reported (Adlekha *et al.*, 1984). Velazquez Cruz (1994) found six genes to be segregating in four moderately resistant to resistant lines (Ginuz, Cugap, Chirya and Subaf) developed at CIMMYT with two or three genes providing good resistance level. The evaluation of large number of genotypes (*T. aestivum* and *T. durum*) and triticale over the year has clearly established that majority of the durum wheat genotypes (AB genome) are highly susceptible to leaf blight pathogen. On the other hand, the common wheat genotypes (ABD genome) fall under the category of susceptible to moderately resistant. Most of the triticale genotypes (ABR genome) seem to be moderately resistant to resistant. These data indicated that in common wheat, resistance is probably located on D genome and in case of triticale on the R genome (Chand *et al.*, 2002). Singh *et al.* (1986) studied 102 genotypes for resistance against spot blotch. Out of 102, 13-showed resistance and the segregation in all crosses followed 1 (resistant): 15 (susceptible) ratio indicating that inheritance of resistance to foliar blight was controlled by two major recessive genes. Information on the inheritance of spot blotch resistance in wheat was determined in field studies conducted in four wheat crosses, each involving a Chinese hexaploid parent with high levels of resistance and a commercial cultivar with low to intermediate levels of resistance (Sharma *et al.*, 1997). Data were recorded in the F_2, F_3 and F_4 generations to estimate heritability, involving 150 lines in each cross. Heritability (h^2) estimates for spot blotch resistance was intermediate to high when measured in terms of HDS (Highest disease score) ($0.47 < h^2 < 0.67$) and AUDPC ($0.58 < h^2 < 0.77$) in both the F_3 and F_4 generations in each of the four crosses. Heritability values were somewhat higher for AUDPC than HDS. The results suggest that selection for resistance to spot blotch could be effective in the segregating populations generated from hexaploid wheat parents having different level of resistance (Sharma *et al.*, 1997).

Breeding Progress

Several attempts have been made to check the spot blotch disease but no single effective control measure has been worked out so far. An integrated approach is considered necessary with host resistance as a major component to control the disease (Joshi and Chand 2002). However, in some instances resistant varieties are the only means of controlling this disease. The slow progress in breeding for resistance to foliar blight has been due to several reasons. The pathogen is highly variable and aggressiveness seems to increase over time (Maraite *et al.*, 1998); no immunity is known against spot blotch in wheat (Van Ginkel and Rajaram 1998) and quantitative and qualitative resistance genes are involved (Sharma *et al.*, 1997), making it difficult to assess disease levels effectively except under strictly standardized experimental conditions (Duveiller *et al.*, 1998). Even the sources of resistance against foliar blight are limited (Chaurasia *et al.*, 1999). However, enhancement of genetic resistance to foliar blight in breeding population remained slow and unsatisfactory (Duveiller *et al.*, 1998). This slow progress is generally attributed to high variability in the pathogen originating from diverse sources (Wood, 1962; Kleine and Nelson, 1963), their shift towards aggressiveness (El Nashaar and Stack, 1989) and quantitative nature of inheritance (Sharma *et al.*, 1997). Nevertheless, there are reports concerning differential interactions between the pathogen population and wheat genotypes (Hetzler *et al.*, 1991). Geographically distinct racial patterns in pathogen populations (Hossain and Azad 1992) indicated a strong influence of specific environmental conditions on the expression of resistance and/or susceptibility. These studies suggested that breeding programs should be planned at regional levels, where environmental conditions are likely to be homogenous. Sharma el al. (1997) investigated four wheat populations involving different Chinese hexaploid parents with high level of resistance and a commercial cultivar moderately resistant to spot blotch to observe the response to selection for low and high area under disease progress curve (AUDPC). Selections were made in the F_3 generation for law and high AUDPC of spot blotch and selected progenies evaluated in a replicated field test. Low AUDPC resulted in higher biomass and grain yield, higher harvest index, and higher 1000-grain weight. Results indicated that selection for low AUDPC of spot blotch in segregating generations would be effective in identifying wheat lines with high levels of resistance and would have positive effects on other characters. Beek (1986) reported that it is possible to increase quantitative resistance to foliar blight and yield potential simultaneously. Transgressive segregants for spot blotch resistance due to the recombination of diverse resistance genes of the parents have been reported in crosses involving moderately resistant genotypes (Sharma *et al.*, 1997; Joshi and Chand 2002), which could be utilized in breeding program. Joshi *et al.* (2002) reported that resistance to spot blotch severity was independent of plant height and days to maturity. Therefore, they suggested that for better evaluation of resistance, breeding population should have susceptible checks of different maturity and height groups. They also suggested that if scoring for disease severity is combined with growth stage, selection for resistance in the segregating generations will be more effective. Joshi and Chand (2002) concluded that genes for spot blotch resistance must be combined with those for erect leaf morphology for better control of spot blotch of wheat. In the last few years, genotypes having good resistance have been obtained through recurrent selection and recombination of the best genotypes (Dubin and Rajaram 1996). CIMMYT strategy for spot blotch is to breed for general resistance based on historically proven stable genes. This non specific resistance can be further diversified by accumulating several minor genes and combine them with different specific genes to provide a certain degree of additional genetic diversity (Van Ginkel and Rajaram 1998).

Marker Assisted Selection for Spot Blotch Resistance

The association of leaf tip necrosis with spot blotch resistance was established by Joshi *et al.* (2004). Later, stay green trait was also showed positive linkage with spot blotch resistance (Joshi *et al.*, 2007). These morphological markers are successfully being used in markers assisted selection of spot blotch resistance germplasm lines. Although, many reports of tagging and mapping of several disease resistance genes and QTLs are available in wheat (Langridge *et al.*, 2001) however, not many reports are available for spot blotch. For this disease, the association of resistance with microsatellite markers in bulks of susceptible and resistant progeny lines was reported (Sharma *et al.*, 2007). The QTLs for spot blotch resistance in the Chinese wheat variety, 'Yangmai 6' were mapped on chromosome 2A, 2B, 5B and 6D (Kumar *et al.*, 2009). However, more information with respect to the identification of QTLs in different genetic background was generated when QTLs were mapped in two other resistance sources ('Ning 8201' and 'Chirya 3') and to compare the chromosomal locations of QTLs with 'Yangmai 6' to identify diagnostic markers that can be used for marker assisted selection and to make an effective breeding program. New biotic stresses like foliar blight has emerged a big constraint for the successful production of wheat in South East Asia (Mathur and Coufer, 1993). This disease is now expending towards non-traditional cooler regions like North West Plain Zone (NWPZ) of India, which is considered as major contributor of wheat in South Asia.

Gene Cataloguing for Disease Resistance Stocks

Since the beginning of Green Revolution in India, tremendous increase in production has occurred due to incorporation of dwarfing genes into indigenous germplasm. As a result, India has been able to successfully avert disastrous shortfall in its food stocks. But with time, yield has become stagnant although population is increasing continuously. In same years insect pests and diseases cause upto 70 per cent loss in yield. To sustain the present level of food production resistance genes in crop plants is therefore very important. Use of genetic resistance is an effective, efficient and economical method to combat different diseases. Although, various genes have been identified from different cultivars it is important to precisely identify diverse genes and to catalogue these. "Identification and characterization of new genes in terms of an internationally accepted system" is referred as cataloguing of genes. The need for search of new genes was highlighted by Dr.B.P.Pal in the early forties where he stated that: The basic steps involved in introduction of new genes into the existing gene pool are **Germplasm collection, Purification and maintenance, Evaluation, Characterization of Novel Genes and Linkage with markers and known genes.**

Germplasm Collection

Exploration and collection of diverse germplasm is the first step towards gene cataloguing. This requires the knowledge of the centres of origin and diversity of a particular plant species, the coexisting species, related wild species and other gene pools. This also requires practical and theoretical knowledge on population sampling. Germplasm collection for biotic stress should be aimed at searching of suitable areas, where stress had been known to occur and the place is having long history of co-existence of host and its various pathogens *e.g.* the geographic distribution of resistance to stem rust in wild emmer wheat. Another important point one should keep in mind while collecting material is that the material should be collected from their hot spots/areas of frequent occurrence of disease. In case a plant has evolved in more than one geographic area, efforts should be made to collect accessions from all such areas and ensure cross evaluation against biotic stresses from both the areas. The evaluation of germplasm is a lengthy process for which sets of collected germplasm have to be prepared for sharing among the collaborators for multiplication and evaluation. This process includes., assigning accession number; processing for medium term conservation; evaluation for resistance against particular stress at hot spots; chemical and physiochemical characterization; regeneration and preliminary evaluation of germplasm accessions using standard descriptors; establishment and maintenance of the germplasm.

Purification and Maintenance

Establishing genetic purity (homogeneity) is an important step towards gene cataloguing. Germplasm should be purified for all the morphological traits, physiological traits and resistance to various diseases and insect pests. Plant germplasm which carries no passport data and descriptor information is only of limited use until this information is gathered. Most of the evaluation information gathered for certain priority characters identified by the National Crop Advisory Committee" (CAC's) is entered into the database using specific software. CAC's identify the descriptors for evaluation. Beyond these descriptors individual scientists can evaluate the germplasm for specific requirement in their area of research. Identification of new disease resistance genes involves steps, such as create artificial epiphytotic conditions in the field by repeatedly spraying a mixture of races of a pathogen to select resistant stocks from the germplasm. Pick up only resistant stocks after evaluation against mixture of races and test each stock for its resistance to individual race of the pathogen. Such evaluation helps to identify race specific and race non-specific resistance and the information can be systematically used by the breeders.

Race Specific and Race Non-Specific Resistance

Race specific resistance is the kind of resistance, which is effective against one race and not effective against other races. It is often short-lived and results in boom and bust cycles. The evaluation for race specific resistance is normally done at seedling stage to economize space and physical resources, however, in some cases, adult plant tests along with seedling tests are to be carried out against individual races. Race nonspecific resistance expresses uniformly against all races of the pathogen and it is never complete. This type of resistance does not force selection pressure on the pathogen, therefore, it is long lasting. From multi-pathotype tests on a set of accessions and their comparison with lines having known resistance genes it is possible to postulate the presence of known genes in a stock and to decide whether a genetic study is worthwhile or not. Putative carriers of novel resistance genes are crossed with a susceptible genotype(s) to determine the nature and number of genes for resistance against a particular pathotype. Intercrossing of resistant lines as well as their crosses with known genes conferring similar resistance is further helpful in establishing novelty of the gene(s) in question.

Characterization of Novel Genes

The novel resistance genes can be further characterized on the basis of following criteria:

Standard Phenotype

Most of the disease resistance genes produce stable and repetitive phenotype over a range of environment conditions but the reaction types for some genes are so characteristic and exact that their genes can be precisely identified only on the basis of their reaction types *e.g.* wheat leaf rust resistance gene *Lr1* produces only 0; reaction types against races avirulent on this gene. The reaction types for some genes are so characteristic and exact, that their genes can be precisely identified only on the basis of reaction types.

Infection Type Matching Technique

The gene-for-gene hypothesis by Flor (1942) states that for every resistance gene in the host, there is a corresponding avirulence gene in the pathogen. Having known the avirulance/virulence pattern of the pathotypes by challenging specific pathotypes on lines with known resistance genes, the resistance gene present in the host can be postulated by comparing its infection types with those observed on lines with known genes for resistance. This technique is useful only if there is differential interaction amongst the genes in NIL's/tester lines against different pathotypes. It is also assumed that the host background in which the known gene is present does not interfere with the expression of infection type.

Sensitivity to Environment

Although resistance genes often show stable reaction over environment but some genes are affected by the environment. Therefore, these can be detected by altering the environment. Therefore, these can be detecting by altering the environment. For example in wheat the adult plant resistance gene *Lr13* conferring resistance to leaf rust expresses best at 20-25°C but fails to express at temperatures below 20°C. Similarly, the gene *Lr11* expresses best at 15 °C and fails to express at 25 °C.

Temperature Sensitivity

Some genes like *Lr11* (Johnston, 1964) and *Lr18* (McIntosh, 1983) express best at low temp. (15-18°C) but fail to express at 25°C. Seedlings carrying *Sr6* conferring resistance to stem rust were tested at temp. ranging from 15-30°C using 3°C interval. Resistance began to break at 18°C and disappeared by 24°C (Luig and Rajaram,1972).

Similarly, other genes like *Lr13* (Pretorius *et al.*, 1984), *Lr3*, *Lr16*, *Lr17* and *Lr23* confer resistance to leaf rust at high temp. (Dyck and Johnston, 1983). This temperature sensitivity may be due to different origin of that cultivar.

Expression over Time

Browder and Eversmeyer (1986) studied NILs of Thatcher carrying *Lr1*, *Lr16* and *Lr17*. These lines were inoculated with an avirulent culture held at overnight, then held at 20°C for 0, 2, 4 and 8 days. In the first environment all three lines were susceptible. Two days were sufficient to induce resistance in *Lr1*, 4 days required for *Lr17* and 8 days for *Lr16*.

Sensitivity to Light

Gene *Sr6* has been extensively studied for sensitivity to light and temp. It was found that the resistance conditioned by *Sr6* broke down between 20° and 25°C. Seedlings carrying *Sr6* were susceptible if grown under a regime of 16 hr of light at 24°C and 8hr of darkness at 16°C. However, if the seedlings were given a 16 hr light period at 16°C between 52 and 91 hr after inoculation, they were resistant. Thus, there is a critical interaction between light and temperature (Knott, 1986).

Linkage with Markers and Known Genes

When resistance genes are tightly linked to a marker, its presence can be inferred simply on the basis of the presence of linked marker *e.g.* the genes *Yr9* for stripe rust resistance in wheat, which is linked to a leaf rust resistance gene *Lr26* confers resistance against all but one yellow rust race occurring in India. This gene can be postulated on the basis of leaf rust resistance genes *Lr26* also. Similarly, the wheat stem rust resistance gene *Sr31* can also be inferred on the basis of the presence of the genes *Yr9* and *Lr26*. The gene *Lr20* for leaf rust resistance and *Sr15*, a stem rust resistance gene are linked together. The wheat stem rust resistance gene *Sr24* and the leaf rust resistance gene *Lr24* are also linked. The gene *Lr24* shows resistance against all races of *P. recondita* in India and it can be identified through tests for the gene *Sr24*. Some resistance genes are also linked to morphological markers *e.g.* the wheat genes *Lr34* and *Yr18* conferring resistance to leaf rust and stripe rust respectively, are linked together and can be identified on the basis of flag leaf tip necrosis. As no avirulence is available against both these genes, the leaf tip necrosis is the only method to look at these two genes. The rust resistance gene *Lr10* and the stem rust resistance gene *Sr2* in wheat can be identified on the basis of the markers tenaceous glumes and black chaff, respectively.

Gene Location of Recessive Resistance Gene

Chromosome location is an important step in establishing the uniqueness and novelty of a gene. The procedure for location of resistance genes varies according to the availability of suitable aneuploid stocks (McIntosh, 1987). Confirmation that the resistance gene if located in the particular chromosome can be established by further association of resistance phenotype and chromosome arms by means of telocentric mapping and by linkage with genes previously known to occur in the chromosome. Only when tests are completed and when uniqueness of a gene from a stock is confirmed, a new gene symbol can be allocated. The gene is then put on the catalogue of a gene symbols for a particular crop.

(i) Hemizygous Effective Recessive Allele

If the recessive allele is hemizygous effective, then analysis in F_1 is sufficient for gene location. In 20 monosomics for chromosome not carrying the resistance gene, the F_1 progenies will be heterozygous and will exhibit the dominant trait uniformly without any segregation and in the critical cross we will get 3 resistants: 1 susceptible ratio.

(ii) Hemizygous Ineffective Recessive Allele

In this case, only the homozygous recessive genotype will show the recessive phenotype. In F_2 generation, similar phenotypic ratios are found in critical as well as in non-critical lines. A detailed cytological study of the recessive phenotypic resistant plants in F_2 allows the identification of the critical line. In the critical line, all recessive individuals will be disomic while in non-critical line the recessive individuals can be disomic, monosomic and nullisomic in a ratio of 24: 73: 3. So absence of monosomic or nullisomic among recessive individuals will suggest a critical line. Hussien *et al.* (1998) conducted monosomic analysis to determine the chromosomal locations of three new leaf rust resistance genes recently transferred from *T. monococcum*. In the case of diploid crops, trisomics are used for the location of gene onto the chromosome. Telocentric mapping is used to locate the gene onto respective chromosome arm. Chromosome substitution lines can also be used for gene location. The procedure can be used to confirm monosomic analysis when suitable materials are available or if the inheritance of resistance is suspected to other than a single dominant gene in a line for which monosomic is not available. All the genes except *Lr47* have been identified through conventional methods. But now a days modern techniques are available which help in quick identification of new genes.

Cytological Markers

If complete karyotype of plant species is known, then this will help in the postulation of genes that have been introduced from alien species by introgression or translocation. For this, careful study of karyotype *i.e.* no. of metacentric, telocentric or satellite chromosomes is required which can be compared with the tester stock. Postulation of some genes like *Lr26/Sr31/Yr9/Pm18* in wheat (1BL/1RS) translocation can be confirmed by microscopic examination of the mitotic chromosomes. The bread wheat chromosomes 1B and 6B are satellite chromosomes but in lines having 1B/1R translocation only one pair of distinct satellite chromosome is observed because the short arm of 1B chromosome is replaced with short arm of 1R chromosome which does not have satellite. Similarly, a gene in Japanese wheat line ST-1 with leaf rust resistance derived from rye was located on chromosome 2A by C-banding to identify wheat chromosomes and genomic or fluorescence in-situ hybridization to detect rye chromatin (Mukai and Gill,1991). On comparing the C-banding patterns of wheat chromosome 2A and rye chromosome 2R with the translocated chromosome, similarity is observed among long arms of 2R and translocated chromosomes. The

GISH pattern shows that the rye segment extends beyond the centromere into the short arm of 2A chromosome. Thus, C-banding permits identification of the translocated chromosome whereas GISH provides an accurate estimation of size of translocated segment.

Molecular Technique

Molecular markers are stable and precise markers which are now being used to establish the uniqueness of the new genes and such markers can circumvent all time-consuming procedures. The technique in specifically useful when such genes are derived by interspecific or intergeneric transfers from alien chromosomes. Dedryver *et al.* (1996) reported linkage between molecular markers and wheat leaf rust resistance gene, *Lr24*. Among the RAPD primers, only one OP-H5 detected an additional band in the resistant line, RL6064. Saturated linkage maps are available for almost all the important crops. Now a day, chromosome specific primers and probes are available which help in location of new gene into the chromosome. Recently, Helguera *et al.* (2000) designated a new leaf rust resistance gene, *Lr47* using RFLP. This gene has been introgressed into *T. aestivum* from *T. speltoides*. This gene is tightly linked to RFLP probe, Xabc 465. For convenience, it was changed into PCR based marker. Translocation line carrying *T. speltoides* segment was crossed with different varieties. Backcross plants were used to test the performance of *T. speltoides* specific primer and to confirm the absence of recombination with 7S chromosome segment. Chinese spring nullisomic-tetrasomic lines were used to assign amplified bands to homeologous chromosomes. PCR amplification of DNA showed a characteristic 282bp fragment in translocation line and absent in the cultivars used. This 282bp PCR fragment was present only in the samples that showed characteristic 6.5 kb *T. speltoides* Bam H1 restriction fragment after hybridization with ABC465. These primers can readily be used for identification of this gene in tester stock. Similarly, Jena *et al.* (2000) introgressed a gene from *O. officinalis* which confers resistance to Indian biotype of brown plant hopper (BPH). Molecular analysis of parents and F_2 plants derived from cross between introgression line and susceptible genotype with several RAPD primers identified a 938bp RAPD product lightly linked to gene for BPH resistance. This locus was cloned, sequenced and PCR primers were developed which could amplify resistance specific STS marker linked to resistance gene. The resistance specific DNA marker was mapped into chromosome 11 by use of primary trisomics and doubled haploid mapping population.

Diseases Caused by Facultative Pathogens

In wheat, such fungal pathogens as *Septoria tritici, S. nodorum, Helminthosporium sativum (Cochliobolus sativus), H. tritici-repentis (Pyrenophora trichostoma), Altemaria triticina,* and *Fusarium* spp. cause various leaf, stem, spike, or grain diseases. Relatively little is known about the genetics of resistance to these diseases, possibly because these pathogens are important only in certain agroclimatic regions where wheat is grown. Wherever work on these pathogens has been done, qualitative and quantitative resistances are reported. Identified a recessive gene, *Crr,* located in chromosome 5BL for resistance to *H. sativum.* Adlakha *et al.* (1984) also found simple inheritance of resistance to this disease. In an inheritance study of wheat cultivar

Carifen 12 from Chile, Lee and Gough (1984) found that resistance to *S. tritici* was conferred by a single dominant gene, whereas resistance to *H. tritici-repentis* was attributed to a single recessive gene. Similarly, Wilson (1979) demonstrated that resistance to *S. tritici* in two wheat cultivars was determined by independent, single dominant genes. Kleijer *et al.* (1977) have shown that the cultivar Altas 66 carries a single dominant gene, located on the wheat chromosome 1 B, for resistance to *S. nodorum*. Some reports indicate that resistance to some of these diseases could be polygenic (Karjalainen, 1983).

Bacteria and Viruses

Very little information is so far available regarding the genetics of resistance to bacteria and viruses. However, genetic variation is known to occur, and various varieties show resistance or tolerance. Tola and Kronstad (1984) have reviewed the situation with Barley Yellow Dwarf Virus (BYDV) and believe that the resistance in various genotypes is quantitative in nature. However, Singh *et al.* (1992) have shown that tolerance to BYDV in cultivar ANZA and various other wheats is based on a partly dominant gene *Bdv1*. ***Mosaic.*** The mosaic disease is incited by a virus. Several forms of the disease have been identified.

Soil-borne mosaic is found in Illinois, Indiana, Missouri, and other states in the soft red winter wheat area. The soil-borne wheat mosaic is characterized by leaf mottling, sometimes through development of a rosette, and by dwarfing of the plant, but the symptoms may vary with the severity of the disease and the variety of wheat. Use of resistant varieties is the only practical control. Two resistant soft red winter wheat varieties, Prairie and Royal, were developed by selecting healthy plants from heavily mosaic-infected areas in fields. Vigo, Saline, Seneca, Chancellor, Thorne, Butler, Knox, Vermillion, and other varieties are also resistant.

Wheat streak mosaic causes considerable damage in the hard red winter wheat area in certain years. The virus causing streak mosaic overwinters in living diseased wheat plants and certain perennial grasses. The virus is carried from one plant to another by the wheat curl mite. If the mite feeds on mosaic-infected plants, the virus is taken into its body. When the mite is moved to a healthy plant and begins to feed upon it, the virus is transmitted to the plant tissue and starts an infection. Symptoms are streaked and mottled leaves, followed by stunting of infected plants to the extent that sometimes there is total crop failure. No highly resistant varieties are known although strains of *Agrotricum (Agropyron* X common wheat) are resistant. However, the genes for resistance in *Agrotricum* strains appear to be on chromosomes that are non-homologous (do not pair) with wheat chromosomes. If this is verified it may be very difficult to transfer the resistance to common wheat.

Catalogue of Symbols

Only when all tests are completed and uniqueness of a gene from a stock is confirmed, the gene is ready for designation. The matter is put forward to the International Committee which then allocates a new gene symbol (McIntosh, 1974, 1983). The gene symbol is then put on the catalogue of symbols for a particular crop. Since crop production is becoming more and more scientific, identification of new genes and their systematic incorporation in developing crop varieties is essential. Uptil 1980s, all the resistance genes has been designated mostly by conventional methods which is very lengthy and cumbersome. But the modern methods of gene identification like C-banding, *in situ* hybridization and molecular markers will help to increase the speed of identification of new resistance genes and their selection in segregating population. Hopefully, this assess to new genes will enhance the yield potential of different crop plants. In naming hereditary function, the use of languages of higher internationality should be given preference. Symbols should be written in italics or in capital Roman letter of distinctive type. All letters and numbers used in symbolization should be written on one line *e.g. Lr13* and not Lr^{13} or Lr_{13}. Temporary symbol designation: Where linkage data are not available, provision has been made for temporary symbols. These shall consist of the basic symbol followed by an abbreviation for the stock and an arabic no. referring to the gene *e.g. Sr.Fr1* (Sr : stem rust; Fr : Federation and 1 : Gene number one in cultivar Federation.).

Organization of catalogue: Information should be given in the order: Gene symbol with principal reference to the particular gene or gene symbol in parentheses. Synonyms (with reference in parenthesis). Chromosome and chromosome arm location, if known, with reference in parenthesis. Stocks that carry the particular gene.

L: N_1. Stock, with no. of backcrosses indicated.

S: Chromosome – substitution stocks, with no. of backcrosses indicated.

V: Cultivated stocks in increasing order of complexity.

So far, the cataloging of genes is systematic and well defined only in wheat. Efforts are being made to bring uniformity in gene cataloguing for various other crop plants also.

Characterization of Genetic Stocks of Leaf Rust Resistance Genes Using Molecular Markers in Wheat

The impact of leaf rust disease caused by *Puccinia triticina* (syn. *P. recondita f. sp. tritici*) on yield reduction in wheat (*Triticum aestivum* L.) is well documented globally, which ranges from 10 per cent under moderate to 65 per cent under intense epidemics. For nearly five decades, breeders have exploited inter-genomic transfer route to transfer resistance to this pathogen into bread wheat through its wild relatives like *Agropyron spp., Aegilops spp.* and *Triticum tauschii* since the discovery of resistance to the disease in the wild relatives of wheat and its transfer to wheat by Sears. Several of these genes (*Lr* genes) have been effective in imparting resistance to the tropical races of *P. triticina* after their transfer into different cultivars. Most of these *Lr* genes are in agronomically poor genetic backgrounds making the genetic stock unsuitable in breeding programmes. But, a concerted effort led to the transfer of the alien *Lr* genes into favourable agronomic backgrounds in India, creating thereby, different near-isogenic lines (NILs). However, during the backcross breeding method employed, the selections were based on the phenotypic assessment of the resistance in general and not on gene-specific virulent races of the pathogen since the same are not available in India. In the absence of *Lr* gene specific resistance expression, unintended and undetectable experimental errors such as wrongly identified selections or wrong genotype being mistaken for the desired donor parent could occur. Such errors committed would perpetuate in the breeding materials where rust resistance wheat breeding is in progress. The molecular markers which are fragments of DNA linked to the gene in question such as random amplified polymorphic DNA (RAPD), sequence characterized amplified regions (SCAR) and microsatellite markers like simple sequence repeats (SSRs) are very effective in detecting such miss-identities and errors in genotyping. Removal of redundant lines and duplicates from genetic stock collection is one of the major aims today in increasing the efficiency of plant breeding and germplasm management. Recently, a SCAR marker developed from one of the RAPD markers linked to the gene *Lr32* also did not distinguish the gene *Lr28* in the NILs. Further, a SSR null allele marker, specific to chromosome 4AL discovered to be associated with the gene *Lr28* in the NIL as well as donor parent, did not distinguish the *Lr32* derivatives. Similarly, the sequence characterized amplified region (SCAR) marker developed earlier by Cherukuri *et al.*, 2003 at this centre as tagged to the gene *Lr19* derived from *Agropyron elongatum* located on chromosome 7DL, which failed to distinguish another alien gene *Lr24* also sourced from *A. elongatum* located on chromosome 3DL. The above observations brought out a possibility of the occurrence of duplicate or misnamed lines in Indian elite genetic stock. Very recently we have been able to identify microsatellite (SSR markers) associated with the gene *Lr19* which enabled us to distinguish the *Lr19* and *Lr24* confusion. The new set of SCAR marker linked to *Lr9* enabled to validate the genetic stocks carrying *Lr9*. The markers thus linked to genes *Lr9, Lr19, Lr24* and *Lr28* were effective in identifying not only the wrongly labeled lines, but also their right identities. Validation of the markers with international reference stock is essential for a common base line communication between laboratories with reference to the markers in question. This was made possible through the near isogenic lines and other lines maintained as international reference stock at Sydney. This process was very useful in detecting redundancies and duplications in the Indian leaf rust resistance genes' genetic. To distinguish between *Lr28* and *Lr32*, the bread wheat material used in the present study comprised of 10 elite near-isogenic line (NIL) pairs, one carrying gene *Lr28* and the other carrying the gene *Lr32* in a common wheat cultivar background. Similar set was employed to distinguish between *Lr19* and *Lr24* in another experiment. The third set described dealt with the NIL pairs listed as isogenic for *Lr9* in different backgrounds. The background cultivars (recurrent parents) making up the NIL pairs for the *Lr* genes were HD 2285, HD 2329, HUW 234, Kalyansona, LOK-1, NI 5439, PBW 226, Sonalika, WH 147 and WH 542 representing wheat varieties released for cultivation in all the agroclimatic regions of the country. All the recurrent parents are highly susceptible to leaf rust disease for the last two decades in India to different races of the pathogen. As a control, the four *Lr* gene donor parents of the NILs, all the four Lr genes were also included for comparison, two concerned donors in each set.

Screening for Leaf Rust Reaction

Nine-day old seedlings raised under controlled environmental conditions at

National Phytotron Facility, IARI, New Delhi were inoculated with the most virulent leaf rust pathotype 77-5 and kept in a humid glass chamber for 36 hrs. at 25°C for incubation. After this, plants were shifted to muslin cloth chamber at the same temperature. Disease reaction was recorded at 10-12 days after inoculation using 0-4 scale described by Stakman *et al.*, where, immune infection type with no visible symptom is scored as '0', fine hypersensitive flecks of dead spots on the leaves are scored as ';' for highly resistant type, small uredial pustules with necrosis scored as '1' for resistance, small to medium size uredial pustules surrounded by necrosis or chlorosis in green islands scored as '2' for moderate resistance, appearance of the medium pustule with mild halo region without chlorosis or necrosis scored as '3' for susceptibility to the disease and appearance of large uredial pustules with a high degree of sporulation without chlorosis often with a ring of secondary uredia around a pustule scored as '4' for highly susceptible type. In the case of susceptibility with high degree of sporulation without the ring of secondary spores, the susceptibility is scored as 33^+.

Lr9

The race specific genes are known to impart seedling resistance and the resistance remain effective throughout the life of a plant. Gene *Lr9* (from *Aegilops umbellulata* Zhuk.) conferred seedling resistance to a number of leaf rust pathootypes and was also effective in adult stage (Sawhney *et al.*, 1977; Tomar and Menon, 1998). A number of cultivars carrying *Lr9* were released in North America, but the pathotypes virulent on *Lr9* were reported from soft winter wheat growing areas. Exotic genes for resistance are no better than any other genes if they are used alone. Tomar and Menon (2001) produced near isogenic lines in different Indian cultivars suited to different agro-climatic regions. India too has relaesed a cultivar Sonali (HP1633) with *Lr9* in north eastern region for late sown irrigated conditions in 1992. Sonali has been postulated to carry stem rust resistance genes, *Sr2, Sr7a, Sr11+* and Yr2+ for stripe rust. A new pathotype virulent on *Lr9* was identified after deploying it in 1992. More over this gene unlike *Lr24* does not combine with yield and hence was not utilized on large scale in Indian wheat Breeding programme.

Lr19

The gene *Lr19* (Agatha) conferred total immunity to Indian leaf rust races at seedling stage (Sawhney *et al.*, 1977) and also exhibited resistance at adult plant stage for a long time at hot spot location, Wellington (India). A wide spectrum of races prevailing. However, virulence(s) on *Lr19* were detected. Two cultivars *viz.*, PBN142 (HD2189/NI917//Agatha) and WH533 (Agatha/Yacora17) were released for general cultivation in 1989 (Maharashtra state) and 1993 (Haryana state), respectively. Sivasamy *et al.* (2009) developed multi-gene lines using two independent DNA segments with two linked genes each namely, *Lr19*/Sr25 and Sr36/Pm6 transferring into 15 bread wheat (*Triticum aestivum* L.) cultivars, susceptible to leaf rust, stem rust and powdery mildew. 7D.7Ag segment with *Lr19*/Sr25 increased grain yield potential by 10-15 per cent in a range of genotypes. The linked genes *Lr19*/Sr25 have been combined in an Australian line Cook with another DNA segment carrying Sr36/Pm6 from *Triticum timopheevi*. India remained free from virulence attacking *Lr19* till 2005, when a new virulence 77-8 (253R31) appearing close to the pathotype 109R31 was detected (Bhardwaj *et al.*, 2005) from Karnataka and Gujarat states. The pathotype 77-8 is avirulent to *Lr9*, Lr23, *Lr24* and *Lr26* and it is still a very useful resistance in Indian wheat breeding programme as it is linked with stem rust resistance gene Sr25.

Lr28

The gene *Lr28* transferred from *Aegilops speltoides* Tausch. has not been used on a large scale, although it exhibited a high degree of resistance in seedling and adult stage for many decades in India and in Europe and other ecological zones. Cultivar Sunland was released in Australia in 1992. Although no detrimental effect appear associated with the presence of *Lr28*, the durability of resistance is likely to be low. Kumar and Raghavaiah (2004) reported that the gene *Lr28* as integrated in the backcross lines developed by Tomar and Menon (1998) neither carried any undesirable linkage drags nor affected yield of the lines. India also released a variety MACS 6145 (HW2034) with *Lr28* (improved version of C306 (Rao *et al.*, 2007) utilizing a translocation CS 2A/2M #4/2 under rainfed condition in North Eastern plains. It does not show any disadvantage in yield. However, a new virulent pathotype 121R60-1 overcoming the resistance of *Lr28*, which appears closely related to the most prevalent pathotype 121R63-1 (77-5) has been reported by Bhardwaj *et al.* (2010). The *Lr28* is still apotential gene which can be utilized in combination of other genes which are effective this particular race. The genes *Lr9*, *Lr19* and *Lr28* have been rendered ineffective by new but different leaf rust pathotypes. However, these genes are potentially useful to provide a high degree of seedling and adult plant resistance if used in combination. To prevent rapid breakdown of seedling resistance genes and combating rust epidemics a strategy to creat pyramids of these genes may provide greater durability of resistance in cultivar as suggested earlier. Samsampour *et al.* (2009) successfully made an effort to pyramid seedling resistnce gene *Lr24* along with Lr48 an adult plantresistance gene through marker assisted breeding. Sivasamy *et al.* (2009) combined *Lr19*/Sr25 and Sr36/Pm6 for leaf rust, stem rust and powdery mildew resistance genes in a number of Indian cultivars through conventional breeding taking advantage of useful linkage and validated the presence of genes through molecular markers.

Validation of the SCAR and SSR Markers Linked to Genes *Lr9*, *Lr24*, *Lr28* and *Lr19*

Among the total known genes of resistance (51 for brown rust, 45 for stem rust and 37 for yellow rust) only a limited number of them (*Lr10*, *Lr13*, *Lr23*, *Lr26*, *Lr24* in brown rust, *Sr2*, *Sr9*, *Sr11*, *Sr31* in stem rust and *Yr2*, *Yr2* (KS), *Yr6* and *Yr9*) were frequent in our breeding programme, so far. Also isolated efforts were made to characterizing the phenomenon of slow rusting (a trait of durability) in Indian wheat germplasm. At this juncture not many genes or their compatible combinations in use, are capable of providing complete protection from prevalent stripe and leaf rust races. Therefore, it becomes imperative to utilize genes from alien sources as well as slow ruster lines (APR-Adult Plant Resistance), to keep the rust epidemics under check in Indian sub- continent. This can be achieved by pyramiding effective resistance genes, which are difficult to monitor in the field for expression of individual resistance genes against the background of other resistance genes. With the advent of molecular marker technology, it is now possible to tackle such complex problems. HI 8498 (Malavshakti) is currently the most popular durum wheat cultivar in Central India due to its high yield with earliness, disease resistance, and excellent grain quality and has proved to be a truly 'landmark' variety in the history of wheat crop improvement in central India. It showed resistance to stem rust pathotypes 40A (62G29) and 40-1(62G29-1); however it is susceptible to 117 group. As the presence of the *Sr36* gene is not documented in any of the known durum genotypes, initial survey was taken up to look for the gene. Hence, 74 durum wheat genotypes representing a cross section of Indian durum wheat germplasm were analyzed using closely linked molecular marker *Xstm773-2,* but none of the durum genotypes tested were found to carry the gene. Australian bread wheat cultivar 'Songlen' which was documented to carry *Sr36* used as donor for this resistance gene. The presence of the gene was also confirmed with the closely linked molecular marker. Stem rust pathotype117-6(37G19) was used for tracking resistance derived from the donors. Cross was performed between Songlen/HI 8498(recurrent parent).

Influence of Resistance Genes in Cultivars in Different Areas

The genes for resistance in cultivars grown in the area leave a marked effect on the rust population of the whole area. When the cultivar HUW206 carrying *Lr26* was released in 1983 in a small area in north eastern region virulence on *Lr26* had not been detected in that region but in different ecological zones falling under central, Southern hills (Nilgiris) and peninsular zones in 1983, 1985 and 1989 respectively. Similarly, a cultivar HP1633 carrying *Lr9* was released for North Eastern region in the year but virulence against this gene was detected in the Nilgiris in 1998. No cultivar carrying *Lr19* has been released in the country, yet there is a virulence attacking this resistance was identified in 2004. These observations indicate that the evolution of a new virulence is a random event and it does not need deployment of cultivars with new resistance genes to create selection pressure. It is very interesting to note that no cultivar carrying *Lr16* has been released in India, but the virulence(s) against this gene have been identified and the stocks possessing *Lr16* have been reported susceptible in adult plant stage. The gene *Lr28* was introduced in North Eastern Plains zone in the year (MACS 6145 = HW2034), however the virulence overcoming this gene was picked up from Peninsular India.

Genetic Linkages

The linkage between useful genes has come handy to the breeders in that attempts to incorporate a particular gene automatically brings along another useful gene. It is happening because more than one desirable genes are linked to have pleiotropic properties. One of the most popular and perhaps most widely used linked gene segment is that of present on short arm of wheat transferred from *Secale cereale* spontaneously. The rye segment carries genes for resistance stem, leaf and stripe rusts and powdery mildew (*Sr31 Lr26 Yr9 Pm8*). This alien segment is presumed to have genes influencing physiological traits leading to higher yield as evidenced by the number of varieties being cultivated world over. The other useful linked genes complexes, namely, *Lr34/Yr18, Lr24/Sr24, Lr37/Sr38/Yr17, Lr16/Sr23, Lr20/Sr15, Lr46/Yr29, Lr21/Sr33, Lr13/Yr27, Sr34/Yr8, Sr36/Pm6, Lr57/Yr40, Lr67/Yr46*. Breeders in India have successfully utilized the *Agropyron elongatum* derived genes *Lr24/Sr24* by releasing several cultivars, which occupied a total of about 14 million ha over a period of 20 years. Another block of genetically linked genes *Lr34/Yr18* providing durable resistance has been in use in commercial cultivars in India. The association of *Lr34* with adult plant stem rust resistance in Canada (Dyck, 1987) and India (Sawhney and Sharma, 1990) and also with stripe rust resistance gene *Yr18* (Singh, 1992; McIntosh, 1992b) has not only aided wheat breeding but assisted the identification of genetic diversity in wheat. The *Lr34* is also associated with a gene manifesting leaf tip necrosis (*Ltn*) (Singh, 1992). Segments carrying *Sr2, Lr34* and *Lr46* have been identified to have linkage with many other genes *viz., Sr2/Yr30/ Lr27/Pbc1, Lr34/Yr18/Sr57/Pm38/Sb1/Ltn1* and *Lr46/Yr29/Sr58/Pm39/Ltn2* providing resistance against multiple diseases. The genes *Lr34/Yr18* conditioning resistance to leaf rust, stripe rust and stem rust along with other diseases such as powdery mildew for a long period. The genotypes carrying *Lr67/Yr46* showed 20 per cent reduction in mean powdery mildew severity. The phenomenon of linkage facilitates simultaneous introgression of genes, permitting postulation of genes in case the matching virulence for other linked gene is not available.

DNA Isolation

Total genomic DNA was isolated from 15-20 days old seedlings using 25-30 mg dry tissue from an uninoculated set using the method.

PCR Amplification with SCAR Markers

The SCAR markers linked to gene *Lr19* namely $SCS73_{719}$ and to gene *Lr32* $SCS421_{572}$ identified earlier as linked to the respective genes were employed to screen the NILs. The SCAR marker linked to gene *Lr9* and SSR marker linked to gene *Lr19* identified recently were employed to work out the genetic identities of the stocks labeled for the respective genes. PCR amplification was carried out in 10mM Tris-HCl (pH 8.3), 50mM KCl, 2mM $MgCl_2$, 0.2mM of each dNTP (Sigma, USA), 200nM Primer and 0.75 unit of Taq polymerase (Bangalore Genei) and 25ng of genomic DNA and 25ng of each of the two SCAR primers per 25µl reaction. PCR was performed in PTC-200 thermal cycler (MJ research) with initial denaturation for 2min at 95°C, followed by 45 cycles of 30s at 94°C, 1min at 60°C, 1min at 72°C with a ramp of 1°/sec, and a final 7min extension at 72°C. Amplification products were separated on 2 per cent agarose gel for minimum 3 hours at 3V/cm constant voltage and viewed on a UV transilluminator after staining with ethidium bromide.

PCR Amplification with Simple Sequence Repeat (SSR) Marker

A null-allele simple sequence repeat microsatellite marker (SSR) Xgwm160 linked to *Lr28* and the SSR marker XGwm 437 linked to *Lr19* identified earlier in the project at the Ludhiana centre as a specific marker co-segregating with the gene *Lr28* was used to confirm the validity of the NILs carrying *Lr32* and NILs carrying *Lr28*. The expressive allele of the SSR marker is known to amplified 196 bp band of (GA)21 in bread wheat genome. In this case, the regions associated with the genes *Lr28* from the *T. speltoides* and *Lr19* from *A. elongatum* lacked the specific respective sequence (null allele) in the donors and the translocated lines in contrast to the chromosome not carrying the segment which expressed an amplification. Amplification was carried out in 10mM Tris-HCl (pH 8.3), 50mM KCl, 2mM $MgCl_2$, 0.2mM of each dNTP, 50ng of each primer, 0.75 unit of Taq polymerase and 80ng of genomic DNA per 25µl reaction. Cycling conditions were as follow: an initial denaturation step of 3min at 94°C followed by 45 cycles of 1min denaturation at 94°C, 1min annealing at 60°C, 2min extension at 72°C and final extension step of 10min at 72°C. PCR products were separated for 4hrs on 4 per cent Metaphore agarose at constant voltage of 5V/cm and viewed on a UV transilluminator after staining with ethidium bromide.

Lr28 and *Lr32*

Phenotypic Assessment of the NILs on Disease Expression: Expression of leaf rust resistance in different NIL pairs displayed consistency between the two NILs carrying genes *Lr32* and *Lr28*. However, the gene donor lines showed distinct difference in the phenotypic expression. The donor parent of the gene *Lr32* recorded up to 2, unlike the case of *Lr28* donor parent which recorded immune reaction 0-. This observation suggested that the transferred trait expression remained more or less comparable to the donor parent in the case of derivatives carrying the gene *Lr28* irrespective of the genetic backgrounds. But the resistance expression was clearly enhanced over that of the donor parent in the derivatives listed as carrying the gene *Lr32* in all the ten genetic backgrounds. Such variations are known to exist when there are more than one resistance genes interacting or background genetic effects in the case of gene transfers involving *Lr32*. But the pathogen virulence used in this study showed clear absence of any other resistance gene in the background, since all the ten recurrent parents were clearly susceptible recording 33^+ – 4 infection type reaction. Thus, the phenotype level differences, which were discernible under the precisely controlled conditions in the phytotron revealed that the expression of disease resistance could be due to a gene other than *Lr32* in the case of *Lr32* derivatives.

Lr34, *Lr46* and other Minor Genes for Durable Resistance to Leaf Rust

The South American cultivar 'Frontana' is considered the best-known source of durable resistance to leaf rust. The Mexican-Rockefeller Program first used this variety in the 1950s. Later its derivatives, such as 'Penjamo 62,' 'Torim 73,' and 'Kalyan/Bluebird,' showed slow-rusting characteristics possibly derived from 'Frontana'. Genetic analysis of 'Frontana' and several CIMMYT wheats possessing excellent slow-rusting resistance to leaf rust worldwide has indicated that such adult plant resistance is based on the additive interaction of *Lr34* and two or three additional slow-rusting genes (Singh and Rajaram, 1992). Leaf-rust severity observed in Mexico on most slow-rusting cultivars could be related to the number of minor genes they carry. Under conditions in which susceptible cultivars display 100 per cent leaf-rust severity, cultivars with only *Lr34* display approximately 40 per cent severity; cultivars with *Lr34* and one or two additional minor genes display 10–15 per cent severity; and cultivars with *Lr34* and two or three additional genes display 1–5 per cent severity. Leaf rust may increase to unacceptable levels if disease continues to increase further on cultivars carrying only Lr34, or Lr34 and one or two additional genes. However, cultivars with Lr34 and two or three additional genes are referred to as "near immune" and showed a stable response in all environments tested so far, with final leaf-rust ratings lower than 10 per cent. The presence of *Lr34* can be indicated by the presence of leaf tip necrosis in adult plants, which is closely linked with it (Singh, 1992a). Slow-rusting resistance to leaf rust is common in spring wheat germplasm and at least 10–12 slow-rusting genes are involved in the adult-plant resistance of CIMMYT wheat germplasm. Lines in which *Lr34* is absent but that still possess high levels of slow-rusting resistance have been identified, indicating that durable resistance is feasible even in the absence of *Lr34.* A second slow-rusting resistance gene *Lr46* was recently identified in wheat cultivar 'Pavon 76' and located in the short arm of chromosome 1BL (William *et al.*, 2003b).

Amplification of the SCAR Markers in the NILs

The SCAR marker was strongly amplified in 8 out of the 10 different varietal backgrounds expected to be carrying the gene *Lr*32 and in 9 out of the same 10 varietal backgrounds carrying gene *Lr*28. The NIL pair of HW 4007 and HW 2035 in the background of variety NI 5439 supposed to possess the genes *Lr32* and *Lr28,* respectively missed the linked marker fragment completely indicating a possibility of the absence of either gene in the lines. Although there is a possibility of recombination between the genes and the markers, in the present set, since common donors were used in one programme at one time and selection practiced identically in the backcross progenies, the chance of selection of the gene with recombination fixed for all the three markers is remote in such a high frequency over repeated backcrosses. But the three markers distinguished the NIL pair HW 4010 and HW 2033, the NILs of variety WH 147, by expressing only in the NIL HW 2033 which possessed the gene *Lr28.* Similarly, the marker fragment got amplified only in the *Lr28* donor parent, CS 2A/2M 4/2 and were completely missing in the *Lr32* donor parent C 86-8/Kalyansona F4. This polymorphism expressed by the molecular markers was comparable to that expressed by the markers when tested for marker validation with other *Lr* genes and leaf rust susceptibility.

SSR Marker Xgwm160

SSR marker Xgwm160 giving null allelic amplification associated with the gene *Lr28* also expressed in the same pattern as that of RAPD and SCAR markers in the 10 derivatives possessing the gene *Lr*32 and *Lr28.* Out of 10 derivatives supposed to be carrying the gene *Lr*32, the presence of the null allele (absence of the DNA fragment) was observed in 8 derivatives while the 196 bp band was present in the *Lr*32 donor parent CS 86-8/Kalyansona F4 as well as in the *Lr32* derivatives HW 4007 and HW 4010. Similarly, the null allele was observed in 9 out of 10 derivatives carrying gene *Lr*28 except the one in NI 5439 background (HW 2035). The *Lr28* donor parent CS 2A/2M 4/2 also displayed the null allele marker establishing the validity of the gene *Lr28.* The results on the 10 pairs of NILs and the two donor parents of the genes *Lr28* and *Lr32* through phenotypic and molecular analyses that include the reliable single band polymorphism of a SCAR and *Lr28* localised chromosome (4AL) specific null allele SSR marker clearly revealed the occurrence of redundancy and miss-identities with reference to the genetic stocks carrying *Lr* genes. The presence of the SCAR and the null allele SSR marker in eight of the ten wheat genetic stock lines; HW 4002, HW 4003, HW 4004, HW 4005, HW 4006, HW 4008, HW 4009 and HW 4011 confirmed that the lines possessed the gene *Lr28* and not the gene *Lr32* as listed earlier. This leads to a situation where, two duplicates of the gene *Lr28* in eight different genetic backgrounds like HD 2285, HD 2329, HUW 234, Kalyansona, Lok 1, PBW 226, Sonalika and WH 542 are being maintained and used in Indian sub-continent of which, one set being maintained mistakenly as *Lr32* genetic stock. From this, it is clear that the RAPD markers mapped earlier to the gene *Lr32* have to be redesignated as linked to the gene *Lr28* since the source *Lr* line HW 4002 employed in the mapping population is now confirmed to be carrying the gene *Lr28* and not *Lr32* as documented. The results also confirm that the derivative HW 2035 in NI 5439 background used to develop the SCAR marker linked to the gene *Lr28*, actually did not contain the gene *Lr28*. The leaf rust resistance in the line HW 2035 is due to an alien gene other than *Lr28*. This inference adds strength to the observation, that the marker identified was not tightly linked to the gene and therefore, could not be reliably used for marker assisted selection for transferring leaf rust resistance. The derivatives HW 4007 and HW 4010 did not carry the gene *Lr28* unlike the other *Lr32* denoted derivatives in the series. We suspect that the two lines also may not be carrying the gene *Lr32* when we consider the higher level of resistance in these lines than the *Lr32* donor parent, but may be carrying some other alien *Lr* gene/s. However, the counterpart NILs expected to carry the gene *Lr28* was confirmed only in the case of HW 2033 unlike the other derivative HW 2035 which clearly did not possess the gene *Lr28*.

Lr19 and Lr24

Molecular Marker Analysis

The marker fragment of the SCAR marker $SCS73_{719}$ developed as linked to the gene *Lr19* was strongly amplified in all the 10 pairs of NILs, each pair documented as possessing the gene *Lr19* and *Lr24*, respectively, in the same genetic background. The marker fragment was also monomorphic between the two donor parents 'Sunstar*6/C80-1' (*Lr19* donor) and 'TR380-14*7/3Ag#14' (*Lr24* donor) for the *Lr* genes in the NIL pairs which was not expected. This indicated that only one gene was involved between the two in each pair of NILs. Presence of the marker in both the donor parents indicated the possibility of both the lines possessing the same gene. In order to distinguish which of the two *Lr* genes was the functional gene in the NIL pairs as well as the donor parents, and also to clarify which of the *Lr* genes was actually tagged, the SCAR marker was tested on the 18 NILs possessing the different alien *Lr* genes in the variety 'Thatcher' background. The marker data revealed that the marker actually tagged the gene *Lr24* and not *Lr19*. The *Lr24* NIL in 'Thatcher' background amplified the critical marker fragment while the NIL for the gene *Lr19* in the same background did not express the marker fragment. Another confirmatory evidence comes from the observation that the SCAR marker fragment was amplified by three varieties carrying the gene *Lr24 viz.* 'Vidisha', 'Vaishali' (Lanes 19 and 20, Figure 4), and 'Agent' (Lane 21), while the critical band was missing in the *Lr19* possessing lines 'Agatha' and 'Cook*6/C80-1' (Lanes 22 and 23). 'Agent' and 'Agatha' are two of the earliest reported source lines of *Lr24* and *Lr19*, respectively. The varieties 'Vidisha' and 'Vaishali' were postulated to possess the gene *Lr24* on the basis of pedigree and lack of virulence on the varieties. While the latter observation confirmed that the SCAR marker $SCS73_{719}$ tagged earlier was for the gene *Lr24*, the observation that the NIL in 'Thatcher' background for gene *Lr42* (Lane 17) also expressed the same band brought to light another possible error in maintenance of the germplasm. The monomorphism observed for the linked genomic regions cannot be expected to exist between the genes *Lr24* (from *A. elongatum* located on chromosome 3D) and *Lr42* (from *Triticum tauschii* located on chromosome 1D). This could be due to one of the many errors during the maintenance breeding activity or port to port handling activities where *Lr24* was inverted or confused for *Lr42*. Since the latter gene is not commonly utilized by breeders, there was no other source available to compare and analyze its genetic status. Further analysis through allelic cross between the NILs *Lr24* and *Lr42* can reveal the sole exception to the clean single band polymorphism shown by the SCAR marker $SCS73_{719}$ in this study against other 17 alien genes.

Test of Allelism

The F_2 population comprising of 350 seedlings of the cross between 'Sunstar*6/C80-1' (*Lr19* donor) and 'TR380-14*7/3Ag#14' (*Lr24* donor) did not show any segregation in leaf rust expression and remained resistant when tested against the pathotype 77-5 of leaf rust. If the two genes were different, a segregation of 15: 1 for resistance: susceptibility should have been obtained. The lack of segregation also confirmed that the same gene (either *Lr19* or *Lr24*) was not located on the two different chromosomes 3DL and 7DL, which was a possibility in a segmental allopolyploid like wheat, however remote, since the marker amplicons had identical sequence of 719 bp. This suggested that the parents did not differ for the *Lr* genes or their location.

Authenticating Stocks Wrongly Identified in India

Using the SCAR or SSR markers linked to the genes *Lr9*, *Lr19*, *Lr24* and *Lr28*, we have been able to authenticate the genetic stocks containing each of these genes and also properly identify the wrongly identified ones with their correct identities since these were the major genes that were being mobilized in India. The conclusions of the marker validated lines are that the genetic stocks documented to be possessing *Lr19* in Indian wheat background DO NOT possess *Lr19* but are only redundant lines carrying *Lr24*, the genetic stocks documented to be possessing *Lr32* in Indian wheat background DO NOT possess *Lr32* but are only redundant lines carrying *Lr28*, the genetic stock HW 2035 does not possess *Lr28*, but possesses *Lr24*, the genetic stock HW 2067 does not possess *Lr9*, but possesses *Lr24* and the genetic stock HW 4010 does not possess *Lr28*, but possesses *Lr9*.

Table 7.19. Seedling reaction to leaf rust pathotype 77-5 of *Puccinia triticina* in wheat NIL pairs for leaf rust resistance genes *Lr32* and *Lr28*, and their recurrent parent background

NILS	*Recurrent Parents*	*Disease Reaction*	
		NILs	*Recurrent Parent*
HW 4002*	HD 2285	;	33^+
HW 4003*	HD 2329	;	33^+
HW 4004*	HUW 234	;	33^+
HW 4005*	Kalyansona	;	33^+
HW 4006*	LOK-1	;	33^+
HW 4007*	NI 5439	;	33^+
HW 4008*	PBW 226	;	33^+
HW 4009*	Sonalika	;	33^+
HW 4010*	WH 147	;	4
HW 4011*	WH 542	;	4
C86-8/Kalyansona F4* (*Lr32* Donor)			
HW2038**		2-	
HW2037**	HD 2285	;	33^+
HW2039**	HD 2329	;	33^+
HW2061**	HUW 234	;	33^+
HW2032**	Kalyansona	;	33^+
HW2035**	LOK-1	;	33^+
HW2040**	NI 5439	;	33^+
HW2031**	PBW 226	;	33^+
HW2033**	Sonalika	;	33^+
HW2062**	WH 147	;	4
CS 2A/2M 4/2**	WH 542	;	4
(*Lr28* Donor)		0-0;	

* The NIL documented as carrying the gene *Lr32* (4, 5).

** The NIL documented as carrying the gene *Lr28* (4, 5).

The utility of the molecular markers in detecting such duplications and miss-identification in the germplasm management has been established. The availability of molecular markers has now made it possible to eliminate redundant genetic stocks and detect mistaken identities as a result of which molecular characterization is recommended as an integral and important aspect of efficient genetic resources management. Where phenotypic characterization is not possible, such unintended duplication and use of wrong parental source lines whose further derivatives could have a manifold enlargement of the wrong identity, the molecular markers are extremely useful. As we observed in the case of the genes *Lr28* and *Lr32*, where, the lack of gene specific virulence to distinguish the two alien genes could have led to the error, there is an urgent need for a molecular marker analysis for fingerprinting the near-isogenic lines and their authentication for the other alien *Lr* genes.

Screening of Backcross Population for Foreground Selection of *Sr36* Gene

In BC_1 population, out of 56 plants, 21 plants showed the presence of the band 155bp *i.e.*, resistant (presence of *Sr36*), whereas, other plants had band 190bpi.e, susceptible (absence of *Sr36*). Plants positive for *Sr36* showed good stem rust resistance in the field along with good expression and were used to develop the BC_2 population by crossing with the recurrent parent HI 8498. In BC_2 population, out of 60 plants, 65 *per cent* of the plants (39 plants) showed the presence of the band 155bp, with good stem rust resistance; whereas, other plants (35 per cent) had band 190 bp *i.e.*, susceptible (absence of *Sr36*). In BC_3 population, out of 92 plants, 27 *per cent* of the plants (24 plants) showed the presence of the band 155 bp *i.e.*, resistant (presence of *Sr36*) with good stem rust resistance and durum wheat 'HI 8498' plant type; whereas, the remaining plants (73 per cent) had band 190 bp *i.e.*, susceptible (absence of *Sr36*). The stem rust resistance in the 'HI 8498' derivatives (backcross populations) carrying *Sr36* individually has improved significantly (terminal disease severity 0 to 5S), compared to the background cultivar HI 8498 (30S – 40S).

Recurrent Parent Genome Recovery (RPG) (per cent) through Marker Assisted Background Selection (MABS)

Based on the information gained through parental polymorphism, polymorphic markers were utilized for MABS to select plants with high recurrent parent genome recovery. In BC_3F_1, RPG ranged from 87.5 to 100 percent in the foreground positive plants, out of which 15 plants showed > 95 per cent with homozygous bands as recipient parent, HI 8498. The rust reactions of these positive plants are in the range of 0 to 5S. Of these 15 plants, 3 plants showed 100 per cent RPG and similar plant type of HI 8498, which were selected for pyramiding of the gene *Sr36*.

Table 7.20. RPG recovery (per cent) of foreground positive plants for *Sr36* by MABS

Plant ID	*Percentage Recovery*	*Plant ID*	*Percentage Recovery*
WE 29	95.4	WE 130	95.4
WE 32	100.0	WE 131	97.2
WE 42	95.4	WE 143	97.2
WE 53	100.0	WE 145	97.2
WE 76	95.4	WE 151	95.4
WE 117	97.2	WE 161	95.4
WE 127	100.0	WE 163	100.0

As many of major and minor genes have been knocked down by stem rust pathogen due to its many evolved variants it is more important to have more genes to fight against this pathogen. The stem rust resistance gene *Sr36* offers near immune reaction to many pathotypes belonging to the 117 group will be helpful in providing resistance to durum wheat on which these pathotypes are highly virulent.

The variety HI 8498 once developed with the gene *Sr36* introgressed, will serve as a good source of resistance for developing stem rust resistant durum wheat varieties and genotypes in near future. Depending on the severity and duration of infection, losses can vary up to 50 per cent depending on the leaf rust resistance genes deployed and the composition of virulence in the region (McIntosh *et al.*, 1995).

Table 7.21. Important released varieties carrying specific leaf rust resistance gene(s)

Name of the Variety	*Gene Postulated*	*Pedigree*
DL784-3 (Vidisha)	Lr24/Sr24	Kalyansona*4/TR380-27* 4/3 #Ag/3/
HW2004 (Amar)	Lr24/Sr24	C306*7/TR380-14*7/3 Ag #14
DL788-2 (Vaishali)	Lr24/Sr24	K7537/HD2160//HD2278/Lr24//K4-14
HW1085 (Bhavani)	Lr24/Sr24	Unnath Kalyansona/CPAN3057
HW2044 (Kurinji)	Lr24/Sr24	PBW226*6/Sunstar*6/C80-1
HW2045 (Kausambi)	Lr24/Sr24	HD2402*6/Sunstar*6/C80-1
HD2781 (Aditya)	Lr24/Sr24	BOW/C306//C591/HW2004
HI1500 (Amrita)	Lr24/Sr24	HW2002*2/HW1042/Strampalli/PNC5
Raj4037	Lr24/Sr24	DL788-2/Raj3717
HD2851 (Pusa Vishesh)	Lr24/Sr24	CPAN3004/WR426//HW2007
WR544 (Pusa Gold)	Lr24/Sr24	Kalyansona/HD1999//HD2204/DW38(Lr24)
HD2833 (Tripti)	Lr24/Sr24	PBW226/HW1042//HD2285
NW1067	Lr24/Sr24	TR380-16-30614/CHAT 'S'
COW(W)-1	Lr24/Sr24	HD2646/HW2002A/CPAN3057
HD2888 (Pusa Wheat-)	Lr24/Sr24	C306/*T.shpaerococcum*//HW2004
AKAW3722 (Vimal)	Lr24/Sr24	WH147/Sunstar*6/C80-1

Growing resistance cultivars is an effective and economical method of reducing losses to leaf rust. To date, 50 leaf rust resistance genes have been identified in wheat and related species (McIntosh *et al.*, 1995). However, the constant search for the novel resistance genes is essential in order to cope with the dynamic and rapidly evolving pathogen population. Development of genetic resistance in new cultivars contributes to the reduction in economic losses, production costs and risk of environmental pollution due to fungicide usage.

Table 7.22. Prevalence of leaf rust resistance genes in old wheat varieties.

Varieties	*Country/Region*	*Genes*
Lerma Rojo	Mexico	*Lr*13, *Lr*17
Champingo 53	Mexico	*Lr*34
Penjamo 62	Mexico	*Lr*14*a*, *Lr*34
Pitic 62	Mexico	*Lr*14*a*
Sonora 64	Mexico	*Lr*1
Mexipak 65, Kalyansona	India, Pakistan	*Lr*14*a*
Sonalika, Bluesilver	India, Pakistan	*Lr*13, *Lr*14*a*
Lyalpur 73	Pakistan	*Lr*1, *Lr*13, *Lr*34

Varieties	*Country/Region*	*Genes*
Bluebird, Yecora 70	Mexico, Pakistan	*Lr1*, *Lr13*, *Lr34*
Ciano 79	Mexico	*Lr16*
Arz	Lebnon	*Lr17*
Pavon F 76, Dollarbird	Mexico, Australia	*Lr1*, *Lr10*, *Lr13*, *Lr46+*
Parula	CIMMYT	*Lr34* and *Lr46+*
Punjab 81	Pakistan	*Lr10*, *Lr13*, *Lr34*
Era	North America	*Lr10*, *Lr13*, *Lr34*

As the cultivars with single resistance genes have been successfully attacked by emergence of virulent pathotypes, pyramiding of rust resistance has been the better approach (Messmer *et al.*, 2000). However, pyramiding of rust resistance genes through traditional phenotypic methods is difficult when different resistance genes produce similar types of infection. Identification of molecular markers linked to rust resistance genes facilitates marker assisted selection (MAS) and gene pyramiding in breeding programs. Additionally, the efficiency of introgression of alien genes to hexaploid wheat can be improved by use of molecular markers. Various molecular markers have been widely used to tag and map resistance genes in wheat, however, simple sequence repeat (SSR) have emerged as the choice of markers in gene mapping studies. In recent efforts, molecular markers linked to rust resistance genes such as *Lr3a*, *Lr19*, *Lr34*, *Lr39*, *Sr2*, *Sr39* have been reported. In our earlier efforts, we have also reported tagging of leaf rust and stem rust resistance genes. Here we have attempted to identify tightly linked flanking markers for *Lr15* and their utility in marker assisted selection (MAS) with gene pyramiding efforts. Stem rust of wheat is a significant disease in south, central, and peninsular India, but epidemics are infrequent and have been confined to local land cultivars. When, such epidemics occur, losses can reach higher percentage in individual fields. The pathogen has a diverse virulence spectrum. Leaf rust has wide spread occurrence in the country and is most dynamic. A broad spectrum of pathogenic races render resistant cultivars susceptible in a short span of time. It causes significant yield losses, when appears in epidemic form. Stripe rust of wheat is important in the northern areas of India, frequent epidemics have occurred which confined to local pockets in northern wheat growing areas. The stripe rust is also recorded in the higher elevations in the south India. It arrives suddenly in some of the areas in Punjab and presents a challenge to breeder because the conditions in December and January are very favourable in northwestern region. Losses of significant magnitude can occur in fields of susceptible cultivars, generally in northwestern regions, northern foothills, and adjacent plains, and in the Nilgiri and Palney hills in the south. The pathogen has a narrow spectrum of virulence on both bread wheat and durum wheat. Under optimum climatic conditions prevailing in the particular locality one or more of these rusts may damage the wheat crop during years of epidemic(s). Losses in yield are generally attributed to several factors, namely, early appearance of rust, congenial environment and load of inoculums and the type of variety grown in the area *etc.* ultimately results in yield reduction (Srivastava *et al.*, 1984).

There are several ways to control the rust infection *viz.*, introduction of varietal diversity, varietal mixtures, multilines, the deployment of cultivars having a combination of oligogenes in a single genetic background and the combination of one major gene with some minor genes providing durable resistance. The breeding rust resistant varieties is most friendly option. Usually the life span of a resistant variety is about 4-5 years. Hence, the search for new genes for resistance and their breeding into cultivars are an ever-continuous process. However, some cultivars remain resistant for fairly long period of time, if the variety carries durable resistance conferred by a single or more than one gene. Deployment of genes for rust resistance in achieving yield stability in wheat production without restoring to potentially harmful chemicals will be environmentally friendly and currently becoming most important. Forty one cultures (PS 830 to PS 876) of wide genetic base and diverse resistant genes are under evaluation programme of various Common Varietal Trails (CVTs). 608 F_1s, 535 F_2s, 687 F_3s, 663F_4s and 1626 advance generation lens were developed. All potential alien genes such as *Lr9, Sr 25+Lr19, Sr 24+Lr 24, Lr28* and *Lr27* were utilized in breeding programme. Against the threat of black rust virulent race *Ug99*, the resistant donor sources such as FLW 2, FLW 4, HW 1085, Raj 4120, Raj 4124 and Raj 4125 were utilized in crossing programme. All possible cross combinations were attempted to introgress diverse genes at single background. Under shuttle breeding programme, 5110 lines of breeding material were grown at Lahaul Spiti and 660 lines at Wellington (2010-2016). Nurseries such as NGSN, EIGN, YCSN, TPN, LBSN, IPPSN and PDSN were evaluated for their disease and ancillary characters. 481 wheat germplasms were maintained for utilization in crossing programme and to provide protection against genetic erosion. Buitre lines and PHR lines *viz.*, PHR 1004, PHR 1006, PHR 1007, PHR 1008, PHR 1009, PHR 1010, PHR 1011, PHR 1012, PHR 1014, PHR 1015, PHR 1021 and PHR 203 were utilized in crossing programme have produced many desirable segregates which are being improved through biparental mating. Chirya-3 and PHR 1001/ PBW 373 were identified as a potential donor against foliar blight (*Helminthosporium* spp). PBW 510 and PBW 579 were utilized as donor source for THT (Terminal Heat Tolerance) at the background of superior agronomic bases. The knowledge regarding the development and annual recurrence of cereal rust was inadequate and was mainly based on the work done by Dr. K.C. Mehta and his associates during early 1920s (Pal *et al.*, 1954). However, the knowledge about the physiological races of three rusts of wheat in India first became available in 1930s. Two stem rust races, 15 and 27 were identified by M.N. Levine from wheat samples collected from Pusa in 1923. Subsequently, races 10, 63, 106, 107 and 108 of leaf rust were identified from 1931 - 1935. The presence of a relatively smaller number of races of the three rusts in India (Mehta and Pal, 1940), the likely hood of their number continuing to remain relatively small and the resistance of a group of physiologic races being controlled by a few genes that are simply inherited for resistance breeding. The work on identification of rust races began in 1931 (Gupta *et al.*, 1989).

Systematic work on race analysis and testing of varieties against wheat rusts in India had started in 1931 and continued since then. Simultaneously screening led to the selection of NP790, NP796, NP797 and NP846. At the same time two Australian varieties Gabo and Ridley, resistant to black rust races. The plant pathological screening nurseries were started in 1969 in hot spot locations. Since 1967, cereal disease surveillance has been conducted by the IARI, New Delhi, using mobile and trap nurseries. The work on rust resistance was started in three stages *viz., first stage*, when no suitable resistant stocks were available for hybridization, rust losses were sought to be minimized by introducing rust tolerant varieties; *second stage*, the attention was devoted in the production of strains possessing resistance to individual rust and third stage attempts were made to achieve a combination of resistance to all the three rusts. Rust resistant sources *viz.*, Khapli, Kenya 184, Kenya Ploughman for black rust, Frontiera, Frontana, Bowie Texas, Mentana and Tremez Molle for brown rust and Ceres Klein, Cometa Klein, Frondosa and Spalding's Prolific for yellow rust and exotics like Gabo, Gaza, Yaqui 53, Timstein (*Lr23*) (resistant to black and brown rusts), Rio Negro, La Prevision (for brown and yellow rusts) were also used in hybridization programme. In the first attempt, rust tolerant varieties like NP710, NP718 and NP761 were developed and released. Breeding for rust resistance continued which led to the evolution of genotypes such as NP783 and NP784 (Democrat x PbC518). Later on the efforts were concentrated to develop genotypes with diverse resistance involving more than two parents. NP809 was released from a cross [(Democrat x PbC518)-F_9 x (Spalding's Prolific x NP114)-F9] x E220, which combined resistance to rust and loose smut (Pal and Sikka, 1956). Further resistance breeding is ongoing. Screening advanced and parental lines for the presence of major genes using molecular or genetic markers are ongoing programme. Major genes functional in India are *Lr9, Lr24, Lr28, Lr48, Lr28, Lr 19, Sr2*, Sr24, Sr25, Sr 31*, Yr9*, Yr5*, Yr10, Yr15, Yr37* (* genetic marker or seedling pathology). Targeting minor genes by adopting molecular markers for rejection of all known Lr, Yr and Sr major genes with the help of molecular markers. This approach was adopted to facilitate the accumulation of minor genes as the major genes can be overcome by the evolving races of the pathogen in a short period after cultivation (Figure 28). Relatively few additive genes, each having small to intermediate effects, required for satisfactory disease control. Near-immunity (trace to 5 per cent severity) can be achieved even under high disease pressure by combining 4-5 additive genes (Singh, 2010).

Development of Backcross Populations

In the last crop season crosses were developed among contrasting parents for heat and drought tolerance. F_1s from twenty advanced in the off-season nursery at Lahaul-Spiti. F_1s were backcrossed with high yielding recurrent parent to generate BC_1F_1 populations. These BC_1F_1s are advanced in the current crop-season to generate BC_2F_1 populations. DNA samples were drawn from the BC_1F_1 for foreground selection of the target trait and recovery of the background genome through background selection. A total of 850 microsatellites markers used and approximately hundred polymorphic markers were observed among them. MABB for rust resistance was utilised in HD 2733, while quantitative rust resistance was also utilized.

Table 7.23. MABB for Rust Resistance in HD 2733

Recipient Parent	*Donor Parents*
HD 2733	HD 2687+Lr19
	HD 2687+Lr24
	HD 2687+Yr15
	HD 2189+Lr34
	HD 2851+Sr26
	HD 2851+Yr10

Table 7.24. Leaf rust severity recorded in parents and selected F_2 plants from their crosses in wheat

Cross (Parent 1 x Parent 2)	*Disease Severity*		
	Parent 1	*Parent 2*	*Selected F_2 Parents**
HS 191 x S 57	30S	30S	10S
HS 191 x S 69	30S	10S	5S
HS 191 x S72	30S	10S	5-10S
HS191 x WBM 145	30S	25S	10S
S 69 x S 72	10S	10S	1-5S
S69 x WBM 145	10S	25S	5S
HB 208 x HS 191	10S	30S	1-5S
HB 208 x S 57	10S	30S	1-5S
HB 208 x S 69	10S	10S	1-5S
HB 208 x S 72	10S	30S	5S

*F_2 progeny in F_3 population have shown most plants with disease severity equal or less than F_2 plants

Yr18, *Yr29* and other Minor Genes for Durable Resistance to Stripe Rust

Singh (1992b) and McIntosh (1992) have indicated that the moderate level of durable adult-plant resistance to stripe rust (caused by *Puccinia striiformis*) in the wheat cultivar 'Anza' and winter wheats, such as 'Bezostaja,' is controlled in part by the *Yr18* gene. This gene is completely linked to the *Lr34* gene; the level of resistance it confers by itself is usually not adequate. However, combinations of *Yr18* and three to four additional slow-rusting genes result in adequate resistance levels in most environments. Genes *Lr34* and *Yr18* occur frequently in germplasm developed. Recently identified slow-rusting gene *Yr29* is completely linked to slow-leaf-rusting gene *Lr46*. Low disease severity to stripe rust is usually associated with at least some reduction in infection type because stripe rust grows systematically in leaf tissues. This phenomenon results in chlorotic or necrotic stripes and therefore creates difficulties in distinguishing slow-rusting resistance from race-specific resistance. Durability and acceptance of adult-plant resistance can be expected if the cultivar's low disease severity is due to the additive interaction of several (four to five) partially effective genes (Navabi *et al.*, 2004).

Recent Advances in Gene Expression Analysis

Analysis of gene expression is important in many fields of biological research. Several techniques are available for studying gene expression at m-RNA level such as Northern blotting, differential display, Serial analysis of gene expression (SAGE) and dot blot analysis. However, these methods have certain limitations. Northern blot analysis only allows limited numbers of mRNAs to be studied at the same time. Differential display allows simultaneous detection of multiple differences in gene expression, but only a limited number of different conditions can be compared. Moreover, the method is not quantitative and screening is based on differences in mRNA length and not identity. Here, serial analysis of gene expression (SAGE) and DNA based microarrays techniques for the analysis of gene expression are discussed.

Serial Analysis of Gene Expression (SAGE)

SAGE is the logical extension of Expressed Sequence Tag (EST) sequencing. This technique is based on two principles: (i) a short nucleotide sequence tag (9 to 10 base pairs) contains sufficient information to uniquely identify a transcript (provided it is isolated from a defined position within the transcript. (ii) Concatenation of short sequence tags allows the efficient analysis of transcripts in a serial manner by sequencing of multiple tags within a single genome. *c*DNA synthesized from mRNA (isolated from specific tissue) are cleaved with a four base cutter restriction endonuclease (anchoring enzyme). The most 3′ portion of the cleaved cDNA is then isolated. The cDNA is then divided in half and ligated to one of the two linkers containing Type II S restriction site (tagging enzyme). Type II S restriction endonuclease cleave at a defined distance upto 20 bp away from their recognition sites. The linkers are designed in such a way that cleavage of ligation products with the tagging enzyme results in release of the linker with a short piece of cDNA. After blunt ends are created, two pools of released tags are ligated to each other and amplified by Polymerase chain reaction (PCR). PCR products are then cleaved with anchoring enzyme, ditags are isolated, concatenated, cloned and sequenced. SAGE is laborious since it involves complex sample preparation procedures, extensive DNA sequencing and is not very sensitive.

DNA Microarray Technology

The technology is based on hybridization of mRNA to a high-density array of immobilized target sequences, each corresponding to a specific gene. Sample mRNAs are labelled as complex mixture and hybridized to the array, where each messenger will quantitatively hybridize to its complementary target sequence. After washing, the fluorescence at each spot on the array is a quantitative measure corresponding to the expression level of the particular gene. The use of two differently labelled mRNA samples allows quantitative comparison of gene expression in both the samples.

Manufacturing of DNA Microarrays

Two main approaches exist. First approach involves direct synthesis of oligonucleotides on a solid surface. Synthesis based on photolithography, (Fodor *et al.*, 1991) is known to be the best. Specific areas of the glass surface, derivatized with linker molecules carrying a photolabile protective group, are selectively illuminated by using a photomask. Subsequently, the surface is incubated with a solution containing a photo protected nucleotide, which will only be coupled to the light activated areas. After removal of the excessive nucleotide, a second photomask is used to deprotect other areas on the surface and subsequently another type of nucleotide is coupled to these areas. By repeating this procedure, a defined set of nucleotides is synthesized on the surface. Second approach involves deposition of small quantities of DNA solution on to a solid surface with the help of micro-dispensing robots. Robots can be divided into two main categories: passive dispensers and active dispensing units based on ink-jet technology. Passive dispensers apply the DNA solution with a pin (or multiple pins) that touches the solid surface. The latter devices use either piezoelectric deliver or delivery by solenoid valves, thus making direct contact redundant. The DNA molecules can be dispensed onto different kinds of surface *e.g.* glass or membranes. In case of glass surfaces, a positively charged layer can be used to bind the negatively charged DNA fragments. Presently, surface-modified glass slides are most commonly used as substrate. Besides glass, other materials are being explored as well.

Hybridization and Scanning

mRNA is purified from the biological sample and labelled fluorescently prior to hybridization *e.g.* by incorporation of a modified nucleotide during *c*DNA synthesis. After labeling of sample mRNA, hybridization to the array is performed. RNA expression level and hybridization signal show a linear correlation. Studies have shown that a four to five-fold difference in expression can be detected with about 95 per cent probability. For imaging, several types of microarray readers are on the market. They can be subdivided into CCD cameras, non-confocal laser scanners and confocal laser scanners.

Screening Methodology and Selection Environment

The probability of identifying resistant parents and resistant progenies is increased by the availability of a reliable screening methodology and an environment favourable for disease development. Depending on the choice of the type of resistance, the methodology may require simple tests in green house on seedling or adult plants, replicated field tests, or even use of protein or DNA markers. Inclusion of check cultivars for resistance and susceptibility is necessary to assess the disease pressure and degree of resistance. Choice of field sites with reliable environmental conditions is crucial for right progress. Use of off-season nurseries is advocated.

Data Analysis

Large amount of complex hybridization data are generated from microarray studies. Software packages are available to assess the quality of data, to find out statistical significance and relative correlations within, and between different arrays and experiments. The applications of microarrays are quantitative changes in the level of gene expression under stress (biotic or abiotic) conditions or at different developmental stages can be compared/monitored; risk assessment in transgenic agricultural products by analysis of altered gene expression; investigation of functional and toxicological effects of food components; unravelling gene function and metabolic pathways by mutant analysis; it is speculated that if whole genome microarrays of inbred parental lines with heterotic hybrids are compared, hybrids would exhibit significant differences in the expression of genes. If this is the case, breeding for certain patterns of gene expression would accelerate progress towards development of heterotic hybrids; microarray studies have revealed extensive cross talk (overlap and interference) among the pathways that are involved in responses to pathogen infection and to various abiotic stresses. Microarray analysis technique does have certain limitations, which are microarray technique is based on *m*RNA analysis. It does not always offer insight into the quantity and quality of the proteins, which are the final gene products. Protein amounts are not always correlated to mRNA amount. Many proteins undergo post-translational modifications (eg removal of signal polypeptides, phosphorylation and glycosylation) that play an important role in their activity and subcellular localization. Such changes can not be predicted from mRNA study. The limitations of microarray analysis can be overcome by studying the proteins themselves that provides information on their real amount and activity at a given time, in a given tissue, or in response to a given treatment. DNA microarray technology and proteomics (large-scale parallel analysis of the proteins that are present in a cell) are complementary which focus on different steps of the same process *i.e.* expression of genetic information into functional molecules, cells and organisms.

Current Status of Resistance Breeding

If India had not experienced frequent rust epidemics, it is because of the cultivation of diverse genotypes carrying different resistance genes in each wheat zone. The efforts are on to develop cultivars with enduring resistance. Presently the cultivars occupying major area in each zone carry diverse genes for rust resistance. Cultivar, HD2967 occupying a large area in NWPZ exhibits a high degree of resistance is likely to carry minor genes, Currently the effective rust resistance genes are being mapped on specific chromosomes. *Secale cereal*e derived gene *Lr45* on 2A were mapped and developed PCR based co-dominant markers. A new gene for leaf rust resistance in *Triticum durum* var. Trinikria has been identified and mapped on 5BS chromosome. The availability of resistance genes in *Triticum aestivum* or *T. durum* is limited and therefore new genes for resistance must be searched from the related and distantly related species. Kuraparthy *et al.* (2009) transferred leaf rust and stripe rusts resistance from *Aegilops geniculata* (UgMg) into common wheat and designated the genes *Lr57* and *Yr40.* Another gene *Lr58* conditioning resistance to leaf rust was transferred from *Aegilops triuncialis* into common wheat and has been mapped on 2BL chromosome. The utilization of this resistance will facilitate wheat improvement. Pyramiding of major genes in agronomically suitable cultivars is also being pursued (Samsampour *et al.*, 2009) to prevent the breakdown of resistance and the pyramided lines are being tested at multi-locations for their yield potential. The apparent effectiveness of some genes is observed in Indian genotypes. Both 'Abha' and 'Vidisha' exhibited seedling infection type '0; 1' (hypersensitive flecks and few type '1' pustules *i.e.* very minute uredia surrounded by distinct necrotic areas), whereas 'Agra Local' displayed infection type '4' (large uredia not accompanied by any necrosis or chlorosis) on 0-4 scale. The two F_1s involvinq susceptible and resistant parents showed infection type similar to that of the resistant parents indicating dominant inheritance of resistance in 'Abha' and 'Vidisha'. Resistance in these two wheat varieties was observed to be under monogenic control since BC_1 and F_2 segregation fitted well to 1 resistant (R): 1 susceptible (S) and 3 R: 1 S ratios, respectively. No susceptible segregants were observed in the F_2 populations of the crosses 'Abha/RL.

Table 7.25. Seedling infection types (ITs)* of wheat varieties and their F_1s to the leaf rust pathotype 77-5

Wheat Varieties and the F_1s	*ITs*
Varieties	
Abha (HI 1454) 01	0: 1
Vidisha (DL 788-2) 0.1	0;1
RL 6064 (Thatcher + *Lr24)*	0;1
Agra Local	4
F_1s	*ITs*
Agra Local/Abha	0;1
Agra Local/Vidisha	0;1
Abha/RL 6064	0;1
Vidisha/RL 6064	0;1
Abha/Vidisha	0;1

* IT '4' indicates susceptibility, other ITs indicate resistance.

The gene *Lr24* has been postulated in bread wheat cultivars 'Amar' (HW 2004) and 'Amrita' (HI 1500) as well. While 'Abha' and 'Vidisha' are recommended for late sown (December) cultivation, 'Amar' (HW 2004) and 'Amrita' (HI 1500) have been released for early (October) planting in central India. Thus, this continuum in time and space of leaf rust resistance in wheat based on a single gene is a cause of concern, particularly in view of the occurrence of virulence for *Lr24* in the neighbouring country Nepal (Mishra, 2001), a focus of leaf rust. Infection for the wheat crop in India. Hence, the leaf rust resistance base of wheat cultivation in central India needs to be diversified by deploying other resistance genes like *Lr19, Lr28, Lr32, Lr39, Lr41, Lr42* and *Lr45* which provide resistance to all the leaf rust pathotypes in India (Nayar, 2004). However, virulence to *Lr19* has recently been reported (Anonymous, 2004), and none of the other genes are available in commercial bread wheat varieties (Anonymous, 2004), except *Lr28* in only one variety 'MACS 6145'. Furthermore, most of the bread wheat cultivars are susceptible to the currently prevalent and virulent leaf rust pathotype 77-5 or 121R63-1. An alternative approach under such a situation is to encourage durum wheat cultivation in central India, 'Since durum wheat varieties are generally resistant to leaf rust race 77-complex pathotypes including 77-5 (Honrao and Rao, 1996). In the past four decades, much has been accomplished in controlling the wheat rusts in India. This collective effort of plant pathologists and plant breeders in developing resistant cultivars and understanding disease epidemiology has gradually reduced the magnitude and frequency of epidemics. Economic gains can be measured over a period since the genetic gain in wheat production is being realized for the past 40 years. Use of knowledge of pathogen variability generated thus far formed strategy for deployment of resistance gene in the field. It is obvious that gene management in field has considerable promise for sustained control of rusts. The popular Indian wheat cultivars HD2329, Lok1, HD2285, HD2189, PBW343, have diverse genes for rust resistance. Mega varieties dominate despite release of many varieties. In India the resistance genes have been fairly and properly managed. The diversification of cultivars in a mosaic pattern in each zone is strategic deployment to contain rust infection. At times, it would be difficult to withdraw susceptible cultivars over an entire epidemiological area because of economic benefits. Also the removing of genes for resistance from particular line in breeding programmes is also not possible. The cooperation between pathologists and breeders is to be well maintained, the rusts should not be a major threat to wheat production in India.

Identification, Screening and Control of Seed Borne Disease in Wheat

Identification and control of seed borne diseases are the main components of seed health testing and quality seed production. Seed health testing involves the assessment and analysis of seed borne pathogens and mycoflora associated with the seed. Seed health testing involves the assessment and analysis of seed borne pathogens and mycoflora associated with the seed. Any infectious agent that is associated with the seed and has a potential to cause the disease has been called a seed borne pathogen and disease caused by it as seed borne disease. The seed borne pathogen includes all pathogenic fungi, bacteria, viruses and nematodes which are carried on, in or with the seed. The first report of a seed borne pathogen came in 1755 when 'Tillet', a French Botanist, established that the causal fungus of hill bunt of wheat, *Tilletia caries*, is seed borne. Since then more than 1000 fungi, 100 bacteria, 126 viruses and about 14 nematodes have been found associated with the seed. Identification and control of seed borne diseases are the main components of seed health testing and quality seed production. The objectives of seed health testing are testing for quarantine, testing for certification schemes, testing for evaluation of planting value, testing for advisability of seed treatment, testing of treated seed, testing for storage quality or feeding value and testing for resistance.

Methods for Detection and Identification of Seed Borne Pathogens

Detection methods for seed borne pathogens have been devised on the basis of their nature and location in relation to seed. If a pathogen is located on the outer surface of the seed, it is called externally seed borne, and if located inside the seed, it is internally seed borne. However, when structures of pathogen *i.e.* ergot sclerotia, earcockle galls *etc.*, are mixed with the seed, then it is called as admixture and the pathogen can be identified by just looking at these structures with a naked eye. Therefore, the detection methods begin with the inspection of dry seed. Seed health testing methods have been standardized by International Seed Testing Association (ISTA, 1993) the account of which is given hereunder:

Detection of Fungal Organisms

Direct Inspection

Inspection of dry seeds: Inspection with naked eye reveals the presence of fruiting structures of fungi and the effect of fungi on the physical appearance of seeds. Fungal fruiting structures such as sclerotia of *Sclerotinia sclerotiorum*, *Claviceps fusiformis* and *C. purpurea* found mixed with the seed; *Melampsora lini* is carried with flax seeds in infected chaff. The resting stages as well as smut balls of many fungi are found associated with the seeds. Observation of seeds under the stereobinocular microscope reveals the presence of chlamydospores, oospores, spores *etc.* on seed surfaces. The seed borne infections of Karnal bunt, hill bunt, dwarf bunt and flag smut of wheat; covered smut of barley; loose smut and covered smut of oat; rice bunt and many other smuts and bunts can be determined by this method. Tentative idea of the seed infection of soybean seeds can be had on the basis of seed discolouration caused by various fungi like *Cercospora kikuchii* (purple), *C. sojina* (Gray or brown), *Macrophomina phaseolina* (indefinite black spots and blemishes) and *Collectotrichum truncatum* (irregular-greyish). Similarly, in rice the seed discolouration of various colours is caused by members of fungi-imperfect.

Examination after Softening or Soaking Seeds

Water Soak Method

The method is useful for detecting fungal infection in which spores are liberated into water after soaking. Infection of *Polyspora lini* in flax seeds can be determined using this method.

Sodium Hydroxide Soak Method

This method is used for detection of rice bunt and sometimes also for karnal bunt of wheat when infection is of very low grade and seed is treated with chemicals. The seeds are soaked in 0.2 per cent NaOH solution for 24 hr at 20-25°C. The solution is decanted-off and the seed is spread over the blotter paper to remove excess moisture. The infected portion of the seeds looks shiny black. To confirm the rice-bunt infection, black seeds may be pressed with finger or punctured with a needle in a drop of water, black spores gush out of the glume which can be examined under the stereobinocular.

Seed Washing Test

Examination of a suspension obtained from washing of seeds popularly been called the washing test. The test is mainly done for recording the presence of those fungi whose inoculum is present on the surface of seeds *i.e.* teliospores of bunts and smuts, oospores of downy mildew fungi and chlamydospores of some fungi. The test consists of shaking seeds in water to which a detergent may be added. The suspension obtained is examined under a compound microscope either directly or after concentrating the released spores by centrifugation and re-suspending them in water or another fluid.

Incubation Methods

Two methods, blotter and agar plate, are recommended by ISTA for routine examination of crop seeds for fungal infection. These are suitable for infections accompanied by hyphae, fruiting structures or spores. The tests are effective for detecting most seed borne fungi. Identification is based on fungal morphology developed during incubation on the seed surface on blotters or on colony characteristics on an agar medium.

Blotter Method

For health testing 400 seeds are tested. Depending on seed size 5,10,25 or 50 seeds are plated in each plate on wet blotters which have been lined in plastic petri dishes. This test can also be conducted by placing the seeds in between water soaked sheets of paper towels. The samples are incubated at 20-22°C for 7 days in alternating cycles of 12 hr darkness and 12 hr light. For identification of internal infections, the seeds may be pre-treated (surface sterilized in 1 per cent free chlorine or 0.1 per cent mercury chloride solution). The observations are recorded under the stereo and binocular microspores.

Agar Plate Method

This method gives an indication of the viable inoculum present in an infected seed sample. This is done by placing seeds on to sterile potato dextrose agar (PDA) or Malt extract agar (MEA) as suggested by ISTA. Prior to plating, the seeds are pre-treated with chlorine or mercury chloride. Usually 10 seeds per 9 cm culture plate are recommended. Incubation conditions and rest of the procedure is similar to that of blotter method. Normally, in order to avoid interference by roots and shoots of one seed to another, the germination of seeds is deliberately suppressed with herbicide 2,4-dichlorophenoxyacetic acid (2,4-D) solution or by **deep freeze method** (Limonard, 1966) where seeds after initial incubation period of 24 hr are transferred overnight to a deep freeze at -20°C and again brought back under normal incubation conditions. This process results in killing of the activated embryos in the seed itself.

Staining Methods

These methods are used where the incubation methods fail. The main standard method is used for loose smut of wheat (*Ustilago tritici*) and barley (*U. segatum* var. nuda) and downy mildew disease of pearl millet (*Sclerospora graminicola*) where the mycelium is located in the embryo of the seed. The embryos of barley or wheat seed (2000-4000 seeds per sample) are extracted over 24 hr at 20°C in a 5 per cent solution of sodium hydroxide and 0.015 per cent trypan blue. The extracted embryos are dehydrated in ethanol (95 per cent) for two minutes and then cleaned in Lactic acid: glycerol: water (1: 2: 1) mixture followed by boiling for 2 minutes in a mixture of lactic acid and glycerol (1: 2) to remove the excess of stain. After cooling for 1 hr the embryos are examined under stereomicroscope for the presence of blue stained mycelium in the scutellum.

Seedling Symptom Tests

Seedling symptom test was evolved by Hiltner (1917) to examine the *Fusarium* in cereals by growing them on sterilized crushed brickstone (Hiltner's brickstone method). Here the seedlings are raised under controlled conditions to allow for symptom development due to the pathogens to predict the field performance of the seed. The seedling symptom tests are conducted for post entry quarantine (PEQ) control, detection of certain obligate pathogens or pathogens that cause symptoms only on seedlings, and testing the efficacy of seed treatment fungicides. Seeds are sown in soil, sand, ground brickstone, blotters and other autoclaved growth media under optimum conditions. Seedling symptom tests are conducted by following methods.

Rolled Paper Towel Method

This method is generally used for measuring seed germination in seed testing laboratories. Pathogens that cause seed decay, seedling blight or seedling abnormalities can be detected. Seeds are placed between layers of water-soaked filter papers, blotters or paper towels, which are rolled and placed flat or in upright position. Moistened porous papers or absorbent cotton or butter papers may be used as a base for the papers. These samples are incubated at prescribed temperature and period and observations recorded.

Blotter Test

Fusarium spp. on cereal seedlings, *M. phaseolina* on sunflower, *Cochliobolus sativus* on barley, *Phoma lingam* on brassica seeds can be detected by this method. Two hundred seeds are placed on filter paper in refrigerator box (10x20 cm, 6 cm high) and incubated in 12 hr light and dark for about 14 days at 6-7°C. Seeds with visible pycnidia then are plated in another box and incubated at 20°C. Infected seedlings are examined for symptom development.

Agar Test in Test Tubes

This method is used for detection and demonstration of seedborne nature of *Colletotrichum graminicola* in sorghum, *Didymella bryoniae* in cucumber and pumpkin, *Drechslera graminea* and *D. teres* in barley and *Septoria nedorum* in wheat. Testing is done on 1 per cent water agar slants in test tubes by placing one seed in each test tube slant with 10 ml agar. The test tube is plugged with a loose cotton and incubated for 4-5 days. Thereafter plug is removed and the tubes are incubated for 14 days under alternate cycles of 12 hr light at 20°C before the disease symptoms are examined.

Soil Tests

Soil tests using multipots are used for detection of *Drechslera* spp., *F. nivale* and *Septoria* spp. and seedling phytotoxicity from fungicide treated seeds. As mentioned earlier that Hiltner's brickstone method is followed by growing the seeds under a 3 cm layer of crushed brickstone (3-4 mm grain size). Seeds are moistured with sterile water. The observation is recorded after incubation of 2 weeks.

Detection of Seed Borne Bacteria and Viruses

Detection of seed borne bacteria and viruses is more difficult than that of seed borne fungi. To have a tentative idea of the pathogen involved, visual observation of seeds may reveal abnormalities such as discolouration, pigmentation, shrivelling, staining, reduced size *etc.* Such abnormalities are specific to few host pathogen combinations.

Growing on Method or Seedling Symptom Test

This is the simplest but one of the reliable methods. Which has been used since a long time. In this method seeds are germinated providing suitable conditions for the expression of the symptoms by the pathogen. It has been found to be efficient in detecting *Xanthomonas Campestris* pv. *Phaseoli* in bean, *X.C.* pv. *malvacearum* in cotton, *Pseudomonas syringae* Pv. *glucinea* in soybean and *X.C.* pv. *Cyamopsidis* in clusterbean. Seedling symptom test by roll towel method is done to detect *X.oryzae* pv. *oryzae* in paddy. For detection of viruses the seeds are planted in blotter papers, sand, soil, vermiculite or other growth media under controlled conditions.

Infectivity or Indicator-Inoculation Test

Indicator or susceptible host plants can be inoculated with seed washings and seed extracts for detection of some bacteria and viruses. The infectivity test is simple and reliable for detecting these pathogens in groups of seeds, but does not provide the percent seed infection unless each seed is tested separately. *X.C.*, pv. *phaseolina* in bean seeds can be detected by inoculating indicator plants with seed washings. In viruses, susceptible hosts which produce local lesions or systemic symptoms are used as indicator plants such as *Chenopodium* spp. are used for cowpea aphid borne mosaic virus in cowpea and mungbean mosaic and cucumber mosaic in mungbeans. *Chenopodium amaranticolor* is used for identification of pea seed borne mosaic virus in pea, cowpea mosaic virus in cowpea and strawberry latent ring-spot virus in celery.

Laboratory Tests for Identification of Bacteria

Basic laboratory methods for detection of seedborne bacteria involve i) their extraction from seeds, ii) isolation into culture and iii) identification by a range of different methods. Bacteria are extracted from seed flour or seeds in a liquid medium; sterile tap water, distilled water, various buffers and enrichments *etc.* Generally, a sterile buffered saline is most commonly used which affords some protection to the bacterial cells. Bacteria are isolated by plating small volumes of the extracted medium onto general or semi-selective agar media. Once isolated in pure culture, plant pathogenic bacteria are identified by means of classic, morphological, physiological and nutritional tests.

Bacteriophage Tests

Bacteriophages are viruses which attack and kill bacteria. The relationship between phage and its host bacterium may be specific and this can be used to detect and identify the bacteria. Suspensions of selected specific phages are spotted on to the plates which have been seeded with pure cultures of bacterial isolates which require identification, where bacteria are killed (lysed) by phases, clear zones develop which are confirmatory of the identity of the pathogen. These methods have been successfully employed for the detection of *Pseudomonas syringae* pv. *phaseolicola* and *Xanthomonas campestris* pv. *phaseoli* in bean seeds.

Immunodiagnostic and Nucleic Acid Methods

Serology

Serological tests for plant pathogenic bacteria have been known since 1918 The methods are based on immunological principle that foreign molecules (*i.e.* immunizing agents or antigens) injected into the blood of mammals stimulate the immune system of those mammals to produce specific antibodies which will recognize and bind to the antigens. Such antibodies recognize many chemical sites, referred to as epitopes, on target antigens; these are polyclonal antibodies. Antisera used against purified pathogens have been effective in the detection of viruses but has a limited value for the detection of bacteria and fungi because of their more complex nature. They contain many non-specific antibodies which may cause cross reaction with related and unrelated species concealing the effects of specific antibodies. The development of monoclonal antibodies (MABs) has greatly increased specificity. Monoclonal antibodies recognize one chemical site (epitope) on target antigens and thus are homogenous and free from the variability as is common in polyclonal antisera. They are helpful in detections at generic, specific, pathovar and strain levels.

Agglutination, Precipitin and Immunodiffusion Methods

In agglutination tests, materials such as bentonite, chloroplast, RBC of sheep or synthetic polysterene latex are used as reaction sites. Here the antigen/antibody reaction results in an agglutination or clumping of particulate antigen. These tests are suitable for elongate virus particles which do not diffuse through a gel medium. The micro-precipitation test is a simple serological test in which drop of a crude or purified seed extracts combine with their specific antibody and result in a macro-or microscopically visible precipitate. The method is helpful for detection of rod shaped or isometric particles. In gel-diffusion method (Ouchterlony double-diffusion method) where the antisesum diffuses from a central well through the agar to precipitate. If similar precipitation lines are formed then the antigens are considered to be identical. Such techniques have been used to detect different viruses like BCMV in urd, BSMV in barley, brome mosaic virus, cowpea mosaic virus, wheat streak mosaic virus *etc.* and bacteria like *P. syringae* Pv. *phaseolicola, X. campestris* pv. *phaseoli* and *P. s.* pv. *pisi*. In addition, methods known as labelled antibodies, immunofluorescence microscopy test and radioisotope-labeled antibodies are also used for detection of some viruses.

Enzyme-Linked Immunosorbent Assay (ELISA)

This method is very useful and now a days used most-widely for detection of viral and bacterial diseases. There are main three types of ELISA i.e direct or double antibody sandwich (DAS-ELISA) method, indirect method and competitive assay (Lange, 1986). The direct method is most often used in seed health assays. Seed testing using, ELISA techniques is based on the use of polystyrenes microtitre plates, each containing up to 96 wells, which are especially suitable for tests involving the indexing of numerous samples and sub-samples of seeds. Seed or seedling extract (*i.e.* antigen) is selectively trapped and immobilized by solid phase-specific antibody in microtitre wells. Enzyme labelled antibody is then reacted with the immobilized antigen and after removing unreacted enzyme labelled antibody, the retained enzyme is assayed by adding a Suitable substrate. The presence of virus is indicated by yellow colour which can be improved with the addition of substrates. However, the quantitative assays can be done by using colorimeter or spectrophotometer. The ELISA have several advantages like the processing large number of samples at a time and it is rapid, sensitive and cost effective. Polyclonal antibodies are effective when the extreme specificity is not needed thus they are commonly applied as trapping antibodies. However, monoclonal antibodies are used for detection of Bacteria and fungi. Highly specific monoclonal antibodies have been developed to *P.syringae* pv. *pisi* (bacterial blight of pea) and have been tested using both indirect and competitive ELISA assays to distinguish between strains of the bacteria, there are monoclonal antibody techniques which now can be used for the detection of seed borne fungi of spruce seeds and certain seed borne fungi) and bacteria of seeds. Commercial kits have been developed and marketed for rapid detection of seed borne pathogens.

Other Immunodiagnostic Methods

There are modifications of ELISA test and also some other immunodiagnostic methods *viz.*, enzyme-linked fluorescent assay (ELFA), Radio immunosorbent assay (RIA), solid phase radioimmuno assay (SPRIA). Immunosorbent electron microscopy (ISEM) and Dot-ELISA (dot blot, dot immunobinding or spot immunodetection) which are employed for detection of various viral and bacterial pathogens.

Nucleic Acid Methods

Nucleic acid hybridization techniques are based on the specificity by which nucleic acids (DNA or RNA) hybridize to form double-stranded molecules. A specific sequence of DNA from a plant pathogen can be selected which only couples (*i.e.* hybridizes) with the identical sequence of DNA in that target organism. this is known as probe and probes are labelled biochemically or using radioisotopes so that the hybridization of the probe and target DNA can be detected visually, target DNA is dotted (dot blots) onto and bound to nylon or nitrocellulose filters. these filters are incubated in the presence of probe DNA. The filters are then auto radiographed and if there is a positive reaction (hybridization of probe and Targetted DNA), the label appears as a spot-on photographic film. The dot blot method is quick, simple and a number of samples can be processed at a time. If the target organisms can be cultured, other forms of hybridization, including colony blotting may be used Nucleic acid probes are used in routine for the detection of plant viruses and viroids but yet have not been used for this purpose in seed pathology. But this technology is widely used in the routine testing of potato samples for the detection of tuber viruses.

Polymerase Chain Reaction (PCR) Technology

The polymerase chain reaction (PCR) is a biochemical method of amplification that multiplies the concentration of target sequence of DNA by upto a millionfold, thereby transforming previously undetectable amounts into detectable quantities. Amplification is done with the help of primers. A further PCR method involves the use of random amplified polymorphic DNA (RAPD) technique. These are very short primers, not having identified DNA sequences and usually with 8-12 bases which will hybridize to large number of arbitrary sites in the genome. Such products are run on the electrophoresis gel to form the banding pattern which are used for distinguishing between species or varieties of fungi. The control of seed borne pathogens and diseases is attained through integrated disease management like selection of disease-free areas, crop management, seed treatment, seed certification, plant quarantine and disease resistance.

Conceptual Framework of Wheat Rust Management

Appropriate pathotypes for evaluation of resistance to stem and leaf rusts in durum and bread wheats identified: Appropriate pathotypes of leaf and stem rusts were identified for evaluating resistance in durum and bread wheats, based on multipathotype seedling tests. In addition to leaf rust pathotypes 12-2 and 104-2, showing virulence to both durum and bread wheats, 77-1, 77-5 and 77-7 can be included in bread wheat evaluation, while 12-4, 106, 108 and 162 are appropriate for evaluating leaf rust resistance in durum wheats, based on the virulence patterns of these pathotypes on the test lines. Among the stem rust pathotypes, 117-6 and 117 A-1 were most virulent on durums, while the pathotypes 40A, 117-5, 122 and 295 can be used for evaluating stem rust resistance in bread wheat. Thus, differences between responses of durum and bread wheats to stem rust pathotypes were reported for the first time. **The** "differential" patterns of resistance to stem and leaf rusts manifested by the two wheat species are potentially useful for disease management **by encouraging the simultaneous cultivation of durum and bread wheats.** Use of appropriate rust pathotypes **for evaluation of resistance will** help in developing durum and bread wheat varieties with stable rust resistance.

Syringe Inoculation in Wheat

Diverse resistance sources need to be used for developing wheat varieties with stable rust resistance. Since conventional genetic analysis aimed at determining diversity among resistance donors is tedious and time consuming, rapid evaluation methods like pedigree-details, field disease scores and seedling infection types are being used as other tools. Studies involving a number of durum and bread wheat genotypes revealed that syringe-inoculation can also be used to rapidly explore the diversity for rust resistance among wheat genotypes, in addition to the above-mentioned tools.

Enhanced Susceptibility to Stem Rust

Enhanced susceptibility to stem rust was observed in the loose smut infected wheat tillers, compared to the non-smutted tillers of the same line. Hence, stem rust infection of any smutted tillers should not be considered for determining the stem rust reaction of a wheat line, to avoid over-rating.

Validity of the Phenotypic Markers in Wheat

Leaf rust resistance gene *Lr34* is known to impart durable resistance to leaf rust worldwide. This gene has been postulated in many Indian bread wheat cultivars including 'C 306', based on the associated leaf tip necrosis. Inheritance studies showed that 'C 306' carried a gene for adult-plant resistance to leaf rust pathotype 77-5, and leaf tip necrosis was associated with this gene. However, susceptible segregants observed in F_2 populations of crosses of C 306 with known *Lr34* carriers confirmed that this gene in C 306 was different from *Lr34*. Thus, leaf tip necrosis in wheat can be associated with other gene/s also, and hence, can not be considered as an exclusive phenotypic marker for *Lr34*. A leaf rust resistance gene, other than *Lr34*, associated with leaf tip necrosis in wheat. *Plant Breeding* 124: 517-519. Recessively inherited gene *Sr2* has provided basis of durable resistance to stem rust in wheat worldwide. This gene has been postulated in many Indian wheat cultivars including 'Lok 1', based on associated pseudo-black chaff in adult-plants, and leaf chlorosis in seedlings. However, dominant inheritance of the resistance factor operating in 'Lok 1', and a 13: 3 (resistant : susceptible) F_2 segregation in the '*Sr2*-line' (Chinese Spring*6/Hope 3B)/'Lok 1' cross confirmed that *Sr2* was absent in 'Lok 1'. Susceptible plants with pseudo-black chaff phenotype were observed in F_2 populations of 'Agra Local' (susceptible)/'Lok 1', and the '*Sr2*-line'/'Lok 1' crosses. Thus, linkage of pseudo-black chaff (pbc) with *Sr2* in wheat can be broken, and hence, caution may be exercised in using 'pbc' as a marker for selecting *Sr2* in breeding programmes. [Ref: Mishra AN, *et al.*, 2005. Linkage between the stem rust resistance gene *Sr2* and pseudo-black chaff in wheat can be broken. *Plant Breeding* 124: 520-522.

Development of Locally Adapted Host Differentials for Indian Pathotypes

Majority of the leaf rust differentials in set 'A' and most varieties in set 'B' are winter types, which require a long photoperiod for flowering. Hence, maintenance of these differentials is difficult, particularly in the plains of India due to their long duration and often there is either seed set failure or production of shriveled grain. Hence, need was felt for developing near-isogenic *Lr* lines in the background of a locally adapted variety. NP 4 was selected as a background parent which is not known to carry any *Lr* gene or suppressor factor for leaf rust resistance, and being early maturing, slow rusting, lodging-tolerant, non-shattering, heat and drought tolerant, and bold seeded variety. Earlier, a total of 9 differentials carrying *Lr1, Lr2a, Lr2c, Lr3a, Lr9, Lr10, Lr15, Lr17a* and *Lr20* were developed in NP4 background and registered. Now, the NILs carrying leaf rust genes *Lr13, Lr18, Lr19* and *Lr26* were developed by backcrossing NP4 line with thatcher lines carrying these genes. A total of six backcrosses were done in succession followed by selection and testing which was completed in 2015-16. The lines *Viz.*, HI KK10 (NP4+*Lr13*) (IC0624491, INGR17034), HI KK11 (NP4+*Lr18*) (IC0624492, INGR17035), HI KK12 (NP4+*Lr19*)

(IC0624493, INGR17036) and HI KK13 (NP4+*Lr26*) (IC0624494, INGR17037) carrying these leaf rust resistance genes were registered with ICAR-NBPGR, New Delhi.

Evaluation of Wheat Genotypes Against Stem and Leaf Rust Resistance at Indore

A total of 584 genotypes of Preliminary disease screening nursery (PDSN) were evaluated for field resistance to stem and leaf rusts under artificial inoculations using mixtures of important pathotypes during *rabi* 2016-17. Of these, 427 entries (~73 per cent of the total) showed resistance (Coefficient of Infection value up to 10.0) to both stem and leaf rusts at Indore. Out of 100 bread wheat entries from Indore, 32 genotypes were found to be resistant to all three rusts. The resistant genotypes are: HAS Nos- 2289, 2292, 2293, 2294, 2295, 2296, 2302, 2305, 2307, 2308, 2309, 2310, 2313,2331, 2335, 2348, 2349, 2352, 2353, 2354, 2361, 2363, 2364,2366, 2367, 2369, 1261, 1515, 1743, 1853, 1878 and 1924. The IARI Common Varietal Trials (CVT) entries (total 350) were evaluated for seedling response to stem rust pathotype 40A. Out of these entries, 305 were observed to be resistant (87.1 per cent), while only 45 were susceptible.

Gene Cloning

The DNA of many cereals is larger and more complex than the human genome, so recent technological advances have been needed to make cloning faster, cheaper, and more practical. The first disease resistant gene to be cloned (HM1) was for resistance to a fungus affecting maize (HCTR). In this case, researchers were already aware of the existence of the fungal toxin and a detoxifying maize enzyme. In contrast, a resistance gene (Xa21) for a bacterial disease in rice was identified without any prior knowledge. Resistance for all Indian and Philippine rice species was provided by a resistant gene from African wild rice, but the process took 10 years. This illustrates the challenges of time and complexity that have been inherent in gene cloning for cereals to date. Cloning most of the major cereal disease resistance genes could be possible in the near future. Further work is needed to understand the most efficient use of resources, looking at how different techniques can be combined and how resistance genes can be more effectively isolated. However, comprehensive breeding programmes are needed to create new crop cultivars with greater disease resistance. New methods for sequencing, gene capture and gene complexity reduction are making the identification, cloning, and effective use of resistance genes more effective:

Faster Trials

Mapping and growing multiple generations of plants with possible resistant traits has typically been a long process. It can now take place considerably more quickly in controlled greenhouse conditions that allow several generations to be grown in a year rather than in seasonal field conditions.

Improved Reference Material

Although genome reference sequences for rice and maize have been available for several years, more complex sequences for wheat and barley have only recently been completed. This will make mapping and isolation of resistant genes much faster. However, gene content differs between different genotypes of the same cereal species, so a resistance gene might not be present in the reference cultivars.

Reduced Complexity

Messenger ribonucleic acid, or *m*RNA, can be used to identify parts of the genome that will not contribute to identifying resistance genes. This allows studies to focus on differences between resistant and susceptible species, with and without the target disease present.

Bulk Sequencing and Mapping

An approach known as MutMap uses whole genome resequencing to quickly isolate genes responsible for specific mutations. Plants with specific characteristics can then be produced by crossing interesting mutations with other cultivars. These can be grown in bulk and compared with the original wild species. The MutMap+ approach doesn't rely on crossing and is especially helpful in identifying mutants that impact reproduction, including sterility or other mutations that prevent crossing. These approaches are successful with rice, which has a relatively small genome, but aren't currently suitable for the more complex genomes of wheat or barley, for example.

Targeted Mapping

Chromosome flow sorting has been used to simplify identification of resistance genes in more complex cereals, although its accuracy can be limited. To overcome this limitation, high-resolution long-range sequence assembly can now be used to allow additional markers to be identified quickly and cost-effectively. When used with other techniques, some time-consuming processes will no longer be needed.

Resistant Gene Enrichment

The natural ability of cereal plants to detect diseases and trigger immune responses is a growing area of interest. Gene enrichment sequencing allows genes that are likely to be associated with diseases to be identified without mapping the whole gene. This is a faster process, although currently limited to certain types of naturally occurring resistance genes.

Wider Diversity

Association genetics and complexity reduction techniques can be used to isolate resistant genes in a wider population without a reference sequence. This can significantly reduce the time needed to clone a resistance gene, although significant time is needed to create a suitably diverse population of plants.

Pedigree-Based Genome Mapping for Marker-Assisted Selection and Recurrent Parent Recovery in Wheat

The pedigree-based genome mapping project will investigate and develop systems for implementing marker assisted selection to improve the efficiency of selection and increase the rate of genetic gain in breeding programs. This is a pilot project to test protocols for implementation of pedigree-based whole genome marker analysis in northern region winter cereal breeding programs. It aims to provide useful information about the use of simple sequence repeat (SSR) and other types of molecular markers for routine genomic analysis, the integration of genotypic, phenotypic and pedigree information for targeted wheat and barley lines, the genomic impacts of strong selection pressure in case study pedigrees, and directions for future pedigree-based marker development and analysis. Pedigree-based whole genome (PBWG) mapping offers significant potential to increase the genetic gain made in plant breeding programs through the development of systems for marker assisted selection. In in-breeding plants, the identification of genomic regions associated with traits of importance traditionally use populations derived from single crosses of inbred lines. This strategy however is frequently sub-optimal for direct application in breeding programs for numerous reasons. The QTL (quantitative trait loci) - trait association strategies frequently use obscure or out-dated parents. This strategy competes directly with breeding programs for significant resources for phenotypic evaluation. It is also limited in coverage because it addresses only the genes segregating in the populations under study. Additionally this strategy provides no information about the frequency or value of the QTLs in the breeding populations and hence requires a further step of validating the usefulness of markers for direct application in the breeding program. Due to the wide-spread adoption of this QTL analysis approach with specific doubled haploid populations, marker application is being impeded through the lack of information about the polymorphism and marker-trait association in genetic material relevant to Australian northern region cereal improvement. Pedigree-based whole genome marker application provides a vehicle for incorporating marker technologies into applied breeding programs by bridging the gap between development and implementation. The PBWG marker concept uses pedigree information to identify markers linked to traits based on identity by descent in breeding populations. The approach makes efficient use of pedigree, phenotypic and genotypic information collected in the normal breeding program, thereby exploiting this valuable resource. The PBWG approach integrates very closely with marker assisted recurrent parent recovery (MARPR), a technology already applied to many crops including barley, maize and rice, that can speed variety development in back-crossing strategies. Effective use of a PBWG-MARPR marker approach requires relevant pedigree and phenotypic data, low-cost, high throughput genotyping and a data management and analysis system that combines pedigree information, genotypic and phenotypic data. Implementation of these components allows linkage to advanced molecular marker and related research, which directly addresses constraints and opportunities arising within the context of the breeding program. The current project is piloting the design and development of a system that integrates these components of PBWG and MARPR marker application for use in applied breeding programs.

Pedigree Based Marker Assisted Selection System including Graphical Genotype with IBD (PBMASS)

Marker data will be analysed in conjunction with both existing and new phenotypic data. The wheat and barley applications will be assessed and compared to determine the effectiveness of PBWG marker analysis in producing relevant information for use in interpreting associations between traits and genomic regions under selection through the pedigrees targeted in each breeding program and for trait selection, parent characterization and MARPR. The analysis of the PBGW mapping approach will allow for an enhanced understanding of the processes of selection within a breeding program, including identification of regions under selection and will act as a basis of validation of markers and QTLs for the more effective implementation of markers, through the confirmation of QTL location, the identification of markers which will work in other, distantly related populations and in the identification of QTLs which are effective in other populations. The expected outcomes from this project will encourage more sophisticated use of markers in targeted breeding populations as selection is applied to conserve and combine critical genomic regions for both environmental adaptation and grain and end-use quality. The use of a focussed set of genotypes means that the marker technology will be quickly available for use with a large proportion of the gene pools in the respective case study breeding programs, however the comparison of the two case studies will also allow for the identification of critical components of this approach for implementation in other crop plants.

New Donors of Stem Rust Resistance in Durum Wheat

Lack of adequate resistance to stem rust in durum wheat is a major constraint in durum wheat improvement, and hence, there is an urgent need to identify additional sources of resistance. Five durum wheat genotypes including B 662, IWP 5019 and Line 1172, interspecific derivatives involving *Triticum aestivum, T. dicoccum* and *T. militinae,* respectively, HG 110, an Indore durum, and ED 2398-A, an Ethiopian durum land race, have been resistant to stem rust since 1996. They are resistant to leaf rust as well, except Line 1172. Differences in seedling reactions of these lines to various pathotypes indicated that they carried diverse genes for stem rust resistance. (Ref: Mishra AN, *et al.*, 2001. New germplasm of durum wheat with stem rust resistance. *Wheat Information Service* 93: 35-37). Genetic studies have confirmed the diversity among these genotypes for adult-plant resistance to stem rust pathotype 117-6 (Mishra AN, unpublished data). Allelic tests identified three diverse genes for seedling resistance to stem rust pathotype 117-6 among four genetic stocks of durum wheat. AKDW 3347 and GW 1139 shared a common gene, which was different from the two independent genes present in B 662 and P 6046. These genes are different from *Sr2, Sr7b, Sr9e* and *Sr11,* the genes commonly postulated in Indian durum wheat genotypes, since the latter are ineffective against the pathotype 117-6 in seedling stage.

Diverse Sources of Adult-Plant Resistance to Leaf Rust in Bread Wheat

The known effective leaf rust resistance base in bread wheat is rather narrow, and hence, identification of new diverse resistance sources is a research priority. Genetic studies showed that a single partially dominant gene in 'CPAN 1842' **conditioned adult plant resistance to leaf rust pathotype 77-5, and was** different from **any of the designated genes for adult-plant resistance to wheat leaf rust** ***viz.***, *Lr12, Lr13, Lr22a, Lr22b, Lr34, Lr35, Lr37, Lr46, Lr48,* **and** *Lr49*. Genetic analysis revealed that the gene/s for adult-plant resistance to leaf rust in bread wheat genotypes- 'CPAN 1676', 'CPAN 1842', 'HD 2135', 'HW 517', 'Pavon 76', and 'WG 138', were different from the resistance component/s of 'Frontana', the Brazilian bread wheat cultivar used worldwide as a source of durable resistance to leaf rust in wheat.

Sources of Rust Resistance in Wheat and Triticale

As a cooperator in the Crop Protection Programme under the All India Coordinated Wheat Improvement Project, I have been actively associated with the evaluation for disease resistance in wheat and triticale through hot spot multilocation tests involving around 2500 lines every year. A number of rust resistance sources in wheat and triticale have been identified based on these tests.

Combining Classical and Novel Resources

Septoria tritici blotch (STB) is one of the most important fungal diseases of wheat in Europe and other wheat-growing areas. Because of the sexual reproduction of the fungus, genetic variability allows fungicide resistant strains to evolve. Many wheat cultivars have already been bred with a specific resistance to the fungus. These can now be cloned and used much more quickly thanks to recently developed techniques such as target interval mapping. Proteins have been identified in this resistant gene that are now recognized as important components of disease resistance in cereals. Another gene has also been discovered that affects the virulence of the disease. It is not yet known whether these genes effectively work together, but they offer valuable options for further development of a plant disease resistance gene. Wheat is the most important winter cereal crop in India and serves as the staple food for more than 1 billion population. It contributes approximately 12 per cent to the world wheat basket and holds the global share of 11 per cent area under cultivation of wheat. This self-sufficiency in food was accomplished due to increased wheat production which must be sustained over the coming years, even though there will be a pressing demand due to increased population pressure and international trade. A moderate growth rate of 2.25 per cent during the past 25 years posed a challenge to wheat workers for enhanced production to meet the targeted need of 109 million tons by 2020 AD. There are several constraints towards realizing the potential yield in wheat, which may hinder timely achievement of the targeted production. Losses incurred due to biotic and abiotic stresses are enormous. Among them, the rust pathogens are prominent and wheat crop is subject to severe depredation of all the three rust endemics to India *viz.*, black or stem rust (*Puccinia garmini* f. sp. *tritici* Erik. and Henn), the brown and leaf rust (*Puccinia triticina* Erik.) and the yellow or stripe rust (*Puccinia striiformis*). Not all the rusts are equally important in all the six agro-ecological zones of the country. Wheat leaf rust, also known as brown rust, is a serious fungal disease affecting wheat and rye caused by *Puccinia triticina*. Leaf rust has potential to cause losses up to 40 per cent whereas up to 100 per cent losses due to stem rust have been reported (Singh *et al.*, 2002). The wheat leaf rust fungus is adapted to a range of different climates and the disease can be found in diverse wheat growing areas throughout the world thereby affecting the productivity and quality of wheat worldwide. Wheat cultivars that are susceptible to leaf rust regularly suffer yield reductions depending on the stage of the crop development.

Gene Pyramiding for Multiple Disease Resistance

Development of resistant varieties for single disease is not enough to save plant product and to feed growing population especially in developing countries. In India, there are number of constraints in the successful production of wheat like leaf and yellow rust, powdery mildew and spot blotch. These all diseases together toll heavy yield losses and put us in the situation to redesign our experiments to develop multiple pest resistant varieties in wheat. The grown varieties on the market do carry some known disease resistance genes against powdery mildew and rusts and low levels of quantitative resistance against leaf spot diseases. However, most of the resistance genes against rusts and powdery mildew are already broken down. In the sustainable agriculture, which is economical both for the farmer and nature, multiple disease resistance is an essential tool against pathogens attack beside cultural practices like crop rotation. With the biotrophic fungi like rusts and powdery mildew, the only solution is the durable disease resistance. In late 1980's Fusarium head blight (FHB) become a world wide problem. Earlier, it was sporadic and localized to China, South America and a few European countries. Varieties resistant to FHB such as Wuhan 1, 2 and 3 and Shangai 7 and 8 and Suzhoe wheats were the product of generous collaboration between various Chinese institutions. Fusarium head blight is a serious disease of wheat (*Triticum aestivum* L.) in humid and semi humid areas of the world. Evaluation of Fusarium head blight resistance is time consuming, laborious and costly as the inheritance of resistance is complex phenomenon. The most recent biotic threat to global wheat production is *Ug99*, a virulent race of stem rust which has emerged from Uganda and has been confirmed at widely distributed testing locations in Kenya and Ethiopia. Yield losses of up to 71 percent have been recorded under experimental conditions (Dixon 2003).

Gene *Yr9*

Although the incorporation of rust resistance genes in to varieties from time to time has prevented large scale damage due to pathogens but the ability of the rust pathogens to produce new virulence has always kept the scientist engaged in searching for new sources. The 1BL.1Rs translocation has been very useful in Indian wheat breeding as it provided multiple disease resistance. Stripe rust is a disease of cooler areas and occurs in the northern hill states and Terai regions. It is also recorded in Nilgiris and Palney hills but the composition of races is different. The stripe rust inoculum is believed to come from the northern hills, Nepal and also exogenously from across the northern border. The information from pathogenicity surveys ensured that the resistant varieties were in the field for cultivation. The gene *Yr9* provided durable resistance for a long period of time in spite of the migration of virulence from across the northern border in the 1990s. It was a significant development and a major concern of the breeders because the *Yr9* resistance was deployed on large area. The resistance to powdery mildew is associated with yellow rust resistance was rapidly overcome by new races at very early stage of their introduction to cultivars. In the UK, the combination of *Yr6* and *Yr9* gave resistance to all the races before 1987, however a combined virulence occurred on these two genes on a cultivar Hornet resulted in a small scale epidemic. Another combination of *Yr7* and *Yr9* present in several CIMMYT cultivars showed resistance in several countries but later on, virulence developed independently in more than one place. Heavy infection of *Yr9* virulence was observed in some of the adjoining countries, where replacement of this gene deployment/variety could not be practiced. In India, the selection for PBW343 at the time of prevalence of *Yr9* virulence proved very useful as it was found resistant. It is not that earlier, races have not been evolved and migrated. The evolution of new races is a continuous phenomenon to occur in nature due to interaction of host and parasite. But this was a major worry for the main wheat productivity and a potential zone. Presently, three *Yr9* virulence(s) have been identified which are spreading in the North and North western parts of the country as a large number of genotypes with *Yr9* released during the past 20 years (PBW343, HD2687, UP2338) have become susceptible and so are the other genotypes carrying *Yr9*.

Lr24/Sr24

The *Agropyron* derived rust resistance genes *Sr24, Sr26* and *SrAgi* conferred effective adult plant resistance to stem rust, whilst *Lr19, Lr24* and *Lr29* conditioned a high degree of resistance to leaf rust for the past several years against the prevailing race flora in the Nilgiris, India. However, the gene *Sr25* was found completely ineffective. The resistance of *Lr24* from *Agropyron elongatum* (Host.) Beauv. has been extensively exploited to develop commercial cultivars. The gene *Lr24* is linked to *Sr24* and both of these genes have been reported ineffective to some of the races of leaf rust and stem rust, but the resistance exhibited by *Lr24* continues to be very effective against Indian leaf rust races in seedling as well as adult stage. Although, Samborski, (1985) expressed his opinion that exotic genes for resistance are not better than other genes if they are used singly. The resistance conferred by linked genes *Lr24/Sr24* available in Agent could not be exploited because it is tighltly linked with a gene producing red kernel colour. The same resistance was utilized from the white seeded induced translocations produced by Sears (1973) and McIntosh (1976). Evaluatioon of different white seeded lines helped in identification of a stock, TR380-14*7/3 Ag#14 with similar resistance as exhibited by Agent. Sawhney (1994) transferred *Lr24* into two cultivars Sonalika and Kalyansona using a different line TR380-27*4/3Ag#3 and developed a genotype DL896-2 identified by All India Wheat Workers Meet in 1990 suitable for cultivation in Peninsular zone. A vigourous programme of transferring the effective alien genes was undertaken at IARI Regional Station, Wellington in 1980. However, DL784-3 (Vaishali) was the first variety released in 1993 by Central Varietal Release Committee for general cultivartion under timely sown, high fertility irrigated conditions in the NWPZ. With a different donor TR380-14*7/7/3 Ag#14 another variety HW2004 was released in 1995 for moisture stress condition in CZ, which became most favoured geneotype with the farmers. Since then about 18 varieties for different ecological conditions ranging from moisture sterss environments to irrigated, high fertiity, under timely and late sown conditions have been relaesed and notified in different zones and states. Tomar and Menon (1998) transferred *Lr24* into 19 Indian cultivars using backcross breeding. All the backcross lines showed high degree of resistance to leaf rust in adult plant stage but recorded low infection to stem rust. It is evident that the release of a large number of varieties with *Lr24* indicate that alien race specific gene *Lr24* has no impediment on yield. HW2004, a back cross derivative carryin *Lr24* significantly yielded higher over recurrent parent C306, particularly under high incidence of leaf rust. The plains of central India are the routr of migration of stem and leaf rusts uredospores to the main wheat growing areas of northern India. Since the rainfed crop in central India is early sown, the inoculum built up onsusceptible cultivars poses a major threat to the timely sown crop of central and northern plains. To check the rust infection in this secondary source, cultivation of HW2004 is the only economic and effective answer. The cultivars carrying *Lr24* are being released since 1993 and the area is also increasing every year. So far about 14 m ha area has been occupied by the cultivars with *Lr24* and no virulence has been reported thus far from India. In 1971, Agent (*Lr24*) was cultivated and virulent strains of leaf rust occurred on it in trace amounts, but races virulent on this gene were not detected. In subsequent year, when Agent occupied 50,600 ha in that state and leaf rust occurred at 1 per cent severity, the pathogenic races were isolated (Samborski, 1972). The isolates of *Puccinia graminis tritici* virulent *Sr24* appeared in South Africa (Le Roux and Rijkenburg, 1987). In India, a virulent biotype 40-1 of race 40 was isolated from the Nilgiris in 1989 (Bhardwaj *et al.*, 1990), well before its deployment in 1993. *Sr24*-virulent culture detected from Wellington has remained confined to this area for more than 20 years. Even within Nilgiris this virulence is not predominant even though present. Therefore, the utility of this gene is undermined and one could use this gene in the areas where such virulences have not been detected/migrated yet. Therefore, it is not necessary that the evolution of a new rust race occur under selection pressure only as it is evident from the ineffectiveness of *Lr26* and *Sr24* when both these genes were overcome by new virulence(s) before they were deployed.

Breeding for Durable Resistance to Rust Diseases

Breeding for durable resistance based on minor additive genes has been challenging and often slow, for several reasons: (1) a sufficient number of minor genes may not be present in a single source genotype, (2) a source genotype may be poorly adapted, (3) there may be confounding effects from the segregation of both major and minor genes in the population, (4) crossing and selection schemes and population sizes are more suitable for selecting major genes, (5) reliable molecular markers for several minor genes are unavailable, and (6) the cost associated with identifying and utilizing multiple markers is high. One suggested approach is to use recurrent selection schemes to accumulate several minor genes in a single genetic background. But, such selection schemes have often been more of a scientific interest than actually being applicable to breeding. Selection for resistance alone will not generate important popular cultivars unless it is simultaneously combined with other traits such as high yield and quality. However, such germplasm carrying combinations of minor genes should be very useful in transferring these genes to modern cultivars. A successful example of breeding for resistance based on minor genes is the development of resistance to leaf and stripe rusts in wheat, which took about 30 years of continuous effort at CIMMYT. In the early 1970s, S. Rajaram, influenced by the concept of slow-rusting resistance in wheat proposed by R. Caldwell (1968), made a strategic decision: to initiate selection for slow-rusting resistance to leaf rust in CIMMYT spring wheat germplasm. In the early phase of breeding he maintained plants and lines in segregating populations that showed 20–30 per cent rust severity with compatible infection type. This strategy led to the release in Mexico and other countries of several successful wheat cultivars including 'Pavon 76' and 'Nacozari 76'. These slow-rusting lines were used heavily in the crossing program and resulted in the wide distribution of minor genes within spring wheat germplasm.

The genetic basis of slow rusting resistance in CIMMYT wheats started to become clear in the early 1990s. High-yielding lines that combine four or five additives, minor genes for both leaf and stripe rusts, and show near-immune levels of resistance were developed in the 1990s. Three or four lines carrying different minor genes were crossed (3-way and 4-way crosses), and plants in large segregating populations were selected under artificially created rust epidemics. Races of pathogens that have virulence for race-specific resistance genes present in the parents were used to create the epidemics. The resulting highly resistant lines are now being used in a planned manner to transfer these minor resistance genes to well-adapted, "farmer's choice" cultivars that are currently grown across large areas but have become susceptible to rust races in Mexico. The crossing and selection scheme described below was developed and applied based on genetic information on the number of additive, minor genes that must be transferred to achieve the desired level of resistance. This strategy has allowed simultaneous transfer not only of resistance genes but also other minor genes with small effects that increase the yield potential or improve the grain quality of an adapted cultivar. To transfer minor gene-based resistance into a susceptible adapted cultivar or any selected genotype, we use a "single backcross-selected bulk" scheme, where the cultivar/genotype is crossed with a group of about 8–10 resistance donors; 20 spikes of the F_1 plants from each cross are then backcrossed to obtain 400–500 BC_1 seeds. Selection is practiced from the BC_1 generation onward for resistance and other agronomic features under high rust pressure. Because additive genes are partially dominant, BC_1 plants carrying most of the genes show inter122 mediate resistance and can be selected visually. About 1,600 plants per cross are space-grown in the F2, whereas about 1,000 plants are maintained in the F_3–F_5 populations. Plants with desirable agronomic features and low to moderate terminal disease severity in early generations (BC_1, F_2, and F_3) and plants with low terminal severity in later generations (F_4 and F_5) are retained. We use a selected-bulk scheme where one spike from each selected plant is harvested as bulk until the F_4 generation, and plants are harvested individually in the F_5. Bulking of selected plants poses no restriction on the number of plants that can be selected in each generation because harvesting and threshing are quick and inexpensive, and the next generation is derived from a sample of the bulked seed. Because high resistance levels require the presence of four to five additive genes, the level of homozygosity from the F_4 generation onward is usually sufficient to identify plants that combine adequate resistance with good agronomic features. Moreover, selecting plants with low terminal disease severity under high disease pressure means that more additive genes may be present in those plants. Selection for seed characteristics is carried out on seeds obtained from individually harvested F_5 plants. Small plots of the F_6 lines are then evaluated for agronomic features, homozygosity of resistance, *etc.*, before conducting yield trials. Resistant derivatives of several cultivars and genotypes were recently developed using the above methodology. In each case we could identify derived lines that not only carry high levels of resistance to leaf rust or yellow rust or both, but also show about 5–15 per cent higher yield potential than the original cultivar. We believe this approach to wheat improvement allows us to maintain the characteristics of the original cultivar while improving its yield potential and rust resistance. It should be noted that having minor gene-based resistance in several backgrounds should ease future selection for these resistance genes.

Breeding for Durable Resistance to Powdery Mildew

Powdery mildew, caused by *Blumeria graminis* f.sp. *tritici,* is an important disease of wheat in highly productive areas. The evolutionary rate for this fungus is even higher than that for rusts because both asexual and sexual mechanisms are prevalent during the crop season. Since natural epidemics of powdery mildew do not occur in its main breeding operations, most of the breeding materials sent from Mexico to other regions were moderately susceptible to susceptible when grown in powdery mildew-prone areas. Because the use of race-specific genes, either singly or in combination, results in a rapid build-up of the new races, the best breeding strategy for a long-term control is the utilization of adult-plant resistance or slow mildewing. Several sources of such durable, adult-plant resistance have been identified. One such source is the North American cultivar 'Knox 62' and its derivatives, in which long-lasting resistance has been attributed to the presence of a few additive, minor genes. Lillemo *et al.* (2006) demonstrated that the high level of adult plant resistance at all tested international sites of CIMMYT spring wheat genotype 'Saar' was due to three or four additive genes. 'Saar' was initially developed for durable resistance to leaf and stripe rusts and evaluation of a common population with two rusts and powdery mildew showed high positive correlations in adult plant responses to these three diseases. Recent studies using the near-isogenic lines of slow-rusting genes *Lr34/Yr18* and *Lr46/Yr29* have shown that these two genes reduce powdery mildew severities by about 50 per cent and 30 per cent, respectively, in Norway under high disease pressure. The effect of *Lr34/Yr18* pleiotropic locus in reducing powdery mildew severity was much higher in Australia (Spielmeyer *et al.*, 2005). The pleiotropic effects of these loci, and perhaps others yet to be discovered, would ease simultaneous selection for resistance to wheat leaf and yellow rusts and powdery mildew. A similar breeding strategy, as discussed earlier, for rust diseases is currently being used at CIMMYT to improve the adult plant resistance to powdery mildew where a systematic transfer is being made from adult plant sources to high-yielding wheats through a single-backcross approach.

Breeding for Resistance to Fusarium Head Blight (Scab)

Several species of the genus *Fusarium* cause fusarium head blight (FHB), or scab, which is a main production constraint in environments where humid and semi-humid conditions coincide with flowering in wheat. The Yangtze River basin of China with about 7 million ha has traditionally been known to be highly prone to scab epidemics. Disease incidences leading to epidemics are now frequent in other developing countries, *e.g.*, Argentina, Brazil, and Uruguay, where residue retention is a common practice for conservation agriculture. Introduction of maize–wheat crop rotation further increases the disease build-up. The most important species implicated in scab epidemics is *Fusarium graminearum* (telomorph: *Giberrella zeae*). In addition to crop losses, the fungus also produces mycotoxins, such as deoxynivalenol (DON) that accumulates in the grain, rendering it unsuitable for human and livestock consumption. Sources of resistance to FHB have been divided into three groups: China and Japan, Argentina and Brazil, and Eastern Europe. More recently, additional sources, including some hexaploid synthetic-derived wheat lines, have been found to carry moderate resistance. Earlier genetic analysis indicated that a few additive genes confer resistance in Chinese and Brazilian wheats, but the genes present in Chinese sources are different sources. Although several genomic regions are now known to contribute quantitative resistance, a gene from Chinese cultivar 'Sumai 3' in the short arm of chromosome 3B has shown the largest and most consistent effect in reducing disease severity and mycotoxin accumulation. The Chinese sources are probably the best resistances currently available and must be combined with other sources of resistance. The Chinese cultivars that best combined with CIMMYT materials to transmit scab resistance are 'Sumai#3,' 'Ning 7,840,' 'Shanghai#5,' 'Yangmai#6,' 'Suzhoe#6,' 'Wuhan#3' and 'Chuanmai 18,'. Further progress in enhancing the level of resistance beyond the current level can come from a breeding strategy that favours the accumulation of multiple minor genes from various sources into a single genotype. This would involve inter-crossing parents with resistance located in different genomic regions, followed by growing large segregating populations and using flanking markers linked to resistance loci for selection in early segregating generations. Conventional field screening must wait until the F_4 or F_5 generations when homozygosity has increased significantly. Progenies that show higher levels of resistance than the parental sources could be used for a targeted transfer to high-yielding cultivars that would already by then carry moderate levels of resistance from the ongoing breeding efforts. This strategy is currently practiced to accumulate resistance genes from different sources.

Table 7.26. Sources of high level of resistance to *Septoria tritici* blotch evaluated at Toluca research station

Cross	*Septoria Score*	*Source*
OASIS/SKAUZ//4*BCN/3/PASTOR/4/KAUZ*2/YACO//KAUZ	42	Bobwhite
OASIS/SKAUZ//4*BCN*2/3/PASTOR	32	Bobwhite
TINAMOU	32	Bobwhite
ALD/CEP75630//CEP75234/PT7219/3/BUC/BJY/4/CBRD/5/TNMU/ PF85487	31	Brazilian
BH1146*3/ALD//BUC/3/DUCULA/4/DUCULA	21	Brazilian
CATBIRD	52	Chinese
N894037 41	41	Chinese

Cross	*Septoria Score*	*Source*
CHUANMAI 107	32	Chinese
MILAN/S87230//BABAX	21	Milan
MINO	31	Milan
MURGA	21	Synthetic
REH/HARE//2*BCN/3/CROC_1/AE.SQUARROSA (213)//PGO/4/ HUITES	21	Synthetic
FINSI/METSO	21	Synthetic
BL 1496/MILAN/3/CROC_1/AE.SQUARROSA (205)//KAUZ	21	Synthetic
ALTAR 84/AE.SQ//OPATA/3/2*WH 542	11	Synthetic

* Disease scores follow a 00–99 two-digit scale and evaluated when susceptible checks had scores between 96 and 98. The first digit represents the relative height of plants infected with the disease and the second digit represents the relative severity within that height.

Breeding for Resistance to *Septoria tritici* Blotch

Breeding for resistance to *Septoria tritici* blotch, caused by *Mycosphaerella graminicola* (anamorph *Septoria tritici*), in semidwarf wheat in early 1970. Steady progress has been made since then. Currently, several high-yielding semidwarf wheats with good resistance are available. Resistance in these wheat genotypes is derived from diverse sources, including synthetic wheats. The main problem encountered in the early breeding work was breaking the association of resistance with lateness and tallness present in the above sources. Efforts are being made to combine these resistances. Two high-rainfall sites, Toluca (Mexico) and Patzcuaro (Michoacan), are used in Mexico for selection. Genetic studies conducted on wheats indicate that five to eight genes confer resistance to *Septoria tritici* blotch, depending on the source population. Two to three genes that have predominantly additive effects are generally needed to confer an acceptable level of resistance. The selection methodology, therefore, is similar to that described for combining minor, additive genes for resistance to leaf and stripe rusts. Some synthetic wheats developed have shown excellent resistance that appears to be leading toward immunity to the disease. These sources offer new genetic diversity of resistance originating from durum wheat and/or *T. tauschii.* A high level of resistance from original synthetic parents was successfully transferred to derived, high-yielding lines. It is not known whether this kind of resistance will remain durable or if a pathogen population can shift to overcome such major gene based resistance.

Breeding for Resistance to Spot Blotch and Tan Spot

The first crosses to incorporate spot blotch (caused by *Bipolaris sorokiniana*) resistance into wheats were made about 20 years ago. These crosses involved moderately resistant cultivars, such as BH1,146, from Brazil. However, the level of resistance in progenies was inadequate when tests were carried out at Poza Rica, Mexico, which is an ME5 testing site. In the mid-1980s, wheat genotypes carrying resistance to scab and obtained from the Yangtze River Valley of China showed varying levels of spot-blotch resistance when tested at Poza Rica. These Chinese lines included Suzhue 1 to 10, Wuhan 1 to 3, Shanghai 1 to 8, and certain Ningmai and Yangmai lines. About the same time, the wide-crossing program produced resistant lines that contain *Thinopyrum curvifolium* in their pedigree. Some of these lines and their derivatives possess good resistance and appear to be promising in Bangladesh, lowland Bolivia, and Nepal. Resistance in these wheats, such as 'Sabuf,' 'Chyria 1,' and 'Cugap,' appears to be controlled by two to three genes, whereas 'Longmai 10' and 'Yangmai 6' may carry polygenic resistance with a relatively high narrow-sense heritability. A few synthetic wheats developed also carry resistance derived from the *T. tauschii* accessions. A key problem with selection for spot-blotch resistance is the negative correlation of disease severity with heading date and plant height. Therefore, care must be taken if short types with early maturity are required. Current strategy is to combine resistances from these diverse sources. Identification of some highly resistant lines from such crosses indicates that resistance is additive. However, a recent study has shown that durable leaf-rust resistance gene *Lr34* also confers a moderate level of resistance to spot blotch, which may be sufficient to prevent losses in areas where disease pressure is usually not too high (Joshi *et al.*, 2004). Tan spot (caused by *Drechslera triticirepentis*) resistance is not widely dispersed in germplasm, but moderate resistance is known to occur. Some newer lines, such as 'Milan,' 'Attila,' 'Corydon,' and 'Tinamou,' and some Chinese wheats and their derivatives, such as 'Luan,' are also reported to carry moderate to high resistance (Diaz de Ackermann and Kohli, 1998). Tan spot is on the increase in areas where reduced tillage practices are being combined with residue retention, so adequate resistance to this disease has become necessary to reduce losses. Therefore, resistance to this disease must be incorporated, especially in germplasm targeted for conservation agriculture.

Challenge by *Ug99* Race of Stem Rust Pathogen

Historically, stem rust of wheat has been the most destructive disease of wheat and barley and has inflicted very heavy losses worldwide some times to the tune of more than 40 per cent loss in America. The organism is a fungus called *Puccinia graminis tritici*. In India, though the occurrence of this rust had been a regular feature before the introduction of semi- dwarf wheat developed through a network of coordinated research programme by the Indian Council of Agricultural Research minimized the loss caused by this fungus through genes which prevent the fungus from causing any harm to the crop. Adjusting to the resistance in the host by throwing up different strains with the capacity to break the resistance is a natural phenomenon, which is the case in stem rust too. Plant breeders constantly keep a vigil for this and keep breeding varieties to upset this phenomenon. However, one unusually potent race was detected in Uganda in 1999 (named *Ug99*) that infected wheat grown in the region. Subsequently, it has quickly spread in the neighbouring countries like Kenya and Ethiopia and has crossed the Red sea as there are reports of its occurrence in Iran at a relatively faster pace than what was expected. Not only it spread quickly, but it also generated newer variants able to break the resistance of the varieties grown in the region. This quick spread and ability to generate variant forms in the African continent and beyond is a concern of global wheat community.

This feature brought the wheat community on high alert prompting Nobel laureate Dr. Norman Borlaug to declare it as a global threat.

The researchers of wheat in India have a position of preparedness in protecting the crop as wheat is accorded top priority in our R and D programmes as it directly impacts our food security. Thus, despite having not faced with any epidemic since last four decades, the strategy of conducting regular countrywide survey and surveillance continued to be in place to monitor the occurrence of all wheat diseases in different regions of the country. Trap-nurseries are also raised to monitor the chance of any stray infection of the disease. So far any presence the *Ug99* or its variants has not been noticed from any part of our country. Not willing to take any chances with the potential threat, the Council has taken necessary steps in wheat research programme to safeguard the wheat crop against destructive pathogen like *Ug99*. In view of the fact that the *Ug99* does not exist in India, the testing of resistance (a plant defense mechanism) against it can only be undertaken in Kenya where the disease is prevalent. Hence the ICAR, in collaboration with International Wheat and Maize Research Centre (CIMMYT), Mexico has stated a testing program of wheat at Njoro in Kenya since 2005. The tests have shown that there are many Indian wheat varieties with resistance to *Ug99* and its variants. The presence of diverse resistance in some of these varieties is very encouraging as it would help wheat to fight against *Ug99* and other future threats of this magnitude. Accordingly, the resistant wheat materials are being multiplied and provided to wheat breeders for further research and the seed of already released varieties with resistance to *Ug99* or its variants are being multiplied. For example, during the years 2007-08 more than 4,000q of breeder seed of 11 wheat varieties having *Ug99* resistance namely, GW 273, GW 322, HI 1500, HD 2781, MP 4010, HUW 510, MACS 2846 (durum), HI 8498 (durum), UP 2338, DL 153-2 and HW 1085 was produced. In the recently concluded national wheat workshop at Hisar in August 2008, another new variety Raj 4120 resistant to *Ug99* has also been identified for release.

Presently, our priority is on extensive surveys in the target areas to detect the first infection of *Ug99* or its variants, if and when it comes, before it acquires alarming proportions. The experience suggests that it takes a few seasons for the race to build up and appear on a large scale to cause economic losses. Thus, even in remotest event of *Ug99* reaching India, the years required for it to take epidemic proportions would be utilized to multiply the already identified resistant varieties, breed new varieties and then aggressively distributing these to the cultivators to avoid any loss in the farmer's field. As an immediate relief measure, chemical control technology is also available to ensure that this race does not spread and cause damage. As a precautionary measure, a special team has been constituted to survey and monitor the summer wheat crop in the higher hills of Himachal Pradesh, Jammu and Kashmir, Southern hills in Tamil Nadu to detect and track down *Ug99* or its variants. A new version of the popular variety PBW 343 which is reported to possess inbuilt resistance to *Ug99* is also being acquired from International Centre for Agricultural Research in Dryland Areas (ICARDA), Syria for its seed multiplication and utilization in the country.

In addition, India is proactively participating in the Borlaug Global Rust Initiative (BGRI) Programme as its core member to be in a state of preparedness against any serious consequences because of this disease. The programmes at the International level are aimed at sensitizing the nations in the migration path of *Ug99* to grow resistant varieties that would help decrease the inoculum build up and its subsequent migration to other countries. The tracking programme for the *Ug99* and its variants is already on to monitor its migration into newer areas. Thus a multi-pronged strategy is already in place to render the rust ineffective even in the most unlikely event of *Ug99* striking Indian territory. We are confident and f eel that such a vibrant technical programme will stand by Indian farming community and will be able to avoid any crisis likely to be caused by this disease. The Indian wheat programme is fully geared to take on the challenge of overcoming the challenge posed by *Ug99*. Since cooperative and collaborative network across geo-political limits are part of a global strategy against the rust, in November, 2008, India will host an **International Conference Wheat Stem Rust Ug99 – A Threat to Food Security** for stressing the importance and also forging a Regional/Global strategy to address this threat.

Stem, or black rust, caused by *Puccinia graminis tritici* (Pgt), is historically known to cause severe losses to wheat production. However, it has been controlled effectively through the use of genetic resistance in cultivars associated with the Green Revolution of the 1960s and 1970s. Over 80 per cent of the spring wheat area in developing countries is currently sown with cultivars either derived directly from germplasm or from germplasm used as parents. For more than 30 years, a major proportion of the CIMMYT wheat germplasm and germplasm developed by other breeding programs has remained resistant to stem rust. Resistance gene *Sr31,* located on rye translocation 1B.1R, has contributed to high levels of resistance in several wheat cultivars developed worldwide in recent years. Consequently, stem rust disease is often not considered important and in many countries is ignored in wheat breeding; research in this area in the last two decades has also declined substantially. Detection in Uganda in 1999 of Pgt race *Ug99* that has a broad virulence, including virulence for *Sr31,* and its migration to Kenya and Ethiopia has been recognized as a highly significant event and led to the launch of the Global Rust Initiative in 2005. All major cultivars currently grown in North Africa, the Middle East, and Asia are moderately or highly susceptible to *Ug99*. Predominant wind patterns or human error are likely to introduce it to these regions and beyond. One of the major challenges in wheat breeding is identifying or developing, and then diffusing, adapted resistant cultivars before this migration occurs. Although some race-specific resistance genes, mostly of alien origin, *viz., Sr22, 24, 25, 26, 27, 29, 32, 33, 35, 36, 39, 40, 44, R* (1A.1R translocation) and *Tmp,* can provide effective control, not all can be used in developing cultivars because some of these alien translocations are associated with negative effects on grain yield or quality. Shortening of these alien segments could make them more useful. Improved wheat germplasm that carry *Sr24, Sr25, Sr26, SrTmp,* and *SrR* genes have already been identified and can be used in breeding. Gene *Sr24* currently occurs with a relatively high frequency in wheat cultivars but it became ineffective in India and South Africa soon after the release of cultivars that carried this resistance gene. The best strategy therefore is to reconstitute durable adult-plant resistance that once protected the Green Revolution and subsequent wheat cultivars. If race-specific genes need to be used, they must be deployed in combination to enhance their longevity.

Durable stem-rust resistance of some older spring wheats is believed to be due to the deployment of *Sr2* in conjunction with other unknown minor, additive genes. McFadden transferred gene *Sr2* to hexaploid wheat in the 1920s from tetraploid emmer wheat cultivar 'Yaroslav.' The slow-rusting gene *Sr2* confers by itself only moderate levels of resistance. Its presence can be detected through its complete linkage with the pseudo-black chaff phenotype. A large number of wheat lines were evaluated in 2005 in Kenya during the stem rust epidemic caused by *Ug99*. Genotypes with a pseudoblack chaff phenotype showed varying degrees of disease severity with a maximum severity reaching about 60 per cent compared with 100 per cent severity for highly susceptible materials. Moreover, the host reaction for these genotypes on the same internode varied from MR to S (moderately resistant to susceptible). These observations clearly indicated that although slow-rusting resistance gene *Sr2* continued to confer at least some resistance, the level of resistance was not sufficient when this gene was present alone under high disease pressure in Kenya. *Sr2* was detected in several highly resistant old, tall Kenyan cultivars, including 'Kenya Plume' (Singh and McIntosh, 1986), and CIMMYT-derived semidwarf cultivar 'Pavon 76.' These cultivars have shown a maximum disease score of 15MR (15 per cent disease severity with moderately resistant reaction). Because 'Pavon 76' is susceptible to race *Ug99* at the seedling stage, its resistance, as speculated earlier (Rajaram *et al.*, 1988), is based on multiple additive genes of which *Sr2* is an important component. Wide testing of improved wheat germplasm also has helped in identifying additional sources of adult-plant resistance. These sources are being used at CIMMYT to incorporate durable stem-rust resistance into high yielding, widely adapted wheat cultivars

Preparedness to Stem Rust Race, *Ug99*

A new race of stem rust known as *Ug99* or TTKS has evolved in Uganda (Africa) in 1999. The most important thing about *Ug99* which is different from other existing races of stem rust elsewhere in the world is that it has rendered *Sr*31 gene ineffective located on 1BL/1RS (*Sr31, Lr26, Yr9, Pm 8*) chromosome along with a group of other important stem rust resistance genes. Since over last four decades the serious epidemics of stem rust have been kept under control with the extensive use of *Sr31* and *Sr2* complex of resistance genes, hence, the defeat of the important stem rust resistance gene *Sr31* and some of the other known *Sr* genes has become a cause of great concern world over. This variant of *Ug99* has been popping up in fields throughout East Africa. Plant breeders, geneticists, plant pathologists and molecular biologists are collaborating internationally to identify and incorporate genetic traits that will protect wheat from the pathogen. The Directorate of Wheat Research under the aegis of ICAR in collaboration with Borlaug Global Rust Initiative (BGRI) took timely steps and evaluated Indian wheat material for *Ug99* at Kenya (2005-2011). This has resulted into identification of resistant varieties in seed chain and genetic stocks.

Development of Wheat Genotypes with Resistance to Black Rust Race *Ug99*

All the released varieties of bread and durum wheats from CVRC for Central Zone and varieties released from individual states *i.e.*, Madhya Pradesh, Chhattisgarh, Gujarat, Southern parts of Rajasthan, Bundelkhand area of U.P. are to be tested for their resistance to black rust race *UG99* at Kenya to have a preliminary knowledge about the available resistance in the genotypes. All the lines included in NIVT (National Initial Varietal Trial), generally numbering around 300 need to be screened in Kenya for stem rust resistance and their responses should be considered for promotion to AVT's (Advanced Varietal Trials). Stable sources of resistance to black rust race *UG 99* selected at Kenya for bread and durum wheats are to be supplied to this centre for use in the crossing programme with the agronomically superior lines of Central Zone. The off-season nursery at Wellington will be used to advance the segregating populations. At the same time, the segregating populations are to be screened and selected at Kenya for resistance to black rust race *Ug99*. Released varieties for Central Zone *viz.*, HI 1500, HI 8498 (durum wheat) and HD 4672 (durum wheat), reported to be resistant to *UG99* by CIMMYT, will be given more emphasis in the crossing programme. In Central Zone, both durum and bread wheats are important for cultivation and exhibit complementary resistance to stem and leaf rusts. This phenomenon will help in reducing the inoculum of rusts in the region and ultimately, reduce the epidemic situation in the North India wheat belt. Hence, emphasis needs to be given on the development of durum genotypes resistant to *Ug99*, at least in the Central and Peninsular Zones, from the beginning itself. Extensive campaign is to be undertaken in the Central Zone to popularize *Ug99* resistant varieties by increasing their breeder seed production, educate the extension workers of the state agricultural department and farmers about their importance, thereby reducing the "**fear psychosis**" with regard to threat of *Ug99* in this region.

Future Prospects

Breeding program for disease resistance still requires strengthening in the perspective of climate change. The effective control of the wheat rusts has been the exchange of improved germplasm, which provides a vital source of genetic variation. The impetus on the high-yielding cultivars must continue and the emphasis on disease resistance may be even higher. Reducing the genetic vulnerability of the crop to rust epidemics will be the prime job of the plant pathologists. Appropriate management of the host resistance genes through deployment, pyramiding genes, *etc.* will require greater attention. Use of durable resistance will be a good strategy for providing agronomically acceptable option to farmers. Wider application of the genes having minor phenotypic effect is advocated. Migration of rust strains from adjoining areas is important. Hence constant visit and monitoring of rust is continuously done. The use of race-specific rust resistance in development of cultivars can routinely be followed with cautious in wheat breeding keeping in mind the continual evolution of new races or biotypes of the pathogen. There are a number of examples that the use of cultivars with single-gene, race-specific resistance became susceptible in a short span of time. At the same time these major genes have provided long lasting resistance and therefore, the use of the race-specific resistance should not be discouraged and these genes need not to be dumped. To achieve strategic deployment of cultivars long-lasting resistance which may provide durable types of resistance may be utilized. In this endeavour both race-specific with non-race-specific resistance could be combined. The basic objective should be to reduce the selection pressure for virulence in the pathogen. A lot remains to be learned in the areas of both applied and basic research.

The past two decades witnessed phenomenal increase in the productivity of wheat and changes in the pest complex of this crop. These changes in the wheat pest complex occurred with the introduction of high yielding semi dwarf varieties, changed cropping pattern, monocultivation of wheat and use of new agrotechniques in wheat cultivation. In the present pest scenario, aphids have emerged as the regular and potential pest of wheat in the major wheat growing areas. The other important pests are termites, shoot fly, brown wheat mite, armyworm and *Heliothis*. In order to avoid the excessive use of chemicals, more emphasis is being laid on the Integrated Pest Management (IPM). Development of varieties resistant to pests is the important component of IPM. In the first stage large number of germplasms is screened under field conditions. The material showing resistance to different pests is then screened in the entomological screening nurseries, after creating epiphytotic conditions for particular pests. The information about, identification, screening techniques and creation of epiphytotics is provided in this article.

Genetic studies have clearly demonstrated that resistance to various insects is governed by specific genes. Thirteen genes are known to confer resistance to hessian fly *(Mayetiola destructor)*. As with fungi, biotypes of hessian fly also occur, and therefore, resistance can break down because of a simple mutation. Specific genes are also known to confer resistance to greenbug *(Toxoptera graminum)*. A single recessive gene *Gb* has been described by Curtis *et al.* (1960) in wheat varieties Cl 9058 and Dickinson Selection 28A. Hollenhorst and Joppa (1983) found a single, dominant gene in chromosome 1A/1R in Amigo wheat. The amphiploid line, Largo, had a single gene for resistance located in chromosome 7D. Studies of resistance to hessian fly and greenbug have clearly demonstrated the existence of a gene-for-gene relationship. Furthermore, these studies imply that resistance to other insects should also follow the same pattern. Nevertheless, as demonstrated with various fungal diseases, non-specific resistance could also occur, if examined carefully.

Identification and Creation of Epiphytotic Conditions for Resistance to Insect Pests in Wheat

Wheat (*Triticum aestivum* L.) is the important grain crop of India, cultivated over an area of 27.2 million hectares with annual production of 74.3 million tonnes of food grains. In Punjab, it is the principal *rabi* crop grown over 34.82 lac hectares with annual production of 146.98 lac tonnes of food grains. The past four decades witnessed phenomenal increase in productivity of wheat and changes in the pest complex of this crop. The changes in the pest complex occurred with the introduction of high yielding semi-dwarf varieties, changing cropping pattern, mono-cultivation of wheat and uses of new agro techniques in wheat cultivation. In the recent past limited epidemics of some pests have been reported from different regions and the incidence of some of these pests was quite common and wide spread. The correct identification of the insect is the key to the management of a particular pest. Similarly, for screening varieties against a particular pest, in the absence of natural population the creation of artificial population pressure is desirable. The information on the identification and creation of epiphytotic conditions for screening is provided in this article.

Termites (*Microtermes obesi* Holmgren; *Odontotermes obesus* Rambur)

Termites are social insects and live in colonies known as termitaria. In mound forming species these termitaria are raised 1.5-2.0 m above ground level. A colony consists of a queen, a king, workers, soldiers, nymphs, colonizing individuals and eggs. Eggs are small and yellowish in colour. The workers are yellowish white and are small in size. The soldiers can be recognized by their large heads. The queen is large (7 cm) with creamy white abdomen having dark brown stripes. The king is much smaller than the queen but slightly bigger than the colonizing individuals. The king and queen live in the "Royal Chamber" which is made of hard and compact mass of soil and located below the termitarium. The colonizing individuals can be seen hovering around lamp posts at night after the first shower in the beginning of rainy season. Termites being the polyphagous pest not much work is being done on screening of varieties against termites. However, for creating population pressures, termites can be attracted to a field by keeping cow dung in the field. The damage by termite can be assessed at the seedling and earhead stage. For this purpose, earmark ten spots of two-meter row length in one-acre field and count the total no. of seedlings three weeks after sowing and data be presented as plant population/m row. *At the seedling stage:* The severity of termite damage can be determined by counting the number. of damaged plants in two-meter row length randomly from ten spots in one acre 3,4 and 5 weeks after sowing. Also the total no. of plants from these spots are counted 3,4 and 5 weeks after sowing. The final data is presented as per cent damaged plant per meter row.

$$\text{Per cent damaged plants} = \frac{\text{No. of damaged plants per m row}}{\text{Total no. of plants per m row}} \quad 100$$

At the Ear Head Stage

When the crop is near maturity but still green, damaged ear heads ("white ears") are counted from already marked ten spots of two-meter row length in one acre. The total no. of tillers are also counted from these spots to work out per cent damaged effective tillers/m row as above.

Aphids (*Sitobion miscanthi* (Takahashi) *S. avenae* Fabricius, *Rhopalosiphum maidis* Fitch, *R. padi*)

These are green to blackish green louse like insects found in small colonies on leaves, leaf sheath, stem, flag leaf and on green ear heads. The important characteristics for the identification of different species. Screening for aphids is carried out in the breeder's material for identifying promising lines showing less damage *i.e.* grade one only (less than 5 aphids/shoot). The promising entries identified from preliminary field screenings along with new lines are sown in two rows, two metre long and replicated thrice. The aphids collected from fields are released in the plots. The leaves having high population of aphids are cut and placed on the plants in the rows. The aphid population is recorded from ten random shoots per replication. Where the population is low actual counts are made and average worked out per shoot. The promising entries identified from the field experiments and from previous screenings along with new lines are sown in two rows, two meter long and replicated thrice. The aphid population is recorded from ten random shoots per replication. The entries which are in grade 2-3 are again evaluated in the screen house under artificial population pressure. The varieties to be screened are sown in plastic cups or wooden trays by planting 4-5 seeds per cup or per hill. After germination two plants are kept uniformly. All these lines are subjected to equal aphid pressure by releasing 10 aphids at each variety. Then these aphids are allowed to multiply for 20-25 days after which aphids are again counted from the seedlings and grades are given accordingly to all the varieties/germplasm. This aphid has been recorded feeding in the root zone of wheat seedlings from the rainfed wheat growing areas. The serious damage of this pest to wheat has been reported from Powarkheda, Madhya Pradesh. The damaged seedlings start withering and initially damage is similar to termites. The development time from first instars to adult is about 20 days. The average longevity of adult ranges from 29.7 days at 30° C. The optimum range of temperature for its development is 20-25° C. Screening is done by counting the aphids clinging to randomly uprooted plants during peak period of infestation.

Table 7.27. Visual grading scale of aphids incidence

Grade	*Approx Number of Aphids/Shoot*	*Rating*
I	0-5	Immune (I)
II	6-10	Resistant (R)
III	11-20	Moderately resistant (MR)
IV	21-50	Susceptible (S)
V	51-75	Highly susceptible (HS)
VI	76 and above	Very Highly susceptible (HS)

The entries found promising in the screening nursery are evaluated in the screen house/glass house where 5 10 young seedings of different entries are planted in pots/trays and exposed to 5 10 aphids collected form field and after 3 4-week entries are rated in different grades as above. Identification of different aphid species

Table 7.28. Differential characteristics of English grain aphid (*Sitobion avenae*), Cereal aphid (*Rhoplalosiphum padi*), Green bug (*Schizaphis graminum*) and Russian wheat aphid (*Diuraphis noxia*).

English Grain Aphid (Sitobion avenae)	*Cereal Aphid (Rhoplalosiphum padi)**	*Green Bug (Schizaphis graminum)*	*Russian Wheat Aphid (Diuraphis noxia)*
Pale to apple green colour occasionally orange or brown	Winged adults blackish green, wingless adults and young ones green	Pale green with dark green mid-dorsal stripe on back	Pale green or lime green except the hind region, with reddish patches
Long dark cornicles	Short cornicles with flanged ends	Short cornicles with black	Very reduced cornicles tip (apex)
Body length upto 3mm	Body length upto 2 mm	Body length upto 2 mm	Elongated, spindle shaped body
Feeds on upper leaves and developing grains	Feeds on stem and may even be found on stem bases below ground level	Feeds on any part of plant, necrotic areas purpling and long white stripes on leaves, rolling of infested leaves	Feeds inside the rolled leaves, leaf rolling, prostrate growth habit, sterile heads
Antenne longer in size and dark in colour	Antenne short	Antenne black (except Ist 3 segments)	Antenne of body colour
Cauda 3/4th of cornicle length	Cauda short	Cauda short	Cauda double 'Double tail'

* Corn leaf aphid, *R. maidis*: Similar to *R. padi*, but slimmer, wingless adults do not have reddish patches on hind region. Favourite feeding site is not yet fully emerged rolled leaves of barley.

Shoot Fly (*Atheriqona naqvii, A. oryzae*)

Shoot fly adult is small insect similar to house fly, but much smaller in size. the flies lay small, elongated, dirty white sculptured eggs, singly on the basal portion of stem or lower side of the leaves just near the ground level. The maggots after hatching creep within the leaf sheaths into the axis of the tiller and cut the central growing shoot causing "dead hearts" formation. The most susceptible stage of the crop is upto 45 days after germination. For shoot fly screening each wheat line is sown in a single row plot of 3m, replicated thrice. One row of Sonalika is planted after every 10 rows. To ensure high pest incidence sowing is done in middle of January. In order to attract the shoot fly, fish meal at the rate of 50 g/ml is broadcasted after 5 and 15 days of germination. In each row one meter row spot is marked and uniform plant stand is maintained at 30 seedlings per meter row. The percent "dead hearts" from one meter row are recorded at 20, 40 and 60 days after germination. The data from all these observations are pooled to denote average infestation index of each row. The mean of the three replications is given as final index. The ranking of the entries is indicated in parenthesis just adjacent to the mean infestation index. All the entries showing higher infestation index than check are rejected and the lines showing lower index are tested again in the following year.

Resistant varieties: K 9025, LOK-1, HUW 234, PDW 215, PDW 175, PBW 377, PDW 243, PDW 232, PBW 226, Raj 1555, WH 147 and UP 1731.

Brown Mite (*Petrobia latens* Muller)

The adults are very active somewhat round in shape brown to metallic in colour. This mite is soil inhibiting and non webbings type. Body is fused into cephalothorax and covered with setae. The nymphs and adults has four pairs of legs. The larvae, nymphs and adults of the mite suck sap from leaves, leaf sheath, green stem and spikes. All the breeders material planted in the rainfed trials is screened against brown mite and germplasm showing least damage is included in the next year screening nursery. For brown mite screening nursery promising entries from the previous year along with new lines are sown in two rows 3 metre long replicated thrice in randomized block design. The sowings are done in last week of October under unirrigated conditions. The variety IWP-72 is planted as a check after every 10 rows. For recording mite population, the infested plants are tapped over 4 glycerine smeared slides held in a thermocole sampler held at ground level close to the plant. The observations are recorded from 3 spots in each plot. The mites sticking on the slides are counted under a binocular (10 x magnification) and computed as mite population/10cm2 area. Population of mite is recorded at 15 20 days interval during peak infestation period of mite attack. The ranking of the entries is denoted along with the mean population. In addition to this visual grading of the same entries is also made during peak infestation period. All the entries are assigned the grades as follows:

Table 7.29. Visual grading scale of mite incidence

Grade	*Infestation*
0	Completely free, no infestation.
1	Few mites seen, few plants showing minor stippling of few leaves.
2	Crop looking green. All plants have one or few leaves showing stippling but no drying.
3	Crop yellowish. All plants showing tip drying of some leaves and stippling symptoms on almost all leaves
4	Crop yellow, leaves in all plants showing drying of tip.
5	Crop drying. All plants showing severe drying of leaves.

Visual grades thus assigned to each entry in the three replications are pooled and the mean infestation grade obtained for each entry, along with the data on mite population at different intervals. Resistant varieties: Hyb 65, UP 2377, UP 2378, WH 611, Raj 3826, WH 627.

Army Worm (*Mythimna separata* Walker)

The adult moths are pale brown in colour and are medium sized with a wing span of 40-50 mm. The eggs are round, grayish white to yellow in colour which changes to dark brown near hatching. The newly hatched larvae are active, light green in colour which changes to dark brown near hatching. The newly hatched larvae are active, light green in colour which change to dark green near maturity. The colour of the larvae is variable but they always have four longitudinal black stripes which run laterally on either side of the mid dorsal stripe. The full-grown larvae measure 40-45 mm in length. The colour of the pupae is light to dark brown with a shinning surface. To assess the damage by armyworm, earmark twenty spots of one-meter row length in one-acre field. Its population in the field can be recorded by counting the no. of armyworm larvae per meter row length on the plants, hiding under clods or plant debris from these spots in one acre. The damage can also be assessed by taking the percentage of tillers showing damaged leaves or ear heads from these spots. The no. of larvae/pupae per sq. meter can be counted from ten spots per acre.

American Pod Borer (*Helicoverepa (Heliothis) armigera* Hub.)

Though American pod borer is a polyphagous pest, its sporadic infestation to wheat crop has been reported from all wheat growing states of India. The moth is stoutly built and is yellowish brown. The caterpillars when full grown are 3.5 cm in length, young larvae are greenish in colour which subsequently change to dark brownish colour with dark brown broken lines along with sides of the body.

Table 7.30. Differential characteristics of Army Worm (*Mythimna separata* (Walker) and American Pod Borer (*Helicoverepa (Heliothis) armigera* Hub)

Army Worm (Mythimna separata (Walker)	*American Pod Borer (Helicoverepa (Heliothis) armigera Hub)*
Young larvae prefer to feed on leaf and the inner surfaces of the leaf sheath. In the later instars, they consume while of the leaf tissues leaving only thick midrib. Mature larvae feed on awns, glumes and developing grains.	Larvae do not consume the awns, only the developing grains are eaten
Newly hatched larvae dull white in colour, changes to green or slightly pink when full grown, two black stripes on either side of the mid-dorsal stripe, out of the lateral stripes dorsal one is thinner than ventral one.	Young larvae are green in early stages, but subsequently turn dark brownish with dark brown broken stripes along the body
Faecal pellets are greenish black in colour and are comparatively large in size than that of *Helicoverpa*	Faecal pellets are white in colour and are smaller in size than faecal pellets of army worm.

Pink Stem Borer (*Sesamia inferens* Walke)

The pink stem borer is a polyphagous insect distributed throughout India. In some parts of the country it is a pest of sugarcane and rice. The caterpillars are pinkish brown, with smooth cylindrical body measuring about 25 mm. The moths are straw coloured having stout body. Severity of damage by pink stemborer can be observed by the 'dead hearts' produced by the larvae in which central shoot of plant dries up. For sampling purpose, earmark ten spots of one meter row length in one acre field and count the no. of 'dead hearts' from these spots. Also count the total no. of tillers/m row length from these spots to determine the per cent dead hearts as below:

$$\text{Per cent dead hearts} = \frac{\text{No. of dead hearts/m row}}{\text{Total no. of tillers/m row}} \times 100$$

Cut Worm (*Cirphis loreyi* Dup., *Euxoa spinifera* Hub. and *Agrotis ypsilon* Rott)

The cutworms comprise of three species *viz.* (*Cirphis loreyi* Dup., *Euxoa spinifera* Hub. *and Agrotis ypsilon* Rott). All these lepidopterous pests cause damage to wheat crop in their larval stages by feeding on leaves and defoliating the plants. The caterpillars are nocturnal in habit and remain hidden during day time in cracks in soil or below the crop debris. Out of the three species the population of C. *loreyi* was more in the fields.

Gujhia Weevil (*Tanymecus indicus*)

It is a pest of wheat in the seeding stage. Both the larvae and adults cause damage, but adults are more serious. They feed on the roots and cut down the young germinating plants at or below ground level. At early stages its damage resembles to those of termites. The pest used to be serious from October-November. In the past, it was a serious pest in northern states of Uttar Pradesh, Rajasthan, Haryana and Punjab. However, since the introduction of high yielding varieties, the damage by this pest has not been reported from these states. This pest species has assumed the status of a non-pest due to changes in the agro-ecosystem of the wheat crop.

Sugarcane Leaf Hopper (*Pyrilla perpusilla*)

The sugarcane pyrilla, *Pyrilla perpusilla* has posed serious problems to wheat crop in some years. This pest migrates from sugarcane to wheat crop. Both nymphs and adults suck plant sap and cause damage to wheat at earing stage.

Flea Beetle (*Chetocnema basilis* Baly)

Flea beetle *Chetocnema basilis* Baly feeds on the foliage causing minute holes, white blotches appear around the damaged portion and finally results in wilting of the leaf. In cases of severe damage the growth of plant is retarded and plants may even die. The damage starts after germination and continues till february. Serious outbreak of flea beetle occurred in 1965-66 in Rajasthan.

Surface Grass Hopper (*Chrotogonus trachypterus*)

This insect is commonly known as *Toka*. Both nymphs and adults cut the germinating plants at ground level and devour tender leaves and growing point. It was a pest of sandy areas with comparatively low rainfall. In the past three decades its damage is on the decline and it is no more a pest now in the state where large area is under irrigated crop.

Leaf Hopper [*Typhlocylbea maculifrons* Motsch and *Zygnidia manaliensis* (Singh)]

Both nymphs and adults suck the sap from leaves and produce small white spots on the leaves. Its damage also remains very low and it had never come in the pest from. Two species *i.e., Typhlocylbea maculifrons* Motsch and *Zygnidia manaliensis* (Singh) has been reported from wheat.

Hessian Fly (*Phytophaga destructor*)

The Hessian fly attacks the wheat crop over a wide area. The principal method of control is to delay planting until after the adult flies have emerged and oviposited. This practice sometimes results in winter injury to the wheat and thereby reduces yields. Furthermore, it must be practiced community-wide in order to be effective. Fly resistance would permit seeding the wheat at an earlier date. Often this practice would result in the production of higher grain yields. It would also make it possible to plant wheat early enough to obtain fall pasture without danger from fly injury. Damage by the Hessian fly may result in a dwarfing of infested plants, reduction in tillering, increased winter injury, and breaking of straw after ripe. The first study of varietal resistance to Hessian fly was made in California in the late 1880's. Varieties used as sources of fly resistance are Dawson, Java, W 38-6, Marquillo, and two durum introductions from Portugal, P. I. 94587 and Ribeiro. Many other strains have been reported to be resistant. Biologic specialization is found in the Hessian fly, just as in many disease-inciting organisms. Three races, each differing in their ability to infect different varieties, have been reported. Several genes for resistance to Hessian fly have been identified in particular varieties. Varietal reaction may be determined either by growing the varieties in the field in a heavily fly-infested nursery or by testing in the greenhouse. Reaction is generally expressed as the percentage of infested plants.

Wheat Stem Sawfly (*Cephus cinctus*)

The wheat stem sawfly is an insect limiting production of wheat in certain spring wheat areas of the United States and Canada. Damage results from the breaking over of infested stems. Yield reductions up to 10 per cent have been reported. In the search for sources of resistance it was originally believed that varieties of wheat with solid stems would have only meagre infestations or escape injury altogether under conditions where hollow-stem varieties would be severely injured. A hybrid line of hard red spring wheat selected in 1942 from a cross between Apex X S-615, the latter a solid-stemmed wheat, was distributed to farmers under the name Rescue. Although inferior to other adapted varieties in quality and yield in the absence of the sawfly, it proved to be highly resistant to this insect pest under most conditions and became widely grown in the sawfly-infested areas. Chinook, distributed more recently, is also resistant. Recent studies have shown that there are other characteristics of the plant, in addition to solid stems, that affect the resistance of

Table 7.31. Genes for Resistance to the Hessian Fly

Class of Wheat	*Variety*	*Genes for Resistance*
White	Dawson	$H_1H_1H_2H_2$ (resistance dominant)
Hard red spring	W38-6	H_3 H_3 (incomplete dominance)
Hard red spring	Java	h_4h_4 (resistance recessive)
Hard red spring	Ribeiro	H_5H_5(incomplete dominance)
Hard red spring	Marquillo	?? (resistance recessive)
Durum	P. I. 94587	?? (2 or more dominant genes for resistance)

varieties to wheat stem sawfly. This work indicates that stem solidness should not be the only guide in selecting sawfly-resistant strains. Tests of the strains should be made with live sawflies to determine actual resistance. In a cross between Rescue and Thatcher, stem solidness was inherited as one major factor pair, with possibly two to four minor modifying factors.

Greenbug (*Toxoptera graminum*)

The green bug causes considerable damage to wheat in the southern Plains area. This insect is, an aphid and feeds on the wheat plant by piercing the leaf tissue with its stylet and injecting saliva which is toxic to the plant. A breakdown of the tissue follows and yellowing or even death of the plant results. Large resistant susceptible areas in a field may be destroyed by heavy infestations. Although the green bug may be partially controlled by spraying, the development of resistant varieties offers a more practical means of control. Observations in Texas and Oklahoma in 1942 during a heavy green bug attack demonstrated that varieties differ in their susceptibility. Most of the resistant strains were selections from the cross Marquillo X Oro. Other wheats showing moderate resistance included Denton, Wichita, Blackhull and Blackhull relatives, and several Chinese and Russian introductions. None of the varieties appeared sufficiently resistant to withstand heavy attacks. In later observations made in Oklahoma, the best tolerance was found in off-type plants in a variety of durum, Dickinson.

Other Insects

Evidence of resistance in varieties of wheat to other insects has been observed. These include resistance to wheat joint worm *(Harmolita tritici)*, grasshoppers *(Melanoplus* spp.), chinch bugs *(Blissus leucopterus)*, and wheat stem maggot *(Meromyza americana)*.

Breeding for Biotic Stress (Insect) Resistance

Global average loss due to insect pests is 14 per cent. Estimated losses in individual crops vary from 5 per cent in wheat to 26.7 per cent in rice and still more in crops like cotton and sugarcane. The ability of a plant to withstand, oppose or overcome the attack of an insect in known as insect resistance. It is the property of a variety or a host crop due to which it is attacked by an insect pest to a significantly lower degree than are other varieties of the same host. Strains of a species of an insect pest, differing in their ability to attack different varieties of the same host species (syn: Physiological races). *Host Habitation* incluides Polyphagy, Oligophagy, Seasonal Oligophagy and Monophagy. *Polyphagy*: Insects feed on a vide range of hosts avoiding few plant species. *e.g.* Scales and moths. *Oligophagy*: Live on one taxonomic unit only. *e.g.* Hessian fly on wheat. *Seasonal oligophagy*: Insects may live on many species in one part of the year and on few in another part of the year. *e.g.*: Aphids. *Monophagy*: Avoid all hosts except one particular species or variety *e.g.* Boll weevil on cotton.

Mechanism of Insect Resistance

Insect resistance is grouped into four categories, *viz.*, Non-preference,

Antibiosis, Tolerance and Avoidance. *Non-preference*: Host Varieties exhibiting this type of resistance are unattractive or unsuitable for colonization, oviposition or both by an insect pest. This type of resistance in also termed as non-acceptance and anti-xenosis. Non-preference involves various morphological and biochemical features of host plants such as - colour, hairiness, leaf angle, taste *etc.* Antibiosis refers to an adverse effect of feeding on a resistant host plant on the development and/or reproduction of the insect pest. In severe cases, it may even lead to the death of the insect pest. Antibiosis may involve morphological, physiological or biochemical features of the host plant; some cases of insect resistance involve a combination of features. *e.g.* Resistance to BPT is due to antibiosis and non-preference. An insect tolerant variety is attacked by the insect pest to the same degree as a susceptible variety. But at the same level of infestation, a tolerant variety produces a higher yield than a susceptible variety. Ability of the host plant to withstand the insect population to a certain extent which might have damaged a more susceptible host. Tolerance is mainly a host character and it may be because of greater recovery from pest damage. *e.g.* Rice varieties tolerant to stem borer/gall midge produce additional tillers to compensate yield losses (as in stem borer in sorghum) or due to the ability of host to suffer less damage by the pest *e.g.* aphid tolerance in Sugar beet and Brassica sps and green bugs tolerance in cereals. Inheritance of tolerance is complex in many cases and is supposed to be governed by polygenes. Pest avoidance is the same as disease escape, and as such it is not a case of true resistance mostly insect avoidance result from the host plants being at a much less susceptible developmental stage when the pest population is at its peak. *e.g.*, Early maturing cotton varieties escape pink boll worm infestation, which occurs late in the season.

Nature of Insect Resistance/Factors for Insect-resistance

Insect resistance may involve: Morphological, Physiological (or); Biochemical features of the host plant. Morphological features: Morphological factors like, hairiness, colour, thickness and toughness of tissues *etc.* are known to confer insect resistance. a) Hairiness of leaves is associated with resistance to many insect pests leaf beetle in cereals, in cotton to Jassids, in turnip to turnip aphid. b) Colour of plant: Colour may contribute to non-preference in some cases. For example: Red cabbage, Red leaved Brussels's sprouts are less favoured than green varieties by butterflies and certain Lepidoptera for oviposition. Boll worms prefer green cotton plants to red ones. c) Thickness and Toughness of plant – Tissues prevent mechanical obstruction to feeding and oviposition and thereby lead to non-preference as well as antibiosis. *e.g.* Thick leaf lamina in cotton contributes to Jassid resistance; Solid stem in wheat confers resistance to wheat stem sawfly; Thick and tough rind of cotton bolls makes it difficult for the boll worm larvae to bore holes and enter the bolls. Other characters: also contribute to insect resistance. *e.g.* 1. *Gossypium arboretum* varieties with narrow lobed and leathery leaves are more resistant to Jassids than are those with broad lobed and succulent leaves. 2. Cotton varieties with longer pedicels are more resistant to boll worms. 2. Physiological Factors: Osmotic concentration of cell sap, various exudates *etc.*; may be associated with insect resistance. *e.g.* Leaf hairs of some *Solanum* sps. secrete gummy exudates. Aphids and Colorado beetles get trapped in these exudates. Exudates from secondary trichomes of Medicago disciformis leaves have antibiotic effects on alfalfa weevil. Cotton- High osmotic concentration of cell sap is associated with Jassid resistance. 3. Biochemical Factors: Several biochemical factors are associated with insect resistance in many crops. It is believed that biochemical factors are more important than morphological and physiological factors in conferring non-preference and antibiosis. *e.g.* High concentrations of gossypol is associated with resistance in several insect pests in cotton. In rice – high silica content in shoots gives resistance to shoot borer

Genetics of Insect Resistance

Insect resistance is governed by - 1. Oligogenes 2. Polygenes 3. Cytoplasmic genes. *Oligogenic Resistance*: Insect resistance is governed by one or few major genes or oligogenes, each gene having a large and identifiable individual effect on resistance. Oligogenic resistance may be conditioned by the dominant or the recessive allele of the concerned gene. The differences between resistant and susceptible plants are generally large and clear-cut. In several cases, resistance is governed by a single gene (monogenic resistance). *e.g.* in wheat to green bugs In cotton to Jassids. In apple to woolly aphis In rice to plant and leaf hopper. *Polygenic Resistance:* It is governed by several genes, each gene producing a small and usually cumulative effect. Such cases of resistance. Involve more than one feature of the host plant, Are much more durable than the cases of oligogenic resistance. Difference between resistance and susceptible plants are not clear cut; Transfer of resistance is much more difficult; Examples for polygenic resistance; In wheat to cereal leaf beetle, Evolution of resistance breaking biotypes is almost rare. *Cytoplasmic Resistance:* governed by plasmagenes *e.g.* Resistance to European corn borer in maize. Resistance to root aphid in lettuce. *Sources of Insect Resistance* are a cultivated variety, germplasm collections, a related wild species and an unrelated organisms.

Breeding Methods for Insect Resistance

Introduction, Selection, Hybridization and Genetic Engineering. Screening Techniques for determining resistance: The most crucial and, perhaps, the most difficult task in breeding for insect resistance is the identification of insect resistant plant during segregation generations. There are two types of screenings. 1. Field Screening 2. Glass house screening. **Field Screening:** The techniques designed to promote uniform infestation by an insect pest in the field are Inter planting a row of known susceptible variety between two rows of testing material, Screening in highly prone areas, in case Soil insect pests to be tested in sick plots only, Testing in a particular season when the infestation is very high. *e.g.* Rice stem borer in off season, Transferring manually equal number of eggs or larvae to each test plant. **Glass house screening:** Result from glass house tests are much more reliable than those from field tests since both the environment and the initial level of infestations are more or less uniform for all the plants being tested.

Problems in Breeding for Insect Resistance

Reduction is quality or make unfit for consumption. Linkage between desirable and undesirable genes. Inter specific varieties are generally low yielding and their produce is often of inferior quality. Screening for resistance is the most critical and difficult step is a breeding programme it necessitates a close co-ordination among scientists belonging to different disciplines. It is a long term programme

Breeding Wheat for Biotic Stress (Nematode) Resistance

Plant parasitic nematode feeds on meristematic zone of the root. In tomato 100 larvae of *Meloidogyne javanica* in 1kg soil is important for causing diseases *i.e.* it is the pathogenic or threshold level. Largest nematode is *Placentonema gigantiformis* (8cm; Spermwhale); Smallest nematode is *Griffiella* (0.1mm). Knot-formation type varies from crop –to-crop although within the same nematode. Only female nematode causes disease. Zone of elongation is preferred by the nematodes. Larvae become male due to food scarcity *i.e.* natural sex reversal. This cause following types of damage: By puncturing the meristematic tissues. It provides the entry site for the other pathogenic fungi, bacteria and viruses. Withdrawal of food (direct damage). Indirect damage by changing the physiology of plants and plants become more susceptible to stresses. When parasitic goes, but it becomes unable to feed and multiplication is restricted. Immunity is the inherent capacity of plant, that, parasitic could not cause disease. Resistance is of two types: Pre-existing: - In tomato; α-Terrenyle (Alkaloid) is scecreted, which is a toxin to *M. incognita*. Come in action after infection: - β- glucosidase by *Radopholous similis*- Amagdalin (alkaloid broken down into HCN) and one more compound. HCN proves to be toxic. Nematode resistance may be due to dominant major genes, recessive genes, or polygenes, for example of the 15 cases considered by Howard and Cotton (1978), no less than 10 had been determined to be due to single dominant genes, one was due to 2 dominant genes, one to two recessive genes, and two to polygenes.

Ear Cockle Mamni and Yellow Ear Rot or Tundu (*Anguina tritci* and *Clavibacter tritici*)

Diseased plants have spreading tendency and swollen base, leaves become crinkled and twisted ear heads contain dark brown, hard and roundish galls (mamni) instead of grains. Several infected plants are stunted and may die at seedling stage. Occurrence of yellow, slimy mass indicates ear rot phase which prevents grain and gall formation. Screening may be done by mixing the galls with the seeds of the varieties to be screened or mixing the galls in the soil where material to be screened is sown. the gall frequency in the grain sample may be recorded for identifying resistance.

Molya Disease (*Heterodera avenae*)

Plants become stunted with yellowing of leaves, reduced tillering, absence of ears on some tillers or small ears with poorly filled grains. The root system of infected plants gets reduced in size and become bunchy with profuse development of thin rootlets. Cysts are seen attached to the roots at the later stage of the crop. Inoculum load of 15 20 cysts/500 cc soil *i.e.*, more than 10 cysts per plant to be screened is used. The plants are dug 70 75 days after sowing, washed in 60 mesh screen for the assessment of root damage and counting of white females and subsequent trans-planting of desirable resistant plants in pots kept in the cage house to raise seeds. The strains upto 4 cysts/plant are rated as resistant and 4.1 to 9 cysts/plant as moderately resistant and more than 9 cysts as susceptible. A new method has been devised to handle a huge population of cross combinations and their progenies. Old method is labour intensive, demanding on space and claims a high mortality of the related transplants. Under the new screening technique, test plants grown in a uniformly infested field are gently dug at early ripening stage and assayed for cyst nematode and subsequently allowed to dry in a cage house to obtain viable seeds. With this method, it is possible to identify even those newly formed cysts which have started turning brown from white.

Nematode Pathotypes/Physiological Races/Biological Races/Biotypes/Trophotypes (Sidhu and Webster, 1981)

Forms of nematodes, which are identical morphologically, but which can be separated by whether different hosts are resistant or susceptible to them.

Pot Screening

Observations are: Gall Index: scales are no gall 0, 1-10,10-30, 30-100, 100>. Number of galls for more precise one. Number of penetrations of Acid fuschin or Cotton Blue (Strain) 0.1 per cent. Destaining with lactophenol. Roots on slide-Total counting of nematode. Number of latrvae/5g soil. Number of eggs. Cobb's sieving and decanting method: Nematode from soil – separated and counted. Final population. Screened in field condition. **DD_{136}:** Nematode formulation as insecticide from beneficial nematode. **Through PTC:** Carliss (1991) obtained protoplasts having nematode research through fusion in potato.

Liskova *et al.* (1991) The reaction of 23 cultivars of the international cereal test assortment to *H. avenae* pathotypes has shown that it is the pathotype Ha 12. The Ha 12 pathotype will serve as the basis for experimental evaluation of cereal cultivar resistance and as the starting point for directing nematode resistance breeding in Czechoslovakia. Mollers *et al.* (1991) As part of the breeding process, the tetraploid potato is converted to the (heterozygous) dihaploid state by anther culture or parthenogenesis, followed by reduction to monohaploidy. This stage is often accompanied by spontaneous reversal to a homozygous dihaploid state. By crossing the heterozygous and homozygous dihaploids, inter-dihaploids are generated, among which the best forms are selected. To achieve maximum yields, the selected genotypes are returned to the tetraploid state by means of meiosis or, increasingly, by protoplast fusion. The latter procedure was tested to determine the extent of inheritance of desirable traits derived from the dihaploid parents. Somatic hybridization between metribuzin or *Globodera rostochiensis* susceptible and resistant parents resulted in the inheritance of herbicide/nematode resistance, indicating that dominant monogenic traits are fully expressed in the fusion progeny. Field trials showed that *Phytophthora infestans* resistance of the hybrids was intermediate between that of the parents, whereas hybrid tuber yield was markedly increased, in some cases exceeding that of the better parent. This improvement in tuber yield was due to a greater tuber weight. Round tuber shape was dominant over elongated shape among the fusion progeny of parents with round and elongated tubers. Skin colour (yellow or red), flesh colour (white or yellow) and starch content tended to be intermediate between the parents.

Nematode Pathotypes

Peng and Cook (1996) Populations of *Heterodera avenae* from wheat fields in 4 provinces of China (Hebei, Beijing, Henan and Hubei) were tested with the International Test Assortment of cereal differentials. The virulence phenotypes of 3 populations which reproduced on some of the cereals were characterized. No population reproduced on any of the six oat differentials. Few of the 12 barleys were susceptible and their responses were different from those to previously described pathotypes. On wheat, the populations showed responses different from each other but similar to previously reported responses of populations of cereal cyst nematodes from other parts of the world. None of the 6 wheat differentials nor 3 local cultivars was resistant to all populations. On the basis of these responses, it seems that there are at least three cereal cyst nematode pathotypes in China and that all have virulence phenotypes not previously described. Muller and Klinke (1996) Crossings between *Beta vulgaris* and 3 wild beet species have been made at different research institutes, resulting in several addition and translocation lines with resistance genes originating from different chromosomes. As resistance is dominantly inherited, the formation of nematode pathotypes, as they exist for example with potato cyst nematodes, was expected. To investigate virulence in H. schachtii, 146 populations were collected and multiplied on rape. A translocation line with homozygous resistance from chromosome 1 of B. procumbens was used to test the nematodes for virulence. In almost all populations at least a few individual new cysts were found, with several populations producing multiplication rates greater than 50 per cent of that on susceptible beets. A population demonstrating extreme virulence was multiplied continuously on the resistant translocation line and was considered to be a pathotype of H. schachtii. In a comprehensive series of tests the three wild species as well as addition and translocation lines with different sources of resistance were inoculated with this pathotype. The pathotype did not break the resistance of the wild species. Difference in reaction by the various breeding lines showed that chromosomes 1 and 7 of B. procumbens and B. webbiana carry identical or very similar resistance genes, that have no effect on the pathotype. Chromosome 7 has additional resistance information, which the pathotype is not able to overcome. In B. patellaris genes for resistance other than the known gene on chromosome 1 can be expected. New symbols are proposed to characterize the genes coding for resistance to *H. schachtii* and for nematode virulence (pathotypes).

Rivoal *et al.* (1993) Lines of wheat with the 6Mv chromosome from *Aegilops ventricosa* display partial resistance to both pathotypes Ha12 and Ha41 of *Heterodera avenae*. With either pathotype, the effect of this alien chromosome on cyst production, size and fecundity was expressed in resistance tests. Partial resistance of 5 6Mv(6D) substitution lines varied according to the intrinsic cyst-forming capacity of the nematode pathotypes and the recipient germplasms. Such partial resistance can be utilized in wheat breeding lines for integrated management of *H. avenae*. Potocek *et al.* (1991) During 1983-1989 about 1200 infested plots in the Czech Republic were investigated for the occurrence of new pathotypes of *Globodera rostochiensis* and *G. pallida*. Only eight populations of the pest were suspected of containing another pathotype than Ro1. Glasshouse pot trials were performed with those eight natural populations, two populations of Ro1, 10 potato cultivars resistant to Ro1 and 10 cultivars resistant to more pathotypes, G. pallida included. The multiplication rates (Pf/Pi) showed that four of the nematode populations were probably mixtures, with pathotypes Ro1 prevailing. Three populations were suspected of containing G. pallida, but their morphological characters corresponded only with *G. rostochiensis*. One population corresponded with pathotype Ro5. The extent of the occurrence of pathotypes or virulence groups different from Ro1 can be estimated at 0.3-0.4 per cent of infested plots.

Nematode Physiological Races

Li *et al.* (1996) A diallel cross was conducted of soyabean varieties differing in resistance to soyabean cyst nematode [*Heterodera glycines*]. The F_1 progeny from 5 resistant and medium resistant X susceptible and highly susceptible crosses, and their reciprocal crosses, were all susceptible. In the F_2, the segregation ratio was 1: 15 between medium resistant and highly susceptible plants. The segregation ratio of the test cross was 1: 3. In reciprocal crosses between plants with resistance and medium resistance, the F1 progeny were resistant. The segregation of resistance to medium resistance in the F_2 was 3: 1, and the segregation ratio in the test cross progeny was 1: 1. It is suggested that a pair of dominant genes and 2 pairs of overlapping genes control *H. glycines* resistance in the resistant varieties Xingxian Huibuzhiheidou and Yingxian Xiaoheidou.

Nematode Biotypes

Sipes *et al.* (1992) Fertility was determined for 7 isolates of *Heterodera glycines* representing three biotypes (B0, B3 and B2,3) collected from North Carolina, Arkansas, and Illinois on soyabean 'Lee 68'. Females produced an average of 300 progeny, 29 per cent of which were deposited in the gelatinous matrix. The covariance between eggs deposited in the matrix and the cyst was small (0.29) and their correlation not significant; therefore, cyst egg production was used to compare fertility among isolates. B2,3 females, virulent on resistant soyabean 'Peking' and PI88788, produced the most progeny - 252 eggs per cyst. B3, virulent on PI88788, produced 217 eggs per cyst. B0, the avirulent biotype, produced 198 eggs per cyst. Females inseminated by B2,3, B0 and B3 males produced 256, 228 and 192 eggs, respectively. Nematode biotypes virulent on resistant soybeans were more fertile than the avirulent ones (k = 100). Isolates from Illinois produced fewer offspring than those from Arkansas or North Carolina (k = 100).

References

Acosta, AC (1963) The transfer of stem rust resistance from rye to wheat. *Diss. Abstr.* 23: 34-35.

Adlakha, KL; Wilcoxson, RD and Raychaudhuri, SP (1984) Resistance of wheat to leaf spot caused by *Bipolaris sorokiniana*. *Plant Disease* **68**: 320-321.

Aguilar-Rincón, VH; Singh, RP and Huerta-Espino, J (2000) Inheritance of resistance to leaf rust in four synthetic hexaploid wheats. *Agrociencia* **34**: 235–246.

Akhavan, A; Strelkov, SE; Kher, SW; Askarian, H; Tucker, JR; Legge, WG; Tekauz, A and Turkington, TK (2017) Resistance to *Pyrenophora teres f. teres* and *P. teres f. maculata* in Canadian barley genotypes. *Crop Science* **57**: 151–160.

Altpeter, F; Diaz, I; McAuslane, H; Gaddour, K; Carbonero, P and Vasil, IK (1999) Increased insect resistance in transgenic wheat stably expressing trypsin inhibitor CMe. *Molecular Breeding* **5**: 53-63.

Anderson, JA; Stack, RW; Liu, S; Waldron, BL; Fjeld, AD; Coyne, C; Moreno-Sevilla, B; Mitchell, FJ; Song, QJ; Cregan, PB and Frohberg, RC (2001) DNA markers for *Fusarium* head blight resistance QTLs in two wheat populations. *Theor. Appl. Genet.* **102**: 1164–1168.

Anderson, JA; Waldron, BL; Moreno-Se-villa, B; Stack, RW and Frohberg, RC (1998) Detection of *Fusarium* head blight resistance QTL in wheat using AFLPs and RFLPs. In: Slinkard AE (ed) Proc 9th Int Wheat Genetics Symp, Saskatoon, Saskatchewan, Canada, 2-7 Aug 1998, vol. 1, pp.135-137.

Anonymous (2004) *Mehtaensis* (Directorate of Wheat Research, Regional Station, Flowerdale, Shimla 171 002, India) 24: 3.

Assefa, S and Fehrmann, H (2000) Resistance to wheat leaf rust in *Aegilops tauschii* Coss. and inheritance of resistance in hexaploid wheat. *Genet. Resour. Crop Evol.* **47**: 135–140.

Atkins, IM and Dahms, RG (1945) Reaction of Small-Grain Varieties to Green Bug Attack. US *Department of Agriculture Technical Bulletin* **901**.

Autrique, E; Singh, RP; Tanksley, SD and Sorrells, ME (1995) Molecular markers for four leaf rust resistance genes introgressed into wheat from wild relatives. *Genome* **38**: 75-83.

Badebo, A; Kema, GHJ;van Ginkel, M and van Silfhout, CH (1997) Genetics of suppressors of resistance to stripe rust in synthetic wheat hexaploids derived from *Triticum turgidum subsp. dicoccoides* and *Aegilops squarrosa. African Crop Sci. Conf. Proc.* **3**: 195–202.

Bahadur, P (2000) Adult plant resistance to leaf rust in 37 Indian wheats. Proc. Int. Conf. on Integrated Plant Disease Management for Sustainable Agriculture, pp.644-649, Indian Phytopathological Society, IARI, New Delhi, India.

Bahadur, P; Nagarajan, S and Nayar. SK (1985) A proposed system of virulence designation in India. II *Puccinia graminis* f. sp. *tritici. Procs of the Indian Academy of Science (Plant Sciences)* **95**: 29-33.

Bahadur, P; Singh, DV and Srivastava, KD (1994) Management of wheat rusts- a revised strategy for gene deployment. *Indian Phytopath.* **47**(1): 41-47.

Bai, D and Knott, DR (1992) Suppression of rust resistance in bread wheat (*Triticum aestivum* L.) by D-genome chromosomes. *Genome* **35**: 276–282.

Baker, EP; Sanghi, AJ; Mcintosh, AA and Luig, NH (1970) Cytogenetical studies in wheat Ill. Studies of a gene conditioning resistance to stem rust strains with unusual genes for avindence. *Aust. J. Biol. Sci.* **23**: 36H7S.

Bakonyi, Y; Aponyi, J and Fischl, G (1998) Disease caused by *Bipolaris sorokiniana* and *Drechslera tritici-repentis* in Hungry. In: Duveiller E, Dubin HJ, Reves J, McNab A (eds), *Helminthosophism* Disease of Wheat: Spot Blotch and Tan Spot. CIMMYT, Mexico, pp. 80-84.

Bansal, M; Kaur, S; Dhaliwal, HS; Bains, NS; Bariana, HS; Chhuneja, P and Bansal, UK (2017) Mapping of *Aegilops umbellulata*-derived leaf rust and stripe rust resistance loci in wheat. *Plant Pathol* **66**: 38–44.

Bansal, U; Hayden, MJ; Venkata, BP; Khanna, R; Saini, RG and Bariana, HS (2008) Genetic mapping of adult plant leaf rust resistance genes *Lr48* and *Lr49* in common wheat. *Theor. Appl. Genet.* **117**: 307-312.

Barcley, A (1892) Rusts and mildews in India. *J. Bot.* **30**: 1-8.

Barloy, D; Lemoine, J; Abelard, P; Tanguy, AM; Rivoal, R and Jahier, J (2006) Marker-assisted pyramiding of two cereal cyst nematode resistance genes from *Aegilops variabilis* in wheat. *Mol Breed* **20**: 31–40.

Beek, MA (1986) Breeding for horizontal resistance to wheat disease. FAO Plant Production and Protection Paper No. 69, pp. 111.

Behera, UK; Mishra, AN and Pandey, HN (2007) Evaluation of heat stress and leaf rust tolerance between very late planted durum and bread wheat cultivars in central India. *Australian Journal of Experimental Agriculture* **47**: 1422-1434.

Bennet, FGA (1984) Resistance to powdery mildew in wheat: a review of its use in agriculture and breeding programmes. *Plant Pathology* **33**: 27HOO.

Berg, LA; Gough, FJ and Williams, ND (1963) Inheritance of stem rust resistance in two wheat varieties Marquis and Kota. *Phytopatho/ogy* **53**: 904-908.

Bettgenhaeuser, J and Krattinger, SG (2019) Mapping and isolation of major resistance genes in cereals, 'Advances in breeding techniques for cereal crops': Frank Ordon and Wolfgang Friedt (eds), Burleigh Dodds Science Publishing, Cambridge, UK *7th Plant Genomics and Gene Editing Congress*: Europe.

Bhardwaj, SC; Nayar, SK; Prashar, M; Kumar, J and Singh., SB (1996) Pathotype distribution of *Puccinia graminis sp. tritici* in India during 1992-93 and 1993-94. *Cereal rust and Powdery Mildews Bulletin* **24**: 85-90.

Bhardwaj, SC; Prashar, M; Jain, SK; Kumar, S; Sharma, YP; Sivasamy, M and Kalappanavar, IK (2010) Virulence of *Puccinia triticina* on *Lr28* and its evolutionary relation to prevalent pathotypes in India. *Cereal. Res. Commun.* **38**: 83-89.

Bhardwaj, SC; Prashar, M; Kumar, S; Jain, SK and Datta, D (2005) *Lr19* resistance in wheat becomes susceptible to *Puccinia triticina* in India. *Disease Notes* **89** (12): 1360.

Bhatta, MR; Pokharel, DR; Devkota, RN; Dution, HJ; Mudwari, A; Bimb, HP; Thapa, BR; Sah, BP and Bhandhari, D (1997) Breeding for *Helminthosporium* blight resistance in Nepal: strategy followed by the national wheat research program and genetic gains. In: Duveiller E, Dubin HJ, Reves J, McNab A (eds), *Helminthosophism* Disease of Wheat: Spot Blotch and Tan Spot. CIMMYT, Mexico, pp. 9-14.

Bhawar, KB; Vinod, SJB; Singh, AK; Sivasamy, M; Singh, M; Prabhu, KV; Tomar, SMS; Sharma, TR and Singh, B (2011) Molecular marker assisted pyramiding of leaf rust resistance genes *Lr19* and *Lr28* in bread wheat (*Triticum aestivum* L.) variety HD2687. *Indian J Genet Plant Breed* **71**: 304–311.

Bhowal, JG and Narkhede, MN (1981) Genetics of pseudo-black chaff in wheat *(Triticum aestivum* L.). *Z. Pflanzenzuchtg.* **86**: 298-304.

Bjarko, ME and Line, RF (1988) Heritability and number of genes controlling leaf rust resistance in four cultivars of wheat. *Phytopathology* **78**: 457-461.

Bliffeld, M; Mundy, J; Potrykus, I and Futterer, J (1999) Genetic engineering of wheat for increased resistance to powdery mildew disease. *Theoretical and Applied Genetics* **98**: 1079-1086.

Bolton, MD; Kolmer, JA and Garvin, DF (2008) Wheat leaf rust caused by *Puccinia triticina*. *Mol Plant Pathol* **9**: 563–575.

Brent, D; McCallum, D; Humphreys, G; Somers, DJ; Dakouri, A and Cloutier, S (2012) Allelic variation for leaf rust resistance gene *Lr34/Yr18* in Canadian wheat cultivars. *Euphytica* **183**: 261-274.

Briceno, FG (1992) Inheritance of resistance to *Septoria* leaf blotch in selected spring bread wheat genotypes. M.Sc. thesis. Oregon State Univ., Corvallis, USA, 86 pp.

Bronnimann, A; Fossati, A and Hani, F (1973) Ausbreitung von *Septoria nodorum* Berk. Und SchAdigung bei Kinstlisch lnduzierten Halmlange-Mutanten der Winterweizensorte 'Zenith *(Triticum aestivum) Z. Pflanzenzuchtg.* **70**: 230-245.

Browder, LE and Eversmeyer, MG (1986) Interactions of temperature and time with some *Pucccinia recondita*: *Triticum* corresponding gene pairs. *Phytopathology.* **76**: 1286-88.

Brown, JKM; Foster, EM and O'Hara, RB (1997) Adaptation of powdery mildew populations to cereal varieties in relation to durable and non-durable resistance. In: IR Crute, EB Holub and JJ Burdon (Eds.), The Gene-for-Gene Relationship in Plant-Parasite Interaction, pp. 119–138. CAB International, UK.

Browning, JA; Simons, MD; Frey, KJ and Murphy, HC (1969) Regional deployment for conservation of oat crown rust resistant genes. *Spec. Rep.–Iowa Agric. Home Econ. Exp. Stn.* **64**: 49–56.

Buerstmayr, H; Lemmens, M; Hartl, L; Doldi, L; Steiner, B; Stierschneider, M and Ruckenbauer, P (2002) Molecular mapping of QTLs for Fusarium head blight resistance in spring wheat. I. Resistance to fungal spread (type II resistance). *Theor. Appl. Genet.* **104**: 84–91.

Bürling, K; Hunsche, M and Noga, G (2010) Quantum yield of non-regulated energy dissipation in psii (y (no) for early detection of leaf rust (*Puccinia triticina*) infection in susceptible and resistant wheat (*Triticum aestivum* L.) cultivars. *Prec. Agric.* **11**: 703–716.

Caldwell, RM; Cartwright, WB and Compton, LE (1946) Inheritance of hessian fly resistance derived from W 38 and *durum* P.i. 94587. *J. Amer. Soc. Agron.* **38**: 398—409.

Caldwell, RM; Roberts, JJ and Eyal, Z (1970) General resistance slow rusting to *Puccinia recondita* in winter wheat and spring wheats. *Phytopathology* **60**: 1287.

Campilan D; TL Lama, SR Ghimire and O Hidalgo (2006) From piloting to scaling up. *LEISA* **22**(3): 16–17.

Campilan, D and Ghimere SR (1998) Mobilizing communities for integrated disease management: comparing experiences in Nepal, Indonesia and the Philippines. Seminar-Workshop on Planning the Implementation of Farmer Field Schools in South Asia, 2–5 June 1998, Kathmandu, Nepal.

Carson, SK; Patterson, FL and Gailun, RL (1978) Inheritance of resistance to hessian fly derived from *Triticum turgidum* L. *Crop Sci.* **18**: 1011-1014.

Chague, V; Fahima, T; Dahan, A; Sun, GL; Korol, AB; Ronin, YI; Grama, A; Roder, MS and Nevo, E (1999) Isolation of microsatellite and RAPD markers flanking the*Yr15* gene of wheat using NILs and bulked segregant analysis. *Genome* **42**: 1050-1056.

Chand, R; Kumar, S; Pandey, SP and Joshi, AK (2002) Occurrence of spot blotch on wheat under rice-wheat cropping system in Eastern India. In: Proc.4th Intl. Wheat Tan Spot and Spot Blotch Workshop, North Dakota State University, USA.

Chand, R; Pandey, SP; Singh, HV; Kumar, S and Joshi, AK (2003) Variability and its probable cause in the natural population of spot blotch pathogen *Bipolaris sorokiniana* of wheat (*T. aestivum* L.). *Journal of Plant Diseases and Protection* **111**(1): 27-35.

Chand, R; Pandey, SP; Singh, HV; Kumar, S and Joshi, AK (2003) Variability and its probable cause in natural populations of spot blotch pathogen (*Bipolaris sorokiniana*) of wheat (*T. aestivum* L.) in India. *J Plant Disease Prod* **110**: 27-35.

Chand, R; Singh, HV; Joshi, AK and Duveiller, E (2002) Physiological and morphological aspects of *Bipolaris sorokiniana* conidia surviving on wheat straw. *Plant Pathol J* **18**: 328-332.

Chang, N and Wn, Y (1998) Incidence and current management of spot blotch of wheat in China. In: Duveiller E, Dubin HJ, Reeves J, McNab A (eds), *Helminthosophism* Diseases of Wheat: Spot Blotch and Tan Spot. CIMMYT, Mexico, pp. 119-125.

Charpe, A; Koul, S; Gupta, SK; Singh, A; Pallavi, JK and Prabhu, KV (2012) Marker assisted gene pyramiding of leaf rust resistance genes *Lr24, Lr28* and *Lr9* in a bread wheat cultivar HD2329. *J Wheat Res* **4**: 20–28.

Chaurasia, S; Joshi, AK; Dhari, R and Chand, R (1999) Resistance to foliar blight of wheat: A search. *Genet Resour Crop Evol* **46**: 469-475.

Chen, DF and Dale, PJ (1992) A comparison of methods for delivering DNA to wheat: the application of wheat dwarf virus DNA to seeds with exposed apical meristems. *Transgenic Research* **1**: 93-100.

Chen, H; Chen, Z; Ni, S; Zuo, S; Pan, X and Zhu, X (2008) Pyramiding three genes with resistance to blast by marker-assisted selection to improve rice blast resistance of jin 23b. *CJRS* **22**: 23–27.

Chhuneja, P; Kaur, S; Goel, RK and Dhaliwal, HS (2006) Mapping of leaf rust and stripe rust resistance QTLs in bread wheat synthetic RIL population under Indian field conditions. *Indian J. Crop Sci.* **1**: 49–54.

Clausen, M; Krauter, R; Schachermayr, G; Potrykus, I and Sautter, C (2000) Antifungal activity of a virally encoded gene in transgenic wheat. *Nature Biotechnology* **18**: 446-449.

Collard, BCY; Grams, RA; Bovill, WD; Percy, CD; Jolley, R; Lehmensiek, A; Wildermuth, G and Sutherland, MW (2005) Development of molecular markers for crown rot resistance in wheat: Mapping of QTLs for seedling resistance in a '2-49' 'Janz' population. *Plant Breeding* **124**: 532–537.

Cox, DJ and Wilcoxson, RD (1982) The relationship of the *Sr6* gene to slow rusting in wheat. *Phytopathology* **72**: 178-181.

Curtis, BC; Schlehuber, AM and Wood, EA (1960) Genetics of greenbug *(Toxoptera graminum* Rond.) resistance in two strains of common wheat. *Agron. J.* **52**: 599-602.

Cuthbert, PA; Somers, DJ; Thomas, J (2006) Fine mapping *Fhb1*, a major gene controlling fusarium head blight resistance in bread wheat (*Triticum aestivum* L.). *Theor Appl Genet* **112**: 1465–1472.

Dale, PJ; Marks, MS; Brown, MM; Woolston, CJ; Gunn, H; Boulton, MI; Mullineaux, PM; Lewis, DM; Kemp, JM; Chen, DF; Gilmour, DM and Flavell, RB (1989) Agroinfection of wheat: Inoculation of *in vitro* grown seedlings and embryos. *Plant Science* **63**: 237-245.

Dedryver, F; Jubier, MF; Thouvenin, J and Goyeau, H (1996) Molecular markers linked to the leaf rust resistance gene *Lr24* in different wheat cultivars. *Genome* **39**: 830-835.

Delibes, A; Delmoral, J; Martinsanchez, JA; Mejias, A; Gallego, M; Casado, D; Sin, E and Lopezbrana, I (1997) Hessian fly-resistance gene transferred from chromosome 4M (V) of *Aegilops ventricosa* to *Triticum aestivum*. *Theor. Appl. Genet.*, **94**: 858-864.

Demeke, T; Laroche, A and Gaudet, DA (1996) A DNA marker for the *Bt-10* common bunt resistance gene in wheat. *Genome* **39**: 51-55.

Dhaliwal, HS; Singh, H; Gill, KS and Randhava, HS (1993) Evaluation and cataloguing of wheat germplasm for disease resistant and quality. In: Damania AB (ed), Biodiversity and wheat improvement. John Wiley and Sons, New York, pp. 123-140.

Dubin HJ; Arun, B; Begum, SN; Bhatta, M; Dhari, R; Goal, LB; Joshi, AK; Khanna, BM; Malaker, PK; Rokhrel, DR; Rahman, MM; Saha, NK; Shaheed, MA; Sharma, RC; Singh, AK; Singh, RM; Singh, RV; Vargas, M and Verma, PC (1998) Results of the South Asia Regional *Helminthosporium* Leaf blight and Yield Experiment, 1993-94. In: Duveiller E, Dubin HJ, Reeves J, McNab A (eds), *Helminthosporium* Diseases of Wheat: Spot Blotch and Tan Spot. CIMMYT, Mexico, pp. 182-187.

Dubin, HJ and Bimb, HP (1994) Soil borne diseases of wheat in warmer areas of South Asia: An update. In: Saunders DA, Hettel GP (eds), Wheat in Heat Stressed Environments: Irrigated, Dry Areas and Rice-Wheat Farming Systems. CIMMYT, Mexico, pp353-359.

Dubin, HJ and Brennan, JP (2009) Combating stem and leaf rust of wheat- historical perspective, impacts and lesson learned. Millions fed: proven success in agricultural development. IFPRI, Washington D.C., IFPRI Discussion paper 00910.

Dubin, HJ and Rajaram, S (1996) Breeding disease-resistant wheats for tropical highlands and lowlands. *Ann Rev Phytopath* **34**: 503-526.

Dubin, HJ and Van Ginkel, M (1991) The status of wheat disease and disease research in warmer areas. In: Sanders DA (ed) Wheat for the non-traditional warm areas. A Proc of the International Conference. CIMMYT, Mexico, pp. 125-145.

Duczek, LJ and Jones-Flory, LL (1993) Relationships between common root rot, tillering and yield loss in spring wheat and barley. *Can J Plant Path* **15**: 153-158.

Duveiller, E and Gilchrist, LI (1994) Production constraints due to *Bipolaris sorokiniana* in wheat: current situation and future prospects. In: Saunders DA, Hettel GP (eds), Wheat in Heat Stressed Environments: Integrated, Dry Areas and Rice-Wheat Farming Systems. CIMMYT, Mexico, pp. 343-352.

Duveiller, E; Garcia, I; Franco, J; Toledo, J; Crossa, J and Lopez, F (1998) Evaluating spot blotch resistance of wheat: Improving disease assessment under controlled condition and in the field. In: Duveiller E, Dubin HJ, Reeves J, McNab A (eds), *Helminthosophism* Diseases of Wheat: Spot Blotch and Tan Spot. CIMMYT, Mexico, pp63-66.

Duveiller, E and Dubin, HJ (2002) *Helminthosporium* leaf blights: Spot blotch and tan spot. In *Bread Wheat Improvement and Production*, edited by B. C. Curtis, S. Rajaram, H. Gomez-Macpherson, Plant Production and Protection Series No. 30. Food and Agriculture Organization of the United Nations, Rome, pp. 285–299.

Dweikat, I; Ohm, H; Mackenzie, S; Patterson, F; Cambron, S and Ratcliffe, R (1994) Association of a DNA marker with Hessian fly resistance gene *H9* in wheat. *Theor Appl Genet* **89**: 964-968.

Dweikat, I; Ohm, H; Patterson, F and Cambron, S (1997) Identification of RAPD markers for 11 Hessian fly resistance genes in wheat. *Theor Appl Genet.***94**(3): 419–423.

Dyck, PL (1987) The association of a gene for leaf rust resistance with the chromosome 7D suppressor of stem rust resistance in common wheat. *Genome*, **29**: 467–469.

Eastwood, RF; Lagudah, ES and Appels, R (1994) A directed search for DNA sequence tightly linked to cereal cyst nematode resistance genes in *Triticum tauschii*. *Genome* **37**: 311-319.

El Bouhssini, M; Ogbonnaya, FC; Chen, M; S. Lhaloui, F. Rihawi, and A. Dabbous. 2012. Sources of resistance in primary synthetic hexaploid wheat to insects pests- Hessian fly, Russian wheat aphid and Sunn pest in the Fertile Crescent. *Genet. Res. Crop Evol.*

El Nashaar, HM and Stack, RW (1989) Effect of long-term conditions cropping on spring wheat on aggressiveness of *Cochliobolus sativus*. *Can J Plant Sci* **69**: 395-400.

Entz, MH; van den Berg, CGJ; Lafond, GP; Stobbe, EH; Rossnagel, BG and Austenson, HM (1990) Effect of late-season fungicide application on grain yield and seed size distribution in wheat and barley. *Canadian Journal of Plant Science* **70**: 699–706.

Evans, IR; Maurice, DC; Penney, DC and Solberg, ED (1993-1994) Wheat diseases and copper nutrition. *Better Crops*, pp. 6-8.

Evdokimova, ZZ (1988) Interspecific hybrids with species of the series Transaequatorialia Buk. - useful source material in breeding for nematode resistance. *Intensifikatsiya-kartofelevodstva-na-Severo-Zapade-RSFSR*. 47-50.

Faris, JD; Anderson, JA; Francl, LJ and Jordahl, JG (1997) RFLP mapping of resistance to chlorosis induction by *Pyrenophora tritici-repentis* in wheat. *Theoretical and Applied Genetics* **94**: 98-103.

Feuillet, C; Messmer, M; Schachermayr, G and Keller, B (1995) Genetic and physical characterization of the *Lr1* leaf rust resistance locus in wheat (*Triticum aestivum* L.). *Molecular and General Genetics* **248**: 553-562.

Flor, HH (1942) Inheritance of pathogenicity in *Melampsora lini*. *Phytopathology* **32**: 653-669.

Flor, HH (1946) The complementary genetic systems in flax and flax rust. *Adv. Genet*. **8**: 29-54.

Flor, HH (1955) Host parasite interaction in flax rust- its genetics and other implications. *Phytopathology* **45**: 680-685.

Flor, HH (1956) Mutation in Flax rust induced by ultraviolet radiation. *Science* **124**: 888-889.

Flor, HH (1971) Current status of the gene-for-gene concept. *Annu. Rev. Phytopathol.* **9**: 275-296.

Flor. HH (1947) Inheritance of reaction to rust in flax. *J. Agric Res*. (Washington DC) **74**: 241-262.

Fowler, RA; Platz, GJ; Bell, KL; Fletcher, SHE; Franckowiak, JD and Hickey, LT (2017) Pathogenic variation of *Pyrenophora teres f. teres* in Australia. *Australasian Plant Pathology* **46**: 115–128.

Fritz, RS and Simms, EL (Eds.) (1992) Plant Resistance to Herbivores and Pathogens: Ecology, Evolution and Genetics. The University of Chicago Press.

Gangwar, OP; Bhardwaj, SC; Khan, H; Prasad, P; Singh, SV and Kumar, S (2013) *Bharat mein gehun ke ratua rog sthiti evam niyantran ranniti. Shodh Bulletin sankhya* 6 31 *prashth* (In Hindi). Directorate of Wheat Research, Regional Station, Flowerdale, Shimla, Himachal Pradesh, India.

Gangwar, OP; Kumar, S; Bhardwaj, SC; Kashyap, PL; Prasad, P and Khan, H (2019) Characterization of three new Yr9-virulences and identification of sources of resistance among recently developed Indian bread wheat germplasm. *J Plant Pathol* **101**: 955–963.

Gao, AL; He, HG and Chen, QZ (2005) Pyramiding wheat powdery mildew resistance genes *Pm2*, *Pm4a* and *Pm21* by molecular marker-assisted selection. *Acta Agron Sin* **31**: 1400–1405.

Gill, BS; Raupp, WJ; Browder, LE; Cox, TS and Sears, RG (1991a) Registration of KS89WGRC7 leaf rust-resistant hard red winter wheat germplasm. *Crop Sci.* **31**: 246.

Gold, J; Harder, D; Townley-Smith, F; Aung, T and Procunier, J (1999) Development of a molecular marker for rust resistance genes *Sr*39 and *Lr*35 in wheat breeding lines. *EJB Elect. J Biotech*, April 15, pp. 1-6.

Goodwin, SB; Hu, X and Shaner, G (1998) An AFLP marker linked to a gene for resistance to *Septoria tritici* blotch in wheat. *In* AE Slinkard, ed. *Proc. 9th Int. Wheat Genetics Symp.*, Saskatoon, Saskatchewan, Canada, 2-7 Aug. 1998, vol. 3, pp108-110.

Grama, AG and Gerechter-Amitai, ZK (1974) Inheritance of resistance to stripe rust (*Puccinia striiformis*) in crosses between wild emmer (*Triticum dicoccoides*) and cultivated tetraploid and hexaploid wheats 11, *Triticum aestivum. Euphytica* **23**: 393-398.

Grumet, R (1994) Development of virus resistant plants via genetic engineering. *Plant Breed. Rev.*, **12**: 47-49.

Gupta, AK and Saini, RG (1987) Frequency and effectiveness of *Lr13* in conferring wheat leaf rust resistance in India. *Curr Sci* **56**: 417-419.

Gupta, PK (2007) Pyramiding genes/QTL for crop improvement using marker-aided selection (MAS). In: Chopra VL, Sharma RP, Bhat SR, Prasanna BM (eds) Search for new genes. Academic Foundation, New Delhi, pp145–171.

Gupta, SK; Charpe, A; Prabhu, KV and Haque, QMR (2006) Identification and validation of molecular markers linked to the leaf rust resistance gene *Lr19* in wheat *Theor Appl Genet* **113**: 1027-1036.

Habgood, RM (1970) Designation of physiological races of plant pathogens. *Nature* (London) **227**: 1268-1269.

Hammerschmidt, R (1999) Phytoalexins: What Have We Learned After 60 Years? *Annu. Rev. Phytopathol.* **37**: 285–306.

Hare, RA and McIntosh, RA (1979) Genetic and cytogenetic studies of durable adult-plant resistances in 'Hope' and related cultivars to wheat rusts. *Zeitschrift für Pflanzenzüch - tung* **83**: 350-367.

Hartl, L; Mohler, V; Zeller, FJ; Hsam, SLK and Schweizer, S (1999) Identification of AFLP markers closely linked to the powdery mildew resistance genes Pm1c and Pm4a in common wheat (*Triticum aestivum* L.). *Genome*. **42**(2): 322–329.

Hartl, L; Mori, S and Schweizer, G (1998) Identification of a diagnostic molecular marker for the powdery mildew resistance gene *Pm4b* based on fluorescently labelled AFLPs. *In* AE Slinkard, (Ed) *Proc. 9th Int. Wheat Genetics Symp.*, Saskatoon, Saskatchewan, Canada, 2-7 Aug. 1998, vol. 3, pp. 111-113.

Hartl, L; Weiss, H; Stephan, U; Zeller, FJ and Jahoor, A (1995) Molecular identification of powdery mildew resistance genes in common wheat (*Triticum aestivum* L). *Theoretical and Applied Genetics* **90**: 601-606.

Hartl, L; Weiss, H; Zeller, FJ and Jahoor, A (1993) Use of RFLP markers for the identification of alleles of the *Pm3* locus conferring powdery mildew resistance in wheat (*Triticum aestivum* L.). *Theoretical and Applied Genetics* **86**: 959-963.

Hatchett, JH; Martin, TJ and Livers, RW (1981) Expression and inheritance of resistance to hessian fly in hexaplold wheats derived from *Triticum tauschii* (Coss) Schmal. *Crop Sci.* **21**: 731-734.

Hayden, MJ; Kuchel, H and Chalmers, KJ (2004) Sequence tagged microsatellites for the *Xgwm533* locus provide new diagnostic markers to select for the presence of stem rust resistance gene *Sr2* in bread wheat (*Triticum aestivum* L). *Theor. Appl. Genet.* **109**: 1641–1647.

Hegde, YR; Kalappanavar, IK and Desai, SA (2006) Effect of Wheat Rusts on Seed Germination and Seedling Vigour. *Seed Research* **34** (1): 110-112.

Helguera, M; Khan, IA; Kolmer, J; Lijavetzky, D; Zhong-qi, L and Dubcovsky, J (2003) PCR assays for the cluster of rust resistance genes and their use to develop isogenic hard red spring wheat lines. *Crop Sci* **43**: 1839.

Helguera, M; Khan, LA and Dubeovsky, J (2000) Development of PCR markers for wheat leaf rust resistance gene Lr47: *Theor. Appl. Genet.*, **100**: 1137-1143.

Hetzler, J; Eyal, Z; Mehta, YR; Campos, LA; Fehrman, H; Kushnir, U; Zekaria-Oren, J and Cohen, L (1991) Interaction between spot blotch (*Cochilobolus sativus*) and wheat cultivars. In: Saunders DA (ed), Wheat for the Non traditional Warm Areas: A Proceeding of the International Symposium. CIMMYT, Mexico, pp146-164.

Heyne, EG and Livers, RW (1953) Monosomic analysis of leaf rust reaction, awnedness, winter Injury and seed colour in Pawnee wheat. *Agron. J.* **45**: 54-58.

Hirschhorn, JN and Daly, MJ (2005) Genome-wide association studies for common diseases and complex traits. *Nat. Rev. Genet*. 6: 95–108.

Hossain, ABS and Azad, AK (1992) Reaction of wheat to *Helminthosporium sativum* in Bangladesh. *Heriditas* **116**: 203-205.

Hu, XY; Ohm, HW and Dweikat, I (1997) Identification of RAPD markers linked to the gene *Pm1* for resistance to powdery mildew in wheat. *Theor Appl Genet* **94**: 832-840.

Huerta-Espino, J and Singh, RP (1994) First report on virulence in wheat with leaf rust resistance gene *Lr19* in Mexico. Plant Dis. 78. In: *Wheat Breeding* (ed. FGH Lupton), University Press, Cambridge, UK.pp.173.

Hulbert, SH; Lyons, PC and Bennetzen, (1991) Reactions of maize lines carrying *Rp* resistance genes to isolates of the common rust pathogen, *Puccinia sorghi. Plant Dis* **75**: 1130–1133.

Hussien, T; Browder, RL; Gill, BS and Cox, TS (1998) Chromosomal locations in common wheat of three new leaf rust resistance genes from *Triticum monococcum. Euphytica*, **101**: 127-31.

Ishii, T and Yonezawa, K (2007a) Optimization of the marker-based procedures for pyramiding genes from multiple donor lines: I. schedule of crossing between the donor lines. *Crop Sci* **47**: 537–546.

Jaiswal, SK; Sweta; Prasad, LC; Sharma, S; Kumar, S; Prasad, R; Pandey, SP; Chand, R and Joshi, AK (2007) Identification of molecular marker and aggressiveness for different groups of *Bipolaris sorokiniana* isolates causing spot blotch disease in wheat (*Triticum aestivum* L.). *Curr Microbiol* **20**: 1-8.

Jha, MP; Sinha, BP; Singh, KMP and Sinha, AK (1991) Scale for recording foliar blights in seedling and adult stages of wheat. *Annual wheat newsletter*.38: Colarado State University: 128-130.

Jha, MP; Sinha, BP; Singh, KMP and Sinha, AK(1992) Pyramiding alien leaf rust resistance genes in cv.UP262. *Annual wheat newsletter*.39: Colorado State University.

Jia, J; Devos, KM; Chao, S; Miller, TE; Reader, SM and Gale, MD (1996) RFLP-based maps of the homoeologous group-6 chromosomes of wheat and their application in the tagging of *Pm12*, a powdery mildew resistance gene transferred from *Aegilops speltoides* to wheat. *Theor Appl Genet* **92**: 559-565.

Jiang, J; Friebe, B and Gill, BS (1994) Recent advances in alien gene transfer in wheat. *Euphytica* **73**: 199-212.

Johnson, R (1992) Reflections of a plant pathologist on breeding for disease resistance, with emphasis on yellow rust and eyespot of wheat. *Plant Pathol* **41**: 239–254.

Johnston, SJ; Sharp, PJ; McIntosh, RA; Guillen-Andrade, H; Singh, RP and Khairallah, M (1998) Molecular markers for the *Sr2* stem rust resistance gene In: Slinkard AE (ed) Proc 9th Int Wheat Genet Symp, Saskatoon, Sas-katchewan, Canada, 2-7 Aug 1998, vol. 3, pp117-119.

Joppa, LR and Williams, ND (1982) Registration of Largo, a greenbug resistant hexaploidy wheat. *Crop Sci.* **22**: 901–902.

Jorgensen, JH and Jensen, CJ (1973) Gene *Pm6* for resistance to powdery mildew in wheat. *Euphytica* **22**: 423.

Joshi, AK and Chand R (2002) Variation and inheritance of leaf angle, and its association with spot blotch (*Bipolaris sorokiniana*) severity in wheat (*Triticum aestivum*). *Euphytica* **124**: 283-291.

Joshi, AK; Chand, R and Arun, B (2002) Relationship of plant height and days to maturity with resistance to spot blotch in wheat. *Euphytica* **123**: 221-228.

Joshi, AK; Chand, R; Kumar, S and Singh, RP (2004) Leaf tip necrosis-a phenotypic marker for resistance to spot blotch disease caused by *Bipolaris sorokiniana* in wheat (*Triticum aestivum* L.). *Crop Science* **44**: 792-796.

Joshi, AK; Chand, R; Prasad, LC; Kumar, S; Pandey, SP and Tripathi, R (2003) Spot blotch disease of wheat caused by Bipolaris sorokiniana: pathogen survival and behaviour of resistance. 10th Int. Wheat Genetics Symposium 1-6 Sept. 2003, Paestum, Italy. pp. 1166-1169.

Joshi, AK; Kumar, S; Chand, R and Ortiz-Ferrara, G (2004). Inheritance of resistance to spot blotch pathogen caused by *Bipolaris sorokiniana* in spring wheat. *Plant Breeding* **123**: 213-219.

Joshi, AK; Kumari, M; Singh, VP; Reddy, CM; Kumar, S; Rane, J and Chand R (2007) Stay green trait: variation, inheritance and its association with spot blotch resistance in spring wheat (*Triticum aestivum* L.). *Euphytica* **153**: 59-71.

Joshi, LM (1986) Monitoring of wheat disease during 1984-85. *Indian J Mycol Plant Pathol* **15**: 135-139.

Joshi, LM; Nagarajan, S and Srivastava, KD (1977) Epidemiology of brown nd yellow rusts of wheat in northern India. I. Time and place of appearance. *Phytopathol.* **90**: 116-122.

Joshi, LM; Singh, DV; Srivastava, KD and Wilcoxson, RD (1983) Karnal bunt-A minor disease that is now threat to wheat. *Bot Rev* **43**: 309-338.

Karjalainen, R (1983) Inheritance of leaf resistance to *Septoria nodorum* Berk in two crosses of spring wheats. *J. Sci. Agril. Soc. Finland* **55**: 425-430.

Kaushal, K; Mishra, AN; Varma, PK; Kapoor, KN and Pandey, HN (2001) Dipteran fly *Mycodiplosis* sp.: a natural bioagent for controlling leaf rust (*Puccinia recondita tritici*) of wheat. *Indian J Agric Sci* **71**: 136-138.

Kazan, K; Schenk, PM; Wilson, I and Manners, JM (2001) DNA microarrays: new tools in the analysis of plant defense responses. *Mol. Plant Pathology* **2**(3): 177-185.

Kema, GHJ; Lange, W and van Silfhout, CH (1995) Differential suppression of stripe rust resistance in synthetic wheat hexaploids derived from *Triticum turgidum subsp. dicoccoides* and *Aegilops squarrosa*. *Phytopathology* **85**: 425–429.

Kerber, ER (1983) Suppression of rust resistance in amphidiploids of *Triticum*. Proc. 6th Int. Wheat Genet. Symp., Kyoto, Japan, pp. 813–817.

Kerber, ER (1987) Resistance to leaf rust in wheat: *Lr32*, a third gene derived from *Triticum tauschii*. *Crop Sci.* **27**: 204-206.

Kerber, ER and Green, GJ (1980) Suppression of stem rust resistance in the hexaploid wheat cv. Canthatch by chromosome 7DL. *Can. J. Bot.* **58**: 1347–1350.

Khairallah, M; Guillén-Andrade, H; Alarcón, J; Rodriguez, C; Ayala, L; Henry, M; Singh, RP; Jiang, C; Sharp, P and Hoisington, D (1998) Mapping of durable resistance to leaf rust and tolerance to BYDV in wheat. *In* AE Slinkard, ed. *Proc. 9th Int. Wheat Genetics Symp.*, Saskatoon, Saskatchewan, Canada, 2-7 Aug. 1998, vol. 3, pp282-284.

Khan, AA; Bergstrom, GC; Nelson, JC and Sorrells, ME (2000) Identification of RFLP markers for resistance to wheat spindle streak mosaic bymo virus (WSSMV) disease. *Genome* **43**: 477-82.

Kleine, DM and Nelson, RP (1963) Pathogenicity of the isolates of *Cochliobolus sativus* from cultivated and wild graminaceous hosts from Western Hemisphere to species of gramineae. *Plant Disease Rep* **47**: 890-894.

Knott, DR (1980) Mutation of a gene for yellow pigment linked to *Lr19* in wheat. *Can. J. Genet. Cytol.* **22**: 651–654.

Knott, DR (1988) Using polygenic resistance to breed for stem rust resistance in wheat. In: Simmonds N W and Rajaram S (eds.) *Breeding strategies for resistance to rusts of wheat.* CIMMYT, Mexico, pp. 39-47.

Knott, DR (2000) Inheritance of resistance to stem rust in Medea durum wheat and the role of suppressors. *Crop Sci.* **40**: 98–102.

Knott, DR (2020) The Wheat Rusts - Breeding for Resistance (Monographs on *Theoretical and Applied Genetics*).

Knox, RE and Howes, NK (1994) A monoclonal antibody chromosome marker analysis used to locate a loose smut resistance gene in wheat chromosome 6A. *Theor Appl Genet* **89**: 787-793.

Kochumadhavan M; Tomar, SMS and Nambisan, PNN (1988) *Agropyron* –derived specific genes in common wheat and their adult plant response to wheat pathogens. *Indian J. Genet.* **48**(3): 383-387.

Kota, R; Spielmeyer, W; McIntosh, RA and Lagudah, ES (2006) Fine genetic mapping fails to dissociate durable stem rust resistance gene *Sr2* from pseudo-black chaff in common wheat (*Triticum aestivum* L.) *Theor. Appl. Genet.* **112**: 492-499.

Kumar, J and Nagarajan, S (1997) Existence of variability in *Tilletia indica* as evident from PAGE test and response on set of differentials. International Conference on Integrated Plant Disease Management for Sustainable Agriculture, Golden Jubilee Celebrations of IPS. Nov. 10-15, 1997, IARI, New Delhi: pp261.

Kumar, J and Nagarajan, S (1998) Role of flag leaf and spike emergence stage on the incidence of Karnal bunt in wheat. *Plant Disease* **82**: 1368-1370.

Kumar, J and Singh, M (1997) Inoculum potential of Karnal bunt pathogen (*Tilletia indica*) as altered by flag leaf surface of wheat genotypes and light conditions: pp116-117 Ibid.

Kumar, J; Mir, RR; Kumar, N; Kumar, A; Mohan, A; Prabhu, KV; Balyan, HS and Gupta, PK (2010) Marker assisted selection for pre-harvest sprouting tolerance and leaf rust resistance in bread wheat. *Plant Breed* **129**: 617–621.

Kumar, J; Nayar, SK; Prashar, M and Nagarajan, S (1996) Mapping the *Pucccinia striiformis* race flora in Ladakh and other valley of the interior Himalayas. *Cereal Rusts and Powdery Mildew Bulletin* **24**: 98-104.

Kumar, J; Nayar, SK; Prashar, M; Bhardwaj, SC and Bhatnagar, R (1994) New variants 47S103 and 102S100 of *Puccinia striiformis* West. *Indian Phytopath.* **47**: 258-259.

Kumar, J; Sharma, AK; Kumar, R and Nagarajan, S (1999) A new technique to develop artificial epiphytotics of foliar blights in wheat. *Indian Journal of Agric. sciences* **69**: 49-52.

Kumar, S (2003) Variation in Spot Blotch pathogen and Genetics of its Resistance in Wheat. PhD. Thesis, BHU, Varanasi, India.

Kumar, S; Chand, R and Joshi, AK (2002) Inheritance of resistance to spot blotch caused by *Bipolaris sorokiniana* in spring wheat. In: Proc.2nd Intl. Group Meeting on Wheat Tech. for Warmer Areas, Pune.

Kumar, S; Ibeagha, AE; Schäfer, P; Langen, G and Kogel, KH (2005) Cytological and molecular studies on wheat resistance to *B. sorokiniana*. 57th International Symposium on Crop Protection, 10th May, 2005, Gent, Belgium pp. 15.

Kumar, S; Joshi, AK; Prasad, LC; Kumar, U and Chand, R (2003) Inheritance and allelic relationship of resistance genes to spot blotch of wheat caused by *Bipolaris sorokiniana*.10th Int. Wheat Genetics Symposium 1-6 Sept. 2003, Paestum, Italy. pp.1184-1186.

Kumar, S; Prasad, LC; Kumar, U; Tyagi, K; Arun, B and Joshi, AK (2007) Inheritance and allelic relationship of resistance genes to spot blotch of wheat caused by *Bipolaris sorokiniana.* HT Buck *et al. (eds.).* Wheat production in Stressed Environments: Developments in Plant Breeding, Springer, AA Dordrecht, The Netherlands, (12): 113-118.

Kumar, S; Schäfer, P; Janson, C; Langen, G; Kogel, KH; Rana, B; Yadav, MK and Joshi, AK (2008) Cellular and molecular characterization of resistance to *B. sorokiniana* in bread wheat. Published on line in eWIS (www.shigen.nig.ac.jp/ewis/article/html/33/article.html;jsessionid).

Kumar, S; Singh, R; Singh, AK; Yadav, SK; Grover, M; Sharma, P; Kumar, S and Kumar, R (Undated) Spot blotch - an emerging issue in South Asia.

Kumar, U; Joshi, AK; Kumar, S; Chand, R and Röder, MS (2010) Quantitative trait loci for resistance to spot blotch caused by *Bipolaris sorokiniana* in wheat (*T. aestivum* L.) lines Ning 8201 and Chirya 3. *Mol Breed* **26**: 477–491.

Kumar, U; Joshi, AK; Kumar, S; Chand, R and Röder, MS (2009) Mapping of resistance to spot blotch disease caused by *Bipolaris sorokiniana* in spring wheat. *Theor. Appl. Genet.* **118**: 783-792.

Labrum, KE (1980) The location of *Yr2* and *Yr6*, genes conferring resistance to yellow rust. In *Proc. 5th Eur. and Med. Cereal Rutsts Cont.*, Bari, Italy, pp. 41-45.

Lagudah, ES; McFadden, H; Singh, RP; Huerta-Espino, J; Bariana, HS and Spielmeyer, W (2006) Molecular genetic characterization of the *Lr34/Yr18* slow rusting resistance gene region in wheat. *Theor Appl Genet* **114**: 21-30.

Leclercq, D and Caubel, G (1991) Varietal resistance of lucerne to the stem nematode Ditylenchus dipsaci (Kuhn) Filipjev: the screening method and its application in selection for resistance. [Resistance varietale de la luzerne au nematode des tiges Ditylenchus dipsaci (Kuhn) Filipjev; test d'evaluation et application en selection.] *Agronomie*. **11**(7): 603-612.

Lee, SC; Choi, HW; Hwang, IS; Choi, DS and Hwang, BK (2006) Functional roles of the pepper pathogen-induced bZIP transcription factor, CAbZIP1, in enhanced resistance to pathogen infection and environmental stresses. *Planta*, **224**: 1209-1225.

Leonard, KJ and Szabo, LJ (2005) Stem rust of small grains and grasses caused by *Puccinia graminis*. *Mol Plant Pathol* **6:** 99–111.

Li, S; Zhang, Z; Wang, B; Zhong, Z and Yao, J (1995) Tagging the *Pm4a* gene in NILs by RAPD analysis. *Acta Genet Sin* **22**: 103-108.

Li, Y; Li, YP; Zhang, XY; Yang, GL; Li, Y; Li, YP; Zhang, XY and Yang, GL (1996) A genetic study on resistance of soyabeans to cyst nematode physiological race No. 4. *Soybean-Science*. **15(**3): 191-196.

Lillemo, M; Chen, XM; He, ZH and Singh, RP (2005) Leaf rust resistance gene *Lr34* is involved in powdery mildew resistance of CIMMYT bread wheat line Saar. In *Abstr. of 7th Int. Wheat Conf.: Wheat Production in Stressed Environments*, Nov 27–Dec. 2, Mar de Plata, Argentina, pp. 17.

Lillemo, M; Skinnes, H; Singh, RP and Van Ginkel, M (2006) Genetic analysis of partial resistance to powdery mildew in bread wheat line Saar. *Plant Dis*. **90**: 225–228.

Liskova, M; Sabova, M; Valocka, B and Vargova, V (1991) Determination of the pathotype Ha12 as the testing basis for evaluation of cereal resistance to *Heterodera avenae*. *Helminthologia*. **28**(2-3): 137-139.

Liu, J; Chang, Z; Zhang, X; Yang, Z; Li, X; Jia, J; Zhan, H; Guo, H and Wang, J (2013) Putative *Thinopyrum intermedium*-derived stripe rust resistance gene Yr50 maps on wheat chromosome arm 4BL. *Theor Appl Genet* **126**: 265–274.

Liu, Z; Sun, Q; Ni, Z and Yang, T (1999) Development of SCAR markers linked to the *Pm21* gene conferring resistance to powdery mildew in common wheat. *Plant Breeding* **118**: 215-219.

Ma, ZQ; Gill, BS; Sorrells, ME and Tanksley, SD (1993) RELP markers linked to two Hessian fly-resistance genes in wheat (*Triticum aestivum* L.) from *Triticum tauschii* (coss.) Schmal. *Theor Appl Genet*. **85**(6–7): 750–754.

Ma, ZQ; Sorrells, ME and Tanksley, SD (1994) RFLP markers linked to powdery mildew resistance genes *Pm1*, *Pm2*, *Pm3* and *Pm4* in wheat. *Genome* **37**: 871-875.

Maddox, DA (1998) Implications of new technologies for seed health testing and the worldwide movement of seed. *Seed Science Research* 8: 277-284.

Mago, R; Bariana, HS; Dundas, IS; Spielmeyer, W; Lawrence, GJ; Prior, AJ and Ellis, JG (2005) Development of PCR markers for the selection of wheat stem rust resistance genes *Sr24* and *Sr26* in diverse wheat germplasm. *Theor. Appl. Genet*. **111**: 496–504.

Mago, R; Tabe, L; McIntosh, RA; Pretorius, Z; Kota, R; Paux, E; Wicker, T; Breen, J; Lagudah, ES; Ellis, JG and Spielmeyer, W (2011) A multiple locus on chromosome arm 3BS in wheat confers resistance to stem rust (*Sr2*), leaf rust (*Lr27*) and powdery mildew. *Theor. Appl. Genet*. **123**: 615-623.

Mahto, BN and Sedhari, GP (1994) Loss assessment of Wheat due to foliar blight. Wheat Procs. NWRP, NARC.

Mahto, BN; Sedhari, GR and Tripathi, JJ (1996) Wheat grain yield losses in farmers' field due to foliar blight infection at outreach site Bankatti, Rupandehi. In: Devkota RN, Saari EE (Eds), Proc of Wheat Research Reports, National Wintercrops Technology Workshop. NARC/CIMMYT, Kathmandu, Nepal, pp. 593-586.

Mallick, N; Vinod, SJB; Tomar, RS; Sivasamy, M and Prabhu, KV (2015) Marker assisted backcross breeding to transfer multiple rust resistance in wheat. *Plant Breed* **134**: 172–177.

Maraite, H; Zinno, TDI; Longree, H; Daumerie, V and Dueveiller, E (1998) Fungi association with foliar blight of wheat areas. In: Duveiller E, Dubin HJ, Reeves J, McNab A (eds), *Helminthosporium* Diseases of Wheat: Spot Blotch and Tan Spot. CIMMYT, Mexico, pp293-300.

Marasas, CN; Smale, M and Singh, RP (2004) The economic impact in developing countries of leaf rust resistance breeding in CIMMYT-related spring bread wheat. Economics Program Paper 04-01. Mexico, DF: CIMMYT. pp. 36.

Markus, KG; Markus, S; Su, JS; Kolukisaoglu, A; Louis, P; Sonia, C; Mark, D; Richter, A; Weder, B; Schulz, B; Martinoia, E (2004) Disruption of AtMRP4, a guard cell plasma membrane ABCC-type ABC transporter, leads to deregulation of stomatal opening and increased drought susceptibility. *Plant J*., **39**: 219-236.

Martens, JW and Dyck, PL (1989) Genetics of resistance to rust in cereals from a Canadian perspective. *Can J Plant Pathol* **11**: 78–85.

Martinez, F; Niks, RE; Singh, RP and Rubiales, D (2001) Characterization of *Lr46*, a gene conferring partial resistance to wheat leaf rust. *Hereditas*, **135**: 111–114.

Mathur, IB and Cunfer BM (1993) Seed borne diseases and seed health testing of wheat. Danish Govt. Institute of seed Pathology for Developing Countries, Copenhegen, Denmark, pp. 168.

Matsumura, H; Reich, S; Ito, A; Saitoh, H; Kamoun, S; Winter, P; Kahl, G; Reuter, M; Kru¨ger, DH and Terauchi, R (2003) Gene expression analysis of plant host–pathogen interactions by SuperSAGE. *Proc Natl Acad Sci USA* **100**: 15718–15723.

Maude, RB (1996) Seed Borne Diseases and Their Control, Principles and Practice, Wallingford, UK, CAB International.

Mavi, GS; Bazzar, SK; Kaur, S; Sohu VS; Prabhu, K and. Chhuneja, P (2018) Marker assisted stacking of drought tolerance QTL and rust resistance genes in an elite Indian wheat genotype. *Indian Journal of Genetics and Plant Breeding*. **78**(1): 1-10.

McDonald, BA and Linde, C (2002) The population genetics of plant pathogens and breeding strategies for durable resistance. *Euphytica* **124**: 163-180.

McIntosh, RA (1988) Catalogue of gene symbols for wheat. p. 1225–1323. Proc. 7th Int. Wheat Genet. Symp, Cambridge, UK.

McIntosh, RA and Brown, GN (1997) Anticipatory breeding for resistance to rust diseases in wheat. *Ann. Rev. Phytopathol*. **35**: 311-326.

McIntosh, RA; Mu, J; Han, D and Kang, Z (2018) Wheat stripe rust resistance gene *Yr24/Yr26*: a retrospective review. *Crop J* **6**: 321–329.

McIntosh, RA; Yamazaki, Y; Dubcovssky, J; Rogers, J; Morris, C; Somers, DJ; Appels, R and Devos, KM (2012) Catalogue of gene symbols for wheat, (Supplement).

Mehta, KC (1929) Annual reoccurrence of rusts on wheat in India. (Presidential address). *Procs, Indian Science Congress*, **16**: 199-223.

Mehta, KC (1940) Further studies on cereal rusts in India. *Imperial Council of Agricultural Research, Monograph*. 14.

Mehta, YR (1981) Identification of races of *Helminthosporium sativum* of wheat in Brazil. *Pesquisa Agropecuaria Brasilia* **16**: 331-336.

Mehta, YR (1985) Breeding wheat for resistance to *Helminthosporium* spot blotch. In: Wheat for the more tropical environments. A proceeding of Mexico, D.F. CIMMYT. pp135-140.

Mehta, YR (1993) Spot blotch (*Bipolaris Sorokiniana*). In: Mathur SB, Cunfer BM (eds), Seed borne disease and seed health testing of wheat. Institute of seed pathology for developing countries, Copenhagen, Denmark, pp. 105-112.

Mehta, YR (1994) Manejo Integrado de Enfermedadas de Trigo-santa Cruz. Bolivia, CIAT/IAPAR, pp. 314.

Messmer, MM; Seyfarth, R; Keller, M; Schachermayr, G; Winzeler, M; Zanetti, S; Feuillet, C and Keller, B (2000) Genetic analysis of durable leaf rust resistance in winter wheat. *Theoretical and Applied Genetics*, **100**: 419–431.

Michelmore, RW; Paran, I and Kesseli, RV (1991) Identification of markers linked to disease-resistance genes by bulked seg-regant analysis: a rapid method to detect markers in specific genomic regions by using segregating populations. *Proc. Natl. Acad. Sci.*, **88**: 9828-9832.

Miedaner, T; Wilde, F; Korzun, V; Ebmeyer, E; Schmolke, M; Hart, L and Schon, CC (2009) Marker selection for *Fusarium* head blight resistance based on quantitative trait loci (QTL) from two European sources compared to phenotypic selection in winter wheat. *Euphytica* **166**: 219–227.

Miller, TE (1984) The homoeologous relationship between the chromosomes of rye and wheat: Current status. *Can. J. Genet. Cytol.* **26**: 578–589.

Minerdi, D; Bossi, S and Maffei, ME (2011) *Fusarium oxysporum* and its bacterial consortium promote lettuce growth and Expansin A5 gene expression through microbial volatile organic compound (MVOC) emission. *FEMS Microbiol Ecol* **76**: 342–351.

Mittler R (2006) Abiotic stress, the field environment and stress combination. *Trends in Plant Science*, **11**: 15-19.

Mohler, V and Jahoor, A (1996) Allele specific amplification of polymorphic sites for the detection of powdery mildew resistance loci in cereals. *Theor Appl Genet* **93**: 1078-1082.

Molnar, I; Gáspár, L; Sárvári, E; Dulai, S; Hoffmann, B; Molnár-Láng, M and Galiba, G (2004) Physiological and morphological responses to water stress in *Aegilops biuncialis* and *Triticum aestivum* genotypes with differing tolerance to drought. *Func Plant Biol* **31**: 1149–1159.

Moshou, D; Bravo, C; Oberti, R; West, J; Bodria, L; McCartney, A and Ramon, H (2005) Plant disease detection based on data fusion of hyperspectral and multispectral fluorescence imaging using kohonen maps. *Real-Time Imaging* **11**: 75–83.

Mujeeb-Kaji, A; Villaveal, RL; Gilchrist, LI and Rajaram, S (1996) Registration of five wheat germplasm lines resistant to *Helminthosporium* leaf blight. *Crop Sci* **36**: 216-217.

Mukai, J and Gill, BS (1991) Detection of barley chromatin added to wheat by genomic *in situ* hybridization. *Genome*, **34**: 448-52.

Mullaney, EJ; Martin, JM and Scharen, AL (1982) Generation mean analysis to Identify and partition the components of genetic resistance to *Septoria nodorum* in wheat. *Euphytica* **31**: 539-545.

Myburg, AA; Cawood, M; Wingfield, BD and Botha, AM (1998) Development of RAPD and SCAR markers linked to the Russian wheat aphid resistance gene *Dn2* in wheat. *Theor Appl Genet* **96**: 1162-1169.

Nagarajan, S and Kumar, J (1997) An review of the increase in importance of research on foliar blights in India: Germplasm improvement and future challenges towards a sustainable high yielding wheat production. International Workshop on *Helminthosporium* blights of wheat Feb.9-14,1997. CIMMYT, Mexico.

Nagarajan, S; Aujla, SS; Nanda, GS; Sharma, I; Goel, LB; Kumar, J and Singh, DV (1997) Karnal bunt (*Tilletia indica*) of wheat - a review. *Review of Plant Pathology* **76**: 1207-1214.

Nagarajan, S; Bahadur, P and Nayar, SK (1984) Contemplating the brown rust resistance (*Lr*) genes to mitigate the spread of *Puccinia recondita* f. sp. *tritici*. *Indian Phytopath* **30**: 490-497.

Nagarajan, S; Joshi, LM; Srivastava, KD and Singh, DV (1980) Epidemiology of brown and yellow rusts of wheat in north India. IV. Disease management recommendations. *Cereal Rusts Bull.* **7**: 15–20.

Nagarajan, S; Nayar, SK and Bahadur, P (1983) The proposed brown rust of wheat (*Puccinia recondita* f. sp. *tritici*) virulence analysis system. *Current Science* **52**: 413-416.

Nagarajan, S; Nayar, SK and Bahadur, P (1985) The proposed systems of virulence analysis III. *Puccinia striiformis* West. *Kavaka* **13** (1): 33-36.

Nagarjan, S and Joshi, LM (1975) A historical account of wheat rust epidemics in India and their significance. *Cereal Rusts Bull* **2**: 29-33.

Navabi, A; Singh, RP; Tewari, JP and Briggs, KG (2004) Inheritance of high levels of adult-plant resistance to stripe rust in five spring wheat genotypes. *Crop Sci.* **44**: 1156–1162.

Nayar, SK (1996) Variation in wheat rusts and their management through resistance. *Indian Phytopath.* **49**: 132-133.

Nayar, SK; Jain, SK; Prashar, M; Bhardwaj, SC; Kumar, S and Menon, MK (2003) Appearance of new pathotype of *Puccinia recondita tritici* virulent on *Lr9* in India. *Indian Phytopath.* **56**(2): 196-198.

Nayar, SK; Nagarajan, S and Singh, DV (1998) Epidemiology and control of wheat rusts in India. "Pathological problems of economic crop plants and their management ". Eds. Paul Khurana, S.M., pp. 29-37.

Nayar, SK; Prashar, M and Bhardwaj, SC (1997) Manual of current techniques in wheat rusts. Technical Bulletin No.1. DWR, Regional Station, Flowerdale, Shimla-171002, India.

Nelson, JC; Autrique, JE; Fuentes-Dávila, G and Sorrells, ME (1998) Chromosomal location of genes for resistance to Karnal bunt in wheat. *Crop Sci.*, **38**: 231-236.

Nelson, JC; Singh, RP; Autrique, JE and Sorrells, ME (1997) Mapping genes conferring and suppressing leaf rust resistance in wheat. *Crop Sci* **37**: 1928-1935.

Nelson, RR (1960) Evolution of sexuality and pathogenicity in interspecific crosses in the genus *Helminthosporium*. *Phytopath* **50**: 375-377.

Nelson, RR; Wilcoxson, RD and Christensen, JJ (1955) Heterocaryosis as a basis for variation in *Puccinia graminis* var. *tritici Phytopathology*, **45**: 639–643.

Ogbonnaya, FC; Moullet, O; Eastwood, RF; Kollmorgen, J; Eagles, H; Appels, R and Lagudah, ES (1998) The use of molecular markers to pyramid cereal cyst nematode resistance genes in wheat. *In* A.E. Slinkard, ed. *Proc. 9th Int. Wheat Genetics Symp.*, Saskatoon, Saskatchewan, Canada, 2-7 Aug. 1998, vol. 3, pp138-139.

Oliveira, AMR; Matsumura, ATS; Prestes, AM; Matos, GS, and van-der-Sand, ST (1998) Morphological variability and pathogenicity in isolates of *Bipolaris sorokiniana*. *Fitopatologia-Brasileira* **23**: 349-353.

Pal, BP; Vasudeva, RS and Kohli, SP (1954) Breeding rust resistant hill wheats in India. Indian Council of Agricultural Research, New Delhi, Research Series No.5. pp52.

Pandey, SP; Joshi, AK; Kumar, S and Chand, R (2003) Sources of genetic resistance against newer epidemic groups of spot blotch pathogen *B sorokiniana* of wheat. 10th Int. Wheat Genetics Symposium 1-6 Sept. 2003, Paestum, Italy. pp792-794.

Pandey, SP; Kumar, S; Kumar, U; Chand, R and Joshi, AK (2005) Sources of inoculum and reappearance of spot blotch of wheat in rice-wheat cropping systems in eastern India. *European Journal of Plant Pathology* **111**: 47: 55.

Pandey, SP; Kumar, S; Prasad, LC; Chand, R and Joshi, AK (2002). Survival of spot blotch pathogen of wheat under rice-wheat cropping system in Eastern India. In: Proc.2nd Intl. Group Meeting on Wheat Tech. for Warmer Areas, Pune.

Park, RF; Bariana, HS; Wellings, CR and Wallwork, H (2002) Detection and occurrence of a new pathotype of *Puccinia triticina* with virulence for *Lr*24 in Australia. *Australia Journal Agricultural Research* **53** (9): 1069-1076.

Parlevliet, JE (1988) Strategies for utilization of partial resistance for the control of cereal rusts. In: Simmonds NW and Rajaram S (eds.). Breeding Strategies for resistance to rust of wheat. CIMMYT, Mexico, pp. 48-62.

Patterson, FL and Gallun, RL (1977) Linkage In wheat of the H3 and HS genetic factors for resistance to Hessian fly. *J. Hered.* **68**: 293-296.

Paull, JG; Pallotta, MA; Langridge, P and The, TT (1995) RFLP markers associated with *Sr22* and recombination between chromosome 7A of bread wheat and the diploid species *Triticum boeoticum*. *Theor Appl Genet* **89**: 1039-1045.

Pearson, C; Samborski, DJ and Rohringer, R (1962) The gene-for-gene concept. *Nature* (London) **194**: 561-562.

Peng, DL and Cook, R (1996) Observations on pathotypes of cereal cyst nematode (*Heterodera avenae*) in China. *Afro Asian Journal of Nematology*. **6**(2): 176-178.

Peterson, RF; Campbell, AB and Hannah, AE (1948) A diagrammatic scale for rust intensity on leaves and stems of cereals. *Canadian Journal of Research*, **26**: 496–500.

Pietrusinska, A; Czembor, JH and Czembor, PC (2011) Pyramiding two genes for leaf rust and powdery mildew resistance in common wheat. *Cereal Res Commun* **39**: 577–588.

Pilon-Smits, EHA; Ebskamp, MJM; Paul, MJ; Jeuken, MJW; Weisbeek, PJ and Smeekens, SCM (1995) Improved performance of transgenic fructanaccumulating tobacco under drought stress. *Plant Physiol.*, **107**: 125-130.

Prasad, P; Gangwar, OP; Bharadwaj, SC and Kumar, S (2019) Monitoring pathotype distribution of *Puccinia* species on wheat and barley. *Mehtaensis* **39**: 5–9.

Prashar, M; Nayar, SK; Bhardwaj, SC and Kumar, J (1996) Colour mutations in brown rust of wheat. *Pl. Dis. Res.* **11**: 163-165.

Pretorius, ZA; Wilcoxson, RD; Long, DL and Schaffer, JF (1984) Detecting wheat leaf rust resistance gene *Lr13* in seedlings. *Plant Disease*. **68**: 585-86.

Priadcencu, A; Miclea, C and Moisescu, L (1964) The first hybrid generation between wheat and rye. An Inst Cere Cereal Pl Teh, *Fundulea*: Ser C **32**: 39–65.

Prins, R; Groenwald, JZ; Marias, GF; Snape, JW and Koebner, RMD (2001) AFLP and STS tagging of *Lr*19, a gene conferring resistance to leaf rust in wheat. *Theor Appl Genet* **103**: 618-624.

Procunier, JD; Knox, RE; Bernier, AM; Gray, MA and Howes, NK (1997) DNA markers linked to a *T10* loose smut resistance gene in wheat (*Triticum aestivum* L.). *Genome* **40**: 176-179.

Puranik, SB and Mathre, DE (1971) Biology and control of ergot on male sterile wheat and barley. *Phytopathology* **61**: 1075-1080.

Qi, L; Cao, M; Chen, P; Li, W and Liu, D (1996) Identification, mapping and application of polymorphic DNA associated with resistance gene *Pm*21 of wheat. *Genome* **39**: 191-197.

Quinones, MA; Larter, EN and Samborski, DJ (1972) The inheritance of resistance to *Puccinia recondita* in hexaploid triticale. *Can. J. Genet. Cytol.* **14**: 495–505.

Rajaram, S; Mann, CE; Ortis Ferrara, G and Muzeeb-Kazi, A (1983) Aaption, stability and high yield potential of certain 1B/1R CIMMYT wheats. In Proc. 6th Wheat Genet. Symp. (Ed. S. Sakamoto), Kyoto, Japan, pp. 613-621.

Randhawa, MS; Bains, NS; Sohu, VS; Chhuneja, P; Trethowan, RM; Bariana, HS and Bansal, U (2019) Marker assisted transfer of stripe rust and stem rust resistance genes into four wheat cultivars. *Agronomy* **9**: 497.

Rani., K; Raghu, BR; Jha, SK, Agrawal, P; Mallick, N; Niranjana, M; Sharma, JB; Singh, AK; Sharma, NK; Rajkumar, S; Tomar, SMS and Vinod (2020) A novel leaf rust resistance gene introgressed from *Aegilops markgrafii* maps on chromosome arm 2AS of wheat. *Theoretical and Applied Genetics*.

Rao, MV (1970) Wheat. Symposium on Crop Plants Evolution, organized by the Indian Society of Genetics and Plant Breeding, March 10-13, 1970, Indian Agricultural Research Institute, New Delhi (Abstr.) pp6-8.

Rasal, PN; Gadekar, DA; Salunke, CB and Gavhane, VN (2008) Breeding for rust resistance in wheat with special reference to *Ug99*. Paper presented in National Symposium held at RWRRS, Mahabaleshwar on Plant Disease Scenario In Agriculture for Eco-Friendly Sustainability, Jan 10-12,2008.O.08, pp22.

Raupp, WJ; Singh, S; Brown-Guedira, GL and Gill, BS (2001) Cytogenetic and molecular mapping of the leaf rust resistance gene *Lr*39 in wheat. *Theor Appl Genet* **102**: 347-352.

Rees, RG and Platz, GJ (1992) Tan spot and its control–some Australian experiments. In *Advances in Tan Spot Research. Proc. of the 2nd Int. Tan Spot Workshop*, edited by

LJ Francl, JM Krupinsky and MP Mc Mullen, Fargo, ND, USA, North Dakota Agric. Expt. Stn, pp1.

Ren, SX; McIntosh, RA and Lu, ZJ (1997) Genetic suppression of the cereal rye-derived gene Pm8 in wheat. *Euphytica* **27**: 353-360.

Ren, SX; McIntosh, RA; Sharp, PJ and The, TT (1996) A storage-protein marker associated with the suppressor of *Pm8* for powdery mildew resistance in wheat. *Theor Appl Genet* **93**: 1054-1060.

Ren, ZL; Lelley, T and Röbbelen, (1990) The use of monosomic rye addition lines for transferring rye chromatin into bread wheat. II. Breeding value of homozygous wheat/rye translocations. *Plant Breed.* **105**: 265–270.

Revathi, P; Tomar, SMS and Vinod, SNK (2010) Marker assisted gene pyramiding of leaf rust resistance genes *Lr24, Lr28* along with stripe rust resistance gene *Yr15* in wheat (*Triticum aestivum* L.). *Indian J Genet Plant Breed* **70**: 349–354.

Riley, R; Chapman, V and Johnson, R (1968) Introduction of yellow rust resistance of *Aegilops comosa* into wheat by genetically-induced homoeologous recombination. *Nature* **217**: 383-384.

Rivoal, R; Jahier, J and Hulle, M (1993) Partial resistance to *Heterodera avenae* in wheat lines with the 6Mv chromosome from *Aegilops ventricosa. Journal-of-Nematology*. **25**(2): 265-269.

Robert, O; Abelard, C and Dedryver, F (1999) Identification of molecular markers for the detection of the yellow rust resistance gene *Yr17* in wheat. *Mol Breed* **5**: 167–175.

Roelfs, AP (1988) Resistance to leaf rust and stem rust in wheat. In Breeding Strategies for Resistance to the Rusts of Wheat, edited by N.W. Simmonds and S. Rajaram, CIMMYT, Mexico, pp10–22.

Roelfs, AP; Singh, RP and Saari, EE (1992) Rust Diseases of Wheat: Concepts and Methods of Disease Management. CIMMYT, Mexico.

Rondon, MA; Gough, FJ and Williams, ND (1966) Inheritance of stem rust resistance In *Triticum aestivum* sp. *vulgare* 'Reliance' and P.I. 94701 of *Triticum durum. Crop Sci.* **6**: 177-179.

Rowland, GG and Kerber, ER (1974) Telocentric mapping in hexaploid wheat of genes for leaf rust resistance and other characters derived from *Aegilops squarosa. Can. J. Genet. Cytol.* **16**: 137-144.

Rubiales, D and Niks, RE (1995) Characterization of *Lr34*, a major gene conferring non-hypersensitive resistance to wheat leaf rust. *Pl Dis* **79**: 1208-1212.

Ruchstuhl, M (1994) Wheat yield losses in farmers fields due to foliar blight infection. In: Saunders DA, Hettel GP (eds), Wheat in Heat-Stressed Environment: Irrigated, Dry Areas and Rice-Wheat Farming Systems. CIMMYT, Mexico, pp. 397-398.

Rutkoski, JE; Heffner, EL and Sorrells, ME (2010) Genomic selection for durable stem rust resistance in wheat. *Euphytica* **179**: 161–173.

Saari, EE (1998) Leaf blight diseases and associated soil borne fungal pathogens of wheat in north and south East Asia. In: Duveiller E, Dubin HJ, Reeves J, McNab A (eds), *Helminthosophism* Diseases of Wheat: Spot Blotch and Tan Spot. CIMMYT, Mexico, pp. 37-51.

Saini, RG; Kaur, M; Singh, B; Sharma, S; Nanda, GS; Nayar, SK; Gupta, AK and Nagarajan, S (2002) Genes *Lr48* and *Lr49* hypersensitive adult plant resistance in wheat (*Triticum aestivum* L.). *Euphytica* **124**: 365-370.

Saiprasad, SV; Verma, DK; Divya, A; Kaushal, K; Prakasha, TL; Singh, JB; Dubey, VG and Mishra, AN (2015) HI-708 (IC061303; INGR 14042), 1 wheat (*Triticum turgidum ssp. durum*) Germplasm with Leaf Rust Resistance, *Indian Journal of Plant Genetic Resources*, **28**(3): 356-357.

Samborski, DJ (1985) Wheat leaf rust. In: The Cereal Rusts (eds. AP Roelfs and WR Bushnell), Vol. 2, Diseases, distribution, epidemiology and control. P.38-56. Orlando, FL, USA, Academic Press.

Samborski, DJ and Dyck, PL (1976) Inheritance and virulence in *Puccinia recondite* on six backcross lines of wheat with single gene for resistance to leaf rust. *Canadian J. Bot.* **54**: 1666-1671.

Samsampour, D; Maleki, ZB; Singh, A; Pallavi, JK and Prabhu, KV (2009) Marker assisted selection to pyramid seedling resistance gene *Lr24* and adult plant resistance gene *Lr48* for leaf rust resistance in wheat. *Indian J. Genet.* **69**(1): 1-9.

Saritha, RK; Sinha, VC; Tomar, SMS and Srivastava, KD (2005) Nature of adult plant resistance in some *Triticum timopheevi* derivatives to pathotype 121R63-1 of leaf rust. *Indian Phytopath.* **58**(2): 129-135.

Sawhney, RN (1992) The role of *Lr34* in imparting durable resistance to wheat leaf rust through gene interaction. *Euphytica* **61**: 9-12.

Sawhney, RN and Joshi, BC (1996) Genetic research as the valid base of strategies for breeding rust resistant wheats. *Genetica* **97**(3): 243-254.

Schachermayr, G; Feuillet, C and Keller, B (1997) Molecular markers for the detection of the wheat leaf rust resistance gene *Lr10* in diverse genetic backgrounds. *Mol Breed* **3**: 65-74.

Schachermayr, GM; Messmer, MM; Feuillet, C; Winzeler, H; Winzeler, M and Keller, B (1995) Identification of molecular markers linked to the *Agropyron elongatum*-derived leaf rust resistance gene *Lr24* in wheat. *Theor Appl Genet* **90**: 982-990.

Schachermayr, GM; Siedler, H; Gale, MD; Winzeler, H; Winzeler, M and Keller, B (1994) Identification and localization of molecular markers linked to the *Lr9* leaf rust resistance gene of wheat. *Theoretical and Applied Genetics* **88**: 110-115.

Scott, PR; Hollins, TW and Muir, P (1978) Pathogenicity of *Pseudocercosporella herpotrichoides* to wheat, barley, oats and rye. *Transactions of the British Mycological Society* **65**: 529-538.

Seo, YW; Johnson, JW and Jarret, RL (1997) A molecular marker associated with the *H21* Hessian fly resistance gene in wheat. *Mol Breed* **3**: 177-181.

Seyfarth, R; Feuillet, C and Keller, B (1998) Development and characterization of molecular markers for the adult leaf rust resistance genes *Lr13* and *Lr35* in wheat. In: Slinkard AE (ed) Proc. 9th Int. Wheat Genetics Symp., Saskatoon, Saskatchewan, Canada, 2-7 Aug. 1998, vol. 3, pp154-155.

Seyfarth, R; Feuillet, C; Schachermayr, G; Winzeler, M and Keller, B (1999) Development of a molecular marker for the adult plant leaf rust resistance gene *Lr35* in wheat. *Theor Appl Genet* **99**: 554-560.

Seyfarth, R; Feuillet, C; Schachermayr, G; Winzeler, M and Keller, B (1999) Development of a molecular marker for the adult plant leaf rust resistance gene *Lr*35 in wheat. *Theoretical and Applied Genetics* **99**: 554-560.

Shah, DM (1997). Genetic engineering for fungal and bacterial diseases. *Current Opinion in Biotechnology* **8**: 208-214.

Shaner, G and Finney, RE (1980) New sources of slow leaf rusting resistance in wheat. *Phytopathology* **70**: 1183-1186.

Shaner, G; Roberts, JJ and Finney, RE (1972) A culture of *Puccinia recondita* virulent to the wheat cultivar Transfer. *Plant Dis. Reorter*. **56**: 827-830.

Sharma, AK (2000) Resistance breeding for new virulent pathotypes of rust. Proc. 39th All India Wheat Workers' Workshop. Aug. 27-30, 2000, BHU, Varanasi, pp1-7.

Sharma, AK; Arun, K; Kumar, J and Nagarajan, S (1998) Worldwide movement of smuts and Bunts. " Bunts and Smuts of wheat: An international Symposium at North American Plant Protection Organization, Ottawa (eds). Malik, VS and Mathre, DE. pp129-136.

Sharma, AK; Kumar, J and Nagarajan, S (1996) Reports of the coordinated Experiments 1995-96: Crop Protection (Pathology and Nematology). All India Coordinated Wheat Improvement Project, Directorate of Wheat Research, Karnal, India, pp. 185.

Sharma, AK; Kumar, J and Nagarajan, S (1998) Disease management strategy in wheat. *Indian Farming* **48**: 51-53 and 64-69.

Sharma, AK; Singh, DP and Singh, AK (1998) Epidemiological studies in biological control of plant pathogens. " Biological suppression of Plant Diseases Phytoparasitic Nematodes and weeds" Eds. S.P. Singh and S.S. Hussaini, Project Directorate of Biological Control, Bangalore pp.110-117.

Sharma, HC; Sharma, KK; Seetharama, N and Ortiz, R (2000) Prospects for using transgenic resistance to insects in crop improvement. *EJB Electronic Journal of Biotechnology* [online].

Sharma, RC; Dubin, HJ; Bhatta, MR and Devkota, RN (1997) Selection for spot blotch resistance in four spring wheat populations. *Crop Sci* **37**: 432-435.

Sharma, RC; Dubin, HJ; Devokota, RN and Bhatta, MR (1997) Heritability estimates of field resistance to spot blotch in four spring wheat crosses. *Plant Breeding* **116**: 64–68.

Sharma, RC; Duveiller, E and Jacquemin, JM (2007) Microsatellite markers associated with spot blotch resistance in spring wheat. *J Phytopathol* **155**: 316–319.

Sharma, SK; Singh, S and Goel, LB (1973) Note on a new record of Sonalika infecting race of yellow rust in India and sources of its resistance. *Indian J. agric. Sci.* **43**: 964-965.

Shi, AN; Leath, S and Murphy, JP (1998) A major gene for powdery mildew resistance transferred to common wheat from wild einkorn wheat. *Phytopathology* **88**: 144-147.

Sibikeev, SN; Krupnov, VA; Voronina, SA and Elisin, VA (1996) First report of leaf rust pathotype virulent to highly effective *Lr* genes transferred from *Agropyron* species to bread wheat. *Plant Breed.* **115**: 276-278.

Siddiq, EA and Sharma, NP (1995) Biotechnological approaches for disease resistance in crop plants. In: B. Sharma *et al.* (eds) Genetic research and education: current trends and the next fifty years. *Indian Society of Genetics and Plant Breeding*, New Delhi pp271-286.

Singh, DP (1998) Effect of gamma irradiation on seed germination and seed borne fungi. *Seed Tech News* **28** : 20.

Singh, DV; Joshi, LM and Srivastava KD (1986) Foliar blight and spot of wheat in India, *Indian J Genet* **46**: 217-245.

Singh, H and Johnson, R (1988) Genetics of yellow rust resistance in Heines VII, Soissonais and Kalyansona. Proc. 7th Intern Wheat Genet. Smp., IPSR, Cambridge, UK (TE Miller and RMD Koebner eds.), pp885-890.

Singh, PJ and Dhaliwal, HS (1993) Resistance to foliar blights of wheat in *Aegilops* and wild *Triticum* species. *Indian Phytopath* **46**: 246-248.

Singh, RP (1992a) Association between gene *Lr34* for leaf rust resistance and leaf tip necrosis in wheat. *Crop Sci.* **32**: 874–878.

Singh, RP (1992b) Genetic association of leaf rust resistance gene Lr34 with adult plant resistance to stripe rust in bread wheat. *Phytopathology*, **82**: 835-38.

Singh, RP and Rajaram, S (2002) Breeding for disease resistance in wheat. In *Bread Wheat Improvement and Production*, edited by B.C. Curtis, S. Rajaram and H. Gomez-Macpherson, FAO, Rome, pp141–156.

Singh, RP; Hodson, DP; Huerta-Espino, J; Jin, Y; Bhavani, S; Njau, P; Herrera-Foessel, S; Singh, PK; Singh, S and Govindan, V (2011) The emergence of *Ug99* races of the stem rust fungus is a threat to world wheat production. *Annu Rev Phytopathol* **49**: 465–481.

Singh, RP; Hodson, DP; Jin, Y; Huerta-Espino, J; Kinyua, MG; Wanyera, R; Njau, P and Ward, RW (2006) Current status, likely migration and strategies to mitigate the threat to wheat production from race *Ug99* (TTKS) of stem rust pathogen. *CAB Reviews* **1**: 54.

Singh, RP; Hodson, DP; Jin, Y; Lagudah, ES; Ayliffe, MA; Bhavani, S; Rouse, MN; Pretorius, ZA; Szabo, LJ; Huerta-Espino, J; Basnet, BR; Lan, C and Hovmøller,

MS (2015) Emergence and spread of new races of wheat stem rust fungus: continued threat to food security and prospects of genetic control. *Phytopathology* **105**: 872–884.

Singh, RP; Kazi-Mujeeb, A and Huerta-Espino, J (1998b) *Lr46*: A gene conferring slow rusting resistance to leaf rust in wheat. *Phytopathology*. **88**: 890–894.

Singh, RP; Ma, H and Rajaram, S (1995) Genetic analysis of resistance to scab in spring wheat cultivar Frontana. *Plant Dis.* **79**: 238–240.

Singh, RP; Mujeeb-Kazi, A and Huerta-Espino, J (1998) *Lr46*: a gene conferring slow rusting resistance to leaf rust in wheat. *Phytopathology*, **88**: 890–894.

Singh, RP; Nelson, JC and Sorrells, ME (2000) Mapping *Yr28* and other genes for resistance to stripe rust in wheat. *Crop Sci.* **40**: 1148-1155.

Singh, RP; Rajaram, S; Miranda, A; Huerta-Espino, J and Autrique, E (1998c) Comparison of two crossing and four selection schemes for yield, yield traits, and slow rusting resistance to leaf rust in wheat. *Euphytica* **100**: 35–43.

Singh, S; Goel, LB; Nayar, SK; Sharma, SK and Chatterjee, SC (1979) Prevalence and distribution of physiologic races of wheat rusts in India during 1976–78 crop seasons. *Indian Phytopathol.* **32**: 417–420.

Singh, SJ and McIntosh, RA (1990) Linkage and expression of genes for resistance to leaf rust and stem rust in triticale. *Genome* 33: 115–118.

Sipes, BS; Schmitt, DP and Barker, KR (1992) Fertility of three parasitic biotypes of *Heterodera glycines. Phytopathology*. **82**(10): 999-1001.

Sivasamy, M; Vinod, Tiwari, S; Tomar, RS; Singh, B; Sharma, JB; Tomar, SMS and Chand, S (2009) Introgression of useful linked genes for resistance to stem rust, leaf rust and powdery mildew and their molecular validation in wheat. *Indian J. Genet*. **69**: 17-27.

Somers, DJ; McCartney, CA and DePauw, RM (2007) Molecular breeding for multiple pest resistance in wheat. *Dev Plant Breed* **12**: 667–676.

Sorty, AM; Meena, KK and Choudhary, K (2016) Effect of plant growth promoting bacteria associated with halophytic weed (*Psoralea corylifolia* L.) on germination and seedling growth of wheat under saline conditions. *Appl Biochem Biotechnol* **180**: 872–882.

Sprague, R (1950) Diseases of Cereals and Grasses in North America. Ronald Press, New York.

Srivastava, OP (1982) Genetics of resistance to leaf blights of wheat. *Indian J Genet* **42**: 140-141.

Srivastava, OP; Luthra, JK and Narula, PN (1971) Inheritance of seedling resistance to leaf blight of wheat. *Indian J Genet Plant Breed* **45**: 941-943.

Stakman, EC and Piemeisel, FJ (1917) A new strain of *Puccinia graminis. Phytopathology* **7**: 73 (Abstr.).

Stakman, EC; Stewart, DM and Loegering, WQ (1962) Identification of physiologic races of *Puccinia graminis* var *tritici*. *US Dep Agric, Agric Res Serv* **E617**: 53.

Stoger, E; Williams, S; Christou, P; Down, RE and Gatehouse, JA (1999) Expression of the insecticidal lectin from snowdrop (*Galanthus nivalis* agglutinin; GNA) in transgenic wheat plants: Effect on predation by the grain aphid *Sitobion avenae*. *Molecular Breeding* **5**: 65-73.

Suenaga, K; Singh, RP and Manilal, HM (2001) Tagging of slow rusting genes for leaf rust, *Lr34* and *Lr46*, using microsatellite markers in wheat. *JIRCAS Research Highlights*, pp8-9.

Sun, GL; Fahima, T; Korol, AB; Turpeinen, T; Grama, A; Ronin, YI and Nevo, E (1997) Identification of molecular markers linked to the *Yr15* stripe rust resistance gene of wheat originated in wild emmer wheat, *Triticum dicoccoides*. *Theor Appl Genet* **95**: 622-628.

Talbert, LE; Bruckner, PL; Smith, LY; Sears, R and Martin, TJ (1996) Development of PCR markers linked to resistance to wheat streak mosaic virus in wheat. *Theor Appl Genet* **93**: 463-467.

Tao, WJ; Liu, JY; Liu, DJ and Chen, PD (1999) Identification of molecular markers linked to the *Pm6* gene conferring powdery mildew resistance in wheat. I *Chuan Hsueh Pao (in Chinese)* **26**: 649-656.

Tomar, SMS; Singh, SK; Sivasamy, M and Vinod (2014) Wheat rusts in India: resistance breeding and gene deployment–a review. *Indian J Genet Plant Breed* **74**: 129–156.

Tomor, SMS and Menon, MK (2001) *In Genes for Resistance to Rusts and Powdery Mildew in Wheat*, Indian Agricultural Research Institute, New Delhi, p. 152.

Tsunewaki, R (1966) Comparative gene analysis of common wheat and its ancestral species. II. Waxiness, growth habit and awnedness. *Jap. J. Bot.* **19**: 175–254.

Tulloch, AP and Hoffman, LL (1974) Epicuticular waxes of *Secale cereale* and *Triticale hexaploide* leaves. *Phytochemistry* **13**: 2535–2540.

Tyagi, S; Mir, RR; Kaur, H; Chhuneja, P; Ramesh, B; Balyan, HS and Gupta PK (2014) Marker-assisted pyramiding of eight QTLs/genes for seven different traits in common wheat (*Triticum aestivum* L.). *Mol Breed* **34**: 167–175.

van der Linden, CG; Wouters, DCAE; Mihalka, V; Kochieva, EZ; Smulders, MJM and Vosmanm, B (2004) Efficient targeting of plant disease resistance loci using NBS profiling. Theor. Appl. Genet. ***109****: 384–393.*

Van der Plank, JE (1978) Genetics and Molecular Basis of Plant Pathogenesis. Springer-Verlag, Berlin, 167 pp.

van Ginkel, M and Rajaram, S (1993) Breeding for durable disease resistance in wheat: an international perspective. *In* T. Jacobs and J.E. Parlevliet, eds. Durability of disease resistance, Dordrecht, Netherlands, Kluwer Academic Press, p. 259-272.

Van Ginkel, M and Rajaram, S (1998) Breeding for resistance to spot blotch in wheat: Global perspective. In: Duveiller E, Dubin HJ, Reeves J, McNab A (eds), *Helminthosophism* Diseases of Wheat: Spot Blotch and Tan Spot. CIMMYT, Mexico, pp. 162-170.

Vanderplank, JE (1968) Disease Resistance in Plants. Academic Press, New York and London.

Varma, PK; Ruwali, KN; Mishra, AN and Pandey, HN (2001) Nodal roots in wheat. In: Bora, K.K., Karan Singh, and Arvind Kumar eds. 2001. Stress and Environmental Plant Physiology. Pointer Publishers, Jaipur, India, pp161-166.

Vasudeva, RS; Prasada, R; Lele, VC; Joshi, LM and Pal, BP (1962) Rust-resistant Varieties of Wheat in India. Indian Council of Agricultural Research, New Delhi.

Venter, E and Botha, AM (2000) Development of markers linked to *Diuraphis noxia* resistance in wheat using a novel PCR-RFLP approach. *Theoretical and Applied Genetics* **100**: 965-970.

Verma, DK; Sai Prasad, SV; Prabhu, KV; Sinha, AK and Chowdhury, S (2019) Integrated Disease Management of Bread Wheat in Bihar, India: Retrospects and Prospects.

Vilareal, RL; Banuelos, O and Muzeeb-Kazi, A (1997) Agronomic performance of released durum wheat (*Triticum turgidum* L.). Stocks possessing the chromosome substitution T1BL.1RS. *Crop Sci.* **37**: 1735-1740.

Villareal, RL; Mujeeb-Kazi, A; Gilchrist, L and Del Toro, E (1995) Yield loss to spot blotch in spring bread wheat in warm non-traditional wheat production areas. *Plant Dis.* **79**: 893–897.

Villareal, RL; Rajaram, S; Gilchrist, L and Mujeeb-Kazi, A (1992) Advanced lines derived from wheat (*Triticum aestivum* L.) and *Thinopyrum currifolium* resistant to *Helminthosporium sativum* In: Abstracts of International Crop Science Congress, Ames, IA, USA, Crop Sci Society of America/Iowa State Univ.

Virmani, SM (1994) Climate resource characterization in stressed tropical environment: Constraints and opportunities for sustainable agriculture. Oxford and IBH Publishing Co. Pvt. Ltd., New Delhi, pp149-160.

Wessels, E and Botes, WC (2014) Accelerating resistance breeding in wheat by integrating marker assisted selection and doubled haploid technology. *South Afr J Plant Soil* **31**(1): 35–43.

Wiese, MV (1987) Compendium of wheat diseases 2nd ed. St. Paul, MN, USA, APS Press. pp112.

William, HM; Hoisington, D; Singh, RP and Gonzalez-de-Leon, D (1997) Detection of quantitative trait loci associated with leaf rust resistance in bread wheat. *Genome* **40**: 253-260.

William, HM; Singh, RP; Huerta-Espino, J; Ortiz-Islas, S and Hoisington, D (2003) Molecular marker mapping of leaf rust resistance gene Lr46 and its association with stripe rust resistance genes Yr29 in wheat. *Phytopathol.* **93**: 153-159.

Williams, KJ; Fisher, JM and Langridge, P (1994) Identification of RFLP markers linked to the cereal cyst nematode resistance gene (*Cre*) in wheat. *Theor Appl Genet* **83**: 919-924.

Williams, KJ; Fisher, JM and Langridge, P (1996) Development of a PCR-based allele-specific assay from an RFLP probe linked to resistance to cereal cyst nematode in wheat. *Genome* **39**: 798-801.

Williams, KJ; Taylor, SP; Bogacki, P; Pallotta, M; Bariana, HS and Wallwork, H (2002) Mapping of the root lesion nematode (*Pratylenchus neglectus*) resistance gene Rlnn1 in wheat. *Theoretical and Applied Genetics* **104**: 874-879.

Williams, ND; Miller, JD and Klindworth, DL (1992) Induced mutations of a genetic suppressor of resistance to wheat stem rust. *Crop Sci.* **32**: 612–616.

Winzeler, M; Winzeler, H and Keller, B (1995) Endopeptidase polymorphism and linkage of the *Ep-Dlc* null allele with the *Lr19* leaf rust resistance gene in hexaploid wheat. *Plant Breed.,* **114**: 24-28.

Wood, LS (1962) Relation of variation in *Helminthosporium sativum* in seeding blight of small grains. *Phytopath* 52: 493-498.

Xu, ZS; Chen, M; Li, LC and Ma, YZ (2011) Functions and application of the AP2/ ERF transcription factor family in crop improvement. *Journal of Integrative Plant Biology*, **53**: 570-585.

Yamamori, M (1994) An N-band marker for gene *Lr18* for resistance to leaf rust in wheat. *Theor Appl Genet* **89**: 643-646.

Yousfi, BE and Ezzahiri, B (2001) Net blotch in semi-arid regions of Morocco l: Epidemiology. *Field Crops Research* **73**: 35–46.

Yousfi, BE and Ezzahiri, B (2002) Net blotch in semi-arid regions of Morocco ll: Yield and yield-loss modelling. *Field Crops Research* **73**: 81–93.

Zeller, FJ (1973) 1B/1R wheat-rye chromosome substitutions and translocations. In Proc.4th Intl. Wheat Genetics Symp. (Eds ER Sears and LMS Sears.) Agricultural Experiment Station, University of Missouri: Columbia, Missouri, USA, pp. 209-221.

Zwart, RS; Thompson, JP and Sheedy, JG and Nelson, JC (2006) Mapping quantitative trait loci for resistance to *Pratylenchus thornei* from synthetic hexaploid wheat in the International Triticeae Mapping Initiative (ITMI) population. *Aust. J. Agr. Res.* **57**: 525–530.

Chapter 8

Breeding Wheat for Quality and Bio-fortification/Micronutrient-Dense Wheat

Outlines

1. Quality Traits
2. Quality Analysis in Cereals
3. Composition and Bioactive Components of Ancient and Modern Wheats
4. Value Added Products of Wheat
5. Wheat Quality Parameters
    a. Market Quality
    b. Milling and Baking Quality
    c. Chapati Making Quality
    d. Industrial Quality
    e. Test Weight or Hectolitre Weight or Weight Per Unit Volume (kg)
    f. Kernel Weight (Thousand Grain Weight) (g)
    g. Kernel Size and Shape
    h. Yellow Berry
    i. Grain Analyser
    j. Grain Appearance Score (Max. 10)
    k. Whole Meal or Flour Sample
    l. Sedimentation Test (SDS Value)
    m. Pelshenke Test (Wheat)

n. Starch and Damaged Starch (AACC 2000)
o. Enzymes
p. Quality Rheological Characteristics
q. Farinograph Method
r. Alveograph Method
s. Amylograph
t. Alveograph
u. Milling Characteristics of Wheat
v. Semolina Yield for Triticum durum wheats, Flour Milling for T. aestivum wheat
w. Spaghetti and Macaroni Properties
x. *Colours and* Provitamin A carotenoids (β-*Carotene) Content*
y. Hardness Index
z. Protein Estimation by Kjeldahl Method

6. Nutritional Quality
   a. Protein Content
   b. Lysine Content
   c. Physico- Chemical Tests
   d. Moisture content (AACC 2000)
   e. Moisture Content in Grain/Flour
   f. Pearling Index (Peter Scott Pearler)
   g. Near-Infrared Reflectance Spectrophotometry
   h. Ogawa Seiki Hardness Tester for single grain and measured as Kg/seed
   i. Visual Screening of Provitamin A Carotenoids
   j. Benzene Sedimentation Test
7. Performance Tests
   a. Cooking Characteristics
   b. Cookies, Chapaties, Bread and Cake Baking; Cooking of End Products from Semolina viz., Noodles, Spaghetti
   c. Extruding Semolina for Pasta Manufacture
   d. Wheat Quality-Future Requirements
8. Breeding for Quality Traits
   a. Breeding for Industrial and Nutritional Quality
   b. Introgression for Grain Yield and Quality Traits from Tetraploid Triticum Species into Triticum durum
   c. Alien introgression as a means to improve yield and quality
   d. Quality Status of Durum Wheat Lines for Evolving Pasta Varieties for Central India

9. Quality Breeding - Problems and Prospects
    a. Genetic Studies
    b. Prospects for Quality Improvement
    c. Quality Evaluation Programme
    d. Marker Assisted Selection for Quality Traits in Wheat
    e. Testing of Chapatti and Bread Making Quality in Wheat
10. Malnutrition - Matter of Concern
    a. Nutritional Improvement – A Human Perspective
    b. Bio-fortification
    c. Wheat and Barley Grain Biofortification
    d. Breeding for Biofortification/Micronutrient-Dense Crops:
    e. Micronutrient Analysis
    f. Contamination in Mineral Analyses
    g. Micronutrient Concentration Versus Content
    h. Estimation of Iron and Zinc by Inductively Coupled Plasma-Optical Emission Spectrometer (ICP-OES)
    i. Genetic Variation–Germplasm Screening
    j. Setting Nutritional Target Levels
    k. Breeding for Increased Bioavailability
    l. Effect of Zn and S application on the productivity of wheat varieties
    m. Popularisation of the CVRC released high yellow pigment variety HI 8663 (Poshan)
11. Problems and Prospects for Quality Improvement
    a. Generation of Material
    b. Success Stories of Food Fortification
    c. Limitations in Quality Breeding
    d. Genetic Control of Quality Breeding
    e. Wide Hybridization for Quality
    f. Biotechnological Approaches for Crop Quality
    g. Enhancement of Iron Content
    h. Chromosomal Location of Genes Controlling Wheat Endosperm Proteins
    i. Location of Genes Including Gliadins in Wheat
    j. Second Generation Transgenics for Quality Improvement
    k. Engineering Fatty Acid Metabolism
    l. A stable high yielding variety HI 8691 *'Pusa Gaurav'* with good content of iron and zinc

Figure 8.1. Biscuit goes best with tea

Figure 8.2. Loaved bread ready to eat

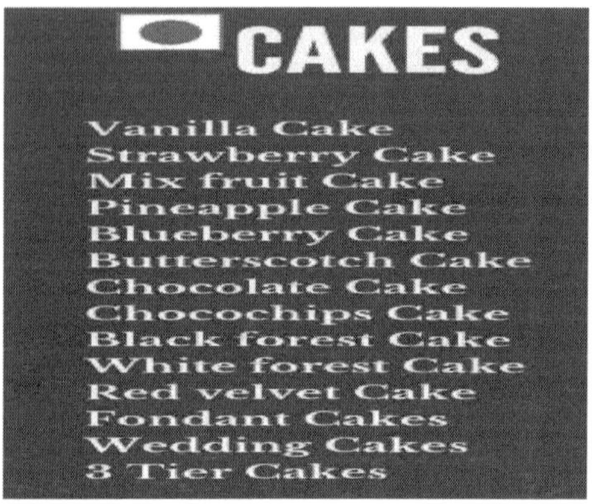

Figure 8.3. Cakes

Figure 8.4. Pastry

Figure 8.5. Yellow berry in wheat

Figure 8.6. Yellow berry affected grains

Figure 8.7. Automatic protein analyzer

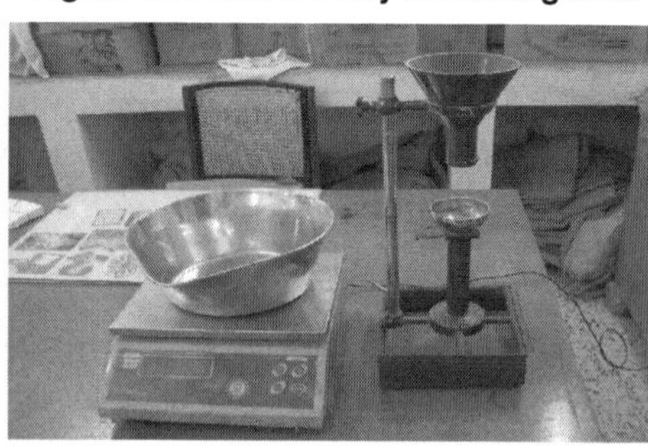

Figure 8.8. Hectolitre machine (kg)

Figure 8.9. Cyclotec Micro Mill (0.5 mm)

Figure 8.10. Sedimentation value (ml)

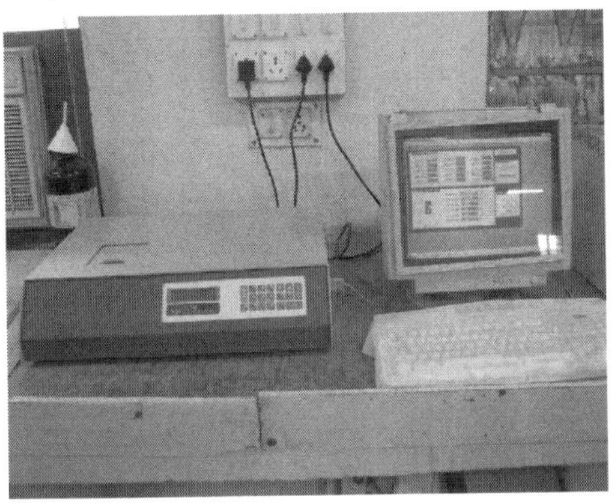

Figure 8.11. Spectrophotometer for -carotene estimation (ppm)

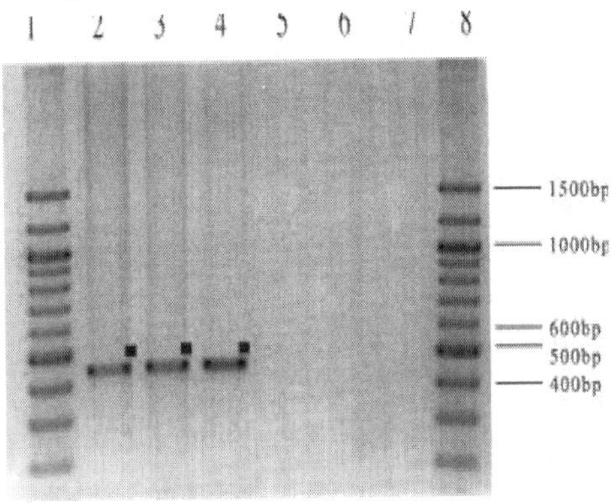

Figure 8.12. Primers (P1 and P2) specific for the 1Dx5 allele were used for PCR reaction on 50 ng of genomic DNA from five bread wheat cultivars

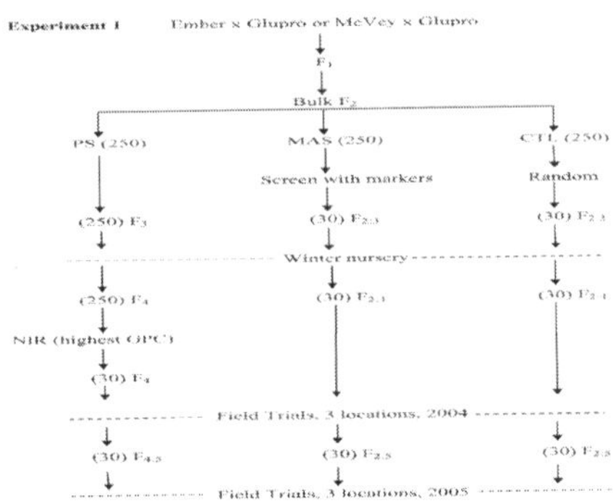

Figure 8.13. Flowchart showing population development and line selection schemes (Adapted from Davies et al. 2006)

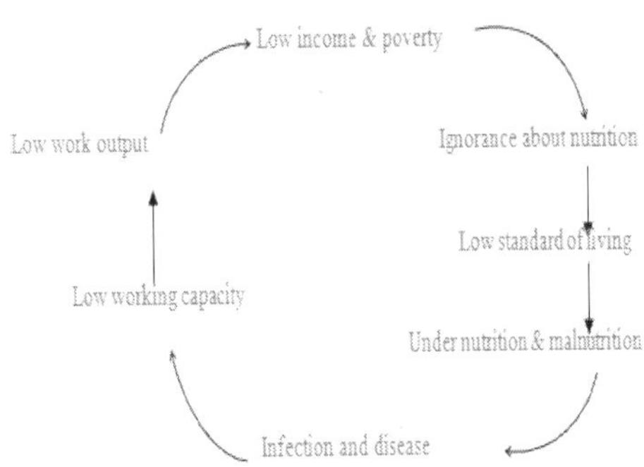

Figure 8.14. Vicious cycle of malnutrition

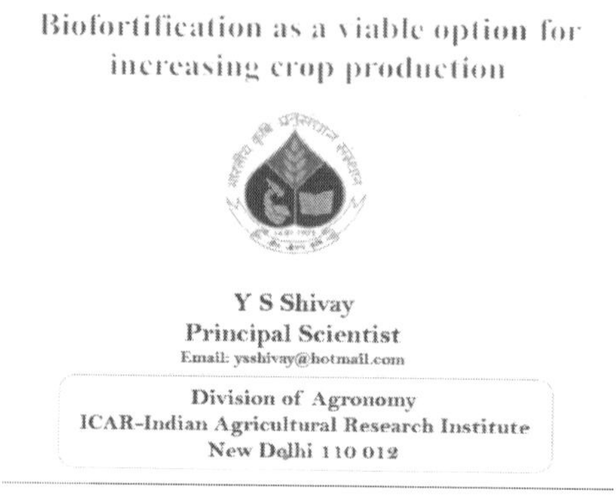

Figure 8.15. Biofortification as a viable option for increasing crop production

Figure 8.16. Digital Release of Commemorative Coin on the occassion of 75 years of FAO dt 16th Oct, 2020

Figure 8.17. Smt Jayshri Govind Jadhav, a woman farmer, ensuring the nutritional security for her family through the cultivation of biofortified wheat variety MACS 4028, in Morgaon, District Pune, Maharashtra

Figure 8.18. Wheat and barley grain biofortification

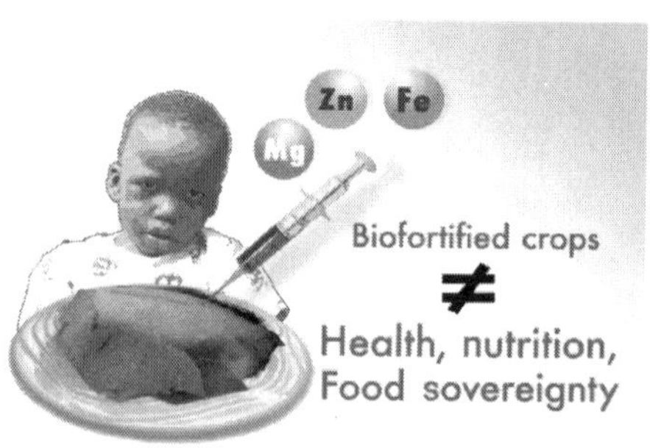

Figure 8.19. Biofortified crops: Health, nutrition, food sovereignty

Figure 8.20. Webinar, entitled, Improving Nutrition Security with Biofortified Crops

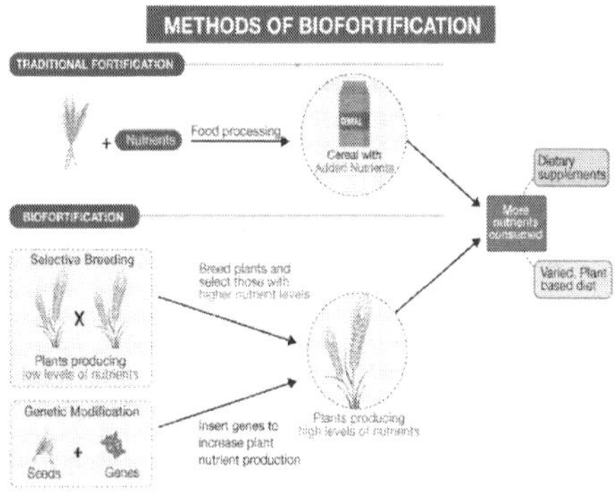

Figure 8.21. Methods of biofortification

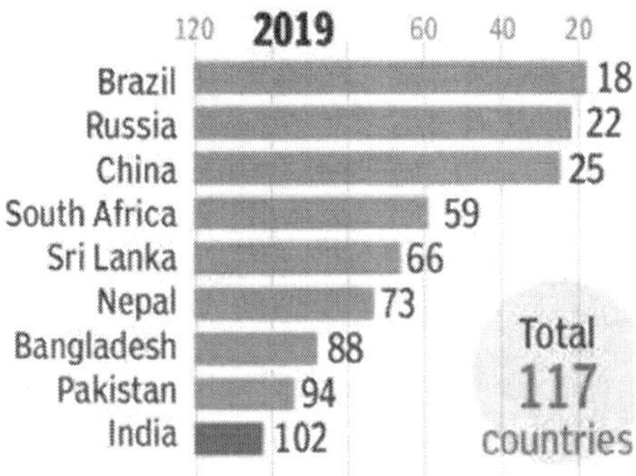

Figure 8.22. Global Hunger Index (GHI)

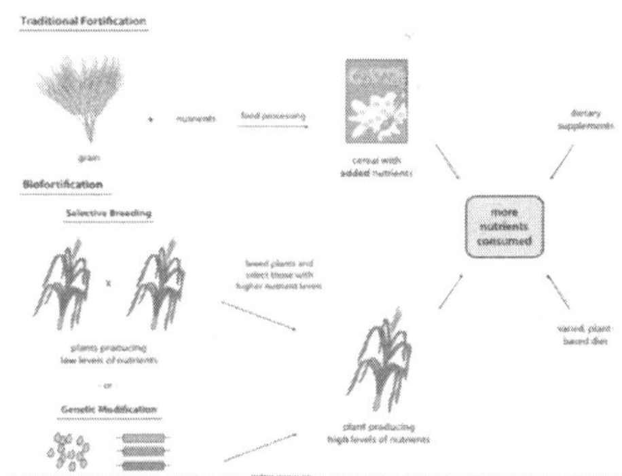

Figure 8.23. Biofortification vs Fortification

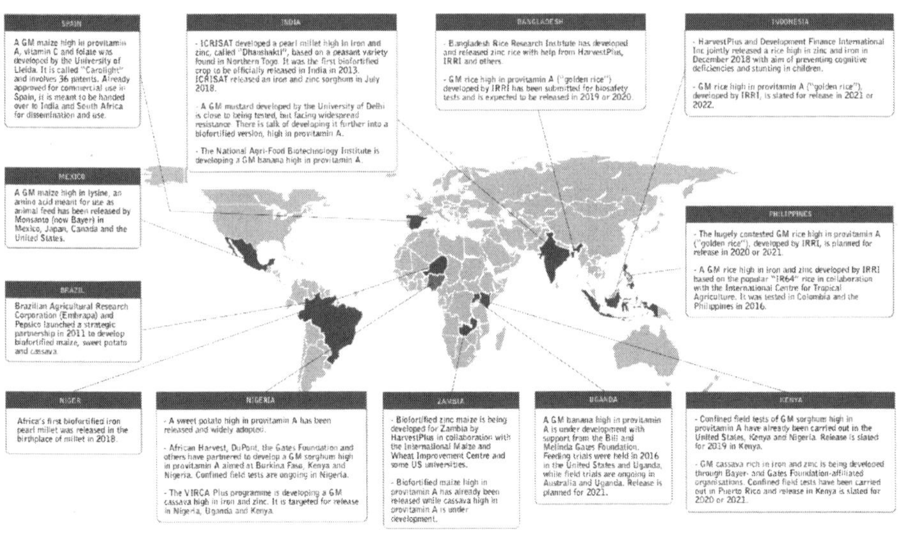

Figure 8.24. Some key biofortified crop research projects

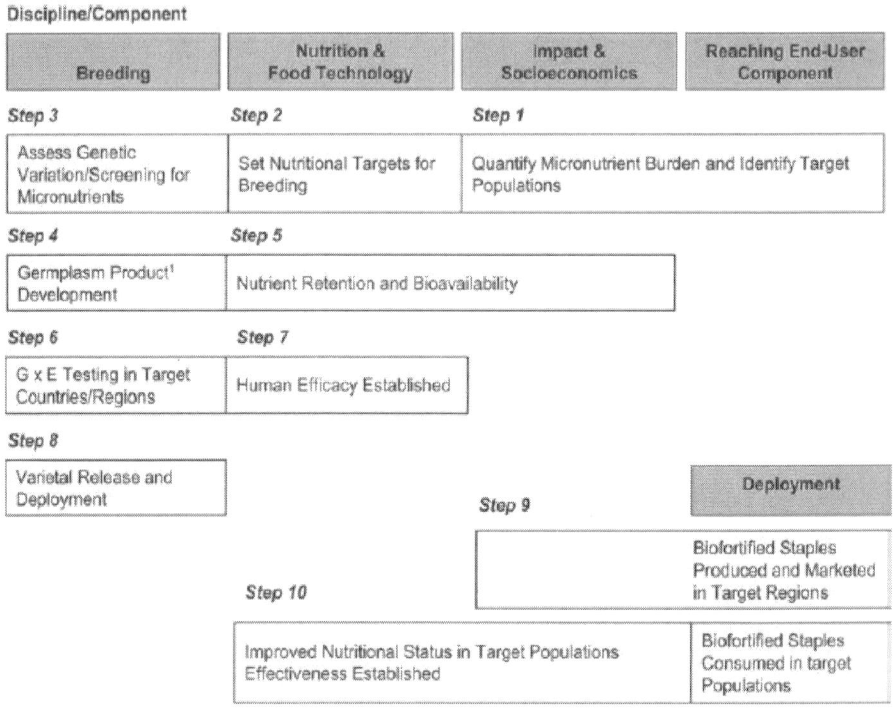

[1] Includes nutrition genomics.
Source: HarvetsPlus Impact Pathway, 2006.

Fig. 8.25. HarvestPlus Impact Pathway

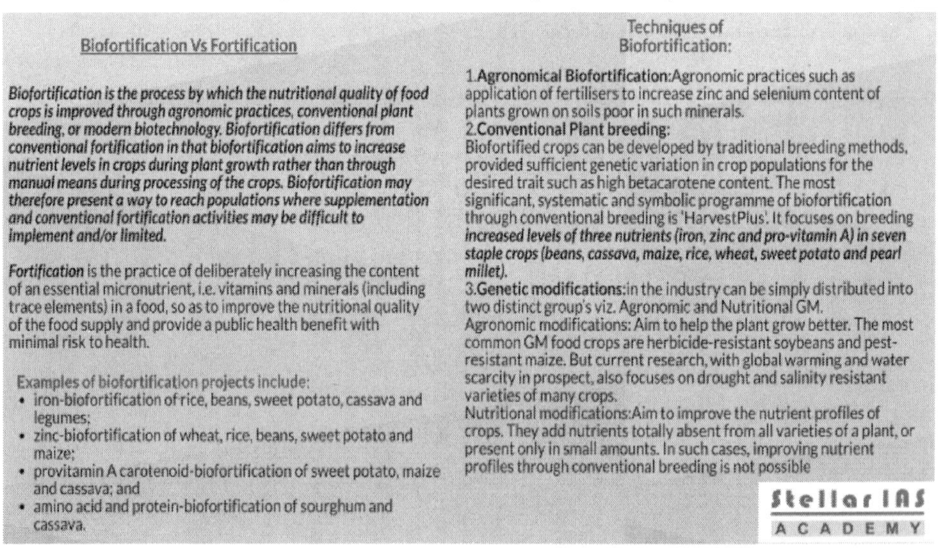

Biofortification Vs Fortification

Biofortification is the process by which the nutritional quality of food crops is improved through agronomic practices, conventional plant breeding, or modern biotechnology. Biofortification differs from conventional fortification in that biofortification aims to increase nutrient levels in crops during plant growth rather than through manual means during processing of the crops. Biofortification may therefore present a way to reach populations where supplementation and conventional fortification activities may be difficult to implement and/or limited.

Fortification is the practice of deliberately increasing the content of an essential micronutrient, i.e. vitamins and minerals (including trace elements) in a food, so as to improve the nutritional quality of the food supply and provide a public health benefit with minimal risk to health.

Examples of biofortification projects include:

- iron-biofortification of rice, beans, sweet potato, cassava and legumes;
- zinc-biofortification of wheat, rice, beans, sweet potato and maize;
- provitamin A carotenoid-biofortification of sweet potato, maize and cassava; and
- amino acid and protein-biofortification of sourghum and cassava.

Techniques of Biofortification:

1.**Agronomical Biofortification:**Agronomic practices such as application of fertilisers to increase zinc and selenium content of plants grown on soils poor in such minerals.
2.**Conventional Plant breeding:**
Biofortified crops can be developed by traditional breeding methods, provided sufficient genetic variation in crop populations for the desired trait such as high betacarotene content. The most significant, systematic and symbolic programme of biofortification through conventional breeding is 'HarvestPlus'. It focuses on breeding ***increased levels of three nutrients (iron, zinc and pro-vitamin A) in seven staple crops (beans, cassava, maize, rice, wheat, sweet potato and pearl millet).***
3.**Genetic modifications:**in the industry can be simply distributed into two distinct group's viz. Agronomic and Nutritional GM.
Agronomic modifications: Aim to help the plant grow better. The most common GM food crops are herbicide-resistant soybeans and pest-resistant maize. But current research, with global warming and water scarcity in prospect, also focuses on drought and salinity resistant varieties of many crops.
Nutritional modifications:Aim to improve the nutrient profiles of crops. They add nutrients totally absent from all varieties of a plant, or present only in small amounts. In such cases, improving nutrient profiles through conventional breeding is not possible

Stellar IAS ACADEMY

Figure 8.26. Biofortification to fight malnutrition

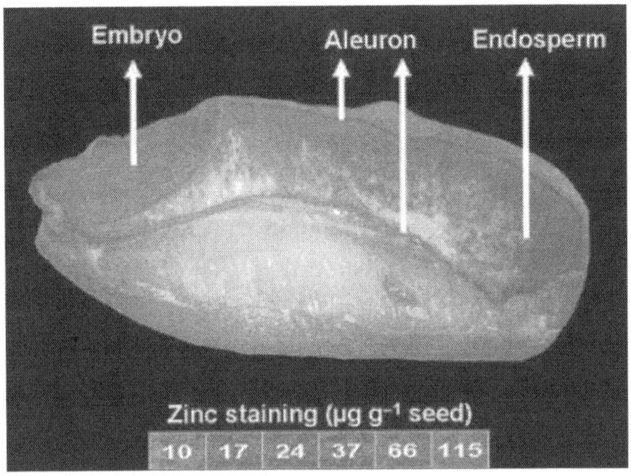

Figure 8.27. Localization and staining of zinc in wheat seed (*Triticum aestivum* L., cv. Balatilla). Seed was stained with Dithizone (DTZ) 52 days after anthesis and Zn content analyzed with ICP-OES (*Source*: Ozturk *et al.*, 2006).

Figure 8.28. HI 8691 'Pusa Gaurav'

Figure 8.29. HI 8691 'Pusa Gaurav'

Wheat is world's most widely cultivated food crop. It is a rabi (winter) season crop. In India it is the second important staple cereal food. It is mostly eaten in the form of chapaties. Wheat is also used for manufacturing bread, flakes, cakes, biscuits *etc.* Wheat straw is a good source of feed for cattle. Wheat contains more protein (8–15 per cent) than in other cereals. Wheat proteins are of special significance. The protein contained in wheat includes albumins, globulins, glutinous and gliadines. Albumins dissolve in water. The other protein forms are insoluble in water and are called gluten. The gluten content in wheat is the highest (16–50 per cent). Because of gluten, wheat flour is used for baking bread. The crop plants which belong to the family graminae and are grown for their edible starchy grains/seed called caryopsis (seed coat + pericarp are fused or united) are called as cereals. The word cereals has been derived from 'Ceres', name of a Roman Goddess, means '**Giver of Grains**'. *Cereals* are cultivated grasses grown for their edible starchy grains (one seeded fruit–caryopsis). Larger grains used as staple food are cereals–rice, wheat, maize, barley, oats *etc.* The word cereal was derived from the word *ceres*, which denotes a goddess who was believed as the giver of grains by Romans. *e.g.* rice, wheat.

Quality Traits

The quality traits are divided in two as (i) evident quality traits and (ii) crytpic quality traits. The grains' evident traits *viz.*, colour, size, shape and hardness of grain determines the consumer preference, marker price and acceptability of food products. The grain constitutes 8 per cent pericarp, 17 per cent germ and 75 per cent endosperm. The colour of the grains depends upon the endosperm and or pericarp colour.

Table 8.1. Nutrient composition of wheat, cereals and small millets (per 100g)

Food Grain	*Protein (g)*	*Carbohy-drates (g)*	*Fat (g)*	*Crude Fibre (g)*	*Mineral Matter (g)*	*Calcium (mg)*	*Phos-phorous (mg)*	*Fe (mg)*
Cereals								
Wheat	11.8	71.2	1.5	1.2	1.5	41	306	5.3
Rice	6.8	78.2	0.5	0.2	0.6	45	160	
Small millet								
Finger millet	7.3	72.0	1.3	3.6	2.7	344	283	3.9
Kodo millet	8.3	65.0	1.4	9.0	2.6	27	188	12.0
Proso millet	12.5	70.4	3.1	7.2	1.9	14	206	10.0
Foxtail millet	12.3	60.9	4.3	8.0	3.3	31	290	5.0
Little millet	7.7	67.0	4.7	7.6	1.5	17	220	6.0
Barnyard millet	6.2	65.5	2.2	9.8	4.4	11	280	15.0

Table 8.2. Essential amino acids contents of small millets and cereals (g/100 g of Proteins)

Amino Acids	*Cereals*		*Finger Millet*	*Kodo Millet*	*Proso Millet*	*Foxtail Millet*	*Barnyard Millet*
	Wheat	*Rice*					
Isoleucine	3.3	3.8	4.4	3.0	8.1	7.6	8.8
Leucine	6.7	8.2	9.5	6.7	12.2	16.7	16.6
Lysine	2.8	3.8	2.9	3.0	3.0	2.2	2.9
Methionine	1.5	2.3	3.1	1.5	2.6	2.8	1.9
Cystine	2.2	1.4	2.2	2.6	1.0	1.6	2.8
Phenylalanine	4.5	5.2	5.2	6.0	4.9	6.7	2.2
Tyrosine	3.0	3.9	3.6	3.5	4.0	2.2	2.4
Threonine	2.8	4.1	3.8	3.2	3.0	2.7	2.2
Tryptophan	1.5	1.4	1.6	0.8	0.8	1.0	1.0
Valine	4.4	5.5	6.6	3.8	6.5	6.9	6.4
Histidine	2.3	2.4	2.2	1.5	1.9	2.1	1.9

Quality Analysis in Cereals

The cereal grains are seeds of the grass family. The principal cereal crops are rice, wheat, maize or corn, jowar, ragi and bajra and millets like samai, varagu *etc.* Percentage composition and nutritive value of different cereals. Most of the population in the world derives its nutrition in terms of protein from cereals. Wheat like other cereals is deficient in tryptophane and other sulphur containing amino acids, hence diverse attempts are made to improve the nutritional quality of cereals by increasing both protein and its quality. The most important food use of wheat is in the main feature of flour for making bread, biscuit and pastry products. Small amounts of wheat are consumed into breakfast cereals liken porridge, flaked products, puffed products, paste products, cake mixes, chapaties, starch, vital gluten, bulger *etc.* Snacks foods like pretzel are also prepared from wheat. Wheat is the king of cereals and its quality means different to different people. The cereals specially wheat and maize starch is used in the paper industries and of course, oil welling, as laundry starches. The residue of wheat as poultry feed in the form of midlings. The deficiencies in essential amino acids in cereals (wheat and maize) are supplemented by each other. Today, world has become self-sufficient in most of the cereals and other crops. The basic definition of quality usually varies from one class of use to another. Farmers are interested in healthy growth of food grain of no disease, miller is concerned with the maximum recovery of milled products (flour, maida, bran, midlinges), baker is interested in getting maximum amount of milled products reduced into baked products like chapaties, biscuits. The ultimate utilizer of cereals is the consumer. The consumer is interested in the quality of the end product and at a reasonable price. A laboratory is required for testing of raw material and its finished products to check the quality of the intermediate products according to the PFA rules which are obligatory. A laboratory is an asset for testing of breeding materials, in milled flour quality and of the finished products. The main emphasis is laid on the quality tests of wheat, maize, barley, pearl millets and of course pulses. The quality testing facilities in the laboratories were further

Table 8.3. Nutritive value of cereals (Per 100 g)

Food	*Energy K. cal*	*Protein g*	*Carbo-hydrates g*	*Calcium mg*	*Thiamine mg*	*Riboflavin mg*	*Niacin mg*
Wheat Flour (whole)	341	12.1	69.4	48	0.49	0.17	4.3
Bajra	361	11.6	67.5	42	0.33	0.25	2.3
Jowar	349	10.4	72.6	25	0.37	0.13	3.1
Maize, dry	342	11.1	66.2	10	0.42	0.10	1.8
Ragi	328	7.3	72.0	344	0.42	0.19	1.1
Rice, parboiled, milled	346	6.4	79.0	9	0.21	0.05	3.8
Rice, raw, milled	345	6.8	78.2	10	0.06	0.06	1.9

Source: Gopalan, C; Rama Sastri, BV and Balasubramanian, SC (1991) Nutritive value of Indian foods, National Institute of Nutrition ICMR Hyderabad.

strengthened with coming into operation of a scheme. The laboratories are now well equipped are capable of carrying out detailed quality analysis of breeding materials for compositional, chapati and bread making properties. Some of the basic quality tests used in evaluating quality in breeding materials, milled and baked products with respect to wheat, barley *etc.* are common for all the cereal crops. In a breeding programme the number of samples is very large and these samples have to be tested in a limited time. In order to get the necessary information on their quality traits, a number of tests have to be limited particularly at the early stages of breeding programme. Among the preliminary tests, protein content, hardness and sedimentation value could be successfully employed on screening tools.

Composition and Bioactive Components of Ancient and Modern Wheats

The composition of wheat grains typically consists of 56–74 per cent available carbohydrates (mainly starch), 11–14 per cent water, 8–12 per cent protein, 2–13 per cent dietary fiber, 2–7 per cent lipids, and 1–3 per cent minerals. Important minor constituents include vitamins such as the B vitamins niacin (15–55 mg/kg), riboflavin (1–2 mg/kg), thiamin (4–7 mg/kg), and pantothenic acid (7–17 mg/kg) (Souci, *et al.*, 2016). Wheat bread provides about 34 per cent of the carbohydrates, 34 per cent

Table 8.4. Representative composition (per 100g) of whole grain flours of bread wheat, durum wheat, spelt wheat, emmer and einkorn

Component	*Bread Wheat*	*Durum Wheat*	*Spelt Wheat*	*Emmer Wheat*	*Einkorn Wheat*
Energy (Kcal)	340	339	324	362	312
Major constituents					
Protein (g)	13.2	13.7	14.7	12.7	12.5
Total lipids (g)	2.5	2.5	2.9	2.1	3.1
Carbohydrates	72.0	71.1	67.7	72.3	53.1
Total dietary fiber	10.7	NA	5.9	10.6	12.5
Minerals					
Calcium (mg)	34.0	34.0	27.0	NA	62.0
Iron (mg)	3.6	3.5	3.2	1.5	4.5
Magnesium (mg)	137.0	144.0	NA	128.0	125.0
Phosphorous (mg)	357.0	508.0	NA	NA	NA
Zinc (mg)	2.6	4.2	NA	4.8	4.7
Vitamins					
B1 (mg)	0.6	0.4	NA	NA	NA
B2 (mg)	0.2	0.1	NA	NA	NA
B3 (mg)	5.0	6.7	NA	8.5	4.4
B5 (mg)	0.6	0.9	NA	NA	NA
B6 (mg)	0.4	0.4	NA	NA	NA

a *Source*: USDA, ARS, Food Data central, 2019 (http: //fdc.nal.usda.gov). b NA: data not available.

Table 8.5. Quality parameters of wheat (aestivum as well as durum)

Varieties	Grain Yield (q/ha)	R	1000 gr. wt (g) Indore	Grain Appear. Score	Hecto. Weight (kg)	Protein (per cent)	SDS Value (ml)	β-carotene (ppm)	Hardness	Yellow Berry (per cent)	Yellow Berry (per cent) Indore
HI 8381	39.2		50.8	6.7	82.8	11.30	29	3.83	88	10.9	3.2
HI 8498	40.6		53.6	6.7	83.7	10.63	35	4.32	84	13.8	7.9
HI 8591	36.3		51.7	6.9	82.3	11.28	19	3.79	86	10.2	4.8
HI 8627	41.8	4	47.9	6.4	83.5	10.79	30	5.73	84	26.6	9.6
HI 8638	40.9		50.6	6.5	83.7	10.76	33	5.49	89	15.7	8.3
HI 8662	43.2	3	51.7	6.6	83.8	10.99	32	5.33	83	21.1	5.6
HI 8663	40.7		45.2	6.8	83.7	10.90	32	5.65	90	15.9	6.2
HI 8664	40.5		54.3	7.1	83.3	11.50	38	5.50	80	9.4	0.2
HI 8668	35.9		47.8	6.9	83.2	11.13	39	5.46	90	10.0	2.4
HI 8670	41.7	5	52.1	6.3	83.0	10.57	31	4.81	80	39.9	33.3
HI 8671	41.1		59.3	7.4	82.1	11.38	28	3.98	78	10.0	4.5
HI 8672	43.7	2	49.3	6.8	83.1	10.66	26	6.23	82	17.9	8.6
HI 8673	46.5	1	43.0	6.2	81.5	10.47	41	5.35	88	10.1	5.7
HD 4672	41.4	7	50.3	6.2	81.7	11.01	27	4.28	85	31.9	20.5
WH 896	40.4		47.7	6.8	84.0	10.91	37	6.33	86	11.0	18.5
PDW 233	38.5		47.9	7.1	83.7	10.76	34	7.77	92	11.2	2.5
NIDW 295	40.9		50.8	6.3	82.5	10.50	30	4.99	86	19.4	15.2
Raj 1555	41.5	6	51.6	6.4	83.0	10.57	34	4.05	78	16.0	5.7
Lok-1 ©	38.9	17	42.8	6.1	81.8	11.06	44	2.27	68	10.6	3.6
DL 788-2 ©	41.3	8	51.2	5.8	83.1	11.01	41	3.27	79	11.4	11.2
Mean	40.7		50.0	6.6	83.0	10.91	33	4.92	84	16.1	8.9
CD	5.09										
CV per cent	11.1										

of the protein, 24 per cent of the dietary fiber, and 13–32 per cent of the B vitamins and minerals of the mean recommended daily nutritional intake in Germany (38) and many other countries with similar dietary habits. Bioactive components in wheat include dietary fiber; phytochemicals such as carotenoids and polyphenols, including phenolic acids, flavonoids and lignans; and vitamins. Flours made from smaller grains typically have higher contents of crude fiber, minerals, and other components that are enriched in the bran. The composition of grain constituents is not only affected by genetic background, but also by environmental factors, such as cultivation region, climatic conditions, and agronomic practices, in particular the type and amount of nitrogen fertilization. Increased nitrogen application leads to higher protein contents but also causes changes in protein composition, with higher proportions of storage proteins gluten), especially gliadins, relative to nongluten proteins (Rossmann, *et al.*, 2019). Climatic conditions also impact grain composition, *e.g.*, with positive correlations between phytochemical contents and the mean temperature during grain development and with negative correlations between water-soluble dietary fiber contents and temperature. Arabinoxylan, tocol, sterol, and alkylresorcinol contents are mainly determined by genetic factors, whereas phenolic acid, B vitamin, and betaine contents are influenced more strongly by environmental factors than by genetic background (Shewry, *et al.*, 2010). The contents of major constituents, as well as minerals and vitamins, differ little between ancient wheats and modern common wheat and durum wheat.

India is expected to emerge as a small, but growing, consistent market for high-protein wheat imports for the growing bakery/confectionery industry and western-style fast food sector," the report said. "The rapidly growing fast food and modernizing bakery/confectionery industries generate demand for specialty flours (used in pizzas and burger buns) that require different wheat classes that are not produced locally. "Wheat is consumed at the household level, local restaurants, and eateries in the form of handmade breads called chapattis or rotis or parathas (unleavened flat bread) using atta (whole wheat flour) estimated at around 75 per cent to 80 per cent of total consumption. Some wheat is used for traditional processed products like raised breads, biscuits (cookies) and other bakery items (about 15 per cent)." The attaché described the organized milling sector as small with 1,250 to 1,300 medium to large flour mills with a total capacity of about 25 million tonnes a year (wheat equivalent) and put capacity utilization at 50 per cent to 60 per cent.

Value Added Products of Wheat

Whole Wheat Flour

It contains the finely ground bran, germ and endosperm of the whole kernel. Whole wheat products have a distinctive flavour and coarser texture than those made from white flour. Because of the higher fat content of the germ, whole wheat flour is more difficult to keep and sometimes becomes rancid in storage under poor conditions.

Maida

The bran and germ are separated in making white flour or maida. Maida bakes more uniformly into a loaf of a greater volume and it is more bland in taste and more easily digested. It can be stored in an air tight container in the refrigerator.

Semolina

It is coarsely ground endosperm and its chemical composition is similar to that of white flour. It is used in the manufacture of macaroni products. It is roasted before storing to save it from insects and worms.

Macaroni Products

These products are also called pasta. These products include macroni, spaghetti, Vermicelli and noodles. The starchy endosperm of wheat is coarsely ground into semolina which is made with water in to a stiff dough. The dough is placed in a cylinder, the lower end of which is fitted with a disc perforated with openings and as the dough is forced through the openings various shapes are formed. Macaroni is a tube form, spaghetti may be either tube or rod, vermicelli is a tiny rod and noodles are flat strips. Pasta products are also available in the shapes of shells and alphabets, usually not less than 5 1/2 percent by weight of egg solids are added to noodles. Popular dishes include vegetable noodles, kesari, upma and vermicelli payasam. Fried noodles are added to soups as thickening agent. Different kinds of wheat are used for different purposes. For cakes and breads soft wheat, for macaroni hard wheat are used. In India Punjab wheat is used for making chapathi and samba wheat is used for rava (upma) purpose. Maida is used for making sweets and snacks.

Malted Wheat

The process of malting consists of the following steps. Good quality grain is steeped in cold water for 36 hours in warm climate with two or three changes of water. The steeped grain is spread on wire mesh trays of 2-3 thickness which are kept in a stand. The germination is allowed to proceed for 3 days in a warm climate. During germination amylases and proteases are formed. Germinated grain is allowed to slow dry during which the amylases act on starch, hydrolysing them. The drying should be at a low temperature to conserve as much of the enzyme activity as possible. During drying, the water soluble carbohydrates and nitrogen (peptones and peptides) increase. The characteristic malt flavour is developed. The malt is dried to a moisture content of about 13 per cent.

Glutamic Acid

It is derived from wheat. A familiar compound of glutamic acid is "mono sodium glutamate", a salt – like product generally available and used to bring out the flavour of other foods or seasonings.

Wheat Bran

It increases the stool weight by increasing the water holding capacity of the bran. Wheat bran prevents constipation and may lower the risk of colon cancer. Bread, biscuits and other baked products and breakfast cereals like wheat flakes and shredded wheat are also considered as value added products from wheat.

Table 8.6. Promising varieties for various wheat products

Chapati	NWPZ	C 306, Raj 3765, PBW 175 and PBW 373
	NEPZ	C 306, K 8027, MACS 6145, UP 262, NW 1014, HUW 533
	CZ	C 306, Sujata, HI 1500, HW 2004, GW 322, GW 366, HI 1544
	PZ	HD 2833 and GW 496
Bread	NWPZ	HD 2285 and PBW 396
	NEPZ	HD 2733 and NW 2036
	CZ	HI 1077, GW 173, GW 496 and Lok-1
	PZ	HI 977, HD 2781, DWR 195, NI 5439
Biscuit	NWPZ	UP 2425
	NHZ	Sonalika
Pasta	NWPZ	WH 896 and PDW 233
	CZ	HI 8498, HD 4672, HI 8627, HI 8663, Raj 1555
	PZ	MACS 2846, HI 8663, DDK 1009

Wheat Quality Parameters

The quality of wheat is determined by a number of factors. The chemical composition of wheat with reference to its nutritional value and milling and baking quality and the physical appearance of the grain are the important parameters that decide the quality of the grain. As far as nutritional value of wheat is concerned, besides starch it also supplies proteins, minerals and important vitamins. The biological value of proteins is dependent upon their amino acid make-up. Protein is an important parameter for making different products of wheat. Loaf volume is known to be the product of the gluten strength and its total amount in a genotype. Protein content around 12 per cent is considered to be the threshold value to bring in a desirable loaf provided the gluten of such genotypes qualify for the standards fixed for determination of its strength. However, low protein, below 10 per cent with high friabilin, a marker for softness are the basic requirements for biscuit making quality. The beta-carotene imparts attractive colour to the pasta products and is considered to be an important quality character of durum wheat in many countries. It is a precursor of Vit A. hence has immense nutritional importance. Durum wheat with more than 7.00 ppm beta-carotene content are generally preferred in international market. For making pasta products from *T. durum*, more than 12 per cent protein is required. The sedimentation value gives an idea about the gluten strength and thus the strength of dough. Higher the sedimentation value the better is predicted to be the bread making potential of a genotype and values of 40 ml or more is required for making good quality pasta products. Time and again medium strong type of wheat has been described to be the most suitable commodity for making quality chapati. Till date all the varieties released for commercial cultivation in the country are good to very good for chapati making but not as good as C 306, a variety released more than three decades ago. The quality of chapati is usually assessed on five subjective criteria, I) Colour, ii) aroma, iii) appearance, iv) taste and v) pliability. Yellow berry (starchiness, mealiness, non-vitreousness) is a physical property and imparts undesirable white spots in dried pasta. Prevalence of yellow berry in durum wheats lowers the cooking quality of pasta products. The accepted limit of yellow berry is less than 10 per cent. Grain appearance score merits consideration when it comes to the grain trade. Chapathi consumers carry the perception that golden amber coloured grains make good quality chapati. Grain size, shape, soundness, colour and lustre are collectively taken into consideration to score the grain appearance. Grain appearance score anything above five is considered good from trade point of view. Hectolitre weight is the another physical parameter. It is a measure of the bulk density of grain *i.e.*,the weight of grain that can be contained in a unit volume packed in a standard manner. It records the plumpness of the grain which is indirectly a determinant of flour recovery. In general, Indian wheats satisfy the international trade requirements for test weight which is more than 78 kg/hl. Perhaps the most targeted trait for genetic engineering in wheat is quality. Seed storage proteins (SSP) are contained in the seed of higher plants. These proteins have been classified as albumins, globulins and glutenins on the basis of their solubility in solvents. The high molecular weight glutenin subunits (HMW-GS) genes in wheat are located on the long arm of the homeologous chromosomes 1A, 1B and 1D. Bread-making properties are particularly associated with variation at the *Glu-D1* and *Glu-A1* loci. The HMW-GS 1Ax1, 1Ax2, 1Dx5 and 1Dx10 have been shown to be associated with stronger dough, better elasticity and, hence, improved bread-making quality. Many elite wheat varieties lack the desired studies have demonstrated that the introduction of one or two HMW-GS genes results in a stepwise increase in dough elasticity. The transgenic lines produced so far have also demonstrated a very high level of expression and stability over several generations. This may imply that native genes are more tolerated by a plant genome subunits and, thus, many research groups are attempting to introduce these via genetic engineering (Shewry *et al.*, 1995; Altpeter *et al.*, 1996; Barro *et al.*, 1997; Vasil and Anderson 1997). In addition to increasing the bread-making quality, altered amino acid composition of the SSP is feasible and could result in improved nutritional properties. For example the insertion of genes for proteins such as zeins or albumins, could lead to an increase in the desired amino acid. Other approaches are also being considered such as reducing the level of anti-nutritional factors and modifying starch and oil composition and content

The increase in domestic demand of baked and pasta products along with economic liberalizations and global trade have offered opportunities for better utilization of wheat. The projected demand for bread by the next decade is 1.4 million tons, biscuits is about 1.9 million tons and cakes/pastry is 0.14 million tons. The annual growth rate of the baking industry is projected to be around 5 per cent. Hence, wheat quality needs uppermost attention so as to meet the domestic and export requirements in the time to come. The quality in wheat may be discussed under the following heads: Grain characteristics, milling quality, Baking and Chapati making quality, Nutritional Quality. Durum wheats are predominantly used for pasta products. They are more suited for pasta products because of their hard nature. They yield higher purified middlings (Semolina) as compared to bread wheats. Semolina is more desirable than flour in preparation of pasta products because it provides a particular al dante (firm bite) to pasta products. Also due to less surface area of semolina less water has to be added for kneading and removed during drying. This results in product economy. A few parameters which are significant from pasta point of view.

Market Quality

It refers to grain that is of desired colour plump, heavy, sound, and disease-free. Selection for these characteristics must always be foremost in the breeding programme, the breeder must ruthlessly discard strains with light, shrivelled, or diseased seeds. The term quality in connection with plant products which serve as food should embrace the parameters *viz.*, pleasing external appearance such as colour, shape, texture and aroma; free from taste and smell of foreign substances, having good nutritional qualities; good cooking qualities; desirable processing and suitability for export purposes. Grain colour is not given importance in the breeding program at CIMMYT because they assume that outer portion of the grain is to be removed. However, in Indian breeding program selection is done for amber grain colour. Red grain colour is not preferred. Bold grains (54 g/1000 grains) with hard texture are preferred. C 273, C 306 *etc.* are very good grain types whereas, Lerma Rajo, Sonora 64, HD 2009, PV 18 *etc.* are average to poor grain types.

Milling and Baking Quality

Milling refers to those properties that permit the economic production of flour. It depends on rate of milling, optimum milling moisture, yield of flour, ash content of flour. In hard textured wheats the starch grains are strongly embedded in thick sheaths of matrix protein whereas adhesion between starch and protein is much less in soft textured wheats. This leads to large difference in their milling properties. In hard grain types the endosperms is readily separated from the bran giving high recoveries of white flour. To develop wheats with appropriate milling properties breeders base their selection on endosperm texture, percentage extraction of white flour and water absorption. A simple, subjective test for endosperm texture is based on amount of endosperm adhering to the bran after removal of the flour from a crushed or ground sample. In soft wheats, considerable endosperm remains attached to bran after milling but with hard wheats there is very little. For flour extraction rates breeders rely on grain configuration in the early generations. Large, plump grains with a thin outer coat are expected to produce more flour. Baking is concerned with the potential of the flour to contribute to the acceptability of the baked product. Baking quality depends on CO_2 gas production and gas retention ability of the flour. Gas production depends on the diastatic activity whereas its retention is determined by the strength, elasticity and extensibility of gluten. These in turn depend on the quality and quantity of gluten. Wheat gluten consists of two major protein types; gliadin which confers extensibility to a dough and gluten which confers elasticity. The type of gluten present have a direct bearing on bread-making quality. Cake volume depends on the kernel hardness.

Chapati Making Quality

Over 90 per cent of wheat produced in the country is consumed in the form of chapatis. Therefore, the grains that cook into good chapati are liked in the market and may bring price advantage. No definite criteria is available to judge the chapati making properties. However, based on the organoleptic tests Austin and Ram (1971) gave the characters, *viz., Water required for making dought* : Water absorption and retention capacity of the flour should be high. This type of flour is called strong. The flours of hard and soft textured endosperms also have different water absorption properties. Those starch grains which are firmly embedded in matrix protein have a greater chance of becoming flattened and damaged during milling. Thus hard textured wheats have greater proportion of damaged starch grains. Damaged starch grains can absorb twice their weight of water when a dough is formed from flour, much more than can be absorbed by undamaged grains and this has an overriding effect on water absorption properties of a flour. *Puffing :* It results from the steam pressure inside the chapati. It could be partial, complete or completely absent. *Appearance :* The surface should be untorn, silky and attractive in its appearance. *Texture :* Soft, smooth, pliable and able to retain pliabilityfor a long time. *Colour :* Creamish is good. Not important from nutritional considerations. *Taste :* Sweetish. *Aroma :* Chapaties should give a pleasing aroma. *Keeping quality :* Longer. To obtain good chapatis following chemical composition is desirable : Protein content : 10-13 per cent, Sedimentation value : 30-37 ml., Pelshenke value : 120-125 minutes, Water absorption : Not less than 60 per cent, Stability, extensibility : Alveograph data and strength of dough, Stability : 100-130 mm, Strength : 15-23 $cm^{2,}$ Extensibility : 20-27 mm, Total sugar : Flour should have a minimum of 2.5 per cent sugar content, Diastatic activity : should be high and not less than 150 mg/10 g of flour. Baking quality of flour is judged from its potentiality to bake into an acceptable baking product. Certain flours which have high water absorption capacity, have medium to medium long mixing requirement, desirable stability and elasticity, on baking produce fine, upstanding loaves of large volume and of good crumb structure. These flours are required by the baking industry for bread making. Other flours have poor water absorption capacity, give lifeless doughs and produce squat, heavy loaves of unsatisfactory texture such flours are called weak flours. These flours are unsuitable for bread making but are suitable for purpose like biscuit, cakes, and the confectionary, foods. The capacity of a flour to produce a large and nicely textured loaf depend upon gas retention capacity of its dough during fermentation. For superior Chapati making properties, a wheat should produce chapaties which are creamish in colour, well puffed, sweet in taste, soft in texture and could be kept for few hours without losing their desirable properties. The studies have revealed that wheat which is amber in colour and produces"atta" of 10-13 per cent protein, with high water absorption, high content of sugars and diastatic activity and is medium strong to strong type, produces very good *'Chapaties'*. The Chapaties produced from a weak type "atta" are leathery, hard to chew and tend to stiffen if kept for a few hours.

Industrial Quality

Three species of wheat namely, *Triticum aestivum*, *T. durum* and *T. dicoccum* are cultivated in the country. Bread wheat is contributing approximately 87 per cent, while around 12 per cent comes from durum wheat and just about 1 per cent is the share of *dicoccum* wheat to the total wheat production. Wheat quality is a complex subject but basically it comes down to four separate factors *i.e.*, protein content, grain hardness, flour dough strength and milling quality (flour yield and flour colour). The quality requirements of wheat for various products like chapati, bread, biscuit and pasta are different. A miller looks for plump and well filled grains that can yield high amount of flour. Chapati consumer wants a variety with bold, lustrous and amber coloured grain and a good dough making quality. Baker's criteria show a wide variability in selection based on the products to be prepared, while pasta producers require altogether a different class of wheat (durum), which is hard with high yellow pigment and capable of making products with excellent cooking quality. Like hard wheat with strong gluten (> 60 ml sedimentation value), > 12.0 per cent protein, 5 + 10 High Molecular Weight Glutenin Subunit with 9 to 10 Glu-1 score are required for making good bread. For biscuit, the quality requirements are weak and soft wheat with < 10.0 per cent protein, < 30 ml sedimentation value and ~ 50 per cent Alkaline Water Retention Capacity (AWRC). For chapati, the quality requirements are in between of those required for bread and biscuit. Interestingly, most of the genotypes, which make good chapati, were found to carry HMWGS '20'. Besides, low response towards phenol showed promise to screen the breeders' material for chapati making quality. For pasta products, hard durum wheat with > 12.0 per cent protein, > 40 ml sedimentation value, below 10 per cent yellow berry incidence, > 7 ppm β-carotene content and γ-gliadin 45 is required.

Table 8.7. Quality parameters in wheat for various end products

Product	*Characters*
Bread	Strong gluten wheat with > 60 ml sedimentation value, > 12.0 per cent protein, 5 + 10 High Molecular Weight Glutenin Subunit with 9 to 10 Glu-1 score
Biscuit	Weak and soft wheat with < 10.0 per cent protein, < 30 ml sedimentation value and ~ 50 per cent Alkaline Water Retention Capacity (AWRC).
Chapati	Protein between 10 to 12 per cent, sedimentation value between 30 – 60 ml, with good dough making quality
Pasta	Hard durum wheat with > 12.0 per cent protein, > 40 ml sedimentation value, below 10 per cent yellow berry incidence, > 7 ppm β-carotene content and γ-gliadin 45 is required.

Table 8.8. Chemical composition (values per 100g)

Nutrient	*Wheat*	*Flour 70 per cent Extraction*	*Bread*
Moisture (g)	12.8	13.3	42.0
Protein (g)	11.8	11.0	7.0
Fat (g)	1.5	0.9	0.7
Fibre (g)	1.2	0.3	0.2
Carbohydrate (g)	71.2	74.1	52.0
Calcium (g)	0.05	0.02	0.01
Phosphorous (g)	0.32	0.09	0.07
Iron (mg)	5.3	1.0	1.0
Calories (Kcal)	348	349	236
Thiamine (mg)	0.45	0.12	0.01

Nutrient	*Wheat*	*Flour 70 per cent Extraction*	*Bread*
Carotene (mg)	64	29	25
Niacin (mg)	5.0	1.2	0.7
Riboflavin (mg)	0.17	0.07	0.04
VitaminB-6(mg)	0.52	0.33	0.10
Pentothenic acid(mg)	1.37	0.54	0.41
Protein (mg)	16.0	1.0	1.1
Inositol (mg)	370	110	51

Table 8.9. Percentile chemical composition of wheat grain

Starch	60 - 68 per cent
Protein	8 per cent
Fat	1.5 -2 per cent
Cellulose	2-2.5 per cent
Minerals	1.5 -2 per cent

The identified varieties shown in Table 8.6 for various wheat products are the result of the breeding programme for high yield and disease resistance, but not from any systematically adopted quality breeding programme. Hence, the urgent need is to improve the quality traits right from the selection of parents and advancement of generation based on the product specific requirements along with yield and rust resistance, so as to develop varieties, which can make various wheat products of international standard. Focussed efforts in germplasm evaluation and working out relationship of end products with different yield and quality components are required. Basic studies on identification of bio-chemical or molecular markers associated with key nutritional components like protein, β-carotene, iron and zinc are further required to assist the bio-fortification in wheat.

Test Weight or Hectolitre Weight or Weight Per Unit Volume (kg)

Test weight is the primary grading factor; it can also, in general, be an indicator of semolina extraction. It is related to variety and environmental conditions and usually determines the plumpness of the grain. Flour yield increases and flour ash decreases with the increase in test weight. A flour yield upto 70 per cent and flour ash of 0.4 per cent would be normal for a wheat with 75 kg hectoliter weight, whereas a flour yield of 62.5 per cent and flour ash of 0.6 per cent would be expected from a wheat with 62 kg hectoliter weight. Kernal size and shape are the two most important features which influence the overall test weight of the grain. Immature and shriveled wheat are usually low in test weight and give correspondingly poor yields of flour. A correlation co-efficient of + 0.75 between test weight and flour yield has been established by previous workers. At lower weights than 71 kg, the milling yield usually falls off rather rapidly with decreasing test weight. Immature wheats or wheat that is badly shriveled as a result of drought or disease is usually low in test weight and give correspondingly poor yield of flour. Hectolitre weight depends mainly on grain shape and density and indirectly indicates the milling yield. In durums, it is strongly related to semolina yield.

The two values of test weights can be inter converted using the expression.

1b/winchester bu = 1.287 kg/ha

For export, durums should have 78 kg/ha as test weight.

The following procedure has been found to be accurate in making hectoliter weight per 1000 ml of grain and has been adopted as the standard method in the determination of test weight. Place the hectoliter machine on a smooth and plain surface so that the machine is in straight position. Pour about 100 – 120 g grain in the hopper of the machine. Make sure that – Opening at bottom of the hopper is round and 11/4 inches in diameter. Bottom of opening in hopper is centered over the kettle and is exactly 2 inches above the top of the kettle. The kettle has the capacity of holding an exactly 100 ml of volume and the inside height of the kettle is of 4 inches. Do not jerk the kettle before or during the stroking operation. Use the valve below the hopper to release the grain from the hopper and see that uniform flow of the grain is there during filling of the kettle. Once the kettle is filled completely, close the lid of the kettle with the slider given at the top of the kettle. Remove all the remaining grains on the top of the kettle. Use a balance, which is sensitive and gives accurate weight. Weight the filled grains of the kettle on the balance. Note the reading of the hectoliter weight (*i.e.*, weight per 100 ml of grain) and represent in Kg. Hectolitre weight depends mainly on grain shape and density and indirectly indicates the milling yield. In durums, it is strongly related to semolina yield. *Vitreous kernel content.* (High percentage desirable). This is a grading factor and is also generally an indicator of semolina extraction. Related to variety and environmental conditions. Bright kernal appearance is desirable. Related to variety and environmental conditions. Dockage: A relatively low amount is acceptable. Weight for 1000 kernels.(High weight desirable). Some people regard this as a better indicator of semolina extraction than test weight. Related to variety and environmental conditions. Kernel size: A uniform kernel size is desirable. The distribution should consist of a high percentage of large and medium sized kernels with a low percentage of small sized kernels. Kernel damage: (Relatively free desirable). Consists of sprouted, heat damaged, frosted, badly ground damaged, badly weather damaged, moldy, diseased, black point, ergot, pink, *etc.* Shrunken and broken kernels: (A low amount is desirable). Foreign material: All matter, other than wheat, which is not separated from the wheat in the dockage determination is classed as foreign material.(Obviously a low amount is desirable). Insect infestation: (Grain should be completely free). Moisture: (Not too low and not too high). This is important, not only from a dry matter stand point, but also for storability and grain handling. Protein content. (A reasonable level of protein desirable). The protein content of the durum wheat produced in North Dakota is usually between 13.3 and 14.3 percent. If the wheat protein is lower than 12.5 percent, it might indicate kernel damage. If much higher than 14.5 percent it might indicate a low-test weight value. Falling number: This is a measure of alpha-amylase. A falling number of 250 units or higher is usually desired related to amount of sprouted wheat kernels in a parcel of grain. Wheat ash: A relatively low value is desirable. In countries that mill to a specific ash level this becomes very important. Grade: To a degree, this is a indicator of quality and the general condition of grain. Numerical grades consist of U.S>I through US No.5 and U.S.Sample grade. Durum wheat is also divided into three subclasses: Hard amber durum wheat (HAD). The subclass Hard Amber durum wheat shall be Durum Wheat with 75 percent or more of hard and vitreous kernels of amber colour. Amber Durum Wheat (AD). The subclass Amber Durum Wheat shall be Durum Wheat with 60 percent or more but less than 75 percent of hard and vitreous kernels of amber colour. Durum Wheat(D). The subclass Durum Wheat shall be Durum Wheat with less than 60 percent of hard and vitreous kernels of amber colour.

Kernel Weight (Thousand Grain Weight) (g)

Kernel weight gives an idea of average grain size and is considered to be a function of kernel size and density. While test weight determines the milling quality of all wheats, kernel weight is decisively superior in predicting the milling quality of hard grain. Both test weight and kernel weight pronounce for the same quality character *i.e.*, milling quality but their relationship has not been conclusively studied. Wide range of variability from 22 – 45 g have been recorded for bread wheat, while for durums, the weight varies from 35 – 55 g.

Kernel Size and Shape

The size and shape of the grain affect the test weight. Large size kernels produce higher flour yield due to high proportion of endosperm. It is also believed that by milling the graded wheat, we can have better milling efficiency. it is measured on visual basis and graded on the basis of 1-10 hedonic scale. Grain appearance gives on indicator of its processing quality. Plump bright grains are likely to give higher extraction rate of flour. Vitrousness is associated with high protein content. Grain weight and appearance varies with site as well as season and can be manipulated by disease control, nitrogen application and its timings. Avoidance of lodging also enhances grain weight.

Yellow Berry

Uneven yellow spots or sometimes the whole grain becomes yellow because of the incidence of yellow berry, which makes the grain unsuitable for pasta products. The central endosperm shows farinous (opacity) instead of vitreousness. Yellow berry has a negative effect on the quantity and quality of semolina and lowers the total protein content of the grain. The pasta products made out of yellow berry affected grains develop stickiness while cooking. Another adverse effect of yellow berry is the loss of the yellow pigment in the grain. It is an important character to be considered before scoring for grain appearance, especially in durum wheat.

Grain Analyser

It is also used for quantitative estimation of various quality parameters including protein. The analyzer used in lab is NIRT- 1241 based a Near Infer Red Reflectance value. This is a non-destructive method for analysis of various quality component in grains of cereals and pulses. For chemical analysis whole grain samples from cultivated and wild accessions were rinsed with N/10 HCl for 1 min to remove any dust particle from seed surface and dried in hot air oven at 80C till constant weight (0.5 g). Grain samples were digested in a mixture of two parts of concentrated nitric acid and one-part perchloric acid as per the standard procedure described by Zarcinas *et al.* (1987). Digestion was continued till white residue was obtained. Required volume was made after the completion of digestion process and digests were analyzed by Atomic Absorption Spectrophotometer (GBC–Avanta Garde M). Concentration of iron and zinc was expressed mg/kg (ppm) on dry weight basis. A minimum of five replications of chemical analysis was made in each of cultivars and wild accessions.

Grain Appearance Score (Max. 10)

Three characters that include grain size, shape and colour are taken into consideration for scoring the grain appearance. Bold grains with attractive shape, amber golden colour and the luster of the grain are the major criteria for scoring. Grains with all these characters fetch high price in the market. These characters help the purchaser of the grain in preliminary surveying of the market. Grain hardness and milling quality are varietal characteristics and not influenced greatly by the environment.

Whole Meal or Flour Sample

These should be prepared by passing wheat through either a Tecator Cyclotec mill fitted with a 0.5 mm screen.

SDS- Sedimentation Test (for Wheat Flour) AACC (2000)

This test is based on the fact that gluten protein of wheat absorbs water and swells considerably when treated with lactic acid under certain conditions. The water absorption capacity depends on the quality and quantity of gluten. This test is superior to protein determination test for predicting baking behavior of flour. Sedimentation value is the level in ml (in a 100 ml cylinder) of suspended flour particles after a specific time in a standard solvent and under standard temperature conditions. 20g sodium dodecy1 sulphates (SDS)/L are added in 20 ml stock diluted lactic acid solution (1: 8v/v). 6g whole meal or 5g flour is added to 50 ml water in 100 ml stoppered cylinder. The material is dispersed by rapid shaking for 15 sec. again shake the contents thrice for 15 sec, 2 min and 4 min with a resting period of few sec in between each shaking. Immediately 50 ml SDS lactic acid reagent is added and mixed by inverting the cylinder four times. Repeat shaking at 2,4 and 6 min. The contents of the cylinder are allowed to settle for 20 min (whole meal) and for 40 min (flour). Read the sediment volume in cubic centimeters. Based on sedimentation values wheat flour is classified as:

Table 8.10. Sedimentation Value and their interpretation

Sedimentation Value	*Interpretation*
Less than 20cc	Weak flour, suitable for making cakes, biscuits, pastries
20-40cc	Medium to strong flour, suitable for Chapati making
More than 40cc	Strong flour, suitable for bread making

Zeleny's test is fast replaced by SDS/lauryl sulphate sedimentation test. As compared to Zeleny's sedimentation values, the SDS values are better correlated with bread making potential of wheat flours. The SDS reagent marks charges on protein and hence swelling is proportional to molecular weight. High molecular weight protein have more swelling, indicating better quality of protein. The value are interpreted as follows

Table 8.11. Macro SDS value, type of gluten and end use

Wheat Type	*Macro SDS value ml*	*Micro SDS value (ml)*	*Type of Gluten*	*End Use*
Bread type	Above 68	Above 18	Strong to very strong	Bread and noodles
	48-68	12-18	Medium	Bread chapatis, cake noodles
	Below 48	Below12	Weak	Biscuit
Durum wheat	Above 10	Above 10	Strong	Noodles and Bread
	7-10	7-10	Medium	Noodles
	Below 7	Below 7	weak	Biscuit

Pelshenke Test (Wheat)

Records the time taken by a whole meal dough ball prepared with yeast suspensions, to disintegrate while being incubated at 30°C. It measures the strength of dough and gluten. This test measure the time required in min, for the initiation of disintegration of dough ball of standard size under specific conditions. (30°C) *i.e.* ball is prepared with yeast suspension. 3.0 g of uniformly ground meal is made into dough with yeast solution (1.8 ml) and allowed to ferment. The dough ball is dropped into breaker containing distilled water maintained at 30°C. The time when the ball is introduced into breaker and the time when it starts disintegrating is noted. This gives Pelshenke value. The Pelshenke value may vary from 50-240 min. The time depends on the strength of gluten (very strong/medium/weak), which gives the dough its particular form and consistency. Based on the Pelshenke value, various groups of wheat are made.

Table 8.12. Groups of wheat based on the Pelshenke value

Group	*Pelshenke Value (min)*	*Suitable for*
Weak flour	less than 80	Making cakes, biscuits, pastries *etc.*
Medium-strong	80-100	Chapati making
Strong	150-200	Bread making
Very strong	> 200	Blending with weak flour group

Starch and Damaged Starch (AACC 2000)

Starch is the most predominant component in flour. It changes from the raw granular form to the gelatinized form on cooking of noodles and causes noodles to set. If the starch on the surface of the noodle is overcooked before the interior becomes cooked, the surface of noodle becomes sticky. If the noodles are properly cooked, the starch on the surface is not sticky but gives a smooth mouth feel. Starch contains two fractions amylose and amylopectin. Both alpha and beeta amylases attack 1-4 alpha linkages but are hindered by the proximity of 1-6 linkages at branching points. The limit dextrins produced from amylopectins result in stickiness. Wheat may be damaged in many different ways and when such damage is of a type and extent that will decrease storage or the processing values, it is considered for assessing the quality of wheat. Different diseases prior to harvest and immaturity at harvest may result in shriveled or green kernels which will reduce the yield and quality of flour produced. The wet weather at maturity but before harvest many cause sprouting of some kernels. The sporting reduces flour yield and the flours have high α-amylase activity which will damage the bread making quality. It may reduce the germination ability also. The threshing operations may cause breakage of wheat kernels which are usually removed in cleaning operations at flour mill and result in economic losses. It addition, the broken kernels are more hygroscopic and under wet storage conditions, they serve as the host for various microorganisms which impart musty and sour odour to the wheat. The musty and sour odour if pronounced is carried over to the flour and baked products. The mould damages and deteriorates baking quality. The level of damaged starch in the flour influences its ability to absorb water. The content of damaged starch in flour is estimated by methods which measure quantity of maltose released from starch by the action of amylase. Only the damaged granules are susceptible to amylase in the ungelatinized condition. Damaged starch may be recognized microscopically by a red coloration with congo- red stain, and have been described, from their microscopically appearance as "ghosts"

Enzymes

The enzymes in flour though present in small amounts considerably affect noodle quality. Excessive amount of alpha amylase (caused either by sprout damage or malt supplementation) in flour causes rapid break down of the noodle structure. A falling number value between 250-300 seconds is considered best for noodles. Polyphenoloxidase is another enzyme important in noodle quality. It oxidises tyrosine (or other polyphenols) to produce dark melanin pigments. This gives the undesirable dark colours in stored fresh noodles.

Now let us take a look at some of the specific tests we perform for a appraisal of wheat quality as summarized below.

Table 8.13. Wheat quality tests for durum

Wheat	*Semolina*	*Spaghetti*
Test weight	Milling characteristics	Processing
Vitreousness	Semolina and Flour yield	Properties
1000 kernel weight	Protein	Colour
Kernel distribution	Ash	Cooked weight

Wheat	*Semolina*	*Spaghetti*
Moisture	Gluten strength	Cooking loss
Protein	Specks	Cooked firmness
Ash		
Falling number		

Alpha-Amylase Activity

It is measured by falling number test. A flour/water suspension in a tube in a boiling water bath is stirred for exactly 60 seconds. A plunger is then allowed to fall freely through the suspension under standardized conditions. Falling time is the time in sec, taken for plunger to fall a standardized distance.

Table 8.14. Falling number test and nature of wheat

Time	*Nature of Wheat*
200-250 sec	Normal wheat (unsprouted). Normal amaylase activity. Good for bread, chapatie
300 sec or above	amylase activity too low. Risk of diminished bread volume and bread crumb is too dry
150 sec or less	High amylase activity (sprouted wheat, Danger that the bread crumb is likely to be sticky.

Falling number test is designed primarily to detect sprouting in wheat caused by rain irrelevant wet weather at the time of harvest of wheat. Falling number when converted to liquefaction number 600 has a linear relation with amylase activity.

Diastatic Activity

This refers to the amount of maltose (mg) produced by flour (5g) when allowed to digest at pH 4.6-4.8 (46 ml buffer) for one hour at 30°C. The results are expressed as mg of maltose produced by 10g flour in our hour. Test indicates production of sugar in dough for use of yeast or sweetness of chapaties. Various other chemical test like the proteolytic activity, damaged starch, crude fat and fibre content are also conducted, at various stages. Various chemical requirements for a bread flour in India are

Table 8.15. Diastatic activity of wheat

Protein per cent (min)	11.0	Granulation (micron)	
Gluten per cent	10.0	105-55	25 per cent
Total ash per cent (min)	0.45	55-35	20 per cent
Damaged starch per cent	7.9	35-25	25 per cent
Acid soluble ash per cent (min)	0.05	25-15	30 per cent
Alcoholic acidity (as H_2SO_4) (min)	0.1		
Sedimentation value (min)	30		
Maltose figure per cent	2-3.5		
Uric acid mg/100g (max)	10		

Quality Rheological Characteristics

Rheology means the science of deformation. Here in the context of wheat quality it is the measure of deformation of the dough during various operations in the process of baking. There are three different stages in the baking of bread *i.e.* mixing, fermentation and baking and there are different equipments available for testing at these stages of processing. The equipments used for these tests are Mixing: Mixograph and Farinograph; Fermentation: Alveograph and Extensograph. The mixograph and farinograph mainly differ on account of the use of water absorption. The mixograph uses a fixed absorption and its basic objective was to assess the mixing requirements of given flour whereas farinograph employs an optimum water level to decipher the changes taking place in the dough. The major use of the farinograph lies in determining the optimum level of water absorption to produce the desired consistency dough. *Mixograph method:* Determine the moisture content in the flour/meal and calculate the weight equivalent to 35g on 14percent moisture basis. Weigh the sample into the mixing bowl and add the calculated quantity of water. The quantity of water is usually based on the 60/68percent absorption for flour/whole meal respectively plus the correction for the moisture content as per the following expression:

Quantity of water = 35x60 or 68/100 + (35 – weight of flour)

Note: The flour water absorption should be adjusted by one percent per percent change in protein content.

Turn on the instrument and lower the mixing head as soon as the recording pen reaches a vertical line. Run the instrument for seven minutes. Clean the bowl. The mixogram recorded is interpreted as: Mixing Time: This is the time taken in minutes by the curve to reach the peak. Maximum height at peak (Strength): This is measured from the center of the curve to the baseline in cm. Rate of dough development: It is the angle enclosed by the line through the ascending curve and the horizontal line through the peak. Rate of dough weakening: It is the angle enclosed by the line through the descending curve and the horizontal line through the peak. Mixing tolerance: It is the angle enclosed by the lines through the ascending and descending curve with centre of the curve at peak as the apex of the angle. Dough development area: It is the area under the curve in cm^2 as enclosed by the vertical line through the centre of the curve at peak. Baking strength: It is the total area under the curve in cm^2. Generally, a lower dough development area and higher baking strength is preferred in baking.

Farinograph Method

Determine the moisture content in the given sample and calculate the quantity of sample equivalent to 50g on 14 percent moisture basis. Place the sample in the mixing bowl and start the instrument in the high speed. Let it run for one minute to equilibrate the temperature and when the pen reaches the zero min line start adding water from the burette. Add a quantity equivalent to the expected value. When the dough begins to form, scrape down sides of the mixer. If the curve goes beyond 500BU line then add more water from the burette so that it is centered at 500BU line. Keep the bowl covered in order to avoid evaporation. Run another titration curve if necessary, until the curve is centred at 500 ± 20BU. Run the final curve for 15min using the standardized quantity of water and interpret the farinogram parameters. Farinographic Water Absorption (FWA): The quantity of water required for centering the curve at 500 BU line. Dough Development time (DDT): The time in minutes required for centering the peak of the farinogram on the 500 BU line. Dough Stability (DS): The time in minutes for which the curve traverses the 500 BU line. Mixing Tolerance index (MTI): The drop in peak consistency of the dough in five minutes as measured from the top of the curve in BU. This is the difference in Brabender units (BU) measured from the top of the curve at the peak and top of the curve 5 min after. Time to Breakdown: It is the time in minutes elapsed from the start of mixing upto a drop of 30 BU in the peak consistency measured from the center of the curve. A higher water absorption and dough stability are preferred for baking bread.

Alveograph Method

Determine the moisture content in the sample and weigh 250g into the mixing bowl. Fill 2.5 per cent salt solution in the burette upto the mark corresponding to the moisture content in the flour. Start the mixer and add the salt solution quickly. Mix the dough for one minute and then scrape the sides of the mixer to bring the material sticking to the sides into the dough mass. Again mix the dough for another six minutes, making the total mixing time to seven minutes. Oil the extrusion plate, relaxation plate, rolling frame and roller *etc.* with peanut or any other equivalent oil. Then start the extrusion process by starting the mixer in the reverse direction. Cut first about one cm dough and discard and then extrude five dough pieces of the specified size. Place these dough pieces on the rolling frame and slide the oiled steel roller 12 times over them. Cut the rolled piece dough with the dough punch to yield patties of the desired size and shape. Put each patty on the relaxing plate and place in the relaxation compartment at 25°C for 20min. Place the patties one by one in the Alveograph blowing section, flatten these to the desired thickness as provided in the mechanism. After flattening each patty, start the Alveograph to blow the patty into a bubble. Stop the Alveograph when the bubble bursts. The resistance offered by the dough patty to the blowing air is recorded in the form of a curve. The following observations are made from the curve:

- P: Tenacity/strength of dough: it is represented by the resistance of dough to deformation and is measured as the height of the curve in mm multiplied by 1.1
- L: Length of the curve in mm and represents the extensibility of the dough
- P/L: This is the curve configuration and is the ratio of the strength and extensibility
- W: It is the baking strength and is measured as the area under the curve in cm^2
- G: It is the index of swelling and is calculated as 2.226 x under root of L

Table 8.16. Farinograph, Extensograph and Amylograph of dough

Water absorption (per cent)	62-64	Extensibility CM	14-18	Paste viscosity (BU)	450-625
Peak time (min)	5-7	Resistance to extension (BU)	450-500	Falling Number	
Stability (min)	6-8			Falling number value (sec)	175-275
Mixing tolerance index (BU)	30-50				

Amylograph

Continuously measure the resistance to stirring of 10 per cent suspension of flour in water while the temperature of the suspension is raised at a constant rate 1.5°C/mm from 20°C – 95°C and then maintained at 95°C. It is of use in testing flour for soups, *etc.* For which purpose the viscosity of the product after gelatinization is an important characteristic, and for adjusting the malt addition to flour for bread making

Alveograph

Measures the ability of dough when subjected to high speed but constant mixing with water at room temperature. It uses air pressure to inflate a bubble of dough and break it, The instrument continuously records air pressure, the time that elapses before the dough breaks. Dough extensibility is measured with Alveograph. It measures the extent to which a gluten can be extended and its ability to retain gases. Basically, extensibility is a varietal character, hence careful selection can result in strains with gluten having good extensibility and gas holding capacity. Alveograph values are interpreter as fellows.

Table 8.17. Alveograph and gluten type

Alveograph Values	*Gluten Type*
W = 200	Weak
W = 200-300	Medium strong
W = above 300	Strong

Milling Quality

Conditioning properties. This is important from the stand point of proper milling. Ease of milling. The wheat should flow through the mill in a uniform manner with an even distribution of the stocks. Durum wheat kernels should not be soft;very rarely will they ever be too hard. Semolina yield. The total semolina extraction should be relatively high with a proper distribution of the semolina streams. Semolina colour. The colour of the semolina should be a deep yellow and should be relatively free from bran, germ, black point or other objectionable specks. Semolina ash. The semolina ash content should be relatively low.

Speck Count

A relatively low number of 'off coloured' specks in a specified area is desirable. Semolina protein. With a normal extraction level, the semolina protein content should not drop more than 1.4 percentage points from the wheat protein. *Gluten*: A relatively high percentage of wet gluten is desirable. *Particle size distribution:* For the production of good quality pasta the particle size of semolina should not be too coarse not too fine. The experimental milling tests are conducted to ascertain yields of straight grade flour or semolina which are correlated to the commercial milling. Experimental milling tests are used to screen varieties with high flour/ semolina yields. The flour and the semolina yield are affected by the grain uniformity and content of yellow berry in a particular wheat. The presence of yellow berry considerably reduce semolina yield and hence is considered undesirable in durum wheat. Rapidly changing environmental conditions, nitrogenous deficiency increase the incidence of yellow berry.

Table 8.18. Particle size distribution for a typical U.S. commercial semolina

U.S. Sieve (Nos.)	*Sieve Opening (mm)*	*Per cent Total Semolina*
On 20	0.86	0.0
On 40	0.38	22.8
On 60	0.23	51.6
On 80	0.18	14.6
On 100	0.14	9.3
Through 100	-	1.7

Semolina Yield for *Triticum durum* Wheats, Flour Milling for *T. aestivum* Wheat

This is another important characteristic of durum kernels. To the layman, semolina is simply the meal that results from grinding the durum kernel after removing the bran layer. But the manufacturer distinguishes between semolina and flour both of which can be milled from durum kernels. Semolina will pass through a 30 mesh sieve but not a 100 mesh sieve. In other words, semolina has larger particles than flour (200-300 μm). The particle size should be uniform. If more particles are fine, they absorb more water due to large surface area and the course ones (says over 500 μm in size) would not properly hydrate causing blotches in the dried pasta. Presence of excess fines also leads to development of thermal stresses during pasta manufacture which may cause protein denaturation. Semolina particles hence should fall between 488-142 μm. Semolina yield is influenced greatly by the kernel size and shape. A large kernel (78 kg/ha, 40 gms/1000 grains) has a higher proportion of endosperm to bran. Thus with large kernels, for each ton of semolina, the miller handles less than he would if he was milling small kernels. Also gives an idea of about the homogeneity of grain in a particular variety. Yellow berry (a pale or white discoloration in kernel resulting from development of floury starch) also influences semolina yield to a considerable extent. The more yellow berry a kernel has, the higher the proportion of flour and the lower the proportion of semolina yield. In order to obtain the maximum yield of semolina and minimum amount of flour, the durum wheat is conditioned so that the endosperm has 15 per cent m.c; the bran about 18 per cent m.c. at the first break. This high moisture differential is important to minimize bran breakage and achieve a good separation of endosperm from bran. The objective of experimental milling are to prepare flour for evaluation of chemical, physical or end use properties. The mills used for milling wheat are quadrumate junior, Cyclotec, Cemotec, Buhller mill and Stone chakki *etc.* Grain is wetted before milling (tempering). Wetting results in wetter exterior and drier interior. Hence under roll pressure, endosperm crumbles to powder whereas bran and germ flatten to flakes and are removed by sieves. For hard wheat usually 11.5-16.5 per cent moisture is required whereas Sami hard and soft wheat require 13.5-15.5 per cent moisture. Various mills used. The water-binding capacity of semolina (absorption) is only important in so far as it varies from the mean. The mixing and extrusion properties are related to a degree to the protein quantity and quality. If the protein of the semolina is at a proper level, very little if any difficulties are experienced. Spaghetti and macaroni products should be dried by carefully removing the water on a falling humidity gradient over a period of time depending upon the product being dried, the time generally will range from about 10 to 20 hours. If the products are dried too fast, they "check" and, conversely, if the drying time is too long, coupled with a high humidity, the products may become "sour" and, in some instances, contain mold growth.

Spaghetti and Macaroni Properties

The general overall appearance of the finished spaghetti product is one of the more important quality criteria. A desirable product should possess a smooth surface of bright amber colour and be free from white, black or brown specks. A properly dried product (long goods) should withstand bending into a bow without breaking. When broken into small pieces, the break should be clean and have a "glossy" and glassy" appearance. In addition, the dried spaghetti or macaroni should be free from "check" or "cracks". Obviously, these problems would adversely affect their packaging, transportation and aesthetic properties.

Milling Characteristics of Wheat

Flour Milling

To separate as completely as possible, the endosperm required for flour from the bran and germ. To reduce the maximum amount of endosperm to flour fineness thereby obtaining the maximum extraction of white flour. The reduced endosperm known as flour; while the germ, bran and residual adhering endosperm are obtained as by products. After the wheat has been conditioned, a succession of grinding statges, using roller mill to open the whole grain in the first stage. These grinding stages grouped into break and reduction system.

Break System

Consists generally of four to five breaks or roller mill grinding stages, each followed by a sieving stage. Ist break open the grain and in subsequent breaks to scrap off the endosperm from the bran coats. Each of the break roller mills is equipped with a pair of rolls usually 10″ (25.4 cm) diameter and 40″ (101.6 cm) length in parallel alignment. The narrow gap 'Nip' between the aligned rolls can be adjusted to allow precision grinding. The rolls rotate in opposite directions at different speed rate between the two generally 2: 5: 1 for break roller mills. The break rolls are corrugated or grooved over their entire surface. The corrugations (flutes) running along the length of the rolls. Every break rolls, have different flutes from 12 per inch (Ist break) to 26-30 per inch in 4th break. In the first break, the flutes sheer open the grains often along the crease and unroll the bran coats. The first break consists of a mixture of particles which differ in size and composition. The largest as the bran coat (break stock), still thickly coated with endosperm, the intermediate sized particles are either semolina, middlings and dunst or bran the smallest are flour *i.e.*, thin particles of endosperm. These particles are separated from one and another by scalping, dusting and grading. The feed to the subsequent break consists of the break stocks scalped from the grind of the previous break; in each grinding stage endosperm is scraped from the bran coats which becomes progressively thinner. The roll gap is made progressively narrower and flutes progressively finer at each successive rate. The break release is dressed over a flour silk to separate the flour which is removed from the milling system as part of the end product. The coarser part of the break release (Semolina, middlings, dunsts) is graded by sieves into a number of fractions differing in particle size, but all consisting mainly of endosperm with some admixture of bran and germ. The semoline and middlings are then purified from bran which is a pure endosperm. The fractions consist the purified semolina middlings are carried to the reduction system.

Sieving

This process is carried out in plansifter which is a machine, consisting of a vertical nest of horizontal sieve. The mixture to be sieved is fed in at the top and falls from the sieve to sieve by gravity. A single plansifter may incorporate sieves of four or five different mesh sizes, thereby delivering or 5 fractions of different particle sizes.

Purifying

The purifier consists of an oscillating sieve set at a slight angle to the horizontal and enclosed by a hood or cover. This comprises four sections which becomes progressively coarser in size of mesh from head to tail. Reduction system: Consists of eight to sixteen grinding stages and differ from the break rolls in two ways *i.e.*, (a) the roll surface are quite smooth or slightly roughened (matt surface); fluted rolls are rarely used and then on one or two particular reductions only (b) the speed differential between two rolls is lower usually 1.25: 1 to 1.5: 1. These characteristics of the reduction roller mill results in the grinding effect being a crushing rather than a shearing action. The feed to the reduction system consists of the purified semolina and break middlings and the dunst. The feeds to A and B rolls despite careful purifying contain a few particles of bran and germ. The purified middlings size goes to C roll. By carefully controlling the roll pressure these bran and germ particles, can be flattened without fragmentation and subsequently sifted off as coarse "cuts" from the grinds. These and corresponding fractions from subsequent reductions stages are fed to certain reduction stages, collectively known as the 'coarse reduction stages' or the 'tailing system'. The other reduction stages *viz.* D,E,G.H>K.and L regrinde the dunst from the preceding grinding stages, although small quantities of new materials of low grade derived from the break or scratch system may be introduced at suitable points from D rolls onwards in order not to contaminate the feeds to A,B and C rolls which yield the purest or patent flour. In the longest reduction system, M is an additional coarse roll, and N, O and P are additional fine rolls.

- **Straight Flour**: Flour containing all wheat herny except the bran and peeds turned 100 per cent flour.
- **Potent Flour**: The clean flour made by grinding the choice protein of the inner protein of wheat.
- **Clear Flour**: Lower grade and higher ash content flour remaining after the patent flour has been separated.

Quadrumat Junior Mill

In the mill 1,2,3,4, roll assembly is used for bread wheat milling. For wheat milling roll assembly is changed to 1,1 (Break assembly with larger flutes). 2,2 (Reduction assembly with smaller flutes). A 6XX siften is used and for semolina an assembly of 18,30 and 130 mesh is used

Flour Mill (Blluher) and Cemotec Mill

Feed inlet in mill is so adjusted that 1 kg of sample passes through rolls in about 10 minutes in triticale and 7-8 minutes in wheat. Mill is made to run for 30 minutes more after all the grain passes through rolls, tapping and cleaning it in between. Milling of durum wheat is done on an experimental semolina mill with aid of small-scale purifier. In order to get the maximum yield of semolina, *T. Durum* wheat is condition ed at 17 per cent moisture content for 49-72 hrs. The semolina produced during milling in Cemotec mill is graded and purified by an elaborated system using purifier so that the light bran particles air lifted and the pure semolina free from bran particles and other smaller light fragments is obtained. Using such a system, a good quality durum wheat yield 70 per cent of semolina and only 5-6 per cent flour. Generally, particles of semolina fall within the particle size range of 486-142, um and preferably between 300-200um. Presence of excess finer leads to the development of thermal stresses during the pasta manufacture which may cause protein denaturation, whereas presence of coarse particles (over 500 um in size) dosen 't absorb water readily, may cause blotches in the dried pasta. The yield obtained on a experimental mill is usually determined by the type of semoina. Lower pigment corresponding to coarser semolina. The semolina produced should be relatively free from bran particles and hence of reasonable ash content. The quality expected in semolina for pasta manufacture are brightness (absence of bran specks), particle size (30-60 mash) semolina ash (0.55-0.75 per cent) grit content and (11-13) protein for adequate gluten formation, yellow pigmentation (4 to 8 ppm), low lipoxidase activity (10-20 units of activity), low microbial counts to ensure long shelf life, deep yellow macaroni colour obtained from 60-65 per cent extraction rate with good baking strength and desirable cooking quality. As the lipoxidase enzyme in semolina may destroy the yellow pigment during subsequent dough making, a low lipoxidase activity in wheat is desirable. The impurities of all types which are normally removed before wheat is milled either into flour or semolina (suji), is a direct economic loss to the buyer

Colours and Provitamin A Carotenoids (β-Carotene) Content

The ratio of semolina pigment to pigment of whole wheat varies from 70-110 per cent. The range of β-carotene pigments in wheat varies from 3 to 8 ppm. Bread wheat semolina exhibit low ppm counts while durum wheat semolina with higher β-carotene pigment counts are considered best. More the pigment content more will be the colour of semolina and the same is reflected in the finished produced like noodles, spaghetti, macaroni products *etc.* The yellow colour is a consumer's preference which gives yellow (golden) colour of finished product. Eight gms of semolina are dipped into 40 ml of saturated butanol for 23 hrs in dark. The intensity of colour present in the supernatant is measured at 440 nm. the OD is then multiplied by 30.1 to get pigment concentration is ppm. Based on bran colour, wheat's can be classified as red or white. Colour is a genetic character but environment also influences it slightly. For bread making any wheat that produces white good quality flour is suitable because bran is removed during the milling process. Passage of bran in flour is undesirable. One can subjectively assess the colour of the flour by visual observation using Petcor test. It should be bright creamy white. Colour can also be determined objectively by using instruments such as Kent Jones and Martin colour grader, Hunter lab Colour Quest Instrument or Actron. Good maida for bread making should have a grade unit of not more than 1.5. But for chapati making, when whole meal is used only the white or amber coloured wheat's are preferred. On the contrary pale yellow or golden coloured durum wheat's are preferred for pasta products. Macaroni manufacture want wheat's with deep yellow colour which is caused by high percentage of β-carotene pigments. This pigment can be destroyed by high level of enzyme lipoxidase. Hence, varieties must have high β-carotene content and low lipoxidase activity. Materials with more than 5 ppm β- carotene are retained during selection process. The durum wheat flour contains two major type of pigments *i.e.* carotenoids and flavanoids. Carotinoid pigments are found in endosperm and give flour its creamy-yellow colour. Durum cultivars vary significantly in carotenoids. These pigments are easily destroyed by oxidizing agents added to flour. Two such agents which are commonly used are benzoyl peroxide and enzyme active soy flour. The other type of pigments are flavanoids which come mostly from bran contamination. These are not bleached by normal bleaching agents. The flavnoids are relatively stable and are colourless at acidic pH but give a yellow colour at high(alkaline) pH. They are the source of yellow colour in Chinese noodles prepared using kansui (sodium and potassium carbonates). The presence of iron salts with the flavanoids at alkaline pH often give rise to green and brown colours. Because of contamination by flavonoid pigments in low grade flours and the resulting problem, noodles are often made from low extraction flour. Thus colour is the single most important quality parameter. Manufacturers want macaroni, noodle and shapes that are deep amber or golden. Durum kernels may range in colour from orange yellow to white. Manufacturers often blend different durums to get the colour they want, but reject durums that are white or grey or that have mottled grains. The desirable yellow colour is caused by a high percentage of carotene. This can be destroyed by a high level of the enzyme lipoxidase. Hence varieties must have high carotene content (minimum 4.5 ppm) and low lipoxidase content. Because of the relationship between yellow colour and consumer acceptance the amount of pigmentation has been used as a selection tool for good quality durum.

Spectrometric Measurement

The quantification of major carotenoids is a challenging task; one of the difficulties results from the variation in the carotenoid composition of different crops (Rodriguez-Amaya and Kimura, 2004; Kimura *et al.*, 2007). Knowing a crop's carotenoid composition is essential for determining which measuring method to use. Due to its sensitivity and selectivity, high performance liquid chromatography (HPLC) is the method of choice to quantify individual carotenoids and their isomers. Other methods may not distinguish between individual carotenoids that differ in their provitamin A activity. However, HPLC equipment is more specialized and more expensive to run. For example, cost per sample can range between US$50 and US$70. The low throughput of 15–45 samples per day makes it unsuitable for rapid screening purposes. Spectrophotometric methods are generally simpler to use. However, they have less capacity to distinguish between different carotenoids and their isomers. For example, a thin-layer chromatography (TLC) method only separates the three different carotenoid groups (β-carotene and α-carotene; β-cryptoxanthin; and lutein and zeaxanthin) but does not detect the presence of trans and cis isomers of β-carotene; cis isomers are known to have lower provitamin A activity and can be present in significant amounts compared with trans isomers. It may be adequate to use a spectrophotometric method, for example, on cassava or orange-fleshed sweet potato, where there is typically one major carotenoid–all trans β-carotene. In maize, however, the major carotenoids are zeaxanthin and lutein, which do not show provitamin A activity, but have much smaller amounts of β-carotene and β-cryptoxanthin. With TLC, lutein and zeaxanthin interfere with the detection of other carotenoids, for example, β-carotene. Hence, for maize, use of HPLC is necessary.

β-Carotene Content

In durum wheat, the yellow color of semolina which imparts golden yellow amber colour to pasta product is desirable and, therefore majority of the pasta consumers prefer the yellow pigment. Xanthophils and specially β-carotene (precursor of vitamin A) contributes to the colour production in the semolina. High lipoxygenase activity has been described to be responsible for an appreciable loss to β-carotene. Linoleic acid acts as a source of substrate for the enzyme for linoleic acid hydro peroxides which ultimately oxidize the pigment. β-carotene acts as a preservative. Durum endosperm contains twice the concentration of β-carotene than that of *aestivum*. Since, β-carotene is highly susceptible to oxidation, and hence, precaution has to be taken for its determination. The yellow colour of semolina is imparted mainly by β-carotene pigment and can be influenced by particle size, enzyme content and processing conditions. Yellow pasta is considered as a mark of quality by many consumers as it indicates that the product is made from durum wheat.

Total amount of β-carotene (ppm) = 0.0105 + (23.566 x O.D. value)

Yellow pigment is essentially a preferable feature of durum wheats. Range of β-carotene is generally 4 to 8 ppm but durums with less than 5 ppm of β-carotene are not acceptable in the international market. Some countries offer little preference to yellow pigment.

Colour Grade

It is performed by kent-jones Martain colour grader, can be used to estimate the degree of contamination of flour with bran particles (30 g flour and 50 ml water). The colour readings have been faced to correlate with the rate of extraction of flour and the corresponding ash contents.

Table 8.19. Classification of flour grade based on colour grade readings

Flour Grade	*Colour Grade Readings*
High grade Patent flour	1.0-1.5
72 per cent extraction flour	2.0-4.5
80 per cent	5.0-7.5
83 per cent	8.0-12.5

Hardness Index

Wheat hardness is a characteristic commonly used in wheat trade and classification. It is to same extent modified by the environment factors, protein, starch and their interaction. The extremely hard wheats require more power for milling. Alternatively, excessive softness interferes with efficient sieving and increases sieving space. Hardness can be measured practically through, Peter, Scott, pearler, Ogawa Satki Hardness tester. Near infra-Red reflectance of ground whet and particle size distribution technique.

Protein Estimation by Kjeldahl Method

The protein content in wheat varies from 6-22 percent. It is influenced both by the genetic and the environmental effects. More of rainfall during the kernel development leads to a low protein content whereas dry spell during this period favours high protein content. It is also influenced by soil N_2 level and timing of its application. It is established fact that the loaf volume is having a linear relation with the protein content between 7-20 per cent. In addition, the higher protein content is associated with a higher water absorption and mixing times. Protein content influences the functional quality of the product and has a greater influence on overall processing quality than any other single factor. While the environment is the major determinant of the actual protein level achieved and the variability between wheat varieties is their capacity to accumulate protein is also important.

Flour dough strength is related to the combination of total protein and also protein quality. It is therefore influence by the environment, but also has a major genetic component. The protein requirements of wheat for various products like chapatti, bread, biscuit and pasta are different. Protein percentage of more than 12 per cent in *aestivum* wheat is needed for good bread, < 10 per cent for biscuit, 10 to 12 per cent for chapatti, and > 12 per cent in durum wheat for good pasta products. Protein quantity and quality are important in noodle making. High amounts (10-14 per cent) of strong protein produce needles with a chewy, elastic texture. Flour with too low a protein content gives noodles with poor cooking tolerance. High gluten quantity (10 per cent dry gluten) and its superior quality as indicated by SDS sedimentation values (68 cc-macro test) result in pasta which is resistant to overcooking, has good flavour and does not break during transportation.

Gluten is a mixture of different proteins of wheat, gliadin and glutenin. Gluten is not present in flour, however when flour is moistened and kneaded into small balls, the development of gluten takes place. It is a rubbery mass made from proteins, lipid, fibrous components of wheat flour. Only wheat flour possesses, such an important feature amongst all the cereals that is why it can be used for the preparation of wide varieties of end products, like, bread cakes, puffed products, pastries, puddings where the inflation of the wheat dough is required. The gluten has elastic properties like rubber band. The interiors of the elastic structure are just mush like work, where the gasses are trapped and help in increase in loaf volume. Twenty five grams of wheat flour is kneaded into balls with 60 per cent moisture content. The balls are left suspended in water for half an hour and then the starch is washed away in running water. The resulting material is a rubbry mass called gluten (wet). It can be dried at 100°C to get dried one. The results are expressed as per cent dry/wet gluten. Often the vital gluten is added to a weak flour to increase its strength. The functional quality of wheat proteins is reflected in amount and strength of gluten. The gluten holds air/gas and is responsible for the puffing of chapaties, rise bread and crispness of cookies and crackers. Gliadin protein provide binder, gluten provides skeleton. Different wheat varieties contain gluten content that varies in quality. The quality and quantity of gluten affects the cooking quality of paste products and also charters ties of bread, chapaties, biscuit, cakes *etc.* The gluten content is largely a varietal character but is also influenced by the climate. Excessively high temperature and low relative humidities during the maturation period as generally encountered in India have a marked deleterious effect on the quality of gluten. The few tests to evaluate gluten or protein quality.

Separation of Gliadin

Polyacrylamide gel electrophoresis (PAGE) is a method of separating dissolved protein components on a polyacrylamid gel support medium according to their molecular size and electric charge. Gliadin proteins are the alcohol soluble proteins of wheat endosperm. The gliadin electrophoregram is a pattern of separated gliadin components at the completion of electrophoresis.

Sample Preparation for Gliadin Analysis

Single seed or half seed was crushed into fine powder using metal crusher or

20 mg flour was used for extraction. 100 µl of 70 per cent ethanol was added to the powder and incubated at 37°C for 30 minutes with brief vortexing at the time interval of 10 minutes. The tubes were centrifuged at 10,000 rpm for 10 minutes and the supernatant was collected. The residue left was used for glutenin extraction. 62.5 µl of the dilution buffer was added to 50 µl of the supernatant obtained. 20 µl of the sample was loaded into the well of the gel for A-PAGE separation of gliadins. The electrophorograms of the gliadin proteins are conventionally evaluated according to Bushuk and Zillman (1978) based on relative mobility. The relative mobility (Rm) values are calculated with reference to a specific band, designated 50, of the variety 'Marquis' (a Canadian hard red spring wheat variety). The Rm value of a given band is obtained by dividing the distance the band migrated from the origin by the distance band 50 migrated from the origin and then multiplying this value by 50. The value obtained is the Rm for that band. Genetic distances were calculated based on presence or absence of bands taking band 50 as reference, using Hierarchical cluster analysis of SPSS programme and dendrograms were drawn using average linkage between groups and Euclidean distance method.

Table 8.20. Recommended uses of different flours on the basis of protein content and nature of gluten

Flour Protein Content (Per cent)	*End Use Suitability*	*Nature of Gluten*	*End Use Suitability*
11.0-12.5	Bread	Strong gluten	Bread, noodles
8.0-9.5	Biscuit	Medium strong gluten	Chapaties
7.0-8.5	Cakes	Weak gluten	Biscuit, cakes
9.0-11.0	Chapaties		

Nutritional Quality

The nutritional quality of proteins depends upon the amount of lysine (first limiting amino acids) present. It is determined by UDY method. The quality, index is an indicative of the lysine status of the grain. Protein content is very important because of all flour properties *e.g.* water absorption and mixing times, mixing tolerance, dough handling, requirements, loaf volume and crumb grain *etc.* are correlated with protein quality and quantity. Protein content is required for wheat flour used in bread making, intermediate for all purpose flour and, low for cookies and lowest for cakes. It is determined by different methods. As high as 35 per cent of world population depends on wheat as the staple diet. Therefore, nutritional adequacy is very essential. Most cereals lack amino acid balance. Lysine is the most limiting amino acid in wheat. No serious efforts have been made to breed varieties with better nutritional quality because a variety can not be accepted primarily due to its better quality in preference to grain yield.

UDY Dye Binding Procedure

In breeding programme involving quality it is very important to scren and further cross materials having high protein quality, as well as quantity in terms of limiting amino acids like lysine. Hence DBC method is used in the initial stages of screening. The dye binding method employs as monosulphanic age dye, acid arrange 12. The amionic buffered dye solution binds with cationic immidozole, guanidine and amino gps which originate from the Histidine, argenine and lysine as well as from free amino acid end groups of protein chain. The resulting insoluble protein dye complex is filtered or centrifuged and unbound dye concentration is colometrically measured. Because argenine and Histidine vry little across lines within most cereal grain, the dye binding procedure is widely used as a rapid screening method for lysine. Lysine determination is performed only on materials where protein quality index has already been worked out:

QI = DBC/per cent protein

Ambati *et al.* (2019) evaluated 600 durum wheat test germplasm and 4 checks, namely HI 8498, HI 8713 and PDW 233 of durum wheat and HI 1544 of bread wheat in an augmented design to identify genetically diverse, agronomically and nutritionally superior germplasm accessions and to develop a core collection. Plant height, days to flowering and maturity, TGW, yellow pigment, protein, starch, and wet gluten content were estimated. Yellow pigment was recorded using AACC standard method 14-50 (AACC 2000). Other quality traits like protein, starch and gluten content for germplasm samples were estimated by near-infrared (NIR) spectroscopy using FOSS Infratec 1241 Grain Analyser.

The composition of wheat is the same as that of flour of 100 percent extraction. It is evident that flour is a poor source of vitamins and has lesser quantities of protein, fat calcium, phosphorous and iron. But the white flour has the advantages like that it produces white bread, has better cooking quality and has lesser phytate phosphorous. The nutrient content of bread on fresh basis is about two third of that of the flour. But if allowance is made for higher water content then the nutritive value of bread is similar to that of the flour from which it has been produced. In general wheat and wheat flour is considered to be low in thiamine, niacin, riboflavin and iron. In view of this, the enrichment of flour has been recommended in many countries for the four nutrients namely thiamine, niacin, riboflavin and iron so that these nutrients could be made available to all income groups through flour. The enrichment has been considered necessary for flours of low extraction rate. But flours of extraction rates of 80 per cent and above have not been found to be detrimental to health of people. The fibre of wheat grains is associated with a substance Phytic acid (inositol hexa- phosphate). This forms insoluble compounds with certain minerals particularly calcium, iron and zinc and impedes their absorption. So high extraction flours which contain higher content of bran are high in Phytic acid and thus tend to immobilize calcium and iron. Phytic acid in animal diets has been reported to decrease the availability of zinc. The effect of Phytic acid can be overcome by the addition of calcium carbonate (chalk). So it has been recommended that flours of 85 per cent extraction and above should be fortified with specified amounts of chalk. It has also been observed that the absorption of zinc bound by phytic acid is increased by autoclaving or vitamin D addition. Wheat contains starch (carbohydrates about 74 per cent) as the principal component and thus basically is an energy giving food. But wheat is a staple food of a large segment of our population. It contains protein as second highest component and the usual proportion of protein is 10-12 percent.

Since it is taken in larger quantities it becomes an important source of dietary protein. According to an estimate, out of the total protein consumed about 22 per cent is provided by wheat. Protein are made up of amino acid. The wheat proteins are found to be deficient in an important essential amino acids lysine followed by threonine and methionine. The flour produced from wheat has still lesser quantities of lysine. When flour is baked into bread, there is losses of certain essential amino acids particularly of lysine which may be to the extent of 10-20 per cent.

Protein Content

The protein content of wheat grains can vary from 6 per cent up to as such as 25 per cent depending upon the growing conditions. Genetic variation for protein content is much less than that due to differences in growing environments, making this a difficult character for improvement by plant breeder. The availability of nitrogen is the major factor determining the protein content of the grain. In the major wheat - exporting countries, such as Canada, USA and Australia, protein content from 12 per cent to 15 per cent is normal, and nitrogen supply is not usually limiting. This is because other factors such as water availability or the length of the growing season restrict carbohydrate production, so there is a high concentration of nitrogen available relative to carbohydrate. Conversely, in a high yielding environment even when high levels of nitrogen are supplied to the crop, relatively low protein contents may still be obtained. Thus, a crop of 2 t/ha at 14 per cent protein is typical of many areas in USA or Canada, compared with a typical UK crop of 7 t/ha at 10 per cent protein. There is a strong negative relationship between grain protein percentage and grain yield. Plant breeders can attempt to offset this relationship in breeding programmes by selecting for the rare combination of higher than average protein content with high grain yield. The rare varieties which have high grain protein without a yield penality may achieve this by more efficient relocation of nitrogen from senescing-tissues to grains or by a more efficient uptake of nitrate and ammonia from the soil. Alternative strategies include the transfer of the high protein trait from some diploid and tetraploid species related to bread wheat.

Lysine Content

Where wheat is the major source of protein for people of developing countries, the nutritional quality of grain protein becomes of importance. As with all the common cereals, the first limiting essential amino acid is lysine, so in relevant breeding programmes major aim is to increase the amount of this amino acid. There is a negative relationship between lysine content and the protein content of a grain (anon., 1980). As the protein increases from 7 per cent to 15 per cent of the total dry matter, the lysine content expressed as a percentage of total protein, falls from 4 per cent to 3 per cent. This is because increasing the protein concentration upto 15 per cent significantly increases the ratio of storage protein to metabolic and structural proteins in the grain the former being lysine deficient and the latter two relatively lysine rich. However, Eggum (1984) has argued that storage proteins are more digestible than structural proteins. So far practical purposes it may be better to opt simply for increased protein content when seeking an improvement in lysine.

From nutritional point of view a better wheat should have :

A balanced composition of protein, fat, carbohydrates *etc.*

Protein

7 per cent — 9 per cent — 12 per cent — 14 per cent — 16 per cent

Biscuits and all bread macoroni *etc.*

crackers purpose making

Protein should be rich in amino acid particularly limiting ones.

Leu-lysine ratio

High diastatic activity and sugar content

Low phytic acid so that it does not interact with metabolism.

High minerals and vitamins.

Physico-Chemical Tests

Various physical and chemical tests used for quality estimation are.

Moisture Content (AACC 2000)

The moisture content is one of the most important factors affecting the quality of wheat. Since the dry matter is inversely related to the moisture content, the same has a direct economic importance. The dry sound wheat can be kept for years if properly stored, but wet wheat may spoil within a few days. Small variation in the moisture content near critical level of 13 per cent may considerably influence the keeping quality. The moisture content determination is important for reporting various results at a particular moisture content. The 14 per cent moisture basis is most commonly used. Result are also expressed on dry matter basis. The moisture content of cereal grain is an important factor for storage, processing and marketing point of view. The less than optimum moisture content may cause injury to the embryo, while excess moisture may lead to the development of moulds and other pathogens, thereby affecting seed storage. The methods commonly used for estimation of seed moisture content. Bring air oven temperature to 130°C. Ground seed sample (2-3g) is put in a moisture dish and kept for 60 min. thereafter, the dish is transferred to a desiccators and final weight of the sample is recorded and expressed as percentage.

Moisture Content in Grain/Flour

The amount of water in grain is also economically significant. The moisture is of utmost importance in the storage of grains. Water is present in free and bound form. Free water conducts electricity while bound water acts as non- conductor. Moisture meter act on the presumption that the ratio of free to bound moisture for grain is constant. OSAW and moto-moco moisture meters are based on this principle.

Pearling Index (Peter Scott Pearler)

Hardness is reflected into greater flour recovery. More coarse semolina and less flour are produced from highly vitreous wheat than from less vitreous wheat. Wheat to qualify as durum must be of specie of *T. durum* and must contain at least 50 per cent vitreous kernels. For the determination of pearling index, the 20 g sample is taken and allowed to be abraded in pearler for 1 min. The grain is separated from bran in a hand sieve in 20 motions. Clear grain is weighed and pearling percent is calculated as.

$$\text{Pearling Index} = \frac{(20\text{-}10)}{20} \times 100 = 50 \text{ per cent}$$

Where, A is the weight of clear grain.

Table 8.21. Classification of grain texture based on pearling index

Pearling Index	*Texture of Grain*
Below 29 per cent	Hard
29-40 per cent	Semi-hard
41-50 per cent	Soft
Above 50 per cent	Very soft

Near-Infrared Reflectance Spectrophotometry

The parameters like protein, hardness water absorption, sedimentation value and loaf volume were tested using Informatic 8100 which uses the light in the Near Infrared region (1400-2400) to test the samples. The reflected light from both the sample and the references is collected in a gold-plated integrating sphere in which the detectors are mounted. The amount of given substance will be propositional to the amount of light absorbed by the sample at given wave lengths. Carotenoids show absorption in the visible and infrared regions of the electromagnetic spectrum, and NIRS has revealed potential for screening a range of crops for total carotenoids and provitamin A. Brenna and Bernardo (2004) applied NIRS to determine carotenoids and the vitamin A precursors β-carotene and β-cryptoxanthin, which are relevant to breeding biofortified maize. Cross-validation procedures indicated close associations between HPLC values and NIRS estimates for major carotenoids. In sweet potato and cassava, β-carotene predominates among total carotenoids, and NIRS screening of the two crops revealed correlations between total carotenoids/β-carotene, as determined by HPLC and NIRS estimates, ranging from >0.80 to >0.90 (Wolfgang Grüneberg, CIP; Thomas zum Felde, personal communications). Based on experience to date, NIRS shows good potential for estimating carotenoids and provitamin A carotenoids with high or medium-to-high precision; determine iron and zinc with the sensitivity required for pre-screening; and predict with high accuracy antinutrients, such as phenolic compounds and promoters, and value-added traits, such as protein, oil, or kernel hardness, which affect fl our yield in milling. Since multiple traits can be determined on the same sample, the costs per compound will be proportionately lower. Hence, NIRS may provide a very inexpensive and rapid method for screening large numbers of genotypes for a wide array of traits.

Table 8.22. Classification of grain texture based on NIR reflectance Index

Proposed Classification	*NIR Reflectance Index*
Extra hard	85-100
Very hard	73-84
Hard	61-72
Medium hard	49-60
Medium soft	37-48
Soft	25-36
Very Soft	13-24
Extra soft	0-12

Ogawa Seiki Hardness Tester for Single Grain and Measured as kg/Seed

Single grain hard nest tester measures the force in kg required to crush the grain. Harder seeds require more weight than softer grain. *i.e.*, if seed coat is hard or soft indicating the texture of grain as either hard or soft.

Visual Screening of Provitamin A Carotenoids

The crop-specific variation in the carotenoids profile reflects differences in the association between provitamin A concentration and visual colour intensity; it also determines the suitability of using colour intensity for visual grading or selection with colour charts when pre-screening for provitamin A. Colour charts can be used for cassava (Chavez *et al.*, 2005) and orange-fleshed sweet potato (Simonne *et al.*, 1993), and for crops in which β-carotene or provitamin A constitutes the major portion of total carotenoids. For maize, visual colour is dominated by the non-provitamin A precursors zeaxanthin and lutein, and inexpensive high throughput visual selection can only be applied to separate white or light yellow maize grains from dark yellow and orange colour types.

Benzene Sedimentation Test

Ash Content

Ash is an indication of milling efficiency. The mineral content of bran is about 20 times that of endosperm. The ash test is fundamentally indicator of purity of flour or the through ness of the separation of bran and gram from the rest of the wheat kernels. Pure endosperm-m yields very less ash content. Ash (mineral content) is determined by charring the weighed amount of grain sample in a muffle furnace maintained at 650°C, for 2-3 h. White powder left in the ash crucibles is measured, which gives percent mineral content. This test can be used to estimate the degree of flour contamination with bran particles. For hard wheat, the temperature at 57.5-59°C is used for charring. One-gram dried grains of two *Ae. kotschyi* accessions, *T. aestivum* cultivars WL711 and CS (PhI), a BC_2 and two BC_1F_2 derivatives each were cleaned thoroughly and kept for incineration at 600C for 10 h. The ash was further processed like the grains for AAS analysis.

Phenol Test

To assess the admixture of *T. aestivum* in *T-durum* wheat lots or contamination of superfirne rice kernels in Basmati type, phenol colour test is done. A representative sample of a test lot is taken by count into a petri dish and dipped in 10 ml of 3 per cent phenol reagent. The pure durum kernels are not affected at all where as *T. aestivum* turn black. Similarly, pure basmati lines are not affected at all.

Flith Test

The rodent hair and insect fragment counts of flour is determined by digesting the flour with acid and adding the cooled digest to petrol in a separatory funnel. The hair and insect fragments are trapped at the petrol/water interface, and can be collected and identified microscopically.

Particle Size Analysis

It may be determined by standard sieving tests, by Simon Funnel method, by means of Coulter Countries, by lasr light scattering *etc.* Particle size index can be used, for assess the wheat hardness rather than by an objective test.

Table 8.23. Particle size analysis by laser light scattering

Hardness Classification	*Psi Per cent*	*Pearling Index*	*SDS Test ml*	*Major End Use*
Hard	Below 20	Below 29	Above 68(S)	Bread, Noodles
Semi Hard	21-26	29-40	48.68(ms)	Cakes
Soft	Above 27	>40	Below 48(ms)	Biscuits

Flag Leaf Analysis

Flag leaves from selected potential donors, recipient parents and their F_1 hybrids were collected between ear emergence and before anthesis, washed thoroughly with N/10 HCl, dried at 80C for 8 h in oven prior to digestion. Dried leaf samples were then digested as a minimum of five replications using mixture of nitric acid and perchloric acid (Zarcinas *et al.,* 1987). Iron and zinc concentrations in the digests were analyzed by AAS.

Performance Tests

Cooking Characteristics

The cooking characteristics of pasta products are considered by many to be number one in importance. In general, cooked spaghetti and macaroni products should be neither "mushy" nor "rubbery", but should retain their shape during cooking and should be firm to the "bite". Cooking time is important both from the standpoint of relative speed of cooking and the tolerance of the product to overcooking. Texture of the cooked product refers to the "finger feel" and "mouth feel" of the product. These are determined subjectively by the customers but objectively in the laboratory by the use of an instrument and are referred to as "tenderness score". Cooking residue determines the amount of material sloughed off during the cooking and can be determined quantitatively. A high percentage of residue is undesirable. Cooked weight or expansion volume is a measure of the water absorbing capacity of the product during cooking. The cooked weight should increase about three times the weight of the dry material. The volume of the cooked product or swelling degree should be about twice that of the dry product. In general, high quality spaghetti and macaroni products should have a good overall appearance. They should possess a deep yellow-amber colour and be free from specks, checks or cracks. When cooked they should expand to about twice their original size but still retain their original shape. When this highly nutritious food is eaten, it should not create the impression of being "mushy" nor rubbery but should be relative "firm" with a pleasant nut-like flavour. Not all grain quality factors relate to the end-use properties of the wheat but are important from an economic standpoint. The correlation studies conducted indicate that the protein is the best single indicator of quality. In a predictive sense, if we picked high test weight, high protein content and quality, and high endosperm colour. We should stand an excellent chance of picking the best quality durum for making pasta.

Cookies, Chapaties, Bread and Cake Baking; Cooking of End Products Prepared from Semolina *viz.*, Noodles, Spaghetti

Flour Test baking procedures are chosen in relation to the ultimate purpose for which it is intended-*viz.* bread making, cake making, biscuit making tests are the final criteria by which the quality of bread, cake and biscuit flour respectively, are judged. However, the condition under which such tests are conducted must be carefully standardized. Moreover the interpretation of the results is not wholly objective, hence simpler, more objective tests have been sought, the result of which will correlate with precast. All the mentioned tests give an idea of flour type. No test has perfect correlation with the actual bread making performance or end product utilization of wheat. The ultimate is the actual performance test or the end use test which gives the true picture of a variety. Since performance tests require larger quantity of materials they can be conducted only on advanced materials. In early generation segregating materials, when sample number is large but sample quantity is limited, only tests which require less material can be conducted. These tests as given above are indicative of ultimate quality of a variety. The chapaties are prepared under controlled conditions in a way almost similar to that of a housewife. The chapateis are than evaluated for puffing, colour, sweetness, flavour and chew ability *etc.* it has been observed that ideal wheat for making chapaties must have protein between 9-12 percent, sedimentation value ranging from 30-39 cc, water absorption of above 68 percent pelshenke value of above 100 minutes, stability value of 100 mm-125 mm combined with strength of 15-24 cm^2, extensibility of 20 -28mm, sugars. The flour with above specifications makes into soft, silky pliable chapaties. Since atta in the form of chapaties accounts for about 80 per cent of wheat consumed in India, study of chapati-making quality of Punjab wheats had been accorded a high priority by the wheat breeders. The characteristics which make chapaties more appealing are puffing, colour, softness and its keeping quality.

Extruding Semolina for Pasta Manufacture

In manufacture of pasta, the semolina is made into a stiff dough using 25-30 per cent of water at 32-38°C semolina weight basis and mixed for 10-15 min in a experimental premier. The material is than transferred into an extruder maintained at 30°C and kneaded in a cylindrical machine equipped with beveled helical blades. The kneading is carried in vacuum because of the absence of air bubbles in the extruded product is advantages. The product is brighter and more transparent. The dough is than extruded through a die of a experimental press. Heavy pressure (1100-1980 1 bs/sq inch) is employed to ensure that the product is translucent and small air bubbles are squeezed out. The extruded product is cut into requirement length and dried at 45-92°C for period of 6-28 hrs. The dried product is then evaluated organoleptically by a panel of judges and scored on 0-9-point hedonic scale. The modern drying process take place in three stages preheating, heating and drying. Drying temp erasure are 45-90°C for 6-28 hours. A three stage drying process using micro wave energy has been described by Kats Kae (1978). Good quality macaroni should be creamy in colour. When broken the fracture should appear glassy. When boiled in water for 10 minutes, the macaroni should swell to twice its original volume, but should retain its shape and firmness without becoming pastry and without disintegrating. Dry pasta product has minimum 11.5 per cent protein, max 12.5 per cent moisture content and 0.60-0.85 per cent ash. Couscous made from durum semolina and water paste, which is dried and ground, has excellent keeping qualities. Durum semolina is also used for making semolina puddings.

Wheat Quality-Future Requirements

This deals with the development of an evaluation system to improve the end-user quality of wheat through cultivar selection. The main focus will be on the completed wheat system, which took almost 10 years to develop. Once it was successfully implemented, the industry followed the same process. This process was, however, far more difficult, as the quality information is consumed as a staple food is not readily available in the world. Towards the end of the development of the criteria for the end quality requirements, the animal feed industry also commenced with a process to identify quality attributes in, which is favourable for users.

Development of a System

The process to define the goals of the breeders and to share the problems that are experienced by the practical millers and bakers in the industry regarding quality characteristics. Millers and bakers were shown the glasshouses to experience first-hand the problems of the breeders. The breeders were taken through the mills and bakeries to experience the results of all their hard work. The role players involved in this process agreed that a set of release criteria should be negotiated to define new goals, which focus on the end quality requirements that are demanded by the consumer, baker and miller. It was decided to identify certain cultivars, which could be used as biological standards, and that all new varieties should be equal or better than the biological standards in an effort to improve the quality of the wheat crop. This process also ensured that no new varieties would be released in the market that would not meet the end requirements of consumers and users.

Table 8.24. Primary attributes and Secondary attributes of wheat

10 Primary Attributes		*7 Secondary Attributes*	
Test weight	80 kg/hl (62 lb)	Baking test 100g	
Falling number	250 sec (min)	Volume	950m^3
Protein flour	12 per cent (m/b) 11 per cent min	Dough characteristic	Very good – non sticky
Extraction	80 per cent	Alveogram	
Colour (KJ)	76 per cent 0,5 per cent	Strength cm^2	40
Mixogram peak time	2.8 min	P/L value	<1
Farinogram absorption	62 per cent	1 000 kernel mass	38 g (1.4 oz)
		Farinogram:	
		Development time	6.0 min
		Stability	7.0 min
		Alveogram	
		Stability	90 mm
		Distensibility	90 mm

Estimation of Micronutrient Content

Micronutrient malnutrition affects more than one-half of the world's population, especially women and preschool children (SCN, 2004). Reaching the Millennium Development Goals (UNDP, 2003) to reduce the under-5 child mortality ratio by two-thirds and the maternal mortality ratio by three-quarters between 1990 and 2015 will require additional technologies and approaches to improving nutritional status, which is an important determinant of these mortalities. The nutritional health of humans is dependent primarily on plant foods either directly or indirectly (DellaPenna 1999). During the past 40–50 years, the major emphasis of plant breeding programs had been to increase crop productivity for meeting the calorific requirements of the world population. More than 2 billion people depending largely on cereals, roots and tubers as staple food suffer from disorders related to micronutrient iron and zinc deficiencies commonly referred to as 'hidden hunger' (FAO 2002). Among the most widespread nutrient deficiencies, the Fe deficiency is estimated to affect more than 2 billion people worldwide and the suboptimal Zn nutrition is more common than previously believed (Stoltzfus and Dreyfuss 1998). Cereal crops vary considerably for micronutrient content in grains and other plant parts. In wheat only 20 per cent of Fe and 70 per cent of Zn, that is taken up by the plant are translocated to the grain, and these are primarily deposited in aleurone layer. Large proportion of micronutrients deposited in the aleurone layer is lost during milling and less than 50 per cent of whatever is consumed, is absorbed by the body due to presence of phytic acid. Some of the wild species of wheat with higher levels of micronutrients might be very efficient in micronutrient uptake and their partitioning to grains as compared to the cultivated germplasm that incorporated little variability for the traits during evolution and domestication. In India alone, the prevalence of anemia ranges from 47 to 98 per cent in different states (FAO 1998).

Conventional and molecular breeding and genetic engineering techniques have been considered to be the most feasible and cost-effective approaches for biofortification of cereals with high iron and zinc content (Bouis, 1999). **Biofortification** is a method of breeding crops to increase their nutritional value. This can be done either through conventional selective breeding, or through genetic engineering. Biofortification differs from ordinary fortification because it focuses on making plant foods more nutritious as the plants are growing, rather than having nutrients added to the foods when they are being processed. Biofortification of staple food crops for enhanced micronutrient content through genetic manipulations is the best option available to alleviate hidden hunger with little recurring costs (Monasterio *et al.*, 2007). Identification of germplasm with high grain Fe and Zn and understanding the genetic basis of their accumulation are the prerequisites for manipulation of these micronutrients.

Breeding for Quality Traits

Since the green revolution era, the breeding programmes to ameliorate grain quality protein malnutrition through improvement of wheat grain protein content and quality have received low priority because of pressing demand for high yielding varieties to feed the growing population of the country. Nevertheless, a small group at IARI have pursued their efforts to develop high protein cultures in a high yielding background. Using the Canadian spring wheats as high protein donors, the high protein potential was transferred to Indian wheat genotypes and Pusa-5-3 was developed. Further crosses with various amber coloured, bold seeded, Indian genotypes and high protein content donors from Plant Breeding Institute, Cambridge in the last decade have resulted in to a number of strains with high protein content, high 1000 kernel weight and good yield potential. Recently emphasis on improving yield and disease resistance in the breeding programme while not loosing sight of the high protein potential have provided some elite strains with 15 per cent protein. At present, the wheat breeders in the country are focussing on incorporating durable resistance to biotic stresses, especially the rusts in a high yielding background, combining one or more quality parameters. In the All India wheat improvement programme all the genotypes entered in the Initial and advanced trials are tested for the quality parameters during evaluation. But for these, there are no directed efforts in India to breed wheat varieties specifically with good quality parameters. The development of molecular tools for targeted breeding for quality traits is envisaged as a futuristic strategy to achieve these goals. All the wheat forms cultivated in the Indian subcontinent are spring types and are grown during the winter season in both plains and hills upto an elevation of 2500 m above sea level. Hexaploid common wheat (*Triticum aestivum*) is by far the most important cereal crop throughout the subcontinent. The durum wheat (*T. durum*) is next in importance and its cultivation is confined to central and peninsular India, primarily under rainfed condition. Emmer wheat (*T. dicoccum*) is grown on a limited scale in south India and Gujarat.

Breeding for Industrial and Nutritional Quality

Wheat provides essential nutrients to human beings and a range of different food is prepared from it. Bread wheat is generally milled into fl our (both refined and whole meal) and made into leavened breads, flat breads, biscuits, and noodles, whereas durum wheat is milled into semolina and consumed as pasta, cooked grits, or couscous, and in some Middle Eastern countries it is baked into bread. The genetics of wheat industrial quality is well understood and our understanding continually increases as more alleles and their effects are discovered. Some attributes, such as protein content and alpha-amylase activity, are influenced by environmental factors, whereas others are not. Protein content tends to be higher when the plant is under stress and lower under well-watered or N-limiting conditions. Protein content affects qualities such as dough strength, dough mixing time, and loaf volume. The largest fraction of total protein is gluten, made up, in turn, of glutenins and gliadins. Gluten influences the viscoelastic properties of wheat fl our and largely determines how a particular variety is used. While a relatively small portion of total variation in protein content across years and locations is genetic, the quality of protein is controlled by known high and low molecular weight glutenins and gliadins. As shown in Figure 5.10, loci that control both high and low molecular weight glutenins can be determined using ID SDS-PAGE (Peña *et al.*, 2002) and the information can be used to make better crosses and ultimately, higher quality cultivars. A number of rapid, indirect quality tests are available to the plant breeders that can be applied in the early generations to increase the probability of identifying progeny with the desired quality profiles. These include near-infrared reflectance (NIR) to measure protein content, grain hardness, and even milling yield (Zeleny, 1947) and SDS-sedimentation (Axford *et al.*, 1979) tests can be used to approximate gluten content and–indirectly–gluten strength. Dough rheological properties can be measured in different ways; some methods are time-consuming but accurate, *e.g.*, the Alveograph, and others are faster and less expensive, but slightly less accurate, *e.g.*, the Mixograph. Peña *et al.* (2002) provided an overview of the application of these and other methods in wheat breeding. The decision about which tests to apply and at what stage in the breeding process they should be applied depends on cost, time, heritability, and program objectives. If the primary objective of a breeding program is yield and industrial quality is secondary, then quality tests should be introduced after yield testing. Trethowan *et al.* (2001) showed that selection for quality, particularly protein content, before yield testing significantly reduced the number of high-yielding lines in a breeding program. A very high emphasis on improvement of leavened and flat bread quality characteristics. Improved wheat materials developed in recent years are more similar to Canadian and Australian wheats, which are known to possess good bread quality characteristics and carry very similar high and low molecular weight glutenin subunits. Millions of people worldwide suffer from micronutrient deficiencies, particularly vitamin A, Fe, and Zn (Graham *et al.*, 2001). In recent years there has been a realization that not only should processing and product quality be high, but so should the nutritional value of wheat products. To date there has been no vitamin A, conditioned by beta-carotene, found in wheat. All yellow pigmented materials have contained only lutein. There is significant variation for Fe and Zn grain concentration, however this variation within the range that most nutritionists agree is required to impact upon human health is known to occur only in unadapted wheat relatives, such as spelt wheat, diploid

Aegilops tauschii and some wild tetraploids. To complicate wheat breeding further, the inheritance of Fe and Zn is quantitative in nature, affected by environment and influenced by seed shrivelling. As much of the Zn and Fe are associated with the aleurone layer of the seed (Graham *et al.*, 2001) there is a dilution effect of these nutrients as yield potential increases. Nevertheless, Trethowan (2006a) showed that it is possible to simultaneously increase yield and Fe content across a broad geographic area; wheat lines developed in Mexico with higher yield and Fe content maintained their performance when grown at seven sites in Pakistan.

Introgression for Grain Yield and Quality Traits from Tetraploid Triticum Species into *Triticum durum*

Wheat is one of the main food crops of India and contributes significantly to the central grain pool. Once a regular importer, today India has become the second largest producer and exporter of wheat in world. Though the area during last 40 years has just doubled the production has gone up more than six times and productivity has increased by three times. Among the cultivated wheats, durum wheat occupies second place both in respect of area and production in the world. It occupies about 10 per cent (≈ 2 million ha area) wheat cropped area in India and contributes almost one tenth of the wheat production. Increasing global demand along with better price in market and resistance to Karnal bunt are the factors which make durum an export commodity. The third *Triticum* species being cultivated in India is *T. dicoccum* or khapali wheat, is also a tetraploid species like *T. durum*. Its cultivation is restricted in the peninsular parts of the country with roughly 50,000 ha of area under its cultivation.

In India durum wheat is grown in Madhya Pradesh Gujarat, Maharashtra, Karnataka and Southern Rajasthan (Central and Peninsular Zone) and in Punjab (North Western Plain Zones). The traditional durum wheats were tall and highly susceptible to yellow rusts leading to their exit from farmers' fields. Durum wheat cultivation was reintroduced in North Western Plain Zone of India including Punjab State in the 1980s after lapse of about half a century. Subsequently, two major pockets of durum wheat cultivation developed in the Punjab state: one in Gurdaspur district and other in areas of Ludhiana, Fatehgarh Sahib, Sangrur and Patiala districts Today durum wheat has emerged as an important alternative crop to diversify the bread wheat monoculture and to combat the menace of Karnal bunt in the submontane regions. Also durum wheat shows field resistance to Karnal bunt and Loose smut diseases of wheat. Under rainfed conditions, durum wheat cultivars are found to out yield their bread wheat counterparts. Durum wheat varieties PDW 233, PDW274 and PDW291 are the best existing cultivars in India and yield at par with the bread wheat cultivars under favorable conditions. Durum wheat is more important for the stringent quality control as special processed products are made from it. Durum wheat endosperm is milled into coarse granular form called as semolina and this is used to make different products like pastas, breads and also used as such by making steamed preparations. Durum has a yellow endosperm, (because of presence of beta-carotene) which gives pasta its color. These specific end uses of durums make them fetch higher prices both nationally and internationally. Therefore, breeding for quality durums keeping in view the specific products made and consumed in different countries is of paramount importance. But durums could only be a crop of choice if the yield levels are higher or at par to the bread wheat. The breeding objectives for durums are breeding for high and stable yield under adverse environments, breeding for improved quality and resistance to disease and pest. Durum wheat genetically is an extremely narrow based species owing to its evolutionary history. The bottlenecks of breaking the yield barrier and improving the quality could be overcome by using the vast reservoir of genetic variability for productivity and quality in immediate progenitor and other cultivated and related species. Indian durums though possess attractive grains generally lag behind in key quality characteristics like protein content, sedimentation value and pasta quality. Two types of storage proteins, gliadins and glutenins have been found to be associated with making quality products (Payne *et al*). These confer extensibility and elasticity to dough which prevent it from over extension and collapse respectively. Glutenins further consists of two major groups known as HMW (high molecular weight) and LMW (low molecular weight). Most of the durum lines do not possess HMW subunits x and y. However, large amount of variations have been found in the related wild progenitor species including *T. dicoccoides* (Randhawa *et al.*, 1995). The potential markets for Indian durums are possibly the Middle East countries, Africa and Mediterranean countries. Therefore, the strategy for the varietal development has to keep in view the needs of the client. Future durum development attempts in India will cover the quality needs to develop the products specifically for catering to the target export countries. As also about 50 per cent of the projected 2.0 million tonnes per year wheat export from India for the next 5 to 8 years can be durum. The consumption in India is mainly in the form of vermicelli and semolina in Southern parts of the country and as noodles along with vermicelli and semolina in Northern parts of the country

Table 8.25. International products of durums for consumption

Products	Per cent
Two-layer bread (Synonyms- khobz, baladi, Shami)	35
Single layer bread (Synonyms- Tannour, Saag, Mountain bread, Markout, Mehrahrah)	18
Burghul (Granulated, boiled sun-dried wheat used as steamed preparation like rice	15
Cous cous	10
Pasta	15
Freekeh (Green, dried wheat ground in coarse particles and steamed like rice)	2
Feed	5

Table 8.26. Desirable quality factors for flat breads, Cous cous and burghul

Product	Desirable Quality factors
Flat breads	Medium protein, high gluten strength, light color
Cous cous	High vitreous kernel count, high protein
Burghul and pasta	High gluten strength, high semolina yield, yellow color

Table 8.27. International quality specifications for durum wheat

Parameter	*Range*
Moisture content (Max)	14 per cent
Protein content (Max)	(12-14 per cent)
Vitreous kernels (min)	80 per cent
Falling Number value	> 300/sec
Test/hectolitre weight	75 kg
Yellow berry	<10 per cent
Foreign material (max)	1 per cent
Broken kernels(max)	4 per cent

Table 8.28. Quality traits of Indian durums

Variety	*Test Weight (kg/hl)*	*Protein (Per cent)*	*Sedimentation Walue (ml)*	*β–carotene (ppm)*
PDW 291	79	11.8	31	6.3
PDW 274	80	11.3	26	6.4
HI 8498	82	12.2	28	6.5
PDW233	78	11.6	38	7.9
WH 896	78	10.7	42	6.2
Raj 1555	80	11.4	31	3.8
HI 8381	82	13.2	25	4.8
PBW 34	79	12.7	33	5.3

Alien Introgression as a Means to Improve Yield and Quality

The target traits to be improved for better yield are grain biomass per day, spikes per meter^2, grains per spike, days to maturity, grain fill duration and tolerance to hot and humid conditions. The quality parameters are hectolitre weight, vitreousness of grains, minimum yellow berry, greater grain hardness, higher protein content (>13 per cent), improved semolina recovery (>60 per cent), beta- carotene content (> 5 ppm), decreased lipoxidase activity. The related species and genera of cultivated crops are excellent reservoirs of traits of economic importance *e.g.* disease and insect pest resistance, tolerance to abiotic stresses, increased biomass, grain yield and quality characteristics (Stalker, 1980). Plant breeders have used unadapted germplasm, almost exclusively, as a source of major genes for disease and insect resistances and have mostly relied on repeated intercrossing of adapted elite genotypes for improvement of yield (Fulton *et al.*, 2000). Studies have shown that wild relatives, though phenotypically inferior, do possess genes for enhanced productivity and quality. During the last decade, convincing evidence at both morphological and molecular level has come forth for utility of wild progenitor species as donors of productivity alleles in some crops. Introgression of yield enhancing genes from unadapted germplasm into adapted cultivars have been reported in oats (Choubey, 1996), chickpea (Singh *et al.*, 2001), rice (Xiao *et al.*, 1996,

1998), and tomato (Bernacchi *et al.*, 1998). Deodikar *et al.* (1979) reported increase in grain yield in the cross between *T. durum* x *T. polonicum*. Cantrell and Joppa (1991) compared 13 disomic substitution lines of durum wheat, each containing a different individual pair of *T. dicoccoides* with each other and with the recipient durum parent. Some genotypes exhibited an average of 39 per cent higher grain yield than the recurrent parent. Similarly, other genotypes had the highest kernel weight. Palve and Raghavaiah (2002) observed some early maturing derivatives of cross of *T. timopheevi* x *T. durum*. Short statured, early flowering derivatives with increased number of spikelets per spike and grains per spike were recovered in the cross *T. aestivum* x *T. durum*.Ciaffi *et al.* (1995) reported some lines derived from durum x dicoccoides crosses having high dough strength and acceptable level for baking. Randhawa *et al.* (1995) identified *T. boeoticum*, *T. dicoccoides* and *T. araraticum* lines carrying both Ax and Ay subunits

Table 8.30. Related and wild species identified for some traits for introgression

Species	*Genome*	*Trait Identified for*
Ae. cylindrica	CCDD	Spike length, spikelet no and size
T. carthalicum	AABB	High tillering, grain weight
T. polonicum	AABB	Drought tolerance, grain weight
T. dicoccoides	AABB	Biomass, grain weight
T. timopheevii	AAGG	Cms source, resistance to rust, powdery mildew, hot and water logging conditions
T. dicoccum	AABB	High protein, 1000- grain weight, High tillering
Ae. kotschyii	UUSS	Heat tolerance

The breeding strategy followed for these introgressions is backcross method along with the help of molecular markers. The number of backcross may vary according to the need of programme but two to three backcross on large number of F_1 and BC progenies supported by selection for recurrent parent is helpful in speedy recovery of the recurrent parent genotype. Pathak (2001) reported the results of a study conducted to determine if related tetraploid *Triticum* species could enhance the grain yield/components of durum wheat. The plant material consisted of *T. durum* cv. PDW 233 as the recurrent parent and several accessions of *T. dicoccum*, *T. dicoccoides*, *T. polonicum*, *T. araraticum* and Extracted Chinese Spring as the donors. All the accessions of *T. dicoccum*, *T. araraticum*, *T. dicoccoides* (except one) had higher tiller number than the recipient parent PDW 233, while *T. polonicum* and *T. durum* did not differ from PDW 233 for this trait. The recurrent parent had 1000- grain weight and number of grains per spike than most of the donors. Selection for increased tiller number and seed size was carried out in BC_1 and BC_2 generations. Selection for longer spikes as the criterion for higher number of grains per spike did not give the desired results. Introgressed lines were evaluated for yield and component traits during *rabi* 2001 in randomized block design. Most of the donors contributed positive QTL/alleles for increased tiller number. The increase of upto 178 per cent for tiller number per plant in the BC_2 progenies on the recurrent parent was recorded in the crosses PDW 233*3/*T. dicoccum* Acc.5366 and PDW 233*3/*T.*

dococcoides Acc. 5259. In the best plants selected from the progenies for higher tiller number, an increase of 100 to 300 per cent in the tiller number was recorded with 21 per cent to 84 per cent increase in grain yield per plant. Similarly, all the donor species except *T. araraticum* and extracted Chinese Spring contributed QTLs/alleles for increased seed size. An increase of up to 25 per cent for seed size was recorded in the crosses, PDW 233*3/*T. dicoccum* cv. DDK 1001, PDW 233*3/*T. dicoccum* cv. DDK 1009 and PDW 233*3/*T. dicoccoides* Acc. 13986. An increase of 17.5 to 65 per cent for seed size over PDW 233 was recorded in the best plants selected from the best progenies having increased grain weight. This was accompanied with an increase of 21 to 174 per cent grain yield per plant. An increase in grain yield upto 58 per cent was observed in BC_2F_2 progenies of the cross PDW 233*3/*T. dicoccum* Acc.5366. Anuradha (2003) evaluated BC_2F_3 and BC_2F_4 generations derived from the crosses of *T durum* cv. PDW 233 with various accessions of *T. dicoccum* and *T. dicoccoides* for grain yield/components and quality traits. 100 plants were selected on the bases of desirable agronomic and quality parameters from each of the BC_2F_3 populations. BC_2F_4 progenies were compared to the recurrent parent for various characters. Many progenies superior to the recurrent parent were recovered. For example, 187 for days to flowering, 18 for reduced plant height, 7 for total biomass, 3 each for grain yield and harvest index, 20 for higher protein content, 2 for higher hectoliter weight, 42 for less yellow berry, 8 for grain hardness and 3 for sedimentation value. Puja (2003) evaluated BC_2F_2 generation derived from the crosses of *T durum* cv. PDW 233 and PDW 274 with *T. dicoccum* and *T. dicoccoides* for agronomic traits. Useful introgression was observed for earliness, plant height, and grains per spike, tiller number and grain yield. Verma (2004) evaluated F_5 progenies derived from *T. aestivum* x *T. durum* crosses for yield components and quality traits. Improvement was observed in the progenies for days to 50 per cent flowering, spike length, plant height, number of tillers per metre, grains per spike and grain weight. Similarly, for quality characteristics, improvement was recorded for protein content, beta carotene content, test weight and sedimentation value.

Quality Status of Durum Wheat Lines for Evolving Pasta Varieties for Central India

Triticum durum Desf. is the second most important wheat species grown in India. Compared to bread wheat, *durum* commands a higher price in the international market, (Feb. 2001 price: durum wheat US $ 195/tonne and bread wheat US $ 130/tonne). Thus, there is scope of exporting durum at premium prices to west European, north African and west Asian countries, if its production is increased and the quality is improved. Recently, consumers in India have started recognising the importance of durum wheat but still it has to go a long way to establish its identity. The main irony in India is that durums are procured at the same price as bread wheat and quite often mixed with it, thereby, totally ignoring its significance. Durum wheat is mainly used in preparation of pasta, suji and other fast food products (Dick, 1981). Semolina milling yield and pasta quality are the most important considerations in durum, which are controlled by protein content and gluten strength in the cultivar. In addition to higher protein, durums contain "vitamin A" (β-carotene) and hence, can help in fighting malnutrition among children and rural population. Vitamin A, a lifesaving micronutrient, acts as a regulator of the growth and activity of epithelial tissues. It plays an essential role in the visual cycle because of its participation in the synthesis of visual purple by the retina. β-carotene is converted to vitamin A in the liver and is stored there. In human beings, vitamin A requirements are only partially covered by the intakes of vitamin A as such. Normally provitamin A (β-carotene), which are converted to vitamin A in the body, provides two third of the requirements. To improve the quality of existing durum wheat cultivars, the donars known for quality are being used in crosses. Three durum varieties released from this centre *i.e.*, Malavshree (HI 8381), Malav Shakti (HI 8498) and Malav Ratna (HD 4672) are gaining popularity in Central India. The area in Malwa plateau under these varieties is more than 20 per cent of the wheat area and production is more than 1 million tonnes, which can be made available for pasta products, nutritional utilisation and export. Donors with superior quality parameters *i.e.*, WH 896, PDW 233, HD 4676, Trinakaria and GW 1139 were crossed with agronomic bases having high yield and disease resistance. A total of 133 crosses involving HI 8498 (in 22 crosses), PDW 233 (17), HD 4676 (7), HD 4672 (9) and MACS 3125 (13) were attempted and subjected for evaluation for making three way and double crosses. About 150 crosses have been attempted during Rabi 2000-01, involving high yielding lines like HI 8498, HI 8591, HD 4672, HD 4692, HI 8653 and other improved lines with good quality like PDW 233, WH 896, HD 4676, Trinakari and GW 1139.

Quality Breeding: Problems and Prospects

Most quality traits are polygenic. Therefore, selection for quality traits during the segregating generations is difficult. Quality traits further enhance the long list of traits and a breeder has to select for in order to develop an acceptable variety. Many quality traits are difficult to estimate and evaluate. Therefore, quality breeding imposes considerable demand on resources, including money. Many quality traits have low heritability and are markedly affected by the environment. This retards the progress under selection. Quality is not a single trait; in fact, it is a conglomerate of traits. Unlike yield, it does not have a single estimate. The various quality traits have to be assayed and then interpreted to define the produce quality. In view of above, a breeder has to familiarize himself with physiological and biochemical aspects of the quality traits; this means additional burden to the breeder. Opaque endosperm of the P-721 opaque high lysine derived sorghum lines was not liked by consumers. However, this problem was eliminated by DES mutagenesis of the P-721 opaque mutant. Most high lysine and/or high protein lines show poor grain filling and lower yields. Even if breeders are able to improve yields of such lines, they will need ~11 per cent and more nitrogen for every increase of 1 per cent in protein content. But nutritional quality is not a basis for pricing of food grains. Therefore, high protein/lysine cultivars are not likely to be popular with farmers unless quality is taken into consideration for pricing of the produce. The definition of quality in bread wheat depends on the purpose for which wheat is being used. It has been defined as the summation effect of soil, climate and the seed stock on the wheat plant and the kernel components particularly the gluten proteins. Old Indian wheats were prized for their quality. The quality characteristics were a part and parcel of selection criteria. But advent of semi-dwarf Mexican strains put all these considerations aside due to their high yield potential. Despite recurrent predictions that world's food resources would run out, international grain trading is still a buyer's market with fierce competition and strong interaction between quality and price. Consumer is very much quality conscious and also the industrial use is fast increasing. It has made the quality aspects much more important. Quality in wheat may be studied from many points. However, we will concern ourselves only for grain quality. The major emphasis in wheat quality research is to develop varieties suitable for various end-use products such as chapati, bread, biscuit and pasta products are made from wheat. For the purpose, following research achievements have been accomplished in wheat quality programme. Under the coordinated and other research programmes, evaluation of thousands of germplasm lines has facilitated identification of genetic resource useful for improvement of processing and nutritional quality of wheat. Allelic diversity of HMW and LMW glutenin subunits has been explored using both protein and DNA based markers and relationship with gluten strength identified useful for the improvement of bread, biscuit and pasta making quality of wheat. Micro-level tests for solvent retention capacity and sedimentation test were developed and applied in evaluating early segregating generations to improve wheat quality. Molecular basis of grain hardness has been studied in detail and found useful in improvement of biscuit making quality of Indian wheats. The major problems are :

Negative Correlation between Yield and Protein Content

It is a common observation, especially in cereals, that as the grain yield increases, the protein content in grains appreciably decreases. Liang *et al.* (1969) established this fact in sorghum. The reduction in yield is much pronounced in maize populations carrying opaque-2 or floury-2 genes, yielding about 8-10 per cent less than their normal counterparts (Sreeramalu *et al.*, 1970).

Negative Correlation between Protein and Lysine Content

Johnson (1976) reported that on the screening of 12,600 common wheat varieties, there occurred a depression of lysine per unit protein as the protein increased from 10 to 15 per cent. A negative association between protein and lysine has also been reported in corn and sorghum. Such a strong negative correlation between lysine, as percentage of protein and total protein, imposes a dilemma to breeders engaged in quality improvement (Wilcke, 1976). Implicit in this is high yield with low protein but of a better quality (more lysine) and vice-versa.

Poor Stability of Total Protein and Lysine Content

The production environment exerts a strong effect on the protein content. Availability of nitrogen in the soil, disease incidence and a close spacing at a single site tend to reduce the environmental variability of protein content. Further, in most cereals, except oats, lysine (per cent of protein) also shows a tendency to decrease as application of N-fertilizer, which raises the total protein amount (Munck, 1972). Similarly, protein content and amino acid fraction in most of the food crops vary over locations as well. This poor stability of protein content dampens quality breeding efforts.

Lack of Proper Field Screening Techniques

Many procedures for a chemical analysis of the protein and amino acid profile have been developed. Some of them are remarkably rapid and reliable. But all these require double screening (first in the field and second in the laboratory) hence the final selection is a time-consuming process. Moreover, the field selection is likely to be attended by rejection of genotypes, which are otherwise highly nutritive. Thus, an absence of suitable and reliable field techniques for mass screening of genotypes, right in the field conditions, leaves the breeder much to chance.

Lack of Price for Quality Produce

Most high lysine and/or high protein lines show poor grain filling and lower yields. Even if breeders are able to improve yields of such lines, they will need approximately 11 per cent more nitrogen for every increase of 1 per cent in protein content. But nutritional quality is not a basis for pricing of food grains. Therefore, high protein/lysine cultivars are not likely to be popular with farmers unless quality is taken into consideration for pricing of the produce.

Genetic Studies

Several lines with excellent grain quality have been used in genetic studies and in quality improvement. notable among these are Atlas 66, 93123 to 93133 series, Chris, Crim, Naphal, WG 357, Tobari 66, *etc.* A world collection (> 4000 lines) has been screened at Nebraska University in USA and the results are : Protein content varies from 9 to 21 per cent and lysine content from 1.77 to 4.15 per cent. Protein content is negatively correlated with grain yield, grain size and lysine content.

Kaulr and Shaikh (1977) evaluated several genotypes for quality, the results are:

Variety	*Protein*	*Lysine Per cent*	*Remarks*
Normal	10	2.5	
Atlas 66	99.9	3.0	
NapHal	20.5	3.2	
A x N	upto 20		
Yugoslavian mutants	upto 20		
Indian mutants	Upto 16	3.0	Sharbati Sanora

Johnson (1974) has reported high level of protein and other acceptable agronomic characters in wheat.

Kernel Size	*Protein*	*Lysine Per cent*
Large	12.7	**3.1**
Medium	12.2	**3.2**
Small	12.5	**3.1**

Size	Configuration	Cerease	Protein Per cent	Lysine Per cent
Large	Plump	Closed	12.0	3.15
		Open	12.6	3.13
		Wrinkled	14.3	3.02
Large	Plump	Closed	11.8	3.31
		Open	12.9	3.24
		Wrinkled	15.4	3.04

D genome contributes to bread making quality (Shebeski, 1958). Morris *et al.*, 1966 4B, 5D and 7B have factors for bread making. Schmidt *et al.*, 1966 1D and 1B also influence bread making. Maystrenko, Troshina and Evamakova (1973) reported chromosomes 28, 1D to carry genes for good flour quality. Morris *et al.*1973. said that 5A, 5B and 5D carry genes for high protein content. Zitelli, Vallega and Bianchi (1983) reported linkage between height and high protein genes. Ht genes to be important for protein differences among genotypes.

Prospects for Quality Improvement

The production and productivity in different crops have reached at high levels in most of the crops. A further jump in production is contemplated by the manipulation of genes using various breeding/biotechnological approaches. Increased emphasis on better nutritive and processing quality will improve the commercial value of a particular crop. New avenues are opening up with the globalisation of international trade. In this context, long slender aromatic Basmati rice varieties with desirable amylose content, better quality durum wheats, high oil double zero and triple zero Brassica, long fibre *hirsutum* and short-course *arboreum* cottons, *etc.* will have good export potential. After golden rice the Dream rice development is to be focussed in IRRI according the Indian Scientist of IRRI S. K. Datta. The dream rice will contain the core vitamins, Iron, beta carotene and lysine. Lines with desirable genes would be developed and they would be crossed with other to make one variety of rice – A fully nutritive rich rice variety. This could be a very perfect diet for use by malnourished women and children and the people who need it the most. While the conventional plant breeding for quality improvement would continue to play its role, recent developments in recombinant DNA technology offer greater opportunities for rapid progress in quality improvement of major crops. Another approach is to engineer microbes to synthesise storage protein for human food. It appears possible that in not too distant further it may be possible to exploit genes for nutritionally improved storage proteins. Thus, there are bright prospects of breeding for quality in various field and horticultural crops. In millets development of hybrids/varieties with fairly high levels, tryptophan and lysine is possible since the protein content and amino acids are governed by both additive and non-additive gene action and have weak or non significant correlation with yield and it is possible to achieve high level of protein and essential amino acids. At present pearl millet does not have much industrial use. Identification suitable genotypes and processing device will make pearl millet an industrially competitive, improving grain appearance, increasing protein and reducing rancid factors could improve quality and accelerate research into alternative uses of pearl millet such as brewing, milling and baking. In pulses the discovery of major genes for nutritional factors would revolutionise breeding for nutritional quality, especially if such genes are easily identified phenotypically. Introduction of genes for better nutritional quality through genetic engineering is the most promising areas of quality improvement in pulses. With the view that the deficiency of sulphur containing amino acids of pulses being balanced by the cereals, much of the research needs to be concerted on elimination of antinutritional factors. In case of oil seeds genetic engineering may be greatly used to increase the stearic acid content in *B. napus*. Similar approach may be employed to increase quality of different oil seeds by making desired change in fatty acid composition. Varieties of sugarcane should possess traits according to their utility in different industries. For sugar industries sugar recovery, sucrose content in juice and purity co efficient of juice, for distillery industries juice volume, higher sugar and for paper industries higher cellulose, lower ash content, decreased pith content or high fibre/ pith ratio are desirable quality traits expected from sugarcane. Several mutants with improved protein quality have been identified in cereals like maize, sorghum and barley. Grain filling has been the major problem in the high lysine mutants of barley, maize and sorghum was successful in combining plump grains and high protein and/or lysine in wheat and barley. In all high Lys mutants, except Hiproly barley, prolamin content has been reduced, *e.g.* in high Lys maize and sorghum mutants prolamin is reduced to a level comparable to that in normal barley, while in high Lys barley it has come down to the level present in rice and oats.

Table 8.31. Varieties released for quality through mutation breeding

Crop/Variety	*Origin*	*Mutant*	*Features*
Wheat			
Sharbati Sonora	India	Dry seeds (12 per cent) ^{60}Co gamma rays (Sonora 64)	Amlur grain colour, Higher protein and lysine content
Pusa Lerma	India	Seeds, gamma irradiation (Lerma Rojo 64-x)	Amlur grained higher extensibility, elasticity and pelshenke value, suggesting stronger gluten
Stadler	USA	Thermal neutrons (Mo W6243)	Strong straw, excellent soft quality and winter hardiness
Novosibirs-kaira 67	USSR	Air-dry seeds 5 Krad ^{60} co gamma rays (Novosibirskaia 7)	Baking quality
Barley	Spontaneous	Hiproly (*lys*)	High protein, Lys
	EMS	Notch-1	High protein Lys
		Notch-2	High protein Lys
	EI	Riso 1508	High lysine
		Canor - Trunpf x Nordal 1976	Good malting quality
Vienna	Austria	Dry seeds, 9400 rad x -rays	High 1000 kernel weight
Pennrad	USA	Thermal neutrons	Increased winter hardiness

Crop/Variety	*Origin*	*Mutant*	*Features*
Jutta	Germany	5 KR X-rays	Increased winter hardiness
Mikns Golden promise	UK	G-24 Krad Gamma rays	Stiffer straw, good grain yield
Midas	UK	RM 759/10 x Milns golden promise	Stiff straw
Diamant	Czechoslovakia	Seeds (10 Krad x-rays)	Very good grain and malting quality, lodging resistance
Pallas	Sweden	Pre-soaked seeds, Y 350 rad X-rays	Stiff strain
Mari	Sweden	Dry seeds, 20 Krad x-rays	Stiffer straw
Kristina	Sweden	Selection from Domen x Mari	Good straw
Gunilla	Sweden	Bringiltax (opal x vega (x "blue lodging - resistant mutant 44/3 from Gull")	Good straw
Bonneville 70	USA	Seeds, 15 Krad gamma - rays	Improved Threshability
Haya shrinriki	Japan	Air-dry seeds 40 Krad ^{60} co gamma rays	Good quality

Table 8.32. Scheme in breeding program for Quality Evaluation Programme (Kaul 1969)

Generation	*Tests*	*Population*	*Seed Needed*
P1 and P2	All possible	Seed from different locations	Severa kgs.
F1	None	Selfed	-
F2	Pelshenke, 1000 k.wt. Prot. per cent	200 plants	5g
F3	Pelshenke, 1000 k.wt. Prot. per cent + sedimentation	100 lines	10-15 g
F4	Pelshenke, 1000 k.wt. Prot. per cent + sedimentation	70 lines	10-15 g
F5	Pelshenke, 1000 k.wt. Prot. per cent + micro	40 lines	500 g
F6,F7,F8	Pelshenke, 1000 k.wt. Prot. per cent + micro bread/biscuit/reducing and non-reducing sugars	20 lines	2-5 kgs
F9	Pelshenke, 1000 k.wt. Prot. per cent + panel tests for chapati bread and biscuits	5 good lines	5-10 kgs
Variety	Pelshenke, 1000 k.wt. Prot. per cent + fat,ash, fibre, thiamine lysine *etc.*	Variety grown over various locations	Several kgs

This scheme is a tentative one and may be modified according to particular requirements. It may be mentioned that this scheme is for bread making quality. In modification to the above Gill (1979) has suggested a scheme for chapati making quality as under:

Table 8.33. Material Tests in different generations

Generation	*Tests*
Germplasm	Colour, size and texture of grains; protein content, stock pelshenke value, Sedimentation value
Promising quality for hybridization	A. grain characters flour, size, texture, pearling B. Milling characters - bran percentage, recovery of atta (whole meal) C. Atta characters - protein content; pel value, damaged starch; diastatic activity; alveographic and micrographic characters; lysine, methionine and theonine content; ash content, phytic acid content; tyrosinase activity. D. chapati making properties - puffing, appearance and texture, aroma and sweetness. Keeping quality, water content of fresh made chapati and water content after fourhour storage.
F1	No test needed
F2	Colour, size, shape and texture of grains
F3 - F5	Colour, size, shape and texture of grains, protein content and pelshenke value.
F6 - F8 *etc.*	Colour, size, shape and texture of grains, protein content and pelshenke value + bran percentage and recovery of atta sugar content, diastatic activity water absorption, progenies in alveographic and mixographic characteristics,
Preliminary replicated yield trials	tyrosniase activity, ash content, and phytic acid content, micro chapati-making quality tests.
Varieties in final trials	Detailed analysis of grain, milling atta and chapati making traits, panel tests for acceptability of chapatis.

Table 8.34. Constitution of major cereals and pearl millet grain (Values per 100g of grain)

	Wheat	*Rice*	*Maize*	*Sorghum*	*Pearl Millet*
A. Constituent					
Moisture (per cent)	12.80	13.00	14.90	11.90	12.40
Protein (per cent)	11.80	6.90	11.10	10.40	11.60
Fat (per cent)	0.90	0.40	3.60	1.90	5.00
Fibre (per cent)	0.30	0.20	2.70	1.60	1.20
Carbohydrates (per cent)	74.10	79.20	66.20	72.40	67.10
Calcium (per cent)	0.02	0.01	0.01	0.03	0.05
Phosphorus (per cent)	0.09	0.28	0.33	0.28	0.35
Iron (mg)	1.00	2.80	2.20	6.20	8.80
Magnesium (mg)	139.00	48.00	144.00	140.00	125.00
Copper (mg)	0.49	0.72	0.19	0.55	0.55
Sodium (mg)	18.00	3.00	6.00	7.00	10.00
Potassium (mg)	349.00	110.00	290.00	321.00	402.00
Oxalic acid (mg)	8.00	3.00	6.00	10.00	14.40
Calories (K cal)	349.00	348.00	342.00	349.00	360.00
Carotene (mg)	29.00	9.00	90.00	47.00	132.00

	Wheat	*Rice*	*Maize*	*Sorghum*	*Pearl Millet*
Thiamine (mg)	0.12	0.27	0.42	0.37	0.33
Riboflavin (mg)	0.07	0.12	0.10	0.13	0.25
Niacin (mg)	1.20	4.00	1.40	1.80	2.30
B. Amino acids+ (g/16 gN)					
Tryptophan	1.20	1.00	0.60	1.10	1.74
Threonine	2.70	3.70	4.00	3.60	3.92
Lysine	2.60	3.80	2.90	2.70	3.01
Methionine	1.40	1.70	1.90	1.70	1.82
Isoleucine	4.10	4.50	4.60	5.40	4.78
Leucine	6.30	8.20	13.40	16.10	10.71
Cysteine	2.10	1.30	1.30	1.70	1.03
Phenylalanine 5.37		4.60	4.80	4.50	5.00
Valine	4.30	6.70	5.10	5.70	5.73
Arginine	4.50	5.50	3.50	3.80	5.14
Histidine	1.90	1.60	2.60	1.90	2.32
Tyrosine	3.50	4.40	6.10	2.80	3.41

Source: Aykroyd *et al.* (1951).

Marker Assisted Selection for Quality Traits in Wheat

Wheat is a major cereal crop of the world which is grown in more than 70 countries across the five continents (Lantican *et al.*, 2005). Bread wheat (*Triticum aestivum* L) and durum wheat (*Triticum turgidum* L) supply nearly 55 per cent of carbohydrates consumed worldwide (Gupta *et al.*, 1999). Wheat quality is a broad term and can mean different things to different people. For a farmer, a healthy and attractive looking crop which can fetch good price in the market is quality. For millers quality is production of white flour per unit grain. For cereal processors quality wheat is one which gives high quality end products like cakes biscuits, pasta products, noodles, flat breads or high-quality animal feeds. As the end product determines the quality of wheat, wheat unsuitable for one may be an excellent choice for another end product. Quality of wheat should be good enough to make high quality end products. In simple words bread wheats are classified as hard or soft on the basis of grain hardiness. Hard wheats are good for bread making and soft wheats are good for cakes and biscuits. While quality in wheat is a very complex trait, it is known that the water-insoluble gluten proteins are responsible for the elasticity and cohesiveness (strength) of dough and are therefore important determinants of breadmaking quality. The water insoluble glutenin proteins encoded by the group 1 homoeologous chromosomes of the wheat A, B and D genomes are mainly responsible for the elasticity and extensibility (visco-elasticity) of dough, which contribute to breadmaking quality. Doughs that have high elasticity and reasonable extensibility are ideal for making bread; doughs which are highly extensible are good for cookies; doughs with intermediate properties are used for making the flat breads in the Middle East and the Indian subcontinent, or noodles in the Far East. Glutenins are composed of both high-molecular-weight (HMW) and low molecular- weight (LMW) subunits based on their mobility on sodium dodecyl sulphate polyacrylamide gel electrophoresis (SDS-PAGE) and are numbered according to their migration in the gel (Payne *et al.*, 1979; Zhen and Mares 1992). The higher molecular-weight glutenin subunits (HMW-GS) are controlled by pairs of tightly linked loci on the long arm of the group 1 homeologous chromosomes. The locus pair contains genes that encode two types of HMW glutenin subunits, one of greater molecular weight, designated the x-type, and the other of lower molecular weight, designated the y-type (Harberd *et al.*, 1986). The A genome y-type gene(s) are silent in hexaploid bread wheats but are active in many diploid and tetraploid wheats. Allelic variation exists at each of the HMW glutenin loci and numerous studies have demonstrated that allelic composition significantly affects dough properties. Most significantly, the HMW glutenin subunits 1Dx5–1Dy10 have been associated with high dough strength and good breadmaking quality, whereas allelic subunits 1Dx2– 1Dy12 are associated with poor breadmaking properties (Payne *et al.*, 1987; Hamer *et al.*, 1992). Another group of proteins termed as gliadins, which are monomers, also interact with glutenins and incorporating themselves in the dough formation with formation of disulfide bonds resulting in protein protein interaction affecting the end quality of dough. The amenability of the starch molecule to enzymatic modifications resulting in waxy (amylose free), partial waxy (low amylose content) and non-waxy (amylose rich) wheats. Waxy wheats, under the influence of three null alleles, and partial waxy wheats, under the influence of one or two null alleles have been developed with biochemical and molecular techniques, the ratio of amylose and amylopectin determines the physiochemical properties of wheat which further determines the end use of the grain.

Phenotypic markers like wheat glume colour controlled by *Rg1* locus is useful as a tool to select *GluB3* alleles encoding LMW glutenin subunits. The correlation between glume colour and gluten strength and the heritability of each trait was estimated in two durum wheat crosses. Brown glume colour appeared to be dominant to white in both crosses. In one cross, glume colour was clearly controlled by one gene while in the other cross it appeared to be controlled by one or two genes with modifiers. The heritability of gluten strength was moderately high. The correlation of F_2 glume colour and F_3 gluten strength was high (r=0.66 and 0.78) indicating that F_2 glume colour was a good predictor of gluten strength in the F_3 generation. Selection for glume colour appears to be an effective breeding strategy for improving gluten strength in those environments where glume colour differences are easily detected (McCulung and Cantrell, 1986). However Yupsanis and Moustakas (2008) reported that in durum wheat gliadin electrophoregrams, protein content and SDS-sedimentation values from the modified SDS-sedimentation test for six Greek cultivars of *Triticum durum* and 98 biotypes possessing white and red glumes indicated that glume colour was not correlated with grain protein content. This highlights the complexity of wheat quality traits and non-reliability on phenotypic markers for quality assessment in wheat. To verify the applicability of this PCR-based method of distinguishing between wheat cultivars possessing the HMW glutenin 1Dx5 gene, Ahmad (2000) analysed DNA samples of five cultivars using primers P1 (GCCTAGCAACCTTCACAATC) and P2 (GAAACCTGCTGCGGACAAG)

(Anderson *et al.*, 1989) specific for alleles Dx2 and Dx5 which produces a 450 bp band. In all the cases the 450-bp amplification product was present only in cultivars known to possess the HMW glutenin 1Dx5 gene (lane 2, Klasic; lane 3, Hartog, lane 4, Neepawa). Cultivars Rata (lane 5) and Chinese Spring (lane 6), encoding subunits 3 and 2 at the 1Dx gene, respectively, did not show the characteristic 450-bp amplification band. At the *Glu-D1* locus of most bread wheat cultivars two allelic pairs, Dx5+Dy10 (good quality) and Dx2+ Dy12 (poor quality) have been detected (Payne *et al.*, 1987). Along with pairs Dx5+Dy10 and Dx2+Dy12, the pairs Dx2+Dy10 and Dx5+Dy12 have also been reported to be present at the *Glu-D1* locus in wheat cultivars. Marker-assisted selection of single alleles, Dx2 or Dx5 assuming their tight linkage with Dy12 (poor quality) and Dy10 (good quality), respectively, may lead plant breeders to select wrong allelic combinations when aiming for bread quality improvement. In this situation, marker assisted selection of a single allele, either Dx2 or Dx5, is not useful for selecting genotypes for bread quality. Therefore, marker-assisted selection of both alleles (Dx2 +Dy12 or Dx5+Dy10) at the same time is a critical factor in wheat breeding for bread quality (Ahmad 2000).

Molecular markers have been developed specifically for the high grain protein QTL. Khan *et al.* (2000) identified flanking microsatellite markers that span a 12.5 cM region of the short arm of chromosome 6B, which contains *QGpc.ndsu-6B1*. Using RFLP markers, Olmos *et al.* (2003) mapped the DIC QTL, now designated *Gpc-B1*, within a 2.7 cM segment. Based on colinearity with rice chromosomes, Distelfeld *et al.* (2004) mapped *Gpc-B1* to a 0.3 cM interval flanked by Cleaved Amplified Polymorphic Sequences (CAPS) markers *Xucw79* and *Xucw71*. As there is a strong relationship between grain yield and protein content molecular markers independent of grain yield must be identified. A significant QTL, *QFpc.agt-6A*, associated with flour protein independent of grain yield has been reported by Kuchel *et al.* (2006). Similarly QTL associated with flour colour have been associated with chromosome 7A. A number of useful molecular marker assisted traits have been reported in wheat however the number of publications detailing the integrated and pragmatic use of molecular markers in quality wheat breeding in limited. In general there has been significant difference in the number of articles with terms like *quantitative trait locus* or *quantitative trait* loci (QTL) and *marker assisted selection* (MAS). A successful application of marker assisted selection for improvement of rust resistance, reduced height and quality in wheat has been demonstrated by Kuchel *et al.* (2007). The target was to produce wheat lines derived from an elite Australian cultivar 'Stylet', with superior dough properties and durable rust resistance donated from 'Annuello'. Molecular markers were used to screen a BC1F1 population produced from a cross between the recurrent parent 'Stylet' and the donor parent 'Annuello' for the presence of rust resistance genes Lr34/Yr18 and Lr46/Yr29. Following this, marker-assisted selection was applied to haploid plants for the rust resistance genes *Lr24/Sr24*, Lr34/Yr18, height reducing genes, and for the grain protein genes Glu-D1 and Glu-A3.

A comparison of marker assisted and phenotypic selection for high grain protein (Davies *et al.* (2006). Population development and line selection schemes are: Selection of high grain protein content (GPC) genotypes by phenotypic selection

(PS) with Near Infrared (NIR) analysis was as effective and under some conditions, more effective than MAS. Perhaps PS was as effective as MAS because selection using NIR accounted for both the large effect of the DIC QTL and expression of other GPC genes. Phenotypic selection was more effective in environments where there was lower overall expression of GPC. In higher GPC environments, the difference in the number of RI lines with high GPC and the mean GPC between MAS and PS was minimal. The reason for this difference could be due to genotype and/or environmental effects (Davies *et al.*, 2006). Although MAS has been successfully applied in cultivar development in the private sector for maize and soybean, there is limited targeted public sector funding to support the large-scale validation, refinement, and application of MAS in field breeding. This is reflected in the annual number of articles with the term "marker-assisted selection," which has consistently lagged behind the number of articles with the term "quantitative trait locus" or "quantitative trait loci" by a factor of three for the past decade, and the gap appears to be widening (Xu and Crouch, 2008) (Figure 1). Moreover, it is likely that only a small proportion of papers with the words "marker-assisted selection" in their text actually report MAS applications as opposed to QTL mapping papers with discussions on the potential MAS application of their research outputs. Most articles on MAS result from either investments from donors with a scientific mandate for, or academic institutions with a specific interest in, demonstrating potential applications of MAS in plant breeding. In contrast, converting promising publications into practical large-scale applications in field breeding requires overcoming many practical, logistical, and genetical constraints.

Testing of Chapatti and Bread Making Quality in Wheat

Wheat is the staple food of majority of the Indian population. It is also an important crop grown extensively in our state and the country. Its breeding is done primarily through bulk method. It is mostly used in the form of "*Chapatties*" as majority of the population prefers these for their staple diet. The wheat to the extent of 80-85 per cent is consumed in the form of chapatties in our country. The consumption of bakery products is also on the increase with improvement in the living standard of the people, increased urbanization and need for convenience foods. The quality of wheat is relative and every consumer has his/her own requirements. It is the satisfaction of these requirements or the quality of the end products produced that determines the quality of wheat. Generally, wheat quality is influenced by its grain size, hardness, protein content, yellow berry incidence and gluten content and quality *etc.*

Table 8.35. Cake and cookie quality characteristics in DBW 14 and its NILs

Biscuit Traits	*Spread Factor (Dia/thickness)*	*Hardness*	*Fracturability (N)*
QBI19-20	11.89	69.67	4.24
QBI19-21	11.44	44.75	5.20
Donor for quality	9.45	51.96	5.49
DBW 14 (RP)	7.30	66.72	7.9
HD 2967	7.1	70.30	8.1

No wheat variety is suitable for producing all sorts of products. The requirements of wheat quality vis-à-vis gluten quality for various products are different, which determine the suitability of different varieties of wheat for various products. The functional quality of wheat is mainly dependent on its inherent quality and quantity of gluten. Most of our wheat varieties produce good to very good chapatties but they do not bake good bread. Our wheat varieties in general do not possess the capacity to produce requisite quantity of elastic gluten, required for baking a good loaf of bread. Keeping in view the needs of consumers and the bakery industry for specific quality of flour, efforts are being made to develop varieties which could meet the needs of the bakery industry and other consumers. The uniqueness of wheat flour for preparation of these products is derived from its gluten forming ability. To achieve the high standard of wheat varieties and to access new markets, cereal chemists devote considerable time and energy in assessing and improving the quality of lines developed by plant breeders. The Indian wheat varieties possess higher proportion of gliadins in their gluten but their gluten/dough exhibit a typical behavior. The Indian wheat gluten is normally strong and non-elastic which rather breaks on stretching. This typical behaviour exhibited by Indian wheat is attributed to the high ambient temperatures prevailing during the ripening period which damage the gluten proteins thereby damaging their elastic properties. It is on this account that the Indian wheat differs from the western wheat. In US or Canada a hard wheat means the wheat containing a higher content of protein and stronger gluten but the hardness in Indian wheat is only physical. In order to improve the baking quality of indigenous wheat, its glutenin content needs to be improved and sensitivity to higher temperatures decreased. The grain quality requirements of each and every product are different. The desirable qualities in good bread wheat flour are high water absorption, a medium long mixing time and good mixing tolerance. The resulting bread should have good colour, grain, and texture and loaf volume. The requirements for producing a good loaf of bread are a high content of elastic and reasonably strong gluten. The balance of two characteristics elasticity and strength is very important. The very strong and less elastic gluten will lead to an insufficient rise of the loaf during baking, leading to the production of a denser loaf with lower volume, on the other hand if the gluten is not strong enough, then it will not be able to resist the pressure of gas and consequently the air cells in the dough will rupture when the gas pressure increases and the loaf may collapse or the crumb may show bigger holes. So, a typical balance between the strength and elasticity is of prime importance. This balance is mainly governed by the proportion of glutenin and gliadin fractions in the gluten.

The quality of chapatties is also influenced greatly by the quality of wheat. The concept of "good chapatties" varies from region to region, as also the preparation. On the whole well filled, bold, hard and lustrous wheat grains having average protein content and medium strong gluten are preferred for baking of quality chapatties. Of these the grain hardness through damaged starch contributes maximum to the chapatti quality. The harder grains when ground to produce atta, suffer a higher degree of starch damage thereby leading to an increase in its damaged starch content. This higher level of the damaged starch in the atta increases its water requirement leading to a higher retention of moisture in the baked chapatties and making them softer in texture. Secondly, the starch splitting enzymes can act only on the damaged starch molecules, the increased availability of such starch leads to an increased production of sugars making the chapatties sweeter and pleasingly flavoured, thereby improving their acceptability. It is mainly due this fact that the wheat grown in the Central India produces better chapatties. Some of you must be knowing the fact that the atta produced on the water mill (Gharat) bakes better chapatties, the reason behind this is the same, the higher content of damaged starch. The content of damaged starch in the atta produced on the water mill is approximately fifteen percent whereas it is only about nine percent in the atta produced on the power chakki. Many screening tests are available for selection of breeding material for different products but final testing is always through the product evaluation. However for screening of breeding material the following parameters can be used.

Table 8.36. Parameters for screening of breeding material for different products

Parameter	*Preferred Value*	
	Chapatti	*Bread*
Grain hardness	Hard	Hard
Protein content (per cent)	9 to 12.0	12.5 and above
SDS Sedimentation value (cc)	35 to 50	50 and above
Gluten content (per cent)	7 to 10	10 and above
Gluten quality	Medium strong	Strong and elastic
Sugar content (per cent)	About four	-
Yellow berry content	Low	Low
Phenol reaction score	Low	-

Malnutrition - Matter of Concern

Malnutrition affects more than 2 billion people across the globe, particularly zinc and iron deficiency causes major health problem in developing world. The biofortified staple food crops such as wheat, is an important channel to contribute to the hidden hunger problem in low income countries. Breeding for enhanced zinc concentration in wheat was initiated by crossing high zinc sources identified among synthetic wheats, *T. dicoccum, T. spelta* and landraces. These crosses have resulted in wheat varieties with competitive yields and enhanced grain zinc were adapted by farmers in South Asia. CIMMYT-derived early-maturity wheat cultivar 'Zinc-Shakti' with about 40 per cent increased zinc (+14 ppm), is now grown in eastern India through public-private partners. The two CIMMYT-derived biofortified varieties: 'WB2' and 'HPBW01' released in 2016 for northwestern plains zone of India. In Pakistan, 'Zincol' was released in 2016. The first high zinc wheat variety (Bari-Gom 33) with better resistance to wheat blast have been released in Bangladesh for commercial cultivation in 2017. Targeted crosses with increased population sizes were used to obtain superior progeny lines that have high zinc levels in combination with other essential traits. This has resulted in the incorporation of several novel alleles for grain zinc and iron in elite, high-yielding germplasm. High zinc and iron are under quantitative genetic control and further progress is possible as multiple

QTL are pyramided in high yielding wheats. High-throughput, non-destructive phenotyping for grain zinc and iron using the X-ray fluorescence (XRF) analysis has facilitated the selection dramatically. Gene discovery and mapping studies leading to the utilization of markers to further improve the breeding efficiency. Rapid adoption of high zinc wheat varieties in South Asia and beyond is expected with the second wave of high zinc wheat lines with superior yield, heat and drought tolerance and resistance to rusts and other foliar diseases. "Micronutrient malnutrition", also known as 'hidden hunger' is a global health problem of our planet and affects the quality of human resources development and productivity of any country. The most important nutrients of public health significance are iron, iodine, vitamin A and zinc. More than 2 billion people of the world are suffering for want of one or more micronutrient deficiency disorders. In India alone, (5-7 per cent) *i.e.* 6.6 – 9.2 million preschool children are suffering from Vitamin A deficiency. Iron deficiency is prevalent among 87.5 per cent of pregnant women and 65 per cent of preschool children. Increasing the death associated with pregnancy and child birth and reduces working capacity. Iodine deficiency has been experienced with reduced working capacity of 167 million people, affiliated with Iodine Deficiency Disorders (IDD). At least 54 million have incidence of goiter or other worst consequences. More than 2 million are cretins and over 65 million have wild neurological disorder. IDD alone causes an estimated 90,000 still births and neonatal deaths each year in India (Begum, 2003). Malnutrition results not only from poverty, but also from absolute ignorance about the nutritional requirement as is evidenced in the vicious cycle.

Nutritional Improvement – A Human Perspective

Plant breeding is an important tool for the improvement of human nutrition for solving the problems of under nutrition and malnutrition. This has become possible as information about genetic variability in respect of the quality traits in various species is now available in most cases. Thus, a strategy for nutritional improvement for different crops can be chalked out. The main idea is that can the plant breeder provide a cheaper solution to the nutritional problem that the nutritionist who can provide the same thing by nutrient fortification or alteration in dietary habits(Axtell,1981). Of course, answers to these questions will vary between various countries or socio-economic groups within the countries. For example, over-population, war and famine have always been the enemies of quality. Quantity conflicts with quality. From the breeder's point of view, new cultivars with improved agronomic traits have been the major factor contributing to increased crop production and there will be a need for continued emphasis in improving these traits. Therefore, compositional traits that are consistent with good nutritional quality can be selected in new plant cultivars. Desired nutritional characteristics must occur in a cultivar that farmers will grow, *i.e.* specific nutritional trait must occur in a cultivar that gives a stable and high yield along with other desirable features. Quality should come as a bonus. Thus, the plant breeder should keep in mind the aspect of quality in the totality of breeding objectives. A character improved in isolation may not be useful to the farmers' *e.g.* opaque maize or high lysine gene in sorghum or barley. Robbelen (1977) proposed three requisites for making effective strategies for modern quality breeding programmes, *viz.*, clear cut information on the nature and priorities of various nutritional criteria; powerful analytical methods for their quantitative determination and sufficient access to genetic variability. Considerable progress has been made in increasing the yield of crop plants. However, the progress in upgrading quality of crop plants has been rather slow.

Quality in verbatum means degree of excellence and is, therefore, a relative term and its improvement is more challenging than of any other trait. Definition of quality varies according to our needs from the viewpoint of seed, crop growth, crop product, post-harvest technology, consumer preferences, cooking quality, keeping quality, transportability *etc.* (Gupta, 2001). The problem of quality improvement is as much of management as of breeding. The methods of breeding may be the same as conventional ones but require a proper understanding of nature of gene control, availability of germplasm, monitoring the manipulation of growth rhythm, developmental patterns and biochemical and physiological processes *etc.* General nutritional goals include improvement of palatability, acceptability, better chemical composition, and energy yield along with high yield. High yield should not be compromised with quality. Also, breeding for quality should safeguard economic interests of the consumer and producer. Quality characters may be classified as: Organolaptic features like taste, smell, aroma; Chemical composition and nutritive value like protein content, amino acid composition, fatty acid composition and antiquality factors; Morphological features, such as, seed size, seed colour (fibre length and strength); Exploitation of genetic variability; Genetic manipulation of growth rhythm; Genetic manipulation of biosynthesis of chemical components. Here we give a brief overview of quality management and genetic improvement towards crop improvement in the near future. For the genetic improvement of chemical traits, we must first study the nutritive value and define parameters to be analysed or manipulated. This should follow the study and evaluation of the germplasm, genetic analyses, and development of sustainable selection criteria. The quality traits are generally influenced by the physiology of growth and development of the organism. It is, therefore, imperative to monitor the growth rhythm and also pinpoint specific parameters to be manipulated by monitoring the growth rhythm. For example, forage quality could be improved by manipulating the growth habit and development of a fodder plant. In grain crops, the present emphasis is on monitoring the biochemical process in the developing grains with a hope to modify the developmental process in a desirable direction, namely, improvement of quality without affecting yield. With the farm income surplus reaching in India to a phenomenal level this year, and the food grain output for India at a targeted level of almost 108.5 million tonnes, the question is, how much of it will reach the consumer in the form in which it has to reach, *viz.*, the healthy food, safety food, quality loaded food and cost effective food. When we look at health, the focus has to be with the nutrition, with to-days world of neutraceuticals coming into play, food is a subject of permanent concern to the consumer. In the developed countries, the question is, "Does the quality meet the consumer needs?" While in the developing countries, it is "Does the quantity meet the consumer demand"? Even though a country produces sufficient quantity of food for its people, whether the people consume food with balanced diet and their food consumed is absorbed and assimilated to meet their average daily requirements is a matter of serious concern.

In addition to its basic calorific value, wheat with its high protein content is an important source of plant protein in the human diet. One of the prominent targets for the application of transgene technology for the nutritional improvement of wheat is targeted at enhancing the grain quality by, i) increasing the protein content, ii) increasing essential amino acids such as lysine, iii) increasing the high molecular weight (HMW) glutenins to improve breadmaking properties of wheat flour, iv) modifying starch composition and v) producing pharma- and neutraceuticals. Amongst the cereals, the flour of bread wheat, *Triticum aestivum*, has a superior capability of forming leavened bread. This superiority stems from the structure and composition of its seed storage proteins, which upon hydration can interact to form gluten, an insoluble, but highly hydrated, visco-elastic aggregate that endows the wheat dough with its unique properties. Although the majority of wheat seed storage proteins participate in gluten formation, biochemical and genetic evidence has demonstrated that high molecular weight glutenin subunit (HMW-GS) plays a major role in determining the visco-elastic properties thereby determining bread making qualities. The HMW glutenins are necessary to create a strong dough, which is essential for making high quality, yeast-raised breads. Strong dough traps tiny bubbles of carbon dioxide gas formed naturally by yeast during mixing and subsequent raising thereby enabling the dough to rise forming leavened breads. Dough strength and the ability to contain gas bubbles is known as visco-elasticity and is an important characteristic of wheat with respect to its end product quality. Increasing understanding of the molecular basis of dough visco-elasticity would thus help in developing strategies for minimizing the effect of unfavourable environmental factors on wheat end-use quality.

Genetic transformation of wheat is a key component in a scheme proposing a complete set of approaches to apply biotechnology to improve wheat quality via direct manipulation of HMW-glutenin genes. To alter the amount and composition of these proteins by genetic engineering, a gene encoding a novel, hybrid-subunit of HMW-GS under the control of native HMW-GS regulatory sequences was inserted into wheat using the biolistic approach. The HMW-GS *1Ax1* gene which is known to be associated with superior bread making quality, was introduced into the cultivar (Bobwhite) lacking this gene. The introduced *1Ax1* gene under the control of HMW-GS promoter was expressed at high levels and stability was maintained for several generations. The results demonstrate the feasibility of manipulating the composition of wheat kernels by genetic engineering and have successfully made changes in both the levels and the types of seed storage proteins. This work also demonstrated the usefulness of tissue-specific promoters for the expression of transgenic proteins in the endosperm tissue of wheat. The prospect of achieving the elusive goal of nutritional improvement of wheat brightened further with the improvement in functional properties of wheat dough due to transformation of wheat with high molecular weight subunit genes. Transformation with one or two subunit genes results in stepwise increase in dough elasticity. The expression of a recombinant protein with x- and y- type HMW-GS subunit in transgenic wheat has resulted in altered gluten polymer assembly and composition. This study thus made it feasible to change the glutenin composition by expressing modified HMW-GS in transgenic plants.

The quality improvement studies were extended to durum wheat by He *et al.* (1999) with the introduction of one of the two subunits of HMW-GS subunit genes (1Ax1 or 1Dx5). Analysis of the expression of the additional subunits in the T2 generation using a mixograph, indicated an increase of dough strength and stability. This study demonstrated the feasibility of manipulating durum wheat for superior bread and pasta making qualities. Similarly, the overexpression of the HMW subunit 1Dx5, in transgenic wheat (*T. aestivum*) resulted in a four-fold increase in this proportion of component in the seed protein and also a corresponding increase in the proportions of the total HMW proteins and glutenins. However, the overexpression of the 1Dx5 gene was found to be associated with a dramatic increase in dough strength by making it too strong, and, therefore, unsuitable for use in conventional breadmaking. Recently, introduced the HMW-GS genes 1Ax1 and 1Dx5 into a commercial cultivar of *T. aestivum* that already expresses five subunits. The overexpression of 1Dx5 gene increases the contribution of the HMW-GS to a level of 22 per cent of the total protein content. In near future, molecular approaches including genetic transformation and marker-assisted selection will provide an opportunity for improving further the wheat processing qualities. Modification of wheat starch is presently targeted in various laboratories to improve its potential utility. Wheat grain is predominantly composed of starch, which is a mixture of two polymers, the almost linear amylose molecules, and the heavily branched amylopectin molecules. The ratio of amylose to amylopectin in starch determines its physico-chemical characteristics and thereby its end-use. Wheat flour, low in amylose content is desirable for noodlemaking as it improves noodle texture. In wheat, scientists are thus working towards increasing the amylopectin content of starch thereby reducing amylose content leading to the formation of a value added, low-amylose flour. The starch branching enzymes catalyse formation of 1, 6-linkages in the glucan polymer and control the amount of amylopectin produced. In recent years, efforts are underway towards detailed characterization of the starch branching enzymes and starch synthases. This increasing information and characterization of the various components of starch biosynthesis will enable researchers to make rational design of novel starches and alteration in starch levels in other crop plants as well.

Bio-fortification

India is ranked 102 in the Global Hunger Index (GHI) out of 117 qualified countries. Nearly 47 million or four out of 10 children in India do not meet their potential because of chronic undernutrition or stunting. This leads to diminished learning capacity, increased chronic diseases, low birth-weight infants from malnourished parents. As per global nutrition report, 614 million women and more than half the women in India aged 15-49 are anemic. Fortification is the practice of deliberately increasing the content of an essential micronutrient, *i.e.* vitamins and minerals (including trace elements) in a food, so as to improve the nutritional quality of the food supply and provide a public health benefit with minimal risk to health. Biofortification is the process by which the nutritional quality of food crops is improved through agronomic practices, conventional plant breeding, or modern biotechnology. Biofortification differs from conventional fortification in that biofortification aims to increase nutrient levels in crops during plant growth rather than through manual means during processing of the crops. Biofortification may therefore present a way to reach populations where supplementation and conventional fortification activities may be difficult to implement and/or limited. Examples of biofortification projects include: iron-biofortification of rice, beans, sweet potato, cassava and legumes; zinc-biofortification of wheat, rice, beans, sweet potato and maize; provitamin A carotenoid-biofortification of sweet potato, maize and cassava; and amino acid and protein-biofortification of sourghum and cassava. It aims at enhancing the nutrient content of staple foods by conventional breeding, opening up a feasible means of reaching malnourished population even in relatively remote rural areas, and delivering naturally fortified foods to people with limited access to supplements or to commercially marketed fortified food. Thus breeding strategy will complement other successful ongoing interventions to reduce the micronutrient malnutrition (Graham *et al.*, 2000). Bio-fortification views agriculture as an instrument for public health and focuses attention or the role of agriculture in delivery of nutrients to humans in balanced amount, that can sustain maximal physical and mental activity of the humans who are simultaneously the drivers of food system and its dependents, aptly confirming the productive, sustainable and nutritious food systems paradigm for agriculture and public health aspects (Welch and Graham, 2000). Such a global challenge program on bio-fortification with proposed funding of US $ 90 million over 10 years has just been approved by the Consultative Group of International Agricultural Research (CGIAR). In this, program, researchers in collaborating agriculture and public health (nutrition) disciplines will apply food systems strategies to deliver more nutritious staple crops to resource poor consumers. Improved grain will be richer in iron, zinc, vitamin A, selenium and iodine as needed. Under this programme, a broadly based survey in the rice germplasm was carried out in Philippines for their iron and zinc concentrations. Though improved cultivars with exceptionally high iron or zinc concentration were not found in the series of screening, several aromatic rice varieties are found among high iron varieties. In a subsequent study of a double haploid population (Azucana, 1964), the linkage between aroma and high iron density was shown to be quite weak, only one of the four QTLs for high iron being located on the same chromosome are as the locus of aroma and at some distance from it (Gregoria *et al.*, 2000). Because of the synergy between these micronutrients in enhancing the absorption and function of each other (Graham *et al.*, 2000), a parallel strategy will be to incorporate high iron high zinc and golden rice traits in that one variety. Such a variety will help overcome not only single deficiencies of iron, zinc or vitamin A but also multiple deficiencies of them and do so more efficiently so that modest expression of each trait (and therefore fewer genetic loci involved), rather than maximal expression may be adequate to solve the problem. This strategy also has the advantage of having an identifiable produce (being yellow) in the market place, and this can be promoted as being healthy for infants, children and pregnant and lactating women. Already the global research on the development of Golden rice is under progress throughout the world. This unique programme aims to fortify the most widely consumed cereal in the whole of Asia, namely rice. This cereal will be enriched with the Vitamin A so that the consumption of this bio fortified food even by the economically weaker sections of the world will largely benefit them against the most widely prevalent malnutritional disorder due to Vitamin A. Besides these, the nutritious cereals as a group are represented by sorghum, pearl millet, finger millet, maize and minor millets like barnyard millet, proso millet, kodo millet and fox tail millet. Over the past three decades, the cropped area of nutritious cereals and their production is decreasing significantly. Of the total cereal consumption, the sorghum, pearl millet, finger millet and other cereals (excluding rice and wheat) contribute 4.32 (3.11 per cent), 2.82 (2.03 per cent), 1.32(1 per cent) and 2.40 (1.72 per cent) per head per annum respectively altogether constituting about 8 percent and the remaining covered by the fine cereals (rice and wheat). Since these coarse cereals are rich in nutrients like starch (56-73 percent in sorghum), proteins, fats, iron and phosphorus and B-complex vitamins and several of the essential amino acids like lysine (71-212 mg per gm of nitrogen in sorghum and 159-380 mg per 100 gm of protein in pearl millet) these nutritious cereals are comparable or even superior than the fine cereals and therefore inclusion of these cereals would definitely ensure the fulfilment of dietary requirements of the malnourished and the healthy people. Research in these coarse cereals should aim in improving the taste and flavour of the grains besides reducing the undesirable effects of the antinutritional factors like tannins and phenols. Development of crop varieties with desirable biochemical properties and increased palatability should be taken up. Due awareness should be given to these nutritious cereals by including these in the regular diet and also popularizing the value-added foods from these through campaigns and adopting measures like including them in the school noon meal schemes. Biofortification is the process of increasing the bioavailable micronutrient density of staple crops through conventional plant breeding and modern biotechnology to achieve a measurable and positive impact on human health. There are marked differences between traditional plant breeding and crop biofortification. Traditional breeding focuses on improving traits of known economic value and developing product concepts for existing markets. Traits are targeted for selection based on whether they can provide better crop and/or utilization options to farmers, but nutritional value as a trait for selection has been largely ignored. Biofortification breeding, on the other hand, seeks to make an impact on human micronutrient status, an endeavour that entails merging breeding with nutrition and socioeconomics research to enhance traits that have measurable value in health outcomes. Biofortification breeding accesses the information it needs to identify such traits by linking directly to the human health and nutrition sectors (Nestel *et al.,* 2006), which have become an integral part of crop improvement and product concept development. At the core of any biofortification breeding program is a product pathway driven by potential impacts of research and nutrition. Collaboration between plant breeding and socioeconomics allows the exchange of information to identify target populations that consume target crops based on their micronutrient burden. The micronutrient burden is the burden imposed on individuals by micronutrient deficiencies and related diseases; it can be quantified using the disability-adjusted life-years (DALYs) approach (Stein *et al.,* 2005). Measuring the effectiveness of biofortified crops in improving human health provides a benchmark for quantifying the ultimate success of biofortification as a cost-effective public health intervention. For biofortification to be successful, micronutrient levels targeted by breeding programs must be derived from nutrition goals set by nutritionists who understand the complexities of making a measurable impact on human health. To set target levels and determine the likely contribution to nutritional status, critical information is needed on the bioconversion/bioavailability of ingested nutrients; retention of the micronutrient after storage, processing, and cooking; human micronutrient requirements; and potential levels of consumption by the target population. Genotypic differences in retention, post-harvest micronutrient deterioration, and concentrations of antinutrients and promoters that inhibit or enhance micronutrient bioavailability have been established. Throughout crop development, nutrition and food technology are engaged in assessing the magnitude of genetic variation and genotypic differences for these traits; this allows breeders to increase bioavailability and determine the effect of micronutrient-dense crops or candidate varieties on micronutrient status via human efficacy trials.

Wheat and Barley Grain Biofortification

Its addresses topics associated with the alleviation of malnutrition in globally diverse populations via wheat and barley biofortification. The book synthesizes the current trends of malnutrition across the globe, the need for wheat and barley nutritional enhancement and how agronomic, microbial and molecular understanding of biofortification can help in devising significant approaches and strategies. In addition, it includes discussions on potential genetic variability available and their efficient utilization in wheat and barley for molecular breeding for nutrients, challenges and opportunities for bioavailability, and technical advancement for analysis of bioavailability.

Breeding for Biofortification/Micronutrient-Dense Crops

"Biofortification" aims to increase a select few nutrients in crops through plant breeding, whether using traditional techniques or through new tools of biotechnology. While there are 40 key nutrients that we must get from our food for good health, biofortified crop research is only focused on three: zinc, iron and vitamin A. Research is currently underway to develop biofortified rice, wheat, sorghum, banana, lentil, potato, sweet potato, cassava, beans and maize in Africa, Asia and Latin America. Part of this research is managed by the Consultative Group on International Agricultural Research divided across three units: the International Rice Research Institute, focusing on genetically modified (GM) rice; the International Potato Centre, focusing on sweet potatoes; and the HarvestPlus programme, which coordinates all the rest. The funding comes from the Bill and Melinda Gates Foundation and USAID, among others. Private research is also being done by PepsiCo, Dupont, Bayer, Nestle and others. Proponents of biofortification argue that it is the cheapest way to address malnutrition: once you breed the plant it can be grown again and again. They often use deceptive language to promote these crops which begins with the very word "biofortified" and suggests that all other foods or plants are inherently weak or deficient. Biofortification is the process by which the nutritional quality of food crops is improved through agronomic practices, conventional plant breeding or modern biotechnology. Biofortification differs from conventional fortification in that biofortification aims to increase nutrient levels in crops during plant growth rather than through manual means during processing of the crops.

Fortification is the addition of key vitamins and minerals such as Iron, Iodine, Zinc, Vitamins A and D to staple foods such as rice, wheat, oil, milk and salt to improve their nutritional content. These nutrients may or may not have been originally present in the food before processing or may have been lost during processing. Micronutrient malnutrition, also known as hidden hunger, is a serious health risk. Sometimes due to lack of consumption of a balanced diet, lack of variety in the diet or unavailability of food one does not get adequate micronutrients. Hence fortification of food is a safe method of improving nutrition among people as the addition of micronutrients to food does not pose a health risk to people. It does not alter the characteristics of the food like the taste, aroma or the texture of the food. According to the National Family Health Survey (NFHS-4) 58.4 percent of children (6-59 months) are anaemic. 53.1 percent of women in the reproductive age group are anaemic. 35.7 percent of children under 5 are underweight. In August, 2018, FSSAI introduced the Food Safety and Standards (Fortification of Foods) Regulations, 2018, to regulate the provisions regarding fortified food. It prescribes the standards of addition of micronutrients for the purpose of food fortification. The manufacturers of the fortified food have to provide a quality assurance undertaking. Packaging and labelling has to state the food fortificant added, +F logo and the tagline "Sampoorna Poshan Swasth Jeevan". It should be in compliance to the Food Safety and Standards (Packaging and Labeling) Regulations, 2011. Fortified salt: In 1950, Indians were among the first countries in Asia to implement mandatory salt iodisation. It is fortified with Iodine. Fortified wheat: The flour is fortified with iron, vitamin A and folic acid. Fortified rice: Iron, Folic Acid, Vitamin B12. Fortified milk: Studies suggest the intake of fortified milk by children not only increased mean serum vitamin D levels but also decreased morbidity rates. It is fortified with Vitamin A, Vitamin D. Fortified oil: Is fortified with vitamins A and D.

Micronutrient Analysis

Mineral micronutrients make up a minuscule fraction of the physical mass of a grain, with concentrations in parts per million (ppm) or even parts per billion range (μgg^{-1} = mg kg^{-1} = ppm). Typical iron and zinc concentrations in major crops range from 5 μgg^{-1} to 150 μgg^{-1}. In crops that show genetic variation for provitamin A or β-carotene, typical concentrations range from >1 μgg^{-1} to >400 μgg^{-1} (for example, in orange-fleshed sweet potato). Although sensitive analytical methods are required to accurately determine micronutrients, the sensitivity requirements in applied breeding may vary greatly for pre-screening large segregating populations and characterizing progenitors for crosses or candidate varieties.

Minerals

Precision analysis methods and state-of-the-art high throughput screening methods (HTMs) that are applied and/or being tested for use in breeding different crops. For precision analysis, the Inductively Coupled Plasma Argon Optical Emission Spectrometer (ICP), Atomic Absorption Spectrometer (AAS),

Table 8.37. Analytical methods for micronutrients.

	ICP-OES: Inductively Coupled Plasma Spectrometer	*AAS: Atomic Absorption Spectrometer*	*XRF: X-Ray Fluorescence Spectrometer*	*NIRS: Near Infrared Reflectance Spectro-photometer*	*Modified Perl's Prussian Blue*	*Modified 2,2 Dipyridal*	*Modified Zincon*
Principle, Throughput, and Practical Considerations							
Principle	Excitation and emissions at various wavelengths	Absorption	X-ray fluorescence	Absorption at wavelengths in the near infra-red	Colour reaction	Colour reaction	Colour reaction
Digestion required	Yes	Yes	No	No	No	Yes	Yes
Sample destructive	Yes	Yes	No	No	Yes	Yes	Yes
Throughput	Up to 2.5 min per sample regardless of number of elements analyzed	~2.5 min per element	5–10 min per sample depending on the number of elements analyzed	~2 min per sample	~4 min per sample	~4 min per sample	~4 min per sample
Pros/comments	Total recovery of nutrient achieved but subject to type of digestion procedure used; good sensitivity	Total recovery of nutrient achieved but subject to type of digestion used; good sensitivity; low cost of purchasing equipment	Total recovery of nutrient achieved; good sensitivity with more expensive models; low cost of bench-type equipment; no digestion step	Low cost	Low cost	Low cost	Low cost

	ICP-OES: Inductively Coupled Plasma Spectrometer	*AAS: Atomic Absorption Spectrometer*	*XRF: X-Ray Fluorescence Spectrometer*	*NIRS: Near Infrared Reflectance Spectro-photometer*	*Modified Perl's Prussian Blue*	*Modified 2,2 Dipyridal*	*Modified Zincon*
Cons/comments	Gas required; high start-up cost; digestion step; destructive	Gas required; digestion step needed; destructive analysis; samples require milling	Problems with Al and Ti sensitivity; samples require milling	Calibration cost can be high; requires ongoing addition of calibrations	Labour-intensive semi-quantitative ;	Semi-quantitative	Semi-quantitative
Analytical Capability Iron	Yes	Yes	Yes	Yes	Yes; separation into high and low groups	Yes; separation into high and low groups	Yes; separation into high and low groups
Zinc	Yes	Yes	Yes	Yes	No	No	Yes; separation into high and low groups
Contamination indicators	Yes	No	Yes, but sensitivity for	Al is inadequate as contaminant indicator	No	No	No
Other elements	Yes	Yes	Yes	Yes, although data limited in this area	No	No	No
Other relevant promoters and inhibitors	No	No	No	For example, protein, compounds–carotenoids, antinutrients	No	No	No
Precision Application	Plant and soil material	Plant and soil material	Plant and soil material	Plant material	Plant material	Plant material	Plant material
Accuracy for Fe	High	High	High	High	High	High	No

	ICP-OES: Inductively Coupled Plasma Spectrometer	*AAS: Atomic Absorption Spectrometer*	*XRF: X-Ray Fluorescence Spectrometer*	*NIRS: Near Infrared Reflectance Spectro-photometer*	*Modified Perl's Prussian Blue*	*Modified 2,2 Dipyridal*	*Modified Zincon*
Accuracy for Zn	High	High	High	High	No	No	High
Accuracy for carotenoids	No	No	No	High to medium-high confirmed for different crops	No	No	No
Accuracy for total carotenoids	No	No	No	High to medium-high confirmed for different crops	No	No	No
Accuracy for β-carotene	No	No	No	High to medium-high confirmed for different crops	No	No	No
Accuracy for minerals	Very accurate	Very accurate	Very accurate	Separates out only high- and low-nutrient genotypes	Separates out only high- and low-Fe genotypes	Separates out only high- and low-Fe genotypes	Separates out only high- and low-Zn genotypes
Economics Start-up costs(equipment)	\$50,000–\$300,000 depending on make and model	\$10,000–\$40,000 depending on make and model	\$50,000–\$350,000 depending on make and model	\$60,000–\$90,000	minimal	minimal	Minimal
Running costs	Varies from lab to lab; between \$4.00 and \$7.00/ sample	Varies from lab to lab; *e.g.*, Argon between\$0.22 and \$2.00/ sample; labour the greatest cost in analysis	\$15–\$25 AUD/ sample for XRF analysis; no cost for gas, just instrument upkeep and labour	\$0.5–\$2.00/ sample; Dramatically decreasing costs/ measurement as more components are measured	0.5–\$1.00/ sample	\$0.5–\$1.00/ sample	\$0.5–\$1.00/ sample

	ICP-OES: Inductively Coupled Plasma Spectrometer	*AAS: Atomic Absorption Spectrometer*	*XRF: X-Ray Fluorescence Spectrometer*	*NIRS: Near Infrared Reflectance Spectro-photometer*	*Modified Perl's Prussian Blue*	*Modified 2,2 Dipyridal*	*Modified Zincon*
Application in Breeding Recommended application in breeding	Pre-screening in population development and validation of Fe- and Zn-dense genotypes identified by rapid screening techniques	Pre-screening in population development and validation of Fe- and Zn-dense genotypes identified by rapid screening techniques	Pre-screening in population development and validation of Fe- and Zn-dense genotypes identified by rapid screening techniques	Pre-screening in population development for minerals and provitamins A; precision analysis for carotenoids and β-carotene for certain crops	Pre-screening in population development	Pre-screening in population development	Pre-screening in population development

Source: James Stangoulis, School of Agriculture and Wine, University of Adelaide, Waite Campus.

and X-Ray Fluorescence Spectrometer (XRF) allow identification of a wide range of micronutrients, including elements, such as phosphorus, which is indicative of the antinutrient phytate. The ICP is the current method of choice for quantifying elements, such as aluminium, which has been proposed as an indicator of contaminant iron. Various HTMs are applied in pre-screening to reduce the large number of samples in populations segregating for micronutrient concentration to a more practical number for subsequent, more expensive high precision analyses. Depending on the method used, an approximate 66 per cent proportion for more qualitative colorimetric methods (Modified Perl's Prussian Blue) to a 75–85 per cent proportion for semi-quantitative methods (2,2 Dipyridal) of "lows" can be discarded. The accuracy of semi-quantitative methods can be increased by using computerized systems with image analyzers. Correlations between iron determined by ICP and the 2,2 Dipyridal colorimetric method in rice, wheat, maize, sweet potato, and cassava ranged between 0.88 and 0.98 (James Stangoulis, personal communication). Colorimetric techniques are simple, fast, and low-cost, but destructive because samples must be milled. Elements indicative of contamination are not determined with colour-staining techniques; they need to be quantified during subsequent precision mineral analysis.

The Near-Infrared Reflectance Spectrophotometry (NIRS) method relates a sample's reflectance of near-infrared light to its chemical composition and covers wavelengths between 730 and 2500 nm emitted by major plant compounds, such as oil, starch, cellulose, water, and protein. In breeding, NIRS is routinely used to determine, for example, grain protein, which it can measure with high accuracy. The latest research has shown that NIRS has potential to predict iron and zinc with sufficient precision for pre-screening, although the causality of the association is not yet understood. Correlations between iron and zinc determined by ICP and NIRS in potato, sweet potato, and beans range between 0.77 and 0.85 (Wolfgang Grüneberg, CIP [International Potato Center], and Steve Beebe, CIAT [International Center for Tropical Agriculture], personal communications), and NIRS has been used to separate high-, medium-, and low-iron barley genotypes (James Stangoulis, personal communication). NIRS is environmentally friendly (no reagents are involved) and easy to operate, but it requires continuing calibration to make sure the calibration set includes samples representative of the genetic variation used in breeding, and it must also account for environmental impact on readings. The section "Provitamin A Carotenoids" later in this chapter elaborates further on NIRS in the context of carotenoid analysis.

Contamination in Mineral Analyses

Detecting contamination while assaying micronutrient concentration is complicated, and references are not yet available to guide researchers. In the past, contamination has, in general, not been addressed in the literature; extremely high iron concentrations that have been reported are likely due to contamination. Contamination–for example, by iron–can result from soil, dust, metal parts or paint in threshing equipment, rubber products (particularly silicon and neoprene), sample preparation, or seed handling. Zinc is less subject to be a contaminant than iron. Research to establish protocols and guidelines for determining approximate thresholds and, in particular, for developing corrective measures, is currently underway. However, diagnostics for contamination cannot substitute for validating micronutrient concentrations of selected genotypes through additional screening. Procedures that can eliminate sources of contamination in experimentation–for example, preventing lodging in cereals or washing tubers before sampling–should be routine. Washing maize grain samples to eliminate iron contamination has been investigated with varying degrees of success (Kevin Pixley, CIMMYT; James Stangoulis, personal communications). Using appropriate spatial experimental designs with replicated standards or checks to estimate error is also warranted, particularly in unreplicated nurseries.

One approach for detecting contamination entails using indicator elements that are: (1) abundantly found in soil, dust, or equipment; (2) uniform in concentration of contaminating fractions (*e.g.*, of soil); and (3) reproducibly released and easily measured. However, contamination-indicator elements must be absent in plants or seed, or present only in trace amounts. Earlier research investigated Al, Ti, and Cr (or a combination of these) for their potential to act as indicator elements, and attempted to establish threshold levels/bands. Other elements may also be suitable, but good indicator elements for soil are, by definition, very hard to determine accurately in plant tissues, if any soil at all is present. To date, research in this area has identified Al as the most suitable indicator element. Effectively correcting for soil/dust contamination is complicated by the large variety of soil types and the varied ratios of Al, Ti, Cr, and Fe and Zn in plant sample analyses, which result, for example, in different recovery rates, depending on particle size/surface area; hence, confidence in such corrections is not high. Statistical approaches use population parameters to identify values that fall outside the range expected for an assumed normal distribution and are probably erroneously high. The accuracy of detection increases when data from replicated trials or replicated check varieties are available. These two approaches are complementary and should be used in combination. Data from mineral analyses conducted at Waite Analytical Laboratories (Adelaide, Australia) and from micronutrient screening of cereal, legume, and tuber crops provided by HarvestPlus crop leaders have been studied to establish tentative Al thresholds for iron contamination. The criteria used to examine contamination consisted of analyzing and comparing data subsets with different Al ranges and using replicated check data to validate results in combination with statistical outlier tests and correlations among elements. Results suggest that Al concentrations of more than 5 to 10 $\mu g g^{-1}$ are frequently associated with contaminant Fe. Analyses also revealed that significant correlations between Fe and Al levels in adapted genotypes generally indicate Fe contamination. Eliminating data for samples with Al values >10 $\mu g g^{-1}$ reduced the average correlation between Fe and Al across datasets/crops from $r = 0.35$ to 0.18; including only data for samples with Al $<$ 5 $\mu g g^{-1}$ further reduced the correlation to $r = 0.11$. These findings coincided with results from statistical analyses.

Micronutrient Concentration Versus Content

Special care in milling is warranted when assessing seed of, for example, wild relatives of crop species, genetic stocks, inbred lines, and unadapted germplasm or genotypes that may have small, shrivelled seed and/or incomplete seed set. Biotic and abiotic stress or other production constraints may have the same effect on seed and seed set in adapted genotypes. If the remaining portions of a seed fraction–for example, the embryo–have high micronutrient concentration, concentration levels can be inflated. Furthermore, shrivelled seed may require a disproportionate degree of milling or polishing to make processed products that satisfy commercial or laboratory standards. Seed shrivelling, wrinkling, and weathering can have dramatic effects on grain micronutrient density, given that micronutrient concentration in the embryo and seed coat is much higher than micronutrient levels in the endosperm. The seed coat to endosperm ratio is high, which can result in elevated micronutrient concentrations, *i.e.*, the "concentration" effect. The concentration effect can result when fewer grains act as the micronutrient sink because few grains per spike have been produced due to sterility or poor seed set. In plump seed, the seed coat to endosperm ratio is much lower, causing a "dilution" effect. Because micronutrient concentration is generally determined on whole grain, concentration levels in shrivelled seed can be overestimated (Imtiaz *et al.*, 2003). The variation in content for narrow concentration ranges (*e.g.*, 48–52 μgg^{-1}) can be crucial in determining micronutrient content (μg seed^{-1} or, in certain cases, μg plant^{-1}) rather than micronutrient concentration (μgg^{-1}) when characterizing germplasm. In hybrid crops, correlations between mineral content and mineral concentration can be significantly affected by the concentration effect (*e.g.*, in inbred or sibbed lines) and warrant consideration in breeding. There are few reports in the literature regarding content within the context of mineral accumulation in conventional germplasm and transgenic materials. Grain yield, agronomic performance, and end-use quality attributes (*e.g.*, protein concentration) of large- or plump-seeded adapted genotypes are often compared with those of non-adapted genotypes with small, shrivelled grain. Since non-adapted genotypes have higher grain protein concentration due to the concentration effect and regularly produce lower grain yields, correlations between these traits and micro nutrient concentration can be overrated. Correlations based on content are usually lower and can partially remove the masking effect of seed size and shrivelling. These factors must be taken into account when comparing different types of germplasm or in germplasm selection.

Estimation of Iron and Zinc by Inductively Coupled Plasma-Optical Emission Spectrometer (ICP-OES)

Processing of Sample

Digestion (Closed) of samples in Microwave Reaction System: - Organic matter of grain sample can be destroyed and oxidize by nitric acid. Grain sample (0.5g) is taken to the tubes of microwave reaction system and 4 ml of distilled nitric acid added to each tube containing samples. In microwave, reaction takes place at 20 bar pressure and 200°C temperature and a clear digested sample is obtained. A blank containing only distilled nitric acid is also run using the same procedure. After digestion rest of the aliquot are transferred to 25 ml volumetric flask and volume is made to 25 ml with double distilled water. This extract is used for the determination of different mineral by using Inductively Coupled Plasma- Optical Emission Spectrometer (ICP-OES).

Standards

1, 2, 4 and 5 ppm of standards prepared by using 16 per cent distilled nitric acid. 2.5 ml of Iron and 2.5 ml of zinc standards are taken in a tube and the volume is make to 50 ml (50ppm) by using 16 per cent Distilled nitric acid. Out of this 50ml 1, 2, 4 and 5 ppm is taken to the tubes and again the volume is make to 50 ml with 16 per cent distilled nitric acid. A blank is also prepared by using only 16 per cent distilled nitric acid and run with these standards during calibration in ICP-OES.

Analysis of Samples

These diluted samples are then subjected to Inductively Coupled Plasma-Optical Emission Spectrometer (ICP-OES, Perkin Elmer 2100 DV) for micronutrient analysis. In ICP-OES various inputs like shear gas (Compressed air), purge gas (nitrogen gas), Plasma gas (ultra-pure argon gas) and coolant circulator are required. Keep the standards as well as the samples to be analyzed in auto sampler. Switch on the plasma. It will show the plasma ignited on the main screen of the computer. Now start the calibration of standards. The instrument will start measuring the standards. If the graph of Fe and Zn comes to be linier and the points lie in one line then click at the analyze the sample position. The sample analysis begun. After completing the analysis, rinse the auto sampler probe in water for few minutes by allowing the plasma on. Switch off the plasma now.

Genetic Variation–Germplasm Screening

The availability of genetic variation for micronutrient density is essential for determining the feasibility of achieving meaningful increments through conventional breeding and high rates of genetic progress under selection (Gs) ($Gs = i\sigma ph^2$, where i is selection intensity, σp is phenotypic standard deviation, and h^2 is heritability). Breeders can capitalize on additive gene effects, transgressive segregation, heterosis, and maternal effects to improve micronutrient density. When the required genetic variation is not available, transgenic approaches can provide novel and additional sources of variation to increase provitamin A or iron–via ferritin–in the endosperm and achieve the target micronutrient density (Khalekuzzaman *et al.,* 2006). In the future, breeding will likely combine both conventional and transgenic approaches. Screening objectives entail assaying representative samples of the genetic diversity for micronutrient density contained in the tactical and strategic gene pools, along with agronomic and end-use quality features of trait-source genotypes. To date, only a relatively small portion of the existing genetic diversity for micronutrients has been evaluated, and an even lower proportion of the genetic diversity for antinutrients and promoters has been assessed. Evaluating all accessions of each relevant species conserved in gene banks is beyond the scope of this chapter, as gene banks at CGIAR centers alone preserve more than 530,000 accessions in-trust (see http: //www. cgiar.org/impact/accessions.html). Because of these large numbers of accessions, current screening by breeding programs under HarvestPlus is focusing on core collections. Screening and research on screening methodology need to be expanded to include generic and crop-specific inhibitors and promoters for bioavailability. Future searches for variation outside core collections may employ state-of- the-art geographical information system (GIS) and molecular tools to enhance screening effectiveness. These modern techniques allow targeting accessions that are most likely to possess untapped genetic variation based on, for example, phytogeography and ancestral, genetic, or functional genomic relationships.

Micronutrient concentrations are affected by micro-environmental variation, G E interaction, germplasm type, and numerous other factors. Consequently, published data on micronutrient genetic diversity reveal significant variation in average values and genotypic variation per se among crops and within crop species. Thus, when interpreting these data, one must consider differences that can produce error such as sampling, milling, analytical protocols, and the type of experimental screening design used. Ranges in micronutrient concentrations reported in the literature reveal there is signify cant genetic variation in wheat (Ortiz-Monasterio and Graham, 2000). White and Broadley (2005) provided a recent review of genetic variation for minerals. Maximum micronutrient levels are frequently present in the strategic gene pool, in genetically distant sources, such as wild relative species, landraces, or germplasm unadapted to the agroecology of the target environment (Zeng *et al.*, 2004). The difficulty is accessing the genetic variation in unadapted sources, and the extent to which variation in the tactical gene pool can be recovered depends on factors such as genetic distance, differences in ploidy levels, and trait linkages. In prebreeding, eliminating unfavourable traits associated with the target trait causes "linkage drag," which, depending on its magnitude, adds to product development time and costs. Since trait recovery from gene sources in the strategic gene pool varies, we used the genetic variation present in the tactical gene pool to predict progress in the shorter term. Figure displays average (baseline) and maximum values for iron and zinc in adapted germplasm. For rice, HarvestPlus used data for polished rice because brown rice is rarely consumed. For both iron and zinc, the short-term exploitable variation in adapted germplasm is of similar magnitude for cereals and legumes, and lower for tuber crops and rice, regardless of the baseline level. The variation for iron in beans is higher than in other crops; HarvestPlus has found an approximate increase in the maximum of 20 μgg-1 resulted from a breeding cycle for high-iron beans. Maximum Fe values reported in the literature for certain crops could be up to 10 times higher than those encountered in later studies and likely coincide with the maximum Al values reported. HarvestPlus has found that maximum values for β-carotene (fresh weight basis) are about 9 $\mu g g^{-1}$ in cassava and >300 $\mu g g^{-1}$ in orange-fleshed sweet potato, whereas for provitamin A in maize (dry matter basis), they are about 15 $\mu g g^{-1}$.

Knowledge of heritability as it relates to genetic progress ($Gs = i\ \sigma p\ h^2$) and associated genetics is crucial for establishing screening, population development, and G E testing strategies, and hence, for effective breeding. All plant-breeding components (such as crossing strategies, breeding methodologies, the operational scale with plot size and number of sites and years required for testing) are based on genetic parameters. Furthermore, the potential for developing molecular markers is closely associated with factors such as the number of genes and their individual contributions. Molecular markers and marker-assisted protocols for applied breeding that can be used at early plant stages to select for micronutrient density can greatly increase breeding efficiency; quantitative trait loci (QTL) for micronutrients have been identified in several crops (Wissuwa, 2005). Growing evidence from

HarvestPlus research supports findings that iron and zinc concentration is controlled by several (2–5) relevant genes, and that mineral heritability is of intermediate magnitude (Cichy *et al.*, 2005). Provitamin A appears to be controlled by a few (~2) major genes, and trait heritability is high (Grüneberg *et al.*, 2005). However, heritability can be overestimated if studies contrast non-provitamin A and high provitamin A genotypes. For both minerals and provitamin A, additive gene action and general combining ability predominate (Long *et al.*, 2004). Heterosis has been found for maize and cassava. Also, studies on temperate maize have revealed significant reciprocal (maternal) effects for provitamin A (Egesel *et al.*, 2003a). Transgressive segregation for provitamin A has been encountered, for example, in cassava crossed with wild relatives, for iron in beans, and for zinc in wheat. Although information on transgressive segregation or heterosis is still incomplete, there is growing evidence that complementary genes are present, particularly in genetically distant sources, such as wild relative species. This is not unexpected, given that in the past, breeders did not select for micronutrients, and latent variation may have been lost. Furthermore, in the past, breeders would often select for white endosperm and, hence, against carotenoids (*e.g.*, in maize and wheat), or for lower ash content, which is associated with lower mineral concentration (*e.g.*, in wheat).

Setting Nutritional Target Levels

The available genetic variation allows the prediction of the magnitude of micronutrient increments that can be added through breeding. However, only a portion of an increment contributes to human micronutrient status; this portion, the bioavailable amount, largely depends on how much nutrient is lost from crop harvest until ingestion and on the bioavailability of a nutrient once ingested. Critical information needed to set nutritional target levels for breeding (for a target country) includes the amount of nutrient retained after storage, processing, and cooking; micronutrient bioconversion/bioavailability in a typical diet once the nutrient is ingested; and nutrient requirements of a target population (Nestel *et al.*, 2006). The daily micronutrient intake supplied by a crop must also be considered when setting target levels. Many of these parameters are interrelated in a highly complex manner, since human micronutrient status, dietary composition, and health status affect bioavailability (for example, for β-carotene: β-carotene absorbed/β-carotene in food) and its components' bioaccessibility (β-carotene freed/micronutrient in food), bioconversion (retinol formed/β-carotene absorbed), and bioefficacy (retinol formed/β-carotene in food). A more detailed discussion of these factors is beyond the scope of this chapter. As a starting point, when initially setting tentative target levels without detailed information at hand for accurate assessment, bioavailability of zinc can be assumed to be 25 per cent and bioavailability of iron to be 5 per cent for legumes (*e.g.*, beans, lentil, cowpea) and cereals with significant phytate concentrations (*e.g.*, wheat, maize, sorghum, pearl millet, barley). Within a context of general assumptions, illustrates the relationship between micronutrient intake and the micronutrient increment from the baseline concentration needed to make a measurable biological impact on women of childbearing age from a public health perspective (for iron, zinc, and β-carotene/provitamin A., 100 per cent retention has been assumed to allow generalizations across crops. Target increment functions are given for iron for 5 per cent, 10 per cent, and 15 per cent bioavailability, for 20 per cent, 25 per cent, and 30 per cent bioavailability for zinc, and for β-carotene/ provitamin A, for bioconversion rates for β-carotene/provitamin A to retinol of 12: 1, 6: 1, and 3: 1.

The feasibility of reaching nutritional target increments through conventional breeding largely depends on bioavailability (provitamin A: retinol bioconversion) and intake. There is scope for increasing micronutrient density in the shorter term via transgressive segregation, heterosis, and maternal effects beyond the variation and for discovering higher levels in the tactical gene pool. Target increments of zinc and provitamin A can likely be reached in most crops. The genetic variation for zinc in varieties, germplasm lines, and parental stocks, especially of cereal and legume crops, is high (White and Broadley, 2005). However, due to the substantially lower bioavailability of iron when compared with zinc, significantly higher micronutrient increments have to be added to reach nutritional target levels and achieve a measurable impact on human health. The bioavailability of iron and zinc is associated with the presence of antinutrients and/or the lack of promoter substances for micronutrients (White and Broadley, 2005). Since an increase in bioavailability translates into a proportional decrease in the nutritional target increment (increasing iron bioavailability from 5 per cent to 10 per cent reduces the target increment by 50 per cent), strategies for breeding micronutrient-dense crops should consider indirect breeding for increased bioavailability, increased retention, or reduced post-harvest micronutrient deterioration, in addition to direct breeding for increased concentration. Although not yet well understood, breeding for increased bioavailability via conventional (Oikeh *et al.*, 2003) or transgenic approaches (Drakakaki *et al.*, 2005) offers tremendous potential (Hambidge *et al.*, 2004).

Breeding for Increased Bioavailability

Breeding for micronutrient bioavailability per se is greatly limited by the lack of diagnostic tools for large-scale, rapid germplasm evaluation, such as adequate in vitro and/or animal bioavailability models. Current studies are exploring the feasibility of dissecting overall bioavailability into its causal components, such as antinutrients and promoters, which can be addressed by breeding. Ongoing exploratory research is investigating the feasibility of breeding for inhibitors/ enhancers from both the crop improvement and human nutrition perspectives. Breeding studies entail determining the genetic variation for antinutrients and promoters, the magnitude of trait expression/stability through G E studies, trait heritability, and associations with agronomic and end-use quality traits. Screening methods are being evaluated in a parallel effort, whereas nutrition research and food science are investigating bioavailability and nutritional impact using in vitro and animal bioavailability models and, subsequently, efficacy and retention studies involving human subjects. Phytate occupied center stage in past and current research on antinutrients (Raboy, 2000). Genetic variation for phytate has been reported in numerous crops: cowpea (Farinu and Ingrao, 1991), beans (Muzquiz *et al.*, 1999), and sorghum (Kayodé, *et al.*, 2006). Low-phytate mutants are available and have been found in barley (Bregitzer and Raboy, 2006), maize (Raboy *et al.*, 2000; Shi *et al.*, 2003;

Shukla *et al.*, 2004), and wheat (Guttierie *et al.*, 2004). Transgenic research focuses on the enzyme phytase (potato, Ulla *et al.*, 2003; rice, Hong *et al.*, 2004; maize, Drakakaki *et al.*, 2005). Introduction of phytase in rice (Lucca *et al.*, 2000), in combination with ferritin in maize (Drakakaki *et al.*, 2005), and co-expression of phytase, ferritin, and a metallothionein-like gene in rice (Lucca *et al.*, 2001a,b) significantly increased iron absorption/bioavailability and demonstrated the potential of transgenic approaches to capitalize on higher micronutrient concentration and increased bioavailability. Lowering phytate concentration can have adverse effects on human health and must be researched by nutrition experts. Safe threshold levels have to be established before addressing the trait in breeding crops for human consumption. The viability of addressing phytate (or heat-stable phytase) in crop improvement via selecting for phytate: zinc or phytate: iron molar ratios also depends on how the crop is consumed. For example, when wheat is milled, there is an over-proportional reduction of phytate (compared with iron and zinc) with decreasing fl our extraction rates; in white fl our (72 per cent extraction according to international trade standards), iron and zinc concentrations are reduced by approximately 50–60 per cent compared with concentrations in the whole grain, whereas phytate concentration is reduced by about 90 per cent. This over-proportional reduction causes a concomitant decrease in the phytate: mineral molar ratio and, hence, in increased bioavailability. Thus, lowering phytate via breeding, may not have the desired effect if wheat is predominantly consumed as white fl our products. Phytate can also be significantly reduced by processing methods (Mamta and Darshan, 2000). Selecting for increased nutrient retention can increase the micronutrient concentration "on the plate" and lower target increments. Significant genotypic differences in retention that could be exploited in breeding have been found, for example, in cassava and yams for provitamin A and, to a lesser extent, for minerals; evaluation of the genetic variation for micronutrient retention in other crops is also warranted. Micronutrient retention has been related to factors associated with fl our extraction rate in wheat (*e.g.*, grain hardness, texture, grain shape) and degree of polishing in rice.

Effect of Zn and S Application on the Productivity of Wheat Varieties

In a field experiment, 18 treatment combinations comprised with two wheat varieties (HI 1544 and HI 8737), three levels of Zinc (0, 5 and 10 kg/ha) and three levels of Sulphur (0, 7.5 and 15 kg/ha) were tried in factorial randomized design with three replications. Results revealed that durum wheat variety HI 8737 gave significantly higher grain and biological yields (5.91 and 14.53 t/ha) than HI 1544 (5.50 and 14.18 t/ha). Whereas, highest dose of Zn application @ 10 kg/ha recorded maximum grain and biological yields (5.33 and 14.87 t/ha), which being at par with 5 kg Zn/ha were significantly higher over control. Interaction effect of variety x Zn on grain and biological yields was non-significant. In case of S application, highest dose of S (15 kg/ha) recorded maximum grain and biological yields (5.83 and 14.78 t/ha) of wheat, which was significantly higher over control (5.48 and 13.71 t/ha); but at par with 7.5 kg S/ha (5.81 and 14.59 t/ha). Response equations were calculated based on mean data of two years, it was recommended that application of Zn @ 6.0 to 8.36 kg/ha and S @ 11.33 to 13.5 kg/ha is better for getting maximum wheat yields in Central India.

Correlations among Minerals and Value-added Traits

Correlations among minerals in cereals (maize and wheat), reveal a generic positive correlation between iron and zinc concentrations, and between iron and zinc and all other nutritionally important minerals and trace elements (Ca, Cu, K, Mg, Mn, P, and S). The substantial positive association among minerals suggests an opportunity for raising levels of a number of micronutrients simultaneously by direct selection for numerous micronutrients or by capitalizing on the in direct selection response (*i.e.*, a parallel increase in micronutrients not targeted in selection). Strain and Cashman (2002) provide an overview of the importance of minerals and trace elements in human nutrition.

Popularisation of the CVRC Released High Yellow Pigment Variety HI 8663 (Poshan)

A total of 112 quintals of breeder seed of HI 8663 (Poshan), produced during *rabi* 2009-10, was distributed to state seed producing agencies and farmers to increase the availability of high-quality durum wheat for improving "nutritional security". There was lot of demand from the farmers and seed growers for this variety as it performed better than to HI 8498 (Malav Shakti), the most popular durum wheat variety in Central India. Until now, 425 quintals of breeder seed of HI 8663 has been distributed to various agencies over last three years, which helped in increasing the area as well as demand from the farmers for this high-quality durum wheat for improving the nutritional intake of the people.

Problems and Prospects for Quality Improvement

People's knowledge, institutions and innovations have influenced the evolution of food basket over the last several millennia. But the expansions of markets, corporalization of agriculture, reduction of biodiversity and emergence of new icons of modernization and cultural progress have reduced the food diversity drastically. The policymakers and food scientists have been always biased against the importance and knowledge about the nutritional benefits like minor/small millets. Further, increase in household income, it has been shown that proportion of processed food increases in the consumption basket. But whether the choice of processed food is nutritionally superior for women, children and workers and others is not an issue of concern and moreover, educational policy only worsens the situation by reinforcing style over substance and nutrition and health over taste. In order to reduce malnutrition among the masses, due awareness should be created by proper campaigning. Moreover, on the research side, the novel aspects of fortification of food needs to be further encouraged and explored to bring out nutritive food at lesser cost. The most successful food fortification campaign was the iodised salt, now reaching 68 per cent of the population in countries with high incidence of goitre. Functional genomics is one area where it is possible to increase greatly the efficiency by which breeders can optimise this bio-fortification programme. A current example is the prebiotic polysaccharides, fructans. The human gut does not have the enzymes to cleave these chains into absorbable sugars so they reach the colon intact where bifid bacteria, present in small numbers normally can use fructans as an energy source, can multiply and by a variety of effects, can increase the absorption of iron, zinc and calcium from the colour. (Another factor is the recognition of phyte as an essential nutrient and establishments of minimum requirement in diets for various functions, including cancer). Thus improved efficiency of genetics and transformation studies is likely to contribute faster, safer and effective ways to improve the nutritional value of cereals. For attaining the objectives of nutrition improvement, close coordination between aspects of nutrition science, analytical and biochemistry and grain processing is required who will support breeders for quality improvement of food crops. Hence channelling research towards attaining accurate requirements of nutrition in food is the need of the hour, because the future of the human race largely depends upon the innovative neutraceuticals and less of pharmaceuticals.

Generation of Material

The material developed showed consistent improvement from 6 ppm to 8 ppm for yellow pigment in the segregating material over generations.

Table 8.38. Filial generations

F_1 generation	Forty crosses were evaluated for improving the yellow pigment content in *rabi* 2009-10. The crosses showing superior performance for the β yellow pigment (> 6.0 ppm) were DBP 01-11 \ HD 4672, HI 8663 \ A-9-30-1, DBP 01-11 \ HI 8381, HI 8663 \ HI 7747, HI 8627 \ A-9-30-1, HI 8627 \ HD 4672 and HI 8627 \ HI 7747. These F_1's were sown to generate the F_2 population, from which the superior plants will be selected.
F_2 generation	Twenty-four crosses were evaluated for improving the yellow pigment content in rabi 2009-10. The crosses showing superior performance for the β yellow pigment (> 6.5 ppm) were PDW 312 \ WH 896, WH 912 \ HI 8663, WH 912 \ WH 896, RS 995 \ WH 896, HI 8692 \ HI 8663, HI 8690 \ WH 896, H: I 8681 \ WH 896, HI 8671 \ WH 896 and A 624 \ HI 8663. A total of 95 plants were selected for advancing to next generation.
F_3 generation	Fifteen crosses were evaluated for improving the yellow pigment content in rabi 2009-10. The crosses showing superior performance for the β yellow pigment (> 7.0 ppm) were HI 8663 \ PDW 233, PDW 233 \ HI 8627, PDW 233 \ HI 8638, HI 8663 \ WH 896 and WH 896 \ HI 8663. A total of 48 plants were selected for advancing to next generation
F_4 generation	Fifteen crosses were evaluated for improving the yellow pigment content in rabi 2009-10. The crosses showing superior performance for the β yellow pigment (> 8.0 ppm) were HI 8638 \ PDW 233, HI 8653 \ HI 8663, PDW 233 \ HI 8627 and PDW 233 \ HI 8663. A total of 42 plants were selected for advancing to next generation.
Advance lines of high yellow pigment in station trials	Twenty one genotypes having high yellow pigment (> 7.0 ppm) were selected in the F_5 to F_6 generations *i.e.*, HG 74, HG 75, HG 77, HG 98, HG 112, HG 114, HG 118, HG 145, HG 149, HG 167, HG 558, HG 575, HG 576, HG 590, HG 591, HG 592, HG 595, HG 597, HG 600, HG 601 and HG 602. All these genotypes were promoted to the station trials for further evaluation for yield, disease resistance and quality traits.
Stable advance lines	Fourteen lines from the station trials were selected, which are having > 6.5 ppm yellow pigment *i.e.*, HG 16, HG 42, HG 43, HG 75, HG 144, HG 150, HG 453, HG 890, HG 948, HG 1945, HG 1034, HG 1156, HG 1264, HI 8662. Of which, four genotypes were promoted to 'IARI – Common Varietal Trials' *viz.*, HG 75 (ID 1017) HG 150 (ID 1022) and HG 42 (ID 1025) for CVT 5B (rainfed and restricted conditions) and HG 1264 (ID 1013) for CVT 4 (Timely sown and high fertility conditions).

Success Stories of Food Fortification

The classic example of a successful food fortification campaign is the iodized salt, now reaching 68 per cent of population in countries with high prevalence of goitre. Food fortification has also produced good results in eradicating such deficiency diseases like rickets (Vitamin D) and pellagra (Niacin). Whereas in US and Europe, the vehicle for multiple nutrient fortification has typically been wheat flour, in Guatemala, sugar has proven to be an effective choice for vitamin A fortification. Iron fortified spices and condiments, such as soy sauce in China and curry powder in India has also been used successfully. Nutritional improvement through food fortification was also possible in Asian countries through the staple food rice. There is a research programme already being conducted in several prominent institutes of the Asian countries with regard to fortifying rice with vitamin A and imparting it golden colour, so as to distinguish it from ordinary rice. The apex institute for this research programme is IRRI. There is also wide scope for fortifying wheat since it is consumed widely in the form of bakery products and chapattis. This is one of the research perspectives in wheat at IARI. Due importance has to be given to the cultivation of pulses, because unless the diet is supplemented with pulses and cereals, the essential amino acids might become unavailable. Moreover, creating awareness about the nutritional importance of the small millets and major millets like sorghum and cumbu and terming them as health foods is highly essential for nutritional improvement of the masses. There are a variety of value-added products, which should gain popularity among the people, since they do supply all the micronutrients and major nutrients required by man.

Quality is a subjective concept and has been defined adequately as conformance to requirements, which are defined by nutritional, processing and specialty needs of consumers. Improvement of grain quality and diversification of output are critical for a continuous and strategic role of agricultural commodities in international economy. Plant breeding has played a major role in enhancing agricultural productivity and improving food quality. This is evident from successes achieved in selective modification of starch, protein and oil storage products. Cereals with modified starch composition are now available to meet the needs of edible or non-edible products without having to resort to expensive and environmentally undesirable chemical modifications. High lysine variants of several cereal crops, especially maize, have helped in improving protein quality. New varieties of soybean, sunflower, rapeseed-mustard now produce oils which have enhanced stability. With the emergence of plant biotechnology as the latest and perhaps most potent tool for plant breeders, stage is now set for achieving higher quality standards with a greater control of grain composition and processing performance. Some aspects of quality are obvious to the consumer and greatly affect the consumer preference for the produce. Such traits are potent factors in determining the popularity of a variety.

Plant breeding is an important tool for the improvement of human nutrition for solving the problems of under nutrition and malnutrition. The necessity of breeding for nutritional quality is well reflected in the general consensus arrived at the coordination meeting of FAO/IAEZ symposium, 1970. It says, "While technological approaches, such as lysine fortification, production of protein from single cells, and manufacture of fish protein concentrates, all have some merits, a major dent on the protein malnutrition problem can be made in the near future only if the basic food grains can be upgraded in their nutritive quality, so that the nutrition of the economically handicapped sections of the population can be improved without any special education efforts or added expense". This has become possible as information about genetic variability in respect of the quality traits in various species is now available is most cases. Thus, a strategy for nutritional improvement for different crops can be chalked out. From the breeder's point of view, new cultivars with improved agronomic traits have been the major factor contributing to increased crop production and there will be a need for continued emphasis in improving these traits. Therefore, compositional traits that are consistent with good nutritional quality can be selected in new plant cultivars.

Improved quality should not be at the cost of yield. Breeding for quality should safeguard the economic interests of both the consumers and the farmers. Improved quality trait must be present in a variety having high and stable yield with good agronomic features. Otherwise, the variety may never be grown by the farmers. A plant breeder must be provided with clear-cut objectives for quality breeding. In case of nutritional quality, human nutritionists must specify the objectives a breeder is expected to achieve. In order to breed for improved nutritional quality, plant breeders must also become familiar with the problems of human nutrition. Incorporation of quality characters should be integrated with the normal breeding procedures for yield. However, traditionally, new lines are developed mainly for enhanced yields and then they are evaluated for quality traits. Increased protein content is usually associated with lower yield. Considerable breeding effort will be required to break this negative association. Findings with wheat, sorghum and barley suggest that this may be achievable. Often higher protein content mutants show poor grain filling. Therefore, protein content should not be considered as per cent; it should be evaluated as total protein yield per seed/per plant/per unit area. A close interaction between nutritionists and breeders will increase the effort of progress in breeding for nutritional quality. The success of a breeding programme for improved quality depends primarily on the following factors: the extent of genetic variability present in the germplasm for the trait, heritability of (or the degree of confusing effects of the environment on) the trait, availability of an analytical tool for a rapid, non-destructive, inexpensive and reliable estimation of the trait and the undesirable associated/pleiotropic effects, if any of the trait. In case, genetic variation for a trait is not present in the germplasm, suitable transgene could be used to bring about the desired alteration in the quality trait.

Limitations in Quality Breeding

The importance of the cereal grains to world nutrition is well recognized. Wheat and rice alone constitute the staple diet of two-third of the world's three billion people. Maize occupies a similar role in tropical and sub-tropical parts of Central and South America. Sorghum is important in several parts of Africa and Asia. Although classified as calorie sources, the cereals are the world's most important protein source. Frey (1977) estimated that on the average, cereal grains provide 70 per cent of the calories consumed in human diet. The world's protein needs come approximately 50 per cent from cereals, 20 per cent from grain legumes and 30 per cent from animal products. In developing countries, animal products contribute only 10 per cent and cereals 70 per cent. Thus, there is need to improve the quality of cereal grains consumed in many parts of Asia, the Near East, Africa and Latin America. Food legumes are particularly important plant sources of proteins, vitamins and minerals. Protein deficiency is generally regarded as the world's most serious nutritional problem, particularly in the young ones. Besides protein, the food deficient countries have a serious calorie problem as well. The two problems cannot be dealt with separately because protein deficiency is difficult or impossible to correct if a serious calorie deficiency exists. Thus, plant protein cannot be improved effectively at the sacrifice of productivity. The probable reasons for the slow progress include priority of quantity over quality. Yield is always the primary consideration. Generally, there is no extra money for improved quality in cereals or pulse crops; lack of funds - very costly equipments are needed. The fact that most quality tests are costly and require several ponds of seed delays their use until later stages of selection in a breeding programme. An unacceptable quality level for any of these traits is sufficient reason to prevent a promising line from being released (Knauff *et al.*, 1987); lack of efficient screening techniques; negative correlation of grain yield and protein content; low heritability of quality traits. Quality traits are greatly influenced by environment; high genotype-environment interaction for quality traits.

Genetic Control of Quality Breeding

Although ample variation exists for many quality traits yet the ease of their manipulation in a breeding programme and eventual incorporation into an improved variety will depend upon the type of gene control, nature and extent of heritable variation and genetic correlation between desirable and undesirable characters. A major distinction is whether the character is controlled by one or a few major genes, which can be easily isolated and manipulated or whether it is under the control of many genes requiring use of more complex biometrical techniques in a breeding programme. Most nutritional characters, however, show continuous variation and are under the control of many genes. Even the traits showing major gene control, may be modified quantitatively. In selecting for such quantitative characters, it is important to know both the amount and nature of gene effects and heritability. The genetic analysis of various germplasms has shown wide range of variation and scope for the improvement of quality traits in pearl millet, Indian mustard rapeseed, groundnut, mung bean, lentil and chickpea. suitable gene pools and genetic stocks have been established and identified for specific quality traits in these crops. Useful correlations have also been observed among quality traits and of these with yield components. However, the correlations vary among the different populations. Also the estimates of the seasonal variation, genetic variability, g x e interaction and heritability estimates of different quality parameters show different trends in different germplasms. The growth rhythm, the developmental and physiological processes, duration of the crop, the span of maturity and stage of harvest also affect the expression of and genetic control of quality traits. Expression of genetic variability and the nature of gene effects also differs over environments as evidenced by our studies in pearl millet, Indian mustard, groundnut, lentil, mungbean and chickpea. Furthermore, the concentration of chemical constituents may vary in different plant parts and the nature and extent of genetic variability may be tissue specific. In general, the heritability is higher in tissues and under environments with higher mean value.

In case of the forage crops, cutting stage greatly influences the quality of green fodder. The gene effects are better expressed in leaves than in stems. The leaf/stem ratio is a good index of quality. In case of grain crops, the longer span of maturity adversely affects the proquality factors as the antiquality factors increase with the advancement of maturity. In durum wheats, the faster rate of dry matter accumulation could be linked with bolder grains and better quality. In pearl millet, most of the desirable chemical components are accumulated by the stage through 2/3 to 3/4th of the usual span of grain development. Unsaturated fatty acids and phytic acid contents increase at later stages. In Indian mustard rate of fatty acid synthesis is maximum upto 30 days after fertilization, the lipid accumulation is 90 per cent by 35 days and the undesirable erucic acid increases after this stage. In lentil and mungbean, trypsin inhibitors have been observed to increase with the advancement of maturity. The effect of environment was significant in lentil but not in mungbean. Study of the biochemical profile at various growth stages has helped in the identification of critical stages for genetic manipulation, biochemical parameters leading to changes in plant behaviour, and inherent potential of plant genotypes in pearl millet, wheat, groundnut, gram, mungbean, lentil, Indian mustard and European rape. Certain biochemical genetic parameters and expression of certain enzymes during different phases of growth could be correlated with the post-harvest quality parameters. Even the combining ability behaviour of the quality traits could be correlated with that of enzyme activity as shown by our studies in Indian mustard.

The environmental stress may also affect the accumulation rate of chemical constituents. Such a response, however, is crop and genotype specific. Our studies with several crops have shown that in crops like cereals, the stress heavily curtails yield but grain quality may not be affected. However, in some Brassica species both yield and quality parameters are affected. Desirable fatty acids decrease while undesirable ones increase under stress with genotype-specific response. In groundnut, quality varies within and between taxonomic groups. Environmental stress reduces oil content but increases protein per cent. Crude protein content in mungbean decreases under all stresses while that of lentil and gram increases under water stress. In lentil, methionine and tryptophan increase under moisture stress. Also the phytohemaglutinins are more but the behaviour of genotypes for this antiquality factor is unpredictable over environments. The protein fractions in gram showed highly significant genotypic differences for water-soluble, salt-soluble and alkali-soluble fractions with highly significant genotype x environment interactions. There were higher genotypic differences for water soluble fractions with less effect of environment. The genotypic effects on the salt-soluble protein fractions were also very high but there was variable response of genotypes to environmental variation. There is, thus enough evidence to show that the biochemical quality parameters are influenced by the growth rhythm, physiology and development of the organism. It would be worthwhile to monitor the pattern of variation for biochemical parameters at different developmental stages, so that the critical stages of development could be identified. This would help in breaking the undesirable linkages and also manipulate the quality in various field crops.

Table 8.39. Gliadin bands, their frequency and quality traits in durum wheat genetic stocks

Land Races, Old Released Varieties and Rust Resistance sources	*Gli - B1*	*Sedimentation Value (ml)*	*Protein Content (per cent)*	*Specific SDS Value (SDS/PC)*
Sarangpur Local	45	40	14.2	2.82
Sawer Local	45	24	15.8	1.52
Dahod Local	45	47	14.2	3.31
Mandsaur Local	45	28	14.3	1.96
Bansi Local	45	31	14.1	2.20
Malvi Local	45	22	14.2	1.55
Motia	45	28	13.4	2.09
Trinakaria	45	33	15.9	2.08
Kathia Red	45	24	13.5	1.78
Bijaga Yellow	45	34	15.0	2.27
IWP 5019	45	34	14.1	2.41
B 146	45	42	13.5	3.11
HG 110	45	31	13.5	2.30
IWP 5004-1	45	32	14.1	2.27
Guji	45	39	15.4	2.53
Kathia 25'	44	29	14.7	1.97
NP 404	44	28	14.6	1.92
N 59	44	34	14.2	2.39
Ju 12	44	40	12.7	3.15
Meghdoot	44	32	14.3	2.24
Bijaga Red	44	30	14.4	2.08
MACS 1967	44	31	14.2	2.18
Karnatakka L	43.5	40	14.8	2.70
A 206	43.5	42	13.8	3.04
A 9-30-1	43.5	42	14.1	2.98
NIDW 15	42	23	13.1	1.76
Yuk	42	45	14.0	3.21
B 662	Null	17	14.6	1.16
ED 2398A	47	32	14.5	2.21
Line 1172	47	23	13.1	1.76

Wide Hybridization for Quality

A number of genes for quality characters have been transferred from alien species. Wild relatives of wheat have been and would be useful sources for improvement in storage proteins. Traditionally, two storage proteins have been recognized in the endosperm: gliadin and glutenin. They are synthesized on the endoplasmic reticulum in the developing endosperm and are deposited in protein bodies. Originally, distinguished by their relative solubilities in aquous alcohols, they are now more generally defined by their molecular size in dissociating solvents. Gliadin molecules are small, ~35,000 mol. wt. and they do not have disulfide bonded subunit structure. When fractionated by gel electrophoresis at low pH, they separate into four groups α, β, and w. Two-dimensional gel electrophoresis suggests that a single variety contains above 45 gliadin components. *Glu*tenins are large, heterogenous molecules built up from some 19 different subunits that are connected by disulfide bonds. The glutenin subunits fall into two unequal groups; the predominant low molecular weight (LMW) subunits and the high molecular weights (HMW) subunits. Gliadins give extensibility to a dough whereas glutenins impart strength and elasticity. They are, therefore, important in determining quality of wheat products. Advances in knowledge of wheat proteins, mainly through the application of various techniques in molecular biology, has made it clear that the glutenin and gliadin are two groups of proteins that belong to the same storage-protein family, the prolamins, as opposed to the globulin storage proteins found principally in leguminous seeds (Payne, 1987). The cytogenetic analyses have shown that: i) genes controlling HMW glutenin subunits are located on the long arm of chromosomes 1A, 1B and 1D at loci designated *Glu* 1. ii) LMW glutenin are encoded by genes present on the short arm of the same chromosomes at loci designated *Glu* 3 closely linked to those encoding gamma and omega-gliadin (designated as *Glu* 1 loci). iii) Genes for α and β gliadin occur on the short arms of chromosomes 6A, 6B and 6D designated as *Gli* 2 loci.

Sodium dodecyl sulphate polyacrylamide electrophoresis (SDS-PAGE) separation of HMW glutenin subunits from bread wheat cultivars has shown the existence of allelic variants at each locus and a numbering system has been proposed to identify different subunits. Molecular analyses have shown that each *Glu*1 locus contains two genes, one encoding for a higher x-type subunit, the other a lower y-type subunit. Usually, the *Glu*-D1 locus encodes both types and the *Glu* A1 can code only the x-type or neither (Payne *et al.*, 1981). The allelic variation at the three *Glu* 1 loci gives different quality scores with respect to their baking performance. However, limited variation is present in HMW glutenin subunits, encoded by *Glu* A1 and *Glu* D1 loci that are principally implicated in influencing the baking characteristics of hexaploid wheat. Cataloguing of wild diploid and tetraploid progenitor species of bread wheat by SDS-PAGE has revealed large intraspecific diversity at the three *Glu*1 loci. Comparison of migration pattern of HMW glutenin subunits of these species with the known subunits has identified three new alleles for Ax subunit and five of Ay subunit in the A genome species. Among the cultivated wheats, most of *T.durum* (AABB) lines do not possess HMW glutenin subunit at *Glu*

A1 locus whereas the *T.aestivum* lines either have a null allele or alleles expressing only the x subunit. All accessions of *T.boeoticum* (A^b) possessed both Ax and Ay subunits whereas T.urartu (A^u) lines either carrying both Ax and Ay subunits or only Ax subunit have been identified. *Aegilops* speltoides (S) was found to be the most variable among the S genome species. In the S genome species, 10 new alleles coding for x subunit and y for 6 subunit have been identified. In *Ae.squarrosa* (D), three alleles for x subunit and six for the y subunit at the *Glu* D1 locus have been found.

The identification of *T.urartu* lines with only Ax or both Ax and Ay HMW glutenin subunits is in agreement with the findings of Waines and Payne, 1987). Identification of lines with both Ax and Ay subunits in *T.urartu* and *T.dicoccoides* (AB) and uniform presence of both the subunits in *T.boeoticum* (A^b) has important implication with respect to their use in improvement of bread making quality of cultivated wheats. One approach to improve the bread making quality of *T.aestivum* could be to transfer or replace the existing gene at *Glu* A1 locus by corresponding gene from *T.urartu, T.boeoticum* o4 *T.dicoccoides* with both Ax and Ay subunits. Transfer of both the Ax and Ay subunits from the related species into *T.durum* would facilitate their further transfer to *T.aestivum*. Also the protein content and quality of various products made from *T.durum* can be improved through transfer of novel HMW subunits at *Glu* A1 locus. At the PAU, Ludhiana, novel genes encoding both x and y subunits at *Glu* A1 locus have already been transferred to *T.durum* background from one accession each of *T.urartu, T.boeoticum* and *T.dicoccoides*. Work is in progress to transfer novel alleles at *Glu* A1 locus from other accessions of these species (Randhawa *et al*,1995). Transfer of novel identified HMW subunits from other species is in progress as well. Effect of introducing the *Glu* A1 alleles from one accession each of the above mentioned three species (carrying A genomes) is being studied. Rogers *et al.* (1989) has transferred *Glu* A1 allele from *T.thaoudar* encoding two HMW glutenin subunits and has been able to demonstrate the positive effect of introducing the *Glu* A1 allele on glutenin strength. Ciaffi *et al.* (1995) observed that rheological and baking properties of *T.durum* lines were strongly associated with the allelic type of proteins encoded by *Glu* A1 and *Glu* B3 loci. In a cross involving an accession of *T.dicoccoides* lines with the best technological performances contained the *Glu* A1 alleles from *T.dicoccoides* (encoding both Ax and Ay subunits and the allelic type LMW-2 (Low molecular weight glutenin subunit) at the *Glu* B3 locus from the parental durum wheat cultivars. Their results indicated that some limes had exceptionally high dough strength and approached acceptable levels for baking with one possessing dough rheological properties and baking performance as good as those of the hexaploid bread wheat cultivars used as control. Quantitative analysis carried by them (Ciaffi *et al.*, 1995) on total protein extracts from 13 F_7 lines showed that the presence of *T.dicoccoides* alleles with both x and y HMW subunits at the *Glu* A1 locus increased the proportion of glutenin by about 4 per cent, when compared with an allele (A327) with only x subunit. Thus, the increase in glutenin strength and bread making properties associated with the presence of both 1Ax and 1 Ay subunits may result from an increase of polymeric proteins due to the greater amount of the high molecular weight glutenin subunits.

Limited work has been done to evaluate and characterize less closely related *Aegilops* species (carrying C, U and M genomes) for HMW glutenin subunit composition. The study of *Ae.umbellulata* (U), *Ae.caudata* (C), *Ae.triuncialis* (UC) and *Ae.ovata* (UM) showed that each of the two diploid species possessed one x and one y subunit specific to the species. The x subunit of both Ae.*umbellulata* (U) and *Ae.caudata* (C) were observed to possess higher molecular weight than the corresponding Ax subunit of *T.aestivum* checks (Singh *et al*,1994). *Ae.triuncialis* possessed x and y subunits of both *Ae.umbellulata* (U) and *Ae.caudata* (C). Ae.ovata also possessed x and y subunits corresponding to that of Ae.*umbellulata*. The variability observed at *Glu*1 locus in these four species is of interest due to two reasons: first, the presence of y subunit and second, the higher molecular weight of x subunit corresponding to *Glu* A1 locus. The replacement of *Glu* A1 by an allele from *Ae.umbellulata*, *Glu* U1, encoding two subunits has shown increase in glutenin strength (Faridi, 1988). *Glu*tenin viscoelasticity, an important factor in pasta cooking quality, is strongly related to some gliadin components and LMW glutenin subunits, encoded by genes at complex loci *Gli* B1 and *Gl* B3, respectively. Only two alleles, 42 and 45 so designated according to their relative mobility of two gamma gliadin bands in polyacrylamide gels with aluminium lactate at pH 3.1, have been detected at these loci in durum wheat collection. They are related to poor and good viscoelasticity, respectively. Study on the variation of the protein components, controlled by genes linked at *Glu* D1 and *Glu* D3 loci, showed also that low variation is present in cultivated wheats at both loci. Patterns of gliadin fraction in wild diploid A genome species had a high level of polymorphism at the *Gli* A1 and *Gli* A2 loci (Ciaffi *et al.* (1992). In particular, 19 and 13 different allele variants were found at *Gli* A1 locus in *T.boeoticum* and *T.urartu*, respectively, whereas 15 and 16 alleles were detected at the *Gli* A2 locus. In a recent study by Ozkan *et al.*, 1995). Two accessions of *Ae.speltoides* showed the band Rm 45 (associated with strong gliadin) whereas *Aegilops umbellulata* showed the based Rm 42 (associated with week gliadin)

Table 8.40. Alien introduction in wheat

Character	*Source*	*Wheat Background or Line*
Increased grain protein	Aegilops comosa	Maris Widgeon
High protein content	Aegilops ovata	Plainsman V
High kernel weight and high protein	Triticum dicoccoides	-
Gliadins	Aegilops longissima	Chinese Spring
Gliadins	Aegilops umbellulata	Chinese Spring
Barley prolamins	Hordeum vulgare	Chinese Spring
Prolamins	Hordeum chilense	Chinese Spring
Rye glutenins	Secale cereale	Gabo

Table 8.41. Quality scores assigned to individual or pairs of HMW subunits of glutenin

Score	Chromosome		
	1A	1B	1D
4	-	-	5-10
3	1: 2*	17-18; 7-8	-
2	-	7-9	2-12; 3-12
1	Null	7: 6-8	4-12

Table 8.42. Variation for high molecular weight (HMW) glutenin subunits in wheat germplasm

Species and Genomes	Number of Accessions Studied	Number of Accessions with New Bands	Percentage of Variants
Triticum urartu (A^u)	31	20	64.5
	Ax	11	35.5
	Ax+Ay	9	29.0
Triticum boeoticum (A^b)	25	24	96.0
Triticum dicoccoides (AB)	24	12	50.0
Triticum araraticum (AG)	5	5	100.0
Aegilops speltoides (S)	20	18	90.0
Aegilops longisimma (S^l)	7	6	85.7
Aegilops searsii (S^s)	1	1	100.0
Aegilops bicornis (S^b)	3	3	100.0
Aegilops squarrosa (D)	42	23	54.8

Biotechnological Approaches for Crop Quality

Plant breeders have been very successful in improving the quality as well as yield of the major crops, while sustaining food supply. This success has been achieved with very little understanding of the biochemical mechanisms of the selected traits. Traditional plant breeding techniques have been used to develop crop plants from ancestors with some desirable characteristics. These techniques are still used to cross crop plants with genetically compatible species in order to obtain hybrids with improved qualities. One limitation of traditional techniques is that they introduce undesirable genes along with desirable ones. As natural resources deteriorate and food needs increase, it is necessary to apply more advanced science and technology to agricultural research and food production. Biotechnology has been employed for many years in the preparation of foods and beverages. From traditional fermentation and tissue cultures the trend is now towards applying DNA technology or the so called Gene Revolution. At present, may genetically modified plants (GMPS) have been released for commercial use especially for manufacture of food products. Much of it is now carried out using (i) plant cell culture (ii) genetic engineering which are two major areas of plant biotechnology.

The enormous advantages of totipotency of plant cells could successfully be translated into standardization of tissue culture techniques namely embryo culture, anther/pollen culture, protoplast culture and cell fusion, meristem culture and callus culture for developing somaclonal variants for economic attributes. Genetic engineering is circumscribed by a non sexual transfer of genetic information coded on the DNA from one organism to the other, thus bringing about genetic transformation of the recipient organism genetic engineering enables the creation of plants and animals incorporating useful traits of other species of microorganisms programmed to manufacture large quantities of valuable chemicals and of bacteria that are "bioengineered" to convert one chemical into another.

Enhancement of Iron Content

India has not only attained self-sufficiency in food grains production through the systematic and concerted efforts of researchers, administrators and farmers within two decades of the ushering in of green revolution but has started exporting wheat and rice. The most widely recognized strategies for reducing micronutrient malnutrition are supplementation with pharmaceutical preparations, food fortification, dietary diversification and disease reduction. For various reasons, none of these have been very successful in reducing the prevalence of iron-deficiency anaemia in researchers have reported the transfer of the entire coding sequence of the soybean ferritin gene driven by the rice seed storage protein glutelins promoter (Gluβ - 1) in to rice by Agrobacterium mediated transformation. The iron content of T1 seeds was three fold higher than that of non-transformed seeds. Luna *et al.* (2001) grain and reported that gene and legumes staples are rich in phytic acid which is a potent inhibitor of iron absorption. The major inhibitor, Phytic acid can be readily degraded in cereal and legume foods by the addition of exogenous phytases either during food processing or during digestion increasing iron absorption dramatically. High phytase rice, with increased iron content and rich in cysteine. Peptide, has the potential to greatly improve iron nutrition in rice-eating population.

The wild progenitors and related cultivated and wild species and genera of our crop plants constitute a rich reservoir of useful genetic variability for various traits including those related to quality aspects which can be transferred to high yielding varieties through wide and somatic hybridization and subsequent chromosomal manipulation and characterization such that the recurrent parent background with the quality trait is recovered in shortest possible time with least linkage drag. Embryo rescue, protoplast fusion, induced homoeologous chromosome pairing, construction of graphical genotypes and *in situ* hybridization are some of the approaches to facilitate gene transfer from related species. Several genes fpr resistance to biotic and abiotic stresses, yield components and quality traits have been transferred from related species and are being exploited commercially. The inherent ability of Totipotency of plant cells enables us to regenerate complete plant from a single cell or protoplast thus making it possible to screen millions of potential plants in a single petri-plate. Regenerable, callus (undifferentiated) and suspension (small cell clusters) cultures can be induced from several juvenile plant parts with appropriate hormonal and tissue culture media modifications. High frequency of genetic and epigenetic changes termed as somaclonal variation occur in callus and suspension cultures. The somaclonal variants for certain traits such as resistance to herbicides, pathotoxins and amino acids and their analogs can be enriched by in vitro selection against these agents in the tissue culture media. The regenerated plants from resistant somaclones can be further tested, evaluated for the heritability and expression of the trait. The HYVs lacking one or two traits can be improved through these approaches. Dr Mehta and his colleagues at IARI, have produced somaclones of *Lathyrus sativus* with as little as 0.01 per cent ODAP (β-N-oxalyl diamino propionic acid) a neurotoxin in the seed of a somaclone as against 0.32 per cent in the parental variety and 50 per cent higher seed yield. Similarly, somaclones with constitutive and higher level of production of some of the essential aminoacid such as lysine and tryptophan have been selected from maize and rice callus culture using the aminoacids lysine and L-threonine or the lysine analog S-(2-aminoethyl)-L-cysteine (AEC) or tryptophan analog 5-methyl-L-tryptophan.

Chromosomal Location of Genes Controlling Wheat Endosperm Proteins

The nutritional value and the unique dough forming and baking properties of wheat flour largely depend on its protein content and composition. Wheat endosperm has poor nutritional quality because it is deficient in certain essential amino acids (especially lysine), but because wheat is largely consumed, it is still the single most important source of protein for much of the world's population. Protein content serves as a good sample of the difficulties encountered in breeding for the desired level of a quality factor exhibiting limited genetic variation, and being extremely sensitive to environmental influences. Although wheat proteins have been extensively investigated by biochemists, geneticists, and breeders for well over the past few decades, our knowledge about them is limited, compared with that of other important sources of dietary proteins, such as meat and milk. Only during the past 15 years have significant advances been made in the biochemical and genetic studies of individual wheat protein components. Most of the genetic studies have been on the aneuploid lines development and made available by sears. Progress in aneuploid analysis of genes controlling proteins has generally depended on advances in protein fractionation and characterization. Osborn (1924) proposed four classes of proteins: albumins, which are soluble in water, globulins, soluble in salt solutions. Prolamins (gliadins in wheat), soluble in aqueous alcohols; and glutelins (glutenin in wheat), insoluble in any of the previous solvents (alkali soluble).

Location of Genes Including Gliadins in Wheat

Gliadins are the prolamins of wheat endosperm. Typical gliadin have been classified into alpha, beta and gamma components based on fractionation by free boundary electro phoresis. Shepherd (1968) observed that changes in chromosomes belonging to homologous groups 1 and 6 cause changes in glaidins. The genes responsible for gliadin pattern changes are located in the Nonstandard arms of groups 1 chromosomes (1-A-short, 1B- short and ID- short) and in the standard arms of groups 6 chromosomes (6A - short, 6B - short and 6D - short). Location of genes encoding glutenin subunits Brown *et al.* (1981) reported that the glutenin subunits were coded by genes located in chromosomes 1A, 1B and 1D. Location of genes affecting total proteins The substitution chromosome 2M for 2D resulted in an increased protein content, whereas the 2M for 2A substitution lines contained less protein (Garciar olmedo *et al.*, 1984). Simultaneous improvement in grain yield and grain protein percentage from breeding is considered difficult because of the negative association usually found between these traits. It has been proposed that wild relatives are useful source of genetic variation for increasing grain protein percentage. Levey and Fieldman (1987) increased the grain protein percentage in high yielding Israeli wheats by using genes from a wild tetraploid wheat *Triticum turgidum* var *dicoccoides* which has been recognised to have exceptionally high grain protein concentration. Cox *et al.* (1990) reported that direct introgression of genes from diploid *Aegilops squarrosa* into bread wheat conferred an improvement in protein percentage. Similarly, high grain protein percentage of a tetraploid (wild) emmer wheat (*Triticum dicoccoides*) has been transferred into bread wheat.

Second Generation Transgenics for Quality Improvement

Agricultural biotechnology is an emerging science. It offers significant advantages including a more efficient system that speeds crop improvement, molecular markers to facilitate the breeding process, and access to new genes that confer significant benefits. The first generation of transgenic crops have already demonstrated that incorporation of input traits has conferred beneficial control of biotic stresses that was not possible with conventional technology. For example, effective and targeted control of specific cotton and maize insect pests as well as potato virus diseases that was not possible through conventional crop improvement programs. Nutritional and quality traits of foods can be altered either enhanced or abridged through transgenic methods. A large number of products are under development and testing Thus the development of nutritive plant-based foods has been a major activity in this generation of transgenics. *e.g.*: Omega 3-fatty acids (Fish oil); Vitamin A; Vitamin C; Natural Vitamin E (Tocopherol); Low saturate canola oil

Improvement of Nutritional Quality

These include "Golden Rice" and "Golden Mustard" which are potential products for the treatment of Vitamin A deficiency (VAD). Vitamin A is an important nutrient and essential for good health, particularly in children. children. Engineering a functional provitamin A (β-carotene) biosynthetic pathway in rice endosperm was accomplished in a single, combined transformation effort the cDNA coding for phytoene synthase (*psy*) and lycopene β-cyclase (β-*lcy*) both from *Narcissus pseudonarcissus* and both under the control of the endosperm-specific glutelin promoter together with a bacterial phytoene desaturase (*crt*I, from *Erwinia uredovora* under constitutive 35S promoter control). This combination covers the requirements for β-carotene synthesis and as hoped, yellow β-carotene-bearing rice endosperm was obtained in the T0-generation. Additional experiments revealed that the presence of β-*lcy* was not necessary, because *psy* and *crt*I alone were able to drive β-carotene synthesis as well as the formation of further downstream xanthophylls.

Metabolic Redesigning of Vitamin E Biosynthesis

Tocotrienols are the most common form of vitamin e in most of the monocot plants including rice and wheat. As potent antioxidants, tocotrienols contribute to the nutritive value of most of the cereal grains in human and livestock grains. cDNAs encoding homogentisic acid geranyl geranyl transferase (HGGT)which catalyses the committed step of tocotrienol biosynthesis were isolated from barley, wheat and rice seeds. Transgenic expression of HGGT in *Arabidopsis thaliana* leaves lead to the increased levels of tocotrienol in leaves which were absent in the non transformed leaves and a 10 to 15-fold increase in the vitamin E antioxidants. Over expression of HGGT in corn seeds showed a six fold increase in tocotrienol and tocopherol content. These results provide an insight into the genetic basis of tocotrienol biosynthesis in plants and demonstrate the ability to enhance the antioxidant content of the plants by the introduction of an enzyme that redirects the metabolic flux.

Engineering Fatty Acid Metabolism

Increasing Lysine Levels

Amino acids represent important targets for metabolic engineering for a number of reasons. Some amino acids (such as proline, arginine, methionine and glutamate) are directly or indirectly involved in the regulation of plant responses to various environmental signals, including light and mineral availability as well as biotic and abiotic stresses (Hofgen *et al.*, 2001). Other amino acids (such as lysine, threonine, methionine and tryptophan) contribute significantly to nutritional quality of plant-based foods (Hesse *et al.*, 2001). These "essential" amino acids cannot be synthesized by humans or monogastric livestock and have to be supplied to them in the diet, which is largely derived from plants. Lysine is synthesized in plants from aspartate via one branch of the aspartate family pathway. Another branch of this pathway leads to the synthesis of two other important essential amino acids namely, methionine and threonine. In plants, the entire aspartate family pathway (except the last step of methionine biosynthesis) is localized in the plastid (Galili, 1995). Lysine biosynthesis in plants is regulated primarily by a lysine-mediated feedback inhibition of dihydro dipicolinate synthase (DHPS) (Galili, 1995). This is in contrast to bacteria in which lysine biosynthesis is primarily regulated by both a lysine mediated repression of gene expression and feedback inhibition of the activity of a lysine-sensitive aspartate kinase isozyme. Developing seeds of transgenic plants expressing the bacterial DHPS under the control of the phaseolin promoter overproduced lysine (Karchi *et al.*, 1994). In cereal seeds, the major storage tissue is the endosperm and lysine catabolism is considered to operate in several specific peripheral cell layers of the endosperm (Arruda *et al.*, 2000). However, expression of the bacterial DHPS in the endosperm caused no increase in seed lysine levels, while its expression in the developing embryo caused a significant increase in seed lysine levels (Mazur *et al.*, 1999).

A Stable High Yielding Variety HI 8691 'Pusa Gaurav' with Good Content of Iron and Zinc

In Madhya Pradesh state, HI 8691 proved to be a widely adapted and high yielding variety, with good content of Iron (31 ppm) and Zinc (31 ppm) with high protein content (> 12 per cent) giving 2.3 per cent to 20.3 per cent higher yield over checks HI 8498, Lok-1 and HI 1544, and identified variety MPO 1215. It has good grain appearance (7.5) for fetching premium market price, and suitable for preparation of good pasta as well. It gave an average 4.0 q/ha and 9.0 q/ha higher yield over HI 8498 and Lok-1, in the co-ordinated trials conducted at 17 and 14 locations, respectively, in Madhya Pradesh during 2007-2010.

HI 8704 with High Levels of Yellow Pigment was Promoted to Second Year Advance Varietal Trial of Central Zone

HI 8704 is a stable and widely adapted genotype having superior yield (49.5 q/ha) compared to HI 8498 (+ 4.0 per cent superiority) with similar quality traits and hence, is equally suitable for chapati and pasta products. Compared to HI 8498, it has higher yellow pigment of 6.2 ppm, along with good grain appearance.

HI 8703 with Good Levels of Yellow Pigment was Promoted to Second Year Advance Varietal Trial of NWP Zone

HI 8703 is a stable and widely adapted genotype having superior yield (47.3 q/ha) compared to PDW 291 (+ 3.5 per cent superiority) with similar quality traits and hence, is equally suitable for chapati and pasta products. It has good yellow pigment level of 5.7 ppm, along with good grain appearance.

HI 8722 with High Levels of Yellow Pigment was Promoted to First Year Advance Varietal Trial in Central Zone

HI 8722 is a stable and widely adapted genotype having higher yield (31.3 q/ha) compared to check variety HD 4672 (28.7 q/ha) under rainfed conditions. It has a yellow pigment content of 6.0 ppm and showed stability over locations.

Evaluation of Durum Germplasm for High β-yellow Pigment Content

Eleven strains showed stable performance for high yellow pigment *i.e.*, > 7.8 ppm *i.e.*, CIMB 1645, E 4291, Plata-1, V 21/23, WH 912, Rolette, CIMB 1648, NIDW 71, JU 130, I 2577 and Modoc, which can be used as stable sources for yellow pigment in future breeding programmes. Of the above lines, V 21/23, CIMB 1645 and CIMB 1648 were developed at this station.

References

Abe, M; Kobayashi, Y; Yamamoto, S; Daimon, Y; Yamaguchi, A; Ikeda, Y; Ichinoki, H; Notaguchi, M; Goto, K and Araki, T (2005) FD, a bZIP protein mediating signals from the floral pathway integrator FT at the shoot apex. *Science*. **309**: 1052–1056.

Ahmad, H (2000) Molecular marker-assisted selection of HMW glutenin alleles related to wheat bread quality by PCR-generated DNA markers. *Theoretical and Applied Genetics*, **101**: 892-896.

Altpeter, F; Vasil, V; Srivastava, V and Vasil, IK (1996) Integration and expression of the high-molecular-weight glutenin subunit 1Ax1 gene into wheat. *Nature Biotechnology* **14**: 1155-1159.

Alvarez, ML; Guelman, S; Halford, NG; Lustig, S; Reggiardo, MI; Ryabushkina, N; Schewry, P; Stein, J and Vallejos, RH (2000) Silencing of HMW glutenins in transgenic wheat expressing extra HMW subunits. *Theoretical and Applied Genetics* **100**: 319-327.

Ambati, D; Phuke, RM; Vani, V; Sai Prasad, SV; Singh, JB; Patidar, CP; Malviya, P; Gautam, A and Dubey, VG (2019) Assessment of genomic diversity and development of core germplasm in durum wheat using agronomic and grain quality traits, *Cereal Research Communications, On-line publication.*

Anderson, DD and Greene, FC (1989) The characterization and comparative analysis of high-molecular-weight glutenin genes from genomes A and B of a hexaploid bread-wheat. *Theoretical and Applied Genetics*, **77**: 689-700.

Anderson, JA; Sorrells, ME and Tanksley, SD (1992) Molecular markers for pre-harvest sprouting resistance in wheat. In: Procs of the 3rd Public workshop of the International Triticeae Mapping Initiative.

Anjum, FM; Lookhart, GL and Walker, CE (2000) High molecular weight glutenin subunit composition of Pakistani hard white spring wheats grown at three locations for two years and its relationship with end-use quality characteristics. *J. Sci. Food Agric.***80**: 1-7.

Anonymous (1980) Genetic Improvement of Productivity and Nutritional quality of Wheat. Final Report. University of Nebraska - Lincoln. USA.

Anonymous (2017) It's a fight against malnutrition in Sub-Sahara - Breeding in Africa for Africa, *Interviews*, 30 May, 2017.

Anonymous (2020) Biofortification, Science, *The Hindu*, 15 May, 2020.

Anuradha, C (2003) Introgression of genes for quality characteristics from *Triticum dicoccoides* and *T. dicoccum* in durum wheat (*T turgidum* var *durum*). M.Sc. Thesis. PAU, Ludhiana.

AOAC (1969) Official methods of analysis. Methods 76-30A. *Cereal Sci. Today* **14**: 320.

AOAC (2000) Official methods of analysis. American, Association of Cereal Chemists Minnesota, Washington D.C., U.S.A.

Aprile, MC; Caputo, V and Nayga, RM (2016) Consumers' preferences and attitudes toward local food products. *J. Food Prod. Market* **22**: 19.

Arude, P (2000) Regulation of lysine catabolism in higher plants. *Trends Plant Sci* 5: 324-333.

Austin, A and Hanslas, VK (1983) Quality characters of Indian wheat cultivars. Wheat Project Directorate, IARI, New Delhi, pp.109.

Avni, R; Nave, M; Barad, O; Baruch, O and Twardziok, SO (2017) Wild emmer genome architecture and diversity elucidate wheat evolution and domestication. *Science* **357**: 93.

Axford, DWE; McDermott, EE and Redman, DG (1979) Note on the sodium dodecyl sulfate test of bread-making quality: Comparison with Pelshenke and Zeleny tests. *Cereal Chemistry* **56**: 582–584.

Axtell, JD (1981) Breeding for improved Nutritional Quality in Plant Breeding II.by K.J. Frey (Ed). Kalyani Publishers, Ludhiana. pp.365-432.

Aykroyd, WR; Patwardhan, VN and Ranganathan, S (1951) The Nutritive Vales of Indian Foods and the Planning of satisfactory Diets. Health Bulletin No.23 4th edition, Ministry of Health, New Delhi. India.

Baga, M; Glaze, S; Mallard, CS and Chibbar, RN (1999b) A starch-branching enzyme gene in wheat produce alternatively spliced transcripts. *Plant Molecular Biology* **40**: 1019-1030.

Baga, M; Nair, RB; Repellin, A; Scoles, GJ and Chibbar, RN (2000) Isolation of a cDNA encoding a granule-bound 152-kilodalton starch-branching enzyme in wheat. *Plant Physiology* **124**: 253-264.

Balyan, HS; Gupta, PK; Kumar, S; Dhariwal, R; Jaiswal, V; Tyagi, S; Agarwal, P; Gahlaut, V and Kumari, S (2013) Genetic improvement of grain protein content and other health-related constituents of wheat grain. *Plant Breed* **132**: 446–457.

Bansal, HC and Mehta, SL (1997) Breeding for protein quality and quantity. In Plant Breeding. Ed. VL Chopra. Oxford and IBH Pub and Co, pp. 309-334.

Barro F, Rooke L, Bekes F, Gras P, Tatham ST, Fido R, Lazzari PA, Shewry PR, Barcelo, P (1997) Transformation of wheat with high molecular weight genes results in improved functional properties. *Nature Biotech* **15**: 1295-1299.

Basu, RN (1990) Seed invigoration for extended storability. *Procs of the International Conference on Seed Science and Technology*, New Delhi.

Bean, SR and Lookhart, GL (1999) Sodium dodecyl sulfate capillary electrophoresis of wheat proteins. 1. Uncoated capillaries. *J. Agric. Food Chem.* **47**: 4246-4255.

Beckmann, JS and Soller M (1986) Restriction fragment length polymorphisms and genetic improvement of agricultural species. *Euphytica*, **35**: 111-124.

Begum, R (1995) Text Book of Foods, Nutrition and Dietetics. Sterling Publ Pvt. Ltd., II Edition, pp95-104.

Belderok, B and Mesdag H and Donner, DA (2000) Bread-Making Quality of Wheat. Springer. pp3.

Bennett, AM (1994) Molecular and functional studies of variant HMW glutenins in wheat. Ph.D. thesis. Univ. Sydney, Australia.

Bhat, NR (2001) Some reflections on breeding for quality with special reference to cash programme for yield of major industrial crops in India. Presidential address Indian society of Plant Breeding and Genetics, Diamond jubilee symposium, pp197-208.

Blanco, A, Degiovanni, C; Laddomada, B; Sciancalepore, A; Simeone, R; Devos, KM and Gale MD (1996) Quantitative trait loci influencing grain protein content in tetraploid wheats. *Plant Breed* **115**: 310-316.

Blatter, RH; Jacomet, S and Schlumbaum, A (2004) About the origin of European spelt (*Triticum spelta* L.): Allelic differentiation of the HMW glutenin B1-1 and A1-2 subunit genes. *Theor. Appl. Genet.* **108**: 360.

Blechl, A and Anderson, O (1996) Expression of novel high-molecule-weight glutenin subunit gene in transgenic wheat. *Nature Biotech.*, **14**: 875-879.

Blechl, AE and Anderson, OD (1996) Expression of a novel high-molecular-weight glutenin subunit gene in transgenic wheat. *Nature Biotechnology* **14**: 875-879.

Bouis, HE (2003) Micronutrient fortification of plants through plant breeding: Can it improve nutrition in man at low cost? *Proc. Nutr. Soc.* **62**: 403–11.

Bouis, HE; Harris, S and Lineback, D (2002) Biotechnology- derived nutritious foods for developing countries: Needs, opportunities, and barriers. *Food and Nutrition Bulletin* **23**(4): 342–383.

Bozzini, A (1966) Mutations in Plant Breeding, FAO-IAEA PL 503/20.

Brandolini, A; Volante, A and Heun, M (2016) Geographic differentiation of domesticated einkorn wheat and possible Neolithic migration routes. *Heredity* **117**: 135.

Brennan, CS and Cleary, LJ (2005) The potential use of cereal (1/3,1/4)-b-D-glucans as functional food ingredients. *J Cereal Sci.* **42** : 1–13.

Brevis, JC and Dubcovsky, J (2010) Effect of chromosome region including the *Gpc-B1* locus on wheat protein and protein yield. *Crop Sci* **50**: 93–104.

Broun, P; Gettner, S and Somerville, C (1999) Genetic engineering of plant lipids. *Annu. Rev. Nutr.* **19**: 197–216.

Brown, JWS and Flavell, RB (1981) Fractionation of wheat gliadin and glutenin subunits by two-dimensional electrophoresis and in gliadin synthesis. *Theor. Appl. Genet.*, **59**: 349-359.

Bryan, GJ; Collins, AJ; Stephenson, P; Orry, A; Smith, JB and Gale, MD (1997) Isolation and characterization of microsatellites from hexaploid bread wheat. *Theor Appl Genet* **94**: 557-563.

Buchanan, BB; Gruissen, W and Jones, RL (2000) In: Biochemistry and molecular Biology of plants American Society of Plant Physiologists, Maryland.

Burton, GW (1981) Meeting human needs through plant breeding: past progress and progress for the future. In K.J. Frey (ed) Plant Breeding II. Iowa State. Univ. Press, Ames, pp. 433-466.

Bushuk W and Zillman, RR (1978) Wheat cultivar identification by gliadin electrophoregrams. I Apparatus, method and nomenclature. *Can. J. Plant Sci.* **58:** 505-515.

Cakmak, I (2005) Role of mineral nutrition in tolerance of crop plants to environmental stress factors. In: Fertigation: optimizing the utilization of water and nutrients; September 20–24; Beijing. China: Fertigation proceedings: Selected papers of the IPI-NATESE-CAUCAAS; 2005. pp. 35–48.

Cakmak, I; Graham, RD and Welch, RM (2002) Agricultural and molecular genetic approaches to improving nutrition and preventing micronutrient malnutrition globally. In *Encyclopedia of Life Support Systems,* UNESCO Publishing, *Vol. 2,* pp. 1075–1099.

Cakmak, I; Kalayc, M; Ekiz, H; Braun, HJ; K l nc, Y and Y lmaz, Y (1990) Zinc deficiency as a practical problem in plant and human nutrition in Turkey: A NATO science-for-stability project. *Field Crops Res.* **60**: 175–188.

Cakmak, I; Ozkan, H; Braun, HJ; Welch, RM and Römheld, V (2000) Zinc and iron concentrations in seeds of wild, primitive, and modern wheats. *Food Nutr. Bull.* **21**: 401–403.

Cantrell, RG and Joppa, LR (1991) Genetic analysis of quantitative traits in wild emmer (*T. turgidum* L. var *dicoccoides*) *Crop Sci.* **31**: 645 – 49.

Capelli, L; Forlani, F; Guerrieri, F; Perini, N; Cerletti, P and Righetti, PG (1998) Wheat cultivar discrimination by capillary electrophoresis of gliadins in isoelectric buffers. *Electrophoresis* **19**: 311-318.

Chao, S; Sharp, PJ; Worland, AJ; Warham, EJ; Koebner, RMD and Gale, MD (1989) RFLP-based genetic maps of wheat homoeologous group 7 chromosomes. *Theor Appl Genet* **78**: 495-504.

Cheryan, M and Rackis, JJ (1980) Phytic acid interactions in food systems. *CRC Crit. Rev. Food Sci. Nutr.* **13**: 297–335.

Chowdhury, RK (2000) Post harvest handling seed Research in India. In advanced course on post-harvest handling and management of seed. 6-15 June 2000. Indian Agricultural Research Institute, Regional Station, Kernal pp. 117 – 125.

Chu, CC; Wang, CC; Sun, CS; Chun, NP; Yin, KC and Hsu, C (1973) Investigation on the induction and morphogenesis of wheat (*Triticum vulgare*) pollen plants. *Acta Botanica Sinica* **15**: 1-11.

Ciaffi, M and Lafindra, D (1995) Bread making potential of durum wheat lines expressing both x and y type subunits at the Glu A 1 locus. *Cereal* **72**: 465 – 69.

Crossby, PB (1979) Quality is free: The art of making quality certain. McGraw Hill, New York.

D'Ovidio, R (1993) Single-seed PCR of LMW glutenin genes to distinguish between durum wheat cultivars with good and poor technological properties. *Plant Molec. Biol.* **22**: 1173-1176.

D'Ovidio, R and Porceddu, E (1996) PCR-based assay for detecting 1B-genes for low molecular weight glutenin subunits related to gluten quality properties in durum wheat. *Plant Breed* **115**: 413-415.

Davies, J; Berzonsky, WA and Leach, GD (2006) A comparison of marker-assisted and phenotypic selection for high grain protein content in spring wheat. *Euphytica* **152**: 117–134.

de la Pena, RC; Murray, TD and Jones, SS (1997) Identification of an RFLP interval containing *Pch2* on chromosome 7AL in wheat. *Genome* **40**: 249-252.

Della-Penna, D (1999) Nutritional genomics: manipulating plant micronutrients to improve human health. *Science.* **285**: 375–379.

Deodikar, GB; Patil, VP and Rao, VSP (1979) Breeding of Indian *durum* and *dicoccum* wheats by interspecific hybridization with 4N *Triticum* species. *Ind J Genet Pl Breed* **39:** 114-25.

Devos, KM; Bryan, GJ; Collins, AJ; Stephenson, P and Gale, MD Application of two microsatellite sequences in wheat storage proteins as molecular markers. *Theor Appl Genet*. **90**(2): 247–252.

Dhaka, V and Khatkar, BS (2015) Effects of gliadin/glutenin and HMW-GS ratio on dough rheological properties and bread making potential of wheat varieties. *J Food Qual* **38**: 71–82.

Dhaliwal, HS; Garg, M; Singh, H; Chhuneja, P and Kaur H (2002) Transfer of HMW-glutenin subunits from wild wheats into *Triticum durum* and improvement of quality. *Cereal Res Commun* **30**: 173–180.

Dharmalingam, C; Lakshmi, S; Malarkodi, K and Anandhi, SK (1998) Halogenation of seeds to prolong shelf life. Department of Seed Science and Technology, TNAU, Coimbatore.

Dick, JW (1981) Basic structure of durum wheat emphasizing its unique chemical, physical and quality characteristics. *Macaroni Journal*. **63**: 35-43.

Duboise, M; Giler, KA; Aaimitton, JK; Rochers, PA and Smith, F (1956) Colorimetric determination of sugars and related substance. *Analytical chemistry*.

Dudley, JW (1984) A method of identifying populations containing favourable alleles not present in elite germplasm. *Crop Sci.*, **24**: 1053-1054.

EBC (2003) Analytica-EBC. EBC analysis committee. Fachverlag Hans Carl, Nurnberg.

Eggum, BO (1984) Evaluation of protein quality of cereal mutants. In Cereal Grain Protein Improvement. (Panel Procs series, 6-10 December 1982 IAFA Vienna. pp347-355.

Erdman, JW Jr (1981) Bioavailability of trace minerals from cereals and legumes. *Cereal Chem*. **58**: 21.

Feil, B and Fossati, D (1995) Mineral composition of triticale grain as related to grain yield and grain protein. *Crop Sci*. **35**: 1426–1431.

Fierce, Y; Morais, VM; Piantedosi, R; Wyss, A; Blaner, WS and Paika, J (2008) *In vitro* and *in vivo* characterization of retinoid synthesis from β-carotene. *Archives of Biochemistry and Biophysics* **472**: 126–138.

Garcia - Olmedo, F; Carbonero, P and Jones, BL (1982) Chromosomal locations of genes that control wheat endosperm proteins. *Advances in Cereal Sci and Technol.*, **V**: 1-47.

Gegas, VC; Nazari, A; Griffiths, S; Simmonds, J; Fish, L; Orford, S; Sayers, L; Doonan, JH and Snape, JW (2010) A genetic frame-work for grain size and shape variation in wheat. *Plant Cell* **22**: 1046–1056.

Geisslitz, S; Longin, CFH; Scherf, KA and Koehler, P (2019) Comparative study on gluten protein composition of ancient (einkorn, emmer and spelt) and modern wheat species (durum and common wheat). *Foods* **8**: 409.

Geisslitz, S; Wieser, H; Scherf, KA and Koehler, P (2018) Gluten protein composition and aggregation properties as predictors for bread volume of common wheat, spelt, durum wheat, emmer and einkorn. *J. Cereal Sci.* **83**: 204.

Giambanelli, E; Ferioli, F; Kocaoglu, B; Jorjadze, M; Alexieva, I; Darbinyan, N and Antuono, F (2013) A comparative study of bioactive compounds in primitive wheat populations from Italy, Turkey, Georgia, Bulgaria and Armenia. *J. Sci. Food Agric.* **93**: 3490.

Gopalan, C; Sastri, BVR and Subramanian, SC (2000) Nutritive value of Indian Foods. National Institute of Nutrition, ICMR, Hyderabad, pp. 27-28.

Graham, RD (2003) Bio-fortification – A global challenge program. *International Rice Research Newsletter*. 4-8.

Graham, RD and Welch, RM (1996) Breeding for staple-food crops with high micronutrient density: Agricultural strategies for micronutrients. Working Paper No. 3; 1996. International Food Policy Research Institute, Washington, DC.

Graham, RD; Humphries, JM and Kilethan, JL (2000) Nutritionally enhanced cereals: a sustainable foundation for a balanced diet. Asia Pacific. *J. Chin. Nutr.* **9 (Suppl.):** S 91 – S 96.

Graham, RD; Senadhira, D; Beebe, SE; Iglesias, C and Ortiz-Monasterio, I (1999) Breeding for micronutrient density inedible portions of staple food crops: Conventional approaches. *Field Crops Res.* **60**: 57–80.

Graham, RD; Welch, RM and Bouis, HE (2001) Addressing micronutrient malnutrition through enhancing the nutritional quality of staple foods: Principles, perspectives and knowledge gaps. *Advances in Agronomy* **70**: 77–142.

Gregg (1967) Seed Processing. National Seed Corporation, New Delhi.

Gupta, VP (1980) Genetics quality improvement. In: Trends in Genetical Research on *Pennisetum*. (VP Gupta and JL Minocha, eds.). Punjab Agricultural University, Ludhiana, India, Pages 91-98.

Gupta, VP (1992) Genetic improvement and management of quality in crop plants. *Indian J. Genet* **52**: 207-212.

Gupta, VP (1996) Strategies in breeding for and management of quality in crop plants, *Crop Improv.* **23**(8): 1-10.

Gupta, VP (2001) Genetic improvement and management of quality in crop plants. Presidential address, Indian society of Genetics and Plant Breeding (Diamond Jubilee celebrations). 287-292.

Guttierie, M; Bowen, D; Dorsch, JA; Raboy, V and Souza, E (2004) Identification and characterization of a low phytic acid wheat. *Crop Sci.* **44**: 418–424.

Hadacek, F (2002) Secondary metabolites as plant Traits: Current Assessment and Future Perspectives. *Cri Rev Plant Sci* **21**: 273-322.

Hamer, RJ; Weegels, WP and Marseille, JP (1992) Prediction of bread making quality of wheat: the use of HMW glutenin A subunit-based quality scoring systems. *Journal of Cereal Science* **15**: 91-102.

Harberd, NP; Bartels, D and Thompson, RD (1986) DNA restriction fragment variation in the gene family encoding high-molecular- weight (HMW) glutenin subunits of wheat. *Biochemical Genetics* **24**: 579-596.

He, GY; Rooke, L; Steele, S; Bekes, F; Gras, P; Tatham, AS; Fido, R; Barcelo, P; Shewry, PR and Lazzeri, PA (1999) Transformation of pasta wheat (*Triticum turgidum* L. var *durum*) with high-molecular weight glutenin subunit genes and modification of dough functionality. *Molecular Breeding* **5**: 377-386.

Heyer, AG; Lloyd, JR and Kossmann, J (1999) Production of modified polymeric carbohydrates. *Curr. Opin. Biotechnol.,* **10**: 169-174.

Horn, MJ; Jones, B and Blum, AE (1946) Colorimetric determination of methionine in protein and food. *J. Biol.Chem.* **116**: 313.

Hostettmann, K; Marston, A and Hostettmann, M (1997) Preparative chromatography techniques: Applications in natural products isolation. Springer, Berlin.

Humphreys, DG; Procunier, JD; Mauthe, W; Howes, NK; Brown, PD and Mac-Kenzie, RIH (1998) Marker-assisted selection for high protein concentration in wheat. In: Fowler DB, Ged-des WE, Johnston AM, Preston KR (eds.) Wheat Protein, Production and Marketing. Proc Wheat Protein Symp, Saskatoon, Saskatchewan, Canada, 9-10 Mar. (1998), pp255-258.

Imtiaz, M; Alloway, JB; Shah, AK; Siddiqui, SH; Memon, MY; Aslam, M and Khan, P (2003) Zinc contents in the seed of some domestic and exotic wheat genotypes. *Asian Journal of Plant Sciences* **2**(15–16): 1118–1120.

Institute of Medicine (2001) Dietary reference intakes for vitamin A, vitamin K, arsenic, boron, chromium, copper, iodine, iron, manganese, molybdenum, nickel, silicon, vanadium, and zinc. National Academy Press, Washington, D.C.

International Atomic Energy Agency (1991) Neutron Irradiation of seeds, Vienna.

Johnson, VA (1976) Wheat protein. Proc Int. Symp. Genetic Control of Diversity in plants, March 1976, Lahore (Pakistan).

Jones, K and Ames, AJ (1967) Modern Cereal Chemistry 6th edition, The Northern Publishing Co. Ltd. Liverpool, England.

Joppa, LR; Khan, K and Williams, ND (1983) Chromosomal location of genes for gliadin polypeptides in durum wheat (*Triticum turgidum L.). Theor. Appl.Genet.* **64**: 289-293.

Jun, L; Fa, C; Anming, D; Chunhua, Z; Xiuqin, W; Lin, W; Yinguang, B; Xiaolei, Q; Xingfeng, L and Jurong, G(2012) QTL detection of seven quality traits in wheat using two related recombinant inbred line populations. *Euphytica.* **183**(2): 207-226.

Kade, M; Barneix, AJ; Olmos, S and Dubcovsky, J (2005) Nitrogen uptake and remobilization in tetraploid 'Langdon' durum wheat and a recombinant substitution line with the high grain protein gene *Gpc-B1*. *Plant Breed* **124**: 343–349.

Khan, IA; Procunier, JD; Humphreys, DG; Tranquilli, G; Schlatter, AR; Marcucci-Poltri, S; Frohberg, R and Dubcovsky, J (2000) Development of PCR-based markers for high grain protein content gene from *Triticum turgidum* ssp. *dicoccoides* transferred to bread wheat. *Crop Science*, **40**: 518-524.

Khardori, A (2001) Miami Group Vs rest of the world. Down to earth, pp36-45.

Khush, GS (2002) The promise of biotechnology in addressing current nutritional problems in developing countries. *Food and Nutrition Bulletin* **23**(4): 354–357.

Khushnir, U and Halloran, GM (1984) Transfer of high kernel weight and high protein from wild tetraploid wheat (*Triticum turgidum dicoccoides*) to bread wheat (*Triticum aestivum*) using homologous and homoeologous recombination. *Euphytica* **33**: 349 - 355.

Kinney, AJ (1996) β-Ketoacyl-ACP synthetase II genes from plants. *US Patent* 5,**500**: 361.

Klepper, LA (1975) Nitrate assimilation and seed protein in wheat.Proc.2nd Int. Wheat Conf. Zagreb.

Koenig, A; Konitzer, K; Wieser, H and Koehler, P (2015) Classification of spelt cultivars based on differences in storage protein compositions from wheat. *Food Chem*. **168**: 176.

Krishnappa, G; Ahlawat, AK; Shukla, RB; Singh, SK; Singh, AM and Singh, GP (2019) Multi-environment Analysis of Grain Quality Traits in Recombinant Inbred Lines of a Biparental Cross in Bread Wheat (*Triticum aestivum* L). *Cereal Res Commun* **47**: 334-344.

Kulwal, P; Kumar, N; Kumar, A; Gupta, RK; Balyan, HS and Gupta, PK (2005) Gene networks in hexaploid wheat: interacting quantitative trait loci for grain protein content. *Funct Integr Genom* **5**: 254–259.

Kumar S, Sohu VS, Singh RP, Gupta SK, Srivastava P, Bains NS (2018) Investigating the role of high molecular weight glutenin subunits (HMW-GS) protein in end use quality of Indian flat breads. *Indian J Biotechnol* **17**: 65–73.

Kumar, D; Kumar, A; Chokar, V; Gangwar, OP; Bhardwaj, SC; Sivasamy, M; Prasad, SVS; Prakash, TL; Khan, H; Singh, R; Sharma, P; Sheoran, S; Iquebal, MA; Jaisawal, S; Angadi, U; Singh, G; Rai, A; Singh, GP; Kumar, D and Tiwari, R (Undated) Genome-wide association studies in diverse spring wheat panel for Stripe, Stem and Leaf rust resistance, *Front Plant Sci*.

Kumar, J; Jaiswal, V; Kumar, A; Kumar, N; Mir, RR; Kumar, S; Kumar, R; Tyagi, S; Khandelwal, M; Prabhu, KV; Prasad, R; Balyan, HS and Gupta, PK (2011) Introgression of a major gene for high grain protein content in some Indian bread wheat cultivars. *Field Crop Res* **123**: 226–233.

Levey, AA and Fieldman, M (1987) Increase in grain protein percentage in high yielding common wheat breeding lines by genes from wild tetraploid wheat. *Euphytica*, **36**: 353.

Li, C and Dubcovsky, J (2008) Wheat FT protein regulates VRN1 transcription through interactions with FDL2. *Plant J.* **55**: 543–554.

Li, Z; Chu, X; Mouille, G; Yan, L; Kosar-Hashemi, B; Hey, S; Napier, J; Shewry, P; Clarke, B; Appels, R; Morell, MK and Rahman, S (1999) The localization and expression of the class II starch synthases of wheat. *Plant Physiology* **120**: 1147-1156.

Li, Z; Mouille, G; Kosar-Hashemi, B; Rahman, S; Clarke, B; Gale, KR; Appels, R and Morell, MK (2000) The structure and expression of the wheat starch synthase III gene. Motifs in the expressed gene define the lineage of the starch synthase III gene family. *Plant Physiology* **123**: 613-624.

Liener, IE and Kakade, ML (1980) Protease inhibitors, in Toxic Constituents of Plant Foodstuffs, 2nd ed., Liener, IE Ed., Academic Press, New York. 7.

Liever, JE (Ed) 1969) Toxic constituents of plant Food stuffs. London, New York, Academic Press.

Lillemo, M; Chen, F; Xia, XC; William, M; Pe~na, RJ; Trethowan, R and He, ZH (2006) Puroindoline grain hardness alleles in CIMMYT bread wheat germplasm. *J. Cereal Sci.* **44**: 86–92.

Longin, CFH; Ziegler, J; Schweiggert, R; Koehler, P; Carle, R and Wuerschum, T (2015) Comparative study of hulled (einkorn, emmer, and spelt) and naked wheats (durum and bread wheat): Agronomic performance and quality traits. *Crop Sci.* **56**: 302.

Lonnerdal, B (2003) Genetically modified plants for improved trace element nutrition. *J Nutr* **133**: 1490S–1493S.

Lowry (1951) Estimation of protein. In: Methods in plant chemistry and economic botany, Kalyani publishers, New Delhi P.32.

Lowry, OH; Rosebrough, NJ; Farr, AL and Randall, RJ (1951) Protein measurement with folin phenol reagent. *J.Biol.chem.* **193**: 265.

Ludvigsson, JF; Leffler, DA; Bai, JC; Biagi, F and Fasano, A (2013) The Oslo definitions for coeliac disease and related terms. *Gut* **62**: 43.

Luo, MC; Yang, ZL; You, FM; Kawahara, T; Waines, JG and Dvorak, J (2007) The structure of wild and domesticated emmer wheat populations, gene flow between them, and the site of emmer domestication. *Theor. Appl. Genet.* **114**: 947.

Maccaferri, M; Harris, NS; Twardziok, SO; Pasam, RK and Gundlach, H (2019) Durum wheat genome highlights past domestication signatures and future improvement targets. *Nat. Genet.* **51**: 885.

Mahalingam, N (2003) Production and Productivity. *Kisan World*, pp5-6.

Mandal, S and Mandal, RK (2000) Seed Storage proteins and approaches for improvement of their nutritional quality by genetic engineering. *Current Science,* **79** (5): 576-589.

Matthews, PD; Luo, RB and Wurtzel, ET (2003) Maize phytoene desaturase and β-carotene desaturase catalyse a poly-Z desaturation pathway: Implications for genetic engineering of carotenoid content among cereal crops. *J. Exp. Bot.* **54**(391): 2215– 2230.

Mc Kanzie, HA and Wallace, HS (1954) The Kjeldahl determination of nitrogen *Aust. J. Chem.* **17**: 55.

McCartney, CA; Somers, DJ and Lukow, OM (2006) QTL analysis of quality traits in the spring wheat cross RL4452 'AC Domain'. *Plant Breed* **125**(6): 565–575.

McCulung AM and Cantrell, RG (1986) Inheritance of glume colour and gluten strength in durum wheat. *Euphytica* **35**: 885-880.

Mclauchlan, A; Ogbonnaya, FC; Hollingsworth, B; Carter, M; Gale, KR; Henry, RJ; Holten, TA; Morell, MK; Rampling, LR; Sharp, PJ; Shariflou, MR; Jones, MGK and Appels, R (2001) Development of PCR-based DNA markers for each homoeo-allele of granule-bound starch synthase and their application in wheat breeding programs. *Aust. J. Agric. Res.* **52**: 1409–1416.

Mehak, K; Singh, N; Singh, V; Kaur A; Ahlawat, AK; Singh, AM and Bajana R (2019) Comparative analysis of native and defatted flour from hard, extraordinarily soft and medium-hard wheat varieties for protein solvation, pasting, mixing, and dough rheological behaviour. *Journal of Food Science* **85**(1).

Miao, S; Duncan, DR and Widholni, J (1988) Selection of regenerable maize callus culture resistant to 5-methyl-DL-tryptophan, S-L2-aminoethyl) L-cysteine and high levels of L-lysine plus L-threonine. *Plant Cell, Tissue Organ Cult.* **14**: 3-14.

Migliorini, P; Spagnolo, S; Torri, L; Arnoulet, M; Lazzerini, G and Ceccarelli, S (2016) Agronomic and quality characteristics of old, modern and mixture wheat varieties and landraces for organic bread chain in diverse environments of northern Italy. *Eur. J. Agron.* **79**: 131.

Misra, BK; Gupta, RK and Sewa Ram (1998) Production of Quality wheats for varied domestic needs. *Indian Farming* **48**: 16-19 and 32.

Monasterio, I; Palacios-Rojas, N; Meng, E; Pixley, K; Trethowan, R and Pena, RJ (2007) Enhancing the mineral and vitamin content of wheat and maize through plant breeding. *J Cereal Sci.* **46**: 293–307.

Mujab-ur Rahman (1999) Variations in dietary pattern and associated factors – A case study. Ph.D. Thesis. Osmania University, Andhra Pradesh.

Mushtari Begum, J (2003) Micronutrient malnutrition. Lecture at International Food Convention held at Mysore, 5-8 December 2003. Souvenir – Abstracts of Plenary and technical lectures, pp43 – 44.

Muyanja, CMBK; Kikafunda, JK; Narvhus, JA; Helgetun, K and Langsrud, T (2003) Production methods and composition of Bushera: a Ugandan traditional fermented cereal beverage. *African Journal of Food, Agriculture, Nutrition and Development*. **3**(1): 10-19.

Nelson, N (1944) A photometric adaption of Sommoggi method for the determination of glucose. *J.Biol. Chem*. **153**: 375-3.

Niranjana, M; Verma, A; Singh, CK; Islam, MA; Jha, SK, Mallick, N; Sharma, JB; Tomar, SMS and Vinod (2019) Plastome micro satellite (SSR) and morphological diversity in *Triticum -Aegilops species. Journal of Genetics.*

Ojakangas, BA (2002) Great Whole Grain Breads. University of Minnesota Press. p48.

Olmos, S; Distelfeld, A; Chicaiza, O; Schlatter, AR; Fahima, T; Echenique, V and Dubcovsky J (2003) Precise mapping of a locus affecting grain protein content in durum wheat. *Theoretical and Applied Genetics* **107**: 1243-1251.

Opienska-Blaunth, J; Charezinski. JM and Berbec. H (1963) A new and rapid method of determining tryptophan. *Anal. Biochem*. **6**: 69.

Ortiz-Monasterio, I and Graham, RD (2000) Breeding for trace minerals in wheat. *Food Nutr. Bull*. **21**: 392–396.

Ozturk, L; Yazici, MA; Yucel, C; Torun, A; Cekic, C; Bagci, A; Ozkan, H; Braun, HJ; Sayers, Z and Cakmak, I (2006) Concentration and localization of zinc during seed development and germination in wheat. *Physiologia Plantarum* **128**: 144–152.

Parker, GD; Chalmers, KJ; Rathjen, AJ and Langridge, P (1998) Mapping loci associated with flour colour in wheat (*Triticum aestivum* L.) *Theor Appl Genet* **97**: 238-245.

Payne, PI (1984) The association between γ-gliadin 45 and gluten strength in durum wheat varieties: a direct causal effect or the result of genetic linkage? *J. Cereal Sci*. **11**: 15-34.

Payne, PI and Lawrence, GJ (1983) Catalogue of alleles for the complex gene loci, *Glu-A1, Glu-B1*, and *Glu-D1* which code for high molecular weight subunits in glutenin in hexaploid wheat. *Cereal Res Communications*, **11**: 29-35.

Pena RJ (2007) Current and future trends of wheat quality needs, in: HT Buck, JE Nisi, N Salomon (Eds.), Wheat Production in Stressed Environments, Springer, Berlin, pp. 411–424.

Pena, RJ; Trethowan, R; Pfeiffer, WH; Ginkel, MV and Van-Ginkel, M (2002) Quality (end-use) improvement in wheat: compositional, genetic, and environmental factors. *J Crop Production*. **5**: 1-2.

Peterson, CJ; Johnson, VA and Mattern, PA (1986) Influence of cultivar and environment on mineral and protein concentrations of wheat flour, bran, and grain. *Cereal Chem*. **63**: 118–126.

Pfeiffer, WH; Trethowan, RM; Ammar, K and Sayre, KD (2005b) Increasing yield potential and yield stability in durum wheat. In C Royo, MM Nachit, N di Fonzo, JL Araus, WH Pfeiffer, and GA Slafer (eds.). Vol. 2. *Durum Wheat Breeding: Current Approaches and Future Strategies*. New York: The Haworth Press, Inc. pp. 531–544.

Philip, TE (1998) Modern cookery for teaching and the trade, orient longman Ltd, Bombay, pp219-221, 269.

Phillips, RW and Freeling. M (1998) Plant Genomics and our food supply: An introduction. *Proc. Natl. Acad. Sci. USA*. **95**: 1969-1970.

Pinstrup-Andersen, P (2000) Improving human nutrition through agricultural research: Overview and objectives. *Food Nutr. Bull.* **21**: 352–355.

Pogna, NE; Autran, JC; Mellini, F; Lafiandra, D and Feillet, P (1990) Chromosome 1B-encoded gliadins and glutenin subunits in durum wheat: genetics and relationship to gluten strength. *J. Cereal Sci* **11**: 15-34.

Punia, S; Sandhu, KS and Siroha AK (2017) Difference in protein content of wheat (*Triticum aestivum* L.): Effect on functional, pasting, color and antioxidant properties. *J Agric Sci* **18**: 378–384.

Rai, A; Singh, AM; Raghunandan, K; Kumar, K; Sharma, TPJ; Ahlawat, AK; Singh, SK; Ganjewala, D; Shukla, RB and Shivasamy, M (2019) Marker assisted transfer of PinaD1a gene to develop soft grain wheat cultivars. *3 Biotech*, **9**: 183-189.

Randhawa, HS; Dhaliwal, HS; Singh, H and Kaur, H (1995) Cataloguing of wheat germplasm for HMW glutenin subunit composition. In: Chopra, VL, Sharma, RP and Swaminathan, MS (eds.) Agricultural Biotechnology, Second Asia Pacific Conference, New Delhi, Calcutta, Oxford and IBH Publishing Co. Pvt. Ltd.

Robbellen (1977) Possibilities and Limitations of Breeding for Nutritional Improvement of Cereals, In nutritional evaluation of cereal mutants. IAIA, Vienna, pp. 47-57.

Rossmann, A; Buchner, P; Savill, GP; Hawkesford, MJ; Scherf, KA and Mühling, KH (2019) Foliar N application at anthesis alters grain protein composition and enhances baking quality in winter wheat only under a low N fertiliser regimen. *Eur. J. Agron.* **109**: 125909.

Roy, A; Archana, S; Jolly, M; Nirupama, T; Johari, RP; Roy, A; Sachdev, A and Tiwari, N (2002) Biochemical characterization of Indian durum wheats. *Indian J of Plant Physiology* **7**: 9-14.

Sankaram, A (1992) Food and Nutrition Security for developing world. *Indian Farming*. **42**: 3-10.

Shepherd, KW (1968) Chromosomal control of endosperm proteins in wheat and rye. In: Proc. 3rd Int. Wheat Genet. Symp. Canberra. KW Finlay and KW Shepherd eds. *Aus. Acad. Sci.*, pp86-96.

Shewry, P; Tatham, A; Barro, F; Barcelo, P and Lazzeri, P (1995) Biotechnology of bread making: unraveling and manipulating the multi-protein gluten complex. *Bio/Technology* **13**: 1185-1190.

Shewry, PR and Hey, S (2015) Do "ancient" wheat species differ from modern bread wheat in their contents of bioactive components? *J. Cereal Sci.* **65**: 236.

Shewry, PR; Piironen, V; Lampi, AM; Edelmann, M and Kariluoto, S (2010) The HEALTHGRAIN wheat diversity screen: Effects of genotype and environment on phytochemicals and dietary fiber components. *J. Agric. Food Chem.* **58**: 9291.

Shintani, D and Della Penna, D (1998) Elevating the vitamin E content of plants through metabolic engineering. *Science* **282**: 2098.

Simmonds NW (1995) The relation between yield and protein in cereal grain. *J Sci Food Agric* **67**: 309–315.

Singh, NK; Shepherd, KW and Cornish, GB (1991) A simplified SDS-PAGE procedure for separating LMW subunits of glutenin. *J Cereal Sci* **14**: 203–208.

Sivasankar, B (2002) Food processing and Preservation. Prentice – Hall of India Pvt Ltd., New Delhi, pp284-295, 308-313.

Sourdille, P; Perretant, MR; Charmet, G; Leroy, P; Gautier, MF; Joudrier, P; Nelson, JC; Sorrells, ME; Bernard, M (1996) Linkage between RFLP markers and genes affecting kernel hardness in wheat. *Theor Appl Genet* **93**: 580-586.

Stoltzfus, RJ and Dreyfuss, ML (1998) Guidelines for the use of iron supplements to prevent and treat iron deficiency anemia. Washington (DC): ILSI Press.

Strain, JJ and Cashman, KD (2002) Minerals and trace elements. In MJ Gibney, HH Voster, and FJ Kok (eds.) *Human Nutrition.* Blackwell Science Ltd. Oxford, U.K. pp177–224.

Sumathi, S and Pattabiram, TN (1976) Natural plant enzyme inhibitors. Part II – Protease inhibitors of seeds. *Indian J. Biochem and Biophysics*, **13**: 52-56.

Swaminathan, M (1995) Essentials of Food and Nutrition. The Bangalore Printing and Publishing Co-Ltd, Bangalore, II, 9-10.

Swaminathan, MS; Naik, MS; Kaul, AK and Austin, A (1970) Improving Plant Proteins by Nuclear Techniques. *IAEA Vienna, pp*165.

Trethowan, RM; Reynolds, M; Sayre, KD and Ortiz-Monasterio. I (2005) Adapting wheat cultivars to resource conserving farming practices and human nutritional needs. *Annals of Applied Biology* **146**(4): 405–413.

Uauy, C; Distelfeld, A; Fhima, T; Blechl, A and Dubcovsky, J (2006) A NAC gene regulating senescence improves grain protein, zinc, and iron content in wheat. *Science* **314**: 1298–1301.

UDY, DC (1954) Dye-binding capacities of wheat flour protein fraction *Cereal Chem.* **31**: 389.

UN SCN (2004) 5th Report on the World Nutrition Situation: Nutrition for Improved Development Outcomes. United Nations System Standing Committee on Nutrition (SCN), Geneva.

Underwood BA and Smitasiri S (1999) Micronutrient malnutrition: policies and programs for control and their implications. *Annu. Rev. Nutr.*, **19**: 303-324.

Underwood, BA (2000) Overcoming micronutrient deficiencies in developing countries: Is there a role for agriculture? *Food Nutr. Bull.* **21**: 356–360.

Vasil, IK and Anderson O (1997) Genetic engineering of wheat gluten. *Trends Plant Sci* **2**: 292-297.

Veenstra, LD; Poland, J; Jannink, JL; Sorrells, ME (2020) Recurrent genomic selection for wheat grain fructans. *Crop Science*.

Velah, RM and Graham, RD (2000) A new paradigm for world agriculture: productive, sustainable, nutritious, healthful food systems. *Food Nutr. Bull.* **21**: 361 – 366.

Velu, G; Crossa, J; Singh, RP; Hao, Y; Dreisigacker, S; Perez-Rodriguez, P; Joshi, AK; Chatrath, R; Gupta, V; Balasubramaniam, A; Tiwari, C; Mishra, VK; Sohu, VS and Mavi, GS (2016) Genomic prediction for grain zinc and iron concentrations in spring wheat. *Theor Appl Genet* **129** (8): 1595-1605.

Weegels, PL; Hamer, RJ and Schofield, JD (1996) Critical Review: Functional properties of wheat glutenin. *Journal of Cereal Science* **23**: 1–18.

Welch, RM and Graham, RD (2004) Breeding for micronutrient in staple food crops from a human nutrition perspective. *J Exp Bot.* **55**: 353–364.

White, PJ and Broadley, MR (2005) Biofortifying crops with essential mineral elements. *Trends in Plant Science* **10**(12): 586–583.

Wilcke, HL (1976) Improving the nutrient quality of cereals. II Rep. 2nd Workshop on Breeding and Fortification, Sept.1976. Colorado, Agency for Intl. Devl., Washington, D.C.

Wissuwa, M (2005) Mapping nutritional traits in crop plants. In: M.R. Broadley and P.J. White (eds.), *Plant Nutritional Genomics.* Ames, Iowa: Blackwell Publishing. pp220–241.

Xu, SS; Khan, K; Klindworth, DL and Nygard, G (2010) Evaluation and characterization of high-molecular weight 1D glutenin subunits from Aegilops tauschii in synthetic hexaploid wheats. *J. Cereal Sci.* **52**: 333–336.

Ye, M and Lou, G (2000) The green revolution strikes gold. *Science*, **287**: 241-243.

Yupsanis, T and Moustakas, M (2008) Relationship between quality, colour of glume and gliadin electrophoregrams in durum wheat. *Plant Breeding* **101:** 30-35.

Zarcinas, BA; Cartwright, B and Spouncer, LR (1987) Nitric acid digestion and multielement analysis of plant material by inductively coupled plasma spectrometry. *Commun Soil Sci Plant Anal* **18**: 131–146.

Zhen, Z and Mares, D (1992) A simple extraction and one-step SDS PAGE system for separating HMW and LMW glutenin subunits of wheat and high molecular weight proteins of rye. *Journal of Cereal Science* **15**: 63-78.

Zitelli, GM Mariani; Bian colatti and J Vallega (1979) A durum wheat of high protein content useful for breeding purposes. *Plant Breeding Abstracts*, **49**: 8774.

Chapter 9

Breeding Wheat for Sustainable, Conservation and Organic Agriculture

Outlines

1. Breeding wheat for Sustainable Agriculture
   a. Sustainable Agriculture/Ecofarming/Alternative Agriculture, Ecological Agriculture and Natural Organic Farming
   b. Sustainable Agriculture - A Must
   c. Nutrient Management for Sustainable Agriculture of Wheat
2. Breeding wheat for Conservation Agriculture
   a. Conservation Agriculture
   b. Challenges in Implementation of CA
   c. Exposition of CA
   d. Availability of CA Machines
   e. Energy Use-Efficiency
   f. Soil Health
   g. Conservation Agriculture for Higher Resource-Use Efficiency
   h. Crop Residue Recycling
   i. Resource-Conserving technologies (RCTs)
   j. Raised Bed Planting
   k. Vibrant Resource Conservation Technologies through Conservation Agriculture

l. Ideotypes for Conservation Agriculture
m. Ideotypes for Resource Conservation Technologies
n. Suitable Wheat Varieties Under Conventional Tillage and Zero Tillage Condition
o. Breeding for Adaptation to Conservation Agriculture

3. Breeding Wheat for Organic Agriculture
   a. Organic Food Production
   b. Potential for Organic Crop Production
   c. Gain the Best Yields from Organic Agriculture
   d. Wheat varieties with special traits for Organic Wheat Production

Breeding Wheat for Sustainable Agriculture

Sustainable Agriculture/Ecofarming/Alternative Agriculture, Ecological Agriculture and Natural Organic Farming is the successful management of resources for agriculture to satisfy changing human needs while maintaining or enhancing the quality of environment and conserving natural resources. **Sustainable agriculture** is also known as ecofarming (as ecological balance is important) or organic farming (as organic matter is the main source of nutrient management) or sometimes as natural farming or permaculture. Some other designated it as regenerative agriculture or alternative farming. Sustainable agriculture is a food and fiber production and distribution system that supports profitable production; protects environmental quality; uses natural resources efficiently; provides consumers with affordable, high-quality products; decreases dependency on nonrenewable resources; enhances the quality of life for farmers and rural communities, and will last for generations to come. Sustainable agriculture is also referred by other names such as alternative agriculture, ecological agriculture and natural organic farming. It is that form of farming which maintains or enhances the flow of its products without damaging its own long term potential. The United States National Research Council (1989) defined alternative agriculture as "those alternative systems incorporating natural processes reducing the use of inputs of off-farm sources, ensuring the long term sustainability of current production levels and conserving soil, water, energy and biological resource: Organic farming is an agricultural production system, which avoids or largely excludes the use of systematically compounded fertilizers and pesticides. To the maximum extent feasible, organic farming systems rely upon crop rotations, crop residues, animal manures, legumes, green manures to maintain soil productivity and tilth to supply plant nutrients. It looks forward to alternative methods of pest-control like pest resistant cultivars, bio-control agents and cultural methods of pest-control. Such ecological farming systems are highly productive and they should not be mistaken for a reversion to inefficient and less productive farming methods. The adoption of ecological farming is not as simple as one may presume. It is highly knowledge intensive, labour-oriented and a complex system integrating several organic recycling processes.

Sustainable Agriculture

Figure 9.1. The 9Rs of Sustainability

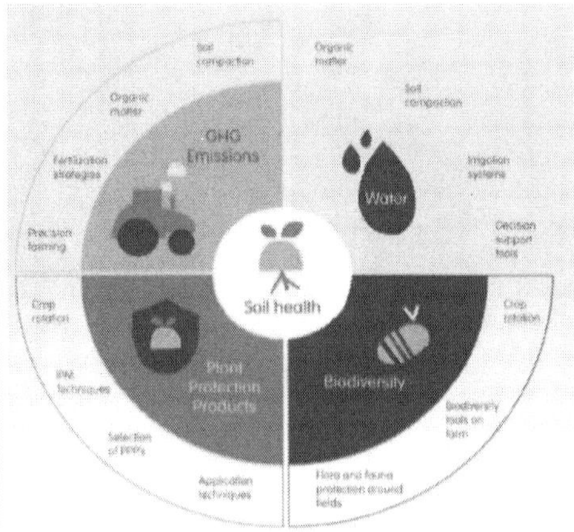

Figure 9.2. Need for Sustainable Agriculture

Figure 9.3. Ecosystem Engineering

Figure 9.4. Dr D K Verma while visiting Petlawad, Dhar, MP came across with the pathetic situation where sustainable agriculture is utmost required in 2019

Figure 9.5. Dr D K Verma, preparing seed packets for sowing of Breeding for Conservation Agriculture Station Trails (CAST), 2018 at ICAR-IARI RS Indore, MP

Figure 9.7. Sustainable Research needs Precision Agruculture

Figure 9.8. Sustainable Research Needs

Figure 9.9. Cover Crops for sustainable agriculture

Figure 9.10. Figure Bed planting in wheat for improved water use efficiency.

Conservation Agriculture

Figure 9.11. Three cardinal principles of conservation agriculture

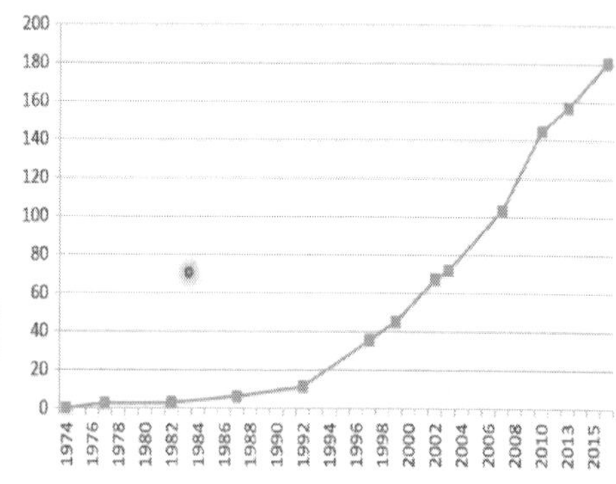

Figure 9.12. Global Adoption of CA in mha of cropland

Figure 9.13. Conservation Agriculture: Challenges and Opportuinties and strategies

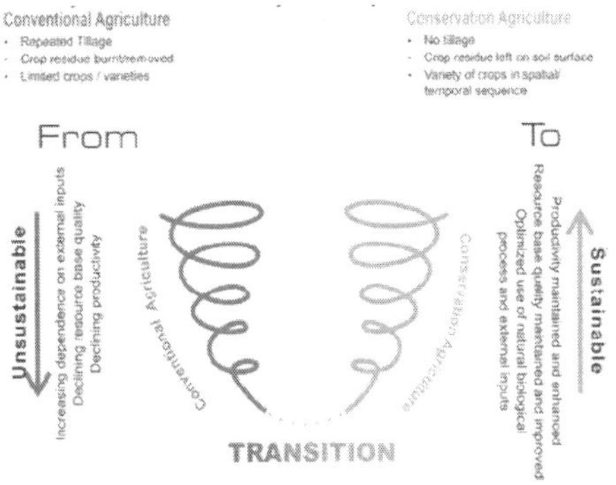

Figure 9.14. Transition of Conventional agriculture to Conservation Agriculture

Figure 9.15a. Wheat sowing under Conservation tillage

Figure 9.15b. Wheat sowing under Conservation tillage

Figure 9.16. Working of crop residue cutter

Figure 9.17. Open burning of crop residue in the field

Figure 9.18a. Wheat crop growing under Conservation Agriculture

Figure 9.18b. Wheat crop growing under Conservation Agriculture

Figure 9.18c. Wheat crop growing under Conservation Agriculture

Figure 9.18d. Wheat crop growing under Conservation Agriculture

Figure 9.18e. Wheat crop growing under Conservation Agriculture

Figure 9.18f. Wheat crop growing under Conservation Agriculture

Figure 9.19. Wheat crop growing as intercrop for Conservation Agriculture

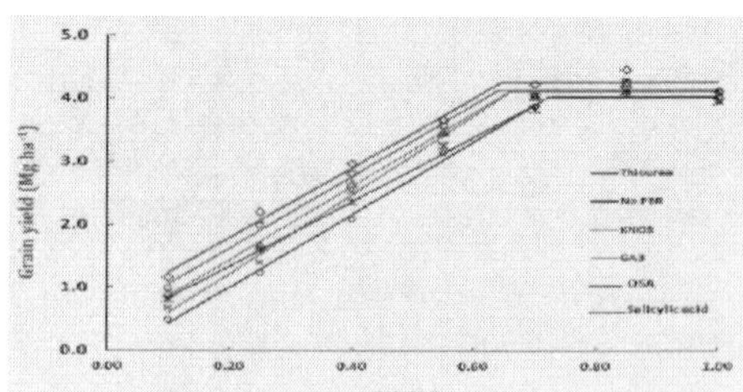

Figure 9.20. Relative response of PBRs to grain yield of wheat

Organic Agriculture

Figure 9.21a. Composting for organic farming an initiative at AIR, Indore

Figure 9.21b. Composting for organic farming an initiative at AIR, Indore

Figure 9.22a. Construction of bio gas plant for organic agriculture

Figure 9.22b. Construction of bio gas plant for organic agriculture

Panchakavya:
wonder pesticide/biofertilizer

Cow Dung	3 kilos
Cow Urine	3 litres
Milk	2 litres
Curds	2 litres
Jaggery	1 kilo
(or)	
Sugarcane juice	2 litres
Bananas	12 nos
Tender Coconut	2 litres
Toddy 2 ltrs or Yeast	100 gm
Groundnut cake	1 kilo

Soak the ground nut cake in water, before mixing all ingredients listed.
Keep container covered to avoid dust and dirt.
Stir the contents of container daily.
The Panchakavya gets ready in one week.
Spray @ 300 ml of concentrate to 10 litres of water.

Figure 9.23. Panchakavya-wonder pesticide/biofertilizer for organic agriculture

Figure 9.24. Compost is an integral part of limited irrigation technology

Figure 9.25. Preparation of Organic Agriculture in the vicinity of IGFRI RRS Srinagar, J&K at Kralapora

Figure 9.26. Worms in vermicompost semi prepared

Figure 9.27. Director, ICAR-IARI visiting the ICAR-IARI RS Pusa, Bihar Organic Wheat Seed Production Programme, 2011

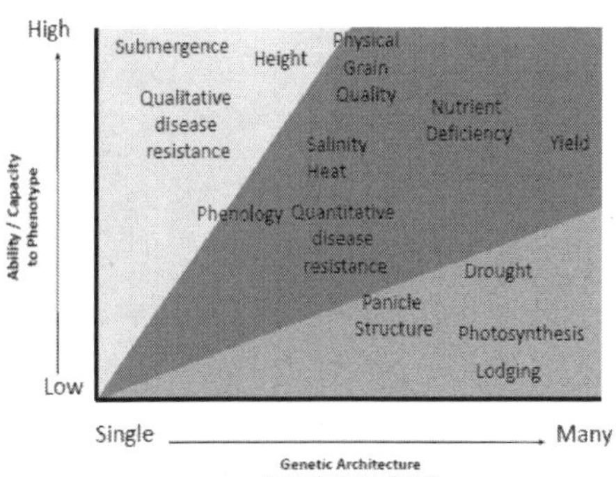

Figure 9.28. Trait selection methodology in organic agriculture

Figure 9.29. Experimental setup and evaluation of PBRs using line source system (LSS)

Indices of Sustainability

Quantification of sustainability is essential to objectively assess the impact of management systems on actual and potential productivity, and on environment. One can assess sustainability or several indices (Lal, 1994). Indices may be simple involving one parameter or complex involving several parameters. Although general principles may be the same, there indices must be fine-tuned and adapted under local environments. Some indices of sustainability include the following:

Coefficient of Sustainability (CS)

It is measure of change in soil properties in relation to production under specific management system (Lal, 1991).

Cs = F(Oi, Od, Om) t,

Where, Cs is coefficient of sustainability, Oi is output per unit that maximizes per capita productivity or profit, Od is output per unit decline in the most limiting or non-renewable resource, Om is the minimum assured output, and t is the time. The time scale is important and must be carefully selected.

Index of Sustainability (Is)

It is a measure of sustainability relating productivity to change in soil and environmental characteristics (Lal, 1993; Lal and Miller, 1993).

Is = f (Pi*Si*Wi*Ci)t,

Where, Is index of sustainability, Si is alteration in soil properties, Wi is change in water resources and quality, Ci is modification in climatic factor and t is time.

Sustainability Coefficient (SC)

It is a complex and a multipurpose index based on a range of parameters, and is similar to As. It is defined as:

Sc + F (Pt * Pd * Pm) t

Sc = d(Pi * Wt* Ct) dt

Where, Pt is productivity per unit input of the limited resource, Pd is productivity per unit decline in soil property, Sc is critical level of soil property, Wt is soil water regime and quality, Ct is climatic factor, and t is time.

Sustainable Agriculture - A Must

A strategy for sustainable living titled 'Caring for the Earth', prepared by IUCN, UNEP and WMF defines sustainable development as "improving the quality of human life while within the carrying capacity of supporting ecosystem". In a predominantly agriculture based country like India, sustainable development is more or less synonymous with sustainable agriculture where sustainable agriculture has been defined as the one that, over the long term, enhances the environmental quality and resource base on which the agriculture depends, provides for human needs of food fibber, fodder and fuel enhances the quality of life for farmers and society as a whole and is economically viable. In the realm of sustainable agricultural development, soil nutrient management has occupied the pivotal position in which "sustainable plant nutrition to increase food production" has been identified as one of the priorities directly linked to land and water management resources in relation to environment. Therefore, the sustained growth the soil health is very important to achieve national food security targets. India is endowed with enormous potential for plant nutrients locked up in these biological wastes. A total quantity of around 875 million tonnes of these so-called waste materials is available in the country, which is equivalent to 18.45 million tonnes of nutrients annually. According to the recent trends available, decline in the productivity of crop, in particular rice-wheat rotation, has been observed in north and north-western Indo-Gangetic alluvial plains despite the application of optimum levels of fertiliser inputs under assured irrigation. Constraint analysis has identified the declining organic matter status as a probable reason. It is in this context that the supplementation of organic along with chemical fertilisers is a decisive viable strategy. Indigenous rock phosphate sources are not of high grade. Phospho-composting prepared out of indigenous low-grade rock phosphate has great promise in India. The future of productive and sustainable agriculture in India lies in increasing the amount, as well as integrated and balanced use of nutrients from inorganic, organic, bio-fertilizers, and biological sources and enhancing their use efficiency.

Nutrient Management for Sustainable Agriculture of Wheat

Soils of India are generally poor in fertility as these are low in organic matter and have consistently been depleted of their finite nutrient resource due to continuous cultivation for many centuries. Because of this, the country had faced food shortages and relied on imports until the mid-sixties. Fortunately, a breakthrough in food production had been achieved in late sixties with the adoption of high yielding varieties (HYV) of cereals which paved the way for "Green revolution" and made the country not only self-sufficient to feed its burgeoning population but also brought it on the list of exporting countries. This could not have been possible without the application of appropriate doses of fertilizers. Dr. Norman E. Borlaugh in accepting Nobel peace prize in 1970 made a strong point of the fact that if the dwarf wheat evolved by him were the vehicle, the fertilizer was the fuel to this vehicle which produced high yields and triggered off green revolution in many developing countries. The Nobel Laureate further stated in 1994 that India and China would have required to put 2-3 times more land under cereal crops to meet the food needs of 1991 if they had not increased the input of fertilizers and continued to use the technology of 1960. Even the HYV or miracle seeds would not have been able to create the miracles of high yield without the fertilizer use. Fertilizers are also the means for land saving through increased land productivity. India's population is expected to cross the one billion mark by year 2000 AD. Estimates show that about 220-240 million tonnes (Mt) of food grains will be required by them. With the current food production of 198 million tonnes in 1996-97, India has the compulsive need to raise food grain output at a rate of more than five million tonnes per annum. This obviously is a daunting task. Technological adoption, using required inputs, especially seeds, plant nutrients, water and energy would be the major determinant to meet the challenges of food growth in future. Among these, most costly inputs would be plant nutrients, whose prudent and need based use would trigger the process of accelerated growth in Indian agriculture in the next decades. As there is not much scope to expand the net cultivated area *i.e.* horizontal expansion of production, which has stabilized around 142.5 m ha with the land to man ratio narrowing rapidly, the future requirement has to be met through vertical growth by intensification of agriculture *i.e.* increasing the cropping intensity (multiple cropping) and production per unit area In the wake of shrinking land and water resources therefore, the only hope for India to meet the future need for food, fodder, fibber, fruit and fuel and ensuring food security lies in enhancing the productivity of the land already under plough. There is ample scope to increase the productivity in the hill ecosystems. One of the ways to achieve this goal is through scientific and need based use of plant nutrient resources through an eco -regional approach.

Breeding Wheat for Conservation Agriculture

Conservation Agriculture

Conservation Agriculture (CA) is a strategy for mitigating climate change and as an adaptive mechanism for alleviating climate change effect. CA practices can help in sequestering atmospheric CO_2 in the form of soil organic matter, as well as reduction in GHG emissions through efficient production system. CA facilitates soil management without excessive disturbance, protection from erosion, compaction, aggregate breakdown, loss in organic matter, leaching of nutrients *etc.* Retaining crop residues as surface mulch, together with minimum soil disturbance and crop rotation and association, forms the basis of conservation agriculture (Kassam *et al.*, 2009). Conservation Agriculture (CA) includes the basic elements of such a sustainable production system, increasing productivity and production while reducing the need for external inputs and the environmental footprint of farming. CA improves the delivery by agriculture of ecosystem services such as water resources, biodiversity and the mitigation of climate change while strengthening the ecological foundation of cropping systems to also adapt to changing climates. Conservation agriculture is practiced on about 156 million ha worldwide. The Argentina, Australia, Breazil, Canada and America are holding around 90 per cent of the total area under CA. The Indo-Gangetic-Plains include four countries in South Asia, India, Pakistan, Nepal and Bangladesh are occupying around 5.0 million ha area under no-tillage system. In India, it is practiced in more than 2.0 million ha area mainly in rice-wheat cropping system of the Indo-Gangetic-Plains. **Benefits of CA** (Jat *et al.*, 2013) are to build-up of organic carbon and arrest decline in factor productivity; saving top fertilized soil from erosion, enhance nutrient-use efficiency by creating favorable environment for microflora and fauna; reduce water requirement of crops by cutting evaporation, check non-point pollution of nearby water bodies; help in sequestering GHG in the soil; improve biological activity and diversity; overall – reduce costs and energy use, efficient input use, stable yields, better use of natural resources. These benefits of CA are needed to be reaped in other input intensive cropping systems such as sugarcane system. Development of technologies/ergonomically designed low power requiring tools for tillage and crop establishment practices with least disturbances to soil/soil cover (*in situ* crop residues). Development of cost-effective technologies for *in-situ* crop residue management and efficient application of different kinds of fertilizers and weedicides. Identification and standardization of new cropping systems appropriate for resource conservation techniques and protected agriculture. Monitoring the dynamics of soil health, insect-pests, diseases and weeds under conservation agriculture based systems. The spectacular growth in food production in India (from merely 50.8 million tonnes in 1950-51 to 235 million tonnes in 2010-11) is mainly attributable to three major factors; namely, development of new superior genotypes in cereal crops, large expansion of irrigated area and fertilizer use. Among these, irrigation may be considered as a backbone of modern intensive agriculture, as it not only ensures physiological water needs of crop plants for desired growth of superior genotypes but also supports efficient utilization of plant nutrients. Agriculture sector is the largest consumer of water (82.8 per cent) but with ever growing population and increased urbanization and industrialization in the country, the requirements of water from competing sectors like domestic and industrial needs are also increasing. In 2025 total demand of water will be 1093 billion cubic meter (BCM), the bulk of which (910 BCM) will be consumed by irrigation and the country will face the stiff competition for water from different sectors (MoWR, GoI).

There are projections that demand for food grains would increase to 345 million tonnes in 2030 and 494 million tonnes by 2050. But, reduction of average size of land holding, deterioration of production environment, declining per capita water availability and climate change are the major concerns to achieve these targets. Per capita per year water resources of the country has also been continuously declining from 5176 m3 in 1951 to 4732 m3 in 1955, 2200 m3 in 1991, 1869 m3 in 2001 and 1703.6 m3 in 2007 (Kumar *et al.*, 2005). To achieve the targeted food production, it is essential to increase the crop productivity from present 2.3 tonnes/ha to 4 tonnes/ha under irrigated conditions and from 1 ton/ha to 1.5 tonnes/ha in rainfed area. This is possible when water use efficiency and productivity would be increased at individual field level. Therefore, a number of improved water management measures are required to be adopted. Improved irrigation water management, water supplies at critical stages of crop growth can alleviate stress and increase yields significantly. With reliability in water supplies the farmers can confidently invest on inputs for higher productivity. Supplementing rainfall to irrigation can increase productivity of water when limited supply is available to crops at critical periods. Precision irrigation using drip and sprinkler technology, proper conveyance system as well as better information and communication technology can reduce non beneficial evaporation, seepage and percolation losses, apply water uniformly to crop and reduce stress. Synergy among water management, high yielding variety, fertilizer and plant protection measures will result in quantum jump in productivity in agriculture. Overexploitation of natural resources during green revolution period has resulted in degradation of underground water resources, soil health and environment. Energy requirement has increased many folds resulting in high input costs. Conservation agriculture technology development for different agro-ecosystems has a golden ray to sustain the natural resources. The concept involves minimum disturbance of soil and perfect land leveling to conserve both soil and water, besides saving labour and fuel. The crop yields increase perceptibly due to enhanced input-use efficiency. The combination of reduced cost and higher production results in better returns to the farmers. This would open up new avenues for farm research in areas such as growing paddy with minimum water and developing new plant strains suited specifically for conservation agriculture. Conservation agriculture (CA) practices are recognized as a powerful tool to address the issues related to land and environmental degradation. CA has great relevance to restore the degraded ecologies where farm income and fatigue in yield have become major concern. Conservation technologies involve minimum soil disturbance, providing a soil cover through crop residues or other cover crops, and crop rotations for achieving higher productivity. This has emerged as way for transition to the sustainability of intensive cropping systems. Hence, several long-term studies have been conducted globally to introduce conservation agriculture technologies in selected cropping systems as a main-ways of improving crop yields, soil health, water use efficiency and income, while reducing requirement of energy and environmental degradation with special emphasis on irrigated cropping system in South Asia and elsewhere have been a success. Conservation tillage systems are systems of managing crop resides on the soil surface with minimum tillage. Retaining crop residues on the soil surface provides a source of plant nutrients, improves organic matter level in the soil and increased soil water content by reducing evaporation and increasing infiltration rate (Chastin *et al.*, 1995). Conservation tillage systems not only reduce erosion and improve the soil environment for crop growth, but also conserve energy and decrease the labour cost of farming (Triplett and Van Doran, 1977). The use of herbicides to control weeds, high costs of fuel in tractor-dominated agriculture and public awareness of the environmental consequences of soil erosion have led to adoption of modified tillage methods for cultivation of different crops even in South Asia. The suitability of soil for sustaining plant growth and biological yield is a function of physical and chemical properties, many of which depend on the quantity and quality of soil organic matter. Soil physical productivity is significantly affected by cropping systems and management practices, including tillage and organic residue management (Bhusan and Sharma, 2005). Approaches to increase soil organic carbon, such as organic manures, minimal tillage and reside management should be encouraged. Returning crop resides to the soil improves soil quality and productivity through favorable effects on soil properties. Crop residue management can have a substantial influence on SOC concentration (Franzluebbers, 2002). There is thus an urgent need for working out the effects of various conservation agriculture practices for enhancing crop productivity, resource-use efficiency, water productivity and physico-chemical properties of the soil under irrigated ecosystems.

Challenges in Implementation of CA

About 15-20 t ha^{-1} trash is left after harvest which is usually burnt because of constraints for fertiliser placement and other intercultural operations. Burning results in loss of organic carbon, plant nutrients such as N and S (Hemwong *et al.*, 2009), soil biota besides the environmental and health hazards due to release of soot particles, smoke and greenhouse gases (Arbex *et al.*, 2007). Several benefits has been reported due to retention of sugarcane trash on the soil surface over the burning such as it prevent weed growth and soil erosion, conserve soil moisture, improve soil structure, increase soil organic matter, cation exchange capacity, nutrient levels (Basanta *et al.*, 2003), carbon sequestration (Hemwong *et al.*, 2008) and nutrients recycling in the agro ecosystems (Fortes *et al.*, 2012) and improve the environmental quality by reducing the emissions of GHGs due to residue burning. Nevertheless, the practice of surface retention of trash is not becoming popular due to non-availability of appropriate machinery particularly for placement of fertilizers under surface retained trash conditions. **Opportunities of conservation agriculture require less tillage operations, *viz.*, a** cropping system based on crop sequence. Often crops remain in field for more than half of the total period of a particular crop rotation, which indicates that cropping systems automatically do away with the requirement of frequent land preparations (primary tillage), normally done under other cropping systems.

Table 9.1. Estimates of the crop residue yield and realizable plant nutrient potential from the residues of principal crops in India

Crop	*Residue Yield ('000 tonnes)*	*Nutrient Concentration (Per cent)*			*Total N, P, K Potential*
		N	*P*	*K*	
Wheat	44987	0.48	0.07	0.98	688.3
Rice	80744	0.61	0.09	1.15	1493.8
Barley	3180	0.52	0.08	1.25	58.8
Maize	6219	0.58	0.09	1.25	119.4
Sorghum	11163	0.52	0.12	1.21	216.2
Pearl millet	8283	0.45	0.07	0.95	121.6
Groundnut	9580	1.65	0.12	1.25	277.3
Sugarcane	15645	0.45	0.08	1.2	270.7
Potato	5062	0.52	0.09	0.85	73.9

Exposition of CA

Zero-tillage

Zero tillage is a technical component used in Conservation Agriculture but not everyone carrying out zero tillage is practicing Conservation Agriculture. Conservation agriculture not only avoids tillage by forcing the seed with appropriate direct drills into the soil, by maintaining a soil cover it also improves the structure of the soil. This facilitates direct planting. Conservation Agriculture uses biological tillage. Zero tillage as stand-alone technique can also be applied in conventional agriculture under certain circumstances.

Conservation Tillage

Conservation tillage is practices that leave crop residues on the surface, which increases water infiltration and reduces erosion. It is a practice used in conventional agriculture to reduce the effects of tillage on soil erosion. However, it still depends on tillage as the structure-forming element in the soil. Nevertheless, conservation tillage practices such as zero tillage can be transition steps towards Conservation Agriculture.

Direct Planting/Seeding

This is only a technique that refers to seeding/planting without preparing a proper seedbed. The same equipment is used in Conservation Agriculture. However, the term direct seeding can also be used for implements, which combine primary and secondary tillage and seeding in one machine/tractor operation like the rotary till drills.

Organic Farming

Conservation Agriculture is not a synonym of organic farming, although it is based on natural processes. CA does not prohibit the use of farm chemical inputs. For example, herbicides are an important component in Conservation Agriculture, particularly in the transition phase, until the new balance of weed populations is managed. However, in view of the importance of the soil life for the system, farm chemicals, including fertilizer, are carefully applied and over the years, quantities applied tend to decline. In some cases organic farming can be practiced within the CA framework.

Table 9.2. Tillage systems and descriptions

Tillage Systems	*Descriptions (Conservation tillage Information Center, 2002)*
Zero-tillage/No-Tillage (NT)	It is a form of conservation tillage, have been used for thousand years of indigenous cultures simply because humans did not have manual force to till a significant area. Soil is left undisturbed from planting harvest except for nutrient injection. Planting or drilling is accomplished in a narrow seed-bed by zero-tillers. Weed control is accomplished primarily herbicides. Cultivation may be used for emergency weed control.
Mulch Tillage	The Soil is disturbed prior to planting. Tillage tools such as chisels, field cultivators, disks, sweeps, or blades are used. Weed control is accomplished with herbicides, cultivation or both.
Reduced Tillage (Minimum tillage)	Full width tillage involving one or more tillage trips leaving residue retained (15-30 per cent) on the field. Reducing tillage operations/trips when compared to conventional systems.
Conventional Tillage	Tillage types that leaves less than 15 per cent residue cover after planting critical wind/water erosion period. Generally, includes ploughing or other intensive tillage operations; residue removed/burning during cultivation. Weed control is accomplished with herbicides, cultivation or both.

Availability of CA Machines

The adoption of conservation agriculture on large scale in rice-wheat cropping system in the Indo-Gangetic Plains of India is due to development/availability of appropriate CA machineries. Therefore, to accelerate the implementation of CA in sugarcane based cropping system, top priority should be given on the development of suitable CA machines. The development of trash shredder/chopper/cutter has brought the revolution in handling the sugarcane trash and opened the avenues for implementation of CA. The trash cutter chopped the trash in small pieces and left on the soil surface, thus only suitable for one operation *i.e.*, trash chopping. These machines are becoming popular (Sri Escorts: Sreenath Sugarcane Trash Cutting Machine, Tamil Nadu; Guru Agro Industries, Haryana and Maharashtra) but those like Happy Seeder are not available where drilling of fertilizers is possible with surface retained straw in rice-wheat cropping system. Recently, efforts have been made in this line and a multi-purpose SORF machine has been developed (Choudhary *et al.*, 2017). The SORF machine performs stubble shaving, offbarring, root pruning and placement of basal doses of fertilizers while retaining the chopped trash on soil surface in a single run. Choudhary *et al.* (2017) have reported that with the use of SORF, cane yield and nitrogen uptake efficiency improved by up to 22 and 11 per cent, respectively over the over the trash burning and broadcast application of fertilizers. Thus, SORF machine has the potential to encourage the farmers to adopt conservation agriculture in cropping systems. However, there is need to develop a machine that could perform all the important operation like trash chopping, stubble shaving, off-barring, root pruning, placement of fertilizers and sowing of crops in a single run for reducing cost and energy requirements.

Energy Use-Efficiency

Farm mechanization plays a vital role for the success of CA in different agro-ecologies and socio-economic farming groups. It ensures timeliness, precision and quality of field operations; reduces production cost; saves labour; reduces weather risk in changing climatic scenarios; improves productivity, environmental quality, sustainability and generates rural employment on on-farm and off-farm activities (Ladha *et al.*, 2009). Reduced labour and machinery costs are economic considerations that are frequently given as additional reasons to use CA practices. Adopting conservation agriculture techniques is a holistic approach for management of soil and water resources, and improving efficiency and productivity per unit of C-based energy consumed. Compared to intensive tilled conventional rice-wheat system, ZT systems require much lesser energy and gives higher energy output; input ratio as well as higher system productivity (Gangwar *et al.*, 2006). For example, continuous ZT with effective weed management using recommended herbicide + 1 hand weeding was more remunerative and energy efficient in Vertisols of Central India, and it was suggested that conventional tillbased rice–wheat system could be replaced with zero-till-based crop establishment method with effective weed control measures to save labor and energy. Similarly, low-cost of cultivation, minimum energy usage, higher water productivity, higher net returns and enhanced energy input: output ratio were reported in ZT maize-wheat cropping system (Ram *et al.*, 2010).

Soil Health

Soil health denotes a state of dynamic equilibrium between flora and fauna and their surrounding soil environment in which all the metabolic activities of the former proceed optimally without any hindrance, stress or impedance from the latter. A healthy soil would ensure proper retention and release of nutrients and water, promote and sustain root growth, maintain soil biotic habitat, respond to management and resist degradation. Soil erosion, organic matter decline, compaction and salinization resulting from the CT based agriculture are the major threats to soil health. Conservation agriculture, which prescribes ZT coupled with crop residue mulching and diversified crop rotation, has come forward as a sustainable management system that could revert physical soil degradation in resource poor farms across very different agro-ecological conditions (FAO 2012). Intensive tillage accelerates soil organic carbon (SOC) loss as CO_2 as a result of physical disruption and enhanced biological oxidation. It is estimated that agriculture has contributed 25 per cent of the historical human-made emissions of CO 2during the past two centuries. Loss of SOC could significantly be reduced by shifting from CT to ZT and other low-disturbance techniques. That's why conservation tillage systems are proposed as a way of achieving SOC sequestration, as relatively higher SOC in the plough layer is noticed under ZT than in CT. It was projected that the conversion of a conventional system to conservation tillage could mitigate approximately 20 per cent of the USA agricultural greenhouse gas emissions (Del Grosso *et al.*, 2005), and could result in a 0.50 MT ha^{-1} yr^{-1} C sequestration rate. The principle of maintaining a permanent soil covers either by planting a cover crop or by using crop residues eventually increases the amount of organic matter and available organic carbon in the soil. The benefit of crop residue recycling is higher when used as mulch on ZT soil than its incorporation under CT system. For example, crop residue treatment in ZT soils showed significantly higher amount of SOC than other treatment combinations in the top 15 cm soil depths. Crop residue served as a source of carbon especially in upper soil layers. Zero-tillage practice minimizes exposure of SOC from oxidation, and thus ensuring higher SOC content in surface soils of ZT with crop residue application.

Table 9.3. Effects of tillage and residue treatments on the SOC content (Ghimire *et al.*, 2008)

Soil Depth (cm)	*Soil Organic Carbon (kg m^{-3})*				*LSD*
	CT		*ZT*		
	M0	*M1*	*M0*	*M1*	
0-5	11.01	12.12	12.73	14.23	1.72
5-10	8.53	10.83	10.08	10.94	1.72
10-15	7.13	9.26	10.11	8.06	1.72
15-30	4.63	5.73	5.80	4.82	1.72
30-50	4.43	4.90	4.69	3.99	1.72
0-50	7.15	8.57	7.81	8.68	0.77

Mo: No crop residue, M1: crop residue @ 4 tonnes ha^{-1} for each crop in the rotation

Conservation agricultural systems are more climate resilient, sustainable, profitable and have low greenhouse gases emissions. Sugarcane has wide potential to adopt conservation agriculture principles and it is the best option we have today for the Sustainable Production of crops. Just, we need to learn how to adapt and apply the principles to farmer conditions and circumstances and sustain the soil health as well as increase the efficiency of inputs. However, adequate and very specific mechanization is needed for its successful implementation and harnessing the potential benefits.

Conservation Agriculture for Higher Resource-Use Efficiency

Indian agriculture is turning in a new phase through various technological advancements and developments. The scientific and technological innovations have been the basis for promoting agricultural research and development. The historical research focus on improved agricultural technologies has been incontestably successful. While during 'Green revolution' period, the major research and developmental efforts was focused on enhancing the production and productivity of selected food grains. This conventional mode of agriculture through intensive agricultural practices leads to increasing scarcity of resources (labour, water, and energy) and cost of production along with climate variability. These are major challenges for the present agricultural sustainability of the country. The traditional practices of intensive agriculture were successful in achieving the goals of crop production, simultaneously, leads to decline in natural resources bases for future generation. However, the recently developed new technologies focused on efficient resource-use and resource conservation which should receive high priority to ensure the past gains that can be sustained and further enhanced to meet the emerging needs. The upcoming issues of resource conservation have assumed importance in view of widespread degradation of future generation natural resources. The growing concerns of sustainable agricultural production have been seen as a positive response to balance between low-input traditional agriculture and intensive modern agricultural practices. It will also help in maintaining the ecological equilibrium and encourage natural regenerative processes such as soil regeneration, nitrogen fixation, nutrient cycling and to provide safeguard for natural enemies of harmful pests and diseases as well as the targeted use of inputs. Modern crop production systems and technologies should rely on such approaches which may not only able to support high productivity requirement for burgeoning population, but also preserve biodiversity and protect the environment. Because of these emerging constraints there is an urgent need to reduce production costs, increase profitability, and make agriculture more competitive and sustainable.

Crop Residue Recycling

Maintaining the soil cover/surface mulch with crop residues is an integral component of the conservation agriculture system. However, sometimes it is very difficult to retain the crop residues in the field particularly in the rainfed areas where crop residues is usually used as source of feed for animals. In contrast to this, availability of the crop residue in cropping systems is plenty. India, produces a bulk amount of crop residues in rice and wheat, which occupies sizable area (Yadav and Srivastava, 2005). Trash production varies with the variety, season of planting, duration of the crop and climatic conditions. This could be used directly in the crop which is occupying around 50 per cent area, while indirectly could be recycled in either plant crop or other crops in the sequence. A vast potential is available to efficiently recycle organic farm wastes/crop residues, especially in rice-wheat belt of Punjab, Haryana and western Uttar Pradesh, where it is burnt *in situ*. It is adversely affecting the ecology in two ways, one by the air pollution and secondly all the macro and micronutrients are lost during burning. The curtailment of this practice seems priority, as straw may form an important component of integrated nutrient management. Scientific management of crop residues (stubbles of the previous crop), on the other hand, helps replenish the soil carbon content and promotes activity of soil microorganisms, land fertility and helps save on fertilizers.

Resource-Conserving Technologies (RCTs)

The terms 'Resource conserving techniques' and 'Conservation agriculture' are used simultaneously and as if their meaning are similar. In general, the RCTs refer to those practices which will sustain or enhance the natural resources as well as input-use efficiency. The zero tillage and permanent bed practices that save fuel, water and improve water productivity at field level and are considered as RCTs. There are so many other practices in intensive farming systems which save water, nutrients, herbicides, energy, *etc.* After more than 35 years of green revolution the Indian agriculture is entering in a new era to overcome the shortfalls of green revolution. The shortfalls of green revolution era mainly occurred because during this period the major research and developmental efforts were focused on enhancing the production and productivity of selected food grains and other crops, the new technologies demand the issues of efficient resource use and resource conservation which receive high priority to ensure that past gains can be sustained and further enhanced to meet the emerging needs. The resource conservation related issues have assumed importance in view of the widespread resource degradation and the need to reduce the agricultural production costs, increase farm profitability, and make agriculture more competitive and beneficial. Resource-conserving technologies (RCTs) such as Zero-tillage (ZT), Un-puddled transplanting, Direct seeded rice (DSR) and Raised Bed Planting have been shown to be beneficial in terms of improving soil health, water use efficiency, crop productivity and farmers' income (Singh *et al.*, 2009). Conservation agriculture based RCTs options has emerged as way for transition to the sustainability of intensive cropping systems as well as improving resource use efficiency, water productivity, soil health and climate mitigation. Direct seeded rice a cost-effective technology which saves 15 per cent water. Zero

Table 9.4. Potential benefits of key-Resource Conservation Technology (RCTs)

Resource Conserving Technology	Potentials benefits relative to conventional practices
Zero Tillage	Reduced water use, C-sequestration, increased yield and income, reduced fuel consumption and GHG emission.
Laser-aided land levelling	Reduced water use, fuel consumption, GHG emission
Direct drill seeding rice	Less requirement of water, saves time, post-harvest condition of field is better for succeeding crop, deeper root growth
Diversification	Efficient use of water, increased income, increased nutritional security, conserved soil fertility, reduced risk
Raised bed planting	Less water use, improved drainage, better residue management, more tolerant to water stress.
Leaf colour chart	Reduces N fertilizer requirement, reduces N loss and environmental pollution, reduces nitrous oxide emission

Source: Wassmann *et al.*, 2009.

tillage offers saving of 15-20 per cent irrigation water. Furrow-irrigated raised bed planting system technology saves 30-40 per cent water as compared to conventional practice. Zero tillage direct seeded rice (ZT-DSR) and reduced – till direct seeded rice (RT-DSR) in rice and bed –drill seeded wheat (Bed-DSW) and zero tillage –drill seeded wheat (ZT-DSW) in wheat are the appropriate tillage and crop establishment options for higher resource use efficiency, water use efficiency, productivity and profitability of rice-wheat system in agro-climatic conditions of Eastern U.P, India. Conservation agriculture-based resource conservation technologies minimize the use of external costly inputs (labour, water, seeds, fertilizer, pesticides and fuel use) reduce cost of production, leading to increased system productivity and profitability at the farm front.

Raised Bed Planting

Bed planting systems have been used in cultivation for centuries. The origin of raised bed cultivation has traditionally been associated with water management issues either by providing opportunities to reduce the impact of excess water in rainfed conditions or to more efficiently deliver irrigation water in high production irrigated systems (Sayre, 2003). In dryland agriculture, bed planting systems are used with small dykes to trap water after a rain so the fields are able to retain more water and store moisture for future crops versus letting it run-off. In bed planting systems, wheat or other crops are planted on the raised beds in ridge- furrow system. This system is often considered more appropriate for growing high value crops that are more sensitive to temporary water logging stress. Farmers often raise crops such as cotton, maïze-soybean and wheat on the raised beds. However, the practice of growing rice, the major water-using crop in rice-wheat systems, on narrow raised beds was introduced only very recently in the Indo-Gangetic Plains (IGP) to reduce water use, conserve rainwater and improve system productivity. Recent work shows that system of raised bed planting of crops may be particularly advantageous in areas where groundwater levels are falling and herbicide-resistant weeds are becoming a problem. This tillage and crop establishment option also facilitates crop diversification and intercropping of wheat, chickpea and Indian-mustard with sugarcane, maize with potato, mint with wheat, rice with soybean, and pigeon pea with sorghum or green gram. Although bed planted wheat in rotation with soybeans covered more than 75 per cent area under wheat by 1994 in Mexico, the South Asian rice-wheat farmers are still experimenting with this system of crop planting in the IGP to address issues of receding water tables, crop losses due to temporary water logging in monsoon season and declining factor productivity and for crop diversification.

In furrow-irrigated raised bed planting systems (FIRB), crops are grown on the raised beds and furrows are used for irrigation. The size of the furrow and raised beds is best dictated by the width of tractor and its size of the tyres. It is always best to use narrow tyres for crop planting in FIRB systems such as to avoid compaction of the raised beds with wide tyres. This system can also be used to plant intercrops in different configurations. Farmers have been planting mint in wheat in wide bed system; chickpea/mungbean with sugarcane or cowpea with maize in narrow bed systems. This technology provides farmers with an opportunity for crop intensification and diversification with substantial saving in irrigation water. The FIRB system is particularly suitable for partially reclaimed alkali soils because beds made with reclaimed soil increases soil depth for rooting. In FIRB system, there is water saving of 30-40 per cent compared to conventional practice (RWC, 2002, Mann and Meisner, 2003). Benefits include (i) improved irrigation water use, (ii) fewer weeds, (iii) facilitates seeding into relatively dry soils and post sowing irrigation, (iv) vigorous and better crop stands, (v) saving of costly seed (vi) reduced crop-lodging and allows N application at grain filling in wheat to improve protein content in grain, (vii) better drainage, improved rainwater conservation; and crop productivity. Permanent bed planting system increased the rice yield by 33 per cent and 60 per cent in wheat along with reduction in the cultivation cost by 26 per cent for rice and 24 per cent for wheat. Compared with conventional farmer's practice, permanent raised bed system improved water and fertilizer use efficiency by 20-25 per cent and reduced the total production cost of R-W system by nearly 30 per cent (Singh *et al.*, 2011).The FIRB technology is gradually becoming popular amongst the R-W farmers as it saves about 30-40 per cent of seed and water, reduces crop lodging, and improves crop productivity and minimizes wilt infestation in crops like pigeon pea and avoids temporary water logging problems. The efficacy of resource conserving technologies further improves when FIRB system is coupled with precision laser land. Results of farmer participatory trials indicate that significant water savings can be obtained by planting rice (major consumer of irrigation water in rice-wheat systems) on raised bed besides improvement in crop yields.

Vibrant Resource Conservation Technologies through Conservation Agriculture

Since a constant supply of water in the soil is necessary for plant survival and growth, the irrigation agronomist/engineer is concerned with how water moves in a given soil, how much water a soil can hold and how much of it is available to plants, and how the water supply can be replenished. The first two are related to size and distribution of the soil pores and to size of the soil particles and their attraction for moisture. The amount of water a soil holds also depends on the amount of organic matter in the soil. Generally, the finer the soil particles and the larger the amount of organic matter, the more water a soil holds. Improving the productivity of irrigation water increasing importance in many developing countries, especially in Asia, where irrigated crop production accounts for a major portion of strategic food supplies for ensuring food security. The water resources available for irrigation are becoming more fragile, especially with respect to increase in soil salinity from poor irrigation management. The use of a conservation agriculture (CA)-based resource conserving technologies RCTs)- zero tillage, furrow irrigated raised bed and direct seeded rice has tremendous potential in significant saving in irrigation for a wide spectrum of crops, over traditional seeding on the flat with flood irrigation. CA involves integration of minimal soil disturbance, residue retention and sensible/profitable cropping/farming system. CA based RCTs options optimize the natural resources and inputs for the sustainability of the system, enhances water productivity and crop diversification. It includes direct seeded rice, surface seeding, unpuddled planted rice, zero tillage, bed planting which saves costly inputs (labour, water, seeds, fertilizer, pesticides and fuel uses) and improves water productivity, profitability and system sustainability. Direct-seeded rice needs only 34 per cent of the total labour requirement and saves 15 per cent water and 29 per cent of total cost of the transplanted crop. Important factors that are forcing a shift from the traditional puddled-transplanting system to unpuddled direct seeded zero till rice are shortages of labour and water and escalating fuel prices. In wheat, ZT reduces irrigation requirements compared with conventional-tillage by using residual water more effectively. In furrow-irrigated raised bed planting systems (FIRB), crops are grown on the raised beds and furrows are used for irrigation. This technology provides farmers with an opportunity for crop intensification and diversification with substantial saving in irrigation water. In FIRB system, there is water saving of 30-40 per cent compared to conventional practice. Permanent bed planting system increased the rice yield by 33 per cent and 60 per cent in wheat along with reduction in the cultivation cost by 26 per cent for rice and 24 per cent for wheat compared with conventional farmer's practice. It also improved water and fertilizer use efficiency by 20-25 percent and reduced the total production cost by nearly 30 percent in RW system. Adopting conservation agriculture techniques is a holistic approach for management of soil and water resources, and improving efficiency and productivity per unit of land area. The compilation of different CA based RCTs studies in IGP shows that these practices help in introduction of new cultivars in system and are more efficient in inputs, improve production and profitability, and address the emerging problems (Saharawat *et al.*, 2009). Ghosh *et al.* (2010) reported that due to higher microbial biomass carbon, dehydrogenase activity and earthworm population in CA practices (residue retention and double no-till) there was good growth of crops like rice, wheat, mustard and linseed, and yield increased by 44 per cent to 63 per cent over conventional tillage. Mishra and Singh (2012) reported that the continuous ZT resulted in significantly higher yield of rice-wheat system compared to continuous conventional tillage. Mishra and Singh (2012) observed that the intensive tilled conventional RW systems require maximum energy (38,187 MJ ha^{-1}), due to intensive field preparations. While, CA based ZT systems requires least energy have higher energy output: input ratio as well as higher system productivity. Several studies were conducted across the India and have projected weeds as the major constraints in adoption of CA technologies. Studies have suggested a small difference in weed populations between conventional and ZT fields, and in some cases, fewer weeds have been observed in ZT conditions (Hobbs and Gupta, 2003). Duxbury *et al.* (2000) estimated that agriculture contributes 25 per cent of the historical human-made emissions of CO_2 during the past two centuries. However, a significant portion of this C can be stored or sequestered by soils managed with no-tillage and other low-soil disturbance techniques. Improved tillage management techniques have shown that scientific agriculture can also be a solution to environmental issues in general, and specifically to mitigating the greenhouse effect (Lal *et al.*, 2007). In the long-term adoption of CA practices, enhanced C sequestration and build-up in SOM, constitute a practical strategy to mitigate GHG emissions and imparting greater resilience to production systems to climate change related aberrations. Saharawat *et al.* (2012) showed that different CA based resource conserving practices in rice-wheat system has pronounced effects on the global warming potential by reducing the GHG emission and sequestering carbon.

Ideotypes for Conservation Agriculture

To achieve 109 million tons target by 2025, we need to concentrate our work on long term and short-term strategies. Among these, the short-term strategies place more emphasis on effective transfer of the available technology to the farmers. While long term strategies are to address the anticipated threats like climate change, water scarcity, need for tillage and planting system specific varieties and improving yield potential of future genotypes. Modern intensive cropping in irrigated agro-ecosystem has brought about substantial increase in the production of food grains. However, it has threatened the environment safety and promoted the degradation and inefficient use of natural resources particularly the land and water. With increasing population, limited resources, uncertain climatic conditions, food security needs become more daunting for the Agricultural Scientists. The technologies are required which can increase the grain yield of the crops and raise te income of the farmers without deteriorating the quality of the environment. Many factors of green revolution like irrigation, input responsive genotypes, fertilizer have almost exhausted in the most part of India. The new technologies are to e developed which aim to conserve, improve, and make more efficient use of natural resources through integrated management of soil, water and biological resources. Conservation agriculture aimed at to optimize the yield and profit rather than maximizing, to achieve a balance of agriculture, economic and environmental benefits. Conservation agriculture refer to as the system of raising crops without tillage along with residue retention. Land preparation through precision land leveling and furrow irrigation raised and furrow configuration for planting crops further improve the resource utilization. The main stay of national food security, rice-wheat cropping is now facing lot of second-generation problems like reduced productivity, increased cost of production, declining water table, development of hard pan, declining organic carbon, increased air, soil and water pollution challenging the sustainability of this system. Conservation agriculture also called as resource efficient or resource effective agriculture. Conservation agriculture has different aspects like 1. Conservation tillage 2. Furrow irrigation raised furrow irrigation raised planting 3. Crop residue mulch 4. Precision land leveling 5. Crop diversification 6. Balanced use of fertilizers and pesticides 7. Improved input use efficiency. Out of all these conservation tillage and furrow irrigation raised furrow irrigation raised planting are gaining popularity among the farmers.

There are conflicting results on the impact of no tillage on wheat yields. No tillage has been reported to reduce, maintain or increase yields relative to conventional tillage practices. There are also conflicting reports that cultivars not differentially affected by tillage system to those that believe that cultivars do differ in their response to tillage. However, under certain case, the higher seed rate may be required due to cool microenvironment and residue retention reduced the plant stand. Early vigour, dry root and shoot weight have been found desirable for no tillage technology. Coleoptiles length, root length, tillering capacity, seedling vigour, nitrogen use efficiency and seedling temperature tolerance will therefore prove important traits for crop establishment and growth under conservation agriculture environment. However, as only few of these characters have been exploited by the breeders for developing the varieties for conventional tillage, there is likely hood that the variability will be limited in the available germplasm for these traits. Two cultivars namely PBW 443 (normal sown variety of medium duration) and HD 2687 (high yielding normal duration) did not perform well under no tillage, whereas varieties, HUW 468, HUW 234 and PBW 343 were found well adaptive under no tillage for both type of tillage environment (Joshi *et al.* (2007).

Table 9.5. Performance of different wheat cultivars under different tillage practices

Genotype	*Grain Yield (kg/ha)*		
	Conventional Tillage	*No Tillage*	*Mean*
HUW 234	4211	4323	4267
HUW 468	5152	5210	5181
HUW 510	4125	4172	4148
HUW 516	5309	5350	5330
PBW 343	5212	5135	5173
PBW 443	4201	3986	4093
HD 2687	5003	4828	4916
HD 2733	4887	4943	4915
UP 2338	4501	4550	4527
NW 1012	4711	4767	4739
DBW 14	4275	4333	4304
Raj 3765	3944	3985	4964
Mean	4628	4614	

Despite limited number of genotypes included in most of the studies, more and more evidences are now accumulating for the importance of tillage x genotype interaction. To realize the genetic gain now breeders from national and international institutes are now targeting the tillage specific varieties. In many parts of the world including north-east India comprising Bihar and Eastern UP, reduced tillage or no tillage has enabled early sowing of wheat. However, resulting in the exposure of the crop to slightly higher temperature and as a consequence of the early termination of vegetative stage. Such conditions are relatively less beneficial for the short duration varieties. Faster growth to cover ground within few days of the germination not only reduces the water evaporation but also smothers the competing weeds and produces high biomass which is of importance in limited water conditions and no tillage conditions. The large variation ranging from prostate to completely erect type exit, which can be utilized for the development of prostate or spreading type initially to smother the weeds and later on become erect to capture the maximum solar radiation. In conventional tillage, weeds were controlled by destroying them at vegetative stage or by burying the seed deep in the soils. No such mechanism exists in no tillage, so weeds may become major factor for yield under these conditions and seed shift is also possible. In no tillage the residue is least disturbed so weeds come out under the diffused light and their composition may be different. So, care should be taken to identify and control theses weeds (Triplett, 1985). The management of residue is of much important, where it not utilized for some other economic use like rice residue. The presence of too much residue of the previous crop under no tillage may be problematic which may reduce the emergence of the crops. Residue degradation is of much importance in the cropping system. The genotypic variation in decomposition of the residue has been observed (Summerel and Burgredd, 1989) in wheat and barley straw, and therefore the variation can be exploited for the development of the cultivars which decompose rapidly. The residue retention has importance in the less available moisture zone or may play big role in reducing soil erosion in medium to heavy rainfall areas. The performance of six recognized varieties of each group, *i.e.*, normal sown and late sown groups, was evaluated under conventional as well as zero tilled sowing conditions. Results revealed that under conventional tillage condition, the order of preference is for HD 2733 (4.5 t/ha) followed by HUW 468 (4.24 t/ha) and HW 2045 (4.21 t/ha) under high yielding group. PBW 373 (3.96 t/ha) performed best, followed by K 9107 and HD 2824 in normal yielding group. Under zero-tillage condition, HD 2733 (4.26 t/ ha) followed by HD 2824 and HUW 468 from the high yielding group, and PBW 373 (3.90 t/ha) followed by HP 1731 and HW 2045 from the normal producing group performed best.

Table 9.6. Interaction between planting systems and genotypes on grain yield and lodging over years

Genotype	*Grain Yield (Ton/ha)*		*Lodging*	
	Flat	*Bed*	*Flat*	*Bed*
PBW 343	8.56	8.39	0	0
UP 2338	8.98	8.38	0	0
Baviacora 92	8.99	9.22	5.0	0
Seri 82	8.73	8.33	12.5	0
STAR	8.08	7.84	16.7	0
MUNIA	8.98	8.85	16.7	0
PASTOR	7.97	8.21	66.7	10.0
SUPER SERI	9.21	9.30	58.3	26.7
WEAWER	9.16	8.43	0	0
WH 542	8.63	8.90	76.7	28.3
CPAN 3004	8.22	8.39	41.7	35.0
HD 2329	7.93	8.08	90.8	78.3
Pavon 76	7.68	7.25	84.2	51.7
Rayon 89	8.11	7.98	95.0	75.8
Bacanora 88	8.81	8.56	50.0	41.7
Oasis 86	9.37	8.84	76.7	21.7
LSD (0.05)	0.41		NS	

The studies conducted by Limon-Ortega *et al.* (2008) showed that the planting system on narrow raised beds is a variety-specific technology. The relative grain yield, stability, and NUE, indicated that Tlaxcala F2000, Nahuatl F2000 and Romoga F96 are the most adequate varieties for the planting system on narrow raised beds in the Central Highlands of Mexico. Sayre and Mareno Ramos (1997) reported that some genotypes such as Oasis 86 and CHIL/2*STAR demonstrate rather large yield depressions when planted on beds. Bacanora 88 and Rayon 89 tend to show more stable yields across both systems. When lodging is a problem, as in the 1995/96 cycle, it is usually more severe in conventional planting, other factors being similar. Thus, genotypes such as VEE/PJN//KAUZ and ROEK//MAYA/NAC/3/TEPOCA lodge more seriously when planted conventionally. It is interesting to note that genotypes that perform well with conventional planting but do poorly under bed-planting are not uncommon. It is rare, however, to find the reverse, unless differential lodging is involved. Of most obvious value as a selection criterion is identifying genotypes that have stable, high yields across different planting systems, especially if farmers in a region are using different planting systems.

Breeding Strategies

Trethoowan *et al.* (2009) have clear breeding plan for different aspects under conservation agriculture (CA).

Screening of the adapted and unadapted germplasm for response to CA

↓

Identifying the donor materials for the characters of value under CA

↓

Combining the key establishment characters with the characters for the target region

↓

Selection under simulated CA environment

↓

Evaluation of fixed CA environment

↓

Modifying the coordinated programme to enable multi-location testing of such material

The genotypes should have quick emergence and early growth, high harvest index, more tillering, kernels/spike more, straw should be easily decomposable in case of conservation tillage. Whereas, the genotypes having early ground cover, profuse genotypes but not erect and upright, better root growth in limited moisture, higher harvest index and bolder grain as well as for long coleoptiles and higher harvest index (HI) are the main characters for the genotypes performed well under furrow irrigated bed planting system.

Ideotypes for Resource Conservation Technologies

The ideotype of a crop is a plant model of a plant community where necessary and beneficial traits for crop performance in a particular environment are combined. (Makela *et al.*, 2008). It is important to consider environment when a developing ideotype since 1. Climate, 2. Edaphic and 3. Agronomic features of the target regions have a major role in determining the crop performance. Generally, no tillage is practiced in wheat crop with rice being cultivated with normal tillage. With the increased acceptance of the framers for these technologies, there is need to for the development of crop cultivars that better adapt to the changing environment. Simultaneously, these cultivars must be able to meet the nutritional requirement of the man. Developing the genotypes adaptive to such conditions can contribute to improvement, if genotype x farming system interaction is exploited. The segregating materials are neither grown nor selected under no tillage, therefore the cultivars developed in normal tillage are also adopted for the no tillage also. There are limited evidences available with the presence of genotype x tillage system interactions in majority of the cases experiments were conducted with the very small number of genotypes that were too released for conventional tillage. The genotype x tillage system or genotype x farming system interaction, once conclusively established with help of breeders identifies the traits on which selection in the segregating material can be placed. Establishment of good crop stand with large crop residue retention particularly in rainfed regions is the biggest challenge faced by the crop agronomist.

Suitable Wheat Varieties Under Conventional Tillage and Zero Tillage Condition

Zero tillage (ZT), as the name implies, is a crop planting system that entails little or no preparatory tillage (plowing, harrowing, *etc.*). The major benefits of zero-till technology include (i) reduced costs due to savings in fuel and labor, (ii) timely planting of *kharif* and winter season (*rabi*) crops, resulting in higher yields, (iii) reduced weed density (iv) saving of irrigation water (up to 15-20 per cent), (vi) improved input use efficiency because of better crop stands due to good seed and fertilizer nutrients placement, and (vii) build-up in soil organic carbon due to reduced burning of crop resides (Singh *et al.*, 2011). Zero tillage is widely adopted in wheat by farmers in the North-western IGP of India, particularly in areas where rice is harvested late. ZT minimizes the loss on account of delayed sowing as it advances the wheat sowing by 10-15 days and also saves the time and cost involved in field preparation (Sharma *et al.*, 2002) However, to get the full benefits of ZT, both rice and wheat need to be grown with a 'double zero-tillage' system (Bhushan *et al.*, 2007). Important factors that are forcing a shift from the traditional puddled-transplanting system to unpuddled direct seeded zero till rice are shortages of labour and water, and escalating fuel prices. In wheat, ZT reduces irrigation requirements compared with conventional-tillage by using residual water more effectively (Erenstein *et al.*, 2007). This was also a 2nd year of testing for six recognized varieties of each group *i.e.* normal and late sown varietal group, where they were evaluated under conventional as well as zero tillage condition of sowing for knowing the suitability of recommended varieties of this zone. Results of this year's trial revealed that under conventional tillage condition in order of preference K 9107 (39.99) followed by HP 1731 (38.10) and HP 1761 (37.57) were at par and formed the higher yielding variety group while PBW 443 (37.49), HUW 234, HD 2824 (35.98) and HD 2733 (35.52) qts/ ha formed the normal producing group. Whereas under zero tillage condition HP 1761 (40.44) followed by HUW 468 (39.54) and HUW 234 (39.08) formed the higher yielding varietal group while HP 1731 (38.63), NW 2036 (37.89) and HW 2045 (37.19) qts/ha formed the normal producing variety group.

Breeding for Adaptation to Conservation Agriculture

Farmers are increasingly adopting conservation agriculture (CA) in many parts of the world. CA reduces soil loss from erosion, conserves water, improves soil organic matter, reduces inputs, and limits CO_2 emissions. This largely involves zero-tillage and the retention of stubble from the previous crop in an appropriate crop rotation. One of the primary impediments to the adoption of CA by farmers has been access to proper machinery; however, recent developments in seeder design and improved access have seen more and more farmers from both developing and developed countries adopting the practice. This movement to CA has implications for wheat breeding because significant genotype tillage interactions (GTI) have been found. Much of the earlier literature was inconclusive regarding GTI, largely because few genotypes were tested and all tested materials were developed under conventional tillage. Nonetheless, adoption and performance of current and new wheat materials has been very successful in many areas of the world; selection under zero-tillage is expected to further improve adaptation to CA. The retention of crop residues has led to a change in the disease spectrum in some areas. Diseases such as tan spot and crown rot (caused by *Fusarium pseudograminearum*) which survive on crop residues can become limiting in some environments. Adequate genetic resistance to most of these diseases exists in the wheat gene pool and among the newly developed synthetic wheats. Genotypes with better emergence and establishment characteristics will also cope better with standing residues (Trethowan *et al.*, 2005b). In many developing countries farmer access to herbicides, so necessary in zero-tillage systems, can be limited. Therefore, genotypes that not only emerge well, but that also have vigorous early growth, will help suppress weeds. The segregating populations are grown under zero-tillage to skew gene frequency in favour of adaptation to this farming system. Preliminary evidence from a study in which populations derived from the same crosses were either always subjected to zero-tillage or always to conventional tillage shows that lines developed under zero-tillage perform better when yield tested under zero-tillage than when tested under regular tillage.

Table 9.7. Guidelines for recording of different traits in Conservation Agriculture Screening Trial (CAST)

Days to Heading	It is calculated as days taken from sowing to emergence of 75 per cent of ears (spikes) in a plot. Observation on off-type plant(s) should not be considered.
Days to Anthesis	Days taken from sowing to emergence of anthers in 75 per cent of plants in a plot.

Days to Maturity	Total days taken from sowing to maturity when all plants in a plot show natural senescence and the grains become hard and fit for harvesting.
Plant height	Measured at the time of maturity in centimetres from the ground level upto the terminal spikelet, excluding the awns. Care should be taken to record the measurement from the most commonly representative plants in the plot.
Biomass	Biomass (g)/net plot at harvest
Productive Tillers	No. of productive tillers/full 3 meter row length
Grain Yield	Total grain yield (g) net plot at harvest
1000 grain weight(g)	Bulk harvest of grains from a test entry should be utilized to draw sample(s) for counting grains (250/500/1000 in number) and their weight is recorded in grams using electronic balance. Grain counter may be used wherever available for increasing efficiency and precision.
GN/Sp	Calculate Grain number/5 main spike during harvest and finally calculate for grain number/spike by dividing it by 5
GW/Sp	Calculate Grain weight/5 main spike during harvest and finally calculate for grain weight/spike by dividing it by 5
CT I	Record only canopy temperature at 2 week after anthesis
CT II	Record only canopy temperature at 3 week after anthesis
Normalised difference vegetation Index (NDVI-I)	Record NDVI at 15 days after sowing
Normalised difference vegetation Index (NDVI-II)	Record NDVI at 21 days after sowing
Chl.Fl.I	Chlorophyll fluorescence at 15 days after anthesis
Chl.Fl.II	Chlorophyll fluorescence at 21 days after anthesis
Ch.Content I	Chlorophyll content at 15 days after anthesis
Ch.Content II	Chlorophyll content at 21 days after anthesis
Weather data	Daily weather data should be provided as per the format
Important	Protect crop from diseases

Breeding Wheat for Organic Agriculture

The concept of organic agriculture has been perceived differently by different people. To most of them, it implies the use of organic manures and natural methods of plant protection instead of using synthetic fertilisers and pesticides. It is regarded by some as farming involving the integrated use of fertilisers and organic manures as well as of chemicals and natural inputs for plant protection. In either case the concept has been understood only partially. Organic agriculture has been defined differently, but the description offered by Lampkin (1990) appears to be most comprehensive one covering all essential features. Organic Agriculture is a possible option for stress mitigation and adaption to climate change. It is an alternative sustainable, eco-friendly production system that assists in biological pest control and crop rotation, supply green manure and composts/FYM to maintain soil fertility. According to an estimate, organic agriculture combined with reduced tillage techniques has the potential to sequester, 500 kg C/ha/yr. This maximum organic scenario would mitigate about 4 Gt CO_2 equivalents/yr or 65 per cent of the agricultural GHG. There are opportunities to use this approach for sustainable and ecofriendly farming models that meet on-farm requirement of organic inputs through biochar, biomass transfer from agroforestry for direct application on farm and composting for farm nutrient recycling. Organic food is as safe to consume as any other kind of food and the organic produces contains significantly lower levels of pesticide residues than conventional produce. Though much work was not carried out on the quality aspects of crop produces, an attempt is made to pool the research works carried out on the quality of crops. Singh and Prasad (1974) observed that, application of FYM equivalent to 100 kg N ha^{-1} increased the starch content upto 74 per cent while the control recorded 71.5 per cent. Further, the Hexose and pentose content in grain was also increased due to the addition of FYM. Besides, the indispensable and semi dispensable amino acids content in grain also increased due to FYM as 100 kg N ha. Dwivedi and Thakur (2000) found that the application of Biogas slurry at 10 t ha^{-1} enhanced the protein content (13.5 per cent) and it was closely followed by the addition of rice straw 5 t ha^{-1} (13.10 per cent).

Table 9.8. Effect of organics on quality parameters of wheat

Treatment	*Starch (Per cent)*	*Pentose (Per cent)*	*Herose (Per cent)*	*Crude Protein (Per cent)*	*Total Amino Acid*
Control	71.5	8.05	0.665	7.08	4.30
FYM equivalent to 100 kg N ha^{-1}	74.0	9.59	0.833	7.40	5.36

Organic Food Production

Organic farming is defined as a production system that avoids or largely excludes the use of synthetic fertilizers, pesticides, growth regulators, and livestock feed additives. To the maximum extent feasible, organic farming systems rely on crop rotations, crop residues, animal manures, legumes, green manures, off-farm organic wastes, mechanical cultivation, mineral-bearing rocks, natural pesticides and biological pest control to maintain soil productivity and to supply plant nutrients, and to control insects, weeds and other pests. All these should be under conditions of sustainable natural resources utilization and a healthy environment. The basic requirements for organic foods are: a) the environment (water, air and arable lands) should not be polluted; b) no synthetic materials should be used during the production; c) the production technologies should not be gene-modified technologies; d) pollution is avoided during processing, transportation, storage and distribution; and e) the necessary procedures for their certification and supervision are followed. Based on the food production methods and the different product standards found in the country, the foods in China can be classified as follows: common food, non-polluted food, green food (A and AA), and organic food. With the exception of the common food, all kinds of foods should follow the required standards. Production and processing should be approved and certified by the relevant authorities. The quality of organic food is the highest among these foods.

Table 9.9. Objectives of organic and conventional farming

Organic Farming	*Conventional Farming*
A. Organization	
Ecological orientation, second economy, efficient labour input.	Economical orientation mechanization, minimising labour input.
Diversification, balanced combination of enterprises.	Specialisation, disproportionate development of enterprises.
Stability due to diversification.	Programme based on market.
B. Production	
Cycle of nutrients within the farm, predominantly farm produced materials.	Supplementing nutrients, predominantly bought in fertilizer.
Weed control by crop rotation and cultural practices.	Weed control by herbicides.
Pest control based on inoffensive substances.	Pest control by pesticides.
Housing of livestock for production and health.	Livestock rarely combined.
C. Mode of influencing life processes	
Production is integrated into environment, building healthy landscapes.	Emancipation of enterprises from their environment by chemical and technical manipulation.
Balanced conditions for plants and animals; few deficiencies need to be corrected.	Excessive fertilisation, necessitating frequent correction of nutrient deficiencies.
D. Social Values	
Optimum input/output ratio.	Low input/output ratio.
No pollution.	Considerable pollution worldwide.
Maximum conservation of soils, water quality and wild life.	Using up soil fertility often resulting in erosion and losses in water quality and wildlife.
Holistic approach.	Economic motivation.

Table 9.10. Differences between organic farming and natural farming (based on natural principles)

Natural Farming	*Organic Farming*
It is not alternative system of farming but part of the philosophy of life involving continuous search to know the true spirit and form of nature.	In many respects close to natural farming, but does not have the philosophical overtone of natural farming.
Totally eliminates all the components of modern farming.	Organic farming does not totally exclude elements of modern farming. It involves limited and essential ploughing, hoeing, weeding, and use of chemicals.
It indicates a 'Do-nothing' approach	It indicates a soil building programme–more intensive style of natural farming. Application of natural plant protection chemicals (which are not inorganic derivatives) use of organic manures (instead of chemical fertilisers) are permitted.
The essential principles are No cultivation, No chemical fertilisers, No weeding, No plant protection.	Principal elements to be considered in practising organic farming are: maintaining a living soil, making available all the essential nutrients, organic mulching.

Potential for Organic Crop Production

The large pollution-free areas facilitates the development of organic crop production. In the past ten years, with the action on pollution-free food and the development of green food, a number of lands (including arable lands and orchard lands) have been used for the production of pollution-free and green food. The development of other kinds of organic food facilitates the development of organic crops. Rotation is necessary for production and organic cropss should rotate with other organic crops, such as maize and rice. In combination with organic crop production, the organic potato production in paddy fields has become more and more common. Covering it with rice straws has become a common way of potato cultivation. It is possible to grow potatoes by just putting seed potatoes on the top of the paddy soil, covering them with rice straw, and picking potatoes from a mulch of rice straw.

Gain the Best Yields from Organic Agriculture

The organic farming methods, in collaboration with organisms says, Nature of operations which the organisms have developed nature again who provided energy plants are organic. The biological system that works in the field of the use of these methods, we will produce more and cereals, also are detoxified microorganisms and organisms of the system of nature again we will be able to cultivate. To understand crop nutrition, we must consider the following facts. 1. Why did the nature of the nitrogen in the air is 78 per cent? 2. Why are all green leaves of plants? 3. Plants need nitrogen from the air, why not take her? 4. The micro-organisms in the soil have done what? 5. The nature of wild trees to nourish what way? Weed how to grow without fertilizers? Different crops require the same ground, which is how the individual meets. Noting that the plant from germination to full age stands on the same spot. The food can not be on the second place. In view of the fact that nature has devised a system for their nutrition. For good growth of plants so much nitrogen in the air is by nature, but the plant can not directly exploited. If they do so, their length becomes so high that other creatures could have trouble staying. Since nitrogen is available to plants through the microorganisms, which convert air nitrogen plants provide. There are many kinds of soil microorganisms. His task is to provide micro elements in soil to plants, dissolved phosphorus available to plants offer. *Rhizobium, Acetobacter* provide nitrogen. Sulfur solvent- potash, iron solvents, *etc.* bacteria plants that provide ingredients. These microbial processes that promote change. The green color of the plants with the help of the sun's energy by photosynthesis for growth of plants that produce carbohydrates. Wild plants in any form nitrogen, phosphorus, potassium is not given, then how they get nutrients. The answer which plants trees shed leaves, they act after the entrance of bacteria or plants become available, either directly or in the ground to cooperate in changing nutritional elements available. This is only possible in the living soil. Neither they nor any of their seed and fertilizer is no. Clearly, the nature of these spells causes a system is created. They are nourished by the same system. As well as a variety of other weeds provide weed nitrogen. Many of them neighboring weeds with chemicals to diseases and insect protection. Its characteristics are mutually cooperating with each other and the nature of the subtle mechanisms of cooperation to complete their life cycles. These microorganisms and plant capacity to produce energy using the sun's power. Making them operate their lifecycle. The seeds which are minerals, they change with germination, plant manganese in calcium, phosphorus, and sulfur are found before the change. Therefore, it is known that bacteria and organisms that nature organic plants are designed by nature to meet their needs. Nature itself is full. It should not get any help from the outside. The same bacteria that work in agriculture.

Wheat Varieties with Special Traits for Organic Wheat Production

The three major constraints on **wheat** production: seed, rust and drought. Most of the potato production regions suffer from late blight and drought. The chemical pesticides to control late blight are strictly forbidden in organic potato production and the late blight resistant varieties could be the best choice in this situation. In the marginal and dry regions with fresh air, healthy soils, but low rainfall, drought tolerant varieties could be the best choice. Rapid economic development will stimulate the consumption of organic foods. Though the potential for organic potato production is remarkable, there are several constraints which affect its development. For food security, the amount of food is more important than the quality of food; this is because of the continuing growth of the country's population and the decrease of its arable land. Also, many people cannot afford to purchase organic potatoes because of the higher prices. The small scale of farming is another constraint on organic potato production. The breakdown products of plant and animal remains, extracted in an alkaline solution, are commonly referred to as humic substances (HS). These can be extracted from a wide variety of sources, including sub-bituminous coals, lignites (brown coals), peat, soil, composts, and raw organic wastes. The application of HS to plants has the potential to improve soil health and plant growth, but the extent of plant-growth promotion is inconsistent and relatively unpredictable when compared to inorganic fertilizers. The goal of this review was to determine the magnitude and likelihood of plant-growth response to HS and to rank the factors contributing to positive growth promotion. These factors included the source of the HS, the environmental growing conditions, the type of plant being treated, and the manner of HS application. Humic substances (HSs) have considerable effects on soil fertility and crop productivity owing to their unique physiochemical and biochemical properties, and play a vital role in establishing biotic and abiotic interactions with in the plant rhizosphere. Humic compounds such as humic acid and fulvic acid have been shown to stimulate plant growth in terms of increasing plant height and dry or fresh weight as well as enhancing nutrient uptake. These effects seem to depend on the concentration) and source of the substance and on the plant species. Humic substances have the ability to hold seven times their volume in water, a greater water holding capacity than clay soils. Water stored within the topsoil enables plant roots to quickly access available nutrients required for plant growth and yield.

Potassium (K) is one of the major plant macronutrients influencing plant growth, development and grain quality and plays a key role in the synthesis of cells, enzymes, proteins, starch, cellulose, and vitamins. Moreover, K not only participates in nutrient transportation and uptake, but also confers resistance to abiotic and biotic stresses, leading to enhanced production of quality crops and provides resistance to plant diseases. Different strains of KMB strains were isolated on Aleksandrov medium and characterized morphologically, following their plant growth promoting activities and potash mobilizing activities. 44 bacterial strains were isolated from the rhizospheric soil from various districts of southern Rajasthan. Among morphological parameters, colony color, shape and size has been considered for this study. Biochemical characteristics of these bacteria included the following: Gram negative, aerobic and catalase positive. All the isolates hydrolyze starch, utilize citrate, and produce H2S. All the strains showed positive reactions for various PGPR activities like production of Indole Acetic Acid, HCN, siderophore, ammonia and phosphate solubilization. All the strains showed zones when inoculated on Aleksandrov medium. The present results concluded the successful isolation of isolates of Potash mobilizing bacteria from the rhizospheric soils having an important role to enhance the soil fertility. Further these strains show PGPR activity and can be used as a biofertilizer. The molecular profiling of the strains was done using ERIC, BOX, REP PCR and 16 S rDNA amplification which exhibited significant diversity amongst the strains.

An experiment was conducted at IFSRP, MPKV, Rahuri to study the evaluation of organic farming package for kharif cropping systems during 2015-16 and 2016-17 in split plot design with three replications. Nine combinations of three crop sequences (onion-sorghum, onion-wheat and onion-chickpea) and three combinations of organic nutrient sources *viz.*,50 per cent N through FYM + 50 per cent N through Vermicompost, 50 per cent N through FYM + 50 per cent N through Neem cake and 50 per cent N through Vermicompost + 50 per cent N through Neem cake were the main plot treatments in kharif season. During rabi season each main plot treatments of residual effect of organic nutrient sources was split into three sub plot treatments of organic nutrient levels *viz.*,100, 75 and 50 per cent N through organic (In equal split of N through 50: 50 per cent FYM: Vermicompost) to rabi season crops (Sorghum, Wheat and Chickpea) resulting in twenty seven treatment combinations. One additional control treatment of GRDF included for comparison not for analysis. The experimental soil was clay loam, pH 8.14, low in nitrogen, medium in phosphorus and very high in potassium respectively. All the biochemical quality parameters were significantly influenced by application of different organic sources. In case of onion, the significantly higher values of quality parameters *viz.*, total soluble solids, reducing sugar, non reducing sugar, total sugar, sulphur per cent, pungency per cent and carbohydrate was recorded with the application of 50 per cent N through FYM + 50 per cent N through VC compared to 50 per cent N through FYM + 50 per cent N through NSC and 50 per cent N through VC + 50 per cent N through NSC. However, the values of non reducing sugar, total sugar and sulphur content were at par over 50 per cent N through VC + 50 per cent N through NSC during both the years and values of reducing sugar during second year. In case of rabi sorghum, the residual effect of 50 per cent N through FYM + 50 per cent N though VC recorded significantly maximum values of quality parameters of sorghum *viz.*, protein content (9.93 and 10.00 per cent), total sugar (1.77 and 1.78 per cent) and voluntary intake of dry fodder (83.99 and 84.10 per cent) over the organic sources of 50 per cent N through FYM + 50 per cent N through NSC.

Application of 100 per cent N through organic (50 per cent N through FYM + 50 per cent N through VC) to rabi sorghum recorded significantly maximum values of quality parameters *viz.*, protein content (9.93 and 10.02 per cent), total sugar (1.76 and 1.76 per cent) and voluntary intake of dry fodder (83.93 and 84.05 per cent) over lower level of treatments during both years of experiments. In case of wheat and chickpea, the residual effect of 50 per cent N though FYM + 50 per cent N though VC was recorded significantly higher protein content (13.13 and 13.89 per cent) and (22.75 and 23.09 per cent) over rest of treatments during 2015-16, 2016-17. Among the different organic nutrient levels the highest grain protein content of wheat was (13.29 and 14.16 per cent) and for chickpea was (22.86 and 23.27 per cent) during 2015-16 and 2016-17, respectively was recorded with 100 per cent N through organic (50 per cent N through FYM + 50 per cent N though VC). In control treatment 100 per cent GRDF, the values of all the quality parameters are lower as compared organic treatments.

References

Abrol, IP and Sunita, S (2005) Conservation agriculture for transition to sustainable agriculture, Meeting report. *Current Science*, **88**(5): 686-687.

Abrol, IP and Sunita, S (2006) Sustaining Indian Agriculture-Conservation agriculture the way forward. *Current Science*, **91 8**: 1020-1024.

Agarwal, PK and Pathak, H (2009) Conservation Cultivation to combat Climate change. *Indian Farming*, 5-10 and 20.

Annon (2019) BAU Desh-Videsh ke logon ko padhayega sanrakshit kheti.

Anonymous (1988) Green manuring in rice. IRRI Pub.

Aquino, P; Carrion, F and Calvo, R (2002) Selected wheat statistics. In *CIMMYT 2000–2001 World Wheat Overview and Outlook: Developing No-till Packages for Small Scale Farmers*, edited by J Ekboir, Mexico, DF, CIMMYT, pp. 52–62.

Arun, B; Joshi, AK; Kumar, S and Singh, BD (2006) Resource conservation technology and somaclonal variants of wheat (*Triticum aestivum* L.) with improved yield traits and disease resistance: Ideal for delayed sowings. 13th Australian Plant Breeding Conference, 18-21 April, 2006 Christchurch (New Zealand).

Bushman, BS; Larson, SR; Mott, IW; Cliften, PF; Wang, RRC; Chatterton, NJ; Hernandez, AG; Ali, S; Kim, RW; Thimmapuram, J; Gong, G; Liu, L and Mikel, MA (2008) Development and annotation of perennial Triticeae ESTs and SSR markers. *Genome* **51**: 779–788.

Carr, PM; Horsley, RD and Polabd, WW (2003) Tillage and seeding rate effects on wheat cultivars: II Yield components. *Crop Sci.* **43**: 210-218.

Chander, K; Goyal, S; Mudra, MC and Kapoor, KK (1997) Organic matter, microbial biomass and enzyme activity of soils under different crop rotation in the tropics. *Biol.Fertil. Soils*. **24**: 306-310.

Chastin, TG; Ward, JK and Wysocki, DJ (1995) Stand establishment response of soft winter wheat to seed bed residues and seed size. *Crop Science* **35**: 213-218.

Chauhan, DS; Sharma, RK and Verma OS (1997) Integrated Weed Management in wheat through Tillage options and new molecules under rice-wheat system. In proceed. IGM on Wheat Research "Wheat Research Needs Beyond 2000 A.D." Aug.12-14,1997, DWR., Karnal. pp. 119.

Chevalier, PM and Ciha, AJ (1986) Influence of tillage on phenology and carbohydrate metabolism of spring wheat. *Agron. J* **78** : 296-300.

Cox, DJ (1991) Performance of hard red winter wheat cultivars under conventional-till and no-till systems. *North Dakota Farm Research* **48**: 17–20.

Curtis, T and Halford, NG (2014) Food security: the challenge of increasing wheat yield and the importance of not compromising food safety. *Ann Appl Biol* **164**: 354–372.

Dahama AK (1996) Organic Farming for Sustainable Agriculture, Agri. Botanical Publishers, Bikaner.

Dahama AK (2002) Organic Farming for Sustainable Agriculture, Agrobios (India), Jodhpur pp.301.

Dao, TH and Nguyen, HT (1989) Growth response of cultivars to conservation tillage in a continuous wheat cropping system. *Agronomy Journal* **81**: 923–929.

Davis, KF; Chhatre, A; Rao, ND; Singh, D; Ghosh-Jerath, S; Mridul, A; Poblete-Cazenave, M; Pradhan, N and DeFries, R (2019) Assessing the sustainability of post-Green Revolution cereals in India. *Procs of the National Academy of Sciences*, **116** (50): 25034.

Derpsch, R; Friedrich, T; Kassam, A and Hongwen, L (2010) Current status of adoption of no-till farming in the world and some of its main benefits. *International Journal of Agricultural and Biological Engineering. Int J Agric and Biol Eng*, **3**(1): 1-25.

Devi, RC (2019) Plant Health and Soil Changes with Application of Humic Acid Substances. Souvenir cum Book of Abstracts, First Global Congress on Biotechnology and Crop Science (BCS-2019), Purnea University, Purnea, Bihar, India.

Dhillon, SS; Prashar, A and Thaman, S (2004) Studies on bed planted wheat (*Triticum aestivum* L.) under different nitrogen levels and tillage methods. *J Curr Sci* **5**: 253-256.

Dhonde, AS; Raskar, BS and Thorat, SD (2019) Biochemical Responses in Onion, Sorghum, Wheat and Chickpea to Effect of Organic Nutrient Management under Onion Based Cropping System. Souvenir cum Book of Abstracts, First Global Congress on Biotechnology and Crop Science (BCS-2019), Purnea University, Purnea, Bihar, India.

Duivedi, DK and Takur, SS (2000) Production potential of wheat crop as influenced by residual organic, direct and residual fertility levels under rice wheat cropping system. *I J Agronomy* **45**(4): 641-647.

Eguchi H and Hirano J (1971) Effect of combinations of tillage and non-tillage, straw mulching and fertilization on weed communities in rice-wheat cropping. *Weed Research* **12**, 36-39.

Erenstein, O; Farooq, U; Malik, RK; Sharif, M (2007) Adoption and impacts of zero-tillage as a resource conserving technology in the irrigated plains of South Asia, Colombo, Sri Lanka, International Water Management Institute, P 55. Comprehensive Assessment of Water Management in Agriculture. Research Report 19.

Fortes, C; Trivelin, PCO and Vitti, AC (2012) Long-term decomposition of sugarcane harvest residues in Sao Paulo state, Brazil. *Biomass and Bioenergy* **42**: 189–198.

Franzluebbers, AJ (2002) Water infiltration and soil structure related to organic matter and its stratification with depth. *Soil and Tillage Research* **66**: 197–205.

Gangwar, KS; Singh, KK; Sharma, SK and Tomar, OK (2006) Alternative tillage and crop residue management in wheat after rice in sandy loam soils of Indo-Gangetic Plains. *Soil and Tillage Research* **88**: 242–252.

Gaur, AC and Sadasivam, KV (1981) Organic manures in aid of fertilizers, *Indian Farming*, **31**.

Gaur, AC; Neelkamal, S and Dargan, KS (1995) Organic manures. ICAR Pub., New Delhi.

Ghimire, R; Shah, SC; Dahal, KR; Duxbury, JM and Lauren, JG (2008) Soil organic carbon sequestration by tillage and crop residue management in rice-wheat cropping system of Nepal. *Journal of Institute of Agriculture and Animal Sciences* **29**: 21–26.

Ghosh, PK; Das A; Saha, R; Kharkrang E; Tripathi AK; Munda GC and Ngachan SV (2010) Conservation agriculture towards achieving food security in North East India. *Current Science* **99**(7): 915-921.

Grover, DK and Sharma, T (2008) Zero tillage – a profitable and resource saving technology, *Indian Farming*, April 25-27.

Gupta RK and Sayre, K (2007) Conservation agriculture in South Asia. *Journal of Agricultural Sciences Cambridge* **145**: 207–214.

Gupta, RK and Seth, A (2007) A review of resource conserving technologies for sustainable management of the rice–wheat cropping systems of the Indo-Gangetic plains, *Crop Protection* **26**: 436–447.

Gupta, RK; Hobbs, PR; Jiaguo, J and Ladha, JK (2003) Sustainability of post-green revolution agriculture. In: JK Ladha *et al.* (Eds.), *Improving the Productivity and Sustainability of Rice–Wheat Systems: Issues and Impacts*, ASA Spec. Publ. ASA, CSSA, and SSSA, Madison, WI, **65**: 1-25.

Gupta, RK; Sayre, K and Ladha, JK (2005) *Conservation Agriculture in South Asia*, In: *Rice Wheat Consortium for the Indo-Gangetic plains*, pp. 1-2.

Hagras, AA; Kishii, M; Sato, K; Tanaka, H and Tsujimoto, H (2005) Extended application of barley EST markers for analysis of alien chromosomes added to wheat genetic background. *Breed Sci* **55**: 335–341.

Hakizimana, F; Haley, SD and Turnipseed, EB (2000) Repeatability and genotype x environment interaction of coleoptile length measurement in winter wheat. *Crop Sci.* **40**: 1233-1237.

Hall, EF and Cholocj, FA (1989) Cultivar x tillage interactions of hard red spring wheat cultivars. *Agron. J.* **81**: 789-792.

Havlin JL; Kissel DE; Maddux LD; Claassen MM and Long JH (1990) Crop rotation and tillage effects on soil organic carbon and nitrogen. *Soil Science Society of America Journal* **54**, 448-452.

Hayward, AC; Tollenaere, R; Dalton MJ Batley J (2015) Molecular marker applications in plants. *Methods Mol Biol*,**1245**: 13-27.

Hemalatha, M; Thirumurugan, V and Balasubramanian, R (2000) Effect of organic sources of nitrogen on productivity quality and soil fertility in single crop wet lands. ***J. Agronomy***, **45**(3): 564-567.

Hemwong, S; Cadisch, G; Toomsan, B; Limpinuntana, V; Vityakon, P and Patanothai, A (2008) Dynamics of residue decomposition and N2 fixation of grain legumes upon sugarcane residue retention as an alternative to burning. *Soil and Tillage Research* **99**: 84–97.

Hobbs, PR and Gupta, RK (2003) Resource conserving technologies for wheat in the rice-wheat system. In: Improving the productivity and sustainability of rice-wheat systems: Issues and impact (eds. Ladha *et al.*), ASA, Spec. Publ. 65, Chapter 7, ASA Madison. pp: 98-107.

Hooker ML; Herron GM and Penas P (1982) Effects of residue burning, removal and incorporation on irrigated cereal crop yields and soil chemical properties. *Soil Science Society of America Journal* 46, 122-126.

IFOAM (2015) *Consolidated Annual Report of IFOAM–Organics International*; IFOAM-Organics International Head Office: Bonn, Germany.

Jat, ML; Gathala, MK; Saharawat, YS; Tetarwal, JP; Guptaa, R and Singh, Y (2013) Double no-till and permanent raised beds in maize–wheat rotation of northwestern Indo-Gangetic plains of India: Effects on crop yields, water productivity, profitability and soil physical properties. *Field Crops Research* **149**: 291–299.

Joshi, AK; Chanr, R; Arun, B; Singh, RP and Ortiz, R (2007) Breeding crops for reduced – tillage management in the intensive rice-wheat system of South Asia. *Euphytica* **151**: 135-151.

Kassam, A; Friedrich, T; Shaxson, F and Pretty, J (2009) The spread of Conservation Agriculture: justification, sustainability and uptake. *Int. J. Agric. Sustain.* **7**: 292–320.

Kumudini, S; Grabau, L; Vanm, SD and Omielan, (2008) analysis of yield-formation processes under no tilla and conventional tillage for soft red winter wheat in the South-Central Region. *Agron. J.* **100**: 1026-1032.

Kushwaha, CP; Tripathi, **S**K and Singh, KP (2001) Soil organic matter and water-stable aggregates under different tillage and residue conditions in a tropical dryland agro-ecosystem. *Applied Soil Ecology* **16**: 229–241.

Ladha, JK; Fischer, KS; Hossain, M; Hobbs, PR and Hardy, B (eds) (2000) Improving the productivity and sustainability of rice-wheat systems of the Indo-Gangetic plains: a synthesis of NARS-IRRI partnership research. IRRI discussion paper series No. 40, IRRI, Los Banos, Phillipines.

Ladha, JK; Kumar, V; Alam, MM; Sharma, S; Gathala, M; Chandna, P; Saharawat, YS and Balasubramanian, V (2009) Integrating crop and resource management technologies for enhanced productivity, profitability, and sustainability of the rice– wheat system in South Asia. (In) Ladha JK, Singh Y, Erenstein O, Hardy B (Eds.) Integrated Crop and Resource Management in the Rice–Wheat System of South Asia, IRRI, Los Banos, Philippines, pp. 69–108.

Lafond, GP and Baker, RJ (1986) Effects of genotype and seed size on speed of emergence and seedling vigour in nine spring wheat cultivars. *Crop Sci.* **26**: 341- 346.

Lal, R (2007) Constraints to adopting no-till farming in developing countries. *Soil and Tillage Research* **94**: 1-3.

Lammerts van Bueren, ET (2002) Organic Plant Breeding and Propagation: Concepts and Strategies. Ph.D. Thesis, Wageningen University, Wageningen, The Netherlands.

Lampkin, N (1990) Organic farming, Ipswich, U.K. Farming Press Books pp.710.

Legzina, L; Skrabule, I (2005) Plant Breeding for Organic Farming: Current Status and Problems in Europe; Compendium ENVIRFOOD: Talsi, Latvia.

Li, C (2001) Conception and method of Biogeochemistry – Development of the DNDC model, *Quat. Sci.* (in Chinese), **21**(2): 89–99.

Limon-Ortega, A; Villase, EM and Espitia-Rangel, E (2008) Nitrogen Management and Wheat Genotype Performance in a Planting System on Narrow Raised Beds. *Cereal Research Communications* **36**: 343–352.

Löve, A (1984) Conspectus of the Triticeae. *Feddes Repert* **95**: 425–521.

Maghirang, RG; Rodulfo, GS; Enicola, EE and Candelaria, R (2011) Organic Breeding and Seed Production in Selected vegetables. In Proceedings of the Organic Seed Preconference of the 17th IFOAM Organic World Congress, Namyangju, Korea, 26–27 September 2011.

Malik, RK; Yadav, A and Singh, S (2004) Resource conservation technologies for maintaining health of natural resources in rice-wheat cropping system. *Proc Workshop on Sustaining Agriculture: Problems and Prospects.* Nov 9-11 Punjab Agricultural University Ludhiana, India.

McGuire, PE and Dvorak, J (1981) High salt-tolerance potential in wheat grasses. *Crop Sci* **21**: 702–705.

Msangi, S; Timothy, B; Sulser, A; Bouis, D; Hawes and Batka M (2010) *Integrated Economic Modeling of Global and Regional Micronutrient Security*, HarvestPlus Working Paper 5 (Washington, DC: HarvestPlus, 2010).

Murugappan (2000) Integrated nutrient management – the concept and overview. S. Kannaiyan *et al.* (eds). Integrated nutrient management, 51-63.

Muthuvel, P and Sivasamy, R (1985) Effect of organic, inorganic and biofertilizer on rainfed redgram. *Madras Agrl. J.* **72**(3): 176-177.

Naidu, R and Mc Laughlin, MJ (1993) Heavy metal contamination of agricultural land and ground water in Tamil Nadu, India and in Australia. Report of ACIAR, p.26.

Neergaard, P (1977) Seed Pathology, Volume 1, London, McMillan Press.

Noble, M and Richardson, MJ (1968) An annotated list of seed borne diseases. *Procs of the International Seed Testing Association* **33**: 1-191.

Ogle, SM; Breidt, FJ; Easter, M; Williams, S and Paustian, K (2007) An empirically-based approach for estimating uncertainty associated with modeling carbon sequestration in soils. *Ecol. Modell.* **205**: 453–463.

Palaniappan, SP and Annadurai, K (1999) Organic Farming: Theory and Practice, Scientific publishers, Jodhpur. 287p.

Pandey, S and Velasco, LE (2002) Economics of direct seeding in Asia: patterns of adoption and research priorities. In: *Pandey, S., Mortimer, M., Wade, L., Tuong, T.P., Lopes, K., Hardy, B. (Eds.), Direct Seeding: Research Strategies and Opportunities*. International Rice Research Institute, Los Banos, Philippines.

Paroda RS (1997) Integrated nutrient management for sustainable agriculture. Keynote address delivered at the inaugural session of the " FAO- IFFCO International Seminar on IPNS for Sustainable Development ", held on 25th November 1999, New Delhi.

Partap, Tej and Saeed M (Edts) (2010) Organic Agriculture and Agribusiness: Innovation and Fundamentals. 2010. e-book. Published by the Asian Productivity Organization, Tokyo, Japan.

Paustian, K; Collins, P and Paul, EA (1997) Characterization of soil organic carbon relative to its stability and turnover. In: E. Paul, K. Paustian, E.T. Elliott and C.V. Cole (eds) Soil Organic Matter in Temperate Agroecosystems: Long-Term Experiments in North America, CC Press, Boca Raton, FL: SI-71.

Paveley, ND; Rennie, WJ; Reeves, JC; Wray, MW; Slawson, DD; Clark, WS; Cockerell V and Mitchell, AG (1997) Cereal seed health strategies in the UK. In: *Seed Health Testing: Progress Toward the 21st Century*, Hutchins JD and Reeves, JC. (eds.), Oxon, UK: CAB International, pp. 95-106.

Pérez-de-Castro, AM; Vilanova, S; Cañizares, J; Pascual, L; Blanca, JM; Díez, MJ; Prohens, J and Picó, B (2012) Application of genomic tools in plant breeding. *Curr Genomics* **13**: 179–195.

Perumal, R; Jayeraman, C; Manikam, TS and Kothandaraman, GV (1984) Studies on management of inorganic and organic nutrients on yield and quality of seed cotton. *Madras Agrl. J.* **71**(11): 724-730.

Phatak, HC (1974) Seed borne plant viruses- identification and diagnosis in seed health testing. *Seed Sci. and Technol.* 2: 3-155.

Preston JC and Kellogg EA. (2007) Conservation and divergence of APETALA1/ FRUITFULL-like gene function in grasses: evidence from gene expression analyses. *Plant J.***52**: 69–81.

Qi, LL; Friebe, B; Zhang, P and Gill, BS (2007) Homoeologous recombination, chromosome engineering and crop improvement. *Chromosome Res* **15**: 3–19.

Rajpal, VR; Rama Rao, S and Raina, SN (2016) *Editors,* Gene Pool Diversity and Crop Improvement, Volume 1. Sustainable Development and Biodiversity 10 Series editor, Kishan Gopal Ramawat, Springer International Publishing Switzerland.

Ram, H; Kler, DS; Singh, Y and Kumar, K (2010) Productivity of maize (*Zea mays*) – wheat (*Triticum aestivum*) system under different tillage and crop establishment practices. *Indian Journal of Agronomy* **55**: 185-190.

Ramamurthy, V and Shivashankar, K (1995) Effect of organic matter and phosphorus on quality and nutrient uptake by soybean. *J. Oil seed res.* **12**(1): L 87-90.

Rebetzke, GJ; Richards, RA; Fisher, VM and Mickelson, BJ (1999) Breeding long coleoptile, reduced height wheats. *Euphytica* **106**: 159-168.

Reeves, JC (1995) Nucleic acid techniques in testing seed borne diseases. In: *New Diagnostics in Crop Sciences,* Skerritt JH and Appels, R (eds.), Oxon, UK: CAB International, pp. 127-151.

Reicosky, DC (2003) Tillage-induced soil properties and chamber mixing effects on gas exchange. In: Proc. International Soil Tillage Research Organisation 16th Triennial Conf. Brisbane, Australia. 14-18 July, 2003, p. 971-976.

Reif, JC; Zhang, P; Dreisigacker, S; Warburton, ML; van Ginkel, M; Hoisington, D; Bohn, M and Melchinger, AE (2005) Wheat genetic diversity trends during domestication and breeding. *Theor Appl Genet* **110**: 859–864.

Richards, RA and Lukacs, Z (2002) Seedling vigour-sources of variation for genetic and agronomic improvement. *Australian J of Agric. Res.* **53** : 41-50.

Richardson, MJ (1996) An annotated list of seed borne diseases, Fourth Edition, Zurich, Switzerland: *International Seed Testing Association.*

Saharawat, YS; Ladha, JK; Pathak, H; Gathala, M; Chaudhary, N and Jat, ML (2012) Simulation of resource-conserving technologies on productivity, income and greenhouse gas (GHG) emission in rice-wheat system. *Journal of Soil Science and Environment Management* **3**(1): 9-22.

Sayre, KD and Mareno, Ramos, OH (1997) Application of raised bed planting system to wheat. *Wheat Programme Special Report* **31**: 7.

Schlotterer, C (2004) The evolution of molecular markers–just a matter of fashion. *Nature Reviews Genetics* 5: 63-69.

Sharma, AK (2002) A Handbook of Organic Farming Agrobios (India), Jodhpur pp. 627.

Sharma, PK and Acharya, CL (2000) Carry-over of residual soil moisture with mulching and conservation practices for sowing of rainfed wheat (*Tricticum aestivum* L.) in north-west India. *Soil and Tillage Research* **57**: 43-52.

Singh, P; Mishra, VK; Sinha, B and Singh, S (2019) Study of correlation of Ppd-D1 gene with the major phenotypic traits in wheat (*Triticum aestivum* L.). Souvenir cum Book of Abstracts, First Global Congress on Biotechnology and Crop Science (BCS-2019), Purnea University, Purnea, Bihar, India.

Singh, RP (1996) Sustainable development of Dryland Agriculture in India. Scientific Publishers, Jodhpur, India.

Singh, UP; Singh, Y; Kumar V and Ladha, JK (2009) Evaluation and promotion of resource conserving tillage and crop establishment technique in rice-wheat system of eastern India. (*In)*: Ladha JK., Singh Yadvinder, Erenstein O,Hardy B, (Eds.), *Integrated crop and resource management in rice-wheat system of south Asia.* International Rice Research Institute, Los Banos (Philippines), pp. 151-176.

Singh, UP; Singh, Y; Singh, HP and Gupta, RK (2005) Cropping system options in no/reduced till - surface residue management system, pp. 79-85. *In*: Abrol, I.P., Gupta, R.K. and Malik, R.K. 2005. Conservation Agriculture - Status and Prospects. Centre for Advancement of Sustainable Agriculture, NASC Complex, New Delhi.

Singh, VN and Prasad, CR (19974) Cumulative effect of manures on carbohydrates, crude protein and amino acids content in wheat on acidic red loams of kanko, *Indian J. Agric. Chem.,* **7**: 93.

Srivastava, AC (2003) Energy savings through reduced tillage and trash mulching in sugarcane *Sugar Crops Newsletter* **8**(1): 2-3.

Summerell, BA and Bugress, LW (1989) Decomposition and chemical composition of cereal straw. *Soil Biol. Biochem* **21**: 551-559.

Taylor, E; Bates, J and Jaccoud, D (2007) Diagnosis of seed borne pathogens. In: *Handbook of Seed Science and Technology*, Basra, AS (ed), Scientific Publishers, Jodhpur, pp. 649-676.

Thampan, PK (1995) Organic Agriculture, Peekay Tree Crops Development Foundation, Cochin. pp.354.

Tilak, KVBR and Singh, G (1994) Biofertilizers research gaps and future needs *Fert. News* (**4**): 11-17.

Trethowan, R; Mannes, Y and Chattha, T (2009) Breeding for improved adaptation to conservation agriculture improves crop yields. In: Innovation for improving efficiency, Equity and Environment. 4th World Congress on Conservation Agriculture. 4-7 Feb., 2009. New Delhi, pp: 207-211.

Trethowan, RM; Reynolds, MP; Sayre, KD and Ortiz-Monasterio, I (2005b) Adapting wheat cultivars to resource conserving farming practices and human nutritional needs. *Annals of Applied Biology* **146**: 404–413.

Tripathi, SC; Sayre, KD and Kaul, JN (2005) Planting systems on lodging behavior, yield components, and yield of irrigated spring bread wheat. *Crop Sci.* **45**: 1448–1455.

Triplett, G and Vand Dorren, DR (1977) Agricultural without tillage. *Scientific American.* **236**: 28-33.

Trivedi, RK and Gunasekran, M (2013) Indian minimum seed certification standards. Ministry of Agriculture (MoA), Government of India (GoI) Delhi.

Unger, PW (1991) Organic matter, nutrient and pH distribution in no- and conventional- tillage semiarid soils. *Agronomy Journal* **83**: 186-189.

Unger, PW (1997) Management-induced aggregation and organic carbon concentrations in the surface layer of a Torrertic Paleustoll. *Soil and Tillage Research* **42**: 185–208.

Ventateshwaralu, B; Sharma, KL and Prasad, JVNS (2009) Conservation agriculture-constraints, issues and opportunities in rainfed areas. *In*: 4th World Congress on Conservation Agriculture (WCCA) Report, Congress held at New Delhi, 4-7 February, 2009.

Vyas, SC; Vyas, S; Vyas, S and Modi, HA (1998) Biofertilizers and Organic farming. Akta Prakashan, Nadiad, pp.252.

Wang, S; Yin, L; Tanaka, H; Tanaka, K and Tsujimoto, H (2010) Identification of wheat alien chromosome addition lines for breeding wheat with high phosphorus efficiency. *Breed Sci* **60**: 372–379.

Wang, YN; Chen, PD; Wang, ZT and Liu, DA (1991) Transfer of useful germplasm from E. giganteus to common wheat. II. Cytogenetics and scab resistance of backcross derivatives. *J Nanjing Agric Univ* **14**: 1–5.

Wilkins, DE; Klepper, B and Rickman, RW (1989) Measuring wheat seedling response to tillage and seeding systems. *Trans ASAE* **32**: 798-800.

Yadav, DV and Dey, P (1998) Integrated nutrient management based on systems approach. *Applied Engineering in Agriculture* **19**(1): 40-45.

Yadav, RL and Srivastava, TK (2005) Conservation Agriculture and Opportunities for Sugarcane Based Cropping Systems, in: Abrol, IP; Gupta, RK; Malik, RK (Eds), Conservation Agriculture – Status and Prospects. Centre for Advancement of Sustainable Agriculture, New Delhi pp242.

Zhang, F; Li, C; Wang, Z and Wu, H (2006) Modelling impacts of management alternatives on soil carbon storage of farmland in Northwest China. *Biogeosciences*, **3**: 451–466.

Chapter 10

Wheat Seed, Maintenance Breeding, Basic Principles, Seed Priming, Electrophoresis, Seed Certification and Farmer Participatory Seed Production

Outlines

1. Seed - A Key Component Among All Inputs
   a. Importance of Quality Seeds
2. Basic Principles of Quality Seed Production
3. Agronomic Principles
4. Different Stages of Rouging
5. Steps in the Production of Breeder Seed
6. Maintenance Breeding, Variety Maintenance and Seed Production
7. Cultivar Maintenance
8. Seed Production Technology
9. Classes of Improved Seed
10. Requirements for Certified Seed
11. Nucleus Seed Production
12. Breeder Seed Production

13. Quality Seed Production
    a. Seed Threshing, Cleaning and Drying
    b. Seed Storage
14. Post-Harvest Managements for Improved Germination and Vigour
15. Inset-Pest and Diseases Management
16. Improving Planting Value of Seeds by Physical and Physiological Methods
17. Seed Processing
18. Seed Priming
    a. Germination (Triphasic Uptake of Water)
    b. Effect of Seed Priming on Seedling Growth and Development
    c. Effect of Priming on Yield and Quality
    d. Physiological and Molecular Basis of Priming
    e. Morphological Changes in Primed Seed
19. Seed Storage Products, Their Biochemistry and Modification Issues: Proteins
20. Gel Electrophoresis of Quality Protein Profiles/Isozyme Profiles of Cultivars
    a. Gel Electrophoresis of Quality Protein Profiles
    b. Separation of Proteins under Denaturing Conditions
    c. Performing Electrophoresis
    d. Preparation of Stock Solutions
    e. Staining
21. Seed Genetic Purity Assessment Using Molecular Markers
22. Prospectus and Constraints in Seed Production
23. Factors Promoting Seed Industry in India
24. Influence of Seed Industry on Economy
25. Rules and Regulations in Seed Act for Quality Seed Production
26. Seed-Certifying Agencies
27. Certification of a Variety
28. How A New Variety Reaches the Farmer
29. Practical Problems in Seed Production

Seed - A Key Component among All Inputs

Seed is a key component among all inputs for sustainable crop production. It is estimated that quality of seed accounts for 20-25 per cent of productivity. The importance of quality seed has been realized by mankind long ago. The need for a good viable seed for prosperity of human race is mentioned in ***Rigveda*** of ancient India. It is mentioned in the **Primeval Manusmriti** as **"Subeejam Sukshetre Jayate Sampadyate"** which literally means **"A good seed in a good field will win**

Figure 10.1. Half of Professional Efficiency

Figure 10.2. Do not judge each day by the harvest you reap, but by the seeds you plant

'Corporates want to dominate food security through seed control'

■ Swadeshi Jagaran Manch, Bharatiya Krishak Samaj, Bharatiya Kisan Sangh to launch joint campaign to make people aware of ill effects of GM crops

■ Staff Reporter

"CORPORATES that are into genetically modified seeds, fertilisers, chemicals, and other agricultural inputs are running various lobbies, as they want to control food security of nations through control over seeds. It is time to see through this conspiracy and protect the interests of farmers and the nation," said Dr Krishan Bir Chaudhary, President of Bharatiya Krishak Samaj.

Dr Chaudhary was talking to 'The Hitavada' on the sidelines of a workshop on Genetically Modified (GM) crops, organised by Swadeshi Jagaran Manch (SJM) at Hindu Dharma Sanskruti Mandir, on Sunday. He said that SJM, Bharatiya Krishak Samaj, and Bharatiya Kisan Sangh were planning to take up a nation-wide campaign to create awareness about the sinister designs of corporate multi-national companies. For the purpose, he said, the three organisations will be bringing out a booklet to be circulated widely.

According to Dr Chaudhary, MNCs like Monsanto are 'exploiting' the Intellectual Property Rights (IPR) regime through seed licensing and patents. "There is a nexus of these corporates, bureaucrats, and pseudo-scientists. Many scientists are on the payrolls of these corporates, and they promote the interests of their pay-masters," he alleged.

Referring to Maharashtra Government's latest decision to allow confined field trials of five GM seeds, and the stand taken by Central Government, Dr Chaudhary said, "The Government also needed to understand things in correct perspective and not fall prey to the malafide propaganda of the corporates." To protect the interests of farmers and the nation's food security, three things are needed -- good price, procurement, and good storage facilities, he said.

"We are not anti-development, but will fight out the attempts to compromise national interest," observed the former Chairman of State Farms Corporation of India. "Further," he said, "we are opposed to bad science and bad technology with bad intentions." Emphasising that farmers should not lose right on seeds, he said that seeds were national resources. Technology should be friendly to human health, environment, ecology, water and soil health, he added.

SJM workshop was attended by delegates from various states of the country. Several scientists and resource persons including Dr Tarak Kate, Manoharrao Parchure, Dilnavaz Variava, who is President of Sahayak Trust, guided the delegates.

Kashmiri Lal, All India Organiser of SJM; Dinesh Kulkarni, All India Organising Secretary of Bharatiya Kisan Sangh; Dr Krishan Bir Chaudhary; Ajay Patki shared the dais at the concluding function.

In his address, Kulkarni said that the aim of the workshop was to prepare well-informed, dedicated workers who would work with conviction and oppose GM crops. "The myth of higher production has been busted. Now the seed companies and GEAC are talking of plant protection while advocating GM crops. However, let me tell you, there are farmers who have taken higher yield with integrated management of agricultural inputs. Traditional seeds and integrated management works wonders without genetic modifications," he said.

Patki said that SJM was no[t] opposed to technology but it wa[s] concerned about national inter[est]. "Careful thought needs to b[e] given to ensure that tradition[al] agricultural wisdom and pra[c]tices are not done away with," [he] added. A convention of agric[ul]ture students would [be] organised soon, he informed t[he] gathering.

Dinesh Kulkarni, All India Organising Secretary of Bharatiya Kisan Sangh, speaks during concluding function of workshop on GM crops. Also seen on the dais are Kashmiri Lal, All India Organising Secretary of SJM; Dr K B Chaudhary, President of Bharatiya Krishak Samaj; and Ajay Patki. (Pic by Anil Futane)

Figure 10.3. Corporates want to domnate food security through seed control

Figure 10.4. ICAR glimpses on the eve of Republic Day, 2020

Figure 10.5. Efforts are better than promises

Figure 10.6. Grow silently

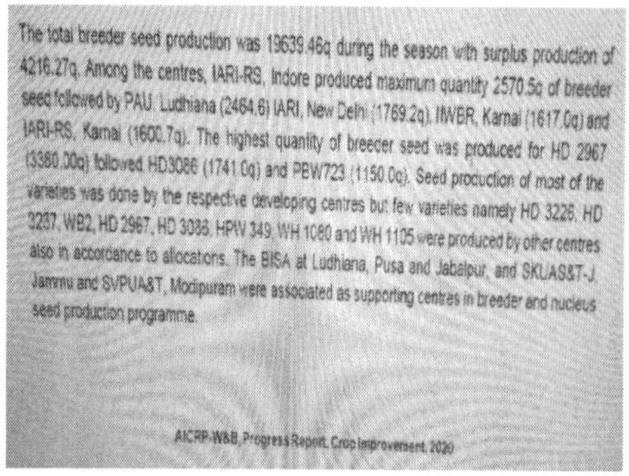

The total breeder seed production was 19639.46q during the season with surplus production of 4216.27q. Among the centres, IARI-RS, Indore produced maximum quantity 2570.5q of breeder seed followed by PAU, Ludhiana (2464.6) IARI, New Delhi (1769.2q), IIWBR, Karnal (1617.0q) and IARI-RS, Karnal (1600.7q). The highest quantity of breeder seed was produced for HD 2967 (3380.00q) followed HD3086 (1741.0q) and PBW723 (1150.0q). Seed production of most of the varieties was done by the respective developing centres but few varieties namely HD 3226, HD 3237, WB2, HD 2967, HD 3086, HPW 349, WH 1080 and WH 1105 were produced by other centres also in accordance to allocations. The BISA at Ludhiana, Pusa and Jabalpur, and SKUAS&T-J, Jammu and SVPUA&T, Modipuram were associated as supporting centres in breeder and nucleus seed production programme.

AICRP-W&B, Progress Report, Crop Improvement, 2020

Figure 10.7. Excerpts from AICRP-W&B Progress Report, Crop Improvement 2020

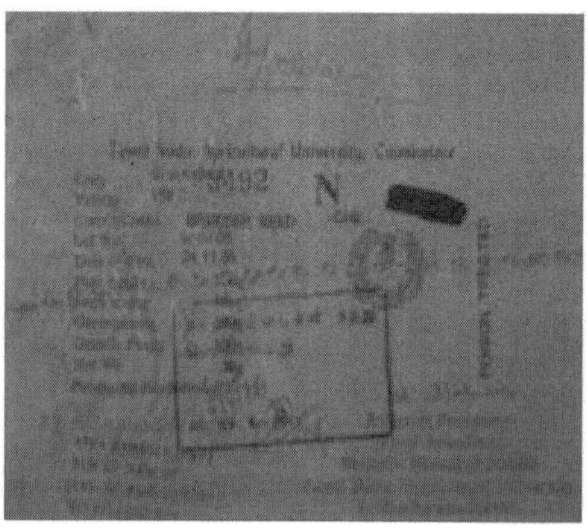

Figure 10.8. Breeder Seed Tag Yellow Colour

Figure 10.9. Foundation Seed Tag; Tag Colour: White

Figure 10.10. Certified Seed Tag; Tag Colour: Azar Blue

Figure 10.11. Registered Seed Tag; Tag Colour: Violet

Figure 1012. ICAR-IARI RS Pusa, Bihar during 2006-2007

Figure 10.13. Display of wheat cultivar developed at IARI RS Pusa, Bihar

Figure 10.14. IARI RS Pusa, Bihar Seed Production, 2009-2010

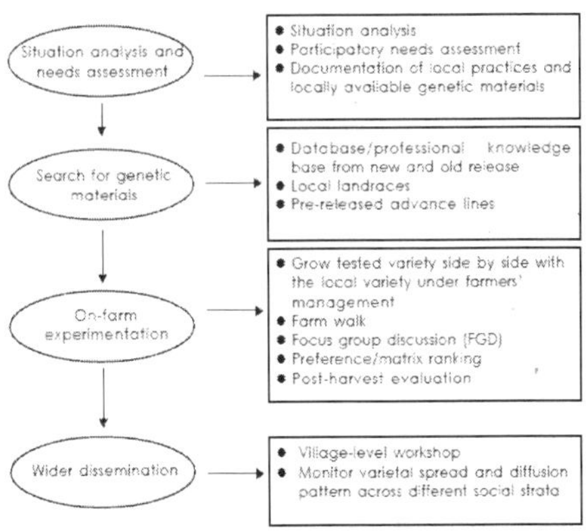

Figure 10.15. PVS process and participatory tools used

Figure 10.16. Elite Material, HD 4672, Seed Production of wheat

Figure 10.17. Elite Material, HD 2987, Seed Production of wheat

Figure 10.18. Elite Material, HD 2932, Seed Production of Wheat

Figure 10.19. Dr D K Verma making observance in Wheat Seed Production, 2019

Figure 10.20. Dr D K Verma addressing the Hoshangabad farmers about Wheat Seed Production, 2019

Figure 10.21. Dr D K Verma addressing the Gujarat farmers about Wheat Seed Production, 2019

Figure 10.22. Scientifc Advisory Committee Meeting at KVK Dhar, MP

Figure 10.23. Dr M S Swaminathan and Pandit jee Dr D K Verma ko tikka lagate hue dated 02nd Feb, 2014

Figure 10.24. Dr D K Verma and Dr A K Singh observing the minute of Wheat Seed Production, 2020

Figure 10.25. Dr A K Singh, presenting lecture to farmers about improved wheat cultivation, 2019

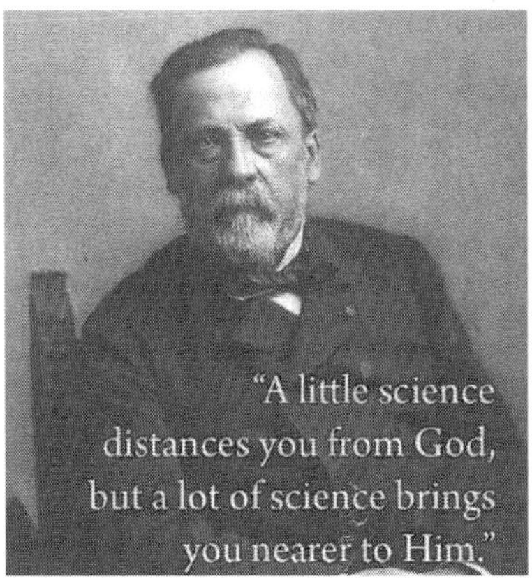

Figure 10.26. Louis Pasteur – A little science distance you from God, but a lot of science brings you nearer to Him

Figure 10.27. Wheat Nucleus Seed, HI 1544 Production

Figure 10.28. Wheat Nucleus Seed, HD 2932 Production

Figure 10.29. Wheat Nucleus Seed, HI 8991 Production

Figure 10.30. Wheat Nucleus Seed, HI 8991 Production space for roguing

Figure 10.31. Wheat Nucleus Seed, HI 1563 Production

Figure 10.32. Wheat Nucleus Seed, HI 8498 Production

Figure 10.33. Breeder seed plot of HI 8663 (Poshan)

Figure 10.34. Breeder seed plot of HI 1500

Figure 10.35. Breeder seed plot of HI 1544 (Purna)

Figure 10.36. Breeder seed plot of HD 2733

Figure 10.37. Breeder seed plot of HD 2824

Figure 10.38. Breeder seed plot of HI 1544 (Rahish Patel Magajpura)

Figure 10.39. Dr D K Verma Dr A K Singh and Dr S R Yadav at farmers seed production training

Figure 10.40a. Dr A K Singh, Dr D K Verma and Dr Bandyopadhyaya with tribal farmers of Dhar District, MP

Figure 10.40b. Dr A K Singh, Dr D K Verma and Dr Bandyopadhyaya with tribal farmers of Dhar District, MP

Figure 10.41. Seed Day at Kusmanya, Dewas, MP

Figure 10.42. Roguing in Seed Production Plot of HI-1544 by Dr D K Verma at CIAE, Bhopal, MP

Figure 10.43. Visit of farmers in Seed Production Plot of HI-1544 at ICAR-IARI RS Indore, MP

Figure 10.44. Dr D K Verma with Dr Indu Sharma in Farmer Participatory Seed Production Field of farmers

Figure 10.45. IARI Indore - Scientist discus with the villagers of Kagjipura, Dhar, MP

Figure 10.46. IARI Indore - Scientist discus with the villagers of Harda, MP

Figure 10.47. Dr K C Sharma, briefing about the Agronomical Package of Wheat Seed Production at ICAR-IARI RS Indore, MP

Figure 10.48. Dr D K Verma with Dr D C Saxena and Dr A K Singh in Farmer Participatory Seed Production Field of durum wheat

Figure 10.49. Seed Production of HI-1544 at CIAE, Bhopal, MP

Figure 10.50. Dr D K Verma and wheat team during Mera gaon Mera Gaurav, delivering lectures about Wheat Seed Production, 2020 at Village level

Figure 10.51. Dr D K Verma, giving a view in Farmer Participatory Trial at Janapaw, Indore, Madhya Pradesh, 2019

Figure 10.52. Wheat farmer in durum wheat seed production, 2013

Figure 10.53. Dr D K Verma briefing about the Wheat Seed Production at ICAR-IARI RS Indore, MP

Figure 10.54. Dr D K Verma with Rajesh and Upendra Singh Chaudhary at Participatory Seed Production, Dewas, 2020

Figure 10.55. Intercropping of Musali with HI 1544 in Participatory Seed Production at Manawar, Dhar, MP

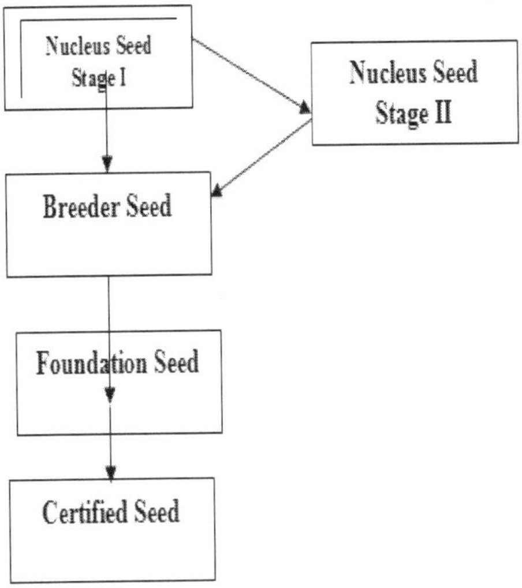

Figure 10.56. Classes of Improved Seed

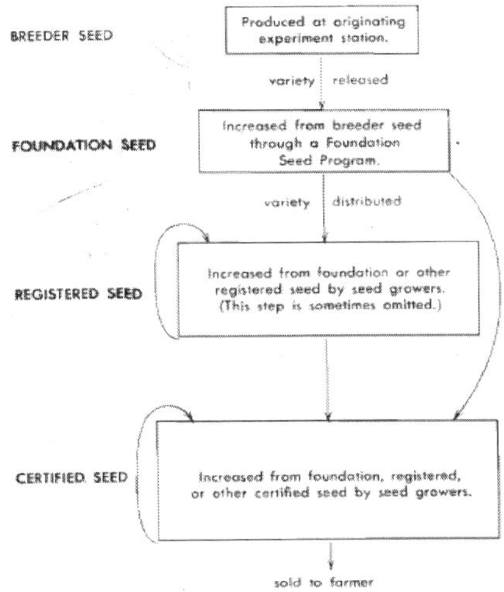

Figure 10.57. Steps in increase and distribution of a variety

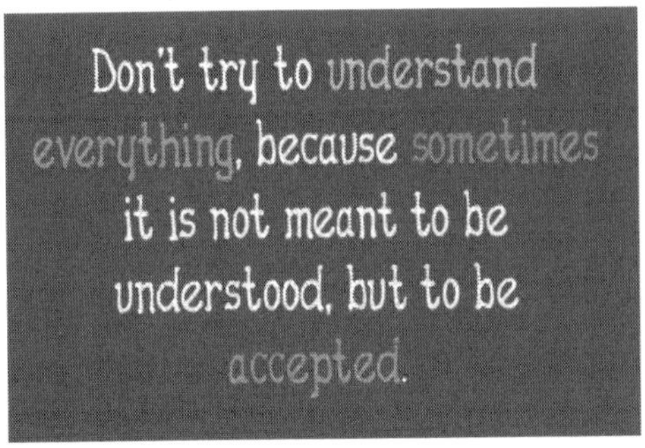

Figure 10.58. Do not try to undderstand everything, because soemtimes it is not meant to be understood, but to be accepted

Figure 10.59. During the pandemic, nutritious food is an urgent requirement for the teeming millions.

Figure 10.60. Display of aestivum and durum varieties in ICAR-IARI RS Indore museum

Figure 10.61. Group of farmers visiting ICAR-IARI RS Pusa, Bihar

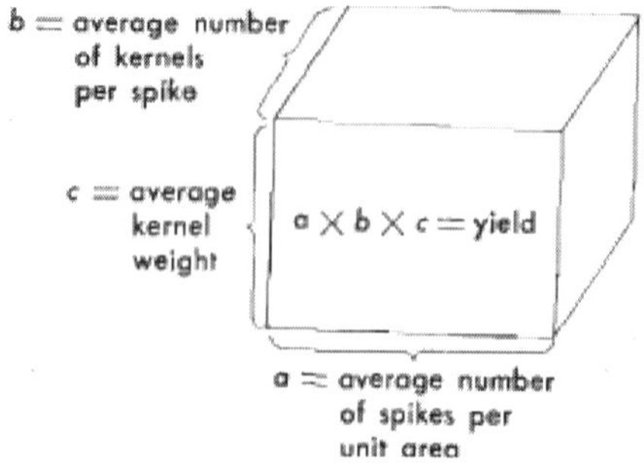

Figure. 12.62. Yield per unit area may be represented geometrically as a box, the volume of which will be dependent upon the number of spikes per unit area, the number of kernels per spike, and the average kernel weight.

Figure 10.63. Rathindra Kumar Singh, Begusarai, Bihar producing seed

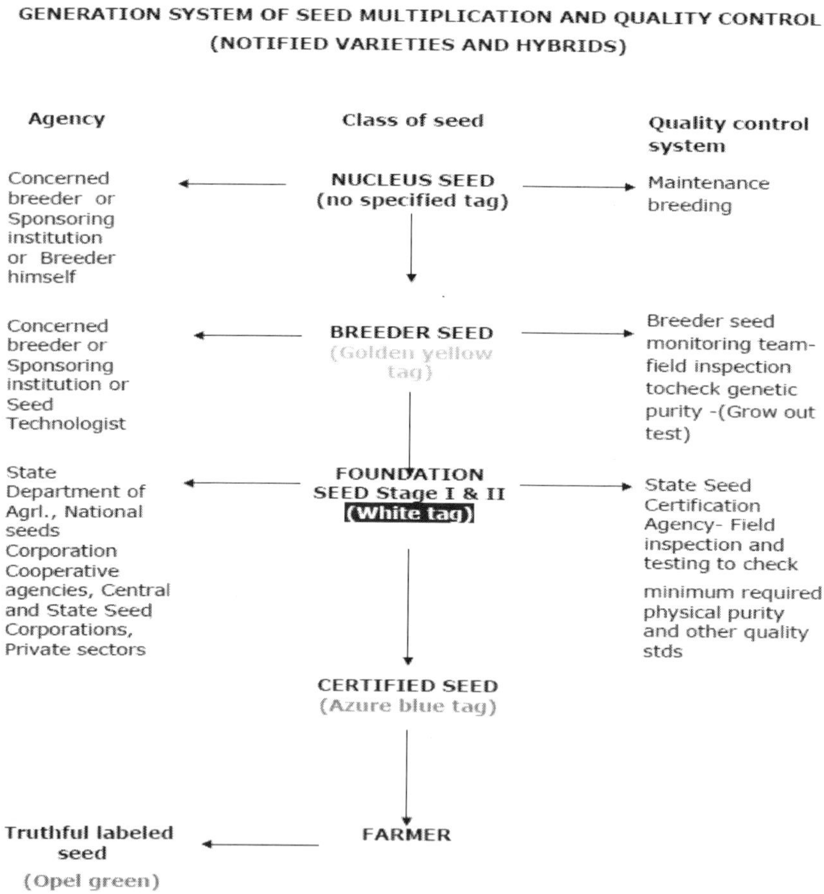

Figure 10.64. Generation System of Wheat Seed Production

Figure 10.65. Seed Production of Durum seeds

Figure 10.66. On farm demonstration at ICAR-IARI RS Pusa, Bihar during 2009-2010.

Figure 10.67 Breeder Seed Production of Wheat, cv. PBW 443 at ICAR-IARI RS Pusa, Bihar, 2009

Figure 10.68. Breeder Seed Production of Wheat, cv. HP 1731 (Rajlaxmi) at ICAR-IARI RS Pusa, Bihar, 2009

Figure 10.69. Breeder Seed Production of Wheat, cv. HP1761 at ICAR-IARI RS Pusa, Bihar, 2009

Figure 10.70. Breeder Seed Production of Wheat, cv. HW 2045 (Kousambi) at ICAR-IARI RS Pusa, Bihar, 2009

Figure 10.71. Breeder Seed Production of Wheat, cv. HD2733 (VSM) at ICAR-IARI RS Pusa, Bihar, 2009

Figure 10.72. Breeder Seed Production of Wheat, cv. HD 2643 at ICAR-IARI RS Pusa, Bihar, 2009

Figure 10.73. Breeder Seed Production of Wheat, cv. HD 2285 at ICAR-IARI RS Pusa, Bihar, 2009

Figure 10.74. Breeder Seed Production of Wheat, cv. HD 2824 at ICAR-IARI RS Pusa, Bihar, 2009

Figure 10.75. Breeder Seed Production of Wheat, cv. HD 2687 at ICAR-IARI RS Pusa, Bihar, 2009

Figure 10.76. Breeder Seed Production of Wheat, cv. HD 2329 at ICAR-IARI RS Pusa, Bihar, 2009

Figure 10.77. Breeder Seed Production of Wheat, cv. HD 2888 at ICAR-IARI RS Pusa, Bihar, 2009

Figure 10.78. Path of 1m to facilitate roguing in seed production plot

Figure 10.79. Breeder Seed Production of Wheat, cv. HI 1563 (Pusa Prachi) at ICAR-IARI RS Pusa, Bihar, 2012

HI 1563 (Pusa Prachi) [MACS 2496*2/MC10], 2011

North Eastern Plains Zone; Late Sown, irrigated

Plant height: 85-90 (cm); Days to heading: 65-70 days; Days to maturity: 105-110 days; Ear character: Shape-Elliptical; Length-Medium long; Density- Medium dense; Leaf colour: Green; Awn attitude and colour: Medium and white, Waxiness-Absent; Grain character: Colour-Amber; Texture-Hard; Shape-Round; 1000-grain weight: 40-45 g; Resistant to: Stripe, Leaf rust, leaf blight, powdery mildew and foot root; Rust resistant genes postulated- *LrR, Yr2+, Sr2+*; Tolerant to- Terminal heat tolerance; Quality traits: Protein (%): 11.82%; Sedimentation value (ml): 47; Hectoliter weight (kg/ha): 77.7; Hardness index: 80; Glu 1 subunit:2*; Glu score: 8/10; Grain Zn : 36.3 & Fe: 43.1 (ppm) content; suitable for making : Chapati and Biscuit ; Grain yield (t/ha): Average : 3.76, Potential: 5.17; Terminal heat tolerance, excellent chapati quality.

Figure 10.80. Breeder Seed Production of Wheat, cv. HD 3117 at ICAR-IARI RS Pusa, Bihar, 2012

Characteristics: It has genetic yield potential of 5.5 t/ha. Average yield of this variety under late sown (after 15 th Dec.) is 4.78 t/ha under tilled condition and 4.79 t/ha under conservation agriculture condition. It is free from brown and yellow rust under natural condition and highly tolerant to Karnal bunt. Grain protein content is 11.7 %. Average yield: 4.7 t/ha, Production conditions: Late sown irrigated conditions under conservation agriculture (CA) environment., Recommended areas: NCR/Delhi state.

Figure 10.81. Breeder Seed Production of Wheat, cv. HD 4728 (Pusa Malvi) at ICAR-IARI RS Pusa, Bihar, 2012

Wheat: HD 4728 (Pusa Malvi), 2015

Characteristics: A semi-dwarf (90 cm), 120 days maturing durum wheat variety with genetic yield potential of 6.8 t/ha. The variety has high degree of resistance to leaf and stem rust diseases. It has bold and lustrous grains and superior quality traits for end-use in semolina-based industry., Average yield: 5.42 t/ha, Production conditions: Timely sown irrigated., Recommended areas: Central Zone.

Figure 10.82. Breeder Seed Production of Wheat, cv. DBW 14 at ICAR-IARI RS Pusa, Bihar, 2009

DBW 14 (Raj 3765 / PBW 343): Released byCVRC in 2003. Early maturing wheat variety for late sown, irrigated condition of North Eastern Plains Zone (NEPZ). This variety was first indigenously developed variety through shuttle breeding approach and is suitable for rice-wheat system of eastern India. A very popular genotype in NEPZ as replacement for HUW 234.

Figure 10.83. Breeder Seed Production of Wheat, cv. DBW 38 at ICAR-IARI RS Pusa, Bihar, 2009

CBW38(*CNDO/R143//ENTE/MEXI_2/3/Ae.sq(Taus) /4/ WEAVER/5/2*PASTOR)*:Released by CVRC in 2008. Wheat varietyfor timely sown, irrigated condition of North Eastern Plains Zone (NEPZ). This variety has yield advantage over all the checks and candidate varieties and possesses APR to yellow rust race *78S84*.

Figure 10.84. Breeder Seed Production of Wheat, cv. WB 2 at ICAR-IARI RS Pusa, Bihar, 2009

WB2(T. dicocccon C19309/Ae.sq(409) /3/MILAN/S87230//BAV92/4/2*MILAN/S87230// BAV92): Released by CVRC in 2017. WB 2 has shown yield superiority over the check variety DPW 621-50. It possesses high Zn & Fe content in grain. It is also resistant against prevalent yellow rust pathotypes 78S84 and 46S119.

Figure 10.85. Dr D K Verma at Farmer Participatory Breeder Seed Production at Dhoot Farm, Kusmanya, Dewas, MP

Figure 10.86. Dr D K Verma during Field Day (Wheat) at Kagdipura, Nalchha, Dhar, 2020

Figure 10.87. Wheat varieties HI 8663 (Poshan) and HI (Purna) and at farmers' field

Figure 10.88. Dr D K Verma delivering lecture at KVK Kasturbagram, Indore during 26.08.2016 to 01.09.2016

Success story

Progressive farmer Mr. Yogendra Kaushik of Village Aslavada, Block Badnagar Dist. Ujjain (MP) had sown HI 8663 (Poshan) breeder seed procured from IARI – Regional Station, Indore and got record yield of 95.32 q/ha as certified by revenue as well as agricultural department personnel in the state crop cutting experiments. He is a recipient of 'Best Farmer" award from President of India.

कृषि कर्मन पुरस्कार 2012-13

बेहतर किसानी के लिए इंजीनियर को मिला सम्मान

● 2010-11 से शुरू हुए थे ये पुरस्कार

● 2010-11 के बाद से कुल खाद्यान्न उत्पादन में दूसरी बार नम्बर एक है मप्र

Figure 10.89. Success story of Mr Yogendra Kaushik, growing Poshan (HI 8663)

Figure 10.90. Field view of HI 8691

Figure 10.91. Field view of HI 1563

Figure 10.92. Amar (HW 2004)

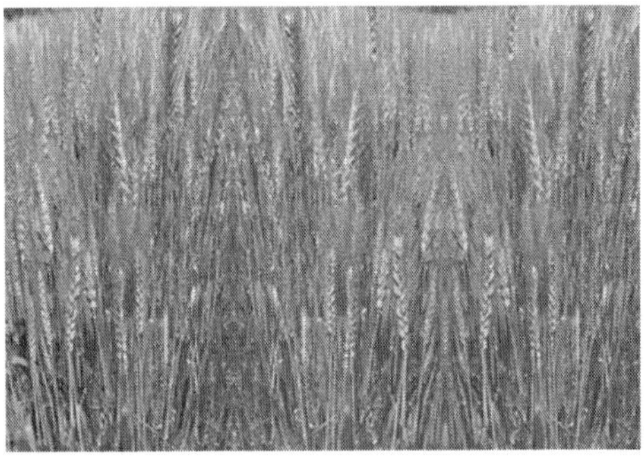

Figure 10.93. Amrita (HI 1500)

Figure 10.94. HI 1531 (Harshita) [HI 1182 / CPAN 1990], 2006

Central Zone, Timely Sown, Rainfed & Restricted irrigation

Plant height: 90-95 (cm); Days to heading: 65-70 days; Days to maturity: 115-120 days; Ear character: Shape-Elliptical; Length-Medium; Density- Medium dense; Leaf colour: Green; Awn attitude and colour: Medium and white, Waxiness-present; Grain character: Colour-Amber; Texture-Semi-Hard; Shape-Round; 1000-grain weight: 40-45 g; Resistant to: Stem, Leaf rust; Rust resistant genes postulated- *Lr24+2+, Yr-, Sr24+*; Tolerant to- Drought & Early and terminal heat tolerance; Quality traits: Protein (%): 9.7%; Sedimentation value (ml): 44; Hectoliter weight (kg/ha): 84; suitable for making : Chapati

Grain yield (t/ha); Rainfed (Average- 2.36, Potential- 3.98); RI (Average- 2.66, Potential- 4.0), Terminal heat tolerance, excellent chapati quality

Figure 10.95. Malav Ratna (HD 4672)

Figure 10.96. Malav Kirti (HI 8627)

Figure 10.97. Malav Kranti (HI 8638), Field View

Figure 10.98. Malav Kranti (HI 8638), Seed View

Figure 10.99. HI 1544 (Purna) [HD 2402/HW3007]

Central Zone, Timely Sown, irrigated

Plant height: 85-90 (cm); Days to heading: 65-70 days; Days to maturity: 110-115 days; Ear character: Shape-Elliptical; Length-Medium long; Density- Medium dense; Leaf colour: Green; Awn attitude and colour: Medium and white, Waxiness-Absent; Grain character: Colour-Amber; Texture-Semi-Hard; Shape-Round; 1000-grain weight: 40-45 g; Resistant to: Stem, Leaf rust; Rust resistant genes postulated- *Lr24+R, Yr2+, Sr24+2+R*; Tolerant to- Terminal heat tolerance; Quality traits: Protein (%): 12.22%; Sedimentation value (ml): 38; Hectoliter weight (kg/ha): 81.6; Hardness index: 97; Glu 1 subunit: N; Glu score: 6/10; Grain Zn : 44.2 & Fe: 35.9 (ppm) content; suitable for making : Chapati

Grain yield (t/ha): Average : 5.14, Potential: 6.82, Terminal heat tolerance, excellent chapati quality

Figure 10.100a. Poshan (HI 8663)

Figure 10.100b. Poshan (HI 8663)

Figure 10.101. HI 1605 (Pusa Ujala) [BOW/VEE/5/ND/VG9144//KAL/BB/3/YACO/4/ CHIL/6/CASKOR/3/CROC_1 /*Ae.squarrosa*(224) // OPATA/7/ PASTOR//MILAN/ KAUZ/3/BAV92]

Peninsular Zone, Timely Sown, Restricted Irrigation

Plant height: 80-85 (cm); Days to heading: 60-65 days; Days to maturity: 105-110 days; Ear character: Shape-parallel; Length – Long; Density- Medium; Leaf colour: Green; Awn attitude and colour: spreading & white; Waxiness : Strong; Grain character: Colour-Amber; Texture-Semi-hard; Shape-Ovate; 1000-grain weight: 35-40 g; Resistant to: Stem rust, Leaf rust, Flag Smut, Karnal Bunt, Leaf blight, Foot rot; Rust resistant genes postulated- *Lr13+, Yr2+, Sr5+11+*; Tolerant to: Drought; Quality traits: Protein (%): 13.1%; Sedimentation value (ml): 55; Hectoliter weight (kg/ha): 78.6; Hardness index: 79; Glu 1 subunit: 2*; Glu score: 8 /10; Grain Zn : 35 & Fe 43 (ppm) content; suitable for making : Chapati, Biscuit & Bread; Grain yield (t/ha): Average : 2.91, Potential: 4.4; Drought and early heat tolerance, Iron rich and excellent chapati quality.

Figure 10.102. HI 1612 (Pusa Wheat 1612) [Kauz//Altar 84/AOS/3/Milan/Kauz/4/Huites]

North Eastern Plains Zone; Timely Sown, Restricted Irrigation

Plant height: 90-95 (cm); Days to heading: 85-90 days; Days to maturity: 125-130 days; Ear character: Shape-Tapering; Length-Long; Density- Medium; Leaf colour: Green; Awn attitude and colour: spreading & white, Waxiness-Strong; Grain character: Colour-Amber; Texture-Semi-hard; Shape-Oblong; 1000-grain weight: 40-45 g; Resistant to: Stripe rust, Leaf rust, leaf blight, Karnal bunt, *Fusarium* head blight and loose smut; Rust resistant genes postulated- *Lr23+, Yr2+, Sr7b+2+*; Tolerant to- Drought and Terminal heat tolerance; Quality traits: Protein (%): 11.54%; Sedimentation value (ml): 58; Hectoliter weight (kg/ha): 78.4; Hardness index: 80; Glu 1 subunit: 2*; Glu score: 8/10; Grain Zn : 35.5 & Fe: 41.5 (ppm) content; suitable for making : Chapati and Biscuit

Grain yield (t/ha): Average: 3.76, Potential: 5.05; Drought & terminal heat tolerance, Iron rich and excellent chapati quality

Figure 10.104. HI 1620 (Pusa Wheat 1620) [NAC/TH.AC//3*PVN/3/MIRLO/BUC/4/2*PASTOR/5/KACHU/6/KACHU]

North Western Plain Zone, Timely Sown, Restricted irrigation

Plant height: 95-100 (cm); Days to heading: 90-95 days; Days to maturity: 145-150 days; Ear character: Shape-Tapering; Length-Long; Density- Medium; Leaf colour: Green; Awn attitude and colour: appressed & white, Waxiness-Strong; Grain character: Colour-Amber; Texture-Semi-hard; Shape-Ovate; 1000-grain weight: 40-45 g; Resistant to: Stripe, Leaf rust, Leaf blight, Karnal bunt, *Fusarium* head blight, loose smut and flag smut; Rust resistant genes postulated- *Lr13+10+3+, YrA+, Sr11+7b+*; Tolerant to- Drought and Terminal heat tolerance; Quality traits: Protein (%): 11.88%; Sedimentation value (ml): 52.2; Hectoliter weight (kg/ha): 78.9; Hardness index: 82; Glu 1 subunit:2*; Glu score: 10/10; Grain Zn : 33.2 & Fe: 35.5 (ppm) content; suitable for making : Chapati and biscuit ; Grain yield (t/ha): Average : 4.91, Potential: 6.18 and Drought & terminal heat tolerance, excellent chapati quality

and prosper". Albeit there have been few private seed industries dealing with production of seeds, the growing of crops especially for seeds in an organized fashion to maintain quality in terms of genetic and physical purity is realized for first time during green revolution period with the establishment of National Seeds Corporation (NSC) in 1963. It is setup by aiming at promoting healthy development of seed industry in India. The principle responsibilities of NSC are establishing an adequate system of quality control inspection for scientific processing, storage and marketing of seeds. Currently there are about 850 seed companies (mostly seed producers) operating in India in 2014, of which about 50 have capacity in crop breeding research. It also undertakes the responsibility of multiplication of seed of pre released varieties and production of foundation seed of varieties. The SRR in crops has gone from 20 percent in eighties to >90 per cent. Out of this total seed market including Open Pollinated varieties is Rs. 4000 crore. Out of 110 hybrids released by AICRP, 60 percent are from Pvt. Sector. Indian seed Industry is currently occupying the 5th position. During the past 5 years the Indian Seed Industry has been growing at a CAGR of 12 per cent compared to global growth of 6.7 per cent. Few mile stone events in seed policy by Govt. of India are 1. The Seeds Act, 1963; 2. National Seeds Project; 3. New Seeds Policy, 1988; 4. PPV and FRA, 2001 and 5. The Seeds Bill, 2004.

They are genetically pure (true to type). The good quality seed has high return per unit area. Less infestation of land with weed seed/other crop seeds. Less disease and insect problem. Uniform in plant population and maturity. The quality seed respond well to the applied fertilizers and nutrients. Good seed prolongs life of a variety. They are vigorous, free from pests and disease. High produce value and their marketability. Crop raised with quality seed are aesthetically pleasing. It is estimated that good quality seeds to improved varieties can contribute about 20-25 per cent increase in yield.

Seed Replacement Rate

Seed Replacement Rate is the percentage of area sown out of total area of crop planted in the season by using certified/quality seeds other than the farm saved seed. Present Seed Replacement Rate for different crops for the year 2019-20, Seed Replacement Rate for Wheat: 13 per cent, Paddy: 19.16 per cent, Maize: 24.41 per cent, Jowar: 26.71 per cent, Bajra: 51.02 per cent, Gram: 7.09 per cent, Urd: 20.48 per cent, Moong: 19.48 per cent, Arhar: 13.60 per cent, Groundnut: 5.5 per cent, Rapeseed and mustard: 66.96 per cent, Soybean: 15.58 per cent, Sunflower: 19.61 per cent, Cotton: 37.25 per cent, Jute: 68.49 per cent for the country.

Seed Multiplication Ratio

It is the number of seeds to be produced from a single seed when it is sown and harvested.

SMR = Seed Yield/Seed Rate.

Table 10.1. Seed multiplication ratio for some crops including wheat

Crop	*Seed Multiplication Ratio*
Wheat	1: 20
Paddy, Maize	1: 80 (Varieties) 1: 100 (Hybrids)
Sorghum, Red gram, Rapesed-Mustard, Jute	1: 100
Bajra	1: 200
Gram, Berseem	1: 10
Blackgram, Greengram, Cowpea	1: 40
Potato	1: 4
Groundnut	1: 8
Soybean	1: 16
Sunflower, Linseed, Cotton, Clusterbean	1: 50
Sesame	1: 250
Lucerne	1: 25
Oats	1: 15
Peas	1: 19

Indian Seed Scenario

For a long time, seed production in the country was mainly depends on the farmers who used to save seeds for their own use as well as for nearby farmers of the locality. The Royal Commission on Agriculture (1928) was given stress on increased breeder seed production and varietal purity and suggested to have a separate organization within agriculture department to deal with seed distribution and testing. In 1945, the private vegetable seed companies had developed facilities for producing seeds of temperate vegetables in Quetta and Kashmir valley; these companies formed an All India Seed Producers Association in 1946. A standing committee on seeds was constituted in 1952 to formulate a programme for strengthening seed production and distribution in the country. Systematic seed production based on scientific principles was initiated by formation of National Seed Corporation (NSC) at central level in 1963 and state seed corporations in state level with the primary responsibility of foundation seed production of varieties of national importance. After the introduction of All India Coordinated Crop Improvement Project (AICRP) in maize during 1957 and sorghum and pearl millet during 1960, led to development of several hybrids which necessitated the enactment of seed legislation. In 1966, Seed Act was passed to regulate the sale of seed in the country. One of the landmarks in the history of seed programme of India has been the launching of AICRP on seed called "National Seed Project" in 1979, with 14 centres in different Agricultural Universities. In 1991, another AICRP on "Seed Borne Diseases" was merged with it and now this project has 35 "Breeder Seed Production" (BSP) centres and 23 "Seed Technology Research" (STR) centres in the country at various Agricultural Universities/ICAR institutes. With upgradation of Project Coordination Unit, the project is now coordinated by Directorate of Seed

Research, Mau. Under AICRP-NSP (Crops), adequate quantity of nucleus and breeder seeds of high quality as per national requirements is being produced. The research on various aspects of seed technology including seed production, drying, processing, storage, value addition *etc.* is being conducted throughout the country. The basic information on seed certification standard including seed health has been generated. The information and technologies on seed production, processing, storage, packing, quality control and seed health were disseminated throughout the country *via.* SAU's and ICAR institutes. The strength of the system lies in the excellent infrastructural facilities, well equipped laboratories, excellent expertise and trained manpower. The progress of Breeder Seed Production (BSP) during 2009-10 was spectacular touching a production level of 115866.81q. This is a record production so far. The mis-match between indent and production has come to a very low level Directorate of Seed Research even in crops like pulses, groundnut, soybean, *etc.* Under Annual Oilseed Scheme, production of breeder seed was 23370.84q against an indent of 17519.80q in groundnut, 7163.75q against 11901.20q in soybean and 68.01q against 18.58q in sunflower, respectively. There is hardly any other similar programme in the world where such diverse activities like breeder seed production, seed technological research, research on morphological, biochemical and molecular characterization of varieties, quality maintenance of nucleus and breeder seed, techniques for seed certification and maximization of hybrid seed production are carried out under same umbrella. The new policy on seed development of 1988 heralded a new era of private enterprise in the seed sector in India. Under 1988 policy, vegetable seeds could be imported freely while seeds of oilseeds, pulses and coarse grains like maize, sorghum and millet could be imported for two years by the companies having collaboration agreement with foreign company. With the view to further strength seed production infrastructure, ICAR launched another mega seed project captioned "Seed production in Agricultural Crops and Fisheries", the country's biggest seed project ever. The project was launched during 2005 -06 of X five year plan (FYP) involving 85 cooperating centers of SAU's and ICAR institutes. The major achievements during last three years in this project in terms of quality seed production is as under:

Table 10.2. Seed replacement rate of major crops

Crop	*SRR Per cent*	
	2007-08	*2008-09*
Wheat	25.23	26.84
Paddy	36.77	30.05
Maize	44.24	48.48
Jowar	19.87	26.16
Bajra	48.47	62.92
Ragi	30.34	-
Barley	16.84	-

In the entire production technology gamut, low availability of quality seed at affordable price in right place and time is a serious constraint to bring about rapid

Table 10.3. Breeder Seed Requirement of Wheat, Rice and Maize in India

Crop	*2009-10 Breeder seed production (in q.)*	*A*	*B*	*C = (AXB)*	*D*	*E = (CXD)*	*F*	*G = (E/F)*	*H = (G/FX1000)*
		Gross Cultivated Area (2008-09) (M ha)	*Seed Rate (Kg/ha)*	*Total certified seed requirement (in 000' tonnes)*	*Target SRR (per cent)*	*Certified seed requirement *(in 000' tonnes) as per targeted SRR*	*SMR (Ratio)*	*Foundation seed Requirement (in 000's tonnes)*	*Breeder seed requirement (in tonnes) (2030) ¥*
Wheat	34893.55	27.2	125	3400	0.5	1700	20	85.0000	4250.0000
Rice	20774.7	45.6	50	2280	0.5	1140	100	11.4000	114.0000
Maize	458.25	8	25	200	0.75	150	80	1.8750	23.4375

Calculated on the basis of 25 per cent replacement in self pollinated and 35 per cent replacement in cross-pollinated OPV's for the year 2010.

transformation in our rural economy. Since "Green Revolution , the productivity in major crops are increasing, however much more increment is essential to feed the people. The Indian agriculture is at the crossroad though it has made significant strides by estimated production of 235.88 million tonnes of food grains during 2010-11. To bring about everlasting second green revolution, quality seeds are deemed to be the major propelling factor. The estimated breeder seed requirement in wheat, rice and maize in India up to 2030.

Basic Principles of Quality Seed Production

Seed production is an exacting task. During seed production strict attention must be given to maintenance of genetic purity and other qualities of seed in order to exploit the full dividends sought to be obtained by introduction of new superior crop plant varieties. It is very difficult for farmer to get quality seed of recommended high yielding newly released wheat variety.

Quality Seed

A quality is one which is genetically pure, physically pure, physiologically active and is healthy as per prescribed seed certification standards. To maintain the quality of seed basic principles may be grouped into genetic and agronomic principles.

Genetic Principles

Genetic purity of a variety can deteriorate due to several factors during production cycles. The important factors responsible for genetic deterioration of varieties as listed by Kadam (1942) are: Developmental variations, Mechanical mixture, Mutations, Nature crossing, Minor genetic variations, Selective influence of diseases, The technique of plant breeder.

Developmental Variations

When the seed crops are grown in difficult environment, under different soil and fertility conditions, or different climatic conditions, or under different photoperiods, or at different elevations for several consecutive generations, the developmental variations may arise sometimes as differential growth responses. To minimize the opportunity for such shifts to occur in varieties, it is advisable to grow them in their areas of adaptation and normal growing seasons.

Mechanical Mixture

Mechanical mixture may take place at the time of sowing, if more than one variety is sown with same drill or through different varieties grown in adjacent field. Two varieties growing alongside each other in the same field are often mixed at the time of harvesting and threshing operation. Threshing equipments (*i.e.* threshing machine) is often contaminated with seeds of other varieties. The gunny bags, seed bines are also quite responsible for mechanical mixture of varieties. To avoid mechanical mixture, it is necessary to rogue the seed fields and care should be taken at the time of harvesting, threshing and handling.

Mutations

Though chances of mutations are very rare (10-6) but if noticed in a seed crop, the plants should be rouged out as soon as visible. *e.g.* 'rabbit ear' in peas.

Natural Crossing

In sexually propagated crops, natural crossing is most important source of varietal deterioration. The deterioration in varieties due to natural crossing may occur due to natural crossing with undesirable type, natural crossing with diseases plants or natural crossing with off-type of plants. According to Bateman (1947) genetic contamination in seed fields due to natural crossing depends upon factors like the breeding system of species, isolation distance, varietal mass and pollinating agents. As the isolation between varieties is increased the contamination decreases. Isolation of seed crop is the primary factor in the seed production of crop plants of cross pollinated by wind or insects and their activities, humidity and temperature at the time of anthesis *etc.*

Minor Genetic Variations

Minor variations exist in phenotypically uniform and homogeneous varieties at the time of their release which are eliminated in later production cycles by the selective influence of environment. Multi-location yield trials before release of the variety in self-fertilized crops (Hann, 1953) and due care during production of nucleus and breeder's seed of cross-fertilized crops will be helpful in minimizing such variations.

Selective Influences of Diseases

New crop varieties often become susceptible to new races of diseases often caused by parasite. Some vegetative propagated stocks deteriorate fast it infected by viral, fungal and bacterial diseases. Therefore, it is very important to produce diseases free-seeds/stocks.

Agronomic Principles

Selection of Agro-Climatic Region

A crop variety to be grown for seed production in an area must be adapted to the photoperiod and temperature conditions prevailing in that area.

Selection of Seed Plot and Land Preparation

The plot selected for seed crop must be free from volunteer plants, weed plants and have good soil texture and fertility. The soil of the seed plot should be comparatively free from soil borne diseases and insects pests. Good land preparation helps in improved germination, good stand establishment and destruction of potential weeds. It also aids in water management and uniform irrigation.

Isolation of Seed Crops

The seed crop must be isolated from other nearby fields of the same crops and other contaminating crops as per requirement of the certification standards. Contamination can be avoided through different methods:

Space Isolation

By providing minimum specified distance from the sources of genetic contamination so that foreign pollens are unable to reach in viable state.

Time Isolation

By adjusting the time of sowing *i.e.* keeping a gap of 15-20 days between sowing of varieties/cross-compatible crops.

Barrier Isolation

Isolation can be achieved by physical barriers such as barrier crop/polythene sheet of optimum height around the seed plot, polyhouse/polytunnels/net houses and mountains which reduce the movement of pollen.

Discarding Border Rows

Foreign pollens mainly pollinate the plants present on the outer periphery of the seed plot. Therefore, contamination can be reduced by discarding the produce of 5-6 lines on the outer periphery.

Different Stages of Rouging

Before Flowering

On the basis of vegetative characters (plant growth, foliage morphology, colour *etc.*) the off types are removed from seed production field.

At Flowering

The early and late varieties can be easily identified on the basis of maturity.

At Maturity

The plants showing late maturity in the early variety and vice versa should be removed immediately from seed production field.

Threshing and Seed Extraction

It varies from crop to crop. Threshing can be done by hand or machines. Threshing machines should be properly cleaned to avoid admixture.

Seed Standards

It refer to the field inspection of the harvested produce as well as the manner of harvesting, transporting, processing and packing. Unless, a seed certification agency keeps track of harvested produce until it is packed and sealed the identity of the lots cannot be assured. It is, therefore, necessary that the seed certification agency should lay down standard for processing plants. In addition, field and seed standards, such as isolation distances, inseparable other crop seeds, weeds, plants affected by seed borne diseases, genetic purity, percentage of pure seed, other crop and weed seeds, inert matter, moisture content, germination and insect damage, should be prescribed for successful accomplishment of the certification.

Steps in the Production of Breeder Seed

Land Requirement

The land that is to be put to seed production should be free from drainage problems and should be well leveled. At higher elevation, specifically in the foothill areas, emergence of volunteer wheat plants is a problem. In such cases, adequate care is to be taken to ensure that such volunteer plants do not come in the way of production of good quality seeds and therefore, the fields with previous crop as wheat should be avoided. Also care should be taken to avoid production of seed in field which has a history of tundu disease infestation.

Methods of Sowing

A low seed rate of 80 kg/ha is recommended with lower nitrogenous fertilizers to avoid lodging of the crop. Seed production plots being large in size are to be sown using a seed drill. Therefore, the following care should be taken before sowing of breeder seed. The seed drill needs a thorough cleaning of the pipes, seed cups and bin. Any left back seed will contaminate the seed production plot. To ensure that physical mixture due to negligence at sowing time does not occur. It is recommended that when seed production of more than two varieties is taken, phenotypically contrasting varieties should be sown alternately. This will help in easy detection of seed drill resultant rogues and can be easily identified and pulled out from the field. In order to walk through the plots and effectively execute rouging, keep one row unsown after every eight-ninth row. One direction sowing always helps in easy cultural operation and monitoring *etc.*

Isolation

Breeder seed plot should be isolated from other wheat plots by a minimum distance of three meters, and no loose smut infected wheat, triticale or rye field should be there within 150 meters. The precaution is necessary to ward-off infection by seed borne pathogen.

Rogueing

Rogueing of seed production plots is done to remove the off-type plants that may arise due to segregation of residual heterozygosity, out-crossing with other varieties, admixtures or mutations *etc.* Aneuploidy is also a common problem in some semi dwarf wheat varieties. Off-types are identified based on the variation noticed for 'DUS' features of a variety like auricle pigmentation, days of flowering, plant height, waxy bloom, ear colour, ear shape, ear density *etc.* It is suggested to perform at least three times rogueing *i.e.* one each at early vegetative growth, 75 per cent ear emergence and maturity. The rogued plants, particularly after seed development, should be removed from the field and disposed at one place far away from the seed production plot to avoid chances of mixing of the pulled out plants with general seed bulk.

Grow Out Test and Monitoring

To monitor the genetic purity of the breeder seed plots, a sample of basic seed lot is drawn. The samples, thus collected from different varieties and from different location are grown at one place and kept under constant supervision for any type of rouge plant emergence or deviation from original variety. Such information is conveyed to the concerned seed producer for initiating appropriate action at their end. The breeder seed plots should be monitored by the monitoring team comprised of producing breeder, breeder from other institute, a member each from NSC and SSCA at two growth stages *i.e.* at vegetative stage and at ripening stage.

Harvesting and Post-harvest Handling of Breeder Seed

Extra care is needed to avoid mechanical mixing that is likely to occur during harvesting, threshing, seed treatment, packing and all through the processing steps. These stages are to be foolproof from potential seed mixture. Threshers, combine harvesters, trailers, processing machinery *etc.*, should be thoroughly cleaned before harvesting. Seed production plots of more than 2 ha are suited for combine harvesting by machine and plots that are smaller can be harvested manually and threshed by *machine*.

Seed Treatment

The seed treatment with Vitavax 75 WP @ 2.5 g/kg or Raxil 2 DS @ 1.25g/kg seed with seed dresser should be done to check loose smut and other seed borne diseases such as hill bunt. If seed production is done in the hills, to check hill bunt seed treatment is strongly recommend.

Storage

The seed moisture of the harvested seed plots should be measured by drawing adequate true samples. Generally, at harvesting time the seed moisture is around 14 to 15 per cent and at such moisture level the seed lots should not go to warehouse. The harvested bags need to be kept in shade at farm till seed moisture comes down to 9-10 per cent. This process may take nearly three weeks time, however, this can be accelerated by the use of seed driers. Low seed moisture helps in increasing seed viability and decreases damage chance due to storage pests and fungal infestation. During storage in ware house/silo, samples must be drawn at periodic intervals to measure grain moisture, inspect for pest/fungi damage. As the situation may demand action must be taken either to reduce seed moisture levels or fumigate seed to suppress insect pests incidence. Occasionally, rodent control also becomes necessary under poor storage conditions. Consult a storage entomologist for precautions, dose and usage of chemicals/fumigants to be used to combat storage pest's infestation.

Seed Purity Standard

The quality of seed is examined through a series of tests. Generally such tests are conducted in seed testing laboratories designated for this purpose. Every state of the country has at least one seed testing laboratories to conduct research on seed testing methods, training of personnel, determining the standards for seed purity and seed quality.

Maintenance Breeding, Variety Maintenance and Seed Production

Breeding programme results in the production of improved cultivars in crops. At the time of variety release a small quantity of seed is available with the breeder. The relatively small amount of seed of improved cultivar needs to be multiplied and made available to farmers as quickly as possible. During seed multiplication, varietal purity and identity needs to be maintained. Each multiplication cycle starts from the 'breeder seed'. If the breeder seed is not of high purity, the contaminants present get multiplied several times in the succeeding generations of foundation and certified seed production. The presence of contaminants may even lead to complete loss of the improved features of the cultivar. Prevention of contamination is at the heart of a successful breeder seed production programme. During seed multiplication process, several factors may reduce the genetic purity of seed. The maintenance of genetic purity is largely dependent on the genetic makeup of the cultivar. The diversity of morphological types is normally greater in predominantly cross pollinated crops *etc.* than self- pollinated crops. Genetic variation may appear within a seed stock for a number of reasons including mechanical contamination, hybrids resulting from undesirable pollination, recombination, residual segregation, and mutation. These factors ensure that no cultivar is likely to retain the precise frequencies established by the breeder without continuous intervention through maintenance breeding techniques. The occurrence of contaminants during seed multiplication process occurs due to:

Out-crossing

Out-crossing crops such as onion, brassicas, cucurbits, onion have substantial outcrossing percentage and even self pollinating crops like common beans show a low percentage of out crossing. In each generation many new genetic combinations appear due to uncontrolled open-pollination. These new genotypes recombine and segregate into a range of new genotypes, thereby enlarging the genetic variations within a cultivar.

Residual Segregation

Some characters have not been selected during breeding or in subsequent purification and production may remain in a population in a range of states. Exposure to new environments may allow expression of states of selected characters not detected previously.

Volunteer Plants

Volunteer plants may arise from vegetative pieces or dormant seed remaining from previous crop grown on the same field. If the volunteer plants are of the same crop to which the cultivar to be maintained belongs will enlarge the genetic variation once harvested together.

Mechanical Mixture

Mechanical mixing with seed from another cultivar of the same crop may occur during sowing, harvesting, threshing, cleaning, storage, bagging, *etc.*

Mutation

During maintenance and production spontaneous mutations do occur and may reduce the agronomic value of the cultivars. The mutation rate is generally low but increases significantly after long periods of seed storage. Since mutations are usually micro mutations and recessive, they are often difficult to detect. However, mutations are probably the least important degrading factor contributing to genetic contamination.

Seed Borne Pathogens

Pathological contaminants occur due to infection with seed borne diseases, which are exclusively transmitted through seeds. Pathogens like viruses in tuber crops or anthracnose in beans can spread rapidly, if not contained. Pathological contaminants result in loss of yield and quality of cultivars and also gradual loss of tolerance of the cultivar to specific seed borne diseases. The above mentioned factors collectively result in genetic variation and these changes accumulate over the years resulting in loss of varietal identity and genetic gain.

Maintenance of a Cultivar

Once a cultivar is released for cultivation, the breeder usually supplies a small quantity of seed for further multiplication and maintenance. The responsibility of breeder seed production centre is to produce breeder seed and varietal maintenance. In order to release seed of an improved cultivar to farmers, it has to be multiplied. Each multiplication cycle has to start from its basic seed stock, 'Nucleus Seed'. Our basic objective of cultivar maintenance is to maintain the purity and identity of a cultivar. The maintenance procedures are in fact the extension of the normal breeding process. The difference is that during maintenance breeding, selection process is relatively mild and our aim is not improvement but to keep the identity unchanged. Selection should maintain the plant type, its uniformity and freedom from diseases. The maintenance procedures for self pollinating crops with a substantial amount of out-crossing are slightly different from cross-pollinating crops. The fields where plants and progenies are to be assessed should be uniform. Essentially these should be grown under optimal growing conditions.

Cultivar Maintenance

Plant to Row Method

The maintenance procedure starts with a small plot raised from the parental material received from the breeder or uniform seed multiplication field in case of established cultivars. A fair number (300-500) of healthy plants typical of the cultivar are selected and marked for progeny testing. The seeds of the marked *true- to- type* plants are harvested separately. The seeds of each plant are planted in a 3m long progeny row. These progeny rows are assessed critically several times during the growing season. Progeny rows that deviate in one or other characteristics are discarded and entire plant progeny rows is rejected. The plant progenies that are uniform and true to type are selected and bulked together as nucleus seed stage-I. This nucleus seed is used for planting larger breeder seed plots. If the breeder seed requirement of a particular cultivar is more, then another cycle of nucleus seed production is followed. In this case the *true- to- type* individual plant progenies are harvested and thrashed separately. Seed of each selected plant- row progeny is now sown in a small plot called plant- row- progeny plot. The second cycle of nucleus seed production provides another opportunity to eliminate any plant row progeny showing segregation or off-type plants. The plots with required uniformity and plant type are bulked together to produce nucleus seed stage-II. The product of this second cycle of progeny testing is sufficient to meet the higher requirements of breeder seed particularly. Many modifications of this scheme are possible. For example in garden pea, where 150 typical plants are selected from seed multiplication plot. These plants are harvested and threshed separately. Next season 150 progeny- rows are planted and critically examined for cultivar characteristic during the entire growing season. The off-type rows and less promising lines about 20 per cent are discarded. Among the remaining 120 progeny rows, the 30 best progeny rows are selected. Ten best plants showing better production are selected from each of the 30 best plant progeny rows. Out of these 10 plants, five best plants showing higher number of seeds per pod are selected. These 5 X 30 plants are used to produce plant progeny rows next year. The remaining seed of 30 best progeny rows is harvested separately and planted as 30 bulk plots next season.

Negative Mass Selection

Part of the basic seed field is given special attention. The undesirable premature bolters and diseased plants are removed. Any doubtful plant not conforming to cultivar description should also be removed. Isolations from related crops should be maintained. The left over plants are harvested and bulked together and is utilized for breeder seed production. In this method, there are chances of cross contamination from offtype pollen grains from undesirable plants. In case of melons nucleus seed is produced by mass selection. A better method is rest seed method.

Rest Seed Method

From the basic seed plot, select 300-500 true-to-type plants based on the cultivar characteristics. The selected plants are harvested and threshed individually. Next year a small part of each progeny (about 200 best plants) is planted in rows and the remaining seed of each progeny is stored. Each plant-row progeny is critically examined for cultivar characteristics and rows showing off type plants are marked and rejected. The remnant seed of the true-to-type plant rows is bulked to form the nucleus seed and is used to raise the breeder seed plots next year. All off-type plants are removed from the progeny rows before flowing until harvest. Again 300-500 best plants are selected for next maintenance cycle. Repeated maintenance, selection results in a continued improvement over time provided the number of progenies is kept fairly large.

Seed Production Technology

Quality seed is one of the important inputs for increasing productivity. Seed is the basic starting point of most of our crops although some are propagated vegetatively from cuttings of tubers. The seed is, therefore, basic for all inputs such as fertilizers, crop protection, irrigation, harvesting and marketing. Thus, it is essential that seed of highest possible quality may be made available to the farmers. Quality of seed includes; genetic purity, physical purity, germination potential and freedom from pest and diseases to produce seed with all these quality attributes, a systematic seed production is must. During the seed multiplication process, the following points are to be kept in mind for obtaining high yields of quality seed with low cost of production.

Environmental Requirements

Crops for seed production should be grown under best possible conditions. Best seed yield and quality is obtained when crop is grown in correct season and in an area where the crops' reaction to appropriate environment can be observed. The regions with abundant sunshine should be selected for seed production. Avoid areas of high rainfall as it may reduce seed viability and increase need for artificial drying. Excessive wind will increase water loss, prevent activity of pollinators, carry wind borne pollen over long distances and increase seed shattering.

Land Requirements

The field selected for raising a seed crop should be free from 'volunteer' plants. Volunteers mean the plants originating from the seed/plant material of the previous market or seed crop. In volunteer plants are seen. The land should be leveled with proper drainage and should have sufficient organic matter. The cultural operations for seed production are similar to the operation on market crop but seed crop is to be maintained beyond the stage of commercial market harvest. Therefore, the control of diseases, pests and weeds should be continued over a longer period.

Pollination Requirement

The natural insect population is normally sufficient under open conditions to ensure satisfactory pollination but high plant population grown for seed in a concentrated area, there is a possibility that natural insect population may be insufficient to ensure proper seed set. Therefore, the introduction of supplementary bee hives will improve pollination and seed set. However, care must be taken to ensure that pest protection chemicals are not used in a way to harm useful pollinating insects. Spray of chemicals should be avoided at peak pollination insect activity and should be done in the late afternoon.

Isolation Requirements

Satisfactory isolation of seed crop helps to maintain purity in three ways, *viz.*, Cross-pollination does not occur between cross compatible crops; During harvesting seeds of different varieties of same crop are not mixed; The transmission of pest and diseases from alternative host crop are minimized

Proper isolation is thus essential to maintain genetic purity and health of a variety. Isolation between cross compatible varieties is achieved as follows.

Isolation by Time

Isolation by time will allow seed of different varieties of the same crop to be produced at the same station each year. If the season is too long enough to allow two production cycles of the cross compatible crops, then they too are isolated by time. For example, early and mid maturity group of cauliflower grown for seed production can be isolated by time.

Isolation by Distance

The mode of pollination is related to the degree of isolation necessary. In case of self-pollinated varieties, the isolation distance is relatively short but, in case of cross-pollinated varieties the distance from other variety should be relatively wide. The isolation distance also depends on the direction of insect flight (in case of insect pollinated varieties) or the direction of winds (in case of wind pollinated varieties).

Roguing

The varieties, which are produced for seed show genetic change over several generations. It is, therefore, necessary to exert control and keep the natural variation within the acceptable limits. This is achieved by inspecting the crops at various growth stages and removing individual plant which do not confirm to the defined limits of that variety. Thus rouging is a technique that is used in seed production to maintain genetic purity of the variety. Rogues or offtypes may occur in a crop due to manifold reasons. The diversity of the morphological types within a crop may be wide. This tendency is greater in predominantly cross-pollinated than self-pollinated crops. This is why varieties of self-pollinated crops are generally more uniform and stable than varieties of cross-pollinated crops. The seeds that result from cross-pollination between the crop for seed production and other compatible varieties or wild plants. These are not always identified in the first generation and show up in the second. Some plants may display deviation from the normal type due to mutation. Seeds of other varieties may have been accidentally mixed in the seed stock during its production, processing or admixture via seed drill. Volunteer plants may arise from vegetative pieces or dormant seed of the previous crop grown in the same field. It is always easier to conduct intensive rouging in breeder seed plots than in large commercial seed production plots. To obtain maximum benefits from rouging operation, we should follow the below mentioned practical points. The crop should be grown in such a way that plants can be seen individually. Paired row system of planting may be followed so that it is easy to walk between rows. This shall facilitate detection of dwarf undesirable plants. Walk systematically through the crop so that each plant is seen. Remove the whole off-type. Do not simply remove the fruits showing undesirable character because the remaining flowers on the off-type plant will still contribute to the material in the next generation. Inspect the crop with the sun behind you as it is difficult to examine plants with the sun on your eyes. Do not delay field inspection. The undesirable plants should be removed before flowering as far as is possible. Remove cross-compatible weeds and wild relatives. Remove all diseased plants and related infected weeds also. Variety description based on morphological characters, generally from a good basis of rouging but some characters like leaf colour, plant height, earliness are affected by environment.

Harvesting, Threshing

The best time of harvesting seed crops is at a stage when the highest yield of best quality seed will be obtained. Seed has to be extracted from dry seed heads, or from dry fruits or from fruits in which the seeds are wet at the time of extraction. Threshing can be done by hand, animal or machines. Care should be taken while transportation of material from the field to threshing floor. Both the trolley and the threshing floor should be clean from the seed/plant parts of the other varieties of the same crop or weeds to avoid admixture at this stage. Threshing machines must be used with care. They should be run at a reduced speed to avoid mechanical damage to the seed. Threshing machines should be properly cleaned to avoid admixture.

Seed Drying

At the time of harvest, the seed contains frequently higher moisture content than the optimum for better germination and storability of seed. Other seeds have relatively low moisture at harvest. Sometimes due to adverse climatic conditions, seed may also have high moisture content. The seed moisture should be reduced to optimum level before storage. For ambient storage seed moisture should be kept under 9-12 per cent and for sealed storage it should be 6-8 per cent. Natural and artificial methods are used in seed drying.

Seed Quality Control

Seed quality reflects the overall value of the seed for intended purpose. Poor quality seed leads to loss of money and potential crop. Seed quality consists of physical purity, germination, potential, genetic purity and seed health.

Genetic Purity

Trueness to type can be assessed by grow-out tests. In addition, biochemical and cytological methods can also be used to ascertain the variety purity. Seed Certification scheme can effectively control genetic purity of seed crops.

Multiplication of Seed of the Released Variety

When an entry is identified by the workshop of the concerned project, the breeder begins seed multiplication of that strain in the following crop season. The seed produced by the breeder after a strain is identified but before it is released as a variety, is termed as the stock or basic seed. The stock seed itself is known as breeder seed once the identified strain is released and notified. Multiplication and distribution of high-quality seed of improved varieties is a crucial step in the agricultural production of any country. In the crop season following notification, foundation seed of the newly released variety is produced. In the second crop season following the release of a variety, certified seed of the variety is produced. Thus, the farmer is able to obtain certified seed of a newly released variety for commercial cultivation only in the third crop season following the release and notification of the variety.

Classes of Improved Seed and Seed Tags

Seed production is a continuous process and each year a huge quantity of seed is required to meet the demands of crop production. Seeds of improved varieties are produced in several stages, each stage generating a particular class of seed. The various classes of improved seeds are recognized in order to: 1) facilitate seed production while maintaining genetic and physical purities, and 2) to ensure a continuous supply of quality seed at a reasonable price. The four generally recognized classes of seeds are: Breeder's seed, Foundation seed, Registered seed and Certified seed. The Association of Official Seed Certifying Agencies (AOSCA) has defined these seed classes. Three basic types of tags can be found on seed bags, *viz.*, Certification tags, Analysis tags, and Treat tags. It is important that the producer retain the tags from each lot of planting seed.

Certification Tags

Good crop production begins with high-quality planting seed. Standards may vary from crop to crop, but a certification tag on a bag of seed is the symbol of quality. It assures the purchaser that the seed inside the bag is the variety stated and it has met the standards for germination and purity. Each class of certified seed is associated with a colored seed tag. Certified seed tags are blue, registered are purple, and foundation tags are white. The certification class refers to the number of generations away from the original variety that was developed by the plant breeder. The first class is Breeder - the seed provided by the breeder of the new plant variety. Progeny from Breeder seed is the Foundation class, or one generation removed from Breeder seed. The Registered class is progeny of Foundation seed, and two generations removed. The final class is Certified, and it may be the progeny of Registered seed or Foundation seed. This is the seed most generally available to farmers. The progeny of Certified seed must enter the commodity market and may not be used for planting seed. This system ensures that planting seed is mostly free of genetic and physical contamination. Information provided on the certification tag includes class of certification, kind of crop, variety, lot number, and name and address of the owner. The certification class is stated and color-coded. The foundation seed "white tag" appears only on seed that has been inspected in the field and meets purity standards for foundation seed. Foundation seed is priced higher than commercial seed since it is a premier grade and intended for increase and not commercial production. Lot number is for identification purposes and is used to trace seed to its originator. Lot number allows seed to be traced back to a specific grower and field. The producer whose name and address appears on the tag may or may not be the actual grower, but is the party responsible for seed contained in the bag.

Analysis Tags

All seed sold must have an analysis tag. Some certification tags also contain the analysis, but others contain only the basic information and there must be a separate analysis tag attached to the bag. The analysis tag that accompanies the certification tag provides information on the quality factors of the crop seed. A grower should read both labels to determine whether the seed meets his purity and germination requirements. Items included on the analysis tag are kind and variety, lot number, percent pure seed, percent other crop seed, percent inert matter, percent weed seeds, germination percentage, and date of test. Other items that may be present on the analysis, depending on crop and situation, are percent noxious weed seed, percent dormant or hard seed, and pure live seed. Kind and variety refer to the crop and stated variety. Lot number should also be identified with the seed lab test used to provide the purity and germination of the seed. Pure seed is the percent of the variety named in the bag. A high pure seed percentage is desirable. Other crop seed is the percent of crop seed not named on the tag. Many times seed may become mechanically mixed during harvesting, storing and processing, or through field contamination. A low other crop percentage is desirable. Inert matter refers to the percent of dirt, sand, stones, sticks, glumes, stems, broken seed, *etc.* Inert matter contributes nothing to crop yield but may create planting problems. Crops such as some grasses contain a relatively high percentage of chaff and empty florets, and thus may be marketed on the basis of pure live seed [PLS].

PLS = pure seed x [germination + dormant or hard seed]/100

Weed seeds in the seed should be very low. Any prohibited noxious weeds in the bag will stop sale of the seed. Germination is important to getting a stand. Test date is important, as normally a reported germination is reliable for six to nine months. The analysis tag will show the percent of pure seed that will produce normal plants under favorable field conditions. Producers should look for high germination and high pure seed when purchasing seed. PLS can be used to determine the seeding rate necessary for an acceptable stand. Hard or dormant seed is live seed that requires a longer time to germinate. Seed count describes the number of seeds per unit. Seed count claims help producers decide how many bags of seed they need to buy for the acres they are going to plant.

Treat Tags

When seeds are treated with poisonous substances such as fungicides or insecticides, the law requires that the bag be labeled to indicate this treatment. The treat tag will note the treatment used. The treat tags shown below are for example only. They do not necessarily imply endorsement of the product.

Basic or Nucleus Seed

It is the original or first seed of a variety available with the producing breeder or any other recognized breeder of the crop. It has 100 per cent genetic and physical purity along with high standards of all other quality parameters. Nucleus seed is multiplied in wheat in two stages *i.e.*, Nucleus Seed Stage I and Nucleus Seed Stage II. In Nucleus Seed Stage I, the true-to-the type ears are selected from a field of nucleus or breeder seed plot of the variety, their seeds are grown in separate progeny rows, and seeds from the true to the type rows are composited to constitute NSS I seed. When NSS II seed is to be produced, seeds from the true-to-the type progeny rows are harvested separately and grown in separate progeny plots. True-to-type plots are selected and their seeds are composited to obtain NSS II. Generally, NSS II is produced only when the demand for nucleus seed is high and cannot be met with NSS I. The current practice for the production of NSS I is based on selection of true-to-the type ear heads.

Breeder Seed

The seed or vegetatively propagated material directly controlled by the originating or the sponsoring breeder or institution which is the basic seed for recurring increase of foundation seed. Breeder seed is the progeny of nucleus seed and is the source of initial and recurring increase of foundation seed. Breeder seed production is directly controlled by the originating plant breeder who developed the variety, or any other institution or qualified breeder recognized by the authorities. Breeder seed is genetically so pure as to guarantee that the subsequent seed class (foundation seed) shall conform to the prescribed standards of genetic purity. Minimum seed standards are less stringent than those of nucleus seed, but they are more stringent than those of foundation seed. Breeder seed tag is golden in colour, which contains the information about seed quality.

Foundation Seed

It is the progeny of breeder seed. The seed stock handled to maintain specific identity and genetic purity, which may be designated or distributed and produced under careful supervision of an agricultural experiment station. This seed is the source of all other certified seed classes either directly or through registered seed. It is the progeny of breeder seed or certified foundation seed itself. When seed is the progeny of breeder seed, it is called foundation seed stage I, while it is called foundation seed stage II, when it is the progeny of the certified foundation seed. In any case, certified foundation seed stage II cannot be used to produce foundation seed; it can be used only for the production of certified seed. Bags of foundation seeds carry tags of white colour, which contain the information about seed quality.

Registered Seed

Registered seed is the progeny of foundation or registered seed. Registered seed maintains satisfactory genetic identity and purity of the variety for the production of certified seed. Registered seed is used as the source of certified seed in some states, or with some crops. In other states, or with certain crops, registered seed is not grown.

Certified Seed

It is the progeny of the foundation seed. Its production is so handled to maintain genetical identity and physical purity according to standards specified for the crop being certified. It should have the minimum genetical purity of 99 per cent. Certified seed may be the progeny of certified seed, provided this reproduction does not exceed two generations beyond foundation seed and provided that if certification agency determines the genetic and physical purity, if not be significantly altered. In case of highly self-pollinated crops certification of one further generation may be permitted. Certified seed produced from certified seed shall be eligible for further seed increase under certification, except in case of highly self-pollinated crops, where certification of one further generation may be permitted. Certification tags issued once for certified seed not eligible for further seed increase under certification. Certified seed is the progeny of foundation seed and its production is so handled as to maintain specific genetic identity and purity standards as prescribed for the crop being certified. The tag of certified seed is of blue colour and carries all the relevant information about the certified seed lot contained in the bag. The various classes of improved seed are recognized to facilitate the maintenance of genetic purity of the variety and to ensure a continuous supply of good quality seed at a reasonable cost. It also helps in the multiplication of the seed rapidly, while maintaining its purity. The production of breeder and foundation seeds is very costly since a very high standard of purity must be maintained. The requirements for certified seed are relatively less rigid than those for foundation seed, hence it is considerably cheaper. For wheat, certified seed produced from certified seed is eligible for certification by NSC up to two generations from foundation seed.

Table 10.4. Differences between certified seed and truthful labelled seed

Certified Seed	*Truthful Labelled Seed*
Certification is voluntary	Truthful labelling is compulsory for notified kind of varieties
Applicable to notified kinds only	Applicable to both notified and released varieties
It should satisfy both minimum field and seed standards	Tested for physical purity and germination
Seed certification officer, seed inspectors can take samples for inspection	Seed inspectors alone can take samples for checking the seed quality

Generation System of Seed Multiplication

Generation system of seed multiplication is nothing but the production of a particular class of seed from specific class of seed up to certified seed stage. The choice of a proper seed multiplication model is the key to further success of a seed programme which basically depends upon, the rate of genetic deterioration, seed multiplication ratio and total seed demand. Based on these factors different seed multiplication models may be derived for each crop and the seed multiplication agency should decide how quickly the farmers can be supplied with the seed of newly released varieties, after the nucleus seed stock has been handed over to the concerned agency, so that it may replace the old varieties. In view of the basic factors, the chain of seed multiplication models could be, Three - Generation model [Breeder seed - Foundation seed - Certified seed]; Four - Generation model [Breeder seed - Foundation seed (I) Foundation seed (II) - Certified seed] and Five - Generation model [Breeder seed - Foundation seed (I)- Foundation seed (II) -Certified seed (I) - Certified seed (II)].

Requirements for Certified Seed

The first and foremost requirement is that the seed must be of an improved variety released by either the central or a state variety release committee for general cultivation and notified to that effect by the Ministry of Agriculture, Government of India; this is essential for the seed to be certified.

Genetic Purity

It means the absence of seeds of other varieties of the same crop species as well as of other crop species. The standard of genetic purity is very high; the amount of contamination permitted ranges from 0 – 0.1 per cent. In most of the cases, contamination by seeds of other crop species is permitted to a small degree (upto 0.1 percent), but contamination by seeds of other varieties of the same crop is not permitted.

Physical Purity

It implies the freedom of seed from inert matter and defective seed. In case of wheat, the total amount of permissible contamination by inert matter and defective seed is about 2 per cent.

Germination

A high percentage of germination is necessary to obtain a good crop *i.e.*, more than 85 per cent in wheat.

Freedom from Weed Seeds

Freedom from weed seeds is necessary to prevent weeds from spreading through seed and to reduce losses caused by weeds. Maximum limit in wheat crop for weed seeds is 5 seeds/kg.

Freedom from Diseases

The certified seed must be free from seed borne diseases and it can be controlled either by effective control of diseases in the standing seed crops and/or by treating seeds with disinfectants and protectants. For example, in wheat, nematode galls of tundu ear cockle or Karnal bunt – 0.5 per cent by number.

Optimum Moisture Content

The seeds must be dried to an optimum moisture level for efficient processing and safe storage *i.e.*, 12 – 13 per cent in wheat.

Table 10.5. Germination and purity standards in wheat crop

Parameter	*Certified Seed*	*Foundation Seed*
Pure seed (Min)	98	98
Other crop seed (Max)	0.1	10/kg
Other varieties seed (Max)	-	-
Total weed seed (Max)	0.1	10/kg
Objectionable weed seed (Max)	5 seeds/kg	2/kg
Total inert matter (Max)	2	2
Moisture (Max)	12	12
Germination (Min)	85	85

Nucleus Seed Production

The maintenance breeding of a variety starts with the maintenance of its nucleus seed by the qualified breeder. The guidelines for the maintenance of nucleus seed are : Procure seed Basic Seed Stock' from variety breeder along with its 'DUS' features. Follow dibbling method for its sowing. Select 300-500 spikes/ear-heads on the basis of its described distinguishable features and thresh them individually. The collection of single ear-heads should be sufficiently large to avoid any type of genetic drift from the original variety. Examine the seeds for uniform colour, shape and size and reject the seeds of those lots where there is variation for these traits appear. The seed from selected ear-heads should be shade dried to bring down the seed moisture to about 11 per cent and packed air tight in small tin foil pouches. For relatively important varieties, it is even suggested to keep adequate quantity of these seeds in pouches at –4°C, to serve as seed bank. Such stored seeds retain viability for 5 to 7 years. This seed sample will constitute the material for Nucleus Seed Stage-I.

Nucleus Seed Stage-I (NSS-I)

In the next cropping season, the seeds obtained from each selected ear-head are space planted in rows of 3 m length in isolation of 5 m from other varieties and examined critically throughout the growing season particularly at sprouting, ear emergence, early dough stage and at maturity for their genetic purity. To enable proper inspection, do the sowing in paired rows by dibbling method. Reject the entire ear row that shows admixture or genetic deviation. To ensure the high-quality seed production ruthlessly reject the row even if a single plant in a row appears to be different. If a progeny has to be rejected after flowering, the both side progenies should also be rejected to avoid the anticipated outcross with the rejected family. Each ear-head progeny is to be harvested and threshed separately and stored in small cloth bag safely. From these ear rows, normally about 75-100 kg seed is obtained which is sufficient to produce NSS II seed in about 0.75 ha and around 25-30 quintals of breeder seed. Select fresh ear-heads (300-400) of uniform feature for successive crop season to continue the NSS- I cycle.

Nucleus Seed Stage-II (NSS-II)

Whenever the demand for breeder seed requirement is more than 25-30 quintals, it is necessary to follow NSS II. Even otherwise for a high-quality seed production, NSS II is followed prior to breeder seed production. Seed harvested from each ear row of NSS I is drilled separately in a plot of 3m x 1.38m size. The production of NSS-II stage may vary depending upon the nucleus seed requirement and availability of NSS I stage seed stock. These plots are re-examined for any type of deviation from its described distinguished features. This second cycle of nucleus seed multiplication provides yet another opportunity to eliminate any off-type in the ear head progeny, which might have escaped during NSS-I. The plots having any mixtures or deviation should be rejected and the retained plots are harvested and threshed separately. The seed of these plots, once again examined for seed colour, shape and size for uniformity. The uniform plots true to their genotype are harvested and bulked to constitute breeder seed stock.

Breeder Seed Production

The nucleus seed is multiplied in isolation under the direct supervision of a qualified plant breeder as breeder seed. The breeder seed production is highly

Table 10.6. Minimum certification requirements for field standard in wheat

Field Selection	*Field Standards*					*Specific Requirement*				*Limits*	
	Field Inspection	*Isolation Distance (m)*				*Off Type Plants (per cent)*		*Inseparable Other Crop Plants (per cent)*		*Seed Born Disease Infested Plants (per cent)*	
		Crop		*Disease*							
		F	*C*	*F*	*C*	*F*	*C*	*F*	*C*	*F*	*C*
Free of volunteer plants	Two	3	3	150	150	0.05	0.2	0.01	0.05	0.1	0.5

F: Foundation Seed; C: Certified Seed.

specialized job and calls for certain precautions and special attention. Although, the breeder seed does not come under the preview of Indian Seed Act (1966), but the breeder has to follow much more stricter minimum seed certification standards for nucleus and breeder seed production than what the Indian Seed Certification Board has prescribed for foundation and certified seed production.

Table 10.7. Minimum certification requirements for seed standard in cereal crops

Factors	*Standard*	
	Foundation	*Certified*
1. Pure seed (minimum)	98 per cent	98 per cent
2. Inert material (Maximum)	2 per cent	2 per cent
3. Other crop seed (Maximum)	10/kg	20/kg
4. Other distinguishable varieties (Maximum)	10/kg	20/kg
5. Total weed seed (Maximum)	10/kg	20/kg
6. Objectionable weed seed (Maximum)	2/kg	5/kg
7. Germination (Minimum)	85 per cent	85 per cent
8. Moisture per cent (Maximum)	12 per cent	12 per cent
9. For vapour-proof containers (Maximum)	8 per cent	8 per cent

Bagging Details

The breeder seed packed 40kg in gunny/cloth bag with necessary variety detail and agency that produced along-with the stitched in golden yellow label (No. 356) size 12x6 cm with following information completes the process. The rear of breeder tag should be rubber stamped with signature, name and designation of the concerned breeder. Packaging of seed in plastic lined gunny bag prevents seed quality deterioration due to dampness and wetness.

Label No.______

Crop

Variety

Lot No.

Date of Test

Physical Purity (Min)

Genetic Purity (Min)

Inert Matter

Treated with

Germination

Valid up to

Producing Institute (Name and Address)

Weight

Table 10.8. Different Class Seed Label colour and size

Class of Seed	*Label Colour*	*Label Size*
Breeder	Golden Yellow	12 cm x 6 cm
Foundation	White	15 cm x 7.5 cm
Certified	Azure Blue	15 cm x 7.5 cm
TL Seed	Opal Green	15 cm x 10 cm

Quality Seed Production

Seed Threshing, Cleaning and Drying

The dry seeds are gently mowed and winnowed to separate the seeds from chaffs. Over drying may damage the brittle seed. Seeds can also be threshed through rolling, threshing machine or combines. The cleaning is achieved with an air screen machine followed by passing through the gravity separator or by floatation. The winnowed and unclean seed be put in a bucket and soaked with clean water and left for not more than 3 minutes in the water, heavy seeds sinks and poor quality seeds and chaffs floats. The pure seed settled in a container should be taken immediately and dried under the morning and late afternoon sun or under shade for 3-4 days; should not be dried in the sun during noon time.

Seed Storage

The final seed lot must be dried down to moisture content not exceeding 12 per cent or lower depending on the method of storage and packaging. When seed has to be packed in porous containers (cloth bag/paper bag) then seed moisture should not be >8 per cent whereas, packaging in moisture proof containers, the seed moisture should be 5- 6 per cent.

Postharvest Managements for Improved Germination and Vigour

High seed germination and seed vigor are directly or indirectly related to the success of crop establishment and its further productivity. The quality seeds having good germination may help in reducing the seed rate per hectare from the existing 7 - 8 kg to 4 - 5 kg. De-Souzaa *et al.*, 2014 investigated the effects of pre-sowing magnetic treatments on germination, emergence, growth and yield of onion (cv.Red Creole) plants under field conditions. In field experiments, the treatments led to a significant increase in seedling emergence, root length, seedling height, seedling dry weight and leaf area per plant. Also, at the vegetative stage, the leaf and root relative growth rates of plants derived from magnetically treated seeds were greater than those of the control plants. At the bulb forming stage, bulb relative growth rates from magnetically exposed seeds were greater than those of controls. At the bulb maturity stage, all magnetic treatments significantly increased mean bulb weight, bulb yield per area, number of tunics per bulb, bulb diameter and dry bulb weight in comparison to the controls. Since the onion seed losses its viability and vigour rapidly thus the role of seed moisture and packing material is more vital for good productivity of crops. *Accordingly, Tripathi and Lawande (2014)* studied the effect of different moisture levels *i.e.* 5, 6, 7 and 8 percent and packing materials *i.e.* cloth bags, polyethylene bags, laminated aluminum bags and laminated aluminum bags with vacuum packing and stored at ambient condition for 27 months. The results indicated that germination percent and seed vigour Index were higher in seed having 5 per cent moisture than seed having 8 per cent moisture. Among the packing materials, lower seed germination and viability was recorded in cotton cloth bags. The seed packed in cloth bags lost their complete viability and vigour within 18 months of storage. The highest seed germination was observed in laminated aluminium bags with vacuum packing. The seed packed in aluminum laminated bags remained viable for 27 months. Among the various treatments combinations seed having 5 per cent moisture and vacuum packed in aluminums laminated bags remain viable for longer period with percent germination was 61.7 per cent germination 27 months of storage. The seed vigour index was also higher in this treatment than other treatments. Abdullah *et al.*, 2011 stored onion seeds under a wide range of temperature (5, 15, 25 and 35°C) and relative humidity (RH) (11.3, 22.5, 32.5, 43.2, 58.4, 75.3 and 84.3 per cent) conditions for various storage periods (1, 3, 6, 9 and 12 months) and recorded seeds stored at 5°C had the highest seed germination percentage (SGP) in shortest mean germination time (MGT). whereas, seeds stored at 35°C had the lowest SGP and the longest MGT. RH up to 58.4 per cent had no significant effect on SGP while higher levels of RH significantly lowered SGP and MGT. The highest RH levels (75 and 84 per cent) showed an obvious decrease in seed quality by lowering SGP and increasing MGT. Gabriel *et al.*, 1997 determine combined effects of the seed size (2.000, 2.375 and 2.750mm dia.) and the gravimetric classification on seed quality and yield components in onion cultivar and observed seed germination increased l in early as seed diameter was greater. Other crop variables such as plant height, commercial yield of bulbs and the percentage of bulb classes 5 and 6 also increased curvilinearly as seed diameter was greater. In the same way, the time from emergence till harvest shortened when seed diameter increased. In addition, the diameter of the bulb neck at harvesting was smaller as the seed diameter increased. According to these results, it is concluded that the performance on onion can be improved classifying the seed lots regarding the seed diameter.

Diseases Management

Among the diseases, purple-blotch, Stemphyllium-blight and Anthracnose causes severe losses, whereas, downy mildew prevalent in temperate regions and, pink-rot and *Fusarium* basal-rot also cause damage in onion growing areas. The severity of the diseases is influenced by the seasonal effect, variety and growing conditions.

Purple Blotch (*Alternaria porri*)

It requires 22-25 0C temperature and 90-100 per cent RH for sporulation and infection. It forms small, water-soaked areas on leaf or seed stalk and turns brown. The enlarged spot becomes zonate and more or less purplish. The margin is a shade of purple or red and surrounded by yellow halo that extends upwards and downwards for some distance. It can cause up to 80 per cent yield losses under Indian conditions.

Stemphylium Leaf Blight (*Stemphylium vesicarium*)

It occurs at same time and even on same plant as of purple blotch and forms the diseases complex. Its symptoms start with small, yellow to brown purple, water soaked lesions. These lesions elongate, become spindle shaped to ovate-elongate and turn into diffusive spot, often extending to leaf tips. The spots frequently coalesce into patches blighting leaves. As conidiophores and conidia develop on the lesions, turn light brown to tan purple at the center, and later dark olive-brown to black.

Anthracnose (*Colletotrichum*)

It mainly shows white sunken-oval lesions, which turns into pale yellow water soaked spots and cover whole leaf in later stage. In severe cases pseudostem twists and show abnormal elongation. The favourable conditions for conidia germination and infection are high humidity and 23-30°C temperature. It is more prevalent in *Kharif* crop.

Downy Mildew (*Pernospora destructor*)

It is highly destructive under cooler and humid regions and develops white to purplish downy growth first on entire surface of the older leaves. The affected parts become pale-green, yellow and collapse. The fungus requires cool moist nights and moderately warm days for best development. The conidia produced in humid atmosphere between 4-25°C with optimum at 13 0C. Cloudy days favour most, as 8 hour of light kills the conidia. Above fungal diseases can be managed by following clean cultivation, crop rotation, raised bed planting and proper drainage. Seed treatment with Bavistin (0.1 per cent) and hot water (50°C for 20 min), high rates of calcium, phosphorus and potassium also reduce diseases infection. Foliar sprays of Mancozeb @ 0.25 per cent, Tricyclazole @ 0.1 per cent and Hexaconazole @ 0.1 per cent effectively controls the diseases.

Inset-Pests

Among the large number of insects including *viz.* thrips, leaf miner, leafhopper, psyllid, army worm, curt worm, gram borer, red spider mite, onion maggot and nematodes that attack onion.

Thrip (*Thrip stabasci*)

It is wide spread and cause damage not only by feeding and oviposition, but as a vector of tospo viruses. Thrips can be managed with regular monitoring of inner-most whorl of the leaves. Thrip nymphs and adults are visible on plant, eggs laid inside the plant, while pre-pupal and pupal stages occur in the soil and debris. Time of planting, crop rotation, barrier crops, irrigation, fertilization, intercropping, deep ploughing, weed control, reflective mulches and clean management followed for cultural control of thrip. Chemical insecticides belonging to organophosphate (malathion), carbamate (Rogor) and pyrethroid (cypermethrin) groups are widely used for controlling the trip in tropical onion.

Improving Planting Value of Seeds by Physical and Physiological Methods

High quality seed is basic input for successful production system. Poor seed quality affects seedling emergence, growth and marketable yield. Quality seed insures better germination, rapid and uniform emergence and vigorous growth. These attributes result in a good stand and increased yield and quality. Poor quality seed results in "skips", excessive thinning or yield reduction due to overcrowding, all of which decrease the profitability of the growers. Seed quality upgradation techniques using physical, physiological and biochemical treatments are performed on seeds after harvest and before sowing to improve its field performance. After harvest seed processing based on screen (size) and density (weight) grading based on physical parameters of seed are useful techniques in improving planting value of seed. In addition, physiological treatments like seed priming and coating are currently used commercially for seed quality enhancement.

Seed Processing

The basic objective seed processing is to achieve maximum physical purity, germination and uniformity of seed size in an economical way. Wide range of seed processing equipment is available to upgrade seed quality based on the principal involving physical differences between good seed and material or seed to be removed.

Pre- cleaning

The most commonly used pre- cleaning machine used is a scalper. The operation is normally used to rough clean various kind of trash or plant debris from the seed lot. It may also be used to break up seed clusters during processing. The seeds which are relatively smaller fall through a vibrating or rotating sieve. This separation is usually accompanied by an air flow that removes dust, chaff and other materials which are lighter than the seed crop undergoing pre- cleaning.

Basic Cleaning

The basic cleaning is the main cleaning of a seed lot. Two main principals involved in basic cleaning are screening and pre- cleaning. The air screen machines combine both these features. It uses a combination of air flow and perforated metal or wire screens to separate seed on the basis of size. Air screen machines have at least two vibrating screens, the upper screen removes impurities larger than the seed while the lower screen separates out any seed or impurities smaller than the optimum seed size of the crop. An aspirating air flow removes impurities and empty seed which are larger than the optimum. Screens used in an air screen machine are constructed in either perforated sheet metal or woven wire mesh on a wooden frame. Metal screen openings are either round, oblong or triangular while openings in a wire mesh screen are either square or rectangular. These screens can be easily changed and almost hundred sizes are available. It is important that appropriate set of screens are available for the specific crop seed to be cleaned.

Gravity Separator

The specific gravity separator removes undesirable seed and inert contaminants that are so similar in size, shape and seed coat characteristics to the crop seed that cannot be removed in any other way. This is the best machine available for upgrading seed quality. For example, deteriorated, mouldy or decayed seed which are usually similar in size and shape to good seed but have a lower specific gravity can be removed by this machine. Insect damaged seed, empty seed or other seeds that have defects that decrease their specific gravity can be separated on this machine. The gravity separator consists of a base, plenum chamber or perforated vibrating deck, a feed hopper and a seed discharge system. Seed are introduced from the feed hopper on the porous metal deck, where the combination of shaking and air flow up through the deck causes them to stratify seed according to specific gravity. Thus heavier particles walk close to the deck surface and move towards the top of the deck where they ultimately fall off and are collected. The lighter particles tend to float on the air cushion above the heavier seeds following the path of least resistance and drift to the lower end of the deck where they fall off and are discharged. Seeds in medium specific gravity ranges called *middlings* can be collected in the middle area.

Indented Cylinder Separator

The indented cylinder separator operates by lifting short seeds out of a mixture containing long and short seed or other particles. This machine is especially used for separating pieces of stem or stalk from lettuce seed.

Electronic Colour Separator

Colour separator makes it possible to separate seeds that cannot be separated by any other method. This type of separation may be necessary on occasions when well graded seed lot contains some discoloured seeds which are known to be of lower potential germination or vigour

Seed Priming

Differences in seed quality within and between seed lots arise from the presence of different proportion of immature, mature and over-mature seeds. Occasionally these differences are associated with colour or some other physiological characteristics such as size, which allow physical techniques like seed processing to separate seed into different viability classes. Another approach is to try to improve seed quality of the poorer seeds in the seed lot by physiological treatments. In the mid- 1970's, Walter Heydecker in U.K invented a technique called priming. Heydecker (1973) defined seed priming as a presowing treatment in which seeds are soaked in an osmotic solution that allows them to imbibe water and go through the first stage of germination, but does not permit radicle protrusion through the seed coat. The seeds then can be dried to their original moisture contents and stored or planted by conventional techniques. Priming has the ability to improve the mean performance of a seedlot and also reduces variation within a seed-lot for a wide range of species.

Germination (Triphasic Uptake of Water)

Seed physiologists recognize three main stages during germination (triphasic uptake of water). *Stage 1:* This stage is recognized with a rapid initial uptake of water that is usually completed in 6- 24h depending on the species. All seeds, even dead ones, take up water rapidly during this stage. *Stage 2:* This is the plateau phase of water uptake during which there is initiation of nucleic acid and protein synthesis in preparation for the emergence of the radicle. This stage may last two or three times as long as stage one. *Stage 3:* This is a stage characterized by the rapid uptake of water, cell expansion and the protrusion of radicle through the seed coat The stage 3 is relatively short and the differences in the times of germination between the seeds of a population are associated with differences in the duration of stage2. Thus 'good' seeds germinate earlier than the 'poorer' seeds. Regulating the availability of water to the seed and preventing it from entering stage 3 can reduce this variation. This enables 'good' seed to be held back, allowing 'poor' seeds in the lot to catch up the development. The hydration of the seeds can be regulated using osmotica (osmopriming), salt (halopriming) and inorganic or organic carriers (solid matrix priming).

Primed seeds emerge faster and grow vigorously. In this process controlled hydration of seed is done to a level that permits pre-germinative metabolic activity to proceed, but prevents actual emergence of the radicle. Different methods are adopted for priming seeds. Prior to sowing, seed is soaked in water for a specified duration depending upon crop and variety. After completion of soaking period, surface dries them either by drying them with cloth or placing in sun. Farmers can hydroprime their own seed if they know the safe limits. These safe limits are calculated for each variety so that germination will not continue once seeds are removed from water. Seeds are osmotically primed by soaking in -0.5 to -1.0MPa Polyethylene glycol 6000 solutions in a test tube or cylinder. Aeration during the priming is provided through a glass tube connected by a rubber pipe to an aquarium pump. Priming is done at a constant temperature (20 to 25°C) for a period ranging from 2 to 7 days. Distilled water is added to test tube as needed to maintain constant volume and thus constant water potential of the solution. After the completion of priming period, seeds are removed from the solution and rinsed with water and surface- dried immediately. Seeds are haloprimed by soaking them in salt solution for a specified duration at constant temperature. Salts like potassium nitrate, calcium nitrate and magnesium nitrate at 10 to 30mM concentration are generally used. After the completion of soaking period, seeds are removed from solution and surface- dried. For solid matrix priming 100g seed is mixed with 200g vermiculite to which 250ml of water is added. The vermiculite and seeds are mixed thoroughly, sealed in a plastic bag and incubated at constant temperature for a specified period. After completion of incubation period, seeds are sieved out and dried to original moisture content. After drying seed to original moisture content, the primed seed can be used for sowing. Occasionally in case sowing is delayed, the primed seed can be stored in dry place for several days.

Effect of Seed Priming on Seedling Growth and Development

The benefits from priming treatments include increased germination, uniform emergence, germination under optimal and sub optimal environments and improved seedling vigor and growth (Pandita and Nagarajan, 2000Khan *et al.*, 1992Penzola and Eira, 1993). Seed priming modify embryonic axis growth and subsequent seedling development. The response varies according to the species and priming conditions. Priming did not modify embryo volume and cell number of leek and onion but under similar conditions, carrot embryo volume increased almost 50 per cent and the number of cells increased by two- fold. Generally, the major effects of seed priming on growth has been observed as early more uniform emergence and not accelerated growth, per se, of the species. Osmopriming of freshly harvested and aged seed of tomato seed improved germination, speed of germination, field emergence and vigor of seedlings (Pandita *et al.*, 2003). Seed priming did not modify the number of basal and lateral roots and taproot length of 14 days old pepper seedlings (Stoffella *et al.*,1992). Root length of primed lettuce seeds germinated at 35°C was greater than that of nonprimed seeds (Wurr and Fellows, 1984). Pill (1986) reported that primed parsley seeds yielded 52 per cent more fresh weight compared to non primed seeds 24 days after sowing. Priming of carrot seeds resulted in 36 per cent increase in dry weight of one- month old seedlings compared to nonprimed seeds (Nagarajan *et al.*, 2003). The differences in root and shoot growth between primed and nonprimed seeds is more evident under stressful conditions. Hydration of bittergourd seed in wet muslin cloth for 48h improved emergence, seedling length and dry weight significantly under low temperatures (Pandita and Nagarajan, 2004). Root growth from perennial rye grass seeds germinated at low temperature was greater in primed seeds than in nonprimed seeds, but no differences were observed at 250 C (Danneberger *et al.*, 1992).

Effect of Priming on Yield and Quality

Seed priming promoted early growth of brinjal, pepper, cucumber, and muskmelon plants, but no differences were detected in early and final yield between primed and nonprimed seeds (Passam *et al.*, 1989). However, Alvarado *et al.* (1987) reported that flowering was early in primed tomato seeds, but fruit maturation, yield or fruit soluble solid content were unaffected. Under stressful conditions, priming increased early seedling growth and marketable yield in tomato (Odell *et al.*, 1992). Beneficial effects of priming on yield and quality have been reported in crops growing under stressful conditions. Pandita *et al.*, 2010 reported that solid matrix priming alone or in combination with *Trichoderma viride* significantly improved final marketable pod yield under sub-optimal temperature but there was no such improvement under optimal temperatures. Priming had no effect on the number of pods per plant and pod yield per plant under either environment. There is no doubt about the beneficial effects of priming on the rate and uniformity of seed germination. However, priming treatments are influenced by complex interaction of factors including plant species, osmoticum, duration, temperature, seed vigor and storage conditions following priming.

Physiological and Molecular Basis of Priming

It is important to understand physiological and molecular basis of seed priming for further refinement of this process to obtain better and more consistent benefits. A number of studies have reported on these effects but the biochemical mechanism of priming remains largely unelucidated. Lettuce seeds primed in PEG 6000 had increased activities of acid phosphatase and esterase and reduced time for RNA and protein synthesis than nonprimed seeds (Khan *et al.*,1978). Coolbear *et al.*(1990) reported large increase in nucleic acid content during priming process. The activities of peroxidase and dehydrogenase markedly increased in osmoprimed seed of carrot (Nagarajan *et al.*, 2003). Solid matrix priming improved emergence in chilli seed under suboptimal temperatures were attributed to increased activity of glyoxylate cycle enzymes (Pandita *et al.*, 2007). Nascimento *et al.*(2000) found that during priming of a thermosensitive lettuce genotype endo-s-mannanase activity was induced after 24 h. After drying and immediately upon reimbibition, endo-s-mannanase levels were high, leading to rapid germination at 35°C. A connection between priming, thermotolerance and endo-s-mannanase activity has finally been established (Cantliffe *et al.*, 2000). Membranes play an active role in seed hydration and dehydration mechanism, but their role during and after priming has not been studied extensively. Basra *et al.*(1988) reported changes in quantity and quantity of membrane phospholipids during and after priming. Parera and Cantliffe (1991) demonstrated that SMP primed sweet corn had less solutes leakage and reduced water uptake rates during early imbibition than nonprimed seeds.

Morphological Changes in Primed Seed

Morphological changes in primed seeds were studied by Guedes *et al.* (1981) under electron microscope. They found that the outer layer of endosperm cells were gradually loosened after 9 h of priming and this loosening weakening of cell wall, possibly is one of the mechanism of priming enhanced seed germination. Osmopriming of tomato seeds showed large free space between embryo and endosperm under X-ray radiography (Pandita *et al.*, 2003, 2007). The occurrence of free space in primed seeds has been suggested to play a role in accelerating germination rate by facilitating uptake of water (Argerich and Bradford, 1989). During priming the embryo expands and compresses endosperm tissue at a location opposite to radicle tip (Liptay and Zariffa, 1993). Both the compression forces of embryo and hydrolytic activities on the endosperm facilitates protrusion of roots upon rehydration (Liu *et al.*, 1996).

Seed Storage Products, their Biochemistry and Modification Issues: Proteins

Proteins are highly complex molecules comprised of linked Amino acids. Amino acids are simple compounds containing carbon, hydrogen Oxygen, nitrogen and occasionally sulfur. There are about 20 different amino acids commonly found in plant and animal proteins. Amino acids link together to form chains called peptides. A typical protein may contain 500 or more amino acids. Each protein has its own unique number and sequence of amino acids which determines its particular structure and function. Proteins are classified on the basis of their solubility properties *viz.* albumins: soluble in water (Heat coagulable), Globulins: soluble in salt solution but insoluble in water, prolamins: soluble in aqueous alcohol and glutelins: soluble in dilute acids and alkalies. In plants and animal tissues broadly, proteins are of three types *viz.*, Structural proteins, metabolic proteins and storage proteins. Storage (reserve) protein possess characteristics features such as, these provide amino acids and nitrogen to young seedling, provide nitrogen to seed nitrogen compounds, possess peculiar amino acid composition, chemically they are model system for gene isolation, characterization and expression and metabolically useful model for the study of storage protein deposition and degradation during seed development and germination respectively.

In cereals, storage proteins are deposited in the endosperm (prolamins and glutelins) of which the outer layer (aleurone cells) are particularly rich in proteins and phosphorus reserves required for lateral growth of seedling. Storage proteins also occur in the embryo mainly in the scutellum, required for the growth of axial organ during beginning of germination. The major storage proteins are alcohol soluble prolamins glutelins and globulins. In Panicodeae family eg: maize, sorghum and millet, 50-60 per cent are prolamins, 35-40 per cent are glutelin. In Triticeae family, wheat, barley (Hordins) and rye (scalins), 40-50 per cent are prolamins (gliadin) and 35-40 per cent are glutelin, while in oat and rice 80 per cent predominant proteins are glutelins, only 8 per cent are prolamin of the total seed storage proteins. In cereals, seed crude protein content varies from 7-15 percent and within storage proteins, prolamin and glutelins are present in the ratio of 1.5 (Panicoideae), 1.15 (Triticeae), 0.15 (oat) and 0.1 (Rice) respectively. From the amino acid composition data of cereal storage proteins, lysine and tryptophan are limiting essential amino acids followed by methionine and basic amino acids. In breeding for cereals protein quantity and quality attempts have been made, by crossing various lines, replacing chromosomes, Mutations, chemical and irradiation methods. Promising results for the higher accumulation of prolamin protein fractions rich in lysine has been obtained for improvement in grain protein quality, quantity and yield. Biochemical means for quality and quantity improvement in cereals proteins are being carried out by elevating nitrate reductase (NR) enzyme activity in roots, vegetative parts and in maturing seeds as well as by efficient translocation of nitrogen from maternal plant to the grain. The higher content of lysine and tryptophan are the limiting amino acids in cereals.

Gel Electrophoresis of Quality Protein Profiles/Isozyme Profiles of Cultivars

Electrophoresis is the migration of charged molecules in solution in response to an electric field. Their rate of migration depends on the strength of the field, on the net charge, size and shape of the molecules and also on the ionic strength, viscosity and temperature of the medium in which the molecules are moving. As an analytical tool, electrophoresis is simple, rapid and highly sensitive. It is used analytically to study the properties of a single charged species, and as a separation technique. Proteins, generally considered as the source of evolution, taxonomy and genetics, play important role in gene expression, mutagenesis, intracellular transport, compartmentation and cell mortality. The seed storage proteins being direct products of genes, are widely used for the identification of crop cultivars. Seed storage proteins are classified as albumins, globulins, prolamins and glutelins depending upon their solubility in different solvents. Further, proteins are composed of a number of polypeptides/subunits. However, the three-dimensional structures of proteins determine their specific metabolic activities. The enzymes, which are essentially proteins, are responsible for the metabolic processes in the living cells. The organisms synthesize an enzyme in several different ioszymic forms and all these isozymic forms catalyze the same enzymatic reaction. However, each isozyme functions in its own distinctive way depending upon the structural and functional requirements of the cells. This diversity in the structure and function of proteins/ ioszymes forms the basis to distinguish crop varieties. The spectra of proteins/ isozymes obtained by electrophoresis are seen as documents of genetic origin.

The migration of charged particles in a liquid in the electric field is termed as electrophoresis. Electrophoretic separation of proteins/enzymes has provided a powerful tool to identify the crop varieties and to verify the genetic identity/purity of seed stocks. The technique is also being used to evaluate varieties for different quality parameters *e.g.* baking quality of wheat cultivars, malting characteristics of barley *etc.* Moreover, the electrophoretic methods are simple, rapid, reliable and less costly than the traditional morphological procedures. In electrophoresis, the separation is achieved mainly due to variation in the molecular weights and isoelectric points of various proteins/isozymes. The electrophoretic separation can be classified as static and dynamic. In static electrophoresis, the molecules move almost up to some end points either on the basis of their molecular weights or isoelectric points. The dynamic separation is obtained due to both size and charge on the molecules. The commonly used supporting media in electrophoresis are agarose, agar, paper, cellulose acetate, starch and acrylamide.The separation of proteins is influenced by several factors including choice of buffers, ionic strength of buffers, electric current and temperature *etc.* Due to difficulty in obtaining starch of reproducible high quality, polyacrlamide gels have superseded in many cases. Polyacrylamide gels are obtained by mixing acrylamide (AA) and N, N′ - methylene bisacrylamide (Bis). The gels of different pore size are prepared by adjusting the proportions of AA and Bis. The separation of proteins can be achieved by using continuous or discontinuous polyacrylamide gel electrophoresis (PAGE). To perform sodium dodecyl sulphate polyacrylamide gel electrophoresis (SDS-PAGE), the samples are heated with SDS and 2-mercaptoethanol so that proteins are dissociated into their polypeptides. The separation of proteins/isozymes can also be obtained by using sieving effect created by the variation in pore size in gel (Poro- PAGE). Similarly, cultivars can be distinguished by performing polyacrylamide gel isoelectric focussing.

PAGE has successfully been used to distinguish potato cultivars on the basis of their protein and esterase patterns (Stegemann, 1976). The electrophoretic band patterns of endosperm proteins obtained by SDS-PAGE have been found to be useful in characterizing barley varieties (Smith and Payne, 1984).The rice varieties have been characterized by the electrophoresis of salt soluble proteins (Sarkar and Bose, 1984). Isolelectric focussing of seed proteins and esterases was utilized for the identification of winter wheat varieties (Pietsch, 1985). The identification of rapeseed cultivars has been achieved by Poro-PAGE (Gupta and Roebbelen, 1986). A method based on SDS-PAGE has been developed to distinguish wheat cultivars (Sreeramulu *et al.*, 1994). To identify different barley cultivars, a procedure using SDS Poro-PAGE hàs been standardized (Ram and Verma, 1998).

Gel electrophoresis enables the separation of macromolecules-either nucleic acids or proteins through a solid matrix -on the basis of size, electric charge, and other physical properties. Therefore, it is one of the most widely used standard tool with numerous applications in nucleic acid/protein biochemistry and molecular biology. The commonly used supporting media in electrophoresis are agarose, agar, paper, cellulose acetate, starch and polyacrylamide. Agarose, a polysaccharide extracted from seaweed, is used at concentrations of 0.5 to 2 per cent. Agarose gels have a large range of separation, but relatively low resolving power. By varying the concentration of agarose, fragments of DNA from about 200 to 50,000 bp can be separated by electrophoresis. Polyacrylamide gels result from the polymerization of a mixture of acrylamide and a cross-linker (*e.g.*; N, N′-Methylene-bis-Acrylamide [BIS, Bis-Acrylamide] or Ethylene Diacrylate). N, N, N′, N′-Tetramethylethylenediamine (TEMED) and Ammonium Per sulphate (APS) are used as accelerators and catalysts for the polymerization reaction. The length of the polymer chains is dictated by the concentration of acrylamide used, which is typically between 3.5 and 20 per cent. The pore size of the gel can be varied by using different proportions of both acrylamide and bis acrylamide. The gels can be prepared in tubes or slabs formed between two glass plates.

Gel Electrophoresis of Quality Protein Profiles

Generally, the sample is run in a support matrix such as paper, cellulose acetate, starch gel, agarose or polyacrylamide gel. The matrix inhibits convective mixing caused by heating and provides a record of the electrophoretic run, at the end of the run the matrix can be stained and used for scanning, autoradiography or storage. In addition, the most commonly used support matrices - agarose and polyacrylamide - provide a means of separating molecules by size, in that they are porous gels. A porous gel may act as a sieve by retarding, or in some cases completely obstructing, the movement of large macromolecules while allowing smaller molecules to migrate freely. Because dilute agarose gels are generally more rigid and easy to handle than polyacrylamide of the same concentration, agarose is used to separate larger macromolecules such as nucleic acids, large proteins and protein complexes. Polyacrylamide, which is easy to make at higher concentrations, is used to separate most proteins and small oligonucleotides that require a small gel pore size for retardation

Separation of Proteins under Denaturing Conditions

Sodium dodecyl sulphate (SDS) is an anionic detergent which denatures proteins by "wrapping around" the polypeptide backbone - and SDS binds to proteins fairly specifically in a mass ratio of 1.4: 1. In so doing, SDS confers a **negative charge** to the polypeptide in proportion to its length i.e, the denatured polypeptides become **"rods" of negative charge cloud** with equal charge or charge densities per unit length. It is usually necessary to reduce disulphide bridges in proteins before they adopt the random-coil configuration necessary for separation by size: this is done with **2- mercaptoethanol or dithiothreitol**. In denaturing SDS-PAGE separations therefore, migration is determined not by intrinsic electrical charge of the polypeptide, but by **molecular weight**.

Determination of Molecular Weight

This is done by SDS-PAGE of proteins - or PAGE or agarose gel electrophoresis of nucleic acids - of known molecular weight along with the protein or nucleic acid to be characterized. A linear relationship exists between the logarithm of the molecular weight of an SDS-denatured polypeptide, or native nucleic acid, and its *Rf*. The Rf is calculated as the **ratio of the distance migrated by the molecule to that migrated by a marker dye-front**. A simple way of determining relative molecular weight by electrophoresis (*M*r) is to plot a standard curve of distance migrated vs. log10MW for known samples, and read off the log*Mr* of the sample after measuring distance migrated on the same gel.

Continuous and Discontinuous Buffer Systems

There are two types of buffer systems in electrophoresis, **continuous and discontinuous**. A **continuous** system has only a single separating gel and uses the same buffer in the tanks and the gel. In a **discontinuous** system, a non-restrictive large pore gel, called a stacking gel, is layered on top of a separating gel called a resolving gel. Each gel is made with a different buffer, and the tank buffers are different from the gel buffers. The resolution obtained in a discontinuous system is much greater than that obtained with a continuous system.

Equipment (Electrophoresis Assembly)

A vertical gel assembly with power pack is generally used for running protein samples. The assembly consists of gel reservoir, glass plates, spacers, clamps, gel cast, gel mould, slot former *etc.*

Acrylamide: **A**

$H_2C=CH-CO-NH_2$

Bis-Acrylamide (N,N'-Methylene-bis-Acrylamide) : **B**

$H_2C=CH-CO-NH-CH_2-NH-CO-CH=CH_2$

TEMED: N,N,N',N'-Tetramethylethylenediamine : $CH_3)_2N-CH_2-CH_2-N(CH_3)_2$

APS: Ammonium Per sulphate : $(NH_4)_2S_20_8$

A part of an acrylamide-bisacrylamide gel

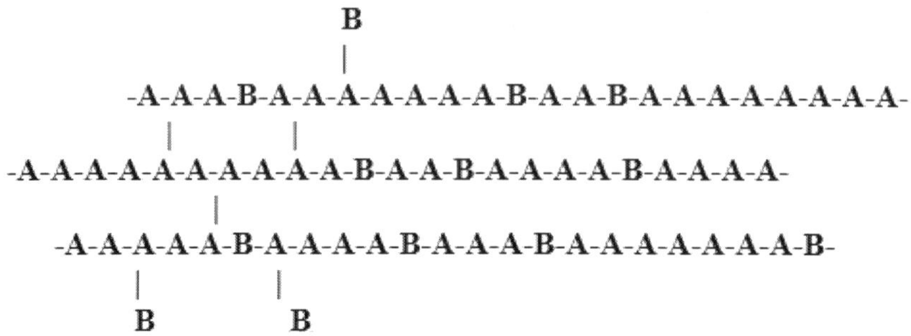

The rate of movement of molecules under influence of an electric field depend size and charge carried by the molecules and resistance of the medium to movement by the molecules.

PAGE

The proteins/isozymes can be separated by continuous or discontinuous polyacrylamide gel electrophoresis. The Continuous Electrophoresis (PAGE) is simple and protein can remain in a more native state as the buffer species and buffer concentrations of the buffer chamber electrode solutions, the sample and the polyacrylamide gel are identical. In case of Discontinuous Electrophoresis (Disc PAGE) better separation and resolution of proteins is achieved by using different buffers for chamber electrode solutions, the sample, the stacking gel and the separating gel. The ions and proteins are stacked closely in the stacking gel which has a relatively larger pore size than the separating gel.

SDS-PAGE

In SDS PAGE the buffers of the electrode solutions, the sample and the gel contain sodium dodecyl sulphate (SDS). The SDS, a reducing agent such as dithiothreitol (DTT) or 2-mercaptoethanol and heat are responsible for the actual denaturation of the proteins. SDS breaks up the two- and three-dimensional structure of the proteins by adding negative charge to the amino acids. Generally, different proteins bind about the same amount of SDS per unit length of protein. Therefore, all proteins acquire about the same charge per unit length, and the same charge per unit mass. As the proteins are separated on the basis of their size their molecular weights can be determined by SDS PAGE.

Poro-PAGE/SDS PAGE

The separation of proteins/isozymes can also be obtained by using sieving effect created by the variation in pore size in gel with (SDS Poro-PAGE) or without SDS (Poro- PAGE). The separating gel is prepared in a manner that its pore size decreases gradually from one end to another. The proteins with large variation in their molecular weights can be separated and resolved on the same gel in Poro-PAGE. This type of electrophoresis is also termed as static electrophoresis as the molecules move almost up to some end point depending upon their molecular weights.

Polyacrylamide Gel Isoelectric Focusing (PAGIF)

It is also static type of electrophoresis as the proteins/isozymes are focused to the points in the gel matrix at which pH of the gel equals to their ioselectric points (pIs). In PAGIF, the protein molecules are subjected to an electric field on a gel in which a pH gradient has been generated using ampholytes (small polymers of MW appoox. 5000 daltons). These ampholytes are composed of thousands of polymers containing both positively and negatively charged groups. PAGE has successfully been used to distinguish potato cultivars on the basis of their protein and esterase patterns (Stegemann and Loeschcke 1976). The electrophoretic band patterns of endosperm proteins obtained by SDS-PAGE have been found to be useful in characterizing barley varieties (Smith and Payne, 1984).The rice varieties have been characterized by the electrophoresis of salt soluble proteins (Sarkar and Bose, 1984). Isolelectric focusing (PAGIF) of seed proteins and esterases was utilized for the identification of winter wheat varieties (Pietsch, 1985). The identification of rapeseed cultivars has been achieved by Poro-PAGE (Gupta and Roebbelen, 1986). A method based on SDS-PAGE has been developed to distinguish wheat cultivars (Sreeramulu *et al.*, 1994). To identify different barley cultivars, a procedure using SDS Poro-PAGE has been standardized (Ram and Verma, 1998). Capelli *et al.* (1998) has used capillary electrophoresis of gliadins in isoelectric buffers for wheat cultivar discrimination. Sodium dodecyl sulphate electrophoresis in uncoated capillaries has also been employed to discriminate wheat proteins (Bean and Lookhart, 1999). Allozymes variability helped in barley cultivar identification (Radovic and Vapa,1999). High molecular weight glutenin subunit composition revealed by electrophoresis has been used to assess end-use quality characteristics of hard white spring wheats (Anjum *et al.*, 2000). Discontinuous (Disc PAGE) gel electrophoresis of peroxidase isozymes has been used to distinguish sugarcane cultivars of south India (Manjunatha *et al.*, 2003).

Performing Electrophoresis

The polyacrylamide gel electrophoresis for the separation of proteins/isozymes is performed by using LKB Multiphor II system as under:

Preparation of Samples

The proteins/enzymes should be extracted from seed or fresh tissue of plants grown under carefully controlled similar conditions. Different solvent/buffers are used to extract albumins, globulins, prolamin or glutelin fractions of proteins. The enzymes are extracted with distilled water or buffers. Sometimes, the powdered seed samples are defatted by acetone or ethyl acetate before extraction of proteins. To inhibit undesirable oxidation reactions during extraction and storage, some reducing agents are added to extracting solvent/buffer. Protein/enzyme extracts containing 20 per cent sucrose as an additive may be stored at + 4°C for two days. For extended storage time the extracts should be stored either at -20°C or preservatives like propylene oxide, sodium azide or thiomersal (an organic mercury salt) can be used. The protein concentration of all samples should be adjusted to some standard concentration prior to mixing with sample buffer, so that the final concentration of protein is the same for all samples. The amount of the sample to be applied on gel is also carefully calculated so that overloading may not result in precipitation and aggregation of proteins, producing streaks and smears in the gel after electrophoresis.

Gel Assembly

Place first glass plate on bench. Inspect upper surface for lint and fingerprints. Clean with alcohol or NaOH solution. Place spacers in position on top. Place second plate in position. Compress outermost plates while placing clamps on edges. Compress outermost plates and move the assembly to vertical position. Align glass plates and spacers so bottom and top edges are flush. Tighten screws in increments to point where they slip on gloved hands. Do not tighten all at once. Place the gray rubber gasket in the bottom of the casting stand, smooth side up. Place assembled plates in the stand. Insert cam levers and rotate to tighten plates into position.

Separating and Stacking Gels

To prepare separating gel mix 7.25 ml of 35 per cent acrylamide, 2.08 ml of 2 per cent bisacrylamide, 5.1 ml distilled water and 9.4 ml Tris (pH 8.8) in a beaker. Add 0.25 ml of SDS, 0.94 ml of freshly prepared 1 per cent Ammonium persulphate and 30 µl of N,N,N',N' tetramethylenediamine (TEMED) to the solution and mix thoroughly. Pour the solution immediately into the gel mould upto the depth of about 2 cm below the position of the sample wells. Over-layer the surface of the polymerizing gel with 4-5 ml of water/isopropanol using a syringe. Leave the separating gel was to polymerize for about 30 minutes. Prepare the stacking gel by mixing 1.75 ml of 35 per cent acrylamide, 0.75 ml of 2 per cent bisacrylamide, 8.0 ml of distilled water and 2.5 ml of Tris (pH 6.8). Add 0.20 ml of SDS, 0.94 ml of freshly prepared 1 per cent Ammonium persulphate and 40 µl of N,N,N',N' tetramethylenediamine (TEMED) to the solution and mix thoroughly. Pour off the water/isopropanol from the top of the separating gel and wash the surface with 4-5 ml of stacking gel solution. Fill the mould with stacking gel solution upto the top of the plates. Place the slot former in between the plates on the top. Let the gel polymerize for about 1 hour. After the gel is polymerized remove the comb was carefully without disturbing the wells and wash the wells twice with the running buffer.

Extraction of Proteins

Grind two to three seeds of each line using plier with smooth surface. Transfer the crushed mixture to an Eppendorf tube. Prepare the extraction solution by mixing 4 ml distilled water, 1.7 ml of 3X sample buffer and 0.3 ml of 2-mercaptoethanol. The amount of extraction solution added to each tube is calculated by multiplying a factor 0.025 with their respective seed weight. After adding extraction solution, the contents are mixed on vortex mixture for 3-4 minutes. Keep the samples at room temperature for about 1 hour.

Sample Loading

Place the samples in boiling water for 10-12 minutes. Centrifuge the tubes with samples at 4000 rpm for 10 minutes. Use the supernatant so obtained for loading.

Place the plates with gel in the electrophoresis apparatus containing the running buffer. Remove the air bubbles from the lower surface with the help of a syringe. Fill the upper reservoir with the running buffer. Load the samples by drawing the supernatant into a micropipette and introduce 10 µl into a sample well. After loading samples, the electrodes are connected to the power pack. The running current is 2-3 mA per sample and the gel is allowed to run until 45 minutes after the tracker dye has disappeared from the bottom of the gel.

Gel Staining and Destaining

After termination of the electrophoresis the glass plates are removed and carefully separated and the stacking gel is removed from the separating gel by cutting the interface with a gel cutter. The separating gel is then placed in staining solution which contains 80 ml methanol, 20 ml GAA, 80 ml distilled water and 0.5 g coomassie brilliant blue-R250 dye. The gel is kept in the staining solution overnight (~12 hrs). Gel destaining is done for 10-12 hour in a solution containing 100 ml methanol, 20 ml GAA and 100 ml distilled water. Again the gel is destained in water for 2-3 days. After destaining the gel is photographed in a gel documentation apparatus using UV light.

Preparation of Stock Solutions

- **Acrylamide (35 per cent):** 35 g of molecular grade acrylamide dissolved in double distilled water and volume made to 100ml. Bis in 75 ml distilled water and stir until solution is clear. Make the volume upto 100 ml and store at + 4°C.
- **Bisacrylamide (2 per cent):** 2 g of molecular grade N, N′-methylene bisacrylamide dissolved in double distilled water and volume made to 100ml.
- Tris-borate buffer (pH 8.9): Dissolve 60.5 g Tris and 4.6 g boric acid in distilled water and make the volume upto 500 ml.
- **SDS (10 per cent):** 10 g of molecular grade SDS dissolved in double distilled water and volume made to 100ml.
- TEMED, 10 per cent V/v.
- Ammonium persulphate,2 per cent w/v
- Coomasie blue,1 per cent w/v
- Staining solution for proteins: mix 50 per cent trichloroacetic acid,96 per cent acetic acid, methanol and distilled water in the ratio of 10: 15: 15: 60 (v/v)
- Detaining solution: mix 50 per cent tricholoacetie acid, 96 per cent acetic acid, methanol and water in the ratio of 5: 20: 40: 335 (v/v)
- Preserving solution: To 50 ml glycerol add 450 ml of detaining solution(G)
- Sodium acetate buffer,0.2M, pH 5.0
- Hydrogen peroxide solution, 2 per cent V/v.

- ☆ Dissolve 8.9 g. $Na_2 HPO_4.2H_20$ in distilled water and make the volume to 100 ml.
- ☆ Dissolve 7.8 g $NaH_2 PO_4. 2H_20$ in distilled water and the make the volume to 100 ml.
- ☆ **1M Tris (pH 6.8):** 12.11 g of Trisma base dissolved in 50 ml of distilled water. The pH adjusted to 7.0 with concentrated HCl and then accurately with 1N HCl to 6.8. The final volume made to 100 ml.
- ☆ **1M Tris (pH 8.8):** 12.11 g of Trisma base dissolved in 50 ml of distilled water. The pH adjusted to 8.8 first with concentrated HCl and then accurately with 1N HCl. The final volume made to 100 ml.
- ☆ **Electrophoresis buffer (pH 8.3):** 75.5 g Glycine, 15 g Tris and 5 g SDS dissolved in distilled water and pH adjusted to 8.3. Final volume made to 5 liters.
- ☆ **3x sample buffer:** 10 ml Glycerol, 6.25 ml 1M Tris (pH 6.8), 2 g SDS, 20 mg bromophenol blue (tracking dye) mixed in distilled water and volume made to 50 ml. Working concentration used is 1X.
- ☆ **Staining solution:** Methanol, Glacial Acetic Acid (GAA) and double distilled water in a ratio of 4: 1: 4. Coomassie Brilliant Blue-R250 is used as staining dye.
- ☆ **Destaining solution:** Methanol, Glacial Acetic Acid (GAA) and double distilled water in a ratio of 5: 1: 5.

Casting of Gels

To perform Horizontal electrophoresis Multiphor II Electrophoresis System (LKB, Sweden) is used. Prepare the gel mould and place on a levelled surface. Prepare the gel (6 per cent PAA) solution by mixing 21 ml of solution A, 8.8 ml solution B, 38.2 ml distilled water and 0.5 ml solution C. Dissolve 25 mg of sodium sulphite in this solution and add 1.5 ml of solution D. Pour the gel solution in the gel mould with the help of a syringe with large needle. Leave the gel to polymerise.

Electrophoresis

When polymerization is complete gel is removed from the mould and placed on the cooling plate of the electrophoresis apparatus. 2pply the prepared protein/ enzyme samples and perform the electrophoresis using 200V current for first 20 min. The electrophoresis is continued at 400V for 2 to 3hr at +4°C.

Staining

Proteins

To 200 ml of solution F add 5 ml of solution E and transfer the gel to this solution. Staining of proteins may take 24 to 48 hr Wash the gel with tap water and destain with solution G for 6 to 8 hr. Preserve the gel in solution H

Peroxidases

Dissolve 0.2g benzidine in 200 ml of buffer I. Boil the contents to get clear solution. Cool and filter. Transfer the gel to this solution and add 10 ml of solution J. Blue bands of peroxidases will appear after ca. 30 min. Destain and preserve the gel in solution H.

Esterases

Remove the gel from the cooling plate and incubate in phosphate buffer PH 7.4 (mix 8.7 ml solution K and 3.8 ml solution L and make volume to 250 ml) for 15 min. at 37°c. Add 250 mg of a-naphthyl acetate (dissolved in 1 ml acetone) and 250 mg of fast blue RR salt. Esterase bands will appear after ca 1 hr. Stop the reaction by adding detaining solution and preserve the gel in solution H.

Table 10.9. SDS-Page Glutenin Electrophoresis

Dye	Bromophenol blue	0.5 g
	Glycerol	50ml
	D.W.	50ml
Extracting Buffer (pH-6.8)	TRIS	0.75g
	SDS	2g
	Mercaptoethanol	5ml
Total Volume		100ml with DW
Stain	Coumassive Brilliant blue (CBB-R 250)	2g
	100 per cent methanol	800ml
	100 per cent TCA (Fixture)	200ml
	Total Volume	2000 ml with D.W.

Table 10.10. Separating/Running Gel

	5 per cent	*7.5 per cent*	*10 per cent*	*12.5 per cent*
Acrylamide	4.5 ml	6.75 ml	9.0 ml	11.25 ml
Bis-Acrylamide	1.17 ml	1.755 ml	2.34 ml	2.925 ml
RG Bufer (8.8 pH)	9.0 ml	9.0 ml	9.0 ml	9.0 ml
Distilled water	21.33 ml	18.495 ml	15.66 ml	12.825 ml
APS*	200/100 µl	200/100 µl	200/100 µl	200/100 µl
Temed	31 µl	31 µl	31 µl	31 µl

Stacking Gel

	6 ml	*7 ml*	*5 ml*
Acrylamide	0.66 ml	0.77 ml	0.55 ml
Bis-Acrylamide	0.36 ml	0.42 ml	0.30
RG Bufer (8.8 pH)	1.5 ml	1.75 ml	1.25 ml

	6 ml	*7 ml*	*5 ml*
Distilled water	3.48 ml	4.06 ml	2.90 ml
APS*	66.66 µl	100/77.77 µl	55.55 µl
Temed	14 µl	21/16.33 µl	11.66 µl

APS: Dissolve 0.06 gm in 600 µl D.W.

Where, SDS: Sodium didoecyl sulfate; TRIS: Hydroxy methyl (Amino methane); APS: Ammonium persulfate; BIS: N,N-Methylene bisacrylamide; TCA: Trichloro acetic acid; TEMED: Tetra methylethylene diamine.

Photography

Photograph the gel by illuminating it with diffused fluorescent light from below, using Agfa Ortho 25 film.

Evaluation

Gels can be scanned by using Ultrascan XL Enhanced laser Densitometer (LKB, Bromma, Sweden) and evaluated for protein/sozyme banding patterns.

Seed Genetic Purity Assessment Using Molecular Markers

The seed is the basic and critical input for enhancing agricultural production and productivity. The seeds sold by seeds companies and grown by farmers are commonly "certified seeds". The certification of seeds is for quality control. The certified seeds have improved traits like better yield, pest resistance, drought tolerance, herbicide tolerance *etc.* Genetic purity of seeds refers to the percentage of contamination by seeds or genetic material of other varieties or species. Genetic purity test is essential as impurity is caused during processing due to pollen shedders and physical admixtures. To detect off-types, Grow out test (GOT) is practiced by researchers for genetic purity. It is time-consuming (takes one full growing season for completion), land intensive, tedious and highly vulnerable to human and environmental error. Crop varieties and hybrids identification is essential for their intellectual property right (IPR) protection, prevention of unauthorised commercial use and misuse of the brand name. A set of qualitative and quantitative characters known as descriptors are currently used for varietal identification in Distinctiveness Uniformity and Stability (DUS) testing. However, quantitative characters interact with the environment and are misleading process for varietal identification. The continued release of new cultivars, particularly F1 hybrid seeds, by public and private breeders in the last 20 years pressed the development of faster and more reliable new technologies for seed genetic purity determination. There are three major types of markers: morphological markers (also called "classical" or "visible" markers) which are phenotypic traits, biochemical markers, which are called isozymes, including allelic variants of enzymes and DNA markers (or molecular markers), which reveal sites of variation in DNA.

Morphological Markers are basically the indicators of phenotype such as seed colour, seed shape, seed size, seed length, *etc.* They generally represent genetic polymorphisms which are easily identified and manipulated.

Biochemical/Protein Markers are based on isozymes. "Isoenzymes" or "isozymes" are the alternative forms or structural variants of an enzyme that have different molecular weights and electrophoretic mobility but have the same catalytic activity or function. Seeds, seedlings or mature leaves *etc.* of a crop plant have a specific mixture of proteins, which are crop and variety specific (genotype-specific). Electrophoresis separates the mixture of proteins extracted from seeds, seedlings or mature leaves into distinct bands. Each variety (or genotype) has a specific "banding pattern" and on the basis of it admixtures of other varieties, differing in "banding pattern" could be detected. Protein banding patterns of analysed samples are compared with the standard banding patterns of a particular variety. This method is being increasingly used for determining the genetic purity of seed samples. The major disadvantages of morphological and biochemical markers are that they may be limited in number and are influenced by environmental factors or the developmental stage of the plant. Besides these limitations, morphological and biochemical markers have been extremely useful to plant breeders.

Molecular Marker are ideal molecular marker should have following characteristics i) highly polymorphic ii) Co-dominant, should be able to discriminate between heterozygous states in diploid organisms iii) frequent occurrence in genome iv) detect multiple independent and reliable loci v) provide adequate resolution of genetic differences vi) selective neutral behaviors – DNA sequences of any organisms are neutral to environmental conditions or management practices and this permits to confer the variations only to a genetic origin vii) easy access and fast assay, it must be simple quick and in expensive viii) high reproducibility to guarantee robust results among different laboratories and equipments ix) require small amounts of tissues and DNA sample x) link to distinct phenotypes xi) require no prior information about genome of an organism

DNA Markers are defined as a fragment of DNA that is associated with a certain location within the genome. They arise from different classes of DNA mutations such as substitution mutations (point mutations), rearrangements (insertions or deletions) or errors in replication of tandemly repeated DNA. Basically, DNA markers are divided into two main categories: PCR-based molecular markers and Hybridization-based molecular markers.

Restriction Fragment Length Polymorphism (RFLP) is a hybridization-based molecular marker. Though it is a co-dominant marker with high reproducibility, it has few limitations like the process is time consuming, expensive, requires large amount of DNA and involves the use of radioactively labelled probes. These disadvantages can be overcome by using PCR-based markers.

Random Amplified Polymorphic DNAs (RAPD) is anonymous stretches of DNA is amplified using arbitrary primers, it is a fast and easy method for detecting polymorphisms but it is a dominant marker along with reproducibility problems.

Amplified Fragment Length Polymorphism (AFLP) is technically demanding, stable and a very reliable method which uses relatively small amount of DNA, it does not require sequence information or probe collection and has a very high multiplex ratio with relative reproducibility. Its limitations include dominant nature, requirement of high quality DNA for complete restriction enzyme digestion, high cost and experience of sequencing gels.

Simple Sequence Repeat (SSRs) is a simple, easy and preferable marker system because of co-dominant nature, vast abundance and uniform distribution throughout the genome. Their limitations are high cost investment in oligonucleotide synthesis and require initial identification and necessary prior sequence information.

Sequence Characterized Amplified Regions (SCARs) are mono locus, less sensitive to reaction conditions, and they can be potentially converted into codominant markers, easy to use and has high reproducibility. Prior sequence information (*i.e.*, sequencing the polymorphic fragments) is required for designing the primers contiguous the polymorphic region.

Sequence Tagged Sites (STS) unlike PCR with arbitrary primers, sequence-tagged sites (STS) are primers that are based on some degree of sequence knowledge. These unique, sequence-specific primers detect variation in allelic, genomic DNA. STS have a particular advantage over RAPDs in that they are co-dominant, that is, they can distinguish between homozygote and heterozygote. They also tend to be more reproducible, because they use longer primer sequences. However, they have the disadvantage of requiring some pre-existing knowledge of the DNA sequence of the region, even if only for a small amount. The investment in effort and cost needed to develop the specific primer pairs for each locus is their primary drawback.

Single Nucleotide Polymorphism (SNPs) is co-dominant markers, often linked to genes and present in the simplest/ultimate form for polymorphism, and thus they have become very attractive and potential genetic markers in genetic study and breeding. Moreover, SNPs can be very easily automated and quickly detected, with a high efficiency for detection of polymorphism. Therefore, it can be expected that SNPs will be increasingly used for various purposes, particularly as whole DNA sequences become available for more and more species (*e.g.*, rice, soybean, maize, *etc.*). However, high costs for start-up or marker development, high-quality DNA required and high technical/equipment demands limit, to some extent, the application of SNPs in some laboratories and practical breeding programs.

Microsatellites (SSRs, STMS or SSRPs) are also called simple sequence repeats (SSRs) and, occasionally, sequence-tagged microsatellite sites (STMS) or simple sequence repeat polymorphisms (SSRPs). They are by far the most widely used type of STS. SSRs are short tandem repeats, their length being 1 to 10 bp, most typically, 2-3 bp. SSRs are highly variable and evenly distributed throughout the genome. This type of repeated DNA is common in eukaryotes, their number of repeated units varying widely among organisms to as high as 50 copies of the repeated unit. These polymorphisms are identified by constructing PCR primers for the DNA flanking the microsatellite region. The flanking regions tend to be conserved within the species, although sometimes they may also be conserved in higher taxonomic levels.

Single Strand Conformation Polymorphism (SSCP) Single-strand conformational polymorphism (SSCP) analysis is a simple and sensitive technique for mutation detection and genotyping. The principle of SSCP analysis is based on the fact that single-stranded DNA has a defined conformation. Altered conformation due to a single base change in the sequence can cause singleAdvances stranded DNA to migrate differently under non denaturing electrophoresis conditions. Therefore wild-type and mutant DNA samples display different band patterns. SSCP analysis involves the following four steps: 1) polymerase chain reaction (PCR) amplification of DNA sequence of interest; 2) denaturation of double-stranded PCR products 3) cooling of the denatured DNA (single-stranded) to maximize self-annealing 4) detection of mobility difference of the single-stranded DNAs by electrophoresis under nondenaturing conditions. Several methods have been developed to visualize the SSCP mobility shifts. These include the incorporation of radioisotope labelling, silver staining, fluorescent dye-labelled PCR primers, and more recently, capillary-based electrophoresis. Silver staining is simple, rapid, and cost-effective, and can be routinely performed in clinical laboratories.

Single polymorphic amplification test (SPLAT) : If sequence tagged site (STS) does not reveal polymorphism it is usually converted into SPLAT. Individual STS products from different genotypes are sequenced any differences revealed can be sequence from nuclear ribosomal DNA. They can be exploited in production of internal primers. Single amplified sequences are dominant markers.

Variable Number of Tandem Repeats (VNTRs) is a location in a genome where a short nucleotide sequence is organized as a **tandem repeat**. These can be found on many chromosomes, and often show variations in length (**number** of **repeats**) among individuals.

RNA- Based Molecular Markers (RBMS) : Markers derived from transcribed/ expressed regions of genomes. Since they are based on expressed portion of genome, it could be also affected by phenological plant stage and environmental conditions. The generated fragments can be easily associated with phenotypic traits (QTLs Genetic mapping studies of qualitative traits loci). Among PCR based markers are – iSNAP inter small RNA polymorphism – based on endogenous non- coding small RNAs consisting of 20-24 nucleotides, highly reproducible and useful in genome mapping and genotyping. Other markers cDNA- SSCP, cDNA- AFLP, cDNA–RFLP and RAPPCR– these are used for differential RNA studies using selective amplification of cDNA. Molecular markers have facilitated in resolving complications of conventional GOT and DUS testing. Molecular markers methods are being adopted by the researchers for fingerprinting cultivars, assessment of genetic purity and varietal identification in seed technology and breeding programmes. In genetic purity assessment, molecular markers detect the degree of contamination due to selfing and out-crossing in a hybrid seed lot, segregation of genotypes in DUS testing and trait confirmation in transgenic. For varietal identification it is desirous to obtain a specific/unique pattern for each variety. Such patterning provides protection to varieties in DUS test and characterization of Essentially Derived Varieties (EDVs).

Retrotransposon-Based Molecular Markers are a class (Class I) of transposable elements (TE) and constitute a major portion of the genome in plants with large genomes. They are important for chromatin modifications and epigenetic reprogramming. Retrotransposons are also an ideal target for developing molecular marker techniques because of their amplification mechanism and sequence characteristics. There are different types of transposon based marker techniques. Some of them are; Inter Retrotransposon Amplified Polymorphism (IRAP), Retrotransposon Microsatellite Amplified Polymorphism (REMAP), inter Primer Binding Site amplification (iPBS), Sequence Specific Amplification Polymorphism (S-SAP), Retrotransposon Based Insertion Polymorphism (RBIP), Inter Sine Amplified Polymorphism (ISAP), RAPD- Retrotransposon Amplified Polymorphism (RRAP), Inverse Sequence Tagged Repeats (ISTR), Inter-MITE Polymorphism (IMP) and Transposable display (TD). These methods are the most widely used for different breeding strategies like; the determination of candidate genes, genetic variability, linkage, genome mapping, the DNA fingerprinting, phylogenetics, systematics, conservation, molecular ecology, somaclonal variation studies, transgenic research, developmental biology, and mutagenesis. Therefore, it is concluded that genetic purity analysis using molecular markers may be used as a potential tool for resolving the problems arise in seed certification program owing to genetic impurity as well as the rapid determination of genetic purity of hybrids with the deployment of molecular makers as compared to GOT. The use of precise molecular marker would definitely replace GOT in the days to come, which is obviously very tedious, time consuming and highly vulnerable to manual mistakes that normally occur during production, processing and marketing.

Prospectus and Constraints in Seed Production

Indian seed industry has been growing awfully in quantity and value over the past fifty years. Both public and private sector corporations/companies are actively involving in quality seed production. The public sector component comprises National Seeds Corporation (NSC), State Farm Corporation of India (SFCI) and 15 State Seeds Corporations (SSCs) (APEDA, 2013 and Nandi *et al.*, 2013), Indian Council of Agricultural Research (ICAR) institutions and State Agricultural Universities. ICAR launched an All India Coordinated Research Improvement project (AICRP) on seed production called National Seed Project in 1979 with 14 centres in different Agricultural Universities (Nandi *et al.*, 2013). AICRP on production of breeder seed in crops is started under National Seed Project in 1994. Twenty two State Seed Certification Agencies and 104 State Seed Testing Laboratories are involving in quality control and certification (Nandi *et al.*, 2013). The private sector comprises around 150 seed companies of national and foreign origin. The Indian public sector seed industry used to dominate the private sector in the very beginning. The order of type of seeds dominating the market in terms of quantity and value has been open-pollinated varieties followed by public hybrids and private hybrids (APEDA, 2013). The situation is quite reversed currently. Seeds of the private hybrids are forming a significant portion of the total seed market. The availability of seeds with NSC as on 2013 is 133.43 t of which 131.68 t of varietal seeds and only 1.75 t of hybrid seeds. Due to new seed policy in 1988, the business of private seed companies taken a new height. Now a day the public sector is mostly confined to certified seeds of high volume, low value segment of high yielding varieties of cereals, pulses and cotton with a limited presence in the high value hybrid sectors of cotton and cereals (APEDA, 2013 and Kumar *et al.*, 2011). In most of the public sector varieties and hybrids are replaced by private sector varieties and hybrids, seed production of which is solely done by the particular manufacturers. Corporate seed firms are mainly concentrating where the seed production of OPVs and hybrids is comparatively easy and more profitable. Private seed corporations are spending 10-12 per cent of their turnover in R and D. Medium sized seed companies annual investment in R and D is growing 20 per cent annually (Anonymous, 2013 and APEDA, 2013). The gargantuan seed manufacturers with multinational base can assemble germplasm from any part of the world which cannot be done by the public sector institutions in easy way. As germplasm serves invaluable resource for any crop improvement programme, this makes the big difference. Moreover, these mega seed giants employ paramount technical personnel with opulence and extravagance, which can never be done by the public sector institutions. At present Indian seed industry is the fifth largest seed market in the world accounting for 4.4 per cent global seed market after US, China, France and Brazil. Indian seed market has grown at 12 per cent rate where the growth rate of global seed market is 5 per cent (Vanitha *et al.*, 2013) while the Indian seed market is growing at a rate of 10-15 per cent in a year. There has been an increase of 194 per cent in Indian hybrid seed market during 1998-2008 (Kumar *et al.*, 2011). The estimated turnover of Indian seed industry (50000 m Rs.) is four percent of the global seed turnover (1250000 m RS.) (Kumar *et al.*, 2011). The seed business in India, at present, amounts for 9000 million Rs. accounting for 18 per cent value wise share of different crops in Indian seed business (Kumar *et al.*, 2011). In future, according to National Seed Association of India (NSAI), the size of Indian seed market is expected to grow at a rate of 11 per cent p.a. to $3.2 bn till FY 2016 on account of favourable global grain supply demand fundamentals, grain productivity well below world's major grain producing regions and government's continued focus on improving seed replacement rate.

Factors Promoting Seed Industry in India

Ever Increasing Demand

The worldwide production has doubled over the past quarter century and the value of global trade now exceeds that of cereals. India is emerging as the second largest producer. Increase in yield is mainly attributed to expanding areas under high yielding varieties and hybrids. Total cultivated area under has been increased from 5.59 m ha in 1991-92 to 8.49 m ha in 2010-11 (Koundinya *et al.*, 2014). Finally, it leads to ever increasing demand for the quality seed. Moreover, the yield of crops are higher when produced from and replaced seeds than own saved seeds. Seed replacement rates are high like to other cereals and oil seeds (Mazumdar, 2012). Total quantity of seeds produced in the country is not sufficient to meet the country's ever increasing demand. Currently quality seeds are met to the extent of 20 per cent only. Framers themselves meet the 75 per cent through own saved seeds (Nandi *et al.*, 2013). India is still importing the seeds from other countries major (Sudha *et al.*, 2006).

Varied Agro Climatic Conditions

India is blessed with assorted agro climatic conditions ranging from tropical to temperate which make possible the cultivation and seed production of all most all belonging to different temperature regimes. Seed production of warm season is possible in Indian plains and Deccan Plateau and seed production (Prasad *et al.*, 2009).

Cheap Labour Availability

Seed production particularly hybrid seed production demands much labour. Labour is needed for performing various cultural operations. Though mechanization reduces the human effort up to some extent, high cost fuel and energy limitations reduce full scale mechanization. Moreover, emasculation and pollination steps during hybrid seed production solely depend on human labour (Sharma, 2011). Smaller flower structure of some need more devotion of time and reduces human efficiency. These operations require specially trained and skilled labour. India is ranked second in hand pollinated seed production in Asia next to China (Prasad *et al.*, 2009). Average number of man-days per acre required for hybrid seed production (Gadwal, 2003). India is having huge human resources availing at reasonably cheaper rates Prasad *et al.*, 2009). This is attracting various corporate sectors of national and international origin to invest in seed business in India.

Rules and Regulations in Seed Act for Quality Seed Production

Seed is the most important input for agricultural production. Efficacy of other agricultural inputs such as fertilizers, pesticides and irrigation is largely determined by the quality of seed. Seed quality accounts for 20-25 per cent of agricultural productivity. It is therefore, important to ensure that the farmers of the country use quality seeds. The seed quality regime in the country is governed by the Seeds Act, 1966, the Seed Rules, 1968 made there under and the Seeds (Control) Order, 1983. In addition to these legal instruments, various guidelines and policy parameters have been laid down in the National Seeds Policy, 2002.

Various committees formed and their function in the seed act are:

Central Seed Committee

The Central Government shall, as soon as may be after the commencement of this Act, constitute a Committee called the Central Seed Committee to advise the Central Government and the State Governments on matters arising out of the administration of this Act and to carry out the other functions assigned to it by or under this Act. The Committee shall consist of the following members, namely: i Chairman to be nominated by the Central Government; ii eight persons to be nominated by the Central Government to represent such interests that Government thinks fit, of whom not less than two persons shall be representatives of growers of seed; iii one person to be nominated by the Government of each of the States. The Committee may, subject to the previous approval of the Central Government, make bye-laws fixing the quorum and regulating its own procedure and the conduct of all business to be transacted by it. The Committee may appoint one or more sub-committees, consisting wholly of members of the Committee or wholly of other persons or partly of members of the Committee and partly of other persons, as it thinks fit, for the purpose of discharging such of its functions as may be delegated to such subcommittee or sub-committees by the Committee. The functions of the

Committee or any sub-committee thereof may be exercised notwithstanding any vacancy therein. The Central Government shall appoint a person to be the secretary of the Committee and shall provide the Committee with such clerical and other staff as the Central Government considers necessary.

Power to Notify Kinds or Varieties of Seeds

If the Central Government, after consultation with the Committee, is of opinion that it is necessary or expedient to regulate the quality of seed of any kind or variety to be sold for purposes of agriculture, it may, by notification in the Official Gazette, declare such kind or variety to be a notified kind or variety for the purposes of this Act and different kinds or varieties may be notified for different States or for different areas thereof.

Power to Specify Minimum Limits of Germination and Purity, *etc.*

The Central Government may, after consultation with the Committee and by notification in the Official Gazette, specify the minimum limits of germination and purity with respect to any seed of any notified kind or variety; the mark or label to indicate that such seed conforms to the minimum limits of germination and purity specified under clause (a) and the particulars which such mark or label may contain.

Regulation of Sale of Seeds of Notified Kinds or Varieties

No person shall, himself or by any other person on his behalf, carry on the business of selling, keeping for sale, offering to sell, bartering or otherwise supplying any seed of any notified kind or variety, unless. such seed is identifiable as to its kind or variety; b. such seed conforms to the minimum limits of germination and purity specified under clause (a) of section 6; c. the container of such seed bears in the prescribed manner, the mark or label containing the correct particulars thereof, specified under clause (b) of section

Central Seed Laboratory and State Seed Laboratory

The Central Government may, by notification in the Official Gazette, establish a Central Seed Laboratory or declare any seed laboratory as the Central Seed Laboratory to carry out the functions entrusted to the Central Seed Laboratory by or under this Act. The State Government may, by notification in the Official Gazette, establish one or more State Seed Laboratories or declare any seed laboratory as a State Seed Laboratory where analysis of seeds of any notified kind or variety shall be carried out by Seed Analysts under this Act in the prescribed manner.

Certification Agency

The State Government or the Central Government in consultation with the State Government may, by notification in the Official Gazette, establish a certification agency for the State to carry out the functions entrusted to the certification agency by or under this Act.

Grant of Certificate by Certification Agency

Any person selling, keeping for sale, offering to sell, bartering or otherwise supplying any seed of any notified kind or variety may, if he desires to have such seed certified by the certification agency, apply to the certification agency for the grant of a certificate for the purpose. Every application under sub-section (1) shall be made in such form, shall contain such particulars and shall be accompanied by such fees as may be prescribed. (3) On receipt of any such application for the grant of a certificate, the certification agency may, after such enquiry as it thinks fit and after satisfying itself that the seed to which the application relates conforms to the minimum limits of germination and purity specified for that seed under clause (a) of section 6, grant a certificate in such form and on such conditions as may be prescribed.

Revocation of Certificate

If the certification agency is satisfied, either on a reference made to it in this behalf or otherwise, that the certificate granted by it under section 9 has been obtained by misrepresentation as to an essential fact; or b. the holder of the certificate has, without reasonable cause, failed to comply with the conditions subject to which the certificate has been granted or has contravened any of the provisions of this Act or the rules made thereunder; then, without prejudice to any other penalty to which the holder of the certificate may be liable under this Act, the certification agency may, after giving the holder of the certificate an opportunity of showing cause, revoke the certificate.

Appeal

Any person aggrieved by a decision of a certification agency under section 9 or section 10, may, within thirty days from the date on which the decision is communicated to him and on payment of such fees as may be prescribed, prefer an appeal to such authority as may be specified by the State Government in this behalf: Provided that the appellate authority may entertain an appeal after the expiry of the said period of thirty days if it is satisfied that the appellate was prevented by sufficient cause from filing the appeal in time. (2) On receipt of an appeal under sub-section (1), the appellate authority shall, after giving the appellant an opportunity of being heard, dispose of the appeal as expeditiously as possible. (3) Every order of the appellate authority under this section shall be final.

Seed Analysts

The State Government may, by notification in the Official Gazette, appoint such persons as it thinks fit, having the prescribed qualifications, to be Seed Analysts and define the areas within which they shall exercise jurisdiction.

Seed Inspectors

The State Government may, by notification in the Official Gazette, appoint such persons as it thinks fit, having the prescribed qualifications, to be Seed Inspectors and define the areas within which they shall exercise jurisdiction.

Powers of Seed Inspector

The Seed Inspector may, a. take samples of any seed of any notified kind or variety fromi. any person selling such seed; or ii. any person who is in the course of conveying, delivering or preparing to deliver such seed to a purchaser or a consignee; or iii. a purchaser or a consignee after delivery of such seed to him; b. send such sample for analysis to the Seed Analyst for the area within which such sample has been taken; c. enter and search at all reasonable times, with such assistance, if any, as he considers necessary, any place in which he has reason to believe that an offence under this Act has been or is being committed and order in writing the person in possession of any seed in respect of which the offence has been or is being committed, not to dispose of any stock of such seed for a specific period not exceeding thirty days or, unless the alleged offence is such that the defect may be removed by the possessor of the seed, seize the stock of such seed; d. examine any record, register, document or any other material object found in any place mentioned in clause (c) and seize the same if he has reason to believe that it may furnish evidence of the commission of an offence punishable under this Act; and e. exercise such other powers as may be necessary for carrying out the purposes of this Act or any rule made thereunder. (2) Where any sample of any seed of any notified kind or variety is taken under clause (a) of sub-section (1), its cost, calculated at the rate at which such seed is usually sold to the public, shall be paid on demand to the person from whom it is taken. (3) The power conferred by this section includes power to break-open any container in which any seed of any notified kind or variety may be contained or to break-open the door of any premises where any such seed may be kept for sale: Provided that the power to break-open the door shall be exercised only after the owner or any other person in occupation of the premises, if he is present therein, refuses to open the door on being called upon to do so. (4) Where the Seed Inspector takes any action under clause (a) of sub-section (1), he shall, as far as possible, call not less than two persons to be present at the time when such action is taken and take their signatures on a memorandum to be prepared in the prescribed form and manner. (5) The provisions of the Code of Criminal Procedure, 1898 (5 of 1898), shall, so far as may be, apply to any search or seizure under this section as they apply to any search or seizure made under the authority of a warrant issued under section 98 of the said Code.

Procedure to be Followed by Seed Inspectors

Whenever a Seed Inspector intends to take sample of any seed of any notified kind or variety for analysis, he shall give notice in writing, then and there, of such intention to the person from whom he intends to take sample; b. except in special cases provided by rules made under this Act, take three representative samples in the prescribed manner and mark and seal or fasten up each sample in such manner as its nature permits. When samples of any seed of any notified kind or variety are taken under subsection (1), the Seed Inspector shalla. deliver one sample to the person from whom it has been taken; b. send in the prescribed manner another sample for analysis to the Seed Analyst for the area within which such sample has been taken; and c. retain the remaining sample in the prescribed manner for production in case any legal proceedings are taken or for analysis by the Central Seed Laboratory under sub-section (2) of section 16, as the case may be. If the person from whom the samples have been taken refuses to accept one of the samples, the Seed

Inspector shall send intimation to the Seed Analyst of such refusal and thereupon the Seed Analyst receiving the sample for analysis shall divide it into two parts and shall seal or fasten up one of those parts and shall cause it, either upon receipt of the sample or when he delivers his report, to be delivered to the Seed Inspector who shall retain it for production in case legal proceedings are taken. Where a Seed Inspector takes any action under clause (c) of sub-section (1) of section 14: a. he shall use all dispatch in ascertaining whether or not the seed contravenes any of the provisions of section 7 and if it is ascertained that the seed does not so contravene, forthwith revoke the order passed under the said clause or, as the case may be, take such action as may be necessary for the return of the stock of the seed seized; b. if he seizes the stock of the seed, he shall, as soon as may be, inform a magistrate and take his orders as to the custody thereof; c. without prejudice to the institution of any prosecution, if the alleged offence is such that the defect may be removed by the possessor of the seed, he shall, on being satisfied that the defect has been so removed, forthwith revoke the order passed under the said clause. (5) Where as Seed Inspector seizes any record, register, document or any other material object under clause (d) of sub-section (1) of section 14, he shall, as soon as may be, inform a magistrate and take his orders as to the custody thereof.

Report of Seed Analyst

The Seed Analyst shall, as soon as may be after the receipt of the sample under sub-section (2) of section 15, analyse the sample at the State Seed Laboratory and deliver, in such form as may be prescribed, one copy of the report of the result of the analysis to the Seed Inspector and another copy thereof to the person from whom the sample has been taken. After the institution of a prosecution under this Act, the accused vendor or the complainant may, on payment of the prescribed fee, make an application to the court for sending any of the samples mentioned in clause (a) or clause (c) of sub-section (2) of section 15 to the Central Seed Laboratory for its report and on receipt of the application, the court shall first ascertain that the mark and the seal or fastening as provided in clause (b) of sub-section (1) of section 15 are intact and may then dispatch the sample under its own seal to the Central Seed Laboratory which shall thereupon send its report to the court in the prescribed form within one month from the date of receipt of the sample, specifying the result of the analysis. The report sent by the Central Seed Laboratory under sub-section (2) shall supersede the report given by the Seed Analyst under sub-section (1). Where the report sent by the Central Seed Laboratory under sub-section (2) is produced in any proceedings under Section 19, it shall not be necessary in such proceedings to produce any sample or part thereof taken for analysis.

During the last four decades, seed production technology has changed and new technologies like transgenic, tissue culture, soil-less agriculture *etc.* have emerged. There is greater emphasis on seed quality assurance particularly to safeguard the interest of the farmers. In addition, significant changes have occurred in the country's socio-economic conditions. The economy has been substantially liberalized and the private sector is playing an increasing role in various spheres including agriculture. India is also increasingly engaged with the rest of the world through Organization for Economic co-operation and development (OECD) and International Seed Testing

Association (ISTA), *etc.* In the liberalized and changed environment, imports and exports of seeds and planting materials into the country have increased. These ongoing changes require upgradation of seed production, quality and regulatory standards. Therefore, a need has arisen for updating the seed quality regulatory regime by enacting a new legislation.

Scope of the Seeds Bill, 2004

Seeds Bill, 2004 covers all the crops and varieties of agricultural, horticultural, forestry, plantation, medicinal and aromatic plants.With respect to fruit nurseries, the Bill proposes compulsory registration of fruit nursery of above one hectare area with State Governments.

Objectives of The Seeds Bill, 2004

The objectives of the Seeds Bill, 2004 are to regulate the quality of seeds and planting materials of all agricultural, horticultural and plantation crops to Ensure availability of true to type seeds for Indian farmers, Curb the sale of spurious and poor quality seeds, Protect the rights of farmers, increased private participation in seed production, Distribution and seed testing, Liberalize imports of seeds and planting materials, *etc.*

Status of Seeds Bill, 2004

The Seeds Bill, 2004 was introduced in the Rajya Sabha in December, 2004. Thereafter, it was referred to the Parliamentary Standing Committee on Agriculture on 16th December, 2004.Parliamentary Standing Committee on Agriculture submitted the report on 12th October, 2006. The Bill is now pending in Rajya Sabha.

The Seeds Bill, 2004 mainly seeks to advance the manifold objectives Enhance the quality of seeds marketed in the country by mandated compulsory registration of varieties; It enhances penalties for offences, compared to the existing Seeds Act 1966; It makes provision for labelling of expected performance of seeds and also provides for compensation to farmers in case of seed failure; It adds seed health as an additional standard for seed quality; It brings conformity with existing National Laws and regulations, especially those relating to the import and export of seeds and use of Genetically Modified seeds; It imposes a ban on the use of Genetic Use Restriction Technology (GURT); It provides for registration of seed producers, seed dealers and seed processing units; Farmers' Rights: The Bill proposes to emphasize the rights of the Farmers in conformity with the PPV and FR Act, 2001. The farmer will have the right to save, use, sow, re-sow, exchange, share or sell his farm seeds and planting materials except when he sells such seeds or planting material under a brand name. However, **the farmers will be exempted from compulsory registration of varieties/nurseries.**

Salient Features

Compulsory Registration of Varieties

No person will be allowed to carry on the business of selling or supplying any seed which is not of a registered kind/variety. Farmers are to be exempted from compulsory registration.

Period of Registration

10 years for annual and biennial crops and 12 years for long duration perennials. Re-registration for a like period after re-establishment of performance in the multi-location trials.

Registration of Transgenic Varieties

Registration of transgenic varieties will be subject to environmental clearance from MOEF and CC under the Environmental (Protection) Act, 1986 and Rules 1988 made there under.

Enhancement of Penalties

It is proposed to enhancement of penalty from Rs. 500/- Rs. 1000/- with or without imprisonment in the Seeds Act 1966 to Rs. 25,000/- Rs. 5,00,000 with or without imprisonment.

Expected Performance

Provision for labelling of expected performance of seeds has been included so that the farmers are assured of quality of seeds purchased by them.

Compensation to the Farmers on Seeds Failure

Provision for compensation to the farmer in case of seed failure has been made.

Farmers' Rights

The bill proposes to emphasise the rights of farmers in conformity with the PPV and FRA (PPV and FR Act, 2001). The farmer will have the right to save, use, sow, re-sow, exchange, share or sell his farm seeds and planting materials except when he sells such seed or planting material under a brand name.

Exclusion of Certain kind of Variety of Seeds from Registration

The Government will have the right to exclude certain kind or variety of seeds from registration to protect public order or public morality or human, animal and plant health or to avoid serious prejudice to the environment.

Seeds Health

Seed health has been included as an additional standard for quality seed.

Cancellation of Registration of Varieties

The Registration of varieties under the Seed Bill will be cancelled on many grounds like Violation of terms and conditions of the registration; Registration obtained by misrepresentation or concealment of essential data; Non-performance of variety and obsolete variety; Commercial exploitation of a variety; Public interest; Protect human beings, animal and plant life and health to avoid serious prejudice to the environment; Provisions to regulate import and export of seeds have been largely incorporated into the Seeds Bill itself are manifold *viz.*, All import of seed shall be subject to Plant Quarantine (Regulation of Import into India) Order, 2003 and other relevant Acts; All imported seeds shall conform to minimum standards of seeds health in addition to other conditions already in existence; All imported seeds shall be subject to registration on the basis of information furnished by the importer on the basis of multi location trials; Exports can be restricted if such exports adversely affect the food security of the country; Genetic Use Restriction Technology (GURT): (**Restriction technology including terminator technology)** has been prohibited. Any person intending to import seed or planting material will declare that such material is, or is not, as the case may be a product of transgenic manipulation or involves Genetic Use Restriction Technology.

Table 10.11. Comparative Analysis of the Seeds Act, 1966 and the Seeds Bill, 2004

The Seeds Act, 1966	*The Seeds Bill, 2004*
Covers seeds of food crops, fruits, vegeTables, fodder, cotton and jute only.	It covers all the crops.
No explicit farmers' exemption.	Explicit farmer's exemptions. The Bill shall not restrict the right of the farmer to save, use, sow, re-sow, exchange, share or sell his farm seeds and planting material except when he sells such seed or planting material under a brand name.
Voluntary notification of variety by Central Government.	Dispensed with the provision of notification of varieties. In place of notification, registration of varieties will be compulsory. Registration of State varieties by the SSC and national varieties by the CSC.
The Central Government notifies the varieties based on the recommendations of the CSC for indefinite period.	The registration Sub-Committee based on the performance in the MLT shall register the varieties for a period of 10 years for annual and biennial crops and 12 year for perennial crops.
The standards to be prescribed will be for germination and purity with respect to notified variety.	The standards to be prescribed will be for germination, physical and genetic purity, seed health and standards for transgenic seeds of all registered varieties sold in the market.
No provision for giving expected performance of variety on the seed label.	Information about expected performance of variety shall be compulsory to be given on the seed label.
No separate provisions for notification of transgenic varieties.	Special provisions for registration of transgenic varieties.
Minimum limits of germination and purity required for import of seeds.	The imported varieties will also be required to be registered if it will meet all the seed quality standards.
Mild penal provision of Rs.500/1000 with or without imprisonment (6 months).	Enhanced penalty for offences *i.e.* Rs 25,000/5,00,000 with or without imprisonment (1 year).
Seeds Inspector may enter into the premises of seed dealer to take samples without approval of any authority.	Seeds Inspector may enter and make search of the premises of seed dealer with the prior approval of Executive Magistrate.
Minimum limits of germination and purity required for import of seeds.	The imported varieties will also be required to be registered if it will meet all the seed quality standards.

The Seeds Act, 1966	*The Seeds Bill, 2004*
Mild penal provision of Rs.500/1000 with or without imprisonment (6 months).	Enhanced penalty for offences *i.e.* Rs 25,000/5,00,000 with or without imprisonment (one year).
Seeds Inspector may enter into the premises of seed dealer to take samples without approval of any authority.	Seeds Inspector may enter and make search of the premises of seed dealer with the prior approval of Executive Magistrate.
Seed Inspector can stop the sale of seeds of suspicious quality upto 30 days for regulating the quality.	Seed Inspector can stop the sale of seeds of suspicious quality up-to 15 days for regulating the quality.
New provisions.	Seed price regulation
Prohibition of GURT including Terminator technology.	Compensation to farmers in case seed failure.
Compulsory registration of Seed producer, Seed Processing Unit and Seed Dealers.	Regulation of fruit nurseries.
Accreditation of Government organizations as Certification Agency	Exemption of Educational, Scientific and Research Organization from all the provisions of this Act.

Public-Private Partnership in Seed Production and Marketing

Agriculture sector accounts for 13.9 per cent to India's gross domestic product (GDP) and employs just a little less than 54.6 per cent of the country's workforce. The Department of Agriculture and Cooperation under the Ministry of Agriculture and farmer's welfare is the nodal organization responsible for the development of the agriculture sector in India. The Immediate challenge to the Ministry of Agriculture, is to sustain the increasing agricultural output of the country. This can be addressed by developing Technology transfer and seed Innovation systems. Indian seed sector is characterized by diverse crops with both public and private sector playing pivotal role. Of the various technological interventions seed is considered as critical basic input for aiming sustained long term growth in agricultural production. Successful production begins with good seed, and many farmers are prepared to pay for such a valuable input. Everything else in production is profoundly affected by the quality of the seed used, the characteristics of the variety, and the potential quantity and quality of its marketable produce. Farmers invest in something that has such significant payoffs, and the private sector invest in marketing such valuable commodities. Because of its value to all parties involved, seed. production and marketing is a key area for successful public-private partnerships for development.

Public-Private Partnership (PPP)

Public private partnership (PPP) is a contractual agreement between a public agency (Government at central or local level) and a private sector entity, in which the private party provides a public service and assumes substantial financial, technical and operational risks and rewards inherent in it. PPP is essentially a business venture which is funded and operated through a partnership of government and one or more private sector companies. The key is partnership between the government and private business sector(s). The structure of the partnership should be designed to allocate risks to the partners who are the best able to manage those risks and thus minimize costs while improving performance. Risk transfer is one of the major components through which PPP projects can generate better value for money. Different types of risks such as public risks, market risk, economic and financial risk, construction risks, environmental risks *etc.* are included in the PPP model (NPC, 2011). Public private partnership reduces public capital investment and improves efficiency due to strong profit incentive Private entity is more accountable than government and results in expedited project completion by grouping multiple responsibilities into a single contract. Public private partnership results in specialized expertise and relieves the government from staffing issues. Further it shares risk/responsibility involved in the project. Government can still step in when private entities are not acting in the desired manners. In public and private partnership, the two players are government and private agencies. The public sector has the social responsibility, whereas, the motive of the private sector is profit oriented. Public sector has extensive infrastructure and higher institutes of learning which can generate knowledge through basic research. In contrary to this private sector is known to have better marketing skills, operational efficiencies, efficient product delivery system, quality service providing capability, and better up-scaling technologies. The main objective of public sector organisation is to conduct research, whereas in private sector research, production and marketing are integrated activities. For developing any model of public private partnership, Identifying and negotiating the common interest is of utmost importance. How the project will be financed should be clear to the partners in the model. All regulatory/ contractual and legal issues must be settled before entering into a contract for its successful completion. So keeping these facts in mind, an appropriate organizational design should be adopted. Provisions for monitoring should also be kept. The basic reason why our policy makers over the years pushed for PPP model was to tap the efficiency and management skills of the Private Sector. The other fundamental reason was to cut red tape and improve flexibility. Government bodies are bound by inflexible rules and regulations which makes decision making very difficult. The PPP model was supposed to change the rules of the game ensuring speed of implementation and avoiding cost overruns. In addition to maximising efficiencies and innovations of Private Enterprise PPPs can provide much needed Capital to finance Government Programs and Projects, thereby freeing public funds for core Economic and Social Sector. In spite of all these noble intentions and so called support from policy makers why is this proving to be a failure? Why the Private sector is now no longer as enthusiastic as it was in being a "Partner in Progress" needs the attention of the governments. Public Private Partnership (PPP) model is supposed to be a collaborative model between Govt. Agency and the Private Sector. The inflexibility of the Govt. partner in a PPP model some time go to silly extremes. Hence the PPP model in its current 'form' needs to be given a decent burial or suitably tweaked to respond to changing Economic and Political environment of the country. Govt. Agency and the Private Sector have to be partners in progress and not adversaries. The bureaucrats manning the Govt. Agencies need to get down from their high horses and should be equally made accountable for the success or failure of the Projects. If Private Sector can fire executives for non-performance what stops Govt. from demanding same from its officers. All this requires a change in mind-set. And will be helpful in achieving the objectives of PPP models in ensuring the economic growth.

Seed Rate

Seed rate is the quantity of seed required for sowing or planting in an unit area. The seed rate for a particular crop would depend not only on its seed size/test weight, but also on its desired population, germination percentage and purity percentage of seed. It is calculated as follows:

$$\text{Seed rate (kg)} = \frac{\text{Area to be sown in m}^2 \times \text{Test weight of the seed} \times 1}{\text{Germination per cent} \times \text{Purity per cent} \times \text{Spacing (m)} \times 1000}$$

Seed Treatment

Seed treatment is a process of application either by mixing or by coating or by soaking in solutions of chemicals or protectants (with fungicidal, insecticidal, bactericidal, nematicidal or biopesticidal properties), nutrients, hormones or growth regulators or subjected to a process of wetting and drying or subjected to reduce, control or repel disease organisms, insects or other pests which attack seeds or seedlings growing there from. Seed treatment also includes control of pests when the seed is in storage and after it has been sown/planted. The seed treatment is done for the reasons *viz.*, to protect from seed borne pests and diseases; to protect from or repel birds and rodents; to supply plant nutrients; to inoculate microorganisms; to supply growth regulators; to supply selective herbicides; to break seed dormancy; to induce drought tolerance; to induce higher germination percentage, early emergence; to obtain polyploids (genetic variation) by treating with X-rays, gamma rays and colchicines; too facilitate mechanized sowing.

Methods of Seed Treatment

1. *Dry treatment:* Mixing of seed with powder form of pesticides/nutrients.
2. *Wet treatment:* Soaking of seed in pesticide/nutrient solutions
3. *Slurry treatment:* Dipping of seeds/seedlings in slurry. Example–rice seedlings are dipped in phosphate slurry.
4. *Pelleting:* It is the coating of solid materials in sufficient quantities to make the seeds larger, heavier and to appear uniform in size for sowing with seed drills. Pelleting with pesticides as a protectant against soil organisms, soil pests and as a repellant against birds and rodents. The seed treatment for different field crops is given in chapter 15.

Public Vs. Private Plant Breeding

Plant breeding in the United States and Canada is conducted both by public and by private resources. The public plant breeding projects in the United States are conducted by tax-supported agencies such as the United States Department of Agriculture, the state agricultural experiment stations, and the agricultural colleges. In Canada they are conducted by the Canada Department of Agriculture, the provincial governments, and the agricultural colleges. Private plant-breeding research is conducted by private industries, such as hybrid seed corn companies, seed cotton firms, and the beet sugar industry. The extent to which plant breeding is done by public or by private agencies varies with the importance of the crop and the resources of the industry. In crops with large recurrent sales of seed, such as hybrid corn, hybrid sorghum, sugar beets, or cotton, private breeders participate to a much greater extent than for crops in which the recurrent seed sales are small. Private industry in the United States and Canada participates very *little* in the breeding of *self-pollin*ated crops such as wheat, oats, soybeans, flax, and tobacco, in which seed sales are relatively small after the initial distribution of a new variety has been made. This is a logical development since seed companies can finance plant breeding programs only in relation to the returns from their seed sales. Some exceptions are found in alfalfa and certain other forage crops where large recurrent seed sales are normal, yet there are no effective private breeding programs in forage crops. Breeding research for all major crops is conducted by various tax-supported agencies. Public support is also used in varying degrees to finance fundamental research which may lead to an advancement in knowledge of plant breeding procedures. At present the latter type of research is supported very little by private means and is therefore almost entirely dependent upon public support. In recent years cooperation between public and private breeders has been increasing. This is a trend that should be encouraged, for it will lead to increased benefits to both groups.

Improved varieties developed by tax-supported institutions, such as state agricultural experiment stations or colleges of agriculture, may be considered public property. Hence it is in the public interest that new varieties developed by these public agencies be increased as rapidly as possible and distributed in an orderly fashion so that seed will become available soon to everyone who may wish to grow the new varieties. It is desirable also that no one be permitted to exploit a new variety developed by a public institution by charging exorbitant prices for seed while the seed supply of the new variety is still scarce. The true benefits of an improved variety should come from the increased production received by the grower, not from excessive profits from the sale of seed that is still in short supply. For the purpose of expediting the orderly and efficient increase of new varieties and efficient increase of new varieties and maintaining pure seed supplies of old varieties, seed improvement organizations have been organized in nearly all the states, in Canada, and in many other countries. Through these organizations, procedures have been developed for the increase, distribution, and certification of new varieties originating at public institutions or, in some instances, by private breeders. Procedures have been established also for maintenance of pure seed and for certification of varieties after distribution. Private seed companies which conduct breeding programs generally have established outlets for marketing the seed of their new varieties. Many of the larger companies, such as those that produce hybrid corn or seed cotton, have efficient advertising departments and sales staffs already in the field. In the case of sugar beets, the beet sugar companies distribute seed of improved varieties directly to the growers who supply them with sugar beets. The increase and distribution of seed of a new variety developed by a private breeder is generally controlled by the seed company financing the breeding program, with the actual operations carried out through the company's production department. In some cases, privately bred varieties may be produced by other growers under a seed certification program. The public generally has little information or knowledge of the problems involved in the increase of privately developed and privately distributed varieties, or how they are handled. However, the technical operations by which pure seed stocks are increased and maintained are similar, whether carried out by a seed improvement association or a commercial seed company.

Before a variety is distributed from a state experiment station, it is generally tested thoroughly in the state, or states, where it originated and from which it is being distributed. Through regional cooperation the tests may be conducted over an area of several states. The results of these tests are available to make possible recommendations about the area of adaptation of the new variety. Generally less information is available to guide the grower regarding the acceptance of a new variety developed by a private plant breeder. For this information the grower must rely almost completely upon the integrity of the company and its advertising. Generally, large seed companies conduct yield trials, comparing the performance of their new strains with those in production, or with standard varieties already grown. In many states corn hybrids and other crops developed by private companies may be included in tests conducted by the state agricultural experiment station. Usually a small fee is paid by the seed company for each strain included in the tests. Results of the tests are then published and made available to the public.

In general, education of the public, seed regulatory laws, and competition have forced reputable seedsmen to set a high standard in the sale of seed of new and established crop varieties. Unfortunately, some unethical seedsmen still foist inferior or unadapted varieties on the public with misleading publicity and pressure sales tactics. One common practice is to sell, at an exorbitant price, seed of an established variety in an area totally unsuited for good performance by that particular variety. This practice could be largely prevented if farmers would purchase seed only of varieties recommended for their area and from seedsmen whose reputation for excellent products and fair dealing has already been established in the community. Use of certified seed, whenever available, is one assurance of obtaining seed of high quality (Figure). It is not an assurance of obtaining an adapted variety, unless the variety has been tested and has been found to be suitable for production in the area where the buyer expects to plant the seed. For example, a variety of oats adapted and certified in Minnesota may be unadapted for production in Missouri or Kansas. Performance trials conducted by the state agricultural experiment stations are the best guides for determining whether a variety is adapted. Recommendations by experiment stations may not extend to privately developed varieties, unless the experiment stations have had an opportunity to test the varieties developed by the private companies. Also, experiment stations may not recommend a privately developed variety unless specific knowledge of the breeding behind the variety is made public, and there is assurance that seed sold by the seed company under the variety name will always be of the same genetic composition.

Seed-Certifying Agencies

The purpose of seed certification as defined by the International Crop Improvement Association is as follows:

'The purpose of seed certification is to maintain and make available to the public sources of high quality seeds and propagating materials of superior varieties so grown and distributed as to insure genetic identity. Only those varieties that contain superior germplasm are eligible for certification. Certified seed is high in varietal purity and of good seeding value.

"Varieties eligible for certification have resulted either from natural selection or through systematic plant breeding. In either case without a planned method for maintaining genetic purity, there is danger of losing varietal identity.

"Varietal purity is the first consideration in seed certification but other factors, such as weeds, diseases, viability, mechanical purity and grading are also important. One of the most effective methods of preventing the wider distribution of weeds is to plant weed-free seed. Adverse effects of plant diseases can be reduced by planting clean seed from disease-free fields. Properly cleaned and graded seed is easier to plant and gives more uniform stands.

"Seed certification is designed, therefore, to maintain not only the genetic purity of superior crop varieties, but also reasonable standards of seed condition and quality."

The plans by which the various state seed-certifying agencies are organized differ, but the general pattern of organization is as follows:

1. The membership comprises seed growers and other parties interested in the production of good seed.
2. The management is vested in a board of directors elected by the members.
3. The agency, through its officers, sets up procedures and standards for inspecting, testing, and certifying seeds in the state in which it is organized.
4. The agency is closely affiliated with the state agricultural experiment station, the state agricultural extension service, and/or the state department of agriculture through membership of staff members on the board of directors of the seed-certifying agency, by staff members serving as officers of the agency, through advisory committees, or by other means.
5. In most the seed-certifying agency has legal status as t.ne official seed certification organization within the state. Legal status may be from direct legislative action, or legal status may be vested in the state agricultural experiment station or the state department of agriculture, who in turn designates the seed-certifying agency as the official seed certification organization.

International Crop Improvement Association

The International Crop Improvement Association was organized in 1919. Its membership includes the seed certifying agencies in the various states, Alaska, and

Canada. Its major effort has been directed toward coordination of the certification programs of the member seed-certifying agencies. This has been accomplished by the establishment of minimum certification standards for the guidance of member agencies. Procedures have also been established for interagency certification of seed samples. For example, a potential certified lot of grass seed grown in one state may be shipped to a seed-processing plant in another state for final cleaning, processing, bagging, tagging, and sealing. By mutual agreement the seed-certifying agencies of the two states may combine efforts, each making the field and processing plant inspections necessary in their respective state to complete final certification on the seed.

Certification of a Variety

Before a variety of a field crop can be certified by a seed-certifying agency in a particular state, it must *first be approved by the certification committee far* that state. Normally, only varieties recommended by the agricultural experiment station in the state are certified. However, additional varieties may be certified if there is a demand for the seed in other states which recommend the variety but which cannot produce the seed satisfactorily. For example, seed of a forage crop variety adapted to an eastern state may he grown and certified. The seed is then shipped back to the eastern state for sale.

Exact certification procedures vary from state to state and with different crops. In general, certification involves the following steps:

1. The grower must plant foundation, registered, or certified seed of an approved variety. In some states only progeny of foundation or registered seed, or, in some cases, first generation increases from certified seed can be certified. Generally, certification must be completed on the total acreage of each variety grown by or belonging to a grower.
2. The seed must be planted on clean ground. The field should not have been planted in the year previous to another variety of the same crop, or to other crops which might volunteer and affect the purity of the crop being certified, such as rye preceding wheat. The ground should be free of noxious weeds that might affect the purity of the crop.
3. In cross-pollinated crops, isolation of the seed producing field is required, either by planting a specified distance from all other fields of the same crop or by planting a specified number of pollinator rows around the border of the field to reduce the opportunity for cross-pollination with other varieties planted in neighbouring fields.
4. Off-type plants are rogued out by the grower before harvest or before flowering, in the case of a cross-pollinated crop. Noxious weeds are removed before harvest, and borders or fence rows are clipped where necessary to maintain seed purity.
5. Field inspections are made by official representatives of the seed-certifying agency to check on the purity of the variety, freedom from other crop plants, freedom from noxious weeds, amount of disease that might affect

certification, and general conformity by the grower to the seed certification rules. Inspections are made at the time that purity and other observations can be seen best. In the case of hybrid corn several visits are usually made to check on thoroughness of de tasseling or on freedom from pollen shedding in *male-sterile lines in the event cytaplasmic male steril*ity is being used to eliminate the de tasseling procedure.

6 Seed inspections are made by official representatives of the seed improvement association as necessary to observe and supervise the harvesting, cleaning, grading, bagging, and other processing operations. Representative samples are drawn by the inspector from the various lots of seed after they have been prepared for sale. The seed samples are tested for bushel weight, moisture, impurities, germination, or other factors affecting seed quality according to the particular crop. Only seed meeting minimum standards of the association in all respects is accepted for certification. An exception to this is that the standards are sometimes temporarily relaxed when it is necessary to maintain an adequate supply of germplasm of the variety.

Official tags of the seed-certifying agency are sealed on the bags of seed accepted for certification. The tag is labelled to show that the seed meets the specific standards for the crop. The production, processing, and marketing of certified seed is exclusively the responsibility of the grower. The responsibility of the seed improvement association is limited to observing whether the grower follows the regulations outlined by the association and to determining whether the seed meets the prescribed standards for certification of that particular crop. The exact regulations and rules for certification of specific crops may *be* obtained by writing to the seed improvement association of a particular state.

How a New Variety Reaches the Farmer

A desirable procedure by which a new variety of a field crop reaches the farmer may be described as follows:

1. A variety is ready for release and distribution when it has been proved to be distinctly superior to existing commercial varieties in at least one or more characteristics, and at least satisfactory in all other important respects. The superiority should have been proved in tests carefully planned and conducted, in comparison with standard commercial varieties in the originating state, and in regional tests which will provide reliable information on the range of the variety adaptation,
2. The decision for release is made by the breeder in consultation with appointed boards of review. The boards of review are set up by appropriate authority within the state agricultural experimental station, to review proposals for release of a new variety. The plant breeder makes a limited increase of the new variety, the amount varying from a few pounds to a few bushels. The breeder seed is then generally turned over to some agency responsible for making the foundation seed increase.

3. Foundation seed is increased from breeder seed. The organization making the increase varies in different states. In some states a foundation seed program has been developed within the agricultural experiment station. In other states foundation seed is produced by a private foundation seed stocks organization closely associated with the agricultural experiment station. Increase may be speeded up in many summer-growing crops by growing a seed crop during the winter in one of the southern states.
4. At least one year before distribution by the originating station, each state experiment station in the region of adaptation of the new variety is normally informed of plans to release the variety, and seed is supplied to them in quantities to permit field plot testing at one or more locations.
5. The variety is named at the originating station, in consultation with representatives of other state agricultural experiment stations within the region, and the United States Department of Agriculture in the case of cooperative state-federal breeding programs.
6. Prior to distribution, either breeder seed or foundation seed is shared within the region with other state agricultural experiment stations who may wish to make a simultaneous increase and distribution of the new variety. In the case of grass or legume seeds, a limited generation increase may be made in states outside the region of adaptation under the supervision of the National Foundation Seed Project, as will be described later. Normally a few hundred to several thousand bushels of seed may be available by the time the variety is ready for release. The amount will depend upon the crop, the facilities for increasing the seed, and the anticipated demand for the new variety. An adequate quantity of foundation seed should be built up to supply immediate needs for the production of registered or certified seed before any release or distribution is made.
7. First distribution of foundation seed is usually made to selected growers who by past experience have proved their ability to produce registered and certified seed with a high standard of quality. Selection of qualified registered and certified seed growers is necessary to insure a rapid increase of the new variety without loss of purity. This distribution, plus a public announcement, constitutes a formal release by the originating station. Seed increased from the foundation seed may be classed as registered seed, or certified seed, according to the crop and the policy of the state concerned. Control over distribution and price of the first increase is sometimes retained by the distributing agency.
8. Distribution of registered seed (or certified seed if certified seed has been produced directly from foundation seed) is made to certified growers throughout the state. The certified seed harvested from this increase then usually becomes available without restriction to any grower within the state insofar as the seed supplies are available.

Exact procedures vary in different states and with different crops. In some states advance release is made of superior inbred lines of corn, without regard to a particular hybrid combination in which they may be used. This makes inbreds with superior characteristics available to private breeders or others who might have a use for them. In some states delayed release of inbred lines of corn is practiced. With delayed release, the agricultural experiment station keeps control of the inbred, and produces and distributes the single crosses in which the inbred is a parent, until the single and double crosses have proved successful. Then the inbred is released. In other states inbreds are never released-only the single crosses in which they are used. Many state seed certifying agencies certify field crops.

Practical Problems in Seed Production

Seed production problems are peculiar to each specific crop. Solutions to these problems have generally been found by long experience with the crop. Small Grains crops are seeded on large acreages. It is common practice for the farmer to buy certified seed of new varieties of these crops as seed supplies of the improved varieties become available. The farmer then harvests seed from his own crop for successive plantings. This practice is satisfactory as long as he can maintain the purity of the variety, keep the crop free from other crop and weed seeds, control seed-borne diseases, and produce seed with good germination. Except for rye, these crops are self-pollinated and isolation is not required to maintain varietal purity. Normally a strip the width of a drill or seeder which may be mowed, uncropped, or planted to some other crop is all that is needed to separate two varieties. With rye, a cross-pollinated crop, isolation of 40 to 60 rods from fields planted with other varieties of rye is required to prevent mixing. In the case of rice, if another variety is seeded by plane parallel to the edge of the seed field, a space not less than 100 feet should separate the varieties. If another variety is seeded by plane at right angles to the seed field, at least one-fourth mile should separate the varieties. Some of the original seed is usually maintained by the organization making the distribution of the variety so that new increases may be made and distributed as needed. If the variety becomes mixed, plant-to-row, or panicle-row, plantings may be used to purify the variety. The procedure is to select several hundred or more plants or panicles and plant each to a row. Only rows which conform to the variety type are harvested. The seed is bulked to start a pure seed increase. If the plant-row method is used to purify a variety, a large number of rows must be grown. Otherwise, if one off-type row is overlooked, the actual percentage of this type in the variety may be increased. Variety purity may be maintained also by planting small seed plots each year and roguing out the off-type plants. By this procedure the percentage mixture will be decreased each year.

Starting Point for Maintenance Breeding (Nucleus Seed Production)

Normally the breeder starts nucleus seed production of a variety only after it gets identified for release with the result that there is no nucleus seed available to take up breeder seed production in the very first year of release of a variety. It is suggested that once the material reaches NIVT (National Initial Varietal Trial) stage, the concerned breeder should initiate the nucleus seed production anticipating that the genotype would get identified in due course of time. This would be the starting point of Maintenance breeding and should be continued once a variety has been identified and released for cultivation. Nucleus seed is produced by growing ear rows (wheat). Select between 400-500 typical ears/panicles/clusters/plants at maturity stage from the seed multiplication plot of a given genotype/variety. Harvest and thresh them separately. Examine the seeds for colour, shape and size and reject the seeds of those ears/panicles *etc.* which are not representative of the variety. During the next cropping season from the seeds of each selected ear/panicle *etc.* grow a row of suitable length. This is known as an Ear/Panicle/Plant row. Grow between 300-400 such rows per variety. Examine these rows periodically throughout the growing season particularly at early growth stage, ear/panicle emergence stage, early dough stage and at maturity. Reject rows that show off type (s). Such rows should be rejected as and when detected. The off types can be with respect to any character such foliage colour, auricle pigmentation, days to flowering, waxy bloom, plant height, ear/panicle colour, shape and density, various glume characters *etc.* Diseased as well as agronomically poor looking rows should also be rejected. When an off type is detected after flowering stage, then the rows on either side of the row showing off type should also be rejected to minimize contamination from off type plants. Rows which are true to type of the original variety for all externally observable characters are selected. From these true to type rows between 400-500 ears? panicles? clusters? plants are selected for the next cycle of nucleus seed production. Harvest and thresh the selected rows separately. Examine the seeds for colour, shape and size and reject the seeds of those rows which show deviation. The bulked seed of the remaining true to type rows is the Nucleus Seed. In some crops like wheat where seed rate is high the nucleus seed produced around 100kg may not be sufficient in some cases when the breeder seed indent is more than 30-35 quintals. In such cases the nucleus seed produced by growing ear rows is referred to as **Nucleus Seed Stage-I (NSS-I).** The seed of true to type ear rows in stage-I is kept separate and used to grow Ear-Row Progeny Plots in the next season. The size of **ear row in progeny** plots can vary depending upon the nucleus seed requirement. This provides another opportunity to reject some of the plots, which escaped detection during state-I. The seed from uniform and true to type ear row progeny plots is bulked to produce **Nucleus Seed Stage-2 (NSS-2).** Where plot-drill facility is not available the NSS-l can be further multiplied in bulk for one generation to produce the desired quantity of NSS-2. Such a nucleus seed production plot should be very carefully and thoroughly rogued.

Breeder' Scientific Language, Farmer Participatory Breeding and Seed Production

The breeders release the variety, farmers has to test the variety (Jahufer and Borovoi, 1992). There should be situation-based planning for seed production programme of cross-pollinated crops in the farmers field, because farmers' are not even sensitive; but technology has to click for the same. There is potential of high harvest in Meghalaya due to favourable climatic pattern in terms of rainfall, relative humidity and sunshine hours. For breeder it is necessary to select by seeing *per se* performance (ideotype) for seed production programme before releasing the same and after release farmers can produce the same at their own field with some introgression of technical know-how from the breeding institutions. The rouging should not be more than 5 per cent in the farmers participatory seed production programme as proved by the same from the breeding institutions. To increase the crop productivity (Reddy *et al.*, 1996) seed production programme should be initiated under "***Farmer participatory seed production programme***" through ICAR to attract the farmers towards the crop and update their knowledge about the latest technologies and know-how so that acceptability and productivity will be increased at the grass root level. The production and income of farmers can be increased further by the adoption of improved seed production programme under hi-tech agriculture. Ramirez and Hacker (1994) found on average, there was a 56 per cent reduction in seed yield a week after maximum seed yield, making identification of optimum harvest date a critical decision.

New Direction for Farmers: Growing Seed as a Crop

There is a buzz among seed farmers about growing seeds. Some growers are discovering the virtues of growing seed for their own use and some are finding out that growing seed can be a great alternative cash crop. In many cases it allows them to diversify and produce something with good market potential that doesn't have the same level of constant care that most farmers demand during the growing season. Many farmers take the seed that they purchase each year for granted, never really considering where it comes from, who produces it and how it's produced. The mail order seed companies that we buy our seed from are usually not producing most of the seed they sell. In general, they are buying it from a few seed companies that perform research, breeding, and large-scale production. Most companies only produce a fraction of their varieties through contracts with independent growers. In fact, the growing of seed is becoming more common across the landscape than it has been in many years. The seed industry is going through a grassroots transformation that is leading to a healthy diversification and blossoming in the number of varieties available to farmers. This is due to an interesting confluence of circumstances that includes consolidations in the seed industry, the burgeoning demand for seed, independent companies specializing in older regional niche varieties, and some new offerings from independent plant breeders who are selecting varieties for new markets. All of these factors have sparked an increased demand for farmer-grown seed (Andrade *et al.*, 1994). The consolidations among the largest players in the seed industry are having a far-reaching impact on the seed that is available to the diversified small-scale grower. First of all, as these companies get bigger they tend to concentrate their energies on their newest, highest profile varieties that also have the highest profit margin. Their older varieties, many of them tried and true "work horse" varieties that have stood the test of time are either dropped or if they are kept on the price list they are relegated to a second-class status whereby they get little or no upkeep or concern for their genetic purity. Many farmers (Trutmann and Kayitare, 1991) have found that these older workhorse varieties are in fact resilient and well adapted to the rigors of cultural practices. As many of these hearty varieties are abandoned by the new mega seed businesses there are more and more farmers who are seeking out new sources for the seed of these varieties (Nordestgaard and Andersen, 1991). In a number of cases they are producing their own seed of important "dropped" varieties that they can't get otherwise. In other instances they are contacting the regional seed companies and requesting the reintroduction of these important open-pollinated varieties. In turn, these companies, which traditionally don't do their own seed production, have been scurrying around in the past few years, trying to find small-scale seed growers to produce these varieties that still have a loyal following.

In the Eye of The Seed Grower

One of the most important benefits of involving growers in the seed growing process is that it can increase the genetic elasticity of the crops through the farmers' selection for resistance under environmental pressures like cold and drought stress or disease and insect attack. While the growers' selection of the plants that are less affected by environmental stresses may seem incremental and slow, many budding plant breeders have been amazed at just how much progress they have made in only a few generations of this type of selection when they subsequently do a side-by-side comparison of their selected variety next to the original stock in a trial (Hare *et al.*, 1989). Farmers' Seeds has an exponential rise in the number of seed varieties that they buy from independent seed growers in the past years. In the seed catalog they have a very telling diagram concerning what they call "Our Own Seed Crops." This diagram shows buying seed of only four varieties from "small-farm grown" sources, contrasted with where they are buying seed of varieties from farmers! The wave of the future is that many of these workhorses from the golden age of plant breeding are going to be dropped. Becoming more philosophical, he says: "Being able to encourage people to grow their own food is a big step in the right direction. It's very different selling someone a packet of seed than selling them a crop (Young, 1993). As this progressive seed movement evolves there will definitely be room for many seed growers to take part as well as a need for farmers to become more savvy in the techniques of growing, rouging, and selecting seed crops so they can effectively improve them for their adaptation to practices and regional climates. There is need to develop a new system of seed production that has the scope and is on the scale of the sustainable market farmers around the country. It needs to be diversified. Just as the market farmers have diversified our diets, then seed growing by an increasing number of farmers is going to increase our seed choices for our different needs. Keep your eyes open for new seed choices. This revolution will not be televised.

Promoting Farmers' Seed: Its Conservation, Enhancement and Effective Utilization

Maintenance of this invaluable resource is crucial to sustainable agriculture. This is especially true for resource poor farmers practicing agriculture under low input conditions in marginal environments. There is also a serious threat of losses of diversity, which requires major inputs to protect and sustainably utilize it. Farmers' varieties, largely represented by indigenous landraces, are among those suffering such losses. In this context, the Seeds of Survival has undertaken a program that promotes the indigenous seed through its conservation, enhancement and effective use. The major emphasis is on promoting enhanced farmers' seed through multiplication and distribution of such material as well as improving/enhancing traditional cultural practices, with a view to promoting sustained production of food crops beyond the subsistence level. This also supports farm community initiatives to enable farmers to benefit from improved market potential of their seed, which will generate the financial resources and other incentives to sustain crop productivity and improved livelihoods with minimal external support.

Farmers: Key Players in Sustaining Seed Production

The traditional farming represents centuries of accumulated experience and skills of peasants who farmed under adverse conditions using locally available resources that often translated into sustained yields. The basis for this is the traditional crops and their landraces which farmers have adapted over centuries of selection and use to meet dynamic and changing needs. Crop management procedures are described for seed production (Cook and Franklin, 1988). Farmers play a key role in creating, maintaining, and promoting crop genetic diversity through a series of other long-standing activities which include intercropping with variety mixtures which result in rapid diversification; identifying and promoting varieties obtained through exchange; promoting the inter-crossing of cultivated crops with wild or weedy relatives, which results in new characteristics; identifying and propagating new, mutant types which occur in their fields, or hybridization between wild and/or cultivated types, or cultivars obtained through exchange; making available their knowledge and skills in identifying, collecting/rescuing and utilizing plants which they have helped develop and maintained for generations (Noland *et al.*, 1988).

Table 10.12. Cultivars details Grown Crops by Farmers Utilized in the Participatory Seed Production Programme

Acc No	*Species*	*Sub Species*	*Race*	*Local Name (Sub-Race)*	*Main Usage*	*Special Characteristics of Local Importance*

Maintenance of Elite Varieties on Farmer's Farms

The materials developed from plant populations subjected to selection based on performance in yield tests under different conditions of environmental stress. Genetic lines (agrotypes) are bulked (mixed) for further selection, multiplication and distribution to farmers.

Table 10.13. Comparative Evaluation of Varieties by Farmers

Factor Considered	*Landraces*	*Improved, High Input Varieties*
Familiarity	Well known for generations	Newly introduced, or been there for some time
Adaptation	Well adapted	Poorly adapted
Need for chemical inputs	Do not require neither fertilizer more chemical inputs	Fertilizer compulsory, some times herbicides needed*
Field performance	Fairly good stands	Mostly sparsely populated
Seed requirements	Less seed per unit of land (76 per cent of that of the improved, high input types)	More seed per unit of land
Lodging	Non-lodging, stand keeps upright	Lodge easily when winds prevail
Grain or kernel weight per unit volume	Heavier; a sack of 100 kgs capacity filled with landraces weighs more (upto 30 kgs in excess)	Lighter, a full sack of 100 kg capacity weighs the same
Disease Resistance	More tolerant	Often unpredictable
Pest Resistance	More tolerant	Often unpredictable
Weed tolerance	More tolerant or competes well	Poor, dominated by weeds
Frost tolerance	Tolerant	Somewhat weak
Yield stability	Fairly stable or within predictable limits	unpredictable, highly variable.

*In the farmer's land," needs to be bribed."

Similarly, on the utilization aspect, evaluation of landraces and IHYVs of wheat revealed that the former had qualities that met the diverse requirements of the farmers who provided the information shown.

Table 10.14. Comparative Evaluation of Materials on Utilization Aspect

Factor Considered	*Landrace Materials*	*Input Improved, High Input Varieties (IHYVs)*
Taste preference	Highly preferred	Less preferred
Nutritional value	High filling ability	Poor filling ability
Utility	Multiple utility (bread, porridge, brewing, *etc.*)	Limited choice
Flour output	Higher, flour lasts longer	Lesser, flour doesn't last long
Baking or dough quality	Good water holding capacity	Poor water holding capacity
Storage quality of grain	Keeps longer time (more than six months)	Attacked by weevils shortly after storage
Use of straw	High feed value	Feed value is inferior
Storage quality of straw	Keeps longer	Doesn't keep for long

The best way to achieve this is probably through community-based seed production or marketing and distribution systems operating in networks. Enhancing or further organizing the traditional networks that were described earlier could possibly develop these. Through this approach, farmers will be able to control the choice of crop types and cultivars to grow, and they will have ready access to planting material adapted to local growing conditions. They will also be in a position to critically evaluate the relative merits of a wide range of cultivars thereby limiting undue expansion of exotic cultivars that are costly and poorly adapted.

Participatory Variety Selection (PVS)

PVS is the selection of released or pre-released advance lines (including landraces) by farmers in their target environments using their own selection criteria. Basically, PVS: provides varietal choices to the targeted farmers under their specific environmental conditions; promotes participatory approaches to test; and select and disseminate the preferred variety. During 2006-2009 at ICAR-IARI RS Pusa, Bihar seeds of wheat varieties *viz.*, NW-2036, HD-2733, HD-2888, HP-1744, PBW-373, HP-1633, HW-2045, HP-1731, PBU-343, HD-2643, HP-1761, WR-544, HD-2824 were produced. Participatory on farm varietal evaluation in large plot and Participatory Seed Production under timely (K-9107, PBW343, HP 1731, HP 1761, PBW 443, HD 2733, UP 262, HUW 468, GW 273, GW 322, and GW 190) and late sown (NW 1014, Raj 3765, DL 788-2, HD 2643, HW 2045, HP 1744, GW 173, DBW 14, PBW 373, Lok –1, HUW 234, HI 1490 and HI 1454) conditions were carried out at ICAR-IARI RS Indore, Madhya Pradesh.

Table 10.15. Performance of Participatory Seed Production

Locations	*Varieties/Yield(Q/ha)*				
District	*Village*	*HD2733*	*HP1731*	*HD2643*	*HUW468*
Nalanda	Kormaha	41.6	32.0	40.0	38.4
	Hardardiock	42.0	30.0	38.0	36.0
Madhubani	Raiyam	43.5	30.6	31.0	29.0
	Rupali	41.6	29/1	32.0	28.1
Darbhanga	Godila	46.8	38.5	37.4	31.9
	Paruri	41.6	41.25	37.4	33.0
Vaishali	Harpur	55.0	26.4	28.1	27.5
	Bajitpur	41.3	*24.8	*24.8	27.5
Muzaffarpur	Pareyan	42.4	25.9	25.8	31.4
	BalraKishan	43.0	42.9	36.9	31.9
	Manika	45.1	36.9	39.6	27.5
Samastipur	Malinagar	49.5	*13.8	38.5	*21.4
	Harsinghpur	46.8	*22.0	30.3	27.5
	R.S.Pusa	50.0	38.0	38.0	36.0

Table 10.16. On farm seed production programme at farmer's field

Locations	*Varieties/Yield (q/ha)*				
Village/District	*HD2733*	*HP1761*	*HUW468*	*HW2045*	*NW1014*
R. S. Pusa	50.61	41.50	40.05	40.86	38.42
Phulhara (Samastipur)	41.61	39.70	38.50	37.00	32.50
Bishanpur (Muzaffarpur)	44.00	38.50	39.50	38.50	35.00
Patepur (Vaishali)	45.10	41.25	38.50	37.50	38.50
Jogiyara (Darbhanga)	44.50	42.00	37.75	36.25	39.00
Sagarpur (Madhubani)	49.50	38.50	39.00	36.00	35.75
Mean	46.88	40.24	38.88	37.68	36.52

Participatory Plant Breeding (PPB)

PPB is a breeding process in which farmers and plant breeders jointly select cultivars from segregating materials under a target environment. The other forms of PPB may also include activities such as germplasm enhancement through pureline or mass selection. PPB could either be consultative or collaborative, based on the typology of farmer participation suggested. Therefore, success of PPB rests on the blending of comparative strengths of farmers, breeder and social scientists involved in the process. The use of consultative or collaborative methods as well as choosing the right level of farmers' participation depends on: crops; capacity of the participant farmers; willingness; and availability of breeder and researcher resources. However, PPB commences only after PVS.

Table 10.17. Consultative PPB vs Collaborative PPB

Consultative PPB	*Collaborative PPB*
farmers are consulted to set breeding goals farmers choose appropriate parents and testing sites	farmers set breeding goals farmers grow segregating genetic material the best plants are selected in their own production environments

Table 10.18. PPB Process and Participatory Tools Used

Process	*Participatory Tools*
1. Need identification and setting breeding goals	
Understating the reasons for growing diverse varieties	RRA/Participatory Rapid Appraisal (PRA): Village level workshop, focus group discussion (FGD), preference matrix ranking, diversity fair, community biodiversity register (CBR)
Setting breeding goals and roles jointly to meet immediate need	FGD
2. Parent selection and generating new diversity	
Identification and use of locally adapted varieties as parent materials	Diversity block, PRA, FGD

Process	*Participatory Tools*
3. Selection of research plots and expert farmer role identification	
Identification and selection of knowledgeable farmers having interest in PPB works	Farmers network analysis (FNA), FGD, expert farmers, CBOs joint farm walk to exchangeable knowledge between breeders and farmers
Farmers assume role of selecting suitable cultivars from the segregating materials	Diversity fairs
Management of research activities under farmers' condition and setting selection criteria	Researcher designed-farmers managed; farmers' selection criteria
4. Selection of segregating lines	
Decentralized selection of segregating lines (variable population) by farmers under target environments breeders	Farm walk, FGD, farm school, preference ranking, selection of lines jointly by population and farmers
Post-harvest evaluation using gender perspective	Micro-milling and visual observation, participatory assessment, preference ranking
5. Variety release and distribution	
Varietal spread through informal seed supply	Ball and string techniques
Release variety on the basis of mother baby trial result, and varietal spread data	Mother and baby trials, PRA monitoring

Limits, Strategy and Scaling Up

Possibilities of PVS have been exhausted, Search process in PVS failed to identify any suitable cultivars for the testing, Farmers identified a new problem in existing cultivars. Demand-led approach, Low cross-high population size strategy, Provide better choice of varieties, At least one parent should be a landrace or locally adapted cultivar, Screened under target habitat utilizing farmers selection criteria and knowledge, Farmers are involved at much earlier stages of breeding process, Mother-baby trials for generating scientific data as well as disseminating PPB results, Linking PPB with formal research systems through collaboration for disease screening and multi-location testing of PPB products. Scaling up of PPB products preferred by farmers through informal seed supply as well as formal variety release systems, Baby trials for dissemination of PPB products.

In Participatory seed production programme were conducted in 14 locations of six district (Nalanda, Madhubani, Vaishali, Muzaffarpur and Samastipur) The performance of HD 2733 was significantly superior over the varieties. The farmers were satisfied with new varieties of wheat. No much variation in respect of yield in order of merit of genotypes under timely and late sown conditions. Out of 15 varieties only HD 2733, HW 2045, HUW 468, NW 1014 and HI 1490 performed better under both the condition. Under evaluation and identification of genotypes for leaf blight very low incidence of brown and black rust was observed but severity of leaf blight in almost all cultures was observed. Out of 23 genotypes screened for heat tolerance under timely sown condition, the genotype 780/4451 showed marked superiority over the check entry HUW 234. Out of 60 cultures evaluation for heat tolerance under late sown condition, 6 cultures showed marked superiority

Table 1.19. Details of Participatory Seed Production at IARI RS Pusa, Bihar, 2006-2009

District	*Villages*	*Name of the Farmers*	*Variety given*
Samastipur	Malinagar	Sh. Shivnath Thakur	HD 2733
		Sh. Omprakash Sharma	HP 1731
		Sh. Jitendra Thakur	HUW 468
		Sh. Sanjay Thakur	HD 2643
	Harsingpur	Sh. Sahdev Rai	HD 2733
		Sh. Manish Rai	HP 1731
		Sh. Ram Kumar Rai	HUW 468
		Sh. Krishna Rai	HD 2643
Vaishali	Harpur	Sh. Jagannath Chaudhary	HD 2733
		Sh. Harishankar Singh	HP 1731
		Sh. Ram Sankar Singh	HUW 468
		Sh. Abhay Kumar	HD 2643
	Bajitpur	Sh. Rajkishore Mishra	HD 2733
		Sh. Ashok Kumar Mishra	HP 1731
		Sh. Sitaram Mishra	HUW 468
		Shibji Mishra	HD 2643
Muzaffarpur	Pareyan	Sh. Surajdev Narayan Kumar	HD 2733
		Sh. Umesh Kumar	HP 1731
		Sh. Satish Chand Chauhan	HUW 468
		Sh. Rakesh Kumar	HD 2643
	Manika	Sh. Subhas Kumar	HD 2733
		Sh. Ghanshyam Pd. Singh	HP 1731
		Sh. Rajkishore Singh	HD 2643
		Sh. Sita Ram Singh	HUW 468
	Balrakishan	Sh. Harihar Saha	HD 2733
		Sh. Ram Kumar Singh	HP 1731
		Sh. Annada Saha	HD 2643
		Sh. Bhola Thakur	HUW 468
Darbhanga	Godhaila	Sh. Sushil Soman	HD 2733
		Sh. Jagannath Prasad	HUW 468
		Sh. Bijoy Kant Lal	HD 2643
		Sh. Dharmendra Kumar	HP 1731
	Patori	Sh. Ajoy Chaudhary	HD 2733
		Sh. Indesh Kumar	HUW 468
		Sh.Sushil Kr.Chaudhary	HD 2643
		Sh. Binay Kr. Chaudhary	HUW 468

District	*Villages*	*Name of the Farmers*	*Variety given*
Nalanda	Kormha	Sh. Shiv Kr. Prasad	HD 2733
		Sh. Pravin Kumar	HD 2643
		Sh. Arvind Kumar	HP 1731
		Sh. Parmanand Kumar	HUW 468
	Hardar diock	Sh. Sunil Kumar	HD 2733
		Sh. Jupendra Kumar	HD 2643
		]Sh. Anil Kumar	HP 1731
		Sh. Srikant	HUW 468
Madhubani	Raiyam	Sh. Masharwar Thakur	HD 2733
		Sh. Kallar Kamath	HP 1731
		Sh. Abhay Jha	HD 2643
		Sh. Lakchinswar Thakur	HUW 468
	Rupoli	Sh. Vivekanand Mishra	HD 2733
		Sh. Mohendra Mishra	HP 1731
		Sh. Tipu Jha	HD 2643
		Sklh. Lalloo Yadav	HUW 468

in yield over the best performing check NW 1014. Again, eleven wheat varieties *viz.* K-9107, PBW343, HP 1731, HP 1761, PBW 443, HD 2733, UP 262, HUW 468, GW 273, GW 322, and GW 190 have been sown in replicated trial. Under late sown condition 13 wheat varieties *viz.* NW 1014, Raj 3765, DL 788-2, HD 2643, HW 2045, HP 1744, GW 173, DBW 14, PBW 373, Lok –1, HUW 234, HI 1490 and HI 1454 were sown in replicated trial. Under seed production programme five best varieties *i.e.* HD 2733, HP 1761, HUW 468, HW 2045 and NW 1014 were chosen and 2 farmer each in Bishanpur (Muzaffarpur), Patepur (Vaishali), Jogiyara (Darbhanga), Sagar (Madhubani) and Phulhara (Samastipur) were selected on farm seed production programme *i.e.* half an acre plot per variety. Among 50 cultures for timely and late sown condition and 45 cultures for very late condition for observing heat tolerance. The varieties HUW 234, NW 1014 and HP 1744 included as a local check. 50 cultures were evaluated and identified for genotypes for salinity tolerance.

Institutional Back Up and Co-Ordination

It is possible to "demystify" seed growing, and which can be returned to the hands of the farmer in the local economy. Regional production of seed by farmers committed to maintaining and improving crop varieties adapted to local environmental conditions and cropping systems is a key component to a truly sustainable agriculture. In this way farmers raising seed for use in their region can gain a better understanding of the inherent traits that contribute to the overall health, resiliency, and productivity of a crop in that climate. Seed production in the area of intended use can provide an opportunity for a selective advantage for bioregional conditions, if the grower understands how to select for the best genetic combinations for the situation. Seed growing can also serve as a good alternative crop with practical use on one's own farm as well as for sale as a cash crop. Certainly, it may also fulfill a sense of self-sufficiency and deep satisfaction that is lost if all inputs are purchased. A characteristic feature of the programme is that it is participatory, developed on the basis of an integrated action plan that unites farmers/local communities, professional institutions and governmental and non-governmental organizations\NGOs\ on a cooperative effort. The role the various partners collaborating in this regard. Peasant farmers are at the center of such a collaborative venture since, apart from being the direct beneficiaries, they play a key role as curators and actively participate in the planning and execution. In a nutshell, this includes managing the plots, selecting, seed multiplication and distribution, on-farm utilization and marketing of material. The scientific and field staff, apart from supporting as well as supervising the on-farm activities, assisted farmers with the selecting and did the enhancement work jointly with scientists. They also participated/assisted with training of farmers and grass roots workers as potential collaborators.

Contribution of IARI RS Pusa, Bihar and IARI RS Indore, Madhya Pradesh

IARI, RS Pusa, Bihar is one of the important centres in NEP zone and **IARI RS Indore, Madhya Pradesh in CZ zone**, which has developed many new high yielding varieties such as HP1102, HP1731 and HP1761 for timely sown irrigated condition while HP1209, HP1633, HP1744 for irrigated late sown condition of NEP zone. HP1209 is a variety for late sown cultivation and was released in 1976. It became very popular as more than 70 percent area of Bihar comes under late sown cultivation. Therefore, demand of the seed of HP1209 was very high and still it persists. HP1633 (Sonali) is grown under late sown condition. This is substitute of popular variety Sonalika. This is the first *Lr9* alien gene introduced variety in the country shows very good tolerance to foliar blight in addition to resistance against brown rust. HP1744 is an outstanding variety for chapatti quality and is most popular among farmers. Another breakthrough was brought when HP1731 and HP1761 wheat varieties were released. Any increase in wheat production of Bihar state was made, the contribution of new high yielding varieties- HP1731 and HP1761 was reckoned as 35 per cent. A number of cultures have been developed possessing resistance against different rust pathogens *viz.* HP1822, 1826, 1832, 1833 showed resistance against black and brown rust, HP1718, 1808, 1811 showed resistance against black rust while HP1529, 1824 and 1825 showed resistance against brown rust. A number of cultures have been developed possessing resistance against different rust pathogens *viz.* HP1822, 1826, 1832, 1833 showed resistance against black and brown rust, HP1718, 1808, 1811 showed resistance against black rust while HP1529, 1824 and 1825 showed resistance against brown rust. HP1531 and HP1731 have been found as reckoned source for resistance against shoot fly. Pusa is a hot spot for the leaf blight. It serves as good platform for screening and developing foliar blight tolerant/resistant varieties. Most popular and leading wheat varieties are HD2733, HD2824 for timely sown and HD2643 and HW 2045 for late sown condition. All these varieties are replacing old varieties and occupying major area under cultivation in NEPZ very fast. During last 3 years the percentage of new Varietal saturation and reduction of old varietal acreage has been noticed. HD2733 is preferred due to high tillering habit, bold grain, good chapatti making quality and high yield. This variety is moderately resistant to leaf blight which is a serious problem in eastern zone. Under late sown, variety HD2643 (Ganga) is very popular due to bold and lustrous grain and there is a heavy demand of seed. Other varieties like PBW 373, HW 2045, NW 2036, DBW14 and PBW 524 are slowly becoming popular under late sown condition. The productivity and production in North Eastern Plain Zone may be improved, or in other words, the yield gap may be reduced significantly if the good quality of seed, mechanized farming and availability of fertilizer are given due importance or assured. Use of seed drills itself enhance yield against traditional method of broadcasting. Loss of produce due to early set of pre monsoon is a serious matter. Use of modern thresher or Mini Combine may help economic return to the farmers. Co-operative farming system is essential due to large number of marginal and poor farmers in this area. The production of wheat will certainly improve if all the constraints are given due consideration.

Seed Village: A Viable Concept for Agricultural and Rural Development

'Seed' is a crucial, critical, vital and basic input to agriculture. Availability of quality seed is a serious impediment in almost all crops in India. The non-availability of good quality wheat seeds to the farmers is a major problem particularly in Bihar due to the non-functioning of State Beej Nigam. It is an emerging concept being coming up; efforts were made through seed village programme on experimental basis. Under this project the 50 farmers from six villages in the jurisdiction of IARI Regional Station, Pusa were selected during Rabi 2003-04. A good quality breeder seeds of best performing five wheat varieties *i.e.* HD 2733, HP 1761, HUW 468, PBW 373 and HW 2045 were chosen and distributed to the farmers as a demonstration-cum-seed production plot. The size of demonstrations was 2.5 acres (1 ha). The seed was supplied at the proper sowing time and demonstrations were monitored at the different stages of crop growth involving wheat expert and representative of seed certification agency. The percent yield gain over check varies from 33.57 to 43.92 per cent at the farmers' field in different location. The quality seed produced by the farmer stored and after proper processing and packaging sold to the other farmers of neighboring villages @ Rs.1000-1200/- per quintal. Farmers have gained additional income @ Rs. 400-600/- per quintal as compare to non-seed. The extent of adoption of recommended practices revealed that 62 per cent farmers adopted the recommendations of wheat seed production in seed village programme at medium level followed by 30 per cent at low level and 8 per cent at high level. The experiences gain from this endeavor suggests that introduction of seed village is an excellent participatory and viable concept for agriculture and rural development.

Agriculture in India is the backbone of the country's economy and seed is considered as the most important, critical, vital and basic input to agriculture. Pure and improved seed itself can increase 20-25 per cent more productivity over the impure seed. The seed replacement rate in wheat is around 15 per cent; it should be at least 30-35 per cent (Dahiya and Srivastava, 2004). Presently National and State Seed Corporation, State Agricultural Universities and ICAR Institutes are producing seeds of different crops. The seeds so produced are not sufficient to meet the national requirement. Due to this reason, private sectors have become more and more powerful in seed market as the farmers are dependent on their seeds, which has led to several problems likes; (i) Availability of seeds at relatively higher price; (ii) Non availability of seeds of recommended varieties for specific agro climatic zone/conditions; (iii) Production of seed and its marketing in separate regions, may lead to partial to complete failure of the crop as seed produced in outside areas of adaptation of a variety generally leads to genetic deterioration of the variety. In Bihar, the network of seed production and marketing of Bihar Rajya Beej Nigam is weak. Today seeds of all classes are being produced by RAU, Pusa at its various centres and Regional Station, IARI, Pusa. Some private seed growers are also involved in seed production and marketing programme. In fact seeds produced in the state is much less than the requirement of the state and consequently the seed replacement rate of different crops is very low which need to be increased to augment the production as well as productivity of different corps. **Seed Village** is to get produced of the appropriate varieties seed of various crops locally by the farmers. It is a very good, an emerging concept being coming up, efforts were made through seed village programme on experimental basis.

The study was conducted in six villages in the jurisdiction of IARI Regional Station, Pusa during Rabi 2003-04. Total 50 farmers were selected group wise from Gamharia (Motihari), Bhoraha (Shivher), Mai (Munger), Kharitiya (Bettiah), Gorhaila (Darbhanga) and Kharia Basti (Araria) for seed village programme. A good quality breeder seeds of best performing five wheat varieties *i.e.* HD 2733, HP 1761, HUW 468, PBW 373 and HW 2045 were chosen and distributed to the farmers as a demonstration-cum-seed production plot. The size of demonstrations was 2.5 acres (1 ha) for five varieties per farmer *i.e.,* half an acre plot per variety. The seed was supplied at the proper sowing time and demonstrations were monitored at the different stages of crop growth involving wheat expert and representative of seed certification agency. The data was collected from each farmer with the help of interview schedule after the completion of the project.

During Rabi 2003-04 the wheat varieties *viz.,* HD 2733, HP 1761, and HUW 468 were timely sown and PBW 373 and HW 2045 were late sown given to the farmers for seed production. During the year wheat crop seriously affected due to terminal heat in the whole Indo-Gangetic plain Zone. But the variety HD 2733 (timely sown) and PBW 373 (late sown) performed undoubtedly superior over rest of the varieties in all the locations.

It is clear from the Table 10.20 that the percent yield gain over checks varies from 33.57 to 43.92 per cent at the farmers' field. It indicates the actual economic gain to the marginal, small and medium farmers. The seed-producing farmers sold most of the seed to the farmers of the neighboring villages @ Rs.1000-1200 per quintal. Similar results stated by Shailesh Kumar and Purushottam (2003). Farmers have gained additional income @ Rs. 400-600/- per quintal as compare to non-seed.

Table 10.20. Yield performance of wheat varieties at farmers field during 2005-06.

Locations	*Varieties/Yield (q/ha)*							
Name of Village	*HD2733*	*HP1761*	*HUW468*	*HW2045*	*PBW373*	*Average Yield (Q/ha)*	*Local Check Q/ha*	*Per cent Gain Pver checks*
Gamharia	55.00	44.00	38.50	40.00	41.25	43.75	30.00	35.83
Bhoraha	54.80	38.50	35.00	33.00	40.00	40.26	30.00	34.20
Mai	42.52	36.50	34.50	35.00	38.50	37.40	28.00	33.57
Kharitiya	45.00	37.00	32.00	32.00	36.00	36.40	27.00	34.81
Gorhaila	41.52	30.20	28.32	26.40	41.63	33.61	25.00	34.44
Kharia Basti	44.50	35.00	30.40	30.20	39.80	35.98	25.00	43.92

Adding individual scores received on seed production of wheat studied the extent of adoption in respect of wheat seed production technology in seed village programme. According to the total score, they were categorized into three categories. The data shows that 62 per cent farmers adopted the recommendations of wheat seed production in seed village programme at medium level followed by 30 per cent farmers adopted the recommendations at low and 8 per cent at high level respectively. Therefore, it may be inferred that there is a vast scope for increasing the adoption of improved seed production practices in seed village programme. Patel *et al.*, also reported similar findings (2003). Seed is considered as the most important, critical, vital and basic input to agriculture, and pure and improved seed itself can increase 20-25 per cent more productivity over the impure seed. Hence, taking into account the serious impediment of quality seed, the seed village programme should be promoted at every institute level for better availability of quality seed to meet the local demand and to improve the social and economic condition of marginal, small and medium farmers. The percent yield gain over check varies from 33.57 to 43.92 per cent at the farmers' field in different location. The extent of adoption of recommended practices revealed that 62 per cent farmers adopted the recommendations of wheat in seed village programme at medium level followed by 30 per cent farmers adopted the recommendations at low level and 8 per cent at high level. The experiences gain from this endeavor suggests that introduction of seed village is an excellent participatory approach and viable concept for agriculture and rural development.

References

Agarwal, RL (1996) Identifying Characteristics of Crop Varieties. Oxford and IBH Publishing Co. Pvt. Ltd. New Delhi.

Agrawal, RL (2013) Seed Technology 2nd ed. Oxford and IBH Publishing Co. Pvt. Ltd. New Delhi. 293p.

Agresti, A (2002) Categorical Data Analysis. 2nd Ed. John Wiley.

Almekinders, CJM; Thiele, G and Danial, DL (2007) Can cultivars from participatory plant breeding improve seed provision to small-scale farmers. *Euphytica* **153**: 363–372.

Annonymous (Undated) The Seeds Act, 1966 and Seeds Rules, 1968.

Anonymous (1999) Annual report national seed project (1998-1999) (Crops) Project Co-ordinator unit. Indian Agricultural Research Institute, New Delhi – 110 012.

Baidu-Forson, J (1997) On-station farmers participatory varietal evaluation: a strategy for client-oriented breeding. *Experimental Agriculture* **33**: 43–50.

Banziger, M and Cooper, M (2001) Breeding for low input conditions and consequences for participatory plant breeding: examples from tropical maize and wheat. *Euphytica* **122**: 503–519.

Basu, RN (1976) Physio-chemical control of seed deterioration. *Seed Research*, **4**: 14-23.

Bateman, AJ (1947) Contamination of Seed Crops I, Insect pollination. *J. Genetics* **48**: 257-75.

Belay, G (2009) Does client-oriented plant breeding work? *CAB Reviews: Perspectives in Agriculture, Veterinary Science, Nutrition and Natural Resources* 4: 1–7.

Bishaw, Z and Turner, M (2008) Linking participatory plant breeding to the seed supply system. *Euphytica* **163**: 31–44.

Borojevis, S (1990) Principles and methods of plant breeding. Elsevier, Amsterdam.

Ceccarelli, S; Grando, S; Singh, M; Michael, M; Shikho, A; Al Issa, M; Al Saleh, A; Kaleonjy, G; Al Ghanem, SM; Al Hassan, AL; Dalla, H; Basha, S and Basha, T (2003) A methodological study on participatory barley breeding. II. Response to selection. *Euphytica* **133**: 185–200.

Chinspeels, MJ and Sadava, DE (1994) Plants, Genes and Agriculture. Jones and Bantleft Publishers, Boston, London. pp. 415.

Chopra, VL (1997) Plant breeding: Theory and Practice. Oxford and IBH Pub. Co. Pvt. Ltd. New Delhi – 2.

Chopra, VL (2000) Plant Breeding: Theory and Practice. Oxford and IBH Publishing Co. Pvt. Ltd., New Delhi.

Chowdhury, S; Kumbhare, NV; Sinha, AK and Prasad, Y (2002) Increasing wheat production and Building up of Research Capabilities in the warmer Areas and Eastern India. Sub Project, National Agricultural Technology Project (NATP) IARI Regional Station, Pusa (Bihar).

Copeland LO and McDonald MB (2001) Seed Science and Technology Kluwer Academic Publishers, The Netherlands.

Copeland, LO and Mc Donald, MB (1985) Principles of Seed Science and Technology. Macmillan Pub Co, New York, Collies Macmillan publishers, London.

Dawson, JC; Murphy, KM and Jones, SS (2008) Decentralized selection and participatory approaches in plant breeding for low-input systems. *Euphytica* **160**: 143–154.

Gadwal, VR (2003) The Indian seed industry: Its history, current status and future. *Current Science*,84(3): 399-406.

Hong, TDS; Linington, S and Ellis, RH (1996) Seed storage behaviour: a compendium. Hand books for Gene banks: No. 4. International Plant Genetic Resources Institute, Rome, Italy.

Joshi, KD; Sthapit, BR and Witcombe, JR (2001) How narrowly adapted are the products of decentralised breeding? The spread of rice varieties from a participatory plant breeding programme in Nepal. *Euphytica* **122**: 589–597.

Kanwar, HS; Sharma, AK; Thakur, KS and Mehta, DK (2017) Advances in quality seed production of vegetable crops (6th to 26th September 2017), Centre of Advanced Faculty Training in Horticulture, YSP University of Horticulture and Forestry Nauni -173 230 Solan, Himachal Pradesh.

Kelly, AF (1988) Seed Production of Agricultural Crops. Longman.

Khare, D and Bhale, MS (2014) Seed Technology 2nd revised ed. Scientific Publishers, India. pp193.

Kumar, S and Purushottam (2003) Seed Village: A means of Enhancing Productivity of Uttaranchal. Abstract. National Seminar on Extension Strategy for promoting development initiatives among farming community, June 18-20, 2003; 130.

Lipton, M and Longhurst, R (1989) New Seeds and Poor People, Johns Hopkins University Press, Baltimore, Md, USA.

McDonald, MB and Copeland, LO (2005) Seed Production: Principles and Practices, Chapman and Hall, New York, pp749.

McGuire, S; Manicad, G and Sperling, L (2003) Technical and Institutional Issues in Participatory Plant Breeding – Done from a Perspective of Farmer Plant Breeding. A Global Analysis of Issues and Current Experience. PPB Monograph No 2. Columbia: Centro International de Agricultura Tropical (CIAT).

Padmavati, S; Prakash, M; Ezhel, S and Kamraj, S (2012) A text book of Seed Science and Technology. New India Publication Agency (NIPA), New Delhi. pp. 254-262.

Patel, MM; Chatterjee, A and Khan, M (2003) Adoption of Wheat Production Technology. *Indian Journal of Extension Education*, XXXIX (1 and 2): 58-62.

Poonia, TC (2013) History of Seed Production and its Key Issues. *Inter. J. Food, Agri. and Vet. Sci*, 3(1): 148-154.

Sperling, L; Loevinsohn, ME and Ntabomvura, B (1993) Rethinking the farmer's role in plant breeding: local bean experts and on-station selection in Rwanda. *Experimental Agriculture* **29**: 509–519.

Sreeramulu, G; Vishnuvardhan, D and Singh, NK (1994) Seed storage protein profiles of seven Indian Wheat (*Triticum aestivum* L.) varieties. *J. Plant Biochemistry and Biotechnology*, **3**: 477-451.

Subedi, A; Gyawali, S; Gautam, R; Sthapit, B; Chaudhary, P and Poudel, D (Undated) Participatory Approaches to Crop Improvement in Nepal, In Conservation and Sustainable Use of Agricultural Biodiversity- A Sourcebook, CIP-UPWARD, in partnership with GTZ GmbH, IDRC of Canada, IPGRI and SEARICE. pp. 354-361.

Tawade, ND; Gonjari, PA and Bhairamkar, MS (1998) Perceived Effectiveness of various mass media used by farmers for seeking agricultural information. *Maharashtra Journal of Extension Education*, **XVII**: 82-85.

Thompson, JR (1979) An Introduction to Seed Technology. Leonard Hill.

Vanangamudi, K (2014) Seed Science and Technology. New India Publication Agency (NIPA), New Delhi. pp. 357-366.

Welch, GB (1954) Seed Processing Equipment. *Mississippi Agricultural Experiment Station Bulletin 520.*

Weltzien, E; Smith, ME; Meitzner, LS and Sperling, L (2003) Technical and Institutional Issues in Participatory Plant Breeding – from the Perspective of Formal Plant Breeding: a Global Analysis of Issues, Results and Current

Experience. CGIAR Systemwide Program on Participatory Research and Gender Analysis for Technology Development and Institutional Innovation. PPB Monograph No 1. Cali, Columbia: Centro Internacional de Agricultura Tropical (CIAT).

Witcombe, JR; Joshi, A and Goyal, SN (2003) Participatory plant breeding in maize: a case study from Gujurat, India. *Euphytica* **130**: 413–422.

Witcombe, JR; Joshi, KD; Gyawali, S; Musa, AM; Johansen, C; Virk, DS and Sthapit, BR (2005) Participatory plant breeding is better described as highly client-oriented plant breeding. I. Four indicators of client-orientation in plant breeding. *Experimental Agriculture* **41**: 299–320.

Appendices

BREAD WHEAT VARIETY :
HI 1620 (Pusa Wheat 1620)
Notification No : 1.4.2019 1498 (E)

Timely sown, Restricted Irrigation conditions

(North Western Plains Zone)

[Punjab, Haryana, Delhi, Rajasthan (Except Kota and Udaipur Divisions), Western Uttar Pradesh (Except Jhansi Division), Jammu & Kathua district of Jammu & Kashmir, Una district & Paonta Valley of Himachal Pradesh and Tarai region of Uttarakhand]

High yielding, good quality Variety

Grain Yield (q/ha)	HI 1620 (PW 1620)	Check varieties			
		WH 1142	HD 3043	PBW 644	WH 1080
Mean yield	49.1	44.8	43.3	45.5	45.8
% increase		9.6*	13.4*	7.8*	7.1*

Average yield : 49.1 q/ha **Potential yield :** 61.8 q/ha

Quality traits	HI 1620
Sedimentation value (ml)	52
Protein content (%)	11.9
Extraction rate (%)	73.0
Grain Hardness index	82
Iron (ppm)	35.5
Zinc (ppm)	33.2
Wet gluten	28.0
Dry gluten	9.3
Gluten index	81
Bread loaf volume (cc)	552
Bread quality (Max. score 10)	6.25
Chapati quality (Max.score 10)	7.52
Biscuit quality (Spread factor)	6.82

- HI 1620 has high levels of field resistance to stripe (ACI: Max.-7.8; Mean–5.8) and leaf (ACI: Max.-5.0; Mean-2.8) rusts
- Good levels of adult plant resistance to prevalent and virulent stripe rust pathotypes 46S119 and 110S119; and leaf rust pathotypes 77-5 and 77-9
- Good levels of resistance to leaf blight, Karnal bunt, *Fusarium* head blight, loose smut and flag smut, compared to all the checks

BREAD WHEAT VARIETY :
HI 1621 (Pusa Wheat 1621)
Notification No : **6.1.2020 91 (E)**

Irrigated and Very late sown conditions

NATIONAL LEVEL

(North Western & North Eastern Plains Zone)

[Punjab, Haryana, Delhi, Rajasthan (Except Kota and Udaipur Divisions), Western Uttar Pradesh (Except Jhansi Division), Jammu & Kathua district of Jammu & Kashmir, Una district & Paonta Valley of Himachal Pradesh and Tarai region of Uttarakhand, Eastern Uttar Pradesh, Bihar, Jharkhand, Orissa, West Bengal, Assam and Plains of North Eastern States]

High yielding, good quality Variety

Grain Yield (q/ha)	HI 1621 (PW 1621)	Check varieties		
		WR 544	DBW 14	DBW 71
Mean yield	32.8	28.1	30.6	31.4
% increase		16.7*	7.2*	4.5*

Average yield : 32.8 q/ha Potential yield : 46.1 q/ha

Quality traits	HI 1621	
	NWPZ	NEPZ
Sedimentation value (ml)	48	55
Protein content (%)	11.9	13.0
Extraction rate (%)	71.8	72.4
Grain Hardness index	78	64
Iron (ppm)	34.1	39.0
Zinc (ppm)	40.5	33.1
Wet gluten	27.9	
Dry gluten	9.2	
Gluten index	87	
Bread loaf volume (cc)	613	
Bread quality (Max. score 10)	7.53	
Chapati quality (Max.score 10)	7.87	
Biscuit quality (Spread factor)	8.52	

- **HI 1621 has high levels of field resistance to stripe (ACI: Max.-6.4; Mean – 4.6) and leaf (ACI: Max.-12.5; Mean-8.0) rusts**
- **Good levels of adult plant resistance to prevalent and virulent stripe rust pathotypes 46S119 and 110S119; and leaf rust pathotypes 77-5 and 77-9**
- **Good levels of resistance to leaf blight, Karnal bunt and head scab; and better resistance to loose smut and flag smut, compared to all the checks**

BREAD WHEAT VARIETY :
HI 1628 (Pusa Wheat 1628)
Notification No : 6.1.2020 91 (E)

Timely sown, Restricted Irrigation conditions

(North Western Plains Zone)

[Punjab, Haryana, Delhi, Rajasthan (Except Kota and Udaipur Divisions), Western Uttar Pradesh (Except Jhansi Division), Jammu & Kathua district of Jammu & Kashmir, Una district & Paonta Valley of Himachal Pradesh and Tarai region of Uttarakhand]

High yielding, good quality Variety

Grain Yield (q/ha)	HI 1628 (PW 1628)	Check varieties			
		WH 1142	HD 3043	PBW 644	WH 1080
Mean yield	50.4	47.9	43.4	45.3	45.3
% increase		5.2*	16.1*	11.3*	11.3*

Average yield : 50.4 q/ha **Potential yield : 65.1 q/ha**

Quality traits	HI 1628
Sedimentation value (ml)	57
Protein content (%)	11.0
Extraction rate (%)	71.4
Grain Hardness index	82
Iron (ppm)	33.6
Zinc (ppm)	35.5
Wet gluten	24.6
Dry gluten	8.0
Gluten index	82
Bread loaf volume (cc)	610
Bread quality (Max. score 10)	7.64
Chapati quality (Max.score 10)	7.56
Biscuit quality (Spread factor)	8.27

- **HI 1628 has high levels of field resistance to stripe (ACI: Max.-11.7; Mean–8.9) and leaf (ACI: Max.-5.8; Mean-2.1) rusts**
- **Good levels of adult plant resistance to prevalent and virulent stripe rust pathotypes 46S119 and 110S119; and leaf rust pathotypes 77-5 and 77-9**
- **Good levels of resistance to leaf blight, Karnal bunt and flag smut, compared to the checks**

DURUM WHEAT VARIETY :
HI 8802 (Pusa Wheat 8802)
Notification No : 6.1.2020 91 (E)

Timely sown, Restricted Irrigation conditions

Peninsular Zone
(Maharashtra, Karnataka & Plains of Tamil Nadu)

High yielding, good quality Variety

Grain Yield (q/ha)	HI 8802 (PW 8802)	Check varieties	
		AKDW 2997-16	UAS 446
Mean yield	29.1	27.2	26.4
% increase		7.0*	10.2*

Average yield : 29.1 q/ha   Potential yield : 36.0 q/ha

Quality traits	HI 8802
Grain appearance	7.4
Test weight (kg/hl)	83.7
Sedimentation value (ml)	40.4
Protein content (%)	13.0
Yellow berry incidence (%)	2.0
Yellow pigment (ppm)	6.2
γ-gliadin	45
Hardness index	82
Iron (ppm)	39.5
Zinc (ppm)	35.9
Cooking time (min./sec.)	10.5
Water uptake ratio	1.25
Gruel solid loss (%)	10.0
Overalll Acceptability	6.2

- **HI 8802 has high levels of field resistance to stem (ACI: Max.-9.6; Mean–6.3) and leaf (ACI: Max.-7.5; Mean-3.4) rusts**
- **Showed good levels of adult plant resistance to prevalent and virulent stem rust pathotypes 117-6 and 40A.**
- **Showed seedling resistance to stem rust pathotypes 15-1, 40-3, 40A; and leaf rust pathotypes 12-2, 104-2 and pathotypes of race 77-group.**
- **Showed good levels of resistance to Karnal bunt, loose smut, flag smut and foot rot.**

ICAR

DURUM WHEAT VARIETY : HI 8805 (Pusa Wheat 8805)

Notification No : 6.1.2020 91 (E)

Timely sown, Restricted Irrigation conditions

Peninsular Zone
(Maharashtra, Karnataka & Plains of Tamil Nadu)

High yielding, good quality Variety

Grain Yield (q/ha)	HI 8805 (PW 8805)	Check varieties	
		AKDW 2997-16	UAS 446
Mean yield	30.4	27.2	26.4
% increase		11.6*	12.9*

Average yield : 30.4 q/ha **Potential yield : 35.4 q/ha**

Quality traits	HI 8805
Grain appearance	7.7
Test weight (kg/hl)	83.7
Sedimentation value (ml)	41.8
Protein content (%)	12.5
Yellow berry incidence (%)	2.0
Yellow pigment (ppm)	4.9
γ-gliadin	45
Hardness index	81
Iron (ppm)	40.4
Zinc (ppm)	33.9
Cooking time (min./sec.)	11.0
Water uptake ratio	121
Gruel solid loss (%)	1.20
Overalll Acceptability	6.0

- **HI 8805 has high levels of field resistance to stem (ACI: Max.-3.7; Mean–2.7) and leaf (ACI: Max.-2.0; Mean-1.3) rusts**
- **Showed good levels of adult plant resistance to prevalent and virulent stem rust pathotypes 117-6 and 40A.**
- **Showed seedling resistance to stem rust pathotypes 15-1, 40-3, 40A, 117-6; and leaf rust pathotypes of race 77-group.**
- **Showed good levels of resistance to Karnal bunt, loose smut, flag smut and foot rot.**

Acronyms and Glossary

β-*Carotene Content*: In durum wheat, the yellow color of semolina which imparts golden yellow amber colour to pasta product is desirable and, therefore majority of the pasta consumers prefer the yellow pigment. Xanthophils and specially β-carotene (precursor of vitamin A) contributes to the colour production in the semolina.

1B/1R translocation (the small arm of the 1 R chromosome of rye trans located on the long arm of the 1 B chromosome of wheat) enhances genetic yield potential.

9Rs of Sustainability.

AAS: Atomic Absorption Spectrometer.

ABA: Abscisic Acid are reported to be associated with drought tolerance.

Abiotic Stresses (Heat, Drought, Waterlogged, Cold) Resistance.

AB-QTL: Advanced Backcross-Quantitative Trait Loci.

Aerator: a fish tank air pump will do) with a rubber tube leading from it, is placed in the beaker.

AFLP (Amplified Fragment Length Polymorphism): is technically demanding, stable and a very reliable method which uses relatively small amount of DNA, it does not require sequence information or probe collection and has a very high multiplex ratio with relative reproducibility. Its limitations include dominant nature, requirement of high quality DNA for complete restriction enzyme digestion, high cost and experience of sequencing gels.

AICW and BIP: All India Coordinated Wheat and Barley Improvement Project.

AKAW 3722 (Vimal): WH147/Sunstar*6/C80-1.

AKDW 3347: Immer 'S'(Jacana*2/RU//Celta/Yav 1).

Allele scoring and Data analysis: Scoring is done on the basis of presence and absence of bands/peaks for all types of markers.

Alpha – Amylase Activity: is measured by falling number test. A flour/water suspension in a tube in a boiling water bath is stirred for exactly 60 seconds. A plunger is then allowed to fall freely through the suspension under standardized conditions. Falling time is the time in sec, taken for plunger to fall a standardized distance.

Alveograph: measures the ability of dough when subjected to high speed but constant mixing with water at room temperature. It uses air pressure to inflate a bubble of dough and break it, The instrument continuously records air pressure, the time that elapses before the dough breaks.

Amber Durum Wheat (AD): the subclass Amber Durum wheat shall be Durum Wheat with 60 percent or more but less than 75 per cent of hard and vitreous kernels of amber colour.

American Pod Borer *(Helicoverepa (Heliothis) armigera Hub)*: is a polyphagous pest, its sporadic infestation to wheat crop has been reported from all wheat growing states of India. The moth is stoutly built and is yellowish brown. The caterpillars when full grown are 3.5 cm in length, young larvae are greenish in colour which subsequently change to dark brownish colour with dark brown broken lines along with sides of the body.

Amino acids (such as lysine, threonine, methionine and tryptophan) contribute significantly to nutritional quality of plant-based foods. These "essential" amino acids cannot be synthesized by humans or monogastric livestock and have to be supplied to them in the diet, which is largely derived from plants. Lysine is synthesized in plants from aspartate via one branch of the aspartate family pathway.

Amylograph: continuously measure the resistance to stirring of 10 per cent suspension of flour in water while the temperature of the suspension is raised at a constant rate 1.5°C/mm from 20°C – 95°C and then maintained at 95°C, is of use in testing flour for soups, *etc.*

Androgenesis is the production of haploids from anther culture or isolated microspore culture.

Aneuploids, such as monosomics, nullisomics, and telosomics, for different wheat chromosomes or chromosome arms.

Anthocyanin pigmentation: accumulation of anthocyanin pigmentation in various plant parts.

Antibiosis may involve morphological, physiological or biochemical features of the host plant; some cases of insect resistance involve a combination of features.

Antibiosis refers to an adverse effect of feeding on a resistant host plant on the development and/or reproduction of the insect pest. In severe cases, it may even lead to the death of the insect pest.

Aphids *(Sitobion miscanthi (Takahashi) S. avenae Fabricius, Rhopalosiphum maidis Fitch, R.padi).*

APR (Adult Plant Resistance): Slow Ruster Lines.

APS: Ammonium persulfate.

Arabinoxylan, tocol, sterol, and alkylresorcinol contents are mainly determined by genetic factors, whereas phenolic acid, B vitamin, and betaine contents are influenced more strongly by environmental factors than by genetic background.

ARI: Agharkar Research Institute, Pune.

ARMP: Asean Regional Maize Programme.

Army Worm (*Mythimna separata* (Walker): the adult moths are pale brown in colour and are medium sized with a wing span of 40-50 mm, the eggs are round, grayish white to yellow in colour which changes to dark brown near hatching, the newly hatched larvae are active, light green in colour which changes to dark brown near hatching.

ASAs: Allele Specific Amplicons.

Ash Content: Ash is an indication of milling efficiency. The mineral content of bran is about 20 times that of endosperm. The ash test is fundamentally indicator of purity of flour or the through ness of the separation of bran and gram from the rest of the wheat kernels.

Atlas 66, 93123 to 93133 series, Chris, Crim, Naphal, WG 357, Tobari 66: excellent grain quality have been used in genetic studies and in quality improvement.

Atlas and Coastal, or varieties such as Bowie, differ greatly in inherent hardiness.

AVT: Advance Varietal Trials will be conducted for two years. Every year, based on the performance in AVT, entries poor in yield and/or disease resistance are rejected. The data from evaluation for two years under Advance Varietal Trials and one year of Agronomical trial (simultaneously conducted for strains promoted to AVT II for one year) is essential for the identification and release of a wheat entry as a variety.

AVT-RF: Advanced Varietal Trial-RF, AVT-RI, AVT-TS-IR-HF, AVT-TS-IR-HF (Bhopal), AVT-LS-IR.

AWRC: Alkaline Water Retention Capacity.

B 662: PBW 34*2/Chuanmai # 18.

B. sorokiniana (*Cochliobolus sativus*) from different geographical regions of Brazil were analyzed for their virulence on wheat cultivars, morphological characteristics, and growth rate on PDA. Variability in cultural characteristics was observed in the morphology and growth rate between the original isolates and the re-isolates.

Baart and Federation.

Baart and Hard Federation varieties.

Baba Mastram, Anandwan, Husru, Badgam, J and K.

BARC Collaborative Trial: Bhabha Atomic Research Centre, Mumbai (derivatives of C 306): Screening for shattering resistance; Replacement of Lok-1 with HI 1544 having hard threshing ability like Lok-1.

Bari-Gom 33: the first high zinc wheat variety with better resistance to wheat blast have been released in Bangladesh for commercial cultivation in 2017.

Baviacora 92: Yield potential about 9 t/ha.

BC: Backcross.

BH-1146 and CNT-1: Older resistant Brazilian commercial varieties.

BHU: Banaras Hindu University Varanasi, Uttar Pradesh.

Bioavailability (provitamin A: retinol bioconversion) and intake. There is scope for increasing micronutrient density in the shorter term via transgressive segregation, heterosis, and maternal effects beyond the variation and for discovering higher levels in the tactical gene pool.

Biochemical/Protein Markers are based on isozymes. "Isoenzymes" or "isozymes" are the alternative forms or structural variants of an enzyme that have different molecular weights and electrophoretic mobility but have the same catalytic activity or function.

Biofortification is a method of breeding crops to increase their nutritional value. This can be done either through conventional selective breeding, or through genetic engineering, biofortification differs from ordinary fortification because it focuses on making plant foods more nutritious as the plants are growing, rather than having nutrients added to the foods when they are being processed. Biofortification of staple food crops for enhanced micronutrient content through genetic manipulations is the best option available to alleviate hidden hunger with little recurring costs (Monasterio *et al.*, 2007), identification of germplasm with high grain Fe and Zn and understanding the genetic basis of their accumulation are the prerequisites for manipulation of these micronutrients.

Biofortified Crops: Health, nutrition, food sovereignty.

Biofortified Variety: Sustainable way to achieve malnutrition.

Biophotometer Method: is used for rapid, simple and convenient measurement of DNA and protein concentration. It measures the absorbance at various wavelengths of 230,280 and 320 nm.

Biotic Stresses (Disease, Insect pests and Nematodes).

BIS: N,N-Methylene bisacrylamide.

BISA: Borlaug Institute of South Asia [Pusa, Samastipur, Bihar; Jabalpur, Madhya Pradesh; Ludhiana, Punjab].

Biscuit traits: QBI19-20, QBI19-21, Donor for quality, DBW 14 (RP), HD 2967.

Black or stem rust (*Puccinia graminis tritici*): Dark brown to black, elongated and large pustules are severe on stalk but also appear on leaf sheaths, leaves and ears.

Blackhull variety originated from three black heads found in a field of Turkey in 1912 by Earl G. Clark of Sedgwick, Kansas.

BM: Breeding Materials.

BMT: Biochemical and Molecular Techniques.

BNI release and dry matter production of some wheat–*L. racemosus* disomic addition lines.

Bonneville 70: Variety with improved threshability.

BRAUSS: Bhim Rao Ambedkar University of Social Sciences, Mhow, Indore, MP.

Bread wheat (Sharbati): (*Triticum vulgare/Triticum aestivum* L.) is an allohexaploid with 21 pairs of chromosomes and three genomes, A, B, and D.

British scientist, *RH Biffen,* discovered that resistance to the stripe rust disease wheat was controlled by single gene. As a result, resistance of many plant diseases was shown to be controlled by single genes.

Brown Mite *(Petrobia latens Muller):* the adults are very active somewhat round in shape brown to metallic in colour. This mite is soil inhibiting and non webbings type. Body is fused into cephalothorax and covered with setae. The nymphs and adults has four pairs of legs, the larvae, nymphs and adults of the mite suck sap from leaves, leaf sheath, green stem and spikes. All the breeders material planted in the rainfed trials is screened against brown mite and germplasm showing least damage is included in the next.

Brown or leaf rust (*Puccinia recondita):* Round, orange pustules, irregularly arranged or in clusters on leaves, less common on leaf sheath and stalk.

BrUTP labeling of nascent RNA revealed that transcription occurs inside the territory, almost exclusively at the surface of condensed chromatin domains.

BSA: Bulk segregant analysis, developed by Michelmore *et al.* (1991) to tag disease resistance genes in lettuce, has been successfully applied in wheat.

BW 11 (Purbali).

BW 925: Wheat Stem Sawfly (*Cephus cinctus*) resistant parent carrying *Sst1.*

BWMRI: Bangla Desh Wheat and Maize Resh Institute, Bangladesh.

C 273, C 306 *etc.* are very good grain types whereas, Lerma Rajo, Sonora 64, HD 2009, PV 18 *etc.* are average to poor grain types.

C306 (*Lr34, Yr2KS, Yr18*).

Cappelle Desprez', French bread wheat, was recognized to possess a measure of resistance to eyespot caused by *Pseudosporella herpotrichoides*.

CAPS (Cleaved Amplified Polymorphic Sequences) markers *Xucw79* and *Xucw71, a*s there is a strong relationship between grain yield and protein content molecular markers independent of grain yield must be identified.

Carleton and Stewart varieties of durum developed by crossing Mindum to Vernal emmer and backcrossing twice to Mindum.

CAST: Conservation Agriculture Screening Trial.

CB: Crossing Block.

CCN : Cereal Cyst Nematode (*Heterodera avenae* Woll.).

cDNA–AFLP: cDNA–Amplified Fragment Length Polymorphism analysis.

Cereal aphid (*Rhoplalosiphum padi*).

Cereals has been derived from 'Ceres', name of a Roman Goddess, means '**Giver of Grains**', are cultivated grasses grown for their edible starchy grains (one seeded fruit–caryopsis), larger grains used as staple food (rice, wheat, maize, barley, oats *etc.*).

CGC: Canada Grain Commission, Canada is answering new and challenging quality assurance needs for exports.

CGIAR-CIAT [International Center for Tropical Agriculture].

CGIAR-CIMMYT, International Maize and Wheat Improvement Institute, Mexico.

CGIAR-CIP [International Potato Center].

CGIAR-ICARDA: International Centre for Agricultural Research in Dryland Areas, Syria.

CGIAR-ICRISAT, International Crops Research Institute, Hyderabad, India.

CGIAR-IRRI, International Rice Research Institute, Manila, Philippines.

CGs: Candidate genes from model plant species, The CGs can be: (a) genes involved in cell protection under drought stress (*e.g.* proteins involved in osmotic adjustment, degradation, repairs, detoxification and structural adaptations) and/or (b) genes involved in regulation of other genes involved in the drought response (protein kinases and transcription factors such as DREB, bZIP, MYB, *etc.*).

CHA: Chemical Hybridising Agent approach indicated certain inherent problems beyond control and, therefore, not found practically feasible to produce hybrid seed.

Chemical composition and nutritive value like protein content, amino acid composition, fatty acid composition and antiquality factors.

Chinch bugs *(Blissus leucopterus).*

Chinese source of resistance used at CIMMYT include Shanghai#4, Suzhoe#8 and Yangmai#6. *Thinophyrum curvifolium* has also been used as alien resistance sources at CIMMYT.

ChIP: Chromatin Immunoprecipitation experiments with a null mutant of MET1.

Chlorophyll Content: As measured by chlorophyll meter (SPAD/CCM) at two stages *i.e.* at anthesis and one week after anthesis.

Chuanmai 42: first documented release of a wheat cultivar derived from SBL by backcrossing SHW with locally adapted Chinese cultivar.

CI-600 root scanner.

CILs: Chromosome Introgression Lines, includes addition, substitution, and translocation lines) that harbor entire or partial chromosomes from the wild species.

CMS: Cytoplasmic Male Sterility System.

Co-dominant informative SSR marker suite for polymorphism frame-work in Indian wheat - A genome, B genome, D genome.

Coefficient of Infection (CI) for rusts of a particular variety is calculated when multilocation testing against rusts is carried out.

Coefficient of Sustainability (CS): It is measure of change in soil properties in relation to production under specific management system (Lal, 1991).

Common wheat, *Triticum aestivum* (2n = 6x = 42), is an allohexaploid comprising three distinct genomic complements, namely A, B, and D, with the genomic structure AABBDD.

Conservation Agriculture (CA) is a strategy for mitigating climate change and as an adaptive mechanism for alleviating climate change effect. CA practices can help in sequestering atmospheric CO_2 in the form of soil organic matter, as well as reduction in GHG emissions through efficient production system, facilitates soil management without excessive disturbance, protection from erosion, compaction, aggregate breakdown, loss in organic matter, leaching of nutrients *etc.*

Conservation tillage: practices that leave crop residues on the surface, which increases water infiltration and reduces erosion, it is a practice used in conventional agriculture to reduce the effects of tillage on soil erosion. However, it still depends on tillage as the structure-forming element in the soil, nevertheless, conservation tillage practices such as zero tillage can be transition steps towards Conservation Agriculture.

Conventional Breeding based on amplified-fragment-length polymorphism (AFLP), single-nucleotide polymorphism (SNP), diversity arrays technology (DArT), simple sequence repeat (SSR) or microsatellite, random DNA marker (RDM), gene targeted marker (GTM), and functional marker (FM).

Conventional Diversity for salinity tolerance has been limiting with just one land race Kharchia 65 playing a major role in salt tolerant varietal development in India where the cultivars KRL1-4 and later KRL 19 emerged.

Conventional Tillage: Tillage types that leaves less than 15 per cent residue cover after planting critical wind/water erosion period. Generally, includes ploughing or other intensive tillage operations; residue removed/burning during cultivation. Weed control is accomplished with herbicides, cultivation or both.

Coordinated Trials at Directorate Level.

Coordinated Trials at Institute Level.

COP: Coefficient of Parentage.

COW(W)-1: HD2646/HW2002A/CPAN3057.

CPAN 1676', 'CPAN 1842', 'HD 2135', 'HW 517', 'Pavon 76', and 'WG 138', were different from the resistance component/s of 'Frontana', the Brazilian bread wheat cultivar used worldwide as a source of durable resistance to leaf rust in wheat.

Creativity Tool: As one puts ideas down on paper, the ideas become clearer and may trigger new associations.

CRFPL: CIMMYT Rainfed Parental Lines.

CRISPR-Cas9.

Crop Growth Stages: There are 5 important stages *viz.*, CRI, Tillering, Jointing stage, Heading, Grain development.

Crossing Block: Maintenance and purposeful utilization for crossing.

CS: Chinese Spring.

CSISA: Cereal Systems Initiative for South Asia.

CT: Canopy Temperature to be directly and easily measured remotely and without interfering with the crop.

CTD : Canopy Temperature Difference, equals to Ambient air temperature-Canopy temperature, measured at two stages *i.e.*, at anthesis and one week after anthesis. measured using handheld infrared thermometer and the CTD will be calculated.

Cut Worm *(Cirphis loreyi Dup., Euxoa spinifera Hub.* and *Agrotis ypsilon Rott):* the cutworms comprise of three species *viz.* (*Cirphis loreyi* Dup., *Euxoa spinifera* Hub. and *Agrotis ypsilon* Rott).

CVT : Coordinated Varietal Trial [CVT-1A, CVT-1B, CVT-2 (Bhopal), CVT-2 (Pune), CVT-3, CVT-4 (Pune), CVT-4 (Bhopal), CVT-4 (Indore), CVT-5A(Res-IR), CVT-5A(RF)].

Cyclotec Micro Mill (0.5 mm).

CYG: Crop Yield Growth.

Cytokinins (Kinetin, BAP).

Cytoplasmic Inheritance: Resistance is determined by cytoplasmic genes or plasma gene(s). *e.g.* The T-male sterility cytoplasm (CMS-T) in maize is extremely susceptible to *Helminthosporium* leaf blight, while the non-T cytoplasm is resistant to this disease.

Cytoplasmic Resistance*:* governed by plasmagenes *e.g.* Resistance to European corn borer in maize. Resistance to root aphid in lettuce. *Sources of Insect Resistance* are a cultivated variety, germplasm collections, a related wild species and an unrelated organism.

CZ: Central Zone.

D genome contributes to bread making quality.

DALYs (Disability-Adjusted Life-Years) approach.

Damaged Ear Heads : white ears.

Date x Variety interaction.

DBT: Department of Biotechnology, Government of India.

DBW 14 (*Raj 3765/PBW 343).*

DBW 17: CMH79A.95/3*CN079//RAJ 3777.

DBW 38 (*CNDO/R143//ENTE/MEXI_2/3/Ae.sq(Taus)/4/WEAVER/5/2*PASTOR).*

DD_{136}: Nematode formulation as insecticide from beneficial nematode.

DDRT-PCR : Differential Display Reverse Transcriptase PCR.

DGE-TAG: Digital Gene Expression TAG, DeepSAGE and RNA-Seq.

DGGE: Denaturing Gradient Gel Electrophoresis for pedigree analysis and fingerprinting.

DH: Doubled Haploidy breeding in wheat following chromosome elimination approach has following advantages over the existing wheat x maize system:.

Dhoot Farm, Kusmanya, Dewas, Madhya Pradesh.

DHPS: dihydro dipicolinate synthase, is in contrast to bacteria in which lysine biosynthesis is primarily regulated by both a lysine mediated repression of gene expression and feedback inhibition of the activity of a lysine-sensitive aspartate kinase isozyme.

DHTSN: Drought and Heat Tolerance Screening Nursery.

Diamant: Very good grain and malting quality, lodging resistance.

Diastatic activity of wheat.

Diastatic Activity: refers to the amount of maltose (mg) produced by flour (5g) when allowed to digest at pH 4.6-4.8 (46 ml buffer) for one hour at 30°C.

Digital Release of Commemorative Coin on the occassion of 75 years of FAO dt 16th Oct, 2020.

Direct drill seeding rice: Less requirement of water, saves time, post-harvest condition of field is better for succeeding crop, deeper root growth.

Direct planting/seeding: this is only a technique that refers to seeding/planting without preparing a proper seedbed. The same equipment is used in Conservation Agriculture. However, the term direct seeding can also be used for implements, which combine primary and secondary tillage and seeding in one machine/tractor operation like the rotary till drills.

Diversification: Efficient use of water, increased income, increased nutritional security, conserved soil fertility, reduced risk.

DL 784-3 (Vaishali): Kalyansona*4/TR380-27* 4/3 #Ag/3/.

DL 788-2 (Vidisha), K 7537/HD 2160 MUT//HD 2278/DL 896-2.

DMSO: dimethyl sulfoxide, six drops (from an eye dropper) of is added to 60 ml of colchicine (1 drop DMSO/10 ml of colchicine).

DNA isolation: Usually CTAB-method (Saghai - Maroof) is used to isolate wheat DNA from 1-2 weeks old plants. Isolated DNA is purified by RNAse treatment and quantified on 0.8 per cent agarose gel/by fluorimeter.

DNA Markers are defined as a fragment of DNA that is associated with a certain location within the genome. They arise from different classes of DNA mutations such as substitution mutations (point mutations), rearrangements (insertions or deletions) or errors in replication of tandemly repeated DNA, basically, DNA markers are divided into two main categories: PCR-based molecular markers and Hybridization-based molecular markers.

Dot-blot hybridization assay in *Ae speltoides* derived lines *viz.*, Sel 2427 and Essay using digoxigenin-labelled *Ae speltoids* DNA as probe.

Dot-ELISA (dot blot, dot immunobinding or spot immunodetection) which are employed for detection of various viral and bacterial pathogens.

Dough Development time (DDT): The time in minutes required for centering the peak of the farinogram on the 500 BU line.

Dough extensibility is measured with Alveograph. It measures the extent to which a gluten can be extended and its ability to retain gases. Basically, extensibility is a varietal character, hence careful selection can result in strains with gluten having good extensibility and gas holding capacity.

Dr Gyanendra Prakash Singh: Wheat man of India, Director, IIWBR, Karnal, India.

Dr Hans-Joachim Braun, DDG (Wheat), CIMMYT, Mexico.

Dr Indu Sharma : Ex-Director, IIWBR, Karnal, India.

Dr MS Swaminathan: Pioneered the Green Revolution in India.

Dr Norman Borlaug (2009): Father of Green Revolution, Noble Prize Winner Peace.

Dr Rajendra Prasad Central University, Pusa, Bihar (erstwhile Rajendra Agricultrual University, Pusa, Bihar).

DRE/CRT: Dehydration Responsive Element/C-repeat is involved in one of the abscisic acid independent pathways.

Drought Avoidance (Dehydration avoidance): is due to the ability of plants to maintain favourable water balance even under stress. The plants which avoid drought retain high moisture content is their tissues and lose less water.

Drought Escapes: is due to ability of a genotype to mature early, before occurrence of drought.

Drought Tolerance (Dehydration tolerance): Ability of plants to produce higher yield even under 'low water potential'.

DS (Dough Stability): The time in minutes for which the curve traverses the 500 BU line.

DSN: Drought Screening Nursery.

DTSN: Drought Tolerant Screening Nursery.

DTT: dithiothreitol.

Durum and red durum: Very hard, translucent, light colored grain used to make semolina flour for pasta.

Durum Wheat (D): the subclass Durum wheat shall be Durum Wheat with less than 60 percent of hard and vitreous kernels of amber colour.

Durum Wheat (Malvi/Kathia): HD 4694, HI 8498, HI 8381, HI 8596, HD 4672, HD 4692, HI 8591, HD 4695, HI 8651 and HI 8653 as lines were crossed with four male parents *viz.*, Bijaga Red, HG 110, IWP 5019 and MPO 215.

DUS: Distinctiveness Uniformity and Stability testing.

Dwarfing (*Rht*12).

DWR 1006: LULA/CREX//AAZ.

DWR 162 carrying *Lr26*.

Ear Cockle Mamni and Yellow Ear Rot or Tundu (*Anguina tritci and Clavibacter tritici):* diseased plants have spreading tendency and swollen base, leaves become crinkled and twisted ear heads contain dark brown, hard and roundish galls (mamni) instead of grains, infected plants are stunted and may die at seedling stage, occurrence of yellow, slimy mass indicates ear rot phase which prevents grain and gall formation, ccreening may be done by mixing the galls with the seeds of the varieties to be screened or mixing the galls in the soil where material to be screened is sown. the gall frequency in the grain sample may be recorded for identifying resistance, the first plant parasitic nematode recorded, also known as seed gall, purples, pepper corns and hard smut.

EBWYT: Elite Bread Wheat Yield Trial.

Ecosystem Engineering.

ED 2398A**:** Ethiopian local variety from the germplasm collection at IARI-Indore**.**

EDTA – Ethylenediamine tetraacetic acid ($Na_2EDTA.2H_2O$); Mol. Wt. = 372.2.

EDVs: Essentially Derived Varieties.

EGPSN: Eastern Gangetic Plains Screening Nursery.

EGS: Early Generation Selection of bulks.

EIGN: Elite International Germplasm Nurseries (EIGN-I for bread wheat and EIGN-II for durum wheat).

Einkorn (*T. monococcum*).

ELFA: Enzyme-Linked Fluorescent Assay.

ELISA (Enzyme-Linked Immunosorbent Assay): It is very useful and now a days used most-widely for detection of viral and bacterial diseases.

Elite Material, HD 2932, HD 2987, HD 4672, to supply the different testing trials conducted all over India.

Emmer wheat (*Triticum dicoccum*).

English grain aphid (*Sitobion avenae*).

Enzymes: in flour though present in small amounts considerably affect noodle quality, excessive amount of alpha amylase (caused either by sprout damage or malt supplementation) in flour causes rapid break down of the noodle structure, a falling number value between 250-300 seconds is considered best for noodles. Polyphenoloxidase is another enzyme important in noodle quality, it oxidises tyrosine (or other polyphenols) to produce dark melanin pigments, this gives the undesirable dark colours in stored fresh noodles.

EST: Expressed Sequence Tags.

ESWYT: Elite Spring Wheat Yield Trial.

Etewah, an Indian variety, used to provide earliness.

Explicit Knowledge is any form of information, expertise, or experience that can be articulated in detail, codified, rendered persistent, and shared.

Exploitation of genetic variability; Genetic manipulation of growth rhythm.

Falling Number test is designed primarily to detect sprouting in wheat caused by rain irrelevant wet weather at the time of harvest of wheat, falling number when converted to liquefaction number 600 has a linear relation with amylase activity.

Farmer Participatory Breeder Seed Production.

Farmer Participatory Trial.

Federation and Rex varieties: Stiffest-strawed strains are found in the white wheats, the white club variety, resulted from a three-way cross involving Fife, Etewah, and Purple straw.

FHB: *Fusarium* head blight (*Fusarium* spp.) resistant parent Carberry carrying *Fhb1* were considered for MAB.

Field Day (Wheat) at Kagdipura, Nalchha, Dhar, 2020.

Field Screening: the techniques designed to promote uniform infestation by an insect pest in the field are Inter planting a row of known susceptible variety between two rows of testing material, Screening in highly prone areas, in case Soil insect pests to be tested in sick plots only, Testing in a particular season when the infestation is very high. *E.g.* Rice stem borer in off season, Transferring manually equal number of eggs or larvae to each test plant.

Fife was introduced from Canada to give quality.

FIRB (Furrow-Irrigated Raised Bed) planting systems, crops are grown on the raised beds and furrows are used for irrigation. The size of the furrow and raised beds is best dictated by the width of tractor and its size of the tyres, is always best to use narrow tyres for crop planting in FIRB systems such as to avoid compaction of the raised beds with wide tyres.

FIRBS: Furrow Irrigated Raised Bed System of wheat planting.

FISH: Fluorescence *in situ* Hybridization has enabled us to map molecular markers, ranging in size from large regions with repetitive elements to small single copy genes.

Flag Leaf (last leaf) appears as this leaf represents about 75 per cent of photosynthesis reactions during the grain filling period and as such should be preserved from disease or insect attacks to ensure a good yield.

Flag Leaf Analysis: Flag leaves from selected potential donors, recipient parents and their F_1 hybrids were collected between ear emergence and before anthesis, washed thoroughly with N/10 HCl, dried at 80C for 8 h in oven prior to digestion. Dried leaf samples were then digested as a minimum of five replications using mixture of nitric acid and perchloric acid.

Flag Smut (*Urocystis agropyri):* Long narrow lead grey or black streaks or stripes running parallel to veins are formed on the leaves, the stripes eventually ruptures and expose black sooty mass of spores, the reduction in crop yield is due to complete loss of productivity of infected plants, screening against flag smut is carried out in the sick plots.

Flea Beetle *(Chetocnema basilis Baly)*: feeds on the foliage causing minute holes, white blotches appear around the damaged portion and finally results in wilting of the leaf, in cases of severe damage the growth of plant is retarded and plants may even die, the damage starts after germination and continues till February.

Flith Test: the rodent hair and insect fragment counts of flour is determined by digesting the flour with acid and adding the cooled digest to petrol in a separatory funnel, the hair and insect fragments are trapped at the petrol/ water interface, and can be collected and identified microscopically.

Flor Hypothesis: gene for gene hypothesis.

Foldscope is an origami-based optical microscope that can be assembled from a flat sheet of paper in less than 10 minutes.

Foliar Blights (*Drechslera, Alternaria, Septoria):* Many pathogens are associated with leaf spots and blights. *Helminthosporium (Drechslera)* and *Alternaria* are the most important pathogens responsible for producing spots, blights and bloches. Elongated straw coloured spots are common, speckled leaf blotch (*Septoria tritici*) also occurs in Punjab, black tip of seeds, black spot of glumes are also associated with these fungi which renders the grains black tipped and shrivelled, the grains loose weight and cause reduction in yield.

Foot Rot/Root Rot/Seedling Blight (*Sclerotinia rolfsii, Drechslera Sorokiniana):* the root and collar show browning when the disease is associated with *Drechslera sorokiniana,* later, elongated brown spots appear on leaves, glumes and spikelets. Ears are poorly filled, grains shrivelled and black tipped, the disease adds to yield losses on establishment of epidemics due to some other major pathogens than *Sclerotinia rolfsii* or may itself acquire the potential to inflict harms above economic threshold.

Fortification is the addition of key vitamins and minerals such as Iron, Iodine, Zinc, Vitamins A and D to staple foods such as rice, wheat, oil, milk and salt to improve their nutritional content.

Four-way (double).

Freeke scale of wheat development.

Frontana: South American cultivar is considered as one of the best sources of durable resistance.

Fulcaster variety of soft red winter wheat produced by a farmer-breeder, SM Schindel, from a cross between Fultz and Lancaster.

Fulcaster, Kawvale, Mercury, Rival, Garnet: are shattering susceptible varieties.

Fulhio and Trumbull, varieties with resistance to the loose smut disease, originated as plant selections from Fultz.

FWA (Farinographic Water Absorption): the quantity of water required for centering the curve at 500 BU line.

Gabo, Lee, and Timstein have two complementary genes governing resistance to race 56.

Gama: Gorham (1990) supposed that salt tolerance gene(s) are located in genomes U, M and D. The D genomes are possessed by 'Gama'.

GBPUA and T Pantnagar, Uttarakhand (Erstwhile Uttar Pradesh).

GBS: Genotype by Sequencing approach, the linkage map of wheat.

GEI: genotype environment interaction analyses.

Gel-Electrophoresis: is carried out depending on the type of marker system used.

Gene Cataloguing for Disease Resistance Stocks: since the beginning of Green Revolution in India, tremendous increase in production has occurred due to incorporation of dwarfing genes into indigenous germplasm.

Gene Density in the euchromatic loops is tenfold higher. The decondensed loops in Arabidopsis might therefore be equivalent to RIDGEs.

Gene Deployment in bread wheat varieties.

Gene Pyramiding for Multiple Disease Resistance: Development of resistant varieties for single disease is not enough to save plant product and to feed growing population especially in developing countries. In India, there are number of constraints in the successful production of wheat like leaf and yellow rust, powdery mildew and spot blotch.

Gene Revolution: may genetically modified plants (GMPS) have been released for commercial use especially for manufacture of food products, using (i) plant cell culture (ii) genetic engineering which are two major areas of plant biotechnology. The enormous advantages of totipotency of plant cells could successfully be translated into standardization of tissue culture techniques namely embryo culture, anther/pollen culture, protoplast culture and cell fusion, meristem culture and callus culture for developing somaclonal variants for economic attributes.

Gene Silencing of HMW glutenins observed in transgenic wheat expressing extra HMW subunits.

Genesis 2354' and 'Genesis 2359: two lines have been released in Uruguay by INIA.

Genetic Engineering is circumscribed by a non sexual transfer of genetic information coded on the DNA from one organism to the other, thus bringing about genetic transformation of the recipient organism genetic engineering enables the creation of plants and animals incorporating useful traits of other species of microorganisms programmed to manufacture large quantities of valuable chemicals and of bacteria that are "bioengineered" to convert one chemical into another.

Genetic manipulation of biosynthesis of chemical components.

Genetic Purity (Homogeneity): An important step towards gene cataloguing.

Genetic stocks, namely, Benno, Kavkaz, Avrora, Clement, Skorospelka 35 carrying *Secale cereale*-derived rust and powdery mildew resistance (1BL.1RS translocation carrying linked genes *Sr31 Lr26 Yr9 Pm8*) were received in late 1970s and their utilization in breeding programme begun.

Genome Annotation, molecular physiological as well as functional genomics studies undertaken in model and/or major crop species provide evidence of the candidate genes (CGs) involved in conferring drought tolerance.

Germination (Triphasic Uptake of Water): Seed physiologists recognize three main stages during germination (triphasic uptake of water). *Stage 1:* This stage is recognized with a rapid initial uptake of water that is usually completed in 6-24h depending on the species. All seeds, even dead ones, take up water rapidly during this stage. *Stage 2:* This is the plateau phase of water uptake during which there is initiation of nucleic acid and protein synthesis in preparation for the emergence of the radicle. This stage may last two or three times as long as stage one. *Stage 3:* This is a stage characterized by the rapid uptake of water, cell expansion and the protrusion of radicle through the seed coat The stage 3 is relatively short and the differences in the times of germination between the seeds of a population are associated with differences in the duration of stage2. Thus 'good' seeds germinate earlier than the 'poorer' seeds. Regulating the availability of water to the seed and preventing it from entering stage 3 can reduce this variation. This enables 'good' seed to be held back, allowing 'poor' seeds in the lot to catch up the development. The hydration of the seeds can be regulated using osmotica (osmopriming), salt (halopriming) and inorganic or organic carriers (solid matrix priming).

GFD : Grain Filling Duration, will be calculated as the number of days required from anthesis to physiological maturity (when 75 per cent glumes turn yellow).

GFLD (Green Flag Leaf Duration), will be calculated as the number of days required from anthesis to physiological maturity (when 75 per cent glumes turn yellow).

GFP (Green Fluorescent Protein) to chromatin proteins enables us to monitor chromatin in living mammalian cells while causing only limited phototoxicity damage.

GHI: Global Hunger Index.

GI : Grain Dormancy.

Gibberellic acid-insensitive dwarfing genes. Rht1 and Rht2 (the Norin 10 dwarfing genes).

Ginuz, Cugap, Chirya and Subaf: are resistant lines developed at CIMMYT with two or three genes providing good resistance level.

GIS: Global Information Systems.

GISH: Genomic *In Situ* Hybridization as per the protocol of Mukai and Gill (1991).

Glaring Disconnect between the academic critique of technological determinism and 'the visions of run-away technology that prevail in society at large'.

Glass House Screening: result from glass house tests are much more reliable than those from field tests since both the environment and the initial level of infestations are more or less uniform for all the plants being tested.

Gliadin analysis: Single seed or half seed was crushed into fine powder using metal crusher or 20 mg flour was used for extraction, 100 µl of 70 per cent ethanol was added to the powder and incubated at 37°C for 30 minutes with brief vortexing at the time interval of 10 minutes, the tubes were centrifuged at 10,000 rpm for 10 minutes and the supernatant was collected, the residue left was used for glutenin extraction. 62.5 µl of the dilution buffer was added to 50 µl of the supernatant obtained. 20 µl of the sample was loaded into the well of the gel for A-PAGE separation of gliadins, the electrophorograms of the gliadin proteins are conventionally evaluated according to Bushuk and Zillman (1978) based on relative mobility, the relative mobility (Rm) values are calculated with reference to a specific band, designated 50, of the variety 'Marquis' (a Canadian hard red spring wheat variety), the Rm value of a given band is obtained by dividing the distance the band migrated from the origin by the distance band 50 migrated from the origin and then multiplying this value by 50, the value obtained is the Rm for that band, genetic distances were calculated based on presence or absence of bands taking band 50 as reference, using Hierarchical cluster analysis of SPSS programme and dendrograms were drawn using average linkage between groups and Euclidean distance method.

Global Warming: Global Threat.

Glutamic Acid: is derived from wheat, a familiar compound of glutamic acid is "mono sodium glutamate", a salt – like product generally available and used to bring out the flavour of other foods or seasonings.

Gluten Complex is composed of two main groups of proteins, gliadins and glutenins constituting about 80-85 per cent of the total grain proteins. The gliadins are responsible for the dough's cohesiveness and along with strong, *Gluten* is a mixture of different proteins of wheat, gliadin and glutenin, is not present in flour, however when flour is moistened and kneaded into small balls, the development of gluten takes place, is a rubbery mass made from proteins, lipid, fibrous components of wheat flour, only wheat flour possesses, such an important feature amongst all the cereals that is why it can be used for the preparation of wide varieties of end products, like, bread cakes, puffed products, pastries, puddings where the inflation of the wheat dough is required, the gluten has elastic properties like rubber band, the interiors of the elastic structure are just mush like work, where the gasses are trapped and help in increase in loaf volume.

Gluten: is made up of glutenins and gliadins, influences the viscoelastic properties of wheat flour and largely determines how a particular variety is used.

Glutenins are extracted and separated by sodiumdodecylsulphate polyacrylamide gel electrophoresis (SDS-PAGE), using the standard procedure.

GMMs: Genic Molecular Markers.

GMOs: Genetically Modified Organisms.

Go Go method is followed for artificial screening of wheat lines against loose smut.

GOT (Grow Out Test) is practiced by researchers for genetic purity.

GPC: Grain Protein Content and gluten quality are the most important factors determining the end-use quality of wheat for pasta-making.

Grain Appearance Score (Max. 10): include grain size, shape and colour are taken into consideration for scoring the grain appearance, bold grains with attractive shape, amber golden colour and the luster of the grain are the major criteria for scoring.

Grain yield/plot: the grain yield of the entire plot will be determined in grams.

Graphic Displays of "concept spaces" visually indicate the relationships and important subsets of concepts, particularly subsets that constitute ontological commitments for representing given phenomena.

Grasshoppers *(Melanoplus* spp.).

GRDC: Grain Research and Development Corporation, Australia.

Green bug (*Schizaphis graminum*): causes considerable damage to wheat, an aphid and feeds on the wheat plant by piercing the leaf tissue with its stylet and injecting saliva which is toxic to the plant, a breakdown of the tissue follows and yellowing or even death of the plant results.

GRT: Germplasm Rejuvenation Trial.

GTI (Genotype Tillage Interactions) inconclusive regarding GTI, largely because few genotypes were tested and all tested materials were developed under conventional tillage, adoption and performance of current and new wheat materials has been very successful in many areas of the world; selection under zero-tillage is expected to further improve adaptation to CA, the retention of crop residues has led to a change in the disease spectrum in some areas. Diseases such as tan spot and crown rot (caused by *Fusarium pseudograminearum*) which survive on crop residues can become limiting in some environments.

Gujhia Weevil *(Tanymecus indicus)*: is a pest of wheat in the seeding stage, both the larvae and adults cause damage, but adults are more serious, they feed on the roots and cut down the young germinating plants at or below ground level, at early stages its damage resembles to those of termites, the pest used to be serious from October-November.

Gunilla: Good strawed variety.

GURT: Genetic Use Restriction Technology.

GW 1139: MACS 2340/IWP 5070.

GW 1189: SU-CULLUGU-7/VD 90-12.

GW 273: TW 275/7/7/10/lok-1.

GW 322: PBW 173/GW 196.

GW 366: DL 802 -3/GW 232.

GWS/GS: Genome-wide selection (GWS) or genomic selection (GS) approach to develop superior germplasm lines with overall excellent performance in a target environment.

GxExM: Genotype x Environment x Management.

HAD (Hard Amber Durum): The subclass Hard Amber durum wheat shall be Durum Wheat with 75 percent or more of hard and vitreous kernels of amber colour.

Haploids developed using the wide-cross method must be subjected to a chromosome doubling treatment.

Hard Red Spring: Hard, brownish, high protein wheat used for bread and hard baked goods. Bread Flour and high gluten flours are commonly made from hard red spring wheat.

Hard Red Winter: Hard, brownish, mellow high protein wheat used for bread, hard baked goods and as an adjunct in other flours to increase protein in pastry flour for pie crusts. Some brands of unbleached all-purpose flours are commonly made from hard red winter wheat alone.

Hard Wheats are harder to process and red wheats may need bleaching, hard, light colored, opaque, chalky, medium protein wheat planted in dry, temperate areas. Used for bread and brewing.

Harvest Index (HI) are the main characters for the genotypes performed well under furrow irrigated bed planting system, the harvest index (Grain yield per plot/ Total biomass produced per plot) will be calculated.

Harvest Plus is focusing on core collections. Screening and research on screening methodology need to be expanded to include generic and crop-specific inhibitors and promoters for bioavailability.

Haya shrinriki: Good quality.

HD 2009: LR 64A/NAI 60.

HD 2135: H 41-3 (HD 1962* (E 4870*K65).

HD 2189 (*Lr13, Lr34, Sr2, Sr8a, Sr11, Yr2* and *Yr18,* showing adult plant resistance for more than 30 years in peninsular India carries race-specific genes, Lr1, Lr13, Lr17 and Lr34.

HD 2189: HD 1963/HD 1931.

HD 2285 (*Lr23, Sr5, Sr9b, Sr11* and *Yr2*).

HD 2285 (*Lr23, Sr9b, Sr11* and *Yr2* genes).

HD 2329 (*Lr13, Lr10* and *Lr34* and *Sr8b, Sr9b* and *Sr11* for stem rust and *Yr2* and *Yr18* against stripe rust.

HD 2329 (*Lr13, Lr10, Lr34, Sr8b, Sr9b, Sr1, Yr2* and *Yr18*).

HD 2380: HD 2255/HD 2257.

HD 2402: HD 2267/HD 2236.

HD 2643 (Ganga), VEE'S'//HD 2407//HD 2329.

HD 2687 (Shrestha).

HD 2733 (VSM).

HD 2733, HD 2824, HI 1500, HI 1544 (Purna), HI 1544 (Rahish Patel Magajpura), HI 8663 (Poshan), DBW 14, DBW 38, HD 2285, HD 2329, HD 2643, HD 2687, HD 2824, HD 2888, HD 3117, HD 4728 (Pusa Malvi), HD2733 (VSM), HI 1563 (Pusa Prachi), HP 1731 (Rajlaxmi), HP1761, HW 2045 (Kousambi), PBW 443, WB 2 : Breeder seed produced.

HD 2781 (Aditya): BOW/C306//C591/HW2004.

HD 2824 (Poorva).

HD 2833 (Tripti): PBW226/HW1042//HD2285.

HD 2851 (Pusa Vishesh): CPAN3004/WR426//HW2007.

HD 2864: DL 509 -2/DL 377-8.

HD 2888 (Pusa Wheat): C306/*T.shpaerococcum*//HW2004.

HD 2932: KAUZ/STAR//HD 2643.

HD 4502 (Malavika).

HD 4672 (Malav Ratna), BRED/PBW 34//ALTAR 84.

HD 4696: AAZ/MORUS.

HD 4728 (Pusa Malvi).

HDD: Heat degree days will be computed with 0°C as base temperature.

Helminthosporium Leaf Blight (HBL): The three important leaf blight diseases reported in South Asia are Spot blotch (*B. sorokiniana*) (Sacc. In sorok) shoem, Tan spot (*Pyrenophora tritici* var. *repentis* (Died.) Drechs.) and Alternaria blight (*Alterania tritici*) (Pras and Prab). However, spot blotch (*B. sorokiniana*) is of increasing concern in South East Asia (Kumar *et al.*, 2010).

Heritability: relates to genetic progress ($Gs = i\ \sigma p\ h^2$) and associated genetics is crucial for establishing screening, population development, and G E testing strategies, and hence, for effective breeding, all plant-breeding components (such as crossing strategies, breeding methodologies, the operational scale with plot size and number of sites and years required for testing) are based on genetic parameters, furthermore, the potential for developing molecular markers is closely associated with factors such as the number of genes and their individual contributions.

Hessian Fly [Mayetiola destructor (Phytophaga destructor)]: attacks the wheat crop over a wide area, the principal method of control is to delay planting until after the adult flies have emerged and oviposited.

HG 110: Sarangpur Local/HI 8185.

Hg*:* hairy glume gene, located in the short arm of chromosome 1A controls the pubescence of the glume.

HGGT (Homogentisic Acid Geranyl Geranyl Transferase): catalyses the committed step of tocotrienol biosynthesis were isolated from barley, wheat and rice seeds.

HI 1418 (Naveen Chandousi), HI 999/HI 601.

HI 1454: HI 1076/CC 505/HI 1136.

HI 1479 (Swarna), HD 2195/HD 2160//WH 283.

HI 1500 (Amrita), barely 10 days early in heading and maturity to "Sujata", HW 2002*2/STREMPALLI/PNC 5.

HI 1531 (Harshita) [HI 1182/CPAN 1990].

HI 1535: BOW/BUC//BUL//CLRP6/HW3034.

HI 1544 (Purna) [HD 2402/HW3007].

HI 1563 (Pusa Prachi) [MACS 2496*2/MC10], 2011.

HI 1605 (Pusa Ujala) [BOW/VEE/5/ND/VG9144//KAL/BB/3/YACO/4/CHIL/6/CASKOR/3/CROC_1/*Ae.squarrosa*(224)//OPATA/7/PASTOR//MILAN/KAUZ/3/BAV92].

HI 1612 (Pusa Wheat 1612) [Kauz//Altar 84/AOS/3/Milan/Kauz/4/Huites].

HI 1620 (Pusa Wheat 1620) [NAC/TH.AC//3*PVN/3/MIRLO/BUC/4/2*PASTOR/5/KACHU/6/KACHU].

HI 1621 (Pusa Wheat 1621).

HI 1628 (Pusa Wheat 1628).

HI 1633 (Pusa Vani).

HI 1634 (Pusa Ahilya).

HI 784: NAPO/TOB 'S'/3/8156/KAL/BB.

HI 8381 (Malavshree), JO 'S'/AA 'S'//FGO 'S'.

HI 8498 (Malavshakti), RAJ 6070/RAJ 911.

HI 8591: HI 8144/NI 8625.

HI 8627 (Malav Kirti), combines a good biomass, harvest index and grain yield and can be very successfully grown in early October yielding 50 q/ha of quality grain.

HI 8627: HD 4672 X PDW 233.

HI 8638 (Malav Kranti), Selection From Buttah.

HI 8663 (Poshan) HI 8177/HI 8185.

HI 8664, HI 8668, HI 8670, HI 8671, HI 8672, HI 8673.

HI 8691 (Pusa Gaurav) [(HG 509/HI 8381)//(Boomer-33/Plata 8)], with good content of iron and zinc.

HI 8704: (HG 822/HI 8498) having good resistance to rusts and distinctively different from HI 8498 with good (>50 g) was proposed during 50th Wheat and Barley Research Worker's Meet, 2011.

HI 8713 (Pusa Mangal).

HI 8737 (Pusa Anmol).

HI 8751 (HD 4685/HI 8634) (IC0623451; INGR 17032).

HI 8759 (Pusa Tejas).

HI 8765 (HI 8504/CPAN 6206//HI 8627), registered at ICAR-NBPGR, New Delhi as genetic stocks with multiple disease resistance, (IC0624495; INGR17038).

HI 8777.

HI 8802 (Pusa Wheat 8802).

HI 8805 (Pusa Wheat 8805).

Hidden Knowledge: knowing the premises, prerequisites and motives influencing our thoughts and action positions.

Hill Bunt or Stinking Smut *(Tilletia caries and Tilletia foetida)* and Dwarf Bunt *(Tilletia controversa)*: is found throughout the wheat-growing areas, is most severe in the Pacific northwest and the intermountain regions, common bunt is both seed borne and soil borne.

HIS: Heat Susceptibility Index.

HMW-GS = High Molecular Weight glutenin subunits.

Hn: hairy node.

Hope : wheat at the same time that McFadden transferred the stem rust resistance gene, currently known as *Sr2*.

HOPE: Half of Professional Efficiency.

Host Habitation includes Polyphagy, Oligophagy, Seasonal Oligophagy and Monophagy.

Hot Spot' for foliar blight.

HP (Hybrid Pusa) No has been adopted at IARI RS Pusa, Bihar.

HP 1102, HP1209, HW741, HW971 carrying Lr13 in combination of Lr34 or Lr23 produced moderate infection type against 77-1, 77-2 and 104 races of leaf rust.

HP 1102: cross 8156 (Blanko)/Nadadores63).

HP 1102: First dwarf variety for NEP zone from the crosses of 8156 (Blanco) and Nadadoses 63. It revolutionized the production of Bihar.

HP 1209 was introduced for cultivation under late sown irrigated condition with an average yield potential upto 33.3 q/ha. This variety is still popular in eastern states and in Nepal as Lumbini.

HP 1633 (Sonali) grown under late sown condition. This is substitute of popular variety Sonalika, which is the first *Lr9* alien gene introduced variety in the country shows very good tolerance to foliar blight in addition to resistance against brown rust. HP1744 is an outstanding variety for chapatti quality and is most popular among farmers.

HP 1731 (Rajlaxmi) [Lira//Parula/Tonichi].

HP 1744 (Rajeshwari) [Ciano/Parula//Chilero/Garuda].

HP 1761 (Jagdish): [RL 6010/6*Inia//3*KAUZ].

HP No allotted to PS series from Pusa, Bihar.

HPI (Host-Pathogen Interaction) in Leaf Rust Infection of Wheat.

HPLC (High Performance Liquid Chromatography) is the method to quantify individual carotenoids and their isomers.

HPT: Hygromycine Phosphotransferase gene is strongly condensed and forms a small neo-chromocenter.

HPW 42: VEE'S'/4/PUN 'S'/CBB//CNO 'S'/#/JAR/ORZ 'S'.

HRD gene, coding for an AP2/ERF-like transcription factor.

HS 295: CQT/AZ//IAS 55/ALD 'S'/3/ALD 'S'/NAFN/4/PIN 'S'/PEZSI – 127.

HS 365: HS 207/SKA.

HSPs: Heat Stress Proteins or chaperones, LEA proteins, osmoprotectants, antifreeze proteins, detoxification enzymes and free-radical scavengers.

HTI: High Temperature Incidence, computed on the basis of daily minimum and maximum temperatures when the crop was exposed to greater than 30°C.

HTMs: High Throughput Screening Methods.

HUW 206 carrying these genes was released for general cultivation in 1983 in the North Eastern Plains Zone, where both stem rust and stripe rust were not a problem.

HUW 234 (*Lr14a, Sr9b, Sr11 Yr2Ks*).

HUW 234, HUW 468, HUW 510, HUW 516, PBW 343, PBW 443, HD 2687, HD 2733, UP 2338, NW 1012, DBW 14, Raj 3765: varieties suitable for No Tillage.

HUW 468: CPAN 1962/TON//LIRA 'S'/PRL 'S'.

HW 1085 (Bhavani): Unnath Kalyansona/CPAN3057.

HW 1085: C 306*7/TR 380-14 *7/3AG 14.

HW 2004 (Amar), C306*7/TR380-14*7/3 Ag #14, had outyielded both the check varieties in production and productivity, proving that early maturity and drought tolerance can be easily combined. Such genotypes will escape terminal drought at maturity and hence, will yield better because of their good grain filling.

HW 2034 (MACS 6145).

HW 2044 (Kurinji): PBW226*6/Sunstar*6/C80-1.

HW 2044: PBW 226 *5/SUNSTAR *6/C801.

HW 2045 (Kausambi): HD2402*6/Sunstar*6/C80-1.

Hybrid 128, was developed from a cross between Jones Winter Fife and Little Club made, a parent in crosses producing Alicel, Hymar, Albit, and other varieties.

Hybridization Programme: performed to create the new genetic variability for further use in the breeding programme.

IARI Common Varietal Trials (CVT) entries.

IBWSN: International Bread Wheat Screening Nursery.

ICAR Research Complex for North Eastern Hill Region, Umiam (Barapani), Meghalaya.

ICAR: IIWBR, Indian Institute of Wheat and Barley Research, Karnal, Haryana (erstwhile DWR, Karnal, Haryana).

ICAR-CIAE: Central Institute of Agricultural Engineering, Bhopal, India.

ICAR-CICR: Central Institute of Cotton Research, Nagpur, Maharashtra.

ICAR-CSSRI: Central Soil Salinity Research Institute, Karnal, Haryana, India.

ICAR-IARI: Indian Agricultural Research Institute, New Delhi, India.

ICAR-IARI: Indian Agricultural Research Institute, Regional Station, Indore, Madhya Pradesh.

ICAR-IARI: Indian Agricultural Research Institute, Regional Station, Pusa, Bihar was the first institution to systematically assess richness of germplasm.

ICAR-IARI: Indian Agricultural Research Institute, Regional Station, Wellington, Tamil Nadu, India.

ICAR-IIMR: Indian Institute of Maize Research (erstwhile Directorate of Maize Research, New Delhi), Ludhiana, Punjab.

ICAR-IIRM: Indian Institute of Rapeseed and Mustard, Bharatpur, Rajasthan, India.

ICAR-IISR: Indian Institute of Soybean Research, Indore (erstwhile National Research Centre on Soybean, Indore; Directorate of Soybean Research).

ICAR-IIWBR: Indian Institute of Wheat and Barley Research, Regional Station, Flowerdale, Shimla, Himachal Pradesh.

ICAR-IIWBR: Indian Institute of Wheat and Barley Research, Regional Station at Dalang Maidan, Lahul-Spiti, Himachal Pradesh, India.

ICAR-Indian Grassland and Fodder Research Institute, Regional Research Station, Srinagar, J and K.

ICP: Inductively Coupled Plasma.

ICP-OES (Inductively Coupled Plasma-Optical Emission Spectrometer): Methods to estimate Iron and Zinc.

ICT: Information and Communication Technology based e-Plant Breeding.

IDD (Iodine Deficiency Disorders): at least 54 million have incidence of goiter or other worst consequences.

IDM: Integrated Disease Management.

IDYN: International Durum Yield Nursery.

ILs: Introgression Lines or Substitutions Lines. This technique offers the potential for further gains in the use of SHW.

Imperata cylindrica hybridization.

Importing Settings: Click button Use in the bottom of LIVE WINDOW to import camera and light settings to the protocol.

Index of sustainability (Is): It is a measure of sustainability relating productivity to change in soil and environmental characteristics.

Indian dwarf wheat (*Triticum spherococcum*).

Indices of Sustainability: Quantification of sustainability is essential to objectively assess the impact of management systems on actual and potential productivity, and on environment, can assess sustainability or several indices, indices may be simple involving one parameter or complex involving several parameters.

Insect Resistance is governed by - 1. Oligogenes 2. Polygenes 3. Cytoplasmic genes.

Institutional Back Up and Co-Ordination: is possible to "demystify" seed growing, and which can be returned to the hands of the farmer in the local economy, but Institutional Back Up and Co-Ordination is must.

International Seed Testing Association (ISTA, 1993).

IP: Identity Preservation systems by automated, portable tests using DNA fingerprinting so that during export, shipments will be able to be certified for a certain desired variety, for the GMO status of grain, or for a variety with a new, valuable trait.

IPM: Integrated Pest Management.

IPPSN: Initial Plant Pathological Screening Nursery.

IPR: intellectual property right.

IRT/SR: Infrared Thermography/Canopy Spectral Reflectance Temperature Trait for Phenotyping.

ISC: Indian Science Congress.

ISEM: Immunosorbent Electron Microscopy.

ISSR : Inter Simple Sequence Repeat Markers:.

ISWYN: International Spring Wheat Yield Nursery.

ITMI: International Triticeae Mapping Initiative, mapping population developed from the cross between the CIMMYT SHW line W-7984 and the hard-red spring wheat (HRSW) 'Opata 85' and it has been extensively used for mapping studies.

ITs: Infection Types.

ITYN: International Triticum Yield Nursery.

Iumillo durum (used in Thatcher)., Hope and Thatcher, or other derivatives from these crosses, have been used in the development of many resistant spring wheat varieties.

IWP 5019: HD 4519*2/NP 200.

Janapaw, Indore, Madhya Pradesh.

Jutta: Increased winter hardiness.

K 9107: K 8101/K68.

K 9465: B 1153/CB 85 = HD 2402/CPAN 1830//VEE 'S'.

Kagjipura, Dhar, Madhya Pradesh, India.

Kalayansona (*Lr13, Lr34, Sr11, Yr2KS, Yr18*).

Kalyan Sona, Safed Lerma, Sharbati Sonara are double gene dwarfs, amber colour wheat varieties.

Kalyansona and Sonalika carrying *Yr2+A* genes provided resistance to the then prevalent races of stripe rust.

Kanred, Tenmarq, and Kawvale: Hard wheat varieties.

Kanred: resistant to some races of stem rust and leaf rust, although otherwise similar to Turkey. It was one of the first rust-resistant varieties. Cheyenne was selected from Crimean in 1922 in Nebraska.

Karnal Bunt (*Neovossia indica*): in an infected plant some of the tillers bear infected ears. In the ear only a few grains are infected and most of the grains get partially infected. The infected grains on pressing give out black powder of spore mass. This powder gives peculiar stinking smell. Screening against Karnal bunt is done using syringe method of inoculation. Sporidial cultures of the pathogen are made in PDA from the teliospores.

Kawvale variety, a semi hard wheat, was selected in Kansas from a soft wheat variety, Indiana Swamp in 1918. It is important because it later became a parent of Pawnee and other varieties. The hard red spring wheat, Haynes Bluestem, was developed from a selection made by LH Haynes in 1895. Mindum, a durum variety, was selected from a field of common durum.

KD: Karewa Damodar Farm, Srinagar, J and K, India.

Kernel Weight (Thousand Grain Weight) (g): gives an idea of average grain size and is considered to be a function of kernel size and density. While test weight determines the milling quality of all wheats, kernel weight is decisively superior in predicting the milling quality of hard grain.

Khapli, Kenya 184, Kenya Ploughman for black rust, Frontiera, Frontana, Bowie Texas, Mentana and Tremez Molle for brown rust and Ceres Klein, Cometa Klein, Frondosa and Spalding's Prolific for yellow rust and exotics like Gabo, Gaza, Yaqui 53, Timstein (*Lr23*) (resistant to black and brown rusts), Rio Negro, La Prevision (for brown and yellow rusts).

Kristina: Good straw.

Kubanka : variety introduced from Russia by the USDA, a leading variety in the durum class for many years.

KVK: Krishi Vigyan Kendra, Kasturbagram, Indore, Madhya Pradesh, India.

KVK: Krishi Vigyan Kendra, Zanskar, Jammu and Kashmir, India.

LBSN: Leaf Blight Screening Nursery.

Leaf Colour Chart: reduces N fertilizer requirement, reduces N loss and environmental pollution, reduces nitrous oxide emission.

Leaf Hopper (*Typhlocylbea maculifrons* Motsch and *Zygnidia manaliensis* (Singh)**:** Both nymphs and adults suck the sap from leaves and produce small white spots on the leaves. Its damage also remains very low and it had never come in the pest from. Two species *i.e., Typhlocylbea maculifrons* Motsch and *Zygnidia manaliensis* (Singh) has been reported from wheat.

Leaf Rust (*Puccinia rubigovera tritici):* Leaf rust is found almost everywhere that wheat is grown. Loss in yield due to leaf rust is generally not so great as the loss caused by a severe stem rust epiphytotic. Yield reductions of 5 to 10 per cent are most common, but severe epiphytotics during early stages of growth may sometimes kill seedling plants.

Leaf spots/Foliar blight caused by *Helminthosporium aestivum* and *Alternaria triticina.*

Leymus racemosus is a wild species belonging to tribe Triticeae (Poaceae), which includes important cereal crops such as bread wheat and barley. This perennial species grows along the coast and in dry lands, and is reportedly tolerant to various biotic and abiotic stresses. Although *L. racemosus* is evolutionarily distant from wheat (*Triticum* spp.) within the tribe, it has been successfully hybridized with wheat, and several wheat–*L. racemosus* chromosome introgression lines have been selected from among the backcrossed progenies of the hybrid. *L. racemosus* is, therefore, a promising wild species for wheat improvement.

LGP: Length of Growing period, depends not only on the rainfall distribution but also on the soil type, depth and water holding capacity.

LI: Lodging Index.

Liberation Theology to Liberation Technology, in which governments seek technological, rather than political, solutions to social problems.

LMW-GS: Low Molecular weight glutenin subunits.

LNMU: Lalit Narayan Mithila University, Darbhanga, Bihar.

Lok 1possess *Lr13, Sr2, Sr9b, Sr11, Yr2Ks*.

Loose Smut (*Ustilago tritici*): is a common disease in most wheat-growing areas.

LOS: Length of Season.

Lr 24 + Lr 28 pyramids using the SCAR marker $SCS421_{640}$ and SCAR marker $scs73_{719}$ in two sub-families PBW 343 background.

Lr 24 + Lr 48 in the F_4 generation.

Lr 28 linked SSR marker Xgwn160 for leaf rust resistance in NIL pairs.

Lr 3 (Democrat), whereas only one pathotype was known to be virulent on this gene.

Lr 34 is associated with leaf tip necrosis (Ltn1), morphological marker.

Lr19 SCAR marker SCS73719 for leaf rust resistance in alien Lr gene NILs of Thatcher.

LSS: Line Source System.

LT (Low temperature): tolerance is a complex quantitative character that is expressed in anticipation of and during exposure of plants to temperatures that approach freezing.

Lysine biosynthesis in plants is regulated primarily by a lysine-mediated feedback inhibition.

Lysine Content: the first limiting essential amino acid is lysine, so in relevant breeding programmes major aim is to increase the amount of this amino acid, there is a negative relationship between lysine content and the protein content of a grain, lysine determination is performed only on materials where protein quality index has already been worked out: QI = DBC/per cent protein.

Maan Annapurna : A symbol of prosperity.

MABB for Rust Resistance in HD 2733.

MABC: Marker-Assisted Backcrossing.

Macaroni Products: these products are also called pasta, include macroni, spaghetti, vermicelli and noodles, the starchy endosperm of wheat is coarsely ground into semolina which is made with water in to a stiff dough.

Macaroni wheat (*Triticum durum*).

MACS 2846: CPAN 6079/MACS 2340.

MACS 6145: HW2034), HD2888 and K307.

Maida: The bran and germ are separated in making white flour or maida. Maida bakes more uniformly into a loaf of a greater volume and it is more bland in taste and more easily digested. It can be stored in an air tight container in the refrigerator.

Mari: Stiffer strawed variety.

Marker Assisted Selection for Quality Traits in Wheat: Wheat quality is a broad term and can mean different things to different people, for a farmer, a healthy and attractive looking crop which can fetch good price in the market is quality, for millers quality is production of white flour per unit grain, for cereal processors quality wheat is one which gives high quality end products like cakes biscuits, pasta products, noodles, flat breads or high-quality animal feeds, as the end product determines the quality of wheat, wheat unsuitable for one may be an excellent choice for another end product.

Markers: morphological markers (also called "classical" or "visible" markers) are phenotypic traits; biochemical markers, are isozymes, including allelic variants of enzymes and DNA markers (or molecular markers), which reveal sites of variation in DNA.

Marquis variety of spring wheat was its early maturity, which enabled it in some seasons to escape stem rust.

MARS: Marker-Assisted Recurrent Selection.

MARs: Matrix Associated Regions.

MAS: Marker assisted selection offers an opportunity to select desirable lines based on genotype rather than phenotype.

ME: Mega-Environments and main targeted traits for wheat improvement in current wheat improvement research [ME1: Irrigated, Temperate, ME2: High rainfall (>500 mm), Temperate, ME4: Low rainfall (<500 mm), Temperate, ME5: Irrigated or High rainfall, Warmer, ME6: Low rainfall, Cool]; was introduced in the early 1990s. MEs are geographical areas where wheat adaptation can be expected to be similar in terms of climatic, disease, or crop management constraints.

Mediterranean, Denton, Wabash, Vigo: soft wheat varieties.

Membrane Stability : assayed in terms of solute leakage from tissues.

Meta Knowledge can be defined as "knowledge about knowledge". Meta-knowledge is both a process and a product.

MGMG: Mera Gaon Mera Gaurav.

MHC: Histocompatibility complex locus appeared to localize outside the chromosome territory depending on cell type and on the activation state of their genes.

Micro Plot Techniques by using small micro plots, small micro plots of size 6 x 3 x 1 m.

Micronutrient Concentrations are affected by micro-environmental variation, G E interaction, germplasm type, and numerous other factors.

Micronutrient Malnutrition (also known as 'hidden hunger') is a global health problem of our planet and affects the quality of human resources development and productivity of any country, the most important nutrients of public health significance are iron, iodine, vitamin A and zinc.

Microsatellites (SSRs, STMS or SSRPs) are also called simple sequence repeats (SSRs) and, occasionally, sequence-tagged microsatellite sites (STMS) or simple sequence repeat polymorphisms (SSRPs).

Microspore Culture involves harvesting immature microspores that are totipotent.

Midas: Stiff strawed variety, an important factor in its replacing Rival, Pilot, Thatcher, and Regent in the hard-red spring wheat area.

Mikns and Golden promise: Stiffer straw, good grain yield.

Milling Quality: is important from the stand point of proper milling, ease of milling, the wheat should flow through the mill in a uniform manner with an even distribution of the stocks.

Mills used for milling wheat are quadrumate junior, Cyclotec, Cemotec, Buhller mill and Stone chakki *etc.*

Mixing tolerance index (BU).

MLHT : Multi Location Heat Tolerant Trial.

MLKS 15 is a multiline mixture of pure lines which are phenotypically similar but genotypically dissimilar.

MLT: Multilocation Trial conducted over seasons to get more reliable results.

MM: Mithila Mahotsav.

Molecular Marker should have the characteristics i) highly polymorphic ii) Co-dominant, should be able to discriminate between heterozygous states in diploid organisms iii) frequent occurrence in genome iv) detect multiple independent and reliable loci v) provide adequate resolution of genetic differences vi) selective neutral behaviors – DNA sequences of any organisms are neutral to environmental conditions or management practices and this permits to confer the variations only to a genetic origin vii) easy access and fast assay, it must be simple quick and in expensive viii) high reproducibility to guarantee robust results among different laboratories and equipments ix) require small amounts of tissues and DNA sample x) link to distinct phenotypes xi) require no prior information about genome of an organism.

Molya Disease *(Heterodera avenae)*: Plants become stunted with yellowing of leaves, reduced tillering, absence of ears on some tillers or small ears with poorly filled grains.

Monophagy: avoid all hosts except one particular species or variety *e.g.* Boll weevil on cotton.

Monosomics : plants with one chromosome less than the normal.

Morphological features, such as, seed size, seed colour (fibre length and strength).

Morphological Markers are basically the indicators of phenotype such as seed colour, seed shape, seed size, seed length, *etc.* They generally represent genetic polymorphisms which are easily identified and manipulated.

MRI : Magnetic Resonance Imaging.

MSD: Multiple Synthetic Derivatives populations by backcrossing elite cultivar and

SHW lines produced using variety of *Ae. tauschii* accessions.

MTI (Mixing Tolerance Index): the drop in peak consistency of the dough in five minutes as measured from the top of the curve in BU, his is the difference in Brabender units (BU) measured from the top of the curve at the peak and top of the curve 5 min after.

Mulch Tillage: The Soil is disturbed prior to planting. Tillage tools such as chisels, field cultivators, disks, sweeps, or blades are used. Weed control is accomplished with herbicides, cultivation or both.

Nanodamnation: focus on technological risk and the potential for unanticipated consequences.

Nanofutures: Provides an introduction to nanotechnology and the shaping of emerging technoscience, describing key narratives and the contestation.

Nano-optimism and nano-pessimism.

Nanosalvation: Assertions based on the belief that nanotechnology is a technological magic bullet.

Nanoscale: The scale of nanometers, or one billionth of a metre.

Nanotechnology is today generating a lot of attention, all around the world, and building great expectations in the academic community, and amongst governments, investors and industry is at three levels: the *macro* (*e.g.* creation of government policy), *meso* (*e.g.* in innovation networks and industry sectors), and *micro* (*e.g.* in research groups).

NATP MM: National Agricultural Technology Project Mission Mode.

Near-Isogenic Lr lines in NP 4 background and genetics of rust resistance in wheat.

Nematode Biotypes/Physiological Races: Nematode biotypes virulent on resistant were more fertile than the avirulent ones (k = 100).

NEPZ: North Eastern Plain Zone.

NGS technology, the sequence-based transcriptome analysis is in many ways considered superior to microarrays, where genome sequence information is lacking, since the sequencing-based method is real time, digital and highly accurate.

NGSN: National Genetic Stock Nursery.

NI 5439 (*Lr34, Sr11, Yr18*).

NIAW 34: CIANO 79/PRL 'S'.

NIDW 15: DOM 50.

NIDW 295: BOOMER 33/PLATA 8.

NILs (Near-Isogenic Lines) differing in the presence or absence of a resistance allele has facilitated the mapping of genes for which such lines exist (*Pm1*, *Pm2* and *Pm3*; *Lr9* and *Lr24*; *Bt-10*; *Yr15*).

NIN (National Institute of Nutrition), Hyderabad, Andhra Pradesh, India.

NIR: Near-Infrared Reflectance measures protein content, grain hardness, and even milling yield and SDS-sedimentation.

NIRS (Near-Infrared Reflectance Spectrophotometry) method relates a sample's reflectance of near-infrared light to its chemical composition and covers wavelengths between 730 and 2500 nm emitted by major plant compounds, such as oil, starch, cellulose, water, and protein. In breeding, NIRS is routinely used to determine, for example, grain protein, which it can measure with high accuracy. The latest research has shown that NIRS has potential to predict iron and zinc with sufficient precision for pre-screening, although the causality of the association is not yet understood. Correlations between iron and zinc determined by ICP and NIRS in potato, sweet potato, and beans range between 0.77 and 0.85, the parameters like protein, hardness water absorption, sedimentation value and loaf volume were tested using Informatic 8100 which uses the light in the Near Infrared region (1400-2400) to test the samples.

NIVT: National Initial Varietal Trial, [NIVT-1A, NIVT-1B, NIVT-2, NIVT-3, NIVT-4, NIVT-5A, NIVT-5B, NIVT-6, NIVT-7, National Initial Varietal Trials, the strains included in the initial evaluation trial for various sowing conditions i.e, Timely sown, Late sown, Rainfed, Restricted Irrigation *etc.*, are evaluated over the zones for evaluating their performance over the zones like North Hills Zone, North Western Plains Zone, North Eastern Plains Zone, Central Zone, Peninsular Zone and South Hill Zone *etc.*, Based on their performance, strains are promoted to Advance Varietal Trials of the respective zone, the pattern of zones for a crop is based on several considerations, including the distribution pattern of the crop and the area under it in the various regions.

NMR: Nuclear Magnetic Resonance.

No. of grains/spike: Ten spikes will be randomly selected from the plot and the number of grains will be counted. Average no. of grains/spike will be calculated.

No. of tillers per meter length: The number of tillers in per meter length of two central rows will be counted and averaged.

NOGAL: variety released by a private company.

Non-Preference: also termed as non-acceptance and anti-xenosis, involves various morphological and biochemical features of host plants such as - colour, hairiness, leaf angle, taste *etc.*

Norin 10, a very short variety introduced from Japan. These and selections made elsewhere from crosses with Norin and other short strains promise to set new standards in the development of short wheat varieties in the United States.

Norin 10: Japanese variety.

Novosibirs-kaira 67: Baking quality.

NP 4, isolated from the local type Mundia was outstanding with regard to maturity, yield, adaptability and quality, among these, Pusa 4 was extensively used as a parent in wheat improvement programme around the world, a famous wheat variety, Pusa 52 (NP 52) was developed as a result of the systematic hybridization, followed by Pusa 80-5 (NP 80-5) and Pusa 165 (NP 165).

NP 4, NP 12, NP 789, NP 824, C 591, C 218, K 68, PKD 4, PV 18 and C 306 exhibiting rust resistance at adult plant stage against prevailing rust pathotypes were cultivated.

Nucleus Seed Stage-I (NSS-I): In the next cropping season, the seeds obtained from each selected ear-head are space planted in rows of 3 m length in isolation of 5 m from other varieties and examined critically throughout the growing season particularly at sprouting, ear emergence, early dough stage and at maturity for their genetic purity. To enable proper inspection, do the sowing in paired rows by dibbling method. Reject the entire ear row that shows admixture or genetic deviation. To ensure the high-quality seed production ruthlessly reject the row even if a single plant in a row appears to be different. If a progeny has to be rejected after flowering, the both side progenies should also be rejected to avoid the anticipated outcross with the rejected family. Each ear-head progeny is to be harvested and threshed separately and stored in small cloth bag safely. From these ear rows, normally about 75-100 kg seed is obtained which is sufficient to produce NSS II seed in about 0.75 ha and around 25-30 quintals of breeder seed. Select fresh ear-heads (300-400) of uniform feature for successive crop season to continue the NSS- I cycle.

Nucleus Seed Stage-II (NSS-II): Whenever the demand for breeder seed requirement is more than 25-30 quintals, it is necessary to follow NSS II. Even otherwise for a high-quality seed production, NSS II is followed prior to breeder seed production. Seed harvested from each ear row of NSS I is drilled separately in a plot of 3m x 1.38m size. The production of NSS-II stage may vary depending upon the nucleus seed requirement and availability of NSS I stage seed stock. These plots are re-examined for any type of deviation from its described distinguished features. This second cycle of nucleus seed multiplication provides yet another opportunity to eliminate any off-type in the ear head progeny, which might have escaped during NSS-I. The plots having any mixtures or deviation should be rejected and the retained plots are harvested and threshed separately. The seed of these plots, once again examined for seed colour, shape and size for uniformity. The uniform plots true to their genotype are harvested and bulked to constitute breeder seed stock.

NUE: N-use efficiency is comprised of two components, N-uptake efficiency (UPE) and N-utilization efficiency (UTE).

Nullisomics (plants with one chromosome pair less than the normal).

NW 1012: PARANA#2/1 JUP/BJY 'S'/3/VEE#5.

NW 1067: TR380-16-30614/CHAT 'S'.

NWPZ: North Western Plain Zone.

October 16: World Food Day.

Oligogenic Resistance: insect resistance is governed by one or few major genes or oligogenes, each gene having a large and identifiable individual effect on resistance, may be conditioned by the dominant or the recessive allele of the concerned gene, the differences between resistant and susceptible plants are generally large and clear-cut.

Oligophagy: Live on one taxonomic unit only. *E.g.* Hessian fly on wheat.

Organic Farming: Conservation Agriculture is not a synonym of organic farming, although it is based on natural processes.

P 52 (NP 52) in 1925 was the result of systematic hybridization work on wheat by Sir Albert Howard followed by P.80-5 (NP 80-5) and P 165 (NP 165).

P 6046: CPAN 1438/3* HD 4519//CBD 274.

P4, P6, P12 are varieties evolved at Pusa are result of pure line selection from local population.

PAGE (Polyacrylamide Gel Electrophoresis) is a method of separating dissolved protein components on a polyacrylamid gel support medium according to their molecular size and electric charge. Gliadin proteins are the alcohol soluble proteins of wheat endosperm,the gliadin electrophoregram is a pattern of separated gliadin components at the completion of electrophoresis.

PAGIF (Polyacrylamide Gel Isoelectric Focusing) is also static type of electrophoresis as the proteins/isozymes are focused to the points in the gel matrix at which pH of the gel equals to their ioselectric points (pIs), the protein molecules are subjected to an electric field on a gel in which a pH gradient has been generated using ampholytes (small polymers of MW appoox. 5000 daltons), these ampholytes are composed of thousands of polymers containing both positively and negatively charged groups. PAGE has successfully been used to distinguish potato cultivars on the basis of their protein and esterase patterns, the electrophoretic band patterns of endosperm proteins obtained by SDS-PAGE have been found to be useful in characterizing barley varieties.

Paleva (pre-sowing irrigation) ensures good germination, uniform crop stand and saves one irrigation, energy of repeated ploughing and 10 to 15 days of time.

Pallas: Stiff strain.

Particle Size Analysis: may be determined by standard sieving tests, by Simon Funnel method, by means of Coulter Countries, by lasr light scattering *etc.* Particle size index can be used, for assess the wheat hardness rather than by an objective test.

Pavon 76′ which is linked with stripe rust resistance gene *Yr29* that had displayed durable resistance.

Pawnee and Triumph are improved over the old Turkey, weak strawed.

Pawnee, Triumph, Knox, and Vermillion can be attributed partly to their short straw as well as to their earliness.

PBI: Plant Breeding Institute.

PBMASS: Pedigree Based Marker Assisted Selection System.

PBR: Plant Breeders' Rights follows distinctness, uniformity and stability (DUS) testing.

PBW 343, UP 2338, Baviacora 92, Seri 82, STAR, MUNIA, PASTOR, SUPER SERI, WEAWER, WH 542, CPAN 3004, HD 2329, Pavon 76, Rayon 89, Bacanora 88, Oasis 86: Lodging Resistant Varieties.

PBW 426: KT BAGE/HP 2449//RAJ 3219/WL 711.

PBW 502: WH 485/PBW 343//RAJ 1482.

PBW 506: KAUZ/PBW 232.

PBW343 (carries *Lr26*).

PBWG (Pedigree-based whole genome) mapping offers significant potential to increase the genetic gain made in plant breeding programs through the development of systems for marker assisted selection.

PCR: Polymerase Chain Reaction methodologies.

PCR-Amplification: is done with the desired primers at required annealing temperature in thermocycler.

PDS: Public Distribution System, Govt of India.

PDSN: Plant Disease Screening Nursery, is being constituted and grown in Delhi, Karnal, Pusa, Dhaula Kuan, Wellington, Indore and Mahabaleshwar.

PDW 233.

Pearling Index (Peter Scott Pearler): Hardness is reflected into greater flour recovery, more coarse semolina and less flour are produced from highly vitreous wheat than from less vitreous wheat, wheat to qualify as durum must be of specie of *T durum* and must contain at least 50 per cent vitreous kernels, for the determination of pearling index, the 20 g sample is taken and allowed to be abraded in pearler for 1 min, the grain is separated from bran in a hand sieve in 20 motions, clear grain is weighed and pearling percent is calculated as Pearling Index = {(20 weight of pearls) x 100}/20.

Pelshenke Test: records the time taken by a whole meal dough ball prepared with yeast suspensions, to disintegrate while being incubated at 30°C, measures the strength of dough and gluten, this test measure the time required in min, for the initiation of disintegration of dough ball of standard size under specific conditions. (30°C) *i.e.* ball is prepared with yeast suspension. 3.0 g of uniformly ground meal is made into dough with yeast solution (1.8 ml) and allowed to ferment, the dough ball is dropped into breaker containing distilled water maintained at 30°C, the time when the ball is introduced into breaker and the time when it starts disintegrating is noted, the Pelshenke value may vary from 50-240 min, the time depends on the strength of gluten (very strong/medium/weak), which gives the dough its particular form and consistency.

Pennrad: Increased winter hardiness.

PET: Positron Emission Tomography.

Phenol Test: to assess the admixture of *T. aestivum* in *T-durum* wheat lots test is done, a representative sample of a test lot is taken by count into a petri dish and dipped in 10 ml of 3 per cent phenol reagent, the pure durum kernels are not affected at all where as *T. aestivum* turn black.

Phenomics provides the opportunity to study previously unexplored areas of plant science, provides the opportunity to bring together genetics and physiology to reveal the molecular genetic basis of a wide range of previously intractable plant processes.

Phenotypic markers like wheat glume colour controlled by *Rg1* locus is useful as a tool to select *GluB3* alleles encoding LMW glutenin subunits.

Phenotyping for leaf rust resistance gene LrS2427.

Phipps Laboratory of Imperial Agricultural Research Institute, Pusa (Bihar) before 1934 Earthquake.

Phipps Laboratory of Imperial Agricultural Research Institute, Pusa (Bihar) at N Delhi.

PHS (Pre-harvest Sprouting) is one of the most important quality constraints in wheat production in environments characterized by summer rainfall and high humidity. PHS resistance is inherited as a quantitative trait controlled by a large number of genes, and is significantly influenced by genetic background, environmental conditions and their interactions.

Phuchka (panipuri) being made of durum wheat is crunchy in taste.

Physico- Chemical Tests: Various physical and chemical tests used for quality estimation are.

Physiological Races: strains of a species of an insect pest, differing in their ability to attack different varieties of the same host species.

Phytate (or heat-stable phytase) in crop improvement via selecting for phytate: zinc or phytate: iron molar ratios also depends on how the crop is consumed.

Phytohemaglutinins are more but the behaviour of genotypes for this antiquality factor is unpredictable over environments.

PI: Plant Introduction.

Pink Stem Borer (*Sesamia inferens Walke*): is a polyphagous insect distributed throughout India, the caterpillars are pinkish brown, with smooth cylindrical body measuring about 25 mm.

PL 480 Scheme for identification of new sources of resistance to wheat rusts.

Plant Fitness.

Plant Phenomics is a simply plant physiology in 'new clothes', but it promises to bring physiology up to speed with genomics by introducing the incredible recent advances made in computing, robotics and image analysis.

Plant Phenotyping.

PMD: Plasma Membrane Damage, known as indicatives of high temperature tolerance, on chromosomes.

Polyacrylamide Gels result from the polymerization of a mixture of acrylamide and a cross-linker (*e.g.* N, N'-Methylene-bis-Acrylamide [BIS, Bis-Acrylamide] or Ethylene Diacrylate); N, N, N', N'-Tetramethylethylenediamine (TEMED) and Ammonium Per sulphate (APS) are used as accelerators and catalysts for the polymerization reaction, the length of the polymer chains is dictated by the concentration of acrylamide used, which is typically between 3.5 and 20 per cent, the pore size of the gel can be varied by using different proportions of both acrylamide and bis acrylamide, the gels can be prepared in tubes or slabs formed between two glass plates.

Polygenic Inheritance*:* disease resistance is governed by many genes with small effects and a continuous variation for disease reaction is produced. The genes show additive and non-additive effects and the environmental effect is also observed. The polygenic resistance does not show pathotype-specificity as against the oligogenic resistance. It is almost same as horizontal resistance, have a oligogenic component, the oligogenes acting in an additive manner *e.g.* bacterial blight resistance in cotton.

Polygenic Resistance: It is governed by several genes, each gene producing a small and usually cumulative effect, involve more than one feature of the host plant, are much more durable than the cases of oligogenic resistance. transfer of resistance is much more difficult;In wheat to cereal leaf beetle, Evolution of resistance breaking biotypes is almost rare.

Polyphagy: insects feed on a vide range of hosts avoiding few plant species. *e.g.* Scales and moths.

Poro-PAGE/SDS PAGE: the separation of proteins/isozymes can also be obtained by using sieving effect created by the variation in pore size in gel with (SDS Poro-PAGE) or without SDS (Poro- PAGE), the separating gel is prepared in a manner that its pore size decreases gradually from one end to another, the proteins with large variation in their molecular weights can be separated and resolved on the same gel in Poro-PAGE, this type of electrophoresis is also termed as static electrophoresis as the molecules move almost up to some end point depending upon their molecular weights.

Pot Screening: Observations are: Gall Index: scales are no gall 0, 1-10,10-30, 30-100, 100>. Number of galls for more precise one. Number of penetrations of Acid fuschin or Cotton Blue (Strain) 0.1 per cent. Destaining with lactophenol. Roots on slide- Total counting of nematode. Number of larvae/5g soil. Number of eggs. Cobb's sieving and decanting method: Nematode from soil – separated and counted. Final population.

Powdery Mildew (*Erysiphe graminis tritici):* The fungus develops numerous superficial white colonies on all the above ground parts of the plant, the white colour of the colonies changes from grey to reddish brown when cleistothecia develop., infected plants become stunted due to reduction in the size and number of leaves, specimens are maintained and multiplied on 7 10 days old seedlings on wheat variety Agra Local or any known susceptible variety in 6″ pots, the potted seedlings are covered with chimneys and the upper side covered with four layers of muslin cloth.

PPB (Participatory Plant Breeding): It is a breeding process in which farmers and plant breeders jointly select cultivars from segregating materials under a target environment, could either be consultative or collaborative, based on the typology of farmer participation suggested.

PPB has been proposed as a strategy to enhance *in-situ* conservation through users. The breeding strategy that employs crossing of landraces with modern cultivars adds value to the landraces. This makes these landraces more attractive to farmers for continued cultivation.

PPFD: Photosynthetically Active Photon Flux Density.

PPP (Precision Phenotyping Platform) established in Jashore in 2018 to screen wheat germplasm for the disease, the platform has been a reliable source of screening for wheat lines around the world and has proven a great success towards taming the wheat blast disease.

Protein Content: vary from 6 per cent up to as such as 25 per cent depending upon the growing conditions, genetic variation for protein content is much less than that due to differences in growing environments, making this a difficult character for improvement by plant breeder, estimated by Kjeldahl Method.

Protein Homogeneity within cultivars is high, only in cultivar Barbara two protein lines (phenotypes) have been detected.

Protein Profiles and the composition of alleles at the *Glu-1* locus of 34 *Triticum aestivum* L. subsp. *aestivum*, 21 *Triticum turgidum* L. subsp.*durum* (Desf.) Husn., 11 *Triticum turgidum* L. subsp.*turgidum*, 13 *Triticum aestivum* L. subsp.*spelta* Thell. accessions have been studied and determined.

Provitamin A carotenoids (β-Carotene) Content: The ratio of semolina pigment to pigment of whole wheat varies from 70-110 per cent, the range of β-carotene pigments in wheat varies from 3 to 8 ppm, bread wheat semolina exhibit low ppm counts while durum wheat semolina with higher β- carotene pigment counts are considered bes, more the pigment content more will be the colour of semolina and the same is reflected in the finished produced like noodles, spaghetti, macaroni products *etc.*, the yellow colour is a consumer's preference which gives yellow (golden) colour of finished product.

PS: Pusa Selection No by earlier breeder at IARI RS Pusa, Bihar.

PSII : Photosynthetic System.

Puccinia pathway.

Punjab Agricultural University, Ludhiana, Punjab.

Purnea University, Purnea, Bihar.

Pusa Lerma: Amlur grained higher extensibility, elasticity and pelshenke value, suggesting stronger gluten.

PVS (Participatory Variety Selection)*:* is the selection of released or pre-released advance lines (including landraces) by farmers in their target environments using their own selection criteria. provides varietal choices to the targeted farmers under their specific environmental conditions; promotes participatory approaches to test; and select and disseminate the preferred variety.

Pyramiding Polygenic Resistance to Leaf and Stripe Rusts in Wheat.

PZ: Peninsular Zone.

QCSN: Quality Screening Nursery.

QK-77 (*T. polonicum* or *T. durum*) A trademarked tetraploid cultivar grown in small quantities that is extensively marketed.

QTL: Quantitative Trait Locus.

Quality in *verbatum* means degree of excellence and is, therefore, a relative term and its improvement is more challenging than of any other trait, definition varies according to needs from the viewpoint of seed, crop growth, crop product, post-harvest technology, consumer preferences, cooking quality, keeping quality, transportability *etc.*, is a subjective concept and has been defined adequately as conformance to requirements, are defined by nutritional, processing and specialty needs of consumers.

Raised Bed Planting: Provides opportunities to reduce the impact of excess water in rainfed conditions or to more efficiently deliver irrigation water in high production irrigated systems, used with small dykes to trap water after a rain so the fields are able to retain more water and store moisture for future crops versus letting it run-off.

Raj 1555, Raj 1972 (HD 2195/HD 2160), Raj 3765(HD 2402/VL 639), Raj 3777(RAJ 3160/HD 2449), Raj 4037(DL788-2/Raj3717).

RAPD (Random Amplified Polymorphic DNA) is anonymous stretches of DNA is amplified using arbitrary primers, it is a fast and easy method for detecting polymorphisms but it is a dominant marker along with reproducibility problems, emerged as a convenient and effective technique for tracing alien chromosome segments in translocation lines, provide a useful alternative to RFLP analysis for screening markers linked to a single trait within near isogenic lines and bulked segregants, provide a useful alternative to RFLP analysis for screening markers linked to a single trait within near isogenic lines and bulked segregants.

RBMS (RNA- Based Molecular Markers)*:* Markers derived from transcribed/ expressed regions of genomes, based on expressed portion of genome, it could be also affected by phenological plant stage and environmental conditions.

RCTs (Resource Conservation Technologies): The terms 'Resource conserving techniques' and 'Conservation agriculture' are used simultaneously and as if their meaning are similar, refer to those practices which will sustain or enhance the natural resources as well as input-use efficiency.

Red Colour of the Grain is controlled by three independent dominant alleles: *Rt, R2,* and *R3,* which are located in chromosomes 30, 3A, and 38. White grain is obtained when recessive alleles *rt, r2,* and *r3* are present together.

Red Rock, had its origin as a single red kernel picked out of a white wheat called Plymouth Rock in Michigan in 1908;.

Reduced Tillage (Minimum tillage): Full width tillage involving one or more tillage trips leaving residue retained (15-30 per cent) on the field.

Regent, Vesta, and the rust-resistant varieties Marquillo and Hope (Spring variety); Clarkan (soft red winter) is resistant to shattering.

Relationship Knowledge is learned by interaction, and could be shared by systematic work in teams complementarily composed, both in the educational system and in the business world.

Retrotransposon-Based Molecular Markers are a class (Class I) of transposable elements (TE) and constitute a major portion of the genome in plants with large genomes, are important for chromatin modifications and epigenetic reprogramming, an ideal target for developing molecular marker techniques because of their amplification mechanism and sequence characteristics, different types of transposon based marker techniques are; Inter Retrotransposon Amplified Polymorphism (IRAP), Retrotransposon Microsatellite Amplified Polymorphism (REMAP), inter Primer Binding Site amplification (iPBS), Sequence Specific Amplification Polymorphism (S-SAP), Retrotransposon Based Insertion Polymorphism (RBIP), Inter Sine Amplified Polymorphism (ISAP), RAPD- Retrotransposon Amplified Polymorphism (RRAP), Inverse Sequence Tagged Repeats (ISTR), Inter-MITE Polymorphism (IMP) and Transposable display (TD).

Rex, Ridit, Triplet, and the club variety Elmar.

RFLP (Restriction Fragment Length Polymorphism) maps provided a more direct method for selecting desirable genes via their linkage to easily detectable markers thereby expediting the movement of desirable genes among varieties, is a hybridization-based molecular marker. Though it is a co-dominant marker with high reproducibility, it has few limitations like the process is time consuming, expensive, requires large amount of DNA and involves the use of radioactively labelled probes. These disadvantages can be overcome by using PCR-based markers.

RHF (Relative Heterochromatin Fraction) the fluorescence intensity of all chromocenters is determined relative to the fluorescence intensity of the entire nucleus.

Rht1 and Rht2: Arguably, increases in grain yields of common wheat over the last several decades came in part from the use of gibberellic acid sensitive dwarfing genes (Rht1 and Rht2), which were widely distributed during the "**green revolution**".

RhtB1 and respectively RhtD1 (previously Rht2).

Seedling Resistance (*Lr28*); APR (*Lr48*).

Seeds Act, 1966.

Seeds Bill, 2004: Enhance the quality of seeds marketed in the country by mandated compulsory registration of varieties; enhances penalties for offences, compared to the existing Seeds Act 1966; makes provision for labelling of expected performance of seeds and also provides for compensation to farmers in case of seed failure; adds seed health as an additional standard for seed quality; brings conformity with existing National Laws and regulations, especially those relating to the import and export of seeds and use of Genetically Modified seeds; it imposes a ban on the use of Genetic Use Restriction Technology (GURT); provides for registration of seed producers, seed dealers and seed processing units; Farmers' Rights: the Bill proposes to emphasize the rights of the Farmers in conformity with the PPV and FR Act, 2001.

Selection Criteria should be based on germination test; growth under stress (measured as plant dry matter accumulation); chlorophyll loss under chilling stress.

Semolina: is coarsely ground endosperm and its chemical composition is similar to that of white flour, is used in the manufacture of macaroni products, is roasted before storing to save it from insects and worms.

Septoria nodorum blotch *(Septoria nodorum),* in which the height of the plant impedes the spread of the inoculum to the upper plant parts.

Septoria tritici blotch evaluated at Toluca research station.

Sharbati Sonora: Amlur grain colour, higher protein and lysine content.

Sharbati, Dara, Safed Pissi, Chandausi, Karachi Choice White, Hard Red Calcutta, Lal Kanak, Lal Pissi, Bansi Kathia, Khandwa, Malvi *etc.* had their origin at Pusa.

Shoot fly *(Atheriqona naqvii, A. oryzae):* is small insect similar to house fly, but much smaller in size. the flies lay small, elongated, dirty white sculptured eggs, singly on the basal portion of stem or lower side of the leaves just near the ground level, the maggots after hatching creep within the leaf sheaths into the axis of the tiller and cut the central growing shoot causing "dead hearts" formation, the most susceptible stage of the crop is upto 45 days after germination.

Shuttle Breeding: Growing of segregating generations in different environments.

SHW: Synthetic Hexaploid Wheat.

SI: Sprouting Index.

Slow Rusting (a trait of durability).

SMR (Seed Multiplication Ratio): It is the number of seeds to be produced from a single seed when it is sown and harvested. SMR = Seed Yield/Seed Rate.

SNPs (Single Nucleotide Polymorphism) is co-dominant markers, often linked to genes and present in the simplest/ultimate form for polymorphism, and thus they have become very attractive and potential genetic markers in genetic study and breeding.

Social Determinism: Society shapes science and technology.

Soft Red Winter: have low protein wheat used for cakes, pie crusts, biscuits, and muffins.

Soft White: Soft, light colored, very low protein wheat grown in temperate moist areas, used for pie crusts and pastry.

Sonalika and Kalyansona : both carries *Lrl3.*

Sonalika: carries *Lr13, Sr2, Sr11, Yr2Sk.*

Sonara 63, Sonara 64 Mayo 64 and Lerma Roja 64: Commercial Spring, semi dwarf wheat from Mexico.

Sonara 63, Sonara 64, Kalyan Sona: triple gene dwarfs.

Spaghetti and Macaroni Properties: The general overall appearance of the finished spaghetti product is one of the more important quality criteria.

Speck Count: a relatively low number of 'off coloured' specks in a specified area is desirable.

Spectophotometer Method*:* measurement of the amount of ultraviolet radiation absorbed by nucleic acid bases offers an accurate means of quantifying DNA concentration in a sample, is the quantification of major carotenoids; one of the difficulties results from the variation in the carotenoid composition of different crops.

Speed Breeding: plant breeding tools that could help to enable the genetic gain.

SPLAT (Single Polymorphic Amplification Test)*:* If sequence tagged site (STS) does not reveal polymorphism, then it is usually converted into SPLAT, individual STS products from different genotypes are sequenced any differences revealed can be sequence from nuclear ribosomal DNA, can be exploited in production of internal primers, Single amplified sequences are dominant markers.

Spot Blotch Resistance: The association of leaf tip necrosis with spot blotch resistance, stay green trait has also positive linkage with spot blotch resistance, these morphological markers are successfully being used in markers assisted selection of spot blotch resistance germplasm lines.

SPRIA: Solid Phase Radioimmuno Assay.

SQI: Soil Quality Index.

SRR (Seed Replacement Rate)*:* It ss the percentage of area sown out of total area of crop planted in the season by using certified/quality seeds other than the farm saved seed, present Seed Replacement Rate for different crops for the year 2019-20 are for Wheat (13 per cent), Paddy (19.16 per cent), Maize (24.41 per cent), Jowar (26.71 per cent), Bajra (51.02 per cent), Gram (7.09 per cent), Urd (20.48 per cent), Moong (19.48 per cent), Arhar (13.60 per cent), Groundnut (5.5 per cent), Rapeseed and mustard (66.96 per cent), Soybean (15.58 per cent), Sunflower (19.61 per cent), Cotton (37.25 per cent), Jute (68.49 per cent) for the India.

SSCP (Single Strand Conformation Polymorphism): is a simple analysis and sensitive technique for mutation detection and genotyping, the principle is based on the fact that single-stranded DNA has a defined conformation, altered conformation due to a single base change in the sequence can cause single Advances stranded DNA to migrate differently under non denaturing electrophoresis conditions.

SSP (Seed Storage Proteins) are contained in the seeds of higher plants, these proteins have been classified as albumins, globulins and glutenins on the basis of their solubility in solvents.

SSRs (Simple Sequence Repeat) is a simple, easy and preferable marker system because of co-dominant nature, vast abundance and uniform distribution throughout the genome, their limitations are high cost investment in oligonucleotide synthesis and require initial identification and necessary prior sequence information.

Stability: ability of a genotype to produce stable and high yield over time in one environment or a variety which shows low genotype *x* environment interaction.

Stadler: strong straw, excellent soft quality and winter hardiness.

Starch and Damaged Starch (AACC 2000): Starch is the most predominant component in flour, changes from the raw granular form to the gelatinized form on cooking of noodles and causes noodles to set, if the starch on the surface of the noodle is overcooked before the interior becomes cooked, the surface of noodle becomes sticky, if the noodles are properly cooked, the starch on the surface is not sticky but gives a smooth mouth feel, contains two fractions amylose and amylopectin, both alpha and beeta amylases attack 1-4 alpha linkages but are hindered by the proximity of 1-6 linkages at branching points, the limit dextrins produced from amylopectins result in stickiness.

Station Trials (Rainfed, High Fertility, Medium Fertility).

STB (*Septoria tritici* blotch) is one of the most important fungal diseases of wheat in Europe and other wheat-growing areas. Because of the sexual reproduction of the fungus, genetic variability allows fungicide resistant strains to evolve.

Stem Rust/Black Rust *(Puccinia graminis tritici):* is one of the most destructive of plant diseases, the fungus inciting stem rust spends part of *its* life cycle on the common barberry plant, also overwinters on growing wheat and *is* spread by wind-borne spores in the spring, stem rust resistance genes were transferred from Yaroslav emmer to the Hope variety, and from Vernal emmer to Carleton and Stewart varieties of durum wheat. Iumillo durum was used in the development of the stem rust-resistant variety of spring wheat, Thatcher.

Storage Proteins of mature seed are used, mainly glutenin and gliadin.

Structural Irresponsibility: producing long-term futures through innovation and technology development and without the capacity for adequate forethought.

STSs (Sequence Tagged Sites) are primers based on some degree of sequence knowledge, these unique, sequence-specific primers detect variation in allelic, genomic DNA. STS have a particular advantage over RAPDs in that they are co-dominant, they can distinguish between homozygote and heterozygote, they also tend to be more reproducible, because they use longer primer sequences, have the disadvantage of requiring some pre-existing knowledge of the DNA sequence of the region, even if only for a small amount.

Sugarcane Leaf Hopper *(Pyrilla perpusilla):* has posed serious problems to wheat crop in some years, migrates from sugarcane to wheat crop, both nymphs and adults suck plant sap and cause damage to wheat at earing stage.

Surface Grass Hopper (Chrotogonus trachypterus): This insect is commonly known as *Toka*. Both nymphs and adults cut the germinating plants at ground level and devour tender leaves and growing point, was a pest of sandy areas with comparatively low rainfall.

Suspercooling: Cooling of water below 0°C without ice crystal formation is called super cooling, In plants water may cool down to -1 to-15°C is herbaceous sps and to -40 to -45°C in hardy trees, becomes possible apparently because internal ice-nucleates are absence in such cases. This is regarded as an important mechanism of freezing avoidance.

Sustainability Coefficient (SC): It is a complex and a multipurpose index based on a range of parameters.

Sustainable Agriculture/Ecofarming/Alternative Agriculture, Ecological Agriculture and Natural Organic Farming is the successful management of resources for agriculture to satisfy changing human needs while maintaining or enhancing the quality of environment and conserving natural resources.

SV: Sedimentation Value.

Tacit Knowledge: is a form of knowledge that is highly personal and context specific and deeply rooted in individual experiences, ideas, values and emotions.

Tag Colour: Breeder Seed Tag - Yellow Colour; Certified Seed Tag Colour - Azar Blue; Foundation Seed Tag Colour - White; Registered Seed Tag Colour - Violet.

Tan Spot (Yellow Leaf Spot).

TCA: Trichloro Acetic Acid.

tcr: tissue culture response.

Technological Determinism: Science and technology shape society.

TEMED: Tetra methylethylene diamine.

Tempering: wetting grain before milling, results in wetter exterior and drier interior, hence under roll pressure, endosperm crumbles to powder whereas bran and germ flatten to flakes and are removed by sieves, for hard wheat usually 11.5-16.5 per cent moisture is required whereas Sami hard and soft wheat require 13.5-15.5 per cent moisture.

Tenmarq parent originated from a cross made between Marquis and P-1066, the latter is a sister selection of Kanred, made from Crimean.

Termites *(Microtermes obesi* Holmgren; *Odontotermes obesus* Rambur*):* are social insects and live in colonies known as termitaria, mound forming species these termitaria are raised 1.5-2.0 m above ground level.

Test Weight or Hectolitre Weight or Weight Per Unit Volume (kg): is the primary grading factor; an indicator of semolina extraction, related to variety and environmental conditions and usually determines the plumpness of the grain, flour yield increases and flour ash decreases with the increase in test weight, depends mainly on grain shape and density and indirectly indicates the milling yield, in durums, it is strongly related to semolina yield.

Thorne, Vigo, Knox, and Vermillion are considerably stiffer strawed. than soft wheat varieties, such as Fulcaster, Mediterranean or Red May.

Thorne: soft red winter varieties, from the cross Portage X Fulcaster.

Three Working Levels: triploid (only one model), tetraploid (three models), and pentaploid (five models).

Threshability: one gene condition non-threshability, Tg from *Ae. tauschii* whereas the additional gene q from the tetraploid parent segregates in crosses between wheat and SHW, single gene segregation results in a higher frequency of threshable progeny.

Tilka Manjhi Bhagalpur University, Bhagalpur, Bihar.

Tillet, a French Botanist, established that the causal fungus of hill bunt of wheat, *Tilletia caries*, is seed borne.

Tilletia caries, T. foetida and *T. controversa*.

TLC (Thin-Layer Chromatography): Separates the different carotenoid groups (β-carotene and α-carotene; β-cryptoxanthin; and lutein and zeaxanthin) but does not detect the presence of trans and cis isomers of β-carotene; cis isomers are known to have lower provitamin A activity and can be present in significant amounts compared with trans isomer.

TMD: Thylakoid Membrane Damage.

Total amount of β-carotene (ppm) = 0.0105 + (23.566 x O.D. value).

Total Biomass Production: harvesting the total plot and determining the total biomass.

TPN: Trap Plot Nursery.

Trait Selection Methodology is employed in breeding for organic agriculture.

Translational Genomics Tool is 'model to crop' translation, that can be contributed to the implementation of genetic and genomics in crop species.

TRIS: Hydroxy methyl (Amino methane).

Trumbull, Fultz, including a leaf rust-resistant strain, were the parents of the Vigo variety, selection are the Nittany variety of soft wheat, which originated as a plant selection from Fulcaster.

TTC: Triphenyl Tetrazolium Chloride.

Turkey, Minturki, Yogo, and Marmin, Varieties are winter wheat.

TWC: Three-Way Cross.

Ug99 race of Stem Rust Pathogen or TTKS: a new race of stem rust evolved in Uganda (Africa) in 1999, is different from other existing races of stem rust, evaluated Indian wheat material for *Ug99* at Kenya, unusually potent race was detected in Uganda in 1999 (named *Ug99*) causing stem rust of wheat, has been the most destructive disease of wheat and barley and has inflicted very heavy losses worldwide, the organism is a fungus called *Puccinia graminis tritici,* Nobel laureate Dr. Norman Borlaug declared it as a global threat; 11 wheat varieties having *Ug99* resistance namely, GW 273, GW 322, HI 1500, HD 2781, MP 4010, HUW 510, MACS 2846, HI 8498 (durum), UP 2338, DL 153-2 and HW 1085, Raj 4120.

UNESCO: United Nations Educational Scientific and Cultural Organisation.

USI (Rotarian), Indore.

Variable Number of Tandem Repeats (VNTRs) *is* a location in a genome where a short nucleotide sequence is organized as a **tandem repeat**. These can be found on many chromosomes, and often show variations in length (**number** of **repeats**) among individuals.

Variety Identification using Bar Code: The variety bar code, **CSIRO ID System,** tests leaf or grain samples using a panel of DNA markers in Australia. Each marker gives a 'yes' or 'no' result. The pattern can be used to develop an individual 'bar code' for each variety. DNA markers are used to develop a variety 'bar code'.

VC: Vice-Chancellor.

Veery selections.

Vernal emmer (used in Carleton and Stewart durum).

Vertical and Horizontal Resistance (Vander plank).

Vertifolia Effect: Vander plank introduced the term Vertifolia effect and refers to epidemic development in a variety carrying vertical resistance genes (oligogenes) leading to heavy economic losses. Total failure of vertical resistance leading to a disease epidemic is known as Vertifolia effect. This failure occurs because of two reasons, *viz.*, the level of horizontal resistance in varieties carrying oligogenes is usually low and the pathogen is able to evolve new virulent pathotypes.

VG: Visual grading scale of aphids' incidence.

Vienna: High 1000 kernel weight.

Vigo: Cross Trumbull X Fultz made.

VL 804: CPAN 3018/CPAN 3004//PBW 65.

Vorobey: synthetic wheat derived from germplasm (trial means ranged from 1 to 8 t/ha).

vrn: Vernalization (*Vrn*1), resistance to preharvest sprouting.

WB2 (*T. dicocccon,* C19309/Ae.sq(409)/3/MILAN/S87230//BAV92/4/2*MILAN/S87230//BAV92).

density (iv) saving of irrigation water (up to 15-20 per cent), (vi) improved input use efficiency because of better crop stands due to good seed and fertilizer nutrients placement, and (vii) build-up in soil organic carbon due to reduced burning of crop resides.

Zincol: Variety released in Pakistan in 2016.

Zinc-Shakti: CIMMYT-derived early-maturity wheat cultivar with about 40 per cent increased zinc (+14 ppm), is now grown in eastern India through public-private partners, the two CIMMYT-derived biofortified varieties: 'WB2' and 'HPBW01' released in 2016 for northwestern plains zone of India.

Index

A

B

C

G

H

M

N

O

R

S

T

U

V

W

Y

Z